T0315061

FIRE IN CALIFORNIA'S ECOSYSTEMS

The publisher gratefully acknowledges the generous support of the August and Susan Frugé Endowment Fund in California Natural History of the University of California Press Foundation.

The publisher and the University of California Press Foundation gratefully acknowledge the generous support of the Ralph and Shirley Shapiro Endowment Fund in Environmental Studies.

Fire in California's Ecosystems

SECOND EDITION

Edited by

JAN W. VAN WAGTENDONK, NEIL G. SUGIHARA,
SCOTT L. STEPHENS, ANDREA E. THODE,
KEVIN E. SHAFFER, and JO ANN FITES-KAUFMAN

UNIVERSITY OF CALIFORNIA PRESS

University of California Press, one of the most distinguished university presses in the United States, enriches lives around the world by advancing scholarship in the humanities, social sciences, and natural sciences. Its activities are supported by the UC Press Foundation and by philanthropic contributions from individuals and institutions. For more information, visit www.ucpress.edu.

University of California Press
Oakland, California

Library of Congress Cataloging-in-Publication Data

Names: van Wagtendonk, Jan W., 1940-editor. | Sugihara, Neil G., 1954-editor. | Stephens, Scott L., 1961-editor. | Thode, Andrea E., 1973-editor. | Shaffer, Kevin E., 1962-editor. | Fites-Kaufman, JoAnn, 1962-editor.

Title: Fire in California's ecosystems / edited by Jan W. van Wagtendonk, Neil G. Sugihara, Scott L. Stephens, Andrea E. Thode, Kevin E. Shaffer, Jo Ann Fites-Kaufman.

Description: Second edition. | Oakland, California: University of California Press, [2018] | Includes bibliographical references and index. |

Identifiers: LCCN 2017044580 (print) | LCCN 2017048227 (ebook) | ISBN 9780520961913 (ebook) | ISBN 9780520286832 (cloth: alk. paper)

Subjects: LCSH: Fire ecology—California. | Fire management—California.

Classification: LCC QH105.C2 (ebook) | LCC QH105.C2 F57 2018 (print) | DDC 577.2/409794—dc23

LC record available at https://lccn.loc.gov/2017044580

To Harold Biswell, who can truly be called the father of fire ecology in California.
His patience, persistence, humor, and devotion to managing wildlands and fire
in harmony with nature have been an inspiration to all of us.

CONTENTS

LIST OF CONTRIBUTORS ix
FOREWORD xi
JAMES K. AGEE

PREFACE xv

1 Introduction: Fire and California Vegetation / 1
NEIL G. SUGIHARA, TODD KEELER-WOLF, AND MICHAEL G. BARBOUR

PART ONE
Introduction to Fire Ecology

2 California Fire Climate / 11
RICHARD A. MINNICH

3 Fire Weather Principles / 27
BRENDA L. BELONGIE AND RICHARD A. MINNICH

4 Fire as a Physical Process / 39
JAN W. VAN WAGTENDONK

5 Fire as an Ecological Process / 57
NEIL G. SUGIHARA, JAN W. VAN WAGTENDONK, AND JO ANN FITES-KAUFMAN

6 Characterizing Fire Regimes / 71
BRANDON M. COLLINS, JAY D. MILLER, JEFFREY M. KANE, DANNY L. FRY, AND ANDREA E. THODE

7 Fire and Physical Environment Interactions: Soil, Water, and Air / 87
PETER M. WOHLGEMUTH, KEN R. HUBBERT, TRENT PROCTER, AND SURAJ AHUJA

8 Fire and Plant Interactions / 103
AMY G. MERRILL, ANDREA E. THODE, ALEXANDRA M. WEILL, JO ANN FITES-KAUFMAN, ANNE F. BRADLEY, AND TADASHI J. MOODY

9 Fire and Animal Interactions / 123
KEVIN E. SHAFFER, SHAULA J. HEDWALL, AND WILLIAM F. LAUDENSLAYER, JR.

PART TWO
The History and Ecology of Fire in California's Bioregions

10 North Coast Bioregion / 149
SCOTT L. STEPHENS, JEFFREY M. KANE, AND JOHN D. STUART

11 Klamath Mountains Bioregion / 171
CARL N. SKINNER, ALAN H. TAYLOR, JAMES K. AGEE, CHRISTY E. BRILES, AND CATHY L. WHITLOCK

12 Southern Cascades Bioregion / 195
CARL N. SKINNER AND ALAN H. TAYLOR

13 Northeastern Plateaus Bioregion / 219
GREGG M. RIEGEL, RICHARD F. MILLER, CARL N. SKINNER, SYDNEY E. SMITH, CALVIN A. FARRIS, AND KYLE E. MERRIAM

14 Sierra Nevada Bioregion / 249
JAN W. VAN WAGTENDONK, JO ANN FITES-KAUFMAN, HUGH D. SAFFORD, MALCOLM P. NORTH, AND BRANDON M. COLLINS

15 Central Valley Bioregion / 279
ROBIN WILLS

16 Central Coast Bioregion / 299
MARK I. BORCHERT AND FRANK W. DAVIS

17 South Coast Bioregion / 319
JON E. KEELEY AND ALEXANDRA D. SYPHARD

18 Southeastern Deserts Bioregion / 353
MATTHEW L. BROOKS, RICHARD A. MINNICH, AND JOHN R. MATCHETT

PART THREE
Fire Management Issues in California's Ecosystems

19 The Use of Fire by Native Americans in California / 381
M. KAT ANDERSON

20 Fire Management and Policy since European Settlement / 399
SCOTT L. STEPHENS AND NEIL G. SUGIHARA

21 Fire and Fuel Management / 411
SCOTT L. STEPHENS, SUSAN J. HUSARI, H. TOM NICHOLS, NEIL G. SUGIHARA, AND BRANDON M. COLLINS

22 Fire, Watershed Resources, and Aquatic Ecosystems / 429
JAN L. BEYERS, ANDREA E. THODE, JEFFREY L. KERSHNER, KEN B. ROBY, AND LYNN M. DECKER

23 Fire, Air Quality, and Greenhouse Gases / 439
SURAJ AHUJA AND TRENT PROCTER

24 Fire and Invasive Plants / 459
ROBERT. C. KLINGER, MATTHEW L. BROOKS, AND JOHN M. RANDALL

25 Fire and At-Risk Species / 477
KEVIN E. SHAFFER AND SHAULA J. HEDWALL

26 Fire and Climate Change / 493
CHRISTINA M. RESTAINO AND HUGH D. SAFFORD

27 Social Dynamics of Wildland Fire in California / 507
SARAH M. MCCAFFREY, GUY L. DUFFNER, AND LYNN M. DECKER

28 The Future of Fire in California's Ecosystems / 517
NEIL G. SUGIHARA, JAN W. VAN WAGTENDONK, SCOTT L. STEPHENS, KEVIN E. SHAFFER, ANDREA E. THODE, AND JO ANN FITES-KAUFMAN

APPENDIX 1 523
INDEX 529

CONTRIBUTORS

JAMES K. AGEE University of Washington (retired), Seattle, WA

SURAJ AHUJA Forest Service (retired), Willows, CA

M. KAT ANDERSON Natural Resource Conservation Service (retired), Davis, CA

MICHAEL G. BARBOUR University of California (retired), Davis, Winters, CA

MICHAEL P. BEAKES Cramer Fish Sciences, West Sacramento, CA

BRENDA L. BELONGIE Forest Service, Redding, CA

JAN L. BEYERS Forest Service, Riverside, CA

MONICA A. BOND Wild Nature Institute, Hanover, NH

MARK I. BORCHERT Forest Service (retired), Fawnskin, CA

ANNE F. BRADLEY The Nature Conservancy, Santa Fe, NM

CHRISTY E. BRILES University of Colorado, Denver, CO

MATTHEW L. BROOKS US Geological Survey, Oakhurst, CA

BRANDON M. COLLINS Forest Service, Davis, CA

FRANK W. DAVIS University of California, Santa Barbara, CA

LYNN M. DECKER The Nature Conservancy, Salt Lake City, UT

TOM L. DUDLEY University of California, Santa Barbara, CA

GUY L. DUFFNER The Nature Conservancy, Salt Lake City, UT

TODD E. ESQUE US Geological Survey, Las Vegas, NV

CALVIN A. FARRIS National Park Service, Klamath Falls, OR

JO ANN FITES-KAUFMAN Forest Service (retired), Vallejo, CA

DANNY L. FRY University of California, Berkeley, CA

SEAN A. HAYES National Marine Fisheries Service, Santa Cruz, CA

SHAULA J. HEDWALL US Fish and Wildlife Service, Flagstaff, AZ

KEN R. HUBBERT Forest Service, Piedmont, CA

SUSAN J. HUSARI Forest Service (retired), San Rafael, CA

JEFFREY M. KANE Humboldt State University, Arcata, CA

TODD KEELER-WOLF California Department of Fish and Wildlife, Sacramento, CA

JON E. KEELEY US Geological Survey, Three Rivers, CA

JEFFREY L. KERSHNER US Geological Survey (retired), Manhattan, MT

ROB C. KLINGER US Geological Survey, Oakhurst, CA

WILLIAM F. LAUDENSLAYER Forest Service (retired), Lake City, CA

JONATHAN W. LONG Forest Service, Davis, CA

JOHN R. MATCHETT US Geological Survey, Oakhurst, CA

SARAH M. MCCAFFREY Forest Service, Fort Collins, CO

KYLE E. MERRIAM Forest Service, Alturus, CA

AMY G. MERRILL Stillwater Science, Berkeley, CA

JAY D. MILLER Forest Service, McClellan, CA

RICHARD F. MILLER Oregon State University, Corvallis, OR

RICHARD A. MINNICH University of California, Riverside, CA

TADASHI J. MOODY California Department of Forestry and Fire Protection, Sacramento, CA

MAX A. MORITZ University of California, Berkeley, Santa Barbara, CA

H. TOM NICHOLS National Park Service (retired), Eagle, ID

MALCOLM P. NORTH Forest Service, Davis, CA

DENNIS C. ODION Southern Oregon University, Ashland, OR

TRENT PROCTER Forest Service, Porterville, CA

JOHN M. RANDALL The Nature Conservancy, San Diego, CA

CHRISTINA M. RESTAINO University of California, Davis, CA

GREGG M. RIEGEL Forest Service, Bend, OR

KEN B. ROBY Spatial Informatics Group, Pleasanton, CA

HUGH D. SAFFORD Forest Service, Vallejo, CA

KEVIN E. SHAFFER California Department of Fish and Wildlife, Sacramento, CA

CARL N. SKINNER Forest Service (retired), Shasta Lake, CA

SYDNEY E. SMITH Forest Service (retired), Cedarville, CA

SCOTT L. STEPHENS University of California, Berkeley, CA

JOHN D. STUART Humboldt State University (retired), Arcata, CA

NEIL G. SUGIHARA Forest Service (retired), El Dorado Hills, CA

ALEXANDRA D. SYPHARD Conservation Biology Institute, La Mesa, CA

ALAN H. TAYLOR Pennsylvania State University, University Park, PA

ANDREA E. THODE Northern Arizona University, Flagstaff, AZ

JAN W. VAN WAGTENDONK National Park Service (retired), El Portal, CA

ALEXANDRA M. WEILL University of California, Davis, CA

CATHY L. WHITLOCK University of Montana, Missoula, MT

KATE M. WILKIN University of California, Berkeley, CA

ROBIN WILLS National Park Service, San Francisco, CA

PETER M. WOHLGEMUTH Forest Service, Riverside, CA

FOREWORD

JAMES K. AGEE

Fire was finally recognized as an important ecological factor in the mid-twentieth century in Rexford Daubenmire's *Plants and Environment*. Before then, it had largely been considered an allogenic factor even by ecologists such as Fredric Clements, who had done some of the first work on fire-dependent lodgepole pine (*Pinus contorta*) in the Rocky Mountains. The volume you have in your hand is the second edition of the most comprehensive work ever on a state's fire ecology, and demonstrates tremendous progress in understanding the role of fire in California wildlands. Although some of the largest wildfires in recorded California history have burned recently, it is the less dramatic truths in this volume that will have a far more lasting effect on wildland fire in California.

Fire and people have interacted for millennia in California. Native Americans burned the landscapes of the state for a variety of purposes, including protection of their villages (the first wildland-urban interface), resources such as basket-weaving materials, the many food plants that were favored by fire, hunting game animals, signaling, and warfare. Their fires, often starting at low elevations, complemented those started from lightning, more common at high elevations. The long dry seasons typical of the Mediterranean climate ensured a prolonged fire season every year. Although fire did visit almost every landscape in California, it did so with a remarkable variety in frequency, intensity, and effects. California has always been and will continue to be a fire environment unmatched in North America.

Institutionalized fire policy in California began early in the twentieth century when the great fires of Idaho and Montana in 1910 galvanized the fledging Forest Service to promulgate a national policy of total fire exclusion. A battle to retain *light burning*, as prescribed fire was called in those days, was fought both in the southern states and in California. Forest industry was leading the charge for light burning in California, not because of altruistic sentiments about natural forests but because they believed it would help protect the old growth forests until they were ready to be harvested.

Aided by the passage of the Federal Clarke-McNary Act that funneled fire protection dollars to the states and by research from the Forest Service, fire exclusion was firmly entrenched in California. Yet some of the same research used to support the fire protection policy, such as that by Bevier Show and Edward Kotok in the Sierra Nevada and Emmanuel Fritz in the coast redwood belt, also showed that fire had played a very significant historical role in forest ecosystem dynamics. The fire exclusion policy remained unchallenged until the 1950s, when it came under attack by a few courageous men, such as Harold Weaver who worked for the Bureau of Indian Affairs and Harold Biswell from the University of California at Berkeley.

Dr. Biswell, more than any other, actually bent the old fire culture in California. The idea of underburning forests to prevent more destructive wildfires was a revolutionary idea in California in the 1950s and 1960s, although fire then was routinely used in some shrublands. It is important to keep in mind that during those times Biswell was widely criticized for the same ideas, presented in the same way, for which he received so much favorable response later in his career, including for his classic integration of science and interpretation, *Prescribed Burning in California Wildland Vegetation Management* (UC Press 1989). Harold was an advocate of fire prevention, but he believed a balance between fire suppression, prevention, and use was critical. Smokey Bear just couldn't say it all in one sentence anymore. One had to be very courageous in those days, and Harold strode on, focusing on spreading the message and taking the high road in terms of his professional demeanor. The logic of that message attracted many of us, including me, to become interested in fire science as a career.

Where have we come since then when we were at least providing lip service to the important role of fire in California wildlands? We have made some great strides in some areas, and seem to be mired in the muck elsewhere. The technology to conduct prescribed burns continues to improve. We now have computer models that incorporate fire behavior information with geographic information systems to predict fire spread across landscapes. It works well. We have more sophisticated fire effects models to predict the ecological outcome of fire. We can tell what size classes of the various tree species on a site are likely to die in fires with various flame lengths. Our understanding of crown fire dynamics is still relatively rudimentary. While the technological fixes for wildland fire are not complete, in comparison to our knowledge about other ecological processes, such as wind, insects, or fungi, fire technology is at the head of the pack. One of the newer systems, the Interagency Fuels Treatment Decision Support System aims to provide access to data and previously stand-alone models in one place, and is now operational.

We use this technology only sparingly, and a strong case remains that we could do much more. The phrase "forest health" emerged in the late 1980s to explain why we see so many trees dead and dying across western landscapes, and high-density, multilayered forests caused by fire exclusion are at the root of the problem. In California, the forest health situation at Lake Arrowhead in the early 2000s was unprecedented, with millions of drought-killed pines helping to fuel wildfires. Insects and disease epidemics are at historic highs, and intense wildfires are expanding like never before on western landscapes. Fuel-laden mixed-conifer forests in the Sierra Nevada, southern Cascades, and Klamath mountains with hundreds of millions of dead trees have made early control of wildfires quite difficult, as evidenced by the 100,000 ha (250,000 ac) Rim Fire in and near Yosemite National Park. California as well as Arizona, Colorado, New Mexico, Oregon, and Washington have experienced their largest-ever wildfires in recorded history since the turn of the millennium.

We continue to despair at the state of our western forests, but the solutions have become mired in political debates. Yet there are some radiant examples of fire use in the state: California state parks are burning in a wide variety of forest and shrub vegetation types; nature organizations such as The Nature Conservancy have been using prescribed fire in prairie restoration and oak woodland maintenance, and are coordinating the pooling of resources of large private landowners to effect landscape burns in the Sacramento Valley; and the National Park Service is continuing to move forward with prescribed fire plans in chaparral and forested portions of national park system lands in the Sierra Nevada and elsewhere. These programs are complex: they require knowledge of plant and animal response to fire, and the effects of varying the frequency of fire, its intensity and extent, the season of burning, and its interaction with other ecosystem processes.

The national forests will see expanded fire programs in the coming years, too. Professor Biswell's idea in the 1950s to use the large Federal emergency firefighting fund upfront to do fuel treatment was recently re-championed by Secretary of the Interior Bruce Babbitt in the 1990s on behalf of all the Federal land management agencies. It made perfect sense 60 years ago, but took more than half a century to become part of the fire culture. A portion of the fund became authorized and available in 1998 for prescribed burning on all Federal lands. Increased funding will result in much more prescribed fire, and reduced threat of wildfire, over millions of acres of the West. The National Fire Plan and the Western Governor's Association are providing consensus policies in the wildland-urban interface. But President Bush's Healthy Forests Initiative in the early 2000s was debated as both a real opportunity to improve forest health or just a way to enable more logging. The National Cohesive Wildland Fire Management Strategy now being developed is intended to result in more resilient post-fire landscapes, communities better adapted to the presence of fire, and safe and effective wildfire response.

The intrusion of residences into wildlands, with its attendant fire problems, was always a major concern of Biswell's, and in his book he warned of impeding catastrophic fire in the Berkeley Hills. His warning was based on precedent, in that one of the most devastating urban-wildland interface fires prior to 1991 occurred in the Berkeley Hills in 1923. A fire started in the hilltop area, and blown by hot, dry autumn winds, it swept down right to the edge of the University of California campus. Fire marshals were considering dynamiting entire residential blocks to save the rest of the town, when fog blew in from the Golden Gate and helped to extinguish the fire. The burned area sprouted back with residences, just as the brush and eucalyptus trees sprouted back, and the residences spread further into the wildlands over the subsequent decades.

The Berkeley Hills are not unique in this regard. They are but one of the innumerable communities where residences are invading wildlands, but Harold lived in the Berkeley Hills, so it was of special interest and concern to him. His late 1980s prediction of a major catastrophic fire there came true in 1991. No one was saddened more than Harold Biswell when the 1991 Tunnel Fire killed 25 people, destroyed more than 3,000 homes, and cost more than $1.5 billion—and it was preventable. Sadly, these property losses were exceeded in the 2003 southern California fires. The 2017 Napa-Sonoma-Mendocino fires burned more than 80,000 ha (200,000 ac), 8,400 structures, will cost more than $100 billion in insured losses, and killed more than 40 people, mostly elders who could not outrun the firestorm. The Ventura-Santa Barbara Counties Thomas Fire in late 2017 is the largest modern wildfire in California history (so far), destroying more than 1,000 homes in the fire and post-fire debris flows.

This growing fire problem in what is called the wildland-urban interface will continue to plague fire managers. Of all the institutional problems with fire, this is the most complex: mostly private land; myriad jurisdictional problems exist for zoning, building codes, fire protection; and continuing attitudes that the disaster will strike somewhere else, or will never strike twice.

This volume will become the secular bible of fire ecology for Californians. But what does the future hold for new knowledge and application? Academic trends over the last decade have disfavored small, technically oriented programs (i.e., fire ecology, forest management) in favor of more general and efficient programs (environmental science) that attract larger numbers of students. California's universities, while not disfavoring fire ecologists, will be hiring general ecologists in order to meet their teaching mandates. Those that can attract research funding in fire ecology may be able to carve a niche for themselves, but few universities will be advertising specifically for fire ecologists.

At the same time, the complex nature of resource management argues for more technically trained managers. The agencies have hired many more doctorate-level fire scientists than have the academic systems, and this demand will continue to grow. But there is a major supply and demand problem emerging: prospects for long-term supply are meager, given the trend in academia to avoid specialist faculty who would guide these students. The typical historical solution to these types of problems has been cooperative programs partly funded by the Federal government at selected universities to maintain viable teaching and research capability in a specific discipline. The Federal Joint Fire Science Program (JFSP) has stepped in and involved more than 90 colleges and universities in fire research, and more than 70% of funding has gone to academic institutions. The JFSP has also directly funded graduate student research, helping to ensure that well-trained scientists continue to emerge from academia. Journals such as *Fire*

Ecology, *International Journal of Wildland Fire*, and *Forest Ecology and Management* have helped to disseminate these research results. However, the $10 million annual budget of JFSP pales in comparison to the almost $2 billion spent by Federal agencies per year on wildland fire suppression.

In wildlands, history does repeat itself. Fire environments of yesterday are those of today, and although altered, will be those of tomorrow. California and the West are fire environments without parallel in North America. Harold Biswell would say that our mountains will always stand majestically, and dry summers and windy spells will always be part of our western heritage. We can only intervene in the fire behavior triangle by managing the vegetation. Biswell and his contemporaries gave us the tools to manage change through controlled fire, integrating it with naturally occurring fire in wilderness and intelligent, cost-effective fire suppression. It is now time for us to recognize that fire is part of our culture, and we need to make good decisions about the use of fire, not just its control. The solutions will be complex, will vary by place, and will occur in a changing environment. This book tells us what we know now, but we have the ability to learn much more as we manage, and we will need to feed this information back into better decision-making. There are many treatments we can apply, in various places with unique land-use histories, and at different scales, in stochastic environments, and perhaps continually changing climates. Fire ecology will inform this debate, with no better place to start than this book. We hope you enjoy this second edition, learn much, and finish with more questions than when you started.

PREFACE

Much has happened in the field of fire ecology since the first edition of *Fire in California's Ecosystems* was published in 2006. The Joint Fire Science Program sponsored by the Department of Agriculture and the Department of the Interior has funded numerous projects, chief among them the Fire and Fire Surrogates program, which had several study sites in California. The surge of funding for fire research has resulted in an exponential increase in scientific articles in journals and technical reports. The second edition contains over 1,300 new citations to articles published in the past 10 years. As a result of this new interest in fire, a new generation of fire ecologists has appeared on the scene. This is evidenced by the inclusion of 26 new authors in the list of contributors and the retirement of 14 of our original authors (including two editors). Should a third edition be published, the transition will be complete.

The second edition has four new chapters. Fire weather has been separated from the chapter on fire climate to form a new chapter. This was done to more fully explain the important factors that affect fire behavior. Advances in the methods of characterizing fire regimes, particularly satellite imagery, warranted a new chapter on that topic. Fire can now be studied as a multifaceted phenomenon rather than just bimodal burn or not burned. The emergence and importance of climate change and the human aspects of fire justified separate chapters.

Alterations to fire regimes have resulted in many changes to the biological communities, including changes in vegetation composition and structure and vegetation type conversions or ecosystem migrations. This text details many of these changes, explains how fire has changed as an ecosystem process, and provides insights for determining the direction that the changes might take in the future. As with introductory treatments of any of the elements of natural ecosystems, we are prone to generalization, simplification, and standardization of processes and interactions that are inherently complex. In describing fire effects and regimes we are by necessity guilty of continuing that trend toward simplification. However, we hope that by communicating the concepts of the role that fire plays as a dynamic ecological process, we can communicate the importance of fire's role in defining what we know as California's ecosystems. This importance has been underscored by the large fires in the fall of 2017, causing the death of dozens of people and the destruction of thousands of structures. If we are to live in fire-prone ecosystems, we must learn how to adapt to this dynamic force.

This book is intended for use both as a text for learning and teaching the basics of fire ecology and as a reference book on fire in California ecosystems. It synthesizes and expands upon our knowledge of fire as an ecological process and facilitates a better understanding of the complex and dynamic interactions between fire and the other physical and biological components of California ecosystems. Modern western society has tended to view ecosystems within narrowly defined ranges of time and space. Focused studies of ecosystems from the standpoint of individual species within their habitats, individual stands of trees, populations, plant communities, fire events, or watersheds allow us to know specific mechanics of ecosystems but, by nature, do not help us develop a broad view of large dynamic landscapes. On the other hand, studies of broad spatial or temporal application are usually limited in their application to specific examples. Understanding fire in ecosystems requires us to greatly expand our spatial and temporal context to include both discrete fire events that occur on finite landscapes and complex multi-scale burning patterns and processes that are dynamic on large landscapes. We intend this text to present an integrated view of fire in California ecosystems from as wide a spectrum of temporal and spatial scales as possible.

This text is divided into three parts. Part One is an introduction to the study of fire ecology that is intended for use in teaching the basics of fire ecology. Part Two is a treatment of the history, ecology, and management of fire by bioregions and is intended for use as a reference and for teaching fire ecology within the various bioregions within California. Part Three is a treatment of fire management issues and is intended for use as a reference and for teaching fire management from a historical, policy, and issue perspective.

Obviously, a book such as this is not written without the help of many people. First, we would like to thank the many authors of all the chapters; they endured structured outlines, tight deadlines, and an authoritarian group of editors. Heath Norton drew the figures, Daniel Rankin prepared the maps, and Lester Thode created the fire regime graphs for both editions of the book. Without their help, the book would have lacked the consistency and attractiveness that add greatly to its readability. The second edition was written by the authors much on their own time, but with some support from their agencies and employers. All of their support was essential.

CHAPTER ONE

Introduction

Fire and California Vegetation

NEIL G. SUGIHARA, TODD KEELER-WOLF,
AND MICHAEL G. BARBOUR

> In California, vegetation is the meeting place of fire and ecosystems. The plants are the fuel and fire is the driver of vegetation change. Fire and vegetation are often so interactive that they can scarcely be considered separately from each other.
>
> BARBOUR et al. (1993)

Fire is a vital, dynamic force of nature in California's ecosystems and shapes the composition and structure of its vegetation. For millennia, human interaction with fire has developed simultaneously around our need to protect ourselves from its harm and the opportunity to use it as a natural resource management tool. As we become more aware of what influences and controls fire, we are gaining an appreciation of the contribution that fire makes to ecosystem complexity and biological diversity. Previous notions that we can, and should, suppress all wildfires to provide for human safety and protect ecological resources are giving way to the realization that we don't have the ability to completely exclude fire, and that many valued attributes of California's ecosystems shouldn't be protected from fire, but actually require fire. To meet both protection and ecological objectives, we must manage wildland fire and adapt our own behavior to coexist with fire. Managing wildland fire is certainly one of the largest and most complex ecosystem management and restoration efforts ever undertaken, and understanding fire and the consequences of its patterns of occurrence and exclusion is essential. This complexity was underscored during the fall of 2017 when over 200,000 ha (500,000 ac) burned, over 9,000 structures destroyed, and over 40 people killed. These fires were driven by high winds under extremely hot and dry conditions into a mixture of development and wildlands.

This revised edition incorporates a better understanding of both fire variability within vegetation patterns and fire-adaptive or non-fire-adaptive life histories of plant species. Our continuously expanding information base is gained through individual studies and through the compilation of information such as the second edition of *A Manual of California Vegetation* (Sawyer and Keeler-Wolf 2009) and online databases (vegetation.cnps.org).

Fire as an Ecological Process

Much of California has a Mediterranean climate conducive to fire (Pyne et al. 1996) with long dry summers and periods of thunderstorms, low relative humidity, and strong winds. These patterns vary through an extremely wide range of climatic zones and complex topography.

Fire is a physical process, and both its direct and indirect effects are vitally important ecological processes. The heat it produces, the rate at which it spreads, and the effects it has on other ecosystem components are all part of that physical process. Human communities, watersheds, soils, air, plants, and animals are affected in one way or another by fire. Water quality and quantity, soil erosion, smoke, and plant and animal mortality are some of the more obvious effects. Other ecosystem effects are less obvious, but perhaps even more important. During the periods between fires, dead biomass accumulates in Mediterranean ecosystems because weather conditions are favorable for growth, but decomposition is active for only a relatively short moist part of the year. Fire complements decomposition in these systems by periodically removing debris through combustion. Fire has a differential effect on plant species mortality and regeneration, allowing those that are best adapted to fire to be perpetuated.

Pyrogenic vegetation has evolved with recurring fire and includes species that tolerate or even require fire in order to complete their life cycles. There is a continuous feedback loop between fire and vegetation. Fire feeds on vegetation as fuel and cannot reoccur without some minimum burnable, continuous biomass; and the vegetation cannot maintain its occupation of a site without recurring fire. Fire and vegetation are often so interactive that they can scarcely be considered separately from each other. Indeed, for a plant species to persist, the attributes of any fire regime—its seasonality, return interval, size, spatial complexity, intensity, severity, and type—elicit specific responses. We consider fire regimes and vegetation types to be mutually dependent with changes to either, facilitating changes to the other.

In California, animal populations and communities have developed in habitats where fire has been a primary dynamic process. The distribution of most animal species on landscapes has been driven by the patterns of fire and vegetation; controlled by climate, weather, and topography, over space and time. Perpetuation of California's biological diversity certainly requires fire to be present as a vital ecological process.

It is difficult to overstate the ecological importance of fire in California's ecosystems. A central theme to this book is that wildfire is a pervasive, natural, environmental factor throughout much of the state, and ignoring its role in ecosystems will seriously limit our ability to manage wildlands in a sustainable, ecologically appropriate, and responsible manner. To better understand the role of fire in California, it is useful to examine the evolution of its vegetation.

California's Floristic Provinces—Evolution of the Vegetation

The State of California is divided into three floristic provinces: California, Desert, and Great Basin (Baldwin et al. 2012). The California Floristic Province corresponds to the Humid Temperate Domain of Bailey (1996) and Miles and Goudey (1997) and comprises the portion of California west of the mountainous crest. Both the Great Basin Floristic Province and the Desert Floristic Province are in Bailey's Dry Domain. The vegetation in the provinces evolved from two different floras. These have been termed as the Arcto-Tertiary Flora (cool climate northern origins) and the Madro-Tertiary Flora (warm temperate-tropical Sierra Madre origins) (Axelrod 1958). The Acrto-Tertiary Flora dominates in the North Coast, Sierra Nevada, Klamath Mountains, and Southern Cascades bioregions. Species from the Madro-Tertiary Flora are most common in the Central Valley, Central Coast, Northeastern Plateaus, Southeastern Deserts, South Coast, and the east portion of the Sierra Nevada bioregions.

The modern array of bioregions is a product of millions of years of plant evolution, geologic upheavals, climate change, and plant community adaptation and migration. Millar (2012) provides a cogent modern perspective of the evolution of California's flora, which we summarize here. Fifty million years ago (Ma), in the Paleocene Epoch, California was low-lying. Increasing temperatures and humidity triggered significant floristic shifts toward tropical species with affinities to taxa now in rain forests of eastern Asia, southern Mexico, and Amazonia. Western California had diverse subtropical plant communities, while upland regions to the east (the proto—Great Basin) supported temperate-adapted species, including many conifers and associates now present in the modern flora.

About 33.5 Ma, abrupt changes in floristic composition and structure took place. Tropical woody angiosperm species disappeared and temperate-adapted species reappeared, especially broad-leaved deciduous trees and conifers. The new plant communities had affinities to modern communities with high geographic variance due to landscape diversity. Highly diverse assemblages with some taxa present in California today and others now native to warmer, and milder climates with year-round rainfall predominated up until about 23 Ma.

Global temperatures rose between 17 Ma and 15 Ma to the highest levels of the past 23 million years. Fossil floras throughout the West reflected adaptations and range shifts in response to these conditions, with increasing latitudinal gradients from coastal environments to inland mountains. Warm dry-adapted species occurred together, alongside now-exotic tropical-subtropical taxa. In the higher ranges of western Nevada and northeast California, fossil assemblages contained diverse conifers and hardwoods.

Fossil floras less than 7 Ma (Pliocene Epoch), showed a Mediterranean-type semiarid climate establishing at low elevations. Grasslands, chaparral, mixed evergreen sclerophyllous forest, and oak woodlands expanded. Elements of the warm-temperate (Madro-Tertiary) flora became dominant throughout low elevations, while cool-temperate Arcto-Tertiary taxa and vegetation retreated to high elevations, riparian areas, or the coastal strip. The Sierra Nevada, Klamath Mountains, and Transverse Ranges were thrust up to ever higher elevations, creating rain shadows to the east that deepened and expanded over time to become today's hot and cold deserts, dominated by drought-tolerant shrubs, succulent cacti, and short-lived ephemeral herbs.

By the end of the Tertiary (2.6 Ma), many vegetation types of modern California were in place. The beginning of the Quaternary Period brought climate variability and cooling leading to the ice ages. Ice-core records show over 40 cycles of glacial (cold) and interglacial (warm) intervals, each with smaller internal variations. The ice ages of the Pleistocene epoch ended about 10,000 years ago with the arrival of our recent warm epoch, the Holocene.

Quaternary glacial advance and retreat forced vegetation types with their associated fire regimes to migrate upslope and downslope, or southward and northward. California's mountain chains are largely oriented north-to-south, montane taxa driven south were able to migrate back north during warmer interglacial periods, including the current period. The most recent glacial retreat was completed about 10,000 years ago, but cold and warm periods continue to alternate and short-term climate fluctuations continually affect the location of ecotones.

A cooling trend about 4,000 years ago distinguished the late Holocene from the warm middle Holocene (6,000 years to 4,000 years ago). Between 700 years and 1,100 years ago two major droughts occurred, each lasting more than a century. These caused many large lakes and rivers to dry, and salinities to increase in those that remained. Tree-ring records for the past several hundred years (Michaelsen et al. 1987) show continued fluctuation in temperature, precipitation, and interannual variation at the time scale of one-to-several decades.

Twelve thousand years ago, before humans arrived in California, vegetation type diversity and distribution were different than they are today or than they were at any time during the modern human eras. However, it is likely that fire regimes played a role in sustaining the same vegetation type in the past. Before the modern Anthropocene, fire regimes within a stable vegetation type probably did not change much over time.

Bioregions and the California Landscape

The diversity of the California landscape is well known; from the mist-shrouded mountains of the north coast to the searing heat of the Southeastern Deserts, and from the sun-drenched beaches of the South Coast to the high Sierra Nevada, the range of climate, geomorphology, and vegetation mirrors this diversity. Similarly, fire's role in each of these bioregions is equally diverse. To give structure to the discussion of fire complexity within these vast and diverse landscapes, we have divided California into a system of bioregions.

If we add up the areas of vegetation types generally regarded as fire-maintained, about 54% of California's 42 million ha (104 million ac) requires the repeated occurrence of fire to persist (Barbour et al. 2007). Fire-adapted in this case requires

TABLE 1.1
Fire-adapted vegetation types as classified through the Wildlife Habitat Relationships model (WHR website: https://www.dfg.ca.gov/biogeodata/cwhr/) with the percentage of the state mapped as of December 2014

WHR habitat name	Percent of state
Annual Grassland	10.5
Mixed Chaparral	5.4
Sierran Mixed Conifer	4.5
Sagebrush	4.0
Montane Hardwood	3.8
Douglas Fir	3.1
Blue Oak Woodland	2.8
Montane Hardwood-Conifer	2.4
Coastal Scrub	1.9
Montane Chaparral	1.7
Klamath Mixed Conifer	1.3
Eastside Pine	1.3
Coastal Oak Woodland	1.3
Chamise-Redshank Chaparral	1.2
Red Fir	1.2
Ponderosa Pine	1.1
Redwood	1.1
White Fir	1.0
Blue Oak-Foothill Pine	1.0
Pasture	0.8
Low Sage	0.8
Jeffrey Pine	0.5
Perennial Grassland	0.4
Lodgepole Pine	0.4
Wet Meadow	0.3
Fresh Emergent Wetland	0.3
Closed-Cone Pine-Cypress	0.2
Valley Oak Woodland	0.1
Aspen	0.1
Marsh	0.1
Eucalyptus	< 0.1
Palm Oasis	< 0.1

major characteristic species of each unit to have adaptions for asexual or sexual regeneration supported by recurring fire. The totals in Table 1.1 are for approximately 55% of the state. A number of natural arid land vegetation types such as desert scrub, alkali desert scrub, Joshua tree woodland, and pinyon-juniper woodland are not well adapted to fire. A number of non-natural anthropogenic land covers such as urban,

TABLE 1.2
Sections from Miles and Goudey (1997) assigned to bioregions used in this book

North Coast Bioregion
Northern California Coast
Northern California Coast Ranges
Klamath Mountain Bioregion
Klamath Mountains
Southern Cascades Bioregion
Southern Cascades
Sierra Nevada Bioregion
Sierra Nevada
Sierra Nevada Foothills
Central Valley Bioregion
Great Valley
Northern California Interior Coast Ranges
Central Coast Bioregion
Central California Coast
South Coast Bioregion
Southern California Coast
Southern California Mountains and Valleys
Southeastern Deserts Bioregion
Mojave Desert
Sonoran Desert
Colorado Desert
Mono
Southeastern Great Basin

orchards, and other crops also make up a large proportion of the non-fire adapted land cover. Only some types of desert scrub, alpine tundra, subalpine woodland, and a few, less widespread, vegetation types are not fire-dependent. Even some wetlands—such as bulrush (*Schoenoplectus* spp.) marsh, riparian forest, and California fan palm (*Washingtonia filifera*) oases—are fire tolerant and experienced fires set both by indigenous human populations and lightning strikes (Anderson 2005). Knowledge of how fire operates as an ecological process within the State's various bioregions is part of the foundation for wise management and conservation of California's natural heritage.

A hierarchical ecosystem classification for ecosystems in the United States was developed based on climate, as affected by latitude, continental position, elevation, and landform (Bailey et al. 1994, Bailey 1996). Miles and Goudey (1997) divided California into 19 sections within Bailey's system (Table 1.2), and we combine these sections into nine bioregions based on relatively consistent patterns of vegetation and fire regimes (Maps 1.1 and 1.2).

The nine bioregions range from the humid northwest corner of the state to the arid southeast. In the North Coast bioregion, numerous valleys and steep coastal and interior mountains create moisture gradients in response to numerous

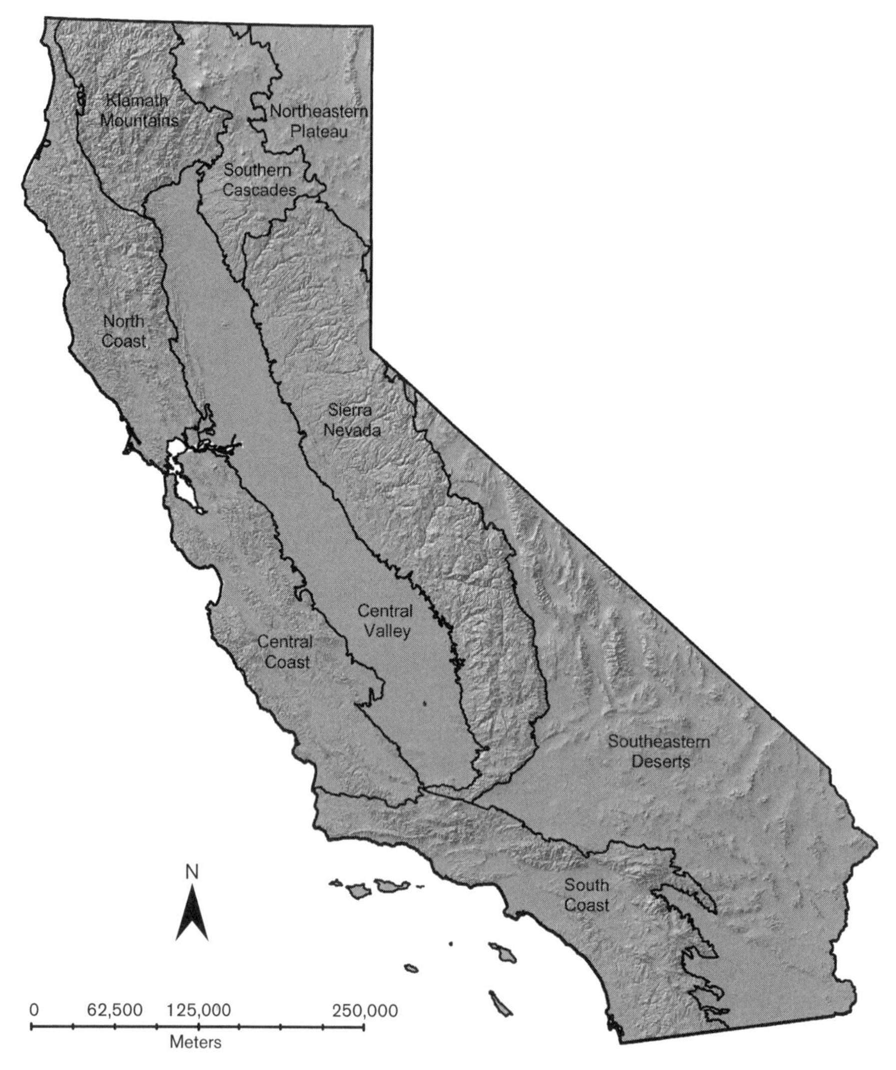

MAP 1.1 Shaded relief map of California bioregions, as defined in this book.

winter storms. The Klamath Mountains bioregion is a complex group of mountain ranges and a diverse flora. Tall volcanoes and extensive lava flows characterize the Southern Cascades and Northeastern Plateau bioregions. Immediately south of the Cascades is the Sierra Nevada bioregion, extending nearly half the length of the state. The Sacramento and San Joaquin rivers flow through broad interior valleys with extensive, nearly flat alluvial floors that constitute the Central Valley bioregion. Coastal valleys and mountains and interior mountains are also typical of the Central Coast bioregion. Southern California with its coastal Valleys and the prominent Transverse and Peninsular Ranges make up the South Coast bioregion. The Mojave, Colorado, and Sonoran deserts along with intervening mountain ranges constitute the Southeastern Desert bioregion.

In Part II of this book, we will examine each bioregion in detail to see how the physical features of the bioregion influence the interactions among fire, vegetation, and other ecosystem components. But first we will take a statewide look at Californians and fire.

Californians and Fire

The arrival of humans in the area that we now know as California had profound influences on fire regimes and the character and distribution of the vegetation. Fire has often been the driving force behind human-induced vegetation change. Native American fire use resulted in many changes to California vegetation, but over several millennia, their activities provided a stabilizing influence for many fire regimes. Relatively speaking, the landscapes of California changed little prior to arrival of Euro-American settlers. Few plant species became extinct, the belts of vegetation in mountains remained nearly the same; the height of mountain peaks, the thickness of sediments beneath meadows and grasslands, the distribution of wetlands, the location of sea level, and the gradients of humidity and aridity from coastal west to interior east shifted little. In Part III of this book we describe how the succession of dominant human cultures that have lived in California view fire, use it, try to suppress it, and deal with its effects on biological and physical resources and on social and political issues.

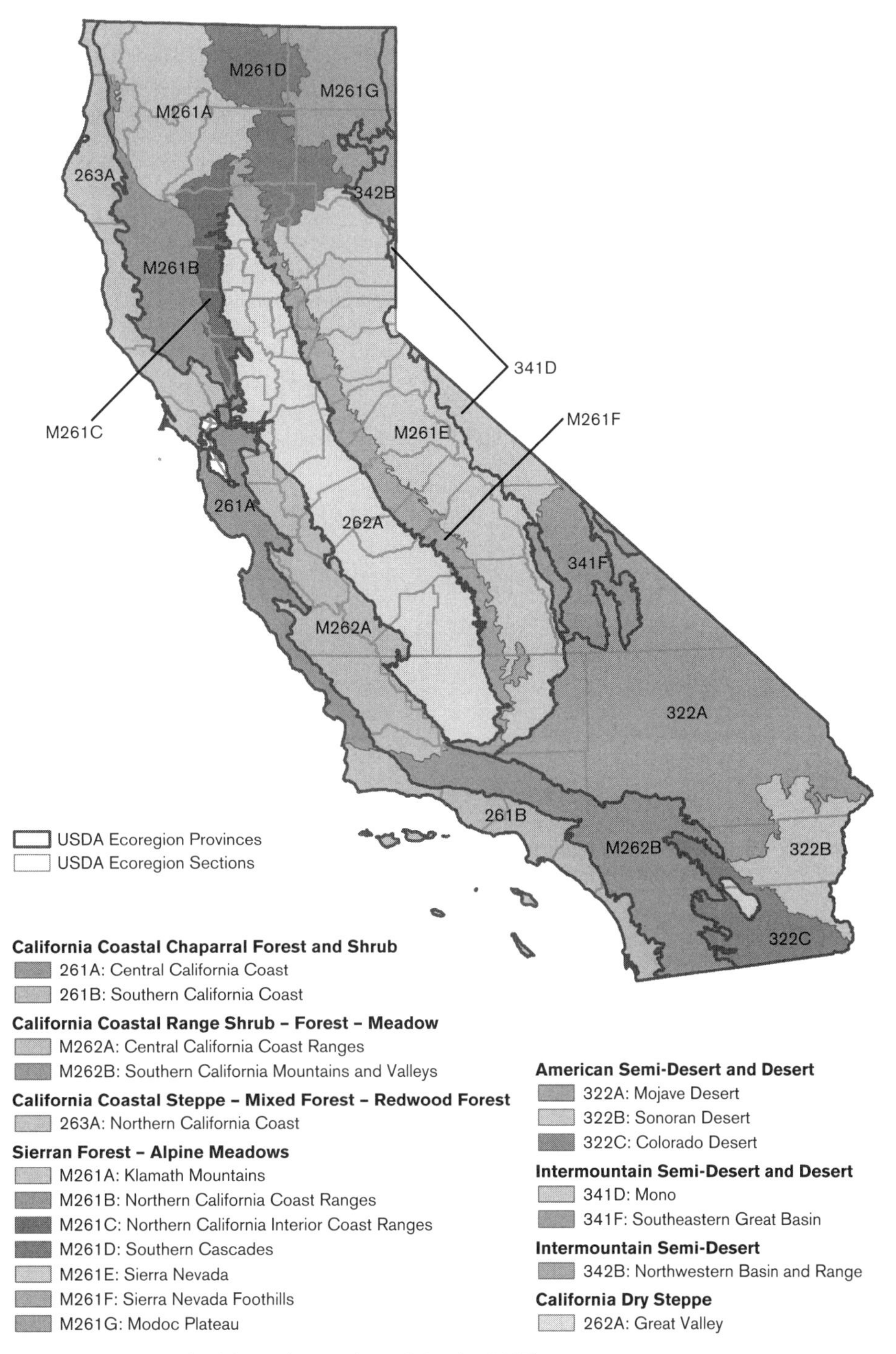

MAP 1.2 Ecoregions of California from Miles and Goudey (1997).

Dramatic change has clearly occurred in the past two centuries due to nonnative and invasive plants, agriculture, fire exclusion, domesticated livestock, and human populations, which are reaching 100 times denser than that of precontact time (Barbour et al. 2007). Much of California's Central Valley and other fertile areas have been converted to agriculture or urbanized, areas of coastal sage scrub have converted to nonnative grasslands, and many mid-elevation conifer forests have been logged and regenerated with species compositions and stand densities that have little historic precedent.

It is not the occurrence of fire in an ecosystem that constitutes an ecological disturbance, but rather our actions that have led to changes to the characteristic fire regimes. Like vegetation, fire regimes have probably never been completely static, but the pace and magnitude of changes to Californian fire regimes accelerated with the arrival of humans from Asia 12,000+ years ago (Rosenthal and Fitzgerald 2012) and again with the arrival of large numbers of settlers of Euro-American origin with the Gold Rush in the mid-1800s. Despite intensive efforts to suppress wildland fires over the past century, fires have continued to burn.

We can identify several historic periods during which fire regimes were altered, and each time they also profoundly altered the ecosystems. Changes to fire regimes occurred while climate, landforms, and species compositions were also changing. The result of these processes defines current California environments and gives us today's California vegetation, much of which is transitional. Although vegetation is always changing, the rates of changes during the past 200 years are far more rapid than in the previous 3,000 or so years.

The Era Prior to Human Settlement

Prior to the arrival of humans in California 12,000 years ago (Rosenthal and Fitzgerald 2012), fire regimes were dependent upon fuel build up to the point at which it would support a spreading flame ignited by lightning. The resulting vegetation distributions were different than they are today or at any time during the human settlement eras. However, it is likely that fire regimes that sustain, and prevent vegetation type conversion for a given vegetation type, also played that role in the past.

Only broad generalizations can be drawn from the evidence that has been documented regarding the history of fire in California prior to human settlement. What is known is general in nature and was reconstructed from paleo-ecological evidence. For example, we do know that in the southern California Coast Ranges, pulses of charcoal deposition in sediments indicate a fire regime dominated by infrequent, large fires (Byrne et al. 1977, Mensing 1998, Mensing et al. 1999). Similarly, charcoal deposition increases during the post-glacial period in the Sierra Nevada indicate that fire was prevalent from 13,700 years to 10,400 years ago (Smith and Anderson 1992).

The Native American Era

Once humans arrived in California and elsewhere in North America, they changed fire patterns with additional ignitions that were focused on their resource needs (Stewart 2002, Anderson 2005). Using fire to manipulate vegetation was universal among Native Americans at the time of European contact. Studies show that vegetation changes occurred following the arrival of Native Americans (Keeter 1995) and again with their demise (Mensing 1998). At the time of European contact, California was already inhabited with large populations of indigenous people living and influencing broad landscapes throughout the state (Cook 1973). As detailed in chapter 19, fire was the most significant, effective, efficient, and widely employed vegetation management tool utilized by California Indian tribes. The pattern of burning was often very specific and focused on particular ecosystem effects (Vale 1998).

Indigenous burning practices defined and maintained the physiognomy of many vegetation types, encouraging particular suites of herbaceous and woody plants. The area that burned during this time period was extensive: Martin and Sapsis (1992) estimate that between 2.2 million ha and 5.2 million ha (5.6 million ac and13 million ac) burned annually from both lightning-caused and human-caused ignitions. Stephens et al. (2007) estimate that 1.8 million ha to 4.8 million ha (4.4 million ac to 11.9 million ac) of California vegetation burned annually. Given the relatively high populations of Native Americans in California, most habitable parts of the state where fuel accumulated and which contained a variety of useful plant or animal species would be expected to have been regularly managed by fire. However, given the diversity of ecosystems, uneven indigenous occupation patterns, and the complex fire characteristics of the vegetation types, the effect of burning by indigenous people was neither uniform nor equally applied across landscapes.

The effects of fires set by Native Americans formed a continuum encompassing a range of human modifications from very little or no influence to fully human-created ecosystems. Vale (2002) concludes that the pre-Euro-American landscape in the American West was a mosaic of areas that were altered by native peoples and areas that were primarily affected by natural processes. However, as Pyne (2003) reminds us, in such ecosystems as Mediterranean California, where fire occurs with or without human influence, it is "infuriatingly difficult" to tease out the originating causes.

The Era of Euro-American and Asian-American Settlement

The early settlement era brought several major changes in the pattern of wildland fire, largely due to the removal of indigenous people and their approach to the land. Newcomers had a very different land-use philosophy and a wider array of tools to modify the landscape. In those ecosystems where burning by Native Americans was a regular practice—California's warm temperate forests and woodlands, montane meadows, coastal prairies—the demographic change led to changes in species composition, invasion of nonnative plants, and even type conversion of the vegetation. Mining and livestock, initially localized, also had widespread influence. An intensive pulse of sheep grazing during the late 1800s greatly changed fire in much of the western United States by breaking the fuel continuity in the herbaceous layer in which fires spread. Fuel continuity in the herbaceous layer and ignition patterns control how and when fires occur because it is the fuel layer in which fire most commonly spreads. In open forest and woodland ecosystems with herbaceous understory, the reduction and fragmentation of the herbaceous layer greatly reduce the ability of a fire to spread. The result is a reduction in the number of days with conditions under which an area will burn, and therefore a drastic reduction in the frequency of fire. This indirect fire exclusion also allowed vegetation to accumulate biomass, developing different fuel structures, changing species mixes, and shifting the geographic distributions of vegetation types.

Introduction of nonnative, invasive plants during this time also had important impacts on fire patterns in many of California's ecosystems (Balch et al. 2012, Brooks et al. 2004, Hunter et al. 2006). In the grasslands and oak woodlands of the Central Valley, replacement of native plants with nonnative species has allowed earlier curing of fuel in the spring and heavier fuel loads in the summer and fall extending the fire season at both ends. In general, exotics have had a greater impact in mesic, lower elevation habitats than in harsher habitats at high elevations, on less productive geologic substrates, or in arid deserts. Exotics can aggressively colonize following high-severity fires.

The impact of humans on fire frequency at this time can be deduced in several ways. As a general rule we have detailed records of recent fire patterns, but as we move back in time, this record becomes more fragmented and less specific. Current records of fire occurrence and vegetation severity are available through land management and firefighting agency records and can be supplemented by remote sensing studies to very fine scales of detail. In contrast, written records often go back to the early 1900s, but they are much generalized and describe only unusually large and destructive fires in the early part of the 1900s. Historic photographs, land survey records, and newspaper accounts from the 1800s can supplement fire records (Egan and Howell 2001). Tree-ring studies extend these records back several centuries (and in some cases for millennia in ecosystems containing long-lived species), but they cannot always reconstruct the details of fire intensity, severity, complexity, or area burned. Studies of charcoal deposits, phytoliths, and pollen in lake sediments allow fire history studies to be extended 10,000 years or more into the past, but the interpretations are more generalized and lack the resolution needed to provide the detailed information derived from other methods. Our knowledge of fire history and fire-related vegetation change is built on information developed from a combination of these methods.

The Fire Suppression Era

The industrialization of America in the early to mid-1900s brought demands for the extraction of wildland resources. Protection of forests and rangelands from "the scourge of fire" was a central part of this era. A fire protection philosophy that was developed in Europe was applied to most plant communities in North America (Wright and Bailey 1982). The primary focus of management in the early National Forest Reserves (1905 to 1910) was protecting the timber supplies and watersheds. Fire was seen as a potential threat to both, and a great deal of effort was committed to the removal of fire in America's forests.

Stephen Pyne (2001) states, "The great fires of 1910 shaped the American fire landscape more than any other fire in any other year throughout the twentieth century." Seventy-eight firefighters were killed and 1.2 million ha (3 million ac) of national forest land burned. These fires instigated the creation of a national system of wildland fire protection that still dominates fire management.

During World War II, the firefighting effort was intensified in the interest of national defense and paired with a public education campaign that included Smokey Bear. The public was now well shielded from the history of human-fire relationships and fighting fires had become "The moral equivalent of war" (Pyne 1997). After World War II, firefighting efforts were intensified. Science and technology allowed important strides to be made in understanding wildfire spread and its control. The study of fire at this time concentrated on fire physics, fire behavior and the relationships between meteorology and fire. Subsequently, the focus broadened into fire effects on vegetation and ecosystems and the development of the field we know today as fire ecology.

The Ecosystem Management Era

Although fire is a natural, recurring process in most California vegetation, the effects of fire are not always what they once were. A number of factors are driving the continuing change of fire patterns in wildlands, including: (1) the increasing density and expansion of humans on the landscape, (2) the technology used for fire suppression continues to become more sophisticated, effective, and efficient, and (3) the intentional, prescribed use of fire as a tool for fuel management, natural resource management, and the protection of communities at the wildland-urban interface.

Starting in the 1960s and continuing today, the emphasis of fire study has focused on natural resource values and the influence of fire as an ecological process. There has been widespread acceptance of the notion that fire is an important part of many ecosystems and that changes in the patterns of occurrence of fire have had many large-scale vegetation effects. Fire ecology and fire management have become central issues in land management and there is greater recognition of the radiating impact of decisions about how to manage fire. Today, the practice of managing wildland fuel to modify future fire behavior has become an important land management activity. The wildland-urban interface has become the contentious focal point for application of fuel management to difficult situations.

In this book, we provide a great deal of information on the ecological role of fire and how our culture has evolved to recognize and appreciate fire as an ecological process. So are we mounting a massive effort to restore fire to all of our wildlands? No. Why? Fire has both positive and negative effects. Our culture also necessarily values other wildland attributes including clean air and water, species and habitats, and living in desirable locations without fire to threaten our health and quality of life.

Fire affected past vegetation patterns, and fire will influence future vegetation patterns. With the current trends in population growth and development and changing climate, fire regimes are likely to remain altered from pre-Euro-American settlement conditions. Wildland fire patterns will continue to change with wildfires pushed to burn mainly during increasingly extreme weather and fuel conditions, while the restoration of fire as an ecological process will occur on a relatively small proportion of California's landscape. Large percentages of the state's natural vegetation can shift over very short periods. Changing climate will work in combination with fire to change fire occurrence and severity. Fire patterns that influence the distributions of forests, woodlands, shrublands, and herbaceous vegetation will change. The impacts of nonnative invasive species will continue to change fuel conditions, and thus fire and vegetation patterns. Simultaneously, understanding of the importance of fire in California vegetation will grow, and the decisions on how to manage fire-adapted vegetation will remain difficult.

This book is an effort to summarize fire's historic and current role in California's ecosystems, and to provide an understanding of fire as an ecological process to facilitate wise management into the future. This story of fire and California's vegetation is an epic adventure played out over millennia in a spectacular setting. This book sets the scenes, introduces the characters and situations and provides you, the future of fire ecology, with concepts and some of the tools to write the next act.

References

Anderson, M.K. 2005. Tending the Wild: Native American Knowledge and the Management of California's Natural Resources. University of California Press, Berkeley, California, USA.

Axelrod, D.I. 1958. Evolution of Madro-Tertiary geoflora. Botanical Review 24: 433–509.

Bailey, R.G. 1996. Ecosystem Geography. Springer, New York, New York, USA.
Bailey, R.G., P.E. Avers, T. King, and W.H. McNab, editors. 1994. Ecoregions and Subregions of the United States (map). U.S. Geological Survey, Washington, D.C., USA.
Baldwin, B.G., D.H. Goldman, D.J. Keil, R. Patterson, T.J. Rosatti, and D.H. Wilken, editors. 2012. The Jepson Manual: Vascular Plants of California, 2nd ed. University of California Press, Berkeley, California, USA.
Balch, J.K., B.A. Bradley, C.M. D'Antonio, and J. Gómez-Dans. 2012. Introduced annual grass increases regional fire activity across the arid western USA (1980–2009). Global Change Biology 19: 173–183.
Barbour, M.G., T. Keeler-Wolf, and A.A. Schoenherr. 2007. Terrestrial Vegetation of California, 3rd ed. University of California Press, Berkeley, California, USA.
Barbour, M.G., B. Pavlik, F. Drysdale, and S. Lindstrom. 1993. California's Changing Landscapes: Diversity and Conservation of California Vegetation. California Native Plant Society, Sacramento, California, USA.
Brooks, M.L., C.M. D'Antonio, D.M. Richardson, J.B. Grace, and J.E. Keeley. 2004. Effects of invasive alien plants on fire regimes. BioScience 54: 677–688.
Byrne, R., J. Michaelsen, and A. Soutar. 1977. Fossil charcoal as a measure of wildfire frequency in Southern California: a preliminary analysis. Pages 361–367 in: H.A. Mooney and C.E. Conrad, technical coordinators. Proceedings of the Symposium on Environmental Consequences of Fire and Fuel Management in Mediterranean Ecosystems. USDA Forest Service General Technical Report WO-3, Washington, D.C., USA.
Cook, S.F. 1973. The aboriginal population of upper California. Pages 66–72 in: R.F. Heizer and M.A. Whipple, editors. The California Indians. University of California Press, Berkeley, California, USA.
Egan, D., and E.A. Howell. 2001. The Historical Ecology Handbook–A Restorationist's Guide to Reference Ecosystems. Island Press, Washington, D.C., USA.
Hunter, M.E., P.N. Omi, E.J. Martinson, and G.W. Chong. 2006. Establishment of non-native plant species after wildfires: effects of fuel treatments, abiotic and biotic factors, and post-fire grass seeding treatment. International Journal of Wildland Fire 15: 271–281.
Keeter, T.S. 1995. Environmental History and Cultural Ecology of the North Fork of the Eel River Basin, California. USDA Forest Service Technical Publication R5-EM-TP-002, Pacific Southwest Region, Vallejo, California, USA.
Martin, R.E., and D.B. Sapsis. 1992. Fires as agents of biodiversity: Pyrodiversity promotes biodiversity. Pages 28–31 in: R.R. Harris, D.E. Erman, and H.M. Kerner, editors. Proceedings of Symposium on Biodiversity of Northwestern California. Wildland Resource Center Report 29, University of California, Berkeley, California, USA.
Mensing, S.A. 1998. 560 years of vegetation change in the region of Santa Barbara, California. Madroño 45: 1–11.
Mensing, S.A., J. Michaelsen, and R. Byrne. 1999. A 560-year record of Santa Ana fires reconstructed from charcoal deposited in the Santa Barbara Basin, California. Quaternary Research 51: 295–305.
Michaelsen, J., L., Haston, and F.W. Davis. 1987. 400 years of California precipitation reconstructed from tree rings. Water Research Bulletin 23: 809–818.
Miles, S.R., and C.B. Goudey. 1997. Ecological Subregions of California: Section and Subsection Descriptions. USDA Forest Service Technical Publication R5-EM-TP-005, Pacific Southwest Region, Vallejo, California, USA.
Millar, C.I. 2012. Geologic, climatic, and vegetation history of California. Pages 49–67 in: B.G. Baldwin, D.H. Goldman, D.J. Keil, R. Patterson, T.J. Rosatti, and D.D. Wilken, editors. The Jepson Manual: Vascular Plants of California, 2nd ed. University of California Press, Berkeley, California, USA.
Pyne, S.J. 1997. America's Fires–Management on Wildlands and Forests. Forest History Society, Durham, North Carolina, USA.
———. 2001. Year of the Fires–The Story of the Great Fires of 1910. Viking Penguin, New York, New York, USA.
———. 2003. Smokechasing. University of Arizona Press, Tucson, Arizona, USA.
Pyne, S.J., P.L. Andrews, and R.D. Laven. 1996. Introduction to Wildland Fire. John Wiley and Sons, New York, New York, USA.
Rosenthal, J.S., and R.T. Fitzgerald. 2012. The Pleistocene Holocene transition in western California. Pages 67–104 in: C.B. Bousman and J. Vierra, editors. From the Pleistocene to the Holocene: Human Organization and Cultural Transformations in Prehistoric North America. Texas A&M University Press, College Station, Texas, USA.
Sawyer, J.O., and T. Keeler-Wolf. 2009. A Manual of California Vegetation, 2nd ed. California Native Plant Society, Sacramento, California, USA.
Smith, S.J., and R.S. Anderson. 1992. Late Wisconsin paleoecologic record from Swamp Lake, Yosemite National Park, California. Quaternary Research 38: 91–102.
Stephens, S.L., R.E. Martin, and N.E. Clinton. 2007. Prehistoric fire area and emissions from California's forests, woodlands, shrublands and grasslands. Forest Ecology and Management 251: 205–221.
Stewart, O.C. 2002. Forgotten Fires–Native Americans and the Transient Wilderness. University of Oklahoma Press, Norman, Oklahoma, USA.
Vale, T.R. 1998. The myth of the humanized landscape: an example from Yosemite National Park. Natural Areas Journal 18: 231–236.
———. editor. 2002. Fire, Native Peoples, and the Natural Landscape. Island Press, Washington, D.C., USA.
Wright, H.R., and A.W. Bailey. 1982. Fire Ecology: United States and Southern Canada. John Wiley and Sons, New York, New York, USA.

PART ONE

INTRODUCTION TO FIRE ECOLOGY

Part One is an introduction to fire ecology. In chapter 2 we describe how climate influence fires and fire regimes. Chapter 3 is a detailed examination of the weather factors that affect fire behavior. We then describe factors that affect fire as a physical process in chapter 4 before discussing fire as an ecological process and fully developing the concept of fire regimes in Chapter 5. Chapter 6 examines the methods used to characterize fire regimes. Next we discuss the interactions of fire with the soil, water, and air components of the physical environment in Chapter 7. Finally, in chapters 8 and 9, we cover the effects of fires on biological communities, first looking at interactions with plants and then with animals. This foundation in fire ecology will provide the basis for understanding fire's varying role in the bioregions of California and the issues confronting fire in today's society.

CHAPTER TWO

California Fire Climate

RICHARD A. MINNICH

> I've lived in good climate, and it bores the hell out of me. I like weather rather than climate.
>
> STEINBECK (1962)

Introduction

To understand fire as an earth surface process in California ecosystems, it is necessary to evaluate how climate (average and predictable weather properties over long time scales) contributes to vegetation flammability, and how short-term weather influences the propagation of flame lines (fire weather). The flammability of vegetation can be envisioned as a "tug of war" between the heat source of the organic energy of plants (carbohydrate), and the heat sink of plant water vital for transport of nutrients and leaf transpiration. Fires occur when fuel energy exceeds the heat capacity of water, i.e., the carbohydrate-to-water energy ratio in vegetation is positive (Rothermel 1972). Climate and weather both affect plant growth and fuel buildup, as well as water/fuel moisture in vegetation and soils, but they reflect different time scales and processes. Climatic parameters include mean precipitation and temperature that affect annual cycles of soil wetting, water loss by evapotranspiration and runoff, plant growth, phenology, desiccation of vegetation, and natural ignitions from thunderstorms. The predictable properties of California's Mediterranean climate are winter precipitation, summer drought, and mild temperatures in most areas of the State. Climate variability at interannual to decadal time scales is believed to correlate with periodic perturbations in the fire regime. Weather, especially relative humidity and wind velocity, affects fuel moisture, heat release in flame lines (fire intensity), and available fuel consumed in pyrolysis. Relative humidity and wind speed influence plant transpiration and water diffusion in dead fuel. Because organic fuels are poor heat conductors and radiative heat transfer from burning vegetation to adjoining unburned vegetation is inversely related to the square of the distance, wind speed (advection) is the dominant mechanism of heat transfer in the propagation of flame lines. Decreasing relative humidity and increasing wind velocity tend to result in higher fire spread rates and intensities, as well as the consumption of ever-coarser fuels.

This chapter reviews California's Mediterranean climate from the standpoint of atmospheric circulation. This is followed by a discussion of climate variability and its possible role on fire regimes. For those interested in the fundamentals of global circulation, see the first edition of this volume. The principles of fire weather are reviewed in chapter 3 and climate change in chapter 26.

Climate

The large variety of climates in California leads to a large diversity in vegetation and associated fire regimes, as well as fire weather conditions. The Mediterranean climate results from seasonal changes in global circulation, including California's location relative to the jet stream and the presence of cold, upwelling ocean waters offshore. During winter, large latitudinal temperature gradients support a strong circumpolar vortex with the mean position of jet stream westerlies extending equatorward to the latitude of northern California, Oregon, and Washington (Jet stream) (Fig. 2.1).

Troughs in the jet stream bring low pressure systems and precipitation. At the surface, low pressure in the Gulf of Alaska advects strong onshore flows of moist westerlies to northern California and the Pacific Northwest and mild temperatures through the state. Annual rainfall is greatest on the windward slope of the mountains due to orographic lift of air masses, and lowest in the deserts due to the rain shadow effect of the mountains. In summer reduced latitudinal temperature gradients weaken the circumpolar vortex, and the mean position of the westerlies and precipitation shifts northward to British Columbia. Surface low pressure in the Gulf of Alaska is replaced by a high pressure that covers most of the North Pacific Ocean. California is dominated by strong onshore flows from the Pacific high to thermal low pressure over the hot desert interior. Air masses are warmer than the ocean, except for a thin layer of cool, moist air, immediately over the ocean surface, called the *coastal marine layer.* The marine layer is also the northeastern boundary of the trade wind layer of the Hadley cell in the North Pacific Ocean. The presence of warm light air overlying the marine layer results in stable air masses that leads to protracted drought.

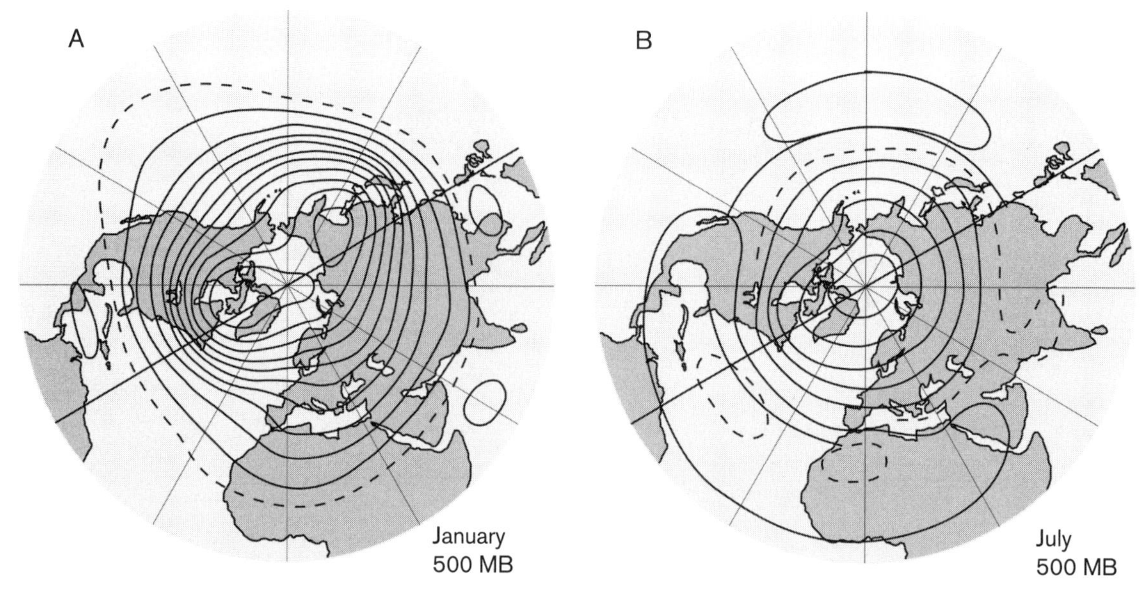

FIGURE 2.1 Mean 500-mbar contours for: (A) January and (B) July. Low pressure is centered near the pole, with westerly winds moving around the hemisphere in a direction parallel to contours (geostrophic wind). The entire counterclockwise flowing system is the circumpolar vortex. The westerlies are strongest in January when the latitudinal temperature gradient is strongest and jet stream (where contours are closest together) arrives at the Pacific Coast in Oregon and Washington. Westerlies are weakest in July when the latitudinal temperature gradient is at a minimum and the mean position of the jet stream retreats to western Canada (redrawn from Palmén and Newton 1969).

Temperature

The position of the jet stream is coupled with global atmospheric latitudinal temperature gradients. Mean winter temperatures increase southward, and in the mountains also reflect atmospheric lapse rates (Map 2.1). Sea level temperatures increase from 9°C (48°F) in coastal northern California to 14°C (57°F) in coastal southern California and 12°C (54°F) in the southeastern deserts. Mean temperatures in the mountains reach freezing at about 1,500 m (4,921 ft) in northern California and 2,200 m (7,218 ft) in southern California. Many low-lying basins contain persistent surface-based inversions resulting from the combination of radiational cooling and low insolation. In the Central Valley, where surface-based inversions are maintained by reflective ground fogs, mean January temperatures average 8°C (46°F). From the high northeastern plateaus to Lake Tahoe, inversions result in mean temperatures as low as −3°C (27°F) on valley floors.

In summer the coastal valleys and mountain slopes are influenced by the coastal marine layer, a steady-state feature that forms from the cooling and moistening of the tropospheric boundary layer overlying the cold California current. The marine layer is associated with extensive coastal low clouds (stratus) over the ocean and is capped by a strong thermal inversion that divides it from warm, subsiding air masses aloft. Northwesterly gradient winds combined with sea breezes and anabatic valley winds and mountain upward slope winds (winds created by local surface heating) transport marine air inland to the coastal ranges usually within 100 km (62 mi) of the coast before the marine layer dissipates from diabatic heating and mixing with warm air aloft (Glendening et al. 1986). Mean July temperatures along the coast reflect gradients in sea surface temperatures, ranging from 14–16°C (57–61°F) in strong upwelling zones north of Point Conception to 18–22°C (64–72°F) in the southern California bight (Map 2.2), the embayment between Point Conception and the Mexican border. Air temperatures increase to 24–28°C (75–82°F) in the inland valleys of southern California and along the Central Valley. The deserts beyond the reach of marine air penetration average 26–35°C (79–95°F) to as warm as 38°C (100°F) in Death Valley.

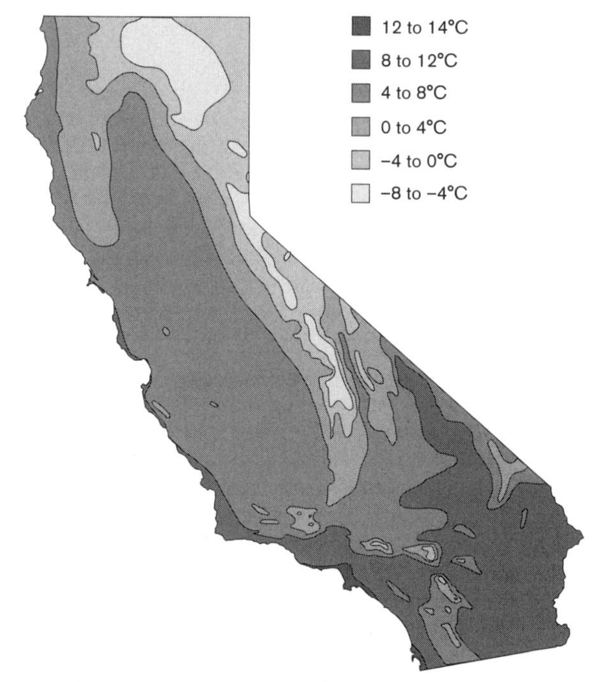

MAP 2.1 Mean temperature (°C) for January.

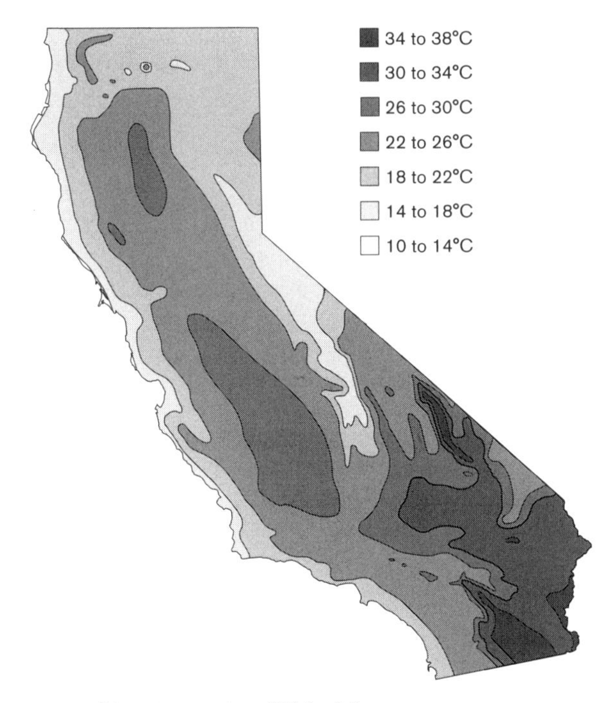

MAP 2.2 Mean temperature (°C) for July.

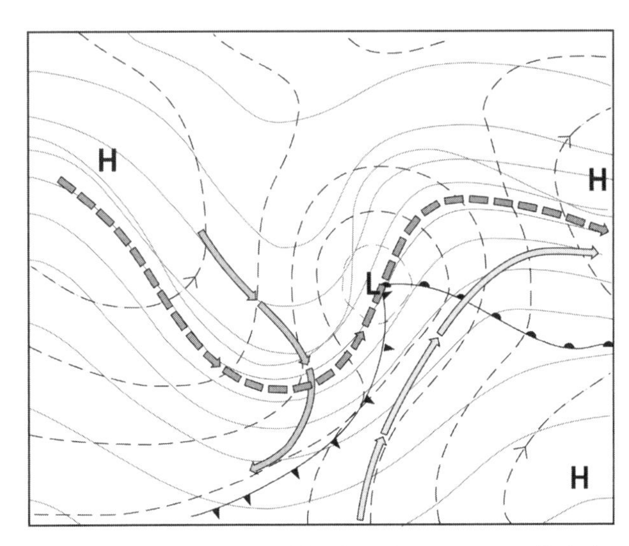

FIGURE 2.2 Schematic of a model cyclone. Pressure pattern (dashed lines) of surface cyclone in middle and anticyclones on right and left shown against wave in the jet stream (solid lines). Frontal boundaries are shown as solid ticked boundaries. Broad arrows show the trajectories of air motion through the system, with sinking air west of the front and rising air east of the front (redrawn from Palmén and Newton 1965).

Temperatures in mountains reflect ambient lapse rates and are isoclinal with latitude, decreasing to 20–24°C (68–75°F) at 1,500 m (4,921 ft) and 14–18°C (57–64°F) at 2,500 m (8,202 ft).

Precipitation

Winter precipitation results from frontal cyclones and associated troughs of the jet stream that move over California from the North Pacific Ocean. Because the mean position of the jet stream lies in northern California (Fig. 2.1), storms decrease in frequency southward through the state. Most precipitation falls in the prefrontal zone (usually from the cold front to 100–300 km [62–186 mi] to the east [Fig. 2.2]). Upward air motions lift cool marine layer, the cold fronts having properties of occluded fronts with long periods of steady rain in stable air. Winds aloft are predominantly southwesterly because frontal zones precede trough axes aloft; low-level winds veer from the south or southeast. Postfrontal precipitation consists of convective showers concentrated over high terrain with upper winds backing from westerly at the surface to southwesterly aloft. Clouds and precipitation decrease rapidly with the onset of subsidence and advection of dry air following the passage of the trough, with winds aloft shifting to westerly to northwesterly.

The interaction between prefrontal circulation and terrain results in strong gradients in mean annual precipitation throughout California. Because storm air masses are stable, the variation in local precipitation is normally a consequence of the dynamic (mechanical) lift of cyclones and attendant physical lift of storm air masses over mountain barriers (orographic effect) rather than from thermal convection. Intense precipitation gradients at local scales in California's mountains are a consequence of stable storm air masses in which the highest precipitation zones reflect windward slope gradients, not altitude. Relatively low, precipitating cloud layers (cloud tops to 3–4 km [2–2.5 mi] msl), fast-moving wind fields and associated high water vapor flux of the jet stream couple mechanical lift in laminar flow with local physiography along windward slopes (high-standing precipitation rates) and descent (low precipitation rates) on leeward slopes. Increasingly weak dynamic (mechanical) lift of storm air masses southward through California paradoxically magnifies mountain orographic precipitation zones because less precipitation is "spent" upwind in the coastal and interior plains.

Orographic lift is most intense on the south- to southwest-facing escarpments that lie at right angles to prefrontal storm winds (Fig. 2.2). Hence, orographic precipitation processes extend southwest to northeast, not west to east. Amounts progressively decrease downwind to inland ranges—regardless of altitude—due to cumulative depletion of storm air mass moisture and descending airflow in rain shadows. The average annual precipitation along the northern California coast varies from 50 cm (20 in) at San Francisco to 100 cm (39 in) north of Ft. Bragg, with locally higher amounts where mountains skirt the coastline (Map 2.3). The highest amounts occur in the coastal mountains north of Eureka and near Cape Mendocino where the annual average is >250 cm (>98 in). Totals in the North Coast Ranges decrease downwind to 150–200 cm (59–79 in) in the Salmon and Siskiyou Ranges, and further decrease southward with the declining general altitude of the mountains to 100 cm (39 in) near Santa Rosa. Rain shadows from the North Coast Ranges produce average annual precipitation of 35–50 cm (14–20 in) in the Sacramento Valley. Orographic lift along the uniformly gentle western slope of the northern Sierra Nevada produces an average annual precipitation of 60 cm (24 in) along the lower foothills to 150–200 cm (59–79 in) at the crest of the range north of Lake Tahoe. East of the Sierra Nevada crest, rain shadows in descending air reduce the average annual precipitation to only 20–60 cm

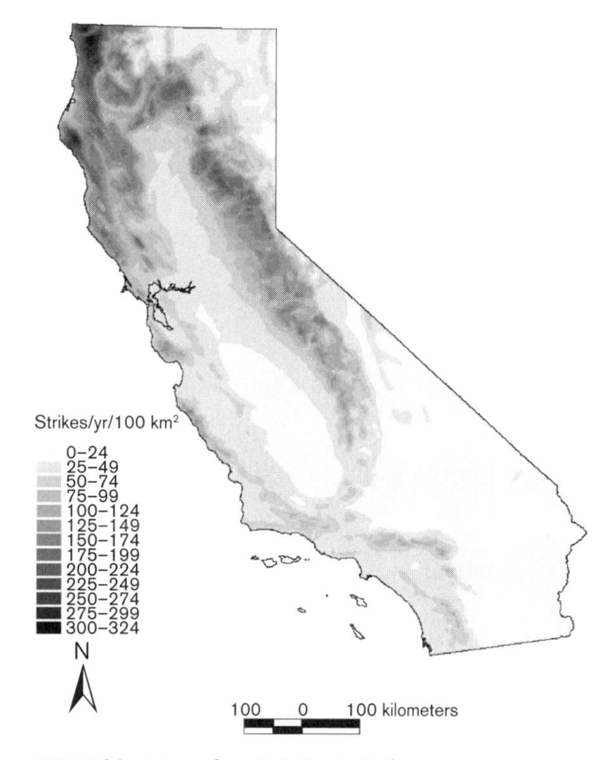

MAP 2.3 Mean annual precipitation in inches.

SOURCE: Western Regional Climate Center, Desert Research Institute, Las Vegas, Nevada.

(8–24 in) in the Modoc Plateau, and 60–80 cm (24–31 in) in the Lake Tahoe Basin.

In the Central Coast Ranges, the average annual precipitation is 100–150 cm (39–59 in) on the steep coastal escarpments of the Santa Cruz and Santa Lucia Mountains but amounts in the rain-shadowed Diablo Ranges seldom exceed 50 cm (20 in). Intervening basins including Salinas Valley receive 30–40 cm (12–16 in). Rain shadows extending from the South Coast Ranges into the San Joaquin Valley produce an average annual precipitation of 30 cm (12 in) near the Sacramento delta, lowering to 15 cm (6 in) near Bakersfield and the Carrizo Plain. Amounts then increase with orographic lift to 35–50 cm (14–20 in) in the Sierra Nevada foothills. The topographic complexity of the southern Sierra Nevada coastal front results in large variability in average annual precipitation. Steep southwestern exposures have averages of 100–150 cm (39–59 in) from Yosemite to Kaiser Ridge, the Great Western Divide, and the Greenhorn Range. Leeward slopes receive 50–100 cm (20–39 in), including the upper Tuolumne River, Mono Creek Basin, the upper Kings River, and the Kern River plateau northward to Mt. Whitney. The average annual precipitation seldom exceeds 50 cm (20 in) in the southernmost Sierra Nevada and Tehachapi Mountains due to their low altitude and leeward position to the western Transverse and South Coast Ranges.

In southern California, precipitation is greatest in the Transverse Ranges (average annual precipitation, 80–110 cm [31–43 in]) due to intense orographic lift along the steep coastal escarpment. Amounts decrease to 60–80 cm (24–31 in) in the "downwind" San Rafael Mountains and Pine Mountain Ridge. Farther inland, relatively high Mt. Pinos, as well as the San Emigdio, Tehachapi and Liebre Mountains, and drainages north and east of Big Bear Basin, receive 35–60 cm (14–24 in). The average annual precipitation is only 40–60 cm (16–24 in) in the Peninsular Ranges because the coastal slopes lie parallel to storm winds. Amounts reach 80–100 cm (31–39 in) on local southern escarpments of the Santa Ana Mountains, Palomar Mountain, and Cuyamaca Peak. Despite their high altitude, the San Jacinto Mountains receive 40–70 cm (16–28 in) and the Santa Rosa Mountains >50 cm (>20 in) due to their leeward position relative to the Santa Ana and Palomar Mountains. The average annual precipitation in the southern California coastal plain varies from 25–35 cm (10–14 in) at the shoreline to 40–50 cm (16–20 in) at the base of the mountains.

The average annual precipitation in the southeastern deserts is mostly 10–15 cm (4–6 in), with totals as low as 5 cm (2 in) in Death Valley and the Coachella-Imperial Valley. Amounts locally reach 25 cm (10 in) in the Panamint and Inyo Mountains, and several ranges in the northeast Mojave Desert.

Snowfall

In most of California, the replenishment of soil water reflects the timing of rainstorms during the winter, which mostly end by March or April. In high mountain watersheds, an accumulating winter snowpack melt serves to increase available soil moisture in summer because melt is delayed until high solar zenith angles in April and May, thereby postponing soil drying and plant desiccation and drying of dead fuels compared to rain-dominated lands below. In effect, the snowpack represents a second layer of water storage on top of moisture storage in the soil and regolith layers (Graham et al. 2010).

The ratio of average frozen precipitation (mostly snow hypothetically melted to a liquid depth) ratioed to the average annual precipitation shows a linear relationship with altitude. In southern California, little snowfall occurs below 1,000 m (3,281 ft). The ratios increase to 25% at 1,750 m (5,741 ft), 75% at 2,750 m (9,022 ft) and 100% at 3,000 m (9,843 ft) (Minnich 1986). Average snow lines are about 200–400 m (656–1,312 ft) lower in the Sierra Nevada (Barbour et al. 1991). With an average annual precipitation of 100 cm (39 in), the Snow Water Equivalent (SWE) in southern California reaches 50 cm (20 in) at 2,300 m (7,546 ft) and 75 cm (30 in) at 2,700 m (8,858 ft). With an average annual precipitation of 150 cm (59 in) near Yosemite, the SWE reaches 50 cm (20 in) at 1,900 m (6,234 ft) and 75 cm (30 in) at 2,200 m (7,218 ft). At Mt. Lassen (average annual precipitation, 200 cm [79 in]), 50 cm (20 in) SWE amounts are reached by 1,400 m (4,593) and 100 cm (39 in) by 2,000 m (6,562 ft). Interannual snow levels in California tend to increase with increasing total annual precipitation due largely to advection of moist subtropical air masses (atmospheric rivers). In southern California, the 50% annual snow line increases from 2,000 m (6,562 ft) with 70% normal precipitation to 2,400 m (7,574 ft) with 140% of normal. The snow line during the floods of January 1969 ranged from 2,700 to 3,000 m (8,858–9,843 ft). The snow line during the New Year's 1997 flood at Yosemite was >2,500 m (>8,202 ft).

The North American Monsoon

From July to early September, the North American monsoon brings occasional afternoon thunderstorms and lightning to

California. For 10–20 days per summer, the western margin of the North American monsoon, a deep layer of moist, unstable tropical air, extends northward from Mexico into the eastern mountains and deserts of California (Tubbs 1972). The monsoon arrives from the tropical Pacific and Gulf of California around an anticyclone in the mid-troposphere centered over the southwestern US desert. The anticyclone is sustained by intense convective heating off the high elevation land surfaces of the Great Basin, Colorado Plateau, and Mexican plateau (Hales 1974, Adams and Comrie 1997, Stensrud et al. 1997).

The occurrence of thunderstorms in California is related to the position of the anticyclone (Minnich et al. 1993). When the jet stream passes over the Pacific Northwest, the anticyclone is typically centered over northern Mexico (Figs. 2.3A and 2.4A). Dry southwesterly flow over California results in mostly clear skies, even over the highest mountains. Monsoon moisture is steered northeastward into northwestern Mexico and Arizona. Deep troughs destabilize Pacific air masses and produce thunderstorms in the mountains of far northern California. When the jet stream is displaced northward into British Columbia, the center of the anticyclone shifts northward over the Colorado Plateau or Great Basin. South to southeasterly winds aloft transport tropical moisture (Figs. 2.3B and 2.4B) into southwestern California and the Sierra Nevada. Moisture arrives at upper levels (3–4 km [1.9–2.5 mi]) as convective debris from afternoon and night thunderstorms and mesoscale convective systems over the Sierra Madre Occidental of northwestern Mexico and Gulf of California. Below 2 km (1.2 mi), low-level moist air masses derived from convective outflows of thunderstorms over Mexico surge into the Salton Sea trough, Colorado River Valley, and occasionally as far north as Owens Valley (Hales 1974, Adams and Comrie 1997, Stensrud et al. 1997). Moisture surges also result from the lift of the trade wind layer overlying the Gulf of California by Mexican Pacific tropical cyclones passing to the south.

Convection tends to be concentrated over high terrain, especially in mountains exposed to low-level Gulf of California air masses, primarily east of a line from the Peninsular Ranges and eastern San Bernardino Mountains to the eastern escarpment of the Sierra Nevada. Average annual summer precipitation (July–September) ranges from 5 to 10 cm (2–6 in) in the most favorable areas of convection. Amounts decrease toward the west because of stable air produced by Pacific marine layer intrusions. Infrequent surges of anomalously moist air encourage outbreaks of thunderstorms throughout California, even along the Pacific Coast. Large potential evapotranspiration (PET) rates compared to summer precipitation limits soil wetting to shallow soils (Franco-Vizcaíno et al. 2002), and monsoon rains seldom interrupt the fire season. Mexican Pacific tropical cyclones enter the State about once a decade, mostly in southern California. These storms are steered northward by southerly upper air winds of the first seasonal troughs or cutoff lows west of California, usually in September, and may produce copious precipitation of 5–20 cm (2–8 in) per day (Smith 1986).

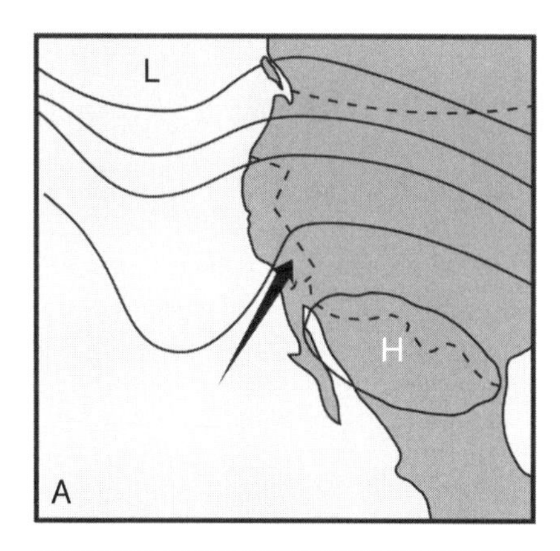

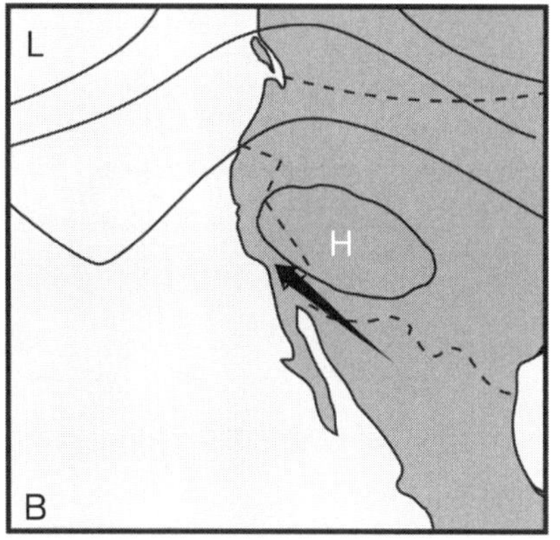

FIGURE 2.3 500-mbar circulation models in California: (A) dry southwesterly flow over California and (B) southeasterly North American monsoon types (from Minnich et al. 1993).

Lightning

In 1985, the Bureau of Land Management installed a system of electromagnetic direction finders in the western United States that record radiation emitted by cloud-to-ground lightning strikes, with lightning located by triangulation within 5 km (3.1 mi) resolution. Over short time scales, the detection of lightning reflects the paths of individual thunderstorm cells. Over time scales of weeks, months, and years, the distribution of lightning strikes reflects a combination of regional circulation and differences in air mass instability with terrain, primarily during the North American monsoon. Lightning detections are highest from Peninsular Ranges of southern California to the eastern crest of the Sierra Nevada and the desert ranges and basins.

During the period from 1985 through 2000, over 1,000,000 lightning strikes were detected in California (van Wagtendonk and Cayan 2008) with nearly half of those occurring in the Southeast Desert bioregion (Table 2.1). On a per area basis, the number varied from 2.96 strikes yr^{-1} 100^{-1} km^{-2} for the Central Coast bioregion to 27.34 strikes yr^{-1} 100^{-1} km^{-2} for the Southeast Deserts.

The spatial distribution of lightning strikes varies within the state. As can be seen in Map 2.4, the Southeastern Deserts, Northeastern Plateaus, and the northern Sierra Nevada have the highest density of lightning strikes.

July and August were the months with the most prevalent lightning in all bioregions except the Central Coast where more lightning was detected during August and September (Table 2.2). The areas with the densest lightning strikes also have the greatest summer precipitation (Minnich et al. 1993). Statewide, the period from noon to 4:00 pm received the greatest number of strikes while the morning hours from 4:00 am to 8:00 am received the least (Table 2.2).

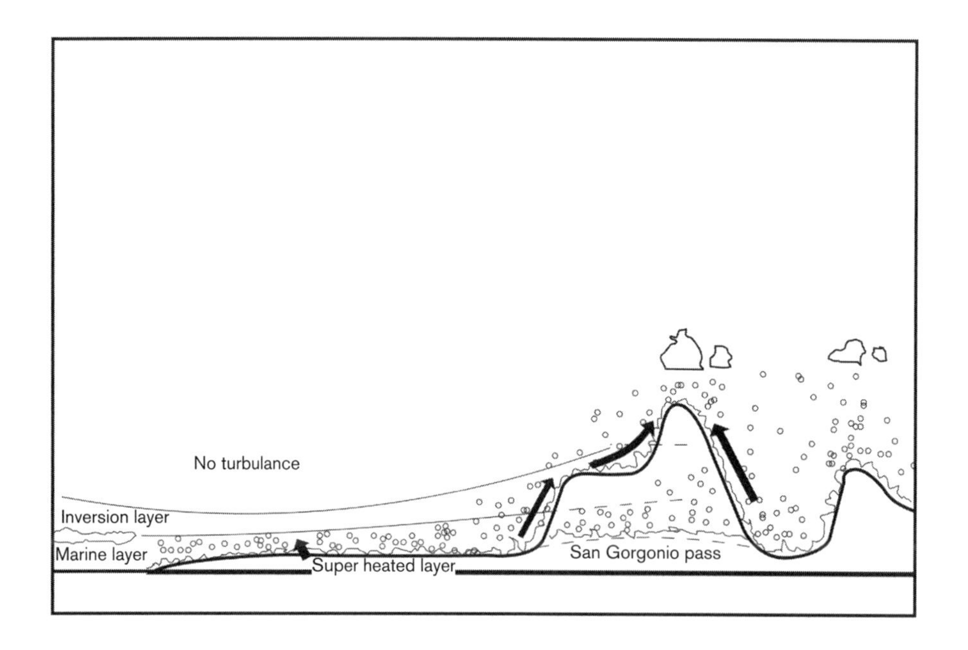

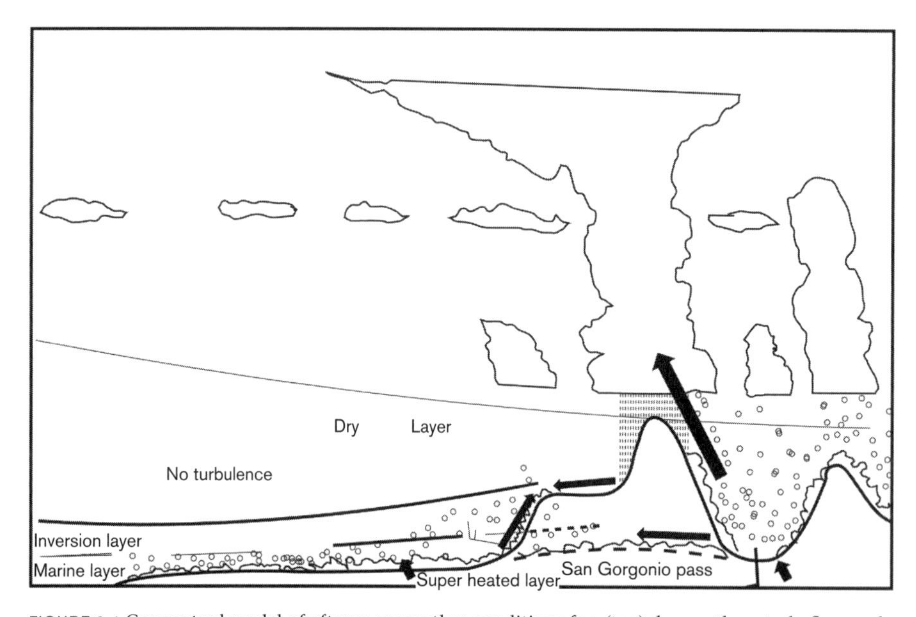

FIGURE 2.4 Conceptual model of afternoon weather conditions for: (top) dry southwesterly flow and (bottom) southeasterly North American monsoon types. MR = hypothetical mixing ratio profiles of atmospheric moisture. Chaotic lines and bubbles represent superheated air updrafts (redrawn from Minnich et al. 1993).

One should not assume that rates of landscape scale burning are proportional to lightning strike rates. The success of ignitions depends on whether they strike flammable targets and weather conditions are conducive to fire spread. Fire record statistics on ignitions are seldom correlated with vegetation successional status of the fuel cycle, and configuration of patch mosaic elements. Clearly, lightning strikes in recently burned areas have less chance of establishing a fire than in areas with high fuel loads.

The frequency of ignitions falling upon a patch of vegetation in ecological time scales requires data for three variables: (1) the ignition flux (per area per time); (2) the time required for patch elements in a mosaic to reach flammability; and (3) the size frequency distribution of the patch mosaic. The patch ignition rate (I_p) is estimated by the equation:

$$I_p = (I_{\text{ltg}} + I_{\text{anth}})\, P\, F$$

where I_{ltg} is the lightning detection rate (number km^{-2} yr^{-1}), I_{anth} is the anthropogenic ignition rate (number km^{-2} yr^{-1}), P is the patch size (km^2), and F is the vegetation fuel threshold, or the time required for vegetation to become flammable (yr).

TABLE 2.1
Lightning strikes by bioregion in California, 1985–2000

	Lightning strikes		
Bioregion	Total	Average	yr^{-1} 100^{-1} km^{-2}
North Coast	15,530	971	2.97
Klamath Mountains	46,187	2,887	12.81
Cascade Range	58,560	3,660	18.81
Northeast Plateaus	73,187	4,574	22.49
Sierra Nevada	210,277	13,142	19.59
Central Valley	35,188	2,199	3.89
Central Coast	18,264	1,142	2.96
South Coast	68,365	4,273	10.17
Southeast Deserts	480,673	30,042	27.34
Total	1,006,231	62,889	16.97

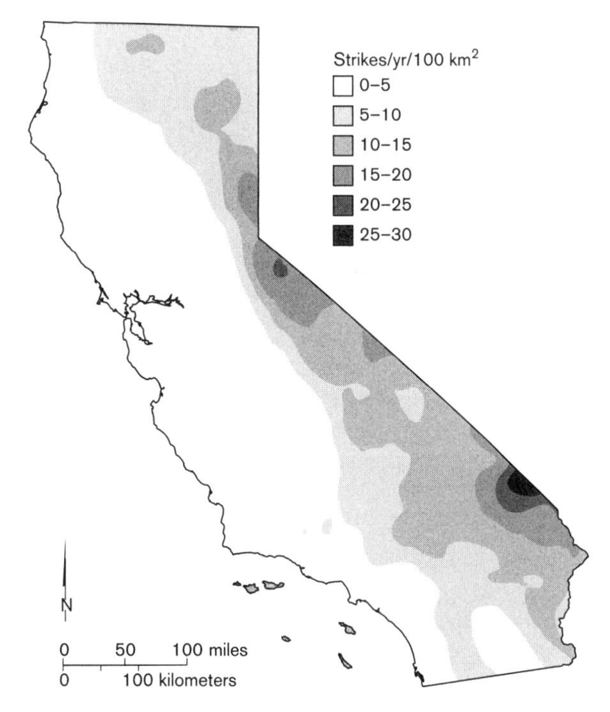

MAP 2.4 Lightning strike density in California (1985–2000) (from van Wagtendonk and Cayan 2005).

The variables contributing to ignition frequency each exhibit a spatial pattern. Lightning (I_{ltg}) occurs periodically in summer, with higher strike densities in the mountains. Anthropogenic ignitions (I_{anth}) are concentrated in heavily populated areas or near roads. With respect to the fuel mosaic, large patches (P) have a greater probability of ignitions than small patches because they are larger targets. The fuel threshold (F) depends on productivity and vegetation status in the fire cycle.

In a fire mosaic, patch elements sustain abundant lightning discharges in a single fire cycle. For example, at detection rates of 1.0 $km^{-2}yr^{-1}$, a 1,000 ha patch of chaparral is struck about 5×10^2 times in a 50-year fire interval. The patch ignition equation holds that the proportion of ignitions that initiate fires is inversely related to the flammability thresholds of ecosystems. Few ignitions establish fires of significant size even in free-burning. In Baja California, <1% of lightning discharges are required to initiate the number of landscape burns observed there (Minnich and Chou 1997).

A universal property of fire regimes is that most ignitions—regardless of source—fail to grow into large fires because of insufficient fuel. Failure rates increase directly with increasing fuel threshold periods and are also the basis of long-tailed fire size frequency distributions observed in all ecosystems, i.e., the distribution comprises a continuum of numerous small burns grading to relatively few large fires that account for most burning (Minnich and Chou 1997, Malamud et al. 1998, Minnich et al. 2000a). The addition of anthropogenic ignitions would be expected to contribute to the same frequency distribution as natural ignitions.

Potential Evaporation and Runoff

During summer, the accumulated soil water from winter storms is pumped out by vegetation through transpiration, leading to depletion of moisture and reduced live fuel moisture. Potential evapotranspiration—the hypothetical moisture returned to the atmosphere from evaporation of wet surfaces and transpiration from plants with unlimited water availability—is largely a function of temperature. Amounts also vary locally in response to wind speeds and relative humidity. Estimated annual rates of 80–100 cm (31–39 in) occur in areas with the coldest summers, notably the crest of the Sierra Nevada and the immediate Pacific Coast. Rates

TABLE 2.2
Temporal distribution of lightning strikes by 2-month and 4-hour periods in California, 1985–2000

Hour	January–February	March–April	May–June	July–August	September–October	November–December	Total
	Month						
	Number of strikes						
0–4	969	987	10,505	25,864	16,258	1,447	56,030
4–8	1,062	637	7,565	23,064	21,999	1,079	55,406
8–12	1,071	5,023	24,236	71,544	22,661	752	125,287
12–16	3,686	19,872	87,523	256,581	71,545	1,683	440,890
16–20	2,912	8,118	51,976	136,918	47,834	2,581	250,339
20–24	1,242	2,047	14,970	38,517	17,850	3,653	78,279
Total	10,942	36,684	196,775	552,488	198,147	11,195	1,006,231

increase with mean annual temperature to 100–150 cm (39–59 in) in the Central Valley and inland valleys of southern California, and 150–250 cm (59–98 in) in the southeastern deserts (Kahrl et al. 1979). Actual evapotranspiration is much less because available water is finite. Potential evapotranspiration exceeds precipitation throughout the State in the summer drought. With respect to runoff and stream flow, the average annual precipitation is greater than the actual evapotranspiration by 100 cm (39 in) only in the northwest coast and in the northern Sierra Nevada. Surpluses of 10–20 cm (4–8 in) occur >1,500 m (>4,921 ft) in the North Coastal Ranges, Santa Lucia Mountains of the Central Coast, the central Sierra Nevada, and >2,000 m (>6,562 ft) in the southern California Coastal Ranges. Runoff is less than 10 cm (4 in) over the rest of the State. Runoff is also dependent on substrate with base flow runoff reduced on porous and deeply weathered bedrock formations such as the Franciscan and Monterey formations of the central coast ranges, as well as on volcanics. Base flow is enhanced in granitic batholiths with opposite properties that dominate the Sierra Nevada and the South Coast Ranges. In the high Sierra Nevada runoff is further enhanced in bedrock-exposed glaciated terrains. Storm runoff under high precipitation intensities occur in all terrains and bedrock units.

Plant Phenology and the Fire Season

The fire season begins in early summer in most parts of California, within two months of the last winter storms. Plant fuel moisture generally declines to >100% dry weight and most species have completed growth flushes. Patterns of plant phenology depend largely on seasonal temperature and precipitation cycles that vary with altitude. At lower elevations >1,000 m (>3,281 ft) covered by nonnative annual grassland, coastal scrub, and desert scrub, and where mean winter temperatures range from 5°C to 10°C (41–50°F), the growth flush begins with the first winter rains and peak productivity is in spring. Above 1,000 m (3,281 ft), the vegetation is dominated by chaparral, woodlands, and conifer forests. Precipitation is heavier, and shrubs and trees have deeper rooting that utilizes accumulated seasonal soil water (Graham 2010). Growth flushes center in late spring after temperatures have warmed. Although evergreen sclerophyllous shrubs possess strong capacity for water regulation (Poole and Miller 1975, 1981), live fuel moisture declines to flammable levels by late spring or early summer. In high mountain conifer forests, the growth flush occurs after snow melt, usually in early summer (Royce and Barbour 2001a, 2001b). The onset of drought is not significantly delayed compared to forests and woodlands at lower elevations because moisture is rapidly depleted in porous soils, especially in glaciated terrain. Fire hazard begins when moisture is depleted, usually by June or July. Snow melt and the onset of the fire season may be postponed to August in subalpine forest after wet winters. Summer thundershowers have limited effect on soil moisture.

Interannual and Decadal Climate Variability

Climatic change is important in fire ecology because variability in temperature and precipitation at scales of years or decades influences regional vegetation productivity, fuel moisture, and fire occurrence. In recent decades atmospheric scientists have recognized that interannual to interdecadal changes in climate that lead to regional floods and droughts are linked to coupled ocean-atmosphere feedbacks that are difficult to interpret because of complex ocean dynamics. Most important are changes in ocean currents and sea surface temperatures, such as in the El Niño/Southern Oscillation and the Pacific Decadal Oscillation (PDO), which influence the amount and distribution of water vapor that is evaporated into the air, condensed into clouds, and precipitated back to earth.

The El Niño/Southern Oscillation

The El Niño/Southern Oscillation (ENSO) comprises the largest source of interannual variability in the troposphere (Lau and Sheu 1991, Diaz and Markgraf 1992). Precipitation variability in California is related to the interannual dislocation of the polar front jet stream along the Pacific Coast by ENSO. During ENSO events the dissipation of trade winds and eastward propagation of warm surface water masses from the western Pacific Ocean, often in discrete Kelvin waves, results in above-normal sea surface temperatures in the central and eastern equatorial Pacific Ocean. The release of latent heat and divergent outflow from enhanced tropical convection in the central equatorial Pacific Ocean results in the dislocation of the jet stream and anomalous, highly variable rainfall patterns in many areas of the Pacific Basin (Philander 1990, Lau and Sheu 1991).

In La Niña events and ENSO neutral years, the climatic norm, the warmest sea surface temperatures cover the western Pacific because the eastern Pacific is influenced by equatorial upwelling and advection of cold water of the Humboldt Current. The center of convection lies along the intertropical convergence zone in the western Pacific and causes ridging of the jet stream in the North Pacific, and troughing of the jet stream of moist surface westerlies over northern California, Pacific Northwest, and Canada.

In El Niño events, the center of convection along the intertropical convergence zone lies east of the International Dateline. This results in subnormal 500-mbar heights in the Gulf of Alaska, high pressure aloft in western Canada, and an enhanced southern branch jet stream across California and the southwestern United States (Renwick and Wallace 1996, Hoerling et al. 1997). The recurrence interval of warm phase El Niño events is variable, but power spectra give a robust period of four years. La Niña events are equally infrequent. However, sea surface temperature "neutral" years resemble La Niña events because the warmest equatorial SSTs remain in the western Pacific.

Statistical studies for California show that annual precipitation correlates with ENSO (McGuirk 1982, Ropelewski and Halpert 1987, Schonher and Nicholson 1989, Cayan and Webb 1992, Minnich et al. 2000b). In moderate and intense La Niña/cold episodes, precipitation departures decrease equatorward along the Pacific Coast. Amounts tend to exceed normal in northern California, approach normal from San Francisco to San Luis Obispo, and fall below normal south of Point Conception. In El Niño events, precipitation departures increase equatorward along the Pacific Coast, with maximum values in southern California and northern Baja California but approach or fall below normal in northern California.

Pacific Decadal Oscillation

The PDO is a rotating sea surface temperature anomaly around the North Pacific gyre. The oscillation affects both precipitation and temperature in the western United States over long time scales. Several models have been put forward to explain the oscillation. Among them, the mechanism of the rotating sea surface temperature anomaly appears to be the spin rate of the gyre that occurs in two phases (Latif and Barnett 1994). In the negative or cool phase of the PDO, sea surface temperatures are above normal from Japan to the Gulf of Alaska, and below normal from California to New Guinea. The reduced latitudinal temperature gradient over the North Pacific results in a northward shift of the jet stream and extratropical cyclones, and a weakening of the Aleutian low (drought in California). The cool phase is associated with a "spin up" of the North Pacific gyre (increasing gyre rotation), with enhanced advection of warm sea waters poleward off Japan. The dominance of high pressure over the entire north Pacific increases frictional wind drag (wind stress curl), and the spin rate of the gyre. After about 10 to 25 years of the "spin-up" mode, the increased gyre spin rate results in rotational advection of the warm ocean pool to the North American coast and then southwestward to the tropics. The increase in the latitudinal water temperature gradient results in the intensification and equatorward displacement of the jet stream and extratropical cyclones, enlargement of the Aleutian low pressure (above normal precipitation in California), and reduced spin rate of the gyre (the positive or warm phase of the PDO). This mode also persists for one or two decades when the "spin-down" phase is reversed by a phase change back to the negative mode. In another model, Psonis et al. (2003), arguing from a thermodynamic perspective, suggest that the decadal oscillations are produced by El Niño cyclicity tied to global temperature change. Given earth-sun geometry, decades of global warming (e.g., 1910–1940, 1977–1999) are primarily expressed in increasing temperature in tropical seas, the enhanced latitudinal temperature gradient in the Pacific generating more frequent El Niño events. Frequent Kelvin wave transfer of warm seas to the central Pacific reduces warm water advection off Japan (a conservation of thermal energy), further increasing latitudinal temperature gradients in a positive feedback. Decades of global cooling (1940–1977) result in a reduced frequency of El Niños (greater frequency of La Niñas), and enhanced advection of warm water off Japan. The hiatus in global warming since 2000 has been correlated with enhanced subduction of cool equatorial East Pacific waters tied to La Niña-like decadal cooling (Meehl et al. 2011, Kosaka and Shang-Ping 2013, England et al. 2014).

Minobe (1997) identifies four shifts (two cycles) in the PDO, with each complete cycle interval spanning about 50 to 70 years. The last two shifts included a negative phase from 1948 to 1976 and a positive phase from 1976 to 1999. The phase transition in the mid-1970s divided a period of protracted drought associated with another global warming hiatus from 1944 to 1968, and above normal precipitation from 1977 to 1998 in California correlated with increased El Niño frequency and global warming. Over the western United States, both negative phase PDO shifts were associated with cold winters, and positive PDO with warm winters.

Analysis of instrumental records (1880–1994) and tree ring data (back to 1700) shows a north-to-south seasaw in precipitation variability in western North America, pivoting at 40°N (Dettinger et al. 1998). Precipitation at California latitudes are positively correlated with the Southern Oscillation Index (SOI) which cause southward displacements of precipitation distribution at interannual and decadal time scales. The overall precipitation amount delivered to the US Pacific Coast has little variation both in the instrumental record and in tree ring records, consistent with the century scale analysis of tree ring data in the Sierra Nevada by Graumlich (1993).

Climate Variability and Fire History

With respect to climatic change, fire ecologists are most concerned with precipitation variability, especially drought that may result in deficits in summer live fuel moisture, primarily in woody fuels. However, it should be recognized that total annual precipitation inclusively is an incomplete predictor of drought because mean annual precipitation greatly exceeds soil field capacities (total soil water available to plants in the soil). In mesic mountainous regions, where subnormal annual precipitation may still saturate soils in dry years, only runoff is reduced (Franco-Vizcaíno et al. 2002). Soil water deficits may occur only in the most extreme droughts, as in Lake Tahoe basin in 1989–1992 and in southern California in 1999–2003 (Minnich et al. 2016).

The relationship between climatic variability and long-term fire history is also dependent on the dynamics and spatial heterogeneity of fuel buildup in ecosystem mosaics. Clearly the effect of drought on fire hazard is expected to be less in a fresh burn than in an old stand that has built-up canopy, accumulated fuels, and increased transpiration over decades or centuries. Young stands may resist fire in spite of drought; old stands may burn preferentially over young stands due to fuels rather than climatic variability.

At regional scales, the effect of climatic variability on regional burning rates is modulated by patch mosaic status in a negative feedback, especially if landscape fire regimes exist in some form of steady state. As such, patch turnover, stand age frequency distribution, and carbohydrate-to-water energy ratios remain relatively constant at broad temporal and spatial scales due to a negative feedback between fire hazard and vegetation status. Multi-year drought may reduce the age of patches subject to fire, thereby increasing burning rates. This may be balanced by large fire outbreaks that would lead to overall reduction in the age and fuel hazard of the landscape mosaic. Likewise, moist periods may temporarily postpone burning but lead to buildup of fuels. Fire weather is unlikely to influence long-term burning rates as long as the occurrence of large events is random relative to fire weather and patch status.

It is often tempting to place "cause" of specific fire outbreaks with short-term events such as a single dry year or a multi-year drought. In a heterogeneous patch mosaic, climatic relationships with fire regimes must integrate production and fuel accumulation dynamics of forest and brushlands at time scales commensurate with ecosystems in question. In short, does the fire outbreak and climate perturbation explain the dynamics of the whole landscape? For example, the denudation of an entire patch mosaic, regardless of age, would suggest that short-term perturbations provide the highest explanatory power (Swetnam and Betancourt 1998). Alternatively, if fires preferentially burn old-growth vegetation, then the explanation may lie in the integrated dynamics of the mosaic, including the fire history linked to growth patterns and cumulative fuel buildup over long time scales. As such, the role of climatic shifts should be examined at time scales of vegetation dynamics, e.g., centuries for forests that turnover in centuries, decades in chaparral that turnover in decades. In summary, the effect of precipitation variability is modulated by patch structure in which changes in regional fire hazard result in only finite portions of stands achieving flammability thresholds. Statistical tests correlating fire and climatic change should therefore involve running averages of the climatic variability scales to mean fire intervals, not specific climate events (Lovell et al. 2002).

In a given weather state, fires spreading from patches with high fuel loads may encounter resistance or reduced fire intensity in adjoining patches with less fuel. Fires still overlap across stands from high to low fuel loads due to the fire's momentum. The flaming front initially retains high energy release rates until it accommodates to energy levels of the new stand (i.e., the fire slows down or eventually dissipates). The magnitude of successive fires at a site fluctuates within the weather window at a modest range of variance because ignitions establish fires at random. The unique weather conditions with each fire causes successive patch mosaic configurations to shift over time (shifting mosaic). Standing mosaics, i.e., fixed patch configurations through multiple fire sequences, occur rarely. The burning of young classes should not be interpreted as evidence against age-dependent flammability. Most fire-overlap zones are "slivers" covering small areas due largely to flame-line momentum. Hence, in the analysis of fire-overlap sequences, it is more instructive to examine the age of vegetation of the entire fire, rather than focusing on edge effects.

Recent studies have correlated climate variability with fire-scar dendrochronology records in which it is assumed that fire intervals can be directly computed from fire scars in tree rings. Studies have correlated fire variability with ENSO (e.g., Norman and Taylor 2003) and at decadal scales (e.g., Swetnam and Betancourt 1998, Westerling and Swetnam 2003). Millennial scale trends reported for California have been obtained from giant sequoia (*Sequoiadendron giganteum*) (Swetnam 1993). Although the scarring frequency documents a site, scaling the scar-year data to regional patterns requires high resolution spatial sampling because fire-size frequency distributions are "long-tailed,"—a property found in all ecosystems. Because most fires are small and collectively add to limited spatial extent, the preponderance of landscape burning is accomplished by relatively few large fires (Minnich et al. 2000a). In western pine forests, abundant small fires over time scales of fire cycles may scar trees as frequently as the few large fires that consume most fuels. Because fire-scar dendrochronology methods cannot directly differentiate mass burns from microburns, regional burning rates in space and time may not covary with scarring rates, i.e., the assumption that fire intervals can be directly computed from the fire scar in tree rings is not constrained. It follows that transient fluctuations in the scarring record may also arise from site-specific microburns (lightning fires, anthropogenic starts), rather than rates of burning at the landscape scale. For example, while the decline of fires in the Sierra San Pedro Mártir from 1790 to 1830 (Stephens et al. 2003) could reflect climate change, it is also plausible that the decline could also reflect the demise of Native Americans as a source of microburns. The San Pedro Mártir mission was established in the 1790s, and lengthening fire intervals due to the introduction of mission livestock (Stephens et al. 2003) is not verified by paleobotanical data. Herbaceous cover is sparse in California forests. Likewise, the extraordinarily short two- to four-year fire intervals in giant sequoia groves of the Sierra Nevada (Swetnam 1993) could be an outcome of cooking fires and patterns of Native American occupation.

In contrast with woody assemblages, the relationship between climatic variability and burn area in herbaceous ecosystems would be expected to correlate over short scales because herbaceous biomass is recycled on a nearly annual

SIDEBAR 2.1 FIRE WEATHER, FIRE SUPPRESSION, AND DYNAMICS OF PATCH MOSAICS IN CHAPARRAL

Richard A. Minnich

At any given site, fires establish over a wide range of weather conditions depending on season, local atmospheric circulation, or time of day. While individual fires are associated with unique combinations of weather and fuel conditions, the cumulative impact of fires scaling over centuries or millennia results in fire occurrence with modal weather states of the local climate over time, if ignitions establish fires at random.

A model that integrates fire weather and fuel-driven patch dynamics is proposed to explain the discontinuity in fire regime from infrequent large burns and coarse mosaics in California and frequent small burns in Baja California (Minnich and Chou 1997) (Fig. 2.1.1). With free-burning in Baja California, numerous fires produce fine-grained mosaics and discrete local-scale fuel heterogeneity linked to stand-replacement fire behavior. Fires consistently establish in old stands and their expansion is constrained in adjoining younger patches with less fuel. Fire suppression in California reduces the number of fires, thereby increasing the size of old-growth patch

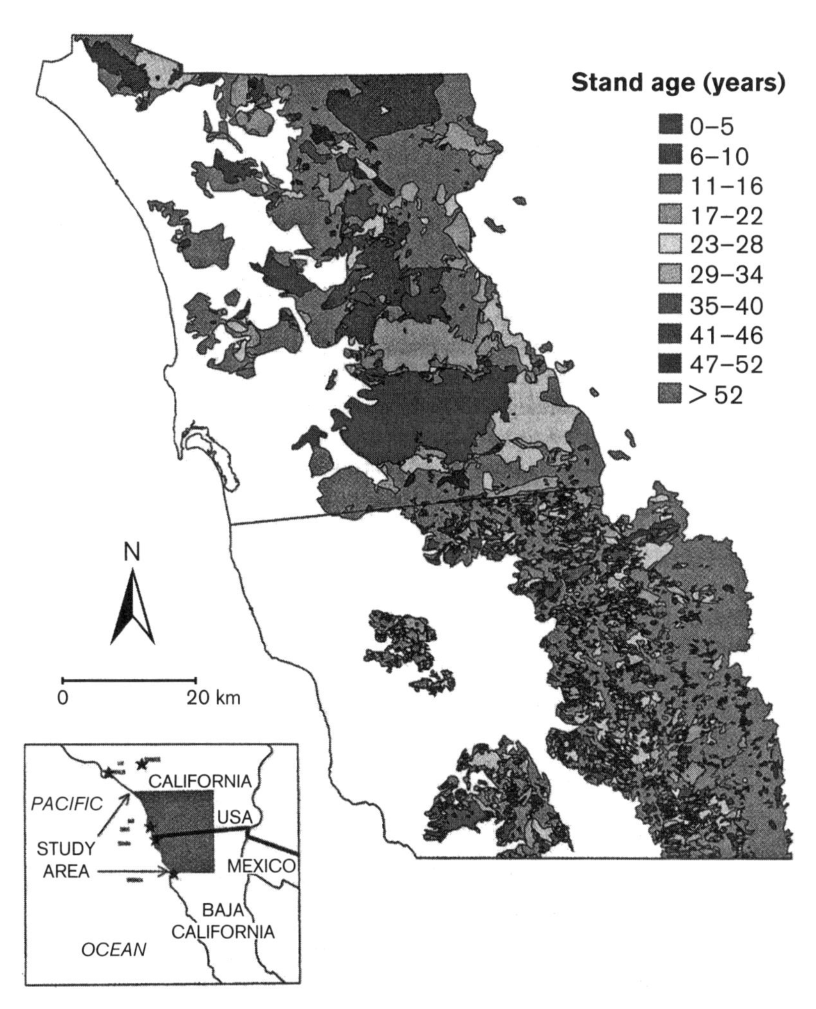

FIGURE 2.1.1 Chaparral patch mosaic in southern California and northern Baja California in 1971.

(continued)

(continued)

elements and fires, but fires still spread largely in old stands. In effect, a fine-grained patch mosaic is well-buffered against megaburns.

Minnich and Chou (1997) found a shift from dominantly Santa Ana wind-driven fires in southern California to fires carried by onshore anabatic flows in Baja California with a discontinuity in fire weather occurrence along the international boundary (Fig. 2.1.1). Satellite imagery documents a disparate seasonality of burning—summer in Baja California and fall in southern California (Minnich 1983). The difference in the weather of fires across the international boundary reflects the nonrandomization of large fire occurrence to extreme-weather states by suppression.

Fig. 2.1.2 is a conceptual drawing that shows how the probability distribution of fire climate and active fire spread combine to produce an expected fire spread rate probability distribution at a landscape scale. Assuming random ignition rates, (A) describes the probability of active fire spread and fire spread rates with both these properties increasing in value as weather turns drier. The dashed curve in (A) represents the probability of active fire spread and is influenced by initial attack suppression of fire starts. In free-burning, fire is not possible in moist weather but involves an increasing proportion of the mosaic as weather becomes ever drier. With fire control, nearly all ignitions are suppressed except in the driest weather. In (B), the fire climate is described as the proportion of time in weather states in a normal statistical distribution. By definition, there is more time in "normal" weather than in the moist or dry extremes. In (C), (A) and (B) are graphically multiplied together and describe the expected proportion of landscape that burns along the climatic continuum.

With or without fire control there is little burning in moist weather (C) because both fire hazard risk in (A) and percent of climate in moist weather in (B) are low. Fire control has little effect because suppressed fire starts would have dissipated at a small size. In normal weather, the expected spread rate frequency distribution in (C) reaches a maximum value without control because modest spread rates (A) are phased with a large portion of time in the fire season (B). In other words, slow burning consumes large landscape area. Fire spread rate distributions in dry weather are low because of the consumption of fuels by slow burning in normal weather; high extreme-weather risk states represent a small proportion of time. With fire control, the nonrandom elimination of fire starts in normal weather reduces burn area in normal weather states. It follows that ignition-stage suppression selectively nonrandomizes the occurrence of large fires to severe-weather risk states compared to chance. The expected spread rate frequency distribution is skewed toward extreme weather, resulting in high average spread rates and flame-front intensities (Fig. 2.1.2). In effect, since fire removes energy responsible for the process, we rob Peter to pay Paul.

The random establishment of fire starts in "normal" weather is a tenable assumption in presuppression California wildlands because of the abundance of natural ignitions and the persistence of fires for months (Minnich 1987, 1988), alternating between expanding flame lines and storage of "live" ignitions as future ignition sources in coarse fuels (logs, snags) through glowing combustion (see lightning detec-

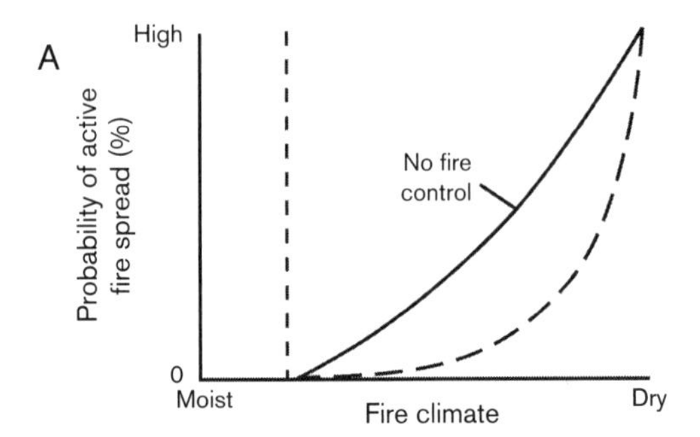

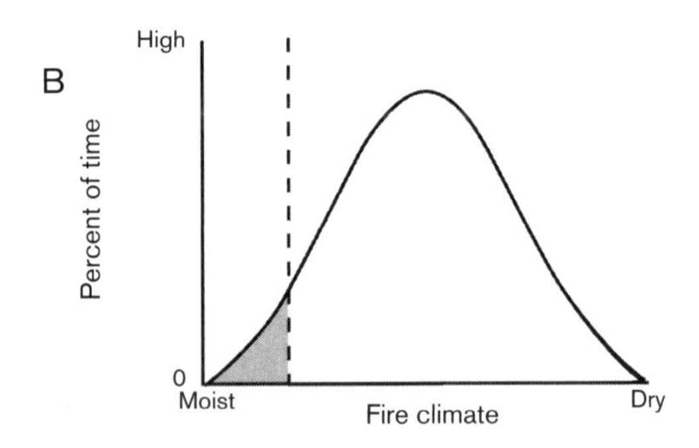

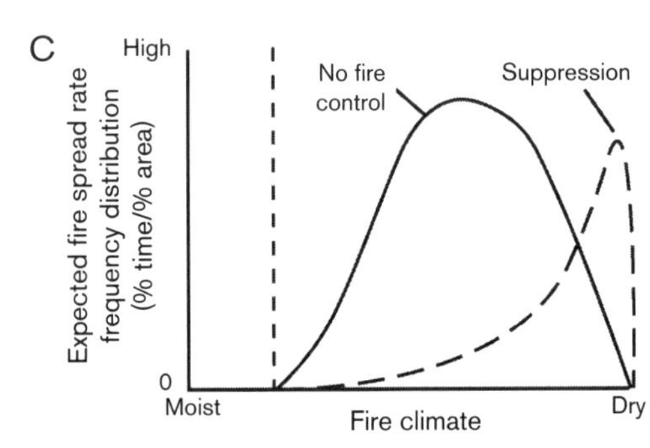

FIGURE 2.1.2 Conceptual model of the expected portion of landscape patch mosaic that burns along a continuum of fire weather risk states of the climate from moist to dry (left to right in all three figures): (A) probability of active fire spread on the landscape; (B) fire climate; and (C) average landscape fire weather and spread rates.

FIGURE 2.1.3 Patchy chaparral burn in the central Sierra Juarez of Baja California.

tion section in chapter 2). It follows that the longer live ignitions persist to trigger flame fronts, the lower the average integrated departure of weather from the long-term mean. Season-long burns still occur in Baja California (Minnich and Chou 1997, Minnich et al. 2000b).

The random establishment of mass fires is assured by the high failure rate of fire starts in a patch mosaic. The rare fire starts that do establish mass burns have a high probability of coinciding with normal weather by chance. Finally, people set fires in independent actions, i.e., there is no "conspiracy" to establish fires in specific weather states. Most such ignitions would also fail.

Suppression would have no effect on the weather of fires only if the firefighting system has a uniform capacity to influence the spread of flame lines during the course of a fire regardless of its size or weather risk state. This premise is of course unrealistic because the effectiveness of suppression actions is dependent on the energy output of fire. All fires begin at an infinitely low rate of energy release (ignition). Energy levels increase exponentially with increasing length of flame lines at the perimeter and flame-front expansion rates. Suppression forces extinguish virtually all ignitions at the ignition stage, but have little effect on escaped large burns because energy release rates exceed the energy of suppression by orders of magnitude.

The weather risk state also influences energy levels at the ignition stage, with ignitions in severe weather (and high-energy release states) having the highest probability of expanding past initial attack efforts. Because the continuity of weather operates in time scales of days (in association with synoptic shifts in atmospheric circulation) and because fires in the flaming stage seldom persist more than a few days under suppression, the weather characteristics of major fires have a high probability of resembling the weather risk state of escaped fire starts. Frequent or long-lived ignition sources in Baja California establish mass burns when weather risk advances pass the "moist threshold" of combustion. Fires periodically rise above or below the moist threshold, and the burning periods are often at marginal to the combustion threshold. On the landscape, this is seen in reticulated (weavy) burning in which separated flaming fronts merge and split with fluctuations in terrain, fuels, and weather, even in head fires. Fig. 2.1.3 shows a patchy chaparral burn in the central Sierra Juárez of Baja California. Hence, complex patch mosaics can also arise from single burns. In California, suppression mass fires begin in extreme risk states due to efficient initial attack extinction of ignitions. Mass fires cease only after the weather risk decreases to the "moist threshold" usually after many days. Fire recurrences are virtually eliminated by the "mopping-up" phase of suppression.

It was argued that differences in transnational fire weather reflects the infrequency of Santa Ana winds in Baja California compared to southern California (Keeley et al. 1999). However, the discontinuity in fire regime along the international boundary indicates that differences in fire weather are related to divergent fire management practices. Offshore "Santa Ana" winds arise from hemispheric circulations operating

(continued)

(continued)

at scales of thousands of kilometers, and the 200-km span of the Peninsular Ranges is too small for large gradients in wind fields. Without distinctive suppression histories, changes in fire weather and patch dynamics should be expressed in a continuum of fire weather, not the discontinuity presently seen along the international boundary (Minnich and Chou 1997). Climatic gradients, especially temperature and precipitation, change largely in response to elevation and distance from the Pacific Ocean, and run east to west, orthogonal to the international boundary.

In the evaluation of presuppression fire regimes, fire ecologists and land managers in California should be careful not to judge "average" fire weather conditions from the history of mass burns in the twentieth century. Observations of fires in Baja California indicate that fires occurred in a broader range of weather environments than in California, including "normal" weather states when ignitions are efficiently suppressed.

The role of weather risk states will shift with the kind of vegetation. For example, in cured grassland, a shift in the moist threshold for active fire spread (to the far left) in (A) would result in more burning under suppression compared to fire control (Fig. 2.1.1). Initial attack is less efficient in suppressing fire starts than in woody assemblages because grass fires enlarge rapidly from ignition point. In vegetation characterized by a shift in the combustion threshold to very dry weather (e.g., pinyon-juniper woodland), active fire spread is possible only in extreme weather, with or without fire control.

basis. Regional fuel patterns and burning rates are independent of fire history. Fire hazard in grasslands tends to increase with total annual precipitation, not drought, due to increased productivity and available cured fuel.

References

Adams, D.K., and A.C. Comrie. 1997. The North American monsoon. Bulletin of the American Meterological Society 78: 2197–2213.

Barbour, M.G., N.H. Berg, T.G.F. Kittel, and M.E. Kunz. 1991. Snowpack and the distribution of a major vegetation ecotone in the Sierra Nevada of California. Journal of Biogeography 18: 141–149.

Cayan, D.R., and R.H. Webb. 1992. El Nino/southern oscillation and stream flow in the western United States. Pages 29–68 in: H.F. Diaz and V. Markgraf, editors. El Niño: Historical and Paleoclimatic Aspects of the Southern Oscillation. Cambridge University Press, Cambridge, UK.

Dettinger, M.D., D.R. Cayan, H.F. Diaz, and D.M. Meko. 1998. North-south precipitation variations in western North America on interannual-to-decadal time scales. Journal of Climate 11: 3095–3111.

Diaz, H.F., and V. Markgraf. 1992. El Niño: Historical and Paleoclimatic Aspects of the Southern Oscillation. Cambridge University Press, Cambridge, UK.

England, M.H., S. McGregor, P. Spence, G.A. Meehl, A. Timmermann, C. Wenju, A. Sengupta, M.J. McPhaden, A. Purich, and A. Santoso. 2014. Recent intensification of wind-driven circulation in the Pacific and the ongoing warming hiatus. Nature Climate Change 4: 222–227.

Franco-Vizcaíno, E., M. Escoto-Rodríguez, J. Sosa-Ramírez, and R.A. Minnich. 2002. Water balance at the southern limit of the Californian mixed-conifer forest and implications for extreme-deficit watersheds. Arid Land Research and Management 16: 133–147.

Glendening, G.W., B.L. Ulrickson, and J.A. Businger. 1986. Mesoscale variability of boundary layer properties in the Los Angeles Basin. Monthly Weather Review 114: 2537–2549.

Graham, R.C., A.M. Rossi, and K.R. Hubbert. 2010. Rock to regolith conversion: producing hospitable substrates for terrestrial ecosystems. GSA Today 20(2): 4–9.

Graumlich, L.J. 1993. A 1000-year record of temperature and precipitation in the Sierra Nevada. Quaternary Research 39: 249–255.

Hales, J.E., Jr. 1974. Southwestern United States summer monsoon source—Gulf of Mexico or Pacific Ocean? Journal of Applied Meteorology 13: 331–342.

Hoerling, M.P., A. Kumar, and M. Zhong. 1997. El Niño, La Niña, and the nonlinearity of their teleconnections. Journal of Climate 10: 1769–1786.

Kahrl, W.L., W.A. Bowen, S. Brand, M.L. Shelton, D.L. Fuller, and D.A. Ryan. 1979. The California Water Atlas. California Department of Water Resources, Sacramento, California, USA.

Keeley, J.E., C.J. Fotheringham, and M.E. Morais. 1999. Reexamining fire suppression impacts on brushland fire regimes. Science 294: 1829–1832.

Kosaka, Y., and X. Shang-Ping. 2013. Recent global-warming hiatus tied to equatorial Pacific surface cooling. Nature 501: 403–407.

Latif, M., and T.P. Barnett. 1994. Causes of decadal climatic variability over the north Pacific and North America. Science 266: 634–637.

Lau, K.A., and P.J. Sheu. 1991. Teleconnections in global rainfall anomalies: seasonal to inter-decadal time scales. Pages 227–246 in: M.H. Glantz,, R.W. Katz, and N. Nicholls, editors. Teleconnections Linking Worldwide Climate Anomalies. Cambridge University Press, Cambridge, UK.

Lovell, C., A. Mandondo, and P. Moriarty. 2002. The question of scale in integration natural resource management. Conservation Ecology 5(2): 25.

Malamud, B.D., G. Morein, and D.L. Turcotte. Forest fires: an example of self-organized critical behavior. Science 281: 1840–1842.

McGuirk, J.P. 1982. A century of precipitation variability along the Pacific coast of North America and its impact. Climate Change 4: 41–56.
Meehl, G.A., J.M. Arblaster, J.T. Fasullo, A. Hu, and K.E. Trenberth. 2011. Model-based evidence of deep-ocean heat intake during surface-temperature hiatus periods. Nature Climate Change 1: 360–364.
Minnich, R.A. 1983. Fire mosaics in southern California and northern Baja California. Science 219: 1287–1294.
———. 1986. Snow levels and amounts in the mountains of southern California. Journal of Hydrology 86: 37–58.
———. 1987. Fire behavior in southern California chaparral before fire control: the Mount Wilson burns at the turn of the century. Annals of the American Association of Geographers 77: 599–618.
———. 1988. The biogeography of fire in the San Bernardino Mountains of California: a historical review. University of California Publications in Geography 28: 1–28.
Minnich, R.A., M.G. Barbour, J.H. Burk, and J. Sosa Ramírez. 2000a. Californian mixed-conifer forests under unmanaged fire regimes in the Sierra San Pedro Mártir, Baja California, Mexico. Journal of Biogeography 27: 105–129.
Minnich, R.A., and Y.H. Chou. 1997. Wildland fire patch dynamics in the chaparral of southern California and northern Baja California. International Journal of Wildland Fire 7: 221–248.
Minnich, R.A., E. Franco-Vizcaíno, and R.J. Dezzani. 2000b. The El Niño/Southern Oscillation and precipitation variability in Baja California, Mexico. Atmosféra 13: 1–20.
Minnich, R.A., E. Franco-Vizcaíno, J. Sosa-Ramírez, and Y.H. Chou. 1993. Lightning detection rates and wildland fire in the mountains of northern Baja California, Mexico. Atmosféra 6: 235–253.
Minnich, R.A., B.R. Goforth, and T.D. Paine. 2016. Follow the water: extreme drought and the conifer forest pandemic of 2002-2003 along the California borderland. Pages 859–890 in: T.D. Paine and F. Lieutier, editors. Insects and Diseases of Mediterranean Forest Systems. Springer International, Cham, Switzerland.
Minobe, S. 1997. A 50–70 year climatic oscillation over the North Pacific and North America. Geophysical Research Letters 24: 683–686.
Norman, S.P., and A.H. Taylor. 2003. Tropical and North Pacific teleconnections influence fire regimes in pine-dominated forests in northeastern California. Journal of Biogeography 30: 1081–1092.
Palmén, E., and C.W. Newton. 1969. Atmospheric circulation systems: their structure and physical interpretation. International Geophysics Series Vol. 13. Academic Press, New York, New York, USA.
Philander, S.G. 1990. El Niño, La Niña, and the Southern Oscillation. Academic Press, New York, New York, USA.
Poole, D.K., and P.C. Miller. 1975. Water relations of selected species of chaparral and coastal sage species. Ecology 56: 1118–1128.
———. 1981. The distribution of plant water stress and vegetation characteristics in southern California chaparral. American Midland Naturalist 105: 32–43.
Psonis, A.A., A.G. Hunt, and J.B. Elsner. 2003. On the relation between ENSO and global climate change. Meteorology and Atmospheric Physics 84: 229–242.
Renwick, J., and J.M. Wallace. 1996. Relationships between North Pacific wintertime blocking, El Niño, and the PNA pattern. Monthly Weather Review 124: 2071–2076.
Ropelewski, C.F., and M.S. Halpert. 1987. Global and regional scale precipitation patterns associated with the El Niño/ Southern Oscillation. Monthly Weather Review 115: 1606–1626.
Rothermel, R.C. 1972. A mathematical model for fire spread predictions in wildland fuels. USDA Forest Service Research Paper INT-RP-115. Intermountain Forest and Range Experiment Station, Ogden, Utah, USA.
Royce, E.B., and M.G. Barbour. 2001a. Mediterranean climate effects I. Conifer water use across a Sierra Nevada ecotone. American Journal of Botany 88: 911–918.
———. 2001b. Mediterranean climate effects II. Conifer growth phenology across a Sierra Nevada ecotone. American Journal of Botany 88: 919–932.
Schonher, T., and S.E. Nicholson. 1989. The relationship between California rainfall and ENSO events. Journal of Climate 2: 1258–1269.
Smith, W. 1986. The effects of eastern North Pacific tropical cyclones on the southwestern United States. U.S. National Oceanic and Atmospheric Administration Technical Memorandum NWS WR-197. U.S. Department of Commerce, Washington, D.C., USA.
Steinbeck, J. 1962. Travels with Charlie. Penguin Press, Cambridge University Press, Cambridge, United Kingdom.
Stensrud, D.J., R.L. Gall., and M.K. Nordquist. 1997. Surges over the Gulf of California during the Mexican monsoon. Monthly Weather Review 125: 417–437.
Stephens, S.L., C.N. Skinner, and S.J. Gill. 2003. A dendrochronology based fire history of Jeffrey pine-mixed conifer forest in the Sierra San Pedro Martir, Mexico. Canadian Journal of Forest Research 33: 1090–1101.
Swetnam, T.W. 1993. Fire history and climatic change in giant Sequoia groves. Science 262: 885–890.
Swetnam, T.W., and J.L. Betancourt. 1998. Mesoscale disturbance and ecological response to decadal climatic variability in the American Southwest. Journal of Climate 11: 3128–3147.
Tubbs, A.M. 1972. Summer thunderstorms over southern California. Monthly Weather Review 100: 799–807.
van Wagtendonk, J.W., and D.R. Cayan. 2008. Temporal and spatial distribution of lightning strikes in California in relationship to large-scale weather patterns. Fire Ecology 4(1): 34–56.
Westerling, A.L., and T.W. Swetnam. 2003. Interannual to decadal drought and wildfire in the western United States. Bulletin of the American Meteorological Society 84: 595–604.

CHAPTER THREE

Fire Weather Principles

BRENDA L. BELONGIE AND RICHARD A. MINNICH

Everybody talks about the weather, but nobody does anything about it.

WARNER (1897)

Introduction

Topography, fuels, and weather are central to the seasonal and daily variations of the fire environment. The aggregate, active interplay between them mainly determines an area's potential fire activity and fire behavior. Topography is a stable, known element of any area as it changes slowly over long, geologic time scales. It can have various direct effects on weather, such as channeling winds in drainages or creating precipitation shadows on the lee side of mountain ranges. Local weather regimes created in part by differing directions slopes face lead to corresponding variations in temperature and moisture that affect fuel conditions. Fuels can also influence weather, such as when a stand of trees blocks wind. Moisture content of smaller fuels is readily affected by diurnal temperature and moisture patterns, whereas moisture content of larger fuels varies more seasonally. These are only a few examples of how topography, fuels, and weather interact.

Weather is the current state of atmospheric conditions making it the most dynamic of the three wildland fire environment elements. It changes frequently and can vary significantly over short distances across a landscape. A combination of individual weather elements such as temperature, sky cover, and barometric pressure are key drivers of fire weather. Core elements of fire weather are atmospheric stability, wind, and temperature/moisture relationships. Weather influences fire outcomes by altering vegetation fuel moisture and the efficiency of heat transfer in combustion (Albini 1993).

The principal generator of weather is the sun, but Earth's rotation is also key to global weather patterns. Most of our weather occurs in the troposphere, which is the lowest layer of the atmosphere approximately 16.1 km (10 mi) deep at California's latitudes. Fire weather is focused on the atmospheric boundary layer that interfaces directly with fuels and governs fire behavior over short time scales. Fire weather is further characterized by individual air masses, or large batches of air that acquire their temperature and moisture characteristics from the nature of the geographic region where they originate. For California, this occurs most often somewhere over the Pacific Ocean. An air mass has fairly uniform temperature and moisture in the horizontal dimension over a large area.

In this chapter, the components of weather that affect fire behavior are first defined, then how fire weather varies seasonally and across California is described. A discussion of the full range of weather conditions capable of propagating fires (the fire weather "window") is at the end.

Components of Fire Weather

Atmospheric Temperature

Atmospheric temperature and moisture vary by region and altitude. In general, the warmest temperatures are found at lower elevations and latitudes. Globally, moisture is mostly concentrated in the lower atmosphere with the highest amounts in equatorial areas and decreasing amounts toward the poles (Trewartha and Horn 1980). These principles apply across California in general, but there are some significant local variations, most noticeably near the coast.

Higher elevations are generally cooler than lower elevations due to the adiabatic process. Adiabatic processes refer to air temperature changes resulting from physical lift or descent of air. The atmosphere's temperature lapse rate as recorded by weather balloons cools with height because of decreasing atmospheric pressure (ever-expanded air parcels with height) and integrated processes of planetary heat transfer. Air at higher elevations is under less atmospheric pressure, so it expands and cools. Conversely, air at lower elevations is under greater atmospheric pressure, is compressed, and warms. This process works because the air does not exchange matter or heat with its surroundings. For example, an average summer day in Sacramento, elevation ~9 m (~30 ft), is much warmer than one at Lake Tahoe, elevation ~1,900 m (near 6,225 ft). Areas near the California coast, however, are generally cooler than similar or higher elevations just a short-distance inland. Temperatures in the cool coastal marine air mass are usually substantially lower than those in nearby areas unaffected by the cool marine air.

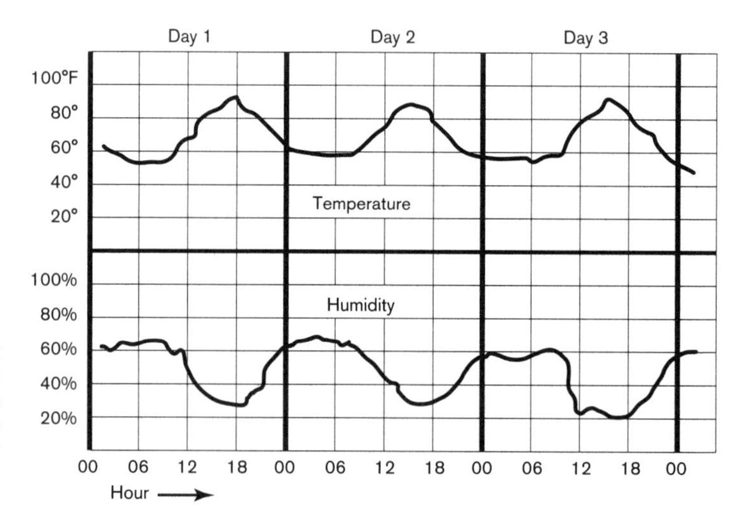

FIGURE 3.1 Simulated hygrothermograph trace that shows common diurnal cycles of temperature versus relative humidity (RH) when there is no large change in the moisture content of an air mass. As temperature rises during the day, RH typically lowers correspondingly and vice versa overnight (adapted from Schroeder and Buck 1970).

Surface conditions modify an area's overlaying air mass to create actual temperature and moisture regimes experienced near the surface (Schroeder and Buck 1970). Earth radiates infrared (heat) energy to space at all times, but during daylight hours, solar radiation delivers more heat than is lost to space. This net gain of energy warms the ground, so air temperatures warm. Daytime surface temperatures on all aspects are cooler when there is less sunlight striking them, such as where ground is shaded by a forest canopy, cloud cover, or a fire's smoke column. When there is no incoming solar radiation overnight, ground temperatures readily cool as terrestrial heat radiates out to space, and air temperatures correspondingly decline. Cloudy nights slow the net effect of terrestrial heat loss and usually keep overnight temperatures warmer than under clear skies. In essence, Earth's surface serves as a heating element during the day and a chilling coil at night for air. This largely accounts for observed, diurnal environmental temperature patterns (Fig. 3.1).

A look around any outdoor environment will reveal a variety of surfaces such as pavement, lawns, shaded forest floor, or water. Sunlight heats each surface differently depending largely on color and specific heat characteristics. Slope steepness and aspect also play an important role in the amount of daily solar radiation the ground receives. Solar rays more perpendicular to the ground will deliver more energy per unit surface area (insolation) and conditions will be warmer there.

In California, southern aspects are generally the warmest and driest as they receive the most intense daily solar radiation. North aspects are generally the coolest and have higher soil moisture as they receive the least amount of insolation. Fuel type and moisture variations across a landscape often reflect these environmental conditions (Schroeder and Buck 1970).

The amount of solar radiation Earth receives has seasonal cycles. Because Earth's axis of rotation is tilted about 23.5 degrees away from the plane of its orbital path around the sun, California's latitudes have a distinct annual solar radiation pattern. Annual temperature patterns across California vary relative to intensity of solar radiation. The warmest yearly temperatures occur in summer when the most intense, longest duration of annual solar radiation creates a net surface temperature gain, and the coolest temperatures come in winter when less insolation and shorter days lead to a net surface temperature loss. In general, maximum daily temperatures in California peak in July about 4 to 5 weeks after summer solstice, except for coastal locations where the peak generally occurs in late August when the seasonal influence of cool marine air is diminished by lighter on shore winds and periods of warmer offshore flow. The coldest maximum daily temperatures of the year commonly occur around winter solstice throughout the state.

Atmospheric Moisture

Atmospheric moisture is a key element of the wildland fire environment, and there are several metrics for the amount present. Dew point and wet-bulb temperatures indicate something about the physical amount of moisture in the atmosphere. Dew point is the temperature at which saturation would occur, and wet-bulb temperature is the coolest temperature achievable by evaporation. Relative humidity (RH) is a commonly observed fire weather element, but it is not a true measure of the absolute amount of atmospheric moisture. RH is defined as:

$$\frac{e_a}{e_s} \times 100 = RH \text{ (in percent)}$$

where e_a is the actual vapor pressure and e_s is the saturation vapor pressure (see sidebar 3.1). In other words, RH is the ratio of atmospheric moisture present to the amount of moisture at saturation at a given temperature and pressure, expressed as a percentage. Changes in the absolute water vapor amount or a change in air temperature will change RH. Because RH depends heavily on air temperature, it commonly responds to diurnal temperature change patterns. Rising temperature typically leads to lower RH and declining temperature to higher RH. The lowest diurnal RH is usually seen around the time of maximum temperature, and the highest RH usually occurs around the time of minimum temperature (Fig. 3.1).

SIDEBAR 3.1 VAPOR PRESSURE, RELATIVE HUMIDITY, AND FUEL MOISTURE

Brenda Belongie

Air pressure is the sum of the pressures contributed by all the individual gases in the air, including water vapor. Each gas's contribution to the total pressure is called the partial pressure of that gas. Since water vapor in the air acts just like any other gas (Schroeder and Buck 1970), we call water vapor's partial pressure *vapor pressure*. Vapor pressure is important in the wildland fire environment because it affects whether condensation or evaporation of water is occurring from fuel.

Imagine a dish of water sitting on a table. There will always be some water molecules energetic enough to break free of the liquid surface to become water vapor in the air above the dish. At the same time, some water (vapor) molecules in the air lose energy and return to the pool of liquid water. When more water molecules leave the dish than return, the net result is evaporation (Fig. 3.1.1A). If more return to the dish than leave, that's condensation (Fig. 3.1.1B). When there is a state of equilibrium where just as many leave as return that is called saturation (Fig. 3.1.1C).

Vapor pressure is temperature dependent. Warmer air makes water vapor molecules more energetic and they are able to remain escaped from a surface and increase the vapor pressure in the air. For any given air temperature, however, there is a maximum vapor pressure that temperature can support. This is the point where equilibrium conditions exist with equal evaporation and condensation. Since that is defined as a state of saturation, that maximum possible vapor pressure at any temperature is called saturation vapor pressure.

Saturation vapor pressure increases nonlinearly with an increase in temperature. This means that at higher air temperatures, saturation vapor pressure can be significantly higher than that for temperatures that are only somewhat cooler. Reaching a high saturation vapor pressure means there would be a lot of water vapor molecules in the air.

When air is quite dry, the actual vapor pressure will be significantly lower than the maximum possible. Since relative humidity (RH) is the ratio of actual vapor pressure to saturation vapor pressure, changes in these elements will lead to a change in RH. A temperature rise increases the saturation vapor pressure, but if the actual vapor pressure does not change, RH lowers and vice versa. For RH to change, either air temperature or amount of atmospheric moisture must change, both of which could happen at the same time since they are independent elements.

Now it is clear how RH affects fuel moisture content. Fuel is a surface that is constantly exchanging moisture with the air. RH affects dead fuel moisture by hygroscopic diffusion. Dead and dormant fuels lose moisture to the air when RH is low and absorb moisture when RH is high. The rate of moisture exchange with the air is related to the difference between the air and fuel moisture contents (vapor pressure is involved!), fuel temperature, and the surface area to volume ratio of the fuels. Moisture exchange continues until equilibrium moisture content is reached, but there is a timelag involved in that process (see chapter 4), so dead fuels often have long periods of being in a state of ongoing moisture content change. Extremely low RH can also affect live fuel moisture content as plants transpire increasingly more water vapor as temperature increases.

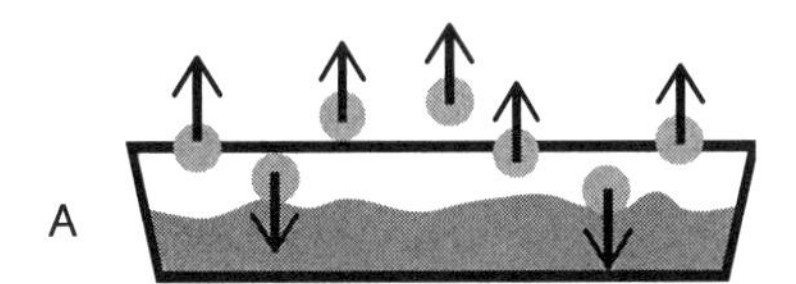

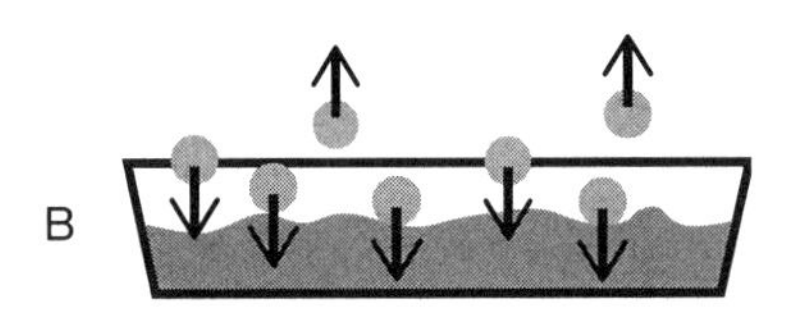

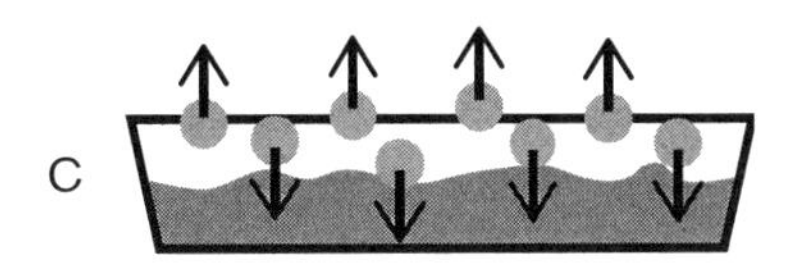

FIGURE 3.1.1 Net motion of water vapor molecules over an open dish of water during states of evaporation (A), condensation (B), and saturation (C) (adapted from Schroeder and Buck 1970).

Trend patterns of fuel moisture content and RH are similar, and sudden changes in RH are reflected in fuel moisture content (Steen 1963). Small fuels have a large surface area to volume ratio and are more reactive to environmental moisture changes. Of course, the bottom line remains that when all fuels become dry enough they will burn, and when they become too moist to sustain combustion, they won't.

Atmospheric pressure also plays a role in saturation vapor pressure, but that was not included in this general discussion. However, psychometric tables appropriately take atmospheric pressure into account which explains why there are elevation-range-specific tables to use when looking up RH and dew point for a given set of dry and wet-bulb temperatures.

Precipitation is another important element of environmental moisture. Its effects on dead or dormant fuel moistures vary with the measured water content, rate, location, and type of precipitation. For example, a rainfall rate of 2.54 cm (1 in) in 48 hours will lead to a more sustained moist environment compared to where that amount falls in an hour. Precipitation type may also affect the duration or location where moisture lingers. For example, a thunderstorm may produce heavy rain for a short duration over a narrow surface path, but snow is likely to linger on a fuelbed and extend the time moisture has an effect.

Diurnal Temperature Ranges and Atmospheric Moisture

Moist environments typically have a smaller daily temperature range than dry ones. Water has high specific heat, so atmospheric moisture absorbs heat during the day and sheds it at night. Moisture acts as a heat sink during the day that keeps temperatures lower than if the air were dry. When cooling overnight, moisture sheds its heat and keeps air warmer than it would be without that moisture. Dry air will have higher daytime temperatures and cooler nighttime ones than moist air will, hence have a larger daily temperature range, and vice versa.

Moisture evaporating from a surface cools it, and the same principle applies when moisture evaporates into air. A good example is water sprayed into a fine mist to cool outdoor patio areas. Conversely, when water vapor condenses in air, the heat shed by the cooling water provides atmospheric warming. This kind of heating is a very important process of cumulus cloud dynamics that helps their vertical development.

Temperature and Moisture Effects on Fuels

Ambient air temperature affects fuel temperature, a key factor determining when fires start and how they spread (Schroeder and Buck 1970). The amount of heat necessary to evaporate fuel moisture and raise fuel temperature to the ignition point is directly related to initial fuel temperature. As air temperatures rise, less additional fuel heating is required for ignition. Smaller fuels react to environmental temperature changes more quickly than larger fuels. Air temperature influences fire effects such as scorch height. Temperature's influence on atmospheric moisture, stability, and wind indirectly affects fire behavior.

Atmospheric moisture directly affects dead, cured, or growth-season dormant fuels, which are the ones most involved in ignition and combustion sustainability. Seasonal precipitation patterns across a region affect vegetation types that make up an area's fuel matrix (Schroeder and Buck 1970). California has a Mediterranean climate characterized by a mild, moist winter and hot, dry summer (Trewartha and Horn 1980). Seasonal precipitation distributions also affect fuel conditions through influences on soil moisture and plant growth cycles. For example, under drought conditions in California, there may be a lack of winter rain to promote annual grass growth. Conversely, a very wet winter may produce an abundance of annual grass growth. Because cured (dry) annual grass is a major carrier of fire, fire intensity is usually lower where there is less grass.

It's easy to understand that precipitation directly affects fuel moistures, but amounts are also significant. Amounts of 2.54 mm (0.10 in) in grass and brush and 6.35–12.7 mm (0.25–0.50 in) in fairly dense timber canopies are considered minimums to bring sufficiently wetting conditions to surface fuels and slow or limit fire activity (NWCG 2007).

Relative humidity serves as an environmental threshold because RH helps drive moisture movement between air and fuels. There are relationships between RH, fuel moisture, and resulting fire activity depending on fuel type and location (Steen 1963). The lower the RH, the more quickly smaller fuels will dry to the point of being able to ignite, sustain combustion, and carry a fire through the landscape. Conversely, when RH rises there will be a point when fuel moisture becomes too high to sustain combustion, and the rate of fire spread will decrease or stop. This relationship often leads to diurnal fire spread patterns of active burning during the warmest, driest portion of the day (usually afternoon and evening) and significant reductions in fire activity during the coolest, moistest time of day (usually overnight). The threshold values of high and low RH significant to fire behavior vary by region and fuel type.

Atmospheric Stability

Atmospheric stability relates to vertical motions within the atmosphere. General definitions of stable and unstable provide insight into the amount of atmospheric motion involved. Stable air moves vertically only a little or reluctantly, and unstable air will support active vertical motions.

A good analogy for effects of stability on wildland fire behavior is a woodstove. With the damper shut in a chimney, air is slow to draw through the firebox up the chimney like that in stable atmospheric conditions, and fire intensity will be relatively low. Fully opening the damper will maximize air draw through the firebox up the chimney similar to air motion in an unstable atmosphere, and the fire will burn actively. Stable conditions tend to limit smoke transport and adversely affect air quality. Unstable conditions encourage and support relatively active fire burning conditions overall, sometimes to the point of extreme fire behavior such as fire whirls or long range spotting.

ATMOSPHERIC TEMPERATURE LAPSE RATES AND STABILITY

Stability is determined by the rate of temperature change with altitude (lapse rate). Environmental lapse rates are measured by weather balloons and range in value along a continuum from very stable to very unstable. Most episodes of instability are due to latent heating from condensation and precipitation under deep convection (conditional instability). Although conditions are normally labeled stable or unstable at any given time, environmental lapse rates vary through the atmospheric column over an area and can be a mixture of stable and unstable layers. Even when all layers are stable or all unstable, there will be varying degrees of stability between layers through the atmospheric column. Stable environmental lapse rates are defined by temperature changes less than 9.8°C km^{-1} (5.5°F $1{,}000^{-1}$ ft^{-1}). Unstable is defined as more than 9.8°C km^{-1} (5.5°F $1{,}000^{-1}$ ft^{-1}), and if the atmospheric lapse rate equals

9.8°C km^{-1} (5.5°F $1,000^{-1}$ ft^{-1}), conditions are referred to as neutral. Lapse rates commonly change from day to day.

When unsaturated air is lifted through the atmosphere, it is assumed to cool adiabatically at 9.8°C km^{-1} (5.5°F $1,000^{-1}$ ft^{-1}). Environmental lapse rates cooling faster than that means the rising air will remain warmer thus lighter than the environment and rise (unstable conditions). For environments that cool more slowly than the lifted air, the lifted air remains cooler thus heavier than the environment and will not be able to rise (stable conditions). If both environmental and lifted air cooling rates are the same, lifted air stays where it's moved (neutral stability) (Schroeder and Buck 1970). If the air cools to the point of saturation, the adiabatic cooling rate slows down to 5.8°C km (3.2°F $1,000^{-1}$ ft^{-1}) due to the latent heat of condensation. This helps make moist air buoyant and contributes to unstable conditions. This is one reason cumulus clouds can build to very high altitudes.

Unstable atmospheric conditions can be supported several ways. Orographic lift is a mechanical process where air flowing along in the atmosphere encounters a mountain range and is forced to lift over it. Frontal zones also have active vertical motions where the leading edge of one air mass is replacing a resident air mass. Because air in a cold front is denser (heavier) than the warmer air it is replacing, the dense air stays near the ground, wedges underneath the warm air, and lifts it over the top of the cold air. When a warm front moves in, warm air lifts (floats) over the denser cold air as it retreats. Low pressure weather systems are a rich source of vertical motions because air flows from high to low pressure areas. Air will converge into low pressure centers from all directions since pressure increases with distance from the center of the low. Although air is a compressible gas, it cannot do so when converging at a given altitude, so it is forced to move upward in the center of lows. A frontal boundary (low pressure extending away from a cyclone's center) develops because of upward vertical motions created by positive vorticity advection in waves of the jet stream in combination with air friction at the Earth's surface. Consequently, upper level air mass divergence exceeds low-level convergence, with the net loss of mass resulting in low pressure. Adiabatic cooling of this rising air leads to clouds and precipitation in the right circumstances, which is why we associate lows with inclement weather. Air can also be lifted thermally such as when near-surface warmed air rises (convection). The jet stream (very strong winds aloft) will sometimes provide a significant amount of atmospheric lift. At California's latitudes the jet stream is generally found at about 9.1 km (30,000 ft) or higher with speeds generally starting around 45 m sec^{-1} (100 mi h^{-1}), although it can be much faster. These strong winds aloft can essentially pull air up from lower altitudes, sometimes quite rapidly. Thunderstorms that form in this environment can develop rapidly.

ATMOSPHERIC INVERSIONS

Under certain conditions, temperature will increase with altitude. This is called an inversion and is a very stable atmospheric condition. Air at the lower altitudes of an inversion layer is cooler than that above it, and this decisively discourages vertical motion. Inversions can occur at any level in the atmosphere. The cooler temperatures associated with surface-level inversions mean higher RH with the resulting increased fuel moisture often becoming high enough to significantly curtail fire activity.

Marine Inversion California's coastal areas frequently experience a marine layer inversion. The cold water of the Pacific Ocean cools the air in contact with it, and when the air over the land is warmed and rises, cool marine air flows ashore to fill in behind it. This is an important summer weather pattern along the California coast and through The Delta of the San Joaquin and Sacramento Rivers.

Overnight Radiation Cooling Inversion Mountain valley and drainage areas can experience a multi-day cycle of overnight surface-based inversions that dissipate during the day. Ground-cooled air will flow downhill at night and pool in valleys and drainage bottoms with the coolest (heaviest) air settling into the lowest points. Even the very large Sacramento and San Joaquin Valleys can develop these. Such inversions will usually dissipate the next day once insolation heats the ground sufficiently to create enough convection to overturn air in the drainage. If smoke is dense in a drainage or clouds cover an area, it takes longer for the sun to heat the ground sufficiently to initiate dissipating convection. Under the right conditions, a rather warm, dry zone develops along the mid to upper slopes in drainages above an inversion. This zone usually forms at night and is call a "thermal belt." Fuels in thermal belts can remain dry enough to burn actively all night.

Subsidence Inversion When a very large, strong high pressure system is over the western states for a week or more, a widespread subsidence inversion can develop. Air slowly subsides in high pressure systems, which leads to warming and drying of the atmosphere overall. Sometimes an inversion layer develops aloft in this pattern where air becomes very warm and dry from the subsidence. These events are so large they clearly show up on routine weather balloon soundings over a region. When the inversion eventually sinks down and comes into contact with Earth's surface (highest elevations first), there will be rather warm temperatures and unusually low RH for several days, even overnight. This fire weather pattern usually happens in summer and leads to very low fuel moistures.

DIURNAL PATTERN OF STABILITY

There is a typical diurnal stability pattern. Afternoons are generally the most unstable time of day due to strong surface heating, and this helps afternoons frequently be the most active burning period of the day. The most stable time of day is generally overnight, and fire activity usually decreases under the more stable conditions occurring then.

Many things can interrupt this typical diurnal pattern, however. For example, if sunlight is blocked by clouds or a large smoke column during the day, surface heating is less than on a sunny day and conditions are likely to be less unstable. A wind pattern that dominates the landscape regardless of the time of day (e.g., a cold front moving through or Santa Ana winds blowing) will also upset the diurnal stability pattern and effects of its cycle.

Wind

Wind is the most variable and least predictable of the three fire weather elements and is the horizontal motion of air.

Wind affects fuel moisture and fire behavior in many ways. By carrying away moisture in the air, wind dries fuels and supplies fresh oxygen to aid or accelerate combustion just as when we gently fan air across a new campfire to get it burning. Rate and direction of fire spread are greatly influenced by wind speed and direction as are smoke and ember transport. Embers deposited downwind can start spot fires ahead of the main fire and increase a fire's lineal and areal rate of growth. Fire spread is also enhanced by heat transfer of convection and wind bending flames toward unburned fuel to preheat it.

Wind is fundamentally caused by uneven heating of Earth's surface. Thermal differences create pressure differences, and air will flow from the resulting areas of higher to lower pressure. The larger the temperature difference across an area, and the stronger the pressure gradient (pressure change over distance), hence the faster the wind blows. Diurnal and seasonal pressure variations create established wind patterns across an area.

Thermal balance across Earth is never achieved, so atmospheric motion is constant. On a global scale, Earth's rotation increases the complexity of general atmospheric motions and provides defined belts of prevailing wind directions (see chapter 2). California is mostly in the belt of westerlies, which explains why most major weather systems come from the west. Seasonal wind patterns occur as well, as when southerly flow around the western edge of summertime high pressure in the desert southwest brings monsoon moisture into California.

WIND DIRECTION

With reference to a compass, wind direction is the direction *from* which the air is coming. A southwest wind would make a fire move toward the northeast. With reference to the ground, i.e., upslope or down drainage, wind direction would be the direction the air is *going*. Near the coast, ground-referenced directions include a sea breeze that blows air onto the land (onshore flow), and a land breeze that moves air from the land out to sea (offshore flow).

GENERAL WINDS

In fire weather, winds aloft over an area are called general winds. They start at about 600–900 m (2,000–3,000 ft) above the ground where the effects of surface friction cease and continue well up into the atmosphere. General wind speed and direction help determine whether that wind might have an effect in the landscape. Exposed locations at higher elevations will have the most interaction with lower altitude general winds. General winds can mix down toward the surface when conditions are unstable, which is fairly common in the afternoon when strong surface heating creates convection.

SLOPE AND VALLEY WINDS

Surface temperatures drive air temperatures; compared to its surroundings, warmer air rises because it is less dense (buoyant), and cooler air sinks because it is more dense (heavier). Duration of sunlight, aspect, and time of day primarily drive local scale, diurnal slope and valley wind dynamics due to differential heating. When clouds or thick smoke block sunlight, daytime diurnal winds are lighter with rather variable directions. General wind can slow down or enhance speeds of local valley winds when the wind is aligned with the layout of the terrain. A synoptic scale weather pattern such as an approaching cold front can provide a dominating wind pattern that will override any local diurnal one.

In California, east aspects are the first to receive sunlight and start warming in the morning. At midday, southern aspects receive the most intense energy per unit surface area and warm readily. By midafternoon, shadows are beginning to form on northeasterly aspects while southwest and west aspects receive their most direct sunlight. South and southwest aspects often have the strongest slope winds of the day do to this timing. North aspects have the lightest winds as they receive the least amount of insolation.

The local wind pattern in an area is typically predictable when the wind is driven by local differential heating of the topography. Such winds tend to be fairly light in the morning and become significantly stronger after noon. Overnight wind speeds are very often calm to 5 km h^{-1} (0–3 mi h^{-1}) on all aspects and flow down a slope or drainage.

6 M (20 FT) WIND

Fire weather observing stations and forecasts are set to measure and provide predictions of the wind 6 m (20 ft) above the average height of the vegetation. Wind at that height is made up of a combination of local slope and valley winds and whatever contribution is provided by the general wind, if any. Locations in the upper third, windward side of the landscape are most affected by the general wind. Locations on the lee (downwind) side of a mountain range tend to be protected from general winds. If conditions are unstable or the general wind is particularly strong, the general wind is more likely to influence the 6 m (20 ft) wind.

FIRE SPREAD AND SLOPE WIND

Fire on a slope will readily spread uphill much faster than it will on flat terrain all else being equal. Slope steepness is considered to have a wind-like effect on fire behavior in that fire spreads readily up a slope as before a wind and fastest up steep slopes (NWCG 2007). When an upslope wind is in place, slope and wind effects combine to create faster fire spread than slope or wind would generate individually.

SEA BREEZE

There is often a marked afternoon sea breeze near the coast driven by land-warmed air rising and the cool marine layer rushing ashore to replace it. This keeps the immediate coastal areas cool and moist compared to just a short-distance inland and has a major influence on coastal fire weather patterns. There is also acceleration of surface winds at midday regardless of direction or individual air mass circulations. Surface solar heating reduces atmospheric viscosity, allowing the momentum of airflow aloft to come down the surface. Nighttime surface cooling reverses this process.

A cool, moist onshore flow can persist for days, and a marine inversion with low clouds or fog and possibly drizzle can develop and remain in place. Persisting fog and low clouds is a common summer pattern near the California coast north of Cape Mendocino. The depth of the marine layer is variable, and when it is relatively thin at around 240–360 m (about 800–1,200 ft), sunny periods regularly develop in the afternoon while deeper marine layers can often keep skies cloudy all day.

TRANSPORT WINDS AND SMOKE DISPERSION

Smoke transport is a key element of air quality and public health near fires. Mixing height is the distance above ground to which smoke can rise. Transport winds are the average wind speed and direction through the depth of that mixing layer. The best smoke dispersion conditions are when smoke rises to a level well above ground before transport winds take effect. In such conditions, transport winds carry smoke away from a fire area with minimal impact on air quality at ground level. However, when overnight local down valley, down drainage winds become the local transport winds, smoke will remain at ground level and degrade air quality in such situations.

Significant Fire Weather Winds

There are a few specific strong wind events or patterns that take over the environment and override any usual local wind. These winds can cause intense burning and rapid-fire spread. Some are short lived like thunderstorm downdraft outflows while others can last for days such as dry, warm Santa Ana winds.

COLD-FRONT WINDS

When a cold front moves into an area, temperature differences between the incoming cold air and the warm air it is replacing drive the strength of wind associated with this weather pattern. The larger the difference in temperature between the two air masses, the higher the wind speeds will be as this dynamic weather change moves through an area. In California, cold fronts typically have moderate to strong southerly winds a day or two ahead of their arrival and much lighter westerly winds within an hour or so of the frontal boundary passing by. Wind speed normally increases as a front approaches and is strongest within the few hours before the front arrives.

As the front approaches and passes, wind direction often goes from rather southerly before its arrival to shifting 90 degrees to become west or northwest once it passes. When a front is strong, this wind shift may take as little as 15 min to occur (sometimes less). These strong, shifting winds can quickly turn a fire's flank into the head of the fire as the front passes through.

Much of California experiences rather weak or no cold frontal passages during the core of summer fire season, and cold fronts in summer seldom reach southern California between late June and early September. Dry cold fronts have little or no precipitation but often have strong, gusty, dry winds and unstable air.

THUNDERSTORMS

Thunderstorms are a major source of fire ignitions. They are essentially an atmospheric convection column produced when sufficient atmospheric instability, moisture, and an air lifting mechanism all work together in just the right way. Common lifting mechanisms in California are orographic and thermal processes as evidenced by the frequency we see thunderstorms form over the mountains. Migratory weather systems provide lift fairly frequently as well. The strength, size, and nature of a thunderstorm depend on the atmospheric conditions at the time of development. If a thunderstorm produces enough rain at ground level, it will help minimize the chance of new fire starts or increase fuel moistures enough to keep fire spread rates rather low.

There is a typical three-stage evolution to thunderstorm development and decay (Fig. 3.2). The stages are called cumulus, mature, and dissipating (Schroeder and Buck 1970). The cumulus stage is when the cloud begins to build vertically and is all updraft. The cloud becomes taller in the mature stage and will begin producing lightning. This stage is a fully developed cumulonimbus cloud that often has an anvil-shaped top. An updraft and a downdraft will be occurring in a mature stage cloud. The downdraft is initiated when precipitation begins. During the dissipating stage, a thunderstorm is essentially a rapidly decaying convection column with a strong downdraft.

The strong, shifting winds associated with thunderstorms are problematic for ongoing fires. Fires near thunderstorms may experience both an indraft wind when the cloud's updraft is strongest followed by a reversal in wind direction when the gusty downdraft winds occur. In California, thunderstorm downdraft winds are typically much stronger than the indraft winds. High-based, dry thunderstorms will often have the strongest gusts of all thunderstorms. Thunderstorm downdraft winds spread out in all directions from the base of the storm cloud once they reach the ground. The length of time these winds significantly affect a fire is often on the order of 10–20 min. A fire within a few kilometers (miles) of the storm base can easily be affected by the downdraft wind, particularly in relatively flat terrain.

Precipitation from high-based thunderstorms is likely to fall as *virga* (rain that evaporates before reaching ground). Virga-cooled air under a thunderstorm is denser (heavier) than its surroundings and sinks to the ground readily. Gravity helps that cool, heavy air gain considerable speed as it travels the rather long distance under a high-based cloud to the ground.

FÖHN (FOEHN) WINDS

Warm, dry, winds that flow down the lee (downwind) side of mountain ranges are called föhn (foehn) winds. They develop when overall atmospheric conditions are stable, and there is strong high pressure and corresponding low pressure on the opposite sides of a mountain range (Schroeder and Buck 1970). The flow of air from the high to the low across the mountain range will create strong downslope winds on the lee side. As the air descends to lower altitudes on the lee side, it warms and dries quite dramatically. Föhn wind speeds can become very strong.

The best-known example of a föhn wind in California is the Santa Ana wind in southern California. Santa Ana winds involve the mountains and coastal plains throughout

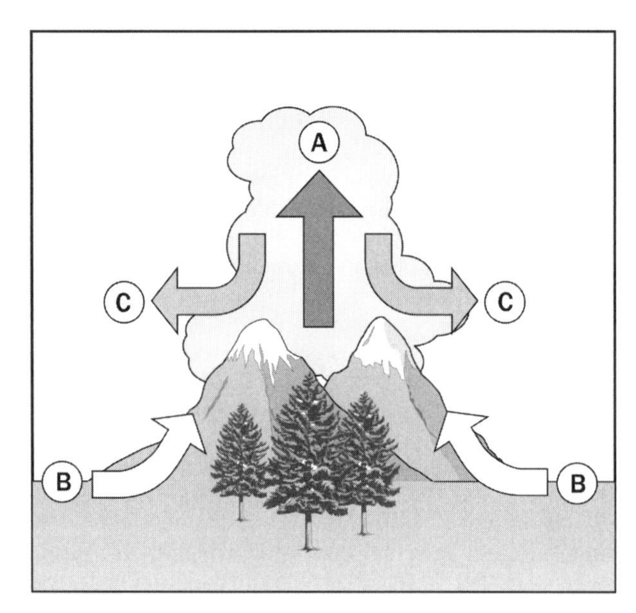

FIGURE 3.2 In a thunderstorm, the cloud begins to build vertically during the cumulus stage (A), causing an updraft of air near the surface (B). During the mature stage, the cloud becomes larger and lightning is produced. The downdraft (C) occurs as long as there is evaporative cooling of the air by precipitation.

southern California. The strongest föhn winds, produced by the Bernoulli effect, run through mountain passes from Palmdale to Oxnard and Santa Monica, Cajon and San Gorgonio Pass through Riverside, Santa Ana Canyon to the Orange County coast, and through the low point in the mountains along the international boundary to San Diego. Under strong upper (general) winds from the east or northeast, mountain waves develop and Santa Ana winds come to the surface throughout southern California, not just through the passes. The Sierra Nevada create a barrier to offshore winds under high surface pressure in the Great Basin, thereby diverting boundary-layer flow southward through southern California. Fires that burn under föhn wind conditions are usually intense and spread quickly. However, because the atmosphere is actually quite stable in this weather pattern, wind-driven fires quickly lose intensity and spread slowly or not at all when föhn wind speeds drop significantly.

Southern California is not the only area of the state to experience föhn winds. The northern Sacramento Valley, its western foothills, and the Bay Area, particularly the East Bay hills, are some other areas that have notable föhn wind events. Most föhn winds in California are easterly, except in the northern and western Sacramento Valley where they are northerly. These winds often last three or more days and can develop a diurnal pattern (Schroeder and Buck 1970). Föhn winds are often strongest overnight and become noticeably lighter during the afternoon hours when local, diurnal wind velocities oppose the föhn winds' velocity.

WHIRLWINDS

Fire whirls and dust devils are relatively short-lived, small diameter events. They are updrafts that spin into a vortex in unstable atmospheric conditions. The updrafts can be caused by strong surface heating (dust devil) or heat from a fire (fire whirl). The stronger the updraft is, the stronger the vortex. In extreme cases, large fire whirls can last a few hours (Schroeder and Buck 1970) and be very large.

The travel path of fire whirls is unpredictable. Some remain fairly stationary during their life span, while others are pushed along by surface winds (Schroeder and Buck 1970) and spread fire along the ground as they move. Fire whirls can also loft burning materials like cones or branches into the fire's convection column that will get carried downwind and start new (spot) fires when they fall down to the ground.

California Dry Season Fire Weather

Each of the components of fire weather varies widely across California. During California's dry season, most areas except near the coast experience weather that can propagate fire virtually every day. In the nineteenth century before fire control, free-burning fires persisted for months in California wildlands (Minnich 1987, 1988). Prehistoric fires that shaped California ecosystems likely burned during "average" weather. It is therefore important to not only describe extremes of fire weather, but also the entire spectrum of fire weather during fire season. Because atmospheric circulation differs fundamentally in summer and fall, the following discussion is divided by season. Emphasis is made on weather conditions at maximum midday fire potential.

Summer

Summer climate occurs in a "steady state" of onshore flow, a persistent marine layer along the coast, and an overlying temperature inversion that divides the marine layer from warm, dry subsiding air aloft. Fire potential increases due to low RH when the marine layer is shallow, but day-to-day wind direction and speed that steer fire movement are predominantly invariant. Prevailing surface winds over California are northwest in response to the East Pacific anticyclone-thermal low pressure configuration over the southwest United States, but local winds are strongly influenced by terrain (Map 3.1). The gap in the Coast Ranges at San Francisco Bay results in divergent low-level wind fields inland, with south winds in the Sacramento Valley and northwest winds in the San Joaquin Valley. In the southern California bight, offshore northwest winds are southwest to southeast and called the southern California eddy. With increasing distance inland, land heating dissipates the marine layer and warm, dry air aloft mixes down to the surface.

Prevailing winds also comprise afternoon sea breezes and anabatic circulations on mountain slopes (Steen 1963, DeMarrias et al. 1965, Hayes et al. 1984, Zack and Minnich 1991). Local winds predominate because upper level winds are less than 10 m s^{-1} (about 22 mi h^{-1}), which disconnect low-level winds from the upper flow (Ryan 1977, 1982). The coupling of surface wind fields with local terrain results in predictable motion of head fires. With anabatic circulations, warm slopes create buoyant surface air layers that move upslope across a valley's axis, and onshore valley flows that move parallel to a valley's axis (Stull 1988). Because the marine layer is stable, airflows are also channeled through complex terrain (Böhm 1992) as documented in the Los Angeles Basin and central California (Fosberg et al. 1966, Schroeder et al. 1967, Edinger et al. 1972, Schultz and Warner 1982, Hayes et al. 1984, Lu and Turco 1995).

MAP 3.1 Prevailing surface winds at 2:30 pm local daylight time in July (data from Ryan 1977, Minnich and Padgett 2003).

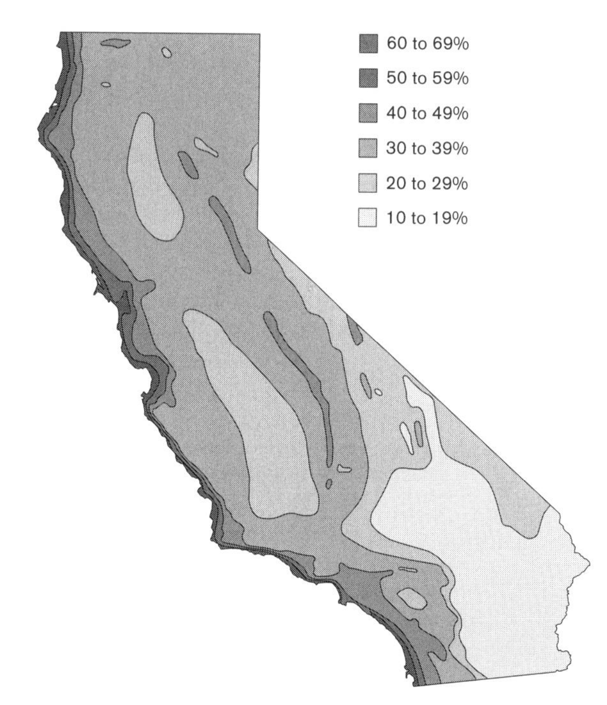

MAP 3.2 Mean relative humidity at 2:30 pm local daylight time in July (data from Ryan 1977, Minnich and Padgett 2003).

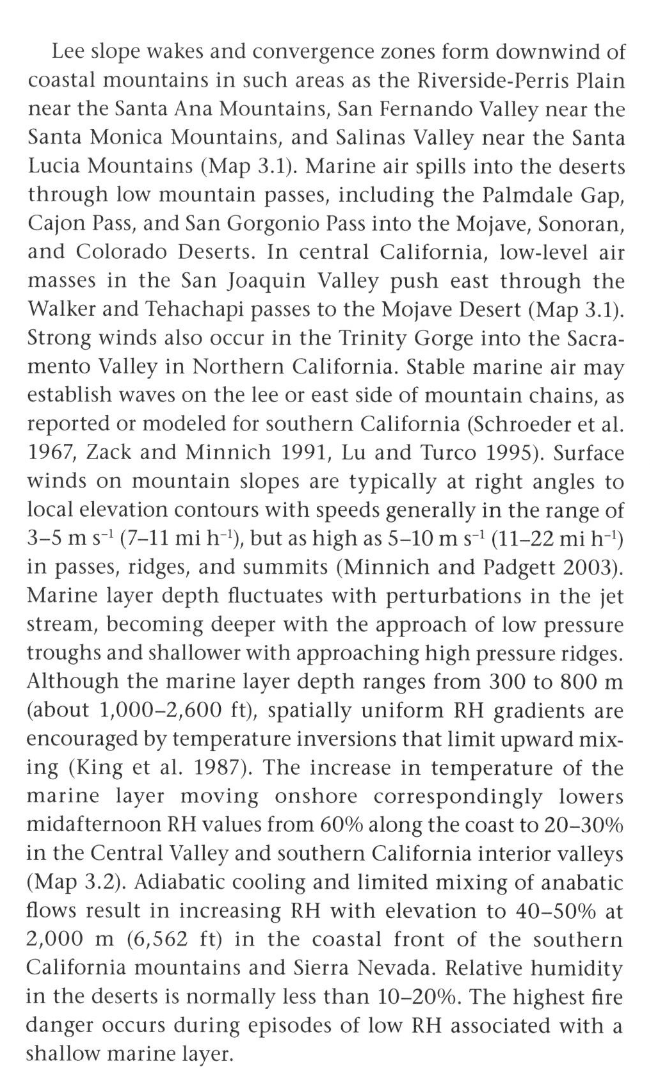

Lee slope wakes and convergence zones form downwind of coastal mountains in such areas as the Riverside-Perris Plain near the Santa Ana Mountains, San Fernando Valley near the Santa Monica Mountains, and Salinas Valley near the Santa Lucia Mountains (Map 3.1). Marine air spills into the deserts through low mountain passes, including the Palmdale Gap, Cajon Pass, and San Gorgonio Pass into the Mojave, Sonoran, and Colorado Deserts. In central California, low-level air masses in the San Joaquin Valley push east through the Walker and Tehachapi passes to the Mojave Desert (Map 3.1). Strong winds also occur in the Trinity Gorge into the Sacramento Valley in Northern California. Stable marine air may establish waves on the lee or east side of mountain chains, as reported or modeled for southern California (Schroeder et al. 1967, Zack and Minnich 1991, Lu and Turco 1995). Surface winds on mountain slopes are typically at right angles to local elevation contours with speeds generally in the range of 3–5 m s^{-1} (7–11 mi h^{-1}), but as high as 5–10 m s^{-1} (11–22 mi h^{-1}) in passes, ridges, and summits (Minnich and Padgett 2003). Marine layer depth fluctuates with perturbations in the jet stream, becoming deeper with the approach of low pressure troughs and shallower with approaching high pressure ridges. Although the marine layer depth ranges from 300 to 800 m (about 1,000–2,600 ft), spatially uniform RH gradients are encouraged by temperature inversions that limit upward mixing (King et al. 1987). The increase in temperature of the marine layer moving onshore correspondingly lowers midafternoon RH values from 60% along the coast to 20–30% in the Central Valley and southern California interior valleys (Map 3.2). Adiabatic cooling and limited mixing of anabatic flows result in increasing RH with elevation to 40–50% at 2,000 m (6,562 ft) in the coastal front of the southern California mountains and Sierra Nevada. Relative humidity in the deserts is normally less than 10–20%. The highest fire danger occurs during episodes of low RH associated with a shallow marine layer.

Wind has less influence than RH on daily fire danger (related to fuel dryness). Regions having the strongest winds, such as mountain passes, have strong winds almost daily because onshore surface pressure gradients are a steady-state feature of the climate. The coastal plains near Santa Barbara experience strong, dry northwest to north winds called "sundowners" that descend the Santa Ynez Mountains in the evening. Sundowners take place under strong north-to-south pressure gradients between San Francisco and Los Angeles. The term refers to the onset of these winds with the dissipation of the sea breeze, usually at sundown.

Leeward mountain slopes and deserts not subjected to marine layer intrusions experience the lowest humidity and greatest fire hazard near the summer solstice (late June–early July), when maximum insolation supports the hottest air masses of the year. The establishment of the North American monsoon, usually by early July, brings slightly cooler tropical maritime air masses to these areas.

Nighttime radiational cooling results in downslope katabatic (cool air drainage) flows in the mountains and creates surface-based inversions in basins. Overnight RH increases to 40% or greater almost everywhere. Another predictable property of summer fire weather is that the diurnal shifts between daytime upslope wind and overnight down drainage wind typically occur around sunrise and sunset, respectively.

Autumn

The southward migration of the jet stream along the Pacific Coast in autumn brings greater day-to-day variability to California's fire weather, including episodes of dry offshore winds from the Great Basin. Fires spread rapidly in offshore wind episodes until the rainy season.

On average, fire danger decreases in autumn due to shorter days and cooling temperatures. Low pressure troughs of the

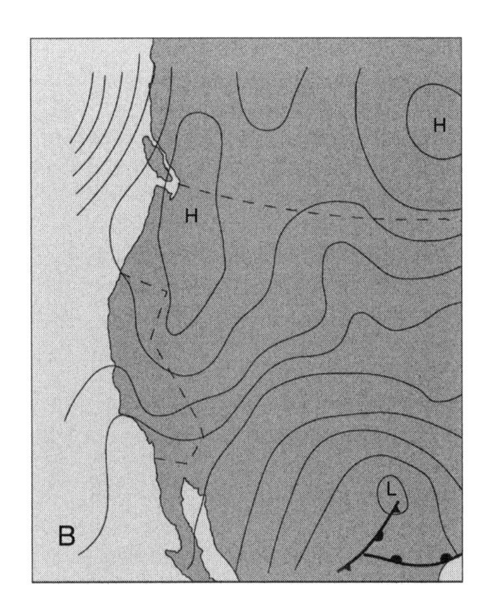

FIGURE 3.3 Weather maps of a strong Santa Ana wind event on November 10, 1970: (left) 500 mbar; (right) surface pressure. At 500 mbar, a trough in New Mexico and a ridge just off the Pacific Coast produce a strong north-northwesterly jet stream over California. Strong surface high pressure centered over the northern Great Basin combined with surface low pressure over Texas and extending west to the California coast cause strong northeasterly winds across southern California (redrawn from NOAA 1970).

jet stream frequently deepen the marine layer, bringing higher RH to the inland valleys and mountains than in summer. Offshore wind flows are generated when high pressure follows the passage of a cold front and strong subsidence develops over the western United States. Strong pressure gradients between high pressure in the Great Basin and low pressure along or just off the California coast bring dry surface air masses to coastal regions normally influenced by the marine layer (Fig. 3.3). Although offshore events are associated with cold air advection, temperatures are warm both in interior valleys and along the coast because of adiabatic warming of the descending air mass.

Because the Sierra Nevada is a high mountain range, post-frontal Great Basin air masses are deflected southward toward southern California, westward through the Trinity River region, or southward through the Sacramento Valley and San Francisco Bay. Southern California Santa Ana winds develop in mountain passes as low-level jets that exceed 20–30 m s^{-1} (about 45–65 mi h^{-1}), especially at Cajon Pass, Santa Ana Canyon, San Gorgonio Pass, the Palmdale Gap, and the Peninsular Range at the international boundary (Fosberg et al. 1966). Offshore winds expand throughout coastal southern California if atmospheric subsidence brings jet stream energy toward the surface, or if mountain waves develop leeward of the physiographic barriers. Southeast winds occasionally surface in the southern San Joaquin Valley and the coastal valleys near Santa Maria. Many areas directly affected by offshore flows experience the strongest winds and lowest humidity of the fire season. In contrast, the Central Valley south of Stockton, normally protected from offshore flows by the Sierra Nevada, experiences persistent ground inversions and tule fog.

The Fire Weather "Window" in Vegetation

A location's fire weather "window" is the full range of weather capable of propagating fires, ranging from a moist threshold to the driest weather of the site's climate. The window varies with changing fuel properties along vegetation gradients. Cured grassland with low fuel moisture has fine fuels that burn in moist conditions (broad window; greater than 50% RH). Woody shrub and forest assemblages with coarse evergreen sclerophyllous foliage and live fuels with high fuel moisture burn in a drier window than grasslands. Open woody assemblages such as pinyon-juniper (*Pinus-Juniperus*) and desert scrub with low fuel continuity in open stand arrangements burn in extreme weather exclusively (narrow window) (chapter 18). Hyper-arid deserts and mountains above the tree line may be fireproof.

In woody vegetation types, the weather window enlarges with time-since-fire due to cumulative fuel buildup (fuel loading), increasing dead fuel in the canopy, and increasing landscape transpiration demand. These factors increase the carbohydrate-to-water energy ratio. In principle, if patch flammability increases with time-since-fire, the resulting differential weather window with age class contributes to nonrandom turnover in patch mosaics (patch heterogeneity dependent on fire history).

Post-fire flammability increases if vegetation recovery is initially dominated by herbaceous fuels, leading to a positive feedback in which fires encourage shorter-interval recurrences, a significant problem with nonnative invasive in wild oat (*Avena*) and brome (*Bromus*) genera (D'Antonio and Vitousek 1992) (see sidebar 18.1). Not all herbaceous species are flammable. Most wildflower species in the California coastal plains and desert, forbs in early recovery of chaparral and coastal sage scrub, plus a few nonnatives such as redstem filaree (*Erodium cicutarium*) disarticulate upon curing, leaving little persistent fuel. Forbs also decline in abundance as they are shaded out by the development of woody overstory vegetation (Minnich 2008).

Summary

Meteorologically speaking, atmospheric temperature and moisture relationships are primarily about relative humidity and what kind, location, and duration of precipitation. From a wildland fire point of view, moisture is all about fuel moisture and whether combustion is sustained. Atmospheric stability largely relates to how well ventilated the atmospheric column is above ground, which for weather means how likely thunderstorms are to form, but to a fire means how active the convection column and fire intensity can be. Wind has a great deal of power over fire and plays a major role in directing its movement, but its role in atmospheric moisture transport is also important.

Weather and climate's influence on the wildland fire environment make them important elements of fire ecology. Weather's daily and seasonal variability may lead to an impression it's not very predictable, but the meteorological principles discussed in this chapter lead to reliably established weather patterns that are useful to fire ecologists.

References

Albini, F.A. 1993. Dynamics and modeling of vegetation fires: observations. Pages 39–53 in: J.J. Crutzen and J.G. Goldammer, editors. Fire in the Environment: The Ecological, Atmospheric, and Climatic Importance of Vegetation Fires. Dahlem Workshop Reports, Environmental Sciences Research Report 13. John Wiley and Sons, Chichester, UK.

Böhm, M. 1992. Air quality and deposition. Pages 64–152 in: R.K. Olson, D. Binkley, and M. Böhm, editors. The response of western forests to air pollution. Ecological Studies, Vol. 97. Springer Verlag, New York, New York, USA.

D'Antonio, C.M., and P.M. Vitousek. 1992. Biological invasions by exotic grasses, the grass/fire cycle, and global change. Annual Review of Ecology and Systematics 23: 63–87.

Edinger, J.G., M.H. McCutchan, P.R. Miller, B.C. Ryan, M.J. Schroeder, and J.V. Behar. 1972. Penetration and duration of oxidant air pollution in the South Coast Air Basin of California. Air Pollution Control Association Journal 2: 882–886.

Fosberg, M.A., C.A. O'Dell, and M.J. Schroeder. 1966. Some characteristics of the three-dimensional structure of Santa Ana winds. USDA Forest Service Research Paper PSW-RP-30. Pacific Southwest Forest and Range Experiment Station, Berkeley, California, USA.

Hayes, T.P., J.J.R. Kinney, and N.J.M. Wheeler. 1984. California surface wind climatology. Aerometric Data Division. California Air Resources Board, Sacramento, California, USA.

King, J.A., F.H. Shair, and D.D. Reible. 1987. The influence of atmospheric stability on pollutant transport by slope winds. Atmospheric Environment 21: 53–59.

Lu, R., and R.P. Turco. 1995. Air pollution transport in a coastal environment-II. Three dimensional simulations over Los Angeles basin. Atmospheric Environment 29: 1499–1518.

Minnich, R.A. 1987. Fire behavior in southern California chaparral before fire control: the Mount Wilson burns at the turn of the century. Annals of the Association of American Geographers 77: 599–618.

———. 1988. The biogeography of fire in the San Bernardino Mountains of California: a historical survey. University of California Publications in Geography 28: 1–120.

———. 2008. California's Fading Wildflowers: Lost Legacy and Biological Invasions. University of California Press, Berkeley, California, USA.

Minnich, R.A., and P.E. Padgett. 2003. Geology, climate and vegetation of the Sierra Nevada and the mixed-conifer zone: an introduction to the ecosystem. Pages 1–31 in: A. Bytnerowicz, M.J. Arbaugh, and R. Alonso, editors. Ozone Air Pollution in the Sierra Nevada: Distribution and Effects on Forests. Elsevier, Amsterdam, The Netherlands.

National Oceanic and Atmospheric Administration. NOAA Central Library. Collections. 1970. U.S. Daily Weather Maps. November 13, 1970. http://library.noaa.gov/Collections/Digital-Documents/US-Daily-Weather-Maps.

NWCG. 2007. Intermediate Wildland Fire Behavior S-290 Student Workbook, NFES 2891. http://www.nwcg.gov/pms/pubs/glossary/w.htm.

Ryan, B.C. 1977. A mathematical model for diagnosis and prediction of surface winds in mountainous terrain. Journal of Applied Meteorology 16: 571–584.

———. 1982. Estimating fire potential in California: atlas and guide for fire management planning. USDA Forest Service, Pacific Southwest Forest and Range Experiment Station. Riverside, California, USA.

Schroeder, M.J., and C.C. Buck. 1970. Fire Weather . . . a Guide for Application of Meteorological Information to Forest Fire Control Operations. USDA Forest Service Agricultural Handbook 360. Washington, D.C., USA.

Schroeder, M.J., M.A. Fosberg, O.P. Cramer, and C.A. O'Dell. 1967. Marine air invasion of the Pacific Coast: a problem analysis. Bulletin of the American Meteorological Society 48: 802–808.

Schultz, P., and T.T. Warner. 1982. Characteristics of summertime circulations and pollutant ventilation in the Los Angeles basin. Journal of Applied Meteorology 21: 672–682.

Steen, H.K. 1963. Relation between moisture content of fine fuels and relative humidity. USDA Forest Service Research Note PNW-RN-4. Pacific Northwest Forest and Range Experiment Station, Portland, Oregon, USA.

Stull, R.B. 1988. An Introduction to Boundary Layer Meteorology. Kluwer Academic Publishers, Dordrecht, The Netherlands.

Trewartha, G.T., and L.H. Horn. 1980. An Introduction to Climate. McGraw-Hill, New York, New York, USA.

Warner, C.D. 1897. Editorial, August 27. Hartford Courant. Hartford, Connecticut, USA.

Zack, J.A., and R.A. Minnich. 1991. Integration of geographic information systems with a diagnostic wind field model for fire management. Forest Science 37: 560–573.

CHAPTER FOUR

Fire as a Physical Process

JAN W. VAN WAGTENDONK

Where there's smoke, there's fire.
ANONYMOUS

In many California ecosystems, the process of decomposition is too slow to completely oxidize accumulated organic material, and another process, fire, steps in to perform that role. The Mediterranean climate in California, with its hot, dry summers and cool, wet winters, is not conducive to decomposition. When it is warm enough for decomposer organisms to be active, it's too dry. Conversely, when it's wet enough, it's too cold. As a result, decomposition is unable to keep up with the deposited material, and organic debris begins to accumulate. This debris becomes fuel available for the inevitable fire that will occur. All that is needed is a sufficient amount of fuel, an ignition source, and weather conditions conducive to burning. In this chapter we will look at fire as a physical process including combustion, fuel characteristics, fuel models, ignition sources, fire behavior and mechanisms for fire spread, and fire effects.

Combustion

Combustion is one of many types of oxidation processes. These processes combine materials that contain hydrocarbons with oxygen and produce carbon dioxide, water, and energy. Additional products are produced if combustion is not complete. Oxidation is the reverse of photosynthesis, where energy is used in combination with carbon dioxide and water to produce organic material. The rate of oxidation can vary from the slow hardening of a coat of linseed oil in a paint film to the instantaneous explosion of a petrochemical. Combustion is a chain reaction that occurs rapidly at high temperatures.

Combustion Chemistry

Byram (1959) presents the chemical equation for combustion using a formula for wood that approximates its carbon, hydrogen, and oxygen contents. Although moisture content affects the amount of fuel available for combustion, water does not take part in the combustion reaction. Nitrogen, which is also a constituent of organic material, has little effect on combustion. If moisture and nitrogen are not included, the combustion equation for four moles of fuel is:

$$4C_6H_9O_4 + 25O_2 \rightarrow 18H_2O + 24CO_2 + 5{,}264{,}729 \text{ kJ}.$$

The energy produced by this reaction is called the *heat of combustion*. The 5,264,729 kJ (4,990,000 Btu) of heat released for the four moles of fuel is equivalent to 20 MJ kg^{-1} (8,600 Btu lb^{-1}) of fuel. For further discussion on different units and their derivation, see Sidebar 4.1. The heat is produced from the fuel once it is ignited. For combustion to occur, fuel, oxygen, and heat must be present. These three factors form the *fire triangle*, and fire control measures are based on breaking the link among them (Fig. 4.1). For example, a fire can be extinguished by removing the fuel, reducing the amount of oxygen, or by lowering the temperature of the fuel.

In most wildland fires, combustion is incomplete and not all fuel is consumed. The amount of heat produced by a fire is less than would occur under conditions of complete combustion. Published heat contents of forest fuels from ideal bomb calorimeter measurements may be almost twice those found under field conditions because of incomplete combustion (Babrauskas 2006). Major heat losses under field conditions come from the vaporization of moisture and because combustion of gases and char are incomplete. Four separate steps are involved in heat loss from water vapor: (1) heat is required to raise the temperature of the water in the fuel to boiling, (2) bound water must be released from the fuel, (3) the water must be vaporized, and (4) the water vapor must be heated to flame temperature. The heat necessary to release the bound water and to vaporize the water can be considered true losses (Byram 1959). The result of subtracting these two values from the heat content is called the *low heat content* or *heat yield*. If there is too much moisture in the fuel, combustion is unable to occur. This threshold level of moisture is called the *moisture of extinction*.

Combustion Phases

In wildland fuels, combustion occurs in three phases: preheating, gaseous, and smoldering (Fig. 4.2). During the

SIDEBAR 4.1 ENERGY UNITS, THEIR DERIVATION, AND THEIR RELATIONSHIPS

We are all familiar with calories—the nutritional content of food and the curse of all dieters. There are several other energy units that are less familiar but equally important. *British thermal units* (Btu) are used to measure the output of furnaces and air conditioners, and *kilowatt hours* (kW h) keep track of our electricity use. There are many types of energy, but all energy can be measured using the same unit, the joule (J), the amount work done to produce the power of 1 watt (w) for 1 second (s). The joule is the preferred metric unit for energy, and the *megajoule* (MJ) is the unit used by fire scientists to measure the amount of heat energy in fuels.

The nutritional calorie is actually 1,000 calories (cal) and is called a *kilocalorie* (kcal). A calorie is the amount of heat necessary to raise the temperature of 1 gram of water from 15°C to 16°C and is equal to 4,184 J. Similarly, the Btu is the amount of heat necessary to raise 1 pound of water one degree Fahrenheit from 60°F to 61°F. It is equivalent to 1,055 J.

The kilowatt hour (kW h) corresponds to 1 kilowatt (kW) of power being used over a period of one hour and is equal to 3.6×10^6 J. Although the kW h is not used to describe fire behavior, the kilowatt (kW) is part of one measure of fire intensity. Fireline intensity, which is described as the amount of energy received per second along 1 meter of the fire front, is measured in units of kW m^{-1}. Regardless of what units are used, the basic concept remains the same. Energy is stored in fuel and released during combustion. The rate of that release is governed by many factors discussed in the remainder of this chapter.

FIGURE 4.1 The fire triangle illustrates the requirements for combustion; heat must be applied to fuel in the presence of oxygen for combustion to occur.

preheating phase, fuels ahead of the fire are heated, and hydration and pyrolysis occur. *Hydration* includes raising the water in the fuel to boiling temperature, evaporating the water, and heating the water vapor to flame temperature. Considerable energy is required to complete hydration, particularly for vaporizing the water. *Pyrolysis* is the thermochemical decomposition of wood at elevated temperatures in the absence of oxygen. During pyrolysis, the temperature of the fuel needs to be raised to the point when gases start to volatize. Then the volatile material in the fuel is vaporized. Once ignition temperature is reached, gaseous combustion begins.

The *gaseous* phase starts with ignition as gases continue to be distilled and active burning begins. During this phase oxidation is initiated and an active flaming front develops. The flames come from the burning distilled gases; both water and carbon dioxide are given off as invisible combustion products. Incomplete combustion results in condensation of some of the gases and water vapor, as small droplets of liquid or solid are suspended over the fire and produce smoke and greatly reduce the energy release from the fire. The amount and rate of fuel consumption as either flaming or smoldering depends on many factors, including the properties of the fuelbed (packing ratio, depth, moisture content, particle sizes, etc.).

During the *smoldering* phase, charcoal and other unburned material remaining after the flaming phase continue to burn leaving a small amount of residual ash (Byram 1959). This phase differs from the gaseous phase in that the surface of the wood is burning rather than the gases above it. During this phase the fuel burns as a solid and oxidation occurs on the surface of the charcoal. The temperature and heat released from a smoldering fire are less than that from an actively flaming fire, and the rate of spread can be as slow as 0.1 mm s^{-1} (0.004 in s^{-1}). Ecologically, low rates of spread can result in a pulse of heat, which, when accumulated over a long time period, can significantly affect plants, seeds, and soil. One of the products of incomplete combustion during the soldering phase is carbon monoxide. Smoldering on a wildland fire can occur for days or weeks, and considerable fuel can be consumed and smoke produced during this time.

Heat Transfer

The heat resulting from combustion is transferred to sustain fire spread and to produce ecological and biological effects by three

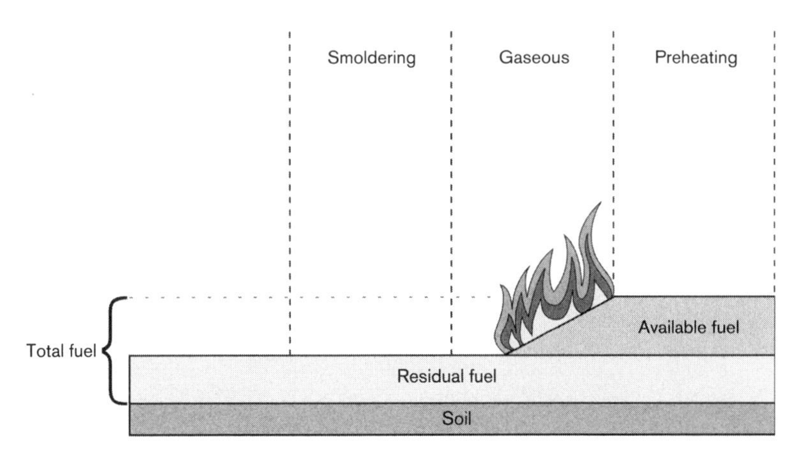

FIGURE 4.2 The three phases of combustion. Available fuel is the amount of fuel actually consumed. Combustion continues during the smoldering phase after passing of the flaming front.

primary mechanisms: conduction, convection, and radiation. *Conduction* occurs when heat moves from molecule to molecule and is the only mechanism that can transfer heat through an opaque solid. Conduction is the reason you are likely to drop a frying pan that has been on the fire when you grab it by its handle. Heat moves by conduction through branches, into the center of logs, and downward into soils. Temperature of the fuel or soil is thus increased and water is driven out.

Convection is the movement of heat in a gas or a liquid. You can feel heat moving by convection when you put your hand above a campfire. When the air above the flames becomes hot it expands, becomes less dense, and rises. Under wind and slope conditions, this convected air can heat unburned fuels ahead of the flaming front. When heated air rises into the canopy of trees it can lead to lethal heating of tree foliage, and eventual torching of individual trees and to crown fires. Heat transfer by convection is enhanced by wind and steep slopes and is instrumental in lofting firebrands that produce new fires, called spot fires, ahead of the flaming front.

Radiation is the process through which heat is transferred via electromagnetic waves through transparent solids, liquids, gases, and even a vacuum. The sun heating the earth is an example of radiant heat. Radiation preheats fuels and can cause spontaneous ignitions of large woody surfaces or compact litter layers. However, radiation heats only the outer surface of wildland fuels where the radiant energy is converted to thermal energy. In addition, heating by radiation is offset by convective cooling of the fine fuel particles (Finney et al. 2015). Whatever heat remains is transferred to the fuel's interior through conduction. Radiant heat is felt while standing in front of a fire and is inversely proportional to the square of the distance away from the source. If you move half the distance toward a campfire, you will receive four times as much heat through radiation.

Fuel

Fuel is the source of heat that sustains the combustion process. Fuels are characterized by physical and chemical properties that affect combustion and fire behavior.

Fuel Characteristics

Fuel characteristics strongly affect the rate of combustion and heat transfer. In general, dry grass supports faster spread than an equivalent weight of woody debris burning under identical conditions. Similarly, fuels distributed as fine particles typical of a tall brush field would burn more intensely than an equivalent mass of fuel arranged in larger particles and less depth (like large downed logs). Fine, porous fuels heat more quickly by convection and burn faster than coarse, compact fuels. The moisture contained in fuels also affects combustion; the drier the fuel, the more rapid the combustion.

SURFACE AREA TO VOLUME RATIO

Fuel coarseness, or fineness, is a function of fuel particle size. Imagine trying to start a log on fire in your wood stove with a single match. The log would not ignite because you would not be able to raise its temperature to ignition. Instead you would split the log into many individual pieces of kindling. Although the total volume of wood has not changed, the surface area of all the kindling is much greater than the surface area of the log (Box 4.1). The smaller the size of a fuel particle, the larger the ratio between its surface and its volume. The surface area to volume ratio is measured in units of $m^2\ m^{-3}$ ($ft^2\ ft^{-3}$) or, for simplicity, m^{-1} (ft^{-1}). For long, cylindrical fuel particles such as conifer needles, twigs, branches, and grasses, the area of the ends can be ignored, and the ratio is determined by dividing the diameter into the number four (Burgan and Rothermel 1984). Leaves from broad-leaved plants have high surface area to volume ratios that can be approximated by dividing the leaf thickness into the number two. For example, an oak leaf with a thickness of 0.0005 m (0.0016 ft) would have a surface area to volume ratio of 4,000 m^{-1} (1,220 ft^{-1}). This ratio is an extremely important fuel characteristic because, as more surface area is available for heat transfer and combustion, heating of the entire particle is quicker, and moisture is driven off more easily. This causes a faster preheating stage allowing the gaseous or smoldering phases to happen more quickly.

PACKING RATIO

Fuelbed compactness is another fuel characteristic that affects fire behavior. Again, imagine compressing all of the kindling you just split into a tight bundle and trying to light the bundle; the kindling would probably not ignite. Remembering all

BOX 4.1. SURFACE AREA TO VOLUME RATIO

This ratio is calculated by dividing the surface area of a fuel particle by its volume.

$$SV = \frac{\pi d l}{\pi (d/2)^2 l}.$$

If you ignore the ends, the equation is simplified to dividing 4 by the diameter of the particle.

$$SV = \frac{4}{d}.$$

The ratio is increased if you split the fuels into smaller parts. For example, when a log with a diameter of 6:

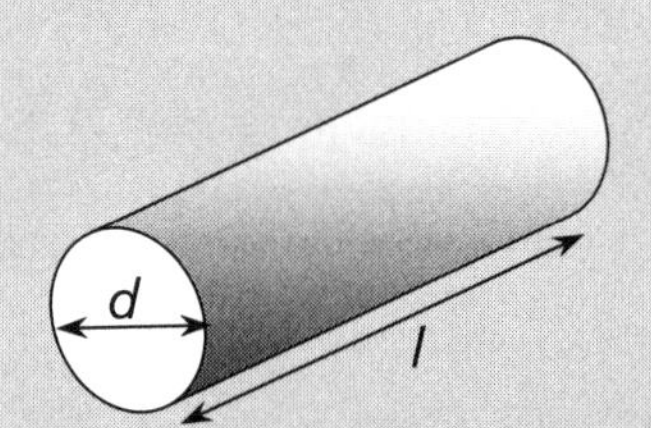

if d = 6,
SV = 0.67.

When split into seven pieces, the ratio increases from 0.67 to 14.

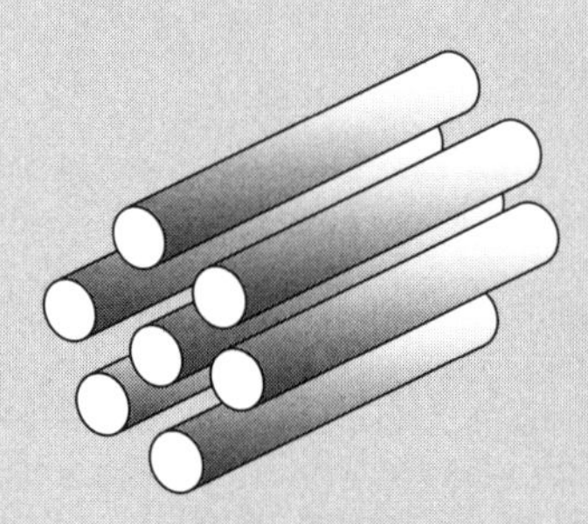

if d = 2,
SV = 2
and
7 × 2 = 14.

BOX 4.2. THE PACKING RATIO IS THE PROPORTION OF FUEL IN A UNIT VOLUME OF FUELBED. THE SAME AMOUNT OF FUEL CAN BE PACKED TIGHTLY WITH ONLY 10% AIR (A) OR LOOSELY WITH 90% AIR (B)

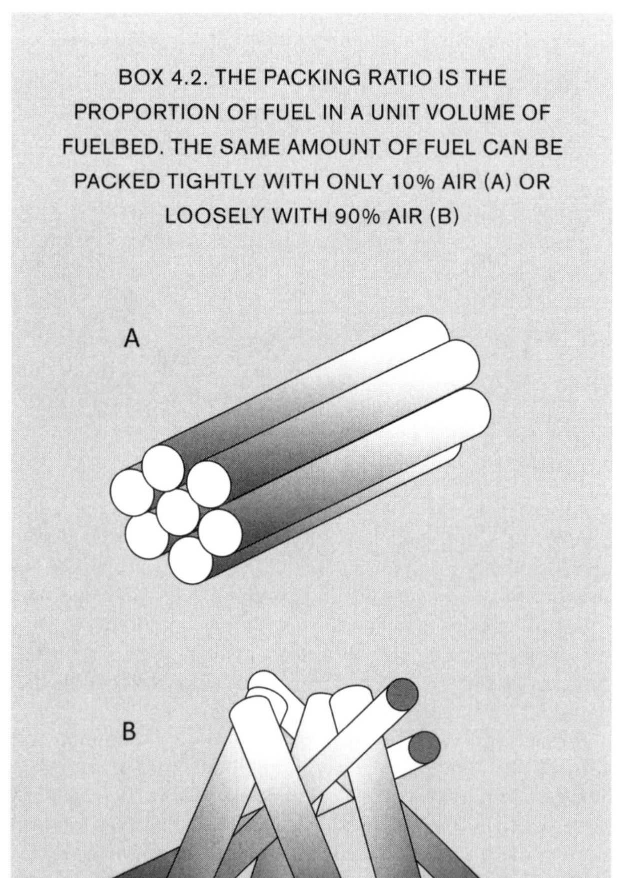

the campfires you have lit, you would instead arrange the kindling in a small log cabin or teepee. The volume of wood has not changed, but the amount of air in the fuelbed has increased (Box 4.2). *Fuelbed bulk density* is the amount of fuel per unit of volume including air and is measured in kg m^{-3} (lb ft^{-3}). The proportion of the fuelbed filled with fuel is called the *packing ratio* and is measured by dividing the bulk density of the fuelbed by the fuel particle density (Burgan and Rothermel 1984). A solid block of wood has a packing ratio of one. If the packing ratio is too high, not enough oxygen can reach the fuel and combustion cannot occur. Conversely, if the packing ratio is too low, the fire has trouble spreading from particle to particle as the distance between particles increases and radiation and convection decrease. The burning rate of a fuelbed varies with compactness and has an optimal range between being too open and too dense. This concept is similar to adjusting the carburetor or fuel injectors on your car to reach the optimum mixture of fuel and air. If the mixture is too rich or too lean, the engine will not burn fuel efficiently.

FUEL LOAD

The amount of fuel that is potentially available for combustion has different effects on fire spread and intensity. As a heat source, the more fuel available, the greater the potential energy stored in the fuel. The amount of this fuel consumed, and whether it is released in flaming or smoldering, depends on many factors besides the physical mass of the fuel. Rate of spread may actually decrease as fuel load increases because the extra fuel also becomes a greater heat sink, and more heat is required to raise it to ignition temperature. Much of the response depends on the size class of the fuel, its packing ratio, and whether it is dead or live. Procedures for inventorying downed woody fuels are found in Brown (1974) and Brown et al. (1982).

LIVE FUEL MOISTURE

Live fuel moisture is the weight of water in live fuels divided by the dry weight of that fuel. The moisture in herbaceous fuels readily responds to changes in relative humidity as opposed to live woody fuels, which hold water longer periods of time. The annual curing of herbaceous fuels can have a great influence on fire behavior. Annual plants can lose all of their moisture by mid-summer and are then considered to be dead fuels. Perennial herbaceous plants can retain their moisture through the fire season, but when their herbage dries or is killed back by frost, they become available for burning, often in late fall or before green-up in the spring. As live fuels age over years, their moisture content decreases. Of more importance, however, is the decrease in the ratio between live and dead fuels with time, making old fuel complexes very flammable. Conifer needles and the leaves of sclerophyllous shrubs and trees have waxy cuticles that inhibit water evaporation and help maintain live fuel moisture. Leaves can vary in their moisture content. Foliar moisture content, the moisture in the needles and leaves, is a critical factor in crown fire initiation and spread. Recent research shows that some of the seasonal changes in moisture content are caused by changing amounts of starch, sugar, and fat in the leaves because of physiological activity of the plant (Jolly et al. 2014). These changes in dry mass of the leaves affect the calculated moisture content by varying the denominator in the dry-weight basis formula even if the amount of water remains the same.

DEAD SURFACE FUEL MOISTURE TIMELAG CLASS

The proportion of a fuel particle that contains moisture is a primary determinant of fire behavior. Dead fuels typically range from 0% to 35% fuel moisture, which is calculated by the wet weight minus the dry weight of the fuel (the weight of the water) divided by the dry weight of the fuel. The interaction of a dead fuel particle with the ambient moisture regime is dependent on its size or its depth in the surface organic layer. The size classes that are traditionally used to categorize fuels correspond to fuel moisture timelag classes (Deeming et al. 1977). *Timelag* is the amount of time necessary for a dead fuel component to reach 63% of its equilibrium moisture content at a given temperature and relative humidity (Lancaster 1970). Or, simply stated, it is how long it takes a fuel to lose or gain moisture to be in equilibrium with the moisture in the air. For example, if there were a change in relative humidity, it would take one hour for a 1-hour fuel to make a 63% adjustment to that change. Table 4.1 lists the various dead fuel timelag classes and the corresponding woody size classes and duff depth classes. Fig. 4.3 shows how the different fuel moisture timelag classes vary as temperature and relative humidity change over a three-day period. In day one, you can see differences in how the 1-hour and 10-hour fuel moistures respond to diurnal changes in relative humidity. The thin solid and dashed diagonal lines indicate the amount of lag of between humidity and the two fuel size classes. The 100-hour fuel moisture responds more slowly and trends downward as relative humidity increases slightly. By day three, the 100-hour fuel moisture has stabilized to the relative humidity regime.

One-hour timelag fuels consist of dead herbaceous plants and small branchwood, as well as the uppermost litter on the forest floor. These fuels react rapidly to changes in relative humidity (the moisture in the air) and show changes from hour to hour. Day-to-day changes in moisture are reflected in the 10-hour fuels. The 100-hour fuels capture moisture trends spanning from several days to weeks, whereas 1,000-hour fuels reflect seasonal changes in moisture. The firewood analogy applies here as well. Your large logs would take several months to dry if left out in the rain for the winter, yet kindling, if brought inside, would dry in a few hours.

HEAT CONTENT

High heat content is the heat of combustion including inorganic ash as determined in a bomb calorimeter. Because these high values are seldom reached in wildland fire conditions, low heat content is used in intensity calculations. Low heat content is calculated by subtracting heat losses from incomplete combustion and the presence of moisture in the fuel from the high heat content. It is this lower value that provides the energy to drive combustion. Rate of spread varies directly with heat content; doubling the heat content results in a twofold increase in rate of spread.

MINERAL CONTENT

The primary chemicals in fuels include inorganic compounds that do not contribute to combustion. The total mineral content includes all of those compounds, whereas effective mineral content is the total mineral content with silica removed. The presence of minerals in fuel inhibits combustion; the higher the mineral content, the less heat is produced.

Fuel Models

Fuel models are used as inputs to the fire spread equation developed by Rothermel (1972) for short-term fire behavior predictions and for long-term fire danger rating. These two uses are distinct and the fuel models developed for one system should not be used for the other. Fuel models for use in the spread equation to predict fire behavior were developed by Albini (1976). These models emphasize the effect of smaller fuels on fire behavior. Deeming et al. (1972, 1977) and Burgan (1988) designed fuel models in the National Fire Danger Rating System (NFDRS) to assess seasonal fire danger used for planning and staffing. For example, the NFDRS is used to initiate the fire season and to determine when areas need to be closed because of extreme fire potential. Because of its seasonal nature, long-term drying and larger fuels are emphasized in the NFDRS.

In both the fire danger rating and fire behavior prediction systems, fuel types that have similar characteristics are grouped into stylized "fuel models" that include the variables important for combustion and fire danger, all of which are still in use. Anderson (1982) provided aids for determining fuel models for estimating fire behavior and showed how the models for fire behavior and fire danger were similar.

Scott and Burgan (2005) described a new set of 40 standard fire behavior fuel models to replace the 13 (Albini 1976) fuel models. Both sets of models split fuels into categories of slash, timber, shrub, and grass models. However, the original 13 models were designed for the severe period of the fire season when dry conditions lead to a fairly uniform fuel complex and were not as well suited for humid conditions or for other purposes under less extreme conditions such as prescribed fire and managed wildfires. In addition, the models were deficient for simulating the effects of fuel treatments on potential fire behavior or for simulating the transition to crown fire

TABLE 4.1
Moisture timelag classes and corresponding woody fuel size and duff fuel depth classes

		Woody fuel size class		Duff fuel depth class	
Timelag class	Time period	cm	in	cm	in
1-hour	Hourly	0.00–0.64	0.00–0.25	0.00–0.64	0.00–0.25
10-hour	Daily	0.25–2.54	0.25–1.00	0.64–1.91	0.25–0.75
100-hour	Weekly	2.54–7.62	1.00–3.00	1.91–10.16	0.75–4.00
1,000-hour	Seasonally	7.62–22.86	3.00–9.00	10.16+	4.00+

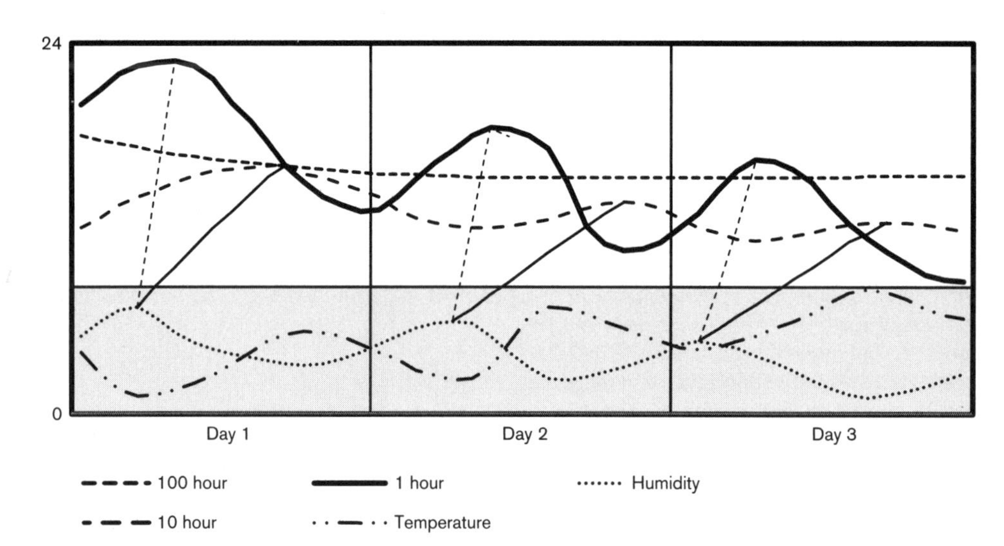

FIGURE 4.3 Timelag moisture content. Different fuel moisture timelag classes vary as temperature and relative humidity change over a three-day period. The thin solid diagonal lines indicate the amount of lag between humidity and 1-hour fuel size classes, and the dashed diagonal lines the lag between humidity and 10-hour fuels. The 100-hour fuel moisture responds more slowly and trends downward as relative humidity increases slightly.

using crown fire initiation models (Scott and Burgan 2005). The new models address these needs and were designed to be used in the Rothermel (1972) surface spread equation.

The Fire Danger Sub-Committee and the Fire Environment Committee of the National Wildfire Coordinating Group are recommending updates to the NFDRS (NWCG 2014). Their proposed fuel models would update live and dead fuel moisture values in the original NFDRS models and consolidate the models into five fuel response types: grass, brush, grass-brush, timber, and slash.

Acknowledging the limitations of existing fuel models, Sandberg et al. (2001) proposed a system of fuel characteristic classes that would provide all the information necessary for calculating potential fire behavior and hazards based on environmental variables. Fuel characteristic classes are defined for a vegetation type and contain data for fuels in up to six strata representing potentially independent combustion environments. A stratum can contain one or more fuelbed components that contribute available biomass and flammable surfaces to the stratum. Fuelbed components combust differently and have a unique influence on fire behavior and effects. Examples of fuelbed components include the different size classes of sound woody fuels and the foliage and twigs of shrubs. Each component is defined by a set of quantitative variables that specify physical, chemical, and structural characteristics.

Ignition

If the weather is conducive for a fire to spread and there is an abundant array of fuel available for burning, all we need is an ignition. Lightning and humans are the primary ignition sources, although volcanoes ignite fires when their infrequent eruptions occur. Spontaneous combustion can occur but is not common. Not all ignitions result in a fire, however, because several conditions must be met before an ignition can become a fire and spread.

Ignition Sources

Virtually all lightning comes from thunderstorms that develop as a result of frontal activity or air mass movements (Schroeder and Buck 1970). Orographic lifting of air masses is the most common cause of thunderstorms and lightning as air moves over mountain ranges. Thunderstorms associated with fronts occur when warm, moist air is forced over a wedge of cold air. Lightning is usually more prevalent with cold-front thunderstorms. As a thunderstorm develops, positive charges accumulate in the top of the cloud and negative charges in the lower portion. Discharges occur within the cloud, between clouds, in the air, and between the cloud and the ground. Both negative and positive charges reach the ground and can start fires. Lightning occurs in thunderstorms when the electrical energy potential builds up to the point that it exceeds the resistance of the atmosphere to a flow of electrons between areas of opposite charge (Schroeder and Buck 1970). Cloud-to-ground lightning accounts for about one-third of all strikes and these strikes are primarily negative (Fuquay 1982). Ignitions occur when the lightning strike has a long continuing current, as do approximately 80% of positive strikes and 10% of the negative strikes. Lightning is pervasive in the state of California. It is most prevalent in the mountains and the southeastern deserts but rare on the coast.

Native Americans have been cited as an ignition source for fires throughout California (Anderson 1996). Although fires set by Native Americans certainly burned areas occupied or used by them, the areal extent of those fires remains uncertain. In chapter 19, Anderson states that their use spanned a gradient from intensive use to no use at all. Vale (1998) suggests that nonhuman processes determined the landscape characteristics of over 60% of Yosemite. In these areas there were scattered Native American camps and few, if any, fires. For the remaining 40%, Vale (1998) states that fires were possibly made more numerous by ignitions by Native Americans and that around village sites, fires were likely to be more frequent as a result of their fires. In all likelihood, ignitions by Native Americans were an addition to lightning ignitions rather than a substitute for them and that the landscape was a mosaic of both natural and cultural characteristics.

Ignition Probability

Ignition can occur as a smoldering ignition, a pilot flaming ignition, and autoignition (spontaneous ignition). All ignitions occur by heat transfer from some external source by radiation, convection, or direct contact with hot solid materials (embers). Although it is tempting to think of ignition as simply raising the temperature of a substance to a critical level, it is more complicated because of the variable temperature of the heat source, the heat flux (kW m^{-2}), and its duration. Engineering studies of ignition commonly refer to a "critical mass flux" for flaming ignitions which means the amount of mass conversion to gas required to support an attached flame (McAllister et al. 2011, Finney et al. 2013). Smoldering ignition can be achieved by widely differing combinations of temperature exposure and duration as well as moisture content. Thus, ignitions sources in the wildlands can vary considerably.

Ignition by a lightning strike or a firebrand occurs in four stages (Deeming et al. 1977). First, contact with a receptive fuel must be made. Once contact is made, the moisture in the fuel must be driven off and a sufficient amount of pyrolysis gasses generated to sustain a flaming ignition. An ignition source may also ignite a smoldering fire which will require days or weeks to develop into flaming spread. The probability that a firebrand will start a fire is a function of the fine dead fuel moisture (1-hour timelag fuel moisture) content, fuel temperature, surface area to volume ratio, packing ratio, and characteristics of the firebrand such as temperature, rate of heat release, length of time it will burn, and whether it is flaming or glowing (Deeming et al. 1977). In the fire behavior prediction system, ignition probability is calculated using fine dead fuel moisture, air temperature, and percent shading (Rothermel 1983). In addition to those three, the ignition component in the NFDRS includes the spread component in order to determine the probability of detecting an ignition that requires suppression action (Deeming et al. 1977).

Using simulated lightning discharges, Latham and Schlieter (1989) found that ignition probabilities for duff of short-needled conifers such as lodgepole pine (*Pinus contorta* subsp. *latifolia*) depend almost entirely on duff depth. Ignition of litter and duff from long-needled conifers including ponderosa pine (*P. ponderosa*) and western white pine (*P. monticola*) was affected primarily by the moisture content. Ignition was also dependent on the duration of the arc, indicating that the length of time of a lightning strike could have an effect on starting a fire.

Arnold (1964) found that only 25% of the lightning discharges of long duration actually started fires. There is considerable variation in strikes and ignitions. For example, van Wagtendonk and Smith (2016) reported that 2,654 strikes started 36 fires in Yosemite in 2014, an ignition rate of only 1.4%. Many discharges might not have resulted in fires that grew large enough to be detected and went out before they could be located. Others discharges might have struck rock, snow, or other noncombustible substances.

Fire Behavior

If there is sufficient fuel quantity and continuity, conducive weather, and an ignition, a fire will occur. We will now look at how these factors, combined with topography, cause a fire to spread. Fires can spread through the ground fuels, surface fuels, crown fuels, or combinations of all three. Spot fires ignited by lofted firebrands can also spread fires. Each method has unique physical mechanisms necessary to sustain fire spread. The fuel stratum that is burned and the method of spread define fire types. Ground fires burn the duff or other organic matter such as peat and usually burn with slow-moving smoldering fires, often after the surface fire as passed. Surface fires burn the litter, woody fuels (up to 7.62 [3 in]), and low vegetation such as shrubs with and active flaming front. Crown fires burn in the canopies of the vegetation with or ahead of the surface fire.

Flaming Front

The flaming front is the area of the fire between its leading edge, where fuels are igniting, and the trailing edge where fuels are finishing the flaming phase of combustion. Properties of the flame zone are the forward rate of spread, residence time, and flaming zone depth. Additional characteristics include the reaction intensity, fireline intensity, flame length, and heat per unit of area. Equations for calculating these characteristics are included in Box 4.3. *Rate of spread* is the speed at which the flaming front moves forward and is measured in units of distance per unit of time. Rate of spread is affected by many fuel, weather, and topographic variables.

BOX 4.3. EQUATIONS FOR THE FLAMING FRONT CHARACTERISTICS

Characteristic		Units	
		Metric	English
Fireline intensity			
FLI = ((Heat/Area) × Rate of Spread)/60	FLI	kW m^{-1}	Btu ft^{-1} s^{-1}
	H/A	kJ m^{-2}	Btu ft^{-2}
	ROS	m min^{-1}	ft min^{-1}
FLI = (Reaction Intensity × Flaming Zone Depth)/60	FLI	kW m^{-1}	Btu ft^{-1} s^{-1}
	RI	kJ m^{-2} min^{-1}	Btu ft^{-2} min^{-1}
	FZD	m	ft
FLI = (Reaction Intensity × Rate of Spread × Residence Time)	FLI	kW m^{-1}	Btu ft^{-1} s^{-1}
	RI	kJ m^{-2} min^{-1}	Btu ft^{-2} min^{-1}
	ROS	m min^{-1}	ft min^{-1}
	RT	min	min
FLI = 258 × (Flame Length)$^{2.17}$	FLI	kW m^{-1}	
	FL	m	
FLI = 5.67 × (Flame Length)$^{2.17}$	FLI		Btu ft^{-1} s^{-1}
	FL		ft
Flame Length			
FL = 0.0775 × (Fireline Intensity)$^{0.46}$	FL	m	
	FLI	kW m^{-1}	
FL = 0.237 × (Fireline Intensity)$^{0.46}$	FL		ft
	FLI		Btu ft^{-1} s^{-1}
Heat per Unit Area			
H/A = (60 × Fireline Intensity)/Rate of Spread	H/A	kJ m^{-2}	Btu ft^{-2}
	FLI	kW m^{-1}	Btu ft^{-1} s^{-1}
	ROS	m min^{-1}	ft min^{-1}

NOTES: ROS = Rate of Spread; RI = Reaction Intensity; FZD = Flaming Zone Depth; RT = Residence Time.

The time the flaming front takes to pass over a point is called the *residence time*. If flame spread is assumed to be one-dimensional (has no vertical dimension) such as very short grass, then the *flame zone depth* is defined as the distance from the front to the back of the active flaming front and is calculated by multiplying the rate of spread by the residence time. Most fuelbeds have some vertical extent which requires additional time to burn downward thereby lengthening the flame zone (Albini 1981a). Anderson (1969) found that the residence time was related to the size of the particles that were being burned, but denser beds tend to take longer to burn for particles of a given size *Heat per unit of area* is the total amount of heat released per unit of area as the flaming front passes.

The rate of energy release is characterized by the reaction intensity and the fireline intensity. *Reaction intensity* is the rate of energy release per unit of flaming zone area (kW m^{-2}). The assumption in the Rothermel spread equation is that the *reaction intensity* is the source of heat that keeps the chain reaction of combustion in motion but it is not easily seen because the fire is moving. A more readily experienced metric of energy release rate, *Fireline intensity*, is the rate of energy release per unit length of fire front (kW m^{-1}) and is likened to the amount of heat you would be exposed to per second while standing immediately in front of a fire. It is equivalent to the product of the available energy (in terms of heat per unit of area) and the forward rate of spread and can also be determined from reaction intensity and flaming zone depth (Box 4.3). Fireline intensity is related to flame length, the average distance from the base of the flame to its highest point. Fig. 4.4 shows flame dimensions with flame length as the hypotenuse and flame height the vertical distance to the highest point.

Byram (1959) provided the approximate relationship between fireline intensity and flame length (Box 4.3). His equations can be reversed to obtain simple expressions for

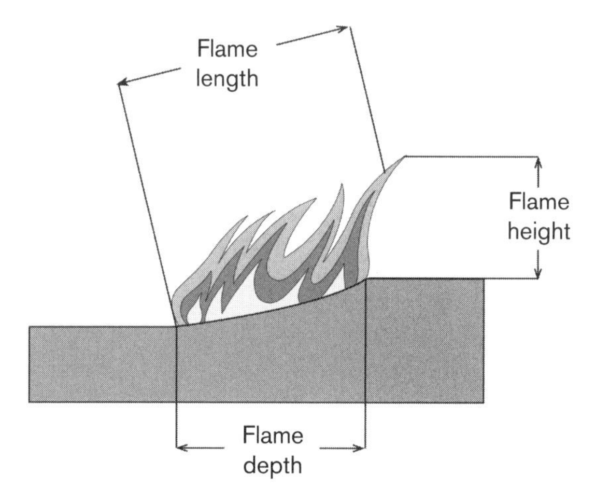

FIGURE 4.4 Flame dimensions for a wind-driven fire. Flame length is related to the fireline intensity and is measured from the base of the flame to its tip.

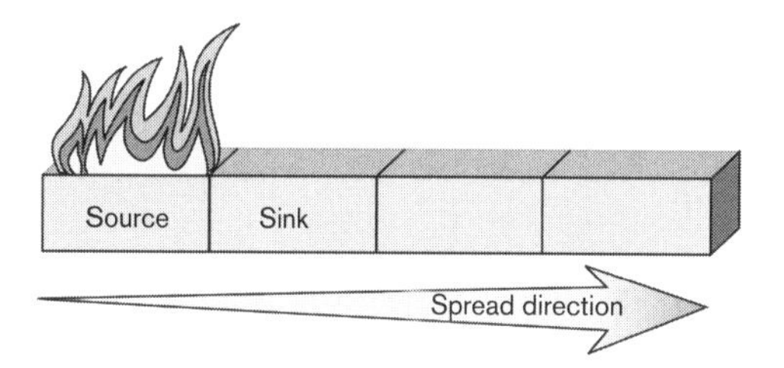

FIGURE 4.5 Rate of spread is the ratio between the heat source and the heat sink. As more heat is generated by the source, the more quickly the heat sink ignites.

fireline intensity in terms of flame length. Byram (1959) cautioned that the equations for fireline intensity based on flame length are better approximations for low-intensity rather than high-intensity fires. Although not without its difficulties, flame length is the only measurement that can be taken easily in the field that is related to fireline intensity (Rothermel and Deeming 1980).

Available fuel energy is the energy that is actually released by the flaming front, whereas *total* fuel energy is the maximum energy that could be released if all the fuel burns. Energy release is measured in heat per unit of area and can be calculated from the fireline intensity and the rate of spread (Box 4.3). Heat per unit area is the primary contributor to fire effects since it is independent of time.

Surface Fire Spread

The first attempt to describe fire spread using a mathematical model was by Fons (1946). He theorized that because sufficient heat is needed at the fire front to ignite adjacent fuels, fire spread is a series of successive ignitions controlled by ignition time and the distance between fuel particles. Conceptually, this is analogous to viewing a fuelbed as an array of units of volume of fuel, each unit being ignited in turn as its adjacent unit produces enough heat to cause ignition. The unit being ignited is the *heat sink*, whereas the unit currently burning is the *heat source* (Fig. 4.5).

Frandsen (1971) took a different theoretical approach to estimating fire spread rate by applying the conservation of energy principle to a unit volume of fuel ahead of a fire front. The unit of fuel that is currently burning serves as the heat source for the unit ahead which acts as a heat sink. Sufficient heat must be generated by the source to ignite the adjacent unit. The rate of spread is determined by the rate at which adjacent fuel units are ignited.

ROTHERMEL'S SPREAD EQUATION

Because some of the terms in Frandsen's (1971) equation contained unknown heat transfer mechanisms, Rothermel (1972) devised experimental and analytical methods to determine these terms using the fire environment triangle of fuel, weather, and topography. His model of surface fire spread has been the most commonly used model in the United States since the mid-1970s. Rothermel's (1972) surface fire spread model was not designed to predict crown fire spread, account for the large fuels that burn after the fire front goes by, or predict fire effects such as duff consumption and smoke production. It predicts rate of spread of the flaming front for surface fire.

The first predictions using Rothermel's (1972) were made with the nomograms contained in Albini (1976). He cautioned that several sources of error could contribute to inaccurate model predictions, including the inapplicability of the model to the situation, the model's inherent inaccuracy, and inaccurate data. He felt that these errors could lead to predictions that were within a factor of two or three. Rothermel (1983) formalized the procedures for training fire behavior analysts to use the spread model for use on wildland fires. The model was subsequently computerized in the Behave and BehavePlus fire prediction and fuel modeling systems (Andrews et al. 2008). The advent of desktop computers made it possible to combine the fire spread model with other fire behavior models into the FARSITE model, which performs spatial simulations of areal and surface fire spread (Finney 1998).

Box 4.4 shows the surface fire spread equation and defines each of its terms. In the numerator, the propagating fluxes were divided into terms that accounted for the total heat release, the proportion of the heat reaching the adjacent fuel unit, and wind and slope effects. Using the campfire analogy, the total heat release is all of the heat produced by the fire, whereas the heat reaching you sitting at the fire's side would be the proportion reaching the adjacent fuel. Imagine how much hotter you would become if you were able to sit while hovering just above the fire. In the denominator, empirical relationships were used to define bulk density, the effective heating number, and the heat of preignition. The final formulation provided an approximate solution to the equation (Burgan and Rothermel 1984). Each of the terms in the surface fire spread equation will be examined individually to gain insight into the complex effects of fuels, weather, and topography on surface fire spread.

Reaction intensity (I_R) is made up of several factors that relate to the rate of energy release (Box 4.5). Reaction velocity (Γ') is a ratio that expresses how efficiently the fuel is consumed compared to the burnout time of the characteristic particle size (Burgan and Rothermel 1984). This ratio is a function of the actual and optimum packing ratios and the surface area to volume ratio. The actual packing ratio is found by dividing the fuelbed bulk density by the ovendry fuel particle density. Albini (1976) specifies a standard value of 51.25 kg m^{-3} (32 lb ft^{-3}) for fuel particle density. The optimum packing ratio is a function of the surface area to volume ratio. Fine

BOX 4.4. THE ROTHERMEL (1972) SPREAD EQUATION

$$R = \frac{I_R \xi (1 + \Phi_w + \Phi_s)}{\rho_b \varepsilon Q_{ig}}$$

where: R is the forward rate spread of the flaming front, measured in m min^{-1} (ft min^{-1}),

I_R is the reaction intensity, a measure of the energy release rate per unit of area of flaming front, in kJ m^{-2} min^{-1} (Btu ft^{-2} min^{-1}),

ξ is the propagating flux ratio, the proportion of the reaction intensity reaching the adjacent fuel, dimensionless,

Φ_w is the wind coefficient which accounts for the effect of wind increasing the propagating flux ratio, dimensionless,

Φ_s is the slope coefficient, a multiplier for the slope effect on the propagating flux ratio, dimensionless,

ρ_b is the fuelbed bulk density, a measure of the amount of fuel per unit of volume of the fuelbed, measured in kg m^{-3} (lb ft^{-3}),

ε is the effective heating number, the proportion of the fuel that is raised to ignition temperature, dimensionless.

Q_{ig} is the heat of preignition, which is the amount of heat necessary to ignite 1 kg (1 lb) of fuel, measured in kJ kg^{-1} (Btu lb^{-1}).

BOX 4.5. THE EQUATION FOR REACTION INTENSITY (FROM ROTHERMEL 1972)

$$I_R = \Gamma' w_n h \, 0_M \, 0_s,$$

where I_R is the reaction intensity, in kJ m^{-2} min^{-1} (Btu ft^{-2} min^{-1}),

Γ' is the optimum reaction velocity, measured in min^{-1},

w_n is the net fuel load, in kg m^{-2} (lb ft^{-2}),

h is the low heat content, in kJ kg^{-1} (Btu lb^{-1}),

0_M is the moisture damping coefficient, dimensionless,

0_s is the mineral damping coefficient, dimensionless.

fuels, such as grass and long-needled pine litter, have near-optimum packing ratios and large surface area to volume ratios. These fuels burn thoroughly in a short period of time and have the highest reaction velocity.

The net fuel load (w_n) is equal to the ovendry fuel load multiplied by one minus the fuel particle total mineral content (Albini 1976). Because minerals do not contribute to combustion, their weight must be removed from the calculation of reaction velocity. Both Albini (1976) and Scott and Burgan (2005) specify a mineral content value of 5.55%.

The low heat content (h) provides the heat necessary to sustain combustion. There is some variation in heat content for fuels of different species. Conifers tend to have higher values than hardwoods because of the presence of resins and higher lignin content. Sclerophyllous shrubs contain oils and waxes in their leaves that increase their heat content. Fuel models in Albini (1976) and Scott and Burgan (2005) use a standard value of 18.61 MJ kg^{-1} (8,000 Btu lb^{-1}). FARSITE (Finney 1998) and BehavePlus (Andrews et al. 2008) allow the user to specify separate live and dead heat content values.

The moisture damping coefficient (0_M) and mineral damping coefficient (0_s) account for the effects that moisture and minerals have in reducing the potential reaction velocity (Rothermel 1972). The moisture damping coefficient is derived from the fuel moisture content and the fuel moisture content of extinction, which is the moisture level at which combustion can no longer be sustained. The mineral damping coefficient is a function of the silica free ash content, termed the *effective* mineral content. A value of 1.00% is used for the 13 standard fuel models (Albini 1976) and the 40 new models (Scott and Burgan 2005).

The proportion of heat reaching adjacent fuel is calculated under the assumption that the fire is burning without any wind and on flat terrain (Burgan and Rothermel 1984). The propagating flux ratio (ξ) is a dimensionless fraction that accounts for the fact that not all of the reaction intensity reaches adjacent fuels. For example, in the no-wind, no-slope situation, most of the heat energy moves upwards by convection while only a smaller proportion is directed at the adjacent fuel by internal convection. Radiation is only a minor component. The minimum value for the flux ratio is zero when no heat reaches adjacent fuels, and the maximum value is one when all the heat reaches adjacent fuels. These extreme values are seldom reached, and a more practical range would be from 0.01 to 0.20 because most of the heat is convected upwards. The surface area to volume ratio and the packing ratio are the determinants of the propagating flux ratio. As these two ratios increase, the flux ratio increases, with fine fuels having the most pronounced effect.

Both the wind coefficient (Φ_w) and the slope coefficient (Φ_s) have the effect of increasing the proportion of heat reaching the adjacent fuel. They act as multipliers of the reaction intensity. In the no-slope case, the wind coefficient increases rapidly with increases in wind speed in loosely packed fine fuels (Burgan and Rothermel 1984). Direct contact, increased convection, and a minor increase in radiation heat transfer occur as the flame tips toward the unburned fuel. Although the smoke might have caused you to move away from the campfire first, if you had remained, you would have felt the added heat from the closer flames. The wind coefficient is affected by surface area to volume ratio, packing ratio, and wind speed. Increasing the surface area to volume ratio increases the wind coefficient, and this effect becomes greater at higher wind speeds as the distance between the flame and the fuel decreases (Burgan and Rothermel 1984). The wind effect is less pronounced as the packing ratio moves beyond the optimum and fuel particles begin to obstruct the convective flow. There is a maximum wind speed beyond which the wind coefficient does not increase (Burgan and Rothermel 1984). At that point the power of the wind forces exceeds the convective forces. This occurs when the wind speed in km h^{-1} is twice the reaction intensity in kJ m^{-2} min^{-1} (or the wind speed in mi h^{-1} is 1/100 of the reaction intensity in Btu ft^{-2} min^{-1}). For typical annual grass fires with sparse fuels, wind speeds in excess of 19 km h^{-1} (12 mi h^{-1}) will not increase the rate of spread. In tall grasses, rate of spread would not increase after wind speeds reach 68 km h^{-1} (42 mi h^{-1}).

Under no-wind conditions, the slope coefficient increases as the slope becomes steeper. The effect is similar, but less pronounced, than that of wind. Although flames are brought closer to the unburned fuels, without wind to bring heated air in contact with the fuel, there is only a slight increase in convection and minimal increase in radiation. If you are standing above a fire on a slope, you will feel much hotter than if you were standing below. The packing ratio and the tangent of the slope are used to calculate the slope coefficient. The packing ratio has a slight influence on the sensitivity of the coefficient to increases in slope steepness (Burgan and Rothermel 1984). This effect is small in comparison to the other effects due to changes in the packing ratio. The wind and slope coefficients do not interact, but their combination can have a dramatic effect on fire behavior.

Now we will take a look at the terms in the denominator of the Rothermel (1972) spread equation that constitute the heat sink. The denominator represents the amount of heat necessary to bring the fuel up to ignition temperature. The first term is *fuelbed bulk density* (ρ_b), the total amount of fuel that is potentially available. It is defined as the ovendry weight of the fuel per unit of fuelbed volume and is calculated by dividing the ovendry fuelbed load by the fuelbed depth. Because bulk density is in the denominator of the spread equation, an increase in density will tend to cause a decrease in spread rate (Burgan and Rothermel 1984). This can happen by either increasing the fuel load or by decreasing the fuelbed depth. However, an increase in load also causes the reaction intensity to increase. In addition, an increase in bulk density can cause the propagating flux and the wind and slope coefficients to go up or down depending on the relative packing ratio.

Not all of the fuel that is available will burn with the passing of the flaming front. Often only the outer portion of a large log or other fuel particle is heated to ignition temperature. The effective heating number (ε) defines the proportion of the fuel that will burn as the flaming front passes and is dependent on fuel particle size as measured by the surface area to volume ratio. Small particles will heat completely through and ignite, whereas decreasing proportions of larger particles will ignite as size increases. Not only do the smaller particles heat all the way through, but their increased surface area allows heating by radiation to occur rapidly. Multiplying the fuelbed bulk density by the effective heating number yields the amount of fuel that must be heated to ignition temperature before ignition can occur (Burgan and Rothermel 1984).

How much heat is required? The heat of preignition quantifies the amount of heat necessary to raise the temperature of a 1 kg (2.2 lb) piece of moist fuel from the ambient temperature to the ignition point. First the moisture must be driven off and then the fuel must be heated. Most of these temperature values are fairly constant and can be calculated in advance (Burgan and Rothermel 1984). Moisture content does vary and is used to calculate the heat required for ignition. As fuel moisture content increases, there is a steady increase in the heat of preignition. The units are in kJ kg^{-1} (Btu lb^{-1}). The product of the fuelbed bulk density, effective heating number, and the heat of preignition is the heat per unit of area in kJ m^{-2} (Btu ft^{-2}) necessary to ignite the adjacent fuel cell.

Rothermel (1972) recognized that his model had limitations but felt that an empirical model could allow fire managers to make reasonable predictions of surface fire spread. Attempts are underway to refine the surface spread model by including heating and combustion processes. Current work is investigating fuel particle heat exchange, live vegetation thermal decomposition processes, nonsteady-state flame structure, convective heat transfer, and flame source burning duration (Finney et al. 2012). Most promising is work done by Finney et al. (2015) on the coupling between flame dynamics induced by buoyancy and fine particle response to convection. They found that heating by radiation of the fine fuel particles was offset by convective cooling until contact with the flames and hot gases occurred. Vorticity and instabilities in the flame zone induced by buoyancy generated by the fire controlled the convective heating needed to ignite the fuel and produce spread (Finney et al. 2015).

Crown Fire Spread

A crown fire occurs when the fire moves from the surface fuels into the canopies. Although shrub canopies can be considered crowns, the models developed for predicting crown fire behavior are specific to trees. Van Wagner (1977) defined three stages of crown fire. The first stage of crowning is a *passive* crown fire, which begins with the torching of trees from a surface fire. If the fire spreads through the crowns in conjunction with the surface fire, it is called an *active* crown fire. A crown fire spreading through the crowns far ahead of or in the absence of the surface fire is an *independent* crown fire.

In a passive crown fire, single trees or groups of trees torch out and there might be some movement of fire into adjacent tree crowns (Fig. 4.6). Torching can occur at low wind speeds with relatively low crown bulk densities if the crown bases are low enough to be ignited by the surface fire. Although a passive crown fire does not spread from crown to crown, embers from torching trees can start fires ahead of the fire front. Transition to passive crowning begins when the fireline intensity of the surface fire exceeds that necessary for igniting the crowns. This point is dependent on the height to the base of the live crowns and the foliar moisture content (Alexander 1988). Ladder fuels are considered in the calculation of the crown base height. Under conditions of low foliar moisture content, the crowns will ignite when the surface fire intensity is great enough to bring the crowns to ignition temperature either through direct contact with flames or through convective heat. Once ignited, the fire in the crowns will spread some, but, as long as the actual rate of spread of the crown fire is less than the threshold for active crown spread, the fire will remain passive. Actual spread rate can be calculated from surface fire spread rate, the proportion of the trees that are involved in the crowning phase, and the maximum crown fire spread rate (Rothermel 1991).

An active crown fire can occur when winds increase to the point that flames from torching trees are driven into the crowns of adjacent trees (Rothermel 1991). The heat generated by the surface fire burning underneath the canopy sustains the fire through the crowns (Fig. 4.7). The fire becomes a solid

FIGURE 4.6 Passive crown fires can occur under conditions of low crown base heights, even with relatively low wind speeds and low crown bulk densities.

FIGURE 4.7 Higher wind speeds and crown bulk densities with low crown base heights lead to active crown fires.

FIGURE 4.8 Very high wind speeds and crown bulk densities can lead to independent crown fires that race ahead of the surface fire.

wall of flame from the surface to the crown and spreads with the surface fire (Scott 1999). Lower crown base heights, higher wind speeds, and higher crown bulk densities than those necessary for passive crowning are required for active crowning. The threshold for transition from passive to active crowning is dependent on the crown bulk density and a constant related to the critical mass flow through the canopy necessary for a continuous flame (Alexander 1988). Active crowning continues as long as the surface fire intensity exceeds the critical intensity for initiation of crown fire and the actual spread rate, as calculated from the Rothermel (1972) equation, is greater than the critical crown fire spread rate. The critical spread rate for active crowning decreases rapidly as crown bulk density increases from 0.01 to 0.05 kg m^{-3} (0.01–0.03 lb ft^{-3}). Consequently, the actual spread rate necessary to initiate crown fire spread becomes less (Scott 1999). As tree canopies become closer and denser, fire is able to spread more easily from tree to tree. After crown bulk densities reach 0.15 kg m^{-3} (0.09 lb ft^{-3}), there is little additional effect on the critical spread rate. Once an active crown fire is initiated, its intensity is calculated using the combined load of the available surface fuels and crown fuels and the crown rate of spread (Finney 1998). The crown fuel load is derived from the crown fraction burned, the mean canopy height, crown base height, and the crown bulk density. The crown fraction burned is dependent on the critical surface spread associated with the critical intensity for initiating a crown fire (Van Wagner 1993).

Independent crown fires burn in aerial fuels substantially ahead of the surface fire and are rare, short-lived phenomena (Fig. 4.8). Steep topography, very high wind speeds, and bulk densities greater than 0.05 kg m^{-3} (0.03 lb ft^{-3}) lend themselves to the extreme behavior of these wind-driven fires. Independent crown fires occur when the surface fire intensity exceeds the critical intensity, the actual rate of spread is greater than the critical rate of spread, and the actual energy flux is less than the critical energy flux for independent crown fires in the advancing direction.

Independent crown fires can also occur under low wind and unstable air conditions. Rothermel (1991) describes fires under those conditions as *plume-dominated* fires. Byram (1959) introduced the concept of energy flow rates in the wind field and in the convection column above a line of fire to explain the behavior of plume-dominated fires. The power of the wind is the rate of flow of kinetic energy through a vertical plane of unit area at a specified height in a neutrally stable atmosphere (Nelson 1993). The wind energy is a function of the air density, the wind speed, the forward rate of spread of the fire, and the acceleration due to gravity. The power of the fire is the rate at which thermal energy is converted to kinetic energy at the same specified height in the convection column. It is calculated from the fireline intensity, the specific heat of air, and the air temperature at the elevation of the fire. When the power of the fire is greater than the power of the wind for a considerable height above the fire, extreme fire behavior can occur (Byram 1959). Both Byram (1959) and Rothermel (1991) give equations for the wind and fire power functions, and Nelson (1993) has generalized the equations for use with any applicable system of units.

Scott and Reinhardt (2001) developed quantitative methods for assessing crown fire hazard. They established links between Rothermel's (1972) fire spread equation and Van Wagner's (1977, 1993) models of crown fire transition and propagation to derive two crown fire hazard indices. The

Torching Index is the open (6.1 m [20 ft]) wind speed at which crown fire activity is initiated. The *Crowning Index* is the open wind speed where active crown fire is possible. Between these two wind speeds the *crown fraction burned*, a measure of the amount of canopy consumed, varies between 0 and 1. The indices incorporate the effects of surface fuel characteristics, dead and live moistures for surface and crown fuels, slope steepness, canopy base height, canopy bulk density, and wind reduction by the canopy.

The indices can be used to evaluate forest stands by their relative susceptibility to crown fire and to compare the effectiveness of treatments designed to mitigate crown fires (Stephens et al. 2009). The hazard indices incorporate the effects of surface fuel characteristics, dead and live moistures of surface and crown fuels, slope steepness, canopy base height, canopy bulk density, and wind reduction by the canopy (Scott and Reinhardt 2001). One problem with using the torching index with the Fire and Fuels Extension of the Forest Vegetation Simulator is that the simulator produces a single value for crown base height, which causes the torching index to vary widely. P-Torch was develop to overcome this problem by predicting the probability of finding small patches in the simulated stand where torching can occur (Rebain 2010).

Cruz and Alexander (2010) found that simulation studies that linked the spread equation with crown transition and propagation models underpredicted potential crown fire behavior when applied to conifer forests of western North America. They attributed this bias to incompatible model linkages, use of spread models that tend to under predict, the use of unsubstantiated crown fraction burned functions, and the use of uncalibrated fuel models.

Spotting

Trees that are ignited during any of the crown fire stages and snags ignited by any fire are sources of firebrands that could ignite spot fires. The spread of a fire is increased dramatically by ignition of numerous spot fires ahead of the flaming front. Albini (1979) developed a model for calculating spot fire distance from a torching tree and enhanced his original model to accommodate embers generated from isolated sources and wind-driven fires (Albini 1981b, 1983a). The model calculates the height to which an ember is lofted, the time it remains burning, and the distance it travels. Characteristics of the torching tree and embers, the intervening area, and the receiving fuelbed all determine the distance and probability of ignition of a new spot fire (Fig. 4.9). Large embers are not lofted as high nor travel as far as small embers and are often still burning by the time they can land and start spot fires. Small embers usually burn out before they can land.

Tree species, height, diameter, and the number of tree simultaneously torching all affect the height and duration of the steady flame above the tree or group of trees. Ember characteristics include size, shape, density, and starting height in the tree. While the embers are in flight, the wind speed and direction, and the evenness and vegetative cover of the intervening terrain influence the distance traveled. If the ember lands on a receptive fuelbed, the fine fuel moisture and temperature determine whether a spot fire is ignited. Chase (1981, 1984) adapted the spot fire distance models for use in programmable pocket calculators and Morris (1987) simplified them. These models have been incorporated in BehavePlus (Andrews et al. 2008) and FARSITE (Finney 1998).

Additional studies have enhanced the ability to predict spot fires. Albini (1983b) conducted further studies to the capability of *strong line thermals* in an unstratified atmosphere to lift firebrand particles. He found that the maximum height of a viable ember is roughly proportional to the square root of thermal strength and was able to calculate the horizontal distance traveled from the point of origin to the point where free descent begins. The downwind drift distance is shown to be both significant and sensitive to the wind-speed profile (Albini 1983b).

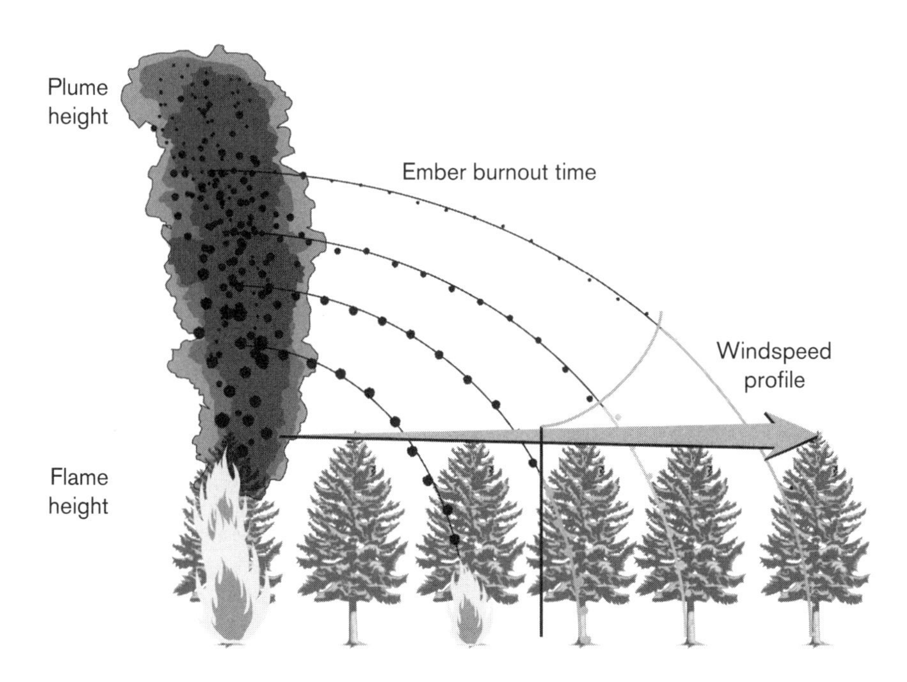

FIGURE 4.9 Larger embers are not lofted as high as small ones but they remain lighted long enough to ignite new spot fires. Small embers are lofted great distances but often are extinguished before they land (redrawn from Finney 1998).

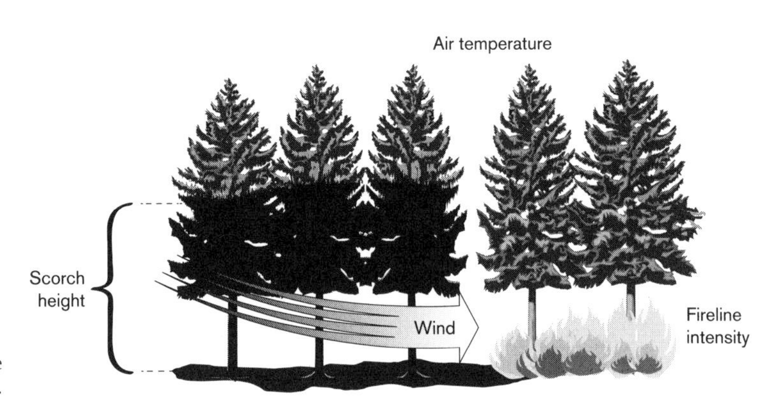

FIGURE 4.10 Scorch height is affected by fireline intensity, wind speed, and air temperature.

Coupled Fire Behavior and Atmospheric Models

A new fire model that includes both fire behavior and atmospheric dynamics was developed by Linn et al. (2002). They coupled an atmospheric hydrodynamics model, HIGRAD, with a wildfire behavior model, FIRETEC, to simulate the relation between a fire and its environment. The model includes the interactions between fire, fuels, atmosphere, and topography over landscape scales from 100 to 1,000 m (330–3,300 ft). The computational requirements preclude the use of the model in the field, but it has great potential as a research tool for predicting fire in complex terrain under varying atmospheric conditions and investigating the causes of extreme changes in fire behavior.

Fire Effects

Once a fire is ignited, it starts to affect other components of the ecosystem. Plants, animals, soil, water, and air all interact in one way or another with fire. In this section we will introduce the physical parameters of fire behavior that affect fire severity, spotting, tree scorch height, plant mortality, biomass consumption, and microclimate. The ecological ramifications of these effects will be discussed in subsequent chapters.

Fire Severity

Fire severity is the magnitude of the effect that the fire has on the environment and is commonly applied to a number of ecosystem components. We include in this definition fire effects that occur while the fire is burning as well as those second-order effects that occur in the postfire environment and recovery after a year. Our definition differs from the definition used by Burn Area Emergency Rehabilitation (BAER) teams who use *fire* severity for the immediate fire environment and *burn* severity for postfire environment (Jain 2004).

Different patterns of fireline intensity and fire duration affect the level of fire severity. For example, a high-intensity fire of short duration could result in the same level of severity as a low-intensity fire of long duration. A high-intensity fire that moves quickly through the crowns may kill all of the trees but have relatively little effect on the soil, whereas a low-intensity fire might leave trees untouched but smolder for days and result in severe soil heating. Precise measures of severity will vary from one ecosystem to the next, depending upon the degree of change to biotic and physical ecosystem components.

Tree Crown Scorch Height

Scorch occurs when the internal temperature of the leaves or needles of a plant are raised to lethal levels. Both the temperature and its duration are important (Davis 1959). Exposure to temperatures of about 49°C (120°F) for an hour can begin to kill tissues, whereas temperatures of approximately 54°C (130°F) can kill within minutes, whereas temperatures over 64°C (147°F) are considered instantaneously lethal. Van Wagner (1973) related crown scorch height to the ambient air temperature, fireline intensity, and wind speed using old metric units. Under conditions of warm air, less intensity is necessary to raise the tissue temperature to the lethal level (Fig. 4.10). For a given fireline intensity, scorch height is reduced sharply as wind speed increases. Winds cool the hot plume as entrained ambient air moves through the canopy. Alexander and Cruz (2012) adapted the Van Wagner (1973) relations to International System (SI) units. They cautioned, however, that fuelbed structure can strongly influence the linkage between fire behavior and fire effects. Scorch height calculations are included in BehavePlus (Andrews et al. 2008) and in FARSITE (Finney 1998).

Plant Mortality

Mortality can occur when plants are either entirely consumed or certain tissues are raised to lethal temperatures for a sufficient duration. However, some species are able to sprout after complete canopy removal or scorch. For other species, if too much cambium or canopy is killed, the plant cannot survive. Ryan et al. (1988) studied long-term fire-caused mortality of mature Douglas-fir (*Pseudotsuga menziesii* var. *menziesii*) and found that the amount of cambium killed was the best predictor of tree mortality and that the percent of the crown scorch was a better predictor than crown scorch height. Peterson and Ryan (1986) used bark thickness, crown scorch height, and Rothermel's (1972) equation to predict cambial damage and mortality for northern Rocky Mountain species. Ryan and Reinhardt (1988) developed a model for predicting percent mortality based on percent volume crown scorch and bark thickness. Bark thickness was derived from species and diameter, while percent crown scorch was calculated from the scorch height, tree height, and crown ratio. Fire-induced mortality calculations based on their work are included in BEHAVE (Andrews et al. 2008). Stephens and Finney (2002) found that mortality of Sierra Nevada mixed conifers was related to percent canopy scorched and local ground fuel consumption. Hood et al. (2007) developed mortality models for western US conifers based on crown scorch and species specific bark thickness. Percent crown killed was determined to be the best overall predictor for mortality of mixed-conifer species in California by Hood et al. (2010).

Biomass Consumption

The amount of biomass consumed by the flaming front can be calculated from the heat per unit of area released by the fire. Fire effects are often more related to the heat given off after passage of the flaming front, however. Van Wagner (1972) provided equations for estimating the amount of the combined litter and fermentation layers that would burn based on the average moisture content of those layers. Similar results for litter and duff layers in the Sierra Nevada were found by Kauffman and Martin (1989). They found that consumption of the litter and fermentation layers was inversely related to the moisture content of the lower duff layer. Albini et al. (1995) modeled burnout of large woody fuels including the influence of smoldering duff. The rate at which these fuels burn is a balance between the rate of heat transfer to the fuel and the amount of heat required to raise the fuel to a hypothetical pyrolysis temperature.

Microclimate

The effect of fire on microclimate can be determined by comparing canopy densities before and after a fire. These secondary effects are manifested through changes in the vegetation. For example, a fire that thins a stand of trees could increase wind speed and air temperature at the ground surface and decrease relative humidity and fuel moisture. A more open canopy allows more sunlight to reach the surface fuels and offers less resistance to winds above the canopy. These changes will, in turn, affect the behavior of subsequent fires.

Bigelow and North (2012) studied the effects of mechanically thinned and group selected stands in the Sierra Nevada on microclimate. They found that wind speeds increased moderately in thinned stands and sharply in group selected openings. Although they found no effect of the treatments on air temperature and relative humidity, duff moisture was reduced in the openings. Simulations of fire behavior in the treated stands showed slight changes in fire behavior in openings but minimal change in thinned stands. It is not clear how well the treatments replicated the effects of fire.

Summary

In this chapter we have covered fire as a physical process. A fire cannot occur unless there is sufficient fuel available to support combustion, weather conditions are conducive for burning, and an ignition source is present. Combustion is an oxidation process that combines hydrocarbons in the form of vegetative fuels with oxygen to produce carbon dioxide, water, and energy, and other products if not completely burned. This basic ecosystem process occurs in three phases: a preheating phase, a gaseous phase, and a smoldering phase. Heat from combustion is transferred through conduction, convection, and radiation. Fuel characteristics determine the amount of fuel energy available and its rate of release. Important fuel characteristics include the size of the fuels, their ability to absorb and release moisture, their compactness, and their total weight. For the purposes of fire behavior prediction and fire danger assessment fuels are categorized into models that include all the various characteristics that affect fire behavior and fire effects.

Once ignited, the fire begins to spread on the surface as the heat generated reaches the temperature necessary to ignite the adjacent fuel. Surface fire spread is accelerated by wind and steep topography. Crown fire spread occurs when the heat from the surface fire crosses the threshold necessary to ignite crowns. Torching trees can loft embers that ignite spot fires downwind from the main fire. Direct fire effects include fire severity, scorch height, plant mortality, and biomass consumption. Microclimate is affected indirectly through the effects of fire on vegetation. Fire plays a dynamic role in natural ecosystems.

Fire plays a dynamic role in natural ecosystems. As a physical process, it reacts to and influences other ecosystem components. In the following chapter, we will see how the physical process of fire is integrated within ecosystems as an ecological process.

References

Albini, F.A. 1976. Estimating wildfire behavior and effects. USDA Forest Service General Technical Report INT-GTR-30. Intermountain Forest and Range Experiment Station, Ogden, Utah, USA.

———. 1979. Spot fire distance from burning trees—a predictive model. USDA Forest Service General Technical Report INT-GTR-56. Intermountain Forest and Range Experiment Station, Ogden, Utah, USA.

———. 1981a. A model for the wind-blown flame from a line fire. Combustion and Flame 43: 155–174.

———. 1981b. Spot fire distance from isolated sources-extensions of a predictive model. USDA Forest Service Research Note INT-RN-309. Intermountain Forest and Range Experiment Station, Ogden, Utah, USA.

———. 1983a. Potential spotting distance from wind-driven surface fires. USDA General Technical Report INT-GTR-309. Intermountain Forest and Range Experiment Station, Ogden, Utah, USA.

———. 1983b. Transport of firebrands by line thermals. Combustion Science and Technology 32: 277–288.

Albini, F.A., J.K. Brown, E.D. Reinhardt, and R.D. Ottmar. 1995. Calibration of a large fuel burnout model. International Journal of Wildland Fire 5: 173–192.

Alexander, M.E. 1988. Help with making crown fire hazard assessments. Pages 147–156 in: W.C. Fischer and S.F. Arno, compilers. Protecting People and Homes from Wildfire in the Interior West. USDA Forest Service General Technical Report INT-GTR-251. Intermountain Forest and Range Experiment Station, Ogden, Utah, USA.

Alexander, M.E., and M.G. Cruz. 2012. Interdependencies between flame length and fireline intensity in predicting crown fire initiation and crown scorch height. International Journal of Wildland Fire 21: 95–113.

Anderson, H.E. 1969. Heat transfer and fire spread. USDA Forest Service Research Paper INT-RP-69. Intermountain Forest and Range Experiment Station, Ogden, Utah, USA.

———. 1982. Aids for determining fuel models for estimating ire behavior. USDA Forest Service General Technical Report

INT-GTR-122. Intermountain Forest and Range Experiment Station, Ogden, Utah, USA.

Anderson, M.K. 1996. Tending the wilderness. Restoration Management Notes 14(2): 154–166.

Andrews, P.L., C.D. Bevins, and R.C. Seli. 2008. BehavePlus fire modeling system, version 4.0: user's guide. USDA Forest Service General Technical Report RMRS-GTR-10. Rocky Mountain Research Station, Fort Collins, Colorado, USA.

Arnold, R.K. 1964. Project skyfire lightning research. Proceedings Tall Timbers Fire Ecology Conference 3: 121–130.

Babrauskas, V. 2006. Effective heat of combustion for flaming combustion of conifers. Canadian Journal of Forest Research. 36: 659–663.

Bigelow, S., and M. North. 2011. Microclimate effects of fuels-reduction and group-selection silviculture: implications for fire behavior in Sierran mixed-conifer forests. Forest Ecology and Management 264: 51–59.

Brown, J.K. 1974. Handbook for inventorying downed woody material. USDA Forest Service General Technical Report INT-GTR-16. Intermountain Forest and Range Experiment Station, Ogden, Utah, USA.

Brown, J.K., R.D. Oberhue, and C.M. Johnston. 1982. Handbook for inventorying surface fuels and biomass in the interior west. USDA Forest Service General Technical Report INT-GTR-129. Intermountain Forest and Range Experiment Station, Ogden, Utah, USA.

Burgan, R.E. 1988. 1988 revisions to the 1978 National Fire-Danger Rating System. USDA Forest Service Research Paper SE-RP-273. Southeastern Forest Experiment Station, Asheville, North Carolina, USA.

Burgan, R.E., and R.C. Rothermel. 1984. BEHAVE: fire behavior prediction and fuel modeling system—FUEL subsystem. USDA Forest Service General Technical Report INT-GTR-167. Intermountain Forest and Range Experiment Station, Ogden, Utah, USA.

Byram, G.M. 1959. Combustion of forest fuels. Pages 61–89 in: K.P. Davis, editor. Forest Fire Control and Use. McGraw-Hill, New York, New York, USA.

Chase, C.H. 1981. Spot fire distance equations for pocket calculators. USDA Forest Service Research Note INT-RN-310. Intermountain Forest and Range Experiment Station, Ogden, Utah, USA.

———. 1984. Spot fire distance from wind-driven fires—extensions of equations for pocket calculators. USDA Forest Service Research Note INT-RN-346. Intermountain Forest and Range Experiment Station, Ogden, Utah, USA.

Cruz, M.G., and M.E. Alexander. 2010. Assessing crown fire potential in coniferous forests of western North America: a critique of current approaches and recent simulation studies. International Journal of Wildland Fire 19: 377–398.

Davis, K.P. 1959. Forest Fire Control and Use. McGraw-Hill, New York, New York, USA.

Deeming, J.E., R.E. Burgan, and J.D. Cohen. 1977. The National Fire Danger Rating System—1978. USDA Forest Service General Technical Report INT-GTR-39. Intermountain Forest and Range Experiment Station, Ogden, Utah, USA.

Deeming, J.E., J.W. Lancaster, M.A. Fosberg, W.A. Furman, and M.J. Schroeder. 1972. National Fire Danger Rating System. USDA Forest Service. Research Paper RM-RP-84. Rocky Mountain Forest and Range Experiment Station, Fort Collins, Colorado, USA.

Finney, M.A. 1998. FARSITE: fire area simulator—model development and evaluation. USDA Forest Service Research Paper RMRS-RP-4. Rocky Mountain Research Station, Fort Collins, Colorado, USA.

Finney, M.A., J.D. Cohen, J.M. Forthofer, S.S. McAllister, M.J. Gollner, D.J. Gorham, K. Saito, N.K. Akafuah, B.A. Adam, and J.D. English. 2015. Role of buoyant flame dynamics in wildfire spread. PNAS 112: 9833–9838.

Finney, M.A., J.D. Cohen, S.S. McAllister, and W.M. Jolly. 2012. On the need for a theory of wildland fire spread. International Journal of Wildland Fire 22: 25–36.

Finney, M.A., T.B. Maynard, S.S. McAllister, and I.J. Grob. 2013. A study of ignition by rifle bullets. USDA Forest Service Research Report RMRS-RP-104. Rocky Mountain Research Station, Fort Collins, Colorado, USA.

Fons, W. 1946. Analysis of fire spread in light forest fuels. Journal of Agricultural Research 7: 93–121.

Frandsen, W.H. 1971. Fire spread through porous fuels from the conservation of energy. Combustion and Flame 16: 9–16.

Fuquay, D.M. 1982. Cloud-to-ground lightning in summer thunderstorms. Journal of Geophysical Research 87: 7131–7140.

Hood, S.M., C.W. McHugh, K.C. Ryan, E.D. Reinhardt, and S.L. Smith. 2007. Evaluation of a post-fire tree mortality model for western USA conifers. International Journal of Wildland Fire 16: 679–689.

Hood, S.M., S.L. Smith, and D.R. Cluck. 2010. Predicting mortality for five California conifers following wildfire. Forest Ecology and Management 260: 750–762.

Jain, T.B. 2004. Tongue-tied. Wildfire Magazine, July–August.

Jolly, W.M., A.M. Hadlow, and K. Huguet. 2014. De-coupling seasonal changes in water content and dry matter to predict live conifer foliar moisture content. International Journal of Wildland Fire 23: 480–489.

Kauffman, J.B., and R.E. Martin. 1989. Fire behavior, fuel consumption, and forest floor changes following prescribed understory fires in Sierra Nevada mixed-conifer forests. Canadian Journal of Forest Research 19: 455–462.

Lancaster, J.W. 1970. Timelag useful in fire danger rating. Fire Control Notes 31(3): 6–8.

Latham, D.J., and J.A. Schlieter. 1989. Ignition probabilities of wildland fuels based on simulated lightning discharges. USDA Forest Service Research Paper INT-RP-411. Intermountain Forest and Range Experiment Station, Ogden, Utah, USA.

Linn, R.R., J. Reisner, J. Colman, and J. Winterkamp. 2002. Studying wildfire using FIRETEC. International Journal of Wildland Fire 11: 1–14.

McAllister, S.S., J-Y. Chen, and A.C. Fernandez-Pello. 2011. Fundamentals of Combustion Processes. Springer, New York, USA.

Morris, G.A. 1987. A simple method for computing spotting distances from wind-driven surface fires. USDA Forest Service Research Note INT-RN-374. Intermountain Forest and Range Experiment Station, Ogden, Utah, USA.

Nelson, R.M. 1993. Byram's energy criterion for wildland fires: units and equations. USDA Forest Service Research Note INT-RN-415. Intermountain Forest and Range Experiment Station, Ogden, Utah, USA.

NWCG (National Wildfire Coordinating Group). 2014. National Fire Danger Rating System (NFDRS). Briefing Paper, September 17, 2014. National Wildfire Coordinating Group, Boise, Idaho, USA.

Peterson, D.L., and K.C. Ryan. 1986. Modeling postfire conifer mortality for long-range planning. Environmental Management 10: 797–808.

Rebain, S.A., compiler. 2010. The fire and fuels extension to the forest vegetation simulator: updated model documentation. USDA Forest Service Internal Report. Forest Management Service Center, Fort Collins, Colorado, USA.

Rothermel, R.C. 1972. A mathematical model for fire spread predictions in wildland fuels. USDA Forest Service Research Paper INT-RP-115. Intermountain Forest and Range Experiment Station, Ogden, Utah, USA.

———. 1983. How to predict the spread and intensity of forest fires. USDA Forest Service General Technical Report INT-GTR-143. Intermountain Forest and Range Experiment Station, Ogden, Utah, USA.

———. 1991. Predicting behavior and size of crown fires in the northern Rocky Mountains. USDA Forest Service Research Paper INT-RP-438. Intermountain Forest and Range Experiment Station, Ogden, Utah, USA.

Rothermel, R.C., and J.E. Deeming. 1980. Measuring and interpreting fire behavior for correlation with fire effects. USDA Forest Service General Technical Report INT-GTR-93. Intermountain Forest and Range Experiment Station, Ogden, Utah, USA.

Ryan, K.C., D.L. Peterson, and E.D. Reinhardt. 1988. Modeling long-term fire-caused mortality of Douglas-fir. Forest Science 34: 190–199.

Ryan, K.C., and E.D. Reinhardt. 1988. Predicting post-fire mortality of seven western conifers. Canadian Journal of Forest Research 18: 1291–1297.

Sandberg, D.V., R.D. Ottmar, and G.H. Cushon. 2001 Characterizing fuels in the 21st century. International Journal of Wildland Fire 10: 381–387.

Schroeder, M.J., and C.C. Buck. 1970. Fire weather . . . a guide for application of meteorological information to forest fire control

operations. USDA Forest Service Agriculture Handbook 360. Washington, D.C., USA.
Scott, J.H. 1999. NEXUS: a spread sheet based crown fire hazard assessment system. Fire Management Notes 59(2): 20–24.
Scott, J.H., and R.E. Burgan. 2005. Standard fire behavior fuel models: a comprehensive set for use with Rothermel's surface fire spread model. USDA Forest Service General Technical Report RMRS-GTR-153. Rocky Mountain Research Station, Fort Collins, Colorado, USA.
Scott, J.H., and E.D. Reinhardt. 2001. Assessing crown fire potential by linking models of surface and crown fire behavior. USDA Forest Service Research Paper RMRS-RP-29. Rocky Mountain Research Station, Fort Collins, Colorado, USA.
Stephens, S.L., and M.A. Finney. 2002. Prescribed fire mortality of Sierra Nevada mixed conifer tree species: effects of crown damage and forest floor consumption. Forest Ecology and Management 162: 261–271.
Stephens, S.L., J.J. Moghaddas, C. Ediminster, C.E. Fiedler, S.M. Haase, M.G. Harrington, J.E. Keeley, J.D. McIver, K. Metlen, C.N. Skinner, and A. Youngblood. 2009. Fire treatment effects on vegetation structure, fuels, and potential fire severity in western U.S. forests. Ecological Applications 19: 305–320.
Vale, T.R. 1998. The myth of the humanized landscape: an example from Yosemite National Park. Natural Areas Journal 18: 231–236.
Van Wagner, C.E. 1972. Duff consumption by fire in eastern pine stands. Canadian Journal of Forest Research 2: 34–39.
———. 1973. Height of crown scorch in forest fires. Canadian Journal of Forest Research 3: 373–378.
———. 1977. Conditions for the start and spread of crownfire. Canadian Journal of Forest Research 7: 23–34.
———. 1993. Prediction of crown fire behavior in two stands of jack pine. Canadian Journal of Forest Research 18: 818–820.
van Wagtendonk, K.A., and D.F. Smith. 2016. Prioritizing lightning ignitions in Yosemite National Park with a biogeophysical and sociopolitically informed decision tool. Pages 129–135 in: S. Weber, editor. Engagement, Education, and Expectations—The Future of Parks and Protected Areas: Proceedings of the 2015 George Wright Society Conference on Parks, Protected Areas, and Cultural Sites. George Wright Society, Hancock, Michigan, USA.

CHAPTER FIVE

Fire as an Ecological Process

NEIL G. SUGIHARA, JAN W. VAN WAGTENDONK,
AND JO ANN FITES-KAUFMAN

> Removing fire from . . . ecosystems would be among the greatest upsets in the environmental system that man could impose—possibly among the most severe stresses since the evolution of the fire-dependent biota evolved. I cannot predict the outcome, but a fundamental reordering of the relationships between all plants and animals and their environments would occur. Many species could be lost through extinction.
>
> HEINSELMAN (1981)

Fire is a defining component of California's ecosystems; for without fire, few native ecosystems and habitats would persist as we know them today. Fire's inherent dynamic nature and complexity are amplified by the state's diverse topography, climate, and vegetation. For millennia, California's ecosystems have developed in tandem with fire. Long-term alterations of fire patterns have occurred with climatic changes and interactions with humans. In the past two centuries, the pace of human-induced alteration has accelerated, resulting in a number of changes in species and ecosystems. Many of these species and ecosystem alterations are still happening, and others are yet to manifest themselves. To understand the importance of the changing ecological role of fire and how to manage fire into the future, it is necessary to understand fire as an ecological process.

Landscapes have repeated patterns of fire occurrence, fire magnitude, and fire type that vary over space and time. When an individual fire is seen as a discrete event, its physical characteristics are vital to understanding fire effects. When fire is considered over centuries or millennia and on large landscapes, this repeated pattern of fire and its properties affect ecosystem function. Compounding the influences of a single fire, the existing patterns during subsequent fires greatly influence the dynamics of fuel patterns, ecosystem properties, species composition, and vegetation structure. Although fire patterns over large expanses of space and long periods of time become extremely complex, they can be distilled into useful summaries known as fire regimes.

There is a continuous feedback loop of fire, fuels, and vegetation within ecosystems. Fire burns the fuel; the vegetation that grows following fire produces the fuel for the next fire. Fire interacts with, and is affected by species composition, vegetation structure, fuel moisture, air temperature, biomass, and many other ecosystem components and processes over several scales of time and space. These ecosystem components are so interdependent that changes to one, including fire, often result in significant changes to others. This dynamic view of ecosystems is the key to understanding fire as an ecological process.

In this chapter we explore fire as a dynamic ecological process by first examining fire in the context of general ecological theory, then discussing the concept of fire regimes, and finally by developing and applying an attribute-based framework for classifying fire regimes that allows us to understand the patterns of fire as processes within ecosystems. This fire regime framework is used in the bioregional chapters that follow in part Two.

Fire in the Context of Ecological Theory

As ecological theory has evolved, so has the manner in which fire along with climate, insects, fungi, floods, erosion, sedimentation, and weather are considered in that theory. We first look at succession theory and then proceed through ecosystem, disturbance, and hierarchical theory. Finally we present our view of fire as an ecological process.

Succession Theory

Classical succession is an ecological concept that was developed and championed by Clements (1916) in the early 1900s. Since it was first published, his framework for viewing plant communities as complex entities that develop over time has served as a basis from which successional ecology theory has developed. Clements (1936) defined *succession* as a predictable, directional, and stepwise progression of plant assemblages that culminates in a self-perpetuating *climax* community controlled by climate. For example, bare ground might first be colonized by grasses, followed by shrubs, and then by a young forest, and finally be covered by a mature forest (Fig. 5.1). According to Clements, the climatic climax is stable, complex, self-perpetuating, and considered to be the final stage in the development of the "complex organism" or plant community.

Clements (1916) considered bare areas created by lightning fires as one of the natural sources for the initiation of succession. He expressed the view that lightning fires were numerous, and often very destructive, in regions with frequent dry thunderstorms. In fact, Clements' views were shaped by some

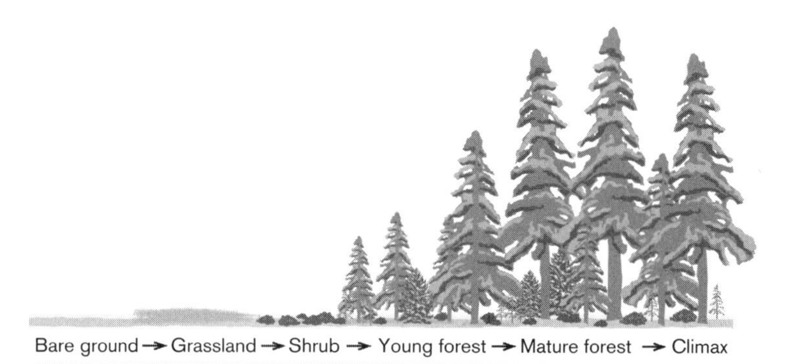

FIGURE 5.1 Clements viewed succession as a stepwise, directional process. As time proceeds, bare ground eventually becomes covered with a mature, climax vegetation.

FIGURE 5.2 Gleason considered plant communities to be distributed according to environmental gradients such as grasses being gradually replaced by oaks as precipitation increases with elevation.

of the most destructive wildland fires known in this country that occurred during the early twentieth century. Clements considered areas where such fires maintained vegetation that differed from the climatic climax to be *subclimax*, because they were continually reset to seral plant assemblages by recurrent fire before they reached climax conditions. He cited chaparral in California and lodgepole pine (*Pinus contorta* subsp. *latifolia*) in Colorado as examples of fire subclimaxes (Clements 1916). Fire was viewed as a retrogressive process that sets back the directional, stepwise progression of succession toward the stable climactic climax. Clements (1936) refined his ideas about the nature and structure of the climax and developed a complex terminology for classifying units of vegetation. Fire subclimaxes were still part of this complex system, and he added California's Monterey pine (*P. radiata*), Bishop pine (*P. muricata*), and knobcone pine (*P. attenuata*) as examples. He used the term *disclimax* for communities that had been degraded by human activities such as logging, grazing, and burning, but seemed to not apply the term to natural fires (Clements 1936).

Gleason (1917) reacted to Clements' theory by proposing the individualistic concept of the plant association. He argued that succession was not inherently directional, but was the result of random immigration of species into a variable environment. As the environment changes, the assemblage of associated species changes based on individual attributes of each species. As an example, he cited the gradual replacement of grasslands by California oak (*Quercus* spp.) forests as one ascends the foothills and precipitation increases (Fig. 5.2). Similarly, Gleason (1917) argued that entirely different plant associations might occupy physiographically and climatically identical environments. For instance, the alpine areas of the Sierra Nevada have essentially the same environment as in the Andes, but their floras are entirely different. Although Gleason (1926) felt that the environment had a strong influence on plant community development, he referred to fire as an unnatural disturbance that limited the duration of the original vegetation.

Daubenmire (1947) was one of the first ecologists to recognize fire as an ecological factor rather than as an allogenic factor. With regard to succession, however, he followed the same terminology as Clements (1916) but considered fire to be one of six different climaxes. *Primary* climaxes included climatic, edaphic, and topographic climaxes, whereas fire and zootic climaxes were termed *secondary* climaxes (Daubenmire 1968). Specific examples included the forests of the Sierra Nevada where episodic fires replaced fire-sensitive species with fire-tolerant pines. Daubenmire (1968) felt that the fire climax could appropriately be called a disclimax because its maintenance depended on continued disturbance.

Whittaker (1953) examined both the organismic (Clements 1916) and individualistic (Gleason 1926) concepts of the climax community and proposed an alternative approach that views the climax as a pattern of vegetation resulting from environmental variables. He postulated that: (1) the climax is a steady state of community productivity, structure, and population, with a dynamic balance determined in relationship to its site; (2) the balance among plant populations shifts with changes in the environment; and (3) the climax composition is determined by all factors of the mature ecosystem. A major contribution that Whittaker (1967) made to ecological theory was his use of gradient analysis to delineate how plant assemblages change in space and time. Whittaker (1953) considered periodic fire to be one of the environmental factors to which some climaxes are adapted. In the absence of fire, the climax plant populations might develop into something entirely different, but that development might never occur. A key point he makes is that burning may cause population fluctuations that make it difficult to distinguish between fire as an environmental factor and fire as a disturbance introduced from outside the ecosystem. For example, in climates between forests and deserts, fire could shift the balance among woodlands, shrublands, and grasslands (Whittaker 1971).

Ecosystem Theory

Tansley (1935) refuted the organism concept of a plant community put forward by Clements (1916) and proposed that succession in a community is a trajectory of a dynamic system with many possible equilibria. That is, depending on the environment, a plant community could develop in one of many different directions and reach a point of equilibrium regardless of which trajectory was followed. He also introduced the term *ecosystem* to describe the entire system to include not only the biotic components but also the abiotic factors that make up the environment. In the ecosystem, these components and factors are in a dynamic equilibrium. Succession leads to a relatively stable phase termed the *climatic climax*. He recognized other climaxes determined by factors such as soil, grazing, and fire. Tansley (1935) considered vegetation that was subjected to constantly recurring fire to be a fire climax, but thought that catastrophic fire was destructive and external to the system.

Odum (1959) defined ecology as the study of structure and function of ecosystems and emphasized that the ecosystem approach had universal applicability. He related the ecosystem concepts of nutrient and energy flow to evolutionary ecological growth and adaptation (Odum 1969). Fire was seen as an important ecological factor in many terrestrial ecosystems, as both a limiting and as a regulatory factor (Odum 1963). He cited examples of fire consuming accumulated undecayed plant material and applying selective pressure favoring the survival and growth of some species at the expense of others.

A systems approach was advocated by Schultz (1968), applying the concepts of energy dissipation to ecosystem function. He described ecosystems as open systems with material being both imported and exported. Rather than reaching equilibrium, an open system attains a steady state with minimum loss of energy. Fire is considered a negative feedback mechanism that prevents the complete destruction of natural ecosystems by returning some of the energy to the system (Schultz 1968).

Disturbance Theory

Traditional theories of natural disturbance considered that disturbance must be a major catastrophic event and that it must originate in the physical environment (Agee 1993). Much discussion has centered on these points and various definitions and thresholds have been applied to distinguish disturbances from processes. Watt (1947) introduced the concept that plant communities were composed of patches in various stages of development that were dynamic in time and space. The patches were initiated by some form of disturbance, be it the death of a single tree or larger factors such as storms, drought, epidemics, or fires. Other than mentioning size differences, he did not distinguish among factors that were internal or external to an ecosystem. Similarly, White (1979) urged that the concept of disturbance not limited to large catastrophic events that originate from within the physical environment but also include external factors. White and Pickett (1985) define disturbance as "any relatively discrete event in time that disrupts ecosystem, community, or population structure and changes resources, substrate availability, or the physical environment." They included disasters and catastrophes as subsets of disturbance. Fire was specified as a source of natural disturbance. Agee (1993) proposed that disturbance comprises a gradient that ranges from minor to major; he did not differentiate between internal and external sources. He did distinguish between fires of natural origin and fires set by Native Americans or Euro-Americans, calling the former natural disturbances.

Walker and Willig (1999) follow the terminology of White and Pickett (1985) and treat fire as a natural disturbance. They go on to state that disturbances that originate inside the system of interest are considered to be endogenous. Fire is driven by interplay of exogenous factors from outside of the system such as climate and endogenous factors such as soil and biota. In this sense, Walker and Willig (1999) consider fire to be an inherent ecological process. They characterize disturbances by their frequency, size, and magnitude. These characteristics are used for grouping disturbances into disturbance regimes.

Turner and Dale (1998) state that large, infrequent disturbances are difficult to define because they occur across a continuum of time and space. One definition they propose is that disturbances should have statistical distributions of extent, intensity, or duration greater than two standard deviations (SDs) of the mean for the period and area of interest. Romme et al. (1998) distinguish large, infrequent disturbances from small, frequent ones by a response threshold—when the force of the disturbance exceeds the capacity of internal mechanisms to resist disturbance or where new means of recovery become involved. For example, an area that burns with a very high-severity fire as a result of unnaturally heavy accumulations of fuels would be qualitatively different from an area that burns with frequent, low-severity fires. However, not all high-severity fires cross the response threshold. Romme et al. (1998) cite the example of jack pine (*Pinus banksiana*), an ecological equivalent of lodgepole pine, that re-establishes itself after stand replacing fires, regardless of size, through the dispersal of seed throughout the area from serotinous cones.

These criteria (Romme et al. 1998, Turner and Dale 1998) form a basis for separating endogenous fires from those arising from outside the environment of the ecosystem.

Hierarchical Theory

O'Neill et al. (1986) proposed a hierarchical concept of the ecosystem to reconcile the species-community and process-function schools of thought. The authors define the ecosystem as being composed of plants, animals, incorporated abiotic components, and the environment. In their view, the ecosystem is a dual organization determined by structural constraints on organisms and functional constraints on processes. These dual hierarchies have both temporal and spatial components.

Disturbances are termed perturbations and are associated with a particular temporal and spatial scale. O'Neill et al. (1986) describe fire as a perturbation that ensures landscape diversity and preserves seed sources for recovery from any major disturbance. They state that viewing ecosystems on the arbitrary scale of the forest stand results in seeing fire as a catastrophic disturbance. If, however, fire is viewed at the scale appropriate to the frequency of occurrence, it can be seen as an essential ecosystem process that retains the spatial diversity of the landscape and permits reaching a dynamic equilibrium after disturbance. O'Neill et al. (1986) consider a perturbation to be *incorporated* if the ecosystem structure exerts control over some aspect of the abiotic environment that is uncontrolled at a lower level of organization.

Systems that are large relative to their perturbations maintain a relatively constant structure (O'Neill et al. 1986). For example, ponderosa pine (*Pinus ponderosa*) forests are usually larger than the fires that burn within them; therefore, the perturbation is incorporated in the sense that the fires do not threaten the survival of the ecosystem but are in fact necessary to perpetuate the spatial diversity of the landscape. This concept is illustrated in Fig. 5.3 (Shugart and West 1981). Above the diagonal line are disequilibrium systems that are the same size or smaller than their characteristic perturbations. Wildland fires would be considered a perturbation in forest stands but would be an incorporated process in large forests.

Pickett et al. (1989) linked the hierarchical organization of ecosystem components with the concept of disturbance. They state that any persistent ecological object such as a tree will have a minimal structure that permits its persistence, and that disturbance is a change in that structure caused by a factor external to the level of interest. Disturbance, then, is identified with specific ecological levels, or hierarchies, of the organization (Pickett et al. 1989). In this view, periodic fire perpetuates a variety of structures that allow the ecosystem to persist.

Our View of Fire

Each of the aforementioned views is based on careful observation and carries something of the truth. Our view of fire synthesizes and builds on these previous theories. We consider fire in its characteristic pattern to be an incorporated ecological process rather than a disturbance. In its natural role, fire is not a disturbance that impacts ecosystems; rather it is an ecological process that is as much a part of the environment as precipitation, wind, flooding, soil development, erosion, predation, herbivory, carbon and nutrient cycling, and energy flow. Fire resets vegetation trajectories, sets up and maintains a dynamic mosaic of different vegetation structures and compositions, and reduces fuel accumulations. Humans have often disrupted these processes, and the result can be that fire behavior and effects are outside of their range of natural variation. At that point, this uncharacteristic pattern of fire is considered an exogenous disturbance factor.

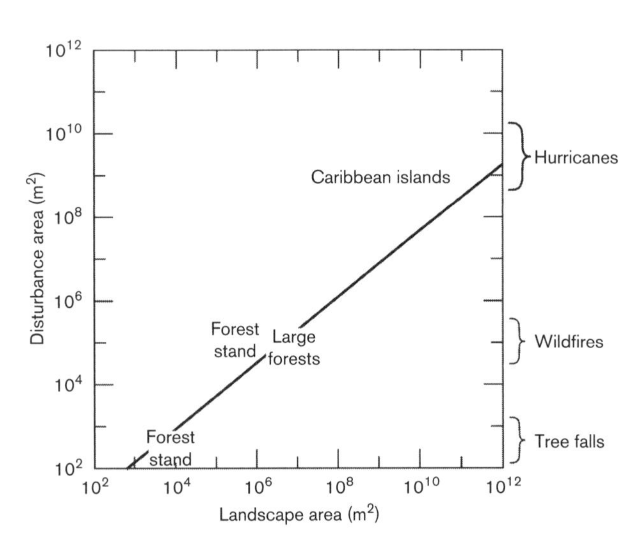

FIGURE 5.3 Relative size of disturbance area and landscape units. Landscapes above the diagonal line are in disequilibrium since they are smaller than the characteristic perturbations (redrawn from Shugart and West 1981).

Fire Regimes

It is relatively simple to understand the influence of a single fire on specific ecosystem properties, but the importance of fire as an ecological process becomes greatly amplified by the complex pattern of fire effects over long time periods, multiple fire events, and numerous ecosystem properties. To synthesize these patterns of fire occurrence, ecologists use the concept of fire regimes. Fire regimes are a convenient and useful way to classify, describe, and categorize the pattern of fire occurrence for scientific and management purposes. Like any classification, a fire regime classification necessarily simplifies complex patterns. Although fire regimes are typically assigned to ecosystems defined by either land areas or vegetation types, or to some combination of area and vegetation, they often vary greatly within a vegetation type and over time on the same piece of land.

Previous Fire Regime Descriptions

Fire regime classification systems have been based on a very small number of attributes that could be described and used to explain basic ecosystem patterns. The classifications offer a variety of information ranging from simple, single-attribute descriptions (e.g., mean fire return interval) to a few attributes, but usually have not provided descriptions of the patterns of fire over time and space. Recent fire history studies have focused on the importance of multi-scaled spatial and temporal variation of fire. As our knowledge of ecosystems and complex processes grows, our need for more sophisticated descriptive tools such as fire regime classifications expands. It is important to recognize that any classification system is an oversimplification, and there is no single "complete" or "right" way to describe fire regimes. The appropriate system to use for classification of fire regimes depends on the character of the ecosystems, the fire regimes, and the intended use of that system.

Heinselman (1981) defined a fire regime as a summary of the fire history that characterizes an ecosystem. He distinguished seven fire regimes based on: (1) fire type and intensity (crown fires or severe surface fires vs. light surface fires), (2) size of typical ecologically significant fires, and (3) frequency or return intervals typical for specific land units. Although these fire regime types described the patterns that he observed in the midwestern United States, this system has served as the basis for fire regime classification throughout the western United States. The classification was not intended to imply mutually exclusive or exhaustive categories; rather it was intended to provide a tool for discussing general fire-occurrence patterns. Heinselman (1981) states, "The purpose here is not to set up a precise classification but to make it possible to discuss important differences in the way fire influences ecosystems." His fire regimes are defined in Box 5.1.

Heinselman (1981) described multiple fire regimes that occur when there are several types of fires in a single ecosystem; each type can be described with its own fire regime. This occurs under the following three conditions: (1) the ecosystem can have more than one type of fire, (2) the types of fires occur under different sets of conditions, and (3) the conditions allow the different types of fires to occur at different frequencies. Multiple fire regimes occur most commonly in vegetation types that have multiple fuel layers that can carry a fire. Heinselman (1981) described red pine (*Pinus resinosa*) forests in the lake states to have both a frequent light surface fire regime carried in the herbaceous layer and a regime of much-less-frequent higher-intensity fire carried in the forests canopy. Many California ecosystems burn with both surface and crown fires that occur at different frequencies and under different weather conditions and can be termed multiple fire regimes (Heinselman 1981).

After applying the Heinselman (1981) fire regimes to the forests and scrublands of the western United States, Kilgore (1981) made a number of observations. There are complex relationships between fire and other attributes of the ecosystem on the variable topography of the western states. Fire acts with different frequencies and intensities, varying with the vegetation, topography, and climate that determine the coincidence of ignitions and burning conditions. Vegetation composition and structure depend on climate, fire frequency, and fire intensity, whereas fire frequency and intensity in turn depend on vegetation structure, topography, and climate. Kilgore (1981) concluded that because of almost annual coincidence of ignitions with suitable burning conditions, western forests, such as some of those found in the Sierra Nevada, have frequent fires of low intensity. Although ignitions are as frequent in many Rocky Mountain forests, they do not coincide as often with dry fuel conditions. These Rocky Mountain forests tend to have less frequent, high-intensity crown fires.

Hardy et al. (2001) modified Heinselman's (1981) original regimes by replacing types of fire with levels of fire severity. They grouped regimes into three levels of frequency and three levels of severity (Box 5.2). These groups have been used to determine natural fire regime condition classes across the

BOX 5.1. HEINSELMAN'S FIRE REGIMES

Seven kinds of fire regimes can be distinguished for forest ecosystems:

0 = No natural fire (or very little)

1 = Infrequent light surface fires (more than 25-year return intervals)

2 = Frequent light surface fires (1–25-year return intervals)

3 = Infrequent, severe surface fires (more than 25-year return intervals)

4 = Short return interval crown fires and severe surface fires in combination (25–100-year return intervals)

5 = Long return interval crown fires and severe surface fires in combination (100–300-year return intervals)

6 = Very long return interval crown fires and severe surface fires in combination (over 300-year return intervals)

BOX 5.2. FIRE REGIME GROUPS USED FOR CONDITION CLASSES

Fire regime group	Frequency	Severity
I	0–35 years	Low (surface fires common) to mixed severity (less than 75% of the dominant overstory vegetation replaced)
II	0–35 years	High (stand replacement) severity (greater than 75% of the dominant overstory vegetation replaced)
III	35–100+ years	Mixed severity (less than 75% of the dominant overstory vegetation replaced)
IV	35–100+ years	High (stand replacement) severity (greater than 75% of the dominant overstory vegetation replaced)
V	200+ years	High (stand replacement) severity

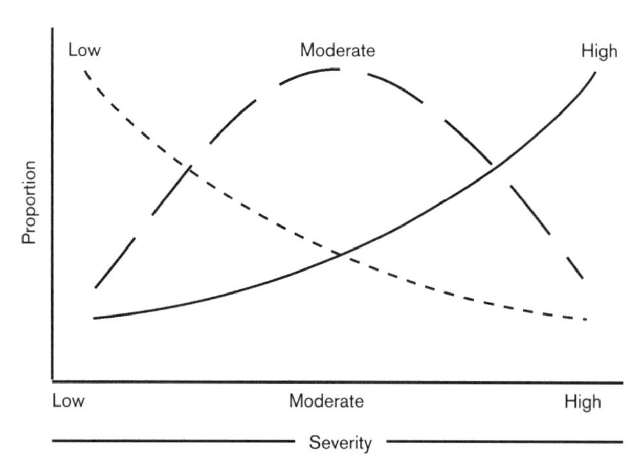

FIGURE 5.4 Variation in fire severity within a general fire regime type (redrawn from Agee 1993). Within a single fire regime type, there could be a combination of low-, moderate-, and high-severity fires.

landscape (Hann and Bunnell 2001). Departures from natural fire regime conditions form the basis for fire- and fuel-management programs.

Agee (1993) describes another system of fire regime classification based on the severity of fire effects on dominant tree species for forests of the Pacific Northwest. To display the variability in fire that occurs on a site, he used a set of distribution curves for illustrating fire severity patterns. The low, moderate, and high fire severity types are presented as distributions composed of different proportions of severity levels (Fig. 5.4). This allows for a range of severity variability within a regime type.

Attribute-Based Fire Regimes

Fire regimes are used to distill and communicate useful information about continuous variation of fire-occurrence patterns into simple categories that help us describe predominant patterns in fire and its effects on ecosystems. As land management objectives evolve to incorporate our expanding understanding of fire as an ecological process, there is a need to re-evaluate what constitutes useful information. Societal objectives for land management have shifted in the past few decades, emphasizing ecosystem and biological values over consumptive uses. The amount and detail of information needed to manage fire to meet these new objectives are greater than ever before.

Attribute-based fire regimes provide a framework that expands on Heinselman's (1981) fire regimes and considers each attribute for every ecosystem. These fire regime descriptions are based on Agee's (1993) treatment of conceptual distributions for severity to include seven fire regime attributes. Temporal attributes include seasonality and fire return interval. Spatial attributes include fire size and spatial complexity of the fires. Magnitude attributes include fireline intensity, fire severity, and fire type. Although there are many other attributes that could be used, these seven include those that are most commonly considered to be important to ecosystem function.

Attributes can be assessed separately, in combinations or as a complete set. This provides a systematic method for describing fire regimes in greater detail and for assessing the change in one or more attributes over time. In altered fire regimes, this system is designed to identify which attributes have changed, the direction of change, and to provide land managers with a tool that can provide direction and focus in determining the effects of the changes and potential restoration of fire regimes.

Sawyer et al. (2009) published a second edition of the manual of California Vegetation, which applied attribute-based fire regime descriptions in tables for each of the more than 300 vegetation alliances in California. This system has served to structure the discussion of fire regimes for restoration and management and to facilitate placing fire regimes and their attributes on landscapes.

Fire regimes are depicted using a set of conceptual distribution curves. For each attribute, there might be several curves with different shapes representing the variability in the distribution of that attribute within different ecosystem types. A fire regime for a particular ecosystem type includes distributions for all seven attributes representing the pattern of variability within that ecosystem.

Fig. 5.5 is an example of fire regime distribution curves for fire return interval. The *x*-axis of each distribution curve represents the range of values for fire return intervals in three different ecosystem types. The *y*-axis always represents the proportion of the burned area with different return interval distributions. The sum of the area underneath each curve is equal to unity and accounts for all of the area that actually burns with that regime type. The three distribution types that are illustrated are short, medium, and long, with each representing a range of short to long, but in different proportions. The conceptual distribution curves allow us to illustrate the features of a fire regime that will affect a specific ecosystem function. For example, if a closed-cone conifer is the only species that distinguishes an ecosystem from the surrounding chaparral ecosystem, the persistence of that closed-cone conifer defines the persistence of the ecosystem. In this case, the distribution of fire return intervals in the two ecosystems may be largely the same, differing only in the presence or absence of the low-frequency events at the extremes of the range of variability (distribution tails) (Fig. 5.6). If the fire return interval extends outside of the range of time (either shorter or longer) when the closed-cone conifer can produce seed, then there is a predicted conversion to the chaparral ecosystem type. For this example, the fire return interval distributions for the two ecosystems have the same general shape, differing only in the absence of the tails of the distribution curve in the conifer type. In this truncated distribution, the tails of the distribution are outside of the range of variability for length of fire return interval within which the conifers can be sustained.

Information for defining and refining the distributions can be obtained using a number of data sources including tree rings with fire scars, charcoal deposits in sediment cores, fire records, and stand-age distributions. These methods require intensive studies and, when used alone, will typically yield only parts of the overall fire regime. Additional information can be obtained through a number of sources that are not currently used in development of fire regime descriptions. The following information should be useful in developing conceptual fire regime attribute distributions that are characteristic for specific ecosystems: (1) geographic location and topography; (2) plant species life history characteristics and fire adaptations; (3) spatial and temporal patterns of fuel quantity, structure, and flammability; and (4) climate and weather patterns. These methodologies and interpretations are detailed in chapter 6.

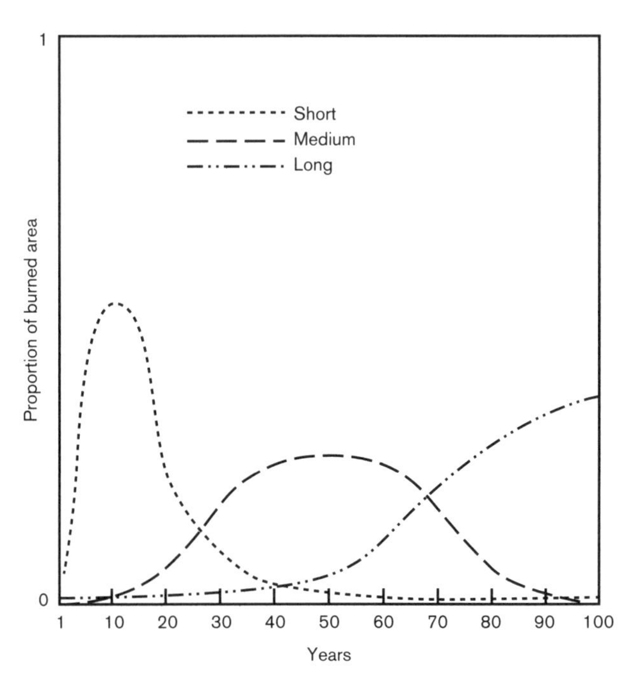

FIGURE 5.5 Example fire regime distribution curve for fire return interval. For short return interval regimes, the majority of the burned area has intervals of only a few years. Medium return intervals regimes range from a few years to several but the majority of the burned area has intervals in the middle range. Similarly, long fire return interval regimes have predominantly long intervals.

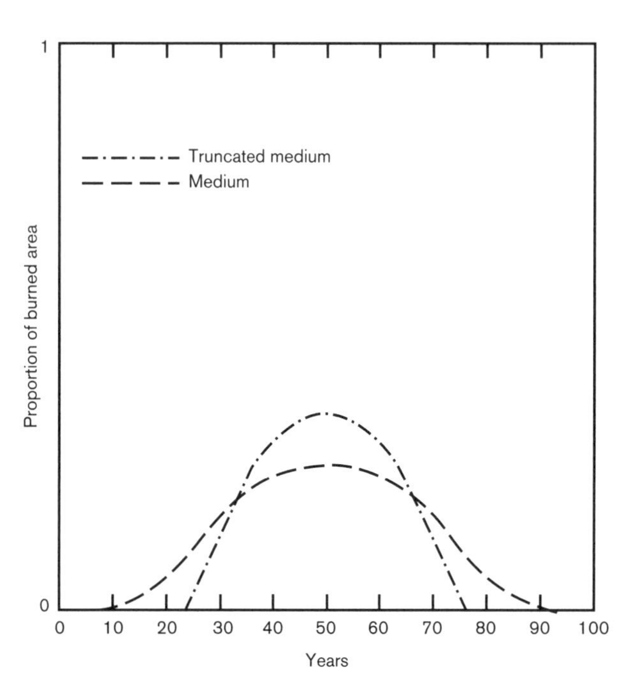

FIGURE 5.6 Example fire regime distribution curves for fire return interval for a closed-cone conifer ecosystem and a surrounding chaparral ecosystem. The curves are the same except for the absence of tails for the closed-cone conifer.

Although there may be no case where we have all of the data needed to know the actual distributions of all seven of the fire regime attributes for any one ecosystem, we can conceptually describe the distributions of these attributes for most ecosystems. These descriptions are based on characteristics of the physical environment and knowledge of the fire relationships of the plant species composing the vegetation types, as well as other vegetation types that interface with it on the landscape. There are different combinations of fire regime attributes that are biologically important and influence stand structure and density, species composition, and distribution and stability of vegetation types with changing fire regimes. Defining the general patterns of fire regimes for ecosystems allows us to gain insight into fire's role in ecosystems.

It is important to keep in mind that the fire regimes that are described in this text are those that are characteristic for ecosystems. That is, when the characteristic fire regime is unchanged, the ecosystem will be maintained. Fire occurrence over the past century or more has changed in many places, and may support the continuance of the former ecosystem, but will often represent changes in specific attributes that will result in tendency of vegetation types to convert. The focus here is to establish the basic relationships between fire and ecosystems, and to better understand and anticipate the results of fire regime changes.

Temporal Fire Regime Attributes

The temporal attributes of fire regimes are described in two ways: seasonality and fire return interval. *Seasonality* is a description of when fires occur during the year; *fire return interval* describes how often fires occur over multiple years. The patterns that are described here for ecosystems are not static on landscapes and can migrate or change in response to changing climate, fuel, continuity, ignition, or species composition. When temporal fire patterns change, there is commonly a change in vegetation type or the distribution of vegetation types.

SEASONALITY

Although California in general can be described as having warm, dry summers and cool, moist winters, season alone does not determine when ecosystems are likely to burn. Other factors, including elevation, coastal influences, topography, characteristics of the vegetation, ignition sources, and seasonal weather patterns, also influence the fire season. Season of burning is especially important biologically because many California ecosystems include species that are only adapted to burning during a fairly limited part of the year. Fig. 5.7 illustrates the four conceptual seasonality patterns that occur in California ecosystems with proportion of the burned area on the *y*-axis and the annual calendar on the *x*-axis.

Spring–Summer–Fall Fire Season The longest fire season type that occurs in California has fire burning well distributed from May to November. It occurs in ecosystems with early spring warming and drying and in which fire is primarily carried in rapidly curing herbaceous layer fuels. The spring–summer–fall fire season type occurs in low elevations and deserts that cure early in the spring and persists until wetting rains occur in the late fall. This fire season type is

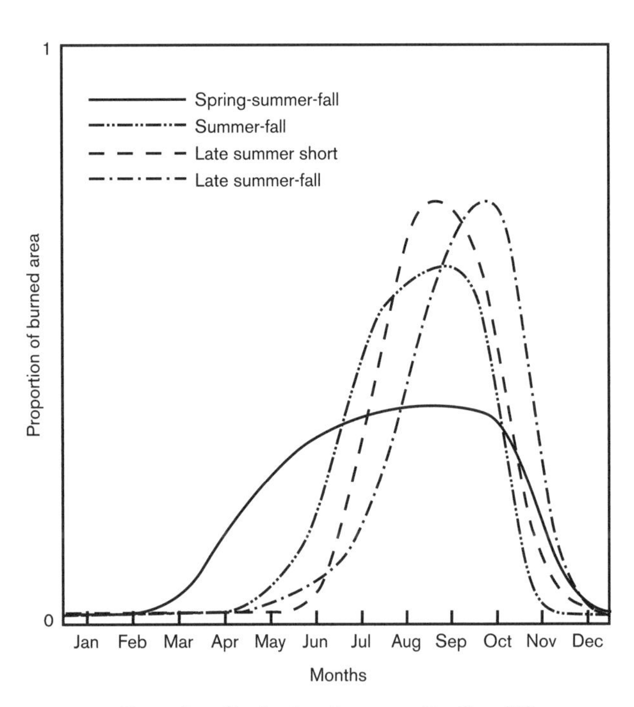

FIGURE 5.7 Fire regime distributions for seasonality. Four different distributions are displayed for fire seasons ranging from spring to fall.

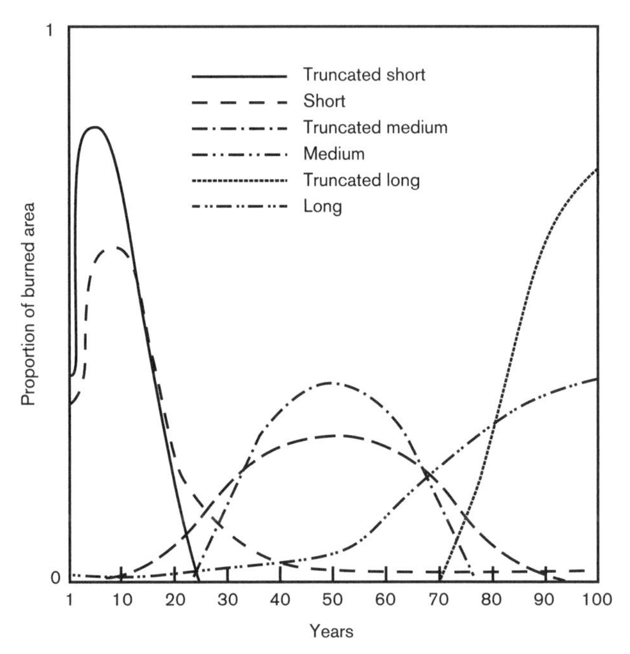

FIGURE 5.8 Fire regime distributions for fire return interval. Six different distribution curves describe the variety of possible return interval regimes.

mostly limited to the Southeastern Deserts bioregion and is characteristic of low-elevation grasslands, shrublands, and woodlands in the Mojave, Colorado, and Sonoran deserts.

Summer–Fall Fire Season This is the characteristic fire season type for most of California, including many low- to middle-elevation grasslands, shrublands, woodlands, and forests. Forest types include coast redwood, Douglas-fir (*Pseudotsuga menziesii* var. *menziesii*), ponderosa pine, mixed conifer, mixed evergreen where fires are primarily carried in herbaceous, duff, and needle layers. Most of the area burns from July to October.

Late Summer-Short Fire Season This is the shortest fire season type that occurs in California. It is characteristic of alpine and subalpine ecosystems where there is a very short period late in the summer when the vegetation is dry enough to burn. The climate excludes fire for the remainder of the year. Although lightning is abundant, fuels are mostly sparse and discontinuous, resulting in few fires. Most of the area burns from July to September.

Late Summer–Fall Fire Season This is the characteristic fire season type for central and south coastal California sage scrub and chaparral ecosystems. Fire occurrence and size are greatly influenced by Santa Ana and north winds that most commonly occur in the late summer and early fall. This is the end of the dry season and live fuel moisture levels are lowest at this time of year. Most of the area burns from September to November.

FIRE RETURN INTERVAL

Fire return interval is the length of time between fires on a particular area of land. Fire rotation (Heinselman 1973) and fire cycle (Van Wagner 1978) are related concepts that display the average time required for fire to burn over an area equivalent to the total area of an ecosystem. Fire return interval distributions illustrate the range and pattern of values that are characteristic of an ecosystem and are critical in determining the mixture of species that will persist as the characteristic vegetation of a given area. A species cannot survive if fire is too frequent, too early, or too infrequent to allow that species to complete its life cycle. For example, survival of a nonsprouting species in a given area may be threatened by fires that occur before there has been time for a seed pool to accumulate or after the plant's longevity has been exceeded and the store of seed is lost (Bond and van Wilgen 1996). The significance of fire return interval in determining the species composition or vegetation structure through time is illustrated when fire burns often enough to prevent Oregon oak (*Quercus garryana*) woodlands from changing to a Douglas-fir forest, which can tolerate a wider range of return intervals (Sugihara and Reed 1987, Engber et al. 2011). Fig. 5.8 illustrates six conceptual fire return interval patterns occurring in California ecosystems with proportion of the burned area on the *y*-axis and the fire return interval on the *x*-axis.

Truncated Short Fire Return Interval All of the area that burns does so with short fire return intervals. Long intervals allow the establishment and growth of species that will convert these ecosystems to another type. Oregon oak and California black oak (*Q. kelloggii*) woodlands, montane meadows, grasslands, and other Native American-maintained ecosystems are typical of this fire return interval pattern.

Short Fire Return Interval Most of the area burns at short fire return intervals, but there is a wide range including a small proportion of longer intervals. Blue oak (*Q. douglasii*) woodlands, ponderosa pine, Douglas-fir and mixed conifer forests, and valley grasslands typify this pattern with the short intervals maintaining the open nature of the stands allowing sufficient light to support continuous herbaceous layer fuelbeds. The occasional low-probability long intervals promote the establishment of a mixture of shade-tolerant and shade-intolerant canopy species in these conifer forests, but do not prevent the shade-intolerant species from maintaining dominance as long as short intervals are typical.

Truncated Medium Fire Return Interval The area that burns does so within a range of fire return intervals that has both upper and lower limits that are defined by the life histories of characteristic species. Intervals outside of that range result in conversion to another ecosystem. This is a variation of the medium fire return intervals with upper and lower boundaries on the length of fire return intervals. Many of the closed-cone pine and cypresses (*Hesperocyparis* spp.) are examples of ecosystems in which fires must occur within a specific range of intervals to facilitate a flush of regeneration allowing these conifers to persist. If fires are too frequent or infrequent, the conifer regeneration does not occur, and the characteristic species cannot persist.

Medium Fire Return Interval Most of the area burns at medium-return intervals, but occasional strong deviation will not usually facilitate conversion to another ecosystem type. This set of fire return interval distributions is typically a few to several decades, but includes a variety of means, ranges, and shapes. Although the distribution on Fig. 5.8 shows a symmetrical shape, this is not always the case. The presence of a relatively wide range of intervals within the regime is characteristic. This pattern includes many chaparral types, live oak forests, and upper-montane forest types including California red fir (*Abies magnifica* var. *magnifica*), and white fir (*A. concolor*) forests.

Truncated Long Fire Return Interval In all of the burned area, intervals are long (typically greater than 70 years), and fires burning over the same area within a few years or even decades do not occur without conversion to another ecosystem type. This return interval pattern is characteristic of ecosystems with discontinuous fuels or very short burning seasons such as most very arid deserts, sand dunes, and alpine and subalpine ecosystems. The characteristic plant species are generally not adapted to fire. Mountain hemlock (*Tsuga mertensiana*), whitebark pine (*Pinus albicaulis*), foxtail pine (*P. balfouriana* subsp. *balfouriana*), bristlecone pine (*P. longaeva*), Sitka spruce (*Picea sitchensis*), and alpine meadows have this return interval pattern.

Long Fire Return Interval In most of the burned area, fire return intervals are long. Infrequently fires burning over the same area at shorter intervals can occur within this ecosystem type but account for only a small proportion of the overall burned area. This pattern is characteristic of ecosystems that are geographically isolated, do not normally have a fuel layer that will typically carry a fire, have discontinuous fuels or very short burning seasons, or lack ignition sources. Ecosystems in which this pattern is typical include some desert scrubs that only develop herbaceous layers in wet years, low-density Jeffrey pine (*P. jeffreyi*), or lodgepole pine on glaciated bedrock that will not support continuous vegetative cover, and singleleaf pinyon pine (*P. monophylla*) and beach pine (*P. contorta* subsp. *contorta*) forests.

Spatial Fire Regime Attributes

The spatial attributes of fire regimes are described in two ways: fire size and spatial complexity. Fire size is the characteristic distribution of total area within the fire perimeter. Spatial complexity describes pattern of burned and unburned areas within the fire perimeter at different levels of severity. Although we have little direct evidence of prefire-suppression-era spatial patterns for most of California's vegetation types, much information can be inferred from the structure of the vegetation and typical burning patterns and conditions.

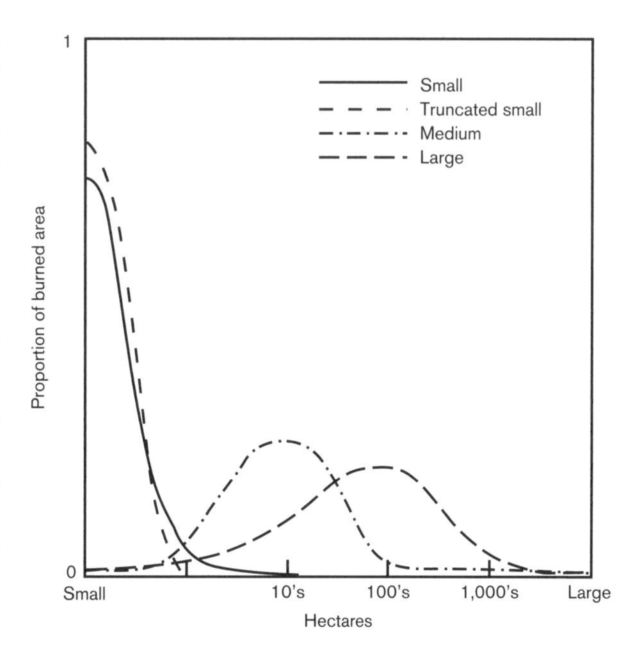

FIGURE 5.9 Fire regime distributions for size. Small, truncated small, medium, and large fire size regimes are displayed.

FIRE SIZE

Fire size is displayed as the distribution of burned area in fires of various sizes. The size of an individual fire is the area contained inside the perimeter of the fire. This is not the same as the total amount of area burned by the fire because the burned area in this case also includes unburned islands and the entire mosaic of burned and unburned areas. The size a fire attains is determined by fuel continuity, site productivity and topography, weather, and fuel conditions at the time of the fire. Fig. 5.9 illustrates four different fire size patterns that occur in California ecosystems, with proportion of the burned area on the *y*-axis and fire size on the *x*-axis. Care should be taken to interpret each curve separately. Although there are usually more small fires than large fires in most ecosystems, small fires do not necessarily burn more total area than large fires. The range of fire sizes is narrower for small fire regimes than for medium or large regimes.

Truncated Small Fire Size All of the burned area is in small fires, usually less than 1 ha (2.5 ac). This is characteristic of ecosystems which are spatially limited such as montane meadows, or with very discontinuous fuels such as whitebark pine, foxtail pine, bristlecone pine, and alpine meadow ecosystems where the potential for fire spread is extremely low.

Small Fire Size Most of the area that burns does so in fires smaller than 10 ha (25 ac) with a few larger fires accounting for much less of the total area burned. Extensive, open Jeffrey pine, western white pine, or Sierra juniper (*Juniperus grandis*) woodlands on glaciated surfaces with sparse, discontinuous fuels are examples.

Medium Fire Size Most of the area that burns does so in medium-sized fires ranging from 10 to 1,000 ha (25–2,500 ac). Smaller and larger fires do occur but account for a small proportion of the total burned area in these ecosystems. This fire size pattern is characteristic of ecosystems that occur with patchy fuel conditions and have limited stand size, limited burning periods, or limited fuel continuity. Many California red fir and white fir forests are examples of this fire size pattern.

Large Fire Size Most of the area that burns is in large fires that are greater than 1,000 ha (2,500 ac) in size with smaller fires accounting for a lower proportion. This pattern is characteristic of ecosystems occurring over extensive areas with fires typically spreading in continuous fuel layers. Many of California's grassland, chaparral, oak woodland, and lower-to middle-elevation conifer forest ecosystems fit into this category.

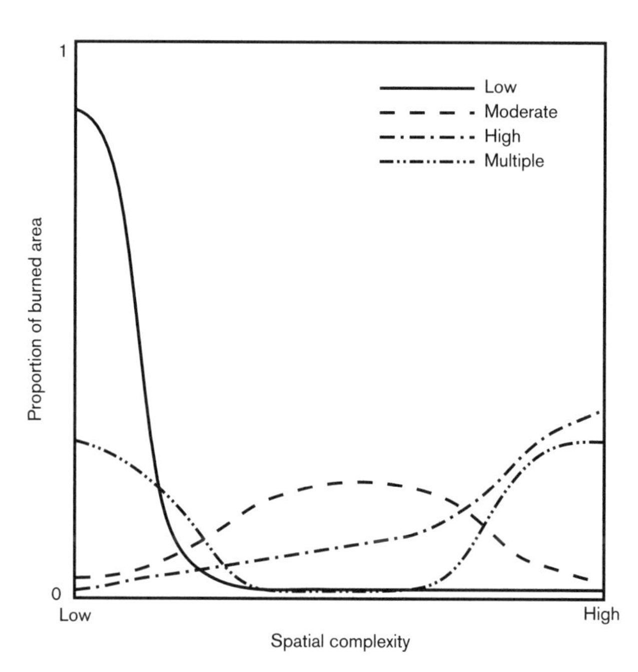

FIGURE 5.10 Fire regime distributions for spatial complexity. Burned areas can have low to high spatial complexity as well as a mixture of multiple complexities.

SPATIAL COMPLEXITY

Spatial complexity, or patchiness, is the spatial variability in fire severity within the fire perimeter. Severity can be a metric of many biotic and abiotic ecosystem components, and spatial complexity can be applied to any of these measures. The pattern of spatial complexity is often the result of fires burning under a wide range of conditions including diurnal change, changing fire weather, seasonal variation, climatic variation, slope, aspect, topography, fuel conditions, fire history, and many others. In this treatment, we address spatial complexity only as a pattern of vegetative severity. Fig. 5.10 illustrates four distribution curves for spatial complexity patterns that occur in California ecosystems, with the proportion of the burned area on the *y*-axis and spatial complexity ranging from low to high on the *x*-axis.

Low Spatial Complexity Most of the area within the perimeter of the fire is homogeneous with few unburned islands and a relatively narrow range of severity producing a course-grained vegetation mosaic. Oak woodlands, grasslands, and chamise (*Adenostoma fasciculatum*) chaparral are often examples of this type of spatial complexity.

Moderate Spatial Complexity Most of the area within the burn perimeter has an intermediate level of complexity. Burned and unburned areas and severity levels produce a mosaic of fine- and coarse-grained vegetation pattern. Douglas-fir and ponderosa pine are examples. This pattern is commonly known as mixed severity and is most common in forests of the interior North Coast and moist portion of the Klamath Mountains bioregions.

High Spatial Complexity Most of the area burns in a highly complex pattern of burned and unburned areas and severity levels producing a fine-grained vegetation mosaic. Mixed conifer and giant sequoia (*Sequoiadendron giganteum*) forests are examples.

Multiple Spatial Complexity Most of the area burns in fires that are of two distinct types: one has a complex burn pattern of burned and unburned areas and severity levels producing a fine-grained vegetation mosaic; the other has a mostly uniform pattern of burned area and severity levels and produces a coarse-grained vegetation mosaic. This is characteristic of ecosystems in which two distinct fire types occur with flaming fronts in two different fuel layers. The characteristic species in multiple spatial complexity ecosystems need to have specific adaptations to survive high-severity fire or reseeding immediately following the fire (see chapter 8). An example is Douglas-fir and White fir mixed forests in the Sierra Nevada, which can have a high frequency of surface fire with high spatial complexity and a low frequency of uniform crown fire with low spatial complexity in the same stands.

Magnitude Fire Regime Attributes

Fire magnitude is separated into three separate attributes: fireline intensity, fire severity, and fire type. Fireline intensity is a description of the fire in terms of energy release pattern. Fire severity is a description of fire effects on the biological and physical components of the ecosystem. Fire type is a description of different types of flaming fronts. Although fire severity is related to fire intensity and fire type, their relationship is very complex depending on which elements of severity are assessed and how they are directly and indirectly influenced by fire intensity and type. Similarly, fire severity is interrelated with fire seasonality and fire return interval through fire intensity.

FIRELINE INTENSITY

Fireline intensity is a measure of energy release rate per unit length of fireline. An ecosystem that burns with a characteristic fireline intensity can burn with higher or lower fireline intensity when rates of spread are higher or lower because more or less area is burned per unit of time for each unit of fireline length. For example, a grass fire that burns with a characteristic fireline intensity will release more energy per second per unit of fireline if the wind is blowing harder and less energy if the winds are calm or still. The range of energy

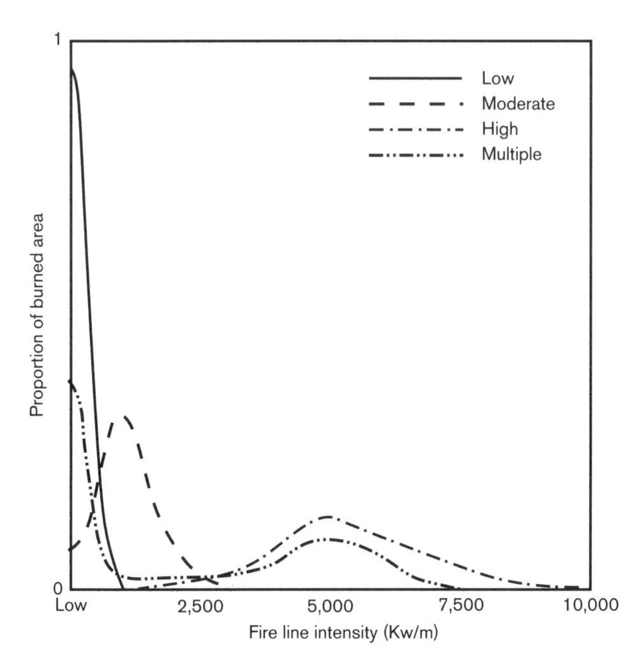

FIGURE 5.11 Fire regime distributions for fireline intensity. Fire regimes include low-, moderate-, high-, and multiple-intensity distribution curves.

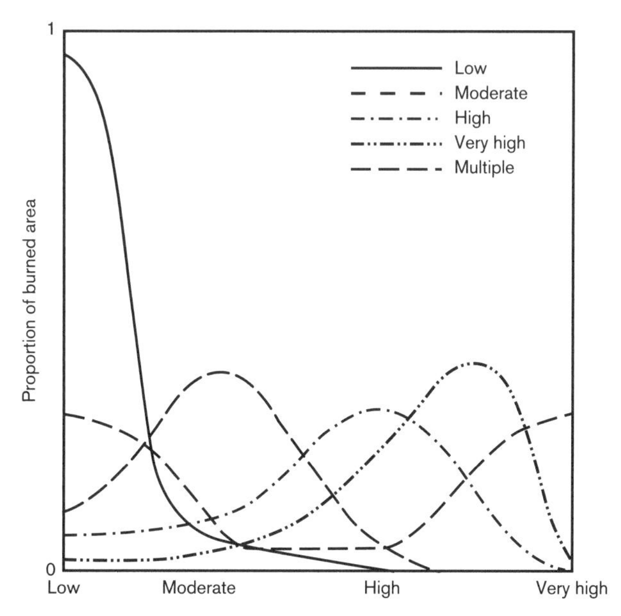

FIGURE 5.12 Fire regime distributions for severity. Five different distribution curves describe the variation in severity for different fire regimes.

release rates is illustrated by the shape of the curve. Intensity is described in detail in chapter 4 and summarized here as it applies to fire regimes. Fig. 5.11 illustrates four different fire intensity distribution patterns that occur in California ecosystems with proportion of the burned area on the *y*-axis and level of intensity on the *x*-axis.

Low Fireline Intensity Most of the area that burns does so in fires that are low intensity with flame lengths less than 1.2 m (4 ft) and fireline intensities less than 346 kW m^{-1} (100 Btu ft^{-1} s^{-1}). A smaller proportion of the area burns at moderate- to high-intensity levels. In practical firefighting terms this means that persons using hand tools can generally attack the fire at the head or flanks. Fire remains on the surface and occasionally consumes understory vegetation. Most annual grasslands, montane meadows, blue oak and many other types of woodlands, ponderosa pine, and Jeffrey pine forests are examples of ecosystems that typically burn with low fireline intensity. Douglas-fir and mixed conifer forests will also burn with low fireline intensity following shorter fire return intervals.

Moderate Fireline Intensity Most of the area burned does so in fires of moderate intensity with flame lengths from 1.2 to 2.4 m (4–8 ft) and fireline intensities between 346 and 1,730 kW m^{-1} (100–500 Btu ft^{-1} s^{-1}). Fire is too intense for direct attack at the head by firefighters using hand tools. Fire usually remains on the surface, although there could be complete consumption of understory vegetation. Moderate fireline intensity is characteristic of several sagebrush ecosystems in the Northeastern Plateaus and Southern Cascades bioregions, middle-elevation shrublands and grasslands in the Southeastern Deserts bioregion, and in coastal sage scrub in the South Coast and Central Coast bioregions. Mixed conifer and giant sequoia forests can burn with this intensity pattern, following fire in the longer end of their characteristic fire return intervals.

High Fireline Intensity Most of the area that burns has fires that are of high to very high intensities greater than 1,730 kW m^{-1} (500 Btu ft^{-1} s^{-1}) with flame lengths over 2.4 m (8 ft). A smaller proportion of the area burns at low to moderate intensity levels. Some crowning, spotting, and major runs are probable. These intensities usually result in complete consumption and mortality of the aboveground vegetation, and consumption of entire individual plants occurs. Chaparral, Sargent cypress (*Hesperocyparis sargentii*), coast live oak, and interior live oak in steep canyons characteristically burn with high fireline intensity.

Multiple Fireline Intensity Most of the burned area has fires that are mostly of two types: low-intensity surface fires and high-intensity crown fires. A smaller proportion of the area burns at moderate or very high-intensity levels. The characteristic species in multiple fireline intensity ecosystems need to have specific adaptations to survive high-intensity fire or reseeding immediately following the fire (see chapter 8). Lodgepole pine, and tanoak–mixed evergreen, canyon live oak (*Quercus chrysolepis*), Bishop pine, and Monterey pine are examples of ecosystems in which this fireline intensity pattern is characteristic.

SEVERITY

Fire severity is the magnitude of the effect that fire has on the environment, and is most commonly used in reference to vegetation or soil effects. Severity is also be applied to a variety of ecosystem components, including geomorphology, watershed, wildlife habitat, and human life and property. Separate, and often very different, distributions are appropriate when severity is displayed for multiple ecosystem characteristics. Fire severity is not always a direct result of fireline intensity, but results from a combination of fireline intensity, residence time, and moisture conditions at the time of burning, and the characteristics of the ecosystem components that are affected. This treatment of severity emphasizes the effect that fire has on the plant communities, especially the species that characterize the ecosystem. Fire severity is becoming a key attribute that is assessed following large fires to aid in assessment and recovery efforts. Thode et al. (2011) have used Landsat TM data to quantify fire severity distributions for 19 vegetation types over a twenty year span in Yosemite National Park. Fig. 5.12 illustrates five severity patterns that occur in California ecosystems with proportion of the burned area on the *y*-axis and severity on the *x*-axis.

Low Fire Severity Most of the area burns in low-severity fires that produce only slight or no modification to vegetation structure; most of the mature individual plants survive. A small proportion of the area can burn at higher severity levels. Many types of grasslands, surface fire in Douglas-fir forests in the Klamath Mountains bioregion, Jeffrey pine, ponderosa pine, and blue oak woodlands are ecosystems characterized by low fire severity.

Moderate Fire Severity Most of the area burns in fires that moderately modify vegetation structure, with most individual mature plants surviving. A small proportion of the area burns at lower and higher severity levels. Some desert shrublands, and mixed conifer and Douglas-fir forests (when following longer fire return intervals) characteristically burn with moderate fire severity.

High Fire Severity Fire kills the aboveground parts of most individual plants over most of the burned area. Most mature individual plants survive below ground and resprout. A small proportion of the area burns at lower and higher severity levels. Chamise and many sprouting chaparral types, some wetlands, quaking aspen, and canyon live oak characteristically burn with high fire severity.

Very High Fire Severity Fires are mostly stand replacing over much of the burned area. All or nearly all of the individual mature plants are killed. A smaller proportion of the area burns at lower severity levels. Knobcone pine, Monterey pine, and many cypress and nonsprouting chaparral types including most manzanita (*Arctostaphylos* spp.) and California lilac (*Ceanothus* spp.) vegetation types are characterized by very high fire severity.

Multiple Fire Severity The area burned is mostly divided between two distinct fire types: low severity and high to very high severity. A smaller proportion of the area burns at moderate severity levels. The characteristic species in multiple fire severity ecosystems need to have specific adaptations to survive high severity fire or reseeding immediately following the fire (see chapter 8). Tanoak (*Notholithocarpus densiflorus* var. *densiflorus*)–mixed evergreen, canyon live oak, Coulter pine, Bishop pine, and Monterey pine are examples of ecosystems in which this fire severity pattern is characteristic.

FIRE TYPE

Fire type is a description of the flaming front patterns that are characteristic of an ecosystem. The types are defined in chapter 4 and include surface, passive crown, active crown, and independent crown fires. Although fire type is a categorical variable, it can be expressed as a continuous variable by using fireline intensity to scale the fire types. Ground fires, although a significant contributor to fire effects, and important in other places, are not a characteristic part of the flaming front in California ecosystems. There are four fire regime types that represent different combinations of the fire types in California. These are the surface-passive crown fire regime, the passive–active crown fire regime, the active-independent crown fire regime, and the multiple-fire-type regime. Fig. 5.13 illustrates the four different patterns for fire-type regimes that occur in California ecosystems. The proportion of the burned area is on the *y*-axis, and the *x*-axis depicts increasing values for fireline intensity with points along the axis for fire type.

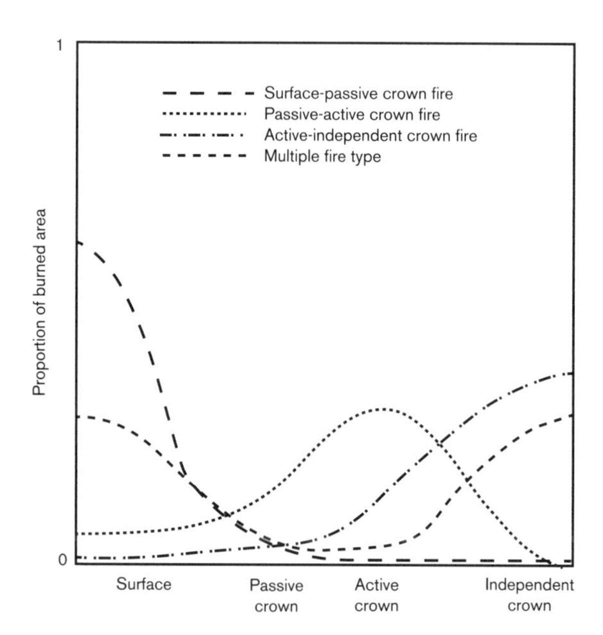

FIGURE 5.13 Fire regime distributions for type. Fire-type regimes include surface fires, crown fires, and multiple-type fires.

Surface-Passive Crown Fire Most of the area that is burned does so with a surface fire. Although as much as 30% of the area may experience torching of individual trees or groups of trees, the flaming front is primarily a surface fire. Organic layers are burned by ground fires, and small amounts of active crowning can burn stands of trees. Grasslands, blue oak woodlands, Jeffrey pine, and low-elevation desert shrublands are typical examples of this fire-type distribution.

Passive–Active Crown Fire Most of the burned area has fire that is a combination of passive and active crown fire supported by surface fire. Active crown fire is dependent on and synchronous with a surface fire and is the most common type of sustained crown fire. This fire regime type occurs in north coastal pine forests, Sitka spruce, knobcone pine, coastal sage (*Salvia* spp.) scrub, and desert riparian woodlands and oases. In recent years, active crown fire is increasingly occurring where mixed conifer, Douglas-fir, Jeffrey pine, and ponderosa pine forests have large, contiguous areas of surface fuel accumulation and continuous crown fuel due to decades of fire exclusion.

Active-Independent Crown Fire Independent crown fire is characteristic of many shrublands. In chaparral ecosystems, independent crown fires are the norm, although some active crowning might occur where there is sufficient surface fuel. In California forests, independent crown fires are rare but do occur occasionally in combination with active crown fires. When they do occur, the crown fire burns independently of the surface fire and advances over a given area ahead of the surface fire. Examples are lodgepole pine in northeastern California and some closed-cone conifer ecosystems. Areas supporting live oak or conifer forests with dense canopies in very steep complex topography can also fit into this fire-type distribution. Examples of vegetation types with a greater preponderance for independent crowning include chamise, several manzanita vegetation types, knobcone pine embedded in chaparral, similarly situated Sargent cypress, and most shrublands.

Multiple Fire Type Both surface fire and crown fire are characteristic of these ecosystems with multiple fire types. Both fire types will usually occur in a complex spatial mosaic within the same fire under different fuel, topographic, and weather conditions. The characteristic plant species in multiple-fire-type ecosystems need to have specific adaptations to survive high-intensity fire or reseeding immediately following the fire (see chapter 8). Lodgepole pine, and tanoak–mixed evergreen, canyon live oak, Bishop pine, and Monterey pine are examples of ecosystems in which this fire-type pattern is characteristic.

COMPREHENSIVE FIRE REGIME

Comprehensive fire regimes are developed for vegetation types by combining the appropriate attribute distribution curves for each attribute. Comprehensive fire regime descriptions are used in the bioregions of this text and were applied to the vegetative alliances of California by Sawyer et al. (2009). Similar combinations could be grouped into fire regime types such as those described by Hardy et al. (2001), but these descriptions would be much more extensive.

Summary

Fire is an important ecological process that occurs regularly and has predictable spatial, temporal, and magnitude patterns that we describe here as fire regimes. In the fire-prone ecosystems of California, fire is inevitable and general patterns are predictable. Species persist in fire-prone ecosystems by having characteristics that make them competitive with recurring fire. Because fire patterns interact with biotic communities and depend on them to provide fuel, the dynamics of many ecosystems are dependent on to fire regimes. Changes to any of the fire regime attributes are alterations to ecosystem function, and inherently affect biological change producing shifts in the composition and distribution of species and ecosystems.

Fire regimes have always been dynamic at multiple scales of time and space. In addition to the scale represented by the distribution curves, fire regimes also operate at larger scales on much larger landscapes over centuries and millennia. Ecosystems and their associated fire regimes have migrated across landscapes with climate changes, human occupation, and geologic and biologic changes. Ecosystems adjust to changes in fire regime by changing composition and structure and by migrating up- and downslope, and north and south.

Although humans have altered fire regimes throughout California for thousands of years, the pace of fire regime change has accelerated over the past 200 years. Recent and current management strategies have imposed directional changes on the pattern of fires in many California ecosystems. For example, fire exclusion from some forests that historically had frequent fires has lengthened fire return intervals, allowing greater fuel accumulations. The long-term statewide trend is to less total area burned, with the reduction mostly in area burned by the characteristic fires of low to moderate intensity and severity. There is also a trend is toward increasing the size of the very largest fires and within them, a greater representation of uncharacteristic large, high-severity patches. Because fire suppression technology has enabled increased human intervention in eliminating the low- to moderate-severity fires, the historic fire regime distributions have now shifted toward a greater proportion of high-severity, large, stand-replacing fires (McKenzie et al. 2004). Although it is unlikely that we will ever universally restore California's ecosystems to any historic condition, it is also apparent that we cannot totally eliminate fire.

In recent decades, ecologists and land managers have become very concerned with mitigating the effects of our changes to historic fire regimes and we have devoted considerable effort to improving our understanding of historic fire regimes and our changes to them. We know that fire's role in ecosystems and fire regime dynamics serve both as mechanisms for maintaining habitats and driving habitat change for many species. An understanding of fire regimes is critical in assessing current conditions and developing strategies for achieving land management objectives. It is also vital in assessing the threat that wildfire poses to people on the urban–wildland interface.

The system used here for describing fire regimes allows description of the attributes involved, comparison of how they differ from those in other ecosystems, and how they change over time. Additionally, fire regime descriptions allow us to view in a structured manner how changing attributes influence fire's role as an ecological process. Knowledge of fire regime–ecosystem interactions allows us to understand mechanisms for ecosystem change due to changing fire regimes. This knowledge further allows prediction of the direction of ecological change that will occur with future planned and unplanned changes to fire regimes.

Today we have the opportunity to manage fire dependent ecosystems and maintain many of their important processes and attributes. We also have the knowledge to recognize that excluding, or not managing fire also has predictable consequences. As our society refines its land management objectives and strategies, managing fire regimes is emerging as a major element of managing ecosystems. We must decide where it is appropriate to manage both natural and altered fire regimes and ecosystems to meet society's desires and demands. The fire regime system described in this chapter is designed to support our decision-making by giving us a tool for assessing fire regime–ecosystem dynamics and to help us to understand the mechanisms and consequences of fire-related ecosystem change. Now our challenges are to define fire management priorities and implement actions that will move us toward long-term fire, land, and resource management goals.

References

Agee, J.K. 1993. Fire Ecology of Pacific Northwest Forests. Island Press, Washington, D.C., USA.

Bond, W.J., and B.W. van Wilgen. 1996. Fire and Plants. Chapman and Hall, London, UK.

Clements, F.E. 1916. Plant succession. Carnegie Institute of Washington Publication 242. Washington, D.C., USA.

———. 1936. Nature and structure of the climax. Journal of Ecology 22: 39–68.

Daubenmire, R.F. 1947. Plants and Environment: A Textbook of Plant Autecology. John Wiley and Sons, New York, New York, USA.

———. 1968. Plant Communities. Harper and Row, New York, New York, USA.

Engber, E.A., J.M. Varner III, L.A. Arguello, and N.G. Sugihara. 2011. The effects of conifer encroachment and overstory structure on

fuels and fire in an oak woodland landscape. Fire Ecology 7(2): 32–50.
Gleason, H. A. 1917. The structure and development of the plant association. Bulletin Torrey Botanical Club 43: 463–481.
———. 1926. The individualistic concept of the plant association. Bulletin Torrey Botanical Club 53: 7–26.
Hann, W. J., and D. L. Bunnell. 2001. Fire and land management planning and implementation across multiple scales. International of Wildland Fire 10: 389–403.
Hardy, C. C., K. M. Schmidt, J. P. Menakis, and R. N. Sampson. 2001. Spatial data for national fire planning and management. International of Journal Wildland Fire 10: 353–372.
Heinselman, M. L. 1973. Fire in the virgin forests of the Boundary Waters Canoe Area, Minnesota. Quaternary Research 3: 329–382.
———, M. L. 1981. Fire intensity and frequency as factors in the distribution and structure of northern ecosystems. Pages 7–57 in: H. A. Mooney, T. M. Bonnicksen, N. L. Christensen, J. E. Lotan, and W. A. Reiners, editors. Fire Regimes and Ecosystem Properties, Proceedings of the Conference. USDA Forest Service General Technical Report WO-26. Washington, D.C., USA.
Kilgore, B. M. 1981. Fire in ecosystem distribution and structure: western forests and scrublands. Pages 58–89 in: H. A. Mooney, T. M. Bonnicksen, N. L. Christensen, J. E. Lotan, and W. A. Reiners, editors. Fire Regimes and Ecosystem Properties, Proceedings of the Conference. USDA Forest Service General Technical Report WO-26. Washington, D.C., USA.
McKenzie, D., Z. Gedalof, D. L. Peterson, and P. Mote. 2004. Climatic change, wildfire, and conservation. Conservation Biology 14: 890–902.
Odum, E. P. 1959. Fundamentals of Ecology. 2nd ed. Saunders, Philadelphia, Pennsylvania, USA.
———. 1963. Ecology. Holt, Reinhart, and Winston, New York, New York, USA.
———. 1969. The strategy of ecosystem development. Science 154: 262–270.
O'Neill, R. V., D. L. DeAngelis, J. B. Waide, and T. F. H. Allen. 1986. A Hierarchical Concept of Ecosystems. Princeton University Press, Princeton, New Jersey, USA.
Pickett, S. T. A., J. Kolasa, J. J. Armesto, and S. L. Collins. 1989. The ecological concept of disturbance and its expression at various hierarchical levels. Oikos 54: 129–136.
Romme, W. H., E. H. Everham, L. E. Frelich, M. A. Moritz, and R. E. Sparks. 1998. Are large, infrequent disturbances qualitatively different from small, frequent disturbances? Ecosystems 1: 524–534.
Sawyer, J. O., T. Keeler-Wolf, and J. M. Evens. 2009. A Manual of California vegetation. 2nd ed. California Native Plant Society, Sacramento, California, USA.
Schultz, A. M. 1968. The ecosystem as a conceptual tool in the management of natural resources. Pages 139–161 in: S. V. Cirancy-Wantrup and J. J. Parsons, editors. Natural Resources: Quality and Quantity. University of California Press, Berkeley, California, USA.
Shugart, H. H., and D. C. West. 1981. Long-term dynamics of forest ecosystems. American Scientist 69: 647–652.
Sugihara, N. G., and L. R. Reed. 1987. Vegetation ecology of the Bald Hills oak woodlands of Redwood National Park. Redwood National Park Research and Development Technical Report 21. Arcata, California, USA.
Tansley, A. G. 1935. The use and abuse of vegetational concepts and terms. Ecology 16: 196–218.
Thode, A. E., J. W. van Wagtendonk, J. D. Miller, and J. F. Quinn. 2011. Quantifying the fire regime distributions for fire severity in Yosemite National Park, California, USA. International Journal of Wildland Fire 20: 223–239.
Turner, M. G., and V. H. Dale. 1998. Comparing large, infrequent disturbances: what have we learned? Ecosystems 1: 493–496.
Van Wagner, C. E. 1978. Age class distribution and the forest fire cycle. Canadian Journal of Forest Research 8: 220–227.
Walker, L. R., and M. R. Willig. 1999. An introduction to terrestrial disturbances. Pages 1–6 in: L. R. Walker, editor. Ecosystems of Disturbed Ground. Elsevier, Amsterdam, The Netherlands.
Watt, A. S. 1947. Pattern and process in the plant community. Journal of Ecology 39: 599–619.
White, P. S. 1979. Pattern, process, and natural disturbance in vegetation. Botanical Review 45: 229–299.
White, P. S., and S. T. A. Pickett. 1985. Natural disturbance and patch dynamics: an introduction. Pages 3–13 in: S. T. A. Pickett and P. S. White, editors. The Ecology of Natural Disturbance and Patch Dynamics. Academic Press, New York, New York, USA.
Whittaker, R. H. 1953. A consideration of climax theory: the climax as a population and pattern. Ecological Monographs 23: 41–78.
———. 1967. Gradient analysis of vegetation. Biological Review 42: 207–264.
———. 1971. Communities and ecosystems. McMillan, New York, New York, USA.

CHAPTER SIX

Characterizing Fire Regimes

BRANDON M. COLLINS, JAY D. MILLER, JEFFREY M. KANE,
DANNY L. FRY, AND ANDREA E. THODE

> The composition, structure, and dynamics of the virgin forests can best be understood if we concentrate on understanding how the system functions with fire as part of that system.
>
> HEINSELMAN (1973)

Patterns of fire effects within individual fire events are driven by fuels, topography, and fire weather. Over time, these fire events have historically formed relatively consistent patterns in many vegetation types, which have allowed for characterization of the fire regime for a given vegetation type/area. Fire regimes vary among ecosystems and geographic regions based on primary productivity, which influences fuel structure and composition, ignition sources, dominant weather patterns, and seasonality of fire occurrence, making fire regime characterization challenging. Despite this, there is great need to describe past and present fire regimes in fire-prone vegetation types. Historical fire regimes, and the coinciding vegetation patterns, are often used as baseline information to compare change in fire and vegetation patterns resulting from modern land management practices. Furthermore, the characterization of historical fire and vegetation patterns often guides the development of desired conditions for ecosystem restoration.

This chapter expands on the fire regime concept introduced in chapter 5, and subsequently referenced in each of the bioregion chapters (10 through 18). We describe the various approaches used to characterize fire regime attributes and discuss the strengths and limitations of these approaches. Because patterns of fire occurrence have been substantially altered by twentieth-century land management practices, we focus on the commonly used methods to characterize historical and contemporary fire regimes. Although these methods apply primarily to forests and woodlands, some can be used to characterize fire regimes in chaparral (sidebar 6.1). There are fundamental differences in the approaches used to determine historical and contemporary fire regimes, owing primarily to differences in data availability. Historical fire regime attributes have to be inferred from both direct and indirect sources including paleoecological data (sedimentary charcoal deposits, tree rings), archived written accounts, and inventory datasets. Contemporary fire regime attributes can be quantified with more direct observations, such as mapped fire perimeters, aerial photographs, and satellite imagery. The datasets created from these observations have allowed for detailed descriptions of fire patterns across landscapes, which are largely unavailable for historical fires. We conclude this chapter by discussing how this information is used to assess change in contemporary fire patterns relative to historical patterns that provide essential, yet imperfect, insights into policy and management decisions in public lands.

Approaches for Reconstructing Historical Fire Regimes and Vegetation Structure

Paleoecological Reconstructions

SEDIMENTARY CHARCOAL AND POLLEN

Wildland fires create charcoal particles from the incomplete combustion of plant matter (see chapter 4). This charcoal is deposited into the soil where it can be burned or transported by wind, overland flow, and erosion processes, and ultimately concentrates into adjacent areas, such as lakes, bogs, wetlands, and alluvial fans. Over time, these charcoal accumulations are deposited within layers of sediment, providing a record of fire activity across broad temporal scales and variable spatial scales (Whitlock and Larsen 2001). Most charcoal studies use lake bed sediments, but bog and soil samples are also used (Power et al. 2008). Unlike other paleoecological methods, sedimentary charcoal and pollen methods can be used to estimate fire regimes for a wide range of vegetation types, including grasslands, shrublands, woodlands, and forests.

Sampling involves taking sediment cores of varying depths. Typically, the deeper the sediment core, the longer the time period that can be examined. Methods to age sediment layers within a core include measuring concentrations of a sedimentary lead isotope ^{210}Pb (Binford 1990) and accelerated mass-spectrometry dating of charcoal or macrofossils present in the samples to estimate radiocarbon dates (Stuiver and Reimer 1993). Radiometric dates are converted to calendar years and age-to-depth models are constructed to estimate sediment accumulation rates for the entire sample. Time periods that exceed an assigned threshold in charcoal accumulation rates relative to the baseline or background levels are identified as

SIDEBAR 6.1 CHARACTERIZING CHAPARRAL FIRE REGIMES

Max A. Moritz, Dennis C. Odion, and Frank W. Davis

Chaparral is typically characterized by crown fires (active-independent fire type), so fire sizes should be expected to vary with gradients in wind patterns and drought. Mapped Santa Ana winds and analysis of historical fire perimeters over several decades support this assertion, as fire sizes are correlated with the severity of fire weather corridors (Moritz et al. 2010). Although analyses of fire perimeter maps (or fire atlases) are increasingly common and sophisticated, it is crucial to note that the smallest events are often routinely omitted. Depending on location, however, the smallest fires may account for very little of the total area burned (Moritz 1997).

Current chaparral fire return intervals in southern California are estimated to range from 30 to 50 years (Keeley and Fotheringham 2001, Moritz et al. 2004, Moritz et al. 2009). Most of these are regional estimates, which can mask substantial site-specific variation, and are based on historical fire perimeter maps. Insights are also provided by a 500-year record of annual fire scars and subsequent growth releases in bigcone Douglas-fir trees (*Pseudotsuga macrocarpa*) embedded in chaparral of the California Transverse Ranges near Santa Barbara (Lombardo et al. 2009). The mean frequency of fires in bigcone Douglas-fir stands was short, only 10 years, prior to settlement (from 1600 to 1864). For multisite events that would have burned more extensive chaparral landscapes, a 90-year frequency was found over this time period. How these tree ring-based estimates relate to natural fire return intervals in other chaparral ecosystems is largely unknown.

Chaparral crown fires naturally exhibit high fireline intensities and relatively low spatial complexity (i.e., stand-replacing and homogeneous), although subdued and patchy burns can occur under more modest fire weather conditions (e.g., Green 1981, Odion and Tyler 2002). A difficulty in using an energy release measure is that fires having very different behavior and effects can have similar fireline intensities (and vice versa). Analysis of heterogeneity within burns (e.g., through dNBR described elsewhere) indicates that remote sensing metrics of fuel consumption do not correlate well with fire severity in chaparral (Keeley et al. 2008).

Physical measures of fire characteristics in chaparral have long included temperature (e.g., Sampson 1944), through the use of compounds with known melting points. In this way, researchers have linked instantaneous maximum temperature to ecological effects in chaparral (Davis et al. 1989). Thermocouples provide an instantaneous measure of temperatures over time and therefore a complete record of heating at a given location. Odion and Davis (2000) found that nearby thermocouples in chaparral fires often had very different temperature-time profiles, revealing that regeneration can be tightly correlated with local species and soil heating patterns.

Other methods for assessing ecological impacts include water loss from containers (Beaufait 1966), which captures the total heat impinging on a container and has been related to regeneration patterns in chaparral (e.g., Moreno and Oechel 1993, Odion and Davis 2000). The minimum diameter of branches remaining on chaparral shrubs after fire has also been used (e.g., Rundel et al. 1987, Rice 1993, Keeley et al. 2008). These diameters, averaged over a small area, may correlate with maximum temperatures, water loss from cans, and regeneration in relatively low-intensity prescribed burns (Moreno and Oechel 1991), but not where fires consume more fuel (Odion and Davis 2000, Keeley et al. 2008).

Seasonality of fire can be important in chaparral. For example, a spring burn in chamise (*Adenostoma fasciculatum*) may have low fire intensity, but unusually high lignotuber mortality due to low resprout reserves (Jones and Laude 1960); seed bank mortality may also be high due to low heat tolerance then (Parker 1987, Odion and Tyler 2002). A low-intensity fire may thus be ecologically severe, largely driven by timing. Although natural timing must always have been influenced by precipitation in months preceding the fire season (Davis and Michaelsen 1995, Dennison and Moritz 2009), or in years prior (Mensing et al. 1999), the availability of ignitions is also a constraint. Native Americans inhabited many chaparral landscapes for thousands of years before Euro-American colonization, and we do not know how precisely ignitions were used in time and space, nor how sensitive fire regimes were to these patterns.

A final caveat is that the above discussion about chaparral fire regimes is based almost entirely on studies of central and southern California shrublands. There are substantial gradients in productivity, fire season characteristics, wind patterns, soils, topography, and species composition across the distribution of chaparral shrublands, most notably in montane areas where shrubs can be a successional component of the forest understory.

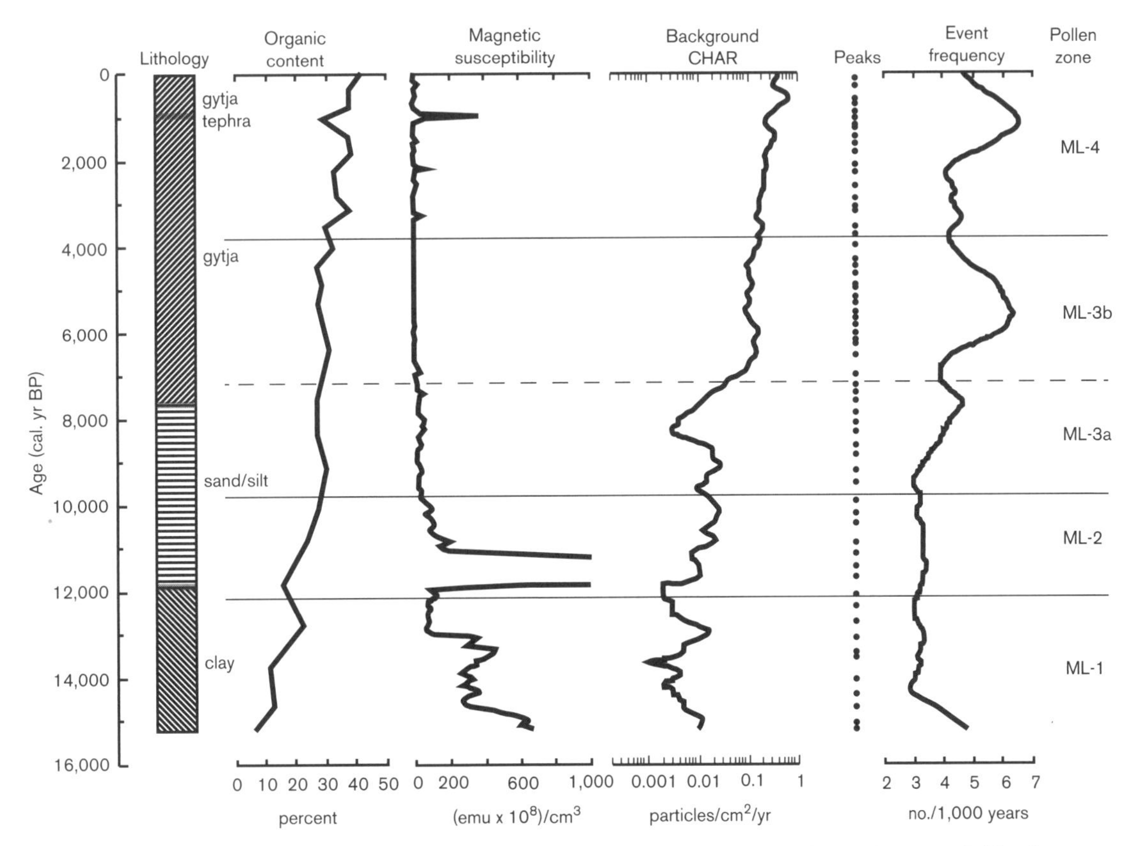

FIGURE 6.1 Example of inferred fire events from sedimentary charcoal record at Mumbo Lake, northwestern California (reproduced with permission from Daniels et al. 2005).

charcoal peaks. Variation in background levels reflects changes in vegetation and/or hydrology, while variations in the size and frequency of peaks reflect changes in the occurrence of fire events or episodes (Long et al. 1998). Charcoal peaks are inferred as a fire event or episode because they may reflect the occurrence of one or more fires within the sample at a particular depth of the sediment record (Whitlock and Larsen 2001). From the charcoal chronology (Fig. 6.1), fire-event frequency and fire-event return interval values are calculated based on the incidence of charcoal peaks over time and time intervals between peaks, respectively.

An advantage of sediment sampling is the additional presence of plant pollen and macrofossils that reflect vegetation changes over time. Pollen grains are visually identified to the lowest taxonomic level based on comparisons to available reference collections, published atlases, and journal articles for the region of interest. Pollen counts for each taxon are converted to percentages based on the sum of all pollen types to assess relative changes in abundance over time. The presence of plant macrofossils such as needles, twigs, and seeds that are embedded into the lake sediment can be used to aid in the identification of pollen specimens and to confirm that a taxon was present within the region.

Sedimentary charcoal and pollen methods provide a long-term chronology (up to 21,000 y) of the fire activity and the vegetation associations present in a given region that can vary from a small to large watershed. Generally, this information can be used in the following ways: (1) to examine changes in fire activity over temporal or spatial scales, (2) to determine the relative importance of climate, topography, and fuel as drivers of fire and vegetation dynamics, (3) to provide insight into modern and prehistoric fire regimes and reference conditions, and (4) to aid the development of ecosystem models that can estimate future fire conditions under different scenarios (Gavin et al. 2007). Additionally, the use of charcoal analyses is particularly helpful in determining fire frequency in regions or forest types with longer fire return intervals (> 100 y) or in high-severity fire regimes (Romme and Despain 1989).

Sedimentary charcoal and pollen methods are not without their limitations. The temporal resolution of sedimentary charcoal methods is generally coarse and typically denotes broader trends, ranging in resolution of approximately 10–1,000 y. Because charcoal peaks can reflect the incidence of multiple fires, fire frequency and return intervals may represent underestimations and may not linearly relate to actual fire frequency. Beyond the coarse estimates of frequency, other fire regime attributes such as fire type, fire size, seasonality, intensity, severity, and spatial complexity are not currently possible, although each of these attributes contribute to the variation in charcoal accumulation rates (Whitlock and Anderson 2003). An additional concern with charcoal methods is that peak detection methods and threshold levels differ among studies (Gavin et al. 2007).

Most charcoal studies in California are primarily located in the Klamath Mountains and the Sierra Nevada, with fewer studies conducted in the Southern Cascades, North Coast, Central Coast, and the Channel Islands in the South Coast bioregion (see specific chapters for more detailed information). The longest charcoal records extend to about 15,000 years BP, a period in the late Pleistocene referred to as the last-glacial-interglacial transition (15,000–10,000 y BP). During

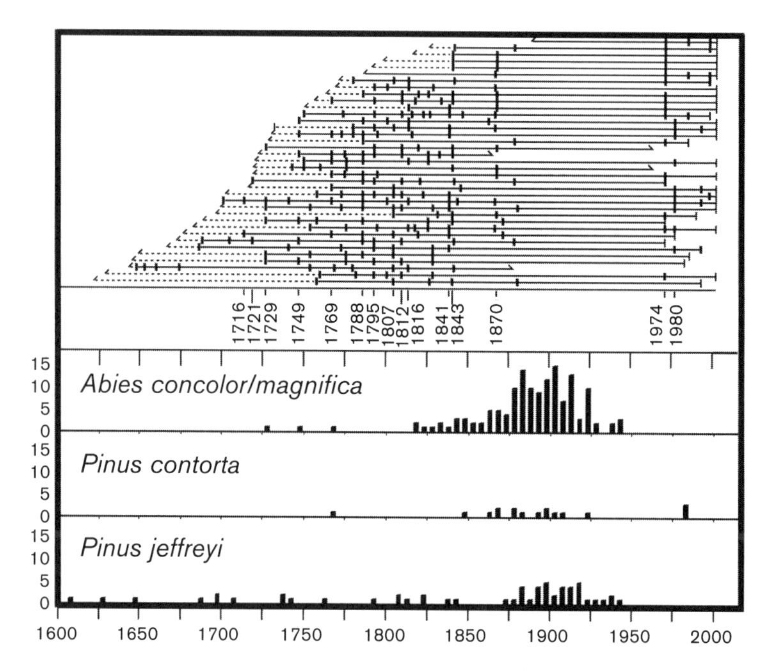

FIGURE 6.2 Dendrochronological reconstruction of fire occurrence and stand age structure from Illilouette Creek Basin, Yosemite National Park. Fire scar and tree establishment dates are from Collins and Stephens (2007b).

this time, many regions were experiencing a cool and moist period with increasing fire activity until the beginning of the Younger-Dryas chronozone (YDC; Marlon et al. 2009). The YDC (12,900–11,700 y BP) is characterized as an abrupt transition to a cool period that resulted in no net change in fire activity in many regions of the western United States, followed by further increases in fire activity at the end of the YDC (Marlon et al. 2009). However, in California these trends were not present across all sites, indicating that some regional scale responses were modulated by local factors. During the early Holocene (~11,700–6,000 y BP), many sites in northern California experienced a warming and drying period that had an increased level of fire activity (Mohr et al. 2000, Daniels et al. 2005); however, this time period was characterized by the detection of relatively low fire activity in montane meadows of the central and southern Sierra Nevada (Anderson and Smith 1997), coastal regions (Anderson et al. 2013), and the Channel Islands (Anderson et al. 2010). During the late Holocene, fire activity across the western United States was generally more variable, with some consistency during warm, dry periods such as the Medieval Warm period (~975–725 y BP) and cool periods such as the Little Ice Age, ~500–300 y BP (Marlon et al. 2012).

DENDROCHRONOLOGY

There are two predominant methods by which past fire evidence can be inferred from tree rings through fire scar and stand age data. The reliance on tree rings limits these methods primarily to forests and woodlands, although information from trees immediately adjacent to shrublands or grasslands has been used to reconstruct fire regimes in those vegetation types (e.g., Lombardo et al. 2009). The fire scar method relies on the occurrence of nonlethal cambial injury created by surface fires burning at the base of trees. These injuries, called fire scars, can be preserved in trees, which can often record numerous fires over their life span (Arno and Sneck 1977). Through cross-dating, matching the pattern of annual tree ring widths with known tree ring chronologies for a particular area, fire scars can be assigned to a calendar year and compiled into a fire history chart (Fig. 6.2). Climate is the common signal driving tree ring widths, thus allowing for relatively consistent patterns across a study area. This method offers a level of temporal resolution and depth that cannot be attained with other reconstruction approaches. Analyzing fire-scarred cross-sections from multiple trees across a study area allows for reconstruction of historical fire occurrence for several hundred years in many dry pine-dominated forest types (e.g., Stephens and Collins 2004, Fry and Stephens 2006, Beaty and Taylor 2008) and up to 3,000 years in giant sequoia-dominated forests (Swetnam 1993, Swetnam et al. 2010). It is worth noting that as the area sampled for fire scars increases, composite estimates of fire frequency decreases due to the greater likelihood of detecting new fire events (Agee 1993). As such, it is important to report spatial extent in conjunction with fire frequencies derived from fire scars. Fire scars also offer the ability to infer seasonality of fire occurrence. This is done by examining the intra-ring position of fire scars, which is commonly assigned to the following classes: earlywood, latewood, and dormant. There is uncertainly, however, as to how these intra-ring positions correspond with actual seasonality given differences in ring phenology among species and regions (Stephens et al. 2018), as well as interannual climate variability.

The second tree ring method uses stand age data, which relies on the formation of distinct postfire cohorts following stand-replacing fire. In order to reconstruct past fire occurrence, tree initiation dates are determined from tree cores to identify "stand-origin dates" (Heinselman 1973, Romme 1982, Fry et al. 2012). This approach can capture several hundred years of fire occurrence over a landscape, but for any particular stand only the most recent fire can be detected since evidence of past fires is consumed by subsequent fires.

Conventionally, fire scar-based reconstructions were used in forest types historically associated with low- to moderate-intensity surface fires (e.g., Dieterich 1980), while stand age reconstructions were used in forest types associated with high-intensity, crown fires (Romme 1982). However, more recent studies, primarily from the former forest types, have coupled fire-occurrence dates from fire scars with extensive tree age structure sampling (e.g., Beaty and Taylor 2008, Scholl and Taylor 2010, Taylor 2010). In addition to being able to directly quantify fire frequency, extent, and seasonality, these studies have allowed for the indirect inference to fire severity through identification of postfire cohorts. These studies have demonstrated that there was a high degree of spatial complexity driven by the interaction between vegetation, fuels, and topography (Nagel and Taylor 2005, Beaty and Taylor 2008, Lauvaux et al. 2016). These interpretations drawn from tree rings have been corroborated with independent historical data (Maxwell et al. 2014, Collins et al. 2015).

There is some uncertainty associated with both tree ring-based fire history reconstructions. Fire scar-based reconstructions have been questioned due to issues associated with sampling methodology and fire frequency calculations. Fire scars are generally sampled by targeting trees with visual signs of multiple wound-wood formations. This nonrandom sampling approach has been criticized as biasing fire frequency toward unrealistically high estimates (Baker and Ehle 2001). Similarly, some have argued that fire scar-based methods of reconstructing fire occurrence "give undue importance to small fires and lead to inaccurate estimates of spatial fire intervals . . .," which also lead to overestimation of "ecologically significant" fire frequency (Minnich et al. 2000, p. 124). Conversely, there can be fires that do not show up in the fire scar record because they either did not scar the collected trees (insufficient heat at the base of the trees) or the evidence was lost (scar consumed by subsequent fires). Both could lead to underestimation of fire frequency. Although logically founded, several studies have robustly refuted these criticisms by conducting intensive fire scar collection (Van Horne and Fulé 2006, Stephens et al. 2010a) and comparing fire scar reconstructions to mapped fire perimeters for contemporary fires (Collins and Stephens 2007a, Farris et al. 2010). The upshot from these studies is that the theoretical biases contended in earlier work were in fact not real. Conducting sampling at an appropriate spatial scale appears to be more important than the actual method used to sample individual fire-scarred trees.

The stand age approach allows for identification of fires that can have large spatial extents and where the majority of evidence is consumed in the event itself. Studies that have employed this approach have been able to explicitly demonstrate that in some forest types, large stand-replacing fires were common prior to Euro-American settlement (e.g., Romme 1982). The stand age approach, however, has a couple key limitations. First, there is often a range in tree establishment dates following a stand-replacing event, which makes identifying actual fire dates difficult. This range can be quite large (1–50 y) due to several factors influencing tree regeneration and establishment, e.g., seed availability, favorable soil moisture, competing vegetation. For example, forests in which dominant tree species have a direct mechanism for regeneration following extensive stand-replacing fire (e.g., Rocky Mountain lodgepole pine [*Pinus contorta* var. *latifolia*] via serotinous cones) can form distinct postfire cohorts rapidly. This is not the case for tree species that rely on wind-blown seed dispersal into stand-replacing patches, which is the primary mechanism for a majority of species in California mixed-conifer forests. One way to address this is to couple tree age sampling with evidence from fire-scarred tree cross-sections, which through cross-dating can yield a calendar year for a particular fire. Where stand-replacing fire is extensive, these fire-scarred cross-sections can be few and far between, but the additional evidence is critical for determining whether or not fire was solely responsible for the initiation of a new cohort (Romme 1982, Fry et al. 2012). There are a number of other events that are stand-replacing, e.g., avalanche, insect outbreaks, windthrow, that can lead to initiation of a new cohort, which is another uncertainty with the stand age approach.

Archived Written Records

PUBLISHED REPORTS AND ARTICLES

There are a number of published sources from the end of the nineteenth century and beginning of the twentieth century with information that can be used to discern some aspects of historical fire regimes. Among the most readily available include a series of reports published by the United States Geological Survey (USGS) documenting surveys conducted after establishment of forest reserves beginning in 1891. The surveys focused primarily on inventorying merchantable timber. However, most surveys conducted in California provide at least some qualitative information on fire effects that can be used to infer severity and ignition sources in forests, and to a lesser extent chaparral systems (Fitch 1900a, 1900b, Leiberg 1900a, 1900b, 1900c, Sudworth 1900, Leiberg 1902). After the Forest Service was established in 1905, publications authored by Forest Service foresters described forest conditions and recent fire patterns in the early part of the twentieth century (Cooper 1906, Greely 1907, Larson and Woodbury 1916, Show and Kotok 1923, 1924, 1925, 1929).

Sudworth (1900), Fitch (1900a, 1900b), Cooper (1906), and Greely (1907) provided only qualitative information on fire effects (i.e., severity) and ignition sources. Sudworth (1900) surveyed an area encompassing what are now the Eldorado and northern half of the Stanislaus National Forests, southern half of the Lake Tahoe Basin Management Unit, and the northwest corner of the Humboldt-Toiyabe National Forest (Fig. 6.3). Fitch (1900a, 1900b) surveyed the southern half of the Stanislaus National Forest and western half of Yosemite National Park (Fig. 6.3). Cooper (1906) reported on sugar pine (*Pinus lambertiana*) and yellow pine (*P. jeffreyi* and *P. ponderosa*) forests statewide, while Greely (1907) focused only on the western Sierra Nevada.

Leiberg (1902) surveyed an area encompassing the southern half of the Plumas National Forest and all of the Tahoe National Forest (Fig. 6.3). Rather than just describing forest types and estimating merchantable timber, Leiberg produced range maps of the most common species, and areas of extensive montane chaparral. In addition, he mapped areas where he estimated 5–100% of timber had been killed by fires in the previous 100 y. His mapped areas of extensive montane chaparral closely, but not precisely, match areas where he mapped 75–100% mortality of timber due to fires in forests dominated by California red fir (*Abies magnifica* var. *magnifica*). Leiberg (1902) also discussed ignition sources.

Leiberg (1900a, 1900b, 1900c) also surveyed three of the southern California forest reserves: San Gabriel which is now part of the Angeles National Forest, San Bernardino, and San

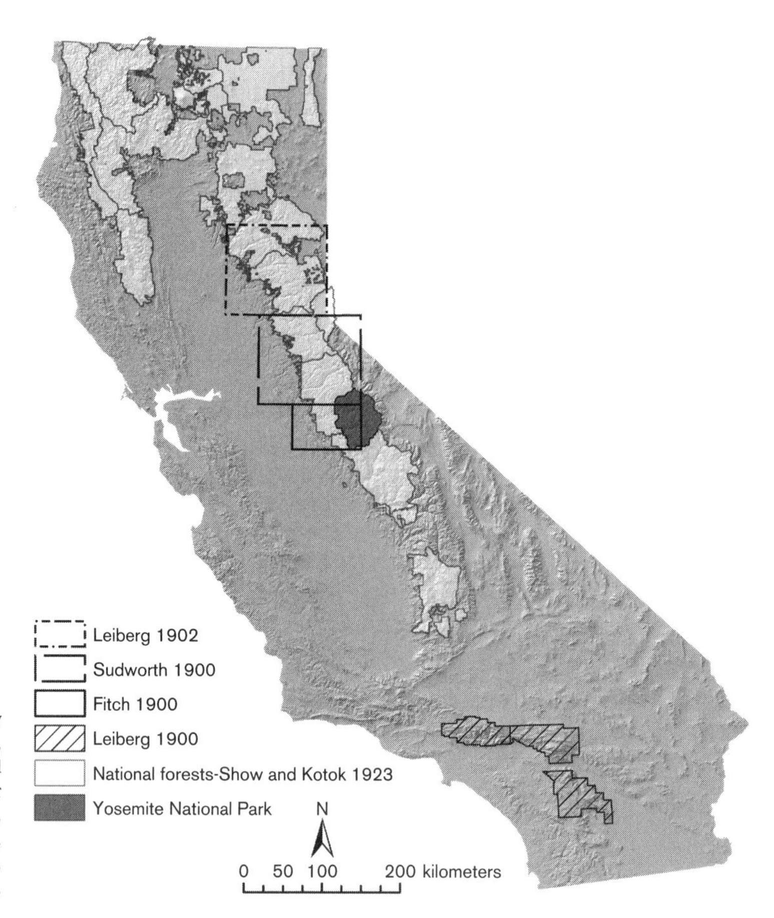

FIGURE 6.3 Areas covered by map sheets surveyed by Fitch (1900a, 1900b), Sudworth (1900), Leiberg (1902), and the 14 National forests for which Show and Kotok (1923) summarized fire information for 1911–1920 (clockwise from left: Mendocino, Six Rivers, Klamath, Shasta-Trinity, Modoc, Lassen, Plumas, Tahoe, Lake Tahoe Basin Management Unit, Eldorado, Stanislaus, Inyo, and Sequoia).

Jacinto which are now both part of the San Bernardino National Forest (Fig. 6.3). His reports followed the general format of the earlier USGS surveys by Fitch (1900a, 1900b) and Sudworth (1900); for conifer forest he only provided some qualitative information on fire effects.

Although Show and Kotok (1923, 1924, 1925, 1929) were advocates for suppressing fire, the data presented in their reports can provide some quantitative comparisons of early twentieth-century fires to modern fires that are, at least in part, a result of the suppression policy. Their reports focus on conifer forests and are based primarily upon data that were collected between 1911 and 1920 in what are now 14 national forests in California (Fig. 6.3). In addition to qualitative accounts of fire effects, their reports provide quantitative summary information on mortality (Show and Kotok 1924), number of ignitions, ignition sources, and fire sizes (Show and Kotok 1923). In addition, they report fire statistics (number of fires, size, ignition source, seasonality, rate of spread, and speed of initial attack) stratified by nine vegetation types (Show and Kotok 1929).

Some reports from the late nineteenth century and early twentieth century mention chaparral systems. The reports do not distinguish between true chaparral and montane chaparral as is commonly done today. However, Sterling (1904a) lists species in northern California and Leiberg (1900a, 1900b, 1900c) lists southern California species by aspect and elevation. All of the reports describe that chaparral fires were stand-replacing (Leiberg 1900a, 1900b, 1900c, Sterling 1904a, Show and Kotok 1924, 1929). Some reported that repeated fires led to a slow attrition of conifer stands, which eventually were converted chaparral fields, but that argument was also used to justify suppressing fires (Sterling 1904a, Show and Kotok 1924). Sterling (1904a) describes that fire maintained, and, in some cases, expanded shrub fields that occupied ridges in northern California forests. Show and Kotok (1924, 1929) reported that chaparral fires were three times larger on average than fires in yellow pine/mixed-conifer forests because they were more difficult to control, and that the fire season in chaparral was one-third of a month longer. Leiberg (1900a, 1900b, 1900c) provides estimates of chaparral area burned.

There are other reports written by early land managers around the turn of the century that also provide at least some qualitative descriptions of fires and forest conditions. But these reports are more difficult to obtain as they are archived only in libraries or land management units (e.g., Flintham 1904, Sterling 1904b, Hodge 1906).

ANECDOTAL INFORMATION

There are a number of written accounts (e.g., personal diaries, expedition journals), historical photos, and Native American oral histories that provide evidence of fire from prior to, or early in the Euro-American settlement period that dramatically changed fire regimes. Relating historical records to contemporary vegetation patterns combined with our current understanding of species' life history traits aids in estimating historical fire regime attributes. Although they provide rich information, they are typically "snapshot" descriptions or images, and, as such, can be limited in spatial and temporal extent. Photographs taken between the mid-1800s and early

1900s (e.g., Gruell 2001) provide information on species composition, physiognomy, and pattern. Oral histories from Native Americans, evidence of fire use, and its effects on ecosystems would be applicable more than 10,000 years ago (Moratto 1984, Anderson 2005). Burning by Native Americans affected successional patterns at multiple organizational and spatial scales, thereby promoting a diversity of habitats and ecological complexity (Anderson 2005). It is important to note that Native American influences on historical fire regime attributes were not ubiquitous across California ecosystems. The influences were likely limited by land use, resource availability, and accessibility (Keeley 2002).

The primary fire regime attributes that can be inferred from historical anecdotal records are fire type, seasonality, and frequency. Most information about historical fire seasonality is likely to come from written documents and oral histories, especially from Native American use of fire. Depending on the vegetation type, fires were set in either spring or late summer/early fall corresponding to the objectives of the resources. For example, Native Americans timed burns in oak (*Quercus* spp.) woodlands to correspond with acorn maturation (September to October), facilitating collection and reducing insect infestations (Lewis and Bean 1973, Timbrook et al. 1982). Approximate fire dates may also be obtained from observations made in dated field journals.

It's generally accepted that burning by Native Americans resulted in an increase in fire frequency. Given the multitude of reasons for using fire (Lewis and Bean 1973), population density throughout California, and supporting information on fire return intervals, the coverage of fire was extensive. In northern California and the Sierra Nevada, annual burning was reported for specific purposes (e.g., promote growth of certain plants, clear forest understory for hunting and safety) (Anderson 2005). Similarly, frequent burning resulted in a type conversion from shrublands/woodlands to grasslands in southern California (Keeley 2002).

DATASETS

Because of the limitations of tree ring-based datasets, researchers have analyzed spatially explicit datasets consisting of forest inventories conducted in the late nineteenth and early twentieth century. These datasets provide detailed information on forest structure and composition prior to the widespread changes associated with fire suppression and timber harvesting (Table 6.1). These datasets have been used to document change in contemporary forest structure and composition, which generally corroborate findings from tree ring-based studies (Scholl and Taylor 2010, Collins et al. 2011, Maxwell et al. 2014). These datasets have also been used to infer historical fire-severity patterns (Baker 2014, Collins et al. 2015, Stephens et al. 2015, Hanson and Odion 2016). However, because these datasets generally lack direct observations of historical fires, they rely on a couple key assumptions to infer fire severity. First, the presence of large trees at sufficient densities is often assumed to be indicative of fairly frequent, low- to moderate-severity fire effects (e.g., Collins et al. 2015, Stephens et al. 2015). Second, areas dominated by non-conifer vegetation (e.g., hardwoods or shrubs) or dense, small conifers are indicative of an early seral condition caused by stand-replacing fire. The former assumption is one that is generally accepted as true, particularly when it is corroborated with tree ring evidence indicting frequent fire. The latter assumption, however, is more problematic in that there are alternate explanations for the presence of those vegetation conditions (e.g., topographic and edaphic control, episodic tree regeneration driven by climate).

LIMITATIONS

Some interpretation is required in order to infer presettlement fire regime characteristics from late nineteenth- and early twentieth-century data sources to pre-Euro-American settlement forests for several reasons. By the end of the nineteenth-century impacts from extensive mining, logging and grazing had already begun to transform the conifer forests of California. For example, prior to his survey Leiberg's (1902) study area had been highly impacted by postsettlement activities. Most of the large-scale hydraulic and placer mining operations in the Sierra Nevada that occurred after 1849 were restricted to the Feather, Yuba, Bear, and American River drainages, all of which are included in Leiberg's study area (Leiberg 1902, Gruell 2001). Leiberg states that montane chaparral, a common early seral condition in Sierra Nevada forests, was primarily contiguous to placer mining camps. Leiberg speculated that many of the fires that created these montane chaparral patches predated the gold rush mining period.

Although lightning ignition patterns probably did not change after Euro-American settlement, human ignition patterns undoubtedly did. Ignitions by Native Americans, which played a large part for thousands of years in eliminating surface fuels and thinning of the lower to middle elevation mixed-conifer and pine forests in at least some locations, had largely been eliminated by the 1870s (Anderson 2005, Morgan 2008). Native Americans had many reasons for using fire, but based upon centuries of experience using fire to manage for specific resources they knew how fire could be used without negatively impacting the forest resources that they were dependent upon (Anderson 2005). After 1849, in addition to escaped camp fires, early settlement period miners, loggers and sheep herders set fires to clear land, assist in the search for gold, remove logging slash, and improve grazing without regard to the possible side effects (Leiberg 1902, Show and Kotok 1924, Cermak 2005).

The reader must also keep in mind the objectives and prevalent attitudes of the period when reading any of the early written accounts. For example, the majority of the early foresters argued that "light burning" (i.e., prescribed fire) was detrimental to timber production and therefore any fire caused tree mortality was unacceptable (e.g., Show and Kotok 1925). Tree mortality was attributed to torching when shrubs were present or through repeated enlargement of fire scar "cat faces." In essence, crown fire, which would constitute the definition of "destructive" fire in modern terms, did not normally occur in the California pine and mixed-conifer forests (Sudworth 1900, Cooper 1906, Greely 1907, Larson and Woodbury 1916, Show and Kotok 1924).

Quantifying Contemporary Fire Regimes

Fire-Occurrence Data

Information on contemporary (post-Euro-American settlement) fire occurrence can be assembled from numerous sources: fire atlases, fire reports (either ignition location and

TABLE 6.1

Historical archived datasets used to derive presettlement forest structure and subsequently infer presettlement fire severity and rotation[1]

Dataset	Information	Geographic extent	Strengths	Limitations	Example studies
Late nineteenth-century General Land Office (GLO) Public Land Survey System (PLSS) survey data	Diameter and species of bearing trees marking quarter corners (two trees) and section corners (four trees) of the PLSS. Surveyor notes on vegetation cover along section lines, including dominant overstory trees and shrubs	Most of the United States; 29 states west of Ohio excluding Kentucky, Tennessee, and Texas	Earliest systematic sample in which the entire data record is still available	There is considerable debate whether these data can be used to estimate forest structure because of the small sample. Methods for pooling the data have been devised (Williams and Baker 2011), but have been shown to overestimate tree density (Levine et al. 2017). This overestimation of tree density has been corroborated by density estimates derived from early twentieth-century forest surveys (Hagmann et al. 2013)	Baker (2014, 2015); Maxwell et al. (2014)
Early twentieth-century Forest Service forest inventories	One to two belt transects, per quarter-quarter section (QQs) in the PLSS grid; trees tallied by species and size class, shrub cover. Notes on site characteristics include logging and fire effects	National Forests	Dense systematic sample. Original transect locations can be relocated	Transect data were only acquired on federal owned lands. Most records have been lost. Transects were not sampled in QQs that did not contain merchantable timber, but summary forms indicate which QQs were not forested (i.e., shrublands)	Scholl and Taylor (2010); Collins et al. (2015, 2017a); Stephens et al. (2015)
1930s Wieslander Vegetation-Type Map (VTM) (Wieslander et al. 1933, Wieslander 1935)	Map has a typical minimum mapping unit of 16 ha, but many map polygons are much larger. Plot data include tree tallies by species and size class, shrub and herb species and cover, and photographs	Predominately national forests with some national parks. Northwestern and northeastern California and southern Sierra Nevada were never completed	First extensive vegetation-type map for California	Plots were located in areas "typical" for the map polygon. Difficult to identify the precise original plot locations	Bouldin (1999); Goforth and Minnich (2008)

NOTES:

1. Inference of high-severity fire in conifer forests relies on the assumption that presence of hardwoods or shrubs is indicative of an early seral condition. However, the potential for conifer forest to occupy the site in question needs to be substantiated for that assumption to be valid.

TABLE 6.2
National fire-severity mapping programs

	Burned Area Emergency Response (BAER)	Rapid Assessment of Vegetation (RAVG)	Monitoring Trends in Burn Severity (MTBS)
Objective	Determine threats from flooding, soil erosion, and instability	Provide information that can assist postfire vegetation management planning	Provide information necessary to monitor the effectiveness of the National Fire Plan and Healthy Forests Restoration Act
Resource focus	Soil burn severity	Tree mortality	Vegetation response
Postfire image acquisition timing	Rapid assessment: postfire image acquired as close to containment as possible; however, the acquisition date is often before containment	Initial assessment: postfire image is acquired within 8 weeks after containment	Primarily extended assessments for forested areas (postfire image is acquired during the first growing season after the fire, normally the summer of the following year). Initial assessments are normally used in grasslands and shrublands and some portions of the southeastern United States
Classification method	dNBR initially classified into four categories by an image analyst. BAER teams adjust the map categories to derive a soil burn severity map based upon field observations of soil burn severity characteristics	RdNBR calibrated to the Composite Burn Index (CBI), percent change in canopy cover, and percent change in tree basal area using field data acquired in California	dNBR classified by an image analyst into five burn severity categories primarily based upon professional experience, but also on field data, which are rarely available

fire size, or mapped perimeter), aerial photographs, and satellite imagery (Morgan et al. 2001). These archived records provide a reference for estimating fire regime characteristics such as frequency, extent, and seasonality at broad temporal and spatial scales (Starrs et al. 2018). Fire-occurrence records have evolved considerably since early fire record keeping began ca. late 1800s. The earliest records often consisted merely of fire size and ignition location identified by Public Land Survey System coordinates. However, many records that remain are only summarized by administrative unit (see Show and Kotok 1923, Westerling et al. 2006, Keeley and Syphard 2015). For larger fires after the establishment of the national forests, hand-drawn perimeters on paper maps may be available at local Forest Service offices. Accuracy and completeness has likely varied because of wildfire size and inaccessibility of wildfire sites, mapping methodologies and precision were not standardized, and some agencies began keeping records at different times (Keeley and Syphard 2015).

While imperfect, fire records and atlases allow for direct assessments of fire regime attributes that are a vast improvement compared to historical records. Excluding human observations, historical fire seasonality can be determined only in woody vegetation types through dendrochronology, and is estimated to a time associated with the growing season. Dates of contemporary fires are attainable in most cases to the day of origin. Similarly, where fire perimeters are available the fire interval and rotation can be quantified, permitting comprehensive analyses of spatial (Morgan et al. 2001) and temporal (Safford and Van de Water 2014) patterns of fire occurrence.

With the advancement of satellite and airborne technology in late twentieth century, the completeness and precision of fire records continue to improve. Although some perimeters are still hand drawn, incident management teams often use infrared data from airborne sensors and Global Positioning Systems (GPS) acquired data to map fire extent (Kolden and Weisberg 2007). Passive satellites, such as LANDSAT, that detect changes in environmental conditions of prefire and postfire landscapes have also been used to develop fire atlases in recent years. The scale of hand-drawn perimeters can be coarse, which may lead to an overestimate of area burned due to a generalization of the perimeter and inclusion of unburned islands (Kolden and Weisberg 2007). Perimeters derived from passive satellites may also underestimate the full extent of a burned area in forested ecosystems where the overstory canopy may obscure the effects of low-intensity surface fire (Morgan et al. 2001).

Fire-Severity Data

Currently, there are three national programs producing severity data from satellite images in the United States: Burned Area Emergency Response (BAER), Monitoring Trends in Burn Severity (MTBS), and Rapid Assessment of Vegetation Condition (RAVG) (Table 6.2). All of the national severity mapping programs normally use satellite images from the LANDSAT program to compute the Normalized Burn Ratio (NBR) index for assessing severity (Key and Benson 2005b). Each of the severity mapping programs has different programmatic objectives, which influence the methods they employ and how they interpret the satellite data. The two primary methodological differences discussed below are the acquisition date of the postfire satellite image and change detection algorithm.

The timing of postfire image acquisition has a significant influence on the type of effects that can be detected. Rapid and initial assessments are best suited for evaluating first-order fire effects. Rapid assessments are based upon images acquired as close to the fire containment date as possible to support postfire response planning, i.e., BAER. Initial assessments use postfire images acquired within eight weeks of fire containment (Key 2006). Rapid or initial assessments are often required to detect fires in areas where vegetation is typically senescent or contains little chlorophyll (e.g., in the Great Basin) or in grasslands, which typically recover after the first growing season. Rapid and initial assessments do not always result in robust severity maps, however. This is particularly the case when containment dates are late in the calendar year (e.g., late September or later) due to low sun angles and greater presence of clouds.

Extended assessments, which use postfire images acquired the first growing season following fire containment (Key 2006), are generally preferred over initial assessments to allow for optimum image acquisition when sun angles are high and visibility is best. However, delayed mortality and/or vegetation recovery may be visible in the postfire image. Extended assessments, therefore, often include some second-order effects which can confound or enhance interpretation depending upon analysis objectives (Key 2006). When postfire management actions, such as salvage logging, commences immediately after containment, the amount of high severity can be over-estimated in extended assessments (Safford et al. 2015).

Change detection algorithms are used to map severity so that areas of little to no vegetation are not misinterpreted as high severity. Developed in the late 1990s, the differenced Normalized Burn Ratio (dNBR) was the change detection algorithm first used by the severity mapping programs. dNBR is an absolute change detection methodology, calculated by subtracting a postfire NBR image from a prefire NBR image (Key and Benson 2005b, Key 2006). More recently, a relativized version of the dNBR (RdNBR), computed by dividing dNBR by a function of the prefire image, has also been used (Miller and Thode 2007, Miller et al. 2009b, Miller and Quayle 2015). Regardless of the change detection methodology, resulting images have similar characteristics.

The most significant characteristic of dNBR data is that they are correlated to the amount of prefire chlorophyll (Miller and Thode 2007). Thus, dNBR values do not always represent the same percent mortality detected by changes in chlorophyll because of differences in species and amounts of vegetative cover within and between fires. As a result, it is difficult to produce a map that consistently rates the percent mortality where vegetation is heterogeneous. Miller and Thode (2007) developed RdNBR to solve the issue with using dNBR to map severity to vegetation in heterogeneous landscapes. Maps derived from RdNBR can depict a more consistent level of percent mortality when there is a broad range of stand density, or vegetation types within a fire perimeter. Image values between multiple fires also represent the same level of severity (Miller and Thode 2007, Miller et al. 2009b, Cansler and McKenzie 2012). Because the same RdNBR values from different fires represent similar severity conditions, calibrations to plot level severity measures can be made from one fire and applied to another fire. Calibrations to RdNBR have been developed for field-measured Composite Burn Index (CBI) (Key 2006), percent change in canopy cover, and percent tree basal area using field data acquired in California (Key and Benson 2005a, Miller and Thode 2007, Miller et al. 2009b, Miller and Quayle 2015). RdNBR extended assessment calibrations based upon CBI plot data acquired in other locations across the western United States are similar to those from California (Holden et al. 2007, Cansler and McKenzie 2012).

BURNED AREA EMERGENCY RESPONSE

BAER teams must produce a plan within seven days of fire containment that assesses postfire watershed conditions such as threats from flooding, soil erosion, and soil instability (Parsons et al. 2010). As part of their assessment of watershed conditions, BAER teams produce soil burn severity maps to identify where postfire management actions should be applied to alleviate the possible effects from erosion. Because the BAER program was the first to produce severity data on a regular basis, some researchers have used those data as a substitute for vegetation severity maps. However, soil burn severity maps were not intended to describe severity to vegetation, as areas with high vegetation mortality do not always signify high soil burn severity (Safford et al. 2008, Parsons et al. 2010).

Initial maps produced for BAER teams, referred to as a Burned Area Reflectance Classification (BARC) maps, are not intended to be final soil severity map products. BARC maps have to be field verified by a BAER team and, if necessary, refined to better represent soil and ground conditions (Parsons et al. 2010). BARC maps are archived and are available for download from the internet (https://fsapps.nwcg.gov/baer). Beginning in 2012 the Forest Service's Geospatial Technology and Applications Center (GTAC) has also made some final soil severity maps available on their website. Soil severity maps from before 2012 are usually only archived at the home units where each fire occurred.

MONITORING TRENDS IN BURN SEVERITY

In 2005 the Wildland Fire Leadership Council sponsored the MTBS program as one element of a strategy for monitoring the effectiveness of the National Fire Plan and Healthy Forests Restoration Act (Eidenshink et al. 2007). MTBS is jointly conducted by GTAC and the US Geological Survey's Earth Resources Observation and Science (EROS) datacenter. The project's charter is to map the location, extent, and severity of both large wildfires and prescribed fire (fires > 400 ha [1,000 ac] in the western United States and fires > 200 ha [500 ac] in the eastern United States) regardless of vegetation type, ownership, or ignition source, throughout the United States beginning in 1984. Severity data are primarily based upon extended assessments. However, initial assessments are often used for grasslands, shrublands, and fires in Florida. Because the MTBS program was conceived by the same organizations that produce BARC maps for BAER teams, the MTBS categorical severity maps are produced using similar methods, e.g., image analyst interpreted severity categories derived from dNBR images (Eidenshink et al. 2007). Along with the categorical maps, continuous dNBR and RdNBR data are also available from MTBS's website which advanced users can use to produce their own categorical maps (http://www.mtbs.gov).

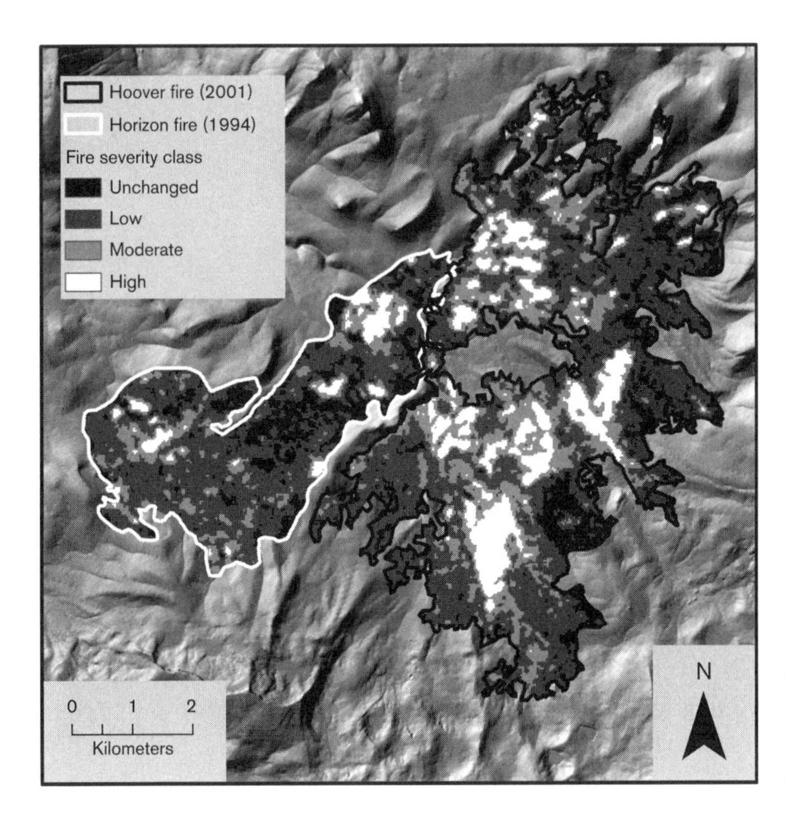

FIGURE 6.4 Fire-severity maps for two fires in the Illilouette Creek Basin, Yosemite National Park. Fire-severity classes are based on the thresholds for the relativized differenced normalized burn ratio described in Miller and Thode (2007).

RAPID ASSESSMENT OF VEGETATION CONDITION

The Forest Service Pacific Southwest Region began the RAVG program in 2006 to support reforestation planning after wildfires > 400 ha (1,000 ac) of forested land. The RAVG program normally uses postfire images acquired within the first 30–45 d after fire containment, i.e., initial assessments. However, "postfire" images can be acquired before official containment for late season fires which are not declared contained until after snowfall. RAVG data products are derived from calibrations of RdNBR to plot data acquired in California. Categorical and continuous map products are calibrated to CBI, percent change in canopy cover, and percent change in basal area based upon field data collected only in California (Miller and Quayle 2015). All data can be downloaded from the internet (https://fsapps.nwcg.gov/ravg).

APPLICATIONS

Spatial data describing fire effects are required to evaluate fire severity and spatial complexity. With the advancement of satellite images to derive spatial severity information on an operational basis there has been a rapid increase in the number of studies quantifying severity and complexity of modern fires within California. The most basic application has been the simple calculation of proportion of area burned within discrete severity categories (e.g., van Wagtendonk et al. 2004, van Wagtendonk and Lutz 2007, Collins et al. 2009, van Wagtendonk et al. 2012, Mallek et al. 2013), or continuous distributions from unburned to high severity (Thode et al. 2011). Some studies, however, have focused on the proportion of high severity (Miller et al. 2009a, Collins and Stephens 2010, Miller and Safford 2012, Miller et al. 2012a, 2012b) or unburned areas (Kolden et al. 2012).

Spatial complexity is described by the patchiness, or spatial variability, of severity within the fire perimeter (Fig. 6.4). Characterizations of high-severity patches have been most often used in describing complexity. Moderate severity often does not form patches *per se*, but rather form rings around high-severity patches as fire transitions from surface to crown fire. Most studies have used basic statistical measures (e.g., mean, minimum, maximum, and variance) to describe high-severity patches (Miller et al. 2009a, Collins and Stephens 2010, Miller and Safford 2012, Miller et al. 2012b). Recently, a new characterization of high severity patch size and shape has been developed (Collins et al. 2017b) and used to analyze fires across land ownerships in California (Stevens et al. 2017). A few studies have used landscape ecology metrics, such as patch density and squareness (van Wagtendonk and Lutz 2007, Kolden et al. 2012). Spatial severity data are a rich source of information that have yet to be fully explored. Studies have contrasted severity and complexity due to fire history, fire type (e.g., prescribed vs. wildfires), vegetation type, and land ownership (van Wagtendonk and Lutz 2007, Collins et al. 2009, Thode et al. 2011, Miller et al. 2012b, van Wagtendonk et al. 2012). Additionally, spatial severity data have been used to understand drivers of contemporary severity patterns (Lydersen et al. 2014, Harris and Taylor 2015, Kane et al. 2015).

LIMITATIONS

The National Wildfire Coordinating Group (NWCG 2014) defines fire severity as the: "degree to which a site has been altered or disrupted by fire; loosely, a product of fire intensity and residence time." In this context, the term "severity" is conceptual in nature that lacks measurement units. Due to the inherent nature of satellites, there are limitations to the types of fire effects that can be interpreted from satellite images. The imaging sensors on satellites are similar to those found in typical consumer digital cameras, i.e., the resolution is limited by size and number of the sensor pixels and optics properties. Various satellites have been used by the three severity mapping programs, with image pixel sizes ranging between 10 and 56 m (32–184 ft); the most common being 30 m (98 ft). Regardless, all are much too large to discern individual trees, and therefore fire effects in the overstory cannot be distinguished from understory when tree canopies are sparse. On the other hand, because the satellite sensors are passive in nature, the satellite cannot "see" under dense tree canopies (Key 2006). Finally, unlike consumer cameras which only record wavelengths visible to humans, satellites that are most often used for severity mapping also record infrared wavelengths. As a result, map products produced by the three national severity mapping programs are primarily sensitive to changes in chlorophyll, but also soil substrate and ash cover (Key and Benson 2005b).

TABLE 6.3
Fire history groups used to estimate Fire Regime Condition Classes

Group	Mean historical fire frequency (y)	Historical fire severity
I	<35	Low to mixed
II	<35	High
III	35–200	Low to mixed
IV	35–200	High
V	>200	High

Fire Regime Departure from Historical Range of Variability

Improving ecosystem conditions and promoting greater resilience to future fires is a primary interest to managers. In the past, much of the efforts have focused on using ecological restoration practices that were informed by the use of reference conditions generated from historical information that preceded presettlement conditions. However, increased recognition that ecosystems are not static and that climate change is altering species distributions and fire regimes has prompted a need to consider future conditions that promote ecosystem resistance and recovery following future fires (Millar et al. 2007, Stephens et al. 2010b). Still, historical fire regime information and current departures from these conditions provide essential details to inform desired future conditions of ecosystems in a period of much uncertainty. Many different methods have been developed to assess vegetation departures from historical conditions. We discuss a few methods commonly used in fire-prone ecosystems, the Fire Regime Condition Class (FRCC), Fire Return Interval Departure (FRID), and Natural Range of Variation (NRV).

Fire Regime Condition Class

FRCC assessment system was first developed as an interagency initiative responding to the 2001 National Fire Plan that emphasized the need to restore and maintain fire-adapted ecosystems (Hardy et al. 2001). Since its inception, FRCC has been updated through the years to address inaccuracies (FRCC 2010) and to better integrate with other systems that model reference conditions (e.g., LANDFIRE; Rollins 2009). FRCC provides a relatively simple tool for managers to evaluate and assess the departure from historical conditions. Historical fire regimes are based on conditions prior to Euro-American settlement and are broadly classified into five groups based on historical fire frequency and dominant fire-severity classes (Table 6.3). Using these groups wildlands are classified into three categories based on the percent departure from historical conditions defined as follows; FRCC 1 is defined as areas with a low departure (<33%), FRCC 2 represents areas with a moderate departure (33–66%), and FRCC 3 indicates areas with a high departure (>66%) from reference conditions (FRCC 2010).

Fire Return Interval Departure

FRID was originally used to assess departures in national parks of the Sierra Nevada (Caprio and Graber 2000), but has recently been expanded more broadly across California (Safford and Van de Water 2014). This method assigns current vegetation types into pre-Euro-American Fire Regime (PFR) groups based on similarities in their historical fire return intervals derived from published fire history studies (Van de Water and Safford 2011). Each PFR is mapped and five FRID metrics are generated including mean, median, minimum, maximum percent FRID (PFRID), and NPS FRID index. PFRID values represent a proportional difference between presettlement and current fire return intervals. Current fire return interval estimates are calculated as the time since last fire based on fire perimeter data. PFRID values can range between –100% and 100%, with negative values indicating that current fire return intervals are more frequent than historical conditions and positive values indicating fire return intervals that are less frequent than historical conditions (Safford and Van de Water 2014). Estimations of FRID metrics can be made at scales ranging from the stand to ecoregions. Note that this FRID analysis only takes fire frequency into account in assessing fire regime departure. Fire-severity distributions and spatial complexity are other important fire regime attributes, but are not included.

Analysis of FRID on national forest lands of California indicates broad differences among northern and southern regions (Safford and Van de Water 2014). In northern California most areas are characterized as missing multiple fire cycles that are likely due to fire exclusion over the past century, while many areas in southern California are characterized as having more frequent fires compared to pre-Euro-American settlement conditions. Variation in FRID was attributed to differences in elevation, precipitation, and temperature. Elevation and temperature had a unimodal relationship with PFRID values, with areas of low and high elevation and temperatures generally having lower departures compared to moderate elevations and temperatures. Precipitation was only found to have a relationship with PFRID values in southern California, in which PFRID values decreased with increasing precipitation.

Natural Range of Variation

NRV assessments use a combination of historical information, vegetation dynamics modeling, and contemporary reference landscape information to identify "natural" fire patterns

(Maxwell et al. 2014, Steel et al. 2015). The primary assumption is that these fire patterns are indicative of a healthy, functional, and resilient ecosystem. NRV assessments address one of the key fire regime attributes that is not explicitly addressed with FRCC or FRID, distributions of fire severity (i.e., proportions of area in different fire-severity classes). This is made possible by the remotely sensed fire-severity products discussed previously in this chapter, which, in addition to the distributions of severity, have allowed for delineation of high-severity patch sizes. These patch sizes are a critical spatial attribute of individual fires and fire regimes, as they affect vegetation recovery patterns and the species associated with recovering vegetation (e.g., wildlife, pollinators). That said, the estimation of "natural" stand-replacing patch sizes can be problematic as this information is very difficult to ascertain from historical reconstructions. This also applies to contemporary reference sites (e.g., Rivera-Huerta et al. 2015, Collins et al. 2016), which may have limited applicability because they tend to be somewhat unique locations. A recent application of the NRV approach evaluated fire-severity distributions and high-severity patch sizes for recent fires that were managed for natural resource benefit (e.g., "managed wildfires"; see chapter 21) relative to those determined through an NRV assessment for mixed-conifer forests in southern Sierra Nevada (Meyer 2015). Results indicated that resource benefit fires were largely within the NRV, suggesting the further expansion of managed fire, under less-than-extreme fire weather conditions may improve the resilience of these forests (Meyer 2015).

Summary

Understanding historical fire regimes is critical for managing fire-prone vegetation types. In this chapter we introduced the common approaches for characterizing historical fire regimes and discussed their applications. We also discussed some limitations of these approaches, which emphasize the need for multiple lines of evidence when attempting to describe the often complex patterns of fire across landscapes. The recent advancements in remote sensing discussed allow for much more explicit descriptions of contemporary fire regimes than are available for historical fire regimes. This mismatch makes for imperfect assessments of current levels of fire regime departure. Despite this we know that contemporary fire patterns in forests historically associated with frequent low- to moderate-intensity fire are considerably departed from historical fire patterns. The degree of departure is less certain for forest types historically associated with greater proportions of stand-replacing fire. If the departed patterns for historically frequent-fire forests continue to prevail, let alone are exacerbated by changing climate, the services they provide (e.g., habitat for wildlife, timber, re-creation) will be diminished. It is incumbent on forest managers, both public and private, to address this departure to insure the integrity of these ecosystems is preserved for future generations.

References

Agee, J.K. 1993. Fire Ecology of Pacific Northwest Forests. Island Press, Washington, D.C., USA.

Anderson, M.K. 2005. Tending the Wild: Native American Knowledge and the Management of California's Natural Resources. University of California Press, Berkeley, California, USA.

Anderson, R.S., A. Ejarque, P.M. Brown, and D.J. Hallett. 2013. Holocene and historical vegetation change and fire history on the north-central coast of California, USA. The Holocene 23: 1797–1810.

Anderson, R.S., and S.J. Smith. 1997. The sedimentary record of fire in montane meadows, Sierra Nevada, California, USA: a preliminary assessment. Pages 313–327 in: J.S. Clark, J.G. Cachier, J.G. Goldammer, and B.J. Stocks, editors. Sediment Records of Biomass Burning and Global Change. Springer-Verlag, Berlin, Germany.

Anderson, R.S., S. Starratt, R.M.B. Jass, and N. Pinter. 2010. Fire and vegetation history on Santa Rosa Island, Channel Islands, and long-term environmental change in southern California. Journal of Quaternary Science 25: 782–797.

Arno, S.F., and K.M. Sneck. 1977. A method for determining fire history in coniferous forests of the mountain west. USDA Forest Service General Technical Report INT-GTR-12. Intermountain Forest and Range Experiment Station, Ogden, Utah, USA.

Baker, W.L. 2014. Historical forest structure and fire in Sierran mixed-conifer forests reconstructed from General Land Office survey data. Ecosphere 5: art79.

———. 2015. Are high-severity fires burning at much higher rates recently than historically in dry-forest landscapes of the Western USA? PLoS One 10 (9): e0136147.

Baker, W.L., and D. Ehle. 2001. Uncertainty in surface-fire history: the case of ponderosa pine forests in the western United States. Canadian Journal of Forest Research 31: 1205–1226.

Beaty, R.M., and A.H. Taylor. 2008. Fire history and the structure and dynamics of a mixed conifer forest landscape in the northern Sierra Nevada, Lake Tahoe Basin, California, USA. Forest Ecology and Management 255: 707–719.

Beaufait, W.R. 1966. An integrating device for evaluating prescribed fire. Forest Science 12: 27–29.

Binford, M.W. 1990. Calculation and uncertainty analysis of 210Pb dates for PIRLA project lake sediment cores. Journal of Paleolimnology 3: 253–267.

Bouldin, J.R. 1999. Twentieth Century changes in forests of the Sierra Nevada Mountains. Dissertation, University of California, Davis, California, USA.

Cansler, C.A., and D. McKenzie. 2012. How robust are burn severity indices when applied in a new region? Evaluation of alternate field-based and remote-sensing methods. Remote Sensing 4: 456–483.

Caprio, A.C., and D.M. Graber. 2000. Returning fire to the mountains: can we successfully restore the ecological role of pre-Euroamerican fire regimes in the Sierra Nevada? Pages 223–241 in: D.N. Cole, S.F. McCool, W.T. Borrie, and J. O'Loughlin, compilers. Wilderness Science in a Time of Change Conference—Volume 5: Wilderness Ecosystems. USDA Forest Service Proceedings RMRS-P-15-VOL-5. Rocky Mountain Research Station, Fort Collins, Colorado, USA.

Cermak, R.W. 2005. Fire in the forest: a history of forest fire control on the national forests in California, 1898–1956. USDA Forest Service Report R5-RF-003. Pacific Southwest Region, Vallejo, California, USA.

Collins, B.M., D.L. Fry, J.M. Lydersen, R. Everett, and S. L. Stephens. 2017a. Impacts of different land management histories on forest change. Ecological Applications 27: 2475–2486.

Collins, B.M., J.M. Lydersen, R.G. Everett, D.L. Fry, and S.L. Stephens. 2015. Novel characterization of landscape-level variability in historical vegetation structure. Ecological Applications 25: 1167–1174.

Collins, B.M., J.M. Lydersen, D.L. Fry, K. Wilkin, T.J. Moody, and S.L. Stephens. 2016. Variability in vegetation and surface fuels across mixed-conifer-dominated landscapes with over 40 years of natural fire. Forest Ecology and Management 381: 74–83.

Collins, B.M., J.D. Miller, A.E. Thode, M. Kelly, J.W. van Wagtendonk, and S.L. Stephens. 2009. Interactions among wildland fires in a long-established Sierra Nevada natural fire area. Ecosystems 12: 114–128.

Collins, B.M., and S.L. Stephens. 2007a. Fire scarring patterns in Sierra Nevada wilderness areas burned by multiple wildland fire use fires. Fire Ecology 3(2): 53–67.

———. 2007b. Managing natural wildfires in Sierra Nevada wilderness areas. Frontiers in Ecology and the Environment 5: 523–527.

———. 2010. Stand-replacing patches within a 'mixed severity' fire regime: quantitative characterization using recent fires in a long-established natural fire area. Landscape Ecology 25: 927–939.
Collins, B. M., J. T. Stevens, J. D. Miller, S. L. Stephens, P. M. Brown, and M. North. 2017b. Alternative characterization of forest fire regimes: incorporating spatial patterns. Landscape Ecology 32: 1543–1552.
Cooper, A.W. 1906. Sugar pine and western yellow pine in California. USDA Forest Service Bulletin 6. Washington, D.C., USA.
Daniels, M.L., S. Anderson, and C. Whitlock. 2005. Vegetation and fire history since the Late Pleistocene from the Trinity Mountains, northwestern California, USA. The Holocene 15: 1062–1071.
Davis, F.W., M.I. Borchert, and D.C. Odion. 1989. Establishment of microscale vegetation pattern in maritime chaparral after fire. Vegetation 84: 53–67.
Davis, F.W., and J. Michaelsen. 1995. Sensitivity of fire regime in chaparral ecosystems to climate change. Pages 435–456 in: J. Moreno and W.C. Oechel, editors. Global Change and Mediterranean-Type Ecosystems. Springer-Verlag, New York, New York, USA.
Dennison, P.E., and M.A. Moritz. 2009. Critical live fuel moisture in chaparral ecosystems: a threshold for fire activity and its relationship to antecedent precipitation. International Journal of Wildland Fire 18: 1021–1027.
Dieterich, J.H. 1980. Chimney spring forest fire history. USDA Forest Service Research Paper RM-RP-220. Rocky Mountain Research Station, Fort Collins, Colorado, USA.
Eidenshink, J., B. Schwind, K. Brewer, Z.-L. Zhu, B. Quayle, and S. Howard. 2007. A project for monitoring trends in burn severity. Fire Ecology 3(1): 3–21.
Farris, C.A., C.H. Baisan, D.A. Falk, S.R. Yool, and T.W. Swetnam. 2010. Spatial and temporal corroboration of a fire-scar-based fire history in a frequently burned ponderosa pine forest. Ecological Applications 20: 1598–1614.
Fitch, C.H. 1900a. Sonora quadrangle, California. Pages 569–570 in: Annual Reports of the Department of the Interior, 21st Annual Report of the U.S. Geological Survey, Part V—Forest Reserves. U.S. Government Printing Office, Washington, D.C., USA.
———. 1900b. Yosemite quadrangle, California. Pages 571–574 in: Annual Reports of the Department of the Interior, 21st Annual Report of the U.S. Geological Survey, Part V—Forest Reserves. U.S. Government Printing Office, Washington, D.C., USA.
Flintham, S.J. 1904. Forest Extension in the Sierra Forest Reserve. Sierra National Forest, Clovis, California, USA.
FRCC. 2010. Interagency Fire Regime Condition Class guidebook, version 3.0. National Interagency Fuels, Fire, and Technology Transfer. National Interagency Fire Center, Boise, Idaho, USA.
Fry, D.L., J. Dawson, and S.L. Stephens. 2012. Age and structure of mature knobcone pine forests in the northern California Coast Range, USA. Fire Ecology 8(1): 46–62.
Fry, D.L., and S.L. Stephens. 2006. Influence of humans and climate on the fire history of a ponderosa pine-mixed conifer forest in the southeastern Klamath Mountains, California. Forest Ecology and Management 223: 428–438.
Gavin, D.G., D.J. Hallett, F.S. Hu, K.P. Lertzman, S.J. Prichard, K.J. Brown, J.A. Lynch, P. Bartlein, and D.L. Peterson. 2007. Forest fire and climate change in western North America: insights from sediment charcoal records. Frontiers in Ecology and the Environment 5: 499–506.
Goforth, B.R., and R.A. Minnich. 2008. Densification, stand-replacement wildfire, and extirpation of mixed conifer forest in Cuyamaca Rancho State Park, southern California. Forest Ecology and Management 256: 36–45.
Greely, W.B. 1907. A rough system of management for reserve lands in the western Sierras. Proceedings of the Society of American Foresters 2: 103–114.
Green, L.R. 1981. Burning by prescription in chaparral. USDA Forest Service General Technical Report PSW-GTR-51. Pacific Southwest Forest and Range Experiment Station, Berkeley, California, USA.
Gruell, G.E. 2001. Fire in Sierra Nevada Forests: A Photographic Interpretation of Ecological Change since 1849. Montana Press Publishing Company, Missoula, Montana, USA.
Hagmann, R.K., J.F. Franklin, and K.N. Johnson. 2013. Historical structure and composition of ponderosa pine and mixed-conifer forests in south-central Oregon. Forest Ecology and Management 304: 492–504.
Hanson, C.T., and D.C. Odion. 2016. Historical forest conditions within the range of the Pacific fisher and spotted owl in the central and southern Sierra Nevada, California, USA. Natural Areas Journal 36: 8–19.
Hardy, C.C., K.M. Schmidt, J.P. Menakis, and R.N. Sampson. 2001. Spatial data for national fire planning and fuel management. International Journal of Wildland Fire 10: 353–372.
Harris, L., and A. Taylor. 2015. Topography, fuels, and fire exclusion drive fire severity of the Rim Fire in an old-growth mixed-conifer forest, Yosemite National Park, USA. Ecosystems 18: 1192–1208.
Heinselman, M.L. 1973. Fire in the virgin forests of the Boundary Waters Canoe Area, Minnesota. Quaternary Research 3: 329–382.
Hodge, W.C. 1906. Forest conditions in the Sierra. U.S. Bureau of Forestry, San Francisco, California. Report on file at the Eldorado National Forest, Supervisor's Office, Placerville, California, USA.
Holden, Z.A., P. Morgan, M.A. Crimmins, R.K. Steinhorst, and A.M. Smith. 2007. Fire season precipitation variability influences fire extent and severity in a large southwestern wilderness area, United States. Geophysical Research Letters 34: L16708.
Jones, M.B., and H.M. Laude. 1960. Relationships between sprouting in chamise and the physiological condition of the plant. Journal of Range Management 13: 210–214.
Kane, V.R., C.A. Cansler, N.A. Povak, J.T. Kane, R.J. McGaughey, J.A. Lutz, D.J. Churchill, and M.P. North. 2015. Mixed severity fire effects within the Rim Fire: relative importance of local climate, fire weather, topography, and forest structure. Forest Ecology and Management 358: 62–79.
Keeley, J.E. 2002. Native American impacts on fire regimes of the California coastal ranges. Journal of Biogeography 29: 303–320.
Keeley, J.E., T. Brennan, and A.H. Pfaff. 2008. Fire severity and ecosystem responses following crown fires in California shrublands. Ecological Applications 18: 1530–1546.
Keeley, J.E., and C.J. Fotheringham. 2001. Historic fire regime in southern California shrublands. Conservation Biology 15: 1536–1548.
Keeley, J.E., and A.D. Syphard. 2015. Different fire–climate relationships on forested and non-forested landscapes in the Sierra Nevada ecoregion. International Journal of Wildland Fire 24: 27–36.
Key, C.H. 2006. Ecological and sampling constraints on defining landscape fire severity. Fire Ecology 2(2): 34–59.
Key, C.H., and N.C. Benson. 2005a. Landscape assessment: ground measure of severity, the Composite Burn Index. In: D.C. Lutes, R.E. Keane, J.F. Caratti, C.H. Key, N.C. Benson, and L.J. Gangi, editors. FIREMON: Fire Effects Monitoring and Inventory System. USDA Forest Service General Technical Report RMRS-GTR-164-CD. Rocky Mountain Research Station, Fort Collins, Colorado, USA.
———. 2005b. Landscape assessment: remote sensing of severity, the Normalized Burn Ratio. In: D.C. Lutes, R.E. Keane, J.F. Caratti, C.H. Key, N.C. Benson, and L.J. Gangi, editors. FIREMON: Fire Effects Monitoring and Inventory System. General Technical Report RMRS-GTR-164-CD. U.S. Department of Agriculture, Forest Service, Rocky Mountain Research Station, Fort Collins, Colorado, USA.
Kolden, C.A., J.A. Lutz, C.H. Key, J.T. Kane, and J.W. van Wagtendonk. 2012. Mapped versus actual burned area within wildfire perimeters: characterizing the unburned. Forest Ecology and Management 286: 38–47.
Kolden, C.A., and P.J. Weisberg. 2007. Assessing accuracy of manually-mapped wildfire perimeters in topographically dissected areas. Fire Ecology 3(1): 22–31.
Larson, L.T., and T.D. Woodbury. 1916. Sugar pine. USDA Forest Service Bulletin 426. Washington, D.C., USA.
Lauvaux, C.A., C.N. Skinner, and A.H. Taylor. 2016. High severity fire and mixed conifer forest-chaparral dynamics in the southern Cascade Range, USA. Forest Ecology and Management 363: 74–85.
Leiberg, J.B. 1900a. San Bernardino Forest Reserve. Pages 429–454 in: Annual Reports of the Department of the Interior, 20th Annual Report of the U.S. Geological Survey, Part V—Forest Reserves. U.S. Government Printing Office, Washington, D.C., USA.
———. 1900b. San Gabriel Forest Reserve. Pages 411–428 in: Annual Reports of the Department of the Interior, 20th Annual Report of

the U.S. Geological Survey, Part V—Forest Reserves. U.S. Government Printing Office, Washington, D.C., USA.

———. 1900c. San Jacinto Forest Reserve. Pages 455–478 in: Annual Reports of the Department of the Interior, 20th Annual Report of the U.S. Geological Survey, Part V—Forest Reserves. U.S. Government Printing Office, Washington, D.C., USA.

———. 1902. Forest conditions in the northern Sierra Nevada. U.S. Geological Survey Professional Paper 8. Department of Interior, Washington, D.C., USA.

Levine, C.R., C.V. Cogbill, B.M. Collins, A.J. Larson, J.A. Lutz, M.P. North, C.M. Restaino, H.D. Safford, S.L. Stephens, and J.J. Battles. 2017. Evaluating a new method for reconstructing forest conditions from General Land Office survey records. Ecological Applications 27: 1498–1513.

Lewis, H.T., and L.J. Bean. 1973. Patterns of Indian Burning in California: Ecology and Ethnohistory. Ballena Press, Ramona, California, USA.

Lombardo, K.J., T.W. Swetnam, C.H. Baisan, and M.I. Borchert. 2009. Using bigcone Douglas-fir fire scars and tree rings to reconstruct interior chaparral fire history. Fire Ecology 5(3): 35–56.

Long, C.J., C. Whitlock, P.J. Bartlein, and S.H. Millspaugh. 1998. A 9000-year fire history from the Oregon Coast Range based on a high-resolution charcoal study. Canadian Journal of Forest Research 28: 774–787.

Lydersen, J.M., M.P. North, and B.M. Collins. 2014. Severity of an uncharacteristically large wildfire, the Rim Fire, in forests with relatively restored frequent fire regimes. Forest Ecology and Management 328: 326–334.

Mallek, C., H. Safford, J.H. Viers, and J. Miller. 2013. Modern departures in fire severity and area vary by forest type, Sierra Nevada and southern Cascades, California, USA. Ecosphere 4: art153.

Marlon, J.R., P.J. Bartlein, D.G. Gavin, C.J. Long, R.S. Anderson, C.E. Briles, K.J. Brown, D. Colombaroli, D.J. Hallett, M.J. Power, E.A. Scharf, and M.K. Walsh. 2012. Long-term perspective on wildfires in the western USA. Proceedings of the National Academy of Sciences of the United States of America 109: E535–E543.

Marlon, J.R., P.J. Bartlein, M.K. Walsh, S.P. Harrison, K.J. Brown, M.E. Edwards, P.E. Higuera, M.J. Power, R.S. Anderson, C.E. Briles, A. Brunelle, C. Carcaillet, M. Daniels, F.S. Hu, M. Lavoie, C. Long, T. Minckley, P.J.H. Richard, A.C. Scott, D.S. Shafer, W. Tinner, C.E. Umbanhowar, and C. Whitlock. 2009. Wildfire responses to abrupt climate change in North America. Proceedings of the National Academy of Sciences of the United States of America 106: 2519–2524.

Maxwell, R.S., A.H. Taylor, C.N. Skinner, H.D. Safford, R.E. Isaacs, C. Airey, and A.B. Young. 2014. Landscape-scale modeling of reference period forest conditions and fire behavior on heavily logged lands. Ecosphere 5: art32.

Mensing, S.A., J. Michaelsen, and R. Byrne. 1999. A 560-year record of Santa Ana fires reconstructed from charcoal deposited in the Santa Barbara Basin, California. Quaternary Research 51: 295–305.

Meyer, M.D. 2015. Forest fire severity patterns of resource objective wildfires in the southern Sierra Nevada. Journal of Forestry 113: 49–56.

Millar, C.I., N.L. Stephenson, and S.L. Stephens. 2007. Climate change and forests of the future: managing in the face of uncertainty. Ecological Applications 17: 2145–2151.

Miller, J.D., B.M. Collins, J.A. Lutz, S.L. Stephens, J.W. van Wagtendonk, and D.A. Yasuda. 2012b. Differences in wildfires among ecoregions and land management agencies in the Sierra Nevada region, California, USA. Ecosphere 3: art80.

Miller, J.D., E.E. Knapp, C.H. Key, C.N. Skinner, C.J. Isbell, R.M. Creasy, and J.W. Sherlock. 2009b. Calibration and validation of the relative differenced Normalized Burn Ratio (RdNBR) to three measures of fire severity in the Sierra Nevada and Klamath Mountains, California, USA. Remote Sensing of Environment 113: 645–646.

Miller, J.D., and B. Quayle. 2015. Calibration and validation of immediate post-fire satellite derived data to three severity metrics. Fire Ecology 11(2): 12–30.

Miller, J.D., and H.D. Safford. 2012. Trends in wildfire severity 1984–2010 in the Sierra Nevada, Modoc Plateau and southern Cascades, California, USA. Fire Ecology 8(3): 41–57.

Miller, J.D., H.D. Safford, M. Crimmins, and A.E. Thode. 2009a. Quantitative evidence for increasing forest fire severity in the Sierra Nevada and southern Cascade Mountains, California and Nevada, USA. Ecosystems 12: 16–32.

Miller, J.D., C.N. Skinner, H.D. Safford, E.E. Knapp, and C.M. Ramirez. 2012a. Trends and causes of severity, size, and number of fires in northwestern California, USA. Ecological Applications 22: 184–203.

Miller, J.D., and A.E. Thode. 2007. Quantifying burn severity in a heterogeneous landscape with a relative version of the delta Normalized Burn Ratio (dNBR). Remote Sensing of Environment 109: 66–80.

Minnich, R.A., M.G. Barbour, J.H. Burk, and J. Sosa-Ramirez. 2000. Californian mixed-conifer forests under unmanaged fire regimes in the Sierra San Pedro Martir, Baja California, Mexico. Journal of Biogeography 27: 105–129.

Mohr, J.A., C. Whitlock, and C.N. Skinner. 2000. Postglacial vegetation and fire history, eastern Klamath Mountains, California, USA. The Holocene 10: 587–601.

Moratto, M.J. 1984. California Archaeology. Academic Press, Orlando, Florida, USA.

Moreno, J.M., and W.C. Oechel. 1991. Fire intensity effects on germination of shrubs and herbs in southern California chaparral. Ecology 72: 1993–2004.

———. 1993. Demography of *Adenostoma fasciculatum* after fires of different intensities in southern California chaparral. Oecologia 96: 95–101.

Moritz, M.A. 1997. Analyzing extreme disturbance events: fire in Los Padres National Forest. Ecological Applications 7: 1252–1262.

Moritz, M.A., J.E. Keeley, E.A. Johnson, and A.A. Schaffner. 2004. Testing a basic assumption of shrubland fire management: how important is fuel age? Frontiers in Ecology and the Environment 2(2): 67–72.

Moritz, M.A., T.J. Moody, M.J. Krawchuk, M. Hughes, and A. Hall. 2010. Spatial variation in extreme winds predicts large wildfire locations in chaparral ecosystems. Geophysical Research Letters, 37: L04801.

Moritz, M.A., T.J. Moody, L.J. Miles, M.M. Smith, and P. de Valpine. 2009. The fire frequency analysis branch of the pyrostatistics tree: sampling decisions and censoring in fire interval data. Environmental and Ecological Statistics, 16: 271–289.

Morgan, C. 2008. Reconstructing prehistoric hunter-gatherer foraging radii: a case study from California's southern Sierra Nevada. Journal of Archaeological Science 35: 247–258.

Morgan, P., C.C. Hardy, T.W. Swetnam, M.G. Rollins, and D.G. Long. 2001. Mapping fire regimes across time and space: understanding coarse and fine-scale fire patterns. International Journal of Wildland Fire 10: 329–342.

Nagel, T.A., and A.H. Taylor. 2005. Fire and persistence of montane chaparral in mixed conifer forest landscapes in the northern Sierra Nevada, Lake Tahoe Basin, California, USA. Journal of the Torrey Botanical Society 132: 442–457.

NWCG. 2014. Glossary of wildland fire terminology. PMS 205, National Wildfire Coordinating Group, Boise, Idaho, USA.

Odion, D.C., and F.W. Davis. 2000. Fire, soil heating, and the formation of vegetation patterns in chaparral. Ecological Monographs 70: 149–169.

Odion, D.C., and C. Tyler. 2002. Are long fire-free periods needed to maintain the endangered, fire-recruiting shrub *Arctostaphylos morroensis* (Ericaceae)? Conservation Ecology 6(2): 4.

Parker, V.T. 1987. Effects of wet-season management burns on chaparral vegetation: implications for rare species. Pages 233–237 in: T.S. Elias, editor. Conservation and Management of Rare and Endangered Plants. California Native Plant Society, Sacramento, California, USA.

Parsons, A., P.R. Robichaud, S.A. Lewis, C. Napper, J. Clark, and T.B. Jain. 2010. Field guide for mapping post-fire soil burn severity. USDA Forest Service General Technical Report RMRS-GTR-243. Rocky Mountain Research Station, Fort Collins, Colorado, USA.

Power, M.J., J. Marlon, N. Ortiz, P. Bartlein, S. Harrison, F. Mayle, A. Ballouche, R. Bradshaw, C. Carcaillet, and C. Cordova. 2008. Changes in fire regimes since the Last Glacial Maximum: an assessment based on a global synthesis and analysis of charcoal data. Climate Dynamics 30: 887–907.

Rice, S.K. 1993. Vegetation establishment in post-fire *Adenostoma* chaparral in relation to fine-scale pattern in fire intensity and soil nutrients. Journal of Vegetation Science 4: 115–124.

Rivera-Huerta, H., H.D. Safford, and J.D. Miller. 2015. Patterns and trends in burned area and fire severity from 1984 to 2010 in the Sierra de San Pedro Mártir, Baja California, Mexico. Fire Ecology 12(1): 52–72.

Rollins, M.G. 2009. LANDFIRE: a nationally consistent vegetation, wildland fire, and fuel assessment. International Journal of Wildland Fire 18: 235–249.

Romme, W.H. 1982. Fire and landscape diversity in subalpine forests of Yellowstone National Park. Ecological Monographs 52: 199–221.

Romme, W.H., and D.G. Despain. 1989. Historical perspective on the Yellowstone fires of 1988. Bioscience 39: 695–699.

Rundel, P.W., G.A. Baker, D.J. Parsons, and T.J. Stohlgren. 1987. Postfire demography of resprouting and seedling establishment by *Adenostoma fasciculatum* in the California chaparral. Pages 575–596 in: J.D. Tenhunen, F.M. Catarino, O.L. Lange, and W.C. Oechel, editors. Plant Response to Stress. Springer-Verlag, Berlin, Germany.

Safford, H.D., J.D. Miller, and B.M. Collins. 2015. Differences in land ownership, fire management objectives, and source data matter: a reply to Hanson and Odion (2014). International Journal of Wildland Fire 24: 286–293.

Safford, H.D., J.D. Miller, D. Schmidt, B. Roath, and A. Parsons. 2008. BAER soil burn severity maps do not measure fire effects to vegetation: a comment on Odion and Hanson (2006). Ecosystems 11: 1–11.

Safford, H.D., and K.M. Van de Water. 2014. Using fire return interval departure (FRID) analysis to map spatial and temporal changes in fire frequency on national forest lands in California. USDA Forest Service Research Paper PSW-RP-266. Pacific Southwest Research Station, Albany, California, USA.

Sampson, A.W. 1944. Plant succession on burned chaparral lands in northern California. University of California Agricultural Experiment Station Bulletin 685. Berkeley, California, USA.

Scholl, A.E., and A.H. Taylor. 2010. Fire regimes, forest change, and self-organization in an old-growth mixed-conifer forest, Yosemite National Park, USA. Ecological Applications 20: 362–380.

Show, S.B., and E.I. Kotok. 1923. Forest fires in California, 1911–1920. An analytical study. U.S. Department of Agriculture Circular 243. Washington, D.C., USA.

———. 1924. The role of fire in the California pine forests. U.S. Department of Agriculture Bulletin 1294. Government Printing Office, Washington, D.C., USA.

———. 1925. Fire and the forest—California pine region. U.S. Department of Agriculture Circular 358. Washington, D.C., USA.

———. 1929. Cover type and fire control in the National Forests of northern California. U.S. Department of Agriculture Bulletin 1495. Government Printing Office, Washington, D.C., USA.

Starrs, C.F., V. Butsic, C.W. Stephens, and W.C. Stewart. 2018. The impact of land ownership, firefighting, and reserve status on fire probability in California. Environmental Research Letters 13: e034025.

Steel, Z.L., H.D. Safford, and J.H. Viers. 2015. The fire frequency-severity relationship and the legacy of fire suppression in California forests. Ecosphere 6: art8.

Stephens, S.L., and B.M. Collins. 2004. Fire regimes of mixed conifer forests in the north-central Sierra Nevada at multiple spatial scales. Northwest Science 78: 12–23.

Stephens, S.L., D.L. Fry, B.M. Collins, C.N. Skinner, E. Franco-Vizcaíno, and T.J. Freed. 2010a. Fire-scar formation in Jeffrey pine—mixed conifer forests in the Sierra San Pedro Mártir, Mexico. Canadian Journal of Forest Research 40: 1497–1505.

Stephens, S.L., J.M. Lydersen, B.M. Collins, D.L. Fry, and M.D. Meyer. 2015. Historical and current landscape-scale ponderosa pine and mixed conifer forest structure in the southern Sierra Nevada. Ecosphere 6: art79.

Stephens, S.L., L. Maier, L. Gonen, J.D. York, B.M. Collins, and D.L. Fry. 2018. Variation in fire scar phenology from mixed conifer trees in the Sierra Nevada. Canadian Journal of Forest Research 48: 101–104.

Stephens, S.L., C.I. Millar, and B.M. Collins. 2010b. Operational approaches to managing forests of the future in Mediterranean regions within a context of changing climates. Environmental Research Letters 5: 024003.

Sterling, E.A. 1904a. Chaparral in northern California. Forestry Quarterly 2: 209–214.

———. 1904b. Fire notes on the Coast Ranges of Monterey County: timber and fires. Forestry Library, University of California, Berkeley, California, USA.

Stevens, J.T., B.M. Collins, J.D. Miller, M.P. North, and S.L. Stephens. 2017. Changing spatial patterns of stand-replacing fire in California mixed-conifer forests. Forest Ecology and Management 406: 28–36.

Stuiver, M., and J. Reimer. 1993. Extended 14C data base and revised CALIB 3.014 C age calibration program. Radiocarbon 35: 215–230.

Sudworth, G.B. 1900. Stanislaus and Lake Tahoe Forest Reserves, California and adjacent territory. 21st Annual Report of the United States Geological Survey, Part V—Forest Reserves. U.S. Government Printing Office, Washington, D.C., USA.

Swetnam, T.W. 1993. Fire history and climate change in giant sequoia groves. Science 262: 885–888.

Swetnam, T.W., C.H. Baisan, A.C. Caprio, P.M. Brown, R. Touchan, R.S. Anderson, and D.J. Hallett. 2010. Multi-millennial fire history of the Giant Forest, Sequoia National Park, California, USA. Fire Ecology 5(3): 120–150.

Taylor, A.H. 2010. Fire disturbance and forest structure in an old-growth *Pinus ponderosa* forest, southern Cascades, USA. Journal of Vegetation Science 21: 561–572.

Thode, A.E., J.W. van Wagtendonk, J.D. Miller, and J.F. Quinn. 2011. Quantifying the fire regime distributions for severity in Yosemite National Park, California, USA. International Journal of Wildland Fire 20: 223–239.

Timbrook, J., J.R. Johnson, and D.D. Earle. 1982. Vegetation burning by the Chumash. Journal of California and Great Basin Anthropology 4: 163–186.

Van de Water, K.M., and H.D. Safford. 2011. A summary of fire frequency estimates for California vegetation before Euro-American settlement. Fire Ecology 7(3): 26–58.

Van Horne, M.L., and P.Z. Fulé. 2006. Comparing methods of reconstructing fire history using fire scars in a southwestern United States ponderosa pine forest. Canadian Journal of Forest Research 36: 855–867.

van Wagtendonk, J.W., and J.A. Lutz. 2007. Fire regime attributes of wildland fires in Yosemite National Park, USA. Fire Ecology 3(2): 34–52.

van Wagtendonk, J.W., R.R. Root, and C.H. Key. 2004. Comparison of AVIRIS and Landsat ETM+ detection capabilities for burn severity. Remote Sensing of Environment 92: 397–408.

van Wagtendonk, J.W., K.A. van Wagtendonk, and A.E. Thode. 2012. Factors Associated with the severity of intersecting fires in Yosemite National Park, California, USA. Fire Ecology 8(1): 11–31.

Westerling, A.L., H.G. Hidalgo, D.R. Cayan, and T.W. Swetnam. 2006. Warming and earlier spring increase western US forest wildfire activity. Science 313: 940–943.

Whitlock, C., and R.S. Anderson. 2003. Fire history reconstructions based on sediment records from lakes and wetlands. Pages 3–31 in T.T. Veblen, W.L. Baker, G. Montenegro, and T.W. Swetnam, editors. Fire and Climatic Change in Temperate Ecosystems of the Western Americas. Springer, New York, New York, USA.

Whitlock, C., and C. Larsen. 2001. Charcoal as a fire proxy. Pages 75–97 in J.P. Smol, H.J.B. Birks, and W.M. Last, editors. Tracking Environmental Change Using Lake Sediments Springer, New York, New York, USA.

Wieslander, A.E. 1935. A vegetation type map for California. Madroño 52: 191–201.

Wieslander, A.E., H.S. Yates, A.E. Jensen, and P.L. Johannsen. 1933. Manual of field instructions for vegetation type map of California. USDA Forest Service Memorandum. California Region, San Francisco, California, USA.

Williams, M.A., W.L. Baker. 2011. Testing the accuracy of new methods for reconstructing historical structure of forest landscapes using GLO survey data. Ecological Monographs 81 (1): 63–88.

CHAPTER SEVEN

Fire and Physical Environment Interactions

Soil, Water, and Air

PETER M. WOHLGEMUTH, KEN R. HUBBERT,
TRENT PROCTER, AND SURAJ AHUJA

Therefore, as fire is to air, so is air to water, water to Earth. And again, as the earth is to the water, so is water to air, and air to fire.

AGRIPPA (1913)

Introduction

Interactions of fire with soil, water, and air play an important role in the ecology of forests, shrublands, rangelands, and grasslands throughout California. Soil is a primary factor in site productivity, and the effects of burning can both enhance and degrade soil quality. Water is especially sensitive to upland environmental changes and is an excellent indicator of ecosystem condition. Fire affects water quantity and water quality both for aquatic and riparian ecosystems and for downstream human consumption. Smoke from fire impacts air quality via its constituent particles and gases, which affect human health and welfare, contribute to the atmospheric pool of greenhouse gases that are changing Earth's climate, and degrade visual range. Smoke can cross state and local jurisdictional boundaries, delivering these impacts and in some cases even affecting subsequent regulatory endpoints some distance from fire origins. By understanding the effects of fire on soil, water, and air resources, fire experts and decision-makers are in a better position to manage the consequences of burning, whether for wildfire mitigation or for developing a program of prescribed fire.

Soils develop in response to geology, topography, climate, organisms, and time. Formed over the years by the weathering of rock and the incorporation of organic material, soil is a dynamic medium that is very sensitive to ecosystem change. Soils provide the natural base for the growth of terrestrial ecosystems and the platform on which fire interactions on the landscape are played.

Hydrology is the cyclic movement of water through the landscape. Precipitation falls to the earth as rain or snow and runs off in surface streams or percolates as groundwater. Some of this water is held in the soil column and is taken up by the local vegetation. Water is returned to the atmosphere by evaporation from land and water bodies and by transpiration from the plant communities. Fire generally changes the water balance at a site by enhancing runoff at the expense of percolation.

Smoke emissions from fire can affect human health, the planetary energy balance, and visibility, though the magnitude of impact depends on the scale, the pollutant, and its concentration. Forest fire smoke is a complex mixture of different gases and particles produced from combustion of forest fuel. The management of smoke is an important aspect of fire management because of the potential effect of elevated concentrations of particulates on human health and visibility, even far removed from the fire's location. Mitigations of interstate and interregional transport of air pollutants require continued coordinated efforts among different management agencies.

This chapter will explicitly address the interactions between fire and the physical and chemical properties of soil; hillslope hydrology and erosion; stream channel hydrology; erosion and sedimentation; and air quality and pollution transport. The biological interactions between fire and soils will be addressed more in chapter 8 (Fire and Plant Interactions). The nuances of the impacts of air pollution on human communities are addressed in chapter 23 (Fire, Air Quality, and Greenhouse Gases).

Fire Interactions with Soils

Effects of fire to soil physical and chemical properties differ with fire temperature, intensity, duration, frequency, season of occurrence, and topographical location. For example, slope aspect can affect both chemical and physical soil properties (Box 7.1). Changes in soil fertility, nutrient availability, organic matter content, water infiltration, and soil mineralogy and color are among the many potential responses to burning that may either benefit or degrade soil quality. Indirectly, fire affects microbial processes such as nutrient immobilization, mineralization, and nitrification by raising soil temperature and reducing soil moisture. Loss of soil nutrients may further result from water and wind erosion, leaching losses of fire-released soil nutrients, and changes in litterfall. On the other hand, fire volatilizes much of the nitrogen deposited as the result of N deposition from urban pollution, potentially benefiting the ecosystem at the stand or the watershed scale by "resetting" the nitrogen accumulation back to very low levels (Johnson et al. 2007).

Soil Physical Properties

SOIL HEATING

Fast propagating, high energy surface fires produce relatively low soil temperatures compared to low intensity, slow spreading or smoldering fires that have a much longer residence time of 10–15 hours (Wells et al. 1979). Less heat is produced in the soil during a fast-moving fire on an uphill run than during a slow-moving backing fire (DeBano 1981).

Fig. 7.1 provides threshold temperatures for physical, chemical, and biological soil processes associated with soil heating. Most of the heat from a wildfire is directed upward, with only 8–10% of the heat produced during a fire radiating downward into the soil (Hungerford 1989). In Sierra Nevada lower montane forests, DeBano et al. (1998) reported typical fire soil surface temperatures of ~210°C (~410°F). Changes in soil physical and chemical properties require higher temperatures that are more typical of severe burning. In fact, the high temperature thresholds for the volatilization of many nutrients ranging from 775°C to 1,960°C (1,427–3,560°F) are rarely, if ever, reached in mineral soil for even the most severe of burns.

Several key biological properties such as microorganism, root, and seed survival are vulnerable at reasonably low soil temperatures. Most studies consider 60°C (140°F) as the lethal soil temperature for root mortality (Varner et al. 2009). Pietikainen et al. (2000) noted that rapid heating of the humus layer to 100°C (212°F) was lethal to most organisms in residence. Therefore, the question often arises, "does wildfire sterilize soils?" If "sterile" is defined as free from living organisms then the answer is no. Studies including Hebel et al. (2009) have shown the ability of some soil microbial species to survive severe burning. Further, the recolonization of the soil by surviving microorganisms, organisms disseminated on the surface by wind or water, and those transported upwards through the soil by ants and other organisms begins almost immediately following fire.

Mineral soil is not a good conductor of heat. Studies show that damaging soil heat, even during severe burning, is usually limited to a fairly thin layer of surface soil (Massman and Frank 2004, Busse et al. 2005), since peak surface temperatures and heat duration rapidly attenuate with depth into the soil column. This is a result of the vast array of soil pores which trap heated air reducing heat penetration. Thus, because thermal conductivity of porous materials is far lower than it is for solid materials, heat transfer is relatively low in soil (Fig. 7.2). Smoldering fires in thick duff layers or large-diameter wood are important exceptions to this rule as trapped heat can lead to considerable soil heat penetration (Hebel et al. 2009). Contiguous macropores can channel heat from fire to deeper soil horizons, and in turn, burning roots can heat soil at depths well below the soil surface. Additionally, compaction can increase maximum soil temperatures and heat duration during burning (Busse et al. 2010) by increasing the thermal contact between soil particles and reducing the air-filled pore space, resulting in a gain in soil thermal conductivity.

Soil water is considered the most important factor controlling heat transfer in soil (Hartford and Frandsen 1992). Moist soils increase the heat capacity of soil, resulting in a high-energy requirement to evaporate water prior to any substantial increase in soil temperature. Therefore, although heat travels faster in moist soil than dry soil based on the principles of thermal conductivity (Jury et al. 1991), its movement is restricted by the energy-dousing effect of water. Temperature of a wet soil will not exceed 100°C (212°F) until the water evaporates or moves deeper into the soil (DeBano et al. 1979). The high water-holding capacity and low thermal conductivity of organic matter can substantially reduce the heat pulse into soil.

BOX 7.1. HOW NORTH AND SOUTH ASPECTS AFFECT SOIL PROPERTIES IN RESPONSE TO FIRE

Canopy and vegetation—Live fuel loads are generally lower and bare ground more prevalent on S-facing slopes than on N-facing slopes. Fuels will dry out earlier in the day on south- and east-facing slopes.

Organic matter—Litter and duff thickness, and organic matter content of A horizon is generally greater on the N-facing slope. The duration of fire at the soil surface and thus soil heating is closely related to the thickness and moisture content of the litter and duff layers.

Depth of the solum (soil profile) and water holding capacity—S-facing soils generally are shallower than those of N-facing slopes. Greater weathering of north-facing slopes results in a deeper solum and finer-textured soils, and thus greater water-holding capacity. Because there is less water-holding capacity, there is a tendency for fuels to dry out quickly on S-facing slopes. But because the biomass is greater on N-facing slopes, more water is generally transpired from these sites. Thus, plants on both the north- and south-facing slopes could dry out quickly during hot, dry summers.

Slope steepness—S-facing slopes tend to be steeper than N-facing slopes (also E-facing slopes tend to be steeper than W-facing slopes). The steeper the slope, the faster the fire will burn upslope, as it preheats soil and fuel above it.

Soil temperature—Soil temperatures are higher on S-facing slopes than on N-facing slopes, and in addition postfire, darker, black ash-mixed soils absorb more heat. Increased soil temperatures can greatly reduce microbial activity.

Soil pH—The soil pH during prefire conditions is usually lower on N-facing (> weathering) than S-facing slope. There is greater change in pH on N-facing slopes postfire (more base cations are released from the litter layer raising the pH); an increase in pH allows for more optimum conditions for nitrifying bacteria, resulting in release of N.

SOIL TEXTURE AND STRUCTURE

The textures of severely burned soils usually become coarser (a decrease in clay content) as a result of the melting and fusion of clay minerals. Heating the soil above 150°C (302°F) aggregates finer clay particles into greater silt and sand-sized particles (Giovaninni and Lucchesi 1997) (Table 7.1). Alternatively, fire can make some soil textures finer by consuming

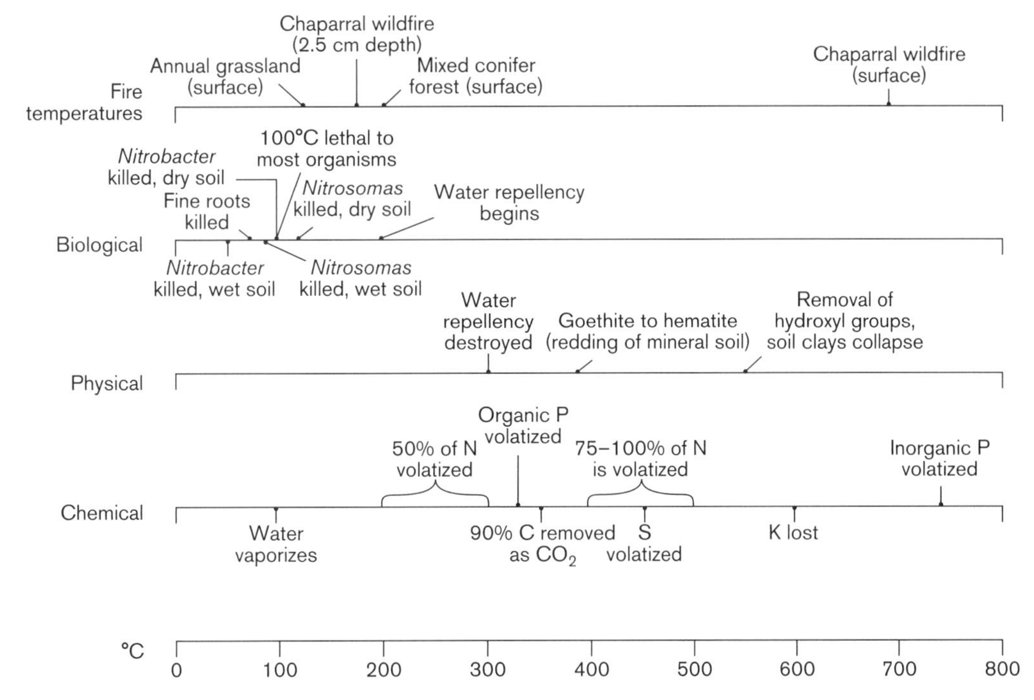

FIGURE 7.1 Wildland fire temperatures and the effects of soil heating on biological, physical, and chemical soil processes.

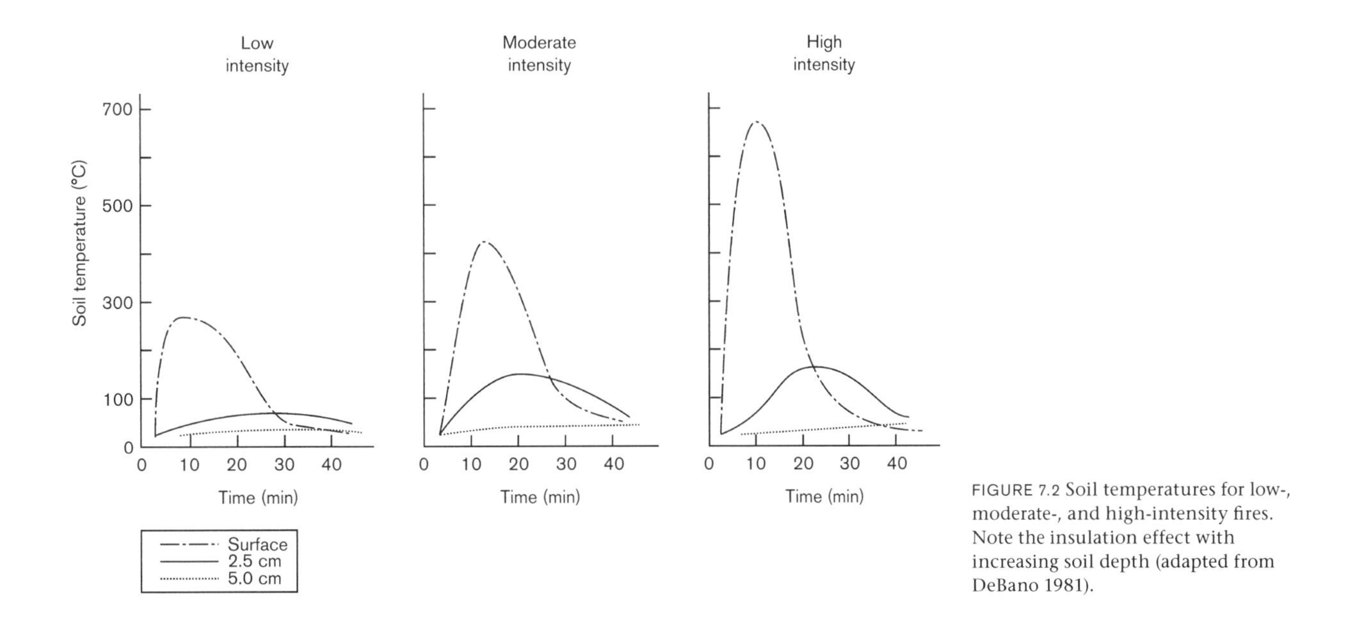

FIGURE 7.2 Soil temperatures for low-, moderate-, and high-intensity fires. Note the insulation effect with increasing soil depth (adapted from DeBano 1981).

the organic matrix that binds primary mineral grains into aggregates, thereby increasing porosity and permeability (Ulery and Graham 1993).

Fire consumes the organic matter and fungi that promote soil aggregation. Soil aggregation is promoted by organic matter, organometallic cementing agents, and microbial organisms. Fungal mycelia can surround the soil particles and intertwine among them to form aggregates with a high degree of stability. Aggregation can reduce erosion by making the soil particles more resistant to soil detachment by the forces of erosion.

SOIL BULK DENSITY, POROSITY, AND PORE SIZE DISTRIBUTION

The consumption of organic glues and fungal mycelia by fire reduces soil aggregation, thus increasing soil bulk density,

TABLE 7.1
Effects of heating by experimental fires on particle size distribution and aggregate stability at 0–2.5 cm depth of mineral soil

Heating temp. °C	Sand %	Silt %	Clay %	Aggregate Sty %
25	34	34	32	42
89	36	33	31	41
184	42	33	25	50
307	45	33	22	50
395	47	33	20	51
558	52	30	17	61

NOTE: Adapted from Giovaninni and Lucchesi (1997).

decreasing soil porosity, and reducing infiltration (Welling et al. 1984, Hubbert et al. 2006). With the loss of soil aggregation, erodibility also increases, as the finer particles are more easily entrained by overland flow (Tiedemann et al. 1979). Coarse textured soils with low carbon content generally show a lack of postfire change in bulk density. Following fire, pore size distribution can be changed as a result of surface sealing or plugging of macropores by smaller particles moving downward into the underlying aggregate soil, effectively reducing infiltration and soil permeability (Welling et al. 1984).

SOIL COLOR

Mineral soil color is influenced by the type and amount of organic matter and iron oxides present before wildfire (Ketterings and Bigham 2000). Low-intensity burns with surface temperatures of 100–250°C (212–482°F) are characterized by black ash and scorched liter and duff. Moderate burning, where surface temperatures reach 300°C to 400°C (572–752°F), consumes most of the litter, depositing gray and black ash. High-intensity burning produces surface soil temperatures in excess of 500°C (932°F), resulting in white ash after the complete combustion of fuel and the reddening of the mineral soil surface (Wells et al. 1979). Reddened soils often occur beneath large downed wood that has been consumed slowly during wildfire (Goforth et al. 2005). Blackened soil surfaces have a lower albedo (percent of incoming solar radiation that is reflected at the soil surface) and thus absorb more heat than unburned litter layers, resulting in higher soil temperatures during the day and greater diurnal temperature ranges (Neal et al. 1965).

SOIL WATER REPELLENCY

Soil water repellency is caused by hydrophobic coatings on soil particles, which reduce the affinity between soil and water, thus decreasing infiltration into the soil (DeBano 1981, Doerr and Thomas 2000). Water-repellent compounds are most commonly associated with trees and shrubs with high foliar resin, wax, or aromatic oil contents, such as evergreen plants (e.g., eucalyptus and pine species) (*Pinus* spp.), and Mediterranean shrublands (Doerr et al. 2000, Hubbert et al. 2006).

The magnitude and persistence of repellency is influenced by soil textural differences. Coarse sandy soils commonly exhibit greater repellency than clayey soils because less hydrophobic substances are needed to coat the small total surface area as compared to finer-sized clay particles that have high total surface area (Woche et al. 2005, Doerr et al. 2006). For example, surface area for a medium-sized sand is 0.0077 $m^2 g^{-1}$ (37.59 $ft^2 lb^{-1}$) (DeBano 1981), whereas clay can have a surface area as large as 800 $m^2 g^{-1}$ (3,905,944 $ft^2 lb^{-1}$) (Jury et al. 1991). Therefore, depth of repellency can extend deeper in coarse-grained soils than in fine-textured soils.

The thickness, depth, intensity, persistence, and spatial distribution of soil water repellency depends on the temperature and duration of wildfire, depth of litter and duff, litter species composition, soil moisture, and soil texture (DeBano 1981). Hydrophobic compounds, when volatilized by fire, move downward through soil along a temperature gradient and condense on cooler soil particles, usually at a soil depth between 0 and 8 cm (Huffman et al. 2001). Water repellency generally intensifies at soil temperatures between 175°C and 200°C (347–392°F) owing to a strengthening of bonding between water repellent substances and soil particles (Doerr et al. 2000).

Water repellent soils typically alternate between repellent and nonrepellent states in response to postfire rainfall patterns (Doerr and Thomas 2000). In most cases, soil water repellency increases in dry soil and either decreases or vanishes following periods of precipitation (Ritsema and Dekker 1994, Dekker et al. 1998, Shakesby et al. 2000) (Fig. 7.3). Several studies show that repellency is greatly reduced when soil moisture content exceeds 10–13% by volume (Dekker et al. 2001, MacDonald and Huffman 2004, Hubbert and Oriol 2005). Continuous layers of the nonwettable substances are generally not formed because of the interspersed nature of hydrophilic soils and rodent, worm, insect, and root activity at the hillslope or catchment scale that continuously disrupts the soil-column structure, and forms conduits for water to enter (Imeson et al. 1992, Hubbert et al. 2006).

Because soil repellency is greatly increased when soils are dry, pronounced overland flow and erosion often occur when storm events follow prolonged dry periods (Doerr et al. 2000). However, as stated by Shakesby et al. (2000), the degree of postfire reduction in water infiltration is dependent on the spatial distribution of hydrophobicity on the landscape. Most studies have only inferred a causal link between water repellency and erosion, and have failed to isolate the erosional impacts of water repellency from the confounding effects of losses in vegetation cover, litter cover, or soil aggregate stability (Doerr et al. 2000, Hubbert et al. 2006). Additionally, surface water ponding due to repellency may increase water availability to deeper rooted chaparral species, when live roots, old root channels, and cracks serve as penetrating points for preferential flow of water to deeper soil layers (Scott and Burgy 1956). Most wildfires produce a mosaic of non-, low-, moderate-, and high-repellency soils across the landscape. The mosaic pattern is a function of nonuniformity of fire temperature and duration, forest floor consumption, vegetation type, and soil moisture and texture (Hubbert et al. 2006). Therefore, one should not expect the spatial distribution of soil water repellency to be uniform across a broad landscape, or even at smaller scales in steep watersheds (Robichaud and Miller 1999).

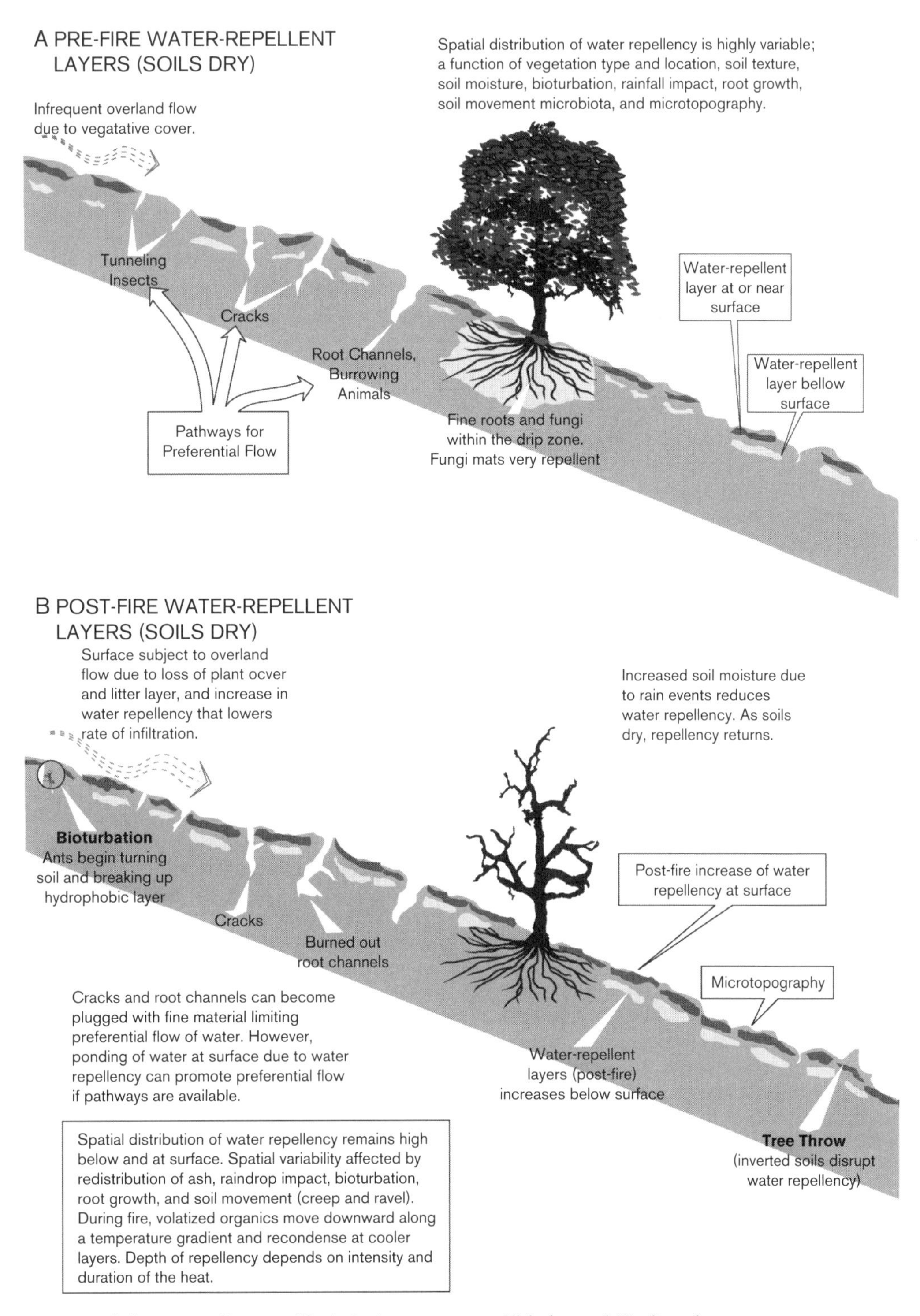

FIGURE 7.3 Soil water repellency and hydrologic consequences (A) before and (B) after a fire.

Soil Chemical Properties

The direct effects of fire on soil chemistry include oxidation of litter and organic horizons, increased mineral mobility caused by ashing, heat-induced changes in soil microflora, and nutrient volatilization, especially nitrogen (N). Indirect fire effects on nutrients include changes in soil moisture and temperature caused by decreased shading and precipitation interception by vegetation and litter; changes in rates of organic matter decomposition and other microbial processes; and selective mortality of soil biota caused by heat. Other indirect effects include changes in N_2 fixation rates, mycorrhizal relationships, and hydrologic export of nutrients by way of mobilization and water movement through the ecosystem (Overby and Perry 1996). Combustion releases nutrients stored in live and dead plant biomass and makes them readily available to the soil (Johnson et al. 2005). However, the majority of the fire generated nutrient flush is transported offsite by the actions of wind, runoff, leaching, and volatilization (Wells et al. 1979).

LITTER, DUFF, SOIL ORGANIC MATTER, AND BLACK CARBON

Litter and duff is important to the health and productivity of shrublands and forested areas because of its nutrient and water content, its influence on physical, chemical, and biological characteristics, and its ability to be a medium for fine roots and mycorrhizae. Fire oxidizes surface organic matter as temperatures exceed 200°C (392°F) (Johnson et al. 2004), releasing carbon dioxide (CO_2) and associated combustion gases such as carbon monoxide (CO), fine particles, charcoal, and ash. Percent loss of carbon is dependent on fire temperature and intensity, amount of organic matter incorporated into the mineral soil, and the nature and structure of preexisting vegetation. Rau et al. (2009) found an increase in soil C following low-severity burning that was likely due to the incorporation of ash, charcoal, and partially burned organic matter into the mineral soil. At 220°C (428°F), Fernandez et al. (1997) reported that 37% of carbon was lost, and at 350°C (662°F), 90% of carbon was lost. Because microorganisms need C for formation of new cells and C is now in short supply, bacteria and fungi (their mass high in N) die and decompose, releasing the N to the soil solution (Donahue et al. 1983).

Fire-derived charcoal (black carbon) can greatly improve soil productivity by providing a reservoir of essential plant nutrients, improving soil water-holding capacity, detoxification of plant- and microbial-inhibiting compounds, and soil-warming capability. Approximately 0.7–3.0% of burning organic matter is converted to charcoal, but spatial distribution in soil charcoal content is high within a fire perimeter (Ohlson et al. 2009), suggesting that charcoal may act primarily at localized "hotspots" to improve soil quality. Charcoal's aromatic ring structure provides high internal porosity and surface area, resulting in a high capacity for water, nutrient, and chemical sorption (Preston and Schmidt 2006). DeLuca et al. (2006) showed that amending soils from ponderosa pine forests with NH_4 and field collected charcoal significantly increased nitrification potential, net nitrification, and gross nitrification.

NITROGEN (N)

Long periods of fire exclusion result in wildlands with N-limiting conditions (Kimmins 1996). Soil N is taken up and held in the organic form resulting in high C/N ratios, and thus only a small portion of total N is available for plant uptake. Soil solution ammonium (NH_4) and nitrate (NO_3) are the primary inorganic N compounds assimilated by plants. Without fire, their concentrations are usually low and vary with seasonal trends in plant uptake, leaching loss, and N mineralization, nitrification, and immobilization rates (Fisher and Binkley 2000).

Wildfire can result in two major changes in soil N: (1) loss of total N to the atmosphere during the consumption of litter and duff layers, and (2) a pulse of plant-available N due to downward movement and condensation of fire volatilized inorganic N (Busse et al. 2014).Simultaneous loss of total N and gain of inorganic N act a double-edge sword, with the two processes linked such that burns resulting in a high loss of total N also yield the greatest increase in plant-available N (Wan et al. 2001). Wildfire acts to jump start the soil inorganic N pool. Ammonium concentrations are usually elevated immediately after burning and are then converted to NO_3 by nitrifying bacteria within several months (Prieto-Fernandez et al. 1998, Busse et al. 2014). In chaparral, Christensen (1973) reported that soluble forms of nitrogen, ammonium (NH_4^+–N) and nitrate (NO_3^-–N), were higher in burned soil as compared to unburned soil. Choromanska and DeLuca (2001) noted initial postfire concentrations of NH_4^+–N for wildfire of 19.5 $\mu g\ g^{-1}$ (ppb) as compared to the unburned control of 1.9 $\mu g\ g^{-1}$ (ppb).

Postfire conditions favor nitrogen fixation by free-living microorganisms (commonly rhizobia or actinomycetes associated with roots of the host plant that supplies the energy for fixing N) (Johnson et al. 2007). Nitrogen fixation is the conversion of molecular N (N_2) to ammonia (NH_3) and, in rapid order, to organic combinations or to forms readily usable in biological processes (Donahue et al. 1983). In the Sierra Nevada of California, whitethorn (*Ceanothus cordulatus*), with nitrogen-fixing symbionts in root nodules, and broadleaf lupine (*Lupinus latifolius*), an herbaceous perennial legume, are both colonizers of recently burned soils and can be expected to add significantly to total soil nitrogen (St. John and Rundel 1976).

Most N lost during fire originates from litter and duff, not mineral soil (Johnson et al. 2007, Busse et al. 2014). For example in the Sierra Nevada, Moghaddas and Stephens (2007) found no significant changes in mineral soil N content even though litter amounts of up to 723 kg ha^{-1} N (645 lbs ac^{-1}) were volatilized from the forest floor. Site conditions that promote high forest floor content, such as (1) high litterfall rates in dense forest stands, (2) slow decomposition rates in cold or dry climates, and (3) a lack of recent forest floor disturbance from burning, contribute to the greatest potential for N loss (Busse et al. 2014). By implementing fuel reduction programs such as prescribed burning, management can mitigate the buildup of litter and duff layers over time and reduce high N loss during high-severity fires.

An exception to the concerns about N loss during burning is in areas saturated with nitrogen from atmospheric deposition. Typically adjacent to urban centers, these areas can actually benefit from wildfires and prescribed burning in reducing critical loads of N from the soil surface. In fact, at the current levels of N critical loads, several ecosystem types (coastal sage scrub, annual grasslands, and desert scrub vegetation) are at risk of type-conversion because N enrichment favors invasion by exotic annual grasses (Gimeno et al. 2009, Fenn et al. 2010).

PHOSPHORUS (P)

The majority of phosphorus is lost during volatilization of the litter and biomass (Giovaninni and Lucchesi 1997). However, volatilized P from the litter and surface organic matter can move down the soil profile along a decreasing temperature gradient, subsequently condensing on the intervening soil particles (Overby and Perry 1996). As soil organic P

TABLE 7.2
The effect of experimental fires on some chemical properties of the 0–2.5 cm depth of soil

Heating temp. °C	CEC meq/100	OM %	N-NH4+ mg/kg	P-Org. mg/kg	P-Avail. mg/kg
25	44	26	40	208	39
89	40	25	34	204	40
184	36	21	60	184	50
210	36	20	33	160	50
312	36	15	50	70	88
395	29	13	52	60	92
457	27	12	52	40	95
558	23	12	54	30	103

NOTE: Adapted from Giovaninni and Lucchesi (1997).

is thermally mineralized, available P increases with the rising heating temperature (Giovaninni and Lucchesi 1997) (Table 7.2).

SOIL pH AND CATION EXCHANGE CAPACITY (CEC)

During combustion, base cations are mineralized from surface fuels leaving behind a nutrient-rich ash layer. Precipitation leaches the cation-laden ash into the soil, providing a flush of nutrients for new growth (Wells et al. 1979). Soil pH increases when the cations displace the hydrogen (H) and aluminum (Al) ions adsorbed on the negative charges of the soil colloids. A postfire increase in pH may be very beneficial for a low pH soil because of its positive effect on nutrient availability, whereas fire-induced changes may be biologically negligible in neutral pH soils (Franklin et al. 2003). Greater fuel consumption leads to greater release of cations in ash and thus greater pH change, whereas highly buffered soils such as clays are capable of absorbing postfire increases in available cations without concomitant changes in pH.

Both clay particles and SOM have high surface areas and negatively charged sites that confer high CEC. Boerner et al. (2009) noted only modest effects of low- to moderate-severity fire on cation availability and which were limited mainly to increased Ca concentrations. St. John and Rundel (1976) reported that burning dramatically increased exchangeable calcium (Ca), as most of the Ca from litter being deposited on the mineral soil. Because ash is highly soluble and most compounds are removed from the soil following the wet season (Tiedemann et al. 1979), postfire increases in pH and CEC are often short-lived, returning to pre-burn levels within 1 to 3 years (Franklin et al. 2003).

Fire Interactions with Water

Fire alters the water balance on burned landscapes and changes the hydrology of the hillslopes, the stream channels, and whole watersheds. These hydrologic changes in turn affect the erosion and sedimentation response of postfire environments at all spatial scales.

Hillslope Hydrology

The disposition of water on hillslopes reflects a balance between the inputs and outputs of the hydrologic cycle (Fig. 7.4). Hillslope hydrology is radically different after a fire, in large part because of altered soil conditions of structure and water repellency mentioned in the foregoing section. These hydrologic changes have serious implications for hillslope erosion and the delivery of sediment and water to the stream channels.

INTERCEPTION, SNOW ACCUMULATION, AND SNOWMELT

Some of the precipitation falling on a watershed never reaches the ground. Interception is that portion of rain (or snow) that is retained in the plant canopy until it evaporates. Interception in unburned vegetation may be up to 11% in shrublands (Rowe and Colman 1951) and as high as 27% in evergreen trees (Xiao and McPherson 2011). Interception reduces raindrop impact, soil water storage, and transpiration. By removing the vegetation, fire reduces interception, thereby potentially increasing water on the hillsides (Baker 1990).

The rates of snowfall accumulation and snowmelt are not uniform across the landscape. Apart from obvious differences in elevation and aspect, snow accumulation is generally inversely proportional to vegetation cover. Thus, snow depths would be enhanced on severely burned sites where the vegetation and the consequent interception have been removed. Fire increased the accumulation of snow on one site in the Sierra Nevada compared to neighboring unburned areas, with the effects lasting over five years (Micheletty et al. 2014). However, accumulation would be less in large bare areas that were subject to snow removal by wind scour (Tiedemann et al. 1979).

Burned landscapes also modify snowmelt. The open areas are subject to more solar radiation that increases melting and reduces snow depths (Gleason et al. 2013), leading to earlier snowmelts. With the removal of vegetation, burned areas are susceptible to extremes of surface temperatures, also affecting soil freezing and snowmelt rates (Baker 1990). Furthermore, scorched ground in burn areas may increase the long-wave solar re-radiation into the snowpack (Tiedemann et al. 1979), thereby enhancing snowmelt.

INFILTRATION

Perhaps the most important hydrologic change in postfire environments is the reduction in soil infiltration. Fire affects infiltration by altering ground cover, soil properties, and thermal regimes. On forested sites, the litter and duff layer can act like a sponge during rainfall, and if it is consumed by fire, the potential for infiltration will be reduced (Wells et al. 1979). Often the changes in infiltration mirror the changes in soil texture and porosity examined previously. The temperature extremes on burn sites can also promote more soil freezing that will reduce infiltration (Tiedemann et al. 1979).

Other soil properties can strongly influence postfire infiltration, as noted above. Water repellent soils will initially sharply reduce infiltration on burned sites, although the water repellent substances are slightly water soluble and will eventually wet up and infiltrate normally (DeBano 1981).

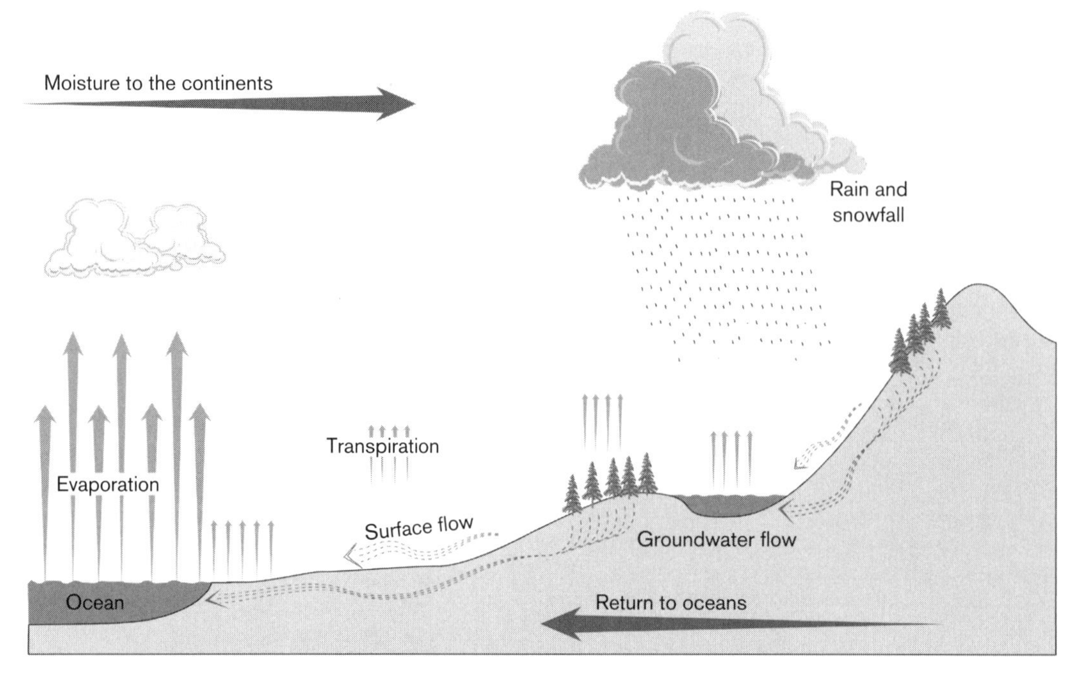

FIGURE 7.4 The hydrologic cycle. Water moves from oceans to continents and back again.

Pore clogging by fine ash particles is another process that reduces infiltration in postfire environments (Woods and Balfour 2008). The ash is easily mobilized by the initial rains following a fire and can plug the preferential flow pathways into the soil. Postfire reduction in infiltration, by whatever mechanism, has been documented in forests (Robichaud 2000), in rangelands (Pierson et al. 2008), and in shrublands (Hubbert et al. 2006).

EVAPOTRANSPIRATION

Generally, site evaporation increases but transpiration by plants is sharply curtailed after a fire. The greater daytime temperature extremes experienced in the open burned areas promote greater evaporation. Postfire evaporation is also enhanced by the loss of the shading and protection from the plant canopy (Ahlgren and Ahlgren 1960). Conversely, water repellent substances can reduce evaporation by disrupting the process of capillary rise (DeBano 1981).

Postfire transpiration is negligible if the vegetation is completely consumed. Partially burning a watershed unit should reduce transpiration in proportion to the percentage of totally burned area (Rowe and Colman 1951). Light and moderately burned sites will have transpiration reduced to the degree that foliage is consumed and roots are killed. Even with initial vegetation recovery, site transpiration rates will be reduced, as leaf area is limited and the herbaceous fire-following plant communities lack great rooting depths to significantly dewater the soil mantle (Rice 1974).

OVERLAND FLOW

Hillslope surface runoff, or overland flow, occurs when the soil becomes saturated or when the precipitation intensity exceeds the rate of infiltration. On unburned areas with a complete canopy and/or litter cover, overland flow is a rare event, produced only during the most intense rains. However, in postfire environments where infiltration and water storage capacities are severely reduced by altered canopy and soil properties, overland flow can be substantial (Tiedemann et al. 1979, DeBano 1981). Rice (1974) estimated that surface runoff could be as much as 40% of the rainfall on a freshly burned chaparral site in southern California. In their review of fire in forested ecosystems, Ahlgren and Ahlgren (1960) reported that burned watersheds in the Sierra Nevada increased overland flow by 31 to 463 times the unburned rate.

Hillslope Erosion

The fire-induced changes in soil conditions and hillslope hydrology discussed above can produce dramatic increases in postfire hillslope erosion. Fire renders the landscape more sensitive to erosional forces, and large storms can result in substantial quantities of rock and soil being stripped off the hillslopes and delivered to the stream channels below.

Hillslope erosion generally increases following a fire, but the amount of accelerated erosion depends on the complex combination of site conditions, fire characteristics, and rainfall patterns. Burning increases surface erosion by removing the protective barriers of vegetation and litter, thereby increasing the erosional effectiveness of rainsplash, wind scour, and overland flow. Fire also alters soil properties (structure, porosity, and water repellency) that enhance hillslope erosional processes (Wells et al. 1979, DeBano 1981).

Accelerated erosion begins while the fire is still burning. As organic barriers on the ground surface are consumed, trapped soil and rock material that have accumulated behind these dams are liberated and move downhill under the influence of gravity. This pulse of dry ravel (Rice 1974) is more pronounced

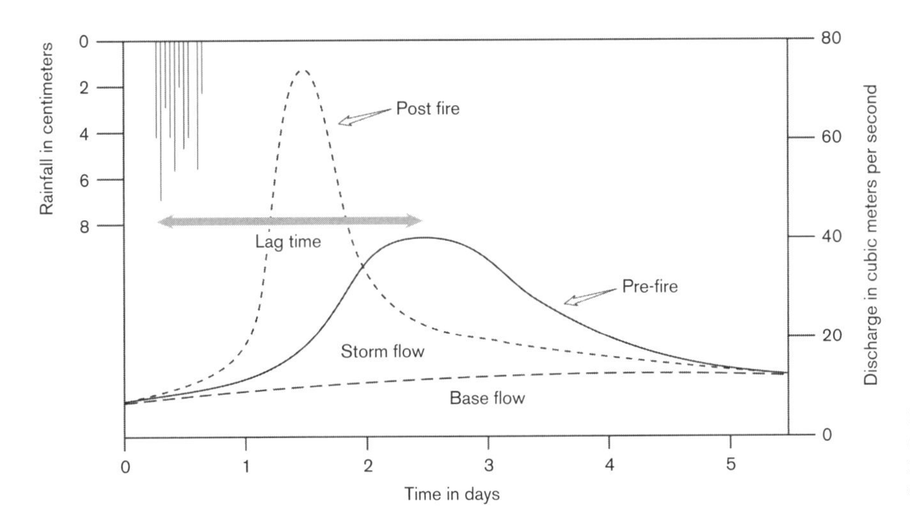

FIGURE 7.5 Streamflow hydrograph (discharge plotted against time) following a rainstorm showing lag time and peak flow changes with fire events.

in steep terrain, and may continue for days or weeks after the fire until a new geomorphic equilibrium is reached.

A second flush of hillslope erosion ensues with the first postfire rains, when gravity is augmented by hydrologic processes. Overland flow, often promoted by restricted infiltration, is an especially effective erosional process. Well-developed networks of micro-channels or rills are a distinctive feature on burned hillslopes (Wells 1981).

Postfire erosion levels can increase by several orders of magnitude over pre-burn rates. Fire increased both wet and dry hillslope erosion by a total of 17 times over pre-burn levels in southern California chaparral (Krammes 1960). Ahlgren and Ahlgren (1960) reported that postfire erosion in the Sierra Nevada was 2 to 239 times greater than pre-burn rates. With vegetation regrowth and the depletion of easily mobilized surface material, hillslope erosion attenuates with time since fire and can return to pre-burn rates within two to three years (Heede et al. 1988, Wohlgemuth et al. 1998).

Hillside slopes are the ultimate source of nearly all sediment produced from a watershed. This accounts for the vast majority of postfire mitigation efforts—primarily covers and barriers—being directed at the hillslopes themselves (Robichaud et al. 2000; see chapter 22). This recognition of the importance of hillslope erosion has led to many modeling efforts to predict postfire hillside sediment production for planning and risk assessment (Robichaud 2005, Moffet et al. 2007, Miller et al. 2011).

Channel Hydrology

Fires can directly impact stream channels by killing the in-stream vegetation and changing riparian habitats. However, the most direct consequence of burning on streams is the extra sediment and water delivered from the hillslopes. The attendant changes in water quantity, water quality, and erosion and sedimentation can drastically alter the character of stream channel systems.

WATER QUANTITY

In the postfire environment there is a change in watershed hydrology. Increased hillslope overland flow—from the reduction in vegetation cover, transpiration, and infiltration—generates more streamflow and downstream flooding. These changes increase the contributing area of hillslope delivery to the channels, and the rill networks quickly convey water off the slopes into the streams (Wells 1981). Thus, in postfire environments, a larger number of storms produce streamflow than before the burn, with greater total water yield, and with intermittent streams often becoming perennial (Tiedemann et al. 1979).

Peak flows are the greatest instantaneous discharge rates experienced in stream channels for individual flood events, while storm flows are the total amounts of water produced by a rainstorm or a related series of storms (Fig. 7.5). Burning has been shown to increase peak flows in California forests and chaparral (Anderson 1949), in Arizona chaparral (Rich 1962), and in Australian scrub forests (Brown 1972). Using comparable storms both before and after a high-severity fire in southern California chaparral, Krammes and Rice (1963) were able to document that postfire peak flows were 200–800 times greater than unburned levels. Tiedemann et al. (1979) reported that extreme increases in postfire peak flows could be up to 10,000 times those in unburned watersheds. Increases in peak flow are generally inversely proportional to watershed area. Thus, the largest increases occur in the smallest catchments. Light and moderate-severity burns can produce peak flows that are 5–100 times greater than from unburned conditions (Tiedemann et al. 1979).

Storm flows also increase after burning. In one of two adjacent watersheds with extensive prefire streamflow records in southern California chaparral, fire increased storm flow by three to five times compared to its unburned companion (Colman 1953). Brown (1972) and Tiedemann et al. (1979) reported similar postfire storm flow responses. Storm flows remain elevated for three years following a fire and may persist for 5–10 years (Baker 1990).

Base flows are the background streamflow levels over which the rainy season storm flows are superimposed (Fig. 7.5). Base flows are typically evaluated at the end of the dry season. Burned watersheds have been shown to increase base flows compared to prefire conditions or to similar unburned catchments (Tiedemann et al. 1979). Base flow increases have been specifically attributed to the reduced transpiration of fire-killed riparian vegetation (Colman 1953), but increased base flow would also be expected to result from the consumption of all vegetation across the burned area.

Flow timing refers to the shape of the streamflow runoff curve (or hydrograph). Fires can alter hydrograph shape by producing multiple storm peaks and reducing the lag time from rainfall initiation to peak flow (see Fig. 7.5). Brown (1972) reported secondary flow peaks in a burned watershed that he attributed directly to the fire. Lag time to peak flows are shorter in burned catchments because water is shed rapidly off the hillslopes and more quickly delivered to the streams. Reduced lag times can also be the result of earlier snowmelts (Tiedemann et al. 1979).

WATER QUALITY

Not only is the amount and timing of streamflow affected by fire, so is the character of the water itself. Changes in turbidity, temperature, and stream chemistry may alter water quality, sometimes to the detriment of aquatic ecosystems and downstream water supplies.

Turbidity is a measure of the suspended solids in streamflow. Stream discharge from unburned watersheds is normally clear, barring any ground disturbing activity. On the other hand, streamflow from burned catchments is typically laden with ash and fine soil grains, giving the water a sooty or muddy appearance. Increased turbidity is the most dramatic postfire water quality response. However, turbidity is difficult to quantify, being highly transient and extremely variable. The source of the stream sediment load is primarily soil stripped off the hillslopes, but also includes material scoured from the channel bed and banks (Tiedemann et al. 1979, Baker 1990).

Stream temperatures are elevated not directly by the fire itself but with the loss of protective shade in the postfire environment. Temperature increases depend on the topography (governing exposure), distance from the water source, and size of the burned area. Measured summer stream temperatures increased by 1.4–2.2°C (2.5–4.0 °F) in response to a fire in a forested landscape in Montana compared with unburned sites, but these changes disappeared within 1.7 km (1.1 mi) downstream from the burned area (Mahlum et al. 2011). However, elevated temperatures are partially offset by increased streamflow (Tiedemann et al. 1979).

Changes in postfire water chemistry can be tremendous. Nutrients and soluble compounds in the ash are flushed off the hillslopes and leached through the soil during the first postfire storms and are incorporated into the streamflow. Although much of the nutrient load is attached to sediment particles and is exported with the suspended sediments (Tiedemann et al. 1979), fires can significantly increase solute concentrations in the stream water as well. Concentrations of these solutes remain elevated long after sediment transport has returned to pre-burn levels (Williams and Melack 1997). Fire caused increased stream nitrate concentrations in a southern California watershed impacted by chronic air pollution compared to a nearby unburned site (Riggan et al. 1994), and these levels remained elevated for 7 to 10 years (Meixner et al. 2006). Fire severity can also affect nitrate mobilization. Annual nitrate loss from a severely burned watershed was 40 times that of an unburned area, while loss from a moderately burned catchment was only seven times the unburned site (Riggan et al. 1994). Compared to prefire rates, postfire concentrations and loads of selected trace metals were up to three orders of magnitude greater in a southern California stream system (Burke et al. 2013). Stephens et al. (2004) reported that stream water calcium concentrations increased in burned watersheds in the Tahoe basin while soluble reactive phosphorus concentrations did not change. Monitoring indicated that the water quality effects in this latter study lasted for approximately three months.

Channel Erosion and Sedimentation

Stream channels are very dynamic systems that erode, transport, and deposit sediment in a quasi-continuous fashion, eventually routing hillslope-derived rock and soil to the watershed outlet. Fire greatly accelerates the activity rate of this process, and can alter riparian structure and geomorphology. Channel stability is often a function of the material making up the bed and banks. Alluvial channels, composed of sand-sized grains, are very mobile and respond quickly to changes in flow regime or sediment load. Conversely, bedrock channels are very stable and are only modified by the highest magnitude floods (Leopold et al. 1964). Channels composed of cobbles or boulders show little morphologic changes in the absence of fire (Doehring 1968).

With the changes in soils, hillslope hydrology, and surface erosion following a fire, the generally observed pattern of channel fill and subsequent downcutting is well documented (Doehring 1968, Heede et al. 1988, Florsheim et al. 1991). Hillslope sediments are delivered to the channels first by dry ravel and second by overland flow with the onset of the rainy season. Initial postfire storms laden with hillslope sediment cause channel filling, while later storms tend to scour the channels, as hillslope sediment supply becomes depleted. However, in alluvial channels and on steeper channel gradients, initial scour can be followed by subsequent filling (Doehring 1968). The increased sediment supply can cause general channel bed aggradation and construction of alluvial fans. Moreover, postfire channel sedimentation will preferentially fill stream pools, smoothing both the longitudinal and transverse profiles (Keller et al. 1997).

Channel scour can also occur by postfire debris flows. Debris flows, a form of mass erosion, are slurries of water and sediment that have incredible erosive power. Some debris flows originate on hillslopes as transformed ground-saturated landslides. But, as noted above, these are not postfire features. Instead, postfire debris flows form and propagate exclusively in the stream channels, as stored sediment is mobilized by the influx of water derived from hillslope overland flow (Wells 1987). Despite their impressive ability to alter channel morphology, debris flows are relatively rare events. Postfire channel flushing is more likely to occur by normal stream transport processes than by high magnitude debris flows (Keller et al. 1997).

Postfire channel sedimentation, subsequently scoured, will eventually migrate downstream out of the burned area. These postfire sediments are periodically flushed by flood flows and temporarily held in storage sites until they reach the outlet of larger watersheds (Heede et al. 1988). This transported sediment can alter riparian habitats and can reduce channel capacity, thereby increasing downstream flood risk (Keller et al. 1997).

Watershed Hydrology and Erosion

The effects of fire on water at a landscape level can be ascertained in studies of whole watersheds. Catchments integrate the local vagaries of fire response into general trends in water yield and sediment yield.

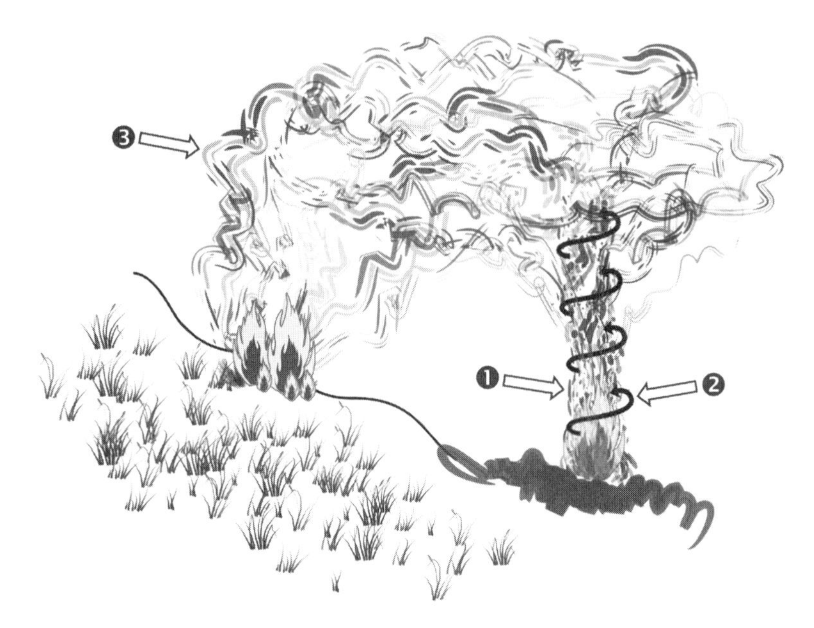

FIGURE 7.6 Illustration of smoke plume dynamics for a hot fire. A central convective column (1) occurs due to heated air becoming more buoyant than the surrounding air. As this air rises it may become a counterrotating vortex (2) that develops strong ground winds as air is pulled up from the ground into the plume. The rapidly developing smoke plume can also form downdrafts (3) that ignite new fires at distant points from the original fire.

WATER YIELD

Fire increases catchment water yield, as might be expected in light of the foregoing discussions of hillslope hydrology and streamflow quantity. Whether comparing the response of adjacent burned and unburned watersheds (Hamilton et al. 1954, Brown 1972) or the response of similar storms in watersheds before and after a fire (Krammes and Rice 1963), burning enhances water yield. Elevated levels of water yield can persist for years after a fire (Rowe et al. 1954, Baker 1990).

SEDIMENT YIELD

Because of accelerated hillslope erosion and enhanced channel transport, fire also increases watershed sediment yield (Anderson 1949, Hamilton et al. 1954, Brown 1972). The source of the sediment comes from both hillslope material and mobilized stream channel deposits. The timing of postfire sediment yield can vary as a function of watershed size. Smaller catchments respond very quickly, but because of temporary storage, there may be a considerable lag time before sediment arrives at the outlet of larger watersheds (Heede et al. 1988).

Documented rates of postfire sediment yield vary considerably, depending on topography, soil characteristics, vegetation type, fire severity, and precipitation patterns. In forested environments, postfire sediment yields can exceed unburned levels by as much as 50 times (Anderson 1949). Burned watersheds in southern California chaparral produce an average of 35 times more sediment than unburned watersheds (Rowe et al. 1954), but can experience 1,000 times more sediment yield than comparably-sized unburned catchments (Kraebel 1934). Postfire sediment yields can remain elevated for 8–10 years after burning (Rowe et al. 1954).

Fire Interactions with Air

Fire affects air quality by introducing smoke and the residues of combustion into the atmosphere. These byproducts of burning can be transported to areas distant from the fire and may affect human health, greenhouse gas (GHG) budgets, and visibility. Wood smoke can increase particulate pollution, contributing to violations of air quality standards and potentially impacting regulatory attainment status. Environmental Protection Agency (EPA) regulations address pollutants as emitted to the atmosphere and not the loading deposited on the landscape. The research community and land management agencies have been working to understand "critical loads" or thresholds of sensitive ecosystems to air pollution exposure or loading.

Air Quality

Smoke is produced as fires burn live and dead vegetation. Wildland fires result in combustion of many types of biomass, and smoke composition varies with topography, weather and fuel density. Emissions from prescribed fires and wildfires contain a number of air pollutants, including particulate matter (PM), carbon monoxide (CO), nitrogen oxides (NO_x), volatile organic compounds (VOC), and certain toxic pollutants (DeBano et al. 1998, Goldammer et al. 2009). Toxic air pollutants include acrolein, benzene and formaldehyde. Climate impacting gases such as carbon dioxide (CO_2), methane (CH_4), N_2O, and black carbon are generated in smoke. The quantity and type of emissions produced depend on the amounts and types of vegetation burned, its moisture content, and the temperature and altitude of combustion (US EPA 1998).

As biomass is consumed, thermal energy is released. This energy creates buoyancy that lifts smoke particles and other pollutants above the fire. In hot fires (Fig. 7.6) buoyancy can result in a fast rising central convective column with counterrotating vortices that entraps the surrounding air. This active column may also produce turbulent downdrafts that ignite new fires distant from the original fire. If fire intensity is lower, then the central column may not form, or it may collapse as fire intensity diminishes. The result is smaller convective cells that do not carry smoke as high into the atmosphere. Low-intensity smoldering fires produce little buoyancy, often resulting in smoke that fills local valleys and basins at night (Sandberg et al. 2002).

Once aloft smoke plumes are subject to complex chemical reactions. These reactions depend on dilution rates, photolysis, position, altitude, and temperature. Dilution occurs when dense smoke mixes with the surrounding "clean" air. Chemical reactions within smoke plumes decrease as smoke gets diluted with this air. For instance, increasing dilution by a factor of 2 will decrease the reaction rate by a factor of 4. Photolysis reactions are chemical reactions that occur in the presence of sunlight, in which molecules are split into their constituent atoms. These reactions occur largely at the top of plumes, especially when thermal inversion layers of air limit plume height. As plumes are diluted with surrounding air, sunlight penetrates deeper in the plume and photolysis reaction rates increase. Photolysis of NO_2 can contribute to ozone development, particularly high in the plume.

Concentration also influences particulate formation. Particles can additionally form as a result of nucleation of gases (molecules of gas combining together) or condensation (entrapment of gases in water droplets). These processes commonly occur as the plume temperature cools, thus plume height, temperature, and concentration may act in combination to control the total particulate and trace gas composition and concentration. Particulates are discussed in more detail in chapter 23.

Downwind air also mixes with the plume, which in California typically includes urban generated air pollution. These urban air pollution sources are predominantly the result of gasoline and diesel combustion. Inorganic compounds including compounds that contain nitrogen (ammonia) are injected into the atmosphere along with light hydrocarbons and particles. Ozone (O_3) is formed by photolytic reaction during transport of these air masses away from urban centers.

Several biomass combustion and secondary products are classified as criteria pollutants (pollutants with established air quality standards) under the Clean Air Act, including O_3, CO, NO_x, and PM (see Table 23.4, chapter 23). Under the current regulatory framework, prescribed fire (planned fire) is generally considered an anthropogenic, as opposed to natural, source of air pollutants by state and regional air quality agencies. However, EPA has developed a policy that recognizes the role of fire in maintaining healthy ecosystems (US EPA 1998). Nevertheless, California Air Pollution Control Districts (APCDs) continue to consider most planned and unplanned fire emissions to be subject to control strategies designed to help meet state and federal air quality standards.

Pollution Transport

In California, prevailing ground-level winds are northwesterly because of the geotropic balance between a subtropical anticyclone over the east Pacific Ocean and thermal low pressure over the desert during the summer (see chapter 2). In gaps along the Coastal Ranges, the onshore winds create air streams flowing into the inland valleys. Warm boundary-layer air masses overlying these valleys in summer are stratified by weak thermal inversions at 1,000–1,300 m (3,280–4,265 ft). Daylight land heating generates local anabatic winds (upslope flows) from the southwest and west along the western slopes, which transport air masses into the interior mountain ranges.

The onshore winds and boundary-layer air masses contribute to complex plume transport characteristics in California. Advection (lateral movement or diffusion) is often easterly for low-elevation fires, but can quickly change with prevailing wind patterns. The inversion layer of air also often limits upward diffusion of the plume and reduces dissipation of the smoke plume in the upper atmosphere. Smoke from higher-elevation fires, such as in the Sierra Nevada, is less affected by thermal inversions but may burn at lower intensities for longer duration. Dispersion is also less predictable in mountainous areas as inversions that normally constrain upward diffusion weaken due to complex high-elevation topography. When this occurs, plumes penetrate to higher-elevation air mass layers and may be transported hundreds or thousands of kilometers before the plume dissipates.

Smoke Effects on Ecosystems

Chemicals generated by smoke affect vegetation in a variety of ways. Smoke may reduce photosynthetic efficiency at low levels, and smoke may also protect vegetation from fungal infection. After plant surfaces were exposed to smoke, germination and growth of several fungi, including annosus root rot (*Heterobasidion annosum*), were reduced in laboratory studies (Parmeter and Uhrenholdt 1976). This suggests that smoke may impact microbial communities that exist in forest and shrub ecosystems.

Smoke has also been shown to induce germination of many annual species in chaparral ecosystems. Smoke and charred wood induced germination in annual and some perennial species (Keeley and Fotheringham 1998). Some responses have been linked to exposure to NO_x, which induces germination in several annual species (Keeley 2000), but others appear to be caused by chemicals present both in smoke and charred wood. In addition, some species have a long-term storage requirement before they will germinate in response to smoke exposure.

Ozone is produced both in smoke and from urban-generated air pollution sources. Ozone is taken into the plant through the stomates, and due to its strong oxidizing capacity it reduces photosynthesis rates, decreases foliar chlorophyll, and accelerates foliar senescence. Field and fumigation chamber studies of conifer species have indicated that ponderosa pine, Jeffrey pine (*Pinus jeffreyi*), white fir (*Abies concolor*), Coulter pine (*P. coulteri*), incense cedar (*Calocedrus decurrens*), bigcone Douglas-fir (*Pseudotsuga macrocarpa*), and sugar pine (*P. lambertiana*) are susceptible to O_3 in California (Miller et al. 1983).

Negative impacts of atmospheric N deposition may also be occurring in California's ecosystems. While small increases of N deposited from smoke act to stimulate growth and regeneration in clean environments, it may have the opposite effect on areas where high deposition of N from urban-generated air pollution has occurred. High levels of N in soils can accelerate community structure changes, and high N deposition has been shown to increase nitrate export through the soil and water systems. Subsequent exposure of high N deposition areas to O_3 has been associated with physiological changes in ponderosa pine species, and can lead to replacement of some pine species with nitrogen, shade, and O_3-tolerant tree species (Takemoto et al. 2001). Issues related to smoke impacts are discussed further in chapter 23.

Under the Regional Haze Regulations of the Clean Air Act of 1977, Congress called for "the remedying of any existing impairment of visibility in 156 national parks and wilderness areas which impairment results from man-made air pollution."

California has 29 such Class I areas (20 wilderness areas and nine national parks). Under the Regional Haze Rule, states must submit their own plan for addressing pollution sources within their jurisdiction that threatens visibility. They must also attain a target for 20% best and 20% worst visibility days by the year 2064 for each Class I area. The state of California submitted its plan in 2009 and will revise it in 2018. CARB produced a midterm progress report of the plan in 2014 that addresses the target visibility values. According to the state, wildfires in northern California are considered major sources for the 20% worst visibility days (California Air Resources Board 2014). Wildfires are on the increase and climate change may aggravate the situation further by increasing wildfire frequency, intensity, and visibility impairment. Issues related to smoke impacts on human communities are discussed further in chapter 23.

Summary

Fire interactions on soil, water, and air resources range from minute changes in soil structure, to alterations in stream water quantity and quality, to potential degradation of air quality across broad regions. These are not isolated effects, however, as fire interactions in one part of the landscape can influence the outcomes in other areas.

Heat from a fire can alter soil structure, chemistry, and soil biota that may radically modify hillslope hydrology, especially infiltration and overland flow. Increased postfire dry ravel and surface runoff greatly accelerate hillslope surface erosion. The delivery of soil and water from the hillslopes can overwhelm the stream channels by increasing the water quantity, changing the water quality, and modifying the patterns of erosion and sedimentation. The net result is an increase in both water yield and sediment yield from upland watersheds that will last for many years. Similarly, smoke from wildland fires can affect the health and well-being of human and biological communities, both adjacent to and at some distance from the actual area of burning. Wildfires also release high amounts of GHGs to the atmosphere that impact climate change, while nitrogen deposition from the atmosphere can cause nutrient enrichment of soils leading to invasive species that could not otherwise thrive in nitrogen-limited systems.

Much is known about the interactions of fire on soil, water, and air; yet many knowledge gaps remain. Although there is an adequate conceptual understanding of these fire interactions, attempts at predictive modeling and risk analysis are woefully inadequate. Only by better understanding the effects of fire on soil, water, and air resources can fire managers and decision-makers be in a better position to manage the consequences of burning.

References

Agrippa, H.C. 1913. The Philosophy of Natural Magic. de Laurence, Scott, and Company. Chicago, Illinois, USA.

Ahlgren, I.F., and C.E. Ahlgren. 1960. Ecological effects of forest fires. Botanical Review 26: 483–533.

Anderson, H.W. 1949. Does burning increase surface runoff? Journal of Forestry 47: 54–57.

Baker, M.B., Jr. 1990. Hydrologic and water quality effects on fire. Pages 31–42 in: J.S. Krammes, technical coordinator. Proceedings of the Symposium on the Effects of Fire Management on Southwestern Natural Resources. USDA Forest Service General Technical Report RM-GTR-191. Rocky Mountain Research Station, Fort Collins, Colorado, USA.

Boerner, R.E.J., J. Huang, and S.C. Hart. 2009. Impacts of fire and fire surrogate treatments on forest soil properties: a meta-analytical approach. Ecological Applications 19: 338–358.

Brown, J.A.H. 1972. Hydrologic effects of a bushfire in a catchment in southeastern New South Wales. Journal of Hydrology 15: 77–96.

Burke, M.P., T.S. Hogue, A.M. Kinoshita, J. Barco, C. Wessel, and E.D. Stein. 2013. Pre- and post-fire pollutant loads in an urban fringe watershed in southern California. Environmental Monitoring and Assessment 185: 10131–10145.

Busse, M.D., K.R. Hubbert, G.O. Fiddler, C.J. Shestak, and R.F. Powers. 2005. Lethal soil temperatures during burning of masticated forest residues. International Journal of Wildland Fire 14: 267–276.

Busse, M.D., K.R. Hubbert, and E.Y. Moghaddas. 2014. Fuel reduction practices and their effects on soil quality. USDA Forest Service General Technical Report PSW-GTR-241. Pacific Southwest Research Station, Albany, California, USA.

Busse, M.D., C.J. Shestak, K.R. Hubbert, and E.E. Knapp. 2010. Soil physical properties regulate lethal heating during burning of woody residues. Soil Science Society of America Journal 74: 947–955.

California Air Resources Board. 2014. California Regional Haze Plan 2014 Progress Report. http://www.arb.ca.gov/planning/reghaze/progress/carhpr2014.pdf. Accessed May 27, 2016.

Choromanska, U., and T.H. DeLuca. 2001. Prescribed fire alters the impact of wildfire on soil biochemical properties in a ponderosa pine forest. Soil Science Society of America Journal 65: 232–238.

Christensen, N.L. 1973. Fire and the nitrogen cycle in California chaparral. Science 181: 66–68.

Colman, E.A. 1953. Fire and water in southern California's mountains. USDA Forest Service Miscellaneous Paper PSW-MP-3. Pacific Southwest Forest and Range Experiment Station, Berkeley, California, USA.

DeBano, L.F. 1981. Water repellent soils: a state-of-the-art. USDA Forest Service General Technical Report PSW-GTR-46. Pacific Southwest Forest and Range Experiment Station, Berkeley, California, USA.

DeBano, L.F., D.G. Neary, and P.E. Ffolliott. 1998. Fire's Effects on Ecosystems. John Wiley and Sons, New York, New York, USA.

DeBano, L.F., R.M. Rice, and C.E. Conrad. 1979. Soil heating in chaparral fires: effects on soil properties, plant nutrients, erosion, and runoff. USDA Forest Service Research Paper PSW-GTR-145. Pacific Southwest Forest and Range Experiment Station, Berkeley, California, USA.

Dekker, L.W., S.H. Doerr, K. Oostindie, A.K. Ziogas, and C.J. Ritsema. 2001. Actual water repellency and critical soil water content in a dune sand. Soil Science Society of America Journal 65: 1667–1675.

Dekker, L.W., C.J. Ritsema, K. Oostindie, and O.H. Boersma. 1998. Effect of drying temperature on the severity of soil water repellency. Soil Science 163: 780–796.

DeLuca, T.H., M.D. MacKenzie, M.J. Gundale, and W.E. Holbe. 2006. Wildfire-produced charcoal directly influences nitrogen cycling in ponderosa pine forests. Soil Science Society of America Journal 70: 448–453.

Doehring, D.O. 1968. The effect of fire on geomorphic processes in the San Gabriel Mountains, California. Pages 43–65 in: R.B. Parker, editor. Contributions to Geology. University of Wyoming, Laramie, Wyoming, USA.

Doerr, S.H., R.A. Shakesby, L.W. Dekker, and C.J. Ritsema. 2006. Occurrence, prediction and hydrological effects of water repellency amongst major soil and land-use types in a humid temperate climate. European Journal of Soil Science 57: 741–754.

Doerr, S.H., R.A. Shakesby, and R.P.D. Walsh. 2000. Soil water repellency: its causes, characteristics and hydro-geomorphological significance. Earth-Science Reviews 51: 33–65.

Doerr, S.H., and A.D. Thomas. 2000. The role of soil moisture in controlling water repellency: new evidence from forest soils in Portugal. Journal of Hydrology 231–232: 134–147.

Donahue, R.L., R.W. Miller, and J.C. Shickluna. 1983. Soils: An Introduction to Soils and Plant Growth. Prentice Hall, Englewood Cliffs, New Jersey, USA.

Fenn, M.E., E.B. Allen, S.B. Weiss, S. Jovan, L.H. Geiser, G.S. Tonnesen, R.F. Johnson, L.E. Rao, B.S. Gimeno, F. Yuan, T. Meixner, and A. Bytnerowicz. 2010. Nitrogen critical loads and

management alternatives for N-impacted ecosystems in California. Journal of Environmental Management 91: 2404–2423.
Fernandez, I., A. Cabaneiro, and T. Carballas. 1997. Organic matter changes immediately after a wildfire in an Atlantic forest soil and comparison with laboratory soil heating. Soil Biology and Biochemistry 29: 1–11.
Fisher, R.F., and D. Binkley. 2000. Ecology and Management of Forest Soils. John Wiley and Sons, Hoboken, New Jersey, USA.
Florsheim, J.L., E.A. Keller, and D.W. Best. 1991. Fluvial sediment transport in response to moderate storm flows following chaparral wildfire, Ventura County, southern California. Geological Society of America Bulletin 103: 504–511.
Franklin, S.B., P.A. Robertson, and J.S. Fralish. 2003. Prescribed burning effects on upland *Quercus* forest structure and function. Forest Ecology and Management 184: 315–335.
Gimeno, B.S., F. Yuan, M.E. Fenn, and T. Meixner. 2009. Management options for mitigating nitrogen (N) losses from N saturated mixed conifer forests in California. Pages 425–455 in: A. Bytnerowicz, M.J. Arbough, A.R. Riebou, and C. Anderson, editors. Wildland Fires and Air Pollution. Developments in Environmental Science 8. Elsevier, Amsterdam, The Netherlands.
Giovaninni, G., and S. Lucchesi. 1997. Modifications induced in soil physico-chemical parameters by experimental fires at different intensities. Soil Science 162: 479–486.
Gleason, K.E., A.W. Nolan, and T.R. Roth. 2013. Charred forests increase snowmelt. Geophysical Research Letters 40: 4654–4661.
Goforth, B.R., R.C. Graham, K.R. Hubbert, W.C. Zanner, and R.A. Minnich. 2005. Properties and spatial distribution of ash and thermally altered soils after high severity forest fire, southern California. International Journal of Wildland Fire 14: 343–354.
Goldammer, J.G., M. Statheropoulos, and M.O. Andreae. 2009. Impacts of vegetation fire emissions on the environment, human health, and security: a global perspective. Pages 3–36 in: A. Bytnerowicz, M.J. Arbough, A.R. Riebou, and C. Anderson, editors. Wildland Fires and Air Pollution. Developments in Environmental Science 8. Elsevier, Amsterdam, The Netherlands.
Hamilton, E.L., J.S. Horton, P.B. Rowe, and L.F. Reimann. 1954. Fire-flood sequences on the San Dimas Experimental Forest. USDA Forest Service Technical Paper PSW-6. California Forest and Range Experiment Station, Berkeley, California, USA.
Hartford, R.A., and W.H. Frandsen. 1992. When it's hot, it's hot . . . or maybe it's not! (Surface flaming may not portend extensive soil heating). International Journal of Wildland Fire 2: 139–144.
Hebel, C.L., J.E. Smith, and K. Cromack, Jr. 2009. Invasive plant species and soil microbial response to wildfire burn severity in the Cascade Range of Oregon. Applied Soil Ecology 42: 150–159.
Heede, B.H., M.D. Harvey, and J.R. Laird. 1988. Sediment delivery linkages in a chaparral watershed following a wildfire. Environmental Management 12: 349–358.
Hubbert, K.R., and V. Oriel. 2005. Temporal fluctuations in soil water repellency following wildfire in chaparral steeplands, southern California. International Journal of Wildland Fire 14: 439–447.
Hubbert, K.R., H.K. Preisler, P.M. Wohlgemuth, R.C. Graham, and M.G. Narog. 2006. Prescribed burning effects on soil physical properties and soil water repellency in a steep chaparral watershed, southern California, USA. Geoderma 130: 284–298.
Huffman, E.L., L.H. MacDonald, and J.D. Stednick. 2001. Strength and persistence of fire-induced soil hydrophobicity under ponderosa and lodgepole pine: Colorado Front Range. Hydrological Processes 15: 2877–2892.
Hungerford, R.D. 1989. Modeling the downward heat pulse from fire in soils and in plant tissue. Pages 148–154 in: Proceeding of the 10th Conference on Fire and Forest Meteorology. Forestry Canada, Ottawa, Ontario, Canada.
Imeson, A.C., J.M. Verstraten, E.J. Van Mullingen, and J. Sevink. 1992. The effects of fire and water repellency on infiltration and runoff under Mediterranean type forests. Catena 19: 345–361.
Johnson, D.W., J.F. Murphy, R.B. Susfalk, T.G. Caldwell, W.W. Miller, R.F. Walker, and R.F. Powers. 2005. The effects of wildfire, salvage logging, and post-fire N-fixation on the nutrient budgets of a Sierran forest. Forest Ecology and Management 220: 155–165.
Johnson, D.W., J.D. Murphy, R.F. Walker, D.W. Glass, and W.W. Miller. 2007. Wildfire effects on forest carbon and nutrient budgets. Ecological Engineering 31: 183–192.
Johnson, D.W., R.B. Susfalk, T.G. Caldwell, J.D. Murphy, W.W. Miller, and R.F. Walker. 2004. Fire effects on carbon and nitrogen budgets in forests. Water, Air, and Soil Pollution 4: 263–275.
Jury, W.A., W.R. Gardner, and W.H. Gardner. 1991. Soil Physics. John Wiley and Sons, New York, New York, USA.
Keeley, J.E. 2000. Chaparral. Pages 203–24 in: M.G. Barbour and W.D. Billings, editors. North American Terrestrial Vegetation, 2nd ed. Cambridge University Press, Cambridge, UK.
Keeley, J.E., and C.J. Fotheringham. 1998. Mechanism of smoke-induced seed germination in Californian chaparral. Journal of Ecology 86: 27–36.
Keller, E.A., D.W. Valentine, and D.R. Gibbs. 1997. Hydrological response of small watersheds following the southern California Painted Cave Fire of June 1990. Hydrological Processes 11: 401–414.
Ketterings, Q.M., and J.M. Bigham. 2000. Soil color as an indicator of slash-and-burn severity and soil fertility in Sumatra, Indonesia. Soil Science Society of America Journal 64: 1826–1833.
Kimmins, J.P. 1996. Importance of soil and role of ecosystem disturbance for sustained productivity of cool temperate and boreal forests. Soil Science Society of America Journal 56: 1643–1654.
Kraebel, C.J. 1934. The La Crescenta flood. American Forests 40: 251–254, 286–287.
Krammes, J.S. 1960. Erosion from mountain side slopes after fire in southern California. USDA Forest Service, Research Note PSW-RN-171. Pacific Southwest Forest and Range Experiment Station, Berkeley, California, USA.
Krammes, J.S., and R.M. Rice. 1963. Effect of fire on the San Dimas experimental forest. Pages 31–34 in: Proceedings of Arizona's 7th Annual Watershed Symposium, September 18, 1963. Arizona Water Commission, Phoenix, Arizona, USA.
Leopold, L.B., M.G. Wolman, and J.P. Miller. 1964. Fluvial Processes in Geomorphology. W.H. Freeman and Company, San Francisco, California, USA.
MacDonald, L.H., and E.L. Huffman. 2004. Post-fire water repellency: persistence and soil moisture thresholds. Soil Science Society of America Journal 68: 1729–1734.
Mahlum, S.K., L.A. Ely, M.A. Young, C.G. Clancy, and M. Jakober. 2011. Effects of wildfire on stream temperatures in the Bitterroot River basin, Montana. International Journal of Wildland Fire 20: 240–247.
Massman, W.J., and J.M. Frank. 2004. Effect of a controlled burn on the thermalphysical properties of a dry soil using a new model of soil heat flow and a new high temperature heat flux sensor. International Journal of Wildland Fire 13: 427–442.
Meixner, T., M.E. Fenn, P.M. Wohlgemuth, M. Oxford, and P. Riggan. 2006. N saturation symptoms in chaparral are not reversed by prescribed fire. Environmental Science and Technology 40: 2887–2894.
Micheletty, P.R., A.M. Kinoshita, and T.S. Hogue. 2014. Application of MODIS snow cover products: wildfire impacts on snow and melt in the Sierra Nevada. Hydrology and Earth System Science 18: 4601–4615.
Miller, P.R., G.J. Longbotham, and C.R. Longbotham. 1983. Sensitivity of selected western conifers to ozone. Plant Disease 67: 1113–1115.
Miller, M.E., L.H. MacDonald, P.R. Robichaud, and W.J. Elliot. 2011. Predicting post-fire hillslope erosion in forest lands of the western United States. International Journal of Wildland Fire 20: 982–999.
Moffet, C.A., F.B. Pierson, P.R. Robichaud, K.E. Spaeth, and S.P. Hardegree. 2007. Modeling soil erosion on steep sagebrush rangeland before and after prescribed fire. Catena 71: 218–228.
Moghaddas, E.Y., and S.L. Stephens. 2007. Thinning, burning, and thin-burn fuel treatment effects on soil properties in a Sierra Nevada mixed conifer forest. Forest Ecology and Management 250: 156–166.
Neal, J.L., E. Wright, and W.B. Bolen. 1965. Burning Douglas-fir slash: physical, chemical and microbial effects in the soil. Forestry Research Laboratory Research Paper 1. Oregon State University, Corvallis, Oregon, USA.
Ohlson, M., B. Dahlberg, T. Okland, K.J. Brown, and R. Halvorsen. 2009. The charcoal carbon pool in boreal forest soils. Nature Geoscience 2: 692–695.
Overby, S.T., and H.M. Perry. 1996. Direct effects of prescribed fire on available nitrogen and phosphorus in an Arizona chaparral watershed. Arid Soil Resources and Rehabilitation 10: 347–357.

Parmeter, J.R., Jr., and B. Uhrenholdt. 1976. Effects of smoke on pathogens and other fungi. Proceedings Tall Timbers Fire Ecology Conference 14: 299–304.

Pierson, F.B., P.R. Robichaud, C.A. Moffet, K.E. Spaeth, S.P. Hardegree, P.E. Clark, and C.J. Williams. 2008. Fire effects on rangeland hydrology and erosion in a steep sagebrush-dominated landscape. Hydrological Processes 32: 2916–2929.

Pietikainen, J., R. Hiukka, and H. Fritze. 2000. Does short-term heating of forest humus change its properties as a substrate for microbes? Soil Biology and Biochemistry 32: 277–288.

Preston, C.M., and M.W.I. Schmidt. 2006. Black (pyrogenic) carbon: a synthesis of current knowledge and uncertainties with special consideration to boreal regions. Biogeosciences 3: 397–420.

Prieto-Fernandez, A., M.J. Acea, and T. Carballas. 1998. Soil microbial and extractable C and N after wildfire. Biology and Fertility of Soils 27: 132–142.

Rau, B.M., D.W. Johnson, R.R. Blank, and J.C. Chambers. 2009. Soil carbon and nitrogen in a Great Basin pinyon-juniper woodland: influence of vegetation, burning, and time. Journal of Arid Environments 73: 472–479.

Rice, R.M. 1974. The hydrology of chaparral watersheds. Pages 27–34 in: M. Rosenthal, editor. Proceedings of a Symposium on Living with the Chaparral, March 30–31, 1973, Riverside, California, Sierra Club, San Francisco, USA.

Rich, L.R. 1962. Erosion and sediment movement following a wildfire in a ponderosa pine forest of central Arizona. USDA Forest Service Research Note RM-RN-76. Rocky Mountain Forest and Range Experiment Station, Fort Collins, Colorado, USA.

Riggan, P.J., R.N. Lockwood, P.J. Jacks, C.G. Colver, F. Weirich, L.F. DeBano, and J.A. Brass. 1994. Effects of fire severity on nitrate mobilization in watersheds subject to chronic atmospheric deposition. Environmental Science and Technology 28: 369–375.

Ritsema, C.J., and L.W. Dekker. 1994. How water moves in a water repellent sandy soil: 2. Dynamics of finger flow. Water Resources Research 30: 2519–2531.

Robichaud, P.R. 2000. Fire effects on infiltration rates after prescribed fire in the northern Rocky Mountains. Journal of Hydrology 231–232: 220–229.

———. 2005. Measurement of post-fire hillslope erosion to evaluate and model rehabilitation treatment effectiveness and recovery. International Journal of Wildland Fire 14: 475–485.

Robichaud, P.R., J.L. Beyers, and D.G. Neary. 2000. Evaluating the effectiveness of postfire rehabilitation treatments. USDA Forest Service General Technical Report RMRS-GTR-63. Rocky Mountain Research Station, Fort Collins, Colorado, USA.

Robichaud, P.R., and S.M. Miller. 1999. Spatial interpolation and simulation of post-burn duff thickness after prescribed fire. International Journal of Wildland Fire 9: 137–143.

Rowe, P.B., and E.A. Colman. 1951. Disposition of rainfall in two mountain areas of California. U.S. Department of Agriculture Technical Bulletin No. 1048. Washington, D.C., USA.

Rowe, P.B., C.M. Countryman, and H.C. Storey. 1954. Hydrologic analysis used to determine effects of fire on peak discharge and erosion rates in southern California watersheds. USDA Forest Service, California Forest and Range Experiment Station, Berkeley, California, USA.

Sandberg, D.V., R.D. Ottmar, J.L. Peterson, and J. Core. 2002. Wildland fire on ecosystems: effects of fire on air. USDA Forest Service, General Technical Report RMRS-GTR-42-VOL-5. Rocky Mountain Research Station, Fort Collins, Colorado, USA.

Scott, V.H. and R.H. Burgy. 1956. Effects of heat and brush burning on the physical properties of certain upland soils that influence infiltration. Soil Science 82: 63–70.

Shakesby, R.A., S.H. Doerr, and R.P.D. Walsh. 2000. The erosional impact of soil hydrophobicity: current problems and future research directions. Journal of Hydrology 231–232: 178–191.

Stephens, S.L., T. Meixner, M. Poth, B. McGurk, and D. Payne. 2004. Prescribed fire, soils, and stream water chemistry in a watershed in the Lake Tahoe Basin, California. International Journal of Wildland Fire 13: 27–35.

St. John, T.V., and P.W. Rundel. 1976. The role of fire as a mineralizing agent in a Sierran coniferous forest. Oecologia 25: 35–45.

Takemoto, B.R., A. Bytnerowicz, and M.E. Fenn. 2001. Current and future effects of ozone and atmospheric nitrogen deposition on California's mixed conifer forests. Forest Ecology and Management 144: 159–173.

Tiedemann, A.R., C.E. Conrad, J.H. Dieterich, J.W. Hornbeck, W.F. Megahan, L.A. Viereck, and D.D. Wade. 1979. Effects of fire on water, a state-of-knowledge-review. USDA Forest Service, General Technical Report WO-10. Washington, D.C., USA.

Ulery, A.L., and R.C. Graham. 1993. Forest fire effects on soil color and texture. Soil Science Society of America Journal 57: 135–140.

US EPA. 1998. Interim Air Quality Policy on Wildland and Prescribed Fires. http://www.epa.gov/ttn/oarpg/t1/memoranda/firefnl.pdf. Accessed May 27, 2016.

Varner, J.M., F.E. Putz, J.J. O'Brien, J.K. Hiers, R.J. Mitchell, and D.R. Gordon. 2009. Post-fire tree stress and growth following smoldering duff fires. Forest Ecology and Management 258: 2467–2474.

Wan, S., D. Hui, and Y. Luo. 2001. Fire effects on nitrogen pools and dynamics in terrestrial ecosystems: a meta-analysis. Ecological Applications 11: 1349–1365.

Welling, R., M. Singer, and P.H. Dunn. 1984. Effects of fire on shrubland soils. Pages 42–50 in: J.J. DeVries, editor. Proceedings of the Chaparral Ecosystems Research Conference. Water Resources Center, Report No. 62. University of California, Davis, California, USA.

Wells, W.G., II. 1981. Some effects of brushfires on erosion processes in coastal southern California. Pages 305–342 in: Proceedings of a Symposium on Erosion and Sediment Transport in Pacific Rim Steeplands. International Association of Hydrological Sciences Publication No. 132. January 25–31, 1981. Christchurch, New Zealand.

———. 1987. The effects of fire on the generation of debris flows in southern California. Geological Society of America, Reviews in Engineering Geology 7: 105–114.

Wells, C.G., R.E. Campbell, L.F. DeBano, C.E. Lewis, R.L. Fredricksen, E.C. Franklin, R.C. Froelich, and P.H. Dunn. 1979. Effects of fire on soil: a state of knowledge review. USDA Forest Service General Technical Report WO-7. Washington, D.C., USA.

Williams, M.R., and J.M. Melack. 1997. Effects of prescribed burning and drought on the solute chemistry of mixed-conifer forest streams of the Sierra Nevada. Biogeochemistry 39: 225–253.

Woche, S.K., M.O. Goebel, M.B. Kirkham, R. Horton, R.R. Van der Ploeg, and J. Bachmann. 2005. Contact angle of soils as affected by depth, texture and land management. European Journal of Soil Science 56: 239–251.

Wohlgemuth, P.M., J.L. Beyers, C.D. Wakeman, and S.G. Conard. 1998. Effects of fire and grass seeding on soil erosion in southern California chaparral. Pages 41–51 in: S. Gray, chair. Proceedings of the 19th Forest Vegetation Management Conference, January 20–22, 1998, Redding, California, USA.

Woods, S.W., and V.N. Balfour. 2008. The effect of ash on runoff and erosion after a severe forest wildfire, Montana, USA. International Journal of Wildland Fire 17: 535–548.

Xiao, Q., and E.G. McPherson. 2011. Rainfall interception of three trees in Oakland, California. Urban Ecosystems 14: 755–769.

CHAPTER EIGHT

Fire and Plant Interactions

AMY G. MERRILL, ANDREA E. THODE, ALEXANDRA M. WEILL,
JO ANN FITES-KAUFMAN, ANNE F. BRADLEY, AND TADASHI J. MOODY

Pyrodiversity creates biodiversity . . .
MARTIN AND SAPSIS (1992)

Introduction

Charcoal records show that fire has been present in California for millions of years (Weide 1968, Keeley and Rundel 2003), creating a long history of fire interacting with plants. In an ecological context, fire is neither "good" nor "bad" for biota—it is an integral ecological process in most plant communities of California. Many plant species in California have responses to fire, and in particular fire regimes that enable them to survive, and even thrive, with fire.

The effects of fire on plants are a result of the interaction between fire's chemical and physical properties and characteristics of the individual plant. Because each fire behaves differently and each species has a unique combination of physiological and physical traits, there is a wide array of resulting fire effects on the plants. Postfire weather conditions, composition and heterogeneity of adjacent unburned vegetation, and survival or immigration of biota also influence these effects.

The heat, smoke, and charcoal that fires produce affect plants and their living tissues in different ways. In biological terms, fire consumes biomass, causes injury and mortality, affects hormone production in growing tissues, alters the physical and chemical characteristics of the soil, and modifies interactions among species.

Species adapt and evolve in context of the entire ecosystem. Fire is a process that is part of an ecosystem, and the effects of fire occur at the local and landscape scale over many generations and fire events. Fire seasonality, frequency, size, spatial complexity, intensity, severity and type, as it occurs over the landscape and over time, is referred to as the ecosystem's fire regime (see chapter 5). Thus, species are not adapted to fire per se, but to a fire regime, and plant species persist because they possess traits that permit them to tolerate or depend on a specific fire regime. To understand the relationship between fire and plants, it is important to understand the relationship between plant species, plant communities, and the fire regimes in which they occur.

Variation in fire regimes and plant species composition combine to generate myriad fire effects on plant communities. In this chapter, we describe how fire interacts with plants at scales ranging from individuals, to populations, to species, to communities, and to landscape patterns (see example in Box 8.1). More information on species and plant community interactions with fire that are specific to different parts of California are described in the bioregional chapters in part Two.

Direct Effects on Individuals

Fire affects individual plants through indirect and direct effects. Indirect effects include changes to the plant's environment, such as changes in water and nutrient availability or available sunlight. For example, a crown fire removes sunlight-blocking canopy, resulting in a postfire period of full sun exposure that supports shade-intolerant species such as buckbrush (*Ceanothus cuneatus* var. *cuneatus*) and manzanitas (*Arctostaphylos* spp.) in the lower montane zone. Similarly, high mortality due to fire temporarily diminishes evapotranspiration, thereby increasing water availability to surviving or newly establishing plants. Direct effects include impacts due to the heat energy of fire. These effects include scorched needles on conifers or portions of a tree trunk in which part or all of the cambium is killed. Direct effects are a function of fire characteristics and the individual's morphological and physiological traits. Morphology refers to plant structure and includes plant size and the location of regenerating tissues such as buds and cambium. Physiology refers to the production, interaction, and regulation of chemicals within living plant tissue that govern plant growth, development, reproduction, and senescence. The amount of heat a plant receives and the duration of that heat are both critical for understanding direct fire effects plants (Martin 1963, Wright and Bailey 1982).

Individual plants or species can be categorized as fire resisters, plants that resist damage from fire, or fire persisters, individual plants or their offspring that regrow after fire such that the local population eventually returns to its prefire state. Plants *resist* fire by having structures that protect them from heat damage, such as thick bark that protects sensitive stem cambium tissue. Plants *persist* by regenerating or sprouting after top-kill, by seeds that survive fire, and by fire-stimulated seed release or germination. In this section, we focus first on plant structures that resist fire and second on vascular plant traits that facilitate persistence and resilience. We then discuss nonvascular plants (mosses and lichens).

Resistance

Heat is the primary output of fire that affects plants; both the absolute amount and the duration of heating are important in determining the degree of damage to individual plants. Heat kills cells by causing structural damage and disrupting cellular metabolism. Most plant cells die when heated up to 50–55°C (122–133°F) (Wright and Bailey 1982). Many plants have morphological structures that enable them to resist heat. These structures and life history responses to fire are summarized in Table 8.1, and some are described in the sections below. The Fire Effects Information System (http://www.feis-crs.org/beta/) provides further detail on the responses of many California plant species to fire.

A plant's lifeform, size, and structure determine how it is affected by fire. Lifeforms of vascular plants include perennial (live over one year) trees and shrubs and perennial and annual (live only one year) grasses and herbs. Each lifeform differs in its response to fire. When the crowns of annual plants are killed by fire, they generally cannot grow new shoots because they lack buds and energy reserves. However, annuals may have died and gone to seed by the time they burn. In order to persist after fire, they rely on seeds present on the site or colonization by seeds raining in from adjacent unburned areas or transported in by wind or water. Depending on exposure to lethal heat, perennials often grow new roots, stems, and leaves following fire. The capacity of buds to sprout and the position of these buds in relation to both heat and the protection of insulating tissue or soil cover are critical to perennial plant survival in a fire.

Plants survive fire by either having insulating tissues (e.g., bark) above ground or by tissues growing underground where they are protected by the insulating soil. Different plant structures and tissues contribute to a plant's response to fire in different ways. The primary structures of a plant include the crown, foliage, and leaves, buds, stems and trunks, roots, and basal meristems. These structures are composed of one or several types of tissue: meristem (found in root and shoot tips, and under bark for lateral growth), vascular tissue (xylem and phloem), ground tissues (found throughout plant for structural support, food storage, and photosynthesis), and dermal (found mostly in leaves, roots for protection, and water loss/absorption).

BUDS

Buds are masses of plant tissue from which new shoots, stems, and leaves form (Fig. 8.1). Plants exposed to fire can lose a lot of their foliage due to heat and because leaves produce energy, plant survival depends upon the rapid regeneration of foliage burned by a fire. Buds occur on many perennial species at different locations on the plant body: aerial, basal, and belowground buds and all provide different kinds of protection (Clarke et al. 2013).

Buds located below the mineral soil surface are most likely to survive fires because they are protected from heat by the soil. Soil is an excellent insulator from heat, depending on soil moisture content (Clarke et al. 2013; chapter 7) (Fig. 8.2). Conversely, buds at or near the soil surface are most sensitive to heat-kill because they are located where active fuel consumption and heat flux is usually greatest.

Buds vary in their degree of protection depending on the amount and type of insulating tissues surrounding them. Located in the plant crown, aerial buds response to fire varies with height above the heat source. Thus, buds in tall tree crowns are less likely to be affected by surface fire heat compared to buds on short-stature trees and low-growing shrubs. For example, many pines have self-pruning stems, large, protected buds, and a crown structure that dissipates heat; as a result, these trees both resist the effects of surface fire and reduce the likelihood of crown fire (Figs. 8.1 and 8.2). Thick bark protects not only vascular tissue and cambium but also epicormic buds, though some fire-resisters, like eucalypts, are notable for having thin bark and deeply embedded meristems (Clarke et al. 2013).

CROWNS AND FOLIAGE

Foliage (leaves and needles) plays a crucial role in fueling plant growth by converting solar energy to chemical energy through photosynthesis. Crown structure and moisture content are primary determinants of fire effects on foliage. The likelihood of crown survival depends on its shape, size, height, and the degree of protection afforded canopy buds. Foliage closer to the flames and heat is more likely to experience tissue death. Plants with crowns that extend from or near the soil surface to the top of the plant usually have greater tissue mortality because branches and foliage near the ground form a "ladder" of vertical fuels that provide a conduit for combustion and heat from the base to the top of the crown. In contrast, many species of pines self-prune their lower branches, reducing the ladder-fuel effect and thereby reducing mortality (Schwilk and Ackerly 2001).

Because of increased heat released, damage also increases with crown density and the proportion of dead branches and leaves in the crown (Schwilk 2003). For example, western junipers (*Juniperus occidentalis*) prevalent in the Northeastern Plateaus bioregion (chapter 13) have crowns that extend to the ground, which can lead to complete crown combustion. Gagnon et al. (2010) suggested that flammable crowns are actually an advantage for resprouting species: because the flammable material burns quickly and well above soil, soil heating is limited and belowground resprouting structures are protected. Many fire-adapted species can resprout even when all of the aboveground plant material has been killed by fire and are discussed in the next subsection.

For nonsprouting species, including most conifers, the ability to recover from foliar damage depends upon the resistance of the aerial buds to heat damage and the proportion of crown retained (Peterson 1985, Fites-Kaufman et al. 2008). Ponderosa pine (*Pinus ponderosa*) buds can resist fir damage, enabling some crown recovery after fire. The proportion of crown scorch a ponderosa pine tree can experience without mortality is variable: some report threshold tolerances range from 60% to 90% (Harrington 1993, McHugh and Kolb 2003), whereas others report that tree mortality and needle scorch varied continuously (Thies et al. 2006). In contrast to ponderosa pine, white fir (*Abies concolor*) and Douglas-fir (*Pseudotsuga menziesii* var. *menziesii*) typically sustain both foliage and bud damage, leaving no means for foliar regeneration. This difference in bud mortality between ponderosa pines and the firs can be largely attributed to differences in morphology. Ponderosa pine and Jeffrey pine (*P. jeffreyi*) buds are relatively

TABLE 8.1
Plant structures and associated definitions, factors associated with fire response, and examples

	Plant structures	Definition	Potential factors associated with fire response	Examples
Vegetative structures	Foliage	Leaves and needles	Phenology, moisture level, leaf thickness, shape, area	Chamise
	Crowns	Sum of all leaves or needles of a plant	Crown height and bulk density	Ponderosa pine
	Bark	Layer of dead, protective cells around outside of stem, phloem	Bark thickness and density; volatile substances	Ponderosa pine giant sequoia
	Roots	Underground structures that absorb water and nutrients, and anchor plant	Amount and duration of heat, depth below surface, concentration and distribution of system	Mountain misery
Sprouting structures	Epicormic buds	Buds in stem capable of sprouting	Fire intensity, fire duration	big-cone Douglas-fir, many hardwoods
	Basal burls, buds at stems	Woody tissue from which roots and stems originate, often covered with buds	Fire intensity, fire duration	manzanita, ceanothus, big-leaf maple
	Caudexes, corms, bulbs	Stem base that stores energy may also produce new leaves/stems	Depth and degree of heat from fire	mule's ears, lily, iris
	Stolons	Aboveground lateral stems form new plants	Depth of litter layer, depth and degree of heat	Strawberry, snowberry
	Rhizomes	Belowground lateral stems form new plants	Depth of heat from fire, soil moisture	Spirea, Oregon grape, mountain misery
	Sprouting roots	Roots that have primordial buds capable of sprouting	Magnitude and duration, soil moisture	Aspen
Sexual reproduction	Fire-stimulated flowering	Plants that flower or flower more with fire	Sensitive to fire intensity: temperature over threshold kills flowering tissue	Mariposa lily
	Serotinous cones	Cones storing seeds: cones only open with high heat	Temperature of fire, longevity of seeds in relation to fire return interval	Knobcone pine, bishop pine, cypresses
	Seed banks	Supply of viable seeds buried in soil	Depth and degree of soil heating, soil moisture	Bigpod ceanothus

large, have greater tissue protection in the form of bracts, and also receive some protection from long needles that extend beyond the terminal buds (Wagener 1961). In contrast, aerial buds in white fir and Douglas-fir are small and relatively unprotected by bracts or foliage.

Foliar moisture content affects flammability and the degree of tissue damage. Although less flammable if a fire does occur, tissue with higher moisture content is more susceptible to heat damage in a shorter time frame than tissue with lower moisture content (Wright and Bailey 1982). In California, foliar moisture is usually highest in spring when actively growing tissues are more susceptible to heat damage. Foliar moisture content tends to be lowest during summer and early fall, when most fires occur (Dell and Philpot 1965).

STEMS AND TRUNKS

Stems and trunks provide structural support for foliage and nutrient and water transport between roots and crowns. Vascular cambium is the layer of tissue that generates vascular tissues called the xylem and phloem (Fig. 8.3). Xylem forms on the inward side of the vascular cambium and transports water from roots to crowns. Phloem forms on the outside of the vascular cambium and transports carbohydrates produced in the foliage down to the roots. Phloem cells die over time, forming part of the stem structural tissue. By continuously generating new xylem and phloem, the vascular cambium tissue is responsible for most stem and diameter growth. Injury or death to the vascular cambium can have a great, even lethal,

FIGURE 8.1 Buds before and after sprouting. Fire can damage buds in either stage.

impact. The vascular cambium of most trees, shrubs, and many herbs is located towards the outside of the stem. Woody plants also have a cork cambium, which is another layer of tissue that continuously generates new cork cells, that make up the bark, to the outside of the vascular cambium and phloem (Fig. 8.3). On the other hand, in monocots—including grasses (Poaceae), lilies (Liliaceae), and palms (Arecaceae)—the vascular cambium, xylem, and phloem tissues are distributed throughout the stem rather than only near the outside. Resistance of stems to fire depends in large part upon how well bark insulates the vascular cambium from heat.

Cambial death is a function of both temperature and duration of heating and can occur by either exposure to high heat or to an extended period of low heat. Accumulation of downed woody fuel or duff at the bases of plants, which burns primarily as smoldering combustion, is a common cause of cambial kill (Ryan and Reinhardt 1980, Sackett and Haase 1992). Girdling occurs when the vascular cambium around the entire circumference of a dicot plant stem is killed. Ultimately, girdling kills a plant by constricting or eliminating water and nutrient transport. However, the mechanism by which individual plants die following heating during a fire is understudied (Midgley et al. 2011). For example, top-kill may be caused by cambium necrosis (Michaletz and Johnson 2007), or plant death may result from damage to the xylem and subsequent water stress (Midgley et al. 2011).

Bark insulates the cambium and buds from heat. The insulating effectiveness of bark is dependent upon its thickness, chemical composition, and flammability. In general, stems with thicker bark provide greater insulation and resistance to heat (Gill 1981). Conifers have thicker bark than hardwoods, and shrubs tend to have thin bark. As trees grow, bark thickness increases so that the likelihood of survival increases with greater age and size (Fig. 8.4) (Peterson and Ryan 1986). Other bark characteristics such as chemical composition and structure (e.g., flakiness) also influence the rate of heat diffusion through the stem (Spalt and Reifsnyder 1962). For example, the flaky bark of ponderosa pine is more flammable than other co-occurring species such as black oak (Plumb 1979). Bark flammability has been little studied in the United States but has been documented as a key factor in effects of fire on Eucalyptus trees in Australia (Gill 1981).

Monocots, such as palm trees, cannot be girdled by fire because they have many independent bundles of vascular tissue distributed throughout the stem. California fan palms (*Washingtonia filifera*) were burned repeatedly by Native Americans to clear oases for human habitation and to enhance palm fruit production (Vogl 1968). Each successive fire burned some of the palm vascular bundles, reducing trunk diameter and crown size, potentially reducing transpiration to the dry desert air, but not necessarily causing mortality.

ROOTS AND UNDERGROUND TISSUES AND ORGANS

Root systems absorb water and nutrients from the soil and confer structural stability to the plant. Feeder roots are closer to the soil surface. Structural roots anchor the plant to the ground and are generally larger and deeper. Death of feeder roots may stress plants even if root mortality does not lead to immediate death after fire (Wade 1993). Some plants also have underground stems, called rhizomes, which can be important for sprouting (Fig. 8.2).

Root mortality is a function of both temperature and heat duration. Smoldering combustion often causes the greatest root mortality because heat lasts longer in a single location compared to heat from a passing flame front (Miller 2000). This is particularly evident where litter has accumulated at the base of plants and where feeder roots are growing into the duff or litter. Smoldering ground fires can exceed 300°C (572°F) for several hours, well exceeding the heat resistance of roots, rhizomes, and underground buds (Ryan 2002). Deeper duff layers increase in burn depth and the likelihood of deeper heat penetration into the soil (Ryan 2002).

The effects of fire on roots and rhizomes are a function of the amount of heat to which tissues are exposed and their physiological state. Deep roots are better protected from heat exposure. The amount of heat that penetrates the soil is a function of soil type, soil moisture, surface duff, and fuel loading (as described in chapter 7). Wet soils conduct heat to plant tissues more readily than do dry soils up to 95–100°C (203–212°F) (DeBano et al. 1998, Debano at al. 2005). At that point, the temperature in wet soils remains the same until water in the soil vaporizes; then heating continues as it does in dry soils (DeBano el al. 1998). In very wet soils, all of the heat from a fire may be required to vaporize the water in the surface soil, so that heat does not get conducted deeper into the soil profile (DeBano et al. 1998). In contrast, fuel consumption, and thus heat output, can be greater when both fuels and soils are dry.

Physiological activity also affects root sensitivity to heat from fire. Feeder roots are most susceptible to heat-kill during the season of active growth (often in early spring) and are more tolerant of heat when dormant. For example, ponderosa pine mortality was higher after spring (30%) than after fall (20%) burning in central Oregon, corresponding to differences in root growth periods (Swezy and Agee 1991). Although the exact mechanism that might explain these observations has not been determined, it is likely that young growing tissues were more sensitive to fire and/or that tree water and nutrient requirements were greater during periods of active growth. In contrast, Kummerow and Lantz (1983)

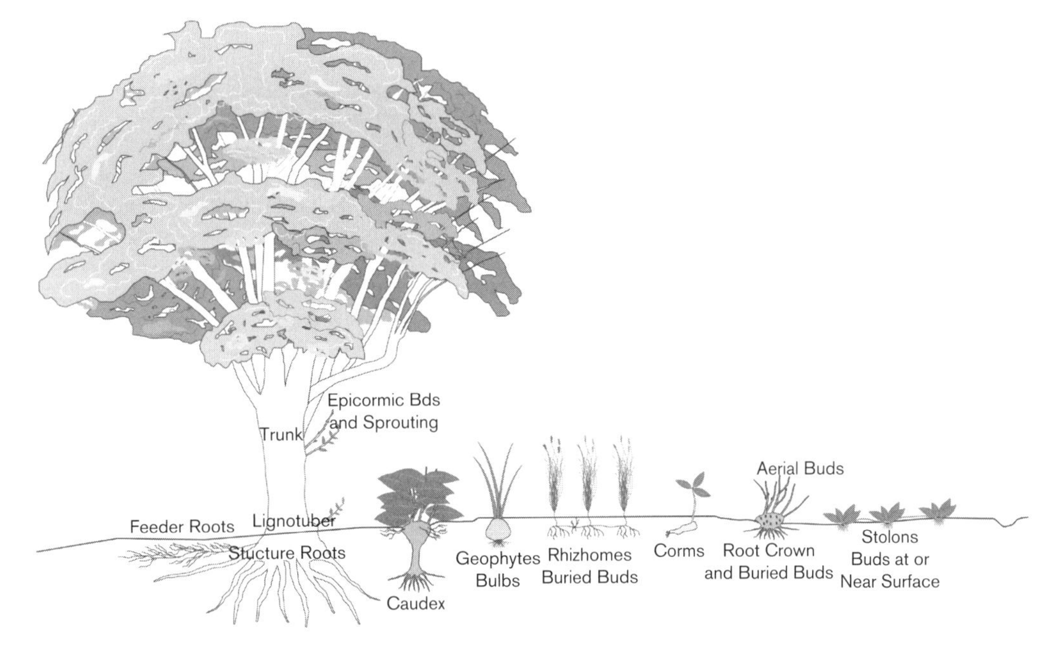

FIGURE 8.2 Variety of plant morphological structures, including those that sprout. The location of the buds relative to the heat source determines the response to fire.

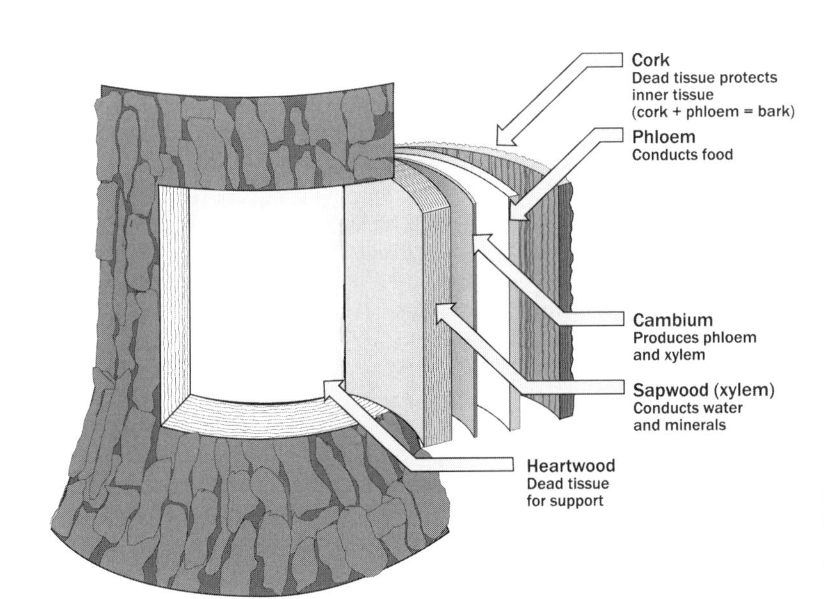

FIGURE 8.3 Cross-section of tree trunk showing protective bark and underlying fire-sensitive tissues. The xylem grows inward of the cambium and forming wood, while the phloem grows outward forming bark.

reported that fine roots of redshank (*Adenostoma sparsifolium*), a chaparral species that grows in southern California, increased following fire. They attributed this increase to enhanced nutrient availability and higher soil moisture following fire.

GRASSES AND GRASS-LIKE PLANTS

Instead of terminal buds, grasses and grass-like plants such as sedges and rushes (hereafter referred to as "grasses") have growing tissues called apical and intercalary (base of leaf) meristems that allow them to survive repeated burning. Unlike the terminal buds on trees, shrubs, and herbs, the intercalary meristems are located at the base of the leaves rather than at the tip, which is why grasses continue to grow after repeated grazing or mowing. The location of intercalary meristems relative to the soil surface, level of decadence, and dead foliage influence how much heat they receive and level of tissue damage or mortality. Just as with tree or shrub crowns, decadent material generates more heat near the foliage and meristems and increases tissue damage and mortality.

Grass lifeforms that are important to fire responses include annuals, perennial bunchgrasses, and perennial stoloniferous or rhizomatous grasses (Miller 2000). Fire typically kills annual grasses outright or completely consumes them if they are cured. Nevertheless, annual grasses can produce many seeds, some of which may survive a fire and reoccupy a site. Seeds of annual and perennial grasses survive a fire depending upon the number produced, how heat resistant they are, and how well insulated the seeds are in the soil (Parsons and

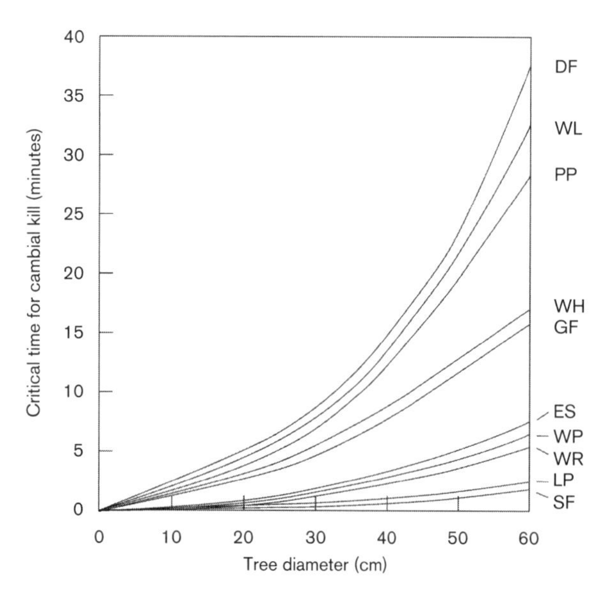

FIGURE 8.4 Variation in time to kill cambium by species and diameter. DF = Douglas-fir, WL = western larch, PP = ponderosa pine, WH = western hemlock, GF = grand fir, ES = Engleman spruce, WP = western white pine, WR = western red cedar, LP = lodgepole pine, SF = subalpine fir (redrawn from data courtesy of D. L. Peterson).

Stohlgren 1989, Kneitel 1997). Perennial grass survival is dependent upon how well their meristems or buds are protected from heat and the amount of decadent material. Stoloniferous grasses are more susceptible to fire damage because aboveground stolons are more exposed to lethal heat (Fig. 8.2). The effect of fire on rhizomatous species and bunch grasses depends on the depth of their buried buds and meristems (Miller 2000). For example, Idaho fescue (*Festuca idahoensis*) is relatively fire sensitive because its meristematic tissue is near the soil surface. In contrast, squirrel tail (*Elymus elymoides*) is more fire-tolerant, because its meristematic tissue is more than 2.5 cm (1 in) below the soil surface.

Individual Plants and Plant Vigor

Plant condition affects its ability to survive a fire. Plants stressed by disease, drought, or herbivory are more likely to die under the additional stress of fire. Moreover, weakened or suppressed plants develop fire resistant characteristics, such as thick bark, more slowly than healthy plants of similar age (Wade 1993). Plants with compromised health are also less likely to generate vigorous sprouts, since underground energy stores are lower than healthy individuals. In some cases, low-severity fire increases individual tree defenses to bark beetles. Hood et al. (2015) tie low-severity fire to an increase in resin duct defenses, which positively links with individual tree survival. They suggest that low-severity fire may actually develop a long-term tree defense against bark beetles whereas moderate to high tree injury from fire may increase probability of tree death.

Persistence and Colonization

Resistance to heat is one means of tolerating fire. Another means of survival is by postfire sprouting or repopulating burned areas through seeds stored in the soil or canopy. Sprouting is the growth of new tissues (e.g., stems, leaves) from buds. Plants known as "sprouters" or "resprouters" respond to fire-induced top-kill through sprouting, improving their chances of crown regrowth and survival. Sprouting can be apical (from terminal buds), epicormic (from boles or branchlets), basal (from collars or lignotubers), or belowground (from rhizomes) (Clarke et al. 2013). Plant species also persist on a site via seeds that are buried in the ground or enclosed in serotinous cones in the canopy; these plants are often called "seeders" or "reseeders." Some species exhibit one, both, or neither of these strategies. Species that depend on only one of the strategies are "obligate resprouters" or "obligate seeders," while species that have seed banks but can also sprout are "facultative resprouters" (Pausas and Keeley 2014). Some species respond to fire by colonization from outside the burn through seed dispersal from adjacent unburned areas. These functional types and examples are detailed in Table 8.2.

SPROUTING

Sprouting is a physiological response stimulated by damage to the top or crown of plants (Miller 2000). Damage can be caused by fire, wind, flooding, and animals. For some species, sprouting can regenerate the entire crown from epicormic buds and energy reserves that were protected underground or in the stem. Temperature and duration of heat relative to bud location affects the amount of bud damage and the sprouting response. Sprouting varies with species, age, size, phenology, bud location, and bud structure.

Fire intensity and duration, together with bud location relative to flames or heat, influence the potential sprouting response (Vesk and Westoby 2004). Sprouting tissues, like other plant tissues, can be killed by either a short exposure of high-intensity fire or a long duration of low-intensity fire. The greatest sprouting response often comes after moderate-intensity fires, where some buds are killed and some are stimulated to sprout (Miller 2000). Long duration fires, even of low intensity, often result in the poorest sprouting response because of the greater heat penetration into the soil and longer exposure of sprouting structures to heat.

Where intense fires remove a large amount of aboveground biomass, such as in chaparral of the South Coast and Sierra Nevada, there is usually a clear distinction between sprouting and nonsprouting species based upon bud tissue locations (Vesk et al. 2004) (also see chapters 14 and 17). Both among and within-species variation in sprouting response occurs in systems with surface-fire dominated regimes, such as displayed by antelope bitterbrush (*Purshia tridentata*) in the Northeastern Plateaus bioregion (Vesk and Westoby 2004) (see chapter 13). This variable response may be partly due to genetic variation and partly due to patchy fire patterns in areas dominated by surface fire regimes.

Sprouting also varies with differences in the type and distribution of buds relative to heat from the fire. A number of different structures contain buds with the potential to resprout after fire, including aerial stems, rhizomes, bulbs, corms, lignotubers, and roots (Fig. 8.2). Most sprouting structures also store energy (e.g., starch) needed by emerging sprouts. For example, geophytes, or plants with bulbs such as death camas (*Toxicoscordion* spp.) often sprout following fire utilizing the starch contained in bulbs (Tyler and Borchert 2002). Sprouting structures that are protected by soil or are heat resistant tissues (e.g., thick bark) are most likely to survive and respond posi-

TABLE 8.2
Modified Bond and van Wilgen (1996; and Pausas and Keeley (2014) classifications of plant reproductive responses to fire for California flora (with examples)
The left side of the table refers to sprouting response, while the right side refers to seeding response

Response direction and degree	Sprouter: not killed by fire (Obligate Sprouters and Facultative Seeder-Sprouters)	Nonsprouter (obligate seeder*)	
		Not killed by fire	Killed by fire
Fire-dependent	Flowering only or almost only after fire (Mariposa lily, Death camas)	Fire-stimulated flowering, germination, seed release (golden-eyes)	Seed release from heat (knobcone and Bishop pines, bigpod ceanothus)
Fire-enhanced	Species increase after fire, but establishment occurs in fire-free interval too (Black oak, aspen)	Seed release and seedling establishment enhanced (ponderosa pine)	Seed germination enhanced (tobacco brush, mountain white thorn)
Fire-neutral	Sprouting recruitment same following fire as in fire-free interval; continuous sprouters (scrub oak, bigleaf maple, cottonwood, sedges)	Seed germination same following fire as in fire-free interval; seed producers survive fire (Douglas-fir, sugar pine)	Long-distance seed dispersal (fireweed, thistle)
Fire-inhibited	Sprouting recruitment less following fire than in fire-free interval (Arizona fescue)	Seed germination less following fire than in fire-free interval (mature firs)	Mature and seedling individuals killed by fire; postfire recruitment low (Sitka spruce, Santa Lucia fir, fir seedlings)

*Although facultative seeders may have some of these traits, they are not specifically represented in this table.

tively to fires. Sprouting from stolons or root crowns is particularly sensitive to and most easily reduced by fire, especially surface fires, since these structures occur on or near the soil surface. Bulbs, rhizomes, and roots are less likely to be damaged by fire, since they are insulated by soil.

Plant size, age, phenology, and time since the last fire also influence sprouting response by affecting levels of starch storage. There is often an increased sprouting response by larger or older plants because of the greater number of buds and/or greater accumulation of stored carbohydrates and nutrients (James 1984), particularly in species with variable sprouting (Vesk et al. 2004). The importance of stored carbohydrates to sprouting has been documented by the decline in carbohydrates in several species following fire-induced sprouting (Bond and Midgley 2001, Tyler and Borchert 2002). It can take two or more years to recover stored starch levels.

Phenology refers to the timing of plant life-cycle events, including growth, reproduction, dormancy, and death. These phases affect a plant's energy reserves and hormonal patterns that govern sprouting (James 1984). Carbohydrate levels in roots or lignotubers are usually highest before the onset of growth. Reserves then decline, reaching their minimum shortly after growth begins. Reserves are lowest after flowering and seed production (Fig. 8.5). Generally, plants are least able to sprout during times of active growth when carbohydrate reserves are lowest. If burning occurs when plants are dormant, sprouting may be delayed until the following growing season. Two other plant conditions that can influence sprouting during the growing season or later in the growing cycle are hormones and water status (James 1984). Hormone levels that suppress sprouting are higher during the summer when temperatures are highest. Sprouting may be greater

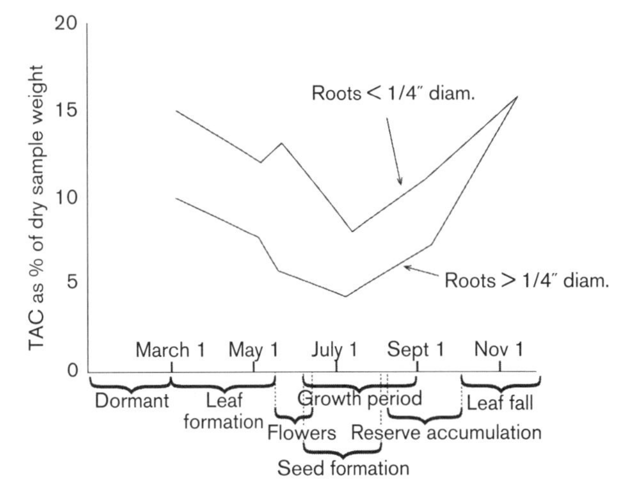

FIGURE 8.5 Seasonal changes in total carbohydrate (TAC) levels in bitterbrush. Levels reach the lowest point after flowering, during the growth period, and during seed formation (redrawn from McConnell and Garrison 1966).

when plants have more available water. Plants stressed by drought or pollution have lower vigor and may have a reduced ability to sprout following fire.

Riparian areas do burn but are typically wetter and less flammable than other portions of the landscape (Dwire and Kauffman 2003). Although riparian plants are thought to be particularly vulnerable to fire, they are often vigorous sprouters. Sprouting permits riparian species to survive not only flooding but also to survive and even flourish after fire (Szalay and Resh 1997). In fact, some riparian species are the first to resprout following fires, often within the same week. The favorable moisture in riparian areas contributes to this rapid response. Observations after wildfires indicate that some widely distributed riparian shrubs, including willow (*Salix* spp.) and mountain alder (*Alnus incana* subsp. *tenuifolia*), and trees such as cottonwood (*Populus* spp.), resprout vigorously after fires. Burning by Native Americans for basketry materials included numerous riparian species such as willow, dogwood (*Cornus* spp.), big-leaf maple (*Acer macrophyllum*), and California hazel (*Corylus cornuta* subsp. *californica*) (see chapter 19).

SEEDS

After fire, seeds are essential to the perpetuation of many plant species, particularly those that do not sprout or that lack heat resistant characteristics. Postfire seed germination includes (1) survival in soil or canopy seed banks, (2) germination or release of seed stimulated by the fire (e.g., serotinous cones), (3) fire-enhanced seedling establishment and survival, and (4) dispersal from outside the burned area. The first two aspects of seed and postfire seed germination response will be discussed in this section and the latter two aspects are discussed in the following section on dispersal and colonization.

The soil seed bank develops by on-site seed accumulation, including both long-lived, persistent seeds as well as transient seeds that are viable for a few weeks or months. The effect of fire on the seed bank depends on the types of seeds present (long-lived or transient), how much heat penetrates the soil, and its duration relative to the distribution of seeds in the soil profile, the heat tolerance of the seeds, and the mechanisms that initiate germination. Species that quickly exploit recently burned sites are able to do so because they have large quantities of long-lived, soil-stored seeds. For example, snowbrush (*Ceanothus velutinus*) seeds may germinate after remaining dormant in the soil for 200 to 300 years (Noste and Bushey 1987) and readily dominate sites after fires. Seed densities can reach 2,000–21,000 seeds m^2 (186–1,951 seeds ft^2) in chaparral seed banks (Zammit and Zedler 1988). Species that have transient seeds do not accumulate large seed banks and therefore depend on other means, such as sprouting, to survive fire, especially frequent fire. Examples of species with transient seeds include toyon (*Heteromeles arbutifolia*) and California coffee berry (*Frangula californica*), both found in California chaparral.

Soil insulates seeds from heat, in the same way it protects roots and rhizomes or other underground sprouting tissues. Seeds on or near the surface are exposed to more heat, which kills them or stimulates germination. Many hard-seeded species have seed coats ruptured by heat from fire or exposure to charcoal-derived chemicals. California-lilac (*Ceanothus*) has numerous species with heat-stimulated germination. Some chaparral seeds require dry heat to germinate but are killed by low temperature if they have imbibed (e.g., absorbed) water (Parker 1986). Species with water imbibing seeds are more likely to sustain high seed mortality if fire occurs when soils are moist. Plant species also vary in the degree of heat-induced germination. For example, seeds of many shrubs species in coastal sage in the South Coast bioregion are "polymorphic" (see chapter 17), meaning some of the seeds are heat-stimulated but not all.

Chaparral in California, particularly in the South Coast and Central Coast bioregions, has many annual and shrub species with germination closely tied to fire (Keeley 1981, 1987, Zammit and Zedler 1988, Borchert and Odion 1995). The most common mechanism for fire-induced germination is heat, which ruptures the seed coat. Seeds of some annuals in chaparral are stimulated not by heat but by chemicals in smoke and charred wood. Generally, large, hard, water impermeable seeds respond to heat, while smaller, water permeable seeds respond to smoke. In some cases, heat and smoke act synergistically to enhance germination (Wilkin et al. 2013).

Though many seeds respond to smoke and/or charcoal, it is unclear which chemicals trigger germination and under what conditions (Nelson et al. 2012). For example, seeds of the annual whispering bells (*Emmenanthe penduliflora*) germinate when exposed to the nitrogen dioxide (NO_2) in smoke (Keeley and Fotheringham 1997). Concentrations of NO_2 sufficient to initiate germination are generated both by the fire and by elevated nitrification in many post-burn soils. NO_2 is also a common air pollutant and deposition of NO_2 in southern California chaparral exceeds levels necessary to elicit germination in the lab. These findings raise questions about air pollution effects on some fire-adapted species (Malakoff 1997). Other smoke-stimulated plants do not respond to nitrogen compounds. A group of compounds known as karrikins, formed from the combustion of plant material, recently have been identified as possible agents in smoke that trigger germination (Nelson et al. 2012). However, karrikins have also been criticized as a fire-specific cue (Bradshaw et al. 2011, Keeley et al. 2011a). Multiple chemicals may contribute to seed germination but are influenced by factors such as temperature and other environmental conditions (Nelson et al. 2012).

In addition to seeds that survive fire in the soil seed bank, many conifer species respond to fire via serotiny. Unlike the cones of most conifers that release seeds when they are mature, serotinous cones remain sealed under normal weather conditions. However, when temperatures are high enough to melt the resin that bonds the cone scales, usually between 45°C and 50°C (113–122°F), serotinous cones open and release seeds (Lotan 1976). These relatively low temperatures allow seeds to remain viable. For example, lodgepole pine (*Pinus contorta* subsp. *murrayana*) cones withstand brief 30–60 s exposures to temperatures as high as 1,000°C (1,832°F) and still produce viable seeds (MacAulay 1975).

In California, serotinous species include knobcone pine (*P. attenuata*), lodgepole pine, Bishop pine (*P. muricata*), Monterey pine (*P. radiata*), Coulter pine (*P. coulteri*), giant sequoia (*Sequoiadendron giganteum*), and many cypresses (e.g., Tecate and Cuyamaca [*Hesperocyparis forbesii* and *H. stephensonii*]). Closed cones can remain on the tree for a decade or more until a fire opens them and releases the stored seeds postfire. Lodgepole pine and Coulter pine produce both serotinous and nonserotinous cones. The degree of serotiny in Coulter pine varied across California amongst plant communities (Borchert 1985). Serotiny is greater in plant communities with fire regimes characterized by uniform, high intensity, crown fires, but serotiny is less well developed in communities where variable surface fires are more common (Radeloff et al. 2004).

SEED DISPERSAL, COLONIZATION, AND ESTABLISHMENT

For some species, seedling establishment on burned sites is essential for site reoccupation. Turner et al. (1998) classified species as residual or colonizer species. Residual species survive fire as individual plants, through the seed bank, or by sprouting. Colonizers do not survive the fire, or were not present at the site before the fire, but traveled to the burned area and became established after the fire.

The ability of plants to colonize burned areas where they were previously absent or where all of the adults were killed depends on how far seeds disperse. Seeds with wings, such as big-leaf maple (*Acer macrophyllum*) or white fir, travel relatively short distances. Other plants such as cottonwood, thistle (*Cirsium* spp.), or willow, have small, light seeds encased in feathery or fluffy fruits that travel long distances, particularly by the wind. Animals carry seeds on their fur that have burs or hooks, such as some grasses and fiddle-necks (*Amsinckia* spp.). Animals also eat fruits, ingest the seeds, and disperse them through their scat while traveling through a burned area. Other animals, particularly rodents, gather and bury nuts, acorns, and seeds. This is important dispersal and recolonization mechanism for Coulter pine (Borchert 1985).

A number of species have enhanced seedling survival in conditions created by fire. For example, seeds of many conifers establish and survive better when they germinate on bare, mineral soil exposed by fire (see giant sequoia sidebar in chapter 14). Endosperm provides immediate food and water for an emerging seedling. Because small seeds, such as conifer seeds, have very little endosperm, seedling survival depends upon rapid root penetration into the soil so that water and nutrients can be absorbed. Thus, when seeds land on bare mineral soil, new roots rapidly reach the water and nutrient resources in the soil and seedlings can survive and grow (Stark 1965). In contrast, when seeds land on accumulated duff and/or litter, the new root may not penetrate far enough through the dry litter and duff into the soil to access the needed water and nutrients and in these conditions seeds with a large endosperm are more likely to survive.

Variation in Fire Response Traits

Fire response traits, including sprouting, serotiny, and fire-triggered germination, vary among closely related species and even within species and populations. The strong sprouters tend to have buds buried deeper in the soil. In the California genus *Ceanothus*, there exists a deep split into two subgenera, the Ceanothus clade, which are mostly sprouters, while the Cerastes clade are primarily nonsprouters. Another common chaparral genus, *Arctostaphylos*, includes 10 resprouting species that have obligate seeding subspecies (Keeley et al. 2011b). Vesk and Westoby (2004) suggested that sprouting response was better characterized as a continuum rather than a dichotomy (sprouts or does not sprout). For example, Vesk et al. (2004) reported that a 79% probability of sprouting for "strong" sprouters after fire in contrast to a 6% probability for "weak" sprouters.

There are within-species differences in the fire response as well. This variable response may be partly genetic variation, partly local environment, and may also be due to the more variable fire patterns in areas dominated by surface fire regimes. Antelope bitterbrush in the Northeastern Plateaus bioregion displays variation in sprouting behavior (see chapter 13). Variation in degree of serotiny within a population has been demonstrated for lodgepole pine (*Pinus contorta* var. *latifolia*) in the Rocky Mountains (Perry and Lotan 1979), jack pine in Quebec (Gauthier et al. 1996) and Wisconsin (Radeloff et al. 2004), and in several other species around the world. In California, there is evidence of variation for fire-stimulated germination for bigpod ceanothus (*Ceanothus megacarpus*) (Carroll et al. 1993), and bark thickness in Monterey pine (Stephens and Libby 2006) when Channel Island populations were compared to mainland populations.

Predicting Species Responses to Fire

Although the response to fire of individual plant structures (e.g., stem resistance to heat) or characteristics (e.g., sprouting) determines whether an individual plant dies, the entire suite of fire response characteristics governs species total response to fire and to fire regimes (Keeley 1998). Various conceptual models of responses to fire have been developed to help characterize and predict fire effects of single plant species or of entire plant communities (e.g., Nobble and Slayter 1980, Rowe 1983, Bond and van Wilgen 1996). Pausas et al. (2004, p.1085) based their model on the "*. . .premise that predicting vegetation change can be accomplished with the use of plant functional types . . . [based on] the combination of life-history traits and [classifying] species into a set of functional groups that best represent the range of strategies present*" [text added].

Most classifications of fire responses have focused on shrublands with crown fire regimes, such as chaparral of California, fynbos of South Africa, and the garrigue and maquis in Mediterranean areas of Europe (Pausas et al. 2004, Ojeda et al. 2010, Keeley et al. 2011b, Enright et al. 2014), though pines have also been studied this way (Schwilk and Ackerly 2001). In these crown fire systems, emphasis has been on suites of adaptive characteristics and trade-offs between sprouting and seed-dependent regeneration. There has been less attention to variation in suites of characteristics associated with other fire regimes (e.g., those dominated by crown fires vs. surface fires or mixed fire types), such as those found throughout the California bioregions.

When comparing species and plant community responses to different fire regimes across California, some general patterns emerge in which the types and proportion of species with different adaptive characteristics align with differences in overall fire regime type. In this section, we describe plant fire response classifications based on those described by Bond and van Wilgen (1996) and more recently by Pausas and Keeley (2014). One way to describe fire response types is based on the overall effect of fire on a plant and its reproduction. Does fire stimulate the plant, harm it, or have no effect? (Bond and van Wilgen 1996). Fire-stimulated responses are those that increase with fire, such as seed germination or sprouting. Fire-stimulated plants are further divided into fire-dependent and fire-enhanced categories, although plants not stimulated by fire are either fire-neutral or fire-inhibited. Fire-dependent refers to responses that only occur with fire, such as seeds that only germinate with heat, smoke, or chemicals from charcoal, though species that are typically described as fire-dependent may have a very limited response in the absence of fire (e.g., bigpod ceanothus germination as described in Carroll et al. [1993]). It should be noted that not all plants fall neatly into those functional types. Fire enhanced refers to responses such as sprouting that are increased by fire but also occur in response to other types of damage to the plant, such as flooding or grazing.

The response that is solicited can be divided into two types based on whether the species is a sprouter or nonsprouter and whether the species exhibits a postfire pulse of seed recruitment or depends on colonization (Pausas et al. 2004). Specific postfire strategies include sprouting as well as storage of seeds in either the canopy (e.g., serotinous cones) or the soil combined with heat or smoke-triggered seed release or germination. Depending on which combination of the above strategies a species displays, it can be classified as an obligate sprouter (no postfire seed recruitment), a facultative seeder (can resprout but also has postfire seedling recruitment), or an obligate reseeder (cannot resprout and depends on postfire recruitment from seeds). The modes of response (e.g., seeder, sprouter, faculative per Pausas and Keeley [2014]) and the direction and degrees of response (dependent, enhanced, neutral, inhibited per Bond and van Wilgen [1996]), are presented in Table 8.2. It is important to note that the degree of response (row vs. columns of Table 8.2) of facultative seeders-sprouters is not well represented in this table because the degree of sprouting response can be different from the seeding response in the same species. For example, chamise (*Adenostoma fasciculatum*) responds to fire by sprouting and seeding, making it a facultative seeder-sprouter. It could be considered as a fire-enhanced sprouter on higher elevations sites and a fire-neutral sprouter on drier, lower elevation sites in southern California (Zedler et al. 1983). However, chamise seed only germinates well with fire (Keeley 1987) placing it in the "fire-enhanced" to "fire-dependent" category for seeding response.

Though most fire-prone regions have species that exhibit a combination of traits and strategies, some of these strategies are associated with particular flora and fire regimes. Obligate seeders with soil-stored seed banks, such as California-lilac species, are common in chaparral regions. In contrast, plants with serotinous cones, such as Monterey pine and Bishop pine, are rare in California relative to those that create soil seed banks.

In communities that have a surface fire dominated fire regime, such as most montane ponderosa pine and mixed-conifer forests in the Sierra Nevada, there tends to be a mix of species with varied responses to fire. Examples include non-sprouters that are fire resistant (e.g., ponderosa pine) with reproduction often enhanced by fire, sprouters that sprout strongly or moderately following fire but are not dependent upon fire (e.g., California black oak [*Quercus kelloggii*], bitterbrush [*Purshia* spp.]), and species with fire-enhanced germination (e.g., deer brush, mountain whitethorn, chaparral whitethorn [*Ceanothus leucodermis*]) or flowering (e.g., Mariposa lilies [*Calochortus* spp.]). Lodgepole pine cones, while serotinous throughout much of the species range, tend to be nonserotinous in the Sierra Nevada (Lotan 1976).

In bioregions with more variable or higher fire severity and intensity and a longer fire return interval, such as the Southeastern Desert and North Coast bioregions, species that do not resist fire well or lack fire-stimulated germination or sprouting are common. However, throughout most of California's bioregions, some species are fire dependent and many species are enhanced by fire or are able to take advantage of postfire conditions for recruitment.

Fire Effects on Nonvascular Plants

In this section, we focus on important groups of organisms other than vascular plants, including mosses, algae, and lichens (a symbiotic combination of algae and fungi). Discussion on ecosystem processes mediated by microorganisms and fungi, such as litter decomposition and nutrient cycling, can be found in chapter 7.

Algae grow on exposed soil and rock surfaces. Mosses are small, nonvascular plants that grow on moist soil, rock, downed wood, and tree boles. Lichens are found on rocks, tree boles, and hanging from tree branches. Lichens are capable of photosynthesis and have the ability to absorb nutrients from the substrate. Most nonvascular plants and lichens reproduce via spores, small seed-like structures that are able to travel great distances via water, animals, or the wind. Although the spores of most nonvascular plants are fire resistant, the plant bodies are not. Nonvascular plants and lichens that survive fire do so by avoiding fire or by relying upon insulation from soil to protect buried spores. Unlike vascular plants, nonvascular plants lack the ability to sprout. Some moss species, however, are known to be "fire-followers" (Viereck and Schandelmeier 1980, Eversman and Horton 2004).

Fire directly affects nonvascular plants through heat and desiccation. The foliage of lichen, algae, and moss species are very sensitive to heat while spore sensitivity varies among species. Interactions between fire and nonvascular plants can be understood in broad categories of fire behavior (crown vs. surface) and habitat: (surface vs. forest canopy or aerial habitats). Nonvascular plants that occur in litter or on downed wood are particularly vulnerable to surface fires. The impact of fire on these organisms depends upon fire intensity and patchiness. The direct effect of fire on nonvascular plants living on tree boles and in the canopy depends upon their proximity to flames, such that surface fires have little effect on arboreal nonvascular plants while crown fires can have a large effect.

Plant Population–Fire Interactions

A *population* is defined as a group of individuals of the same species that potentially interbreed and together form a gene pool. In order for a population to persist, individuals must survive and reproduce. The size of the area occupied by a population and the distribution of individuals within that area are both important characteristics of a population. A metapopulation is composed of multiple populations that infrequently interbreed within a portion of the entire extent of a species (Levins 1969). While some populations persist, others in a metapopulation may go extinct. The overall distribution and trend of the populations that comprise the metapopulation provides important indicators of species genetic diversity and are useful in characterizing the pattern and significance of population and species change in space and time (Alexander et al. 2012). In this section, we discuss interactions between plant populations with different fire-adaptive characteristics and seven fire regime attributes (as described in chapter 5).

Plant Populations and Fire Regime Characteristics

Plant population responses to fire are based on the entire fire regime rather than a single attribute. However, we describe general response patterns for species adapted to different fire regime attributes. In the bioregional chapters, examples of integrated species responses to whole fire regimes are discussed.

SEASONALITY

The season of fire influences plant populations at two life history stages: (1) young established plants: through the increased susceptibility to injury and fire damage for cohorts at a young, sensitive stage of development; and (2) seeds and seedlings: through reduction in mature individual seed production and in seedling recruitment for the year of the fire. These individual effects can influence population size and age distribution, which in turn can influence population growth and reproduction. These fire effects can be near-term, such as destruction of the fire year's progeny recruitment or reduction of particular age-classes, or long-term, such as depressed reproductive capacity or changes in the population's competitive advantage in the community.

Fires that burn during flowering or seed development result in greater mortality and population reduction than fires that burn during other times of the year. Even if reproduction is not directly impacted, energy reserves may be reduced or utilized for responding to fire effects. As a result, population recovery can be slow and seed production for the year can be reduced. Fire effect on a population is also a function of plant life cycle or lifeform. Annual species, dependent on annual seed production and recruitment, are more likely affected by variation in fire seasonality than are perennial species.

Populations that rely on seed banks for regeneration also can be affected by fire timing. Seasonal fluctuations in soil moisture influences heat transfer through the soil and therefore seed mortality (Parker 1986). A species seed bank may be heavily impacted if fire occurs when soils are wet enough to conduct heat deeply but not wet enough to absorb the heat. This is particularly important for species populations with seeds that imbibe water and for species that are easily stem-killed.

Populations of sprouting species can be enhanced by an early growing season fire even though there may be high aboveground mortality. During this time newly formed foliage is sensitive to heat; but individuals also have high below ground energy reserves to produce a vigorous sprouting response.

FIRE SIZE

The spatial overlap between a fire and population extent determines the effect that fire has on a population. A small or large fraction of a plant population can be affected by a fire, depending upon the species distribution in relation to fire extent. Interactions between fire size and population distribution influence populations at the landscape scale.

Rarely does a single fire extend across the entire range of a species but the distribution of populations within its range can affect the degree and type of impact fire can have on the species. Plant populations occur in three general patterns: (1) endemic; (2) patchy; and (3) continuous. A population with an endemic distribution is one in which the plant is limited to a small geographic area (Cox and Moore 2000). Examples include Brewer's spruce in the Klamath Mountains bioregion and bristlecone fir (*Abies bracteata*) in the Central Coast bioregion. A patchy distribution refers to a species that is widespread, but within that broad area, individual occurrences are clustered. Examples include mountain dogwood (*Cornus nuttallii)* or trail plant (*Adenocaulon bicolor*), both of which are found in discontinuous areas of the North Coast, Southern Cascades, Klamath Mountains, and Sierra Nevada bioregions. A continuous distribution pattern refers to an extent that is widespread and roughly evenly distributed across the area. Ponderosa pine exhibits a continuous distribution.

Population distributions usually refer to the distribution of plants. However, the distribution of viable seeds in the seed bank is as important and does not necessarily overlap with the living, aboveground population. Determining the potential distribution of plants that have long-lived seeds accumulated in the soil can be difficult (Parker 1986).

The importance of an individual plant population varies with the distribution pattern of the species. For example, a single fire can have a large effect on an endemic population, because a single fire could encompass most or all of its populations. Conversely, a single fire is unlikely to have a significant effect on a wide-ranging species, because the fire would only affect a small number of populations. The exception would be if the fire event fragmented the population by creating a barrier to dispersal and colonization.

The interaction between a fire and a species population depends on whether the species is fire-dependent, fire-enhanced, fire-neutral, or fire-inhibited. For fire-dependent species, a single fire often has positive effects on abundance and distribution within the extent of the fire. For example, the rare geophyte, Pleasant Valley mariposa lily (*Calochortus clavatus* var. *avius*), was thought to be very limited in distribution in the central Sierra Nevada until a major wildfire occurred. The Cleveland fire in Eldorado County stimulated flowering of the plant, and the reported distribution and abundance increased significantly after the fire; although a fire stimulated response is likely, at least part of this increased number of observations could be due to increased visibility (Forest Service 1997). For fire-dependent species, the greater the spatial overlap between a fire and the population, the greater the positive effect on the overall population. Fire-enhanced species respond similarly to fire-dependent species but may be more influenced by the severity of the fire and how it changes the physical environment for the plant, and its competitive ability relative to other species. Fire effects on fire-intolerant species populations are typically negative and vary with the amount of population burned by the fire.

SPATIAL COMPLEXITY

Plant populations are affected by the pattern of fire intensity and severity as well as overall fire size. The effects of fire spatial patterns on populations are influenced by (1) the population's spatial distribution: whether the species is widespread or restricted in extent, continuous or patchy in distribution; (2) the dispersal characteristics of the species; and (3) fire magnitude fire intensity and severity. The pattern of fire intensity (chapter 5) influences the effect of fire on the portion of a plant population within a fire (Fig. 8.6). At one extreme, a fire with uniform intensity will have uniform effects, either positive or negative, on the survival, age-class distribution, abundance and distribution of individuals in a plant population. At the other extreme, a complex fire of different intensities will have varied effects on plant populations within the area burned. Typically, crown fires are considered to be more uniform, whereas surface fires are considered to be more complex, particularly at fine spatial scales. Mixed severity fires are also complex but on a coarser, often patchy scale. However, these common observations have never been verified with rigorous scientific testing.

As long as fire characteristics match the fire regime, fire-dependent and fire-enhanced species are more likely to respond positively to a uniform fire than to a variable one, because more individuals will be burned or otherwise positively affected. For fire-inhibited or fire-intolerant species, uniform intensity fires may result in large decreases in local abundance, including local extirpation, compared to the

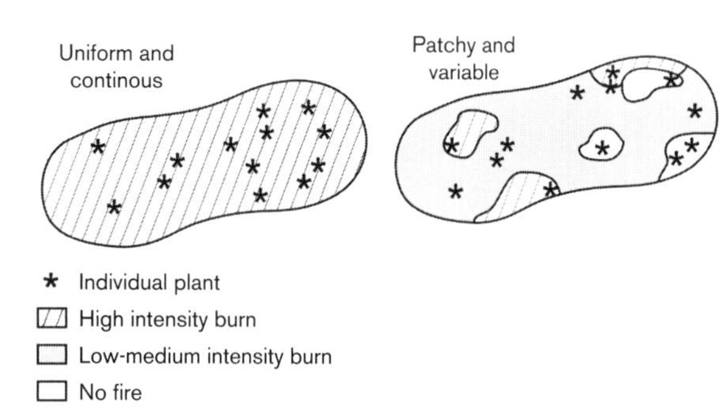

FIGURE 8.6 Interaction of fire spatial complexity and effects to plant populations. A fire of uniform intensity will have uniform effects, while a fire with variable intensities will have variable effects.

effects of a more discontinuous fire. Fires with greater spatial complexity would result in a higher probability of at least some individuals of each fire response-type surviving. However, this hypothesis has not been tested.

Studies from southern California chaparral illustrate interactions between fire pattern and plant population distributions. In chamise communities, variation in the amount and distribution of dead branches on the ground can cause variation in surface heating, which in turn can affect the distribution of surviving seeds and sprouting tissues. High surface heating in dense shrub clumps can reduce regeneration of most species in those clumps (Odion and Davis 2000), and can result in small-scale patterns of germinable seeds and seedlings (Borchert and Odion 1995). Patchiness of plant populations in postfire chaparral has also been attributed to uneven prefire seed distributions, such as those of annuals that are concentrated in canopy gaps (Davis et al. 1989).

For many fire-dependent species, distribution patterns remain relatively stable through repeated burns due to at least two mechanisms (Odion and Davis 2000). First, seeds of many fire-dependent species have thick, heavy coats, and therefore do not disperse widely. Secondly, sprouting species retain roughly constant spatial distributions since they regenerate "in place." As a result, many fire-dependent populations retain fairly fixed spatial distributions.

Plant species' prefire distributions and a fire's spatial pattern can affect postfire plant dispersal and recolonization (Hanski 1995), thereby affecting the plant population as a whole. For example, large burned areas are more likely to be recolonized by species with continuous distributions because the seeds are usually proximate to burned areas. However, if a complex fire overlaps with a species with a patchy distribution, fire could significantly reduce the ability of that species to survive and recolonize after fire. Similarly, a fire that results in mortality and extirpation of a subpopulation or a significant proportion of the entire population can create a gap in species distribution that extends beyond typical dispersal distances between populations; consequently, the likelihood of recolonization is reduced (Hamrick and Nason 1996).

Large, high-severity patches in low-severity fire regimes can have huge impacts on populations and community's ability to regenerate. If high-severity patches are larger than seed dispersal distances, there is a potential for the area to convert to a new vegetation type, particularly with subsequent fires. The distances seeds travel are affected by differences in travel speed, mode (i.e., air, water, animal), and timing. Species that have high dispersal distances, such as fireweed, will be more capable of recolonizing large burned areas than those that have low dispersal capabilities. Ponderosa pine, for example, has seed dispersal of only 200–450 m (650–1,475 ft) so that recolonization of large high-severity patches may take hundreds of years, assuming there is no recurring fire (Haire et al. 2013). These large, high-severity patches could convert from forests to shrublands and grasslands (Barton 2002, Savage and Mast 2005).

FIRE RETURN INTERVAL

The consequences of fire return intervals on plant populations depend on fire interval length relative to age at which plants reach reproductive maturity and on the ability of plants to survive between fires. The effects of fire return interval are most pronounced with fire-dependent species. Both length of time between fires and regularity of these intervals influences population dynamics. Fire return intervals have shifted in response to fire suppression, human-caused ignitions, climate change, and other factors, with consequences for plants that are adapted to a particular fire interval.

Zedler (1995) described two main risks to fire-dependent or fire-enhanced plants that arise from varying intervals between fires. Senescence risk is present when intervals between fires are so long that the capacity of a species to recover from fire is diminished by a combination of the reduction or loss of input of fresh seeds owing to death and malaise of reproductive individuals and age-related loss of remaining stored seeds viability. Immaturity risk is present when the reproductive capacity of a species is reduced because the fire return interval is too short. For obligate seeding species, this will be when few or no plants have reached reproductive age and size and therefore no seed bank is present. The few sprouting species that can suffer significant mortality from fire and that reestablish abundantly from stored seeds after fire will likewise risk significant population declines if few or no resprouting plants have not reached reproductive age and size.

Some species minimize senescence risk by producing seeds that remain viable for long periods. For example, the longevity of Gregg's ceanothus is less than the average intervals between fires (Zedler 1995), and the perpetuation of the species is dependent upon seed bank accumulations. Some California-lilac seeds maintain viability for hundreds of years (Noste 1985), so the populations may be able to overcome senescence risk through reliance on seed banks. Geophytes, such as lilies (*Lillium* spp.), irises (*Iris* spp.), and onions (*Allium* spp., *Brodiaea* spp.), persist in a dormant state for decades with little growth or flowering (Keeley 1991). Thus, geophytes overcome senescence risk by living longer through these underground reserves. Long-lived species, such as giant sequoia, can overcome senescence risk by individuals surviving extremely long fire free periods. In summary, three strategies to minimize senescence risk rely upon seed banks, underground reserves, and long life spans.

Giant sequoia reproduction and recruitment are heavily dependent upon fire; the cones are serotinous and seeds require mineral soil for optimum establishment and survival (Rundel 1969). However, because individuals live for thousands of years, the population can persist through widely varied fire return intervals, still successfully reproduce, and recruit new cohorts to perpetuate the population when fires do occur. Many chaparral species, especially nonsprouters, face immaturity risk in southern California due to shortened fire intervals and invasive grasses (the "grass-fire cycle")

(D'Antonio and Vitousek 1992, Enright et al. 2014). If a second fire occurs before seedlings reach reproductive maturity or before the population can build up a sufficient seed bank, the population will have difficulty recovering and type conversion may occur.

Variability in intervals between fires can also influence populations. Although the focus in fire regime literature is often on the average fire return interval, the distribution of intervals as discussed in chapter 5 may be more important in understanding plant–fire interactions (Buma et al. 2013). Regular, short-return fire intervals tend to favor abundant populations of fire-resistant or fire-tolerant species. For example, populations of annual plants or sprouters may be little affected by recurrent frequent fire because they are able to reproduce or survive following fire and maintain abundant population levels. In contrast, populations of species that are not fire-resistant or fire-tolerant may persist with fire return intervals that are on average frequent but that include occasional long intervals between fires, but this also depends on fire intensity. During long nonfire intervals, such species can establish larger and more widely distributed populations that have a greater likelihood of surviving a subsequent fire. For example, white fir achieves fire tolerance at maturity but is vulnerable to fire before then. Populations of white fir can exist in an area when there are at least occasional long intervals between fires. Overall abundance of the nonfire resistant population would be higher in a landscape with a long fire return interval, but because it is long-lived, infrequent and irregularly occurring long return intervals can provide for a smaller but persistent population of a nonfire resistant species.

FIRE INTENSITY AND TYPE

Fire intensity and fire type, such as ground, surface or crown fire, interact with species characteristics to produce different fire severities. Fire intensity variably affects plant populations, depending on patterns of survival, reproduction, and recruitment in response to fire. Depending upon the species response and fire characteristics, fires of a given intensity can increase population size by stimulating recruitment of more individuals than originally existed, or reduce population size by killing mature individuals and seedlings, destroying the seed bank, or not significantly stimulating regeneration. Crown fires result in high mortality above ground and species with populations that persist tend to do so with sprouting, regenerating from the seed bank or colonizing. Surface fires, which are usually of low to moderate intensity, result in lower mortality and more varied individual responses within a population than do crown fires. Individuals may be killed, have aboveground mortality, or be unaffected or only partially consumed.

Most fire-resistant plant species have minor population fluctuations with low- to moderate-intensity fires but are significantly affected by very high-intensity fires. For example, mature individuals of ponderosa pine and Douglas-fir are slightly affected by low- to moderate-intensity fires because of their fire resistant characteristics, but experience high mortality with high-intensity fires. For some species, younger cohorts of a population may be more susceptible than older individuals to fire intensity. Ponderosa pine seedlings are more vulnerable to fire than older trees and are more likely to die from moderate- or even low-intensity fire. This means that low-intensity fires change age distributions, which has implications for future forest structure over the coming centuries, even if only younger trees are currently affected. Conversely, lack of fire may result in large changes in population structure (size and age distribution) and in species composition. Ponderosa pine can reach greater densities in the absence of fire, leading to greater mortality from tree competition. In species that lack fire resistant characteristics, fires of any intensity may reduce mature plant survival and population levels. For example, Brewer spruce (*Picea breweriana*) in the Klamath Mountains bioregion has thin bark, and fire of any intensity causes high mortality for young to old cohorts.

Fire intensity affects population recruitment from sprouts and the seed bank. Low-intensity fires enhance seedling establishment of some species, including pines (Vale 1979), providing that mineral soil is exposed. However, the response of a population to fire intensity is not always straightforward. In one low-intensity fire, high regeneration from root sprouts occurred in snowbrush but did not induce seed germination of the heat-dependent seeds (Noste 1985). After a more intense burn, snowbrush root sprouting was very low but seed germination was high. Thus, intensity did not affect the size or extent of the population, but did affect its age structure.

Plant Community–Fire Interactions

In this section, we focus on community characteristics and landscape distribution of community types that change with fire. California is host to an amazing array of plant communities and associated fire regimes. Plant community–fire interactions are often specific to each community type and fire regime. Many characteristic, unique, and dominant plant community–fire interactions are presented in more comprehensive discussions in part Two of this book. In this section, we outline the types and broad implications of plant community–fire interactions.

Communities and Their Components

Plant communities are comprised of all species that occur together and interact on a particular site. Three major aspects of plant communities include: (1) species composition, (2) community structure, and (3) interactions among plants and between plants and their environment. Plant composition refers to the number and proportion of different species in a given community. The size, arrangement, density, and plant lifeforms (i.e., shrub, tree, grass, herb, moss), make up the plant community structure.

Researchers have observed reciprocal effects between fire and vegetation (Agee 1996, Chang 1996). These are most evident at the community level and reveal themselves in landscape patterns of vegetation. The influence of vegetation structure and community composition on fire was illustrated in chapter 4 by detailing the effect of vegetation on fuel type, loading and arrangement, and on fire behavior. Fire, along with other factors such as climate fluctuations, influences plant community dynamics. Animals also influence plant communities by themselves or in concert with fire (chapter 9). Here, we discuss the relationship of fire and plant composition, fire and community structure, and indirect environment and climate effects on fire and community interactions.

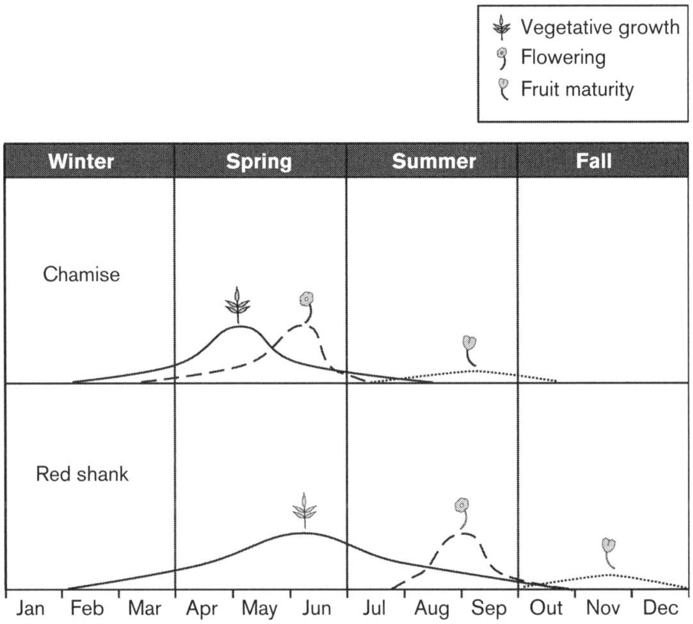

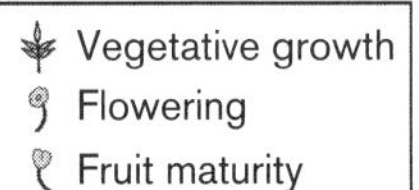

FIGURE 8.7 Different phenological patterns for two co-occurring southern California chaparral species. An early spring fire is likely to cause more damage to chamise than redshank because the chamise would be flowering at that time (redrawn from Hanes 1965).

Fire and Species Composition

The overall plant community response to fire is a function of the responses of its constituent species: the greater the response differences among several species, the greater the effect on the community composition and/or structure. Variation in species responses to the fire regime attributes of season, fire interval, and fireline intensity determine the effects of fire on the community.

SEASONALITY

Each species within a community responds differently to seasonal patterns of fire based on that species' phenology. For example, chamise and red shank are two common shrub species that often co-occur in southern California chaparral but have different phenologies (Fig. 8.7). Fires that occur in the early spring are likely to result in greater mortality of chamise than red shank because the chamise flowers earlier and is more sensitive to fire. Thus, following a spring fire red shank could become more dominant than chamise, thereby changing the community composition. This is just one example of how the season in which a fire occurs can change the plant species composition of a community.

Species composition can also affect the season when fires burn most readily. In areas of the Great Basin and the Southeastern Deserts where composition has shifted to annual grasses, fires tend to occur earlier in the year and earlier relative to the life cycles of other species. Annuals typically die early in the year and, if present in sufficient densities, provide a readily available fuel source. A clear example is a community that has become dominated by cheat grass (*Bromus tectorum*) (see chapters 13, 18, and 25 for details). Cheat grass dies back by early to mid-summer and creates a carpet of very flammable, available fuel. Ignitions spread easily during this time of year, and the fire season shifts toward early rather than late summer. This creates a positive feedback mechanism with the earlier fire season favoring cheat grass because cheat grass has already produced seed. In contrast, reestablishment of perennials that generally flower and produce seeds later in the year is diminished.

FIRE RETURN INTERVAL

Both the length and regularity of the fire interval are important in determining community composition. Longer fire intervals allow establishment and persistence of species that are fire-inhibited, or fire-neutral. Shorter fire intervals prevent fire-inhibited species from establishing and favor fire-enhanced and fire-dependent species. For example, ponderosa pine, a fire-enhanced species, often dominates communities where fire intervals are short. When occasional long fire intervals occur or when average fire intervals are longer, fire-inhibited young individuals such as Douglas-fir or white fir co-dominate communities with ponderosa pine or dominate communities at the expense of ponderosa pine (for examples, see sidebar 4.1 and the fire regime–plant community interactions section of chapter 14). Grasses and herbs are often able to out-compete young shrubs and saplings for water, nutrients, and light. In grassland systems, there is often a similar pattern of plant community responses to fire that retards or halts encroachment of woodland species.

Community species composition can also influence fire return interval (Mutch 1970). Communities dominated by ponderosa pine, with high annual production of flammable litter, are more likely to burn frequently because of accumulated of surface fuels. Similarly, communities dominated by species with naturally high concentrations of volatile oils, such as California sagebrush (*Artemisia californica*) (Montgomery and Cheo 1969, Rothermel 1976) or mountain misery (*Chamaebatia foliolosa*) (McDonald et al. 2004), may have higher fire frequencies because of greater inflammability. A similar shift in fire regime can occur in riparian zones invaded by giant reed (*Arundo donax*) because a positive feedback in fire and postfire establishment of giant reed favors the nonnative over native riparian species (Coffman et al. 2010).

FIRE INTENSITY

High-intensity fires often lead to plant communities with lower diversity and increased dominance of a few species. High-intensity fires favor species that have some mechanism for resistance to fire-related mortality or that establish or recruit in response to fire through sprouting, fire-stimulated seed banks, or wide spread seed dispersal. These reproductive strategies enable rapid growth following fire and can allow one or several species to quickly dominate a site where high-intensity fire has killed most other species. Chamise is a good example of this type of postfire domination by a single species.

If the fire is intense, the forest or shrubland may convert to herbs or grasses, at least temporarily, until shrubs and trees recolonize. The time needed for shrubs and trees to reestablish depends on burn intensity, patch size, topography, the species of herbaceous plants immediately taking over the site, and the presence of surviving seed on-site or adjacent vegetation supplying seed. Often, only a few years are needed for some shrub communities to reestablish, such as those dominated by chamise, some manzanitas, and California-lilac species. Tree-dominated communities take much longer to repopulate sites, but the transition can begin within 10 to 20 years if trees quickly establish and out-compete shrubs.

Community response to low-intensity fires differs from response to high-intensity fires. More species with low fire resistance or low regeneration response to fire can survive low-intensity fires. However, with increased species survival, competition among all species can be high in post-burn, low-intensity fire communities. In these situations, fire-dependent or fire-enhanced species may have a competitive advantage.

Fire and Community Structure

Shifts in forest or shrubland structure can occur after fires. Canopy closure can be reduced, and vertical or horizontal continuity can be altered, thereby influencing subsequent fire behavior. Fires affect the density and vertical structure of forests by reducing overall biomass, including dead and live fuel (Kilgore and Sando 1975). In general, stands thin differentially leaving some areas intact and others with fewer live plants immediately following fire. The result is a mosaic of unburned areas resembling the prefire condition, and burned areas with more open surface and canopy. Formation of such a mosaic increases the variety of habitats and extent of ecotones within an area.

Recurrent fire generally results in reduced small tree density (Kilgore and Sando 1975, Lathrop and Martin 1982, Keifer 1998) and a shift in tree diameter distribution toward larger trees (Gordon 1967). Frequent fire also reduces recruitment since younger, more vulnerable individuals are killed by fire. As a result, each cohort of individuals is less abundant, reducing not only the number of younger individuals, but also the recruitment pool for older individuals over the long run. With more frequent fires, it is also more likely that a plant community will either be composed of small (young) individuals or have an open mid-story. In forests, frequent fires can lead to crowns further above the ground due to frequent scorching of low branches and needles, allowing for fewer ladder fuels. Long fire intervals allow dense vegetation to develop.

Not only do fires affect plant community structure, but the resulting structure can also influence subsequent fire behavior and effects. Forests with high representation of young or understory trees have more ladder fuels, which can increase fire intensity and severity. Similarly, crown fires occur in coniferous forests when trees are densely packed, have low moisture contents, and high litter accumulation (van Wagtendonk 1996).

Disease and reduced plant vigor due to drought or other stressors can increase the fuel load but do not consistently increase likelihood and severity of fire. For example, some forests types infested with sudden oak death were found to be more susceptible to fire than noninfested forests, while susceptibility in other forest types was not increased with sudden oak death (Moritz et al. 2008, Kuljian and Varner 2009). Similarly, elevated wildfire activity was not observed to co-occur in areas infested with mountain pine beetles in the western United States over a three-year period (Hart et al. 2015).

The relationship between community structure and landscape position can also have important effects on fire. For example, Coulter pine patches surrounded by chaparral burned at higher severity than similar stands surrounded by grassland (Vale 1979). The Coulter pines were short in stature and formed an open canopy. The heavy fuel load in the adjacent chaparral created a high-intensity fire that was able to sweep into the Coulter pine stand and burn trunks high into the canopy. In contrast, fuel loads in the grasslands were not as high, so only a low-intensity surface fire moved underneath the adjacent pine stand. Riparian areas can also play an important role in fire movement across the landscape. The relation between riparian corridors and fire and fuel management are discussed in sidebar 21.2.

Indirect Environment and Climate Effects on Community–Fire Interactions

In addition to the direct, physical effects fires have on plant communities; fires also change the physical and chemical environment of the vegetation community. These changes can influence weather in the short term and climate in the long term and can affect changes in the fire regime, plant community, and fire–plant community interactions.

SHORT-TERM ENVIRONMENT–FIRE–PLANT COMMUNITY INTERACTIONS

Fires alter availability of nutrients, water, light, soil surface substrate and chemistry, and microclimate of plant communities (see chapter 7). These changes can differentially affect sprouting, growth, colonization, and establishment of varied species in a community. Vegetation mortality increases light, water, and nutrients available to surviving or establishing plants by decreasing competition among plants for these resources. Also, increased solar radiation and loss of litter cover increases soil temperature fluctuations. Warmer soil and increased light levels after a fire have been found to enhance the sprouting response. In annual grass communities, late spring fires have been shown to induce prolific flowering and seed production. Fires remove thatch and expose the mineral soil, which provides greater light and fluctuation

BOX 8.1. EXAMPLE OF BIOLOGICAL AND ECOLOGICAL LEVELS OF ORGANIZATION FOR BISHOP PINE

Level	Example	Example response to fire
Tissue	Bark	Weak protection of interior tissues from heat damage
Individual organism	Bishop pine	Individual dies; releases seeds from serotinous cones
Population	Bishop pines in Point Reyes National Seashore	High-severity, low-frequency fire results in near complete mortality of mature population with very high seed release from serotinous cones, followed by dense regeneration
Community	Mature forest overstory composed of bishop pine with patches of madrone, tanoak, coast live oak, and California bay; huckleberry, salal, and western swordfern	Succession following stand-replacing fire results in dense stands of regenerating pines interspersed with dense stands of blue blossom ceanothus and Marin Manzanita; over time stand thins and mid-story, shrub, and ground cover species of mature forest increase in percent cover
Landscape	Bishop pine communities occur in discontinuous pockets along Coastal California with heavy summer fog and granitic quartz-diorite soils; bounded by coastal prairie grassland, beach, Douglas-fir forest, pasture, and residential lands	Stands of different ages and successional stages are distributed along the coast in clusters of available and appropriate habitat. Age distribution among separate stands is a function of fire history

in ground temperatures which in turn promotes forb flowering and seed germination. Fire also kills seeds of most European grasses, thereby creating a temporary reduction in the annual grass population (Stone 1951, Reiner 2007).

The litter layer is often completely consumed by surface fires, leaving bare mineral soil. Many plants have greater establishment and seedling survival on mineral soil than on litter. After fire, species that sprout, have fire-stimulated seeds in the soil, colonize from external seed sources, or have enhanced germination on mineral soil often have a competitive advantage.

Weather proceeding and following fire can influence plant community response to fire. If conditions preceding fire were dry, plants may be drought-stressed and less able to respond to increased stress imposed by fire. After a fire, weather can play an important role in determining sprouting and germination levels, and seedling survival. Moisture following fire provides plants with the conditions necessary for successful seedling survival and successful sprouting. Moderate to intense rains after a fire can also increase erosion of topsoil and ash-derived nutrients, particularly in steep chaparral lands. Loss of topsoil and nutrient can affect short-term plant survival, growth, and species interactions.

LONG-TERM CLIMATE CHANGES

Changes in climate can shift fire–plant community interactions by affecting plant community characteristics and distributions. Climate change influences plant communities directly by changing the conditions favorable for individual plant species to establish and survive in different parts of the landscape, which, over long time periods, results in development of new plant communities. Over shorter time scales, climate changes can also affect stand age structure through effects on germination and establishment. Climate change can affect species composition by altering competitive species advantages under changing climate conditions, such as heat and moisture levels and seasonality.

Climate also affects community structure and fire patterns through changes in fuel loads resulting from increased stress-induced mortality and altered litter decomposition rates. These shifts in plant community composition and structure affect flammability, fire-season interactions, and community response to periodic fires. Climate change could result in a shift in the average fire return interval or just increase variation in the fire return interval, with one or a few fires occurring well outside of the historically normal range for a particular location (Buma et al. 2013). One example of this is the potential extirpation of local serotinous populations with a single fire that occurs before the population seed bank has matured (increase in variation) or with a more gradual shift in fire return interval until the period is outside of the viable range for the species (change in average return interval) (Buma et al. 2013). Periods of no fire that extend beyond the duration of their serotinous seed bank (e.g., seeds are lost to predation or decay) could also result in a decline or disappearance of a local population.

Compounding situations resulting from a changing climate, such as changes in precipitation or summer fog frequency, or increases in seed predation as wildlife habitat ranges expand and contract, could limit postfire seedling recruitment, and further stress local populations of serotinous species. At the landscape scale, loss of local populations due to shifts in the average or variance in the return interval could have larger effects on the metapopulation and resilience of the species in that region (Buma et al. 2013). Changes in climate that directly affect fire frequency and intensity include changes in duration and frequency of drought, changes in precipitation and changes in temperature, lightning ignitions, and winds. Much more work needs to be done to understand and appropriately respond to these changes from landscape, community, and species population perspectives.

Evidence of climatically driven shifts in fire regimes has been inferred from correlations between fluctuations in charcoal levels and climate, as interpreted from buried pollen, over centuries and millennia (Anderson and Smith 1997). These findings indicate that fire frequencies and intensities varied over long time periods. Over the scale of thousands of years, Swetnam (1993) found variation in both fire intervals and synchronicity between cohorts of giant sequoias based on fire scars and apparent changes in climate, deduced through tree ring analysis. Millar and Woolfenden (1999) took a more detailed look at interactions between climate, fire regimes, and plant communities on the eastside of the southern Sierra Nevada. These communities include long-lived species such as California red fir (*Abies magnifica* var. *magnifica*), whitebark pine (*Pinus albicaulis*) and lodgepole pine, all of which can live for well over 400 years. The authors found that fire return intervals, recorded as fire scars on trees, varied substantially on the scale of centuries. Some centuries, including the most recent, had little evidence of fire. Other centuries had more regular and frequent fires, and others had longer, more irregular intervals. The plant community compositions and age structures varied with changes in fire return interval, with shifts in dominance of lodgepole pine or red fir in various cohorts, depending upon the century. These studies illustrate that plant community–fire interactions may vary over time, even at the relatively short temporal scale of centuries.

Fire and Geographic or Landscape Patterns of Communities

Plant communities also respond to fire by changes in the geographic distribution of the community. Communities expand, contract, are converted to different types, or change their community landscape patterns in response to changes in the fire regime.

Repeated low-intensity fire or occasional high-intensity fire can result in contraction or expansion of adjacent plant communities in relation to each other. For example, some grassland communities in various bioregions of California are maintained or can expand with recurrent fire (Reiner 2007). Without recurrent fire, trees from adjacent forests or woodlands become established in grasslands and eventually shade out the grass. This results in an expansion of the forest or woodland communities into the grassland. In contrast, with recurrent fire in the grasslands, less fire resistant tree seedlings are unable to survive repeated fire and fail to establish or survive. Examples of this are discussed in the bioregional chapters in part Two of this text. Often the expansions or retractions are influenced not only by fire, but also by fluctuations in climate and/or soil moisture conditions.

Bioregional boundaries, such as between the east slopes of the Sierra Nevada and the Great Basin, are characterized by differences in precipitation and temperature that result in changes in the suite of plant communities present. As precipitation increases, forests or woodlands replace grasslands or open shrublands. Climate at the boundaries may represent the limit of what plant species can tolerate to survive. For example, the boundary might represent the area where precipitation becomes insufficient or surface temperatures become too high for survival. Sometimes, plant seedlings at bioregional boundaries depend on shade protection from overstory trees to establish and survive. With high-intensity fire that kills the protective overstory, seedlings of those species that were present only because of the ameliorating effects of the canopy on the plant microclimate can no longer survive and establish in the burned area. Other species that can tolerate the site conditions dominate, and the plant community composition changes from prefire conditions. This change or conversion to a new type of community may persist for decades or centuries. Type conversions also occur following intense fire when nonnative species invade and dominate an area after fire, excluding future recolonization by native species. Examples include conversions to nonnative annual grasslands from many scrub and sagebrush communities in the state.

Summary and Conclusions

Fire directly affects individual plants through heat, smoke, and charcoal, and indirectly affects plants through fire-induced changes in nutrient and light availability and germination conditions. Plants respond to these direct and indirect effects in myriad ways. Many species have physical characteristics, such as thick bark, that enable them to survive fires. Fires, depending on fire intensity, distribution, and season, also enhance reproduction in some species. Other species are negatively affected by fire and proliferate during long fire-free periods. Fire seasonality, return interval, size, spatial complexity, intensity, severity, and type all affect individual plant survival and reproduction, plant population extent and age structure, and plant community structure and species composition. Plant community structure and composition have feedback effects on fire frequency, intensity, extent, and season. Shifts in fire regime characteristics, such as fire size, severity, frequency, and seasonality, can occur due to climate change and human influence; these changes can also affect the relationship between plants and fire. As we learn more about fire interactions among plant, animal, and microbial communities, we will increase our understanding of how fire functions as an ecological process and how we might best manage fire in California's dynamic ecosystems.

References

Agee, J.K. 1996. The influence of forest structure on fire behavior. Pages 52–68 in: J. Sherlock, editor. Proceedings: Forest Vegetation Management Conference. January 16–18, 1996, Redding, California, USA.

Alexander, H.M., B.L. Foster, F. Ballantyne, IV, C.D. Collins, J. Antonovics, and R.D. Holt. 2012. Metapopulations and metacommunities: combining spatial and temporal perspectives in plant ecology. Journal of Ecology 100: 88–103.

Anderson, R.S., and S.J. Smith. 1997. The sedimentary record of fire in montane meadows, Sierra Nevada, California, USA: a

preliminary assessment. Pages 313–327 in: J.S. Clark, H. Cachier, J.G. Goldammer, and B.J. Stocks, editors. Sediment Records of Biomass Burning and Global Change. NATO ASI Series 51. Springer, Berlin, Germany.
Barton, A.M. 2002. Intense wildfire in southeastern Arizona: transformation of a Madrean oak–pine forest to oak woodland. Forest Ecology and Management 165: 205–212.
Bond, W.J., and J.J. Midgley. 2001. Ecology of sprouting in woody plants: the persistence niche. Trends in Ecology and Evolution 16: 45–51.
Bond, W.J., and B.W. van Wilgen. 1996. Fire and Plants. Chapman and Hall, New York, New York, USA.
Borchert, M.I. 1985. Serotiny and cone-habitat variation in populations of *Pinus coulteri* (Pinaceae) in the southern coast ranges of California. Madroño 32: 29–48.
Borchert, M.I., and D.C. Odion. 1995. Fire intensity and vegetation recovery in chaparral: a review. Pages 91–100 in: J.E. Keeley and T. Scott, editors. Brushfires in California Wildlands: Ecology and Resource Management. International Association of Wildland Fire, Fairfield, Washington, USA.
Bradshaw, S.D., K. Kingsly, S.D. Hopper, H. Lambers, and S.R. Turner. 2011. Little evidence for fire-adapted plant traits in Mediterranean climate regions. Trends in Plant Science 16: 69–76.
Buma, B., C.D. Brown, D.C. Donato, J.B. Fontaine, and J.F. Johnstone. 2013. The Impacts of changing disturbance regimes on serotinous plant populations and communities. BioScience 63: 866–876.
Carroll, M.C., L.L. Laughrin, and A.C. Bromfield. 1993. Fire on the California Islands: does it play a role in chaparral and closed cone pine forest habitats? Pages 73–88 in: F. Hochberg, editor. Third California Islands Symposium: Recent Advances in Research on the California Islands. Santa Barbara Museum of Natural History, Santa Barbara, California, USA.
Chang, C. 1996. Ecosystem responses to fire and variations in fire regimes. Pages1071–1099 in: D.C. Erman, general editor. Sierra Nevada Ecosystem Project: Final Report to Congress, Volume II. Wildland Resources Center Report 37. University of California, Davis, California, USA.
Clarke, P.J., M.J. Lawes, J.J. Midgley, B.B. Lamont, F. Ojeda, G.E. Burrows, N.J. Enright, and K.J. Knoxl. 2013. Resprouting as a key functional trait: how buds, protection and resources drive persistence after fire. New Phytologist 197: 19–35.
Coffman, G.C., R.F. Ambrose, and P.W. Rundel. 2010. Wildfire promotes giant reed (*Arundo donax*) invasion in riparian ecosystems. Biological Invasions 12: 2723–2734.
Cox, C.B., and P.D. Moore. 2000. Biogeography: An Ecological and Evolutionary Approach. 6th ed. Blackwell Science, Oxford, UK.
D'Antonio, C.M., and P.M. Vitousek. 1992. Biological invasions by exotic grasses, the grass/fire cycle, and global change. Annual Review of Ecology and Systematics 23: 63–87.
Davis, F.W., M.I. Borchert, and D.C. Odion. 1989. Establishment of microscale vegetation pattern in maritime chaparral after fire. Vegetatio 84: 53–67.
DeBano, L.F., D.G. Neary, and P.F. Ffolliott. 1998. Fire Effects on Ecosystems. John Wiley and Sons. New York, New York, USA.
———. 2005. Soil physical properties. Pages 29–51 in: D.G. Neary, K.C. Ryan, and L.F. DeBano, editors. Wildland Fire in Ecosystems: Effects of Fire on Soil and Water. USDA Forest Service General Technical Report RMRS-GTR-42-VOL. 4. Rocky Mountain Research Station, Fort Collins, Colorado, USA.
Dell, J.D., and C.W. Philpot. 1965. Variations in the moisture content of several fuel size components of live and dead chamise. USDA Forest Service Research Note PSW-RN-83. Pacific Southwest Forest and Range Experiment Station, Berkeley, California, USA.
Dwire, K.A., and J.B. Kauffman. 2003. Fire and riparian ecosystems in landscapes of the western USA. Forest Ecology and Management 178: 61–74.
Enright, N.J., J.B. Fontaine, B.B. Lamont, B.P. Miller, and V.C. Westcott. 2014. Resistance and resilience to changing climate and fire regime depend on plant functional traits. Journal of Ecology 102: 1572–1581.
Eversman, S., and D. Horton. 2004. Recolonization of burned substrates by lichens and mosses in Yellowstone National Park. Northwest Science 78: 85–92.
Fites-Kaufman, J., D.A. Weixelman, and A.G. Merrill. 2008. One-year post fire mortality of large trees in low- and moderate-severity portions of the Star Fire in the Sierra Nevada. USDA Forest Service General Technical Report PSW-GTR-189. Pacific Southwest Research Station, Albany, California, USA.
Forest Service 1997. Whale Rock Forest Health Multi-Resource Project: Final Environmental Impact Statement. Eldorado National Forest, Pacific Ranger District, Pollock Pines, California, USA.
Gagnon, P.R., H.A. Passmore, W.J. Platt, J.A. Myers, C.E.T. Paine, and K.E. Harms. 2010. Does pyrogenicity protect burning plants? Ecology 91: 3481–3486.
Gauthier, S., Y. Bergeron, and J-P. Simon. 1996. Effects of fire regime on the serotiny level of jack pine. Journal of Ecology 84: 539–548.
Gill, A.M. 1981. Fire adaptive traits of vascular plants. Pages 208–230 in H.A. Mooney, T.M. Bonnicksen, N.L. Christensen, J.E. Lotan, and W.A. Reiners, editors. Proceedings of the Conference Fire Regimes and Ecosystem Properties. USDA Forest Service General Technical Report WO-26. Washington, D.C., USA.
Gordon, D.T. 1967. Prescribed burning in the interior ponderosa pine type of northeastern California. USDA Forest Service Research Paper PSW-RP-45. Pacific Southwest Forest and Range Experiment Station, Berkeley, California, USA.
Haire, S.L., K. McGarigal, and C. Miller. 2013. Wilderness shapes contemporary fire size distributions across landscapes of the western United States. Ecosphere 4(1): 15.
Hamrick, J.L., and J.D. Nason. 1996. Consequences of dispersal in plants. Pages 203–236 in: O.E. Rhodes, R.K. Chesser, and M.H. Smith, editors. Population Dynamics in Ecological Space and Time. University of Chicago Press, Chicago, Illinois, USA.
Hanes, T. D. 1965. Ecological studies on two closely related chaparral shrubs in southern California. Ecological Monographs 35(2): 213–235.
Hanski, I. 1995. Effect of landscape pattern on competitive interactions. Pages 203–224 in: L. Hansson, L. Fahrig, and G. Merriam, editors. Mosaic Landscapes and Ecological Processes. Chapman and Hall, New York, New York, USA.
Harrington, M.G. 1993. Predicting *Pinus ponderosa* mortality from dormant season and growing-season fire injury. International Journal of Wildland Fire 3: 65–72.
Hart, S.J., T. Schoennagel, T.T. Veblen, and T.B. Chapman. 2015. Area burned in the western United States is unaffected by recent mountain pine beetle outbreaks. Proceedings of the National Academy of Sciences 112: 4375–4380.
Hood, S.M., A. Sala, E.K. Heyerdahl, and M. Boutin. 2015. Low-severity fire increases tree defense against bark beetle attacks. Ecology 96: 1846–1855.
James, S. 1984. Lignotubers and burls—their structure, function and ecological significance in Mediterranean ecosystems. Botanical Review 50: 225–266.
Keeley, J.E. 1981. Reproductive cycles and fire regimes. Pages 231–277 in: H.A. Mooney, T.M. Bonnicksen, N.L. Christensen, J.E. Lotan, and W.A. Reiners, editors. Proceedings of the Conference Fire Regimes and Ecosystem Properties. USDA Forest Service General Technical Report WO-26. Washington, D.C., USA.
———. 1987. Role of fire in seed germination of woody taxa in California chaparral. Ecology 68: 434–443.
———. 1991. Seed germination and life history syndromes in the California chaparral. The Botanical Review 57: 81–116.
———. 1998. Coupling demography, physiology and evolution in chaparral shrubs. Pages 257–264 in: P.W. Rundel, G. Montenegro, and F.M. Jaksic, editors. Landscape Disturbance and Biodiversity in Mediterranean-Type Ecosystems. Springer, New York, New York, USA.
Keeley, J.E., W.J. Bond, R.A. Bradstock, and J.G. Pausas. 2011b. Fire in Mediterranean Ecosystems: Ecology, Evolution and Management. Cambridge University Press, Cambridge, UK.
Keeley, J.E., and C.J. Fotheringham. 1997. Trace gas emissions and smoke-induced seed germination. Science 276: 1248–1250.
Keeley, J.E., and S.C. Keeley. 1987. Role of fire in the germination of chaparral herbs and suffrutescents. Madroño 34: 240–249.
Keeley, J.E., J.G. Pausas, P.W. Rundel, W.J. Bond, and R.A. Bradstock. 2011a. Fire as an evolutionary pressure shaping plant traits. Trends in Plant Science 16: 406–411.
Keeley, J.E., and P.W. Rundel. 2003. Evolution of CAM and C_4 carbon-concentrating mechanisms. International Journal of Plant Science 164(3 suppl.): S55–S77.

Keifer, M. B. 1998. Fuel load and tree density changes following prescribed fire in the giant sequoia-mixed conifer forest: the first 14 years of fire effects monitoring. Proceedings Tall Timbers Fire Ecology Conference 20: 306–309.

Kilgore, B. M., and R. W. Sando. 1975. Crown-fire potential in a sequoia forest after prescribed burning. Forest Science 21: 83–87.

Kneitel, J. M. 1997. The effects of fire and pocket gopher (*Thomomys bottae*) disturbances on California valley grassland. Thesis, California State University, North Ridge, California, USA.

Kuljian, H., and J. M. Varner. 2009. Effects of sudden oak death on the crown fire ignition potential of tanoak (*Lithocarpus densiflorus*). Proceedings of the Sudden Oak Death Fourth Science Symposium. USDA Forest Service General Technical Report PSW-GTR-229. Pacific Southwest Research Station, Albany, California, USA.

Kummerow, J., and R. K. Lantz. 1983. Effect of fire on fine root density in red shank (*Adenostoma sparsifolium* Torr.) chaparral. Plant and Soil 70: 236–347.

Lathrop, E. W., and B. D. Martin. 1982. Response of understory vegetation to prescribed burning in yellow pine forests on Cuyamaca Rancho State Park, California. Aliso 10: 329–343.

Lotan, J. E. 1976. Cone serotiny–fire relationships in lodgepole pine. Proceedings Tall Timbers Fire Ecology Conference 14: 267–278.

Levins, R. 1969. Some demographic and genetic consequences of environmental heterogeneity for biological control. Bulletin of the Entomological Society of America 15: 237–240.

MacAulay, J. D. 1975. Mechanical seed extraction of lodgepole pine. Dissertation, University of British Columbia, Vancouver, British Columbia, Canada.

Malakoff, D. A. 1997. Nitrogen oxide pollution may spark seeds' growth. Science 276: 1199.

Martin, R. E. 1963. A basic approach to fire injury of tree stems. Proceedings Tall Timbers Fire Ecology Conference 2: 151–162.

Martin, R. E., and D. B. Sapsis. 1992. Fires as agents of biodiversity: pyrodiversity promotes biodiversity. Pages 150–157 in: R. R. Harris and D. C. Erman, editors. Proceedings of the Symposium on Biodiversity of Northwestern California. Wildland Resources Center, Report 29. University of California, Berkeley, USA.

McDonald, P. M., G. O. Fiddler, and D. A. Potter. 2004. Ecology and manipulation of bearclover (*Chamaebatia foliolosa*) in northern and central California: the status of our knowledge. USDA Forest Service General Technical Report PSW-GTR-190. Pacific Southwest Research Station, Albany, California, USA.

McHugh, C. W., and T. E. Kolb. 2003. Ponderosa pine mortality following fire in northern Arizona. International Journal of Wildland Fire 12: 7–22.

Michaletz, S. T., and E. A. Johnson. 2007. How forest fires kill trees: a review of the fundamental biophysical processes. Scandinavian Journal of Forest Research 22: 500–515.

Midgley, J. J., L. M. Kruger, and R. Skelton. 2011. How do fires kill plants? The hydraulic death hypothesis and Cape Proteaceae "fire-resisters". South African Journal of Botany 77: 381–386.

Millar, C. I., and W. Woolfenden. 1999. The role of climate change in interpreting historical variability. Ecological Applications 9: 1207–1216.

Miller, M. 2000. Fire autecology. Pages 9–34 in: J. K. Brown and J. K., Smith, editors. Wildland Fire in Ecosystems. USDA Forest Service RMRS-GTR-42 VOL 2. Rocky Mountain Research Station, Fort Collins, Colorado, USA.

Montgomery, K. R., and P. C. Cheo. 1969. Moisture and salt effects on fire retardance in plants. American Journal of Botany 56: 1028–1032.

Moritz, M. A., T. J. Moody, B. Ramage, and A. Forrestel. 2008. Spatial distribution and impacts of *Phytophthora ramorum* and sudden oak death in Point Reyes National Seashore. California Cooperative Ecosystems Study Unit Report Task Agreement No. J8C07050015. Berkeley, California, USA.

Mutch, R. W. 1970. Wildland fires and ecosystems: a hypothesis. Ecology 51: 1046–1051.

Nelson, D. C., G. R. Flematti, E. L. Ghisalberti, K. W. Dixon, and S. M. Smith. 2012. Regulation of seed germination and seedling growth by chemical signals from burning vegetation. Plant Biology 63: 107–130.

Nobble, I. R., and R. O. Slayter. 1980. The use of vital attributes to predict successional changes in plant communities subject to recurrent disturbances. Vegetatio 43: 5–21.

Noste, N. V. 1985. Influence of fire severity on response of evergreen ceanothus. Pages 91–96 in: J. E. Lotan and J. K. Brown, editors. Fire's Effects on Wildlife Habitat—Symposium Proceedings. USDA Forest Service General Technical Report INT-GTR-18. Intermountain Research Station, Ogden, Utah, USA.

Noste, N. V., and C. L. Bushey. 1987. Fire response of shrubs of dry forest habitat types in Montana and Idaho. USDA Forest Service General Technical Report INT-GTR-239. Intermountain Research Station, Ogden, Utah, USA.

Odion, D. C., and F. W. Davis. 2000. Fire, soil heating, and the formation of vegetation patterns in chaparral. Ecological Monographs 70: 149–169.

Ojeda, F., J. G. Pausas, and M. Verdú. 2010. Soil shapes community structure through fire. Oecologia 163: 729–735.

Parker, V. T. 1986. Evaluation of the effect of off-season prescribed burning on chaparral in the Marin municipal water district watershed. Marin Municipal Water District Report. San Rafael, California, USA.

Parsons, D. J., and T. J. Stohlgren. 1989. Effects of varying fire regimes on annual grasslands in the southern Sierra Nevada of California. Madroño 36: 154–168.

Pausas, J. G., R. A. Bradstock, D. A. Keith, and J. E. Keeley. 2004. Plant functional trait analysis in relation to crown-fire ecosystems. Ecology 85: 1085–1100.

Pausas, J. G., and J. E. Keeley. 2014. Evolutionary ecology of resprouting and seeding in fire-prone ecosystems. New Phytologist 204: 55–65.

Perry, D. A., and J. E. Lotan. 1979. A model of fire selection for serotiny in lodgepole pine. Evolution 33: 958–968.

Peterson, D. L. 1985. Crown scorch volume and scorch height: estimates of post-fire tree condition. Canadian Journal of Forest Research 15: 596–598.

Peterson, D. L., and K. C. Ryan. 1986. Modeling postfire conifer mortality for long-range planning. Environmental Management 10: 797–808.

Plumb, T. R. 1979. Responses of oaks to fire. Pages 202–215 in: T. R. Plumb, technical coordinator. Ecology, Management, and Utilization of California Oaks. USDA Forest Service General Technical Report PSW-GTR-4. Pacific Southwest Forest and Range Experiment Station, Berkeley, California, USA.

Radeloff, V. C., D. J. Mladenoff, R. P. Guries, and M. S. Boyce. 2004. Spatial patterns of cone serotiny in *Pinus banksiana* in relation to fire disturbance. Forest Ecology and Management 189: 133–141.

Reiner, R. J. 2007. Fire in California grasslands. Pages 207–217 in: M. R. Stromberg, J. D. Corbin, and C. M. D'Antonio, editors. California Grasslands: Ecology and Management. University of California Press, Berkeley, California, USA.

Rothermel, R. C. 1976. Forest fires and the chemistry of forest fuels. Pages 245–259 in: F. Shafizadeh, K. V. Sarkanen, and D. A. Tillman, editors. Thermal Uses and Properties of Carbohydrates and Lignins. Academic Press, New York, New York, USA.

Rowe, J. S. 1983. Concepts of fire effects on plant individuals and species. Pages 135–154 in: R. W. Wein and D. A. MacLean, editors. The Role of Fire in Northern Circumpolar Ecosystems. Scope 18. John Wiley and Sons, New York, New York, USA.

Rundel, P. W. 1969. The distribution and ecology of the giant sequoia ecosystem in the Sierra Nevada, California. Dissertation, Duke University, Durham, North Carolina, USA.

Ryan, K. C. 2002. Dynamic interactions between forest structure and fire behavior in boreal ecosystems. Silva Fennica 36: 13–39.

Ryan, K. C., and E. D. Reinhardt. 1980. Predicting postfire mortality of seven western conifers. Canadian Journal of Forest Research 18: 1291–1297.

Sackett, S. S., and S. M. Haase. 1992. Measuring soil and tree temperatures during prescribed fires with thermocouple probes. USDA Forest Service General Technical Report PSW-GTR-131. Pacific Southwest Research Station, Albany, California, USA.

Savage, M., and J. N. Mast. 2005. How resilient are southwestern ponderosa pine forests after crown fires? Canadian Journal of Forest Research 35: 967–977.

Schwilk, D. W. 2003. Flammability is a niche construction trait: canopy architecture affects fire intensity. American Naturalist 162: 725–733.

Schwilk, D. W., and D. D. Ackerly. 2001. Flammability and serotiny as strategies: correlated evolution in pines. Oikos 94: 326–336.

Spalt, K. W., and W. E. Reifsnyder. 1962. Bark characteristics and fire resistance: a literature survey. USDA Forest Service Occasional

Paper SO-OP-193. Southern Forest Experiment Station, New Orleans, Louisiana, USA.
Stark, N. 1965. Natural regeneration of Sierra Nevada mixed conifers after logging. Journal of Forestry 63: 456–457.
Stephens, S.L., and W.J. Libby. 2006. Anthropogenic fire and bark thickness in coastal and island pine populations from Alta and Baja California. Journal of Biogeography 33: 648–652.
Stone, E.C. 1951. The stimulative effect of fire on flowering of the golden brodiaea (*Brodiaea ixioides* Wats. var. *lugens* Jeps.). Ecology 32: 534–537.
Swetnam, T.W. 1993. Fire history and climate change in giant sequoia groves. Science 262: 885–889.
Swezy, D.M., and J.K. Agee 1991. Prescribed fire effects on fine root and tree mortality in old growth ponderosa pine. Canadian Journal of Forest Research 21: 626–634.
Szalay, F.A., and V.H. Resh. 1997. Responses of wetland invertebrates and plants important in waterfowl diets to burning and mowing of emergent vegetation. Wetlands 17: 149–156.
Thies, W.G., D.J. Westlind, M. Loewen, and G. Brenner. 2006. Prediction of delayed mortality of fire-damaged ponderosa pine following prescribed fires in eastern Oregon, USA. International Journal of Wildland Fire 15: 19–29.
Turner, M.G., W.L. Baker, C.J. Peterson, and R.K. Peet. 1998. Factors influencing succession: lessons from large, infrequent natural disturbances. Ecosystems 1: 511–523.
Tyler, C., and M.I. Borchert. 2002. Reproduction and growth of the chaparral geophyte, *Zigadenus fremontii* (Liliaceae), in relation to fire. Plant Ecology 165: 11–20.
Vale, T.R. 1979. *Pinus coulteri* and wildfire on Mount Diablo, California. Madroño 26: 135–140.
van Wagtendonk, J.W. 1996. Use of a deterministic fire growth model to test fuel treatments. Pages 1155–1166 in: D.C. Erman, general editor. Sierra Nevada Ecosystem Project: Final Report to Congress, Volume II. Wildland Resources Center Report 37. University of California, Davis, California, USA.
Vesk, P.A., D.I. Warton, and M. Westoby. 2004. Sprouting by semi-arid plants: testing a dichotomy and predictive traits. Oikos 107: 72–89.
Vesk, P.A., and M. Westoby. 2004. Sprouting ability across diverse disturbances and vegetation types worldwide. Journal of Ecology 92: 310–320.
Viereck, L.A., and L.A. Schandelmeier. 1980. Effects of fire in Alaska and adjacent Canada—a literature review. Technical Report 6; BLM/AK/TR-80/06. U.S. Department of the Interior, Bureau of Land Management, Alaska State Office, Anchorage, Alaska, USA.
Vogl, R.J. 1968. Fire adaptations of some southern California plants. Proceedings Tall Timbers Fire Ecology Conference 7: 79–109.
Wade, D.D. 1993. Thinning young loblolly pine stands with fire. International Journal of Wildland Fire 3: 169–178.
Wagener, W.W. 1961. Guidelines for estimating the survival of fire-damaged trees in California. USDA Forest Service Miscellaneous Paper PSW-MP-60. Pacific Southwest Forest and Range Experiment Station, Berkeley, California, USA.
Weide, D.L. 1968. The geography of fire in the Santa Monica Mountains. Thesis, California State University, Los Angeles, California, USA.
Wilkin, K.M., V.L. Holland, D. Keil, and A. Schaffner. 2013. Mimicking fire for successful chaparral restoration. Madroño 60: 165–172.
Wright, H.A., A.W. Bailey, and W. Arthur. 1982. Fire Ecology: United States and Southern Canada. John Wiley and Sons, New York, New York, USA.
Zammit, C.A., and P.H. Zedler. 1988. The influence of dominant shrubs, fire, and time since fire on soil seed banks in mixed chaparral. Vegetatio 75: 175–187.
Zedler, P.H. 1995. Fire frequency in southern California shrublands: biological effects and management options. Pages 101–112 in: J.E. Keeley and T. Scott, editors. Brushfires in California Wildlands: Ecology and Resource Management. International Association of Wildland Fire, Fairfield, Washington, USA.
Zedler, P.H., C.R. Gautier, and G.S. McMaster. 1983. Vegetation change in response to extreme events: the effect of a short interval between fires in California chaparral and coastal scrub. Ecology 64: 809–818.

CHAPTER NINE

Fire and Animal Interactions

KEVIN E. SHAFFER, SHAULA J. HEDWALL,
AND WILLIAM F. LAUDENSLAYER, JR.

> Big game creatures do not eat mature trees—they feed on sun-drenched browse, new grass, and lush regrowth, all of which rely on fire to rewind their biotic clocks.
>
> PYNE (1995)

Introduction

Fire effects on animals may include direct fatality, changes in physiology and behavior, shifts or displacements in populations and communities, and immediate and longer-term alterations to habitat structure, composition, and function. In this chapter we explore how fire affects terrestrial and aquatic animals, animal populations and communities, and habitat structure, composition, and function. Specific fires affect individual animals and discrete habitat, but the historic and potential fire regime can have long-term effects on animal populations and communities.

The last three decades have seen great advances in understanding how fire and fire regimes affect animals (e.g., McIver et al. 2012, Latif et al. 2016). The study of animal–fire interactions has historically been focused on effects to a single or few species and the response of animals, as individual organisms, to specific fire events. Some research and syntheses focus on the broader relation between fire and animal communities, usually addressing the longer-term effects from fire-induced changes to habitat (Lyon et al. 1978, Smith 2000). Fire effects investigations in many regions of North America, including California, had generally focused on high-profile species, such as the spotted owl (*Strix occidentalis*) (Bond et al. 2009, Lee et al. 2012, Lee and Bond 2015a, Jones et al. 2016), and more generally on birds (Mendelsohn et al. 2008, Russell et al. 2009), and mammals (Roberts et al 2008, Meyer et al. 2008).

Fire affects animal habitat, including cover and shelter, the physical dimensions of an area, plant species, food and forage, and other biological attributes important to a species, including nutritional needs, life history stages, and reproduction and rearing. Fire affects the presence and condition of various habitat elements and temporal and spatial characteristics of habitat critical to animal populations, including patch size, juxtaposition, and connectivity (Cooperrider et al. 2000). There are complex, ecological relations between fire, animals, and specific environmental elements (Saab and Vierling 2001, Cunningham et al. 2002, Cunningham and Ballard 2004, Jenness et al. 2004, Converse et al. 2006, Saab et al. 2007, Mahlum et al. 2011). These interactions reveal the multifaceted relation of fire to California's diverse, wildlife diversity.

Fire Effects on Animals

Each of the fire regime attributes is significant to animals. The interaction between animals and fire, including the implications of altered fire regimes or fire management practices, may be best understood when viewing individual fire regime attributes and their particular effects on animals.

Temporal Attributes of Fire

Of all of the fire regime attributes, seasonality of fire may have the greatest potential for affecting individual animals and animal populations. Most animal species have specific temporal behaviors and resource needs associated with nesting, mating, raising young, foraging, and periods of inactivity. In many regions of California, late summer to autumn represents the period when fire is most frequent, although fire can and does occur in other seasons of the year. For most animals, the historic fire season occurs after breeding and rearing of young is completed. Animal species are best adapted to historic conditions, and timing of animal reproduction and rearing, migration, and other behavioral and physiologic traits are ecologically tied with historic fire regimes. Singular fires burning in other seasons have implications for plant species and communities, and thus, at least short-term implications for animals. Multiple or continued fires out of season could have considerable long-term implications. Key factors are the frequency and regularity of change, the uniformity of the change compared to historic condition, and the immediate effect of initial out-of-season fire(s) on animal species or communities. These ideas will be explored further in the section on direct and indirect effects of fire to populations and communities.

Like seasonality, a given bioregion has a number of characteristic fire return intervals that occur within its ecological zones. Changes in fire return intervals can have long-term effects on vegetation communities, animal habitat, and animal communities. A reduction in the interval between fires can impact plant populations and can even contribute to habitat modification by plant community type conversion

(Zedler et al. 1983, Haidinger and Keeley 1993). Increased fire frequency can disrupt behavior essential to reproducing, rearing, defending territory, and acquiring resources. Too-frequent fire could have the additive effect of removing enough individuals from a population to impact reproduction or survival or to cause animals to emigrate or cease to immigrate. On the other hand, infrequent fire could eventually render an animal's habitat to be unsuitable due to changes in composition and structure (Jones et al. 2016).

Departures in Fire Return Interval (FRID) have occurred in and been broadly investigated across California (Safford and Van de Water 2014) and discussed in chapter 6. Fire return intervals in northern California national forests have expanded, while fires have become more frequent in southern California (Safford and Van de Water 2014). Both of these changes to historic fire regimes could be render wildlife habitat unsuitable. Fire exclusion from northern forests has been a long-standing issue for management of several animal species, including fisher, pine marten, and California spotted owl (*Strix occidentalis occidentalis*). In southern California, where the fire return intervals are decreasing, suitability of habitat to animals is also being impacted. Whether return intervals are increasing or decreasing, the net results to habitat in both cases are decreased food and prey, decreased nesting and cover, and obstruction of travel and migration paths. Nonetheless, frequency, as a single element of fire regime, does not determine effects to animals. In southern California's San Gabriel Mountains, frequent of large fires have been positively tied to the availability and distribution of high-quality habitat for bighorn sheep (*Ovis canadensis nelson*) and mule deer (*Odocoileus hemionus*) (Bleich et al. 2008, Holl and Bleich 2010, Holl et al. 2012). Changes in vegetation composition and structure of habitat resulted in increased preferred forage, nutrient levels, and openness of habitat. The latter is needed both for sheep and deer movement and ready recognition of predators.

Spatial Attributes of Fire

The spatial complexity and size of a fire affect animals in several ways. The most obvious may be that differential burning in an area results in a varied, more complex vegetation mosaic comprised of different plant species and habitat elements (e.g., food, roosts, nesting sites and materials, cover) in different conditions of suitability to different species. The significance to animals varies with size and scale of patches within the mosaic. For animals such as salamanders, relevant patches are of square meters within centimeters above the ground, but for larger animals such as bears and large ungulates, they are of square kilometers and meters above the ground. Spatial complexity can be caused by both variation within one fire and by the pattern of multiple fires at various times in a landscape.

Magnitude Attributes of Fire

Fire intensity is a measure of the heat released by fire (see chapter 4) and is typically responsible for most direct fatality and injury to wildlife. Temperatures above 100°C (212°F) can be considered lethal for most living organisms (DeBano et al. 1998), whereas much lower temperatures (40–60°C [104–140°F]) are lethal for animals. Longer residence times can result in increased mortality rates at the lower end of the lethal temperature range for animals.

Fire intensity determines whether animals are injured or killed by a fire, whereas fire severity provides an indication of the degree of habitat alteration. Fire severity is a measure of the aboveground and belowground organic matter consumed by a fire. Alterations to the plant and soil communities can affect the shelter, nesting, and forage components of animal habitat. The extent and depth of heating in soil and woody debris is an important factor to many animal species. Soil and woody material can reduce the lethal effects of heat on animals through their insulating properties. Consumption of vegetation also affects animals through changes in their habitat. Animals associated with more open habitats generally benefit from fire severe enough to remove overstory vegetation, eliminate decadent plants, and maintain open conditions. These changes in vegetation have been shown to benefit some species (Bleich et al. 2008, Holl et al. 2012) and have differing effects across an animal community in California (Mendelsohn et al. 2008, Roberts et al. 2008, 2015). Animals associated with more homogenous habitats benefit from fires of more uniform severity and low spatial complexity. Low-severity fire allows existing species to remain or quickly recolonize, while moderate to high-severity fires allow new species to establish themselves into the resulting, altered landscape. Animals associated with plant community ecotones benefit from fires of varying severity that leave a relatively diverse patch work of unburned and burned areas and a range of habitat conditions.

Crown fires reduce or eliminate trees and shrub canopies, stressing or killing them. The response of animals to these changes can be varied. Canopy reduction can substantially reduce habitat for invertebrates and nesting vertebrates by reducing nesting sites and material, forage, cover, and escape routes from predators (Kotliar et al. 2007). Conversely, the removal of vegetation structure associated with crown fires creates or enhances habitat for species preferring more open canopy or exposed surfaces (Saab et al. 2007, Latif et al. 2016).

Direct Effects of Fire on Individuals

A properly functioning ecosystem, complete with some degree of mortality and habitat change for resident species, should be the focus for understanding animal–fire interactions. The immediate, direct effects to animals and animal populations are complex and difficult to quantify. Direct fatality counts or estimates likely do not completely reveal either the immediate or long-term effects on animal populations. Animals can move in from adjacent areas or be displaced, reproductive rates may decline or increase, and fewer inter- or intra-specific competitors may benefit those individuals that survive fire.

Direct Fatality of Animals from Fire

Much attention has been given to how animals respond to actual fire events. Mortality rates are difficult to quantify, because scavengers often consume the majority of carcasses within 24 hours of a fire (Brooks, U.S. Geological Survey, Oakhurst, California, USA, pers. comm.). Perhaps the best studies on the direct effects of wildfire on wildlife species have occurred at Yellowstone National Park following the

1988 fires. In many cases, direct fatality was minimal (i.e., large mammals, fishes; Singer et al. 1989, Romme et al. 2011). Knowledge concerning animal response to fire events is often based on inferences from observations of animal behavior before and after fires, and what is thought to be happening where observations cannot be conducted (i.e., in burrows, under tree or brush canopy). Studies with prefire data reveal important information on how fire, both wildfire and prescribed fire, affects wildlife (Gamradt and Kats 1997, Mendelsohn et al. 2008, Romme et al. 2011).

Escaping the fire's heat is the most common way for animals to escape direct fatality. Animals can either flee or take shelter from fire. Slower moving animals (e.g., small mammals, reptiles, and amphibians), by necessity, have to take shelter from fire. More mobile animals (e.g., deer, birds) will flee burning areas. If escape from fire is not possible, their size or particular habitat use, such as ground or tree nesting, can become a liability. Animals that are considered dispersal limited (e.g., reptiles, amphibians, flightless invertebrates, small mammals, some bird species) have shown greater direct and indirect impacts from fire in southwestern North America (Friggens et al. 2013). These same relations likely exist for animal species in California.

The vertical zones occupied by a species determine its behavioral response to fire. Where and how an animal moves about, forages, finds shelter, and nests is dependent on an animal's morphology, physiology, and behavioral abilities as well as the characteristics of the habitat at the time of a fire. These factors govern animal response to fire. We will discuss animal reaction to fire by generalizing the landscape into four vertical zones: subterranean, ground level, arboreal, and aquatic (Fig. 9.1).

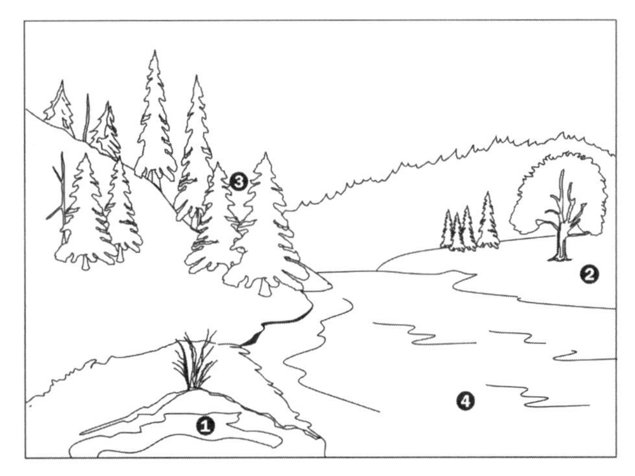

FIGURE 9.1 Animal habitats zones: (1) subterranean, (2) surface-dwelling, (3) arboreal, and (4) aquatic.

SUBTERRANEAN FAUNA

Subterranean fauna include species that live in or under the ground and species that either spend a considerable amount of time underground or the underground habitat serves an essential function(s) (Fig. 9.2). Subterranean animals create burrows, or utilize burrows made by other species (Box 9.1).

Subterranean species typically respond to fire by taking shelter underground. The insulating properties of soil generally protect animals under most fire conditions. For species that find protection underground, direct fatality due to fire is often avoided (Harestad 1985, Yensen et al.1992, Koehler and Hornocker 1997) (Box 9.2). Though some of these species can be fast-moving, most are relatively slow or immobile. For some species (e.g., earthworms), there is no option for escaping the effects of fire other than the thermal protection of the soil. Conversely, mobile species such as rabbits can leave the immediate area, returning once the fire has passed. Heated soil and smoke can also cause mortality in underground fauna. As depth increases, the insulating properties of soil reduce the heat penetration beneath the surface. However, the insulating properties of soils are altered depending on the residence time of the fire coupled with soil texture and soil moisture. Longer residence time will drive the heat deeper into the soil. Smoke, the fire's consumption of oxygen, and the production of carbon monoxide also threaten underground-dwelling animals. In addition to providing multiple escape routes, multiple entrances may have value for burrowing mammals during a fire by allowing for a continual supply of breathable air thereby avoiding asphyxiation (Bendell 1974, Sullivan 1996).

SURFACE-DWELLING FAUNA

Animals living above soil and on the ground constitute the surface-dwelling fauna (Fig. 9.3). For these animals, the ground is where the majority of activities occur (e.g., nesting, rearing of young, foraging). The surface faunal community has the widest variation in size, ranging from small snails and beetles to large mammals and have a wide range of mobility. Some move extremely slowly, and, if also small, cannot move sufficient distances quickly enough to escape fire. Other animals, both small and larger, are much more mobile. Size and mobility contribute to how animals can respond to fire (Box 9.3).

Many surface fauna try to escape fire by sequestering themselves inside or under surface features, such as the litter layer, downed and decaying woody matter, ground nests, dens, talus, and rocky outcrops. Animal morphological characteristics and mobility determine what form of shelter they seek. Smaller, slower species such as woodrats (*Neotoma* spp.) take shelter by retreating to dens or cavities in rocks or talus piles (Howard et al. 1959, Wright and Bailey, 1982), whereas toads hide underneath logs or under wet litter in depressions (Komarek 1969). Some animals den in dense vegetation or create their dens from dead vegetation and sticks (e.g., woodrats). These materials are highly flammable, and direct fatality of species seeking shelter from fire in such places has been recorded (Williams et al. 1992). Many surface inhabitants utilize or nest in downed wood, which is differentially affected by fire (see below). In Arizona and Utah, research has revealed that many surface species (i.e., snakes, lizards, rodents, lagomorphs, ground birds) occurred in burned areas shortly after fire, suggesting they avoided the fire in some way (Esque et al. 1995).

Larger animals, such as black bears (*Ursus americanus*), mule deer, and pronghorn (*Antilocapra americana*), often evade a fire and are rarely killed during fires (National Park Service 1966). Other species for which escape may be the primary response, such as bighorn sheep and American pine martens (*Martes americana*), data regarding direct effects to and responses of these species to fire are lacking (Tesky 1993, Koehler and Hornocker 1997). However, even species that may be thought as being relatively slow movers, such as toads, have been observed moving ahead of oncoming fires (Sullivan 1994).

FIGURE 9.2 Subterranean fauna habitat zone, showing burrowing owls (*Athene cunicularia*) in burrow. (Photo by Gary Kunkel.)

BOX 9.1. SUBTERRANEAN ANIMAL COMMUNITY

Inhabits soil	Mites, earthworms, insects, moles
Creates burrows or nests	Ants, trapdoor spiders, wasps, bees, gophers, ground squirrels, rabbits, badgers, kangaroo rats, spadefoot toads
Utilizes existing burrows	Many arachnids, snakes, frogs, and toads

BOX 9.2. AVOIDING THE HEAT: EXAMPLES OF ANIMALS INSULATING THEMSELVES FROM HEAT, FLAME, AND SMOKE

Living beneath the soil	Various insects, earthworms, moles
Sheltering or retreating to burrows	Many amphibians and reptiles, ground squirrels, kangaroo rats, rabbits, badgers
Sheltering under rocks and talus	Woodrats, some snakes and lizards
Sheltering in logs and trees	Many types of invertebrates, cavity-nesting birds, bats, arboreal mammals

ARBOREAL FAUNA

Animals living in vegetation above the ground, whether in main stems, branches, or crowns, constitute arboreal fauna (Fig. 9.4). Some of these species spend their time primarily above the ground, nesting and foraging in the canopy of shrubs and trees (e.g., tree squirrels, porcupine [*Erethizon dorsatum*], tree frogs) but also traverse the ground. Others are able to either glide or fly, perhaps leaving branches or the canopy to forage but otherwise remaining above the ground (e.g., many insects, arboreal nesting birds, northern flying squirrel [*Glaucomys sabrinus*], bats).

Some arboreal animals respond to fire by taking shelter whereas others take flight. Again, there is little direct evidence to confirm how a species responds to fire. Animals that construct or occupy existing nest cavities within trunks may find suitable shelter from fire. However, unlike the protection provided by a soil layer for benthic fauna, the protection of plant matter may be more tenuous for arboreal animals. Heat convection is greater and the material providing the protection itself is flammable. Thus, moderate to severe fires may pose a threat to animals sheltering within standing trees, especially if the trees are already dead.

Observations of birds suggest that they are not frightened by fire (Lyon et al. 1978), perhaps because of their ability to easily and quickly avoid fire and its effects. Instead, many flying insects and birds respond quickly to fires to take advantage of injured plants, whereas other bird species immigrate to feed on species attracted to the burned area. Bark beetles (e.g., *Dendroctonus* spp., *Scolytus* spp.) and other beetles that forage on the cambial tissues (e.g., Cerambycidae and Buprestidae) quickly arrive to lay their eggs under the bark of stressed trees, and the resulting larvae on the cambium, contributing to tree mortality. In turn, predaceous insects and insectivorous birds, such as woodpeckers and nuthatches, quickly take advantage of new, prolific food supplies, whereas raptors congregate in burned areas to prey on rodents.

FIGURE 9.3 Ground-dwelling fauna habitat zone, showing tule elk (*Cervus canadensis nannodes*) in grass and scrub vegetation. (Photo by Mark Salvestrin.)

BOX 9.3. GROUND-DWELLING ANIMAL COMMUNITY: RELATIVE MOBILITY AND RESPONSE TO FIRE

Limited mobility and likely cannot flee fire, taking shelter or seeking refuge:

Snails, slugs, arachnids, centipedes, millipedes, beetles and other insects, tortoises, toads

Moderate mobility, and fleeing versus taking shelter likely a function of rate of approaching fire front:

Amphibians (e.g., western toad, *Bufo boreas*), many salamanders, reptiles (e.g., rattlesnakes, lizards), some ground nesting birds (e.g., quail, turkeys), many rodent species (e.g., mice, voles, chipmunks, woodrats)

Highly mobile and likely fleeing fire:

Flying insects (e.g., grasshoppers, some ground nesting birds (e.g., meadowlark, wild turkeys) mid- to large-sized mammals (e.g., many carnivores, ungulates)

FIGURE 9.4 Arboreal fauna habitat zone, showing acorn woodpeckers (*Melanerpes formicivorus*) in tree cavity nest. (Photo by Lyle Madeson.)

AQUATIC FAUNA

Animals that live entirely in water (i.e., fishes, some mollusks, and insects) have an aquatic life stage (i.e., amphibians, some insects), or primarily occupy or use aquatic, wetland, or riparian ecosystems for nesting and shelter, reproduction or foraging (e.g., many amphibians and reptiles, river otter, beaver, many waterfowl and wading birds) constitute aquatic fauna (Fig. 9.5).

The effects of wildfire on aquatic systems include direct or immediate influence of the fire on water quality and the indirect or subsequent effects on watershed characteristics and processes that influence water quality and quantity, and stream channels. Wildfire has been perceived as a destructive force in streams (Rieman and Clayton 1997), but not all fires result in detrimental impacts to streams, and subsequently to the aquatic species occupying the stream. The fire effects,

FIGURE 9.5. Aquatic fauna habitat zone, showing adult Kokanee [land-locked sockeye] salmon (*Oncorhynchus nerka*) and gravel substrate and submerged vegetation. (Photo by Matt Elyash.)

both direct and indirect, depend on the extent and severity of fire, and size of the watershed (Gresswell 1999, Rieman et al. 2012). Potential direct and immediate effects to aquatic systems include increased temperature, changes in stream chemistry, and infusion of debris and sediment delivery from the first rain events following the fire (Kershner et al. 2003, Spencer et al. 2003). Longer-term watershed effects include increased repeated, increased sediment delivery, increased temperature regime, habitat fragmentation, and interrupted wood delivery and cycling, and altered aquatic invertebrate communities (Bisson et al. 2003).

Species or specific life stages of animals that inhabit aquatic ecosystems are likely buffered from most direct effects of fires. However, this is conditional. For example, direct effects to more vulnerable life stages (e.g., nonmobile amphibian egg masses, aquatic insect larvae) would be greater than to more mobile life stages (e.g., fully developed amphibian tadpoles or riparian-dependent birds). Studies have shown that amphibians inhabiting aquatic environments suffer little to no fatality during fire, although responses were variable (Ford et al. 1999, Russell et al. 1999, Hossack et al. 2013). Similarly, temporal characteristics of fire indirectly affect the aquatic environment by affecting terrestrial elements, such as soil and vegetation, which then change aquatic environments. Many bird and mammal species occupy marshes, wetland, riparian, and other aquatic habitats. They are also often highly mobile and likely to avoid the direct effects of fire by escaping. Animals living along a body of water likely retreat to the water, as they would when faced with other threats. Nonflying vertebrates (e.g., amphibians and reptiles, many ducks, and mammals able to swim), occupying the shores of a pond or lake or the banks of a stream or river, take refuge in the water.

Increased water temperature and decreased oxygen due to fire are potential direct threats to aquatic life and can result in fatality of aquatic species (Rieman et al. 1997, Gresswell 1999). If a severe fire burns near or across a stream, water temperature can increase substantially (Hitt 2003). However, temperatures in multiple streams impacted by high-severity burns were found to have no apparent increases in maximum water temperature during the fires (Mahlum et al. 2011). As smoke, ash, and volatile compounds dissolve, the pH may be altered as well as concentrations of trace metals, nutrients, and other chemical constituents in streams (Gresswell 1999). The spatial pattern of direct effects seems to depend on the extent and severity of fire and size of the watershed (Gresswell 1999). In addition, the size of the watershed likely also influences the ability of a fire to influence temperature in the stream. Fire is not likely able to influence temperature in third order or larger streams (Minshall and Brock 1991).

Indirect fire effects to the aquatic environment influence aquatic animals, populations, and communities. These effects on watershed characteristics and processes can influence water quality and quantity, stream channels, and aquatic biota. The indirect effects of wildfire on streams are influenced and limited by direct effects, the supply of water, sediment, woody debris, nutrients, and other materials; energy (insolation); stochastic events such as postfire rain events; and, the interactions of these factors on physical and biological succession of the stream. Fires of higher magnitude or greater extent likely would have more effect on aquatic systems than smaller fires or those of lesser magnitude; both of these regime attributes could impact water temperature and oxygen concentration and affect terrestrial features influencing the aquatic environment.

The amount of heat necessary to change liquid water to gas is considerable and water provides good insulation for animals. The volume of the water mass inhabited by aquatic animals would determine the level of protection afforded them. Lakes, large ponds, rivers, and perennial streams provide considerable protection from fire. Interestingly, California has a unique aquatic ecosystem, vernal pools, which are both ephemeral and relatively shallow. This would seem to be one aquatic environment that potentially could be negatively impacted by fire. But as with other ecosystems in California, vernal pools, and the native fauna, appear to be in synchrony with the fire regimes where they occur (see chapter 15).

Direct Effects of Fire on Populations and Communities

The same factors that are important to individual animals are important to populations and communities. Fires over large spatial and temporal scales influence the availability, condition, and spatial distribution of key habitat features such as migration corridors and refugia. In fact, unburned areas adjacent to burned areas function as biological refugia for animals during and immediately after fire and during the period of recovery from the fire. In addition to direct fatality, fire directly influences intra- and inter-specific behavior including population residency and displacement, migration, colonization, foraging and predation, and reproduction.

Only in situations where populations are in dangerously poor condition, such as low abundance, restricted age-class distribution or already-established decline, or where the fire regime is drastically altered, does a single fire or series of fires pose a real threat to an animal species or community (e.g., desert tortoise [*Gopherus agassizii*], see sidebar 18.3). Alteration of fire regimes and vegetation communities, along with the fragmentation of animal populations, create situations in which fire can significantly impact a particular aquatic or terrestrial species population or community.

There is a growing understanding of the effects of fire on animal communities and the role of different fire regimes on multiple animal species in a given ecosystem. Within California, there have been investigations on fire effects on southern California rodent, bird, and ungulate populations and communities, the Stephen's kangaroo rat (*Dipodomys stephensi*), California gnatcatcher (*Polioptila californica*), and mule deer. Each study has shown that different species have varied responses to a given fire regime (e.g., one rodent species benefits from more open vegetation while another may tolerate the change or emigrate until vegetation recovers), and single species' response is based on specific attributes of the fire (e.g., bighorn sheep benefiting from high-intensity fire depending on how much area is burned under that intensity).

Mendelsohn et al. (2008) found variable shifts in bird species diversity, relative abundance, and community composition two and three years following the 2003 Otay and Cedar fires in southern California. The fires burned more than 1,299 km^2 (502 mi^2), primarily affecting chaparral and Coastal Sage Scrub Communities (CSS). The changes to vegetation structure, biomass, and composition had different consequences to the bird communities, supporting the complex relationship between fire, vegetation, and animal response. Table 9.1 depicts the variable responses of the bird communities to the 2003 wildfires. In CSS, species diversity at lower elevation increased, while at higher elevation, there was no detectable change in diversity. For community structure, there were detectable changes low-elevation CSS and low-elevation chaparral, but no detectable changes in community structure in any other vegetation community. At low-elevation chaparral and CSS, three species increased in abundance postfire while Anna's hummingbird (*Calypte anna*) decreased. In contrast, only one species, horned lark (*Eremophila alpestris*) increased in abundance in higher-elevation vegetation communities.

The variation of bird community response (changes in abundance, community structure, distribution) suggests that bird communities across different vegetation communities in southern California are relatively resilient to changes created by fire. It is likely that bird communities and animal communities in general are able to respond well to wildfire when fires are low to moderate with only some areas experiencing high severity and where unburned habitat is proximate to burned areas.

Residence and Displacement

Animal are disrupted by fire, often dispersed from nests, shelters, and foraging areas. Resident species are often displaced only during the actual fire, but others are displaced for varying periods. In contrast, some species exploit the burned area, even to point of colonization and establishment where

TABLE 9.1
Changes in bird species diversity, community structure, and relative abundance in low- and high-elevation chaparral, coastal sage scrub, grassland, oak woodland, and riparian vegetation communities following the 2003 Otay and Cedar fires, southern California
Data from Mendelsohn et al. (2008)

Prefire condition	Postfire condition (2005 and 2006)
Species diversity	
Low-elevation chaparral	No detectable change
Low-elevation coastal sage scrub	Increase
High-elevation chaparral	No detectable change
High-elevation coastal sage scrub	No detectable change
Oak woodland	No detectable change
Riparian	No detectable change
Community structure	
Low-elevation chaparral	Detectable change
Low-elevation coastal sage scrub	Detectable change
High-elevation chaparral	No detectable change
High-elevation coastal sage scrub	No detectable change
Low-elevation grassland	No detectable change
High-elevation grassland	No detectable change
Oak woodland	No detectable change
Riparian	No detectable change
Relative abundance	
Increased abundance postfire	Decreased abundance postfire
Low-elevation chaparral	Low-elevation chaparral
lazuli bunting, *Passerina amoena*	Anna's hummingbird, *Calypte anna*
spotted towhee, *Pipilo maculatus*	Low-elevation coastal sage scrub
Phainopepla, *Phainopepla nitens*	spotted towhee
High-elevation grassland	Wrentit, *Chamaea fasciata*
horned lark, *Eremophila alpestris*	Bushtit, *Psaltriparus minimus*

postfire habitat conditions are suitable. Each fire regime attribute has the potential to influence the residency or displacement of animal populations and by extension, the entire animal community (Table 9.2). The season of the fire dictates which behaviors of animals might be altered, and when changes in fire frequency occur, animal populations and entire community are commonly altered. The size and complexity of fire determines how much of the range of a population is affected and how each species' essential habitat is distributed. Likewise, the magnitude of the fire controls whether habitat elements remain and how they are distributed across the area, dictating whether and where a species can continue to reside.

Rodent communities in southern California illustrate how fire affects populations of different species and how these populations, and the resulting biotic community continues to adjust as the burned area continues to change following fire (Price et al. 1995, Wirtz 1995). In grasslands and coastal sage scrub, rodent communities with different vegetation structure and forage requirements differentially occupy and respond to burned and unburned areas. The mixture and juxtaposition of patches of unburned and burned areas affect species composition and distribution of the rodent species, and this effect is evident for up to 10 years. Fires created a wide variety of habitats, both opening up some areas, an essential condition for rodents such as kangaroo rats (*Dipodomys* spp.), and provided for a wider array of plants species, used as both shelter and forage by animals. In unburned areas, rodent diversity was lower, and some species abundances declined due to lack and gradual loss of habitat.

Table 9.3 displays a comparison between rodent species showing survival and response following fire in coastal sage scrub and southern chaparral. Species were first detected at different periods following fire, from less than four months for Pacific kangaroo rats (*Dipodomys agilis*) in coastal sage scrub to up to 11 months for dusky-footed woodrats (*Neotoma fuscipes*). Peak abundance ranged from as early as immediately after fire (e.g., deer mouse [*Peromyscus maniculatus*] in both plant communities) to several years following fire (e.g., cactus mouse [*P. eremicus*] in coastal sage scrub). The presence and period of occupation of each rodent species was related to the habitat condition over time after fire. At first, the area had the most contrasting mosaic of unburned and burned patches–open space and sparse vegetation in burned patches, with abundant and accessible seed from the surviving and flourishing plant species. Over time, the open areas regained vegetation structure and density, closing in the open areas and changing available forage as plant community composition changes.

The particular life stage of aquatic species impacted by the fire determines the response of the species postfire. For example, after a 1993 chaparral fire in the Santa Monica Mountains, stream habitat occupied by the California newt (*Taricha torosa*) was altered, resulting in loss of over 50% of pool and run habitat types. Three years after the fire, adult density did not change, but the number of egg masses decreased to one-third of what was surveyed prefire conditions (Gamradt and Kats 1997). Postfire, all life stages of fish can be displaced depending upon the level of habitat modification (Rinne 1996, Rieman et al. 1997, 2003, 2012).

Fire's Influence on Animal Migration

Migration involves unique behavior and considerable energy expenditure, and contributes to the crucial genetic exchange between populations and affects animal community health and composition. Each attribute of fire regimes can affect migration. Fires occurring during active migration times can cause migration routes to be interrupted. The size, complex-

TABLE 9.2

Potential change in wildlife residence and displacement with changes in fire regime attributes

Temporal	Seasonality	Changes in the seasonality of fire can cause shifts in which species are disturbed or displaced. Critical activities, such as reproduction and nesting, can be disturbed for the season. Slow-moving young or newborn nestlings can be killed by unusually early fire
	Frequency	Increased frequency could cause displacement or behavioral change frequently enough that species do not continue to reside in that area. Decreased or increased frequency each over time could alter habitat structure and other resources so that the area is no longer inhabitable for some species but perhaps suitable for other species
Spatial	Size	Large fires could potentially affect more habitat and could displace animals for a season or until enough area becomes suitable again for residency
	Complexity	Heterogeneity can determine whether suitable habitat persists and the quantity and distribution of that habitat. Thus, the presence, number, and distribution of resident species could change with the degree of a fire's heterogeneity. Species benefiting from a more uniform environment would benefit more from a more homogeneous fire
Magnitude	Severity	Severity can either create or destroy the habitat attributes required by resident species. Too severe fire could pose a greater threat for direct fatality, potentially impacting the future reproductive success of populations. And, it could destroy habitat structure and function, displacing species for variable periods after fire. On the other hand, low-severity fire could displace species due to inadequate habitat conditions in the future that would have been initiated by increased fire severity
	Type	Surface fires might not displace species occupying habitat above the ground any longer than the time it takes for smoke to dissipate. Crown fires greatly alter habitat in all three dimensions, displacing species for as long as it takes to regain the habitat structure. However, crown fires kill more trees, allowing species that require dead trees and openings

ity, and magnitude of the fire could alter the path of subsequent migrations because of the juxtaposition of altered habitat gaps with past migration routes, adequacy of maintenance of migration corridors, and behavioral response of other species (e.g., predators) to changes in migration corridors (Table 9.4).

Ungulate migration is one of many animal population life history activities studied following the 1988 Yellowstone fires that consumed both forest and range vegetation. The more open postfire forests were found to allow pronghorn different routes to summer foraging areas (Scott and Geisser 1996). For migration to successfully occur through a particular area, the essential food and cover needed by the species in question must be present. Fire can affect such characteristics, either positively or negatively, thus changing the quality, or perhaps use of burned corridors. Indigenous Californians were aware of the vegetation changes that fire caused and used fire to create or manipulate animal habitat to both benefit preferred wildlife and create easier hunting opportunities (see chapter 19).

Annual movement and occupation between seasonal habitats (i.e., foraging, mating, rearing) following fires are influenced by the size and intensity of fire. Postfire movement, occupation and use of habitat, distribution, and abundance have been demonstrated in a variety of California wildlife, including rodents, birds, ungulates, amphibians, mustelids, and arthropods (Wirtz 1995, Beyers and Wirtz 1997, Gamradt and Katz 1997, Ferrenberg et al. 2006, Bleich et al. 2008, Mendelsohn et al. 2008, Holl and Bleich 2010).

Fire's Influence on Animal Reproduction

For a population to remain viable, a sufficient number of individuals must survive and reproduce in the course of short- and long-term changes that fire brings to their environment. Fire affects survival, mating and rearing activity, and can affect the amount and food available for newborn, juvenile, and adult animals alike. Environmental conditions can both benefit and impact populations. For example, abundant forage after high rainfall that allows for greater reproductive success in a population may also fuel fires in subsequent years just as immature individuals are rearing on the landscape (see sidebar 18.3). If enough animals in a population are affected, the dynamics and viability of that population will change.

Animals subjected to fire may leave the burned area temporarily or be displaced for longer periods. Animal populations might benefit from newly made-available food, reduced predators, or reduced competition. Any of these changes following a fire could contribute to more adults surviving to produce and raise offspring or more offspring simply surviving. But fire could have equally negative impacts on the reproductive success of animal populations. Adults, or more likely their slowly moving or immobile young, could be killed by fire. And essential forage or prey could be impacted, decreasing fecundity or survival of young. As with other characteristics of animals, the attributes of fire regime affect reproduction in different ways (Table 9.5).

Seasonality of fire and fire return interval can have a dramatic effect on reproductive viability. Many times, diverse

TABLE 9.3

Survival, time of appearance, and postfire abundance of rodent species following fire in coast sage scrub and southern chaparral communities of southern California

Rodent species	Survive fire?	Time to appear at burned sites	Postfire peak abundance	Study
Coastal sage scrub				
Pacific kangaroo rat *Dipodomys agilis*	Unknown	4 months or less	2–4 years	Price et al. 1995
San Diego pocket mouse *Chaetodipus fallax*	Unknown	4 months or less	3–7 years	Price et al. 1995
Deer mouse *Peromyscus maniculatus*	Yes	6–9 months	Postfire	Wirtz 1995
Cactus mouse *Peromyscus eremicus*	Unknown	Unknown	3–7 years	Price et al. 1995
Southern Chaparral				
California pocket mouse *Perognathus californicus*	Yes	6–7 months	Unknown	Wirtz 1995
Brush mouse *Peromyscus boylii*	Yes	6–9 months	Unknown	Wirtz 1995
Parasitic mouse *Peromyscus californicus*	Yes	6–9 months	Unknown	Wirtz 1995
Deer mouse *Peromyscus maniculatus*	Yes	6–9 months	Postfire	Wirtz 1995
Pacific kangaroo rat *Dipodomys agilis*	Unknown	8–9 months	Unknown	Wirtz 1995
Dusky-footed wood rat *Neotoma fuscipes*	No	8–11 months	Unknown	Wirtz 1995

TABLE 9.4

Potential change in wildlife migration with changes in fire regime attributes

Temporal	Seasonality	Fires burning during migration season could alter the path migrant species take
	Frequency	Change in frequency causes changes in habitat that could, depending on the species, either increase or decrease its use during the migration period
Spatial	Size	The size of a fire might (1) dictate whether migration routes are kept sufficiently open or (2) create gaps too wide for animals to cross or hide from predators
	Complexity	Independent of size, complexity probably has little effect on migration. But increased spatial complexity could mean more variability in migration routes, benefiting more species. Decreased complexity in combination with large fires can significantly shift migration routes
Magnitude	Severity	High-severity fire could establish or maintain migration routes necessary for larger species while destroying the cover or finer scale of routes important to smaller species. Low-severity fires might not be sufficient to maintain the open habitat structure needed by migrating ungulates/larger(?) animals
	Type	Crown fires could cause the most change in habitat structure that may be improve or decrease habitat quality for migrating species. For example, crown fires could remove the arboreal structure used by migrating birds

TABLE 9.5
Potential change in wildlife reproduction with changes in fire regime attributes

Temporal	Seasonality	In-season fire could allow resident wildlife to maintain reproductive and rearing behaviors and activities, access high-quality habitat and forage. Out-of-season fire could impact those same behaviors or activities or result in the death of nestlings or recently fledged or slow-moving young
	Frequency	Repeated disturbance of reproduction or longer-term alterations in an area's habitat could have successive influences on reproductive success. This could result in a decline in a population. If the impacted wildlife species are of greater significance (e.g., keystone, integral in food web) to the ecosystem, the entire community can be affected
Spatial	Size	Larger fires have the potential of affecting a larger portion of reproductively active members or offspring of a population, more mating pairs' territories, greater foraging or hunting range of females with offspring, and/or creation of openings for foraging and better predator detection
	Complexity	Independent of other fire regime attributes, complexity probably has little effect on reproduction. Increased fire complexity could result in more diverse or increased number of prey species, which could benefit predators. Species that use more homogeneous habitats likely would benefit from a less complex fire
Magnitude	Severity	Severity probably has unique effects to reproduction. High-severity fire may drive adult animals from nests and borrows, abandoning their young. In addition, higher severity often results in increased smoke, resulting in increased fatality and increased fleeing behaviors. Fire of any magnitude that disturbed reproductive or rearing activities would have the same, net effect. And, heat from a low to moderate fire could pose the same threat to fatality of young as high intensity. However, severity would come into play with forage for lactating mammals and species that collect food for their young
	Type	As with other animal activities, fire type has differential effect on animals living in different zones. Ground fires would be a threat to ground-nesting species but not to species high in the canopy or main stem of plants

groups within a population are differentially affected. Entire age-classes (e.g., newborn) or life stages (e.g., eggs) can be impacted once (e.g., an out-of-season fire) or repeatedly (e.g., more frequent fires than the return interval would indicate over a period of time), affecting not just the number of individuals within a population but also an entire or substantial proportion of a brood year. For example, nesting of ground and tree-nesting birds, fawning of ungulates, emergence from or initiation of aestivation (e.g., reptiles, amphibians, mammals) are closely linked to season, bioregion, latitude, and elevation. Invertebrate populations are equally susceptible. In southern California chaparral, some life stages of butterflies (i.e., eggs, larvae, pupae) are quite vulnerable to fire, and even adults may suffer direct fatality due to low flight speed and high site fidelity (Longcore 2003). Thus, fire may affect the age structure of these butterfly populations. However, increased postfire plant production, changes in plant community composition, and increased structural complexity resulted in overall beneficial conditions for butterflies. In a broader geographic and ecologic context, butterfly diversity increased two to three times in open areas created by fire in forest communities and doubled in riparian communities (Huntzinger 2003).

The effects of seasonality are not always clear. Research in the San Gabriel Mountains (Bleich et al. 2008, Holl and Bleich 2010, Holl et al. 2012) found that differences in severity and size of summer and fall fires had important consequences for ungulates. Fall fires burned at higher intensity and greater area, and these regime attributes resulted in better quality and more open habitat. In contrast, little effect between early and late-season prescribed burns was detected on rodents (e.g., deer mice, chipmunks) in mixed-conifer communities in the southern Sierra Nevada (Monroe and Converse 2006). These differences could be related to the difference in animal ecology, fire effects on preferred elements of habitat, or the vegetation communities themselves.

Change in vegetation structure, biomass, and nutritional content contribute to the reproductive success of each animal population inhabiting or using a burned area. Because fire severity and fire type alter vegetation differently, some populations' young of the year will benefit while others will not. Bird and mammal adults searching for food will have differing success both in nesting and gaining food for their young depending on the changes to vegetation brought on by the severity, spatial complexity, and size of the fire. Both surface fires and crown fires may provide access to more food for such animals as rodents and woodpeckers. Maturation of young may be enhanced because fires can increase the amount and nutritional value of forage and access to that forage. Fires also can create nesting structures (e.g., large snags, cavities or deformities in live trees) for bird species such as the spotted owl.

Reproduction and survival of animal species with extended breeding and foraging seasons, such as chipmunks and bark beetles, may be significantly impacted by a long fire season. Fire affects availability of food necessary for overwintering or migrating birds and mammals putting on fat and energy stores. Fire, coupled with one or more years of low primary production, could reduce health and reproduction in many animals, including many insect and bird species, ungulate mammals, and food cachers such as rodents and several species of birds.

Fire's Influence on Animal Foraging and Predation

Fire's alteration of the vegetation community affects reproductive success of animals by influencing the amount and nutritional value of available food. Fire can result in alterations to available animal prey and plant food, including forage, fruits, and seed. Following fire, trophic structure of the prefire animal community can be altered, and the changed landscape may attract different species which feed on new food sources, including plant species that germinate or sprout only after fire, regrowth of previously occurring plants, and newly available seed. New herbivores attract predators, some new and some not, to the burned area. The result is an altered food web, including some animals from the prefire community, as well as new species stimulated by the fire itself. Over decades, the response of the vegetation community to subsequent fires continues to affect the animal population–food web relations.

The effect of smoke is an immediate, usually short-lived, change in animal community composition and structure. Asphyxiation due to smoke is a threat to individual terrestrial animals. As with humans trapped in structures, smoke may pose a greater threat to life than heat. Many species escape the threat of smoke by fleeing fires, while others find shelter beneath materials such as logs and rocks.

Some wildlife, including insects, birds, and mammals, are attracted to a fire front, preying upon other animals fleeing or attracted to the fire and scavenging freshly killed animals or infecting injured trees. However, it is unclear if the smoke, heat, or other physical-chemical attributes of fire attract them. Bark-infesting beetles (e.g., *Dendroctonus* spp., *Melanophila* spp.) may be attracted to smoke (Álvarez et al. 2015), but a more likely means of detection is infrared (Evans 2005, Israelowitz et al. 2011). In California, ravens (*Coryus* spp.; Wirtz 1979), screech owls (*Megascops* spp.; Elliott 1985), woodpeckers (Moriarty et al. 1985), raptors and scavengers (Dodd 1988), golden eagles (*Aquila chrysaetos*), and peregrine falcons (*Falco peregrinus*; Lehman and Allendorf 1989) increase in abundance while the fires are active and for days and weeks following fire. Of particular interest is the community change related to woodpeckers. Stand-replacing fires kill large numbers of trees, leaving many others dying or susceptible to insects or diseases that can eventually kill them. Many species of insects are attracted to the dead and dying trees, followed by predaceous insects and insectivorous birds. In these environments, some species of woodpeckers are known to be the primary or exclusive occupant, or to dramatically increase following fire where dead and dying trees are abundant.

Some mammals are attracted to active or recently active fires. Deer are attracted to newly available forage and in turn attract mountain lions (*Puma concolor*) (Quinn 1990). French and French (1996) documented bears, raptors, and scavenging birds foraging along actively burning fires fronts of the 1988 Yellowstone fires. Grizzly and black bears, in particular, were not inhibited by either smoke or fire while foraging on ungulate carcasses resulting from fire mortality. Fire also can affect the type and quantity of food and forage available to animals but this topic is more properly discussed below under the indirect effects of fire.

Indirect Effects of Fire on Terrestrial Animals

Indirect fire effects are among the most important to animals, because both the animals' biotic and abiotic environments may be affected for a considerable period of time. Influences include alterations to plant community composition and structure, soil and litter, and aquatic habitat and ecosystems. These habitat changes determine the short-term conditions for survival of individuals residing in the area before the fire, newly arriving individuals, and the long-term consequences for entire animal populations and communities. In recent years, effects of climate change on wildlife habitat have become a topic of key concern, with regard to wildlife management, species and biodiversity conservation, and vulnerability assessments (Haak et al. 2010, Klausmeyer et al. 2011, Point Reyes Bird Observatory Conservation Science 2011, Romme et al. 2011, Moyle et al. 2012, 2015, Neville et al. 2012, Friggens et al. 2013, Homel et al. 2015). Wildfire has both immediate and longer-term effects on habitat, as does climate change. How climate change alters or affects wildfire itself has consequence to wildlife habitat changes (Friggens et al. 2013).

Fire alters both the structure and composition of vegetation, temporarily or permanently reduces or eliminates some food types while increasing others, and opens up areas previously closed by vegetation. Where essential physical features or food items are lost, animal species that depend on them will leave the area until those features or food items return, if they do. This is also the case for cover needed for shelter, protection, or migration, and even thermal requirements of animals (Esque et al. 1995, Rinne 1996, Brooks and Esque 2002). On the other hand, changes can result in habitat benefits for species dependent upon early successional habitat. For example, species requiring more openings, downed or dying vegetation, or types of food enhanced by fire may establish themselves in postfire environments until the resources that attracted them are exhausted or are lost to vegetation growth following fire.

Habitat Structure

One of the most significant roles of fire with regard to animal populations and communities is the alteration of vegetation and the subsequent shift in available habitat and habitat quality. These alterations are temporal, and some changes, such as the elimination of forbs and grasses, may be relatively short-lived (i.e., 1–10 years). Other changes, such as the reestablishment of large snags and mature, larger trees, or mature shrubs, can take decades to over a century. Interestingly, the same wildfires that remove more mature habitat elements (e.g., large snags, mature trees, roosting platforms) create new snags and downed wood. Fire alters individual plants, the density of vegetation, the age-class distribution of plant species, the three-dimensional structure of habitat, and the composition of plant species. Each of these changes can potentially influence animal occupation and use of a burned area for months, years, or decades. Postfire changes to habitat, both from wildfire and prescribed fire, have been documented to affect many types of wildlife in California, including arthropods (Longcore 2003, Ferrenberg et al. 2006), birds (Wirtz 1979, Beyers and Wirtz 1997, Mendelsohn et al. 2008), rodents (Price et al. 1995; Wirtz et al. 1995; Roberts et al. 2008, 2015; Brehme et al. 2011), spotted owls (Roberts et al. 2011, Jones et al. 2016), and ungulates (Holl and Bleich 2010, Holl et al. 2012).

WOODY PLANTS

Woody plants constituting the canopy structure are altered in several ways that affect animals. Fires kill mature trees creating new snags, burn down snags creating new downed logs, and consume old logs creating ash beds for new tree seedlings. Some microhabitat elements are created or facilitated by fire such as fire scars, cat faces, and hollowing of trees or by scorching.

Fire can repeatedly scar individual, mature trees, eventually resulting in "cat faces," deeply burned sections of the bole. Cat faces permit the invasion of insects and fungi into the exposed cambium, sapwood, and heartwood of trees, accelerating tree decline but providing food sources for invertebrates and vertebrates such as wood boring insects, insectivorous birds, and flying squirrels. Scorching can either kill trees or induce physiological stress, permitting invasion by insects and fungi. As with scarring, scorching provides habitat for insects and other invertebrates, and abundant food (e.g., bark beetles) for vertebrates such as woodpeckers and nuthatches. Decay sequences can last from years in moist, mild climates, to decades, or even centuries for large wood in very cold, dry subalpine environments and supports an everlasting but changing array of organisms associated with decay.

Fire can facilitate the development of hollow trees by either a fire entering through existing fire scars and cat faces or facilitating decay in heart and sapwood. Hollow trees provide shelter, nests, and food for many species of animals, including woodpeckers, birds of prey, bats, rodents, mustelids, and black bears.

PLANT DENSITY AND SIZE

Fires alter the configuration and degree of canopy closure and juxtaposition and density of plants. These changes affect the condition of habitat both in the canopy as well as of that of the forest or woodland floor. Opening up of dense forest canopies improves the habitat for some animals (e.g., chipmunk [*Tamias* spp.], ground squirrels [*Spermophilus* spp.]). For other animal species, fires that open up the canopy degrade or remove essential habitat, such as creating excessively large openings, destroying nesting sites, or eliminating snags. Species such as northern flying squirrels and chipmunk (e.g., Townsend's chipmunk [*Tamias townsendii*]), species associated with denser forest habitats, would be expected to be reduced by either greater canopy opening or lesser tree density. Similarly, the desert tortoise, which requires shrub cover to escape intense, desert sun, typically display population declines in areas where shrub cover is lost due to fire (Brooks and Esque 2002; see sidebar 18.3).

Decreasing plant density can also improve primary productivity by decreasing competition and increasing growth rate, which benefits animal species associated with larger diameter or taller shrubs and trees. Large trees, in particular, appear to be of great value to animals in woodlands and forests. Their dimension and volume provide habitat for animals that cannot be provided by smaller trees. Large animals, such as nesting birds of prey, bears, and porcupines, may find it impossible to use small trees because of dimensional constraints alone. Smaller animals, such as passerine birds or rodents, may prefer larger trees to avoid predators. If large trees are missing, these animals have to find alternatives or are forced to move to areas where habitat conditions are appropriate. Large trees also provide a greater abundance of cambium for invertebrate food and habitat than do smaller trees. It is possible that some of the larger boring insects can only find sufficient habitat in larger trees. Small trees do perform some important animal habitat functions. Some insects prefer habitat in smaller, dying trees or in the branches of larger trees. The larger density of smaller trees provides a constant background of habitat for invertebrates and fungi that serve as food for predators.

Like trees, changes in shrub size and density have important consequences for animals. Fire often consumes a very large proportion of the plant biomass in shrubland communities. The effects on animals vary and are related to the effects on individual plant species, the entire plant community, and to the ecological needs of the animals. Often, shrub canopy reduction benefits animals that forage in more open habitat or on newly available food (i.e., kangaroo rats) are more mobile in the more open habitat (e.g., mule deer). For example, opening up shrub communities initially benefits animals consuming seeds (Brehme et al. 2011). Other animals require a more mature shrub community or one with greater canopy cover. Beyers and Wirtz (1997) found that the California gnatcatcher inhabited coastal sage scrub with 50% cover or more, with plants typically taller than one meter. When fire burned occupied scrub communities, gnatcatchers typically did not recolonize the area for four to five years.

SOIL AND LITTER

Fires generally consume much of the litter, duff, and small dead wood, and may also consume a large proportion of the organic matter in the upper soil layers. Consumption of litter fuels generally releases a flush of nutrients to plants, encouraging growth. While reducing litter, fire also initiates deposition of a new pulse of organic matter which will initiate the new litter and duff layers.

Deposition of new organic material has two immediate wildlife-related results. First, deposition provides the only available habitat for species whose habitat was, at least partially, consumed. Secondly, the deposition of seeds (e.g., cones and acorns) or sprouting provide substantial readily available food for ungulates, rodents, especially squirrels, ants, and seed-eating birds.

SNAGS AND DOWNED WOOD

Dead and dying woody plants, snags, and downed wood and logs (coarse woody debris) provide important habitat for hundreds of vertebrate (CDFG 1999) and invertebrate species in shrub, woodland, and forest ecosystems. Fire, alone and in combination with other agents such as rusts, fungi, and insects, facilitates the production of new snags, downed wood and coarse woody debris, and other deadwood. Fire regime attributes of severity, extent, and spatial complexity result in creation and distribution of snags and coarse woody debris and consumption of dead wood existing before fire. Often single trees and large shrubs are killed, but large patches of shrubs and trees, and even entire stands may be killed depending on the size, intensity, and severity of the fire. During the deadwood cycling process, a wide range of animals utilizes the wood in various ways, including food, shelter, and nesting (Box 9.4).

BOX 9.4. EXAMPLES OF ANIMAL USE OF SNAGS, DEAD WOOD, AND DOWNED WOOD

Dead wood habitat element	Animal use
Dying, standing trees	Insects (those that forage on trees and their predators), insectivorous insects and birds, arboreal rodents
- Loosened bark	Creepers, bats, insects
- Excavated galleries and holes	Squirrels, woodpeckers, nuthatches, chickadees, black bears, insects
Standing snags	Owls, murrelets, goshawk, osprey, woodpeckers
Downed, coarse debris	Cover for invertebrates, salamanders, reptiles, rodents
Logs	Habitat for insects and other invertebrates; cover for amphibians, reptiles, rodents; perches for flycatchers; food source for some woodpeckers

Fire can also consume existing snags and coarse woody debris. Because these habitat elements are essential to animals, uniform loss over large extensive geographic areas could threaten the continued occurrence or recolonization of animals after a fire. However, the distribution of these elements and fire behavior may not result in complete consumption of snags and logs postfire. The effects to wildlife from loss of snags and woody debris are dependent on how much biomass is consumed and geographic pattern of loss, and these effects are not uniform or consistent across all fires (Horton and Mannan 1988, Harrington 1996, Laudenslayer 2002).

Habitat Distribution

In addition to changing the structural nature of habitat, fire also alters the spatial relation of distinct patches of habitat—the number and size of the patches, the juxtaposition of the suitable habitat for any given animal species, and the connectivity between these habitat patches. The spatial distribution of burned and unburned patches of animal habitat dictates animal community composition and carrying capacity for each population in an area. Unburned areas can function as refugia for resident animals during fires, sites for immediate occupation or recolonization following the fire, temporary shelter, or provide forage for migrating and early succession species (Esque et al. 1995). Burned areas can function for foraging or hunting of resident or immigrant species or newly available habitat for some resident or immigrant species (Ferrenberg et al. 2006, Mendelsohn et al. 2008, Roberts et al. 2008, Holl et al. 2012). Roberts et al. (2008, 2011, 2015) found that the effects of low- to moderate-severity fires have different effects on habitat patchiness and fire refugia, and benefit different species of the late-successional forest ecosystems. Different fire regimes alter habitat differently, influencing if and how burned areas will be colonized or exploited by animals (Table 9.6).

Responses to the mosaics created by fire elicit differ by species. Small habitat patches are necessary for animals that require either small openings or denser vegetation patches to be scattered in an otherwise homogeneous landscape. Some rodents need open areas to forage or to move within their range, while other rodent species and birds need cover for shelter and nesting. For example, the California gnatcatcher occupies and nests in coastal sage communities with shrub cover of 50% or more (Beyers and Wirtz 1997). Similarly, for many rodents (e.g., *Microtus* spp.) shorter grasses or more open situations, following fire, make it difficult to conceal their movements from potential predators. When fires open up this plant community, some bird and rodent species cease to occupy the habitat.

The interface between severely burned areas and the adjacent unburned or very lightly burned areas forms a complex mosaic of edges and openings. The shapes of these mosaics depend on a complex mix of variables affecting fire behavior, including the presence and types of fuels, topography, and weather during the burn. Openings at the edges of most closed forests are utilized by browsing and predatory animals, which take advantage of newly available forage and exposed prey, respectively. Larger openings dominated by herbaceous plants provide habitat for species including pocket gophers, voles, and ground foraging birds. The edges of larger openings can also provide forage for deer. Norland et al. (1996) concluded that fire effects on ungulates are still not well-understood.

The relations between fire regime, habitat alteration, and species and community response are complicated. Fire's modification of individual plants, canopy, and habitat has a cumulative effect. Spatial and temporal variation amongst fires creates a complex, varied environment for animals. In California, the interaction of fire and the northern goshawk (*Accipiter gentilis*) illustrates the complex relation of fire regime and animal habitat (Graham et al. 1997, Russell et al. 1999). In California, northern goshawks occupy forests with various fire regimes, ranging from surface fires in ponderosa pine (*Pinus ponderosa*) communities to stand-replacing fires of coastal Douglas-fir (*Pseudotsuga menziesii*). Smaller openings within the forests are preferred by many prey species of the northern goshawk, both birds and mammals. The goshawks themselves hunt at the edges of larger openings. Additionally, openings of adequate size adjacent to suitable nest sites are needed for adequate clearance for flight to and from nests, and for detection of predators. The northern goshawk requires a spectrum of opening sizes in order to prey, breed, and successfully rear young. An inadequate diversity of opening sizes likely would reduce population viability for the northern goshawk.

Plant Community Composition

Fire has even larger-scale, indirect effects on animals by altering plant community composition and condition. These longer-term, larger-scale changes affect available, suitable habitat for animal populations and communities. As the

TABLE 9.6
Potential change in wildlife colonization and exploitation with changes in fire regime attributes

Temporal	Seasonality	In-season fire would allow resident wildlife to exploit resources (e.g., food, nesting sites) normally, while out-of-season could either impact those same resources or fail to affect them in a manner necessary for a segment of the resident animal community
	Frequency	Changes in frequency alter both habitat structure and composition, which in turn could lead to new species being able to exploit an area. Long-term shifts might mean colonization and establishment of invading animal species; shorter-term shifts might mean relatively shorter periods of exploitation
Spatial	Size	Larger fires have the potential to reduce or increase habitat quality for different species across larger areas, depending upon the fire severity; newly burned and open space may facilitate or deter different species' use or occupation
	Complexity	Independent of size, complexity probably has little effect on colonization. But increased spatial complexity could mean more variability in habitat quality affecting more species positively or negatively. Decreased complexity in combination with large fires could significantly change the capability of the habitat to support some species
Magnitude	Severity	High severity could establish or maintain habitat necessary for larger species while removing the cover or finer scale of habitat patches important to smaller species. Low-severity fires might not be sufficient to maintain the habitat structure needed by some species, but may result in limited effects to wildlife currently using the area
	Type	Crown fires would cause the most obvious change in habitat structure, benefiting some resident species while impacting others. Crown fires could remove the arboreal structure used by birds and insects and other arboreal animals

plant community develops between fires, animal populations respond accordingly and changes in animal community composition and structure occur.

FOOD AVAILABILITY

Fire alters the type, quantity, and nutritional value of plant matter, including foliage, fruit, and nuts. At a broader level, the species composition and age-class distribution of plant species in the vegetation community dictates the composition and structure of the animal community. In California, fire in chaparral ecosystems may benefit mule deer and upland bird species (Biswell 1957, 1969, Kie 1984) by increasing forage and forage value as well as opening up areas for access to foliage.

The type and quantity of food and forage available to animals during years immediately following fire depends on both the fire type and plant community condition. Food for granivorous and insectivorous animals may be greatly increased after a fire. This effect may remain for two to five years following a fire in a grass or shrub community, and many more years in forests, depending on the ever-changing patterns and the types of plant communities. Newly available or newly depleted food may mean an introduction or displacement of animal species (Price et al. 1995, Van Dyke et al. 1996).

Fire affects the nutritional value of plant materials for animals. For grasslands, fire has been shown to increase forage productivity and quality (Norland et al. 1996, Tracy and McNaughton 1996, Van Dyke et al. 1996); however, studies indicate that this is not always the case. Fire has been shown to reduce nutritional value of elk winter forage in Yellowstone National Park (DelGiudice and Singer 1996) and to have no effect (positive or negative) on mule deer in pine-oak scrub of Trinity County (Kie 1984). Biswell (1957, 1969) suggested that fire in various types of California chaparral communities increased both deer utilization and deer fecundity. Though much is still unknown about the specific effects of fire to food values for wildlife, it is clear effects are variable and different species of animals may benefit or suffer from nutritional changes to plant matter.

PLANT COMMUNITY SHIFTS

Plant communities are in a continuous state of change. If the fire regime is altered by the occurrence of one uncharacteristic fire, an uncharacteristic sequence of fires, or fire exclusion, plant communities can change on a given site. Forests can be converted to another forest type, shrublands or herbaceous vegetation, and shrublands may be converted to herbaceous vegetation. The time needed for shrubs and trees to become reestablished depends on variables including the severity of the burn, topography, reproductive strategies of the species making up the plant community, the species of plants colonizing the site, the seed bank and adjacent vegetation that can supply seed.

Changing from a forest or shrub community to herbaceous-dominated community will immediately shift the fauna from those associated with trees and shrubs to those relying on herbaceous plants if suitable animals are nearby and the size of the opening is sufficient. This shift in species composition will, upon establishment of shrubs and trees, begin to revert to the fauna present before the fire. Often only a few years are needed for some rapidly developing shrub communities like those dominated by chamise (*Adenostoma fasciculatum*) and California lilac (*Ceanothus* spp.), to become reestablished. Tree-dominated communities take much longer to reestablish themselves, but the transition can begin immediately, and within 10–20 years, tree heights can exceed those of the shrubs and the shade provided by the trees begins to cause the loss of the shrubs. In grassland systems, the plant community itself often retards or halts encroachment of woody shrubs and trees, at least for a time. Grasses and forbs often outcompete young shrubs and saplings for nutrients, light, and especially water. Many animal species, including solitary bees, rodents, and birds, are dependent on native grasses for food. Frequent fire in grasslands maintain suitable habitat by maintaining grassland ecosystems (Roberts et al. 2012).

Shifts in forest or shrubland structure also occur after fire. Stands are differentially thinned by fire, leaving some areas intact and other areas with many fewer plants alive, immediately following fire. Additionally, canopy closure and canopy layering is reduced. The result is a mosaic of unburned areas, resembling the prefire condition, and areas of more open surface and canopy. This creates both a greater variety of habitats and ecotones, or edge effect, which many animals exploit.

Changes in plant species composition can have a profound effect on the composition of the fauna associated with the site. Invertebrates are often closely tied to specific species or genera of plants and can be lost until required plant species, consumed by fire, recolonize the burned area. Animals that are resource generalists (e.g., squirrels, chipmunks, jackrabbits [*Lepus* spp.], and jays [*Aphelocoma* and *Cyanocitta* spp.]) tend to be more tolerant of changes in vegetation composition because they do not require specific plant species for food, shelter, or nesting.

Indirect Effects of Fire on Aquatic Animals

Aquatic and amphibian communities and habitat can be indirectly affected by fire in several ways. Among these effects are changes to water flow, increased sedimentation, changes in supply and distribution of woody debris, and changes in composition and structure of riparian vegetation. In California, one of the most important questions about fire is its relation to native aquatic species and communities. Knowledge learned from other western states, and likely to be applicable to California, indicates that: fire has an important, ecological function for aquatic ecosystems and the adjacent, mesic vegetation communities and fire is a threat to aquatic systems only when either natural fire regime has been compromised, or native populations are highly fragmented, disjunct, or depressed. Postfire flooding that results in highly sedimented and ash and debris-laden runoff of water from the burned forest into streams that pose a risk to aquatic communities or species when one or both of these factors exist.

Delivery of Large Woody Debris, Sediment, and Water

Indirect fire effects to river and lake systems primarily involve shorter-term increase in water discharge and longer-term increases in sediment, nutrients, and large woody debris influx to the aquatic habitat. The annual, seasonal, and spatial distribution of each influences aquatic habitat and its quality. The materials delivered to aquatic systems can have beneficial or detrimental effects on aquatic animals and their habitat. Animals that inhabit riverine environments can benefit from increased water flow as long as pulse or sustained flow is not so strong as to degrade aquatic habitat or displace species downstream.

Large woody debris delivered from uplands is an essential aquatic habitat element for many species, such as salamanders, fishes, and benthic invertebrates, creating pools and backwaters, increasing stream structural complexity, and stabilizing lake and stream banks and substrate. Large moderate- to high-severity fires cause mortality in woody plants, resulting in significant quantities of woody debris being transported into water bodies over many years. In-stream wood forms log jams, and these log jams form barriers to some aquatic species moving up- or downstream. Log jams rarely form barriers that last long or prove to be complete blockage to upstream or downstream migration. In addition, the water held behind the log jams is habitat for other species. Fire has a crucial role in the ecological cycling of large woody debris, both creating new sources of debris and decreasing the stability of existing debris (Young and Bozek 1996).

Large, severe fires also can result in extreme erosion conditions, producing large quantities of sediment moving into water bodies. Sediment and organic material deposition can impact the benthic environment by filling in gravel substrate essential for spawning fishes and benthic habitat for macroinvertebrates, covering benthic plants or habitat, or creating a shallower environment. Animals inhabiting lacustrine systems may be especially susceptible, because the filling in of ponds and lakes alters the benthic environment, creates a shallower water body, and impacts aquatic plants. Suspended sediment can also impact aquatic animals by damaging gills and increasing respiratory stress and by temporarily decreasing photosynthesis by reducing light penetration. Decreased light reduces aquatic plants and algae, affecting herbivorous and predatory, aquatic species.

The alterations to aquatic habitat due to fire, and the runoff from postfire, successive storms can cause fatality of aquatic animals and impact animal community density and diversity. However, aquatic species appear to be resilient, if not adapted, to the interrelated cycles of fire, water, and transport of upslope sediment (Rinne 1996, Rieman et al. 1997, 2003, 2012, Gresswell 1999, Romme et al. 2011).

Thermal Regime

Heat influx from fire to aquatic ecosystems has not been documented and is likely not typically significant due to water's thermal and insulating properties. The potential effect fire can have on the aquatic temperature regime is due to fires that are severe enough to remove the vegetation that shades streams and lakes. Elevated water temperature can lower available dissolved oxygen, increase metabolism, and increase susceptibility of fishes to pathogens. For Coho Salmon (*Oncorhynchus kisutch*) in British Columbia, Holtby (1988) demonstrated that increased water temperature contributed to early emergence of fry, early, less favorable smolt migration, and lowered abundance and diversity of prey species. In California, elevated water temperatures following decreased riparian cover have been documented for salmonid streams, but neither fish mortality nor reduced population viability has been observed (Cafferata 1990).

Riparian Communities

Fire is a common ecological process in riparian communities, as it is in other natural communities. In many instances, fire removes much of the dead wood, litter and duff deposits, as well as trees from riparian areas. Effects on animals would result from the interrelation of fire severity, vegetation conditions, and soil moisture, and whether standing or moving water was present. Despite these potential effects on animals residing in riparian areas, it is generally believed that animals that inhabit riparian areas are rarely affected by fire (Smith 2000).

Fire and Ecosystem Interactions

The relation of fire to animals, especially populations and communities, is of great importance in California. The more significant relation, however, is the interaction of fire within an ecosystem. In this context, fire is one of several, critical ecological processes (e.g., energy, water, nutrient, geomorphic, weather, and climate) that interact with the entire biotic community. Interactions between vegetation, wildlife, soil microbes, and fire can play an important part in determining how the entire system responds to single burns or changes in the fire regime. In this section, we discuss the kinds of interactions that can occur and possible ramifications at the broader, ecological levels where there are important management implications as well as questions in which further research is needed.

Ecosystem Feedbacks from Animal and Fungi Activity

Biotic interactions between plants, herbivores, insects, fungi, or other soil biota all influence plant community responses. Changing from a forest or shrub to herbaceous vegetation will immediately shift the fauna from those associated with trees and shrubs to those relying on herbaceous plants, if a suitable reservoir of those animals is within a reasonable distance to immigrate. Herbivores, ranging from insects to ungulates, can browse newly sprouted shoots following a fire, depleting the plant's energy reserves and reducing its ability to become established. Selective herbivory can favor regeneration of some plant species while suppressing regeneration of others, thereby influencing the postfire succession trajectory. Fire's variable effects in different vegetation communities (i.e., plant species, vegetation structure, soil and litter) have been shown to have consequences on both arthropod and fungal communities in California (Ferrenberg et al. 2006, Meyer et al. 2008). Changes in invertebrate and fungal community structure, biomass, diversity, and distribution have immediate and longer-term consequences on entire animal communities.

Insect or fungus attack of already-weakened species can also influence plant species composition. A cause of plant mortality associated with fire can come from attack by insects attracted to the burned area or that were already present before the burn. The level of the insect population at the time of the fire, as well as whether the trees were already physiologically stressed from drought at the time of the fire, all interact to cause additional stress and secondary mortality. Secondary mortality can also favor reestablishment of some plant species while suppressing establishment of others, thereby affecting the overall composition and succession pathway of the ecosystem.

Animals and other biota can also have reciprocal effects on an area's fire regime. Grazing and pathogens can greatly alter vegetation structure and composition. Fire can also increase the dominance of nonnative plants (see chapter 24). Such changes in vegetation structure, affect litter and woody debris inputs, litter chemistry and decomposition rates, and therefore fuel loading. These changes directly affect fire conditions. For example, increased grazing in the African Serengeti decreased overall fuel levels and reduced fire frequency and intensity over several decades (Norton-Griffiths 1979). In North America, intensive grazing has also contributed to the reduction of grasses and forbs (that become fine fuels after drying) that can reduce fire frequency but has also led toward increases dense stands of advanced tree reproduction (Rummell 1951, Biswell 1972, Wright 1974, Savage 1991) that can alter fire behavior and severity.

Ecosystem Relations between Animals and other Environmental Processes

Fire does not simply affect a single animal species or animals alone. Across California, fires and different fire regimes affect biotic communities and ecological cycles (e.g., hydrology, nutrient cycles). Plant species dominance and distribution, which affects animal population occurrence, density, and distribution, is altered by fire. Those same animal species have different population demographics after the fire, and new animal species may either temporarily thrive or colonize the area. In chaparral and grassland ecosystems of the Central Valley and South Coast bioregions, rodent and insect herbivores may take advantage of new growth or abundant seed, even aiding in transporting seed to or within the burned area. Predators may take advantage of less cover, birds or mammals of altered nesting structure in the vegetation. Loss of vegetation affects available water, affecting the resident and colonizing animals alike, and products of the fire affect nutrient cycling, influencing plant response as well as detritus-consuming animals. All of this, in turn, affects the very vertebrates inhabiting or using a burned area.

In forest-dominated systems of the Sierra Nevada, the relation of fire's effects on forest structure and density also alter the distribution, density, and viability of many prominent vertebrates, including California spotted owl, northern goshawk, northern flying squirrel, pine marten, and fisher (*Martes pennanti pacifica*). Fire affects vegetation, fungi, and animal prey species, which in turn affects each other and other animal populations both individually and synergistically. Often, researchers attempt to investigate the relation of fire to one species, the vegetation forest composition and structure it needs, and possibly some animal species that make up the prey or food base (Meyer et al. 2007, 2008, Roberts et al. 2008, 2011, 2015, Scheller et al. 2011). Sidebar 9.1 provides details on the California spotted owl and fire-prone forests.

Fire influences carnivores, herbivores, omnivores, and scavengers simultaneously. Research at Black Mountain Experimental Forest in northern California has revealed some of the ecological relations between forest trees, insects, and insectivorous birds during and after fire, but little is known about the interactions or relations of the broader faunal communities (Jenkins et al. 2014). As fire regimes continue to be altered in the Sierra Nevada and other forest-dominated bioregions such as the Southern Cascades and Klamath Mountains, little is known about the ecological relations of fire, animal communities, the forests, and other ecosystem processes such as water and nutrient cycling or how changes in fire regimes may affect the health of animals. Research in oak woodland communities of the Central Coast bioregion has investigated the response of oak species and communities to fire and the response of more than 30 species of small vertebrates, including mammals, birds, and reptiles, to both fire and the change oak community (Vreeland and Tietje 2002, Tietje et al. 2008). Only one species showed initial lower abundance (e.g., pinyon deer mouse [*Peromyscus truei*]) following fire, and some species (e.g., deer mouse and California pocket mouse [*Chaetodipus californicus*]) showed higher abundance in areas of higher vegetative. Low- to moderate-intensity fires in coastal oak woodlands likely do not impact animals or do so for only short periods, and vegetation regeneration and increased seed production benefit herbivores and granivores, respectively, and healthy small animal populations benefit carnivores.

Overall, there is a lack of understanding of the interrelations of fire, vegetation, and other ecological processes to animal communities. Such understanding requires both time, as measured in successive fires and animal population generations interacting, and interdisciplinary research, focused on crucial information about those same fires' effects on hydrology, vegetation, and nutrient cycling.

Climate Change and Animal Response

Climate change and its effects and interactions with California vegetation and fire regimes are discussed in detail in chapter 26. The issues, discussions, and studies regarding the relation of current and predicted changes in climate and animals is now very similar to where such investigations were regarding animal–fire interactions 60 years ago. Three significant differences now are that climate change is considered to be of issue to all vegetation and animals communities, the investigations and concerns about animal response and resilience are joined by concerns for all natural resources, land management, and human communities, and the focus of examinations in California and other western States is acute, given the social and scientific concerns regarding climate change (Haak et al. 2010, Klausmeyer et al. 2011, Point Reyes Bird Observatory Conservation Science 2011, Romme et al. 2011, Luce et al. 2012, Friggens et al. 2013, Podhora 2015).

Climate change is resulting in variable and complicated environmental and habitat changes to all ecosystems and animal communities, aquatic and terrestrial; invertebrate and vertebrate. Already, fire ecologists and wildlife biologists are tying changes in climate to changes in stream flow timing, quantity, and temperature at different temporal and spatial scales (Field et al. 2007, Haak et al. 2010, Friggens et al. 2013) and migration corridors and habitat (Wormworth and Mallon 2006, Friggens et al. 2013) across entire ecosystems (Palmer et al. 2009, Luce et al. 2012, Friggens et al. 2013).

Regardless of the uncertainty surrounding climate change projections, developing adaptation strategies to promote habitat resiliency is an important emphasis in current wildlife and fisheries management (Haak et al. 2010, Friggens et al. 2013). Some of the first climate change assessments being conducted at state (Moyle et al. 2013), regional (Luce et al. 2012, Friggens et al. 2013), and national scales (Heinz Center 2008) are animal vulnerability assessments. To date, vulnerability assessments conducted in California have been focused at the species level, and assessments suggest a majority of native fauna are at moderate to extreme risk from habitat and ecological alterations from climate change. A total of 83% of native fishes (Moyle et al. 2013) and 36% of birds (Gardali et al. 2012) are considered at high risk due to climate change, while less than 20% of California's reptiles and amphibians are considered to be at greater than moderate to climate change (Wright et al. 2013). Wright et al. (2013) did find that there was more risk to species in the Central Valley and South Coast ecoregions.

California species and communities more dependent on cooler ambient air or water temperatures may be more affected, as many species with unique and narrow suitable or optimal ecological and habitat conditions. Such species could be the southern extent of Coho Salmon, in coastal watersheds south and north of San Francisco Bay, Chinook salmon (*Oncorhynchus tshawytscha*) in the Central Valley floor, wolverine (*Gulo gulo*), and pika (*Ochotona princeps*) in higher elevations of the Sierra Nevada. Interestingly, different phylogenetic groups may have broad adaptions that give them different advantages to climate change. These might include aerial dispersal and travel (i.e., many insects and birds), variability in form, size, and ecology (i.e., mammals), and physiology adaptations to dry and hot climate (i.e., most reptiles) (Friggens et al. 2013). Ecosystems in California and western North American have evolved and adapted with fire being a common, significant, and essential ecosystem driver. Therefore, even phylogenetic groups (i.e., fishes, amphibians) and communities (i.e., riparian, riverine) that may be seen as being the most vulnerable are adapted and have thrived in fire-influenced ecosystems. Climate change alterations to habitat and fire regimes is complex, and the influence on animals and their response will only add to the complexity of changes in the future.

Summary and Conclusions

To remain viable, animal species must adapt to the conditions of the regions they inhabit. Fire has been historically and still is an essential ecological process in California's ecosystems, and is a dominant force for many of the plant communities. Therefore, many animal populations and communities in California have adapted and persist in a fire-prone environment. Though individuals perish in fires and, in some cases, populations may be reduced or absent for a time, animal community health and interrelationships are tied to fire fulfilling its ecological functions. Fire maintains habitat complexity and ecotones, recycles and makes available nutrients, water, and other ecological elements, and changes the trophic relations between the various animal species in a given community or area.

We now have an understanding beyond fire's effects on animals or animals' response to fire. The relation of insects, birds, fire, and vegetation structure has been studied at Klamath National Forest and at Black Mountain Experimental Forest in northeast California. Further south in the Sierra Nevada (e.g., Kings Canyon and Yosemite national parks), natural resource managers have a growing understanding of fire's effects on several species of the biotic community. In oak woodlands of the Central Coast bioregion, there has been a land management focus on the relationship of oak species, small vertebrates, in the South Coast the focus has been small mammals and bird species of coastal sage scrub, and in the Southeast Desert bioregion, investigators have investigated the relation of fire, vegetation, precipitation, and ground-dwelling animals and ungulates.

SIDEBAR 9.1 SPOTTED OWLS IN FIRE-PRONE FORESTS OF CALIFORNIA

Monica L. Bond, Shaula Hedwall, and Kevin E. Shaffer

The California spotted owl (*Strix occidentalis occidentalis*) is generally associated with mid- to late-seral stage conifer forests in California. As a result of the owl's reliance for nesting and roosting habitat in older forests and its status as a key forest management species, attention is focused on understanding the effects of wildfire on the owl's survival, reproduction, territory occupancy, habitat use, and the long-term sustainability of late-seral forest habitat at the landscape scale.

The California spotted owl inhabits forests with high levels of canopy cover from overhead foliage (>70%), and large (>61 cm [24 in] diameter-at-breast height) live trees and snags are important components of nesting and roosting stands (Bond et al. 2004, Blakesley et al. 2005); thus managers have presumed that high-severity, landscape level fire, which kills all or most of the dominant vegetation within an area, posed a risk to owls. Over the past several decades, forest management within the range of the species has been guided by efforts to protect owl nest and roost habitat from high-severity, landscape-scale wildfire, under the assumption that severe fire is harmful to the maintenance and persistence of owl nest and roost habitat.

However, the California spotted owl occurs in fire-adapted forest systems, and research has revealed that fire extent, severity, size of high-severity patches, and prefire quality of the territory dictate the overall effect of fire on the subspecies. Studies over the past 15 years demonstrate that these birds can occupy and successfully reproduce in landscapes that experience a mosaic of fire severities, including a component of high-severity fire, and forage in high-severity burn areas.

Bond et al. (2002) examined survival and occupancy of spotted owls one year after mixed-severity wildfires in southern and northern California. All nest and roost areas within the fire perimeters were burned, and no postfire (salvage) logging had occurred before owls were surveyed the year after fire. The authors found 18 of 21 (86%) individual owls were resighted after fire and 16 of the 18 (89%) were in same sites immediately postfire. Reproductive rates were higher in sites that contained burned areas compared to long-unburned sites during the study period (one year). Roberts et al. (2011) compared longer-term effects of mixed-severity wildfires on occupancy and reproduction of spotted owls in randomly selected burned (>15 years since fire) and long-unburned surveyed areas in Yosemite National Park and found no difference in occupancy rates or densities of spotted owls. In an 11-year study of spotted owl occupancy at 41 sites located within the burn perimeters of six large fires and 145 long-unburned control sites on National Forest System lands in the Sierra Nevada, Lee et al. (2012) concluded that fires had no significant effect on owl occupancy 5 to 7 years postfire. In 2013, the Rim Fire burned over 100,000 ha (250,000 ac) in the Stanislaus National Forest and Yosemite National Park. Single-season postfire occupancy rates in the forest following the Rim Fire were higher than other previously published occupancy rates in both burned and long-unburned forests (Lee and Bond 2015a). Further, Lee and Bond (2015b) found in higher-quality owl sites in southern California that were consistently occupied and reproductive, the amount of severe fire had negligible effect on occupancy or reproduction.

Large patches of high-severity fire in nesting and roosting stands may adversely affect occupancy of lower-quality sites. For example, increasing amounts of severe fire surrounding nest and roost sites decreased occupancy probability in the Rim Fire, but did not affect occupancy by pairs of owls (Lee and Bond 2015a). In lower-quality sites in southern California that were often vacant and nonreproductive, occupancy was negatively correlated with increasing amounts of severe forest fire in the site's core (Lee and Bond 2015b). One year after the King Fire in the central Sierra Nevada, some California spotted owl sites were extirpated in a very large high-severity patch which experienced some postfire logging (Jones et al. 2016), exacerbating a long-term, 20-year population decline in this heavily managed region.

Why do spotted owls often remain and even thrive in many territories containing burned habitat? Fires that provide a mosaic of habitats and maintain nest and roost tree structure result in increased biodiversity and improved prey habitat. For example, Bond et al. (2009, 2013, 2016) found California spotted owls hunted (mostly for gophers and woodrats) in stands recently burned by severe fire.

In all of these studies, spotted owls still generally preferred to nest and roost in green forests with high canopy cover, underscoring the importance of unburned and low-severity refugia within the larger landscape mosaic of mixed-severity fire. However, moderate- and high-severity fires are part of the fire regimes of most California forests. These fires can increase habitat diversity that results in increased foraging opportunities.

Multiple years of investigation are needed to understand the relations of fire regime and fire cycles, animal populations, vegetation changes, climate, hydrology, and nutrient cycling. It will be these types of investigations, studies that focus on broader ecological processes and multiple animal populations, which will give the best insight into the ecological role of fire on animals and California and provide the best options to conserving both the ecosystems and the animal populations, particularly as climate continues to change.

Improved knowledge will be Continued progress is important, because expanded knowledge will be essential tool in managing and conserving the State's remaining wildlands and the animal communities that inhabit these lands. The confounding, environmental factors tied to climate change, weather, vegetation, and habitat make it vital to better understand the complex relation of fire to animal communities and the ecological processes affecting species diversity, distribution, habitat selection, and long-term, population viabilities.

As Biswell (1989) noted, fire has been a crucial process in many California forests, woodlands, and grasslands for thousands of years. As a process, fire affects the myriad of animals both directly and indirectly through effects on their habitat. Recent alterations in fire regimes have resulted in changes in California landscapes. In some cases, fires have become more intense, directly impacting larger numbers of animals more severely in a single event than most fires likely did historically. Fire regime changes also have indirectly affected California's wildlife through the alteration of habitat composition and structure; habitat quality for some of these species has improved but for others it has declined. It is not clear whether these current conditions will be sustained through time. Although we know that contemporary fire regimes differ from historical regimes in many bioregions in California, the effects of these changes to animal species and communities are not yet understood, and the changes in populations or communities likely would take generations of animals to be detectable. This change in fire regimes has resulted in changes in vegetation composition and structure that have altered the habitats for many animals that reside in these areas. For some, the changes have improved the quality of their habitat, whereas for others, the changes have reduced the quality of their habitat. Yet, in many of these habitats, fire will, and in many cases should, eventually return regardless of attempts to suppress it. The question is: what kind of fire will it be?

References

Álvarez, G., B. Ammagarahalli, D.R. Hall, J.A. Pajares, and C. Gemeno. 2015. Smoke, pheromone and kairomone olfactory receptor neurons in males and females of the pine sawyer *Monochamus galloprovincialis* (Olivier) (Coleoptera: Cerambycidae). Journal of insect Physiology 82: 46–55.

Bendell, J.F. 1974. Effects of fire on birds and mammals. Pages 73–138 in: T.T. Kozlowski and C.E. Ahlgren, editors. Fire and Ecosystems. Academic Press, New York, New York, USA.

Beyers, J.L., and W.O. Wirtz, II. 1997. Vegetative characteristics of coastal sage scrub sites used by California gnatcatchers: implications for management in a fire-prone ecosystem. Pages 81–90 in: J.M. Greenlee, editor. Fire Effects on Rare and Endangered Species and Habitats. International Association for Wildland Fire, Fairfield, Washington, USA.

Bisson, P.A., B.R. Rieman, C. Luce, P.F. Hessburg, D.C. Lee, J.L. Kershner, G.H. Reeves, and R.E. Gresswell. 2003. Fire and aquatic ecosystems: current knowledge and key questions. Forest Ecology and Management 178: 213–229.

Biswell, H.H. 1957. The use of fire in California chaparral for game habitat improvement. Pages 151–155 in: Proceedings of the Society of American Foresters. Bethesda, Maryland, USA.

———. 1969. Prescribed burning for wildlife in California brushlands. Pages 438–446 in: Transactions of the 34th North American Wildlife and Natural Resources Conference. Wildlife Management Institute, Washington, D.C., USA.

———. 1972. Fire ecology in ponderosa pine-grassland. Proceedings Tall Timbers Fire Ecology Conference 12: 69–96.

———. 1989. Prescribed Burning in California Wildlands Vegetation Management. University of California Press, Berkeley, California, USA.

Blakesley, J.A., B.R. Noon, and D.R. Anderson. 2005. Site occupancy, apparent survival, and reproduction of California spotted owls in relation to forest stand characteristics. Journal of Wildlife Management 69: 1554–1564.

Bleich, V.C., H.E. Johnson, S.A. Hall, L. Konde, S.G. Torres, and P.R. Krausman. 2008. Fire history in a chaparral ecosystem: implications for conservation of a native ungulate. Rangeland Ecology and Management 61: 571–579.

Bond, M.L., C. Bradley, and D.E. Lee. 2016. Foraging habitat selection by California spotted owls after fire. Journal of Wildlife Management 80: 1290–1300.

Bond, M.L., R.J. Gutiérrez, A.B. Franklin, W.S. LaHaye, C.A. May, and M.E. Seamans. 2002. Short-term effects of wildfires on spotted owl survival, site fidelity, mate fidelity, and reproductive success. Wildlife Society Bulletin 30: 1022–1028.

Bond, M.L., R.J. Gutiérrez, and M.E. Seamans. 2004. Modeling nesting habitat selection of California spotted owls (*Strix occidentalis occidentalis*) in the central Sierra Nevada using standard forest inventory metrics. Forest Science 50: 773–780.

Bond, M.L., D.E. Lee, R.B. Siegel, and M.W. Tingley. 2013. Diet and home-range size of California spotted owls in a burned forest. Western Birds 44: 114–126.

Bond, M.L., D.E. Lee, R.B. Siegel, and J.P. Ward. 2009. Habitat use and selection by California spotted owls in a postfire landscape. Journal of Wildlife Management 73: 1116–1124.

Brehme, C.S., D.R. Clark, C.J. Rochester, and R.N. Fisher. 2011. Wildfires alter rodent community structure across four vegetation types in southern California, USA. Fire ecology 7(2): 81–97.

Brooks, M.L., and T. Esque. 2002. Alien plants and fire in desert tortoise (*Gopherus agassizii*) habitat of the Mojave and Colorado deserts. Chelonian Conservation and Biology 4: 330–340.

Cafferata, P. 1990. Temperature regimes of small streams along the Mendocino coast. Jackson Demonstration State Forest Newsletter 39. California Department of Forestry and Fire Protection, Fort Bragg, California, USA.

CDFG (California Department of Fish and Game). 1999. California Wildlife Habitat Relationships System. California Department of Fish and Game, Sacramento, California, USA.

Converse, S.J., W.M. Block, and G.C. White. 2006. Small mammal population and habitat responses to forest thinning and prescribed fire. Forest Ecology and Management 228: 263–273.

Cooperrider, A., R.F. Noss, H.H. Welsh, Jr., C. Carroll, W. Zielinski, D. Olson, S.K. Nelson, and B.G. Marcot. 2000. Terrestrial fauna of redwood forests. Pages 119–163 in: R.F. Noss, editor. The Redwood Forest: History, Ecology, and Conservation of the Coast Redwoods. Save the Redwood League. Island Press, Washington, D.C., USA.

Cunningham, S.C., R.D. Babb, T.R. Jones, B.D. Taubert, and R. Vega. 2002. Reaction of lizard populations to a catastrophic wildfire in a central Arizona mountain range. Biological Conservation 107: 193–201.

Cunningham, S.C., and W.B. Ballard. 2004. Effects of wildfire on black bear demographics in central Arizona. Wildlife Society Bulletin 32: 928–937.

DeBano, L.F., D.G. Neary, and P.F. Ffolliiott. 1998. Fire's Effects on Ecosystems. Wiley and Sons, New York, New York, USA.

DelGiudice, G.D., and F.J. Singer. 1996. Physiological responses of Yellowstone elk to winter nutritional restriction before and after the 1988 fires: a preliminary examination. Pages 133–135 in: J.M. Greenlee, editor. Ecological Implications of Fire in Greater Yellowstone. International Association for Wildland Fire, Fairfield, Washington, USA.

Dodd, N.L. 1988. Fire management and southwestern raptors. Pages 341–347 in: R.L. Glinski, B.G. Pendleton, M.B. Moss,

M.N. LeFranc, Jr., B.A. Milsap, and S.W. Hoffman, editors. Proceedings of the Southwest Raptor Symposium and Workshop. National Wildlife Federation Scientific and Technology Series 11. National Wildlife Federation, Washington, D.C., USA.

Elliott, B. 1985. Changes in distribution of owl species subsequent to habitat alteration by fire. Western Birds 16: 25–28.

Esque, T.C., T. Hughes, L.A. DeFalco, B.E. Hatfield, and R.B. Duncan. 1995. Effects of wildfire on desert tortoises and their habitats. Pages 153–154 in: A. Fletcher-Jones, editor. 19th Proceedings of the Desert Council, Tucson, Arizona, USA.

Evans, W.G. 2005. Infrared radiation sensors of *Melanophila acuminata* (Coleoptera: Buprestidae): a thermopneumatic model. Annals of the Entomological Society of America 98: 738–746.

Ferrenberg, S.M., D.W. Schwilk, E.E. Knapp, E. Groth, and J.E. Keeley. 2006. Fire decreases arthropod abundance but increases diversity: early and late season prescribed fire effects in a Sierra Nevada mixed-conifer forest. Fire Ecology 2(2): 79–102.

Field, C.B., L.D. Mortsch, M. Brklacich, D.L. Forbes, P. Kovacs, J.A. Patz, S.W. Running, and M.J. Scott. 2007. North America. Pages 616–652 in: M.L. Parry, O.F. Canziani, J.P. Palutikof, P.J. van der Linden, and C.E. Hanson, editors. Climate Change 2007: Impacts, Adaptation and Vulnerability. Contribution of Working Group II to the Fourth Assessment Report of the Intergovernmental Panel on Climate Change. Cambridge University Press, Cambridge, UK.

Ford, W.M., M.A. Menzel, D.W. McGill, J. Laerm, and T.S. McCay. 1999. Effects of a community restoration fire on small mammals and herpetofauna in the southern Appalachians. Forest Ecology and Management 114: 233–243.

French, M.G., and S.P. French. 1996. Large mammal mortality in the 1988 Yellowstone fires. Pages113–116 in: J.M. Greenlee, editor. Ecological Implications of Fire in Greater Yellowstone. International Association for Wildland Fire, Fairfield, Washington, USA.

Friggens, M.M., D.M. Finch, K.E. Bagne, S.J. Coe, and D.L. Hawkworth. 2013. Vulnerability of species to climate change in the southwest: terrestrial species of the middle Rio Grande. USDA Forest Service General Technical Report RMRS-GTR-306. Rocky Mountain Research Station, Fort Collins, Colorado, USA.

Gamradt, S.C., and L.B. Kats. 1997. Impact of chaparral wildfire-induced sedimentation on oviposition of stream-breeding California newts (*Taricha torosa*). Oecologia 110: 546–549.

Gardali, T., N.E. Seavy, R.T. DiGaudio, and L.A. Comrack. 2012. A climate change vulnerability assessment of California's at-risk birds. PLoS ONE 7(3): e29507.

Graham, R.T., T.B. Jain, R.T. Reynolds, and D.A. Boyce. 1997. The role of fire in sustaining northern goshawk habitat in Rocky Mountain forests. Pages 69–76 in: J.M. Greenlee, editor. Fire Effects on Rare and Endangered Species and Habitats. International Association for Wildland Fire, Fairfield, Washington, USA.

Gresswell, R.E. 1999. Fire and aquatic ecosystems in forested biomes of North America. Transactions of the American Fisheries Society 128: 193–221.

Haak, A.L., J.E. Williams, D. Isaak, A. Todd, C. Muhlfel, J.L. Kershner, R. Gresswell, S. Hostetler, and H.M. Neville. 2010. The potential influence of changing climate on the persistence of salmonids of the inland west. U.S. Geological Survey Open-File Report OF-2010-1236. U.S. Geological Survey, Reston, Virginia, USA.

Haidinger, T.L., and J.E. Keeley. 1993. Role of high fire frequency in destruction of mixed chaparral. Madroño 40: 141–147.

Harestad, A.S. 1985. *Scaphiopus intermontanus* (Great Basin spadefoot toad) mortality. Herpetology Review 16: 24.

Harrington, M.G. 1996. Fall rates of prescribed fire-killed ponderosa pine. USDA Forest Service Research Paper INT-RP-489. Intermountain Research Station, Ogden, Utah, USA.

Heinz Center. 2008. Strategies for Managing the Effects of Climate Change on Wildlife and Ecosystems. Washington, D.C., USA.

Hitt, N.P. 2003. Immediate effects of wildfire on stream temperature. Journal of Freshwater Ecology 18: 171–173.

Holl, S.A., and V.C. Bleich. 2010. Responses of bighorn sheep and mule deer to fire and rain in the San Gabriel Mountains, California. Proceedings of the Northern Wild Sheep and Goat Council 17: 139–156.

Holl, S.A., V.C. Bleich, B.W. Callenberger, and B. Bahro. 2012. Simulated effects of two fire regimes on bighorn sheep: the San Gabriel Mountains, California USA. Fire Ecology 8(3): 88–103.

Holtby, L.B. 1988. Effects of logging on stream temperatures in Carnation Creek, British Columbia, and associated impacts on the coho salmon. Canadian Journal of Fish Aquatic Science 45: 502–515.

Homel, K.M., R.E. Gresswell, and J.L. Kershner. 2015. Life history diversity of Snake River finespotted cutthroat trout: managing for persistence in a rapidly changing environment. North American Journal of Fisheries Management 35: 789–801.

Horton, S.P., and R.W. Mannan. 1988. Effects of prescribed fire on snags and cavity-nesting birds in southeastern Arizona pine forests. Wildlife Society Bulletin 16: 3744.

Hossack, B.R., W.H. Lowe, R.K. Honeycutt, S.A. Parks, and P.S. Corn. 2013. Interactive effects of wildfire, forest management, and isolation on amphibian and parasite abundance. Ecological Applications 23: 479–492.

Howard, W.E., R.L. Fenner, and H.E. Childs, Jr. 1959. Wildlife survival in brush burns. Journal of Range Management 12: 230–234.

Huntzinger, P.M. 2003. Effects of fire management practices on butterfly diversity in the forested western United States. Biological Conservation 113: 1–12.

Israelowitz, M., J. Kwon, S.W.H. Rizvi, C. Gille, and H.P. von Schroeder. 2011. Mechanism of infrared detection and transduction by beetle *Melanophila acuminata* In memory of Jerry Wolken. Journal of Bionic Engineering 8(2): 129–139.

Jenness, J.S., P. Beier, and J.L. Ganey. 2004. Associations between forest fire and Mexican spotted owls. Forest Science 50: 765–772.

Jenkins, M.J., J.B. Runyon, C.J. Fettig, W.G. Page, and B.J. Bentz. 2014. Interactions among the mountain pine beetle, fires, and fuels. Forest Science 60: 489–501.

Jones, G., R.J. Gutiérrez, D.J Tempel, S.A. Whitmore, W.J. Berigan, and M.Z. Peery. 2016. Megafires: an emerging threat to old-forest species. Frontiers in Ecology and the Environment 14: 300–306.

Kershner, J.L., L.H. MacDonald, M. Decker, D. Winters, and Z. Libohova 2003. Fire-induced changes in aquatic ecosystems. Pages 232–243 in: R.T. Graham, technical editor. Hayman Fire Case Study. USDA Forest Service General Technical Report RMRS-GTR 114. Rocky Mountain Research Station, Fort Collins, Colorado, USA.

Kie, J.G. 1984. Deer habitat use after prescribed burning in northern California. USDA Forest Service Research Note PSW-RN-369. Pacific Southwest Forest and Range Experiment Station, Berkeley, California, USA.

Klausmeyer, K.R., M.R. Shaw, J.B. MacKenzie, and D.R. Cameron. 2011. Landscape-scale indicators of biodiversity's vulnerability to climate change. Ecosphere 2(8): art88.

Koehler, G.M., and M.G. Hornocker. 1977. Fire effects on marten habitat in the Selway-Bitterrroot Wilderness. Journal of Wildlife Management 41: 500–505.

Komarek, E.V. 1969. Fire and animal behavior. Proceedings Tall Timbers Fire Ecology Conference 9: 161–207.

Kotliar, N.B., P.L. Kennedy, and K. Ferree 2007. Avifaunal responses to fire in southwestern montane forests along a burn severity gradient. Ecological Applications 17: 491–507.

Latif, Q.S., J.S. Sanderlin, V.A. Saab, W.M. Block, and J.G. Dudley. 2006. Avian relationships with wildfire at two dry forest locations with different historical fire regimes. Ecosphere 7(5): e01346.

Laudenslayer, W.F., Jr. 2002. Effects of prescribed fire on live trees and snags in Eastside Pine Forests in California. Pages 256–262 in: N.G. Sugihara, M.E. Morales, and T.J. Morales, editors. Proceedings of the Symposium: Fire in California Ecosystems, Integrating Ecology, Prevention, and Management. Association for Fire Ecology. Miscellaneous Publication 1. Davis, California, USA.

Lehman, R.N., and J.W. Allendorf. 1989. The effects of fire, fire exclusion, and fire management on raptor habitats in the western United States. Pages 236–244 in: B.G. Pendleton, editor. Proceedings of the Western Raptor Management Symposium and Workshop. National Wildlife Federation Scientific and Technical Series 12. National Wildlife Federation, Washington, D.C., USA.

Lee, D.E., and M.L. Bond. 2015a. Occupancy of California spotted owl sites following a large fire in the Sierra Nevada. The Condor 117: 228–236.

———. 2015b. Previous year's reproductive state affects spotted owl site occupancy and reproduction responses to natural and anthropogenic disturbances. The Condor 117: 307–319.

Lee, D.E., M.L. Bond, and R.B. Siegel. 2012. Dynamics of breeding-season site occupancy of the California spotted owl in burned forests. The Condor 114: 792–802.
Longcore, T. 2003. Ecological effects of fuel modification on arthropods and other wildlife in an urbanizing wildland. Pages 111–117 in: K.E.M. Galley, R.C. Klinger, and N.G. Sugihara, editors. Proceedings of Fire Conference 2000: The First National Congress on Fire Ecology, Prevention, and Management. Miscellaneous Publication No. 13. Tall Timbers Research Station, Tallahassee, Florida, USA.
Luce, C., P. Morgan, K. Dwire, D. Isaak, Z. Holden, and B. Rieman. 2012. Climate change, forests, fire, water, and fish: building resilient landscapes, streams, and managers. USDA Forest Service General Technical Report RMRS-GTR-290. Rocky Mountain Research Station, Fort Collins, Colorado, USA.
Lyon, J.L., H.S. Crawford, E. Czuhai, R.L. Fredricksen, R.F. Harlow, L.J. Metz, and H.A. Pearson. 1978. Effects of Fire on Fauna: A State-of-the-Knowledge Review. USDA Forest Service General Technical Repot GTR-WO-6. Washington, D.C., USA.
Mahlum, S.K., L.A. Eby, M.K. Young, C.G. Clancy, and M. Jakober. 2011. Effects of wildfire on stream temperatures in the Bitterroot River Basin, Montana. International Journal of Wildland Fire 20: 240–247.
McIver, J.D., S.L. Stephens, J.K. Agee, J. Barbour, R.E.J. Boerner, C.B. Edminster, K.L. Erickson, K.L. Farris, C.J. Fettig, C.E. Fiedler, S.M. Haase, S.C. Hart, J.E. Keeley, E.E. Knapp, J.F. Lehmkuhl, J.J. Moghaddas, W. Otrosina, K.W. Outcalt, D.W. Schwilk, C.N. Skinner, T.A. Waldrop, C.P. Weatherspoon, D.A. Yaussy, A. Youngblood, and S. Zac. 2012. Ecological effects of alternative fuel-reduction treatments: highlights of the National Fire and Fire Surrogate study (FFS). International Journal of Wildland Fire 22: 63–82.
Mendelsohn, M.B., C.S. Brehme, C.J. Rochester, D.C. Stokes, S.A. Hathaway, and R.N. Fisher. 2008. Responses in bird communities to wildlife fires in southern California. Fire Ecology 4(2): 63–82.
Meyer, M.D., D.A. Kelt, and M.P. North. 2007. Microhabitat associations of northern flying squirrels in burned and thinned forest stands of the Sierra Nevada. American Midland Naturalist 157: 202–211.
Meyer, M.D., M.P. North, and S.L. Roberts. 2008. Truffle abundance in recently prescribed burned and unburned forests in Yosemite National Park: implications for mycophagous mammals. Fire Ecology 4(2): 105–114.
Minshall, G.W., and J.T. Brock. 1991. Observed and anticipated effects of forest fire on Yellowstone stream ecosystems. Pages 123–135 in: R.B. Keiter and M.S. Boyce, editors. The Greater Yellowstone Ecosystem: Redefining America's Wilderness Heritage. Yale University Press, New Haven, Connecticut, USA.
Monroe, M.E., and S.J. Converse. 2006. The effects of early season and late season prescribed fires on small mammals in a Sierra Nevada mixed conifer forest. Forest Ecology and Management 236: 229–240.
Moriaty, D.J., R.E. Farris, D.K. Noda, and P.A. Stanton. 1985. Effects of fire on a coastal sage scrub bird community. Southwestern Naturalist. 30(3): 452–453.
Moyle, P., J.D. Kiernan, P.K. Crain, and R.M. Quinones. 2013. Climate change vulnerability of native and alien freshwater fishes of California: a systematic assessment approach. PLoS ONE 8(5): e63883.
Moyle P.B., R.M. Quiñones, J.V.E. Katz, and J. Weaver. 2015. Fish species of special concern in California. California Department of Fish and Wildlife, Sacramento, California, USA. https://www.wildlife.ca.gov/Conservation/Fishes/Special-Concern.
Moyle, P., R.M. Quiñones, and J. Kiernan. 2012. Effects of climate change on the inland fishes of California with emphasis on the San Francisco Estuary region. Our Hanging Climate. California Energy Commission Publication CEC-500-2012-029. Sacramento, California, USA.
National Park Service. 1966. Conservation Plan: Wind Cave National Park, Custer County Conservation District. Department of the Interior, National Park Service, Wind Cave National Park, South Dakota, USA.
Neville, H.M., R.E. Gresswell, and J.B. Durham. 2012. Genetic variation reveals influence of landscape connectivity on population dynamics and resiliency of western trout in disturbance-prone habitats. Pages 177–186 in: C. Luce, P. Morgan, K. Dwire, D. Isaak, Z. Holden, and B. Rieman. Climate Change, Forests, Fire, Water, and Fish: Building Resilient Landscapes, Streams, and Managers. USDA Forest Service RMRS-GTR-290. Rock Mountain Research Station, Fort Collins, Colorado, USA.
Norland, J.E., F.J. Singer, and L. Mack. 1996. Effects of the Yellowstone fires of 1988 on elk habitat. Pages 223–232 in: J.M. Greenlee, editor. Ecological Implications of Fire in Greater Yellowstone. International Association for Wildland Fire, Fairfield, Washington, USA.
Norton-Griffiths, M. 1979. The influence of grazing, browsing, and fire on vegetation dynamics in the Serengeti, Tanzania, Kenya. Pages 310–352 in: A.R.E. Sinclair and M. Norton-Griffiths, editors. Serengeti: Dynamics of an Ecosystem. University of Chicago Press, Chicago, Illinois, USA.
Palmer, M.A., D.P. Lettenmaier, N.L. Poff, S.L. Postel, R. Richter, and R. Warner. 2009. Climate change and river ecosystems: protection and adaptation options. Environmental Management 44: 1053–1068.
Podhora, E. 2015. Lessons for climate change reform from environmental history: 19th century wildlife protection and the 20th century environmental movement. Journal of Environmental Law and Litigation 30: 1–55.
Point Reyes Bird Observatory Conservation Science. 2011. Projected Effects of Climate Change in California: Ecoregional Summaries Emphasizing Consequences for Wildlife. Version 1.0. http://data.prbo.org/apps/bssc/climatechange.
Price, M.V., N.M. Waser, K.E. Taylor, and K.L. Pluff. 1995. Fire as a management tool for Stephens' kangaroo rat and other small mammal species. Pages 51–61 in: J.E. Keeley and T. Scott. Brushfires in California Wildlands: Ecological and Resource Management. International Association of Wildland Fire. Fairfield, Washington, USA.
Pyne, S.J. 1995. World Fire: The Culture of Fire on Earth. University of Washington Press, Seattle, Washington, USA.
Quinn, R.D. 1990. Habitat preferences and distribution of mammals. In California Chaparral. USDA Forest Service Research Paper PSW-RP-202. Pacific Southwest Research Station, Berkeley, California, USA.
Rieman, B.E., and J. Clayton. 1997. Wildfire and native fish: Issues of forest health and conservation of sensitive species. Fisheries 22: 6–15
Rieman, B., R. Gresswell, and J. Rinne. 2012. Fire and fish: a synthesis of observation and experience. Pages 159–177 in: C. Luce; P. Morgan, K. Dwire, D. Isaak, Z. Holden, and B. Rieman. Climate change, forests, fire, water, and fish: Building resilient landscapes, streams, and managers. USDA Forest Service RMRS-GTR-290. Rock Mountain Research Station, Fort Collins, Colorado, USA.
Rieman, B., D. Lee, D. Burns, R. Gresswell, M. Young, R. Stowell, J. Rinne, and P. Howell. 2003. Status of native fishes in the western United States and issues for fire and fuels management. Forest Ecology and Management 178: 197–211.
Rieman, B.E., D. Lee, G. Chandler, and D. Myers. 1997. Does wildfire threaten extinction for salmonids: responses of redband trout and bull trout following recent large fires on the Boise National Forest. Pages 47–57 in: J.M. Greenlee, editor. Fire Effects on Threatened and Endangered Species and Habitats. International Association of Wildland Fire, Fairfield, Washington, USA.
Rinne, J.N. 1996. Short-term effects of wildfire on fishes and aquatic macroinvertebrates in the southwestern United States. North American Journal of Fisheries Management 16: 653–658.
Roberts, S.L., D.A. Kelt, J.W. van Wagtendonk, A.K. Miles, and M.D. Meyer. 2015. Effects of fire on small mammal communities in frequent-fire forests in California. Journal of Mammalogy 96: 107–119.
Roberts, S.L., J.W. van Wagtendonk, A.K. Miles, and D.A. Kelt. 2011. Effects of fire on spotted owl site occupancy in a late-successional forest. Biological Conservation 144: 610–619.
Roberts, S.L., J.W. van Wagtendonk, A.K. Miles, D.A. Kelt, and J.A. Lutz. 2008. Modeling the effects of fire severity and spatial complexity on small mammals in Yosemite National Park, California. Fire Ecology 4(2): 83–104.
Romme, W.H., M.S. Boyce, R. Gresswell, E.H. Merrill, G.W. Minshall, C. Whitlock, and M.G. Turner. 2011. Twenty years after the 1988 Yellowstone fires: lessons about disturbance and ecosystems. Ecosystems 14: 1196–1215.
Rummell, R.S. 1951. Some effects of livestock grazing on ponderosa pine forest and range in central Washington. Ecology 32: 594–607.

Russell, K.R., D.H. van Lear, and D.C. Guynn, Jr. 1999. Prescribed fire effects on herpetofauna: review and management implications. Wildlife Society Bulletin 27: 374–384.
Russell, R.E., J.A. Royle, V.A. Saab J.F. Lehmkuhl, W.M. Block, and J.R. Sauer. 2009. Modeling the effects of environmental disturbance on wildlife communities: avian responses to prescribed fire. Ecological Applications 19: 1253–1263.
Saab, V., W.M. Block, R.E. Russell, J. Lehmkuhl, L. Bate, and R. White. 2007. Birds and burns of the interior West: descriptions, habitats and management in western forests. USDA Forest Service General Technical Report PNW-GTR-712. Pacific Northwest Research Station, Portland, Oregon, USA.
Saab, V.A., and K.T. Vierling. 2001. Reproductive success of Lewis' woodpecker in burned pine and cottonwood riparian forests. Condor 103: 491–501.
Safford, H.D., and K.M. Van de Water. 2014. Using fire return interval departure (FRID) analysis to map spatial and temporal changes in fire frequency on national forest lands in California. USDA Forest Service Research Paper PSW-RP-266. Pacific Southwest Research Station, Albany, California, USA.
Savage, M. 1991. Structural dynamics of a southwestern pine forest under chronic human influence. Annals of the Association of American Geographers 81: 271–289.
Scheller, R.M., W.D. Spencer, H. Rustigian-Romsos, A.D. Syphard, B.C. Ward, and J.R. Strittholt. 2011. Using stochastic simulation to elevate competing risks of wildfires and fuels management on an isolated forest carnivore. Landscape Ecology 26: 1491–1505.
Scott, M.D., and H. Geisser. 1996. Pronghorn migration and habitat use following the 1988 Yellowstone fires. Pages 123–132 in: J.M. Greenlee, editor. Ecological Implications of Fire in Greater Yellowstone. International Association for Wildland Fire, Fairfield, Washington, USA.
Singer, F.J., W. Schreier, J. Oppenheim, and E.O. Garton. 1989. Drought, fires, and large mammals. BioScience 39: 716–722.
Smith, J.K., editor. 2000. Wildland fire in ecosystems: effects of fire on fauna. USDA Forest Service General Technical Report RMRS-GTR-42-VOL 1. Rocky Mountain Research Station, Ogden, Utah, USA.
Spencer, C.N., F.R. Hauer, and K.O. Gabel. 2003. Wildlife effects on stream food webs and nutrient dynamics in Glacier National Park, USA. Forest Ecology and Management 178: 141–153.
Sullivan, J. 1994. *Bufo boreas*. In: W.C. Fisher, compiler. The Fire Effects Information System [data base]. USDA, Forest Service, Intermountain Research Station, Ogden, Utah, USA. http://www.fs.fed.us/database/feis.
———. 1996. *Taxidea taxus*. In: W.C. Fisher, compiler. The Fire Effects Information System [data base]. USDA, Intermountain Research Station, Ogden, Utah, USA. http://www.fs.fed.us/database/feis.
Tesky, J.L. 1993. *Ovis canadensis*. In: W.C. Fisher, compiler. The Fire Effects Information System [data base]. USDA, Intermountain Research Station, Ogden, Utah, USA. http://www.fs.fed.us/database/feis.
Tietje, W.D., D.E. Lee, and J.K. Vreeland. 2008. Survival and abundance of three species of mice in relation to density of shrubs and prescribed fire in understory of an oak woodland in California. Southwestern Naturalist 53: 357–369.
Tracy, B.F., and S.J. McNaughton. 1996. Comparative ecosystem properties in summer and winter ungulate ranges following the 1988 fires in Yellowstone National Park. Pages 181–191 in: J.M. Greenlee, editor. Ecological Implications of Fire in Greater Yellowstone. International Association for Wildland Fire, Fairfield, Washington, USA.
Van Dyke, F., M.J. DeBoer, and G.M. Van Beek. 1996. Winter range plant production and elk use following prescribed burning. Pages 193–200 in: J.M. Greenlee, editor. Ecological Implications of Fire in Greater Yellowstone. International Association for Wildland Fire, Fairfield, Washington, USA.
Vreeland, J.K., and W.D. Tietje. 2002. Numerical response of small vertebrates to prescribed fire in a California oak woodland. Pages 269–279 in: R.B. Standiford, D. McCreary, and K.L. Purcell, technical coordinators. Proceedings of the Fifth Symposium on Oak Woodland: Oaks in California's Changing Landscape. USDA Forest Service, General Technical Report PSW-GTR-184. Pacific Southwest Research Station, Albany, California, USA.
Williams, D.F., J. Verner, H.F. Sakai, and J.R. Waters. 1992. General biology of major prey species of the California spotted owl. Pages 207–221 in: J. Verner, K.S. McKelvey, K.S. Noon, R.J. Guti Rez, G.L. Gould, Jr., and T.W. Beck, technical coordinators. The California Spotted Owl: A Technical Assessment of Its Current Status. USDA Forest Service General Technical Report PSW-GTR-133. Pacific Southwest Research Station, Albany, California, USA.
Wirtz, W.O., II. 1979. Effects of fire on birds in chaparral. Cal-Neva Wildlife Transactions 1979: 114–124.
———. 1995. Responses of rodent populations to wildfire and prescribed fire in southern California chaparral. Pages 63–67 in: J.E. Keeley and T. Scott. Brushfires in California Wildlands: Ecological and Resource Management. International Association of Wildland Fire, Fairfield, Washington, USA.
Wormworth, J., and K. Mallon. 2006. Bird species and climate change: the global status report Version 1.1. World Wildlife Fund, Gland, Switzerland.
Wright, H.A. 1974. Range burning. Journal of Range Management 27: 5–11.
Wright, H.A., and A.W. Bailey. 1982. Fire Ecology: United States and Southern Canada. Wiley and Sons, New York, New York, USA.
Wright, A.N., R.J. Hijmans, M.W. Schwartz, and H.B. Shaffer. 2013. California amphibian and reptile species of future concern: conservation and climate change. University of California, Davis, Final Report Contract No. P0685904. California Department of Fish and Wildlife, Sacramento, California, USA.
Yensen, E., D.L. Quinney, K. Johnson, K. Timmerman, and K. Steenhof. 1992. Fire, vegetation changes, and population fluctuations of Townsend's ground squirrels. The American Midland Naturalist 128: 299–312.
Young, M.K., and M.A. Bozek. 1996. Post-fire effects on coarse woody debris and adult trout in northwestern Wyoming streams. Page 43 in: J.M. Greenlee, editor. Ecological Implications of Fire in Greater Yellowstone. International Association of Wildland Fire, Fairfield, Washington, USA.
Zedler, P.H., C.R. Gautier, and G.S. McMaster. 1983. Vegetation changes in response to extreme events: the effects of short interval fire in California chaparral and coastal scrub. Ecology 64: 809–818.

PART TWO

THE HISTORY AND ECOLOGY OF FIRE IN CALIFORNIA'S BIOREGIONS

Part Two of this book describes nine bioregions and their fire regimes, beginning in the humid northwest and ending in the arid southeast. We start in chapter 10 with the North Coast bioregion in northwestern portion of the state where numerous valleys and steep coastal and interior mountains create moisture gradients in response to numerous winter storms. In chapter 11 we describe the Klamath Mountains bioregion, a complex group of mountain ranges and a diverse flora. Tall volcanoes and extensive lava flows characterize the Southern Cascades (chapter 12) and Northeastern Plateau (chapter 13) bioregions. Immediately south of the Cascades is the Sierra Nevada bioregion (chapter 14), extending nearly half the length of the state. The Sacramento and San Joaquin rivers flow through broad interior valleys with extensive, nearly flat alluvial floors that constitute the Central Valley bioregion (chapter 15). Coastal valleys and mountains and interior mountains are also typical of the Central Coast bioregion (chapter 16). Southern California with its coastal valleys and the prominent Transverse and Peninsular Ranges are included in chapter 17 on the South Coast bioregion. Finally, in chapter 18, we discuss the vast southeast corner of California that constitutes the Southeastern Desert bioregion.

CHAPTER TEN

North Coast Bioregion

SCOTT L. STEPHENS, JEFFREY M. KANE, AND JOHN D. STUART

> In the early days of forestry we were altogether too dogmatic about fire and never inquired into the influence of fire on shaping the kind of forests we inherited.
>
> FRITZ (1951)

Description of Bioregion

Physical Geography

The North Coast bioregion is classified as being within the California Coastal Steppe, Mixed Forest, and Redwood Forest Province of the Mediterranean Division of the Humid Temperate Domain (Bailey 1995). Specifically, it is composed of the Northern California Coast and the Northern California Coast Ranges Sections (Map 10.1) (Miles and Goudey 1997).

Mesozoic sedimentary bedrock from the Franciscan Formation is the dominant type in the bioregion. Sandstone, shale, and mudstone are most common with lesser amounts of chert, limestone, and ultramafic rocks. Basalt, andesite, rhyolite, and obsidian can be found in the volcanic fields of Sonoma, Napa, and Lake Counties, while granitic rocks similar to those found in the Sierra Nevada are located west of the San Andreas Fault near Point Reyes and Bodega Bay (Harden 1997). Soils in the northwestern California ecological units have been classified as Alfisols, Entisols, Inceptisols, Mollisols, Spodosols, Ultisols, and Vertisols (Miles and Goudey 1997).

The bioregion is topographically diverse. Elevations range from sea level to around 1,000 m (3,280 ft) in the Northern California Coast Section and from around 100 to 2,470 m (328–8,100 ft) in the Northern California Coast Ranges (Miles and Goudey 1997). Slope gradients vary from flat valley bottoms to steep mountain slopes that are commonly greater than 50%. Numerous mountain ranges (e.g., Kings Range, South Fork Mountain, Yolla Bolly Mountains, and the Mayacamas Mountains) and rivers (e.g., Smith, Klamath, Mad, Van Duzen, Mattole, Eel, Noyo, Navarro, Big, Russian) are located within the north coast region.

Climatic Patterns

Three predominant climatic gradients help determine the vegetation patterns in northwestern California: (1) a west–east gradient extending from a moist, cool coastal summer climate to a drier, warmer interior summer climate; (2) a north–south latitudinal gradient of decreasing winter precipitation and increasing summer temperatures; and (3) a montane elevational gradient of decreasing temperature and increasing precipitation. These gradients, while important individually, interact in a complex fashion, especially away from the coast.

The bioregion experiences a Mediterranean climate with cool, wet winters and cool to warm, dry summers. Over 90% of the annual precipitation falls between October and April (Elford and McDonough 1964). Annual precipitation varies from 500 to 3,000 mm (20–118 in) (Miles and Goudey 1997). The Pacific Ocean greatly moderates temperature, resulting in a sharp west to east temperature gradient. The mean maximum monthly temperature at Fort Ross, for example, varies from 13.8°C (56.4°F) in January to 20.2°C (68.4°F) in September, a difference of only 6.4°C (43.5°F). In contrast the mean maximum monthly temperature at Angwin, near the Napa Valley, varies from 11.2°C (52.2°F) in January to 30.5°C (86.9°F) in July (Western Region Climate Center 2001). Most coastal forests experience summer fog, an important water source that increases soil moisture and reduces plant moisture stress (Dawson 1998). Summer relative humidity and temperature is strongly influenced by proximity to the Pacific Ocean and the presence of summer fog. Whereas lightning does occur along the North Coast bioregion during the summer fire season, it is much less prevalent than on the higher ridges and mountains to the east (Keeley 1981). van Wagtendonk and Cayan (2008) found that lightning strike density ranged from 0.9 to 9.3 yr^{-1} 100^{-1} km^{-2}, with density increasing with distance from the Pacific Ocean and increasing elevation for the period between 1985 and 2000. Notwithstanding the lightning fire potential in northwestern California, ignitions by Native Americans likely accounted for most prehistoric fires (Fritz 1931, Lewis 1993, Stephens and Fry 2005).

Synoptic weather systems in northwestern California influence fire activity (Hull et al. 1966). Gripp (1976), in a study of critical fire weather in northwestern California, found that 37.5% of fires larger than 120 ha (300 ac) were associated with the Pacific High (Postfrontal) Type. The Great Basin High Type accounted for 29.7% of the fires, the Subtropical High Aloft Pattern was linked with 21.9%, and other miscellaneous

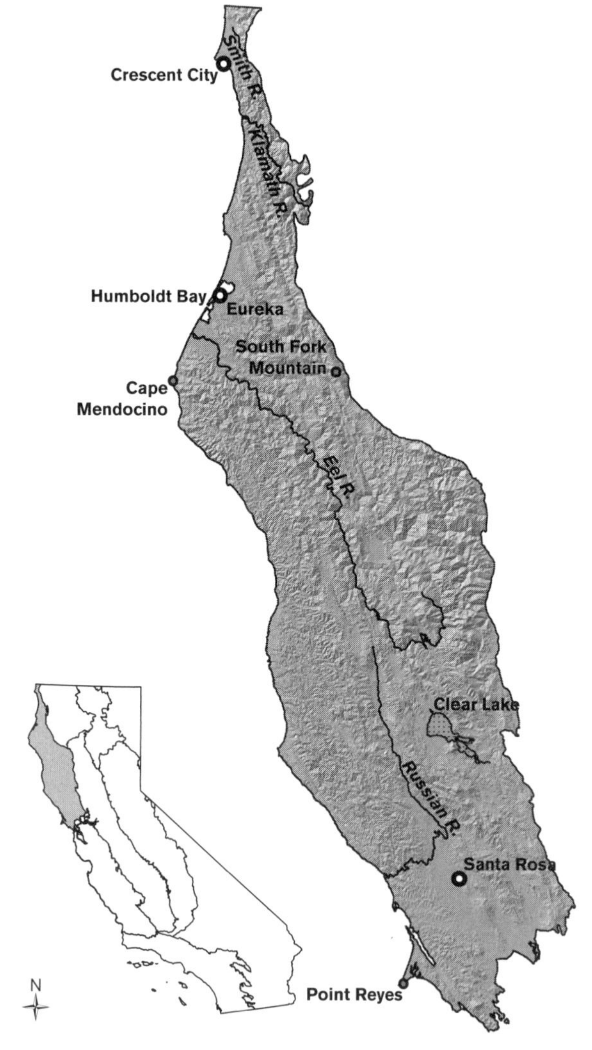

MAP 10.1 North Coast and North Coast Ranges Ecological Zones.

types were associated with 10.9%. The Pacific High (Postfrontal) and the Great Basin High Types produce warm, dry east winds (foehn winds) that displace the marine air mass off the coast (Hull et al. 1966).

Ecological Zones

The Northern Coastal Scrub and Prairie Zone is found in the fog belt along the California coast in a discontinuous band below 1,000 m (3,280 ft) elevation from Santa Cruz north to the Oregon border (Heady et al. 1988). Northern Coastal Scrub was described by Munz and Keck (1959) and is variously dominated by species such as coyote brush (*Baccharis pilularis*), yellow bush lupine (*Lupinus arboreus*), salal (*Gaultheria shallon*), and California huckleberry (*Vaccinium ovatum*). Scrubs dominated by salal, California huckleberry, ferns, and blackberry (*Rubus* spp.) are more common in the northern part of the bioregion. Coastal prairies are interspersed within the Northern Coastal Scrub and are often dominated by grass and herbaceous species, with one study recording an average of 23 species m^{-2} (Stromberg et al. 2001)

The Northern Chaparral Zone is typically made up of chamise (*Adenostoma fasciculatum*) dominated chaparral that occurs in the inland portion of the north coast (Fig. 10.2). Other common shrubs include manzanita (*Arctostaphylos* spp.), California-lilac (*Ceanothus* spp.), and oak (*Quercus* spp.). Large areas of chaparral are found near Clear Lake and on the lower elevations of the interior northern mountains.

The Northern Coastal Pine Forest Zone (Fig. 10.3) is made up of isolated stands along the north coast (Barbour 2007). Principal species include shore pine (*Pinus contorta* ssp. *contorta*), Bishop pine (*P. muricata*), Bolander pine (*P. contorta* ssp. *bolanderi*), knobcone pine (*P. attenuata*), and pygmy cypress (*Hesperocyparis pygmaea*).

The Sitka Spruce (*Picea sitchensis*) Forest Zone (Fig. 10.4) is generally found inland of the Northern Coastal Scrub and Prairie Zone in a narrow strip approximately 1–2 km (0.6–1.2 mi) wide (Zinke 1988) extending south from the Oregon border and terminating near Fort Bragg. Along rivers and in the Wildcat Hills south of Ferndale, Sitka spruce forests can extend inland as far as 25 km (15.5 mi) (Zinke 1988).

The Redwood (*Sequoia sempervirens*) Forest Zone (Fig. 10.5) is inland of the Sitka Spruce Forest Zone due to intolerance of salt spray and strong, desiccating winds (Olson et al. 1990). The distribution of redwood forests to the north, east, and south is mostly limited by inadequate soil moisture and excessive evapotranspiration (Mahony and Stuart 2001). Redwood forests occur in an irregular narrow strip, ranging in width from 8 to 56 km (5–35 mi) (Olson et al. 1990). Stands in Napa County are 68 km (42 mi) from the coast (Griffin and Critchfield 1972).

Increased evapotranspiration inland limits the coastal conifers allowing for complex mixtures of Douglas-fir (*Pseudotsuga menziesii* var. *menziesii*) and a variety of evergreen and deciduous broad-leaved trees defining the Douglas-fir–Tanoak (*Notholithocarpus densiflorus*) Forest Zone (Fig. 10.6). Notable among the tree species present are tanoak, Pacific madrone (*Arbutus menziesii*), Oregon oak (*Quercus garryana*), and California black oak (*Q. kelloggii*) (Stuart and Sawyer 2001). Douglas-fir and tanoak forests dominate inland lower montane forests. Montane forests characteristically have Douglas-fir mixed with ponderosa pine (*P. ponderosa*) and white fir (*Abies concolor*) at higher elevations.

Low-elevation riparian forests are interspersed throughout much of the Sitka spruce, Redwood, and Douglas-fir–Tanoak Forest zones. These forests are typically represented by the presence of red alder (*Alnus rubra*) and big-leaf maple (*Acer macrophyllum*) in addition to the common conifer species present in the respective zones. Further inland other species are associated with riparian forests including California bay (*Umbellularia californica*) and California buckeye (*Aesculus californica*). We do not consider these forests as a separate zone, but note that there are likely subtle but distinct differences in the fire ecology of these forests relative to their adjacent upland forest zones.

The Oregon Oak Woodland Zone (Fig. 10.7) occurs sporadically throughout the North Coast and North Coast Ranges. In the Redwood Forest Zone and the Douglas-fir–Tanoak Forest Zone, Oregon oak woodlands often occur in patches of a few to hundreds of hectares in size, usually near south and west facing ridges. In the warmer, drier parts of the North Coast and North Coast Ranges, Oregon oak can form open savannas or be interspersed with other tree species.

A few of the higher mountains support red fir forests and on the highest peaks, foxtail pine (*P. balfouriana*) and Jeffrey pine (*P. jeffreyi*) forests can be found. These are discussed in the Klamath Mountains chapter (chapter 11). Blue oak (*Q. douglasii*) woodlands and grasslands are found in the interior lowlands

FIGURE 10.1 North Coastal Scrub and Prairie at Sea Ranch, Sonoma County (photograph by Rand Evett).

FIGURE 10.2 North coastal chaparral zone (photograph by Scott Stephens).

FIGURE 10.3 North coastal pine forest zone. Bishop pine forest regeneration following the 1995 Mount Vision fire (photograph by Scott Stephens).

FIGURE 10.4 Sitka spruce forest zone. Young Sitka spruce forest at Patrick's Point State Park (photograph by John Stuart).

FIGURE 10.5 Redwood forest zone. Old growth redwood forest in Redwood National Park (photograph by John Stuart).

on the eastern border of the region (Stuart and Sawyer 2001) and are discussed in the Central Valley chapter (chapter 15).

Overview of Historic Fire Occurrence

Prehistoric Period

Holocene fire history reconstructions from lake sediments in western Oregon (Long et al. 1998, Long and Whitlock 2002) and the Klamath Mountains (Mohr et al. 2000) indicate relatively frequent fire during the warm, dry early to mid-Holocene and less frequent fire as the climate became cooler and wetter. Pollen analyses reveal increased levels of fire-adapted vegetation concomitant with thicker charcoal deposits and a warm, dry climate (Long and Whitlock 2002). Similar patterns were detected by Anderson et al. (2013) on the Point Reyes peninsula and presumably apply more generally to North Coast and North Coast Ranges.

The North Coast Ranges have experienced three major climatic periods since the end of the Pleistocene: a cool, somewhat continental climate from the early Holocene to about 8,500 years BP; a warmer period with presumably drier summers from 8,500 years BP to about 3,000 years BP; and a cool,

FIGURE 10.6 Douglas-fir–tanoak forest zone. Young Douglas-fir–tanoak forest in Humboldt County (photograph by John Stuart).

FIGURE 10.7 Oregon oak woodland zone. Prescribed fire in Oregon oak woodland at Copper Creek (photograph by John McClelland).

moist climate since about 3,000 years BP (Keter 1995). Native Americans are known to have lived in the region since around 8,200 years BP to 8,600 years BP, and by the middle of the Holocene, humans lived throughout the North Coast (Sawyer et al. 2000). During the early Holocene pine forests with sparse shrubs and herbs dominated parts of the North Coast Ranges (West 1993). Redwood forests were present in relatively low abundance at this time but peaked in the mid-Holocene (~5,500 BP). As the climate warmed and dried, pine pollen counts remained high and oak counts increased while Douglas-fir pollen was reduced. The cool, moist climate in the late Holocene enabled Douglas-fir, tanoak, and true fir pollen counts to increase and oak counts to decrease (Keter 1995). Fire was presumably more frequent during the warm, dry period and less frequent during the cool, moist period.

Fire history studies from the last 1,000 years reveal a variable pattern of fire frequencies throughout northwestern California. The most frequently burned landscapes were ignited on a near annual basis by Native Americans (Lewis 1993) and were generally in close proximity to villages or in areas cultured for food and basketry materials such as in grasslands and oak woodlands. Vegetation adjacent to Native American use areas experienced more frequent fire than would be found in the same vegetation type further away (Whitlock and Knox 2002). Lightning fires were generally less frequent, but relatively more numerous at higher elevations in the North Coast Ranges than the coastal regions. In general, the

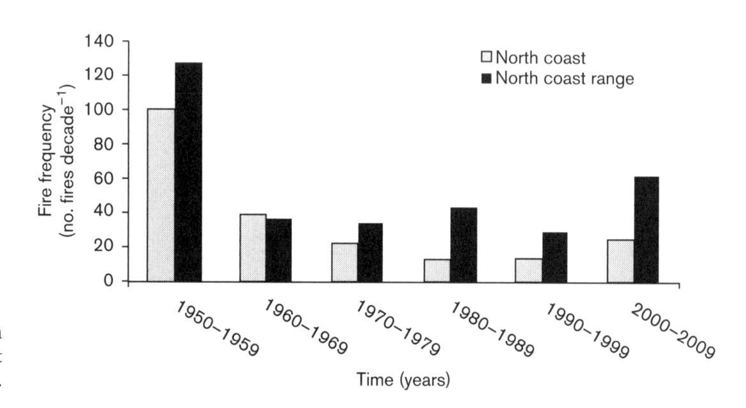

FIGURE 10.8 Fire frequency by decade for fires larger than 120 ha (300 ac) in the North Coast and North Coast Ranges Ecological Sections.

most frequent fire occurred in grasslands and oak woodlands, with less frequent fire in chaparral, mixed evergreen, and montane mixed conifer. The least frequent fire occurred in moist, coastal conifer forests (e.g., Sitka spruce forests).

Historic Period

The removal and prevention of Native American ignited fires, the introduction of cattle and sheep by ranchers, and intensive logging altered fire regimes during the historic period. Changes in the fire regimes occurred as Euro-American settlement moved north from the San Francisco Bay area in the early nineteenth century to the northern counties in the mid- to late-nineteenth century. The earliest known Euro-American settlement in the region was established at Fort Ross in 1812 by Russian-American fur traders (Lightfoot et al. 1991). Burning by Native Americans of the North Coast was likely interrupted starting in the early to mid-1800s. After which, traditional burning practices were increasingly curtailed as indigenous populations were decimated by disease and warfare and as survivors were relocated to reservations (Keter 1995).

Early settler fires originated either from escaped campfires or were deliberately set to improve forage for livestock (Barrett 1935). Ranchers primarily grazed cattle during the early to mid-nineteenth century with sheep grazing increasing in the mid- to late-nineteenth century. Sheepherders were notorious as indiscriminate users of fire (Barrett 1935). Rather than burning for a single reason, Native Americans skillfully employed burning for multiple purposes at different times of the year (Lewis 1993).

During the mid- to late-nineteenth century fire frequency, intensity, and severity were generally high throughout this period. Loggers regularly burned recently cut lands to remove downed fuels and to facilitate log extraction by draft animals. The potential for fire to escape and burn into unlogged forests was high. By the late nineteenth century and through the 1920s, mechanical yarding systems and railroads enabled logging of whole watersheds in coastal and interior drainages. Following logging, some timberland owners attempted to convert forestland to grassland by repeatedly burning the logging slash and sowing grass seed (O'Dell 1996). Large fires were frequent in northwestern California during the historic period. For example, Gripp (1976) conducted an extensive review of northwestern California newspapers and various other documents and found that large fires in Humboldt and Del Norte counties from about 1880 to 1945 had an average three years between severe fires.

Fire suppression began on national forest lands in northwestern California in 1905 (Keter 1995). Fire suppression on private and state land, in the latter part of the nineteenth century through the early twentieth century, was largely the responsibility of the counties and various landowner associations. During the 1920s fire wardens used their power of conscription to recruit fire fighters and by the early 1930s, the California Division of Forestry assumed the role of fire suppression (Clar 1969). Effective fire suppression on private, state, and federal land did not begin until after 1945 when an increased number of returning soldiers that became firefighters had access to technologies used during World War II.

Current Period

Fire records dating to around 1915 exist for the bioregion, although reasonably complete records for large fires are only available since 1945 and complete records in digital format for fires of all sizes are available for only the past few decades. The records for fires larger than approximately 120 ha (300 ac) (CDF-FRAP 2015) reveal that two to three times as many fires occurred in the 1950s as in subsequent decades. In addition, there were consistently more fires in the North Coast Ranges than the North Coast (Fig. 10.8). The vast majority of fire records did not identify fire cause, although the large number of fires in the 1950s coincided with a period of increased logging. Fewer large fires in ensuing decades can be attributed to more effective fire prevention and suppression.

The North Coast has experienced a consistent decrease in cumulative area burned since the 1950s, and the North Coast Ranges similarly experienced progressively smaller areas burned from 1950 through the 1970s, but in the last few decades cumulative fire size is potentially trending upward (Fig. 10.9), possibly due to increased fuel load and continuity associated with logging and fire suppression (Stuart and Salazar 2000) and increasing global temperatures (Westerling et al. 2006). This trend of increased fire size in the bioregion appears to be continuing as demonstrated by the spate of destructive fires in the Northern Bay area during the fall of 2017.

Major Ecological Zones

Northern Coastal Scrub and Coastal Prairie Zone

Northern coastal scrub extends from Monterey County into Oregon in a narrow strip generally less than a few hundred

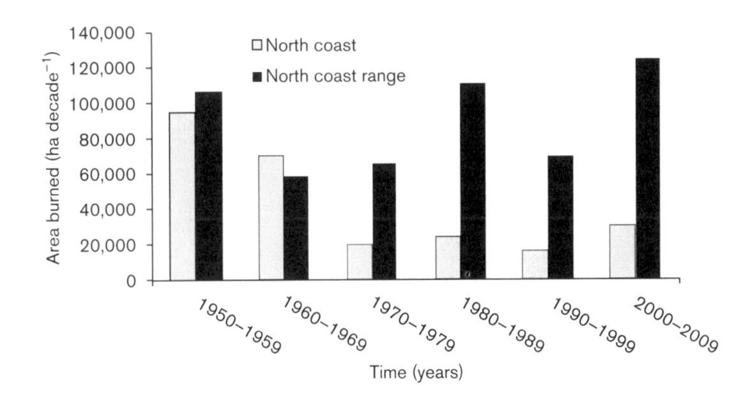

FIGURE 10.9 Area burned by decade for fires larger than 120 ha (300 ac) in the North Coast and North Coast Ranges Ecological Sections.

meters wide. Salal in combination with blackberry species and a rich mix of other shrubs, subshrubs, and herbaceous species often forms thickets and brambles. A transition zone between northern coastal scrub and coastal sage scrub lies in Marin and San Mateo Counties.

Prior to Euro-American settlement, the Northern Coastal Prairie was probably dominated by native perennial grasses, including California oat grass (*Danthonia californica*), purple needlegrass (*Stipa pulchra*), Idaho fescue (*Festuca idahoensis*), and tufted hair grass (*Deschampsia cespitosa*) (Heady et al. 1988). Highly susceptible to invasion, coastal prairie now includes many Mediterranean annual species as well as non-native perennial grasses, notably common velvet grass (*Holcus lanatus*) and sweet vernal grass (*Anthoxanthum odoratum*) (Hektner and Foin 1977).

FIRE RESPONSES OF IMPORTANT SPECIES

Salal, California huckleberry, coyote brush, thimbleberry (*Rubus parviflorus*), salmonberry (*R. spectabilis*), and California blackberry (*R. ursinus*) are all fire neutral, facultative sprouters (Table 10.1). While regeneration is not fire dependent, these species have the capability to aggressively recolonize burned landscapes through sprouting, seeding, or germination from buried seed.

Dominant grasses in the Northern Coastal Prairie are fire neutral, facultative sprouters. One study of coastal prairie native bunchgrass fire responses showed no significant changes in foliar cover or frequency for California oat grass, purple needle grass, and foothill needle grass (*Stipa lepida*) after burning (Hatch et al. 1999). Tufted hairgrass in other plant communities is resistant to all but the highest intensity fires and recovers to pre-burn levels in a few years (Walsh 1995). Idaho fescue is resistant to low-intensity burning but can be killed at higher intensities (Zouhar 2000). A small unpublished study at The Sea Ranch in Sonoma County showed temporary decreases in the cover of velvet grass and sweet vernal grass following prescribed burns with little effect on tufted hairgrass (Evett 2002).

FIRE REGIME–PLANT COMMUNITY INTERACTIONS

Fire is relatively uncommon in Northern Coastal Scrub but Native Americans would most likely have ignited any fires that did burn, and fire spread probably would have been dependent on warm, dry east winds. Pre-Euro-American fire intervals are unknown, but Greenlee (1983) estimated variable fire return intervals ranging from 1 to 100 years depending on the vegetation type and proximity to Native American villages. Northern Coastal Scrub is capable of self-perpetuating with or without fire. Ten years following the 1995 Vision fire in Point Reyes National Seashore where 5,000 ha (12,355 ac) burned, coastal scrub communities decreased by 27% and transitioned to either blue blossom scrub or Bishop pine communities (Forrestel et al. 2011).

The role of fire in maintaining coastal prairie prior to Euro-American settlement is poorly documented but likely included widespread burning (Blackburn and Anderson 1993). Soil and phytolith evidence suggests that many coastal prairie sites have been grassland for thousands of years (Bicknell et al. 1992). In the absence of burning or grazing, many of these sites with high grass phytolith content in the soil have been invaded by shrubs and trees, suggesting that regular aboriginal burning was required to maintain the coastal prairie (Bicknell et al. 1993, Evett 2000). Following displacement of the Native American populations, many ranchers practiced deliberate burning to maintain the prairie and promote understory forage (Bicknell et al. 1993). For the past 150 years, continuous livestock grazing has replaced frequent burning, which collectively resulted in the increase in dominance of non-native Mediterranean annuals. Whether the proliferation of non-native annuals was caused by the transition to heavy grazing, removal of burning, or a combination of the two is not well understood. Aboriginal fires were probably small and of low intensity due to the discontinuous nature of the coastal prairie and lack of fuel accumulation under a high-frequency fire regime. Seasonality of aboriginal fires is unknown; fires were probably more likely in the dry season of summer or early fall.

The removal of fire and livestock grazing from the coastal prairie has profound effects on plant community composition. In most cases, the absence of fire or grazing results in conversion to Northern Coastal Scrub after 15 to 25 years (Ford and Hayes 2007). However, sites at The Sea Ranch in northern Sonoma County, where livestock grazing was removed in the late 1960s and fires were excluded, show a shift from mixed annual and perennial grasses to overwhelming dominance of non-native perennial grasses that reduced biodiversity and substantially increased fuel loads (Foin and Hektner 1986).

Small-scale experiments to reintroduce fire to coastal prairie sites to reduce nonnative species and restore native species dominance had mixed success. Burning at The Sea Ranch sites reduced cover of the nonnative grasses (velvet grass and sweet vernal grass) but increased cover of another nonnative grass,

TABLE 10.1
Fire response types for important species in the North Coast Bioregion

Lifeform	Type of fire response: Sprouting	Seeding	Individual	Species
Tree	None	Neutral	Killed	Shore pine, Sitka spruce, western hemlock, western red cedar
	None	Neutral	Survive	Grand fir, Port Orford cedar
	None	Stimulated (establishment)	Survive	Douglas-fir
	None	Stimulated (release)	Killed	Bishop pine, Bolander pine, pygmy cypress
	Stimulated	Neutral	Survive/top-killed	Big-leaf maple, California bay, canyon live oak, giant chinquapin, tanoak
	Stimulated	Stimulated (establishment)	Survive/top-killed	Redwood, California black oak, Oregon oak, Pacific madrone
Shrub	Stimulated	Neutral	Survive/top-killed	California huckleberry, salal
	Stimulated	Stimulated (establishment)	Survive/top-killed	Chamise, California blackberry, coyote brush, salmonberry, thimbleberry
	None	Stimulated (establishment)		Wedgeleaf ceanothus, blue blossom, common manzanita
Grass	Stimulated	Stimulated (establishment)	Survive/top-killed	California oat grass, Idaho fescue, purple needlegrass, tufted hairgrass

hairy oat grass (*Rytidosperma penicillatum*) (Evett 2002). Biological diversity increased because of increased cover of nonnative forbs and annual grasses. Native grass species were mostly unaffected by burning and this was confirmed in a separate study in coastal San Mateo County (Hatch et al. 1999).

Northern Chaparral Zone

Chaparral constitutes 7% of the land area in California, but it hosts more than one-quarter of the state's endemic flora and fauna—nearly half of which are endemic to chaparral (Keeley and Davis 2007). Chaparral ecosystems have been extensively studied in the southern and central coasts of California (see chapters 16 and 17 of this text) but some work has also occurred in northern California, particularly in the coastal interior mountains (Potts and Stephens 2009, Potts et al. 2010). In many areas chaparral is located on steep slopes with thin soils and high solar radiation (Wilkin et al. 2017, Newman et al. 2018). Mature chaparral is commonly 1.5–2 m (4.9–6.6 ft) in height and nearly 100% shrub cover but cover varies with site productivity. Understory native herbaceous plants are uncommon in mature chaparral because of low light levels under the canopy and limited moisture. Chaparral was probably not as dense as it is today and Native Americans likely burned some areas to facilitate grasslands and oak woodlands that were higher valued resources.

Crown fires are the typical fire type in chaparral, characterized by high flame lengths and rapid rates of spread (Stephens et al. 2008). Chaparral has been burned for decades in northern California to increase range resources for livestock and to improve forage for black-tailed deer (*Odocoileus hemionus columbianus*) (Biswell 1989). Because of its high fire hazards chaparral has also been burned to reduce fuel loads. In southern California, the impacts of high velocity, dry foehn winds have been shown to be a dominant factor in chaparral fire regimes (Moritz et al. 2004).

FIRE RESPONSES OF IMPORTANT SPECIES

Chaparral plant species can be grouped into three functional groups, depending upon their post-disturbance regeneration response: obligate seeders, obligate sprouters, and facultative seeders. Obligate seeders (e.g., *Ceanothus cuneatus*, *Arctostaphylos manzanita*), primarily reproduce from long-lived fire stimulated soil-stored seed banks and lack any sprouting ability (Table 10.1). In contrast, obligate sprouting species regenerate well after fire primarily by sprouting vigorously from adventitious buds (e.g., Eastwood's manzanita [*A. glandulosa*], scrub oak [*Quercus berberidifolia*]). Obligate sprouters can also reproduce from seed, but most species only produce a few seeds that are intolerant of heat exposure (e.g., oaks), and a few species (e.g., Eastwood's manzanita) produce seeds that require fire cues (Keeley 1991). Facultative seeders reproduce both sexually by seed and vegetatively by sprouting (e.g., chamise). The seed bank of facultative seeders is often polymorphic, containing both fire-stimulated and fire sensitive seeds to ensure germination after and between fire events

(Zammit and Zedler 1988). Facultative seeder species should have an advantage in most chaparral ecosystems since they can successfully regenerate after diverse environmental cues.

FIRE REGIME–PLANT COMMUNITY INTERACTIONS

Chaparral, like most Mediterranean shrublands, is highly fire resilient and historically burned with high-severity, stand-replacing events every 30 to 100 years (Stephens et al. 2007). Though adapted to infrequent fires, chaparral plant communities can be exterminated by frequent fires ignited by humans (Keeley 2005). Historically, periodic summer or fall high-intensity wildfires were considered necessary to maintain a full suite of native chaparral plants because many species depend on fire cues for germination; however, many native chaparral plants can also germinate without fire (Potts et al. 2010, Wilken et al. 2017).

Northern Coastal Pine Forests Zone

Bishop, shore, Bolander, and knobcone pine forests are sporadically arranged along coastal bluffs and marine terraces from the Oregon border to the San Francisco Bay Area. Bishop pine occurs in disjunct populations in coastal California from Humboldt County to Santa Barbara County. Shore pine is widely distributed along the Pacific Coast from Yakutat Bay, Alaska, south to Mendocino County (Little 1971), primarily occurring along stabilized dunes and low lying deflation plains with poor site conditions and high water tables (Green 1999). Bolander pine and pygmy cypress are endemic to the pygmy forests of western Mendocino County. Knobcone pine forests primarily occur on dry steep slopes and ridgetops in the North Coast and Klamath ranges (Griffin and Critchfield 1972).

FIRE RESPONSES OF IMPORTANT SPECIES

Bishop pine is a fire dependent, obligate seeder, with postfire regeneration from crown-stored seed bank (Table 10.1). Cones can remain closed for years and primarily open after fire or on hot, dry days (Van Dersal 1938). Most Bishop pine stands are very dense and commonly have stand-replacing crown fires, although older trees have thick bark that enables survival following multiple surface fires (Stephens and Libby 2006). Cone serotiny is somewhat variable, with northern populations less serotinous than southern populations (Zedler 1986). Morphological differences in cones have been observed but were not associated with differences in cone opening temperatures (Ostoja and Klinger 1999).

Shore pine can be considered a fire neutral or fire-dependent obligate seeder, while Bolander pine and pygmy cypress are fire-dependent, obligate seeders (Table 10.1). Most shore pine have open cones (Lanner 1999), though serotiny has been observed (Griffin 1994). Bolander pine and pygmy cypresses are mostly serotinous and cones typically open after fire or by desiccation (Vogl et al. 1977, Lotan and Critchfield 1990). Detached or desiccated cones commonly open but deep litter beds and low light conditions limit seedling establishment.

Knobcone pine is a fire-dependent, obligate seeder (Table 10.1). Mature trees have thin bark and are often killed following fire. However, serotinous cones are retained in the canopy for up to 50 years and fires with temperatures of 160°C (320°F) or greater open cone scales and allow seed dispersal (Vogl 1973). Maturity in knobcone pine trees is reach by seven years, with 90% of individuals bearing an average of nine cones per tree (Keeley et al. 1999). Seed content of cones varies slightly with age but generally ranges from 60 to 100 seeds per cone with approximately 80% seed viability (Fry and Stephens 2013).

FIRE REGIME–PLANT COMMUNITY INTERACTIONS

Bishop pine is adapted to stand-replacing high-severity fires (Appendix 1). Prior to Euro-American settlement, the majority of these fires probably occurred in the late summer and fall when fuel moisture contents were low. High-severity fires were probably mostly associated with warm, dry east winds. Fire history data are not available to estimate the spatial extent of fires, but isolated Bishop pine groves surrounded by shrubland probably had moderate to large fires (> 100 ha [> 250 ac]).

The majority of Bishop pine forests have not burned in the last 40–90 years because of fire exclusion. However, in 1995 the Vision fire burned through 423 ha (1,045 ac) of Bishop pine forests in Point Reyes National Seashore. Following the fire, Bishop pine prolifically reseeded and expanded in area by 85% (Forrestel et al. 2011) (Fig. 10.3). Expansion occurred in areas that were previously Northern Coastal Scrub communities adjacent to preexisting Bishop pine stands (Harvey et al. 2011). Fourteen years following the Vision fire, two distinct Bishop pine community pathways were apparent, a low-density, open-canopy pathway with high shrub cover on gentle slopes and a high-density, closed-canopy pathway with low shrub cover on steep slopes (Harvey and Holzman 2014). The lack of fire could threaten the long-term existence of Bishop pine because it relies on high-severity crown fires that prepare seed beds, enable the release of large quantities of seed, and increase light availability to seedlings. Introduction of nonnative grasses into recently burned areas would increase the risk of extirpation because the grasses could produce a highly continuous fuelbed in 1 to 2 years. Little is known about the role of fire in Bolander pine and pygmy cypress forests. Fires are likely infrequent and of high severity when they occur.

Knobcone pine typically experiences a high-severity stand-replacing fire. In fires between 1930 and 1960 in the Mayacamas Mountains of the North Coast Range, 67% of stands examined had even-aged distributions while the remaining stands had multiage distributions (Fry et al. 2012). Multiage distributions were attributed to areas that experienced multiple fires or a low-intensity fire that promoted regeneration but retained some of the prefire cohort (Fry et al. 2012).

Sitka Spruce Forest Zone

Sitka spruce forests are variously dominated by Sitka spruce, western hemlock (*Tsuga heterophylla*), Douglas-fir, western red cedar (*Thuja plicata*), Port Orford cedar (*Chamaecyparis lawsoniana*), grand fir (*Abies grandis*), and red alder. Forests immediately along the coast are primarily composed of Sitka spruce.

As the distance from the coast increases the other co-occurring species become more common and Sitka spruce reduces in abundance.

FIRE RESPONSES OF IMPORTANT SPECIES

In general, Douglas-fir and red alder regenerate well following fire, whereas western hemlock, grand fir, western red cedar, and Port Orford cedar regenerate well in either undisturbed or disturbed forests. Douglas-fir is a fire-enhanced, obligate seeder and red alder is a fire neutral, facultative sprouter (Table 10.1). Both species require full sunlight and mineral soil to regenerate. Sitka spruce can establish well on organic seedbeds but require light gaps from small disturbances (Taylor 1990). Young red alder trees sprout vigorously, but older trees rarely sprout (Harrington 1990). Western hemlock, grand fir, western red cedar, and Port Orford cedar are all fire inhibited, obligate seeders and regenerate well on organic seedbeds in shade or partial shade (Burns and Honkala 1990).

With the exception of Douglas-fir and larger Port Orford cedar (Zobel et al. 1985), other potential canopy dominants and codominants in these forests are not fire resistant. Western hemlock, Sitka spruce, western red cedar, and red alder are shallow rooted and have thin bark (Burns and Honkala 1990). Grand fir is fire sensitive when small, but can develop bark thick enough to resist light surface fires (Howard and Aleksoff 2000).

FIRE REGIME–PLANT COMMUNITY INTERACTION

Fires were generally uncommon in Sitka spruce forests due to high fuel moisture and low lightning densities. Fuel moisture is almost always too high to support fire in these coastal forests with the exception of late summer and early fall after summer fog has dissipated. Even then, temperature and humidity are usually not conducive to burning so close to the Pacific Ocean. With the exception of ignitions by Native American in prairies or near villages, fire intervals were long to very long. Lower-intensity fires with long residence times and the thin bark of Sitka spruce resulted in a stand-replacing, high-severity fire in this ecosystem. Fires were small to moderate in size with moderate spatial complexity (Appendix 1).

Fire history in Sitka spruce forests is not well documented. Based on multiple sources, Van de Water and Safford (2011) estimated a range in fire return intervals for spruce-hemlock forests of California to range from 180 to 550 years. Agee (1993) reported fire intervals of around 200 years for the southern Oregon coast, 400 years from the northern Oregon coast, and 1,146 years in Sitka spruce forests of western Washington (Fahnestock and Agee 1983). Red alder is known as an early seral species that aggressively colonizes moist, mineral soils in full sunlight. Early growth is rapid and it usually dominates competitors for the first 25 years or so until it is overtopped by other conifers (Uchytil 1989). Port Orford cedar and western hemlock can regenerate in the shade of a canopy, while Sitka spruce and grand fir, though not as shade tolerant as Port Orford cedar or western hemlock, can self-perpetuate following windthrow or pockets of overstory mortality (Franklin and Dyrness 1973). There has been little change in fire regime from pre-European time to the current time period. The 70 to 90 years of effective fire suppression is much shorter than the pre-Euro-American settlement fire intervals.

Redwood Forest Zone

FIRE RESPONSES OF IMPORTANT SPECIES

Redwood is a fire-enhanced, facultative sprouter (Table 10.1). Seedling establishment is problematic in the absence of fire, windthrow, or flooding because of low seed viability (Olson et al. 1990) and unsuitable seedbeds. It is rare to find redwood seedlings that have established on their own litter because of the combination of damping-off fungi (Davidson 1971) and low light intensities commonly found beneath redwood canopies (Jacobs et al. 1985). However, exposure of mineral soil and canopy openings following fire, windthrow, or flooding often results in successful seedling establishment. Following recent wildfires in the Santa Cruz Mountains of the Central Coast, redwood seedling regeneration was highly variable and ranged from 0 to 450 seedlings ha^{-1} across three sites (Lazzeri-Aerts and Russell 2014).

Sprouting in redwood can occur from lignotubers at the root crown, induced lignotubers on layered branches, trunk burls, and from adventitious buds on tree trunks (Del Tredici 1998). The development of redwood lignotubers and axillary meristems are a normal part of seedling development (Del Tredici 1998) and may represent an evolutionary response to fire, windthrow, or flooding. Sufficiently intense surface fires can stimulate basal lignotubers to sprout, while crown fires often result in redwood "fire columns" (Jepson 1910, Fritz 1931) whose denuded trunks sprout new leaves and eventually develop new branches. In general, larger redwood trees (12 in [>30 cm] diameter at breast height) that experienced full crown scorching or consumption by wildfire were more likely to regenerate through epicormic sprouting than smaller redwood trees (Lazzeri-Aerts and Russell 2014). Postfire basal sprouting (number, area, and height) per tree decreased with tree size (Ramage et al. 2010, Lazzeri-Aerts and Russell 2014). However, fire intensity (measured as bole char height) had a parabolic relationship with basal sprout area and height, where intermediate fire intensity resulted in the tallest and most numerous sprouts per tree (Ramage et al. 2010).

Redwood bark serves as either a resister or enabler of fire damage to the cambium layer depending on its thickness and the water content of its loosely packed, sponge like fibers. Bark 15–30 cm (6–12 in) thick (Fritz 1931) protects the cambium from heat damage, especially when moist. Small, thin barked trees, however, are susceptible to fire damage and are readily top-killed (Finney and Martin 1993). Dead cambium acts as an infection court for sapwood and heartwood rots. Surface fires burn into the rotten sapwood or heartwood forming basal hollows often referred to as "goose pens." Subsequent fires in goose pens continue to expand the cavity with the result that many trees are consumed standing (Fritz 1931) but cavity formation can also be assisted by the large radial growth rates of redwood (Finney 1996). Fire scars have been recorded to be as tall as 70 m (230 ft) (Sawyer et al. 2000). Redwood stands are among the most productive in the world (Fritz 1945) and consequently produce impressive fuel loads. Litter load of old-growth redwood forests ranged from 9.6 to 16.7 Mg ha^{-1} (26.2–45.4 ton ac^{-1}) with depth ranging from 5.0 to 9.6 cm (2–3.8 in) (Graham 2009). Rapid litter decomposition helps to keep the litter loads relatively low. Pillers and Stuart (1993) found that the time to decompose 95% of the weight of oven dried litter ranged from approximately 7 to 11 years on four sites in Humboldt County.

Redwood litter is highly flammable in comparison to many other western conifer species (Fonda et al. 1998) because leaves do not abscise individually but fall as sprays of leaves. The resultant fuelbed is made up of loosely compacted fuel enabling the litter to quickly respond to moisture changes and greater oxygen permeability during combustion (Pyne et al. 1996) but when fog is present redwood litter will not readily burn. Litter is the predominant surface fuel that carries fire, but understory plants can either inhibit or intensify fire behavior. Coastal redwood forests in Del Norte County, for example, include many plants (Mahony 1999) with a moisture content high enough to retard fire spread. Inland redwood forests, in contrast, have a higher proportion of sclerophyllous understory trees and shrubs that can exacerbate fire behavior.

FIRE REGIME–PLANT COMMUNITY INTERACTIONS

Redwood forests have a complex fire history, typically burning across a wide range of conditions over space and time (Stephens and Fry 2005). This variation mainly reflects differences and changes in ignition types and climatic gradients. Many fires were ignited by Native Americans with less common ignitions occurring from lightning. Thus, stands in closer proximity to Native American village sites or resource use areas usually burned more frequently (Brown and Swetnam 1994, Norman 2007). Three main climatic gradients in the redwood region contribute to variation in redwood forest fire regimes (Lorimer et al. 2009). Forests with more moisture and lower temperatures (coastal and northern forests) burned less frequently than forests with less moisture and higher temperatures (inland and southern forests). Forests with less fog or during greater fog-free periods burned more frequently than forests with more fog. Also, climate–fire relationships may be muted by the presence of Native American burning due to burning across wider climatic conditions.

Historical information on fire intensity and severity in redwood forests is not directly ascertainable, but redwood forests generally experienced low- to moderate-intensity surface fires (Appendix 1). Stuart (1986) found that during prescribed burns in Humboldt County, flame lengths on redwood litter and duff were 10–90 cm (4–35 in) long, yet beneath California huckleberry the flames were 300–500 cm (9–16 ft) long. In September 2003, lightning ignited the Canoe fire, which burned 5,554 ha (13,744 ac) of predominantly old-growth redwood forest in Humboldt Redwoods State Park. Fire behavior was generally characterized as a low-intensity surface fire with flame lengths ranging from 15 to 30 cm (6–12 in) and rates of spread of 7.9 km hr^{-1} (4.9 mi hr^{-1}) (Scanlon 2007). Occasional passive crown fires occurred, especially along the southern and eastern edges of the range. On average, fire severities were lowest in the coolest, wettest regions and highest in the warmer, drier areas.

Numerous fire history studies in redwood forests indicate that the fire return interval was quite variable. Fire return intervals from 125 to 500 years have been reported for coastal forests in Del Norte County where redwood grows in association with Sitka spruce, western hemlock, and western red cedar (Veirs 1982, Mahony and Stuart 2001). Norman (2007) estimated that the fire return interval for warmer redwood forests in eastern Del Norte County were relatively frequent (mean = 23 years, range 5 to 86 years); fire intervals in its southern range generally ranged between 6 and 44 years (Table 10.2). Some redwood stands had fire intervals of 1 or 2 years due to regular Native American burning in prairies surrounded by redwood forest, near villages, and along travel corridors (Fritz 1931, Lewis 1993). Redwood tree ring growth patterns are often complacent (i.e., uniform) with high numbers of missing and false rings (Brown and Swetnam 1994), which makes cross-dating and assigning years difficult. While numerous fire history studies have been conducted in redwood forests throughout much of its range, only two studies have successfully cross-dated chronologies (Brown and Swetnam 1994, Norman 2007). However, a recent study by Carroll et al. (2014) has provided the most comprehensive and extensive cross-dated chronologies in redwood to date, with the potential to develop more detailed fire histories in redwood forests.

Fire season in redwood forests was also historically variable across sites. Most studies indicate that the fire season occurred between the late summer or early fall, based on the fire scar position in the latewood or between annual rings. For instance, Brown and Swetnam (1994) and Brown and Baxter (2003) both determined that 91–95% of fires occurred in the late summer or early fall. In Del Norte Coast Redwoods State Park, however, Norman (2007) found greater occurrence (32% of fire scars) of early season fires.

Few studies have attempted to estimate fire size in redwood forests. Fire size is generally thought to be small, though occasionally larger fires would occur, especially during droughts or warm, dry east wind events. Stuart (1987) estimated fire sizes in old-growth redwood stands in Humboldt Redwoods State Park to be approximately 786 ha (1,940 ac) for the presettlement period, 1,097 ha (2,710 ac) for the settlement period, and 918 ha (2,270 ac) for the postsettlement period. Stuart and Fox (1993) examined fire records between 1940 and 1993 for Humboldt Redwoods State Park and nearby areas and found 30 fires larger than 120 ha (300 ac) burned (average of 505 ha [1,250 ac] and range from 100 to 1787 ha [250 –4,400 ac]).

Succession and the climax status of redwood forests have been discussed since the early 1900s. Some authors contend that redwood is a seral species that, in the absence of fire, windthrow, or flooding, would eventually be succeeded by more shade-tolerant species as individual redwoods die (Cooper 1965, Osburn and Lowell 1972). Most ecologists, however, have argued that redwood is self-perpetuating and should be considered as a climax or fire subclimax species (Fisher et al. 1903, Weaver and Clements 1929, Veirs 1982, Olson et al. 1990, Agee 1993). While redwoods are generally highly resistant and resilient to numerous forest disturbances, climate-change induced warming and reduction in fog may have important effects on redwood growth and survival in the near future. A recent study determined that fog frequencies in the redwood region have declined by 33% since the 1900s (Johnstone and Dawson 2010). However, the growth response to this warming and reduced fog has been shown to differ across the redwood region. Carroll et al. (2014) found that redwood growth was negatively correlated to dry summer conditions in the southernmost portion of the range and positively with decreased summer cloudiness (i.e., fog) in the northern portion of the range.

Whether redwood is dependent on fire is an open question and should be answered on a case-by-case basis with each redwood community assigned a place on a fire dependency continuum. A redwood-western hemlock/salmonberry association (Mahony and Stuart 2001) in coastal Del Norte County, for example, is fire tolerant but not fire dependent (Veirs 1982). Low to moderate surface fires kill fire susceptible species such as western hemlock, western red cedar, and Sitka

TABLE 10.2

Fire intervals in redwood forests

Location	Mean fire interval (years)	Composite area (ha)	Source
Del Norte and northern Humboldt Counties	50–500; 12–26	1; variable	Veirs 1982; Norman 2007
East of Prairie Creek Redwoods State Park	8	0.25–3	Brown and Swetnam 1994
Humboldt Redwoods State Park	11–44	7	Stuart 1987
Southern Humboldt County	25	12	Fritz 1931
Jackson State Forest	6–20	4–20	Brown and Baxter 2003
Salt Point State Park	6–9	~200	Finney and Martin 1989
Annadel State Park	6–23	14 trees[a]	Finney and Martin 1992
Near Muir Woods National Monument	22–27	75	Jacobs et al. 1985
Western Marin County	8–13	5–10; 10 trees in one stand	Finney 1990; Brown et al. 1999
Jasper Ridge Preserve, San Mateo County[b]	9–16	1–3	Stephens and Fry 2005
Big Basin Redwoods State Park[b]	~50	Variable, dependent on estimated fire area	Greenlee 1983
Santa Cruz Mountains[b]	61	51 trees[a]	Jones and Russell 2015

NOTES:

a. Point data.

b. study sites located south of North Coast bioregion.

spruce. These shade-tolerant species quickly regenerate, grow, and have the capacity to develop into important ecosystem structural elements. A redwood-tanoak/round-fruited sedge (*Carex globosa*)–Douglas iris (*Iris douglasiana*) association (Borchert et al. 1988) near the drier southern limits of the range, in contrast, could be considered as fire dependent (Veirs 1982). In these areas, reproduction from seed may be limited due to drier soil conditions, but redwoods can persist through postfire resprouting (Noss 2000). The vast majority of redwood's range is found on sites intermediate to those described above with individual stands occupying intermediate positions on the fire dependency continuum. While the redwood species may not be dependent on fire to perpetuate in these forests, the associated ecosystem structures and functions could provide critical habitat for a variety of wildlife species. Additionally, fire can increase the availability and cycling of soil nutrients, increase understory plant diversity, and create natural edges.

Fire regime changes have not been consistent across the redwood range. Post-World War II (1950–2003) natural fire rotations for the northern, central, and southern redwood zones for large fires (>331 ac [134 ha]) were 1,083 years, 717 years, and 551 years, respectively (Oneal et al. 2006). The current fire intervals in the central and southern redwoods zones greatly exceed the 6- to 44-year presuppression intervals (Table 10.2). It is possible that a fire regime change in the northern range of redwood will eventually occur because of modifications in fuel loads and structure following logging and increasing global temperatures. Only about 10% of the redwood range is currently old-growth forest, with the remaining forest composed of young-growth (Fox 1988). Some of the young-growth forests are over 100 years old and dominated by large redwoods with fuel complexes similar to old growth, but most young forests are composed of small diameter, dense complexes of conifers intermixed with broad-leaved trees and shrubs. Other fuel complex alterations include: the presence of large, persistent redwood stumps and logging slash, as well as greater shrub and herbaceous plant cover. In general, fuel load and continuity have increased. Eventually, fire size, intensity, and severity may increase, while fire complexity may change and the fire type may have a greater occurrence of passive/active crown fire.

Douglas-Fir–Tanoak Forest Zone

Douglas-fir–tanoak forests are widely distributed in areas inland from the redwood belt. Both tanoak and Douglas-fir are major components of northern mixed evergreen forests and Douglas-fir–hardwood forests (Sawyer et al. 1988). Douglas-fir dominates the overstory, while tanoak dominates a lower, secondary canopy. Other important tree associates can include: Pacific madrone, giant chinquapin (*Chrysolepis chrysophylla*), big-leaf maple, California bay, canyon live oak (*Quercus chrysolepis*), ponderosa pine, sugar pine (*Pinus lambertiana*), incense-cedar (*Calocedrus decurrens*), California black oak, and Oregon oak.

FIRE RESPONSES OF IMPORTANT SPECIES

Douglas-fir regeneration is often episodic in part due to irregular seed crops (Strothmann and Roy 1984), and establishment is enhanced when good seed crop years are followed by low to moderate surface fire. If, however, there was a poor seed crop other species would become established and inhibit Douglas-fir seedlings in ensuing years. Seedlings are most likely to become established on moist, mineral soil, while relatively few seedlings are found on thick organic seedbeds (Hermann and Lavender 1990). Douglas-fir is considered moderately shade tolerant in northwestern California (Sawyer et al. 1988). Young seedlings are better able to tolerate shade than older seedlings, although Douglas-fir seedlings growing on dry sites need more shade. In northwestern California, optimum seedling survival occurs with about 50% shade, but optimum seedling growth occurs with 75% full sunlight (Sawyer et al. 1988). Mature Douglas-fir has thick corky bark and is fire resistant. Other fire adaptations include: great height with branches on tall trees over 30 m (100 ft) above the ground, rapid growth, longevity (up to 700 to 1,000 years old), and the ability to form adventitious roots (Hermann and Lavender 1990).

Douglas-fir and tanoak litter decomposes to 5% of original dry weight in 7 years and 9 years, respectively (Pillers 1989). Because of this, fuelbeds are typically thin. Douglas-fir leaves are short and fall individually resulting in compact litter (Engber et al. 2011). Tanoak litter, however, is deeper and less dense promoting good flammability when dry (Engber and Varner 2012a).

Tanoak, Pacific madrone, chinquapin, canyon live oak, bigleaf maple, and California bay are all fire neutral facultative sprouters (Table 10.1). Tanoak generally dominates its associated broad-leaved trees and reproduces from seed under most light conditions, but fares best in full sunlight (Sawyer et al. 1988). While tanoak is very tolerant of shade, regeneration from seed in dense shade is limited. Tanoak can be easily top-killed by fire but vigorously sprouts from dormant buds located on burls or lignotubers (Plumb and McDonald 1981). Stored carbohydrates and an extensive root system aid in a rapid and aggressive postfire recovery (McDonald and Tappeiner 1987). Resistance to low-intensity surface fires increases with size because of increased bark thickness (Plumb and McDonald 1981). Dried tanoak leaves are very flammable in comparison to most California oaks (Engber and Varner 2012a).

FIRE REGIME–PLANT COMMUNITY INTERACTIONS

Pre-Euro-American settlement fires were relatively frequent in Douglas-fir–tanoak forests due to their warmer, drier, inland locations and increased lightning activity at more interior and higher elevations (Keeley 1981). In the North Fork of the Eel River, for example, an average of about 25 lightning strikes occur per year (Keter 1995). Native Americans were the primary ignition source in the North Coast Ranges, as they regularly burned to promote food plants and basketry materials (Keter 1995). Similar to other regional forest types, fires most likely occurred during the months of July through September (Keeley 1981).

There is little literature describing pre-Euro-American settlement fire size, intensity, and severity in Douglas-fir–tanoak forests. Ethnographic data on indigenous populations and fire history data suggest that fires sizes, intensities, and severities were highly variable. In areas subject to frequent burning by Native Americans, fire intensities and severities were low (Lewis 1993). Other areas experienced fire intensities and severities that varied spatially and temporally across the landscape resulting in a complex mosaic of mostly multiaged stands of varying sizes (Rice 1985, Wills and Stuart 1994). Fires in interior sites spread more extensively than those closer to redwood forests. Surface fires were common and were intermixed with areas that supported passive/active crown fires.

The few fire history studies of Douglas-fir–tanoak forests indicate that average presuppression fire return intervals varied from 10 to 16 years (Rice 1985, Wills and Stuart 1994). Further inland in the North Coast Range, dry forest sites containing Douglas-fir and other species, without tanoak, had shorter fire return intervals between 4 and 6 years with slightly higher and more variable fire return intervals on more mesic sites (Skinner et al. 2009). These high-frequency fires likely promoted much more open forests with greater cover of understory plant species. Presuppression fire sizes were undoubtedly variable. In addition to environmental factors, such as soil type and topography, successional trajectories depend on fire severity, seed availability, and sprout density. The climax forest is characterized by an overstory of Douglas-fir with tanoak dominating the lower, secondary canopy. Fire suppression and past logging have increased the density of shade-tolerant tanoak in many Douglas-fir–tanoak forests (Hunter et al. 1997). However, Douglas-fir is often able to maintain its dominance because of its large size and longevity.

Sawyer et al. (1988) describe several successional pathways for Douglas-fir–tanoak forests. Following a severe, extensive stand-replacing fire, seed producing Douglas-fir are killed leaving the sprouting tanoak or other sprouting hardwoods to dominate during early succession. Salvage logging and broadcast burning can enhance sprouting hardwood dominance; while the absence of postfire treatment will promote the persistence of shrubs and sprouting hardwoods (Stuart et al. 1993). Eventually Douglas-fir will slowly reinvade as growing space is created by maturing, self-thinning hardwoods. A moderate- to low-severity surface fire, in contrast, would not kill many Douglas-fir but would kill Douglas-fir seedlings and saplings (Rice 1985), allowing Douglas-fir trees to dominate the overstory and tanoak to perpetuate in the understory and secondary canopy. A third scenario in more open mixed stands would allow both Douglas-fir and tanoak to persist.

Since Euro-American settlement, Douglas-fir–tanoak forests have been modified through the absence of fire. The most significant change in old growth is the greater density of understory shrubs and trees creating greater vertical fuel continuity, increasing the probability that a surface fire could burn into the crown. Many Douglas-fir–tanoak forests, however, have been logged or have experienced a stand altering wildfire. The current regime in these forests can be characterized as having longer fire return intervals due to effective and aggressive fire suppression (Skinner et al. 2009), greater intensity because of increased fuel loads from slash and increased densities of understory shrubs and trees, and greater severity because of the accumulated ladder fuels and increased dead surface fuel loads. Seasonality of fire occurrence is probably unchanged and fire size changes are uncertain.

Oregon Oak Woodland Zone

Oregon oak is distributed from southwestern British Columbia through western Washington and Oregon into California in the North Coast Ranges and Sierra Nevada (Little 1971). Oregon oak woodlands often occur in the margin between conifer forest and prairies in the North Coast Ranges. Oregon oak can also be found in open savannas, closed-canopy stands, mixed stands with conifers, or other broadleaved trees (Burns and Honkala 1990). Common associates of Oregon oak woodlands are Pacific madrone, California black oak, California bay, and Douglas-fir.

FIRE RESPONSES OF IMPORTANT SPECIES

Oregon oak and California black oak are fire-enhanced, facultative sprouters (Table 10.1). Seedling establishment for these oak species is relatively less common than sprouting, but may be enhanced by the removal of the litter layer (Arno and Hammerly 1977). Both oak species have moderately thick bark that can often withstand low- to moderate-intensity surface fires. During higher-intensity fires, Oregon oak and California black oak are frequently top-killed by fire but vigorously sprout from the bole, root crown, and roots (Sugihara et al. 1987). Sprouting has been reported to decrease with age (Sugihara and Reed 1987) and increase with higher severity fire (Cocking et al. 2012, 2014).

FIRE REGIME–PLANT COMMUNITY INTERACTIONS

Prior to Euro-American settlement, Oregon oak woodlands experienced frequent, low-intensity surface fires, many of which were ignited by Native Americans. Mean fire return intervals varied from 7 to 13 years in Oregon oak woodlands in Humboldt County (Reed and Sugihara 1987). Fires probably spread in cured herbaceous fuels and oak litter. The litter of Oregon oak and California black oak are among the most flammable of California oak species (Engber and Varner 2012a). There is no information on fire sizes in this vegetation type but they probably were diverse depending on annual variation in climate. Fires ignited by Native Americans under moister conditions probably would not spread far into adjoining conifer forests (Gilligan 1966), those ignited in the summer and fall could have been extensive because of continuous dry herbaceous fuels. Frequent surface fires produced open savannas in the Bald Hills in Humboldt County (Sugihara et al. 1987) (Appendix 1).

Fire exclusion has resulted in an estimated 29–44% loss in the spatial extent of prairies and oak woodlands in the Bald Hills region of Redwood National Park (Sugihara and Reed 1987, Fritschle 2008). Frequent surface fires once inhibited seedling establishment and reduced the density of Douglas-fir and other competing conifers (Barnhardt et al. 1996). In the absence of fire conifers can overtop, suppress, and eventually kill the shade-intolerant Oregon oak (Hunter and Barbour 2001). In a study of Oregon oak and California black oak stands in Humboldt and Mendocino counties, most oaks (>80%) established before 1905, while the majority (73%) of Douglas-fir established after 1950 (Schriver and Sherriff 2015).

Future Direction and Climate Change

Climate Change

The combination of fire exclusion, logging, and increased global temperatures has altered the fire characteristics of the North Coast bioregion. Fire size and frequency has increased over the past few decades due in part to increasing temperatures and earlier springs (Westerling et al. 2006, Miller and Safford 2012). This trend of increased burned area is predicted to continue over the next century (Westerling et al. 2011); however, some models indicate fire size may decrease for some areas along the coast (Lenihan et al. 2008). While the amount of area burned each year continues to increase, the rates of burning are still considered much lower than presettlement levels (Safford and Van de Water 2014). Research assessing changes in fire severity for the North Coast bioregion has not been conducted and findings from other bioregions have been mixed (Miller et al. 2009, Miller et al. 2012, Mallek et al. 2013). Return intervals for some ecosystems closer to the coast are generally thought to be longer and may not have deviated substantially from presettlement fire severity. However, more inland forest types may experience increases in fire severity.

Management Issues

Native American burning was interrupted in the mid-1860s largely because of the impacts of introduced diseases. Managers in this region are increasingly incorporating the influence of past Native American ignitions into restoration and management objectives because Native Americans were an integral component of this region for thousands of years (Underwood et al. 2003; Crawford et al. 2015) and continue to live in the region today.

Fire has been an essential part of coastal prairie ecology and fire exclusion has led to undesirable consequences. Livestock grazing can mitigate some of these consequences (Hayes and Holl 2003) but effective management should include prescribed burning. The sparse published data on burning of coastal prairie in California combined with anecdotal evidence and evidence from outside the region suggest regular burning on a 3- to 5-year rotation will slow the invasion of some nonnative species without reducing cover of existing native species. If the seed bank contains viable seeds of displaced native species, burning may stimulate increased cover of these species. To achieve restoration objectives on most sites, extensive reseeding of desirable native species is required, followed by a regular prescribed burning and/or grazing program (Evett 2002). See sidebar 10.1 for information on chaparral management in the north coast.

The majority of closed-cone pine forests have not burned in the last 40 to 80 years because of fire exclusion (except for areas burned in the 1995 Mt. Vision and the 2017 Northern Bay Area fires). The lack of fire could threaten the long-term existence of closed-cone pines (e.g., Bishop, knobcone) because they rely on high- and mixed-severity fires to prepare seed beds, enable the release of large quantities of seed, and remove the canopy thereby increasing the light reaching the forest floor. Where possible, wildfire or use of higher intensity prescribed fires could perpetuate these forests types.

The majority of the remaining redwood forests are relatively young and more homogeneous than their old-growth forest counterparts, due to a legacy of logging that occurred throughout the region. As such, many land managers are interested in restoring young-growth redwood forests to promote old-growth forest structures and improve wildlife habitat and greater connectivity (Porter et al. 2007). Much of this restoration has focused on the use of mechanical thinning treatments to reduce stand density, increase growth, and increase structural complexity (O'Hara et al. 2010, Teraoka and Keyes 2011, Plummer et al. 2012). One potential drawback of this approach is the increased accumulation of surface fuels (activity fuels) following thinning treatments (Agee and Skinner 2005) and may require subsequent fire or other fuel reduction treatments to reduce potential risks to wildfire.

Redwood forests once experienced relatively frequent fire at varying intervals depending on their geographic location. Redwood sites should use fall-ignited prescribed fires to

SIDEBAR 10.1 NORTHERN CALIFORNIA CHAPARRAL

Kate M. Wilkin

Chaparral typically burns in high-severity stand-replacing events every 30 to 100 years (Quinn and Keeley 2006). These fires are quite dangerous for the people living in the wildland-urban interface and fire hazard reduction treatments, such as prescribed fire and mastication, are often implemented to proactively protect people and infrastructure (Keeley 2002, Dicus and Scott 2006). These treatments help reduce wildfire hazard, yet can have drawbacks for ecosystems adapted to infrequent crown fire (Keeley 2002, Schoennagel et al. 2004). In California chaparral, non-native plants are known to invade after fuel treatments, but mainly persist in areas with the most shrub cover reduction (Merriam et al. 2006, Potts and Stephens 2009).

Many are concerned that fuel reduction treatments altered chaparral ecosystems through changes to the historic fire regime (Parker 1987, Keeley 2002, Wilkin et al. 2017). Though adapted to lower frequency fires, increased fire frequency can convert chaparral to a non-native annual grassland and reduce species diversity, especially under global-change-type drought (Syphard et al. 2007, Pratt et al. 2013). Despite wide spread application of fuel treatments in chaparral, there have been few large-scale, long-term experiments to determine how fuel hazards, non-native species invasion, and shrub composition change through time.

A recent study addressed these research gaps by examining the effects of fuel treatment type (fire or mastication) and season (fall, winter, spring) over a 13-year timespan in chamise-dominated chaparral of northern California (Fig. 10.1.1) (Wilkin et al. 2015). Both treatment type and season of application had distinct influences on plant communities and fire hazards (Fig. 10.1.2). In contrast to prescribed fire treatments, masticated areas had consistently lower shrub cover, higher nonnative plant abundance, and more non-native grasses. However, mastication surprisingly increased buckbrush (*Ceanothus cuneatus*) cover, an obligate seeder and a preferred deer browse, compared to some fire treatments (Fig. 10.1.2). Fall treatments had consistently lower shrub cover, greater non-native plant abundance, non-native annual grasses, and greater buckbrush cover than spring or winter treatments. Ten years after treatment, fall mastication had the lowest shrub fuel load, but highest annual grass cover. These results highlight ecological and fire hazard reduction trade-offs among treatment types and season of treatment in chaparral shrublands.

FIGURE 10.1.1 Fuel reduction treatments included prescribed fire (top), a control (center), and mastication (bottom) (photo credits: S. Stephens, K. Wilkin, and D. Fry).

(continued)

(continued)

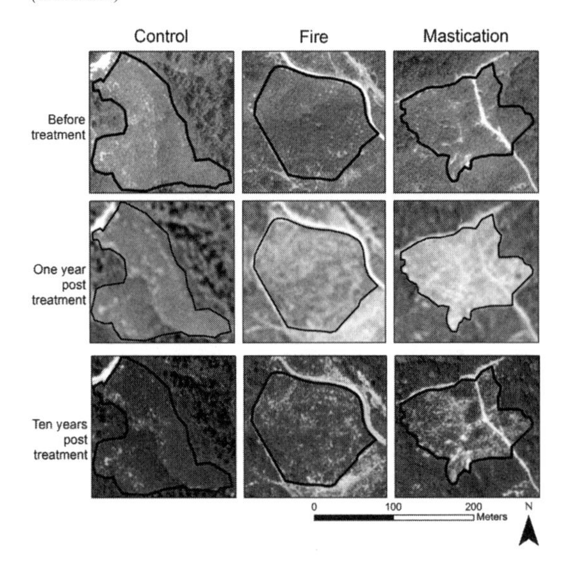

FIGURE 10.1.2 Repeated imagery of northern California chaparral fire hazard reduction study, demonstrating different long-term recovery between fire and mastication treatments (Wilkin et al. 2017). Shrub cover was persistent and continuous through time in the control unit. One year after treatment, the masticated unit had fewer shrubs, whereas fire treatment had rapid shrub recovery. A decade after treatment, differences in shrub recovery were maintained, where the fire unit had a nearly continuous shrub canopy and the masticated unit had a shrub-grass mosaic. The before treatment photos are black and white aerial photos from a USGS Digital Ortho Quarter Quad. The posttreatment photos are aerial images from National Agricultural Imagery Products for Mendocino County California.

Management Implications

Prescribed fire

- Generally fosters long-term native plant diversity and community structure
- Reduces fire hazard for a shorter time than mastication
- Decreases certain native shrubs

Mastication

- Fosters some native shrub species, but nonnative understory species invade and persist
- Reduces fire hazard more than fire, but may also increase fire frequency due to an increase amount of highly flammable annual grasses
- Increases certain native shrubs, such as the obligate seeder buckbrush

Season of treatment

- Fall treatments slow shrub recovery more than winter and, even more so, spring treatments
- Fall treatments promote greater nonnative plants and nonnative annual grass density than other seasons
- Fall fire treatments increase the preferred dear browse buckbrush. Within mastication treatments, season did not influence buckbrush outcomes

reintroduce fire as an ecosystem process. Some prescribed fire is presently occurring in parks in this area but this should be expanded. While redwood trees may not be solely dependent on fire to perpetuate in these forests, many ecosystem structures and functions can be enhanced or maintained.

Pre-Euro-American settlement fires were relatively frequent in Douglas-fir–tanoak forests due to their warmer, drier, inland locations and increased lightning activity at higher elevations; fire use by Native Americans was also important in areas with tanoak since their acorns were an important food. Fire suppression has reduced fire frequency, particularly in young-growth forests. The most significant change in old growth is the greater density of understory shrubs and trees that has increased vertical fuel continuity, which can increase the probability of high-severity crown fires.

A major concern for forests containing tanoak in the North Coast bioregion is sudden oak death, a disease caused by the pathogen, *Phytophthora ramorum* (Rizzo and Garbelotto 2003). The pathogen has reached epidemic levels in the north and central coast of California and is projected to continue to spread over the coming decades (Meentemeyer et al. 2011). In its lethal form the pathogen can kill native tree species, including tanoak, leading to rapid ecosystem changes.

Sudden oak death alters the distribution, availability, and abundance of fuels. In some locations more than 95% of the trees in a stand can be killed (USDA Forest Service 2009). Once dead, the tanoak leaves are often retained in the canopy for two or more years. Kuljian and Varner (2010) found that infected tanoak leaves had 5–10% lower foliar moisture content than leaves of uninfected tanoak, while the foliar moisture content of dead leaves in the canopy of sudden oak death killed tanoak ranged between 5.9% and 26.4% over the fire season. Increased availability and abundance of canopy fuels

FIGURE 10.10 Chaparral, mixed evergreen forest, and oak woodlands burned in the Bouverie Preserve near Glen Ellen by the 2017 Nuns Fire (photo by Scott Stephens).

increases the potential for crown ignition and torching in these forests, which have been substantiated by laboratory experiments (Kuljian and Varner 2013) and modeling studies (Valachovic et al. 2011, Forrestel et al. 2015). Increases in the surface and crown fire behavior in sudden oak death infected stands can increase fire severity. In a recent study, researchers found that the density of standing dead tanoak was positively related to fire severity, leading to increased redwood mortality (Metz et al. 2011, 2013) following the 2008 wildfires in the Big Sur area. The possible impact of sudden oak death on the 2017 Northern Bay Area fires is an important question.

Removal of fire from coastal grasslands and Oregon oak woodlands has dramatically changed these ecosystems. The establishment of conifers and loss of Oregon oak have had cascading effects in Oregon oak woodlands. The reduction of Oregon oak populations reduces the amount of food and habitat important to wildlife in the region. Douglas-fir canopies reduce light penetration, which also reduces native herbaceous species resulting in a net loss of biodiversity in invaded areas (Livingston et al. 2016). The increased presence of Douglas-fir and the reduction of herbaceous plants also reduce the flammability of these ecosystems by increasing fuelbed bulk density and increasing fuel moisture (Engber et al. 2011). Lessened flammability results in reduced surface temperatures that are essential in killing young Douglas-fir trees (Engber and Varner 2012b) and contributes to a positive feedback that further promotes Douglas-fir invasion and continued loss of Oregon oak woodlands and prairies.

Frequent surface fires historically maintained these open oak woodlands and prairies in the Bald Hills in Humboldt County. The use of fall ignited prescribed fires should be used to enhance and maintain these ecosystems. Management ignited prescribed fires have been ineffective in reducing the density of competing conifers when they are over 10 ft (3 m) in height (Sugihara and Reed 1987). Hand removal of small Douglas-fir has been used in some areas and larger trees can be girdled to increase the dominance of the oak (Hastings et al. 1997). To maintain oak dominance, a minimum fire frequency of 3 to 5 years has been recommended (Sugihara and Reed 1987).

Fire research on the ecosystems of the North Coast bioregion should be expanded through adaptive management and experimentation including in riparian and wetland habitats. This area includes significant amounts of federal and state land along with the highest proportion of industrial forests in California. Site specific questions on fire's role in North Coast ecosystems need to be addressed on the full spectrum of private and public lands. Most fire-related research has been on public land and relatively little has been done on the extensive private lands found in this bioregion. Fire has been an essential ecological process throughout most of the North Coast and has greatly contributed to the biological diversity of the region. Fire management in this area could focus on improving the resilience of these ecosystems to maintain and enhance the biodiversity of the region. The recent 2017 Northern Bay Area fires were tragic in terms of lives and property lost but initial burn patterns may have coarsely followed vegetation types with high-severity fire in chaparral, mixed severity in mixed evergreen forests (coast live oak [*Quercus agrifolia*], California bay, Pacific madrone, Douglas-fir), and low severity in oak woodlands (Fig. 10.10). Much more information on the drivers and effects of these fires is needed to better understand their impacts.

References

Agee, J.K. 1993. Fire Ecology of Pacific Northwest Forests. Island Press, Washington, D.C., USA.

Agee, J.K., and C.N. Skinner. 2005. Basic principles of forest fuel reduction treatments. Forest Ecology and Management 211: 83–96.

Anderson, R.S., A. Ejarque, P.M. Brown, and D.J. Hallett. 2013. Holocene and historical vegetation change and fire history on the north-central coast of California, USA. The Holocene 23: 1797–1810.

Arno, S.F., and R.P. Hammerly. 1977. Northwest Trees. The Mountaineers, Seattle, Washington, USA.

Bailey, R.G. 1995. Description of the Ecoregions of the United States, 2nd ed. USDA Forest Service, Washington, D.C., USA.

Barbour, M.G. 2007. Closed-cone pine and cypress forests. Pages 296–312 in: M.G. Barbour, T. Keeler-Wolf, and A.A. Schoenherr, editors. Terrestrial Vegetation of California. University of California Press, Berkeley, California, USA.

Barnhardt, S.J., J.R. McBride, and P. Warner. 1996. Invasion of northern oak woodlands by *Pseudotsuga menziesii* (Mirb.) Franco in the Sonoma Mountains of California. Madroño 43: 28–45.

Barrett, L.A. 1935. A record of forest and field fires in California from the days of the early explorers to the creation of the forest reserves. Report to the Regional Forester. October 14, 1935. USDA Forest Service, California Region, San Francisco, California, USA.

Bicknell, S.H., A.T. Austin, D.J. Bigg, and R.P. Godar. 1992. Late prehistoric vegetation patterns at six sites in coastal California. Bulletin of the Ecological Society of America 73: 112.

Bicknell, S.H., R.P. Godar, D.J. Bigg, and A.T. Austin. 1993. Salt Point State Park prehistoric vegetation. Final Report, Interagency Agreement 88-11-013. California Department of Parks and Recreation, Arcata, California, USA.

Biswell, H.H. 1989. Prescribed Burning in California Wildlands Vegetation Management. University of California Press, Berkeley, California, USA.

Blackburn, T.C., and K. Anderson, editors. 1993. Before the Wilderness: Environmental Management by Native Californians. Ballena Press, Menlo Park, California, USA.

Borchert, M.I., D. Segotta, and M.D. Purser. 1988. Coast redwood ecological types of southern Monterey County, California. USDA Forest Service General Technical Report PSW-GTR-107. Pacific Southwest Forest and Range Experiment Station, Berkeley, California, USA.

Brown, P.M., and W.T. Baxter. 2003. Fire history in coast redwood forests of Jackson Demonstration State Forest, Mendocino coast, California. Northwest Science 77: 147–158.

Brown, P.M., M.W. Kaye, and D. Buckley. 1999. Fire history in Douglas-fir and coast redwood forests at Point Reyes National Seashore, California. Northwest Science 73: 205–216.

Brown, P.M., and T.W. Swetnam. 1994. A cross-dated fire history from coast redwood near Redwood National Park, California. Canadian Journal of Forest Research 24: 21–31.

Burns, R.M., and B.H. Honkala. 1990. Silvics of North America. Agriculture Handbook 654. U.S. Department of Agriculture, Forest Service, Washington, D.C., USA.

Carroll, A.L., S.C. Sillett, and R.D. Kramer. 2014. Millennium-scale crossdating and inter-annual climate sensitivities of standing California redwoods. PLoS ONE 9: e102545.

CDF-FRAP. 2001. Fire perimeters. Fire and Resource Assessment Program. California Department of Forestry and Fire Protection. http://frap.fire.ca.gov/data/frapgisdata-sw-fireperimeters_download. Accessed December 12, 2015.

Clar, C.R. 1969. Evolution of California's Wildland Fire Protection System. California State Board of Forestry, Sacramento, California, USA.

Cocking, M.I., J.M. Varner, and E.E. Knapp. 2014. Long-term effects of fire severity on oak-conifer dynamics in the southern Cascades. Ecological Applications 24: 94–107.

Cocking, M.I., J.M. Varner, and R.L. Sherriff, 2012. California black oak responses to fire severity and native conifer encroachment in the Klamath Mountains. Forest Ecology and Management 270: 25–34.

Cooper, D.W. 1965. The Coast Redwood and Its Ecology. University of California, Agricultural Extension Service, Berkeley, California, USA.

Crawford, J.N., S.A. Mensing, F.K. Lake, and S.R. Zimmerman. 2015. Late Holocene fire and vegetation reconstruction from the western Klamath Mountains, California, USA: a multi-disciplinary approach for examining potential human land-use impacts. The Holocene 25: 1341–1357.

Davidson, J.G.N. 1971. Pathological problems in redwood regeneration from seed. Dissertation, University of California, Berkeley, California, USA.

Dawson, T.E. 1998. Fog in the California redwood forest: ecosystem inputs and use by plants. Oecologia 117: 476–485.

Del Tredici, P. 1998. Lignotubers in *Sequoia sempervirens*: development and ecological significance. Madroño 45: 255–260.

Dicus, C.A., and M.E. Scott. 2006. Reduction of potential fire behavior in wildland-urban interface communities in southern California: a collaborative approach. Pages 729–738 in: P.L. Andrews and B.W. Butler, compilers. Fuels Management—How to Measure Success: Conference Proceedings. USDA Forest Service Proceedings RMRS-P-41. Rocky Mountain Research Station, Fort Collins Colorado, USA.

Elford, R.C., and M.R. McDonough. 1964. The climate of Humboldt and Del Norte Counties. Agricultural Extension Service. University of California, Eureka, California, USA.

Engber, E.A., and J.M. Varner 2012a. Patterns of flammability of the California oaks: the role of leaf traits. Canadian Journal of Forest Research 42: 1965–1975.

———. 2012b. Predicting Douglas-fir sapling mortality following prescribed fire in an encroached grassland. Restoration Ecology 20: 665–668.

Engber, E.A., J.M. Varner, L.A. Arguello, and N.G. Sugihara. 2011. The effects of conifer encroachment and overstory structure on fuels and fire in an oak woodland landscape. Fire Ecology 7(2): 32–50.

Evett, R.R. 2000. Research on the pre-European settlement vegetation of the Sea Ranch coastal terraces. Unpublished Final Report. Sea Ranch Association, Sea Ranch, California, USA.

———. 2002. Sea Ranch grassland vegetation inventory and monitoring program associated with the prescribed burn program. Unpublished Final Report. Sea Ranch Association, Sea Ranch, California, USA.

Fahnestock, G.R., and J.K. Agee. 1983. Biomass consumption and smoke production by prehistoric fires in western Washington. Journal of Forestry 81: 653–657.

Finney, M.A. 1990. Fire history from the redwood forests of Bolinas Ridge and Kent Lake Basin in the Marin Municipal Water District. In: Vegetation and Fire Management Baseline Studies: The Marin Municipal Water District and the Marin County Open Space District (Northridge Lands) Unpublished Report. Leonard Charles and Associates and Wildland Resource Management, Marin County, California, USA.

———. 1996. Development of fire-scar cavities on old growth coast redwood. Pages 96–98 in: J. Leblanc, editor. Coast Redwood Forest Ecology and Management. University of California, Berkeley, California, USA.

Finney, M.A., and R.E. Martin. 1989. Fire history in a Sequoia sempervirens forest at Salt Point State Park, California. Canadian Journal of Forest Research 19: 1451–1457.

———. 1992. Short fire intervals recorded by redwoods at Annadel State Park, California. Madroño 39: 251–262.

———. 1993. Modeling effects of prescribed fire on young-growth coast redwood trees. Canadian Journal of Forest Research 23: 1125–1135.

Fisher, R.T., H. von Schrenk, and A.D. Hopkins. 1903. The redwood. USDA Bureau of Forestry Bulletin 38. Washington, D.C., USA.

Foin, T.C., and M.M. Hektner. 1986. Secondary succession and the fate of native species in a California coastal prairie community. Madroño 33: 189–206.

Fonda, R.W., L.A. Bellanger, and L.L. Burley. 1998. Burning characteristics of western conifer needles. Northwest Science 72: 1–9.

Ford, L.D., and G.F. Hayes. 2007. Northern coastal scrub and coastal Prairie. Pages 180–207 in: M.G. Barbour, T. Keeler-Wolf, and A.A. Schoenherr, editors. Terrestrial Vegetation of California. University of California Press, Los Angeles, California, USA.

Forrestel, A.B., M.A. Moritz, and S.L. Stephens. 2011. Landscape-scale vegetation change following fire in Point Reyes, California, USA. Fire Ecology 7(2): 114–128.

Forrestel, A.B., B.S. Ramage, T. Moody, M.A. Moritz, and S.L. Stephens. 2015. Disease, fuels and potential fire behavior: impacts of sudden oak death in two coastal California forest types. Forest Ecology and Management 348: 23–30.

Fox, L. 1988. A classification, map, and volume estimate for coast redwood forest in California. The Forest and Rangeland Resources Assessment Program. California Department of Forestry and Fire Protection, Sacramento, California, USA.

Franklin, J.F., and C.T. Dyrness. 1973. Natural vegetation of Oregon and Washington. USDA Forest Service General Technical Report PNW-GTR-8. Pacific Northwest Forest and Range Experiment Station, Portland, Oregon, USA.

Fritschle, J.A. 2008. Reconstructing historic ecotones using the public land survey: the lost prairies of Redwood National Park. Annals of the Association of American Geographers 98: 24–39.

Fritz, E. 1931. The role of fire in the redwood region. Journal of Forestry 29: 939–950.

———. 1945. Twenty years' growth on a redwood sample plot. Journal of Forestry 43: 30–36.

———. 1951. Letter to Harold Weaver quoted in D. Carle. 2002: 62–63. Burning questions: America's fight with nature's fire. Praeger Publishers, Santa Barbara, California, USA.

Fry, D.L., J. Dawson, and S.L. Stephens. 2012. Age and structure of mature knobcone pine forests in the northern California Coast Range, USA. Fire Ecology 8(1): 49–62.

Fry, D.L., and S.L. Stephens. 2013. Seed viability and female cone characteristics of mature knobcone pine trees. Western Journal of Applied Forestry 28: 46–48.

Gilligan, J.P. 1966. Land use history of the Bull Creek Basin. Pages 42–57 in: Proceedings of the Symposium, Management for Park Preservation: A Case Study at Bull Creek, Humboldt Redwoods State Park, California. School of Forestry, University of California, Berkeley, California, USA.

Graham, B.D. 2009. Structure of downed woody and vegetative detritus in old-growth *Sequoia sempervirens* forests. Thesis, Humboldt State University, Arcata, California, USA.

Green, S. 1999. Structure and dynamics of a coastal dune forest at Humboldt Bay, California. M.A. thesis, Humboldt State University, Arcata, California, USA.

Greenlee, J.M. 1983. Vegetation, fire history and fire potential of Big Basin Redwoods State Park, California. Dissertation, University of California, Santa Cruz, California, USA.

Griffin, J.R. 1994. Pinaceae. Pages 115–120 in: J.C. Hickman, editor. The Jepson Manual. University of California Press, Berkeley, California, USA.

Griffin, J.R., and W.B. Critchfield. 1972. The distribution of forest trees in California. USDA Forest Service Research Paper PSW-RP-82. Pacific Southwest Forest and Range Experiment Station, Berkeley, California, USA.

Gripp, R.A. 1976. An appraisal of critical fire weather in northwestern California. Thesis, Humboldt State University, Arcata, California, USA.

Harden, D.R. 1997. The Coast Ranges: mountains of complexity. Pages 252–288 in: D.R. Harden, editor. California Geology. Pearson Prentice Hall, Upper Saddle River, New Jersey, USA.

Harrington, C.A. 1990. Red alder. Pages 116–123 in: R.M. Burns and B.H. Honkala, technical coordinators. Silvics of North America. Vol. 2. Hardwoods. USDA Agricultural Handbook. Washington, D.C., USA.

Harvey, B.J., and B.A. Holzman. 2014. Divergent successional pathways of stand development following fire in a California closed-cone pine forest. Journal of Vegetation Science 25: 88–99.

Harvey, B.J., B.A. Holzman, and J.D. Davis. 2011. Spatial variability in stand structure and density-dependent mortality in newly established post-fire stands of a California closed-cone pine forest. Forest Ecology and Management 262: 2042–2051.

Hastings, M.S., S.J. Barnhardt, and J.R. McBride. 1997. Restoration management of northern oak woodlands. USDA Forest Service General Technical Report PSW-GTR-160. Pacific Southwest Research Station, Albany, California, USA.

Hatch, D.A., J.W. Bartolome, J.S. Fehmi, and D.S. Hillyard. 1999. Effects of burning and grazing on a coastal California grassland. Restoration Ecology 7: 376–381.

Hayes, G.F., and K.D. Holl. 2003. Cattle grazing impacts on annual forbs and vegetation composition of mesic grasslands in California. Conservation Biology 17: 1694–1702.

Heady, H.F., T.C. Foin, M.M. Hektner, M.G. Barbour, D.W. Taylor, and W.J. Barry. 1988. Coastal prairie and northern coastal scrub. Pages 733–760 in: M.G. Barbour and J. Major, editors. Terrestrial Vegetation of California. California Native Plant Society, Sacramento, California, USA.

Hektner, M.M., and T.C. Foin. 1977. Vegetation analysis of a northern California prairie: Sea Ranch, Sonoma County, California. Madroño 24: 83–103.

Herman, R.K., and D.P. Lavender. 1990. *Pseudotsuga menziesii* (Mirb.) Franco. Pages 527–554 in: R.M. Burns and B.H. Honkala, technical coordinators. Silvics of North America. Volume 1. Conifers. USDA Agricultural Handbook 654. Washington, D.C., USA.

Howard, J.L., and K.C. Aleksoff. 2000. *Abies grandis*. Fire Effects Information System. U.S. Department of Agriculture, Forest Service, Rocky Mountain Research Station, Fire Sciences Laboratory. http://www.feis-crs.org/beta/. Accessed December 12, 2015.

Hull, M.K., C.A. O'Dell, and M.J. Schroeder. 1966. Critical fire weather patterns—their frequency and levels of fire danger. USDA Forest Service, Pacific Southwest Forest and Range Experiment Station, Berkeley, California, USA.

Hunter, J.C. 1997. Fourteen years mortality in two old-growth *Pseudotsuga-Lithocarpus* forests in northern California. Journal of the Torrey Botanical Society 124: 273–279.

Hunter, J.C., and M.G. Barbour. 2001. Through-growth by *Pseudotsuga menziesii*: a mechanism for change in forest composition without canopy gaps. Journal of Vegetation Science 12: 445–452.

Hunter, J.C., V.T. Parker, and M.G. Barbour. 1999. Understory light and gap dynamics in an old-growth forested watershed in coastal California. Madroño 46: 1–6.

Jacobs, D.F., D.W. Cole, and J.R. McBride. 1985. Fire history and perpetuation of natural coast redwood ecosystems. Journal of Forestry 83: 494–497.

Jepson, W.L. 1910. The Silva of California. The University Press, Berkeley, California, USA.

Johnstone, J.A., and T.E. Dawson. 2010. Climatic context and ecological implications of summer fog decline in the coast redwood region. Proceedings of the National Academy of Sciences 107: 4533–4538.

Jones, G., and W. Russell. 2015. Approximation of fire-return intervals with point samples in the southern range of the coast redwood forest, California, USA. Fire Ecology 11(3): 80–94.

Keeley, J.E. 1981. Distribution of lightning and man caused wildfires in California. Pages 431–437 in: C.E. Conrad and W.C. Oechel, editors. Proceedings of the Symposium on Dynamics and Management of Mediterranean-Type Ecosystems. USDA Forest Service General Technical Report PSW-GTR-58. Pacific Southwest Forest and Range Experiment Station, Berkeley, California, USA.

———. 1991. Seed germination and life history syndromes in the California chaparral. The Botanical Review 57: 81–116.

———. 2002. Fire management of California shrubland landscapes. Environmental Management 29: 395–408.

———. 2005. Fire as a threat to biodiversity in fire-type shrublands. Pages 97–106 in: B.E. Kus and J.L. Beyers, technical coordinators. Planning for Biodiversity: Bringing Research and Management Together. USDA Forest Service General Technical Report

PSW-GTR-195. Pacific Southwest Research Station, Albany, California, USA.
Keeley, J.E., M. Baer-Keeley, and C. Fotheringham. 2005. Alien plant dynamics following fire in Mediterranean-climate California shrublands. Ecological Applications 15: 2109–2125.
Keeley, J.E., and F.W. Davis. 2007. Chaparral. Pages 339–366 in: M.G. Barbour, T. Keeler-Wolf, and A.A. Schoenherr, editors. Terrestrial Vegetation of California. University of California Press, Los Angeles, California, USA.
Keeley, J.E., G. Ne'eman, and C.J. Fotheringham. 1999. Immaturity risk in a fire-dependent pine. Journal of Mediterranean Ecology 1: 41–48.
Keter, T.S. 1995. Environmental history and cultural ecology of the North Fork of the Eel River Basin, California. USDA Forest Service Technical Publication R5-EM-TP-002. Pacific Southwest Region, Vallejo, California, USA.
Kuljian, H., and J.M. Varner. 2010. The effects of sudden oak death on foliar moisture content and crown fire potential in tanoak. Forest Ecology and Management 259: 2103–2110.
———. 2013. Foliar consumption across a sudden oak death chronosequence in laboratory fires. Fire Ecology 9(3): 33–44.
Lanner, R.M. 1999. Conifers of California. Cachuma Press, Los Olivos, California, USA.
Lazzeri-Aerts, R., and W. Russell. 2014. Survival and recovery following wildfire in the southern range of the coast redwood forest. Fire Ecology 10(1): 43–55.
Lenihan, J.M., D. Bachelet, R.P. Neilson, and R. Drapek. 2008. Response of vegetation distribution, ecosystem productivity, and fire to climate change scenarios for California. Climatic Change 87: 215–230.
Lewis, H.T. 1993. Patterns of Indian burning in California: ecology and ethnohistory. Pages 55–116 in: T.C. Blackburn and K. Anderson, editors. Before the Wilderness: Environmental Management by Native Californians. Ballena Press, Menlo Park, California, USA.
Lightfoot, K.G., T.A. Wake, and A.M. Schiff. 1991. The archaeology and ethnohistory of Fort Ross, California. Volume 1, Introduction. Contributions of the University of California Archaeological Research Facility 49. University of California, Berkeley, USA.
Little, E.L. 1971. Atlas of United States trees: Volume I. Conifers and important hardwoods. USDA Forest Service Miscellaneous Publication 1146. Washington, D.C., USA.
Livingston, A.C., J.M. Varner, E.S. Jules, J.M. Kane, and L.A. Arguello. 2016. Prescribed fire and conifer removal promote positive understorey vegetation responses in oak woodlands. Journal of Applied Ecology 53: 1604–1612.
Long, C.J., and C. Whitlock. 2002. Fire and vegetation history from the coastal rain forest of the western Oregon Coast Range. Quaternary Research 58: 215–225.
Long, C.J., C. Whitlock, P.J. Bartlein, and S.H. Millspaugh. 1998. A 9000-year fire history from the Oregon Coast Range, based on a high-resolution charcoal study. Canadian Journal of Forest Research 28: 774–787.
Lorimer, C.G., D.J. Porter, M.A. Madej, J.D. Stuart, S.D. Veirs, S.P. Norman, K.L. O'Hara, and W.J. Libby. 2009. Presettlement and modern disturbance regimes in coast redwood forests: implications for the conservation of old-growth stands. Forest Ecology and Management 258: 1038–1054.
Lotan, J.E., and W.B. Critchfield. 1990. *Pinus contorta* Dougl. ex Loud. lodgepole pine. Pages 302–315 in: R.M. Burns and B.H. Honkala, editors. Silvics of North America. Volume I. Conifers. Agriculture Handbook 654. U.S. Department of Agriculture, Forest Service, Washington, D.C., USA.
Mahony, T.M. 1999. Old-growth forest associations in the northern range of redwood. Thesis, Humboldt State University, Arcata, California, USA.
Mahony, T.M., and J.D. Stuart. 2001. Old-growth forest associations in the northern range of coastal redwood. Madroño 47: 53–60.
Mallek, C., H. Safford, J. Viers, and J. Miller. 2013. Modern departures in fire severity and area vary by forest type, Sierra Nevada and southern Cascades, California, USA. Ecosphere (12): art153.
McDonald, P.M., and J.C. Tappeiner. 1987. Silviculture, ecology, and management of tanoak in northern California. Pages 64–70 in: T.R. Plumb and N.H. Pillsbury, editors. Proceedings of the Symposium on Multiple-Use Management of California's Hardwood Resources. USDA Forest Service General Technical Report PSW-GTR-100. Pacific Southwest Forest and Range Experiment Station, Berkeley, California, USA.
Meentemeyer, R.K., N.J. Cunniffe, A.R. Cook, J.A.N. Filipe, R.D. Hunter, D.M. Rizzo, and C.A. Gilligan. 2011. Epidemiological modeling of invasion in heterogeneous landscapes: spread of sudden oak death in California (1990–2030). Ecosphere (2): art17.
Merriam, K.E., J.E. Keeley, and J.L. Beyers. 2006. Fuel breaks affect nonnative species abundance in Californian plant communities. Ecological Applications 16: 515–527.
Metz, M.R., K.M. Frangioso, R.K. Meentemeyer, and D.M. Rizzo. 2011. Interacting disturbances: wildfire severity affected by stage of forest disease invasion. Ecological Applications 21: 313–320.
Metz, M.R., J.M. Varner, K.M. Frangioso, R.K. Meentemeyer, and D.M. Rizzo. 2013. Unexpected redwood mortality from synergies between wildfire and an emerging infectious disease. Ecology 94: 2152–2159.
Miles, S.R., and C.B. Goudey. 1997. Ecological subregions of California: section and subsection descriptions. USDA Forest Service Technical Publication R5-EM-TP-005. Pacific Southwest Region, San Francisco, California, USA.
Miller, J.D., and H.D. Safford. 2012. Trends in wildfire severity: 1984 to 2010 in the Sierra Nevada, Modoc Plateau, and southern Cascades, California, USA. Fire Ecology 8(3): 41–57.
Miller, J.D., H.D. Safford, M. Crimmins, and A.E. Thode. 2009. Quantitative evidence for increasing forest fire severity in the Sierra Nevada and Southern Cascade Mountains, California and Nevada, USA. Ecosystems 12: 16–32.
Miller, J.D., C.N. Skinner, H.D. Safford, E.E. Knapp, and C.M. Ramirez. 2012. Trends and causes of severity, size, and number of fires in northwestern California, USA. Ecological Applications 22: 184–203.
Mohr, J.A., C. Whitlock, and C.N. Skinner. 2000. Postglacial vegetation and fire history, eastern Klamath Mountains, California, USA. The Holocene 10: 587–601.
Moritz, M.A., J.E. Keeley, E.A. Johnson, and A.A. Schaffner. 2004. Testing a basic assumption of shrubland fire management: how important is fuel age? Frontiers in Ecology and the Environment 2: 67–72.
Munz, P.A., and D.D. Keck. 1959. A California Flora. University of California Press, Berkeley, California, USA.
Newman, E.A., J.B. Potts, M.W. Tingley, C. Vaughn, and S.L. Stephens. 2018. Chaparral bird community responses to prescribed fire and shrub removal in three management seasons. Journal of Applied Ecology. doi: 10.1111/1365-2664.13099.
Norman, S.P., 2007. A 500-year record of fire from a humid coast redwood forest. Final Project Report. Save the Redwoods League, San Francisco, California, USA.
Noss, R.F. 2000. The Redwood Forest: History, Ecology, and Conservation of the Coast Redwoods. Island Press, Washington, D.C., USA.
O'Dell, T.E. 1996. Silviculture in the Redwood Region: an historical perspective. Pages 15–17 in: J. LeBlanc, editor. Proceedings of the Conference on Coast Redwood Forest Ecology and Management. University of California Cooperative Extension, Humboldt State University, Arcata, California, USA.
O'Hara, K.L., J.C.B. Nesmith, L. Leonard, and D.J. Porter. 2010. Restoration of old forest features in coast redwood forests using early-stage variable-density thinning. Restoration Ecology 18: 125–135.
Olson, D.F., D.F. Roy, and G.A. Walters. 1990. *Sequoia sempervirens* (D. Don) Endl. Redwood. Pages 541–551 in: R.M. Burns and B.H. Honkala, editors. Silvics of North America, Volume 1, Conifers. U.S. Government Printing Office, Washington, D.C., USA.
Oneal, C.B., J.D. Stuart, S.J. Steinberg, and L. Fox, III. 2006. Geographic analysis of natural fire rotation in the California redwood forest during the suppression era. Fire Ecology 2(1): 73–99.
Osburn, V.R., and P. Lowell. 1972. A review of redwood harvesting. State of California, The Resources Agency, Department of Conservation, Division of Forestry, Sacramento, California, USA.
Ostoja, S.M., and R.C. Klinger. 1999. The relationship of bishop pine cone morphology to serotiny on Santa Cruz Island, California. Pages 167–171 in: D.R. Brown, K.L. Mitchell, and H.W. Chaney, editors. Proceedings of the 5th Channel Islands Symposium. Santa

Barbara Museum of Natural History, Santa Barbara, California, USA.
Parker, V.T. 1987. Can native flora survive prescribed burns? Fremontia 15(2): 3–6.
Pillers, M.D. 1989. Fine fuel dynamics of old-growth redwood forests. Thesis, Humboldt State University, Arcata, California, USA.
Pillers, M.D., and J.D. Stuart. 1993. Leaf-litter accretion and decomposition in interior and coastal old-growth redwood stands. Canadian Journal of Forest Research 23: 552–557.
Plumb, T.R., and P.M. McDonald. 1981. Oak management in California. USDA Forest Service General Technical Report PSW-GTR-54. Pacific Southwest Forest and Range Experiment Station, Berkeley, California, USA.
Plummer, J.F., C.R. Keyes, and J.M. Varner. 2012. Early-stage thinning for the restoration of young redwood—Douglas-fir forests in northern coastal California, USA. ISRN Ecology 2012: 1–9.
Potts, J.B., E. Marino, and S.L. Stephens. 2010. Chaparral shrub recovery after fuel reduction: a comparison of prescribed fire and mastication techniques. Plant Ecology 210: 303–315.
Potts, J.B., and S.L. Stephens. 2009. Invasive and native plant responses to shrubland fuel reduction: comparing prescribed fire, mastication, and treatment season. Biological Conservation 142: 1657–1654.
Porter, D., V. Gizinski, R. Hartley, and S.H. Kramer. 2007. Restoring complexity to industrially managed timberlands: the Mill Creek interim management recommendations and early restoration thinning treatments. Pages 283–294 in: R.B. Standiford, G.A. Giutsi, Y. Valachovic, W.J. Zielinski, and M.J. Furniss, technical editors. Proceedings of the Redwood Region Forest Science Symposium: What does the Future Hold? USDA Forest Service General Technical Report PSW-GTR-194. Pacific Southwest Research Station, Albany, California, USA.
Pratt, R.B., A.L. Jacobsen, A.R. Ramirez, A.M. Helms, C.A. Traugh, M.F. Tobin, M.S. Heffner, and S.D. Davis. 2013. Mortality of resprouting chaparral shrubs after a fire and during a record drought: physiological mechanisms and demographic consequences. Global Change Biology 20: 893–907.
Pyne, S.J., P.L. Andrews, and R.D. Laven. 1996. Introduction to Wildland Fire. 2nd ed. John Wiley and Sons, Inc., New York, New York, USA.
Quinn, R.D., and S.C. Keeley. 2006. Introduction to California Chaparral. University of California Press, Berkeley, California, USA.
Ramage, B.S., K.L. O'Hara, and B.T. Caldwell. 2010. The role of fire in the competitive dynamics of coast redwood forests. Ecosphere (6): art20.
Reed, L.J., and N.G. Sugihara. 1987. Northern oak woodlands—ecosystem in jeopardy or is it already too late? Pages 59–63 in: T.R. Plumb and N.H. Pillsbury, editors. Symposium on Multiple-Use Management of California's Hardwood Resources. USDA Forest Service General Technical Report PSW-GTR-100. Pacific Southwest Forest and Range Experiment Station, Berkeley, California, USA.
Rice, C.L. 1985. Fire history and ecology of the North Coast Range Preserve. Pages 367–372 in: J.E. Lotan, B.M. Kilgore, W.C. Fischer, and R.W. Mutch, editors. Wilderness Fire Symposium. USDA Forest Service General Technical Report INT-GTR-320. Intermountain Forest and Range Experiment Station Intermountain Forest and Range Experiment Station, Ogden Utah, USA.
Rizzo, D.M., and M. Garbelotto. 2003. Sudden oak death: endangering California and Oregon forest ecosystems. Frontiers in Ecology and the Environment 1: 197–204.
Safford, H.D., and K.M. Van de Water. 2014. Using fire return interval departure (FRID) analysis to map spatial and temporal changes in fire frequency on national forest lands in California. USDA Forest Service Research Paper PSW-RP-266. Pacific Southwest Forest and Range Experiment Station, Berkeley, California, USA.
Sawyer, J.O., J. Gray, G.J. West, D.A. Thornburgh, R.F. Noss, J.H. Engbeck, Jr., B.G. Marcot, and R. Raymond. 2000. History of redwood and redwood forests. Pages 7–38 in: R.F. Noss, editor. The Redwood Forest: History, Ecology, and Conservation of the Coast Redwoods. Island Press, Washington, D.C., USA.
Sawyer, J.O., D.A. Thornburgh, and J.R. Griffin. 1988. Mixed evergreen forest. Pages 359–381 in: M.G. Barbour and J. Major, editors. Terrestrial Vegetation of California, new expanded edition. California Native Plant Society, Sacramento, California, USA.
Scanlon, H. 2007. Progression and behavior of the Canoe Fire in coast redwood. Pages 223–232 in: R.B. Standiford, G.A. Giusti, Y. Valachovi, W.J. Zielinski, and M.J. Furniss, technical editors. Proceedings of the Redwood Region Forest Science Symposium: What does the Future Hold? USDA Forest Service General Technical Report PSW-GTR-194. Pacific Southwest Research Station, Albany, California, USA.
Schoennagel, T., T.T. Veblen, and W.H. Romme. 2004. The interaction of fire, fuels, and climate across Rocky Mountain forests. Bioscience 54: 661–676.
Schriver, M., and R. Sherriff. 2015. Establishment patterns of Oregon white oak and California black oak woodlands in northwestern California. Pages 529–539 in: R.B. Standiford and K.L. Purcell, technical coordinators. Proceedings of the Seventh California Oak Symposium: Managing Oak Woodlands in a Dynamic World. USDA Forest Service General Technical Report PSW-GTR-251. Pacific Southwest Research Station, Berkeley, California, USA.
Skinner, C.N., C.S. Abbott, D.L. Fry, S.L. Stephens, A.H. Taylor, and V. Trouet. 2009. Human and climatic influences of fire occurrence in California's North Coast Range, USA. Fire Ecology 5(3): 76–99.
Stephens, S.L., and D.L. Fry. 2005. Fire history in coast redwood stands in the northeastern Santa Cruz Mountains, California. Fire Ecology 1(1): 2–19.
Stephens, S.L., and W.J. Libby. 2006. Anthropogenic fire and bark thickness in coastal and island pine populations from Alta and Baja California. Journal of Biogeography 33: 648–652.
Stephens, S.L., R.E. Martin, and N.E. Clinton. 2007. Prehistoric fire area and emissions from California's forests, woodlands, shrublands and grasslands. Forest Ecology and Management 251: 205–216.
Stephens, S.L., D.R. Weise, D.L. Fry, R.J. Keiffer, J. Dawson, E. Koo, J. Potts, and P. Pagni. 2008. Measuring the rate of spread of chaparral prescribed fires in northern California. Fire Ecology 4(1): 74–86.
Stromberg, M.R., P. Kephart, and V. Yadon. 2001. Composition, invisibility, and diversity in coastal California grasslands. Madroño 48: 236–252.
Strothmann, R.O., and D.F. Roy. 1984. Regeneration of Douglas-fir in the Klamath Mountains Region, California and Oregon. USDA Forest Service General Technical Report PSW-GTR-81. Pacific Southwest Forest and Range Experiment Station, Berkeley, California, USA.
Stuart, J.D. 1986. Redwood fire ecology. Final Report. California Department of Parks and Recreation, Weott, California, USA.
———. 1987. Fire history of an old-growth forest of *Sequoia sempervirens* (Taxodiaceae) in Humboldt Redwoods State Park, California. Madroño 34: 128–141.
Stuart, J.D., and L. Fox. 1993. Humboldt Redwoods State Park Unit Prescribed Fire Management Plan. California Department of Parks and Recreation, Arcata, California, USA.
Stuart, J.D., M.C. Grifantini, and L. Fox. 1993. Early successional pathways following wildfire and subsequent silvicultural treatment in Douglas-fir/hardwood forests, NW California. Forest Science 39: 561–572.
Stuart, J.D., and L.A. Salazar. 2000. Fire history of white fir forests in the coastal mountains of northwestern California. Northwest Science 74: 280–285.
Stuart, J.D., and J.O. Sawyer. 2001. Trees and Shrubs of California. University of California Press, Berkeley, California, USA.
Sugihara, N.G., and L.J. Reed. 1987. Prescribed fire for restoration and maintenance on Bald Hills oak woodlands. Pages 446–451 in: T.R. Plumb and N.H. Pillsbury, editors. Symposium on Multiple-Use Management of California's Hardwood Resources. USDA Forest Service General Technical Report PSW-GTR-100. Pacific Southwest Forest and Range Experiment Station, Berkeley, California, USA.
Sugihara, N.G., L.J. Reed, and J.M. Lenihan. 1987. Vegetation of the Bald Hills oak woodlands, Redwood National Park. Madroño 34: 193–208.
Syphard, A.D., V.C. Radeloff, J.E. Keeley, T.J. Hawbaker, M.K. Clayton, S.I. Stewart, and R.B. Hammer. 2007. Human influence on California fire regimes. Ecological Applications 17: 1388–1402.
Taylor, A.H. 1990. Disturbance and persistence of Sitka spruce (*Picea sitchensis* (Bong) Carr.) in coastal forests of the Pacific Northwest, North America. Journal of Biogeography 17: 47–58.

Teraoka, J.R., and C.R. Keyes. 2011. Low thinning as a forest restoration tool at Redwood National Park. Western Journal of Applied Forestry 26: 91–93.

Uchytil, R.J. 1989. *Alnus rubra*. Fire Effects Information System. USDA Forest Service, Rocky Mountain Research Station, Fire Sciences Laboratory. http://www.feis-crs.org/beta/. Accessed December 12, 2015.

Underwood, S., L. Arguello, and N. Siefkin. 2003. Restoring ethnographic landscapes and natural elements in Redwood National Park. Ecological Restoration 21: 278–283.

USDA Forest Service. 2009. Major Forest Insect and Disease Conditions in the United States 2007. Department of Agriculture, Washington, D.C., USA.

Valachovic, Y.S., C.A. Lee, H. Scanlon, J.M. Varner, R. Glebocki, B.D. Graham, and D.M. Rizzo. 2011. Sudden oak death-caused changes to surface fuel loading and potential fire behavior in Douglas-fir-tanoak forests. Forest Ecology and Management 261: 1973–1986.

Van Dersal, W.R. 1938. Native Woody Plants of the United States, Their Erosion Control and Wildlife Values. U.S. Department of Agriculture, Washington, D.C., USA.

Van de Water, K.M., and H.D. Safford. 2011. A summary of fire frequency estimates for California vegetation before Euro-American settlement. Fire Ecology 7(3): 26–58.

van Wagtendonk, J.W., and D.R. Cayan. 2008. Temporal and spatial distribution of lightning strikes in California in relationship to large-scale weather patterns. Fire Ecology 4(1): 34–56.

Veirs, S.D. 1982. Coast redwood forest: stand dynamics, successional status, and the role of fire. Pages 119–141 in: J.E. Means, editor. Proceedings of the Symposium, Forest Succession and Stand Development Research in the Northwest. Forest Research Laboratory, Oregon State University, Corvallis, Oregon, USA.

Vogl, R.J. 1973. Ecology of knobcone pine in the Santa Ana Mountains, California. Ecological Monographs 43: 125–143.

Vogl, R.J., W.P. Armstrong, K.L. White, and K.L. Cole. 1977. The closed-cone pine forest. Pages 295–358 in: M.G. Barbour and J. Major, editors. Terrestrial Vegetation of California. University of California Press, Berkeley, California, USA.

Walsh, R.A. 1995. *Deschampsia cespitosa*. Fire Effects Information System. USDA Forest Service, Rocky Mountain Research Station, Fire Sciences Laboratory. http://www.feis-crs.org/beta/. Accessed December 12, 2015.

Weaver, J.E., and F.C. Clements. 1929. Plant Ecology. McGraw-Hill Book Company, Inc., New York, New York, USA.

West, G.J. 1993. The late Pleistocene-Holocene pollen record and prehistory of California's North Coast Ranges. Pages 65–80 in: G. White, P. Mikkelsen, M.E. Hildebrandt, and M.E. Basgall, editors. There Grows a Green Tree: Papers in Honor of David A. Fredrickson. CARD Publication 11. University of California, Davis, California, USA.

Westerling, A.L., H.D. Hidalgo, D.R. Cayan, and T.W. Swetnam. 2006. Warming and earlier spring increase western U.S. forest wildfire activity. Science 313: 940–943.

Westerling, A.L., M.G. Turner, E.A.H. Smithwick, W.H. Romme, and M.G. Ryan. 2011. Continued warming could transform Greater Yellowstone fire regimes by mid-21st century. Proceedings of the National Academy of Sciences 108: 13165–13170.

Western Region Climate Center. 2001. Western U.S. Climate Historical Summaries: Crescent City and Angwin Pacific Union College. Desert Research Institute. http://www.dri.edu/. Accessed December 12, 2015.

Whitlock, C., and M.A. Knox. 2002. Prehistoric burning in the Pacific Northwest: human versus climatic influences. Pages 195–231 in: T.R. Vale, editor. Fire, Native Peoples, and the Natural Landscape. Island Press, Washington, D.C., USA.

Wilkin, K.M., L.C. Ponisio, D.L. Fry, C.L. Tubbesing, J.B. Potts, and S.L. Stephens. 2017. Decade-long plant community responses to shrubland fuel hazard reduction. Fire Ecology 13(2): 105–136.

Wills, R.D., and J.D. Stuart. 1994. Fire history and stand development of a Douglas-fir/hardwood forest in northern California. Northwest Science 68: 205–212.

Zammit, C.A., and P.H. Zedler. 1988. The influence of dominant shrubs, fire and time since fire on soil seed banks in mixed chaparral. Vegetatio 75: 175–187.

Zedler, P.H. 1986. Closed-cone conifers of the chaparral. Fremontia 14(3): 14–17.

Zinke, P.J. 1988. The redwood forest and associated north coast forests. Pages 679–698 in: M.G. Barbour and J. Major, editors. Terrestrial Vegetation of California. Native Plant Society, Sacramento, California, USA.

Zobel, D.B., L.F. Roth, and G.M. Hawk. 1985. Ecology, pathology, and management of Port Orford-cedar (*Chamaecyparis lawsoniana*). USDA Forest Service General Technical Report PNW-GTR-184. Pacific Northwest Forest and Range Experiment Station, Portland, Oregon, USA.

Zouhar, K.L. 2000. *Festuca idahoensis*. Fire Effects Information System. USDA Forest Service, Rocky Mountain Research Station, Fire Sciences Laboratory. http://www.feis-crs.org/beta/. Accessed December 12, 2015.

CHAPTER ELEVEN

Klamath Mountains Bioregion

CARL N. SKINNER, ALAN H. TAYLOR, JAMES K. AGEE,
CHRISTY E. BRILES, AND CATHY L. WHITLOCK

Fires . . . have been ground fires, and easily controlled. A trail will sometimes stop them.

WILSON (1904)

Description of the Bioregion

The Klamath Mountains bioregion occupies much of northwestern California continuing into southwestern Oregon. In California, the bioregion lies between the Northern California Coast bioregion on the west and the southern Cascade Range to the east. The southern boundary is the Northern California Coast Ranges and Northern California Interior Coast Ranges (Miles and Goudey 1997). This bioregion covers approximately 22,500 km^2 (8,690 mi^2), or 6% of California. It includes the Klamath and Trinity River systems, the headwaters of the Sacramento River, the most extensive exposure of ultramafic rocks in North America, and some of the most diverse conifer forests in North America (Sawyer 2006) (Map 11.1).

Physical Geography

The Klamath Mountains have been deeply dissected by the Klamath, McCloud, Sacramento, Salmon, and Trinity Rivers with no consistent directional trends. Only two sizable alluvial valleys, Scott and Hayfork, occur here (Oakeshott 1971). Elevations range from 30 m (100 ft) to 2,755 m (9,038 ft). Several prominent ranges or ridge systems comprise the bioregion with Mt. Eddy being the highest peak (Oakeshott 1971). The crests of these ridge systems are usually between 1,500 m (4,900 ft) and 2,200 m (7,200 ft) (Irwin 1966).

The complexity of the geology and terrain has a strong influence on the structure, composition, and productivity of vegetation (Whittaker 1960, Sawyer 2006) and thus influence fire regimes. Spatial variation in soil productivity combined with steep gradients of elevation and changes in slope aspect controls the connectivity, structure, and rates of fuel accumulation.

Climate

Climate of the bioregion is Mediterranean. However, the local expression of climate is remarkably variable due to strong west-east moisture and temperature gradients caused by proximity to the Pacific Ocean and steep elevation gradients that influence temperature and the spatial pattern of precipitation via orographic effects. The contemporary climatic phase appears to have become established ~3,000 to ~4,000 years ago (Mohr et al. 2000, Briles et al. 2011).

Table 11.1 shows climatic data for selected stations from west to east. Notably, temperature records in the Klamath Mountains are only from valleys or canyon bottoms because no regularly reporting stations are located above 1,000 m (3,280 ft).

There is considerable local and regional variation of annual precipitation. Generally, less precipitation falls in valleys and canyons than in the surrounding uplands with strong gradients over short distances. The driest areas occur adjacent to the Shasta and Sacramento Valleys. However, the watersheds of the Sacramento, McCloud, and Pit Rivers in the eastern Klamath Mountains are noted for high annual precipitation. At higher elevations, most precipitation falls as snow. The average annual early April snowpack depth and water content for snow courses are shown in Table 11.2.

WEATHER SYSTEMS

Critical fire weather in the Klamath Mountains is associated with any weather condition that creates sustained periods of high velocity winds with low humidity. Three weather patterns are important (Hull et al. 1966): (1) Pacific High—Postfrontal (Postfrontal), (2) Pacific High—Prefrontal (Prefrontal), and (3) Subtropical High Aloft (Subtropical High).

Postfrontal conditions occur when high pressure following the passage of a cold front causes strong winds from the north and northeast. Temperatures rise and humidity declines with these winds. Examples of fires fanned by Postfrontal conditions occurred in 1999 when the Megram fire burned over 57,000 ha (141,000 ac) east of Hoopa and the Jones fire, northeast of Redding near Lake Shasta, consumed over 900 structures while burning over 10,000 ha (25,000 ac).

Prefrontal conditions occur when strong, southwesterly or westerly winds are generated by the dry, southern tail of a rapidly moving cold front. Strong winds are the key here because

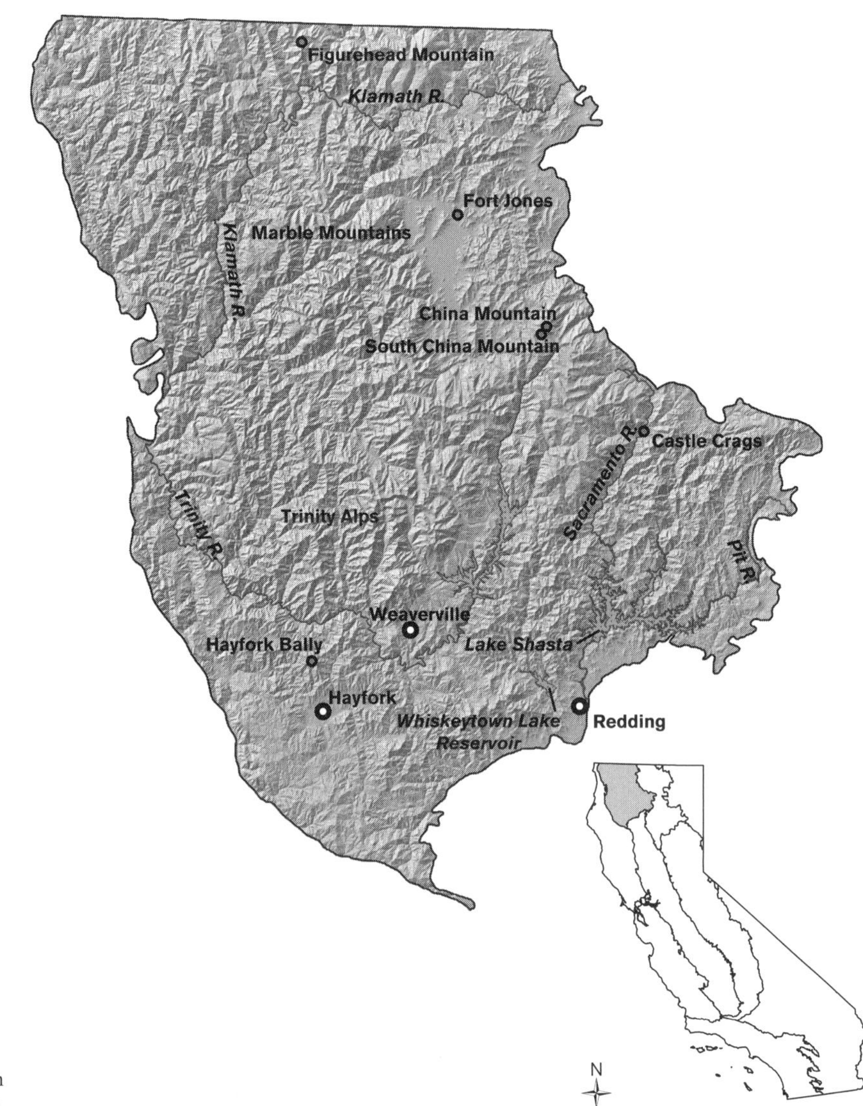

MAP 11.1 Klamath Mountains bioregion in California.

temperatures usually drop and relative humidity rises. These strong Prefrontal winds are able to spread fires rapidly as happened when the Oregon fire (2001) west of Weaverville burned over 650 ha (1,600 ac) and 13 homes and the Panther fire (2008) south of Happy Camp, already over 20,200 ha (50,000 ac) burned an additional 5,260 ha (13,000 ac) at high severity in a single run.

Subtropical High conditions occur when the region is under the influence of high pressure that causes temperatures to rise and humidity to drop. In this bioregion, these conditions lead to fires controlled mostly by local topography. Subtropical High conditions promote the development of strong temperature inversions that inhibit smoke from venting out of the canyons and valley bottoms (Robock 1988, 1991) leaving only the ridgetops in full sun. Smoke trapped under the thermal inversions, especially following initiation of widespread lightning-caused fires, shades the surface from solar heat, thus amplifying differentials in temperature, humidity, and fuel moisture between the canyon bottoms and the ridgetops (Schroeder and Buck 1970, Robock 1988, 1991, Estes et al. 2017) reducing fire intensity below the inversions. Examples of recent major fire episodes burning under these conditions include the Hayfork fires (1987) where 70% of the burned area sustained low-moderate-severity fire effects (Weatherspoon and Skinner 1995) and the widespread lightning fires of 2008 (Miller et al. 2012a). Fires burning above the inversion layer and immediately after dissipation of the inversion, especially when accompanied by strong winds, can produce large areas of high severity (Weatherspoon and Skinner 1995). Examples of such fires are the Megram fire in 1999 (Jimerson and Jones 2003), the Biscuit fire in 2002 (USDA Forest Service 2003), and the Motion and Panther fires of 2008 (Miller et al. 2012b).

LIGHTNING

Lightning is common in the Klamath Mountains with 12.8 strikes (range 6.4 to 26.4) yr^{-1} 100^{-1} km^{-2} (33.7 strikes [range 16.8 to 69.4] yr^{-1} 100^{-1} mi^{-2}) (van Wagtendonk and Cayan 2008). Lightning-caused fires have accounted for most area burned in recent decades (for example, 1977, 1987, 1999, 2002, 2006, 2008, 2012, and 2014). Indeed, from 1984 to

TABLE 11.1

Average annual, January, and July precipitation and normal daily January and July maxima and minima temperatures for representative stations in the Klamath Mountains

Location—elevation	Average precipitation cm (in)	Normal daily maximum temperature °C (°F)	Normal daily minimum temperature °C (°F)
Willow Creek—141 m			
Annual	143.5 (56.5)		
January	24.3 (9.6)	11.1 (52)	1.5 (35)
July	0.4 (0.2)	34.7 (95)	11.5 (53)
Sawyers Bar—659 m			
Annual	117.6 (46.3)		
January	21.6 (8.5)	9.1 (48)	–2.9 (27)
July	2.3 (0.9)	32.8 (91)	10.9 (52)
Fort Jones—830 m			
Annual	57.6 (22.3)		
January	10.8 (4.3)	6.6 (44)	–5.1 (23)
July	0.9 (0.4)	32.9 (91)	8.6 (48)
Weaverville—610 m			
Annual	101.2 (39.8)		
January	18.8 (7.4)	8.3 (47)	–2.8 (27)
July	0.5 (0.2)	34.2 (94)	9.6 (49)
Whiskeytown—367 m			
Annual	160.4 (63.1)		
January	30.0 (11.8)	12.0 (54)	2.1 (36)
July	0.7 (0.3)	35.3 (96)	17.4 (63)
Dunsmuir—703 m			
Annual	163.6 (64.4)		
January	29.7 (11.7)	9.9 (50)	–0.9 (30)
July	0.7 (0.3)	31.8 (89)	12.1 (54)

2008 over 87% of area burned in northwestern California was from lightning-caused fires (Miller et al. 2012a). Lightning may ignite hundreds of fires in a 24-hour period such as on June 21, 2008. The large number of simultaneous ignitions combined with poor access for fire-suppression forces, steep topography, and extensive strong canyon inversions (see above) generate widespread lightning fires that often burn for weeks to months over large areas (e.g., Estes et al. 2017).

For example, on June 21, 2008 lightning ignited widespread fires across northern California following an unusually early and dry spring (Hayasaka and Skinner 2009). These ultimately burned >175,000 ha (>432,000 ac) in the bioregion (Miller et al. 2012b). Many burned until extinguished by rain and snow in the fall (Miller et al. 2012a).

Lightning occurrence increases with distance from the coast and increasing elevation (van Wagtendonk and Cayan 2008). Interestingly, the two years with relatively few lightning strikes recorded—1987 and 1999—were years with some of the greatest amount of area burned by lightning-caused fires during the period of lightning strike data (Fig. 11.1).

Though counterintuitive, the number of lightning-caused fires and total area burned in a region are not necessarily related to the number of lightning strikes. More important than number of strikes is the ratio of ignitions to strikes (Hayasaka and Skinner 2009). Storms producing lightning-caused fires are associated with higher instability and drier air than storms that produce the most lightning strikes (Rorig and Ferguson 1999, 2002). Additionally, in each of 1987, 1999,

TABLE 11.2

Average April 1 snowpack data for representative courses ordered from north to south (CCSS 2002)

Course name	Elevation m (ft)	Snow depth cm (in)	Water content cm (in)
Etna Mountain	1,798 (5,900)	190.2 (74.9)	76.2 (30.0)
Sweetwater	1,783 (5,850)	94.2 (37.1)	34.5 (13.6)
Parks Creek	2,042 (6,700)	231.9 (91.3)	92.5 (36.4)
Deadfall Lakes	2,195 (7,200)	174.5 (68.7)	72.6 (28.6)
North Fork Sacramento R	2,103 (6,900)	153.9 (60.6)	59.9 (23.6)
Gray Rock Lakes	1,890 (6,200)	246.6 (53.8)	57.2 (22.5)
Middle Boulder 3	1,890 (6,200)	136.7 (97.1)	103.4 (40.7)
Wolford Cabin	1,875 (6,150)	218.7 (86.1)	91.4 (36.0)
Mumbo Basin	1,737 (5,700)	145.0 (57.1)	59.9 (23.6)
Whalan	1,646 (5,400)	124.5 (49.0)	53.1 (20.9)
Highland Lakes	1,829 (6,000)	172.5 (67.9)	74.9 (29.5)
Slate Creek	1,737 (5,700)	163.6 (64.4)	73.9 (29.1)
Red Rock Mountain	2,042 (6,700)	259.8 (102.3)	111.8 (44.0)
Bear Basin	1,981 (6,500)	197.6 (77.8)	84.8 (33.4)

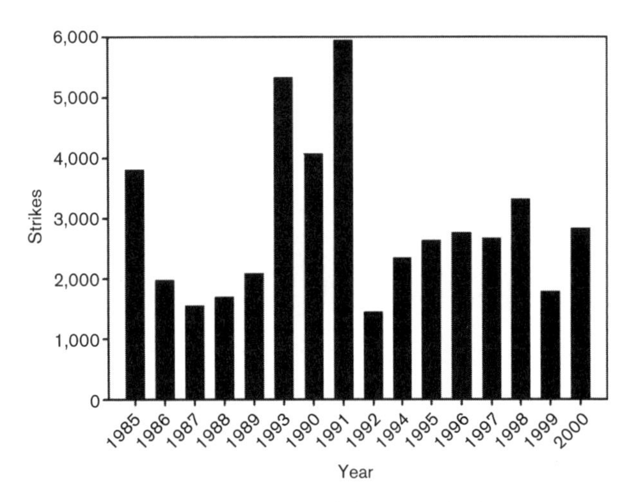

FIGURE 11.1 Variation in number of lightning strikes by year.

2008, and 2012 a single storm episode was responsible for nearly all of the area burned by lightning-caused fires.

The contribution of lightning-caused fires to total area burned has increased from 42% to 87% over the last century while the annual area and sizes of fires have significantly increased (Miller et al. 2012a). Further, the annual area burned appears to have changed over this time from being initially associated with drought to being inversely associated with summer rainfall. As the region receives very little summer rain, this further attests to the importance of thunderstorm characteristics—i.e., are they wet or dry. With improved access for suppression forces in more recent decades, thunderstorms accompanied by rain inhibit fire spread long enough for crews to arrive and suppress them. When little or no rain accompanies thunderstorms, fires ignite easily and quickly spread to become widespread fires that exceed capacity of suppression forces (Miller et al. 2012a).

It appears likely the incidence of lightning fires will increase as climate warms and fire season lengthens (Westerling et al. 2006, Yang et al. 2015). Lengthening fire seasons are characterized by earlier onset to spring warming and later fall cooling. Thus, fires can more easily ignite and spread earlier in the year and burn over a longer period of the year with increased drying of fuels (Yang et al. 2015). The widespread outbreak of lightning-caused fires in 2008 were set up by the early, dry spring (Hayasaka and Skinner 2009) which may become more common under a warming climate (Yang et al. 2015).

FIRE CLIMATE

To date, no studies have found a consistent association of fire activity with the El Niño Southern Oscillation (ENSO). This appears largely due to the location of the bioregion in the pivot zone (40–45°N) of the north/south dipole pattern of ENSO (Dettinger et al. 1998, Wise 2010). Resulting in the fire climate in this bioregion acting like the Southwest in some years and the Northwest in others. Examples are the droughts in the El Niños of 1965–1966 and 1991–1992 (Northwest pattern) and the extremely wet years of the strong El Niños of 1982–1983 and 1997–1998 (Southwest pattern).

Years with widespread and larger fires are usually dryer and warmer than the norm (Trouet et al. 2006, Taylor et al. 2008, Trouet et al. 2009). Trouet et al. (2009) found that these anomalous warm/dry years are associated with positive phases of both the Pacific North American Pattern (PNA) and the Pacific Decadal Oscillation (PDO). The development of a positive PNA + PDO produces persistent high-pressure ridges along the Pacific Coast in winter (Wallace and Gutzler 1981, Mantua et al. 1997) which reduces the amount and alters timing of annual precipitation (Trouet and Taylor 2010) contributing to fire seasons beginning earlier and being generally warmer, dryer, and longer (Trouet et al. 2009).

Ecological Zones

The Klamath Mountains have exceptional floristic diversity and complexity in vegetative patterns (Whittaker 1960, Sawyer 2006). The diverse patterns of climate, topography, and parent materials create heterogeneous vegetation patterns more complex than found in the Sierra Nevada or the Cascade Range (Sawyer and Thornburgh 1977, Sawyer 2006). The bioregion is thought to be of central importance in the evolution and development of western forest vegetation because of this diversity and the mixing of floras from the Cascade/Sierra Nevada axis and the Oregon/California coastal mountains that intersect here (Whittaker 1961, Sawyer 2006). Vegetation and species diversity generally increases with distance from the coast and species diversity is highest in woodlands with highly developed herb strata (Whittaker 1960). Conifer forests and woodlands are found in all elevational zones throughout the bioregion (Sawyer 2006).

The rugged, complex topography and resulting intermixing of vegetation in this bioregion defies a simple classification of ecological zones by elevation (Sawyer 2006). Nevertheless, this chapter will discuss three general zones: (1) a diverse lower montane zone of mixed-conifer and hardwood forests, woodlands, and shrublands, (2) a mid-upper montane zone where white fir (*Abies concolor*) is abundant and hardwoods are less important, and (3) a subalpine zone where white fir, Douglas-fir (*Pseudotsuga menziesii* var. *menziesii*), sugar pine (*Pinus lambertiana*), and ponderosa pine (*P. ponderosa*) drop out and are replaced by upper montane and subalpine species such as Shasta red fir (*A. magnifica* var. *shastensis*), mountain hemlock (*Tsuga mertensiana*), western white pine (*P. monticola*), Jeffrey pine (*P. jeffreyi*), whitebark pine (*P. albicaulis*), lodgepole pine (*P. contorta* subsp. *murrayana*), foxtail pine (*P. balfouriana* subsp. *balfouriana*), and curl-leaf mountain-mahogany (*Cercocarpus ledifolius*).

LOWER MONTANE

The lower and mid-montane zone is characterized by a very complex and diverse intermixing of vegetation assemblages (Fig. 11.2). This heterogeneity is caused by rugged complex terrain, diverse lithology, and a diversity of fire regimes.

Grasslands are most extensive in the two alluvial valleys (Scott and Hayfork). Shrublands are found throughout the Klamath Mountains. Lower elevation shrublands are found on warm or rocky, dry sites and on ultramafic and limestone derived soils. Species that commonly dominate lower montane shrublands are whiteleaf manzanita (*Arctostaphylos viscida*), greenleaf manzanita (*A. patula*), Brewer oak (*Quercus garryana* var. *breweri*), and deer brush (*Ceanothus integerrimus*). Shrublands also occupy extensive areas around historic mining districts (e.g., near Lake Shasta and Whiskeytown Reservoirs) where the combination of heavy cutting to support mining and air pollution from smelters caused drastic soil erosion and reduced site quality. The northern-most stands of chamise (*Adenostoma fasciculatum*) are found in the Whiskeytown area. Douglas-fir dominated and mixed evergreen forests are found throughout this zone.

MID TO UPPER MONTANE

In the western Klamath Mountains are areas on upper slopes and ridgetops locally known as prairies supporting dense perennial grasses. Grasslands also occur on shallow ultramafic soils and on cemented glacial till, while wet montane meadows are scattered throughout the upper montane and subalpine areas. Shrublands occur at higher elevations on poor sites and where severe fires inhibit or have removed tree cover. Important shrubs here are tobacco brush (known locally as snowbrush) (*Ceanothus velutinus*), shrub tanoak (*Notholithocarpus densiflorus* var. *echinoides*), giant chinquapin (*Chrysolepis chrysophylla*), bush chinquapin (*C. sempervirens*), huckleberry oak (*Quercus vacciniifolia*), and greenleaf manzanita.

Woodlands dominated by any combination of blue oak (*Q. douglasii*), Oregon oak (*Q. garryana*), California black oak (*Q. kelloggii*), gray pine (*Pinus sabiniana*), or ponderosa pine are found on sites similar to grasslands. Woodlands are also found on steep, dry, south- and west-facing slopes such as those along the Trinity River west of Junction City. Dry woodlands of ponderosa pine, western juniper (*Juniperus occidentalis*), Douglas-fir, Oregon oak, and incense cedar dominate sites around the Scott and Shasta Valleys. Woodlands are also common on harsh sites in the upper montane zones where they may be dominated by western white pine, Jeffrey pine, incense cedar (*Calocedrus decurrens*), Shasta red fir, or curl-leaf mountain-mahogany.

The montane conifer forests can be quite diverse with up to 17 conifer species have been identified in some watersheds in the north central Klamath Mountains (Keeler-Wolf 1990). However, stands typically have Douglas-fir in combination with any of five other conifer species: sugar pine, ponderosa pine, incense cedar, Jeffrey pine, and white fir (Fig. 11.3). Areas of ultramafic soils are an exception to this vegetation pattern, and stands here are usually dominated by Jeffrey pine or gray pine. Douglas-fir is the dominant conifer in the western portion of the bioregion. Ponderosa pine becomes an important associate on drier sites and may codominate or even dominate sites in the eastern part of the bioregion. White fir is of significant importance throughout except on ultramafics where Jeffrey pine is more important. With increasing elevation, white fir generally gives way to Shasta red fir and then mountain hemlock. Western white pine is commonly an important species throughout the upper montane areas.

The hardwood component of Klamath montane forests is equally diverse and distinguishes them from montane forests in other bioregions. Hardwoods commonly present in the subcanopy include: giant chinquapin, big-leaf maple (*Acer macrophyllum*), Pacific madrone (*Arbutus menziesii*), tanoak (*N. densiflorus* var. *densiflorus*), California black oak, and canyon live oak (*Q. chrysolepis*). Tanoak and giant chinquapin, dominant hardwoods in the west, are replaced by California black oak in the central and eastern Klamath Mountains.

Several species of fire sensitive conifers (Brewer spruce (*Picea breweriana*), Engelmann spruce (*P. engelmannii*), subalpine fir (*Abies lasiocarpa*)) are found in scattered, disjunct, and

FIGURE 11.2 Lower montane zone. Photo from the McCloud River canyon shows the diversity typical of the lower montane zone. Left side of photo shows Douglas-fir stands, California black oak stands, and Brewer oak intermixed on soils derived from weathered metasediments. Right side of photo shows gray pine woodland, buck brush, and Brewer oak on soils derived from limestone (photo credits: Carl Skinner, USDA Forest Service).

FIGURE 11.3 Mid-montane zone. Douglas-fir usually dominates conifer forests in this zone. Jud Creek looking north toward Hayfork Bally (photo credits: Carl Skinner, USDA Forest Service).

isolated locations that are fireless refugia or places that escape fire for long periods (Sawyer and Thornburgh 1977, Thornburgh 1990, Ledig et al. 2005). These species are sometimes found in watersheds that support a high diversity of conifer species such as the Sugar Creek Research Natural Area (Cheng 2004, Sawyer 2006, Agee 2007).

SUBALPINE

Subalpine woodlands and forests dominate the highest elevations in the Klamath Mountains (Fig. 11.4). There is no upper limit to this zone as trees are able to grow to the tops of the highest peaks (Sawyer and Thornburgh 1977, Sawyer 2006). The alpine character of the higher elevations is primarily due to shallow soils (Sharp 1960) or soils derived from strongly ultramafic parent materials and not due to low temperatures that prevent forest growth (Sawyer and Thornburgh 1977). Forests in the subalpine zone are generally open, patchy woodlands of widely spaced trees with a discontinuous understory of shrubs and herbs. Extensive bare areas are common (Sawyer and Thornburgh 1977). However, occasionally dense stands can be found on deeper soils.

Stands on mesic sites are dominated by mountain hemlock while xeric sites are usually occupied by Shasta red fir (Sawyer and Thornburgh 1977, Keeler-Wolf 1990). Woodlands are also common on harsh sites in the upper montane and subalpine zones where any mixture of western white pine, Jeffrey pine, whitebark pine, foxtail pine, mountain hemlock, or curl-leaf mountain-mahogany may occur.

Historic Fire Occurrence

Holocene Fire, Vegetation, and Climate History

Our understanding of the prehistoric vegetation and fire regimes come primarily from pollen and charcoal preserved in the sediments of lakes occupying cirques or landslide depressions with some records extending into the last ice age

FIGURE 11.4 Subalpine landscape looking toward the Trinity Alps from Mount Eddy. Most trees in the foreground and middle ground are foxtail pines (photo credits: Carl Skinner, USDA Forest Service).

(>20 ka [20,000 years before present]; West 1990, Mohr et al. 2000, Daniels et al. 2005, Briles et al. 2005, 2008, 2011). Besides the slow variations in climate related to the seasonal cycle of insolation and long-term droughts significantly influencing vegetation and fire regimes, nonclimatic factors also have shaped the region's vegetation and fire history. For example, lags in plant response to climate change among sites have been attributed to a forest's relative distance from the coast and the penetration of the coast fog zone (Briles et al. 2008). Local topographic gradients have also influenced the density of forest cover and through time accounted for sharp gradients in vegetation and fire patterns. Additionally, the geology and specifically nutrient limitations and lower moisture due to increased evapotranspiration of ultramafic substrates in the bioregion have had a strong influence on forest composition and structure through time (Briles et al. 2011).

Eight published pollen and charcoal records are available from the Klamath bioregion, mostly from montane and subalpine forests. Bolan Lake (1,637 m elevation [5,371 ft]) and Sanger Lake (1,547 m [5,075 ft]) occur in the Siskiyou Mountains, and their watersheds are underlain by diorite substrates (Briles et al. 2005, 2008). Campbell Lake (1,750 m [1,741 ft]) in the Marble Mountains lies on metamorphic substrates, and Taylor Lake (1,979 m [6,493 ft]) in the Russian Mountains is located on granodiorite substrates (Briles et al. 2011). Bluff Lake (1,926 m [1,819 ft]), Crater Lake (2,288 m [7,507 ft]), Mumbo Lake (1,860 m [6,102 ft]), and Cedar Lake (1,742 m [5,715 ft]) are located on the extensive Trinity Ultramafic Sheet (Mohr et al. 2000, Daniels et al. 2005, Briles et al. 2011).

The pollen and charcoal records document major plant community and fire regime changes since the last glacial period (Fig. 11.5). The glacial period was characterized by cold dry conditions due to significantly lower summer insolation than today. The Bolan Lake record between ~17 and 15 ka suggests that Klamath forests were initially open with sagebrush, grasses, and scattered white pines (Haploxylon pines) and mountain hemlock (Briles et al. 2005). Only the highest peaks in the Klamath Mountains support similar plant communities today. Fire activity was very low and fires did not burn much biomass.

As summer insolation increased in the late-glacial period, warm mesic conditions prevailed between 15 and 11.5 ka, conifers including white pines, fir (*Abies* spp.), and mountain hemlock increased their ranges upslope while forests on nonultramafic substrates became more closed throughout the bioregion. However, forests in the central bioregion on ultramafic substrates remained open and supported primarily yellow pines (Diploxylon pines) likely Jeffrey pine. Fires occurred infrequently at all locations during the late-glacial period. Although forests became more closed on nonultramafic locations, fire activity was low, suggesting cool moist conditions limited ignition and fire spread.

A period of warm dry conditions occurred from 11.5 to 7 ka as summer insolation increased and drought became more intense. Forests on nonultramafic substrates transitioned to an open forest of xerophytic species including white pines, cedars, shrub oaks, and roses, and on ultramafic substrates, open forests were dominated by yellow pines, cedars, and shrub oaks. Douglas-fir and mountain hemlock were limited by dry conditions in the central bioregion and absent completely on ultramafic substrates. They were also significantly reduced in abundance in the Siskiyou Mountains. At all sites, fire occurrence was the highest during the early Holocene.

Summer radiation declined after 7 ka through present, and conditions became cooler and wetter. During this period, firs replaced white pines as the dominant tree species on nonultramafic soils through the Klamath Mountains, and Douglas-fir became more abundant in the Siskiyous. Forests on nonultramafic soils were generally closed. On ultramafic substrates, yellow pines and firs increased, while shrub oaks and cedars declined, but forests remained open. Fire frequency decreased after 7 ka, but was higher than during the late-glacial period.

The modern vegetation developed at approximately 3–4 ka as summers cooled and became moister due to declining summer radiation. Douglas-fir increased in the Siskiyous after 2 ka and dominated the forests along with firs, and increased to low levels on nonultramafic soils and remained nonexistent on ultramafic sites in central forests. Mountain hemlock, which was significantly reduced at all locations after 11.5 ka, increased in abundance after 4 ka (and especially after 2 ka) except on ultramafic substrates. It was particularly sensitive to submillennial climate variations, decreasing in abundance during the Medieval Climate Anomaly and increasing during the Little Ice Age. In central forests on ultramafic substrates, yellow pines increased while shrub oaks declined during the last 4 ka.

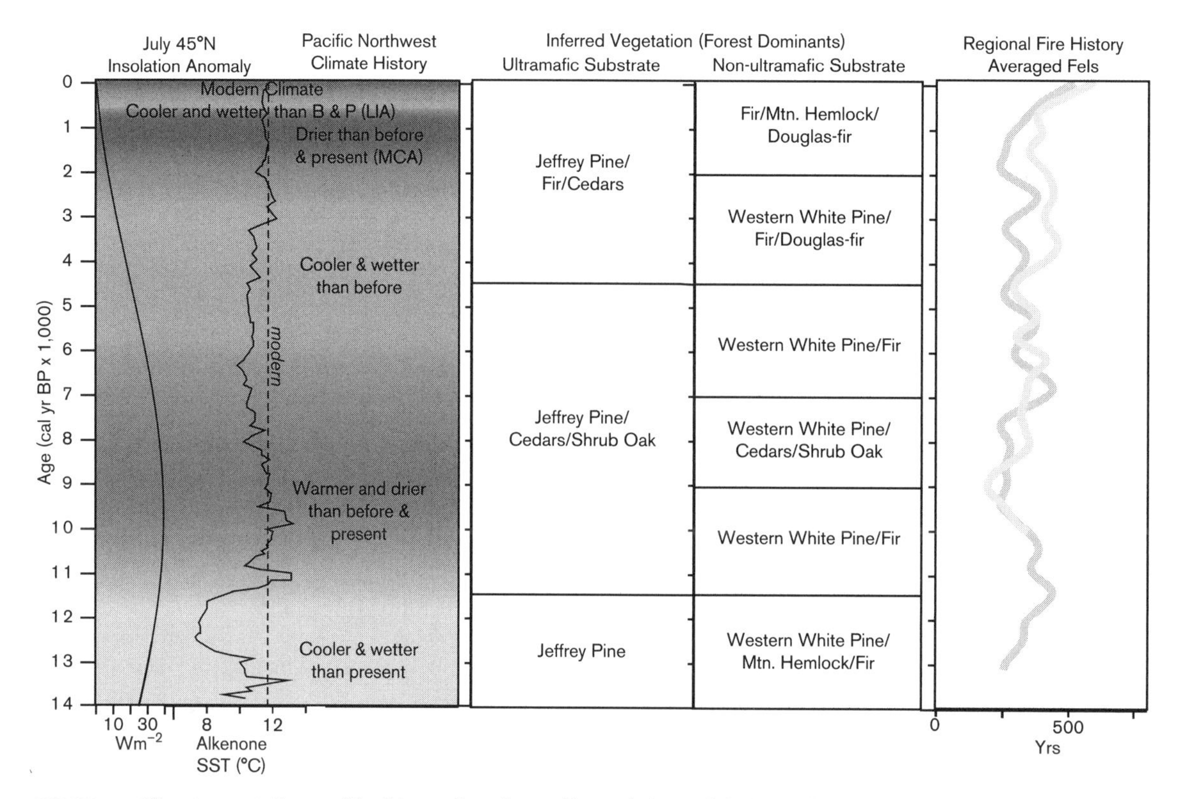

FIGURE 11.5 Climate, vegetation, and fire history (based on pollen and charcoal data). Pacific Northwest climate history summarized based on July 45°N insolation anomaly, alkenone d18O-derived sea surface temperatures from ODP 1019, and modeled climate (Alder and Hostetler 2015, Bartlein et al. 1998, Barron et al. 2003). LIA = Little Ice Age; MCA = Medieval Climatic Anomaly.

Fire activity in the last 4 ka remained moderately high on nonultramafic sites, but declined on ultramafic substrates possibly due to decline in the shrub oak understory that helps to carry fires. At all locations, fire activity decreased to unprecedented low levels after 1 ka, likely due to a combination of cool moist conditions during the Little Ice Age followed by fire suppression in recent times.

In summary, vegetation and fire regimes have continually changed in the Klamath Mountains since the last ice age and have been significantly influenced by slow variations in summer insolation and substrate differences. Past climate change has resulted in elevational adjustments in plant species, with important conifer species, such as Douglas-fir and mountain hemlock, showing particular sensitivity to warm/dry conditions. Climate has been the dominant driver of fires in the montane and subalpine Klamath forests, except on ultramafic substrates when understory shrubs were sparse.

Fire History Reconstructions of the Last 500 Years

Several fire history studies describe fire regimes of the Klamath Mountains over the last few centuries (Agee 1991, Wills and Stuart 1994, Taylor and Skinner 1998, 2003, Stuart and Salazar 2000, Skinner 2003a, 2003b, Fry and Stephens 2006). These studies indicate there are two periods with distinctly different fire regimes: (1) the Native American period, which usually includes both the prehistoric and European settlement periods, and (2) the fire-suppression period. Native people of the bioregion used fire in many ways: (1) to promote production of plants for food and fiber; (2) for ceremonial purposes; and (3) to improve hunting conditions (Long et al. 2016). Though ignitions by natives appear to have been widespread, the extent of their influence on fire regimes and vegetation is not known. Though there is variation among sites in when fire suppression became effective, before fire suppression began it appears most stands experienced at least several fires each century. This suggests a general fire regime of frequent, low-moderate-intensity fires.

Historic

Europeans began to explore the bioregion by the 1820s (Sullivan 1992). Following the 1848 discovery of gold along the Trinity River (Jackson 1964, Hoopes 1971) nonnative people began to enter the bioregion in large numbers and permanently settle the area. Settlers are reported to have set fires to make travel easier, to clear ground for prospecting, to drive game, and to encourage forage production for sheep and cattle (Whittaker 1960). However, no increases in fire occurrence during the settlement period are evident in fire-scar studies (Agee 1991, Wills and Stuart 1994, Taylor and Skinner 1998, 2003, Stuart and Salazar 2000, Fry and Stephens 2006). It may be that fires caused by settlers, either intentional or accidental, replaced fires ignited by Native Americans as the latter populations declined (Taylor et al. 2016). One study found that fire frequency in the Whiskeytown mining district declined following influx of miners in the mid-1800s (Fry and Stephens 2006). Fires went from frequent and local to less frequent and more extensive. This was similar to findings by Skinner et al. (2009) in the North Coast Range where they suggested it may be due to a decline in Native American burning accompanied by diminished herbaceous fuels from livestock grazing. In any case, many areas in the Klamath Mountains did not experience a major change in the fire regime until fire suppression became effective sometime after establishment of the Forest Reserve system in 1905 (Shrader 1965). Fire suppression had become effective in more accessible areas by the 1920s (Agee 1991, Stuart and Salazar 2000, Skinner 2003a, 2003b, Taylor and Skinner 2003, Fry and Stephens 2006), while it did not become effective in more remote areas until after 1945 (Wills and Stuart 1994, Taylor and Skinner 1998, Stuart and Salazar 2000).

Twentieth-Century Fire Activity

Fire occurrence declined dramatically with onset of fire suppression. Over the 400 years preceding effective fire suppression, there are no comparable fire-free periods when large landscapes experienced decades without fires simultaneously across the bioregion (Agee 1991, Wills and Stuart 1994, Stuart and Salazar 2000, Taylor and Skinner 1998, 2003, Skinner 2003a, 2003b, Fry and Stephens 2006).

These changes in the fire regimes are accompanied by changes in landscape vegetation patterns. Before fire suppression, fires of higher spatial complexity created openings of variable size within a matrix of forest that was generally more open than today (Taylor and Skinner 1998). This heterogeneous pattern has been replaced by a more homogenous pattern of smaller openings in a matrix of denser forests (Skinner 1995a). The ecological consequences of these changes are likely to be regional in scope but are not yet well understood.

The annual maximum fire size and total area burned have been increasing since the onset of fire suppression in the early twentieth century even as number of fires has declined (Miller et al. 2012a). When modern fires burn under relatively stable atmospheric conditions conducive to thermal inversions in narrow canyons as described above, patterns of severity appear to be similar to historical patterns (Weatherspoon and Skinner 1995, Taylor and Skinner 1998, Miller et al. 2012a).

Although there is no clear trend in overall proportion of high-severity burn area, the size of fires and the size of high-severity burn patches have been increasing over the last several decades—the larger the fire, the larger the maximum high-severity burn patches (Miller et al. 2012a). The extent of the recent high-severity burn patches appears to exceed historic patch size patterns (Skinner 1995a, Taylor and Skinner 1998). We suggest this is related, in part, to higher quantities and more continuous, homogeneous fuels caused by accumulation during the fire-suppression period.

Fire severity, though counterintuitive, is inversely associated with the total area burned in a year—greater area burned is associated with lower proportion of high severity (Miller et al. 2012a). This is largely a result of lightning-caused fires accounting for most area burned and having a lower proportion of high-severity than human-caused fires. The large lightning-caused fires are generally in rugged, remote landscapes where fires burn for weeks to months under variable and often less than severe conditions. In contrast, human-caused fires are generally near communities and travel corridors that provide better access. Such human-caused fires escape mostly under very severe burning conditions. Since they do not come in swarms of ignitions like lightning, human-caused fires are usually isolated events and get sufficient resources to be contained quickly when weather and burning conditions moderate. Thus, much of the area burned by human-caused fires burns under severe conditions leads to greater proportions of high severity (Miller et al. 2012a).

Major Ecological Zones

Fire Regimes

The steep and complex topography of the Klamath Mountains makes it difficult to separate fire regimes by ecological zones. The most widespread fire regime in the bioregion is found from the lower montane into the upper montane and it crosses ecological zones. Indeed, the patterns we present run the elevational gradient from the lowest canyon bottoms to nearly 2,000 m (6,250 ft). Generally, the steep, continuous slopes that run from low to higher elevations interact with changes in slope aspect and the dominating influence of summer drought, to create conditions for frequent mostly low- and moderate-intensity fires in most ecological zones of the bioregion. Given the importance of topographical controls on fire regimes, we will discuss the fire regimes more generally rather than assign them to specific ecological zones as in other bioregions.

The Klamath Mountains are often described as characterized by complex, mixed-severity fire regimes (Halofsky et al. 2011, Perry et al. 2011). Thus, this bioregion generally had neither completely low-intensity, surface-fire regimes nor totally stand-replacement, crown-fire regimes (Halofsky et al. 2011, Perry et al. 2011). Yet, to be useful this label (mixed severity) needs to be accompanied by specific descriptors (e.g., patch scales, proportions of severity levels) as by itself it is simply stating that burned patches lie somewhere on the gradient between being mostly alive to mostly dead. Importantly, though fires were largely quite heterogeneous, the central tendency for fire effects was to generally fall on the end of the gradient with mostly low-moderate-severity surface fires with variation influenced by topography resulting in a complex mosaic of fire effects (Taylor and Skinner 1998, Halofsky et al. 2011, Perry et al. 2011). This interpretation of the fire regime is largely supported by tree-ring derived fire histories (Wills and Stuart 1994, Stuart and Salazar 2000, Taylor and Skinner 1998, 2003, Skinner 2003a, 2003b, Fry and Stephens 2006).

TOPOGRAPHY

Topography strongly influences Klamath Mountain fire regimes. The long-term record of fire occurrence suggests spatial patterns were related to differences in timing of fires from place to place and differences in fire severity rather than fire frequency. With the exception of riparian zones (Skinner 2003b), only small differences in median Fire Return Intervals (FRIs) have been found within watersheds of several thousand hectares despite considerable variability in elevation, slope aspect, and species composition (Taylor and Skinner 1998, 2003).

Areas that burned with similar timing were found to be of several hundred hectares and bounded by topographic fea-

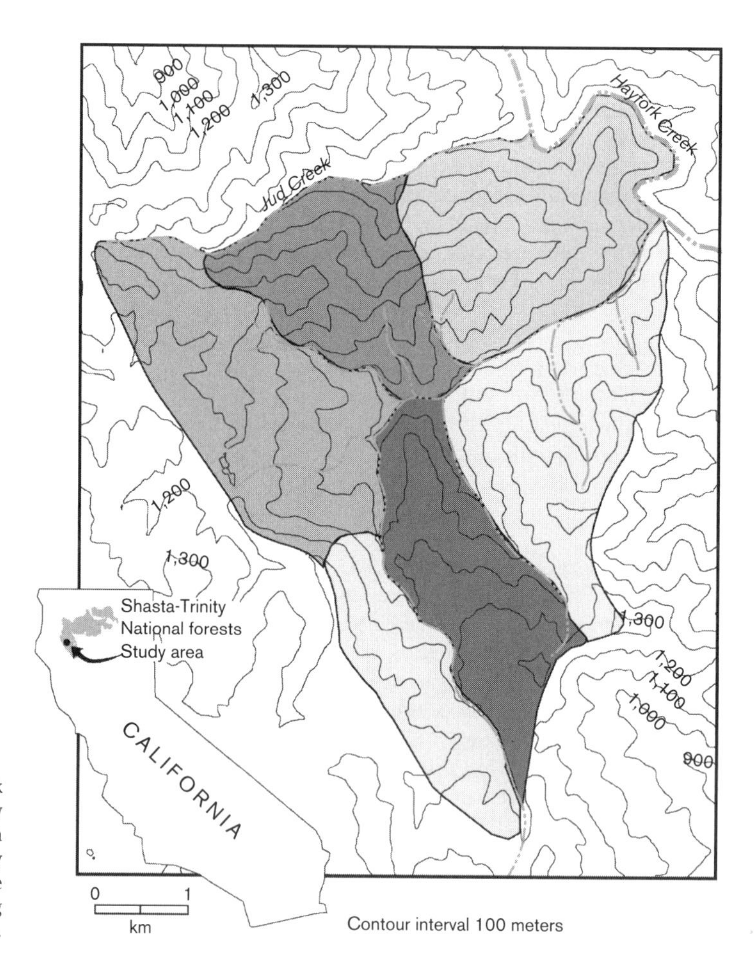

FIGURE 11.6 Map of FOAs in the Rusch/Jud creek watersheds near Hayfork. Figure illustrates how topographic features limited the spread of fires in most years. Though fire frequency did not vary significantly from area to area, the year of fire occurrence was often different from neighboring FOAs (adapted from Taylor and Skinner 2003).

tures (e.g., ridgetops, aspect changes, riparian zones, lithologic units) that affect fuel structure, fuel moisture, and fire spread (Taylor and Skinner 2003). We refer to such areas as Fire-Occurrence Areas (FOA). It is likely that the sizes of FOAs vary from landscape to landscape depending upon topographic complexity and are more localized than what is implied by the term fireshed. Although FOAs separated by topographic boundaries commonly had similar FRI distributions, they often experienced fires in different years—thus, different timing of fires.

These topographic boundaries between FOAs were not simple barriers to fire spread, but acted more like filters. In many years these features contained fires within the FOAs, but in others, especially unusually dry years such as 1829 or more recently 2008, fires would spread across boundaries and burn large portions of the greater landscape (Taylor and Skinner 2003) (Fig. 11.6). Thus, although nearby FOAs may each experience frequent fires, topographic factors set them up to generally burn in different years contributing to landscape heterogeneity.

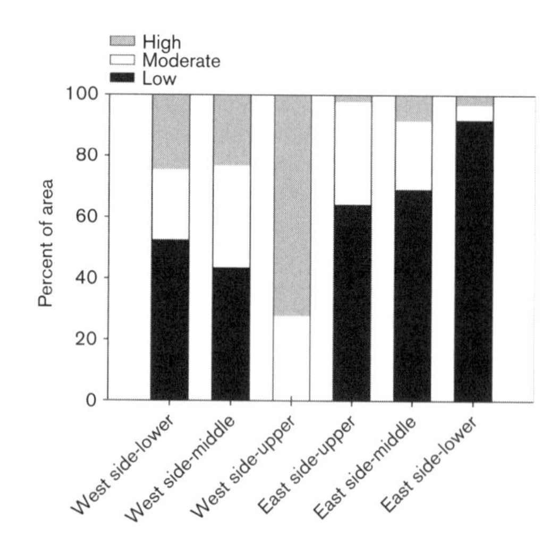

FIGURE 11.7 Chart depicting the distribution of cumulative fire severity patterns on Thompson Ridge near Happy Camp (Taylor and Skinner 1998).

FIRE SEVERITY

Patterns of fire severity, an important determinant of stand and landscape structural diversity, have been associated with topographic position in both the prefire-suppression and contemporary periods (Weatherspoon and Skinner 1995, Taylor and Skinner 1998, Jimerson and Jones 2003). A typical pattern of fire severity is illustrated in Fig. 11.7. Generally, the upper third of slopes and the ridgetops, especially south- and west-facing aspects, experience the highest proportion of high-severity burn. This is seen as larger patches of shrubs, young even-aged conifer stands, or stands of knobcone pine

FIGURE 11.8 Photo of Figurehead Mountain in the Thompson Creek watershed illustrates how patch size varies with topographic position in response to variation in fire intensity. The largest patches of high severity are on the upper thirds of the slopes, intermediate patches are in the middle third slope position, with the lower third slope position exhibiting a fine grain pattern dominated by large, old trees indicating fires burned primarily as low-intensity surface fires in these locations (Taylor and Skinner 1998). Photo taken in 1992, five years after entire landscape burned in the 1987 fires (photo credits: Carl Skinner, USDA Forest Service).

(*Pinus attenuata*). The lower third of slopes and north- and east-facing aspects experience mainly low-severity fires. Thus, more extensive stands of multiaged conifers with higher densities of old trees are found in these lower slope positions. Severity patterns of middle slope positions are intermediate between the other two positions.

Mid- and upper-slope positions, especially on south- and west-facing aspects, are more likely to experience higher fire intensities than other slope positions due to factors that affect fire behavior. These slopes generally experience greater drying and heating of fuels which contributes to greater fire intensity and makes it more likely to experience higher-intensity burns (Rothermel 1983). The common occurrence of strong thermal inversions that trap smoke in the steep, narrow canyons amplify differentials in temperature, humidity, and fuel moisture and thus differentials in fire behavior between the canyon bottoms and the ridgetops (Schroeder and Buck 1970, Robock 1988, 1991). Diurnal patterns of local sun exposure and wind flow combine with slope steepness to affect fire behavior (Schroeder and Buck 1970, Rothermel 1983). Thus, upper slopes tend to support higher-intensity fires running uphill through them compared to the tendency for lower slopes to support lower-intensity backing fires. Exposure to wind and solar insulation combine with position on steep slopes to create conditions where upper slopes experience higher-intensity fires more often than do lower slopes. The cumulative effects of the interaction of these factors on landscape patterns are depicted in Fig. 11.8.

It is important to note that this interpretation/description of severity patterns assumes that relatively even-aged cohorts of regeneration are exclusively the result of gaps and openings created by a preceding fire (Taylor and Skinner 1998). Other pathways to such patches exist (Brown 2006, Brown et al. 2008) with two likely being important in this bioregion. First scenario: when a climatic anomaly (e.g., cool/wet period) induced a hiatus in fire occurrence, even for only an additional decade or so, that provided sufficient time for enough regeneration to grow sufficiently to survive subsequent fires (Brown 2006). This would be especially so if the climatic anomaly was accompanied by a good seed crop resulting in abundant tree recruitment. Second scenario: the abundant regeneration documented throughout the western United States that originated coincidently with the onset of the fire-suppression era. This widespread, synchronous recruitment that regenerated in gaps or openings was simply the result of fire exclusion rather than a preceding high-severity fire. It is likely that many of these gaps and openings were created and maintained by chronic, frequent, low-moderate-severity fires. Thus, our description of the historical proportion high-severity patterns in Klamath Mountains landscapes is likely inflated, and possibly greatly so in some landscapes.

RIPARIAN ZONES

Few fire history data are available from riparian zones, but available data suggest that FRIs, and possibly fire behavior, are more variable within riparian zones than in adjacent uplands (Skinner 2003b). Median FRIs were generally twice as long on riparian sites as on neighboring uplands. However, the range of FRIs did not differ between riparian zone and adjacent upland sites (Skinner 2003b). Importantly, these data are from riparian sites adjacent to perennial streams and not ephemeral or intermittent streams. Riparian areas associated with ephemeral and intermittent streams dry out over the warm summers and probably have a fire regime similar to the surrounding uplands (Skinner 2003b). Additionally, the later types of streams are more likely to be above the smoke inversions where the streams initiate. Thus, they would tend to burn more severely than the perennial streams located lower in the watersheds that would be more shaded by the persistent smoke inversions.

Riparian areas along perennial watercourses often served as effective barriers to low-intensity and some moderate-intensity fires and strongly influenced patterns of fire occurrence beyond their immediate vicinity. The ability to be effective barriers would be enhanced by the effects of shading from inversion-trapped smoke. Consequently, by affecting fire spread, riparian areas are a key topographic feature that not only constitute a unique habitat, but also contribute to the structure and dynamics of upland forest landscapes (Skinner 2003b, Taylor and Skinner 2003).

Lower Montane

Fire regime information of common forest, woodland, and shrubland alliances of the Klamath Mountains is summarized in Appendix 1. Generally, before the fire-suppression era, mostly low-moderate-intensity surface fires characterized forest and woodland area fire regimes while mostly high-intensity crown fires were characteristic of shrubland fire regimes.

FIRE RESPONSES OF IMPORTANT SPECIES

More fire ecology information is available for alliances with Douglas-fir an important species than any others in this bioregion. Douglas-fir, once mature, is very resistant to low-moderate-intensity surface fires due to a variety of characteristics. When mature, Douglas-fir has very thick bark, a deep rooting habit, high crowns (Agee 1993), short needles, heals fire wounds rapidly, and does not slough bark. In fact, Douglas-fir is the most fire-resistant tree species in the Klamath Mountains. Its common conifer associates, ponderosa, Jeffrey, and sugar pines are also fire resistant as they too have thick bark, root deeply, and have high, open crowns. The pines, however, have longer needles and slough bark which forms a less compact litter bed that is better aerated so surface fires are more intense at the base of pine trees. It is not unusual for these three pine species to exhibit open fire wounds (commonly referred to as cat faces). In contrast, Douglas-fir rarely maintains open fire wounds. Wounds generally heal rapidly and are bark covered after only a few years. Moreover, Douglas-fir has shorter needles than the pines and does not slough bark so litter beds beneath the trees become more compact reducing fire intensity at the base of the tree. Thus, Douglas-fir has advantages in this bioregion following occasional extended periods (20 years to 30 years) without fire when Douglas-fir is less likely than the pines to incur basal bole damage.

Canyon live oak, generally considered sensitive to fire, is common in the lower montane zone of the Klamath Mountains. They may be easily top-killed by fire due to dense canopy and thin bark that makes them highly susceptible to crown scorch and cambium damage. As most oaks, if the top is killed, canyon live oak sprouts vigorously from the root crown (Tollefson 2008).

California black oaks, common throughout lower and mid-montane forests in the bioregion, have thin bark and are fire sensitive compared to the conifer associates that invade oak stands during longer fire-free periods. The size and shape of California black oak leaves provide for a well-aerated litter bed that can burn rapidly (Engber and Varner 2012). However, oak litter beds decompose rapidly contributing to low accumulations of fuel so fires that burn in oak litter are low intensity compared to fires in pine litter and rarely damage mature stems. Moreover, California black oak crowns are open and rarely support crown fires. With regular burning, the understory fuels are light, generally composed of grasses, forbs, scattered shrubs, and oak litter. Additionally, if black oaks are top-killed they sprout vigorously from the root crown and are able to maintain their presence on a site (Cocking et al. 2012).

Stands with a major component of buck brush (*Ceanothus cuneatus* var. *cuneatus*) are found scattered throughout the lower to mid-montane zones in the Klamath Mountains on xeric sites with shallow soils on limestone, ultramafic, or granitic bedrock. Often associated with buck brush are birch-leaf mountain-mahogany (*Cercocarpus betuloides* var. *betuloides*), hollyleaf redberry (*Rhamnus ilicifolia*), California buckeye (*Aesculus californica*), and the trees California bay (*Umbellularia californica*) and California black oak. It is interesting that buck brush does not sprout but establishes from seed that germinates following fires, while its associates are all strong sprouters.

Dense stands of shrubs dominated by Brewer oak (*Quercus garryana* var. *breweri*) are common and often support a diverse association of woody species. Brewer oak stands are found well into the mid-montane areas. Common associates are deer brush (*C. integerrimus*), poison oak (*Toxicodendron diversilobum*), snowdrop bush (*Styrax redivivus*), California ash (*Fraxinus dipetala*), birch-leaf mountain-mahogany, wild mock orange (*Philadelphus lewisii*), redbud (*Cercis occidentalis*), and California buckeye. Brewer oak is generally more flammable than other shrub oaks and other shrub associates. The leaf morphology creates a less compact more flammable litter bed than its associates (Engber and Varner 2012). Thus, where Brewer oak is a major component of shrubfields, it is usually the primary carrier of fire. All of these species sprout vigorously following fires (Skinner 1995b).

FIRE REGIME–PLANT COMMUNITY INTERACTIONS

The fire regimes of forests dominated by Douglas-fir were discussed at length in the section describing the common fire regimes of the Klamath Mountains and so are not repeated here. Here we concentrate on alliances more common in this zone than in others. Fire responses of important species in the lower montane zone are presented in Table 11.3.

In the lower- to mid-montane zone, canyon live oaks commonly achieve tree stature and dominate steep, xeric slopes in landscapes that experienced frequent, low-moderate-intensity fires. Canyon live oaks on these sites sometimes have open wounds with fire scars evident. However, the fire record is generally undatable due to decay. Fire-scar records collected from ponderosa pines, sugar pines, and Douglas-firs scattered in five canyon live oak stands near Hayfork had median FRIs of 6 years to 22 years (Taylor and Skinner 2003).

Sites where canyon live oak makes up a major portion of the canopy are often rocky, unproductive (Lanspa, n.d.) and have sparse, discontinuous surface fuels that do not carry fire well (Skinner and Chang 1996). Canyon live oak is less flammable than California black oak due to small leaf size producing compact litter beds (Engber and Varner 2012). Slopes with canyon live oak are often so steep that surface fuels collect mainly in draws, on small benches, and the upslope side of trees. Fires on these slopes would likely follow the draws and burn in a discontinuous manner. The fire-scar record comes from trees located near the head of ephemeral draws on the upper third of slopes. The presence of fire scars in the canyon live oaks suggest they were scarred by very light fires that burned in fuel that collected on the uphill side of the stem.

Stands dominated by California black oak (CBO) are common throughout lower and mid-montane areas especially in the central and eastern Klamath Mountains. Their highly nutritious acorns were an important food source for the native people of the bioregion. In order to perpetuate this food source, the native people promoted and maintained CBO stands by regular burning (Long et al. 2016). The cultural use of regular fire to manage CBO was promoted by the

TABLE 11.3
Fire responses of important species in the Lower Montane Zone of the Klamath Bioregion

	Type of fire response			
Lifeform	Sprouting	Seeding	Individual	Species
Conifer	None	Stimulated (establishment)	Resistant/killed	Douglas-fir, ponderosa pine
	None	Stimulated (seed release)	Resistant/killed	Gray pine
	None	Fire stimulate (seed release)	Killed	Knobcone pine
Hardwood	Fire stimulated	Stimulated (establishment)	Top-killed/survive	California black oak
	Fire stimulated	None known	Top-killed/survive	Brewer oak, tanoak, foothill ash, Oregon ash, Fremont cottonwood, white alder
Shrub	None	Fire stimulated	Killed	Whiteleaf manzanita
	Fire stimulated	Stimulated (germination)	Top-killed/survive	Chamise, deer brush, greenleaf manzanita, mahala mat
	Fire stimulated	None	Top-killed/survive	California buckeye, Lemmon's ceanothus, shrub tan oak, birch-leaf mountain-mahogany, wild mock orange, California storax, poison oak

species having the more flammable foliage among oaks in the bioregion due to leaf morphology (Engber and Varner 2012). Since the onset of fire suppression, conifers have invaded many of these stands and are poised to overtop and replace the oaks on many sites.

California black oak usually suffers the greatest fire damage when moderate-intensity fires burn in stands that have a significant component of conifers. Greater fuel accumulates under conifers due to slower decomposition. Moreover, CBO in mixed stands often have lower vigor due to competition from the conifers, making the trees more susceptible to fire damage. Conifers often survive fires in these mixed stands because they have thicker bark while many CBO may be top-killed. Where much of the conifer canopy remains, the oaks then sprout in the shade of the conifers and are unlikely to reach the main canopy as they would in an open environment or under other oaks.

Where conifers have overtopped CBO and begun to dominate the stands, low-moderate-intensity fires often enhance the process of succession to conifers by damaging the older, now weakened, understory oaks. Once stands have succeeded to conifer dominance, a high-intensity fire can kill the conifers outright while only top-killing the oaks. After such fires, CBO can sprout from root crowns to quickly regain dominance (Cocking et al. 2012). Though this will allow the area to return oak-dominance, it will take many decades before CBO develop the large, mature condition necessary to produce significant acorn crops, denning sites, and other important habitat qualities (Long et al. 2016).

California black oak (as well as tanoak and Pacific madrone) seedlings can survive for many years in the shaded understory of conifers. During this time, they are able to develop a large root system with a long taproot with limited top growth. Top growth on the seedlings may die back to the root crown and resprout several times waiting to quickly put on height growth following formation of a canopy gap. Thusly, CBO are able to survive for long periods as isolated trees in relatively dense conifer stands. Then, when a high-intensity fire kills much of the conifer overstory, existing CBO seedlings can quickly grow and reclaim dominance of the site (McDonald and Tappeiner 2002). An example can be seen near Volmers along Interstate 5 where a severe fire in 1986 killed several hundred hectares of mixed-conifer forest. The burned area is now dominated by fast-growing California black oak on mesic sites, and knobcone pine on xeric sites.

Extensive stands of California black oaks survived the ~12,000 ha (29,600 ac) High Complex in 1999. Even though this fire burned in the driest time of the year, August and early September, the light fuelbeds under oak stands supported mostly low-intensity surface fire.

Mid to Upper Montane

Forests in this zone are differentiated from lower elevation forests by the increased importance of white fir throughout and Shasta red fir in higher portions and the decreased importance of hardwoods. Specified in this way, the lower extent of

TABLE 11.4

Fire response of important species in the mid- to upper montane zones of the Klamath Bioregion

	Type of fire response			
Lifeform	Sprouting	Seeding	Individual	Species
Conifer	None	Fire stimulated (seed release)	Killed	Knobcone pine
	None	Fire stimulated (establishment)	Resistant/killed	Douglas-fir, ponderosa pine, Jeffrey pine
	None	None	Resistant/killed	Incense cedar, Port Orford cedar, sugar pine, western white pine, red fir, white fir, western juniper
	None	None	Killed	Brewer's spruce, lodgepole pine
Hardwood	Fire stimulated	None	Top-killed/survive	Big-leaf maple, tanoak, canyon live oak, Pacific dogwood, white alder, Oregon ash, western birch
	Fire stimulated	None	Resistant/top-killed/survive	California black oak, blue oak, Pacific madrone, giant chinquapin
	Fire stimulated	Stimulated (establishment)	Resistant/top-killed/survive	Oregon oak
	None	None	Killed	Curl-leaf mountain-mahogany
Shrub	Fire stimulated	Stimulated (germination)	Top-killed/survive	Tobacco brush, greenleaf manzanita, mahala mat
	Fire stimulated	None	Top-killed/survive	Bush chinquapin, shrub tanoak, huckleberry oak, California buckeye, wild mock orange, vine maple, mountain maple

the zone varies from approximately 600 m (2,000 ft) in the west to ~1,300 m (4,250 ft) in the eastern portion of the range (Sawyer and Thornburgh 1977). The fire regime information for vegetation alliances common in this part of the bioregion is summarized in Appendix 1.

FIRE RESPONSES OF IMPORTANT SPECIES

Fire responses for important species in the mid- to upper montane ecological zone are presented in Table 11.4.

White fir has thin bark when young, but its bark is not shed and thickens with age, making it more fire-tolerant when mature. Shasta red fir is similar but appears to be more sensitive than white fir at all ages. Knobcone pine, a serotinous cone pine with relatively thin bark, is common in the Klamath Mountains in areas that tend to burn intensely.

Port Orford cedar (POC) (*Chamaecyparis lawsoniana*), commonly associated with mesic conditions on soils derived from ultramafic material, is found in two disjunct areas of the bioregion. The largest stands of POC occur in the western Klamath Mountains, especially in the Siskiyous. Inland, POC stands are primarily found in riparian settings in the Trinity Pluton ultramafic formation (Jimerson et al. 1999) mostly in the Trinity and Sacramento River watersheds. POC stands often include trees over 300 years of age with open, charred wounds indicating they survived low-moderate-intensity surface fires.

Stands dominated by Jeffrey pine are found primarily on soils derived from ultramafic rock and they occur from the lower montane through the subalpine zones (Sawyer and Thornburgh 1977). Incense cedar is a common associate with huckleberry oak (*Quercus vaccinifolia*) and California coffeeberry (*Frangula californica*) common understory shrubs. Jeffrey pine is similar to ponderosa pine in that it develops thick bark relatively early in life rendering it resistant to most low- and moderate-intensity fires. Incense cedar becomes very resistant to low- and moderate-intensity fires as it approaches maturity due to thick bark and high crowns. Incense cedar has also been found to withstand high levels of crown scorch (Stephens and Finney 2002).

Important shrub-dominated alliances of the upper montane Klamath Mountains are greenleaf manzanita, deer brush, tobacco brush, and huckleberry oak. All of the dominant shrubs in these alliances sprout vigorously following fire. Moreover, manzanitas (*Arctostaphylos* spp.) and most California-lilacs (*Ceanothus* spp.) also establish after fire from long-lived seeds stored in soil seed banks (Knapp et al. 2012).

FIRE REGIME–PLANT COMMUNITY INTERACTIONS

Most information on fire regimes and fire effects in white fir forests comes from the edges of the Klamath Mountains. On the western edge, Stuart and Salazar (2000) found median FRIs of 40 years in the white fir alliances, and shorter 26- and 15-year median intervals where white fir was found in the Douglas-fir and incense cedar associations. Atzet and Martin (1992) reported a 25-year FRI for white fir, and Agee (1991) found a range from 43 years to 64 years from dry to moist white fir forest in the Siskiyous. Thornburgh (1995) reported a 29-year FRI for the centrally located Marble Mountains.

Before the fire-suppression era, the modal severity of most fires was not high, due to the fire tolerance of mature white fir and generally low to moderate fire intensities. Generally, more fires are dated from fire scars than from fire-initiated cohorts of regeneration (Taylor and Skinner 1998, Stuart and Salazar 2000), allowing inference that most fires were underburns. The natural forest structure is patchy, and this structure was maintained by fire. Areas that burn with high severity usually are young stands of pure white fir, open stands of white fir with a shrub understory, or montane chaparral that may contain a few white fir (Thornburgh 1995) and these patches create coarse scale heterogeneity.

Our understanding of the frequency and extent of high-severity fire and its role in stand and landscape dynamics in the white fir zone is limited. Fire likely interacted with wind to influence dead fuel accumulations. In the Klamaths and in the Cascade Range, white fir stand structure is often all-aged and sites often have pit and mound topography created by windthrow, suggesting that wind is an important disturbance that creates gaps (Agee 1991, Taylor and Halpern 1991, Taylor and Skinner 2003). Moreover, at these higher elevations winter snowfall is common, and, when followed by high winds, can cause substantial snapping of treetops. Wind in 1996 generated stem snap and windthrow that created the high fuel accumulations producing the higher-severity burn patterns in the 1999 Megram fire (Jimerson and Jones 2003). The degree to which higher stand densities and surface fuel accumulations due to fire exclusion stimulated this synergistic effect is not clear.

Forest density in white fir forests has tended to increase with fire suppression, and the shade-tolerant white fir generally shows the largest increases (Stuart and Salazar 2000). Fire-tolerant species such as ponderosa pine, sugar pine, and California black oak are declining and this should continue as long as fire exclusion is effective.

In the Klamath Mountains, fire history has been documented in only one Shasta red fir stand near Mumbo Lakes (Whitlock et al. 2004). The composite median FRI was 10 years and the median FRI for individual trees ranged from 9 years to 30 years. No fires were detected after 1901. Though Shasta red fir is common throughout this bioregion in upper montane and subalpine environments (Sawyer and Thornburgh 1977), Shasta red fir has been studied most extensively in the Cascades so more detail is presented in chapter 12.

Though Port Orford cedar is resistant to low- and moderate-intensity fires, extensive mortality to POC has occurred in stands where high-intensity fires have burned in recent years. For example, high mortality in the No Mans Creek drainage (a proposed Research Natural Area) from the Bear Fire (1994) led to concern that POC would disappear locally due to lack of seed source (Creasy and Williams 1994). Since POC stands are located on mesic, riparian sites and their wood is highly resistant to decay, these stands can produce heavy fuel loadings, especially following unusually long periods without fire. When such areas finally burn in inevitably dry years, high-intensity burns should be expected with accompanying mortality of the Port Orford cedars.

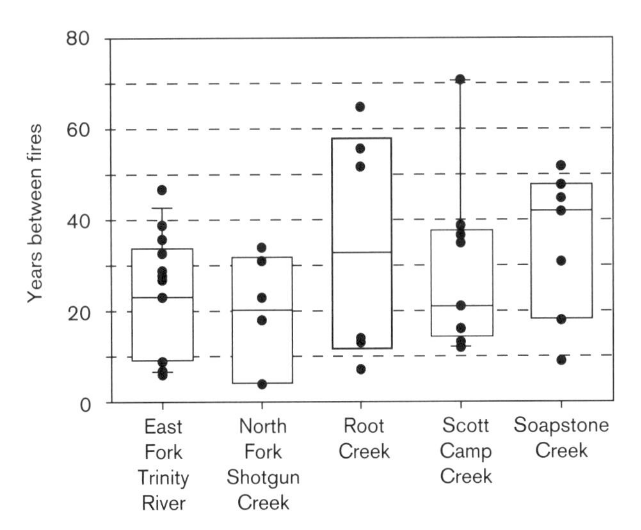

FIGURE 11.9 Distribution of FRIs in inland stands of Port Orford cedar (Skinner 2003a).

The previous discussion on the influence of riparian areas on Klamath Mountain fire regimes was based on information from inland sites in the Trinity and Sacramento River watersheds dominated by POC. These fire-scar data for the inland POC stands indicate that fires burned with median FRIs of 16 years to 42 years (Fig. 11.9). In each case, the median FRI in the POC stands was at least twice that of forests in the surrounding uplands (Skinner 2003b). Though there was considerable variation in length of fire intervals, the time since the last fire (the fire-suppression period) exceeds the longest interval previously recorded in each of the sampled stands.

Knobcone pine in this bioregion exhibits a unique bimodal geographical distribution in higher and lower elevation areas that experience high-intensity fires. As in other bioregions, knobcone pine is found at low elevations intermixed with chaparral as around Lake Shasta and Whiskeytown Lake reservoirs. However, in this bioregion, knobcone pine is also found in the upper montane zone on upper slopes and ridgetop positions, especially on south- and west-facing slopes that tend to burn more severely than the surrounding landscape. An example can be seen in Fig. 11.8 where much of the severely burned upper slope is occupied by knobcone pine.

Jeffrey pine forests are generally thought to have fire regimes similar to ponderosa pine forests—frequent, low-moderate-intensity fires—with more variability (Skinner and Chang 1996, Taylor 2000, Stephens et al. 2003). This variation is probably due to the combination of nutrient poor soils and shorter growing seasons, especially at higher elevation, that increases variability in fuel production compared to typical ponderosa pine sites.

Fire-scar data from ultramafic sites with Jeffrey pine are available from seven sites in the vicinity of Mt. Eddy (Skinner 2003a) and 10 sites near Hayfork (Taylor and Skinner 2003). The median FRI for the sites near Mt. Eddy ranged from 8 years to 30 years and 8 years to 15 years for those near Hayfork. For the sites near Mt. Eddy, the current fire-free periods of 57 to 129 years exceed the 95th percentile prefire-suppression fire-free interval for five of seven sampled stands. At

Hayfork, the current fire-free periods of 68 to 125 years exceed the longest prefire-suppression intervals recorded on all 10 sites.

The fire sensitive species in the region—Brewer spruce, Engelmann spruce, subalpine fir—are at risk of being extirpated from their scattered locations by high-severity fire. For example, the entire Rock Creek Butte Research Natural Area (RCBRNA) burned severely in one of the large lightning-caused fires of 2008. This resulted in all Brewer spruce, the target species of the research natural area (Cheng 2004), being killed by the fire leaving no local seed source. This effectively extirpated Brewer spruce from RCBRNA due to inability to regenerate (Cole et al. 2010). Brewer spruce appears to be able to regenerate under a tree overstory but may have problems in open conditions especially after severe fires. Thus, a warming climate may work to make it more difficult for young trees to become established in open patches of severe burn even when there is a local seed source nearby (Ledig et al. 2012).

The occurrence of montane shrub stands may be associated with either edaphic conditions unsuitable for tree growth or high-intensity fires. Once established, because of the nature of shrub fuels, fires that burn in these communities are more likely to be high-intensity events. Thus, where shrub communities become established, recurring fire plays a key role in the maintenance of these communities by inhibiting succession from shrubs to trees (Nagel and Taylor 2005, Lauvaux et al. 2016). More information on fire regimes of shrub-dominated alliances can be found in chapter 12.

The only known herbaceous alliance with fire ecology information in the bioregion is the California pitcher plant (*Darlingtonia californica*). Pitcher plant seeps are common in open habitats saturated with flowing water, usually on ultramafic substrate (Sawyer and Keeler-Wolf 1995). The continuous presence of flowing water through these herbaceous communities would seem to limit opportunities for fires and little is known of their fire ecology (Crane 1990). However, Port Orford cedars, incense cedars, western white pines, or Jeffrey pines with charred catfaces are commonly scattered in and adjacent to the seeps.

A prescribed burn in September 1997 in the Cedar Log Flat Research Natural Area on the Siskiyou National Forest was found to spread easily through dead herbaceous material in sedge and pitcher plant seeps under conditions easily achieved, or exceeded, in the summer where pitcher plant is found. Only limited effects of the burn were detected three years post-burn (Borgias et al. 2001).

Fire histories from fire-scarred trees in pitcher plant seeps have been documented in two tributaries to the Sacramento River. Median fire return intervals for these sites were 18 and 42 years, respectively (sites 2 and 5 in Fig. 11.9) (Skinner 2003b). Differences in the length of FRI in these seeps are probably related to conditions in the surrounding forests. One is located in the upper third of a steep, southeast-facing slope surrounded by mixed stands of Jeffrey pine, white fir, incense cedar, sugar pine, and Douglas-fir. Consequently, it would be expected to have shorter FRIs. The other is near the bottom of a u-shaped canyon on a gentle slope surrounded by mixed stands of Shasta red fir, white fir, western white pine, Jeffrey pine, sugar pine, and Douglas-fir. Thus, it would be expected to have longer FRIs. Port Orford cedar is the most common tree on both sites, however.

Fires burning in these environments probably occur very late in the season when water is low, in very dry years, or possibly after an early frost has killed much of the herbaceous material aboveground as in Fig. 11.10.

FIGURE 11.10 Pitcher plant seep in September following an early hard frost in the Scott Camp Creek watershed near Castle Lake (photo credits: Carl Skinner, USDA Forest Service).

Subalpine

FIRE RESPONSES OF IMPORTANT SPECIES

Tree species in the subalpine zone including mountain hemlock, Shasta red fir, whitebark pine, western white pine, foxtail pine, lodgepole pine, and curl-leaf mountain-mahogany have thinner bark than species of lower elevations and are easily damaged or killed by moderate-intensity fire or the consumption of heavy surface fuels at the base of the tree (Table 11.5). Appendix 1 summarizes the fire regime information for vegetation discussed in the subalpine ecological zone.

FIRE REGIME–PLANT COMMUNITY INTERACTIONS

Landscapes of this zone are a heterogeneous mosaic of stands, rock outcrops, talus, morainal lakes, and riparian areas so fuels are discontinuous. Moreover, deep snow packs generally persist into late June or July so the fire season is very short. Fuelbeds from the short-needled species are compact and promote slow spreading, mostly smoldering surface fires. Fuel buildup tends to be slow because of the short growing season. Higher-intensity fires that burn in subalpine forests primarily occur in areas of locally heavy fuel accumulations during periods of extreme fire weather.

The only fire history data for the subalpine zone in the Klamath Mountains are from stands on China Mountain (Mohr et al. 2000, Skinner 2003a). Species present in these stands are mountain hemlock, Shasta red fir, whitebark pine, western white pine, foxtail pine, and lodgepole pine.

Fire-scar samples were collected from 14 trees on three 1 ha (2.5 ac) sites in the Crater Creek watershed. Over the period spanned by the fire-scar record (1404 to 1941), the median fire return intervals for these sites were 11.5, 12, and 13 years. However, 44 of 51 fires were detected on only single trees.

TABLE 11.5
Fire response of important species in the subalpine zone of the Klamath Bioregion

	Type of fire response			
Lifeform	Sprouting	Seeding	Individual	Species
Conifer	None	None	Resistant/killed	Red fir, mountain hemlock, Jeffrey pine, foxtail pine, western white pine, whitebark pine
	None	None	Killed	Lodgepole pine
Hardwood	None	None	Killed	Curl-leaf mountain-mahogany

Thus, fires in this subalpine basin were mainly low intensity and small. Ranges of individual-tree median FRIs were 9 years to 276 years with a grand median of 24.5 years. No fires were detected after 1941.

Management Issues

Managers face several fire-related challenges in the Klamath Mountains with several issues—wildlife habitat, wildland-urban interface, and smoke—standing out in this bioregion.

Wildlife Habitat

Management objectives often include the desire to maintain resilient forest ecosystems similar to that of their Historic Range of Variability (HRV) (e.g., Swanson et al. 1994, Long et al. 2014a) using ecological processes (FEMAT 1993, USDA-USDI 1994, Long et al. 2014b) in order to sustain a mix of desirable wildlife habitats. Recent studies suggest that vegetation patterns and conditions generated by prefire-suppression fire regimes (Taylor and Skinner 1998, 2003) may be advantageous for wildlife species of concern such as the northern spotted owl (Franklin et al. 2000, Agee 2007) and several species of butterflies (Huntzinger 2003). Of all management activities, fire suppression alone has been ubiquitously applied throughout the bioregion. Consequently, there is a need to better understand the role of frequent low- and moderate-severity and mixed-severity fires on development of the forest landscape mosaic (Agee 1998, 2003, Perry et al. 2011). This understanding will help managers better assess risks associated with different management alternatives (e.g., Agee 2003, Hessburg et al. 2016).

More recent management activities such as logging, replacement of multiaged old-growth forests with even-aged forest plantations, and continued fire suppression have reduced forest heterogeneity, increased the proportion of even-aged forests, and altered habitat conditions for forest-dwelling species compared to the prefire-suppression landscape (USDA-USDI 1994). Large wildfires with large patches of high severity have burned in the Klamath Mountains in the last several decades (1977, 1987, 1995, 1996, 2002, 2008, 2012, 2014, 2017). These fires have reduced the extent, in some places dramatically, of multiaged, old-growth stands. Indeed, wildfire, among all causes, is responsible for the largest proportion of loss of nesting and roosting habitat for the northern spotted owl, especially in reserves, since implementation of the Northwest Forest Plan (Davis and Dugger 2011). Areas burned by these fires become occupied by plantations, even-aged hardwood stands, or brushfields, and, in some watersheds (e.g., north and south forks of the Salmon River), these vegetation types are now the landscape matrix. Moreover, some large areas that burned intensely in 1977 burned intensely again in 1987, while large areas that burned intensely in 1987 did so again in 2008 (Perry et al. 2011).

The increasing size of high-severity patches results in reduced structural heterogeneity across landscapes. Positive feedbacks between management (i.e., fire suppression, plantations), stand conditions in the new even-aged vegetation matrix, and intense fire have the potential to initiate and expand persistent broad scale changes from forest to shrubland (Coppoletta et al. 2016, Lauvaux et al. 2016).

Forests in the lower- and mid-montane zone that historically burned at low to moderate severity are now particularly susceptible to high-severity fire because of high surface and canopy fuel loads from fire suppression (Taylor and Skinner 2003). A shift to larger patches of high-severity fire can induce a vegetation switch if fire effects or climate change favor postfire establishment of different species (Collins and Roller 2013, Lauvaux et al. 2016). High fire severity in lower and mid-montane forests initiates a period of dominance by fire-dependent shrubs (montane chaparral) that establish from sprouts or buried seedbank (Knapp et al. 2012). Montane chaparral impedes tree seedling establishment and growth, and shrubs can dominate severely burned sites capable of supporting trees for decades or even a century or more (Coppoletta et al. 2016, Lauvaux et al. 2016). Consequently, postfire rates of forest development following severe fire can be slow and may be further exacerbated by interactions with subsequent fire.

Accumulating evidence indicates that fire initiated montane chaparral exhibits self-reinforcing fire behavior with shrublands often burning severely again in subsequent fires (Perry et al. 2011, Lauvaux et al. 2016). Similar self-reinforcing severe fire effects have been observed where plantations were established in the footprint of severe fires to facilitate rapid forest development and then burned severely again in subsequent fires before reaching maturity (Thompson et al. 2007, Thompson and Spies 2010). Old shrubfields that converted to forest during the fire-suppression period have also been observed to burn severely again suggesting that shrubfields may impart a long-term ecological memory of fire effects in a landscape (Taylor et al. 2013, Coppoletta et al., 2016, Lauvaux et al. 2016).

A warming climate with more extreme fire weather (e.g., Lenihan et al. 2008, Collins 2014) and large areas burning with high severity may increase the likelihood of vegetation shifts. Interactions between fire, vegetation, and climate change could lead to a "landscape trap" where a vegetation switch and with positive feedback processes maintain the new vegetation regime across entire landscapes (Lindenmayer et al. 2011, Coppoletta et al. 2016, Lauvaux et al. 2016). This would have significant implications for the type and heterogeneity of habitat for forest-dwelling species in these ecosystems.

Wildland-Urban Interface

With an average of less than 1.2 people km^{-2} (3 people mi^{-2}), the Klamath Mountains have a low human population compared with California as a whole (USCB 2002). Yet, a large proportion of the bioregion is classified as mixed interface (CDF 2002a) because of the dispersed nature of dwellings in small, scattered communities in flammable, wildland vegetation. As a result, hundreds of homes have been lost to wildfires that originated in the bioregion in just the last several decades (CDF 2002b). Examples of major suppression efforts in WUIs in the bioregion include fires near Hayfork and Happy Camp (1987), Redding and Lakehead (1999), Weaverville (2001), Jones Valley and French Gulch (2004), and Orleans (2008, 2013). The Jones fire (1999) alone burned over 900 structures (CDF 2002b) including nearly 200 homes in and around Redding. The fire problem at the wildland-urban interface will continue to grow as more people move into low-density housing at the edges of communities throughout the bioregion.

Smoke

Most people living in the Klamath Mountains reside in the lower reaches of canyons where smoke collects under the thermal inversions. The periodic outbreaks of widespread fires in this bioregion create profound smoke episodes that can go on for weeks to months leading to health problems for susceptible people (Mott et al. 2002) and impairment of visibility inconveniencing tourists and businesses that rely on tourism. As a result of these effects, smoke is regulated and managed as primarily a human health hazard and secondarily as a nuisance. Smoke is discussed in this regard in chapter 23.

Historically, the presence of smoke was pervasive though varying from year to year and the clear vistas expected and desired today are largely an artifact of fire suppression. Yet, smoke is rarely discussed in a context of being part of the ecological process of fire. Further, though it is generally recognized that there is a fire deficit today, this is usually not accompanied by a recognition of a corresponding smoke deficit (although see Stephens et al. 2007, Lake and Long 2014). Smoke has always been a major part of the ecosystem in the productive, summer dry forests of this bioregion. The continuous record of charcoal from lake sediment data attests to the persistent presence of smoke (Whitlock et al. 2004, Briles et al. 2011). Indeed, before fire suppression became the rule, rather than having clear vistas of the mountains in the summer, C. Hart Merriam, chief of the Biological Survey, noted in 1899, ". . . few see more than the immediate foreground and a haze of smoke that even the strongest glass cannot penetrate (Agee 2007)." If "pristine" means original and unaltered then pristine summer skies would have been characterized by considerable smoke and haze from the many fires that would be burning in most years (Stephens et al. 2007).

Rarely discussed are the benefits of smoke. Though a number of benefits have been noted (Lake and Long 2014), we limit our discussion to the potential affect of smoke on maintenance of cooler stream water temperature (Mahlum et al. 2011, Lake and Long 2014).

Stream temperature measurements acquired during ongoing fires in the Klamath Mountains (F. Lake, Forest Service, Pacific Southwest Research Station, Redding, California, USA, pers. comm.) and in the Rocky Mountains (Mahlum et al. 2011) often remain either unchanged or lower than in areas outside of the influence of smoke. It appears the cooling under the thermal inversions not only dampens fire behavior reducing fire severity to vegetation, but may also promote cooler stream water temperatures. It is hypothesized that before fire suppression, smoke trapped in canyons would be a common occurrence, especially in late summer during warm, dry years. Typically streamflow would be low during these warm/dry periods and in the absence of smoke, water temperatures could become elevated and stressful to cool water fish and other aquatic life. Having smoke a common occurrence during these periods would help to keep streamwater temperature lower and less stressful to aquatic life. It may be too that reduction of solar radiation and cooling would lower the need for water by vegetation, reducing transpiration, leaving more water available for streamflow.

Because the degree of stream warming in a landscape following fire is largely due to the level and pattern of fire severity across the watershed (Gresswell 1999, Rieman et al. 2003), we hypothesize that a more indirect, yet longer lasting effect of smoke dampening fire behavior would likely be less area burned severely resulting in more vegetative shading along riparian areas during periods without smoke. Halofsky and Hibbs (2008) found that the fire severity in the riparian zone along fish bearing streams, measured as basal area mortality, was most strongly associated with fire severity in adjacent uplands. Amaranthus et al. (1989) found that though some headwater riparian areas that burned severely had considerable water temperature increases after the Silver fire (1987), less than 5% of headwater riparian zones in their study area had burned severely, while considerable shading remained along lower reaches with net effect of maintaining cool temperatures in the lower reaches. From this it is reasonable to surmise that smoke, by reducing fire intensity and resulting fire severity on the lower slopes of canyons in presuppression times, helped to promote cooler stream temperatures both during and indirectly long after fires and continues to do so today.

Future Directions

There is a critical need to better understand the synergistic relationships between low-, moderate-, and high-intensity fire and prefire-suppression vegetation patterns. There is a particular need for quantitative estimates of the proportion of landscapes in different stand types (i.e., old-growth, young even-aged, hardwood, etc.) and how they were patterned on the landscape to provide a stronger foundation for applying concepts of historical range of variability to forest management. There is great potential for fire and landscape ecologists to work with wildlife ecologists to examine wildlife responses to landscape dynamics across a range of spatial and temporal scales.

Hardwoods, especially oaks, provide important habitat elements for many species of wildlife. As a result, managers may use prescribed fire to inhibit conifer encroachment into oak stands as well as to improve acorn crops (Long et al. 2016). Oak woodlands have also been associated with rich vegetation diversity. Yet, their ecology is little studied in these montane forest environments. We need to better understand the ramifications of the potential loss of large areas of hardwoods to conifers for associated vegetative diversity and wildlife habitat.

The temporal and spatial dynamics of large, dead woody material in areas where presuppression fire regimes were characterized by frequent, low-mixed-severity fires is not known but it is probably very different than those identified by current standards and guidelines used by both federal and state agencies. Current standards and guidelines were generally developed from contemporary old-growth forests that had experienced many decades of fire suppression. These quantities of woody material were probably unusually high compared to typical prefire-suppression values. Consequently, a management emphasis on meeting or exceeding current standards and guidelines for dead woody material has and will increase fire hazard over time and threatens the very habitat the standards and guidelines were designed to protect (Skinner 2002, Knapp 2015).

Managing Wildfire

In 2009, recognizing the increasing wildland fire problem, Congress enacted the FLAME Act, which mandated a national cohesive wildland fire management strategy. Its vision was to safely and effectively extinguish fire, when needed; use fire where allowable; manage our natural resources, and as a nation, live with wildland fire (Wildland Fire Leadership Council 2014). Implementing this vision in the Klamath Mountains requires a comprehensive change in current policy. Indeed, current policies relying primarily on fuels treatments and fire suppression will continue to create an increasing backlog of area outside of historical range of variation (North et al. 2012). Large wildfires that burn for weeks to months dominated by mostly low- and moderate-severity effects have become common over the last several decades in this bioregion. The results of these long-burning fires demonstrate that the bioregion is well set up to benefit from a planned, purposeful use of managed wildfire. We believe this can be successfully accomplished by pursuing two general types of strategies: strategies for remote backcountry areas and strategies for front country areas. They are linked because the success or failure of one will affect the other.

In backcountry areas, including wilderness, more active management of wildland fire, not simply fire suppression, would better support achieving wilderness and broader management goals. In many wilderness areas, annual area burned by natural fires remains far below historical estimates (Parsons and Landres 1998). This is true even in the remote and rugged Klamath Mountains (Miller et al. 2012a), and in the face of large fires of the past few decades, where fire-suppression efforts involve long-term commitments of resources, overhead teams, and expenditures (Lewis 2006). Aggressive fire suppression can also compromise firefighter safety, as happened in 2008 on the Iron 44 fire with a helicopter crash that killed nine people and occurred 15 miles into the Trinity Alps wilderness with no structures at risk. Consideration needs to be given to managing all wildland fire, both natural and human-caused, with an appropriate management response. An example of this was the 2005 Woolley fire in the Marble Mountain wilderness. The source of ignition could not be confirmed, but it was managed primarily with monitoring of fire behavior and effects, with minimal suppression action (Lewis 2006).

The probability of success for this strategy depends on a much more aggressive fuel reduction strategy in and around communities. This would provide for firesafe communities that could better coexist with fire in the broader landscape. Due to the complexity of ownership and responsibility for fire protection, partnerships involving local, tribal, state, and Federal entities are needed to plan the appropriate mix of fuel reduction strategies (Harling 2014a). Leadership can come from local nonprofit groups such as the Watershed Research and Training Center, located in Hayfork, the Mid-Klamath Watershed Council located in Orleans, or national groups such as The Nature Conservancy. Prescribed fire can play a much larger role than it has in the past, but locally it faces the same impediments as in most of northern California: narrow burn windows, regulations, lack of adequate personnel, and environmental laws (Quinn-Davidson and Varner 2012). On the Salmon River, where prescribed fire has been used to create firesafe zones around the community, local residents are now advocating for backcountry wildfires to burn around them to clear away fuels created by earlier burns and reduce the threat of future fires moving into the community (Harling 2014b). Where no structures are present, prescribed fire can be used to manage fuels at the boundary of the front and backcountry (Agee 1996).

Mechanical and manual fuel reduction can be used where topography is gentle and access is present. Near Weaverville, understory clearing along Highway 3 has created a linear zone where wildfire behavior will be reduced, making suppression efforts more effective. Such treatments can be followed by light prescribed fires to further reduce remaining undesired residues and for maintenance. A new biomass cogeneration facility in Weaverville will have the capability to accept clean thinning residues created from fuel reduction treatments, although its 5 megawatt output will primarily be fueled by mill residues (North State Resources 2014). Utilizing a cogeneration facility will reduce the impacts created by smoke if such residues had to be pile burned.

The lessons of the last several decades indicate that the complex, rugged, and often remote terrain of the bioregion precludes treatment of sufficient area with either mechanical treatments or prescribed fire, either alone or in combination, without incorporating managing wildfires to accomplish goals of creating firesafe communities while restoring fire resilient forest landscapes (North et al. 2012, 2014).

The Warming Climate

Long-term fire histories suggest that fire has been an important part of Klamath ecosystems since their evolution, and that climate has been the primary driver of fire occurrence, both directly and through changes in vegetation. At millennial scales severe fires are associated with periods of fir-dominated forest in high-elevation locations, but fire frequency was highest during times when the forest structure was relatively open and conditions warmer and drier than today. Another observation is that while fire history is similar on ultramafic and nonultramafic sites, the consequences were quite different. Ultramafic sites have shown little change in vegetation over the last 10,000 years with only subtle shifts in pine and oak, whereas nonultramafic sites showed more dynamic responses. Conservation planners and resource managers should consider that nonultramafic regions may likely experience far greater change in vegetation with projected climate change than ultramafic sites, despite the high level of endemism in the latter ecosystems. Future shifts on nonultramafic substrates will be hindered by the altered landscapes and in some cases human intervention may be needed to assure plant establishment in the new environment. Economically important species, such as Douglas-fir, and species that are at their ecological limits, such as mountain hemlock, or endemics, such as Brewers spruce, will require special monitoring under warmer future conditions.

Summary

Primarily due to the annual summer drought and ample winter precipitation, fires were historically frequent and generally of low-moderate and mixed severity in most vegetation assemblages, especially those that cover large portions of the bioregion. Fire exclusion and other management activities have led to considerable changes in Klamath Mountain ecosystems over the last century. Of all management activities that have contributed to altered ecosystems in the bioregion, fire suppression has been the most pervasive since it alone has been ubiquitously applied. Though there is much current discussion of the need for restoring fire as an ecological process, or at least creating stand structures that would help reduce the general intensity of fires to more historical levels, there are many competing social/political concerns and objectives (for example, fine filter approaches to managing wildlife habitat and air quality) that make doing anything problematic (Agee 2003). Regardless of how these controversies are resolved, the ecosystems of the Klamath Mountains will continue to change in response to climate and social/political choices for the use of forest resources and their associated fire management alternatives.

References

Agee, J.K. 1991. Fire history along an elevational gradient in the Siskiyou Mountains, Oregon. Northwest Science 65: 188–199.

———. 1993. Fire Ecology of Pacific Northwest Forests. Island Press, Washington, D.C., USA.

———. 1996. Alternatives for implementing fire policy. Pages 107–112 in: J.K. Brown, R.W. Mutch, C.W. Spoon, and R.H. Wakimoto, technical coordinators. Proceedings: Symposium on Fire in Wilderness and Park Management. USDA Forest Service General Technical Report INT-GTR-320. Intermountain Forest and Range Experiment Station, Ogden, Utah, USA.

———. 1998. The landscape ecology of western forest fire regimes. Northwest Science 72: 24–34.

———. 2003. Burning issues in fire: will we let the coarse-filter operate? Tall Timbers Research Station Miscellaneous Publication 13: 7–13.

———. 2007. The Steward's Fork: A Sustainable Future for the Klamath Mountains. University of California Press, Berkeley, California, USA.

Alder, J.R., and S.W. Hostetler. 2015. Global climate simulations at 3000-year intervals for the last 21,000 years with the GENMOM coupled atmosphere-ocean model. Climate of the Past 11: 449–471.

Amaranthus, M., H. Jubas, and D. Arthur. 1989. Stream shading, summer streamflow and maximum water temperature following intense wildfire in headwater streams. Pages 75–87 in: N.H. Berg, technical coordinator. Proceedings of the Symposium on Fire and Watershed Management. USDA Forest Service General Technical Report PSW-GTR-109. Pacific Southwest Research Station, Berkeley, California, USA.

Atzet, T., and R. Martin. 1992. Natural disturbance regimes in the Klamath Province. Pages 40–48 in: R.R. Harris, D.C. Erman, and H.M. Kerner, editors. Symposium on Biodiversity of Northwestern California. Wildland Resources Center Report 29. University of California, Berkeley, California, USA.

Barron, J.A., L. Heusser, T. Herbert, and M. Lyle. 2003. High-resolution climatic evolution of coastal northern California during the past 16,000 years. Paleoceanography 18(1): doi:10.1029/2002PA000768.

Bartlein, P.J., K.H. Anderson, P.M. Anderson, M.E. Edwards, C.J. Mock, R.S. Thompson, R.S. Webb, T. Webb, III, and C. Whitlock. 1998. Paleoclimate simulations for North America over the past 21,000 years: features of the simulated climate and comparisons with paleoenvironmental data. Quaternary Science Reviews 17: 549–585.

Borgias, D., R. Huddleston, and N. Rudd. 2001. Third year post-fire vegetation response in serpentine savanna and fen communities, Cedar Log Flat Research Natural Area, Siskiyou National Forest. Unpublished Report Agreement 00-11061100-010, The Nature Conservancy. On file at the Siskiyou National Forest, Grants Pass, Oregon, USA.

Briles, C.E., C. Whitlock, and P.J. Bartlein. 2005. Postglacial vegetation, fire, and climate history of the Siskiyou Mountains, Oregon, USA. Quaternary Research 64: 44–56.

Briles, C.E., C. Whitlock, P.J. Bartlein, and P. Higuera. 2008. Regional and local controls on postglacial vegetation and fire in the Siskiyou Mountains, northern California, USA. Palaeogeography, Palaeoclimatology, Palaeoecology 265: 159–169.

Briles, C.E., C. Whitlock, C.N. Skinner, and J. Mohr. 2011. Holocene forest development and maintenance on different substrates in the Klamath Mountains, northern California, USA. Ecology 92: 590–601.

Brown, P.M. 2006. Climate effects on fire regimes and tree recruitment in Black Hills ponderosa pine forests. Ecology 87: 2500–2510.

Brown, P.M., C.L. Wiene, and A.J. Symsted. 2008. Fire and forest history at Mount Rushmore. Ecological Applications 18: 1984–1999.

CCSS. 2002. Historical Course Data. California Resources Agency, Department of Water Resources, Division of Flood Management, California Cooperative Snow Surveys. http://cdec.water.ca.gov/snow/.

CDF. 2002a. Historical Statistics in Fire and Emergency Response. California Department of Forestry and Fire Protection, Sacramento. http:/www.fire.ca.gov/FireEmergencyResponse/HistoricalStatistics/HistoricalStatistics.asp.

———. 2002b. Information and data center. In: Fire and Resource Assessment Program. California Department of Forestry and Fire Protection, Sacramento. http://cdfdata.fire.ca.gov/incidents/incidents_statsevents.

Cheng, S. 2004. Forest service research natural areas in California. USDA Forest Service General Technical Report PSW-GTR-188. Pacific Southwest Research Station, Albany, California, USA.

Cocking, M.I., J.M. Varner, and R.L. Sherriff. 2012. California black oak responses to fire severity and native conifer encroachment in the Klamath Mountains. Forest Ecology and Management 270: 25–34.

Cole, D.N., C.I. Millar, and N.L. Stephenson. 2010. Responding to climate change: a toolbox of management strategies. Pages 179–196 in: D.N. Cole and L. Yung, editors. Beyond naturalness: rethinking park and wilderness stewardship in an era of rapid climate change. Island Press, Washington, D.C., USA.

Collins, B.M. 2014. Fire weather and large fire potential in the northern Sierra Nevada. Agricultural and Forest Meteorology 189–190: 30–35.

Collins, B.M., and G.B. Roller. 2013. Early forest dynamics in stand-replacing fire patches in the northern Sierra Nevada, California, USA. Landscape Ecology 28: 1801–1813.

Coppoletta, M., K.E. Merriam, and B.M. Collins. 2016. Post-fire vegetation and fuel development influences fire severity patterns in reburns. Ecological Applications 26: 686–699.

Crane, M.F. 1990. *Darlingtonia californica*. In: Fire Effects Information System. U.S. Department of Agriculture, Forest Service, Rocky

Mountain Research Station, Fire Sciences Laboratory (Producer). http://www.fs.fed.us/database/feis/. Accessed January 20, 2016.
Creasy, M., and B. Williams. 1994. Bear fire botanical resources report. In: Dillon Complex Burn Area Rehabilitation Report. USDA Forest Service, Klamath National Forest, Yreka, California, USA.
Daniels, M.L., R.S. Anderson, and C. Whitlock. 2005. Vegetation and fire history since the late Pleistocene from the Trinity Mountains, northwestern California, USA. The Holocene 15: 1062–1071.
Davis, R.J., and K.M. Dugger. 2011. Habitat status and trends. Pages 21–61 in: R.J. Davis, K.M. Dugger, S. Mohoric, L. Evers, and W.C. Aney, editors. Northwest Forest Plan—The First 15 Years (1994–2008): Status and Trend of Northern Spotted Owl Populations and Habitats. USDA Forest Service General Technical Report PNW-GTR-850. Pacific Northwest Research Station, Portland, Oregon, USA.
Dettinger, M.D., D.R. Cayan, H.F. Diaz, and D.M. Meko. 1998. North-south precipitation patterns in western North America on interannual-to-decadal time scales. Journal of Climate 11: 3095–3111.
Engber, E.A., and J.M. Varner. 2012. Patterns of flammability of the California oaks: the role of leaf traits. Canadian Journal of Forest Research 42: 1965–1975.
Estes, B.L., E.E. Knapp, C.N. Skinner, and J.D. Miller. 2017. Factors influencing fire severity under moderate burning conditions in the Klamath Mountains, northern California, USA. Ecosphere 8(5): e01794.
FEMAT. 1993. Forest ecosystem management: an ecological, economic, and social assessment. Report of Forest Ecosystem Management Assessment Team. USDA Forest Service, Portland, Oregon, USA.
Franklin, A.B., D.R. Anderson, R.J. Gutierrez, and K.P. Burnham. 2000. Climate, habitat quality, and fitness in northern spotted owl populations in northwestern California. Ecological Monographs 70: 539–590.
Fry, D.L., and S.L. Stephens. 2006. Influence of humans and climate on the fire history of a ponderosa pine-mixed conifer forest in the southeastern Klamath Mountains, California. Forest Ecology and Management 223: 428–438.
Gresswell, R.E. 1999. Fire and aquatic ecosystems in forested biomes of North America. Transactions of the American Fisheries Society 128: 193–221.
Halofsky, J.E., D.E. Donato, D.E. Hibbs, J.L. Campbell, M. Donaghy Cannon, J.B. Fontaine, J.R. Thompson, R.G. Anthony, B.T. Bormann, L.J. Kayes, B.E. Laws, D.L. Peterson, and T.A. Spies. 2011. Mixed-severity fire regimes: lessons and hypotheses from the Klamath-Siskiyou Ecoregion. Ecosphere 2(4): zrt40.
Halofsky, J.E., and D.E. Hibbs. 2008. Determinants of riparian fire severity in two Oregon fires, USA. Canadian Journal of Forest Research 38: 1959–1973.
Harling, W. 2014a. 2013 wildfires highlight advances in community and agency fire management, and where work is still needed. Mid-Klamath Watershed Council Newsletter 2014: 13–15.
———. 2014b. 2013 wildfires: a success story. Mid-Klamath Watershed Council Newsletter 2014: 10–12.
Hayasaka, H., and C.N. Skinner 2009. 2008 forest fires in northern California, USA (extended abstract). In: B.E. Potter and T.J. Brown, technical coordinators. The Eighth Symposium on Fire and Forest Meteorology, October 13–15, 2009. American Meteorological Society, Kalispell, Montana. http://ams.confex.com/ams/8Fire/techprogram/paper_155842.htm.
Hessburg, P.F., T.F. Spies, D.A. Perry, C.N. Skinner, A.H. Taylor, P.M. Brown, S.L. Stephens, A.J. Larson, D.J. Churchill, P.H. Singleton, B. McComb, W.J. Zielinski, B.M. Collins, N.A. Povak, R.B. Salter, J.J. Keane, J.F. Franklin, and G. Riegel. 2016. Management of mixed-severity fire regimes in forests in Oregon, Washington, and northern California. Forest Ecology and Management 366: 221–250.
Hoopes, C.L. 1971. Lure of Humboldt Bay Region. Kendall/Hunt Publishing, Dubuque, Iowa, USA.
Hull, M.K., C.A. O'Dell, and M.J. Schroeder. 1966. Critical fire weather patterns: their frequency and levels of fire danger. USDA Forest Service, Pacific Southwest Forest and Range Experiment Station, Berkeley, California, USA.
Huntzinger, M. 2003. Effects of fire management practices on butterfly diversity in the forested western United States. Biological Conservation 113: 1–12.
Irwin, W.P. 1966. Geology of the Klamath Mountains province. Pages 19–28 in E.H. Bailey, editor. Geology of Northern California. California Division of Mines and Geology Bulletin 190. Sacramento, California, USA.
Jackson, J. 1964. Tales from the Mountaineer. The Rotary Club of Weaverville, Weaverville, California, USA.
Jimerson, T.M., S.L. Daniel, E.A. McGee, and G. DeNitto. 1999. A field guide to Port Orford cedar plant associations in northwest California and Supplement. USDA Forest Service Technical Report R5-ECOL-TP-002. Pacific Southwest Region, Vallejo, California, USA.
Jimerson, T.M., and D.W. Jones. 2003. Megram: blowdown, wildfire, and the effects of fuel treatment. Tall Timbers Research Station, Miscellaneous Report 13: 55–59.
Keeler-Wolf, T., editor. 1990. Ecological surveys of Forest Service Research Natural Areas in California. USDA Forest Service General Technical Report PSW-GTR-125. Pacific Southwest Research Station, Berkeley, California, USA.
Knapp, E.E. 2015. Long-term dead wood changes in a Sierra Nevada mixed conifer forest: habitat and fire hazard implications. Forest Ecology and Management 330: 87–95.
Knapp, E.E., C.P. Weatherspoon, and C.N. Skinner. 2012. Shrub seed banks in mixed conifer forests of northern California and the role of fire in regulating abundance. Fire Ecology 8(1): 32–48.
Lake, F.K., and J.W. Long. 2014. Fire and tribal cultural resources. Pages 173–186 in: J.W. Long, L. Quinn-Davidson, and C.N. Skinner, editors. Science Synthesis to Support Socioecological Resilience in the Sierra Nevada and Southern Cascade Range. USDA Forest Service General Technical Report PSW-GTR-247. Pacific Southwest Research Station, Albany, California, USA.
Lanspa, K.E. n.d. Soil survey of Shasta-Trinity Forest area, California. USDA Forest Service, Pacific Southwest Region, National Cooperative Soil Survey. San Francisco, California, USA.
Lauvaux, C.A., C.N. Skinner, and A.H. Taylor. 2016. High severity fire and mixed conifer forest-chaparral dynamics in the southern Cascade Range, USA. Forest Ecology and Management 363: 74–85.
Ledig, F.T. 2012. Climate change and conservation. Acta Silvatica and Lignaria Hungarica 8: 57–74.
Ledig, F.T., P.D. Hodgskiss, and D.R. Johnson. 2005. Genic diversity, genetic structure, and mating system of Brewer spruce (*Pinaceae*), a relict of Arcto-Tertiary forest. American Journal of Botany 92: 1975–1986.
Lenihan, J.M., D. Bachelet, R.P. Neilson, and R. Drapek. 2008. Response of vegetation distribution, ecosystem productivity, and fire to climate change scenarios for California. Climatic Change 87(Suppl 1): S215–S230.
Lewis, G.E. 2006. Management action on the Woolley Fire is the appropriate one. Fire Management Today 66(4): 33–35.
Lindenmayer, D.B., R.J. Hobbs, G.E. Likens, C.J. Krebs, and S.C. Banks. 2011. Newly discovered landscape traps produce regime shifts in wet forests. Proceedings of the National Academy of Sciences of the United States of America 108: 15887–15891.
Long, J.W., M.K. Anderson, L. Quinn-Davidson, R.W. Goode, F.K. Lake, and C.N. Skinner. 2016. Restoring California black oak ecosystems to promote tribal values and wildlife. USDA Forest Service General Technical Report PSW-GTR-252. Pacific Southwest Research Station, Albany, California, USA.
Long, J.W., C. Skinner, M. North, C.T. Hunsaker, and L. Quinn-Davidson. 2014b. Integrative approaches: promoting socioecological resilience. Pages 17–54 in: J.W. Long, L. Quinn-Davidson, and C.N. Skinner, editors. Science Synthesis to Support Socioecological Resilience in the Sierra Nevada and Southern Cascade Range. USDA Forest Service General Technical Report PSW-GTR-247. Albany, CA, USA.
Long, J.W., C. Skinner, H. Safford, and S.L. Charnley. 2014a. Introduction. Pages 3–16 in J.W. Long, L. Quinn-Davidson, and C.N. Skinner, editors. Science Synthesis to Support Socioecological Resilience in the Sierra Nevada and Southern Cascade Range. USDA Forest Service General Technical Report PSW-GTR-247. Albany, California, USA.
Mahlum, S.K., L.A. Eby, M.K. Young, C.G. Clancy, and M. Jakober. 2011. Effects of wildfire on stream temperatures in the Bitterroot River Basin, Montana. International Journal of Wildland Fire 20: 240–247.
Mantua, N.J., S.R. Harre, Y. Zhang, J.M. Wallace, and R.C. Francis. 1997. A Pacific interdecadal climate oscillation with impacts on

salmon production. Bulletin of the American Meteorological Society 78: 1069–1079.
McDonald, P.M., and J.C. Tappeiner, II. 2002. California's hardwood resource: seeds, seedlings, and sprouts of three important forest-zone species. USDA Forest Service General Technical Report PSW-GTR-185. Pacific Southwest Research Station, Albany, California, USA.
Miles, S.R., and C.B. Goudey, editors. 1997. Ecological subregions of California: section and subsection descriptions. USDA Forest Service Technical Publication R5-3M-TP-005. Pacific Southwest Region, San Francisco, California, USA.
Miller, J.D., C.N. Skinner, H.D. Safford, E.E. Knapp, and C.M. Ramirez. 2012a. Trends and causes of severity, size, and number of fires in northwestern California, USA. Ecological Applications 22: 184–203.
———. 2012b. Northwestern California national forests fire severity monitoring 1987-2008. USDA Forest Service Technical Publication R5-TP-035. Pacific Southwest Region, Vallejo, California, USA.
Mohr, J.A., C. Whitlock, and C.N. Skinner. 2000. Postglacial vegetation and fire history, eastern Klamath Mountains, California, USA. The Holocene 10: 587–601.
Mott, J.A., P. Meyer, D. Mannio, S.C. Redd, E.M. Smith, C. Gotway-Crawford, and E. Chase. 2002. Wildland forest fire smoke: health effects and intervention evaluation, Hoopa, California, 1999. Western Journal of Medicine 176: 157–162.
Nagel, N., and A.H. Taylor. 2005. Fire and persistence of montane chaparral in mixed conifer forest landscapes in the northern Sierra Nevada, Lake Tahoe Basin, California, USA. Journal of the Torrey Botanical Society 98: 96–105.
North, M., B. Collins, J. Keane, J. Long, C. Skinner, and W. Zielinski. 2014. Synopsis of emergent approaches. Pages 55–70 in: J.W. Long, L. Quinn-Davidson, and C.N. Skinner, editors. Science Synthesis to Support Socioecological Resilience in the Sierra Nevada and Southern Cascade Range. USDA Forest Service General Technical Report PSW-GTR-247. Pacific Southwest Research Station, Albany, California, USA.
North, M., B.M. Collins, and S. Stephens. 2012. Using fire to increase the scale, benefits, and future maintenance of fuels treatments. Journal of Forestry 110: 392–401.
North State Resources. 2014. Trinity River Lumber Biomass Energy Project. Trinity County Planning Department, Weaverville, California, USA.
Oakeshott, G.B. 1971. California's Changing Landscapes: A Guide to the Geology of the State. McGraw-Hill, San Francisco, California, USA.
Parsons, D.J., and P.B. Landres. 1998. Restoring natural fire to wilderness: how are we doing? Proceedings Tall timbers fire Ecology Conference 20: 366–373.
Perry, D.A., P.F. Hessburg, C.N. Skinner, T.A. Spies, S.L. Stephens, A.H. Taylor, J.F. Franklin, B. McComb, and G. Riegel. 2011. The ecology of mixed severity fire regimes in Washington, Oregon, and northern California. Forest Ecology and Management 262: 703–717.
Quinn-Davidson, L.N., and J.M. Varner. 2012. Impediments to prescribed fire across agency, landscape and manager: an example from northern California. International Journal of Wildland Fire 21: 210–218.
Rieman, B., D. Lee, D. Burns, R. Gresswell, M. Young, R. Stowell, J. Rinne, and P. Howell. 2003. Status of native fishes in the western United States and issues for fire and fuels management. Forest Ecology and Management 178: 197–211.
Robock, A. 1988. Enhancement of surface cooling due to forest fire smoke. Science 242: 911–913.
———. 1991. Surface cooling due to forest fire smoke. Journal of Geophysical Research 96(D11): 20869–20878.
Rorig, M.L., and S.A. Ferguson. 1999. Characteristics of lightning and wildland fire ignition in the Pacific Northwest. Journal of Applied Meteorology 38: 1565–1575.
———. 2002. The 2000 fire season: lightning-caused fires. Journal of Applied Meteorology 41: 786–791.
Rothermel, R.C. 1983. How to predict the spread and intensity of forest and range fires. USDA Forest Service General Technical Report INT-GTR-143. Intermountain Research Station, Ogden, Utah, USA.
Sawyer, J.O. 2006. Northwest California: A Natural History. University of California Press, Berkeley, California, USA.
Sawyer, J.O., and T. Keeler-Wolf. 1995. A Manual of California Vegetation. California Native Plant Society, Sacramento, California, USA.
Sawyer, J.O., and D.A. Thornburgh. 1977. Montane and subalpine vegetation of the Klamath Mountains. Pages 699–732 in: M.G. Barbour and J. Major, editors. Terrestrial Vegetation of California. John Wiley and Sons, New York, New York, USA.
Schroeder, M.J., and C.C. Buck. 1970. Fire weather—a guide for application of meteorological information to forest fire control operations. USDA Agricultural Handbook 360. Washington, D.C., USA.
Sharp, R.P. 1960. Pleistocene glaciation in the Trinity Alps of northern California. American Journal of Science 258: 305–340.
Shrader, G. 1965. Trinity forest. Pages 37–40 in: Yearbook of the Trinity County Historical Society. Weaverville, California, USA.
Skinner, C.N. 1995a. Change in spatial characteristics of forest openings in the Klamath Mountains of northwestern California, USA. Landscape Ecology 10: 219–228.
———. 1995b. Using prescribed fire to improve wildlife habitat near Shasta Lake. Unpublished File Report. USDA Forest Service, Shasta-Trinity National Forest, Shasta Lake R.D., Redding, California, USA.
———. 2002. Influence of fire on dead woody material in forests of California and southwestern Oregon. Pages 445–454 in: W.F. Laudenslayer, Jr., P.J. Shea, B.E. Valentine, C.P. Weatherspoon, and T.E. Lisle, editors. Proceedings of the Symposium on the Ecology and Management of Dead Wood in Western Forests. USDA Forest Service General Technical Report PSW-GTR-181. Pacific Southwest Research Station, Albany, California, USA.
———. 2003a. Fire regimes of upper montane and subalpine glacial basins in the Klamath Mountains of northern California. Tall Timbers Research Station Miscellaneous Publication 13: 145–151.
———. 2003b. A tree-ring based fire history of riparian reserves in the Klamath Mountains. Pages 116–119 in: P.M. Farber, editor. California Riparian Systems: Processes and Floodplains Management, Ecology, and Restoration. Riparian Habitat and Floodplains Conference Proceedings. Riparian Habitat Joint Venture, Sacramento, California, USA.
Skinner, C.N., C.S. Abbott, D.L. Fry, S.L. Stephens, A.H. Taylor, and V. Trouet. 2009. Human and climatic influences on fire occurrence in California's North Coast Range. Fire Ecology 5(3): 76–99.
Skinner, C.N., and C. Chang. 1996. Fire regimes, past and present. Pages 1041–1069 in: D.C. Erman, general editor. Sierra Nevada Ecosystem Project: Final Report to Congress, Volume II. Wildland Resources Center Report 37. University of California, Davis, California, USA.
Stephens, S.L., and M.A. Finney. 2002. Prescribed fire mortality of Sierra Nevada mixed conifer tree species: effects of crown damage and forest floor combustion. Forest Ecology and Management 162: 261–271.
Stephens, S.L., R.E. Martin, and N.E. Clinton. 2007. Prehistoric fire area and emissions from California's forests, woodlands, shrublands, and grasslands. Forest Ecology and Management 251: 205–216.
Stephens, S.L., C.N. Skinner, and S.J. Gill. 2003. A dendrochronology based fire history of Jeffrey pine-mixed conifer forests in the Sierra San Pedro Martir, Mexico. Canadian Journal of Forest Research 33: 1090–1101.
Stuart, J.D., and L.A. Salazar. 2000. Fire history of white fir forests in the coastal mountains of northwestern California. Northwest Science 74: 280–285.
Sullivan, M.S. 1992. The Travels of Jedediah Smith. University of Nebraska Press, Lincoln, Nebraska, USA.
Swanson, F.J., J.A. Jones, D.O. Wallin, and J.H. Cissel. 1994. Natural variability—implications for ecosystem management. Pages 80–94 in: M.E. Jensen and P.S. Bourgeron, editors. Eastside Forest Ecosystem Health Assessment. USDA Forest Service General Technical Report PNW-GTR-318. Pacific Northwest Research Station, Portland, Oregon, USA.
Taylor, A.H. 2000. Fire regimes and forest changes along a montane forest gradient, Lassen Volcanic National Park, southern Cascade Mountains, USA. Journal of Biogeography 27: 87–104.
Taylor, A.H., and C.B. Halpern. 1991. The structure and dynamics of *Abies magnifica* forests in the southern Cascade Range, USA. Journal of Vegetation Science 2: 189–200.
Taylor, A.H., and C.N. Skinner. 1998. Fire history and landscape dynamics in a late-successional reserve in the Klamath Mountains, California, USA. Forest Ecology and Management 111: 285–301.

———. 2003. Spatial patterns and controls on historical fire regimes and forest structure in the Klamath Mountains. Ecological Applications 13: 704–719.
Taylor, A.H., C.N. Skinner, and B.L. Estes. 2013. A comparison of fire severity in the late 19th and early 21st Century in a mixed conifer forest landscape in the southern Cascades. Final report for USDI/USDA Joint Fire Science Program Project Number 09-1-10-7. https://www.firescience.gov/projects/09-1-10-7/project/09-1-10-7_final_report.pdf.
Taylor, A.H., V. Trouet, and C.N. Skinner. 2008. Climatic influences on fire regimes in montane forests of the southern Cascades, California, USA. International Journal of Wildland Fire 17: 60–71.
Taylor, A.H., V. Trouet, C.N. Skinner, and S. Stephens. 2016. Socioecological transitions trigger fire regime shifts and modulate fire-climate interactions in the Sierra Nevada, USA, 1600-2015 CE. Proceedings of the National Academy of Sciences of the United States of America 113: 13684–13689.
Thompson, J.R., and T.A. Spies. 2010. Factors associated with crown damage following recurring mixed-severity wildfires and post-fire management in southwestern Oregon. Landscape Ecology 25: 775–789.
Thompson, J.R., T.A. Spies, and L.M. Ganio. 2007. Reburn severity in managed and unmanaged vegetation in a wildfire. Proceedings of the National Academy of Sciences of the United States of America 104: 10743–10748.
Thornburgh, D.A. 1990. *Picea breweriana* Wats. Brewer spruce. Pages 345–357 in: R.M. Burns and B.H. Honkala, editors. Silvics of North America, Volume 1, Conifers. USDA Agriculture Handbook 654. Washington, D.C., USA.
———. 1995. The natural role of fire in the Marble Mountain Wilderness. Pages 273–274 in: J.K. Brown, R.W. Mutch, C.W. Spoon, and R.H. Wakimoto, editors. Proceedings: Symposium on Fire in Wilderness and Park Management. USDA Forest Service General Technical Report INT-GTR-320. Intermountain Forest and Range Experiment Station, Ogden, Utah, USA.
Tollefson, Jennifer E. 2008. Quercus chrysolepis. Fire Effects Information System. U.S. Department of Agriculture, Forest Service, Rocky Mountain Research Station, Fire Sciences Laboratory (Producer). http://www.fs.fed.us/database/feis/ (2018, January 21).
Trouet, V., and A.H. Taylor. 2010. Multi-century variability in the Pacific North American (PNA) circulation pattern reconstructed from tree rings. Climate Dynamics 35: 953–963.
Trouet, V., A.H. Taylor, A.M. Carleton, and C.N. Skinner. 2006. Fire-climate interactions in forests of the American Pacific Coast. Geophysical Research Letters 33: L18704.
———. 2009. Interannual variations in fire weather, fire extent, and synoptic-scale circulation patterns in northern California and Oregon. Theoretical and Applied Climatology 95: 349–360.
USCB. 2002. Census 2000 data for the state of California. In United States Census 2000. US Census Bureau, Washington, D.C., USA. http://www.census.gov/census2000/states/ca.html.
USDA Forest Service. 2003. Biscuit post-fire assessment—Rogue River and Siskiyou national forests: Josephine and Curry counties, Oregon. Siskiyou National Forest, Grants Pass, Oregon, USA.
USDA-USDI. 1994. Record of decision for amendments to Forest Service and Bureau of Land Management planning documents within the range of the northern spotted owl; standard and guidelines for management of habitat for late-successional and old-growth forest related species within the range of the northern spotted owl. USDA Forest Service and USDI Bureau of Land Management, Portland, Oregon, USA.
van Wagtendonk, J.W., and D.R. Cayan. 2008. Temporal and spatial distribution of lightning strikes in California in relation to large-scale weather patterns. Fire Ecology 41(1): 34–56.
Wallace, J.M., and D.S. Gutzler. 1981. Teleconnections in the geopotential height field during the Northern Hemisphere winter. Monthly Weather Review 109: 784–812.
Weatherspoon, C.P., and C.N. Skinner. 1995. An assessment of factors associated with damage to tree crowns from the 1987 wildfires in northern California. Forest Science 41: 430–451.
West, G.J. 1990. Holocene fossil pollen records of Douglas fir in northwestern California: reconstruction of past climate. Pages 119–122 in: J.L. Betancourt and A.M. MacKay, editors. Proceedings of the Sixth Annual Pacific Climate (PACLIM) Workshop. California Department of Water Resources, Interagency Ecological Studies Program Technical Report 23. Sacramento, California, USA.
Westerling, A.L., H.G. Hildalgo, D.R. Cayan, and T.W. Swetnam. 2006. Warming and earlier spring increase western U.S. forest wildfire activity. Science 313: 940–943.
Whitlock, C., C.N. Skinner, T. Minckley, and J.A. Mohr. 2004. Comparison of charcoal and tree-ring records of recent fires in the eastern Klamath Mountains. Canadian Journal of Forest Research 34: 2110–2121.
Whittaker, R.H. 1960. Vegetation of the Siskiyou Mountains, Oregon and California. Ecological Monographs 30: 279–338.
———. 1961. Vegetation history of the Pacific Coast States and the "central" significance of the Klamath region. Madroño 16: 5–23.
Wildland Fire Leadership Council. 2014. The National Strategy: The Final Phase in the Development of the National Cohesive Wildland Fire Management Strategy. Department of the Interior and Department of Agriculture, Washington, D.C., USA. http://www.forestsandrangelands.gov. Accessed February 25, 2015.
Wills, R.D., and J.D. Stuart. 1994. Fire history and stand development of a Douglas-fir/hardwood forest in northern California. Northwest Science 68: 205–212.
Wilson, R.B. 1904. Township Descriptions of the Lands Examined for the Proposed Trinity Forest Reserve, California. US Department of Agriculture, Bureau of Forestry, Washington, D.C., USA.
Wise, E.K. 2010. Spatiotemporal variability of the precipitation dipole transition zone in the western United States. Geophysical Research Letters 37: L07706.
Yang, J., P.J. Weisberg, T.E. Dilts, E.L. Loudermilk, R.M. Scheller, A. Stanton, and C. Skinner. 2015. Predicting wildfire occurrence distribution with spatial point process models and its uncertainty assessment: a case study in the Lake Tahoe Basin, USA. International Journal of Wildland Fire 24: 380–390.

CHAPTER TWELVE

Southern Cascades Bioregion

CARL N. SKINNER AND ALAN H. TAYLOR

> In . . . the southern portions of . . . the Cascades . . . where the forests are largely or mainly of yellow pine in open growth, with very little litter or underbrush, destructive fires have been few and small, although throughout these regions there are few trees which are not marked by fire, without, however, doing them any serious damage.
>
> GANNETT (1902)

Description of the Bioregion

Physical Geography

The Cascade Range extends from British Columbia, Canada south to northern California where it meets the Sierra Nevada. The Southern Cascades bioregion in California is bounded on the west by the Sacramento Valley and the Klamath Mountains, and on the east by the Modoc Plateau and Great Basin. The bioregion encompasses the Southern Cascades section of Miles and Goudey (1997) and covers approximately 4% (16,740 km^2 [6,460 mi^2]) of the area of California.

The Cascades are geologically young and characterized by prominent volcanic peaks (some recently active) that stand above an extensive mainly basaltic plateau. In parts of the central and southern Cascades, volcanics overlie granitic and metamorphic rocks similar to those of the Klamath Mountains and Sierra Nevada (Oakeshott 1971). Soils are derived from volcanic material and are classified as Alfisols, Entisols, Inceptisols, Mollisols, Ultisols, and Vertisols (Miles and Goudey 1997).

Overall, topography in the southern Cascades is gentler than in the Klamath Mountains or the Sierra Nevada. Elevations range from ca. 60 m (196 ft) in the southwestern foothills adjacent to the Sacramento Valley to 4,317 m (14,162 ft) at the summit of Mt. Shasta. Other notable topographic features include Mt. Lassen, the Medicine Lake Highlands, Butte Valley, Hat Creek Valley, Burney Falls, Shasta Valley, and the Pit River canyon. The headwaters of both the Klamath and Pit rivers originate east of the Cascade crest and breach the range as they flow westward toward the Pacific Ocean (Map. 12.1).

Climate

The climate of the southern Cascade Range is Mediterranean. The expression of this climate regime is mediated by location along three gradients: (1) a west-east gradient in precipitation and winter temperature where wetter and warmer conditions prevail on the west side of the range south of Mt. Shasta; (2) a north-south gradient where annual precipitation is lower on the west side of the range north of Mt. Shasta due to the rain shadow of the Klamath Mountains; and (3) decreasing temperatures and increasing annual precipitation with increasing elevation (Table 12.1). The driest areas are Butte and Shasta Valleys to the north of Mt. Shasta.

Most precipitation falls as snow at higher elevations. The average early April snowpack depth and water content for selected sites in the Cascade Range are shown in Table 12.2.

The west to east gradient in precipitation and temperature help to create very different environments at similar elevations on the west side of the crest compared to the east side albeit not as dramatically as in the Sierra Nevada. Nevertheless, distinctly different vegetation develops in response to the different climatic regimes. Therefore, the term "westside" is used to refer to environments typical of the west side of the crest and "eastside" for those typical of the east side of the range.

CLIMATE AND FIRE

Years with widespread and larger fires are dryer and warmer than the norm (Trouet et al. 2006, Taylor et al. 2008, Trouet et al. 2009). These anomalous warm/dry years are associated with positive phases of the Pacific North American Pattern (PNA) and the Pacific Decadal Oscillation (PDO) (Trouet et al. 2009). The development of persistent high-pressure ridges along the Pacific Coast in winter under the positive PNA (Wallace and Gutzler 1981) and PDO (Mantua et al. 1997) reduces the amount and timing of annual precipitation (Trouet and Taylor 2010) contributing to earlier onset and warmer/dryer fire seasons (Trouet et al. 2009).

In this bioregion the El Niño Southern Oscillation (ENSO) is not consistently associated with fire activity (Taylor et al. 2008). This is largely due to the location of the bioregion in the pivot zone (40–45°N) of the north/south dipole pattern of ENSO (Dettinger et al. 1998, Wise 2010). Thus, the fire-climate in the Southern Cascades bioregion responds like the Southwest in some years and the Northwest in others. Examples are the droughts in the El Niños of 1965–1966 and

MAP 12.1 Southern Cascade Range in California.

1991–1992 (Northwest pattern) and the extremely wet years of the strong El Niños of 1982–1983 and 1997–1998 (Southwest pattern).

WEATHER SYSTEMS

Three types of fire weather conditions that occur during fire season are important in the southern Cascades (Hull et al. 1966): (1) Pacific High—Postfrontal (Postfrontal); (2) Pacific High—Prefrontal (Prefrontal); and (3) Subtropical High Aloft (Subtropical High).

Postfrontal conditions occur when high pressure follows the passage of a cold front and causes strong foehn winds from the north and northeast. The Cone fire in September 2002 that burned 812 ha (2,006 ac), mostly in the Blacks Mountain Experimental Forest (BMEF), is an example of a major fire that burned under Postfrontal conditions.

The Prefrontal type occurs when the southern, dry tail of a cold front crosses the area and generates strong southwest or west winds. The strong winds are the key fire weather component in this type because temperatures usually drop and relative humidity rises. The Lost fire (1987) in the Hat Creek Valley that burned over 9,700 ha (24,000 ac) and the Fountain fire (1992) that burned over 26,000 ha (64,250 ac) on Hatchet Mountain are examples of fires driven by Prefrontal conditions.

The Subtropical High type occurs when stagnant high pressure produces high temperatures and low relative humidity for extended periods. These conditions are often accompanied by periods of high atmospheric instability (Schroeder and Buck 1970) with high values of the Haines Index, which are associated with widespread burning (Trouet et al. 2009). The Ponderosa and Reading fires (2012) are examples of fires burning and spreading for several days under these conditions after initial wind driven runs.

LIGHTNING

The Cascade Range averages 18.8 lightning strikes (range 5.5–33.8) yr^{-1} 100^{-1} km^{-2} (49.5 strikes [range 14.5–89.1] yr^{-1}

TABLE 12.1
Normal annual, January, and July precipitation and normal maxima and minima January and July temperatures for representative stations of the southern Cascade Range (WRCC 2002)

Location—elevation	Average precipitation cm (in)	Normal daily maximum temperature °C (F)	Normal daily minimum temperature °C (F)
Paradise (533 m)—west side			
Annual	139.7 (55.0)		
January	28.2 (11.1)	12.1 (53.7)	3.1 (37.5)
July	0.3 (0.1)	33.1 (91.6)	17.6 (63.6)
Mt. Shasta City (1,204m)—west side			
Annual	99.1 (39.0)		
January	18.8 (7.4)	5.9 (42.6)	–3.5 (25.7)
July	0.8 (0.3)	29.2 (84.6)	10.1 (50.2)
Mineral (1,486 m)—west side			
Annual	141.0 (55.5)		
January	25.7 (10.1)	4.9 (40.8)	–5.8 (21.5)
July	1.0 (0.2)	27.1 (80.7)	5.9 (42.7)
Burney (956 m)—east side			
Annual	70.6 (27.8)		
January	12.4 (4.9)	6.4 (43.5)	–7.7 (18.1)
July	0.5 (0.2)	30.8 (87.4)	6.2 (43.2)
Mt. Hebron (1,295 m)—east side			
Annual	30.0 (11.8)		
January	3.3 (1.3)	3.9 (39.1)	–9.2 (15.5)
July	1.0 (0.4)	28.3 (83.0)	5.9 (42.6)
Chester (1,379 m)—east side			
Annual	86.1 (33.9)		
January	16.5 (6.5)	5.5 (41.9)	–6.9 (19.6)
July	0.8 (0.3)	29.4 (85.0)	6.7 (44.0)

100^{-1} mi^{-2}) and they are a common source of ignition. The density of lightning strikes increases from south to north (van Wagtendonk and Cayan 2008). Occasionally, incursions of subtropical moisture moving north from the eastern Pacific and the Gulf of California produce widespread thunderstorms resulting in numerous fires. Hundreds of lightning fires can be ignited over short periods during these events. The occurrence of widespread, simultaneous, lightning ignitions has contributed to fires that burn for weeks and cover very large areas as in 1977, 1987, 1990, 1999, 2008, 2009, 2012, and 2014.

Ecological Zones

South of the latitude of Mt. Shasta, vegetation composition and species dominance in the lower and mid-montane zones is similar to that in the northern Sierra Nevada while the upper montane and subalpine zones are more similar to the Klamath Mountains and Cascades of southern Oregon (Parker 1991). When compared to the Sierra Nevada, vegetation composition in the Cascades is more strongly controlled by local topography and substrate and less so by elevation (Parker 1995) likely due to less dramatic elevation differences and youthful geology with young soils. Open woodlands, shrublands, and areas of sparse vegetation occur over wide areas on harsh sites. These conditions are common where young lava flows or other young volcanic materials with shallow, poorly developed soils inhibit vegetative growth regardless of ecological zone. North of Mt. Shasta, in the rain shadow of the Klamath Mountains, the vegetation of the west side of the Cascades resembles vegetation more characteristic of the drier east side of the Cascades. Lower elevations on both sides of the Cascades are dominated by grasslands, shrublands, and woodlands. However, there is considerable variability in the composition and physiognomy of these vegetation types.

SOUTHWESTERN FOOTHILLS

The southwestern foothills of the Cascades, along the northeastern edge of the Sacramento Valley, is a low elevation, dissected volcanic plateau with vegetation similar to that of the Sierra Nevada foothills (Fig. 12.1). Common alliances in this portion of the bioregion are blue oak (*Quercus douglasii*), gray pine (*Pinus sabiniana*), interior live oak (*Q. wislizeni*), valley oak (*Q. lobata*), buck brush (*Ceanothus cuneatus* var. *cuneatus*), annual grassland, and vernal pools (Miles and Goudey 1997). Two other important alliances in this zone are ponderosa pine (*P. ponderosa*) and California black oak (*Q. kelloggii*). This ecological zone corresponds to subsection M261Fa of Miles and Goudey (1997). Three major creeks (Mill, Antelope, and Deer) flow from the Cascades westward to the Sacramento Valley through this area.

The southwestern foothills have a complex pattern of vegetation primarily influenced by the depth of the underlying soil (Biswell and Gilman 1961). Blue oak woodland is the most common alliance and it varies from forest to savanna in terms of tree density.

NORTHWESTERN FOOTHILLS

The northwestern foothills are north of Mt. Shasta and include the Shasta Valley and adjacent foothills. They generally correspond to subsections M261Db and M261Dc of Miles and Goudey (1997). Being in the rain shadow of the Klamath Mountains and above 790 m (2,600 ft) in elevation, the area is more of an eastside type (Fig. 12.2). Common alliances include big sagebrush (*Artemisia tridentata* subsp. *tridentata*), western juniper (*Juniperus occidentalis*) woodlands, meadows (mostly sedges [*Carex* spp.]), buck brush, California black oak, and ponderosa pine (Miles and Goudey 1997). Introduced grasses, especially cheat grass (*Bromus tectorum*), dominate extensive areas. See chapter 13 on the Northeastern Plateau for more on vegetation alliances common to this zone.

TABLE 12.2
Average April 1 snowpack data for representative snow courses in the southern Cascade Range ordered from north to south (CCSS 2002)

Snow course	Elevation m (ft)	West/east side	Snow depth cm (in)	Water content cm (in)
Medicine Lake	2,042 (6,700)	East	202.7 (79.8)	81.3 (32.0)
Dead Horse Canyon	1,372 (4,500)	East	65.3 (25.7)	29.0 (11.4)
Burney Springs	1,433 (4,700)	East	12.7 (5.0)	5.1 (2.0)
Blacks Mountain	2,042 (6,700)	East	56.9 (22.4)	20.3 (8.0)
Chester Flat	1,402 (4,600)	East	40.6 (16.0)	16.5 (6.5)
Mount Shasta	2,408 (7,900)	West	307.3 (121.0)	106.2 (41.8)
Stouts Meadow	1,646 (5,400)	West	206.5 (81.3)	95.8 (37.7)
Snow Mountain	1,859 (6,100)	West	161.8 (63.7)	81.8 (32.2)
Thousand Lakes	1,981 (6,500)	West	206.2 (81.2)	86.4 (34.0)
Lower Lassen Peak	2,515 (8,250)	West	454.2 (178.8)	202.7 (79.8)
Mill Creek Flat	1,798 (5,900)	West	234.7 (92.4)	98.0 (38.6)

FIGURE 12.1 Southwestern Foothills occur on a dissected volcanic plateau. The thin, rocky soils often support blue oak and gray pine as shown here in Bidwell Park, Chico (photo credits: Carl Skinner, USDA Forest Service).

FIGURE 12.2 Northwestern Foothills. Shrublands on north slopes of Mount Shasta looking northwest into the Shasta Valley. Shrubs shown are mostly greenleaf manzanita, tobacco brush, and bitterbrush (photo credits: Carl Skinner, USDA Forest Service).

FIGURE 12.3 Lower montane eastside. Gray pine mixed with western juniper on rim of Pit River Canyon near Fall River Mills depicting the mixing of characteristically westside vegetation with that more characteristic of the eastside (photo credits: Carl Skinner, USDA Forest Service).

FIGURE 12.4 Mid-montane westside. Mixed stand of Douglas-fir, ponderosa pine, sugar pine, incense cedar, California black oak, white fir, and Pacific dogwood in the Flatwoods near Big Bend. Stand was mechanically thinned for fire hazard reduction 10 years before taking the photograph (photo credits: Derrick B. Skinner).

LOW-ELEVATION EASTSIDE

Lower elevations of the east side of the Cascades occur primarily along the Pit River and lower reaches of Hat Creek. This zone corresponds roughly to the lower portions of Miles and Goudey's (1997) subsection M261Dj. The low elevation of the Pit River gorge may have allowed it to serve as a corridor for plant species characteristic of westside habitats to migrate into eastside environments. For example, between Burney and Fall River gray pine, California black oak, Oregon oak (*Quercus garryana*), and buck brush, usually associated with westside environments, co-occur with western juniper and bitterbrush (*Purshia tridentata*) (Fig. 12.3). See chapter 13 on the Northeastern Plateau for more on vegetation alliances common to this zone.

The Pit River corridor near Burney Falls has disjunct stands of mountain misery (*Chamaebatia foliolosa*) that grow in association with ponderosa pine, California black oak, Douglas-fir (*Pseudotsuga menziesii* var. *menziesii*), Oregon oak, and other conifers. For more information on mountain misery see chapter 14.

MID-MONTANE WESTSIDE/EASTSIDE

Conifer forests dominate the mid-montane zone on both sides of the range and they are intermixed with woodlands and shrublands. Few natural meadows or grasslands occur on the west side of the range. Species composition varies from west to east over the crest but less so than in the Sierra Nevada as crest elevation and mountain passes are generally lower (Griffin 1967). Nevertheless, the mid-montane zone of the bioregion is quite different on the east compared to the west side of the crest because of the rain shadow effect and differences in temperature. Consequently, our discussion will be split between westside and eastside alliances. The Mid-Montane Westside corresponds to subsections M261Dg and M261Di of Miles and Goudey (1997). The Mid-Montane Eastside corresponds to subsections M261Da and M261Dj of Miles and Goudey (1997).

Mixed-species conifer forests dominate the mid-montane zone west of the Cascade crest (Fig. 12.4). Any of six conifer

FIGURE 12.5 Upper montane. Shrubfields intermixed with patches of conifers on Snow Mountain. Dominant shrubs are tobacco brush, greenleaf manzanita, bitter cherry, and bush chinquapin (photo credits: Carl Skinner, USDA Forest Service).

species (ponderosa pine, Douglas-fir, incense cedar [*Calocedrus decurrens*], sugar pine [*Pinus lambertiana*], Jeffrey pine [*P. jeffreyi*], and white fir [*Abies concolor*]) may co-occur and share dominance (Parker 1995, Beaty and Taylor 2001). A subcanopy of the deciduous hardwoods California black oak (*Quercus kelloggii*), big-leaf maple (*Acer macrophyllum*), and mountain dogwood (*Cornus nuttallii*) and the evergreen canyon live oak (*Q. chrysolepis*) may occur beneath the conifer canopy. Stand composition is influenced by elevation, slope aspect, soil moisture, and substrate (Griffin 1967). Mixed white fir and California red fir (*A. magnifica* var, *magnifica*) forests occur above the elevation of the mixed-conifer zone in the Cascades (Parker 1995). Forest cover is often interrupted by stands of montane chaparral. The most common shrubs of the Cascade Range montane chaparral are greenleaf manzanita (*Arctostaphylos patula*), California-lilac (*Ceanothus* spp.), and huckleberry oak (Q. vacciniifolia). Montane chaparral occupies sites that are unable to support trees due to shallow soils, exposed slopes where cold, high winds and ice damage are common, or have a history of severe fires (Beaty and Taylor 2001, Lauvaux et al. 2016).

Extensive areas on the east side are dominated by ponderosa pine, Jeffrey pine, or a combination of both. Other conifers, such as white fir and incense cedar, may be locally important but do not usually attain dominance, especially on the drier sites (Rundel et al. 1977). Western juniper (*Juniperus occidentalis*) and curl-leaf mountain-mahogany (*Cercocarpus ledifolius* var. *ledifolius*) may be associates on drier and rockier sites. Widely scattered, small stands of quaking aspen (*Populus tremuloides*) occur around seeps, on meadow edges, and young exposed basalt.

UPPER MONTANE

Upper montane zone conifer forests and shrublands have similar species composition on both sides of the crest. However, species dominance varies widely and is influenced by annual precipitation, topography, and substrate. The more common conifers in this zone are Jeffrey pine, ponderosa pine, white fir, red fir, lodgepole pine (*Pinus contorta*), and western white pine (*P. monticola*). Common shrubs include bush chinquapin (*Chrysolepis sempervirens*), greenleaf manzanita, pine-mat manzanita (*Arctostaphylos nevadensis* subsp. *nevadensis*), mountain whitethorn (*Ceanothus cordulatus*), tobacco brush (more commonly known as snowbrush in this region) (*Ceanothus velutinus*), huckleberry oak, Parry's goldenbush (*Ericameria parryi*), rubber rabbitbrush (*E. nauseosa*), and big sagebrush (*Artemisia tridentata*). As in the mid-montane zone, forest cover can be interrupted by stands of shrubs on harsh sites or where there have been severe fires (Fig. 12.5). The dryer portions of this zone are found on the eastern edge of the bioregion and correspond to subsections M261Dd and M261Dh of Miles and Goudey (1997). See chapter 13 for more on herbaceous and shrub vegetation alliances common to the eastern part of this zone.

Large, seasonally wet, montane meadows are characteristic of the upper montane zone east of the crest. The remainder of the zone corresponds to the portions of subsections M261Df, M261Di, and M261Dm of Miles and Goudey (1997) that are at elevations lower than the subalpine zone.

SUBALPINE

Subalpine and alpine vegetation is generally limited to the highest peaks in the southern Cascades such as Mt. Shasta, Mt. Lassen, and Crater Peak. Common alliances are mountain hemlock (*Tsuga mertensiana*), whitebark pine (*Pinus albicaulis*), and California red fir. These areas are the higher elevations of subsections M261Df, M261Di, and M261Dm of Miles and Goudey (1997) (Fig. 12.6).

Overview of Historic Fire Occurrence

There are generally two periods with distinctly different fire regimes in the Cascades. First was a Native American period, before 1905, when fires were generally frequent. This period

FIGURE 12.6 Subalpine woodland of mountain hemlock near Reading Peak in Lassen Volcanic National Park. Foreground is near the origin of the 2012 Reading Fire (photo credits: Carl Skinner, USDA Forest Service).

includes both the prehistoric and European-settlement periods. The Native American period was followed by the fire-suppression period after the establishment of the national forest reserves in 1905 when fire occurrence commonly decreases dramatically (Taylor 1990a, 1993, 2000, 2010, Skinner and Chang 1996, Beaty and Taylor 2001, Bekker and Taylor 2001, Norman and Taylor 2005, Lauvaux et al. 2016).

Prehistoric Fire Occurrence

As discussed in chapter 19, native people of the southern Cascade Range used fire to promote production of food and basketry materials, to help gather grasshoppers and other insects, to improve hunting conditions, and for ceremonial purposes (Long et al. 2016). Though the use of fire by native people appears to have been widespread, the extent of its influence on vegetation at broad scales is unknown.

The Mediterranean climate, the commonality of lightning ignitions, and the widespread use of fire by native people in the Cascades promoted frequent surface fires of mostly low–moderate intensity with frequency decreasing with elevation. Pronounced local variations in fire frequency also occur due to interruptions in fuel connectivity caused by volcanics (i.e., lava flows, scoria depositions, debris flows) (Taylor 2000). More mesic environments, especially with increasing elevation, promoted mixed-severity fires with severity tending toward the low–moderate type (Perry et al. 2011, Hessburg et al. 2016).

Historic

Parts of the bioregion began to experience a decrease in fire occurrence as early as the late nineteenth century (Norman and Taylor 2005, Gill and Taylor 2009, Taylor 2010). The decrease was pronounced near meadows, coinciding well with a documented period of heavy sheep grazing on the eastside of the range (Taylor 1990a, Norman and Taylor 2005). Most areas did not experience a fire frequency decline until the beginnings of organized fire suppression. Even then, rural residents would often continue burning to maintain forage for livestock. The earliest accounts of wildland fire suppression are from 1887 for fires burning along the railroad lines near what is now the city of Mt. Shasta (Morford 1984). The first recorded organized fire protection in wildland areas was by the Central Pacific Railroad in 1898 which supported mounted patrols to suppress fires in the McCloud flats east of Mount Shasta (Morford 1984).

Many long-time residents of the area describe how farmers and ranchers regularly burned large areas well into the early decades of the twentieth century to promote forage for livestock and keep forest understories open. They described how they would burn in the fall while moving herds from summer to winter pasture. The fires were ignited by using long ropes soaked at the end in kerosene and then lit and dragged behind horses on their way to winter pasture. Others described how regular underburning of forest stands allowed them to increase forage while avoiding higher taxes from an increase in land values that would result from converting forest to pasture (C. N. Skinner, Shasta Lake California, pers. inform.).

Current

Area burned by fires was greatly reduced in the twentieth century compared to prehistoric levels, due to highly effective fire suppression in the relatively gentle terrain. This terrain also permitted early logging of the large fire resistant trees which, combined with the effects of fire suppression, have left forests heavily stocked with smaller, young trees. Unlogged forests in parks and wilderness areas have also developed high densities of young trees due to fire suppression (Dolph et al. 1995, Taylor 2000). Thus, where surface fires were frequent and extensive (i.e., foothill through mid-montane) and mostly low–moderate intensity, fires are now either suppressed while very small or escape initial fire-suppression efforts and become large, mostly high-intensity fires. Examples of these recent large, severe fires include the Pondosa

(1977, 6,000 ha [~15,000 ac]), Lost (1987, 9,700 ha [24,000 ac]), Campbell (1990, >48,500 ha [>120,000 ac]), Fountain (1992, ~26,000 ha [64,250 ac]), Gunn (1999, >24,000 ha [59,000 ac]), Storrie (2000, >21,000 ha [52,000 ac]), Chips (2012, >30,500 ha [75,400 ac]), Ponderosa (2012, >11,200 ha [27,600 ac]), Reading (2012, >11,300 ha [29,000 ac]), Eiler (2014, >13,100 ha [32,400 ac]), and Bald (2014, >16,000 ha [39,700 ac]). All of these fires except the Fountain, Storrie, and Chips Fires were lightning caused.

The southwestern foothills have the greatest proportion of area burned during the period 1910–1993 of all major watersheds assessed by the Sierra Nevada Ecosystem Project (SNEP) (McKelvey and Busse 1996). Despite the frequent occurrence of large fires in the twentieth century, Moyle and Randall (1996) found the biological integrity of the three major tributaries to the Sacramento River (Mill, Antelope, and Deer Creeks) to be unusually high compared to the rest of the SNEP assessment area with much of the native fish and amphibian faunas intact ". . . and the biotic communities . . . still largely governed by natural processes." Thus, relatively frequent, large twentieth century fires had not had a detrimental effect on aquatic communities at least in this part of the Cascades.

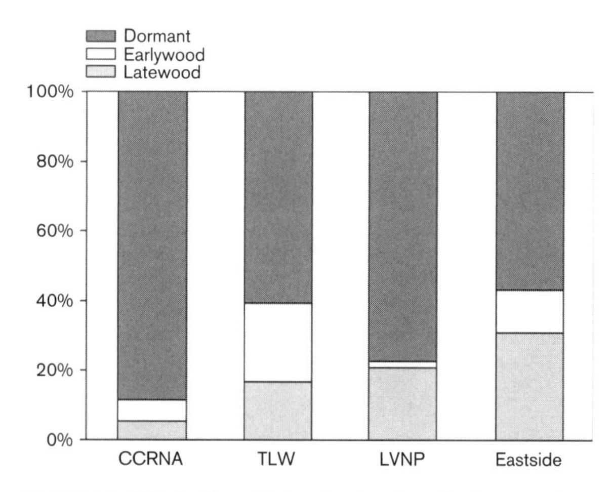

FIGURE 12.7 Distribution of intra-ring locations for fire scars in this bioregion (Taylor 2000, Beaty and Taylor 2001, Bekker and Taylor 2001, Norman and Taylor 2003).

Major Ecological Zones

Fire Patterns

General fire regime characteristics are widespread; thus, we will break with the protocol of other bioregional chapters and discuss the fire regimes more generally, without assigning them to a specific ecological zone. Assigning them to a specific ecological zone could convey a misleading impression that these characteristics are specific to a particular zone. The pre-fire-suppression fire regimes for forested areas and woodlands in the foothills, lower and mid-montane on both the west and east sides of the range were characterized by mostly frequent fires of low–moderate intensity. Fire frequency generally declined with increasing elevation but remained generally frequent throughout these areas while varying locally with aspect and topographic position (Taylor 2000, Beaty and Taylor 2001, Norman and Taylor 2002, 2003, Taylor et al. 2008, Taylor 2010, Lauvaux et al. 2016). Shrubfields and upper montane forests and woodlands had less frequent fires with forests tending toward mixed-severity fires and shrubfields tending toward high-severity fires (Taylor and Halpern 1991, Taylor 1993, Beaty and Taylor 2001, Bekker and Taylor 2001, Taylor and Solem 2001, Nagel and Taylor 2005, Lauvaux et al. 2016).

SEASONALITY

Fire history studies have shown that most fire scars occurred in latewood or at the ring boundary (Fig. 12.7) indicating that fires burned mostly in mid-summer through fall (Skinner 2002). Both dead and live fuels reach their lowest moisture levels at that time (Estes et al. 2012) and ignite and burn easily. However, there is temporal and spatial variation in the seasonal timing of fires suggesting a complex set of interactions between slope aspect, elevation, climate variation, and the seasonal occurrence of fire (Taylor 2000, Beaty and Taylor 2001, Bekker and Taylor 2001, Norman and Taylor 2003, 2005). Fires occurred earlier in the season on dryer pine dominated sites (i.e., south- and west-facing aspects) and at lower elevation compared to more mesic sites (i.e., north- and east-facing aspects) and higher elevation sites (Taylor 2000, Bekker and Taylor 2001, Norman and Taylor 2005, Taylor 2010). Seasonality has also varied over time, with fires that burned in years with normal or above normal precipitation burning mainly in the late summer and fall.

TOPOGRAPHY

As in the Klamath Mountains (see chapter 11), fire history studies demonstrate that variability in Fire Return Intervals (FRIs) and fire severity are strongly influenced by topography, especially slope aspect, elevation, and slope position. Differences in FRIs have been identified within the same forest type on different slope aspects and FRIs increase with elevation from montane to the upper montane zone (Taylor 2000, Beaty and Taylor 2001, Bekker and Taylor 2001). FRIs have also been found to vary with slope position. Upper slopes had shorter median FRIs than middle and lower slopes (Norman and Taylor 2002). FRIs also vary with other terrain characteristics. Forests interspersed with lava flows, scoria fields, or other features with little or no vegetation that impede fire spread have longer FRIs than areas with continuous forest cover (Taylor 2000). Riparian zones affected landscape patterns of fire occurrence by acting as barriers to the spread of some fires as in the Klamath Mountains (Beaty and Taylor 2001, Norman and Taylor 2002).

FIRE SEVERITY

Patterns of fire severity during the prefire-suppression period were associated with topographic position at least in deeply incised terrain. In the Cub Creek Research Natural Area (CCRNA), Beaty and Taylor (2001) found that the upper thirds of slopes and ridgetops, especially south- and west-facing aspects, experienced the highest proportion of high-severity burns. Lower thirds of slopes, north-, and east-facing aspects experienced the smallest proportion of high severity and the highest proportion of low-severity burns. Middle slope positions were intermediate (Fig. 12.8). Severity patterns of the

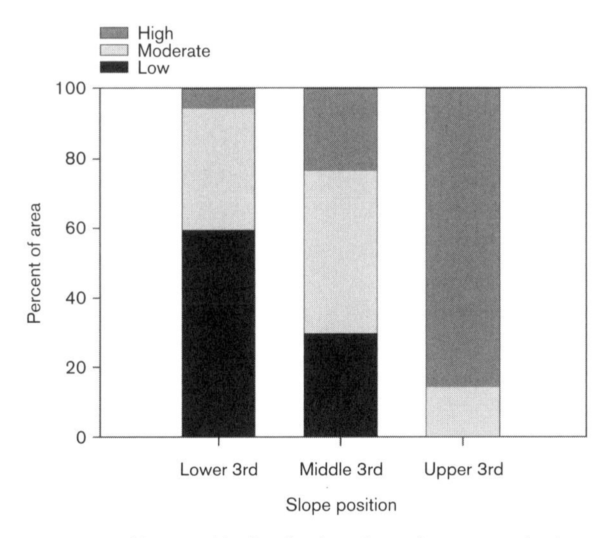

FIGURE 12.8 Topographic distribution of severity patterns in the Cub Creek Research Natural Area (Beaty and Taylor 2001).

Cub Fire, which burned the entire CCRNA in 2008, were similar to the presuppression patterns. Slope position and vegetation type in 1941 were the most important variables explaining severity (Taylor et al. 2013). This pattern, strongly pronounced in the Klamath Mountains, is discussed further in chapter 11.

Southwestern Foothills

FIRE RESPONSES OF IMPORTANT SPECIES

Blue oak woodland is the most common alliance in this subregion and tree densities vary from forest to savanna. Most dominant shrubs and hardwood trees in the foothills sprout following fire (Table 12.3). These include scrub oak (*Quercus berberidifolia*), interior live oak (*Q. wislizeni*), birch-leaf mountain-mahogany (*Cercocarpus betuloides* var. *betuloides*), California yerba santa (*Eriodictyon californicum*), bearbrush (*Garrya fremontii*) (commonly called silktassel in this bioregion), flannelbush (*Fremontodendron californicum*), and California bay (*Umbellularia californica*). Germination after fire from a soil seed bank is characteristic of three dominant shrubs—buck brush, whiteleaf manzanita (*Arctostaphylos viscida*), and common manzanita (*A. manzanita*) (Biswell and Gilman 1961). California-nutmeg (*Torreya californica*), a sprouting conifer, is commonly found in the steep canyons, especially on north-facing slopes (Griffin and Critchfield 1976).

Additional important species in this zone are gray pine, ponderosa pine, California black oak, and California juniper (*Juniperus californica*). Each of these species survives surface fires of low–moderate intensity. Among these species, ponderosa pine is most fire resistant and California black oak least. Moreover, both gray pine and California black oak have characteristics that promote regeneration after more severe fire. When top-killed, California black oak sprouts vigorously from the root crown. Gray pine is semi-serotinous, and its heavy cones protect the seeds from heat and cones open slowly over several years. California junipers have dense crowns that inhibit herbaceous understory growth so fires mostly scorch the edges of individual crowns. However, California juniper is not resistant to fire and is killed by moderate-intensity surface fires.

FIRE REGIME–PLANT COMMUNITY INTERACTIONS

The southwestern foothills have broad, flat ridgetops that support open stands of ponderosa pine and California black oak. An example is the Beaver Creek Pinery (BCP) in the Deer Creek watershed (Fig. 12.9). BCP is approximately 250 ha (615 ac) and burned at least five times in the twentieth century (Taylor 2010). BCP most recently burned in the large Campbell (1990) and Barkley (1994) fires, suffering only patchy tree mortality. Interestingly, diameter distributions for unlogged pine stands of BMEF in 1938 were similar to current diameter distributions in BCP (Oliver 2001). Thus, it is thought that BCP may broadly represent fine-grained patchy stand conditions similar to prefire-suppression ponderosa pine forests (Taylor 2010) (Fig. 12.10).

Since the 1994 fire, three major fires have been suppressed shortly before entering the BCP. These fires would have thinned the large crop of pine seedlings consequent of the 1998 El Niño that now occupy much of the understory (Fig. 12.11). Thus, there is concern that a continued lack of fire will soon cause the BCP to suffer similar stand structural changes that have occurred elsewhere in westside forests due to fire suppression.

Chapter 14 discusses fire regimes of the Sierra Nevada foothills that are likely to be similar to those of this ecological zone (Appendix 1).

Northwestern Foothills

Chapter 13 discusses fire regimes of the Northeastern Plateau which are likely similar to those of this area since many species and alliances are common to both areas.

Mid-Montane–Westside

Most information available on fire ecology and fire history in the mid-montane zone is for woody plants in conifer-dominated landscapes (Appendix 1).

FIRE RESPONSES OF IMPORTANT SPECIES

All of the more common conifer species (ponderosa pine, Douglas-fir, incense cedar, sugar pine, Jeffrey pine, and white fir) survive frequent surface fires of low–moderate intensity when mature (Table 12.4). The primary difference is how early in life they become resistant to these fires. All of the common deciduous hardwoods, California black oak, big-leaf maple, and mountain dogwood, and the evergreen canyon live oak are able to survive low-intensity surface fires and they sprout vigorously when top-killed. Stand composition is influenced by elevation, slope aspect, soil moisture conditions, and substrate (Griffin 1967). With few exceptions, the more common shrubs such as greenleaf manzanita, California-lilac species, and shrub-like oaks sprout vigorously after being top-killed. One exception is whiteleaf manzanita, which is easily top-killed by even low-intensity fires and relies on soil seed banks to germinate following fires. Manzanitas and California-lilacs generally germinate from seed following fires. An exception is Lemmon's ceanothus (*Ceanothus lemmonii*), which relies entirely on sprouting after being top-killed.

TABLE 12.3
Fire response types for important species in the Southwestern Foothills Zone of the South Cascades bioregion

	Type of fire response			
Lifeform	Sprouting	Seeding	Individual	Species
Conifer	None	Fire-stimulated (seed dispersal)	Resistant/killed	Gray pine
	Fire stimulated	None	Resistant/top-killed/survive	California nutmeg
	None	None	Resistant/killed	Ponderosa pine
Hardwood	Fire stimulated	None	Resistant/top-killed/survive	California black oak, blue oak
	Fire stimulated	None	Top-killed/survive	Interior live oak, valley oak, California bay, big-leaf maple, Pacific dogwood, foothill ash, Fremont cottonwood
Shrub	None	Stimulated (germination)	Killed	Buck brush
	Fire stimulated	None	Top-killed/survive	California buckeye, Lemmon's ceanothus, flannel bush, birch-leaf mountain-mahogany, toyon, yerba santa, poison oak

FIGURE 12.9 The Beaver Creek Pinery sits on this flat ridgetop in the Ishi Wilderness Area. Entire landscape shown burned in both the Campbell (1990) and Barkley (1994) fires. Photo taken in 1998 (photo credits: Carl Skinner, USDA Forest Service).

FIGURE 12.10 Typical stand structure in BCP in 2000, six years following last fire. Note two people standing near large ponderosa pine at lower, right of center (photo credits: Carl Skinner, USDA Forest Service).

FIGURE 12.11 Abundant seedlings in understory of BCP in 2008, only 14 years after last fire. In absence of fire, seedlings will grow to alter stand structure (photo credits: Carl Skinner, USDA Forest Service).

TABLE 12.4
Fire response types for important species in the Mid-Montane Westside Zone of the South Cascades bioregion

	Type of fire response			
Lifeform	Sprouting	Seeding	Individual	Species
Conifer	None	Fire-stimulated (establishment)	Resistant/killed	Ponderosa pine, Douglas-fir, Jeffrey pine
	None	None	Resistant/killed	Incense cedar, sugar pine, white fir, western juniper
	None	Fire-stimulated (seed dispersal)	Killed	Knobcone pine, McNabb cypress
	None	None	Killed	Pacific yew
Hardwood	Fire-stimulated	None	Resistant/top-killed/survive	California black oak
	Fire-stimulated	None	Top-killed/survive	Big-leaf maple, canyon live oak, white alder, Oregon ash, quaking aspen
Shrub	Fire-stimulated	Stimulated (germination)	Top-killed/killed	Greenleaf manzanita
	None	Stimulated (establishment)	Killed	Curl-leaf mountain-mahogany
	Fire-stimulated	Stimulated	Top-killed/survive	Deerbrush, tobacco brush, mahala mat
	None	Stimulated (germination)	Killed	Buck brush
	Fire-stimulated	None	Top-killed/survive	Birch-leaf mountain-mahogany, shrub tanoak, vine maple, mountain maple, bush chinquapin

FIRE REGIME–PLANT COMMUNITY INTERACTION

Fire regimes in mixed-species, mid-montane forests have been documented in the CCRNA in the Deer Creek drainage as discussed above under topography and severity. They varied with forest species composition and environment (i.e., slope aspect, slope position, elevation) (Beaty and Taylor 2001). Most fires burned only one or two slopes but occasionally burned more extensively and these were in very dry years (Taylor et al. 2008). This suggests that topography is an important control on fire severity and forest structure at landscape scales in complex terrain. Similar landscape patterns have been described in the steep, complex terrain of the Klamath Mountains (Taylor and Skinner 1998) and are discussed in more detail in chapter 11.

TABLE 12.5
Fire response types for important species in the Low-Elevation Eastside Zone of the South Cascades bioregion

Lifeform	Type of fire response			Species
	Sprouting	Seeding	Individual	
Conifer	None	Fire-stimulated (seed dispersal)	Resistant/killed	Gray pine
	None	None	Resistant/killed	Ponderosa pine, western juniper
Hardwood	Fire-stimulated	None	Resistant/top-killed	California black oak
	Fire- stimulated	Stimulated (establishment)	Survive/ top-killed	Oregon oak
Shrub	None	Stimulated (germination)	Killed	Buck brush
	Fire-stimulated when young	None	Top-killed/sometimes survive when young	Bitterbrush
	Fire-stimulated	Fire-stimulated (germination)	Top-killed/survive	Mahala mat
	Fire-stimulated	None	Top-killed/survive	Mountain misery

Fire frequency in CCRNA also varied by time period. Mean composite FRIs were similar in the Native American (6.6 years) and settlement (7.7 years) periods but longer (30.7 years) in the fire-suppression period. There was no evidence in CCRNA, or in other mid-montane forests on the west slope of the Cascades (Norman and Taylor 2002), of a fire frequency change during the settlement period like that reported for mid-montane forests in the Sierra Nevada (Gill and Taylor 2009, Taylor et al. 2016) and north Coast Range (Skinner et al. 2009).

The mixed-species conifer forests on the west side of the Cascades have changed since the onset of fire suppression (Norman and Taylor 2002). Forest density has increased along with a shift in species composition toward increasing density of fire-sensitive white fir. Moreover, many areas that were montane chaparral early in the twentieth century have been invaded by trees, especially white fir, and are now closed forests. These areas today appear to have a tendency to burn at high-severity setting succession back to montane chaparral (Taylor et al. 2013). Thus, it is likely that fire suppression has been an important factor in vegetation changes that have reduced the structural diversity of both forest stands and landscape patterns in the mid-montane zone.

Of concern is the increase in large fires containing unusually large patches (100s ha to 1,000s ha [200s–2,000s ac]) of high severity that can affect postfire succession and the associated trajectory of wildlife habitat (Long et al. 2014a). Succession in such fires usually begins with forests initially replaced with a combination of snags, herbaceous plants, and shrubs. Where no management is undertaken, snags decline over a relatively short time (Ritchie et al. 2013, Ritchie and Knapp 2014), there is only limited conifer regeneration, and shrubs tend to come to dominate (Zhang et al. 2008, Collins and Roller 2013, Crotteau et al. 2013, Coppoletta et al. 2016). Generally, the larger the high-severity burn patch, the fewer the available seed trees, with increasingly less conifer regeneration except along edges of the patches (Crotteau et al. 2013, 2014, Coppoletta et al. 2016). It is likely to take many decades to over a century without a repeat severe fire to reforest and develop old-growth-like conditions in large, high-severity burn patches. Moreover, areas that burned at high severity initially and converted to chaparral, especially those with dense snags, tend to burn again at high severity—e.g., the Storrie Fire (2000) followed by the Chips Fire (2012) (Collins et al. 2009, Coppoletta et al. 2016). Consequently, there is concern that large high-severity burn patches may cause a more permanent shift of vegetation from forest to shrubland (Long et al. 2014a, Hessburg et al. 2016, Lauvaux et al. 2016).

Mid-Montane–Eastside

FIRE RESPONSES OF IMPORTANT SPECIES

Most of the more common conifer species (Jeffrey pine, ponderosa pine, incense cedar, and white fir) survive frequent surface fires of low–moderate intensity when mature (Table 12.5). Again, the primary difference is how early in life each species becomes resistant to these fires. Western juniper is more easily killed by fires that other conifers would survive and it invades open sites from rocky refugia during longer fire-free periods. Three conifers in this zone, knobcone pine (*Pinus attenuata*), Macnab cypress (*Hesperocyparis macnabiana*), and Baker cypress (*H. bakeri*) have serotinous cones (Rentz and Merriam 2011, Milich et al. 2012). These species rely on occasional severe crown fires to induce regeneration.

The more common hardwoods of this zone are Oregon oak, curl-leaf mountain-mahogany, quaking aspen, and California black oak. Oregon oak is resistant to low–moderate-intensity surface fires and is able to sprout vigorously if top-killed by fire. Curl-leaf mountain-mahogany is killed by even low-intensity surface fires. It reinvades burned sites from rocky fire refugia during periods of reduced fire activity. The fire ecology of quaking aspen has not been well studied in this bioregion. However, quaking aspen is fire sensitive but sprouts vigorously after it has been top-killed by fire. Mahala mat (*Ceanothus prostratus*) is a common understory shrub in the mid- to upper montane forests of the range on both sides of the crest. It germinates from stored seed banks following fires.

FIRE REGIME–PLANT COMMUNITY INTERACTIONS

Of the three species of serotinous-cone conifers, Baker cypress and Macnab cypress occur in small, widely scattered groves on rocky, shallow soils (Griffin and Critchfield 1976). More information on the fire ecology of these cypresses can be found in chapter 13. The most widespread of the serotinous-cone species is knobcone pine found mostly on the westside, but also on the eastside. Knobcone pine has not been well studied in this bioregion. The most extensive stands occur on the south- and east-facing slopes of Mt. Shasta. On the eastside, knobcone pine extends to the lava beds north of Fall River Valley and into the Northeast Plateau along the base of the Big Valley Mountains. The widespread distribution of this serotinous cone species suggests that severe fires were an important component of mid-montane fire regimes at least in some locations. Knobcone pines are short-lived and usually begin dying after 50 or more years without fire (Vogl et al. 1977). Thus, after decades of fire suppression and the associated low fire occurrence, many stands are experiencing widespread mortality (e.g., Imper 1991). Severe crown fires are advantageous to mature, standing live trees because these fires move quickly through the stands and melt heavy resin coating the cones. This promotes dispersal onto mineral soil soon after the passage of the fire. Conversely, as time since fire increases and stands die, the dead, fallen trees accumulate as dry fuel leaving the seedlings and cones susceptible to intense fires that consume cones and kill seeds and seedlings. Consequently, contemporary stands may lose the ability to regenerate and this species may experience a range contraction in the Cascades (Merriam and Rentz 2010). More information on the fire ecology of knobcone pine can be found in chapters 11 and 14.

The low growth habit of mahala mat situates green, relatively moist leaves on and near soil and litter surfaces. This low growing shrub often does not burn well except under extreme burning conditions. Unburned patches of the shrub on the forest floor provide refuge for small, fire-susceptible seedlings and saplings of conifers and other shrubs (e.g., white fir, bitterbrush, curl-leaf mountain-mahogany) to establish and survive. By giving protection to young woody plants, mahala mat may have played an important role in long-term stand dynamics and vegetative diversity (Dunning 1923).

The historical fire regimes of yellow pine (Jeffrey pine and ponderosa pine) dominated alliances, especially in the area transitioning from mid-montane to upper montane environments in the southern portion of the bioregion, have been described by Taylor (2000) and Norman (2002). Fires were frequent in yellow pine forests until late in the nineteenth century or the onset of the fire-suppression period (Table 12.5).

Norman (2002) identified prefire-suppression fire regimes and stand structure across a ~800 km^2 (~300 mi^2) landscape between Eagle Lake and Butte Creek. Both Open Pine Forest sites (OPF) and Closed Pine Forest sites (CPF) were studied. Median FRIs were 12 years (range 6–17 years) for OPF sites and 14.4 years (range 7–22 years) for CPF sites. Notably, widespread fires, detected on multiple sites across the large study area, had a median return interval of 20.5 years (range 7–49 years).

These conifer forests have changed, often considerably, with shifts in species composition toward more fire-sensitive white fir and increasing density of pines coincident with the period of fire suppression (Norman 2002). Moreover, many areas that were montane meadows early in the twentieth century have been invaded mostly by Jeffrey or lodgepole pine, and are becoming forests (Norman and Taylor 2005). Thus, fire suppression, interacting with climatic variation and more locally with other land uses such as grazing and logging, has contributed to vegetation changes that have reduced the structural diversity of both forest stands and landscapes.

It is likely there was only limited accumulation of dead woody material in these forests due to the frequency of low–moderate-intensity fires (Skinner 2002). Accumulations today appear to be in excess of that expected under the original fire regime (Stephens et al. 2007a, Uzoh and Skinner 2009). Greater accumulations of dead woody material can contribute to increased fire intensity and more severe fire effects (Brown et al. 2003, Uzoh and Skinner 2009). Research has shown that decomposing woody material is more likely to be consumed in prescribed fires, and presumably wildfires, than sound wood (Stephens and Moghaddas 2005, Uzoh and Skinner 2009). Thus, given the frequency of fires before fire suppression, it appears unlikely that woody material would have been able to fully decompose before being consumed by fire. Thus, the dead wood on the forest floor likely consisted mostly of recently fallen sound wood with little in advanced stages of decay (Skinner 2002).

Small stands of quaking aspen are scattered throughout the eastside montane and upper montane zones. Recent research and years of observation suggest several important relationships between quaking aspen and fire. In quaking aspen stands with few conifers, the high fuel moisture of the herbaceous understory and typically low fuel loads reduce fire intensity as fire enters stands, even under more severe burning conditions. Severe passive crown fires in adjacent conifer stands have been observed to change to low-intensity surface fires upon entering quaking aspen stands. However, if the quaking aspen stand has been heavily invaded by conifers, the fires will continue to burn severely.

Conifers are capable of replacing quaking aspen and fires were probably a key process that maintained such stands before the fire-suppression period (Arno and Fiedler 2005, Jones et al. 2005, 2011). Since the onset of fire suppression, many quaking aspen stands have been invaded and overtopped by conifers (Pierce and Taylor 2010). Where heavy conifer invasion has significantly reduced the vigor of quaking aspen clones, the clones may have difficulty sprouting following severe fires. In some areas, conifers have already replaced quaking aspen and contemporary quaking aspen stands probably cover less area than before the fire-suppression period. Restoration work on the Lassen National Forest has shown that selective removal of conifers helps release and reinvigorate aspen stands (Jones et al. 2005, 2011). Application of prescribed fire to small mixed aspen-conifer stands in Lassen Volcanic National Park (LVNP) promoted aspen regeneration, which was lowest in stands with the greatest basal area of conifers. However, browsing of regeneration by deer was high, indicating that interaction of fire effects and biotic factors (browsing, competition) may benefit or contribute to decline in small aspen stands (Margolis and Farris 2014).

Young basalt flows with little or no soil cover large areas such as in the Hat Creek Valley. These harsh sites limit fuel accumulations and fires are unable to burn well except under severe weather conditions. These harsh sites seem to serve as

TABLE 12.6
Fire response types for important species in the Upper Montane Zone of the South Cascades bioregion

Lifeform	Type of fire response: Sprouting	Seeding	Individual	Species
Conifer	None	None	Resistant/killed	Jeffrey pine, ponderosa pine, white fir, red fir, western white pine
	None	None	Killed	Lodgepole pine
Hardwood	Fire-stimulated	None	Top-killed/survive	Quaking aspen, willows, black cottonwood
Shrub	Fire-stimulated	Stimulated (seed production)	Top-killed/survive	Bush chinquapin, mountain whitethorn
	Fire-stimulated	Stimulated (germination)	Top-killed/survive	Greenleaf manzanita, tobacco brush
	None	Stimulated (germination)	Killed	Pinemat manzanita
	Fire-stimulated	None	Top-killed/survive	Huckleberry oak, rubber rabbitbrush
	None	None	Killed	Big sagebrush

fire refugia for many fire sensitive species including curl-leaf mountain-mahogany, western juniper, bitterbrush, and quaking aspen.

Many yellow pine dominated forests and woodlands today have a relatively continuous understory of bitterbrush such as those surrounding Butte Valley. Bitterbrush in this bioregion is fire sensitive and easily killed by even low-intensity fires. It does not sprout well unless young and vigorous. Where mature and robust it is highly flammable, burns with high intensity, and plants are usually killed outright. Given these traits, it is unlikely the contemporary widespread continuity of bitterbrush, especially the mature robust type, was typical of these environments under prehistoric fire regimes. It is likely bitterbrush took advantage of refugia provided by basalt flows and mahala mat to maintain itself in this fire-prone ecosystem.

Upper Montane

FIRE RESPONSES OF IMPORTANT SPECIES

Fire ecology information is available for montane shrubs and four tree-dominated alliances in this bioregion—white fir, red fir, lodgepole pine, and Jeffrey pine (Table 12.6). Of the more common conifer species found in the upper montane, Jeffrey pine is most fire resistant, followed by white fir, red fir, western white pine, and lodgepole pine, respectively (Agee 1993).

Shrub species dominance varies with substrate, soils, and other conditions. However, dominant woody species in these shrubfields sprout following fires including bush chinquapin, huckleberry oak, and bitter cherry (*Prunus emarginata*). Important species that both sprout and germinate from long-lived soil seed banks include tobacco brush, greenleaf manzanita, mountain whitethorn, and deer brush.

FIRE REGIME–PLANT COMMUNITY INTERACTIONS

Due to their ecological importance and potential as competition for commercial conifers, there has long been interest in shrub soil seed banks. As noted, many of the important shrubs in the bioregion can germinate following fires from seed stored in the soil (Knapp et al. 2012). The most common seeds found in soil seed banks were California-lilac species (deer brush, tobacco brush, whitethorn, mahala mat) followed by manzanitas (greenleaf, whiteleaf, common, pine-mat) and bitter cherry, respectively. Seeds were found generally throughout the soil profile on sites not recently burned. Sites recently burned in late summer or fall (wildfire or prescribed fire) generally had far fewer seeds in the upper several inches of soil. Sites burned in spring or early summer (prescribed fire only) maintained seed profiles similar to unburned conditions. Results of Knapp et al. (2012) suggest that recurring, low–moderate-intensity consumptive fires under dry, late summer conditions probably served to limit seed stores of montane shrub species.

Truffle abundance was evaluated 10 years after thinning and nine years after burning in the same study area (Waters et al. 1994). Interestingly, no difference in the frequency or biomass of truffles was found between thinning treatments with or without the use of prescribed fire (Waters et al. 1994). However, the composition of the truffle species assemblage differed between the thinned and unthinned environments. Since this was a one-time measurement after 10 years, there was no way to determine if the two treatments started the same and shifted or if succession differed from time of treatments.

There are few data on the fire history of montane shrubfields because the severe nature of fires in shrubfields consumes most evidence needed to reconstruct fire histories (Skinner and Chang 1996). However, fire scar data collected by Lauvaux et al. (2016) in LVNP from trees along the edges and scattered within stands of montane chaparral (n = 30) had a median

FIGURE 12.12 Paired photos showing representative changes in structure of white fir/ Jeffrey pine forests in LVNP. Photo A taken in 1925. Photo B taken at same location in 1993 (top photo by A.E. Wieslander. Bottom photo by Alan Taylor [from Taylor 2000—used here with permission of Blackwell Publishing]).

point fire return interval of 26 years (range 17.2–52 years). In contrast, surrounding forests (n = 10) had a median point fire return interval of 12.7 years (range 9–28.8 years). This suggests that FRIs in montane chaparral may be longer and more variable than in surrounding conifer forests. Fuel structure, fuel moisture, and other conditions needed for burning, and rates of fuel recovery in montane chaparral are different than in adjacent conifer forests and are probably responsible for the longer and more variable FRIs. The absence of fire, primarily due to fire suppression, has also led to widespread conversion of many montane shrublands to forest (Beaty and Taylor 2001, Bekker and Taylor 2001, 2010, Lauvaux et al. 2016).

White fir-dominated stands in the Cascades are common in the transition area from mid-montane to upper montane environments especially on mesic sites. Stands are usually mixed with either Jeffrey pine or red fir at lower and higher elevations, respectively. Fire regimes generally vary between the characteristics of these other two species depending upon which is more prevalent. Fig. 12.12 shows how conditions have changed in stands where white fir is a major component over the twentieth century in LVNP.

Stands where red fir is dominant or codominant are common, especially on mesic sites, in the upper montane zone (Laacke 1990, Parker 1991). Where this zone transitions to subalpine environments, red fir is found most frequently on xeric sites (Taylor 1990b). Common red fir associates include white fir, western white pine, Jeffrey pine, lodgepole pine, and mountain hemlock (Laacke 1990).

Pine-mat manzanita is a common, prostrate understory shrub in upper montane forests and woodlands. Its low growth habit situates green, relatively moist leaves on and near soil and litter surfaces. Pine-mat manzanita burns well only under extreme burning conditions and burns are usually patchy. Islands of pine-mat manzanita serve as fire refugia for small, fire-susceptible conifer seedlings and saplings similar to that described for mahala mat. The authors observed many

FIGURE 12.13 Top, October 1925, on west side of Prospect Peak. Photo shows red fir-western white pine stand with few seedlings and saplings. Understory has patches of tobacco brush and bush chinquapin with little herbaceous cover. Fire scars indicate the stand last burned in 1883. Bottom, July 1993. Trees have increased in height and diameter with new seedlings and saplings established since 1925. Seedlings and saplings have established in the opening created by the loss of tree in 1925 foreground left. Surface fuels (needles, twigs, branches, boles) have increased on the forest floor (top photo by A.E. Wieslander. Bottom photo by Alan Taylor [from Taylor 2000—used here with permission of Blackwell Publishing]).

examples of unburned islands of pinemat manzanita with surviving conifer seedlings and saplings following the Huffer (1997) and Reading (2012) fires in LVNP.

Fire occurrence in red fir forests has diminished considerably since the onset of fire suppression (Taylor 1993, McKelvey et al. 1996, Taylor 2000). Along with the ever-increasing time between fires, there has been an increase in forest density, similar to forest changes that are occurring in lower and mid-montane forests (Taylor 2000). Fig. 12.13 is representative of the ongoing changes in stand conditions that are representative of upper elevation red fir forests in the southern Cascades.

Fire regimes in red fir dominated landscapes are characterized by mixed-severity burns (Agee 1993, Perry et al. 2011) that occur mainly in the late summer or fall after trees have stopped growth for the year (Taylor 2000). Mean FRIs for points (individual trees) in forests dominated by red fir range from 25 to 110 years with the median being 70 years (Taylor and Halpern 1991, Taylor 1993, 2000, Bekker and Taylor 2010). This is similar to mean values (i.e., 65–76 years, range 25–175 years) reported for high-elevation California red fir-western white pine forests in the northern and southern Sierra Nevada (Pitcher 1987, Scholl and Taylor 2006). However, most studies in red fir forests have used fire scars extracted from open wounds and did not have access to stumps. Open wounds on trees of these species tend to rot. Access to stumps allows the discovery of hidden scars in trees that successfully healed following the wound. Without access to stumps the estimates of fire occurrence may be under estimated.

Stands dominated by lodgepole pine are more similar to those of central Oregon (e.g., Stuart et al. 1989, Parker 1991, 1993, Agee 1993) than to stands in the Sierra Nevada (e.g., Parker 1986, Potter 1998). In the Cascades, lodgepole pine dominates topographic settings conducive to cold-air ponding, high water tables, or sites with infertile, coarse volcanic material (Parker 1991, 1993, Agee 1994, Bekker and Taylor 2001, Taylor and Solem 2001). Red fir and white fir are often important associates, especially on more upland sites (Parker 1991, 1993) where the firs may eventually replace lodgepole

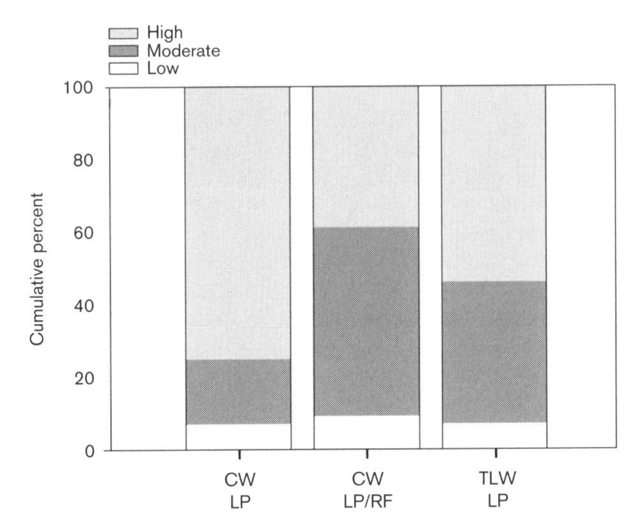

FIGURE 12.14 Distribution of fire severity classes in lodgepole pine, Caribou (Taylor and Solem 2001) and Thousand Lakes (Bekker and Taylor 2001) wilderness areas.

pine after long periods without fire (Bekker and Taylor 2001, Taylor and Solem 2001).

Lodgepole pines are considered fire sensitive since they have thin bark and are easily damaged or killed by moderate intensity, surface fires (Agee 1993, Cope 1993). In this bioregion, average point (individual tree) FRIs were 63 years (range 59–67 years) in the Caribou Wilderness (Taylor and Solem 2001) and 47 years (range 38–54 years) in the Thousand Lakes Wilderness (Bekker and Taylor 2001). These are more comparable to upper montane, red fir forests than montane yellow pine forests. FRIs in this bioregion are shorter than those reported for lodgepole pine in central Oregon (Agee 1993). However, lodgepole stands north and east of Mt. Shasta may be more similar to those of central Oregon since they too are most extensive on pumice and andesite flats.

A mixed-severity fire regime (Perry et al. 2011, Hessburg et al. 2016) consisting of small, low-severity fires and larger fires with significant portions of moderate- and high-severity, stand-replacing effects appears to have been common in lodgepole pine forests (Fig. 12.14). The 1984 Badger Fire (563 ha [1,391 ac]) burned with mixed effects through lodgepole pine forest in LVNP. Twenty-four years postfire, tree regeneration was highest in areas burned with moderate-severity effects. Regeneration in patches burned at high severity is dependent on proximity to seed sources. Large high-severity patches still have areas devoid of seedlings because of long distances to the forest edge (Pierce and Taylor 2011). Unlike the Rocky Mountain variant of lodgepole pine, Sierra Nevada lodgepole pines do not have serotinous cones. Recurring mixed-severity fires, interacting with beetles and wind events, have created variable multiaged forests at both plot and landscape scales (Bekker and Taylor 2001, 2010, Taylor and Solem 2001). Average extent of prefire-suppression fires in lodgepole pine forests were 150 ha (370 acres) in the Caribou Wilderness (Taylor and Solem 2001) and 405 ha (1,000 acres) in the Thousand Lakes Wilderness (Bekker and Taylor 2001, 2010).

Fire rotations in lodgepole pine forests varied by time period. In the Caribou Wilderness, the fire rotation has increased from 76 years in the Native American period to 577 years (Taylor and Solem 2001) and from 46 years to no fires detected in fire scars in lodgepole pine forests in the Thousand Lakes Wilderness since the beginning of fire suppression (Bekker and Taylor 2001); however, some lodgepole pine stands in the Thousand Lakes Wilderness burned in the 2014 Eiler fire.

Stand density in eastside yellow pine forests has increased during the twentieth century (Dolph et al. 1995) and the onset of the increase corresponds with the beginning of the fire-suppression period (Taylor 2000, Norman 2002). Repeat photography of eastside pine forests in LVNP that have never been logged (Weislander 1935) illustrate changes in forest conditions between the 1920s and 2000. These commonly seen changes are unlikely due to climate change. They are due mainly to the dramatic reduction of fire as a result of fire suppression (Fig. 12.15).

Jeffrey pine-dominated forests occupy sites generally less mesic than ponderosa pine in eastside environments—e.g., dry ridgetops and cold valley bottoms. In LVNP they occupy the lower elevations on south-, east-, and west-facing slopes. There the median point FRIs (n = 36) for this type of forest was 16 years (range 13–32 years). Longer FRIs on slopes that dry earlier each year (i.e., south, west) and hence are more fire prone is counterintuitive and appears to be due to lava flows and scoria fields that interrupts fuel continuity on these slopes (Taylor 2000). This underscores the importance of local topography and geology as a control on fire regimes.

Twentieth-century changes in stand density and species composition have been documented in unlogged stands of BMEF (Dolph et al. 1995, Ritchie et al. 2008). Dolph et al. (1995) found the density of small diameter trees (<38 cm [15 in]) in unlogged stands increased four or more fold between 1938 and 1991, while the density of large diameter trees (> 70 cm [27.5 in]) declined by 3.5 ha^{-1} (1.4 ac^{-1}). Ritchie et al. (2008) studied changes in stand conditions between the 1934 BMEF tree census to 1999 within the Blacks Mountain Research Natural Area (BMRNA). They too found a four or more fold increase in small diameter trees (<30 cm [12 in]) with the density of large trees (>60 cm [24 in]) having declined by 58% from an average of ~49.3 ha^{-1} (20 ac^{-1}) to ~20.8 ha^{-1} (8.4 ac^{-1}). The remaining large trees also showed greater signs of low vigor and declining health compared to large trees in nearby stands treated in the Blacks Mountain Ecological Research Project (BMERP) (Oliver 2000). The low vigor explains why the use of prescribed fire in the BMRNA appears to have accelerated the rate of mortality of the large, old trees, especially when compared to the nearby stands thinned of the smaller trees. Their modeling suggests that the declines in the BMRNA stands will continue in the absence of stand density management.

It has been hypothesized that the loss of larger trees has been at least partly due to increased competition from the smaller trees that have established since the onset of fire suppression. Increased competition may have created stressful conditions for the larger trees making them more susceptible to fatal insect attack, especially during dry periods. Intense competition between small- and intermediate-sized trees also reduces growth of intermediate size trees and inhibits their recruitment into larger size-classes as existing larger trees die (Dolph et al. 1995, Ritchie et al. 2008).

Subalpine

Lightning is frequent in the subalpine zone of the Cascade Range (van Wagtendonk and Cayan 2008). Yet, fire does not appear to play a significant role in this zone. Trees in this zone

FIGURE 12.15 Top, October 1925, south side of Prospect Peak. Photo shows mixed Jeffrey pine-white fir stands with few small trees. A small patch of tobacco brush is in the foreground (right) with little herbaceous cover. Note the charred bark on the large diameter stems. Fire scars indicate the stand last burned in 1892. Bottom, June 1993, a dense stand of mostly white fir now dominates the understory and surface fuels have accumulated on the forest floor. Stand density is now 820 trees (>4.0 cm dbh) ha^{-1} (top photo by A.E. Wieslander. Bottom photo by Alan Taylor in 1993 [from Taylor 2000—used here with permission of Blackwell Publishing]).

are generally very sensitive to even low-intensity fires. Though lightning is frequent, the few fires that ignite rarely spread beyond individuals or small groups of trees. This is the result of generally late melting snowpacks and sparse fuels that are not conducive to fire spread. Historic fire regimes were likely similar to that described for the adjacent Sierra Nevada bioregion (see chapter 14).

Management Issues

Four management issues—wildlife habitat, wildland-urban interface, changing stand structures, and compounding fire deficit—stand out in this bioregion. The gentle terrain with limited topographic breaks is unique to the bioregion when compared to the Sierra Nevada and the Klamath Mountains. Thus, when conditions are prime for large fires, fuel patterns, more than topography, determine spread potential (Collins and Skinner 2014). Effects of the ongoing fire deficit combined with the spread of wildland-urban interface leads to large areas of continuous, similar vegetation making management of wildfire more difficult.

Wildlife Habitat

There is a need to better understand the role frequent low- and moderate-intensity fires once had on the development of wildlife habitat at watershed and landscape scales. Especially important is an improved understanding of the influence of fire on development of old forest ecosystems and patterns of large, dead woody material in the relatively gentle terrain of the Cascade Range. The topography of the Cascades, more conducive to extensive fires (Norman and Taylor 2003) than in the Klamath Mountains (Taylor and Skinner 2003), indicates relationships with these habitat factors may be different than those described for other bioregions. Describing these patterns would have important implications for management of such species as spotted owls (Keane 2014), fur-bearers (Zielinski 2014), and northern goshawks (*Accipiter gentilis*). Managers would be better able to assess risks associated with different management alternatives that inadequately consider fire as a process in fire-prone ecosystems (e.g., Arno and Allison-Bunnell 2002).

Wildland-Urban Interface

A large proportion of the bioregion is classified as mixed interface (CDF 2002a) because dwellings are widely dispersed in small, scattered communities that are embedded in wildland vegetation. This pattern has resulted in the loss of hundreds of homes to wildfires in just the last several decades (CDF 2002b). The Fountain fire alone burned over 300 homes in 1992 (CDF 2002b). More recent fires that burned significant numbers of homes were the Ponderosa (2012) and Eiler (2014) fires. The wildland-urban interface fire problem will continue to grow as development inexorably expands throughout the bioregion.

Changing Stand Structures

Up to a century of mostly successful fire exclusion in forests that once experienced frequent fires has greatly altered compositional and structural diversity in forest stands and forested landscapes. The result has been an increase in stand density, shift from fire-tolerant to fire-intolerant species, and reduced structural heterogeneity in both forest stands and across forested landscapes similarly to many parts of the western United States (Agee 1993, Arno and Allison-Bunnell 2002). Such changes (Taylor 2000, Ritchie et al. 2008) in this bioregion have resulted in increasing area burned by severe wildfires, especially in lower and mid-elevation forests (McKelvey et al. 1996, Weatherspoon and Skinner 1996). Miller and Safford (2012) determined that the number of large wildfires (≥400 ha [1,000 ac]) has been increasing since 1950. Additionally, over the period of satellite imagery (1984–2010), the proportion of area burned with high severity has increased significantly. This increase is consistent with a fire exclusion-tree densification model of vegetation change. It has been suggested that a shift to more severe fires may likely occur in previously less fire-prone, higher elevation forests as stand density and time since last fire increases with continuing fire suppression (Taylor 2000) and a warming climate.

The Compounding Fire Deficit

Though managers and society often state goals of restoring fire resilience and ecological integrity to California forests (Long et al. 2014b), there remains a significant and growing deficit in area burned needed to sufficiently achieve those goals compared to the annual rates of burned area before fire exclusion (Stephens et al. 2007b, North et al. 2012). The gentle terrain of the southern Cascade Range has facilitated access to large areas and promoted the use of mechanical means to manage stand structure and fuels. Managers and society favor the use of mechanical means wherever possible to avoid short-term risks associated with the use of fire for vegetation and fuels management (Quinn-Davidson and Varner 2012). Yet, research and experience from recent fires (Ritchie et al. 2007, Symons et al. 2008) demonstrate that reliance on mechanical means without the additional use of prescribed fire to manage surface fuels falls short of ecological restoration (Stephens et al. 2012, McIver et al. 2013) and gives a false impression of having achieved fire resilience (Schmidt et al. 2008, Stephens et al. 2009, Knapp et al. 2011).

Several wilderness areas are found in this bioregion (Caribou, Ishi, Lassen, Mount Shasta, Thousand Lakes) where managed fire (either prescribed or use of wildfire [see North et al. 2012]) is the primary tool available for manipulating vegetation and fuels. Yet, only in LVNP, the largest of the wilderness areas, has there been a major program to use and manage fires. Whereas the National Park Service has been successfully managing fires in large areas of Yosemite and Sequoia-Kings Canyon National Parks for decades (Miller et al. 2012), the experience in LVNP has been mixed. We believe this is largely due to lower physical complexity of the landscapes with higher fuel continuity.

The southern Sierra Nevada, where Yosemite and Sequoia-Kings Canyon National Parks are located, has considerable topographic relief, a long elevation gradient, and extensive fuel-free geologic substrate. Thus, areas where fires have been permitted to spread naturally occur within compartments of vegetation and fuels where fires can burn but not easily spread into other landscapes due to interruptions of fuel and changes in burning conditions. In contrast, the wilderness areas in the more gentle terrain of this bioregion are elevated islands connected to expanses of mostly continuous fuels with few interruptions to fire spread. These qualities make the broader Cascade Range landscapes particularly vulnerable to large fires under severe conditions. Thus, with fire the primary tool available to manage these landscapes, it becomes necessary to use fire under desired conditions whenever possible in conjunction with managing the broader landscape. The alternative is more fires like the Eiler Fire (2014) that burn out from the wilderness and do considerable damage over the broader landscape.

That the gentle landscapes of the southern Cascade Range are particularly vulnerable to large, severe fires suggests the expanded use of managed fire is perhaps more imperative here than in other forested bioregions of California. It appears that without a greater use of managed fire it will remain difficult to achieve the oft-stated goals of improving fire resilience and restoring ecological integrity in this bioregion.

Future Directions

Ecological Effects of Fuels Treatments and Altering Stand Structures

The changing stand and fuel conditions in fire-prone lower elevation forests has led many to conclude that widespread treatment of forest fuels is necessary to restore ecological integrity and to reduce the high risk of destructive, uncharacteristically severe fires in forests that have been highly altered by fire exclusion, timber harvesting, and grazing (e.g., Hardy and Arno 1996, Turhune 2002). Treatments that are being discussed for widespread application include selective tree removal, mechanical manipulations of fuels, and prescribed fire but the effectiveness of these treatments and especially long-term ecological effects are poorly known (Weatherspoon and Skinner 2002). Resource managers need better information about the long-term ecological consequences of using different methods to alter stand structures to reduce fire hazard (Zack et al. 1999, Oliver 2000, Weatherspoon and Skinner 2002).

To address this need, three large stand-manipulation studies were initiated in the Cascade Range by the Forest Service, Pacific Southwest Research Station's Redding Laboratory. The three large studies are: (1) the BMERP (Oliver 2000, Ritchie and Skinner 2014); (2) the Little Horse Peak Research Project (GAMA [Goosenest Adaptive Management Area]) (Zack et al. 1999, Ritchie 2005); and (3) the Southern Cascades site of the National Fire and Fire Surrogates Study (FFS) (Boerner et al. 2008). These studies are interdisciplinary, complimentary, and designed to assess both the short- and long-term ecological effects of altering stand structures using mechanical thinning, prescribed fire, or both. With the completion of the FFS treatments in November 2002, all initial treatments for each study are in place. One study area, BMERP, was tested by the Cone wildfire burning under severe conditions in September of 2002 and demonstrated the ability of the stand structure treatments to significantly reduce fire severity and associated effects. Though a number of papers describing the short-term findings of these studies have been published, continued measurements of changing conditions over several more decades is necessary to provide an understanding of their long-term ecological effects.

It would take too much space here to summarize the findings on the short-term effects of treatments from these studies. However, syntheses of this work can be found in several publications: BMERP in a special issue of the *Canadian Journal of Forest Research* (2008, Vol. 38, pp.909–980), GAMA in a conference proceedings (Ritchie et al. 2005), and FFS in Stephens et al. (2012), McIver et al. (2013), and special issues of *Ecological Applications* (2009, Vol. 19, pp.283–358), *Forest Ecology and Management* (2008, Vol. 255, pp.3075–3211), and *Forest Science* (2010, Vol. 56, pp.2–138). Assessment of the BMERP influence on the severity and subsequent effects of the Cone fire can be found in Ritchie et al. (2007, 2008), Symons et al. (2008), and Ritchie and Knapp (2014).

Climate Change and Fire Regimes

The increasing number of large fires combined with the increasing proportion of high-severity burn (Miller and Safford 2012) is occurring during a period of rapid global climatic change (Fried et al. 2004, IPCC 2013, Jardine and Long 2014). We suggest that this trend, combined with a warming climate and longer fire seasons (Westerling et al. 2006, Safford et al. 2012, Jardine and Long 2014), may serve as a catalyst to more permanent shifts in vegetation from forests to shrublands (Collins and Skinner 2014, Lauvaux et al. 2016). Large forest areas of severe burn are expected to recover initially to shrubfields. In the absence of intensive reforestation following large, severe burns that severely limit seed sources, it would likely take many decades for trees to reinvade the landscape and grow to sufficient size to withstand a subsequent fire (Zhang et al. 2008). Under a warming climate we suggest it would be more likely for areas like this to reburn frequently enough to maintain many of these areas as shrubfields rather than allowing them to return to forest (Coppoletta et al. 2016, Lauvaux et al. 2016). An example can be seen in the recent >30,300 ha (>75,000 ac) Chips Fire (2012) that reburned much of the >21,000 ha (>52,000 ac) Storrie Fire (2000) (Long et al. 2014a, Coppoletta et al. 2016). Thus, old-growth or late-seral conditions that take many decades to centuries to develop are likely to become much more limited on the landscape.

Summary

Annual summer drought, ample winter precipitation, and abundant lightning combined to make fire frequent and generally low or moderate in severity in most of the southern Cascade Range during the Native American period. Fire exclusion, grazing, and logging have dramatically changed fire regimes in Cascade ecosystems over the last century. Fire suppression has likely been the most pervasive management activity since it alone has been ubiquitously applied. There is considerable desire to reduce fire hazard by manipulating fuels and stand structure either mechanically or with the use of managed fire. However, accomplishing this is often problematic due to competing social/political objectives for Cascade Range forests (e.g., wood production, preservation for wildlife, maintenance of air quality). Regardless of how these controversies are resolved, the ecosystems of the Cascade Range will continue to change in response to historic and contemporary management activities and the warming climate in expected and unexpected ways.

References

Agee, J.K. 1993. Fire Ecology of Pacific Northwest Forests. Island Press, Washington, D.C., USA.

———. 1994. Fire and weather disturbances in terrestrial ecosystems of the eastern Cascades. USDA Forest Service General Technical Report PNW-GTR-320. Pacific Northwest Research Station, Portland, Oregon, USA.

Arno, S.F., and S. Allison-Bunnell. 2002. Flames in Our Forest: Disaster or Renewal? Island Press, Washington, D.C., USA.

Arno, S.F., and C.E. Fiedler. 2005. Mimicking Nature's Fire. Island Press, Washington, D.C., USA.

Beaty, R.M., and A.H. Taylor. 2001. Spatial and temporal variation of fire regimes in a mixed conifer forest landscape, southern Cascades, California, USA. Journal of Biogeography 28: 955–966.

Bekker, M.F., and A.H. Taylor. 2001. Gradient analysis of fire regimes in montane forests of the southern Cascade Range, Thousand Lakes Wilderness, California, USA. Plant Ecology 155: 15–28.

———. 2010. Fire disturbance, forest structure, and stand dynamics in montane forests of the southern Cascades, Thousand Lakes Wilderness, California, USA. Ecoscience 17(1): 59–72.

Biswell, H.H., and J.H. Gilman. 1961. Brush management in relation to fire and other environmental factors on the Tehama deer winter range. California Fish and Game 47: 357–389.

Boerner, R.E.J., S.C. Hart, and J.D. McIver. 2008. The national fire and fire surrogates study: ecological consequences of alternative fuel reduction methods in seasonally dry forests. Forest Ecology and Management 255: 3075–3080.

Brown, J.K., E.D. Reinhardt, and K.A. Kramer. 2003. Coarse woody debris: managing benefits and fire hazard in the recovering forest. USDA Forest Service General Technical Report RMRS-GTR-105. Rocky Mountain Research Station, Fort Collins, Colorado, USA.

CCSS. 2002. Historical Course Data. California Resources Agency, Department of Water Resources, Division of Flood Management, California Cooperative Snow Surveys. http://cdec.water.ca.gov/snow/.

CDF. 2002a. Information and Data Center, Fire and Resource Assessment Program (FRAP). California Department of Forestry and Fire Protection, Sacramento, California, USA. http://frap.cdf.ca.gov/infocenter.html.

———. 2002b. Historical Statistics, Fire and Emergency Response. California Department of Forestry and Fire Protection, Sacramento, California, USA. http:/www.fire.ca.gov/FireEmergencyResponse/HistoricalStatistics/HistoricalStatistics.asp.

Collins, B.M., J.D. Miller, A.E. Thode, M. Kelly, J.W. van Wagtendonk, and S.L. Stephens. 2009. Interactions among wildland fires in a long-established Sierra Nevada natural fire area. Ecosystems 12: 114–128.
Collins, B.M., and G.B. Roller. 2013. Early forest dynamics in stand-replacing fire patches in the northern Sierra Nevada, California, USA. Landscape Ecology 28: 1801–1813.
Collins, B., and C. Skinner. 2014. Fire and fuels. Pages 143–172 in: J.W. Long, L. Quinn-Davidson, and C.N. Skinner, editors. Science Synthesis to Support Socioecological Resilience in the Sierra Nevada and Southern Cascade Range. USDA Forest Service General Technical Report PSW-GTR-247. Pacific Southwest Research Station, Albany, California, USA.
Cope, A.B. 1993. *Pinus contorta* var. *murrayana*. In: Fire Effects Information System [Data base]. USDA Forest Service, Intermountain Research Station, Intermountain Fire Sciences Laboratory, Missoula, Montana, USA.
Coppoletta, M., K.E. Merriam, and B.M. Collins. 2016. Post-fire vegetation and fuel development influences fire severity patterns in reburns. Ecological Applications 26: 686–699.
Crotteau, J.S., M.W. Ritchie, and J.M. Varner. 2013. Post-fire regeneration across a fire severity gradient in the southern Cascades. Forest Ecology and Management 287: 103–112.
———. 2014. A mixed-effects heterogeneous negative binomial model for post-fire conifer regeneration in northeastern California, USA. Forest Science 60: 275–287.
Dettinger, M.D., D.R. Cayan, H.F. Diaz, and D.M. Meko. 1998. North-south precipitation patterns in western North America on interannual-to-decadal timescales. Journal of Climate 11: 3095–3111.
Dolph, K.L., S.R. Mori, and W.W. Oliver. 1995. Long-term response of old-growth stands to varying levels of partial cutting in the eastside pine type. Western Journal of Applied Forestry 10: 101–108.
Dunning, D. 1923. Some results of cutting in the Sierra forests of California. U.S. Department of Agriculture Bulletin 1176. Washington, D.C., USA.
Estes, B.L., E.E. Knapp, C.N. Skinner, and F.C.C. Uzoh. 2012. Seasonal variation in surface fuel moisture between unthinned and thinned mixed conifer forest, northern California, USA. International Journal of Wildland Fire 21: 428–435.
Fried, J.S., M.S. Tom, and E. Mills. 2004. The impact of climate change on wildfire severity: a regional forecast for northern California. Climatic Change 64: 169–191.
Gannett, H. 1902. The forests of Oregon. US Geological Survey Professional Paper 4, Series H, Forestry, 1. Washington, D.C., USA.
Gill, L., and A.H. Taylor. 2009. Top-down and bottom-up controls on fire regimes along an elevation gradient on the east slope of the Sierra Nevada, California, USA. Fire Ecology 5(3): 57–75.
Griffin, J.R. 1967. Soil moisture and vegetation patterns in northern California forests. USDA Forest Service Research Paper PSW-RP-46. Pacific Southwest Research Station, Berkeley, California, USA.
Griffin, J.R., and W.B. Critchfield. 1976. The distribution of forest trees in California, with supplement. USDA Forest Service Research Paper PSW-RP-82. Pacific Southwest Research Station, Berkeley, California, USA.
Hardy, C.G., and S.F. Arno. 1996. The use of fire in forest restoration. USDA Forest Service General Technical Report INT-GTR-341. Intermountain Research Station, Ogden, Utah, USA.
Hessburg, P.F., T.A. Spies, D.A. Perry, C.N. Skinner, A.H. Taylor, P.M. Brown, S.L. Stephens, A.J. Larson, D.J. Churchill, P.H. Singleton, B. McComb, W.J. Zielinski, B.M. Collins, N.A. Povak, R.B. Salter, J.J. Keane, J.F. Franklin, and G. Riegel. 2016. Tamm review: management of mixed-severity fire regimes in forests in Oregon, Washington, and northern California. Forest Ecology and Management 366: 221–250.
Hull, M.K., C.A. O'Dell, and M.J. Schroeder. 1966. Critical fire weather patterns: their frequency and levels of fire danger. USDA Forest Service, Pacific Southwest Forest and Range Experiment Station, Berkeley, CA.
Imper, D.K. 1991. Ecological survey of the proposed Mayfield Research Natural Area, SAF Type 248 (Knobcone Pine), Lassen National Forest. Unpublished Report PO #40-9AD6-0409 on file. USDA Forest Service, Pacific Southwest Research Station, Albany, California, USA.
IPCC. 2013. Climate change 2013: the physical basis. In: T.F. Stocker, D. Qin, G.-K. Plattner, M. Tignor, S.K. Allen, J. Boschung, A. Nauels, Y. Xia, V. Bex, and P.M. Midgley, editors. Contribution of Working Group I to the Fifth Assessment Report of the Intergovernmental Panel on Climate Change. Cambridge University Press, Cambridge, UK.
Jardine, A., and J. Long. 2014. Synopsis of climate change. Pages 71–81 in: J.W. Long, L. Quinn-Davidson, and C.N. Skinner, editors. Science Synthesis to Support Socioecological Resilience in the Sierra Nevada and Southern Cascade Range. USDA Forest Service Pacific General Technical Report PSW-GTR-247. Pacific Southwest Research Station, Albany, California, USA.
Jones, B.E., D.F. Lile, and K.W. Tate. 2011. Cattle selection for aspen and meadow vegetation: implications for management. Rangeland Ecology and Management 64: 625–632.
Jones, B.E., T.H. Rickman, A. Vazquez, Y. Sado, and K.W. Tate. 2005. Removal of encroaching conifers to regenerate degraded aspen stands in the Sierra Nevada. Restoration Ecology 13: 373–379.
Keane, J.J. 2014. California spotted owl: scientific considerations for forest planning. Pages 437–467 in: J.W. Long, L. Quinn-Davidson, and C.N. Skinner, editors. Science Synthesis to Support Socioecological Resilience in the Sierra Nevada and Southern Cascade Range. USDA Forest Service Pacific General Technical Report PSW-GTR-247. Pacific Southwest Research Station, Albany, California, USA.
Knapp, E.E., J.M. Varner, M.D. Busse, C.N. Skinner, and C.J. Shestak. 2011. Behavior and effects of prescribed fire in masticated fuelbeds. International Journal of Wildland Fire 20: 932–945.
Knapp, E.E., C.P. Weatherspoon, and C.N. Skinner. 2012. Shrub seed banks in mixed conifer forests of northern California and the role of fire in regulating abundance. Fire Ecology 8(1): 32–48.
Laacke, R.J. 1990. *Abies magnifica* A. Murr. California red fir. Pages 71–79 in R.M. Burns and B.H. Honkala, technical coordinators. Silvics of North America. Volume 1, Conifers. USDA Forest Service, Washington, D.C., USA.
Lauvaux, C.A., C.N. Skinner, and A.H. Taylor. 2016. High severity fire and mixed conifer forest-chaparral dynamics in the southern Cascade Range, USA. Forest Ecology and Management 363: 74–85.
Long, J.W., M.K. Anderson, L. Quinn-Davidson, R.W. Goode, F.K. Lake, and C.N. Skinner. 2016. Restoring California black oak ecosystems to promote tribal values and wildlife. USDA Forest Service, General Technical Report PSW-GTR-252. Pacific Southwest Research Station, Albany, California, USA.
Long, J.W., C. Skinner, S. Charnley, K. Hubbert, L. Quinn-Davidson, and M. Meyer. 2014a. Post-wildfire management. Pages 187–220 in: J.W. Long, L. Quinn-Davidson, and C.N. Skinner, editors. Science Synthesis to Support Socioecological Resilience in the Sierra Nevada and Southern Cascade Range. USDA Forest Service General Technical Report PSW-GTR-247. Pacific Southwest Research Station, Albany, California, USA.
Long, J.W., C. Skinner, H. Safford, S. Charnley, and P.L. Winter. 2014b. Introduction. Pages 3–16 in J.W. Long, L. Quinn-Davidson, and C.N. Skinner, editors. Science Synthesis to Support Socioecological Resilience in the Sierra Nevada and Southern Cascade Range. USDA Forest Service General Technical Report PSW-GTR-247. Pacific Southwest Research Station, Albany, California, USA.
Mantua, N.J., S.R. Hare, Y. Zhang, J.M. Wallace, and R.C. Francis. 1997. A Pacific interdecadal climate oscillation with impacts on salmon production. Bulletin of the American Meteorological Society 78: 1069–1079.
Margolis, E.Q., and C.A. Farris. 2014. Quaking aspen regeneration following prescribed fire in Lassen Volcanic National Park, California, USA. Fire Ecology 10(3): 14–26.
McIver, J.D., S.L. Stephens, J.K. Agee, J. Barbour, R.E.J. Boerner, C.B. Edminster, K.L. Farris, C.J. Fettig, C.E. Fiedler, S. Haase, S.C. Hart, J.E. Keeley, J.F. Lehmkuhl, J.J. Moghaddas, W. Otrosina, K.W. Outcalt, D.W. Schwilk, C.N. Skinner, T.A. Waldrop, C.P. Weatherspoon, D.A. Yaussy, and A. Youngblood. 2013. Ecological effects of alternative fuel reduction treatments: principal findings of the national fire and fire surrogates study (FFS). International Journal of Wildland Fire 22: 63–82.
McKelvey, K.S., and K.K. Busse. 1996. Twentieth-century fire patterns on Forest Service lands. Pages 1119–1138 in: D.C. Erman, general editor. Sierra Nevada Ecosystem Project: Final Report to Congress, Volume II. Wildland Resources Center Report 37. University of California, Davis, California, USA.

McKelvey, K.S., C.N. Skinner, C. Chang, D.C. Erman, S.J. Husari, D.J. Parsons, J.W. van Wagtendonk, and C.P. Weatherspoon. 1996. An overview of fire in the Sierra Nevada. Pages 1033–1040 in: D.C. Erman, general editor. Sierra Nevada Ecosystem Project: Final Report to Congress, Volume II. Wildland Resources Center Report 37. University of California, Davis, California, USA.
Merriam, K., and E. Rentz. 2010. Restoring fire to endemic cypress populations in northern California. Final Report for JFSP Project 06-2-1-17. USDI-USDA Joint Fire Science Program, Boise, Idaho, USA.
Miles, S.R., and C.B. Goudey, editors. 1997. Ecological subregions of California: section and subsection descriptions. USDA Forest Service Technical Publication R5-3M-TP-005. Pacific Southwest Region, San Francisco, California, USA.
Milich, K.L., J.D. Stuart, J.M. Varner, and K.E. Merriam. 2012. Seed viability and fire-related temperature treatments in serotinous California native *Hesperocyparis* species. Fire Ecology 8(2): 107–124.
Miller, J.D., B.M. Collins, J.A. Lutz, S.L. Stephens, J.W. van Wagtendonk, and D.A. Yasuda. 2012. Differences in wildfires among ecoregions and land management agencies in the Sierra Nevada region, California, USA. Ecosphere 3(9): art80.
Miller, J.D., and H. Safford. 2012. Trends in wildfire severity: 1984 to 2010 in the Sierra Nevada, Modoc Plateau, and southern Cascades, California, USA. Fire Ecology 8(3): 41–57.
Morford, L. 1984. 100 years of wildland fires in Siskiyou County. Siskiyou County Library Reference No. R 634.9618 M. Yreka, California, USA.
Moyle, P.B., and P.J. Randall. 1996. Biotic integrity of watersheds. Pages 975–985 in: D.C. Erman, general editor. Sierra Nevada Ecosystem Project: Final Report to Congress, Volume II. Wildland Resources Center Report 37. University of California, Davis, California, USA.
Nagel, N., and A.H. Taylor. 2005. Fire and persistence of montane chaparral in mixed conifer forest landscapes in the northern Sierra Nevada, Lake Tahoe Basin, California, USA. Journal of the Torrey Botanical Society 132: 442–457.
Norman, S.P. 2002. Legacies of anthropogenic and climate change in fire prone pine and mixed conifer forests of northeastern California. Dissertation, The Pennsylvania State University, State College, Pennsylvania, USA.
Norman, S., and A.H. Taylor. 2002. Variation in fire-return intervals across a mixed-conifer forest landscape. Pages 170–179 in: N. Sugihara, M. Morales, and T. Morales, editors. Symposium on Fire in California Ecosystems: Integrating Ecology, Prevention, and Management. Association for Fire Ecology Miscellaneous Publication No. 1, Davis, California, USA.
———. 2003. Tropical and north Pacific teleconnections influence fire regimes in pine-dominated forests of north-eastern California, USA. Journal of Biogeography 30: 1081–1092.
———. 2005. Pine forest expansion along a forest-meadow ecotone in northeastern California, USA. Forest Ecology and Management 215: 51–68.
North, M.P., B.M. Collins, and S.L. Stephens. 2012. Using fire to increase the scale, benefits, and future maintenance of fuels treatments. Journal of Forestry 110: 392–401.
Oakeshott, G.B. 1971. California's Changing Landscapes: A Guide to the Geology of the State. McGraw-Hill Book Co., San Francisco, California, USA.
Oliver, W.W. 2000. Ecological research at the Blacks Mountain Experimental Forest in northeastern California. USDA Forest Service General Technical Report PSW-GTR-179. Pacific Southwest Research Station, Albany, California, USA.
———. 2001. Can we create and sustain late successional attributes in interior ponderosa pine stands? Large-scale ecological research studies in northeastern California. Pages 99–103 in: R.K. Vance, C.B. Edminster, W.W. Covington, and J.A. Blake, editors. Proceedings-Ponderosa Pine Ecosystems Restoration and Conservation: Steps toward Stewardship. USDA Forest Service Proceedings RMRS-P-22. Rocky Mountain Research Station, Ogden, Utah, USA.
Parker, A.J. 1986. Persistence of lodgepole pine forests in the central Sierra Nevada. Ecology 67: 1560–1567.
———. 1991. Forest/environment relationships in Lassen Volcanic National Park, California, U.S.A. Journal of Biogeography 18: 543–552.
———. 1993. Structural variation and dynamics of lodgepole pine in forests of Lassen Volcanic National Park, California. Annals of the Association of American Geographers 83: 613–629.
———. 1995. Comparative gradient structure and forest cover types in Lassen Volcanic and Yosemite National Parks, California. Bulletin of the Torrey Botanical Club 122: 58–68.
Perry, D.A., P.F. Hessburg, C.N. Skinner, T.A. Spies, S.L. Stephens, A.H. Taylor, J.F. Franklin, B. McComb, and G. Riegel. 2011. The ecology of mixed severity fire regimes in Washington, Oregon, and northern California. Forest Ecology and Management 262: 703–717.
Pierce, A.D., and A.H. Taylor. 2010. Competition and regeneration in quaking aspen-white fir (*Populus tremuloides-Abies concolor*) forests in the northern Sierra Nevada, USA. Journal of Vegetation Science 21: 507–519.
———. 2011. Fire severity and seed source influence lodgepole pine (*Pinus contorta* var. *murrayana*) regeneration in the southern Cascades, Lassen Volcanic National Park, California. Landscape Ecology 26: 225–237.
Pitcher, D.C. 1987. Fire history and age structure in red fir forests of Sequoia National Park. Canadian Journal of Forest Research 17: 582–587.
Potter, D.A. 1998. Forested communities of the upper montane in the central and southern Sierra Nevada. USDA Forest Service General Technical Report PSW-GTR-169. Pacific Southwest Research Station, Albany, California, USA.
Quinn-Davidson, L.N., and J.M. Varner. 2012. Impediments to prescribed fire across agency, landscape, and manager: an example from northern California. International Journal of Wildland Fire 21: 210–218.
Rentz, E., and K. Merriam. 2011. Restoration and management of Baker cypress in northern California and southern Oregon. Pages 282–289 in: J.W. Willoughby, B.K. Orr, K.A. Schierenbeck, and N. Jensen, editors, Proceedings of the CNPS Conservation Conference: Strategies and Solutions, January 17–19, 2009. California Native Plant Society, Sacramento, California, USA.
Ritchie, M.W. 2005. Ecological research at the Goosenest Adaptive Management Area in northeastern California. USDA Forest Service General Technical Report PSW-GTR-192. Pacific Southwest Research Station, Albany, California, USA.
Ritchie, M.W., and E.E. Knapp. 2014. Establishment of a long-term fire salvage study in an interior ponderosa pine forest. Journal of Forestry 112: 395–400.
Ritchie, M.W., E.E. Knapp, and C.N. Skinner. 2013. Snag longevity and surface fuel accumulation following post-fire logging in a ponderosa pine dominated forest. Forest Ecology and Management 287: 113–122.
Ritchie, M.W., D.A. Maguire, and A. Youngblood, editors. 2005. Proceedings of the symposium on ponderosa pine: Issues, trends, and management, 2004 October 18-21, Klamath Falls, OR. USDA Forest Service General Technical Report PSW-GTR-198. Pacific Southwest Research Station, Albany, California, USA.
Ritchie, M.W., and C.N. Skinner. 2014. Interdisciplinary research on the Blacks Mountain Experimental Forest. Pages 129–146 in: D.C. Hayes, S.L. Stout, R.H. Crawford, and A.P. Hoover, editors. USDA Forest Service Experimental Forests and Ranges: Research for the Long Term. Springer, New York, New York, USA.
Ritchie, M.W., C.N. Skinner, and T.A. Hamilton. 2007. Probability of wildfire-induced tree mortality in an interior pine forest of northern California: effects of thinning and prescribed fire. Forest Ecology and Management 247: 200–208.
Ritchie, M.W., B.M. Wing, and T.A. Hamilton. 2008. Stability of the large tree component in treated and untreated late-seral interior ponderosa pine stands. Canadian Journal of Forest Research 38: 919–923.
Rundel, P.W., D.J. Parsons, and D.T. Gordon. 1977. Montane and subalpine vegetation of the Sierra Nevada and Cascade ranges. Pages 559–599 in: M.G. Barbour and J. Major, editors. Terrestrial Vegetation of California. John Wiley and Sons, New York, New York, USA.
Safford, H.D., M.P. North, and M.D. Meyer. 2012. Climate change and the relevance of historical forest conditions. Pages 23–45 in M.P. North, editor, Managing Sierra Nevada Forests. USDA Forest Service General Technical Report PSW-GTR-237. Pacific Southwest Research Station, Albany, California, USA.
Schmidt, D.A., A.H. Taylor, and C.N. Skinner. 2008. The influence of fuels treatment and landscape arrangement on simulated fire

behavior, southern Cascade Range, California. Forest Ecology and Management 255: 3170–3184.
Scholl, A.E., and A.H. Taylor. 2006. Fire regimes, forest change, and self-organization in old-growth red fir-western white pine forests in the northern Sierra Nevada, Lake Tahoe, USA. Forest Ecology and Management 235: 143–154.
Schroeder, M.J., and C.C. Buck. 1970. Fire weather: a guide for application of meteorological information to forest fire control operations. U.S. Department of Agriculture Handbook 360. Washington, D.C., USA.
Skinner, C.N. 2002. Influence of fire on dead woody material in forests of California and southwestern Oregon. Pages 445–454 in: W.F. Laudenslayer, Jr., P.J. Shea, B.E. Valentine, C.P. Weatherspoon, and T.E. Lisle, editors. Proceedings of the Symposium on the Ecology and Management of Dead Wood in Western Forests. USDA Forest Service General Technical Report PSW-GTR-181. Pacific Southwest Research Station, Albany, California, USA.
Skinner, C.N., C.S. Abbott, D.L. Fry, S.L. Stephens, A.H. Taylor, and V. Trouet. 2009. Human and climatic influences on fire occurrence in California's North Coast Range. Fire Ecology 5(3): 76–99.
Skinner, C.N., and C. Chang. 1996. Fire regimes, past and present. Pages 1041–1069 in: D.C. Erman, general editor. Sierra Nevada Ecosystem Project: Final Report to Congress, Volume II. Wildland Resources Center Report 37. University of California, Davis, California, USA.
Stephens, S.L., D.L. Fry, E. Franco-Vizcaíno, B.M. Collins, and J.J. Moghaddas. 2007a. Coarse woody debris and canopy cover in an old-growth Jeffrey pine-mixed conifer forest from the Sierra San Pedro Mártir, Mexico. Forest Ecology and Management 240: 87–95.
Stephens, S.L., R.E. Martin, and N.E. Clinton. 2007b. Prehistoric fire area and emissions from California's forests, woodlands, shrublands, and grasslands. Forest Ecology and Management 251: 205–216.
Stephens, S.L., J.D. McIver, R.E.J. Boerner, C.J. Fettig, J.B. Fontaine, B.R. Hartsough, P.L. Kennedy, and D.W. Schwilk. 2012. The effects of forest fuel-reduction treatments in the United States. BioScience 62: 549–560.
Stephens, S.L., and J.J. Moghaddas. 2005. Fuel treatment effects on snags and coarse woody debris in a Sierra Nevada mixed conifer forest. Forest Ecology and Management 214: 53–64.
Stephens, S.L., J.J. Moghaddas, C. Edminster, C.E. Fiedler, S. Haase, M. Harrington, J.E. Keeley, E.E. Knapp, J.D. McIver, K. Metlen, C.N. Skinner, and A. Youngblood. 2009. Fire treatment effects on vegetation structure, fuels, and potential fire behavior and severity in western U.S. forests. Ecological Applications 19: 305–320.
Stuart, J.D., J.K. Agee, and R.I. Gara. 1989. Lodgepole pine regeneration in an old, self-perpetuating forest in south central Oregon. Canadian Journal of Forest Research 19: 1096–1104.
Symons, J.N., D.H.K. Fairbanks, and C.N. Skinner. 2008. Influences of stand structure and fuel treatments on wildfire severity at Blacks Mountain Experimental Forest, northeastern California. The California Geographer 48: 61–82.
Taylor, A.H. 1990a. Tree invasion in meadows of Lassen Volcanic National Park. Professional Geographer 53: 457–470.
———. 1990b. Habitat segregation and regeneration patterns of red fir and mountain hemlock in ecotonal forests, Lassen Volcanic National Park, California. Physical Geography 11: 36–48.
———. 1993. Fire history and structure of red fir (*Abies magnifica*) forests, Swain Mountain Experimental Forest, Cascade Range, northeastern California. Canadian Journal of Forest Research 23: 1672–1678.
———. 2000. Fire regimes and forest changes along a montane forest gradient, Lassen Volcanic National Park, southern Cascade Mountains, USA. Journal of Biogeography 27: 87–104.
———. 2010. Fire disturbance and forest structure in an old-growth *Pinus ponderosa* forest, southern Cascades, USA. Journal of Vegetation Science 21: 561–572.
Taylor, A.H., and C.B. Halpern. 1991. The structure and dynamics of *Abies magnifica* forests in the southern Cascade Range, USA. Journal of Vegetation Science 2: 189–200.
Taylor, A.H., and C.N. Skinner. 1998. Fire history and landscape dynamics in a late-successional reserve in the Klamath Mountains, California, USA. Forest Ecology and Management 111: 285–301.
———. 2003. Spatial patterns and controls on historical fire regimes and forest structure in the Klamath Mountains. Ecological Applications 13: 704–719.
Taylor, A.H., C.N. Skinner, B.L. Estes. 2013. A comparison of fire severity patterns in the late 19th and early 21st century in a mixed conifer forest landscape in the southern Cascades. Final Report for JFSP Project 09-1-10-7. USDI-USDA Joint Fire Science Program, Boise, Idaho, USA.
Taylor, A.H., and M.N. Solem. 2001. Fire regimes and stand dynamics in an upper montane forest landscape in the southern Cascades, Caribou Wilderness, California. Journal of the Torrey Botanical Club 128: 350–361.
Taylor, A.H., V. Trouet, and C.N. Skinner. 2008. Climatic influences on fire regimes in montane forests of the southern Cascades, California, USA. International Journal of Wildland Fire 17: 60–71.
Taylor, A.H., V. Trouet, C.N. Skinner, and S. Stephens. 2016. Socioecological transitions trigger fire regime shifts and modulate fire-climate interactions in the Sierra Nevada, USA, 1600-2015 CE. Proceedings of the National Academy of Sciences 113(48): 13684–13689.
Trouet, V., and A.H. Taylor. 2010. Multi-century variability in the Pacific North American (PNA) circulation pattern reconstructed from tree rings. Climate Dynamics 35: 953–963.
Trouet, V., A.H. Taylor, A.M. Carleton, and C.N. Skinner. 2006. Fire-climate interactions in forests of the American Pacific Coast. Geophysical Research Letters 33: L18704.
———. 2009. Interannul variations in fire weather, fire extent, and synoptic-scale circulation patterns in northern California and Oregon. Theoretical and Applied Climatology 95: 349–360.
Turhune, G. 2002. The QLG defensible fuelbreak strategy. Pages 253–255 in: N. Sugihara, M. Morales, and T. Morales, editors. Proceedings of the Symposium: Fire in California Ecosystems: Integrating Ecology, Prevention and Management. Miscellaneous Publication 1. Association for Fire Ecology, Davis, California, USA.
Uzoh, F.C.C., and C.N. Skinner. 2009. Effects of creating two forest structures and using prescribed fire on coarse woody debris in northeastern California, USA. Fire Ecology 5(2): 1–13.
van Wagtendonk, J.W., and D.R. Cayan. 2008. Temporal and spatial distribution of lightning strikes in California in relation to large-scale weather patterns. Fire Ecology 41(1): 34–56.
Vogl, R.J., W.P. Armstrong, K.L. White, and K.L. Cole. 1977. The closed-cone pines and cypresses. Pages 295–358 in: M.G. Barbour and J. Major, editors. Terrestrial Vegetation of California. John Wiley and Sons, New York, New York, USA.
Wallace, J.M., and D.S. Gutzler. 1981. Teleconnections in the geopotential height field during the Northern Hemisphere winter. Monthly Weather Review 109: 784–812.
Waters, J.R., K.S. McKelvey, C.J. Zabel, and W.W. Oliver. 1994. The effects of thinning and broadcast burning on sporocarp production of hypogeous fungi. Canadian Journal of Forest Research 24: 1516–1522.
Weatherspoon, C.P., and C.N. Skinner. 1996. Landscape-level strategies for forest fuel management. Pages 1471–1492 in: D.C. Erman, general editor. Sierra Nevada Ecosystem Project: Final Report to Congress, Volume II. Wildland Resources Center Report 37. University of California, Davis, California, USA.
———. 2002. An ecological comparison of fire and fire surrogates for reducing wildfire hazard and improving forest health. Pages 239–245 in: N. Sugihara, M. Morales, and T. Morales, editors. Proceedings of the Symposium: Fire in California Ecosystems: Integrating Ecology, Prevention and Management. Miscellaneous Publication 1. Association for Fire Ecology. Davis, California, USA.
Weislander, A.E. 1935. First steps of the forest surveys in California. Journal of Forestry 3: 877–884.
Westerling, A.L., H.G. Hildalgo, D.R. Cayan, and T.W. Swetnam. 2006. Warming and earlier spring increase western U.S. forest wildfire activity. Science 313: 940–943.
Wise, E.K. 2010. Spatiotemporal variability of the precipitation dipole transition zone in the western United States. Geophysical Research Letters 37: LO7706.
WRCC. 2002. Historical Climate Data. Western Regional Climate Center, Desert Research Institute, Reno, Nevada, USA. https://wrcc.dri.edu/Climate/summaries.php.
Zack, S., W.F. Laudenslayer, Jr., T.L. George, C. Skinner, and W. Oliver. 1999. A prospectus on restoring late-successional forest structure to eastside pine ecosystems through large-scale,

interdisciplinary research. Pages 343–355 in: J. E. Cook and B. P. Oswald, editors. First Biennial North American Forest Ecology Workshop. Forest Ecology Working Group, Society of American Foresters, Bethesda, Maryland, USA.
Zhang, J., J. Webster, R. F. Powers, and J. Mills. 2008. Reforestation after the Fountain Fire in northern California: an untold success story. Journal of Forestry 106: 425–430.
Zielinski, W. J. 2014. The forest carnivores: fisher and marten. Pages 395–435 in J. W. Long, L. Quinn-Davidson, and C. N. Skinner, editors. Science Synthesis to Support Socioecological Resilience in the Sierra Nevada and Southern Cascade Range. USDA Forest Service General Technical Report PSW-GTR-247. Pacific Southwest Research Station, Albany, California, USA.

CHAPTER THIRTEEN

Northeastern Plateaus Bioregion

GREGG M. RIEGEL, RICHARD F. MILLER, CARL N. SKINNER,
SYDNEY E. SMITH, CALVIN A. FARRIS, AND KYLE E. MERRIAM

The only means for taking deer wholesale was by firing. This method was called kupi't (firing) and was practiced in late summer, about the middle of August. When deer were sighted on a hill, a group of hunters hastened there, some on either side of the mountain. They started fires, working them around until the band was completely encircled. This accomplished, the fires were brought closer, constricting the circle until the animals were bunched on the crest on a hill where they could be shot conveniently. Deer firing was ordinarily executed without undue noise, but if a bear were accidentally caught in the fire, a great hue and cry was set up to warn the others and to inform them in which direction the animal was headed.

KELLY (1932)

Description of the Bioregion

Northeastern California landscapes are a mixture of vast arid basins and uplands, forested mountain ranges interspersed with both fresh water and alkaline wetlands. The entire bioregion is significantly influenced by the rain shadow effect of the Cascade Range to the west. Three ecological unit subsections are treated in this chapter: (1) Modoc Plateau Section (M261G), (2) northwestern Basin and Range Section (342B), and (3) extreme northern portion of the Mono Section (341Dk) (Miles and Goudey 1997). The Cascade Range defines the western edge of the Modoc Plateau and the Sierra Nevada defines the western boundary for the southern portion of the northwestern Basin and Range and Mono Sections. There is no one overriding feature that cleanly depicts this bioregion to the north and east except the political borders of the states of Oregon and Nevada, respectively (Map 13.1).

Physical Geography

The Northeastern Plateaus bioregion extends east of the southern Cascade Range and south along the east slopes of the Sierra Nevada. This region represents the northern extent of California's portion of the Great Basin. The climate is dry, cold, and continental. Volcanism dominates the geologic, topographic, and geomorphic processes of the area. The most conspicuous feature of the region is the Modoc Plateau, a high, flat terrain characterized by basalt plains and volcanic shields (MacDonald and Gay 1966). The topography can be extremely abrupt with elevations, ranging from 1,204 m (3,950 ft) to 3,016 m (9,892 ft). The Pit River traverses the Modoc Plateau from east to west where it then flows through the Cascade Range to finally enter the Sacramento River within the Klamath Mountains (Pease 1965).

The volcanism that shaped the area is fairly recent, dating from the Pliocene to Pleistocene eras, and includes volcanic ash deposits, lava flows, talus jumbles, obsidian deposits, and lava rims. Soils that have weathered in place, particularly at the lowest elevations of the region, are often poorly developed, shallow, and rocky. As a result, vegetation and fuels are sparse. Clays weathered from basalt are an important soil texture component in many locations. River deposits include Mollisols and Aridisols, and lakebed deposits include silt, sandstones, and diatomite (Young et al. 1988). At higher elevations in the bioregion, most residual soils are Mollisols, often weathered from basalt parent materials. Soils transported to lower-lying areas via alluvial processes are often Vertisols. Soil temperature regimes are mesic (warm), frigid (cool), and cryic (cold), and soil moisture regimes are mostly xeric (>300 mm) and aridic (<300 mm). The quantity of available moisture from precipitation or soil (soil moisture regime) and temperature (soil temperature regime) are the primary determinants of vegetation and fire regimes in the Northeastern Plateaus bioregion.

Climate Patterns

FIRE CLIMATE VARIABLES

The bioregion is buffered from Pacific storms by being in the rain shadow of the Cascade Range and Sierra Nevada. Most of the precipitation occurs between October and May, with the majority coming between November and April as snow. Summer thunderstorms can be locally significant and are the source of lightning ignitions; they also account for 12–19% of total annual precipitation. The range of annual precipitation in the Northeastern Plateaus bioregion is

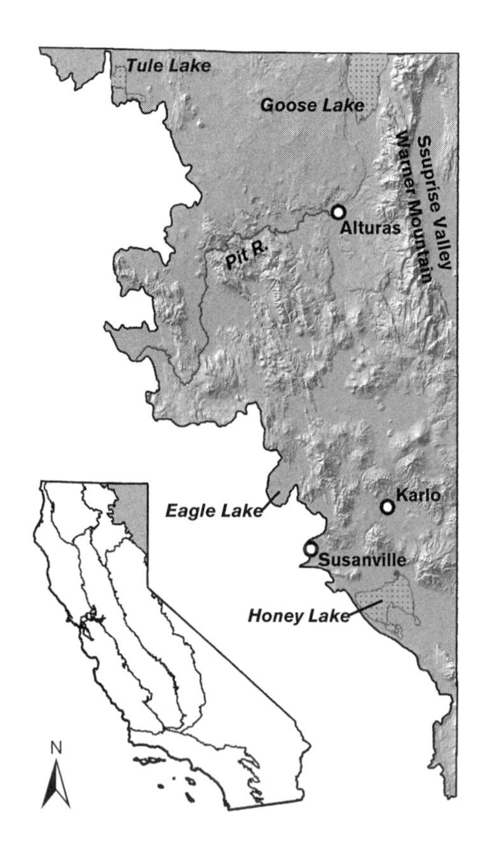

MAP 13.1 Northeastern Plateaus bioregion map. Note the inclusion of Carson Valley and the Pine Nut Mountains in the northern Mono sections of the Great Basin, southeast of Lake Tahoe.

17.8–121.9 cm (7–48 in) (Daly et al. 1994). The overall bioregional average is 43.2 cm (17 in). However, local orographic enhancement boosts precipitation from 91.4 to 121.9 cm (36–48 in) in the northern Warner Mountains. Temperatures range from January minimums of –11°C to –2°C (12–28 °F) to July maximums of 21°C to 32°C (69–89°F). The majority of fires occur from June to September, and certain fire weather conditions can affect fire ignitions and fire behavior during those months.

WEATHER SYSTEMS

There are three primary fire weather patterns that can significantly affect fire behavior and natural ignitions in northeastern California during the May-to-October fire season: (1) Prefrontal Winds, (2) Lightning with Low Precipitation, and (3) Strong Subsidence/Low Relative Humidity patterns. A fourth pattern, Moist Monsoon, is very rare in northeastern California. The prefrontal and subsidence patterns mostly affect fire behavior and spread, while the lightning with low precipitation and moist monsoon patterns are important for their potential to ignite many simultaneous widespread fires.

Prefrontal Winds Prefrontal wind events are frequent in springtime and again in late summer and fall. They are of most consequence in the latter period, when both live and dead fuel moistures are low. The prefrontal weather pattern occurs in advance of spring or fall cold fronts, or upper low-pressure troughs. Tight pressure gradients aloft produce moderate to strong South to West winds, which can surface strongly on the lee side of the Cascade Range and Sierra Nevada. As the downward moving air masses lose elevation, they gain temperature and lose humidity due to adiabatic compression. These phenomena are combined with moderate to strong wind speeds of 24–56 km h^{-1}, (15–35 mi h^{-1}) with gusts as high as 80 km h^{-1} (50 mi h^{-1}). This pattern usually occurs between 5 and 10 times a year, with one or two significant events during the fall season of most years. These conditions can lead to rapid fire spread and extreme fire behavior.

Lightning with Low Precipitation This weather pattern typically affects northeastern California to varying degrees, once or twice per fire season, although it can occur as many as four or five times in a single year. This pattern includes episodes of thunderstorms that are more likely to cause fires than others and is most common in July and August, but can occur from June through mid-September. It sets up when the dominant high pressure aloft becomes oriented from roughly northern Nevada down through Arizona. When relatively cool disturbances aloft are caught up in this larger-scale flow pattern they can act as "triggers," destabilizing the mid- and upper layers enough to generate high-based thunderstorms. Because the resulting cells have high bases, much of the precipitation associated with them evaporates before reaching the ground. These events usually result in many fire ignitions over a relatively short time, a situation that can be rapidly compounded by the gusty erratic downdraft winds associated with the thunderstorms (Rorig and Ferguson 1999, 2002).

Strong Subsidence/Low Relative Humidity A third pattern, strong subsidence/low relative humidity (RH) can, with enough duration, cause a significant increase in northeastern California fire potentials, even without much wind. The pattern occurs when a strong mid- and/or upper level high-pressure area is centered to the west of northeastern California for a period of at least several days. High pressure over the area leads to diverging air at the surface. The surface divergence induces large-scale slow sinking of the mid-upper air mass over northeastern California. This phenomenon is called subsidence and it leads to adiabatic compression that warms and dries the lower atmosphere. An important effect is the hindrance of normal nighttime RH recovery. Nonsubsidence pattern recovery occurs on lower slopes and in drainage bottoms. The RH recovery ranges from 35% to 45%, and may reach 55%. In a subsidence situation, which is most common in August or September and only pertains to mid-slopes and above, the daytime minimum RH usually drops to 4–12%, but nighttime recovery is very low, reaching only the 15–30% range. Dead fuel moistures drop, live fuels become more stressed, and fires ignite, spread, and spot more easily. Extended periods like this are often noted for periods of high atmospheric instability.

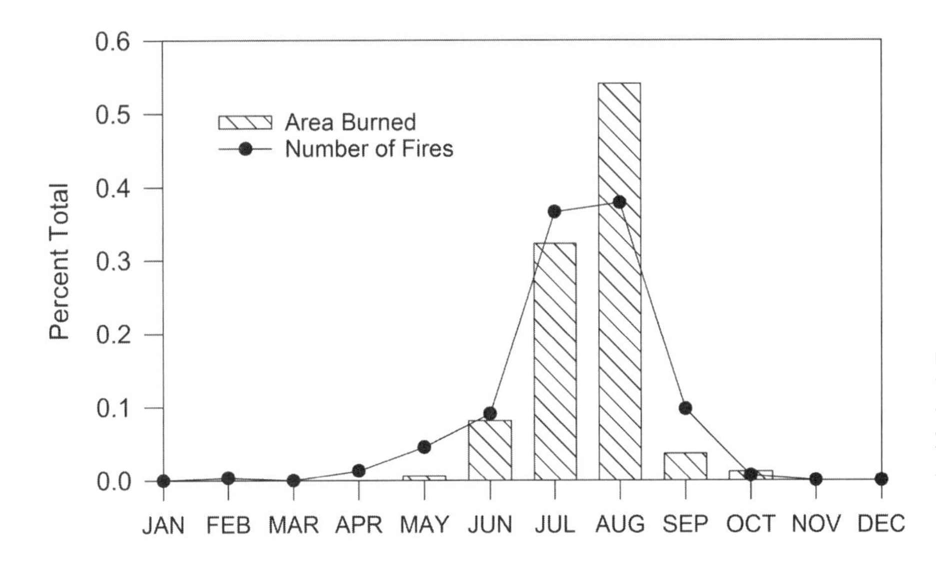

FIGURE 13.1 Monthly distribution of the number of lightning-ignited fires and total area burned between 1910 and 2013 in the northeast California bioregion. Area burned was assigned to the month in which the fire started (fire source data from: http://frap.cdf.ca.gov/data/frapgisdata-subset.php).

LIGHTNING

Lightning strikes are most common from June through August and peak in late July through August, though they may occur from May through September (van Wagtendonk and Cayan 2008). Total number of lightning strikes in the Northeast Plateau Bioregion from 1985 to 2000 was 73,187 covering 2,034,103 km^2 with 22.49 strikes yr^{-1}.

Ignition probability, number of lightning-ignited fires, and total amount of area burned in the bioregion, is highest in late July through September as fuel moistures are typically at their lowest and abundant lightning coincide (Fig. 13.1). More than half of the total acreage burned occurs in August alone. Lightning occurrence is not linearly related to an increase in elevation in this bioregion. The distance lightning can travel in this area ranges from 180 km (112 mi), which is the longest at the lower end range of all bioregions, to 340 km (211 mi), which is second to the Southeast Desert bioregion.

Ecological Zones

The quantity of available moisture from precipitation or soil (soil moisture regime) and temperature (soil temperature regime) are the primary determinants of vegetation and fire regimes in the Northeastern Plateaus bioregion. Major zones include: Sagebrush Steppe, Lower Montane, Mid-Montane, Upper Montane, and Subalpine Zones.

Overview of Historical Fire Occurrence

Prehistoric Period

The Northeastern Plateaus bioregion appears to have been a dry region for more than a million years since the uplifting of the Cascade Range began to block moisture from Pacific winter storms. Over this period the assemblages of vegetation across the bioregion appear to have remained relatively stable, though oscillating from grasslands to sagebrush and juniper to pine woodlands and forests depending upon the trends in effective moisture as the climate changed between glacial and interglacial periods (Adam et al. 1989). The last cold periods appear to have occurred between 18.9 and 17.6 ka (thousands of years before present) (Hakala and Adam 2004). This coincides with glacial advances in several areas of the south Warner Mountains where there is extensive evidence of glacial activity. What is now Paterson Lake was then covered with glacial ice (Osborn and Bevis 2001).

Only a few paleoecological studies have looked at the association of climate, fire, and vegetation in this bioregion. Wigand et al. (1995) and Calo (2014) conducted studies in the Eagle Lake area on the southwestern edge of the bioregion, and Minckley (2003) and Minckley et al. (2007) evaluated paleoecological records at Patterson and Lily Lakes in the Warner Mountains.

In the southeastern portion of the bioregion around Eagle Lake, it appears that most of the early and mid-Holocene (11.5–4 ka) was characterized by warm and very dry conditions. Vegetation appears to have oscillated between grasslands and shrublands during this period. Fire frequency appears to have been variable at this time with little charcoal produced, possibly due to limited fuel productivity. Juniper woodlands had their greatest presettlement expansion during the Neoglacial period of 4–2 ka. This was accompanied by changes in the fire regime that indicated the greater available biomass would promote occasional extensive fires that would be followed by short-term expansion of grasslands that would then give way again to juniper woodlands. Eventually, a general increase in effective moisture over the last millennium appears to have increased vegetation productivity and promoted more frequent fires. As a result, the juniper woodlands gave way to the eastside yellow pine (Jeffrey pine [*Pinus jeffreyi*], ponderosa pine [*P. ponderosa*], and Washoe pine [*P. ponderosa* var. *washoensis*]) dominated forests and woodlands that are characteristic of the area today (Wigand et al. 1995).

The Warner Mountains show a warming and drying trend from 11.5 to 9 ka with sagebrush expansion similar to the rest of the Great Basin. However, white fir (*Abies concolor*) expansion began between 9 and 8 ka, indicating greater effective moisture in contrast to other Great Basin locales. This also coincides with increased fire activity, as evidenced by charcoal influx to lakes in the Warner Mountains. Fir and pine continued to increase and modern forest conditions appear to have started to assemble between 4.5 and 3 ka. Two other periods of high fire activity took place between 4 and 3 ka and from 2 to 1 ka. Interestingly, the mixed-conifer vegetation at Lily Lake in the north Warners has been unusually stable in spite of the oscillations in fire activity. Here, variation in fire activity appears to have been independent of trends in vegetation. However, variation in fire activity, as indicated in charcoal influx, has been similar to variation in fire activity across the region (Minckley 2003). Recent work suggests that background levels of charcoal from lake sediments are an indicator of extralocal or regional fire activity (Whitlock et al. 2004). It may be that Lily Lake is capturing a more regional picture of fire activity and that local fires were of low intensity or patchy enough for locally produced charcoal to not overwhelm the regional picture (Minckley 2003).

Historic Period

Many early accounts of settlers, soldiers, surveyors, and others mention stands composed of large, widely scattered ponderosa and Jeffrey pine (Laudenslayer et al. 1989). Fire was almost certainly one of the main factors that affected the forest structures observed in those early accounts. Historic period disturbances of logging, grazing, mining, agriculture, rail networks, highways, etc., began in the region in the 1860s, and were in full swing by the 1920s.

Timber harvesting of ponderosa and Jeffrey pine was an important activity in the forested areas of this bioregion. Fire suppression was actively pursued and promoted as a means of protecting the economic resource provided by the region's forests, and was probably effective in the flatter areas of the bioregion (Laudenslayer et al. 1989).

The number of fires and area burned from 1910 to 2017 in the Modoc National Forest are depicted in Table 13.1. Though only two fires were recorded to have burned more than 800 ha (2,000 ac) from 1910 to 1919, the area burned was the second largest in the 80-year record. In the early 1900s, many small fires were started by trains, escaped warming fires, and fires set to clear land for agriculture. As railroad and highway traffic increased, so did the number of ignitions along travel corridors in this bioregion.

Information on the use of fire by Native Americans in this bioregion, either prehistoric or during initial contact with settlers, is rare. Deer firing (kupi't) was an effective tool used by the Northern Paiute to move and corral frightened mule deer (*Odocoileus hemionus*) into a waiting party of hunters (see full quotation at the beginning of this chapter, Kelly 1932). In the Lower Montane zone, Native American fire use coupled with lightning ignitions would have favored herbaceous species over sensitive shrubs such as bitterbrush (*Purshia tridentata*), which is a key browse plant of mule deer (*O. hemionus*) and a major food source of the Native Americans. However, in the Mid-Montane Zone, repeated fire may have been used to improve human travel within canopy gaps and persistent shrubfields where open conditions facilitate fire-adapted shrub species such as tobacco bush (commonly referred to as snowbrush in this region) ceanothus (*Ceanothus velutinus*) and manzanita (*Arctostaphylos* spp.) but are not preferred browse species of mule deer.

Current Period

Since the turn of the century, fire regimes have changed among many of the semiarid plant associations in northeastern California. These changes are largely attributed to the reduction of fine fuels through livestock grazing, fire suppression, the introduction of highly flammable, non-native annual herbs, and climate change. Since the late 1870s, some plant associations within ecological zones burn more frequently and others less frequently than prior to this period. Where fires were historically more frequent and have currently become less frequent, fire severity has increased. Fire perimeters, total area burned, and numbers of fires in the northeast California bioregion between 1910 and 2014 are depicted in Fig. 13.2. The most striking feature is the very large amount of area burned since 2010 across all vegetation communities relative to previous time periods. Despite the fact that this period contains only five years (2010 to 2014 inclusive), the total area burned is 3 to 14 times higher than any previous 10-year decade period (4 to 26 times higher for shrublands, and 1.2 to 9 times higher for conifer). The period since 2010 has been characterized by warmer temperatures, low precipitation, and longer fire seasons than usual, which is part of a broader trend throughout much of the western United States. The largest fire year in the region was 2012 and the third largest was in 2014; both years had among the driest and warmest combination on record, and live fuel moistures were very low. Accumulation of forest fuels and widespread cheat grass (*Bromus tectorum*) cover also likely contributed.

TABLE 13.1
Number and area of fires recorded on the Modoc National Forest (N.F.) from 1910 to 2017. These values include human-caused and lightning ignitions (data on file, Modoc N.F., Alturas, CA)

Years	Total number of fires	Fires ≥ 2000 acres	Total acres burned
1910–1919	64	2	212,652
1920–1929	65	6	57,847
1930–1939	73	7	74,847
1940–1949	78	7	91,733
1950–1959	49	12	162,487
1960–1969	34	2	13,087
1970–1979	65	10	282,736
1980–1989	47	6	37,058
1990–1999	123	10	109,381
2000–2009	74	6	75,593
2010–2015	35	2	31,933
2016	2	1	3,028
2017	28	3	93,037

Major Ecological Zones

Sagebrush Steppe Zone

The most widespread zone is the Sagebrush Steppe, which occurs across the driest areas in the bioregion, typically on the extensive flatter, lower elevation areas. This zone is predominately shrub steppe, shrub desert, western juniper (*Juniperus occidentalis*) woodlands and singleleaf pinyon pine-Sierra juniper (*Pinus monophylla-J. grandis*) woodlands in the northern

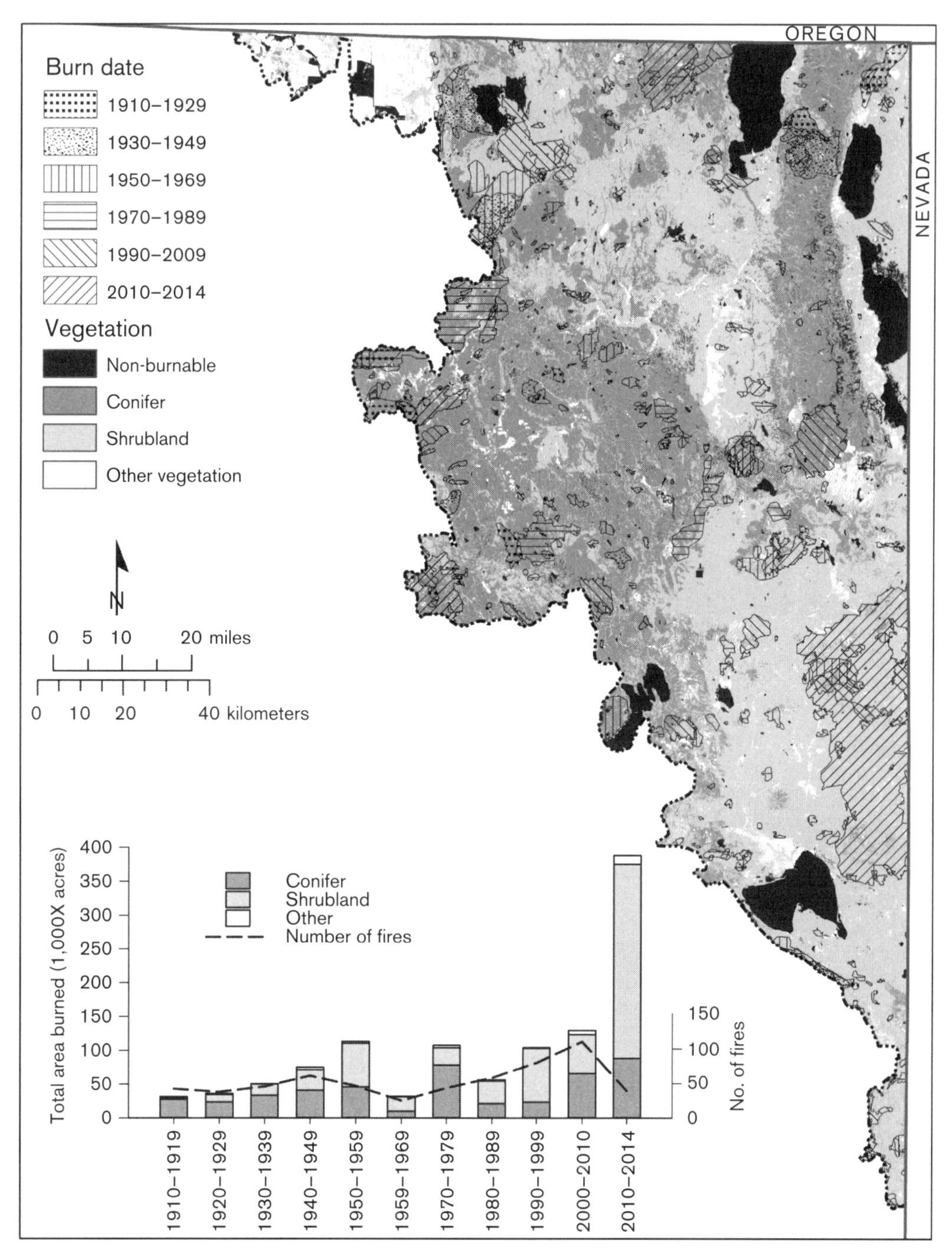

FIGURE 13.2 Fire perimeter locations and trends in total area burned and numbers of fires between 1910 and 2014 for the northeast California bioregion. Vegetation groupings were derived from LANDFIRE Biophysical Settings classification, which is intended to represent reference vegetation communities that may have existed prior to Euro-American settlement (http://www.landfire.gov/NationalProductDescriptions20.php). Fire perimeters are shown for fires >10 acres. Fire perimeter data were compiled from all federal and state agencies by the Fire and Resource Assessment Program of the California Department of Forestry and Fire Protection (http://frap.cdf.ca.gov/data/frapgisdata-subset.php).

Mono section (Vasek and Thorne 1988, Young et al. 1988). The predominant sagebrush species in the Modoc region are big sagebrush (*Artemisia tridentata*), which includes subspecies Wyoming big (*A. tridentata* subsp. *wyomingensis*), big (*A. tridentata* subsp. *tridentata*), mountain big (*A. tridentata* subsp. *vaseyana*), and low sage (*A. arbuscular*) (Rosentreter 2005) (Fig. 13.3). Black sagebrush (*A. nova*) occurs as very minor component of the bioregion. Alliances associated with the sagebrush steppe are western juniper woodlands, quaking aspen (*Populus tremuloides*), riparian, and associations dominated by bitterbrush and curl-leaf mountain-mahogany (*Cercocarpus ledifolius* var. *ledifolius*). Pre-Euro-American settlement juniper communities were primarily shrub savannas, which usually occupied shallow rocky soils associated with low sagebrush. However, 80% of present day juniper woodlands in northeastern California are <150 years old (Bolsinger 1989) often establishing in deeper soils replacing big sagebrush steppe communities where precipitation is sufficient to support western juniper. Western juniper

FIGURE 13.3 Sagebrush steppe zone. This mountain big sagebrush site had recently burned. Rabbitbrush has resprouted and bitterbrush appears to be growing in the background (photo by Rick Miller).

pollen has increased fivefold from the late 1800s to 1980 (Mehringer and Wigand 1990). Generally this occurs in areas that receive a minimum of 30 cm (12 in) of precipitation. The warmer-drier (mesic-aridic) sagebrush communities have been invaded by annual grasses.

Annual grasses and forbs such as cheat grass, medusa head (*Elymus caput-medusae*), tumble mustard (*Sisymbrium altissimum*), and the native tansy mustard (*Descurainia pinnata*) have successfully invaded and resulted in a shift of shrub steppe communities throughout the region to annual grasslands.

The pluvial valley bottoms, primarily found in the Surprise Valley and Honey Lake areas, are occupied by the greasewood (*Sarcobatus vermiculatus*) and shadscale (*Atriplex confertifolia*) alliances. Greasewood is also found in the lowlands around Alturas and in the Klamath Basin around Tule Lake and Klamath Marsh (Young et al. 1988). These environments have soils that are too saline or precipitation that is too low for most species of sagebrush to be a dominant component. Predominant shrub species are four-wing saltbush (*Atriplex canescens*), greasewood, spiny hop-sage (*Grayia spinosa*), and winter fat (*Krascheninnikovia lanata*).

TABLE 13.2
Relative response of common shrubs in the sagebrush biome and salt deserts to fire

Tolerant	Moderately tolerant	Intolerant
Yellow rabbitbrush (s)	Rubber rabbitbrush (s)	Low sagebrush
Wax currant (s)		Big sagebrush
Desert gooseberry (s)		Shadscale
Wood rose (s)		Curl-leaf mountain-mahogany
Greasewood (s)		Hop-sage
Mountain snowberry (s)		Bitterbrush (ws)
Horsebrush (s)		

NOTE: s = sprouter, ws = weak sprouter.
SOURCE: Derived from Blaisdell (1953) and Wright et al. (1979).

FIRE RESPONSES OF IMPORTANT SPECIES

One of the largest changes in community composition following fire is the immediate reduction of the shrub layer. Shrubs associated with sagebrush steppe plant communities across the bioregion are composed of fire tolerant and intolerant species (Table 13.2). The composition of fire tolerant and intolerant shrubs prior to the burn influences shrub composition during the early years following fire.

Species of sagebrush found in the Northeastern Plateaus bioregion are easily killed by fire except for silver sagebrush (*Artemisia cana* subsp. *bolanderi*), which resprouts (Winward 1985) (Table 13.2). Reestablishment of the big sagebrush subspecies and the two forms of low sagebrush is entirely dependent upon unburned seed or seed dissemination from adjacent unburned areas. The potential for large inputs of sagebrush seed following a fire is limited and dependent on the amount of unburned edge and the amount and distribution of unburned sagebrush shrubs in the burn. Sagebrush seed is mainly distributed by wind, with no evidence of seed caching by animals. Seed movement from adjacent unburned areas has been reported to take many years to move into the interior of a burn (Mueggler 1956, Johnson and Payne 1968, Rick Miller, Eastern Oregon Agricultural Research Center [EOARC], unpublished data). The majority of sagebrush seed (>90%) is disseminated within 9 m (30 ft) of the parent plant and nearly 100% within 30 m (98 ft) (Johnson and Payne 1968, Harniss and McDonough 1974). This suggests the rate of recovery is highly dependent upon soil seed pools immediately following the fire, particularly for large fires with few unburned islands.

Sagebrush seed matures and disperses in the autumn and winter (Young and Evans 1989, Meyer and Monsen 1991). The majority of wild and prescribed fires occurs during the summer and early fall, prior to maturity of the current year's seed crop. Thus, soil seed banks are determined by seed crops in previous years.

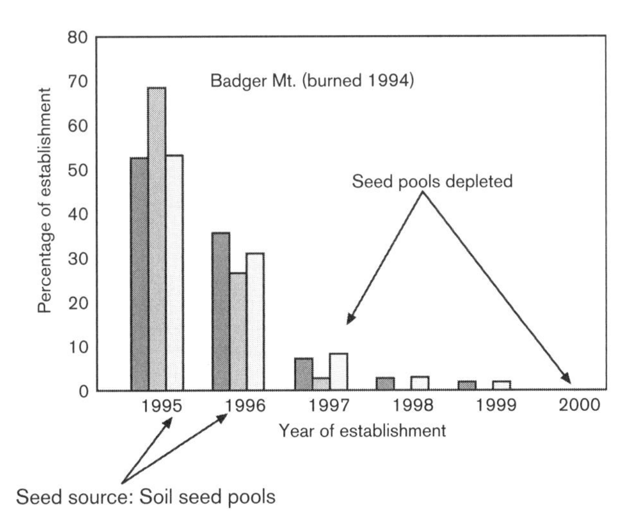

FIGURE 13.4 Mountain big sagebrush seedlings became established from the soil seed bank only in the first two years postfire. The seed bank decline was precipitous in the first year and by the third year the pools were exhausted, emphasizing the short-lived seeds of sagebrush. Beyond the second year, recruitment occurs from new seed that is wind dispersed.

Although seed crop production is highly variable from year to year, seed production is usually not a limiting factor for big sagebrush establishment (Harniss and McDonough 1974). Goodwin (1956) reported that sagebrush produced 350,000 seeds m^{-2} (3,252 ft^{-2}) of canopy annually. In the first year after fire, sagebrush seedling density can be high (Blaisdell 1953, Mueggler 1956). However, with little seed input, seed pool reserves will decline through germination, loss of viability, and predation within the first two years following fire. Hassan and West (1986) reported Wyoming big sagebrush seed densities following a summer fire were 0.7 viable seeds m^{-2} (0.07 ft^{-2}) compared to 3.7 viable seeds m^{-2} (0.34 ft^{-2}) on an adjacent unburned site. In the second fall after the fire, viable seed densities declined to 0.3 m^{-2} (0.03 ft^{-2}). In northwestern Nevada, a 4,000 ha (9,900 ac) wildfire occurred in 1994, following an extremely wet year when large seed crops were produced. The fire burned prior to seed maturation and left few surviving sagebrush plants. During the first two growing seasons following the fire, rapid establishment of mountain big sagebrush and bitterbrush occurred, and 0.3 and 0.9 shrubs m^{-2} (0.03 and 0.08 ft^{-2}) were successfully established (Ziegenhagen and Miller 2009) (Fig. 13.4). In mountain big sagebrush associations, shrub densities in fully established stands ranged between 0.8 and 1.2 shrubs m^{-2} (0.07 and 0.11 ft^{-2}) (Ziegenhagen and Miller 2009, EOARC data). The decline in seedling establishment in 1997 and 1998, both wetter than average years, was attributed to depleted seed pools due to germination of seed, loss of seed viability, granivory, and limited input from newly established plants. In several other fires, seedling establishment began to increase 8 to 10 years after the fire as previously established plants matured and began producing adequate seed. Wijayratne and Pyke (2012) reported that 40–60% of Wyoming big and mountain big sagebrush seed remained viable for two years when covered with a thin layer of soil. Viability of seed sharply declined if seed was exposed. If buried more than a few millimeters, the small sagebrush seeds are unable to germinate (Chambers 2000).

Rate of big sagebrush recovery following fire is highly variable but tends to be slower in the more arid sites (Bunting 1984a, Miller and Heyerdahl 2008). Within 15 years to 30 years following the fire mountain big sagebrush canopy cover can be 15–25% (Bunting et al. 1987, Ziegenhagen and Miller 2009, Nelson et al. 2014). In northeastern California, mountain big sagebrush was the dominant plant cover 30 to 60 years following fire but then declined within 80 to 100 years to near 0% cover as a result of increasing juniper dominance (Miller and Heyerdahl 2008). Rate of recovery is highly dependent on soil seed pools and moisture availability in growing seasons immediately following the fire. Compared to mountain big sagebrush, postfire recolonization of Wyoming big sagebrush is very slow (Hosten and West 1994, Beck et al. 2008, Miller et al. 2013). Soil moisture availability is a principle variable in determining establishment of sagebrush (Boltz 1994, Board et al. 2011) and if present, cheat grass is a more effective competitor for early growing season moisture.

Bitterbrush is a weak sprouter in the Northeastern Plateaus bioregion and its response to fire is highly variable (Table 13.2). Blaisdell and Mueggler (1956) reported survival from resprouting in eastern Idaho was 49% in a light burn, 43% in a moderate burn, and 19% in a hot burn. They also reported greater survival of resprouting in plants <15 years old. Driscoll (1963) concluded survival of bitterbrush resprouts was related more to soil surface texture than fire intensity in central Oregon. In California, 5–25% of the bitterbrush successfully resprouted in 5 of 13 fires (Nord 1965). Survival from resprouting was also greater in spring burns than fall burns (Clark et al. 1982).

In northwestern Nevada, <1% of the bitterbrush survived by resprouting (Ziegenhagen and Miller 2009). The majority of shrub establishment occurred from the soil seed bank. However, unlike sagebrush, an important vector of bitterbrush seed dispersal was small mammals (Vander Wall 1994). When bitterbrush does successfully resprout following fire, the species can recover to levels near those in unburned stands within 9 years to 10 years after fire (Wright et al. 1979). However, on sagebrush associated sites evaluated several years after fire in northern California, southeastern Oregon, and northwestern Nevada, reestablishment was primarily from seed and occurred at the same rate as sagebrush. On a heavily browsed deer winter range in south central Oregon, bitterbrush had not recovered to unburned levels 40 years after the fire (EOARC data). Bitterbrush seedlings also compete poorly with cheat grass, which can severely limit reestablishment (Holmgren 1956).

Rubber and yellow rabbitbrush (*Ericameria nauseosa*, *Chrysothamnus viscidiflorus*) and horsebrush (*Tetradymia* spp.) are capable of sprouting and more rapidly recovering immediately following fire than big sagebrush (Table 13.2). However, rubber rabbitbrush is more sensitive to fire than green rabbitbrush (Wright et al. 1979). In some areas establishment is from both seeds and shoots (Young and Evans 1978). In eastern Idaho and northeastern California establishment was primarily from crown and root shoots with little from seed (Bunting et al. 1987, EOARC data). Although percent composition is usually higher for these sprouting species during the early years following fire, percent cover of these sprouters often does not exceed pre-burn levels on good condition sites (Bunting et al. 1987, EOARC data). The abundance of these sprouting species also usually declines over time as sagebrush abundance increases and the intervals between fires increases (Young and Evans 1978, Whisenant 1990). However, density and cover of these species can exceed unburned levels, especially on degraded sites (Chadwick and Dalke 1965, Young and Evans 1978). Abundance of horsebrush remained higher than pre-burn levels 30 years after the fire (Harniss and Murray 1973). Heavy grazing following fire can also increase the abundance of rabbitbrush and horsebrush.

Curl-leaf mountain-mahogany, a weak sprouter, is highly susceptible to fire (Wright et al. 1979). Pre-Euro-American settlement aged plants are usually found on rocky ridges, which are fire protected (Dealy 1975, Gruell et al. 1984, Davis and Brotherson 1991). Postsettlement stands are commonly found in communities where presettlement fire return intervals were <20 years but have significantly increased since the late 1800s (Dealy 1975, Gruell et al. 1984, Miller and Rose 1999). Demographic analysis indicates a large increase of curl-leaf mountain-mahogany since the late 1800s.

Reestablishment following fire is largely dependent on seedling establishment (Wright et al. 1979); thus a nearby seed source is important. Curl-leaf mountain-mahogany has regenerated in several large burns in northeastern California (Tule Mountain 1959 and Three Peaks 1957) where mature plants, often located on rocky micro sites, survived the fire. However, little recruitment has occurred in a large curl-leaf mountain-mahogany stand burned in a high-severity fire in the 1973 Ninemile Fire south of Likely Mountain where few plants survived.

Ground cover of native herbaceous vegetation can quickly recover and exceed pre-burn levels in some plant associations and decrease or equal pre-burn levels in others (Blaisdell 1953). Rate of recovery (or resilience) and composition following a fire is largely determined by soil moisture and temperature regimes, plant composition prior to the burn, soil seed reserves, fire tolerance of species on the site, fire intensity, weather conditions, and postfire management (Miller et al. 2013). In general, ground cover of native grasses and forbs growing in the cool-wet sagebrush communities; characterized by mountain big sagebrush, Lemmon's needle grass (*Stipa lemmonii*), Idaho fescue (*Festuca idahoensis*), and blue bunch wheat grass (*Elymus spicatus*); recover rapidly and often exceeds pre-burn levels within 2 years to 3 years after fire (Miller et al. 2013). However, in the warmer and more arid (mesic-aridic) plant associations that are lower in resilience and resistance to invasive annuals or contain fire-sensitive grasses and forbs, recovery is slower and cover may not exceed pre-burn levels for many years. If density of native woody species is less than 3 plants m^{-2} (0.28 ft^{-2}) (Bates et al. 2000), and cheat grass is abundant in the understory, burning will likely convert the site to introduced annual grassland.

In general, broad leaf grasses such as squirreltail (*E. elymoides*), blue bunch wheat grass, and Lemmon's needle grass are relatively resistant to fire, recovering quickly and often producing greater amounts of biomass following a fire (Blaisdell 1953, Wright 1971, Bunting et al. 1987, EOARC data). In contrast, fine leaf grasses such as Idaho fescue and Thurber's needle grass (*S. thurberiana*) are more sensitive to fire, suffering greater crown mortality and slower recovery rates than broad leaf grasses (Blaisdell 1953, Wright 1971). Fine leaf grasses usually accumulate more dead material in the crown, which causes the plant to burn more slowly, transferring more heat to the growing points (Wright 1971).

Conrad and Poulton (1966) reported grazed Idaho fescue and blue bunch wheat grass plants were less damaged by fire than ungrazed plants. In a Wyoming big sagebrush plant association in south central Oregon, blue bunch wheat grass was severely damaged in an ungrazed exclosure where considerable dead leaf material had accumulated in the crowns (Davies et al. 2010). Outside of the exclosure on the grazed portion of the study area, blue bunch wheat grass biomass and ground cover was near pre-burn levels in the first year following fire.

Although the majority of literature reports Idaho fescue is fire-sensitive and declines in the first year following fire (Blaisdell 1953, Countryman and Cornelius 1957, Conrad and Poulton 1966), this species usually recovers, and biomass and cover can exceed pre-burn levels within 3 to 5 years after the fire (Miller et al. 2014, EOARC data). In Lassen County, Idaho fescue crown area decreased by one-third in the first year following fire but was 35% greater than pre-burn levels in the third growing season. Contradictory results in the literature regarding this species may partially result from differing intensities of postfire herbivory (Miller et al. 2013).

Forb species that resprout below ground from a caudex, corm, bulb, rhizome, or rootstock usually exhibit rapid recovery following fire. Table 13.3 shows the relative negative response of common perennial forbs in the sagebrush steppe to fire. The majority of these forbs are dormant at the time of the fire and their growing points are protected from burning. However, forbs that are suffrutescent or mat forming, such as sandwort (*Arenaria* spp.) and wild buckwheat (*Eriogonum* spp.), have their growing points above ground and can be severely damaged by fire, resulting in crown area reduction or mortality. Perennial forb production usually increases two- to threefold following fire in the more moist sagebrush plant associations (Blaisdell 1953, Bates et al. 2016). However, perennial forb response is usually less in the drier plant associations (Blaisdell 1953, Bunting et al. 1987, Fischer et al. 1996, Miller et al. 2013). Postfire increases can persist from 3 to >10 years (Bates et al. 2016).

In relatively good condition sites, the largest increases in vegetation during the first several years following fire are often the native annuals if sufficient moisture is available. Most species have completed their life cycle by early summer, prior to most fire events. During the first growing season following fire, annuals are able to take advantage of increased nutrient availability and decreased competition from perennials. In several fires in northeastern California and northwestern Nevada, annuals increase three- to fivefold in the first and second year following fire (Bates et al. 2016). However, annual response typically lasts only two to five growing seasons following a fire event. Their response can also be greatly limited by dry conditions in the spring. In heavily disturbed or warmer sites (i.e., Wyoming big sagebrush) the native annual response is replaced by exotic annuals and biennials, which dominate the site.

FIRE REGIME–PLANT COMMUNITY INTERACTIONS

Proxy information must be used to develop fire regimes for much of the Sagebrush Steppe Zone since little direct information is available. The lack of large pre-Euro-American settlement wood or trees that repeatedly scar, such as ponderosa pine, across most of the sagebrush biome limit our ability to date pre-Euro-American settlement fires and determine mean fire return intervals. Presettlement fire return intervals for some of the wetter mountain big sagebrush associations adjacent to forested communities have been described (Houston 1973, Miller and Rose 1999, Miller and Heyerdahl 2008). But descriptions of fire regimes for the majority of plant associations in the sagebrush steppe are lacking. Here we use available literature and ecological–fire–climate relationships to describe presettlement fire regimes for shrub steppe, desert shrub, and woodland associations (Appendix 1). This includes: (1) fire scars and charred wood, (2) fine fuel load potentials, and (3) juniper age structure to describe presettlement fire regimes in several mountain big sagebrush plant associations.

Over 90% of the presettlement fires dated in this region occurred between mid-summer and early fall (Miller and Rose 1999, EOARC data). The likely ignition source was lightning,

TABLE 13.3
Relative negative response of common perennial forbs in the sagebrush biome to fire

None to slight	Moderate to severe
Yarrow	Pussy-toes
Agoseris spp	Sandwort
Onion	Matted buckwheat
Aster	Parsnipflower buckwheat
Milkvetch	Douglas' buckwheat
Woollypod milkvetch	Slender buckwheat
Balsam-root	Sulfur flower
Indian paintbrush	Spiny phlox
Hawksbeard	
Fleabane daisy	
Geranium	
Avens	
Prickly lettuce	
Lomatium spp.	
Lupine	
Bluebells	
Slender phlox	
Penstemon	
Longleaf phlox	
Cinquefoil	
Lambstongue ragwort	
Goldenrod	
Dandelion	
Goat's beard	
Largehead clover	
Foothill death camas	

NOTE: Derived from Blaisdell (1953), Pechanec et al. (1954), Lyon & Stickney (1976), Klebenow & Beall (1977), Wright et al. (1979), Volland & Dell (1981), and Bradley et al. (1992).

which usually occurs in dry thunderstorms during July and August. Although there is evidence of burning by Native Americans in this region, there is no consensus as to the magnitude of human-caused fire starts compared to lightning.

Juniper trees <50 years old are easily killed by fire (Burkhardt and Tisdale 1969, 1976, Bunting 1984b, Miller and Rose 1999). Although there was considerable variation in fire return intervals, the mean intervals were probably less than 50 years necessary to limit woodland encroachment. As fire return intervals increase in length (>50 years) the potential for some trees to survive fire increases. In mountain big sagebrush it is not uncommon to see scattered presettlement trees on less productive south slopes. However, on the opposing north aspect evidence of presettlement trees is typically absent unless present on fuel-limited micro sites. In these arid land plant communities, the potential for shorter fire return intervals decreases along an increasing moisture gradient resulting in more abundant fuels and great fuel continuity.

Frost (1998) estimated a very general presettlement fire return interval of 13 years to 25 years for the sagebrush region in the Intermountain West. However, this was based on fire return intervals reported for the more cool-moist mountain big sagebrush communities in Yellowstone National Park (Houston 1973). Brown (2000) estimated the presettlement fire return interval for this same region as varying between 35 and 100 years which probably better characterizes the more arid sagebrush associations, which occupy a larger portion of this zone. Using the combined range of both authors, 13 years to 100 years probably best captures the range of fire return intervals that occurred across much of the sagebrush steppe, with return intervals getting increasingly longer going from cool-moist to warm-arid big sagebrush communities. In the fuel limited hot-dry big sagebrush, low sagebrush, and black sagebrush communities, fire return intervals were more commonly >150 years (Young and Evans 1981, Miller and Rose 1999).

Pre- and postsettlement fire regimes among and within desert shrub, salt desert shrub, shrub steppe, and juniper woodlands of northeastern California are spatially and temporally complex. Presettlement fire regimes varied from low-intensity fires occurring at intervals of less than 20 years in some more grass-dominated mountain big sagebrush associations to events rarely occurring in desert salt shrub associations. Even within the mountain big sagebrush alliance presettlement fire return intervals on the Lava Beds National Monument range from <20 years to >200 years among different plant associations (Miller and Heyerdahl 2008).

Amount and continuity of fuels are often limiting among many plant associations in the Sagebrush Steppe Zone (Bunting et al. 1987) and associated plant communities. Fire models that describe rate of spread and behavior are usually inadequate for most of these plant associations due to the patchiness of fuels and presence of bare ground (Brown 1982). Minimum levels of fuels required to carry fire in sagebrush steppe communities are 20% shrub cover, 228–336 kg ha^{-1} (200–300 lb ac^{-1}) of herbaceous fuel, 13–24 km h^{-1} (8–15 mi h^{-1}) winds, 15–20% RH, and 21–27°C (70–80°F) temperatures (Britton et al. 1981). Brown (1982) reported shrub canopies began influencing fire spread at 30–40% cover. For fire to carry through woodlands under relatively dry conditions, Klebenow and Bruner (1976) reported a minimum of 40–60% pinyon, juniper, and shrub cover were required. Wright et al. (1979) concluded that about 270–320 kg (600–700 lb) of fine fuel were necessary to carry a fire through pinyon-juniper woodland. However, wildfires that burn under severe-weather conditions are capable of burning across plant communities with limited fuels.

Areas in this zone with fire regimes characterized by short return intervals were probably grass dominated with a low density of shrubs. As Median Fire Return Interval (MFRI) increased, the physiognomy shifted from grassland to shrub steppe. Sparse stands of western juniper probably became established as MFRI became >10 years (Miller and Rose 1999).

Two stands comprise the northernmost populations of pinyon pine are found on the edge of the Northeastern Plateaus bioregion and the eastern slopes of the Sierra Nevada (David Charlet, Community College Southern Nevada, Henderson, Nevada, pers. comm.). Both stands consist of only a few scattered trees and have been spared by several wildfires that have burned in the vicinity. Much more expansive stands of

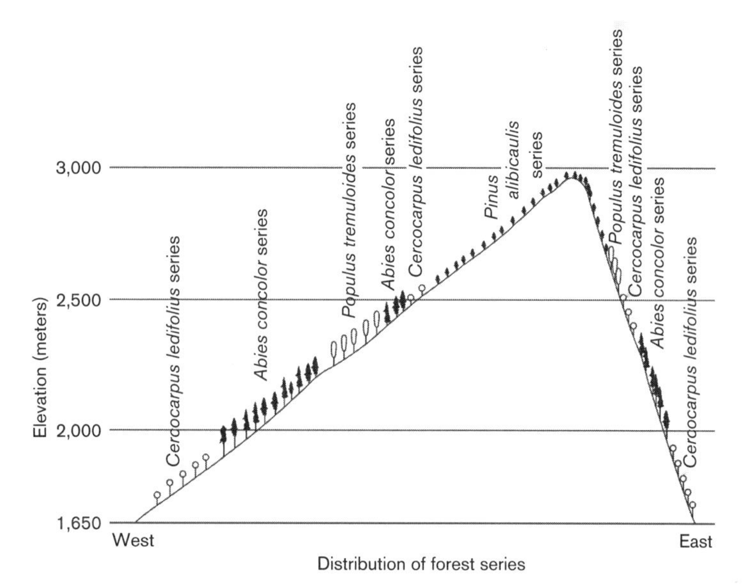

FIGURE 13.5 Distribution of vegetation types and elevation relationships in the South Warner Mountains. This graphic depicts a cross-section of the bioregion that includes precipitation patterns with vegetation and elevation in two cross-sections: (1) presettlement-past vegetation distribution prior to fire exclusion and alien weed introductions, and (2) current climatic potential vegetation distribution (Riegel et al. 1990).

FIGURE 13.6 Lower Montane zone. Landscape view of ponderosa and Jeffrey pines, California black oak, western juniper, and bitterbrush dispersed between large expanses of low and mountain big sagebrush on the Big Valley Ranger District, Modoc National Forest (photo by Sydney Smith).

pinyon pine occur in the northern Mono section of the bioregion. Their fire ecology is discussed in chapter 18.

Since the late 1800s, fire regimes have significantly changed among many of the plant associations in the Sagebrush Steppe Zone of the bioregion. In some areas, the reduction of fine fuels through grazing, fire suppression, and the removal of anthropogenic burning have significantly altered fire interval, intensity, severity, and season (Burkhardt and Tisdale 1969, 1976, Whisenant 1990, Miller and Rose 1999). Fire events in these areas have generally decreased across the mountain big sagebrush alliance, especially those where presettlement MFRI were <100 years. In northern California, the result has been an increase in shrub cover and density and decline in the herb layer, and often a shift from shrub steppe communities to western juniper woodlands.

In contrast, other portions of the Sagebrush Steppe Zone are now burning much more frequently as a result of cheat grass invasion. Over 25,000 ha (62,000 ac) have been mapped as dominated by cheat grass in the region. Due to its early-season growth, cheat grass can outcompete native grasses, forbs, and shrubs by reducing moisture and nutrients in surface soils (Norton et al. 2004). Once established, cheat grass alters fire regimes by creating continuous fine fuels that promote frequent, high-severity fires (Zouhar 2003). Sagebrush and other plant species that are slow to recolonize following fire, like bitterbrush and curl-leaf mountain-mahogany, decline with increased fire frequencies. The invasion of cheat grass has converted thousands of hectares of sagebrush, bitterbrush, and curl-leaf mountain-mahogany plant communities in the region to low diversity annual grasslands.

Lower Montane Zone

The Lower Montane Zone is wetter and cooler than the Sagebrush Steppe, and generally occurs where precipitation levels range from approximately 40 cm (16 in) to about 64 cm (25 in) (Figs 13.5 and 13.6). Vegetation within the Lower Montane Zone is characterized by ponderosa and Jeffrey pine forests growing on sites with moderately deep to deep well-drained soils. The Lower Montane Zone in this bioregion occurs on the lower slopes of the Warner, Adin, and Big Valley Mountains, at the base of the Diamond Mountains, and in large patches on the flat, expansive Devil's Garden area. Slopes are generally less than 30%. At the north end of the bioregion, ponderosa pine is the sole native yellow pine, although Jeffrey pine can be found in older tree plantations such as Sugar Hill (reforested after the 1929 fire) in the north Warner Mountains. South of a latitude line corresponding with Alturas, California, Jeffrey pine mixes with ponderosa pine.

This zone was heavily affected in the twentieth century by logging, grazing, fire exclusion, and possibly climate change. In general, stands are much denser than in the late nineteenth century, and trees are much smaller, larger ones having been removed mostly by logging. Fuel loadings are likely quite high compared to pre-twentieth century levels because of dense trees, dead trees, and accumulations of logging slash (Sydney Smith, Modoc National Forest, Alturas, California, personal observations).

FIRE RESPONSES OF IMPORTANT SPECIES

Fire response types for important species in the Lower Montane Zone are shown in Table 13.4. Before the twentieth century, frequent surface fires of low–moderate intensity removed needle and branch litter and created exposed surface mineral soil where yellow pines, both ponderosa and Jeffrey, and incense cedar seeds can germinate and grow. Seeds of these species are wind dispersed in the late summer through fall. Depending on the time at which fire returns to a site these seeds may be eaten or cached by animals, consumed by fire, or fall on mineral surface soil.

Jeffrey pine occurs on the coldest sites, higher elevations, steeper slopes, and shallower soils relative to ponderosa pine (Smith 1994). Thus, Jeffrey pine sites are generally lower in productivity compared to ponderosa pine sites (Jenkinson 1990).

Bitterbrush is the most widespread understory dominant associated with Jeffrey and ponderosa pine forests, and is adapted to open canopy (<40% cover) environments that burned with low intensity and severity. Historically, frequent fires (<10 years) limited the ability of bitterbrush to occupy a site and favored fire resilient graminoids. The ability of bitterbrush to persist in fire prone environments is primarily a function of seed caches buried deep enough to avoid lethal surface fire temperatures. Post fire sprouting also contributes to bitterbrush persistence, but successful sprouting is highly variable in ponderosa pine forests, ranging from <10% (Driscoll 1963, Nord 1965, Martin 1982, 1983, Martin and Driver 1983, Busse and Riegel 2009) to 25% of top-killed plants (Busse et al. 2000). Fire free intervals of >11 years are needed to maintain bitterbrush seed crop, browse, and ladder fuel functions (Busse and Riegel 2009).

Utah service-berry (*Amelanchier utahensis*) may be slightly injured by fire, depending on moisture conditions and flame lengths but is generally considered to be fire tolerant (Crane 1982, Bradley et al. 1992). However, mortality does occur if a thick litter layer is present at the time of the fire. Utah serviceberry sprouts from adventitious-buds at the root crown and it is also a prolific seed producer with persistent animal dispersed berries.

Fire often kills aboveground western choke cherry stems and foliage, but it quickly resprouts from root crowns and rhizomes (Volland and Dell 1981, Young 1983). Fire-induced sprouts can be produced within the growing year following a spring burn, or, if a fall burn, by the next spring. It can double its stem density within one growing season prior to being burned. Large, animal-dispersed seeds aid in postfire regeneration. Scarification of the seed coat by the heat from fire improves germination.

Hardwood species in the bioregion, including Oregon oak (*Quercus garryana*) and California black oak, have a number of adaptations that allow them to tolerate fire, most importantly the ability to sprout from the root crown after mortality of aboveground stems (McDonald and Tappeiner 1996). Sprouting can be vigorous, with up to 100 sprouts observed emerging from a single stump. Growth rates of sprouts can also be high, allowing hardwood species to quickly reoccupy sites after fire. Even first year seedlings and saplings can resprout after being top-killed by fire or other disturbances (Standiford et al. 1996). Mature California black oak and Oregon oak can survive low- to moderate-intensity fire (Plumb 1979). However, the bark of these oak species is thinner than that of ponderosa and Jeffrey pines of similar size, making them more sensitive to moderate-severity fire than associated conifers in lower montane forests.

The primary response of herbaceous plants five to six years post-prescribed fire on the Fremont National Forest, in south-central Oregon, was a slight increase in diversity and a change in the relative dominance of graminoid species, Ross' sedge (*Carex rossii*), and Idaho fescue (Busse et al. 2000). Squirreltail and Ross' sedge both increased in biomass and cover, while fescue cover declined. Fire stimulates squirreltail root and shoot biomass of surviving plants and increases seedling recruitment (Young and Miller 1985, Vose and White 1991). Ross' sedge has shallow rhizomes and is capable of responding rapidly depending on fire severity. In comparison, fescue, a tufted bunchgrass, varies from moderate to severe susceptibility to fire, depending on the amount of detritus in the basal tufts that cause fire to linger and kill the perennating buds (Johnson 1998) (see further discussion in the Sagebrush Steppe Zone above).

FIRE REGIME–PLANT COMMUNITY INTERACTIONS

Though no fire history studies have been conducted in the Jeffrey pines forests in northeastern California, Smith (1994) speculates that return intervals were short, probably ranging from 5 to 20 years (Appendix 1). In the eastern Sierra Nevada adjacent to our bioregion, Jeffrey pine mean site fire intervals (C1) (Grissino-Mayer 2001) varied from 6.8 years to 8.2 years and mean fire return intervals for fires scarring two or more trees (C10) varied from 11.7 years to 17.8 years in Bridgeport and Sierraville, respectively (North et al. 2007). In the ponderosa pine/bitterbrush-snowbrush ceanothus/needlegrass association on Pringle Butte in Central Oregon (Deschutes National Forest), between 1362 and 1900, the MFRI ranged from 7 years to 20 years where mean annual precipitation was 61 cm (24 in) (Bork 1984). However, Jeffrey pine, which reaches its eastern geographic range limit in Modoc County (Little 1971, Griffin and Critchfield 1972), is absent from these central Oregon sites. Thus, the fire return intervals may be shorter than would be expected for Jeffrey pine. The harsher conditions usually associated with Jeffrey pine sites, as noted above, usually leads to slower fuel accumulation, typically resulting in longer fire return intervals than found on sites dominated by ponderosa pine (Skinner and Chang 1996, Taylor 2000, Stephens 2001, Skinner 2003, Stephens et al. 2003) (see chapters 11 and 12).

Within the Lower Montane Zone, fire return intervals were probably longer at the drier and/or lower elevation Sagebrush Steppe Zone ecotone than in the mid- and upper elevations,

TABLE 13.4
Fire response types for important species in the Lower Montane Zone

	Type of fire response			
Lifeform	Sprouting	Seeding	Individual	Species
Conifer	None	None	Resistant/killed	Ponderosa pine, Jeffrey pine, incense cedar, western juniper
Hardwood	Fire stimulated	None	Resistant/killed	California black oak
Shrub	Fire stimulated	Fire stimulated	Top-killed	Mahala mat, greenleaf manzanita, western choke-cherry
	Fire stimulated	None	Top-killed	Utah service berry, bitter cherry, Modoc plum
	+/–Fire stimulated	None	Killed	Bitterbrush
	None	None	None	Mountain big sagebrush
Forb	Fire stimulated	None	Top-killed	Woolly mule's ears, arrowleaf balsam-root, lambs tongue ragwort, lupine, wild pea, lomatium, peony, Indian paintbrush
Graminoids	+/–Tillering	None	Top-killed	Idaho fescue
	Tillering	None	Top-killed	Squirreltail, Ross' sedge, Canby bluegrass
	None	None	Killed	Cheat grass

regardless of dominance Jeffrey pine or ponderosa pine. Bork (1984) reported an MFRI of 16 years to 38 years, from 1460 to 1970, in a ponderosa pine/bitterbrush-mountain big sagebrush/Idaho fescue plant association where mean annual precipitation is 24 cm (9 in). Fire history was reconstructed from scattered ponderosa pine growing within mountain big and low sagebrush communities in the upper Chewaucan River basin near Paisley, Oregon, where precipitation was approximately 40 cm (16 in). MFRI ranged from 3 years to 38 years between the years 1601 and 1897. Fire return intervals from a relatively small sample size were reported along an elevation and vegetation gradient in the Buck Creek watershed on the west side of the north Warner Mountains (Goheen 1998). For the period between 1650 and 1879, return intervals were four times longer (MFRI ranged from 2 years to 56 years) on a lower elevation (1,567 m [5,140 ft]) juniper and ponderosa pine site than in the middle elevation (1,735–1,905 m [5,690–6,250 ft]) ponderosa pine sites (MFRI ranged from 1 year to 11 years). Vegetation spatial pattern is more heterogeneous on these lower elevation sites as limited soil moisture regulates plant density and biomass production. Both limited fuel quantities and lack of continuity result in longer intervals between fires. Additionally, slopes are not as steep on the lower elevation sites and would be less likely to produce fires intense enough to scar trees as on steeper slopes.

Busse et al. (2000) demonstrated that low-severity prescribed fire following a prolonged period of wildfire exclusion produced a significant, yet slight reduction in tree growth in thinned ponderosa pine stands. Basal area growth reduction was proportional to the level of crown scorch and O horizon reduction during burning. No evidence was found to suggest that prescribed fire will have a long-term impact on stand productivity. Specifically, soil resources, including nutrient content and exposed mineral soil, were unaffected by fire.

Hardwood species are relatively rare in northeastern California due to extremely cold minimum temperatures and aridity. Oregon oak achieves its most interior occurrence primarily in the western portion of the bioregion in the Big Valley and Adin Mountains. Forests and woodlands with a significant Oregon oak or California black oak component have fire return intervals of between 5 and 25 years have been reported, depending on aspect, soil type, and elevation (Taylor and Skinner 2003). Shorter fire return intervals restricted conifer encroachment and allowed Oregon and California black oak to persist prior to Euro-American settlement. Patchy, stand-replacement fire occurred occasionally in low and mid-elevation forests (Vankat and Major 1978, Parsons 1981, Stephenson et al. 1991). These small patches were also important in maintaining oak stands where conifers can outcompete hardwood species without fire (McDonald 1969). Fire return intervals in fractured basalt flows were likely less frequent because ignition must be initiated on site due to the discontinuity of vegetation in basalt flows and shallow soils matrix.

The largest population of Baker cypress (*Hesperocyparis bakeri*) occurs on the northwestern edge of the bioregion (Griffin and Critchfield 1972). Fire is the only effective process that opens the serotinous cones and creates the conditions required for successful regeneration (Vogl et al. 1988). Fire return intervals largely determine the size of the canopy seed bank for Baker cypress (Ne'eman et al. 1999). In more open grown stands, canopy seed storage significantly increases in populations over 50 years of age, and remains constant across older age classes suggesting a 50 year minimum fire return interval (Rentz and Merriam 2009). Knobcone pine (*Pinus attenuata*), another closed-cone fire adapted species that occurs at the western edge within our bioregion, is discussed in chapters 11, 12, 16, and 17.

FIGURE 13.7 Mid-Montane Zone. The Mid-Montane Zone provides habitat for many woody species. In the foreground, ponderosa pine is regenerating at the edge of a mountain big sagebrush and woolly mule's ears site. Sagebrush is edaphically controlled by higher soil clay content, which limits plant-available water. Quaking aspen occurs, as both small clones comprised of several trees growing within conifer stands or as larger stands that are topographically controlled where the water table is closer to the surface. Historically, discontinuous fuel types and varying biomass productivity contributed to complex burn patterns and fire intensity. With fire exclusion, many smaller quaking aspen clones are being lost to confer succession. In the background, Mid-Montane through Subalpine Zones are visible in the South Warner Wilderness Area (photo by Gregg Riegel).

Mid-Montane Zone

The Mid-Montane Zone for this bioregion is defined as the area where the climate supports the lower elevation white fir forests (Fig. 13.7). The Mid-Montane Zone occurs in the Big Valley Mountains, the Adin Mountains, the northern part of the Devil's Garden, and in the Warner Mountains. This zone is typically colder and wetter than the Lower Montane Zone. Average annual precipitation in this zone ranges from about 64 cm (25 in) to 82 cm (32 in) and elevations range from about 1,430 to 1,980 m (4,700–6,500 ft) in the Big Valley Mountains and about 1,615 m (5,300 ft) to as high as 2,225 m (7,300 ft) in the Warner Mountains (Smith and Davidson 2003). White fir occurs mixed with ponderosa pine, Jeffrey pine, incense cedar, and as more-or-less pure stands.

This zone presents unusual fuels management challenges because of: (1) changes in stand species composition and structure in the twentieth century, and (2) susceptibility of trees to mortality from drought, disease, and insects. Because of preferential logging of ponderosa pine, fire exclusion, and climate changes, white fir has generally increased while pine species have decreased. These changes have dramatically altered fire regimes in this zone. Stand physiognomy has generally changed from less dense or relatively open stands of large trees to much denser stands of small trees. Extensive tree mortality from insects and diseases has accompanied periodic droughts. Thus, the potential for high-intensity fires increase as these mortality episodes lead to increased fuel accumulations as the dead trees fall (S. Smith, personal observation).

FIRE RESPONSES OF IMPORTANT SPECIES

Jeffrey pine, ponderosa pine, incense cedar, and white fir all develop thick bark as they age which allows for their survival in characteristic frequent and low-intensity fire (Table 13.5). Jeffrey and ponderosa pine are more resistant to fire when young as compared to white fir or incense cedar (van Wagtendonk 1983, Agee 1993). White fir sapling and pole-sized trees are very susceptible to fire-caused mortality due to: (1) numerous resin blisters on the bark that increase potential for flammability, (2) flat, lateral branches that often extend to the ground especially in trees that are in canopy gaps and isolated "wolf trees," and (3) comparatively shallow roots that increase susceptibility to lethal temperatures when surface fires slowly burn through high fuel loads in deep and compact litter (Agee 1993, Miller 2000).

Fire profoundly affects individual species responses which in turn greatly influence successional dynamics in the Mid-Montane zone. Surface fires burn litter temporarily leaving an ashy mineral soil surface that ponderosa and Jeffrey pines seeds require for successful germination and establishment, whereas incense cedar tolerates a variety of surface substrates (Powers and Oliver 1990). White fir is the only tree species in this bioregion that can germinate on top of litter, which typifies sites with longer fire return intervals.

Following fire, greenleaf manzanita prolifically sprouts from root crowns. Observations of the authors suggest that the occurrence of burls or lignotubers decreases with increasing latitude and may be rare or absent in the farther northeastern populations of our bioregion. Long-lived seed bank reserves are also released from dormancy ensuring the establishment or perpetuation of seral montane shrubfields (Knapp et al. 2012). Mountain snowberry (*Symphoricarpos oreophilus*) response to fire is quite variable. Most plants survive fire and typically resprout from basal buds at the root crown. However, mountain snowberry is considered a weak sprouter especially after severe fire (Young 1983) (Table 13.5). Low-severity fire can kill branches as well as all the above-crown portions of a plant making it difficult to predict individual plant survival. Even following severe fires, mountain snowberry root crowns usually survive (Pechanec et al. 1954, Kuntz 1982). Sprouting can occur the first year but may be initially limited, eventually reaching pre-burn levels within 15 years (Pechanec et al. 1954, Kuntz 1982).

FIRE REGIME–PLANT COMMUNITY INTERACTIONS

Historic fire regimes are similar to those described for the East Montane zone of the Sierra Nevada and the Mid-Montane Eastside in the southern Cascade Range. Fire return intervals were short and intensity and severity were low (Appendix 1). Intensive logging of ponderosa pine, coupled with fire suppression and exclusion, has shifted species composition from

TABLE 13.5
Fire response types for important species in the Mid-Montane Zone

	Type of fire response			
Lifeform	Sprouting	Seeding	Individual	Species
Conifer	None	None	Resistant, killed	Ponderosa pine, Jeffrey pine, white fir, incense cedar
Shrub	Fire stimulated	Fire stimulated	Top-killed	Mahala mat, greenleaf manzanita, snowbrush ceanothus
	Fire stimulated	None	Top-killed	Mountain snowberry, creeping barberry
	None	None	None	Mountain big sagebrush
Forbs	Fire stimulated	None	Top-killed	Heartleaf arnica, tuber starwort, hawkweeds, lupines, wild peas, sweet cicely
Graminoids	Tillering	None	Top-killed	Squirreltail, Ross' sedge, Wheeler and Canby bluegrass needlegrasses, orcutt brome

pine dominated and codominated stands to overstories where white fir is the dominant species. Fire helped to regulate stand density and suppressed the continual recruitment of white fir. The exclusion of fire has helped promote the shift to white fir dominance. White fir seedlings, which grow best in shade, can survive under the canopy of montane shrubs and dense forests but once established, grows best in full sunlight (Laacke 1990). Shade-tolerance ranking for tree regeneration (seedlings and saplings) from most to least tolerant: white fir > incense cedar > ponderosa pine = Jeffrey pine (Minore 1979). White fir is also a prolific tree, producing up to 1.5 million seeds ha^{-1} (6 million seeds ac^{-1}) generally every 2 years to 5 years, beginning at age 40 and continuing beyond 300 years (Laacke 1990).

The only cross-dated fire scar record from this zone comes from the Adin Mountains just west of Adin Pass and south of Stone Coal Valley (Table 13.6). These data are from three stands dominated by ponderosa pine with incense cedar and white fir as associates. Each stand is sufficiently moist to have populations of mountain lady's-slipper (*Cypripedium montanum*). The fire scar record was developed from stumps of mostly ponderosa pine and a few incense cedar. The record of fires extends from 1654 to 1894. Median FRI's ranged from 4.5 years to 8 years. Intra-ring position of fires scars was 24.1% ring boundary, 56.6% latewood, 19.3% in earlywood (unpublished data on file USDA Forest Service, Pacific Southwest Research Station, Redding, California, USA). Interestingly, the last fire that scarred more than one tree on these sites occurred in 1864, around the time of introduction of large herds of livestock to northeastern California (Pease 1965). The cessation of fire on these sites predates organized fire suppression by many decades. This lack of fire may be due to a lack of fuel continuity resulting from livestock consuming much of the available herbaceous fuel that was likely the primary carrier of fire as has been suggested from findings in other ponderosa pine dominated forests in northeastern California (Norman and Taylor 2005) and the southwest (Savage and Swetnam 1990, Grissino-Mayer and Swetnam 1995).

When precipitation is greater than or equal to the annual mean for a site, white fir can be very successful with rapid growth and continual reproduction and recruitment. However, in the 64 cm (25 in) precipitation zone in northeastern California, due to recurring droughts, white fir tends to become stressed, especially where the denser stands have developed, and eventually succumbs to insects and diseases, often before a fire occurs. White fir's stomatal regulation, Water Use Efficiency (WUE), ability to control transpiration/unit of fixed carbon is far less than ponderosa pine (Hinckley et al. 1982). If white fir is under moisture stress (–2.0 Mpa [290 lb in^{-2}]) for several summers, they are more susceptible to successful fir engraver beetle (*Scolytus ventralis*) attacks than trees under less stress (Ferrel 1978). Recurring drought is the common denominator that regulates a site's ability to produce more leaf area and stand density. Cochran (1998) advises that if prolonged droughts are forecast, removal of white fir on drier sites is recommended. Fire would have regulated white fir regeneration under a functioning historic fire regime on these sites.

Despite extended drought in the southeastern Warner Mountains, Vale (1975) found that white fire invaded the mountain big sagebrush ecotone between 1915 and 1944. He concluded that heavy livestock grazing of competing vegetation was the main factor and that successful fire suppression, which did not occur until after the trees had been established, was secondary. Eventually these trees on the ecotone will be eliminated by: (1) drought, (2) stress, (3) insect interactions, or (4) fire.

Outbreaks of Modoc budworm (*Choristoneura viridis*) defoliating white firs occurred during the years 1959 to 1962 and 1973 to 1975 in the Warner Mountains and nearby ranges of northeastern California (Ferrel 1980). These outbreaks were associated with periods of three years or more when precipitation averaged at least 25% below the historic mean annual precipitation. From 1959 to 1962, a Douglas-fir Tussock Moth (*Orgyia pseudotsugata*) infestation occurred on 182 ha (450 ac) from 1964 to1965, and defoliated second growth white fir populations in the Warner Mountains between 1,700 and 1,900 m (5,600–6,200 ft). Wickman (1978) interpreted from the existing number and size of stumps that the prelogged forest had been a ponderosa and Washoe pine stand with large trees, 91–102 cm dbh (36–40 in) before being logged around the early 1900s. Though the area had been logged a second time for white fir overstory trees in 1954, Wickman (1978) noted

TABLE 13.6
Adin Mountains fire history summary; west of Adin Pass, south of Stone Coal Valley, Modoc County, California
Mean scars per tree = 10.2

Site	Latitude	Longitude	No. trees	No. scars	Earliest scar	Early 2+	Latest scar	Latest 2+	Median FRI	Mean FRI	Min FRI
ACM	41.340763	−121.00855	13	119	1691	1733	1894	1864	4.5	5.97	1
APM	41.336605	−121.00337	5	52	1657	1691	1864	1864	8	9	2
APR	41.338062	−120.99318	20	218	1654	1654	1865	1864	5	5.02	1
Total			38	389	1654	1654	1894	1864			

that no fires had burned in the area for at least 50 years and more likely 75 years. The effect of logging the pine and suppressing and excluding fire has been to allow white fir to become the dominant tree species in all structural and age classes. The consequence of this change is that the current forest is more susceptible to high-intensity fire because of large amounts of live and dead fuels. The ensuing high-intensity fires have produced large landscape-level stand-replacing events such as the Scarface Fire (1977), the Crank Fire (1987), Blue Fire (2001), and the Barry Point Fire (2012).

After a forest is removed by fire or logging, the developing shrub and herbaceous understory often suppresses conifer regeneration as a result of competition for soil moisture (Conard and Radosevich 1981, 1982, McDonald 1983a, 1983b, Roy 1983, Radosevich 1984, Lanini and Radosevich 1986, Shainsky and Radosevich 1986, Parker and Yoder-Williams 1989, Collins and Roller 2013, Knapp et al. 2013). Though there are many pathways succession that can take following disturbance in Mid-Montane pine forests, a generalized summary includes: (1) early seral-woolly mule's ears dominance (Riegel et al. 2002) with associated herbaceous species and some conifer regeneration; (2) mid-seral-greenleaf manzanita, snowbrush ceanothus, mahala mat (*Ceanothus prostratus*), and bitterbrush shrub dominance with pole-sized conifers; and (3) and late seral-conifer dominance with some shade-tolerant herbs and shrubs. A shrub canopy also provides a shaded microenvironment, however, that reduces evaporative demand and improves the water balance of trees, which may be less physiologically stressful than competition for soil resources on hot, dry sites (Conard and Radosevich 1981, Lanini and Radosevich 1986). Soil moisture availability on shrub-dominated sites can be significantly greater than on woolly mule's ears-dominated sites during the latter part of the growing season (Williams 1995).

Shrub dominance in the mid-seral phase should increase soil nutrient concentrations with time between fires and logging disturbance intervals that temporarily reduce the cover of nitrogen fixing shrubs such as snowbrush, mahala mat, and bitterbrush (Conard et al. 1985, Johnson 1995, Busse et al. 1996, Busse 2000a, 2000b). Nitrogen fixation occurs from actinorhizal symbionts associated with snowbrush ceanothus and mahala mat (Delwiche et al. 1965, Conard et al. 1985, Busse 2000a, 2000b), and bitterbrush (Webster et al. 1967, Busse 2000a, 2000b). Annual nitrogen fixation in forests east of the Sierra Nevada and Cascade Range crest varies from 5 to 15 kg ha^{-1} (4.5–13.4 lb ac^{-1}) for snowbrush ceanothus and 1 kg ha^{-1} (0.9 lb ac^{-1}) for mahala mat and bitterbrush (Busse 2000a, 2000b). Busse et al. (1996) found increased soil carbon, nitrogen, and microbial biomass in the upper horizon of a ponderosa pine forest due to long-term retention of shrubs from longer fire return intervals due to fire exclusion. Johnson (1995) also found improved soil nitrogen concentration at pine sites with a dominance of snowbrush ceanothus. After nearly 50 years of fire exclusion in naturally regenerated ponderosa pine stands in Central Oregon, nitrogen fixing shrubs that would have been periodically consumed in historic fire regimes increased radial-increment growth (Cochran and Hopkins 1991). Succession in these forests may be more influenced by competition for soil moisture and changes in microclimate.

Upper Montane Zone

The Upper Montane Zone in this bioregion is defined as the area where the climate supports the higher elevation white fir forests (Figs. 13.8 and 13.9). The Upper Montane Zone occurs at elevations of about 1,830–2,225 m (6,000–7,300 ft) in the Big Valley Mountains, and at elevations between about 1,800 and 2,430 m (5,900–8,000 ft) in the Warner Mountains (Smith and Davidson 2003). This zone is colder and wetter than the Mid-Montane Zone. Average annual precipitation in this zone exceeds 81.5 cm (32 in). The vegetation of the zone is quite patchy, responding to soil moisture, edaphic variation, and winter wind patterns. The patches are of several general types: (1) conifer-dominated, (2) shrub-dominated, (3) herbaceous-dominated, and (4) rocky, relatively barren areas primarily along exposed ridges. White fir can occur in pure stands, as well as in mixed stands with western white pine, lodgepole pine, quaking aspen, occasional yellow pines (ponderosa, Jeffrey, or Washoe pines), and whitebark pine (Riegel et al. 1990, Smith 1994). Western juniper at these higher elevations can still be found on exposed sites with shallow soils. The bioregion is too dry and cold to support California red fir (*Abies magnifica* var. *magnifica*) and mountain hemlock (*Tsuga mertensiana*). There is little fire research conducted in this zone. Most of the following discussion is based on observations and discussions with local fire and resource managers.

Although this zone has been affected by twentieth century logging, grazing, and climate change, the fire regime, as indicated by conifer and shrub species composition, has likely been affected the least of the montane zones in this bioregion. One major change, however, is the decrease in the extent and occurrence of quaking aspen (described in more detail in the Nonzonal Vegetation section below).

FIGURE 13.8 Upper Montane Zone. Landscape perspective of the Owl Creek watershed on the east side of the Warner Mountains showing the patchy, discontinuous nature of forest in this zone. Photo was taken during an August thunderstorm (photo by Carl Skinner).

FIGURE 13.9 Upper Montane Zone. Photo shows the pattern of exposed slopes dominated by shrubs and herbaceous plants with trees confined mostly to more protected sites such as the draw in the foreground and the patch of trees in the upper center on the lee side of the crest. Photo taken near High Grade Spring north of Mt. Vida (photo by Carl Skinner).

FIRE RESPONSES OF IMPORTANT SPECIES

Fire responses of Upper Montane species are listed in Table 13.7. Scouler's willow (*Salix scouleriana*) is fire sensitive but crown mortality is highly variable (0–100%) and depends on fire severity (Owens 1982). Severe fires may result in 100% aboveground mortality (Lyon and Stickney 1976). Following low- to moderate-severity fire, Scouler's willow will quickly resprout. High-severity fire that kills live foliage but does not kill the vascular cambium will cause vigorous epicormic sprouting from the root crown (Weixelman et al. 1998). Sprouting typically occurs within days following a fire and it is not uncommon for a plant to reach and exceed pre-burn frequency and cover in 5 years or less. Scouler's willow also produces wind borne seeds that can travel for considerable distances off site.

Bitter cherry response to fire is variable. It can be killed if burned while actively growing or if fire severity is high (Young 1983). Resprouting occurs from the root crown. It also establishes from large buried seed or seed dispersed from off site by animals, primarily birds. Creeping snowberry is usually top-killed by fire; however, vegetative regrowth does occur from rhizomes (Volland and Dell 1981). Fire will kill rhizomes if surface litter layers are thick and/or organic matter constitutes a significant portion of the upper soil horizon (Neuenschwander, n.d., cited in FEIS). Sticky currant is considered only moderately resistant to fire (Volland and Dell 1981). Low- to moderate-severity fire can stimulate rapid resprouting from basal stems. Seeds are relatively heavy and animal dispersed. Heat scarification increases germination rates. Pinemat manzanita is killed by fire but reestablishes from seed during the first postfire growing season. Kruckeberg (1977) speculated that it may be an obligate seeder, requiring fire and/or leachate from smoke or charred wood to break seed dormancy.

FIRE REGIME–PLANT COMMUNITY INTERACTIONS

Fuel accumulation rates, not ignitions, are probably the limiting factor for fire occurrence in the Upper Montane Zone. The generally dry climate coupled with cold winters contributes

TABLE 13.7
Fire response types for important species in the Upper Montane Zone

Lifeform	Type of fire response: Sprouting	Seeding	Individual	Species
Conifer	None	None	Resistant/killed	White fir, Washoe, ponderosa, Jeffrey, and western white pine
	None	None	Killed	Lodgepole pine
Hardwood	Fire stimulated	None	Top-killed	Quaking aspen
Shrub	Fire stimulated	Fire stimulated	Top-killed	Bush chinquapin
	Fire stimulated	None	Top-killed	Creeping snowberry, mountain snowberry, sticky currant, snowfield sagebrush
	None	Fire stimulated	Top-killed	Pinemat manzanita
	None	None	Killed	Mountain big sagebrush
Forbs	Fire stimulated	None	Top-killed	Heartleaf arnica, tuber starwort, hawkweeds, lupines, slender penstemon, whiteveined wintergreen, sweet cicely
Graminoids	Tillering	None	Top-killed	Squirreltail, Ross' sedge, Wheeler bluegrass, needlegrasses, orcutt brome, Brainerd's sedge

to slow fuel accumulation. Although episodes of widespread lightning are common in summer, ignitions are not likely a limiting factor for fire occurrence. The predominance of short-needle conifers and the winter snow compact the fuelbed. Thus, fires started by thunderstorms are often slow spreading; patchy, low-intensity surface fires and become more intense only where concentrations of fuels from dead trees or small thickets, usually of white fir regeneration, have developed (Appendix 1). Accumulations of dead fuels are generally caused: (1) by the stem exclusion phase of white fir regeneration patches, (2) beetle-killed patches, or (3) deaths of individual large trees.

Some areas in this zone that were once dominated by very large yellow pines have converted to lodgepole pine and white fir (Vale 1977). Examples of this are found: (1) on the north facing slopes near Lily Lake where large stumps of both yellow pines and incense cedar are found in what are now dense thickets of mostly lodgepole pine, and (2) along the west side of the main crest of the Warner Mountains in the vicinity of the headwaters of the south fork of Davis Creek where large stumps of yellow pines are found in lodgepole/white fir stands. In both of these cases no yellow pine regeneration is evident. It is likely the original fire regimes for these specific sites have changed dramatically with the change in species composition. The South Warner Wilderness Area provides an outstanding example of succession from yellow pines to white fir without logging but with fire exclusion (Riegel et al. 1990).

There are three yellow (i.e., three needle pines; subgenus *Diploxylon* of *Pinus*): that occur in the Upper Montane Zone in the South Warner Mountains: ponderosa, Jeffrey, and Washoe (Haller 1961, Critchfield and Allenbaugh 1969, Griffin and Critchfield 1972, Riegel et al. 1990, Smith 1994). Washoe pine, which grows at elevations from 1,890 to 2,195 m (6,199–7,200 ft) has distinctive characteristics that favor its survival in more severe climate and environments at higher elevations and slow growth rates (Rehfeldt 1999). The primary threats to Washoe pine are potential destruction by stand-replacing fires where white fir succession has altered the fire regime, and loss of habitat due to climate change.

The only cross-dated fire scar record in the Warner Mountains is from the Horse Creek Watershed in this zone (Table 13.8). These data are from stands dominated by white fir with lodgepole pine and western white pine as associates. The fire scar record was developed from stumps of mostly white fir, lodgepole pine, and white pine. The record of fires extends from 1746 to 1957. Median FRIs ranged from 3.5 years to 40 years. Intra-ring position of fires scars was 90.6% ring boundary, 7.8% latewood, 1.6% in middle-earlywood (S. Smith, unpublished data). This indicates that seasonality was mostly late summer to early fall and was similar to other upper montane forests in northern California (Taylor 2000, Skinner 2003) (see chapters 11 and 12).

Western juniper often occupies exposed, almost barren ridgelines in this zone. These are areas where fire rarely spreads due to very limited fuels. The barren nature of these ridgelines often limits the spread of fires from one watershed to the next. Many of the larger trees in these locations appear to be quite old, although no formal aging has been done.

Curl-leaf mountain-mahogany is often found on sites exposed to both the winter winds that remove snow and more intense summer insolation. Curl-leaf mountain-mahogany plants appear to initiate in small areas protected from wind and sun by rock outcrops. These sites are often surrounded by

TABLE 13.8
Fire history data derived from cross-dated fire scars at six sites of <10 ha each in the Horse Creek watershed of the Warner Mountains

Site	Number of trees	Number of scars	Median FRI	Mean FRI	Minimum FRI	Maximum FRI
Horse A	8	21	7.5	13.3	1	51
Horse B	9	17	3.5	8.4	1	43
Horse C	10	12	37.0	33.0	8	55
Horse D	6	6	40.0	35.3	1	65
Horse F	14	20	5.5	7.1	1	20
Horse H	14	19	15.5	20.6	4	56

open, sparse stands of low shrubs and herbaceous plants. They can then survive for long periods as these areas appear to not carry fire well. Larger curl-leaf mountain-mahogany plants in one location of this type north of Fandango Pass have been aged at over 400 years (unpublished data on file Forest Service Pacific Southwest Research Station, Redding, California, USA).

Conifer stands in this zone likely have similar responses to fire as found in other bioregions and the reader should reference chapters 11, 12, and 14. The character of lodgepole pine stands varies from more open stands of larger trees and scattered individuals in the south Warner Mountains to more extensive and denser stands north of Highway 299, such as around Dismal Swamp, continuing through the Warner Mountains into Oregon. It is likely the fire regimes will vary with the nature of the stands, being more like the Cascades in the north (chapter 12) and the Sierra Nevada in the south (chapter 14).

Subalpine Zone

The Subalpine Zone is limited in this bioregion to the Warner Mountains with the majority of the zone found in the South Warner Wilderness Area at elevations starting at about 2,253 m (7,390 ft) (Riegel et al. 1990). Whitebark pine forms nearly pure stands in this zone (Fig. 13.10). White fir can either be rare or common at the lower ecotonal edge; however, lodgepole pine, western white pine, and quaking aspen are relatively rare (Vale 1977, Riegel et al. 1990, Figura 1997).

Fire responses for the Subalpine Zone species are listed in Table 13.9. The fire regimes are similar to that described for the Subalpine Zone of the Sierra Nevada and Klamath Mountains, characterized by infrequent, small, low-intensity fires due to the lack and discontinuity of fuels (Appendix 1). The largest and most contiguous stands of whitebark pine are found in South Warner Wilderness Area from Mill Creek to the southern slopes of Eagle Peak where relatively high-elevation ridges occur on a gentle west slope. Heavy sheep grazing in the late nineteenth century greatly reduced perennial grasses and forbs, caused erosion of surface soils and reduction fuel loads and potential fire spread which facilitated downslope expansion of whitebark pine into mountain big sagebrush communities (Figura 1997). These lower elevation stands below approximately 2,470 m (8,102 ft) are characterized as relatively young (most trees established between 1915 and 1965) with a few older trees (200 years to 300 years), yielding low stem density and basal area. Figura (1997) noted the lack of standing snags, large fallen logs, burnt coniferous wood, and downed wood in the lower elevation stands, which was consistent with his hypothesis of a downslope migration. High-elevation forests are old (400 years to 800 years), self-perpetuating stands. Fire scarred trees were found in the upper elevation stands, though all trees cored by Figura (1997) were rotten, precluding dating historic fires. Lightning strikes are very common at these elevations yet fuel continuity and cool moist conditions limit the ability of fire to spread beyond one to several multistemmed trees. We are unaware of evidence that suggests large fires have burned sizeable whitebark pine stands in the Warner Mountains. Millar et al. (2014) estimated that 38% of whitebark pine in the Warner Mountains has succumbed to drought facilitated mountain pine beetle (*Dendroctonus ponderosae*) mortality which could affect how fire behaves in the subalpine zone. Small, old wind and snow-shaped krummholz whitebark pine grow to the summit of Eagle Peak (3,016 m [9,892 ft]), which is the highest peak in the bioregion (Riegel et al. 1990). There is no alpine zone (land above tree line) in this bioregion (Riegel et al. 1990).

Nonzonal Vegetation

Riparian areas, including wetlands, meadows, and riparian forests are dispersed throughout the bioregion in all ecological zones due to both topographic and edaphic influences of surface and subsurface water. To our knowledge no fire ecology research has been done in these hydric vegetation types. Fire regimes in riparian areas are dependent on the size of stream and whether it is perennial or intermittent (Olson 2000, Van de Water and North 2010, 2011). Smaller-sized streams and intermittent streams typically have the same fire frequency as the adjacent uplands. We speculate that pre-Euro-American settlement fire return intervals were longer and were typically more spatially complex than adjacent uplands. When these sites did burn, fire intensity was moderate to high with mixed severity due to microtopographic and surface/water table interactions. The removal of fine fuels in the form of forage by livestock has altered these fire regimes since settlement.

Meadows in the Northeastern Plateau bioregion have been dramatically affected by tree invasion, primarily by lodgepole pine. The reasons for invasion are complex but the combination of reduced competition from herbaceous species and fine fuel removal due to livestock grazing coupled with longer fire

FIGURE 13.10 Subalpine Zone. Photo looking south from Warren Peak toward Eagle Peak along the main crest of the Warner Mountains. All of the forests (foreground, middle ground, and background) shown in this photo are nearly 100% whitebark pine. The patches of whitebark pine are interspersed with patches of shrubs and herbaceous plants dominated mostly by snowfield sagebrush. Photo was taken during an August thunderstorm (photo by Carl Skinner).

TABLE 13.9
Fire response types for important species in the Subalpine Zone

	Type of fire response			
Lifeform	Sprouting	Seeding	Individual	Species
Conifer	None	None	Resistant/killed	Whitebark pine
Shrub	Fire stimulated	None	Top-killed	Mountain snowberry, snowfield sagebrush
	+/−Fire stimulated	Fire stimulated	Top-killed	Gooseberry currant
Forbs	Fire stimulated	None	Top-killed	Slender penstemon, prickly sandwort, Davidson's penstemon, phlox
Graminoids	Tillering	None	Top-killed	Western needlegrass, Wheeler's bluegrass, Ross' sedge

intervals due to fire exclusion has reduced the number and size of meadows in this bioregion. Since lodgepole pine is an evergreen conifer it continues to transpire as long as the soil within the rooting zone is not frozen (Running 1980, Running and Reid 1980, Sparks et al. 2001). This increased, nearly yearlong transpiring leaf area significantly reduces water to the meadow and is a net water loss to a watershed.

Common riparian hardwood trees and shrubs comprised mostly of black cottonwood (*Populus trichocarpa* subsp. *trichocarpa*), willows (*Salix* spp.), silver sagebrush, interior rose (*Rosa woodsii* subsp. *ultramontana*), and scattered populations of mountain (*Alnus incana* subsp. *tenuifolia*) and white alder (*A. rhombifolia*), dwarf birch (*Betula nana*), and water birch, (*B. occidentalis*) can sprout from stumps, stems, and root crowns following low and sometimes moderate-intensity fires (Dwire and Kauffman 2003).

Quaking aspen, considered a facultative or facultative upland wetland species, occurs in different environmental conditions in the Northeastern Plateaus bioregion, depending on ecological zone and n topographic or geographic position. In the Sagebrush Steppe Zone, quaking aspen trees are somewhat rare, and if they occur they are confined to sites such as streams, springs, high water tables, and talus rock "reservoirs" that have perennial moisture. In the Lower and Mid-Montane zones, quaking aspen stands are more common and are similarly associated with sites containing perennial water. The most extensive occurrences of quaking aspen in the Northeastern Plateaus bioregion are in the Upper Montane Zone, mainly in the Warner Mountains. In this zone quaking aspen stands are found on sites with added moisture from near surface water such as rocky outcrops, streams, meadows, and springs, as well as sites with added moisture from snowdrifts. In the Subalpine Zone, quaking aspen occurrence declines with increasing elevation. In areas where snow accumulations can persist well into the growing season (typically June through July), nivial or snow morphed stands occur (Riegel et al. 1990), which are very important to many wildlife species in this zone.

Aspen in the Northeastern Plateaus bioregion are on the western edge of their range and occupy only one percent or less of the landscape. Despite their rarity, aspen communities provide many critical ecological services and functions in the bioregion. Unfortunately, declines in the health and distribution of aspen communities have been observed across most of the western United States over the past century (Rehfeldt et al. 2009). A comparison of aerial photographs from 1946 and 1994 in the Warner Mountains found that the total area occupied by aspen declined by 24% during this 48-year period, and that modern aspen groves are smaller and more fragmented (Di Orio et al. 2005). Concerns about the decline and deterioration of aspen have prompted a number of restoration efforts in the bioregion.

Quaking aspen stands in this bioregion occur in two broad themes: meadow quaking aspen and upland quaking aspen. Meadow quaking aspen that grow on wetter sites on stream sides and meadow edges are often more resistant to conifer succession. If the aboveground stems senesce and are not replaced by younger stems, or if younger stems are repeatedly browsed, the clones will often thin out or disappear and the sites will be occupied by forb communities such as corn lily in the Upper Montane Zone, or even mountain big sagebrush if the sites are also dewatered. Upland quaking aspen grow in drier sites and are susceptible to conifer succession in the absence of fire or other disturbances that remove or kill the conifers. Quaking aspen in both environments have decreased in the last 150 years. Meadow quaking aspen have been more affected by livestock grazing impacts and browsing on aspen than changes in fire regimes in the twentieth century. Repeated browsing of young aspen suckers can incite aspen decline and could eventually eliminate an entire aspen stand. Upland quaking aspen stands have been profoundly affected by changed fire regimes, and have largely become dominated by conifers. In healthy aspen stands, fire would often be low-intensity surface fire due to green herbaceous understory and little dead fuel buildup. The trees in larger quaking aspen clones affected by the Blue Fire often survived even though the surrounding conifer stands sustained considerable mortality. The fire burning intensely in adjacent conifer stands appears to have dropped to the ground when entering the larger quaking aspen groves and burned through as a low-intensity surface fire. This change in fire behavior was likely due to higher moisture in the live and dead fuels, higher humidity, and the open nature of the crowns. None of these stands in the Blue Fire had significant encroachment of conifers.

Today, fires will often burn more intensely where conifers have invaded aspen stands and altered fuel structure by increasing litter and dead woody material. Quaking aspen clones can respond vigorously to stand-replacing fire by producing thousands of suckers per hectare. Often the extent of a degraded clone with a few surviving stems per hectare is not apparent until a stand-replacing fire event releases the clone and multitudes of quaking aspen suckers appear where there were remnant aspen within a conifer dominated stand.

Management Issues

Altered Fire Regimes

Altered fire regimes and introduction of invasive nonnative species have caused large changes in plant composition and structure across many plant communities. These changes affect important ecological processes including the water, nutrient, and energy cycles, in addition to wildlife habitat and forage production.

The reintroduction of fire has been proposed to restore some of these arid-land communities. However, there is considerable debate over the use of fire as a tool to restore these habitats. Much of the debate is due to a lack of research on fire from this bioregion and the wide variation of resilience and resistance to invasion among plant communities (Chambers et al. 2014). Concerns related to the use of fire include the rapid expansion of exotic annual communities, further loss of habitat for sagebrush-obligates, potential damage to watersheds, and liability for damage to personal property or public harm caused by fire. Some of these concerns are fueled primarily by a lack of local information (see sidebar 13.1). The lengthening periods without fire in some plant communities in the quaking aspen and mountain big sagebrush alliance will likely lead to their loss. On the other hand, concerns about restoring fire to areas invaded by cheat grass are legitimate, and in some areas fire suppression may be the only effective tool for preventing the extirpation of sagebrush steppe. A developing concern is the rapid expansion of invasive annual North Africa grass (*Ventenata dubia*) which is replacing medusa head at lower elevations and is expanding into recently distributed ponderosa pine forests increasing fine fuel continuity in these systems. Besides altered fire regimes, the primary issues in the Northeastern Plateaus bioregion include grazing effects, wildlife needs, and impacts of invasive plant species, especially cheat grass (see chapter 24). However, due to the lack of investigation and interest, this region is one of the most poorly understood with regard to fire ecology. In the future, more attention and investigation need to be given to the fire ecology of the various shrub and woodland communities that occur in this region.

Grazing

Both the presence and level of grazing have implications for many areas in the Northeastern Plateaus bioregion and have affected a wide variety of species, including Idaho fescue, blue bunch wheat grass, Wyoming big sagebrush, bitterbrush, rabbitbrush, white fir, and whitebark pine. In some instances, grazing maintains conditions suitable for native vegetation by keeping areas more open and preventing excessive accumulation of dead material. On the other hand, direct impacts to plants occur from foraging and trampling, and indirect impacts occur when both native and invasive plant species are promoted by livestock grazing, reducing competition for resources, primarily soil moisture.

Because grazing affects vegetation composition and condition, fire regimes are related to grazing regimes (Miller et al. 1994, 2013). Prescribed burning is used to improve range conditions and can have dramatic effects on sagebrush and grass species, as well as junipers. The timing, intensity, and seasonality of grazing, as well as prescribed burning conducted for range improvement, need to integrate the unique fire ecology and the resilience and resistance to invasive annuals of the various plant communities in the bioregion to minimize both impacts to native vegetation and alterations to future fire regimes.

Wildlife

Many plant species in the shrub and woodland communities of the Northeastern Plateau bioregion that are important to wildlife are greatly affected by fire. For example, bitterbrush, rabbitbrush, and juniper provide significant habitat and food

Long-term monitoring is an integral component of adaptive management (Gitzen et al. 2012). This is especially true in mountain big sagebrush communities where every fire management decision has the potential to significantly degrade or improve habitat conditions depending on the specific treatment characteristics and site conditions. Local monitoring data allow managers to determine if treatments are achieving management objectives, to modify future prescriptions or objectives based on local trends, and to identify unintended consequences of different management scenarios. A recent review by Miller et al. (2013) indicates that continuous, long-term postfire monitoring data (>5 years) are rare in sagebrush ecosystems.

One example of long-term monitoring from the northeast California bioregion is Lava Beds National Monument, an 18,842 ha (46,560 ac) sagebrush-dominated landscape where systematic fire effects monitoring began in the 1980s (National Park Service 2003). Lava Beds is a useful case study because it encompasses a very wide array of mountain big sagebrush habitat conditions and fire effects due to (a) its biogeography along a sagebrush-coniferous forest ecotone, (b) an abundant history of lightning and prescribed fires over the past several decades (0 to 7 overlapping fires), and (c) complex geology that contributes to fine-scale variation in productivity and fire regimes (Miller and Heyerdahl 2008) (Fig. 13.1.1). Such variability

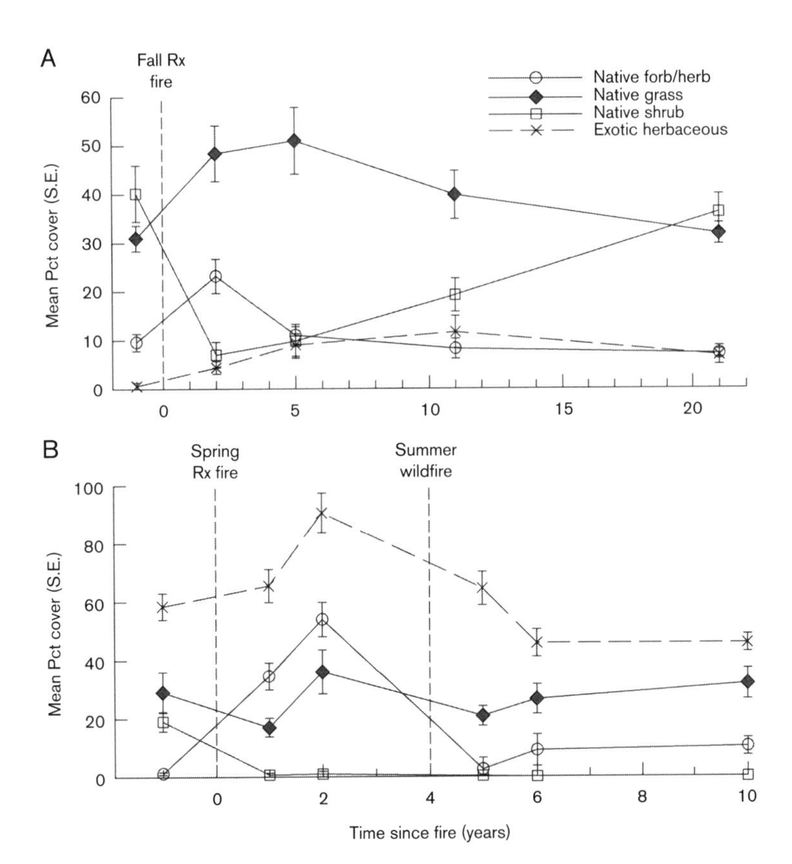

FIGURE 13.1.1 Representative examples of long-term fire effects monitoring in Lava Beds National Monument. Panel (a) shows 21-year composite results for three fall prescribed burns (n = 10 plots) conducted between 1989 and 1991 in Zone III. Native grass and forb cover increased significantly; shrub cover was initially reduced but recovered to prefire levels in just 21 years. Panel (b) shows composite results for a spring prescribed fire and a subsequent short-interval summer wildfire (n = 8 plots) in Zone I. Native grass and forb cover was relatively resilient to the two short-interval fires, but shrub cover has been reduced to near 0. Exotic cover remains high 10 years after the first burn, although it decreased slightly after the second burn (due to a decrease in tumble mustard).

(continued)

(continued)

reinforces the importance of local monitoring data to address management challenges.

When prescribed burning was initiated in the 1970s the overarching goal was to reduce fuels and restore ecosystem structure and composition by restoring pre-Euro-American fire return intervals (i.e., a process-oriented restoration approach consistent with National Park Service policies at the time). As burning expanded to different parts of the park it became apparent that fire effects varied considerably and were not always compatible with ancillary resource objectives such as maintaining shrub cover or reducing exotic plant cover. Fire managers were able to utilize systematic monitoring results to modify the timing, frequency, and location of prescribed burns (including areas where no burning was best) to maximize positive benefits of fire and reduce negative impacts. This information was also important for refining and coordinating resource and fire management objectives across the monument.

At higher elevations in the southern part of the monument, the most immediate threat to sagebrush is extensive encroachment by western juniper and other woody species that reduce shrub and native herbaceous cover, and increase fire intensity and potential postfire exotic cover. Cool season prescribed fires have successfully increased ecosystem resilience in these areas by reducing tree cover and increasing native herbaceous cover, thereby preventing irreversible thresholds from being reached that may lead to permanent loss of mountain sagebrush (Fig. 13.1.2). In contrast, the drier, hotter low elevation areas where grazing occurred more recently, and which have experienced higher contemporary wildfire frequencies, tend to recover to shrub much more slowly. Although native herbaceous cover in many of these dry sites has been relatively resilient following fire (in contrast to Wyoming big sagebrush), exotic cover is much higher following fire and may increase the size and homogeneity of subsequent fires. Successive short-interval fires can lead to extensive loss of shrub cover and dominance by cheat grass (Fig. 13.1.2). The intermediate elevations represent transition zones where fire effects have been largely beneficial, although exotic cover expansion is an increasing concern. Managers must carefully weigh

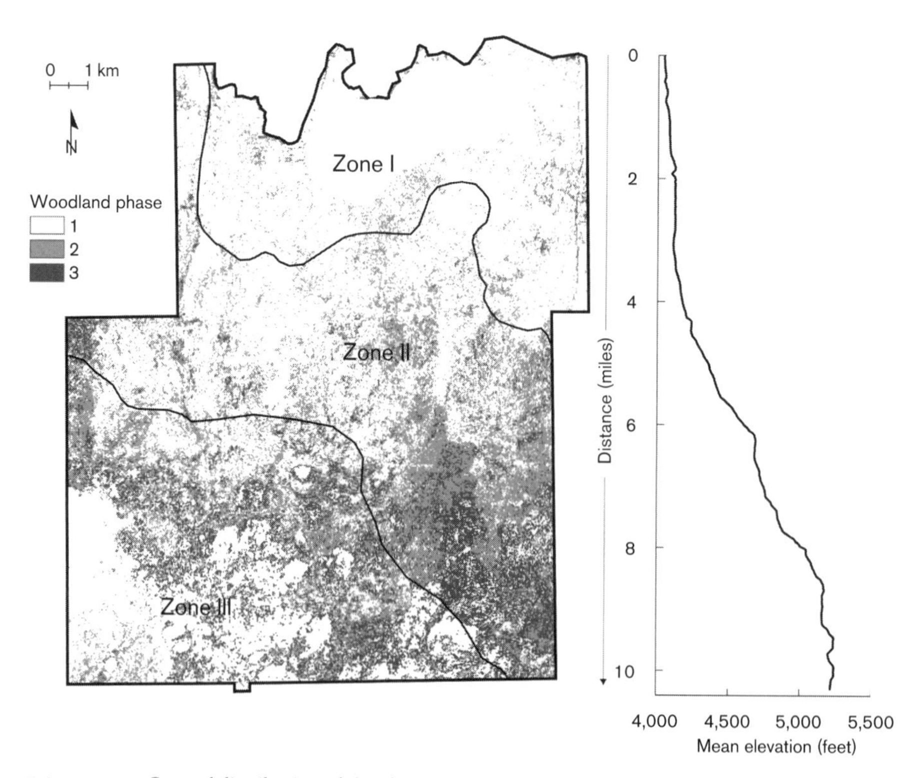

FIGURE 13.1.2 General distribution of the elevation gradient along the sagebrush-conifer ecotone, and woodland development phases of juniper in Lava Beds National Monument (phases based on Miller et al. (2005). Three general fire management zones are shown, which are based on a combination of vegetation characteristics, postfire response and successional rates (from monitoring data), contemporary fire frequency, special habitats and resource objectives, and operational considerations. They correlate closely with the general elevation gradient.

FIGURE 13.1.3 Variation in postfire community development and condition across the elevation gradient in Lava Beds National Monument. Panel (a) shows a lower elevation area (1,244 m [4,081 ft]) area in Fire Management Zone I four years after a hot summer wildfire in old mountain big sagebrush (which had not burned previously in >80 years). Nonnative cover is high and shrub cover is low over a large area. Panel (b) shows a transitional site in Zone II (1,383 m [4,539 ft]) 15 years following a patchy fall prescribed burn, and panel (c) shows an area in Zone III (~1,527 m [~5,010 ft]) five years following a fall prescribed fire. Residual shrub cover remains in both (b) and (c), native grass and forb cover increased significantly, and exotic plant cover is relatively low. Stands of young, expanding juniper can be seen in the Zone II photo (panel (b)).

the risks and benefits of fire in these stands to maintain resilience of the sagebrush mosaic; too much burning may facilitate cheat grass expansion, but very long periods without disturbance will lead ultimately to loss of habitat via woody vegetation expansion, loss of the herbaceous understory, and increased fire intensity. Fire effects monitoring and adaptive management are critical here (Fig. 13.1.3).

As environmental conditions continue to change at Lava Beds National Monument due to climate change and exotic plant hybridization and expansion, fire occurrence patterns and fire effects will continue to change as well. Decades of monitoring data will allow managers to anticipate and identify these changes, and develop the best adaptive management strategies to deal with them.

SIDEBAR 13.2 SAGE-GROUSE

Sage-grouse (*Centrocercus urophasianus*) are a sagebrush obligate species dependent upon fire sensitive plants for both food and cover. During late summer and fall sagebrush leaves comprise the majority of their diet and constitute 100% of their diet during the winter (Crawford et al. 2004). From the early spring and through summer forbs are important parts of their diet, especially for chicks. Prior to the twentieth century fire was a variable but important ecosystem process across their range (Miller and Heyerdahl 2008, Littell et al. 2009). Since Eurasian settlement, fire regimes have drastically changed across much of the sagebrush biome, resulting in large losses of sage-grouse habitat. Fire return intervals have lengthened in portions of the mountain big sagebrush series and greatly shortened in the Wyoming big sagebrush series (Miller and Rose 1999, Miller and Tausch 2001, Wright and Bailey 1982, Miller and Heyerdahl 2008). In northeast California, exotic annuals have replaced large areas of warmer drier sagebrush communities, which are potentially important winter habitat. On cooler more mesic sagebrush communities western juniper woodlands are rapidly increasing resulting in the decline of sagebrush and forbs. These contrasting scenarios of altered fire regimes resulting in conifer and exotic weed encroachment have caused a significant loss in sage-grouse habitat (Miller and Eddleman 2000, Crawford et al. 2004). Thus, a primary question related to sage-grouse is "How do we manage fire?"

There is considerable debate on the pros and cons of using fire to enhance sage-grouse habitat. The decision to use or suppress fire for the enhancement of sage-grouse habitat must be made on a site-by-site basis. Five factors that determine the negative or positive outcomes of fire on sage-grouse habitat are: (1) site resiliency and resistance to invasive annual grasses, (2) plant composition at the time of the fire, (3) functional plant group(s) that is the limiting food resource for sage grouse, (4) pattern and size of the burn, and (5) plant community composition at the landscape scale (Miller et al. 2014a, 2015). Because sage-grouse have extensive home ranges habitat limitations must be considered at large scales.

In the cool-moist (frigid-xeric soil temperature moisture regimes) mountain big sagebrush plant associations fire can be a useful tool to enhance the availability of native perennial forbs and grasses (Wrobleski and Kaufman 2003, Miller et al. 2013, 2014b). This is primarily in communities where sagebrush cover exceeds 30%, good populations of native herbs are present, and exotic species are limited. Perennial forb abundance can increase two- to threefold following fire (Pyle and Crawford 1996). Fire also enhances nutrient quality of forbs, especially protein content (McDowell 2000), and lengthens the growing season. The effects of fire on beetles and ants (important chick food) are mixed. Pyle and Crawford (1996) reported fire did not affect beetle populations while Fischer et al. (1996) reported decreases in beetle populations following fire. In Nevada, sage-grouse have been reported to be attracted to burn areas during the summer (Klebenow and Beall 1977, Martin 1990). Small burns with adjacent stands of sagebrush have also been used as leks. Possibly the most important justification for the use of fire for the restoration or maintenance of sage-grouse habitat is limiting the development of juniper woodlands across many of these shrub steppe communities.

The use of fire in warm-dry (mesic-aridic) sagebrush plant associations for improving sage-grouse habitat is probably very limited. Burning in Wyoming big sagebrush communities or low sagebrush/Sandberg bluegrass association usually does not increase desirable forbs used as sage-grouse food (Fischer et al. 1996, Miller et al. 2014b). Many Wyoming big sagebrush sites also have depleted understories and have a low resistance to invasion of nonnative plant species. With little to gain and the potential to increase populations of nonnative weeds, the use of fire as a tool to enhance sage-grouse habitat in these warm-dry sagebrush communities is limited and very risky (Miller et al. 2013).

for wildlife species. Mule deer, pronghorn antelope (*Antilocapra americana*), hares and rabbits (*Lepus* spp.), and sage-grouse (*Centrocercus urophasianus*) and other ground-nesting birds are dependent on the cover, shelter, and food that the scattered native grass, brush, and woodland vegetation provide.

Greatly altered fire regimes and the spread of exotic plant species have impacted this bioregion's wildlife. As in other areas of the State, the use of prescribed fire has become a topic of interest amongst land managers and agencies responsible for the conservation of wildlife. One example of this debate is the sage-grouse, a species that in recent years has shown a decline in numbers sufficient to raise concerns about its health and viability. Whether prescribed fire has a conservation role for the sage-grouse or other wildlife species is still unknown (see sidebar 13.2).

Summary

This bioregion has a rich history of fire throughout the vast arid basins and uplands, forested mountain ranges interspersed with fresh and alkaline wetlands. Both lightning and Native American burning patterns were abruptly changed when fuel continuity was disrupted and nonnative invasive species were introduced with the arrival of large numbers of domestic livestock in the last two decades of the 1800s. Fire suppression, logging, and agriculture have further altered historic fire regimes. Changes to fire regimes have caused changes in plant community composition and structure and wildlife habitat in many plant communities. Whereas fire once limited woody plants, juniper now takes advantage of longer fire return intervals and is expanding its range into the sagebrush steppe. Conversely, widespread cheat grass invasion has promoted more frequent fire return intervals in the sagebrush steppe, converting much of the Sagebrush Steppe Zone to annual grasslands. Fire has also become less frequent throughout the montane zones, with the Mid-Montane more changed than Lower Montane and Upper Montane Zones the least changed of the three. Fire continues to be an important ecological process throughout the Northeastern Plateaus bioregion, and land management in the region must consider its historical and changing role.

References

Adam, D.P., A.M. Sarna-Wojcicki, H.J. Rieck, J.P. Bradbury, W.E. Dean, and R.M. Forester. 1989. Tulelake, California: the last 3 million years. Palaeogeography, Palaeoclimatology, Palaeoecology 72: 89–103.

Agee, J.K. 1993. Fire Ecology of Pacific Northwest Forests. Island Press, Washington, D.C., USA.

Bates, J.D., R.F. Miller, and T.S. Svejcar. 2000. Understory dynamics in cut and uncut western jumper woodlands. Journal of Range Management 53: 119–126.

Bates, J.D., T. Svejcar, R.F. Miller, and K.W. Davies. 2016. Plant community dynamics spanning 25 years after control. Journal of Range Ecology and Management 70: 106–115.

Beck, J.L., J.G. Klein, J. Wright, and K.P. Wolfley. 2008. Potential and pitfalls of prescribed burning big sagebrush to enhance nesting and early brood-rearing habitats for greater sage-grouse. Natural Resources and Environmental Issues 16: 17–19.

Blaisdell, J.P. 1953. Ecological effects of planned burning of sagebrush-grass range on the upper Snake River Plains. USDA Technical Bulletin 1075. Washington, D.C., USA.

Blaisdell, J.P., and W.F. Mueggler. 1956. Sprouting of bitterbrush (*Purshia tridentata*) following burning or top removal. Ecology 37: 365–370.

Board, D.I., J.C. Chambers, and J.G. Wright. 2011. Effects of spring prescribed fire in expanding pinyon-juniper woodlands on seedling establishment of sagebrush species. Natural Resources and Environmental Issues 16: 1–10.

Bolsinger, C.L. 1989. California's western juniper and pinyon juniper woodlands: area, stand characteristics, wood volume, and fence posts. USDA Forest Service Resource Bulletin PNW-RB-166. Pacific Northwest Research Station, Portland, Oregon, USA.

Boltz, M. 1994. Factors influencing postfire sagebrush regeneration in south central Idaho. Pages 281–290 in: S.B. Monsen and S.G. Kitchen, editors. Proceedings-Ecology and Management of Annual Rangelands. USDA Forest Service General Technical Report INT-GTR-313. Intermountain Forest and Range Experiment Station, Ogden, Utah, USA.

Bork, J.L. 1984. Fire history in three vegetation types on the east side of the Oregon Cascades. Dissertation, Oregon State University, Corvallis, Oregon, USA.

Bradley, A.F., N.V. Noste, and W.C. Fischer. 1992. Fire ecology of forests and woodlands in Utah. USDA Forest Service General Technical Report INT-287. Intermountain Research Station, Ogden, Utah, USA.

Britton, C.M., R.G. Clark, and F.A. Sneva. 1981. Will your sagebrush range burn? Rangelands 3: 207–208.

Brown, J.K. 1982. Fuel and fire behavior predication in big sagebrush. USDA Forest Service Research Paper INT-RP-290. Intermountain Forest and Range Experiment Station, Ogden, Utah, USA.

———. 2000. Introduction and fire regimes. Pages 1–7 in: J.K. Brown and J.K. Smith, editors. Wildland Fire in Ecosystems: Effects of Fire on Flora. USDA Forest Service General Technical Report RMRS-GTR-42-VOL 2. Rocky Mountain Research Station, Fort Collins, Colorado, USA.

Bunting, S.C. 1984a. Fire in sagebrush grass ecosystems: successional changes. Pages 7–11 in: K. Sanders and J. Durham, editors. Rangeland Fire Effects. USDI Bureau of Land Management Idaho State Office, Boise, Idaho, USA.

———. 1984b. Prescribed burning of live standing western juniper and post-burning succession. Pages 69–73 in: T.E. Bedell, compiler. Oregon State University Extension Service Proceedings, Western Juniper Short Course. Oregon State University, Corvallis, Oregon, USA.

Bunting, S.C., B.M. Kilgore, and C.L. Busbey. 1987. Guidelines for prescribed burning sagebrush-grass rangelands in the northern Great Basin. USDA Forest Service General Technical Report INT-GTR-231. Intermountain Forest and Range Experiment Station, Ogden, Utah, USA.

Burkhardt, J.W., and E.W. Tisdale. 1969. Nature and successional status of western juniper vegetation in Idaho. Journal of Range Management 22: 264–270.

———. 1976. Causes of juniper invasion in southwestern Idaho. Ecology 76: 472–484.

Busse, M.D. 2000a. Ecological significance of nitrogen fixation by actinorhizal shrubs in interior forests of California and Oregon. Pages 23–41 in: R.F. Powers, D.L. Hauxwell, and G.M. Nakamura, technical coordinators. Proceedings of the California Forest Soils Council Conference on Forest Soil Biology and Forest Management. USDA Forest Service General Technical Report GTR-PSW-178. Pacific Southwest Research Station, Albany, California, USA.

———. 2000b. Suitability and use of the 15N-istope dilution method to estimate nitrogen fixation by actinorhizal shrubs. Forest Ecology and Management 136: 85–95.

Busse, M.D., and G.M. Riegel. 2009. Response of antelope bitterbrush to repeated prescribed burning in central Oregon ponderosa pine forests. Forest Ecology and Management 257: 904–910.

Busse, M.D., P.H. Cochran, and J.W. Barrett. 1996. Changes in ponderosa pine site productivity following removal of understory vegetation. Soil Science Society of America Journal 60: 1614–1621.

Busse, M.D., S.A. Simon, and G.M. Riegel. 2000. Tree-growth and understory responses to low-severity prescribed burning in thinned *Pinus ponderosa* forests of Central Oregon. Forest Science 46: 258–268.

Calo, D.B. 2014. Sedimentary analysis of fire history and paleohydrology, Eagle Lake, California. Thesis, California State University, Chico, California, USA.
Chadwick, H.W., and P.D. Dalke. 1965. Plant succession on dune sands in Fremont County, Idaho. Ecology 46: 765–780.
Chambers, J.C. 2000. Seed movements and seedling fates in disturbed sagebrush steppe ecosystems: implications for restoration. Ecological Applications 10: 1400–1413.
Chambers, J.C., B.A. Bradley, C.A. Brown, C. D'Antonio, M.J. Germino, S.P. Hardegree, J.B. Grace, R.F. Miller, and D.A. Pyke. 2014. Resilience to stress and disturbance, and resistance to *Bromus tectorum* L. invasion in the cold desert shrublands of western North America. Ecosystems 17: 360–375.
Clark, R.G., C.M. Britton, and R.A. Sneva. 1982. Mortality of bitterbrush after burning and clipping in eastern Oregon. Journal of Range Management 35: 711–714.
Cochran, P.H. 1998. Examples of mortality and reduced annual increments of white fir induced by drought, insects, and diseases at different stand densities. USDA Forest Service Research Note PNW-RN-525. Pacific Northwest Research Station, Portland, Oregon, USA.
Cochran, P.H., and W.E. Hopkins. 1991. Does fire exclusion increase productivity of ponderosa pine? Pages 224–228 in A.E. Harvey and L.F. Neuenschwander, compilers. Proceedings: Management and Productivity of Western Montane Forest Soils. USDA Forest Service Research General Technical Report INT-GTR-280. Intermountain Research Station, Ogden, Utah, USA.
Collins, B.M., and G.B. Roller. 2013. Early forest dynamics in stand replacing fire patches in the northern Sierra Nevada, California, USA. Landscape Ecology 28: 1801–1813.
Conard, S.G., A.E. Jaramillo, K. Cromack, Jr., and S. Rose. 1985. The role of *Ceanothus* in western forest ecosystems. USDA Forest Service General Technical Report PNW-GTR-182. Pacific Northwest Forest and Range Experiment Station, Portland, Oregon, USA.
Conrad, C.E., and C.E. Poulton. 1966. Effect of a wildfire on Idaho fescue and bluebunch wheatgrass. Journal of Range Management 19: 148–141.
Conard, S.G., and S.R. Radosevich. 1981. Photosynthesis, xylem pressure potential, and leaf conductance of three montane chaparral species in California. Forest Science 27: 627–639.
———. 1982. Growth responses of white fir to decreased shading and root competition by montane chaparral shrubs. Forest Science 28: 309–320.
Countryman, C.M., and D.R. Cornelius. 1957. Some effects of fire on perennial range type. Journal of Range Management 10: 39–41.
Crane, M.F. 1982. Fire ecology of Rocky Mountain Region forest habitat types. Final Report Contract No. 43-83X9-1-884. Missoula, Montana: USDA Forest Service, Region 1. USDA Forest Service, Intermountain Research Station, Fire Sciences Laboratory, Missoula, Montana, USA.
Crawford, J.C., R.A. Olson, N.E. West, J.C. Mosley, M.A. Schroeder, T.D. Whitson, R.F. Miller, M.A. Gregg, and C.S. Boyd. 2004. Ecology and management of sage-grouse and sage grouse habitat. Journal of Range Management 57: 2–19.
Critchfield, W.B., and G.L. Allenbaugh. 1969. The distribution of the Pinaceae in and near northern Nevada. Madroño 19: 12–26.
Daly, C., R.P. Neilson, and D.L. Phillips. 1994. A statistical topographic model for mapping climatological precipitation over mountainous terrain. Journal of Applied Meteorology 33: 140–158.
Davies, K.W., J.D. Bates, T. Svejcar, and C.S. Boyd. 2010. Effects of long-term livestock grazing on fuel characteristics in rangelands: an example from the sagebrush steppe. Rangeland Ecology and Management 63: 662–669.
Davis, J.N., and J.D. Brotherson. 1991. Ecological characteristics of curl-leaf mountain-mahogany (*Cercocarpus ledifolius* Nutt.) communities in Utah and implications for management. Great Basin Naturalist 51: 153–166.
Dealy, J.E. 1975. Ecology of curl-leaf mountain-mahogany (*Cercocarpus ledifolius*) in eastern Oregon and adjacent area. Dissertation, Oregon State University, Corvallis, Oregon, USA.
Delwiche, C.C., P.J. Zinke, and C.M. Johnson. 1965. Nitrogen fixation by *Ceanothus*. Plant Physiology 40: 1045–1047.
Di Orio, A.P., R. Callas, and R.J. Schaefer. 2005. Forty-eight year decline and fragmentation of aspen (*Populus tremuloides*) in the South Warner Mountains of California. Forest Ecology and Management 206: 307–313.
Driscoll, R.S. 1963. Sprouting bitterbrush in central Oregon. Ecology 44: 820–821.
Dwire, K.A., and J.B. Kauffman. 2003. Fire and riparian ecosystems in landscapes of the western USA. Forest Ecology and Management 178: 61–74.
Ferrel, G.T. 1978. Moisture stress threshold of susceptibility to fir engraver beetles in pole-sized white firs. Forest Science 24: 85–92.
———. 1980. Growth of white firs defoliated by Modoc budworm in northeastern California. USDA Forest Service Research Paper PSW-RP-154. Pacific Southwest Forest and Range Experiment Station, Berkeley, California, USA.
Figura, P.J. 1997. Structure and dynamics of whitebark pine forests in the South Warner Wilderness Area. Thesis, Humboldt State University, Arcata, California, USA.
Fischer, R.A., K.P. Reese, and J.W. Connelly. 1996. An investigation on fire effects within xeric sage grouse brood habitat. Journal of Range Management 49: 194–198.
Frost, C.C. 1998. Presettlement fire frequency regimes of the United States: a first approximation. Proceedings Tall Timbers Fire Ecology Conference 20: 70–81.
Glitzen, R.A., J.J. Millspaugh, A.B. Cooper, and D. Licht. 2012. Design and Analysis of Long-term Ecological Monitoring Studies. Cambridge University Press, Cambridge, UK.
Goheen, A. 1998. Buck Creek Drainage fire analysis: the ecological consequences of aggressive fire suppression. Technical Fire Management 11. Washington Institute, Colorado State University, Fort Collins, Colorado, USA.
Goodwin, D.L. 1956. Autecological studies of *Artemisia tridentata* Nutt. Dissertation, State College of Washington, Pullman, Washington, USA.
Griffin, J.R., and W.B. Critchfield. 1972. The distribution of forest trees in California. USDA Forest Service Research Paper PSW-RP-82. Pacific Southwest Forest and Range Experiment Station, Berkeley, California, USA.
Grissino-Mayer, H.D. 2001. FHX2—software for analyzing temporal and spatial patterns in fire regimes from tree rings. Tree-Ring Research 57: 115–124.
Grissino-Mayer, H.D., and T.W. Swetnam. 1995. Effects of habitat diversity on fire regimes in El Malpais National Monument, New Mexico. Pages 195–200 in: J.K. Brown, R.W. Mutch, C.W. Spoon, and R.H. Wakimoto, editors. Proceedings: Symposium on Fire in Wilderness and Park Management. USDA Forest Service General Technical Report INT-GTR-320. Intermountain Forest and Range Experiment Station, Ogden, Utah, USA.
Gruell, G.E., S.C. Bunting, and L. Neuenschwander. 1984. Influence of fire on curl-leaf mountain-mahogany in the Intermountain West. Pages 58–72 in: J.K. Brown and J. Lotan, editors. Proceedings: Fire's Effects on Wildlife Habitat. USDA Forest Service, General Technical Report INT-GTR-186. Intermountain Research Station, Ogden, Utah, USA.
Hakala, K.J., and D.P. Adam. 2004. Late Pleistocene vegetation and climate in the southern Cascade Range and the Modoc Plateau region. Journal of Paleolimnology 31: 189–215.
Haller, J.R. 1961. Some recent observations on ponderosa, Jeffrey and Washoe pines in northeastern California. Madroño 16: 26–132.
Harniss, R.O., and W.T. McDonough. 1974. Yearly variation in germination in the subspecies of big sagebrush. Journal of Range Management 29: 167–168.
Harniss, R.O., and R.B. Murray. 1973. 30 years of vegetal change following burning of sagebrush-grass range. Journal of Range Management 29: 322–325.
Hassan, M.A., and N.E. West. 1986. Dynamics of soil seed pools in burned and unburned sagebrush semi-deserts. Ecology 67: 269–272.
Hinckley, T.M., R.O. Teskey, R.H. Waring, and Y. Morikawa.1982. The water relations of true firs. Pages 85–94 in: C.D. Oliver and R.M. Kenady, editors. Proceedings of the Biology and Management of True Fir in the Pacific Northwest Symposium. University of Washington, College of Forest Resources Contribution, No. 45. Seattle, Washington, USA.
Holmgren, R.C. 1956. Competition between annuals and young bitterbrush (*Purshia tridentata*) in Idaho. Ecology 37: 370–377.

Hosten, P.E., and N.E. West. 1994. Cheatgrass dynamics following wildfire on a sagebrush semidesert site in central Utah. Pages 56–62 in: S.B. Monsen and S.G. Kitchen, editors. Proceedings—Ecology and Management of Annual Rangelands. USDA Forest Service General Technical Report INT-GTR-313. Intermountain Research Station, Ogden, Utah, USA.
Houston, D.B. 1973. Wildfires in northern Yellowstone National Park. Ecology 54: 1109–1117.
Jenkinson, J.L. 1990. *Pinus jeffreyi* Grev. & Balf. Jeffrey Pine. Pages 359–369 in: R.M. Burns and B.H. Honkala, technical coordinators. Silvics of North America: Volume I. Conifers. USDA Forest Service Agricultural Handbook 654. Washington, D.C., USA.
Johnson, C.G., Jr. 1998. Vegetation responses after wildfire in national forests of northeastern Oregon. USDA Forest Service Technical Publication R6-NR-ECOL-TP-06-98. Pacific Northwest Region, Portland, Oregon, USA.
Johnson, D.W. 1995. Soil properties beneath *Ceanothus* and pine stands in the eastern Sierra Nevada. Soil Science Society of America Journal 59: 918–924.
Johnson, J.R., and G.F. Payne. 1968. Sagebrush reinvasion as affected by some environmental influences. Journal of Range Management 21: 209–213.
Kelly, I.T. 1932. Ethnography of the Surprise Valley Paiute. University of California Publications in American Archaeology and Ethnology 31: 67–209.
Klebenow, D.A., and R.C. Beall. 1977. Fire impacts on birds and mammals on Great Basin rangelands. Pages 1–13 in Proceedings, Joint Intermountain Rocky Mountain Fire Research Council, Casper, Wyoming, USA.
Klebenow, D.A., and A. Bruner. 1976. Determining factors necessary for prescribed burning. Pages 69–47 in: F.E. Busby and E. Storey, editors. Use of Prescribed Burning in Western Woodland and Range Ecosystems: Proceedings of a Symposium. Utah State University, Logan, Utah, USA.
Knapp, E.E., C.N. Skinner, M.P. North, and B.L. Estes. 2013. Long-term overstory and understory change following logging and fire exclusion in a Sierra Nevada mixed-conifer forest. Forest Ecology and Management 310: 903–914.
Knapp, E.E., C.P. Weatherspoon, and C.N. Skinner. 2012. Shrub seed banks in mixed Conifer forests of Northern California and the role of fire in regulating abundance. Fire Ecology 8(1): 32–48.
Kruckeberg, A.R. 1977. Manzanita (*Arctostaphylos*) hybrids in the Pacific Northwest: effects of human and natural disturbance. Systematic Botany 2: 233–250.
Kuntz, D.E. 1982. Plant response following spring burning in an *Artemisia tridentata* subsp. *vaseyana*/*Festuca idahoensis* habitat type. Thesis, University of Idaho, Moscow, Idaho, USA.
Laacke, R.J. 1990. *Abies concolor* (Gord. & Glend.) Lindl. ex Hildebr. White fir. Pages 36–46 in: R.M. Burns and B.H. Honkala, technical coordinators. Silvics of North America: Volume I, Conifers. USDA Forest Service Agricultural Handbook 654. Washington, D.C., USA.
Lanini, W.T., and S.R. Radosevich. 1986. Response of three conifer species to site preparation and shrub control. Forest Science 32: 61–77.
Laudenslayer, W.F., Jr., H.H. Darr, and S. Smith. 1989. Historical effects of forest management practices on eastside pine communities in northeastern California. Pages 26–34 in: A. Tecle, W.W. Covington, and R.H. Hamre, technical coordinators. Multiresource Management of Ponderosa Pine Forests: Proceedings of the Symposium. USDA Forest Service General Technical Report RM-GTR-185. Rocky Mountain Forest and Range Experiment Station, Fort Collins, Colorado, USA.
Littell, J.S., D. McKenzie, D.L. Peterson, and A.L. Westrerling. 2009. Climate and wildfire area burned in western U.S. ecoprovinces, 1916–2003. Ecological Applications 19: 1003–1021.
Little, E.L., Jr. 1971. Conifers and important hardwoods. Vol. 1. Atlas of the United States trees. USDA Forest Service Miscellaneous Publication 1146. Washington, D.C., USA.
Lyon, L.J., and P.F. Stickney. 1976. Early vegetal succession following large northern Rocky mountain wildfires. Proceedings Tall Timbers Fire Ecology Conference 14: 355–375.
MacDonald, G.A., and T.E. Gay, Jr. 1966. Geology of the southern Cascade Range, Modoc Plateau, and Great Basin areas in northeastern California. Pages 43–48 in: Mineral Resources of California. California Division of Mines and Geology Bulletin 191. Sacramento, California, USA.
Martin, R.C. 1990. Sage Grouse Responses to Wildfire in Spring and Summer Habitats. Thesis, University of Idaho, Moscow, Idaho, USA.
Martin, R.E. 1983. Antelope bitterbrush seeding establishment following prescribed fire in the pumice zone of the southern Cascade Mountains. Pages 92–99 in: A.R. Tiedemann and K.L. Johnson, compilers. Proceedings-research and management of bitterbrush and cliffrose in western North America. USDA Forest Service General Technical Report INT-GTR-152. Intermountain Forest and Range Experiment Station, Ogden, Utah, USA.
Martin, R.E., and C.H. Driver. 1983. Factors affecting antelope bitterbrush reestablishment following fire. Pages 266–279 in: A.R. Tiedemann and K.L. Johnson, compilers. Proceedings-research and management of bitterbrush and cliffrose in western North America. USDA Forest Service General Technical Report INT-GTR-152. Intermountain Forest and Range Experiment Station, Ogden, Utah, USA.
McDonald, P.M. 1969. Silvical characteristics of California black oak (*Quercus kelloggii* Newb.). USDA Forest Service Research Paper PSW-RP-73. Pacific Southwest Forest and Range Experiment Station, Berkeley, California, USA.
———. 1983a. Climate, history, and vegetation of the eastside pine type in California. Pages 1–16 in: T.F. Robson and R.B. Standiford, editors. Management of the Eastside Pine Type Northeastern California: Proceedings of a Symposium. Northern California Society of American Foresters SAF. 83-06. Arcata, California, USA.
———. 1983b. Weeds in conifer plantations of northeastern California—management implications. Pages 70–78 in: T.F. Robson and R.B. Standiford, editors. Management of the Eastside Pine Type Northeastern California: Proceedings of a Symposium. Northern California Section, Society of American Foresters SAF. 83-06. Arcata, California, USA.
McDonald, P.M., and J.C. Tappeiner. 1996. Silviculture-ecology of forest-zone hardwoods in the Sierra Nevada. Pages 621–636 in: D.C. Erman, general editor. Sierra Nevada Ecosystem Project: Final Report to Congress, Volume II. Wildland Resources Center Report 37. University of California, Davis, California, USA.
McDowell, M.K.D. 2000. The effects of burning in mountain big sagebrush on key sage grouse habitat characteristics in southeastern Oregon. Thesis, Oregon State University, Corvallis, Oregon, USA.
Mehringer, P.J., and P.E. Wigand. 1990. Comparison of late Holocene environments from woodrat middens and pollen. Pages 294–325 in: J.L. Betancourt, T.R. Van Devender, and P.S. Martin, editors. Packrat Middens: The Last 40,000 Years of Biotic Change. University of Arizona Press, Tucson, Arizona, USA.
Meyer, S.E., and S.B. Monsen. 1991. Habitat-correlated variation in mountain big sagebrush (*Artemisia tridentata* subsp. *vaseyana*) seed germination patterns. Ecology 72: 739–732.
Miles, S.R., and C.B. Goudey. 1997. Ecological subregions of California: section and subsection descriptions. USDA Forest Service Technical Publication R5-EM-TP-005. Pacific Southwest Region, San Francisco, California, USA.
Millar, C.I., R.D. Westfall, D.L. Delany, M.J. Bokach, A.L. Flint, and L.E. Flint. 2012. Forest mortality in high-elevation whitebark pine (*Pinus albicaulis*) forests of eastern California, USA; influence of environmental context, bark beetles, climatic water deficit, and warming. Canadian Journal of Forest Research 42(4): 749–765.
Miller, M. 2000. Fire autecology. Pages 9–34 in: J.K. Brown and J.K. Smith, editors. Wildland Fire in Ecosystems: Effects of Fire on Flora. USDA Forest Service General Technical Report RMRS-GTR-42-VOL-2. Rocky Mountain Research Station, Fort Collins, Colorado, USA.
Miller, R.F., J.D. Bates, T. Svejcar, F.B. Pierson, and F.E. Eddleman. 2005. Biology, ecology, and management of western juniper (*Juniperus occidentalis*). Agriculture Experiment Station Technical Bulletin 152. Oregon State University, Corvallis, Oregon, USA.
Miller, R.F., J.C. Chambers, and M. Pellant. 2014b. A Field Guide for Selecting the Most Appropriate Treatment in Sagebrush and Piñon-Juniper Ecosystems in the Great Basin – Evaluating Resilience to Disturbance and Resistance to Invasive Annual Grasses, and Predicting Vegetation Response. USDA Forest Service

response. USDA Forest Service General Technical Report RMRS-GTR-322-rev. Rock Mountain Research Station, Fort Collins, Colorado, USA.
Miller, R.F., J.C. Chambers, and M. Pellant. 2015. A Field Guide for Rapid Assessment of Post-Wildfire Recovery Potential in Sagebrush and Piñon-Juniper Ecosystems in the Great Basin – Evaluating Resilience to Disturbance and Resistance to Invasive Annual Grasses and Predicting Vegetation Response. USDA Forest Service, General Technical Report RMRS-GTR-338. Rocky Mountain Research Station, Fort Collins, Colorado, USA.
Miller, R.F., J.C. Chambers, D. Pyke, F.B. Pierson, and C.J. Williams. 2013. A review of fire effects on vegetation and soils in the Great Basin Region: response and ecological site characteristics. USDA Forest Service Technical Report RMRS-GTR-308. Rocky Mountain Research Station, Fort Collins, Colorado, USA.
Miller, R.F., and L.E. Eddleman. 2000. Spatial and temporal changes of sage grouse habitat in the sagebrush biome. Oregon State University Agricultural Experiment Station, Technical Bulletin 151. Corvallis, Oregon, USA.
Miller, R.F., and E.K. Heyerdahl. 2008. Fine-scale variation of historical fire regimes in sagebrush-steppe and juniper woodland: an example from California, USA. International Journal of Wildland Fire 17: 245–254.
Miller, R.F., J. Ratchford, B.A. Roundy, R.J. Tausch, A. Hulet, and J.C. Chambers. 2014a. Response of conifer-encroached shrublands in the Great Basin to prescribed fire and mechanical treatments. Rangeland Ecology and Management 67: 468–481.
Miller, R.F., and J.A. Rose. 1999. Fire history and western juniper encroachment in sagebrush steppe. Journal of Range Management 52: 520–559.
Miller, R.F., T. Svejcar, and N.E. West. 1994. Implications of livestock grazing in the intermountain sagebrush region: plant composition. Pages 101–146 in: M. Vavra, W.A. Laycock, and R.D. Pieper, editors. Ecological Implications of Livestock Herbivory in the West. Western Society for Range Management, Denver, Colorado, USA.
Minckley, T.A. 2003. Holocene environmental history of the northwestern Great Basin and the analysis of modern pollen analogues in western North America. Dissertation, University of Oregon, Eugene, Oregon, USA.
Minckley, T.A., C.P. Whitlock, and P.J. Bartlein. 2007. Vegetation, fire, and climate history of the northwestern Great Basin during the last 14,000 years. Quaternary Science Reviews 26(17): 2167–2184.
Minore, D. 1979. Comparative autecological characteristics of northwestern tree species—a literature review. USDA Forest Service General Technical Report PNW-GTR-87. Pacific Northwest Forest and Range Experiment Station, Portland, Oregon, USA.
Mueggler, W.F. 1956. Is sagebrush seed residual in the soil of burns or is it wind borne? USDA Forest Service Research Note INT-RN-35. Intermountain Research Station, Ogden, Utah, USA.
National Park Service. 2003. Fire Monitoring Handbook. Fire Management Program Center, National Interagency Fire Center, Boise, Idaho, USA.
Ne'eman, G., H.J. Fotheringham, and J.E. Keeley. 1999. Patch to landscape patterns in post fire recruitment of a serotinous conifer. Plant Ecology 145: 235–242.
Nelson, Z., P.J. Weisberg, and S.G. Kitchen. 2014. Influence of climate and environment on post-fire recovery of mountain big sagebrush. International Journal of Wildland Fire 23: 131–142.
Neuenschwander, L.F. (n.d.). The Fire Induced Autecology of Selected Shrubs of the Cold Desert and Surrounding Forests: A State-of-the-Art-Review. College of Forestry, Wildlife and Range Sciences, University of Idaho, Moscow, Idaho, USA.
Nord, E.C. 1965. Autecology of bitterbrush in California. Ecological Monographs 35: 307–334.
Norman, S.P., and A.H. Taylor. 2005. Pine forest expansion along a forest-meadow ecotone in northeastern California, USA. Forest Ecology and Management 215: 51–68.
North, M., J. Innes, and H. Zald. 2007. Comparison of thinning and prescribed fire restoration treatments to Sierran mixed-conifer historic conditions. Canadian Journal of Forest Research 37: 331–342.
Norton, J. B., T. A. Monaco, M. Norton, D. A. Johnson, and T. A. Jones. 2004. Soil morphology and organic matter dynamics under cheatgrass and sagebrush-steppe plant communities. Journal of Arid Environments 57: 445–466.
Olson, D.L. 2000. Fire in riparian zones: a comparison of historical fire occurrence in riparian and upslope forests in the Blue Mountains and southern Cascades of Oregon. Thesis, University of Washington, Seattle, Washington, USA.
Osborn, G., and K. Bevis. 2001. Glaciation in the Great Basin of the western United States. Quaternary Science Reviews 20: 1377–1410.
Owens, T.E. 1982. Postburn regrowth of shrubs related to canopy mortality. Northwest Science 56: 34–40.
Parker, V.T., and M.P. Yoder-Williams. 1989. Reduction of survival and growth of young *Pinus jeffreyi* by an herbaceous perennial, *Wyethia mollis*. American Midland Naturalist 121: 105–111.
Parsons, D.J. 1981. The historical role of fire in the foothill communities of Sequoia National Park. Madroño: 111–120.
Pease, R.W. 1965. Modoc County: a geographic time continuum on the California volcanic tableland. University of California Publications in Geography Vol. 17. University of California Press, Berkeley, California, USA.
Pechanec, J.F., G. Steward, and J.P. Blaisdell. 1954. Sagebrush burning—good and bad. U.S. Department of Agriculture Farmer's Bulletin 1948. Washington, D.C., USA.
Plumb, T.R. 1979. Response of oaks to fire. Pages 205–215 in: T.R. Plumb and N.H. Pillsbury, technical coordinators. Proceedings of the Symposium on the Ecology, Management, and Utilization of California Oaks. USDA Forest Service General Technical Report PSW-GTR-217. Pacific Southwest Research Station, Albany, California, USA.
Powers, R.F., and W.W. Oliver. 1990. *Libocedrus decurrens* Torr. Incense-cedar. Pages 173–180 in: R.M. Burns and B.H. Honkala, technical coordinators. Silvics of North America: Volume I, Conifers. USDA Forest Service Agricultural Handbook 654. Washington, D.C., USA.
Pyle, W.H., and J.A. Crawford. 1996. Availability of foods of sage grouse chicks following prescribed fire in sagebrush-bitterbrush. Journal of Range Management 49: 320–324.
Radosevich, S.R. 1984. Interference between greenleaf manzanita (*Arctostaphylos patula*) and ponderosa pine (*Pinus ponderosa*). Pages 259–270 in: M.L. Dureya and G.N. Brown, editors. Seedling Physiology and Reforestation Success. Martinus Nijhoff/Dr. W. Junk Publishers, Dordrecht, The Netherlands.
Rehfeldt, G.E. 1999. Systematics and genetic structure of Washoe pine: applications in conservation genetics. Silvae Genetica 48: 167–173.
Rehfeldt, G.E., D.E. Ferguson, and N.L. Crookston. 2009. Aspen, climate, and sudden decline in western USA. Forest Ecology and Management 258: 2353–2364.
Rentz, E., and K.E. Merriam. 2009. Restoration and management of Baker Cypress in Northern California and Southern Oregon. Pages 282–289 in: Proceedings of the CNPS Conservation Conference, January 17–19, 2009. Sacramento, California, USA.
Riegel, G.M., T.J. Svejcar, and M.D. Busse. 2002. Does the presence of *Wyethia mollis* limit growth of affect growth of *Pinus jeffreyi* seedlings? Western North American Naturalist 62: 141–150.
Riegel, G.M., D.A. Thornburgh, and J.O. Sawyer. 1990. Forest habitat types of the South Warner Mountains, Modoc County, California. Madroño 37: 88–112.
Rorig, M.L., and S.A. Ferguson. 1999. Characteristics of lightning and wildland fire ignition in the Pacific Northwest. Journal of Applied Meteorology 38: 1565–1575.
———. 2002. The 2000 fire season: lightning-caused fires. Journal of Applied Meteorology 41: 786–791.
Rosentreter, R. 2005. Sagebrush identification, ecology, and palatability relative to sage-grouse. Pages 3–16 in: N.L. Shaw, M. Pellant, and S.B. Monsen, compilers. Sage Grouse Habitat Restoration Symposium Proceedings. USDA Forest Service Proceedings RMRS-P-38. Rocky Mountain Research Station, Fort Collins, Colorado, USA.
Roy, D.F. 1983. Natural regeneration. Pages 87–102 in: T.F. Robson and R.B. Standiford, editors. Management of the Eastside Pine Type in Northeastern California: Proceedings of a Symposium. Northern California Society of American Foresters SAF 83-06. Arcata, California, USA.
Running, S.W. 1980. Environmental and physiological control of water flux through *Pinus contorta*. Canadian Journal of Forest Research 10: 82–91.

Running, S.W., and C.P. Reid. 1980. Soil temperature influences on root resistance of *Pinus contorta*. Plant Physiology 65: 635–640.

Savage, M., and T.W. Swetnam. 1990. Early 19th-century fire decline following sheep pasturing in a Navajo ponderosa pine forest. Ecology 71: 2374–2378.

Shainsky, L.J., and S.R. Radosevich. 1986. Growth and water relations of *Pinus ponderosa* seedlings in competitive regimes with *Arctostaphylos patula* seedlings. Journal of Applied Ecology 23: 957–966.

Skinner, C.N. 2003. Fire regimes of upper montane and subalpine glacial basins in the Klamath Mountains of northern California. Tall Timbers Research Station Miscellaneous Publication 13: 145–151.

Skinner, C.N., and C. Chang. 1996. Fire regimes, past and present. Pages 1041–1069 in: D.C. Erman, general editor. Sierra Nevada Ecosystem Project: Final Report to Congress, Volume II. Wildland Resources Center Report 37. University of California, Davis, California, USA.

Smith, S. 1994. Ecological guide to eastside pine plant associations, northeastern California: Modoc, Lassen, Klamath, Shasta-Trinity, Plumas, and Tahoe National Forests. USDA Forest Service Technical Publication R5-ECOL-TP-004. Pacific Southwest Region, San Francisco, California, USA.

Smith, S., and B. Davidson. 2003. User's manual: terrestrial ecological unit inventory (TEUI), land type associations, Modoc National Forest. USDA Forest Service Technical Publication R5-TP-015, Version 1.0. Pacific Southwest Region, San Francisco, California, USA.

Sparks, J.P., G.S. Campbell, and R.A. Black. 2001. Water content, hydraulic conductivity, and ice formation in winter stems of *Pinus contorta:* a TDR case study. Oecolgia 27: 469–475.

Standiford, R.B., J. Klein, and B. Garrison. 1996. Sustainability of Sierra Nevada hardwood rangelands. Pages 637–680 in: D.C. Erman, general editor. Sierra Nevada Ecosystem Project: Final Report to Congress, Vol. III, Assessments and Scientific Basis for Management Options. Centers for Water and Wildland Resources Report 37. University of California, Davis, California, USA.

Stephens, S.L. 2001. Fire history differences in adjacent Jeffrey pine and upper montane forests in the Sierra Nevada. International Journal of Wildland Fire 10: 161–167.

Stephens, S.L., C.N. Skinner, and S.J. Gill. 2003. A dendrochronology based fire history of Jeffrey pine-mixed conifer forests in the Sierra San Pedro Martir, Mexico. Canadian Journal of Forest Research 33: 1090–1101.

Stephenson, N.L., D.J. Parsons, and T.W. Swetnam. 1991. Restoring natural fire to the sequoia-mixed conifer forest: should intense fire play a role? Proceedings Tall Timbers Fire Ecology Conference 17: 321–337.

Taylor, A.H. 2000. Fire regimes and forest changes along a montane forest gradient, Lassen Volcanic National Park, southern Cascade Mountains, USA. Journal of Biogeography 27: 87–104.

Taylor, A.H., and C.N. Skinner. 2003. Spatial patterns and controls on historical fire regimes and forest structure in the Klamath Mountains. Ecological Applications 13(3): 704–719.

Vale, T.R. 1975. Invasion of big sagebrush (*Artemisia tridentata*) by white fir (*Abies concolor*) on the southeastern slopes of the Warner Mountains, California. Great Basin Naturalist 35: 319–324.

———. 1977. Forest changes in the Warner Mountains, California. Annals of the Association of American Geographers 67: 28–45.

Vander Wall, S.B. 1994. Seed fate pathways of antelope bitterbrush: dispersed by seed-caching yellow pine chipmunks. Ecology 75: 1911–1926.

Van de Water, K., and M. North. 2010. Fire history of coniferous riparian forests in the Sierra Nevada. Forest Ecology and Management 260: 384–395.

Van de Water, K., and M. North. 2011. Stand structure, fuel loads, and fire behavior in riparian and upland forests, Sierra Nevada Mountains, USA; a comparison of current and reconstructed conditions. Forest Ecology and Management 262: 215–228.

Vankat, J.L., and J. Major. 1978. Vegetation changes in Sequoia National Park, California. Journal of Biogeography 5: 377–402.

van Wagtendonk, J.W. 1983. Prescribed fire effects on understory mortality. Fire and Forest Meteorology Conference 7: 136–138.

van Wagtendonk, J.W., and D. Cayan. 2008. Temporal and spatial distribution of lightning strikes in California in relationship to large-scale weather patterns. Fire Ecology 4(1): 34–57.

Vasek, F.C., and R.F. Thorne. 1988. Transmontane coniferous vegetation. Pages 797–832 in: M.G. Barbour and J. Major, editors. Terrestrial Vegetation of California. California Native Plant Society Special Publication Number 9. Sacramento, California, USA.

Vogl, R.J., W.P. Armstrong, D.L. White, and K.L. Cole. 1988. The closed-cone pines and cypress. Pages 295–358 in: M.G. Barbour and J. Major, editors. Terrestrial Vegetation of California. California Native Plant Society Special Publication Number 9. Sacramento, California, USA.

Volland, L.A., and J.D. Dell. 1981. Fire effects on Pacific Northwest forest and range vegetation. USDA Forest Service Technical Publication R6 RM 067 1981. Range Management and Aviation and Fire Management, Pacific Northwest Region, Portland, Oregon, USA.

Vose, J.M., and A.S. White. 1991. Biomass response mechanisms of understory species the first year after prescribed burning in an Arizona ponderosa-pine community. Forest Ecology and Management 40: 175–187.

Webster, S.R., C.T. Youngberg, and A.G. Wollum. 1967. Fixation of nitrogen by bitterbrush (*Purshia tridentata* [Pursh] D.C.). Nature 216: 392–393.

Weixelman, D.A., T.R. Bowyer, and V. Van Ballenberghe. 1998. Diet selection by Alaskan moose during winter: effects of fire and forest succession. Pages 213–238 in: W.B. Ballard and A.R.J. Rodgers, editors. Proceedings, 33rd North American Moose Conference and Workshop: 4th International Moose Symposium, Alces 34. Lakehead University, Thunder Bay, Ontario, Canada.

Whisenant, S.G. 1990. Changing fire frequencies on Idaho's Snake River plains: ecological and management implications. Pages 4–10 in: E.D. McArthur, E.M. Romney, S.D. Smith, and P.T. Tueller, compilers. Proceedings: Symposium on Cheatgrass Invasion, Shrub Die-off, and Other Aspects of Shrub Biology and Management. USDA Forest Service General Technical Report INT-GTR-276. Intermountain Forest and Range Experiment Station, Ogden, Utah, USA.

Whitlock, C., C.N. Skinner, P.J. Bartlein, T. Minckley, and J.A. Mohr. 2004. Comparison of charcoal and tree-ring records of recent fires in the eastern Klamath Mountains. Canadian Journal of Forest Research 34: 2110–2121.

Wickman, B.E. 1978. A case study of a Douglas-fir tussock moth outbreak and stand conditions 10 years later. USDA Forest Service, Research Paper PNW-RP-244. Pacific Northwest Forest and Range Experiment Station, Portland, Oregon, USA.

Wigand, P.E., M.L. Hemphill, S. Sharpe, and S. Patra (Manna). 1995. Eagle Lake Basin, Northern California, Paleoecological Study: Semi-Arid Woodland and Montane Forest Dynamics during the Late Quaternary in the Northern Great Basin and Adjacent Sierras. University and Community College System of Nevada, Reno, Nevada, USA.

Wijayratne, U.C., and D.A. Pyke. 2012. Burial increases seed longevity of two *Artemisia tridentata* (Asteraceae) subspecies. American Journal of Botany 99: 438–447.

Williams, M.P. 1995. Inhibition of conifer regeneration by an herbaceous perennial, *Wyethia mollis*, in the eastern Sierra Nevada, California. Dissertation, University of Washington, Seattle, Washington, USA.

Winward, A.H. 1985. Fire in the sagebrush-grass ecosystem—the ecological setting. Pages 2–6 in: K. Sanders and J. Durham, editors. Rangeland Fire Effects. USDI Bureau of Land Management Idaho State Office, Boise, Idaho, USA.

Wright, H.A. 1971. Why squirreltail is more tolerant to burning than needle and thread. Journal of Range Management 24: 277–284.

Wright, H.A., L.F. Neuenschwander, and C.M. Britton. 1979. The role and use of fire in sagebrush-grass and pinyon-juniper plant communities: a state of the art review. USDA Forest Service General Technical Report INT-GTR-58. Intermountain Forest and Range Experiment Station, Ogden, Utah, USA.

Wrobleski, D.W., and J.B. Kauffman. 2003. Initial effects of prescribed fire on morphology, abundance, and phenology of

forbs in big sagebrush communities in southeastern Oregon. Restoration Ecology 11: 82–90.
Young, R.P. 1983. Fire as a vegetation management tool in rangelands of the intermountain region. Pages 18–31 in: S.B. Monsen and N. Shaw, compilers. Managing Intermountain Rangelands—Improvement of Range and Wildlife Habitats: Proceedings. USDA Forest Service General Technical Report INT-GTR-157. Intermountain Forest and Range Experiment Station, Ogden, Utah, USA.
Young, J.A., and R.A. Evans. 1978. Population dynamics after wildfires in sagebrush grasslands. Journal of Range Management 31: 283–289.
———. 1981. Demography and fire history of a western juniper stand. Journal of Range Management 34: 501–505.
———. 1989. Dispersal and germination of big sagebrush (*Artemisia tridentata*) seeds. Weed Science 37: 201–206.
Young, J.A., R.A. Evans, and J. Major. 1988. Sagebrush steppe. Pages 763–796 in: M.G. Barbour and J. Major, editors. Terrestrial Vegetation of California. California Native Plant Society Special Publication Number 9. Sacramento, California, USA.
Young, R.P., and R.F. Miller. 1985. Response of *Sitanion hystrix* (Nutt.) J.G. to prescribed burning. American Midland Naturalist 113: 182–187.
Ziegenhagen, L.L., and R.F. Miller. 2009. Postfire recovery of two shrubs in the interiors of large burns in the Intermountain West, USA. Western North American Naturalist 69: 195–205.
Zouhar, K. 2003. *Bromus tectorum*. In: Fire Effects Information System. U.S. Department of Agriculture, Forest Service, Rocky Mountain Research Station, Fire Sciences Laboratory (Producer). http://www.fs.fed.us/database/feis/plants/graminoid/brotec/all.html. Accessed March 23, 2014.

CHAPTER FOURTEEN

Sierra Nevada Bioregion

JAN W. VAN WAGTENDONK, JO ANN FITES-KAUFMAN,
HUGH D. SAFFORD, MALCOLM P. NORTH, AND
BRANDON M. COLLINS

> In the main forest belt of California, fires seldom or never sweep from tree to tree in broad all-enveloping sheets . . . Here the fires creep from tree to tree, nibbling their way on the needle-strewn ground, attacking the giant trees at the base, killing the young, and consuming the fertilizing humus and leaves.
>
> MUIR (1895)

The Sierra Nevada is one of the most striking features of the state of California, extending from the southern Cascade Mountains in the north to the Tehachapi Mountains and Mojave Desert 620 km (435 mi) to the south (Map14.1). The Central Valley forms the western boundary of the Sierra Nevada bioregion, and the Great Basin is on the east. The bioregion includes the central mountains and foothills of the Sierra Nevada Section and the Sierra Nevada Foothills Section of Miles and Goudey (1997). The area of the bioregion is 69,560 km^2 (26,442 mi^2), approximately 17% of the state. Significant features along the length of the range include Lake Tahoe, Yosemite Valley, and Mount Whitney.

Description of the Bioregion

The natural environment of the Sierra Nevada is a function of the physical factors of geomorphology, geology, and regional climate interacting with the resident biota. These factors are inextricably linked to the abiotic and biotic ecosystem components including local climate, hydrology, soils, plants, and animals. The nature and the distribution of the Sierra Nevada's ecological zones are directly influenced by these interactions. The ecological role of fire in the bioregion varies with changes in the natural environment.

Physical Geography

The Sierra Nevada is a massive block mountain range that tilts to the west-southwest and has a steep eastern escarpment that culminates in the highest peaks. Elevations range from 150 m (492 ft) on the American River near Sacramento to 4,421 m (14,505 ft) at Mount Whitney. The moderately inclined western slope of the Sierra Nevada is incised with a series of steep river canyons from the Feather River in the north to the Kern River in the south. The western foothills are gently rolling with relatively narrow valleys. At the mid-elevations, landforms include deep canyons and broad ridges that run primarily from east-northeast to west-southwest. Rugged mountainous terrain dominates the landscape at the higher elevations. The eastern slope is steep and in some places precipitous, with the relative height of the mountain front increasing greatly as one moves south.

The oldest rocks of the Sierra Nevada were metamorphosed from sediments deposited on the sea floor during the early Paleozoic Era (Huber 1987). Granitoid rocks began to form 225 million years ago with the beginnings of the Nevada Orogeny, and pulses of magma continued for more than 125 million years, forming the granite and grandiorite core of the range (Schweickert 1981). During the Cretaceous Period, mountains were uplifted and erosion stripped the metamorphic rocks from above the granite and exposed large expanses of the core throughout the range. Meandering streams became deeply incised as gradients became steeper. By the Eocene Epoch, about 55 million years ago, this high "proto-Sierra Nevada" had been eroded into a chain of low mountains. Volcanic eruptions during the second half of the Tertiary Period blanketed much of the subdued landscape of the northern Sierra Nevada and portions of the central Sierra Nevada with lava and ash (Hill 1975). Today volcanic and metamorphic rocks occur primarily in the northern and central Sierra Nevada, although outcrops can be seen throughout the range. The sharp relief and high altitude of the modern Sierra Nevada are the products of Late Cenozoic uplift, and periodic earthquakes are a sign that the uplift continues today.

During the Pleistocene Epoch, snow and ice often covered most of the high country, and glaciers filled many of the river valleys. Several glaciations are recognized to have occurred in the Sierra Nevada. These glaciers further deepened valleys and scoured ridges, leaving the exposed granite landscape so prevalent today. Glaciers appear to have more or less disappeared from the Sierra Nevada by the Early Holocene. Modern glaciers scattered on high peaks between Yosemite and Sequoia national parks grew as the climate cooled and water balance rose over the last 4,000 years (Konrad and Clark 1998).

Geology primarily influences fire activity through its influence on soil formation, which in turn influences vegetation. Hard rock types such as granites and quartzites weather slowly, especially where glaciation has stripped soils and removed vegetation. As a result, glaciated ridges in the Sierra Nevada often support relatively little biomass. Valleys and catchments in canyons accumulate sediment, and deep soils

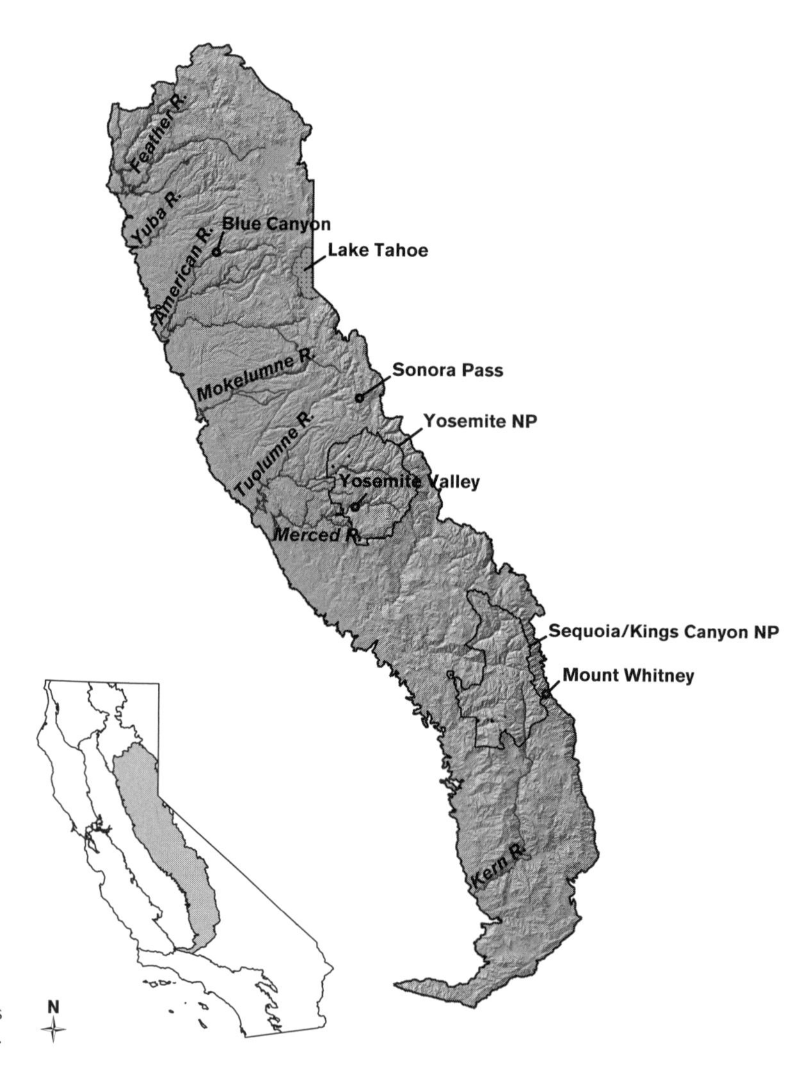

MAP 14.1 The Sierra Nevada bioregion. Locations mentioned in the text are shown on the map.

can develop and support dense forests and grasslands. Differences in topography and glacial history and subtle changes in rock chemistry affect erodibility and nutrient availability, creating a patchwork of forested swales and rocky outcrops and peaks (Hahm et al. 2014). This highly heterogeneous patchwork of vegetation, fuels, and natural barriers affects fire spread. North of Yosemite, rock lithology becomes progressively more dominated by metamorphic and volcanic rocks, and the time for soil formation has been much longer. Forest cover thus tends to be denser and more continuous, and fires can run more easily through the landscape.

Climatic Patterns

The pattern of weather in the Sierra Nevada is influenced by its topography and geographic position relative to the Central Valley, the Coast Ranges, the Pacific Ocean, and the Southeastern Deserts. Winters are dominated by low pressure in the northern Pacific Ocean while summer weather is influenced by high pressure in the same area. The climate type is primarily Mediterranean, with a wet winter and dry summer, but the influence of the Southeastern Deserts is important east of the Sierra Nevada crest, where rainfall from summer monsoons originating in the Gulf of California can be substantial.

FIRE CLIMATE VARIABLES

The primary sources of precipitation are winter storms that move from the north Pacific and cross the Coast Ranges and Central Valley before reaching the Sierra Nevada. Over the long term 30–40% of all precipitation comes in the form of "atmospheric rivers," which arrive as long, linear storm systems carrying large amounts of atmospheric water, often from subtropical sources (Dettinger et al. 2011). Lower coastal mountains catch some of the arriving moisture, but the gap in the mountains near San Francisco Bay allows storms to pass with relatively little orographic loss, producing heavy precipitation in the Sierra Nevada to the east and north. As air masses move up the western slope, precipitation increases and, at the higher elevations, falls as snow. Once across the crest, precipitation decreases sharply. This decrease is less prominent in the northern Sierra Nevada where the crest is not as high as in the central and southern parts of the range. Precipitation also decreases from north to south with nearly twice as much falling in the northern Sierra Nevada as does in the south. Mean annual precipitation ranges from a low of 25 cm (10 in) at the western edge of the southern foothills to over 200 cm (79 in) north of Lake Tahoe. More than half of the total precipitation falls in January, February, and March, much of it as snow. Summer precipitation is associated with afternoon thunder-

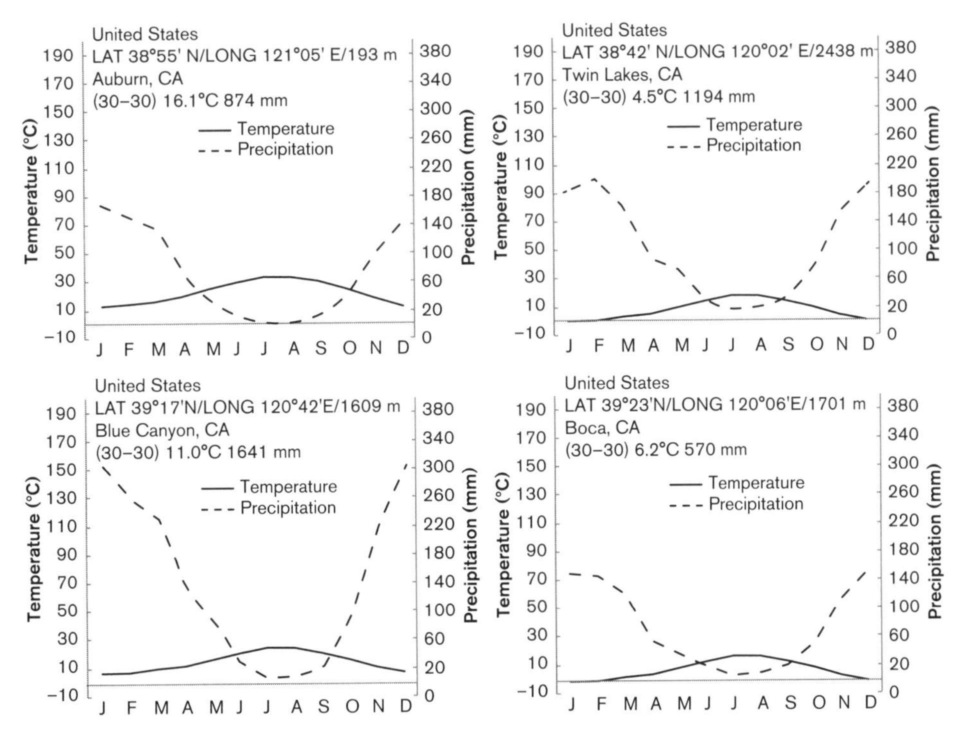

FIGURE 14.1 Annual precipitation and temperature patterns for four stations along a transect from the foothills to the eastern side of the Sierra Nevada.

storms and subtropical storms moving up from the Gulf of California.

Sierra Nevada temperatures are warm in the summer and cool in the winter. Temperatures decrease as latitude and elevation increase, with a temperature lapse rate of approximately 4.5°C with each 1,000 m of elevation (2.5°F in 1,000 ft) (Major 1988). Normal 10:00 am relative humidity is highest in January and lowest in July. Extremely low relative humidity is common in the summer. Sustained wind speeds are variable, averaging up to 10 km hr^{-1} (6 mi hr^{-1}) but have been recorded as high as 100 km hr^{-1} (70 mi hr^{-1}) or more out of the north during October or across high mountain ridges during winter storms.

Fig. 14.1 shows the annual pattern of precipitation and temperature for four stations along a transect in the northern Sierra Nevada. Auburn is located at the upper edge of the foothills and has the hottest summer temperatures. Midway up the slope, Blue Canyon receives the greatest amount of precipitation and slightly cooler temperatures. Twin Lakes is located near the crest with the coldest temperatures and decreasing amounts of precipitation. On the eastern side of the crest, Boca has cool temperatures and scant precipitation.

Monthly precipitation and temperature, in conjunction with soil water holding capacity and snow cover, can be used to calculate annual potential evapotranspiration, actual evapotranspiration, and annual climatic water deficit (Stephenson 1998). Deficits occur when the potential evapotranspiration demand exceeds the water supply (rain plus snowmelt). Both actual evapotranspiration and deficit have been shown to affect the distribution of forest species in the Sierra Nevada (Lutz et al. 2010), and deficits during the summer months can affect fire ignitions and fire severity (Lutz et al. 2009, van Wagtendonk and Smith 2016). Fig. 14.2 shows the water balance for three locations in Yosemite National Park at 1,680 m (5,512 ft), 2,426 m (7,959 ft), and 3,018 m (9,902 ft) elevation. In the ponderosa pine (*Pinus ponderosa*) forest, where temperatures are mostly above 0°C (32°F), a large deficit occurs during the summer and provides opportunities for fire ignitions. In the red fir (*Abies magnifica* var. *magnifica*) forest, precipitation peaks in May, mean monthly temperature is below 0°C (32°F) from mid-November to March, and the resulting deficit is relatively small. Abundant snowmelt and low temperatures limit ignitions in the whitebark pine (*Pinus albicaulis*) forest.

Lightning is pervasive in the Sierra Nevada, occurring in every month (van Wagtendonk and Cayan 2008). There is, however, important spatial and temporal variation. Most strikes occur between noon and 4:00 pm during July and August. Map 14.2 shows the spatial distribution of the average annual number of lightning strikes for the 16-year period from1985 through 2000. The highest concentration of lightning strikes occurs across the crest east of Sonora Pass. In the Sierra Nevada, there is a strong correlation between the number of lightning strikes and elevation, with strikes increasing with elevation up to about 3,200 m (10,500 ft) (Fig. 14.3). Nearly a third of strikes occur in mid-afternoon during July and August.

WEATHER SYSTEMS

Fires tend to be associated with critical fire weather patterns that occur with regularity during the summer (Hull et al. 1966). The Pacific High Postfrontal type is a surface type where air from the Pacific moves in behind a cold front and causes northerly to northwesterly winds in northern and central California. A föhn effect is produced by steep

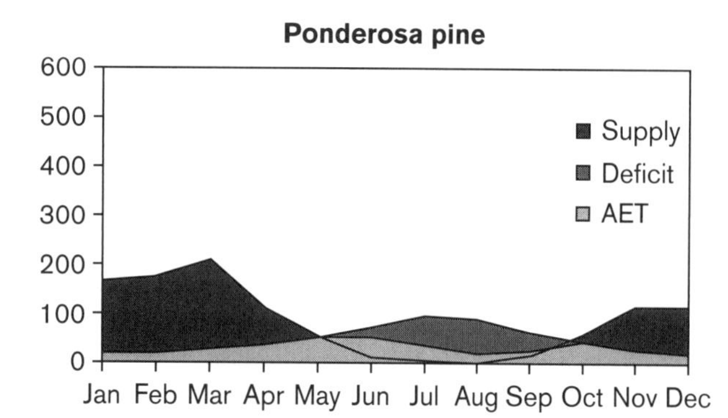

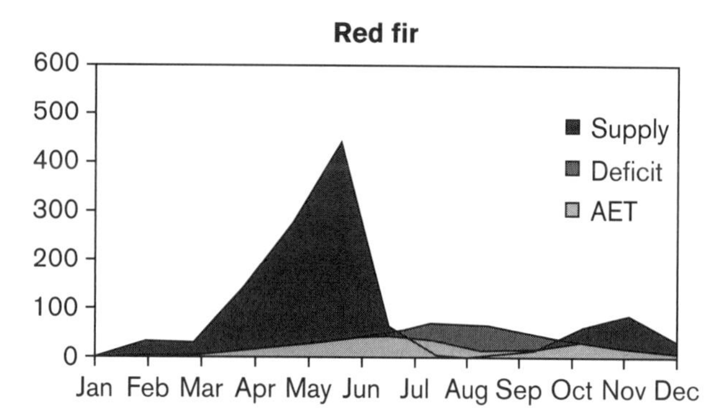

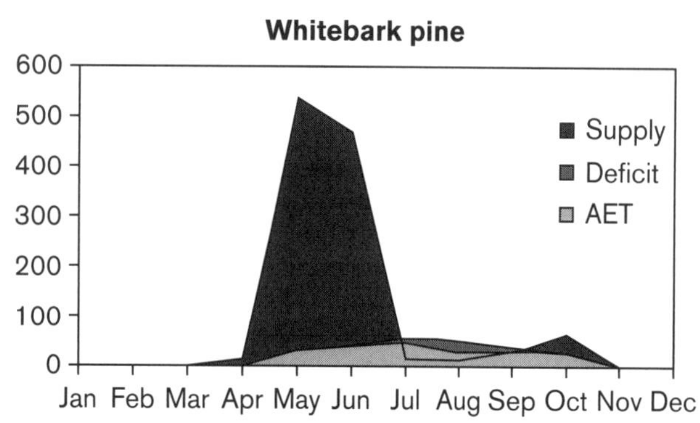

FIGURE 14.2 Water balance relationships for three species in Yosemite National Park in the central Sierra Nevada. Winter temperatures, spring water supply, and soil water capacity combine to determine Deficit and AET. Species water balance curves were based on the means of the plots where these species were present. Among species at similar elevations, increased soil water capacity results in higher AET. Deficits occur when evaporative demand has exhausted stored water and exceeds water supply (from Lutz et al. 2010).

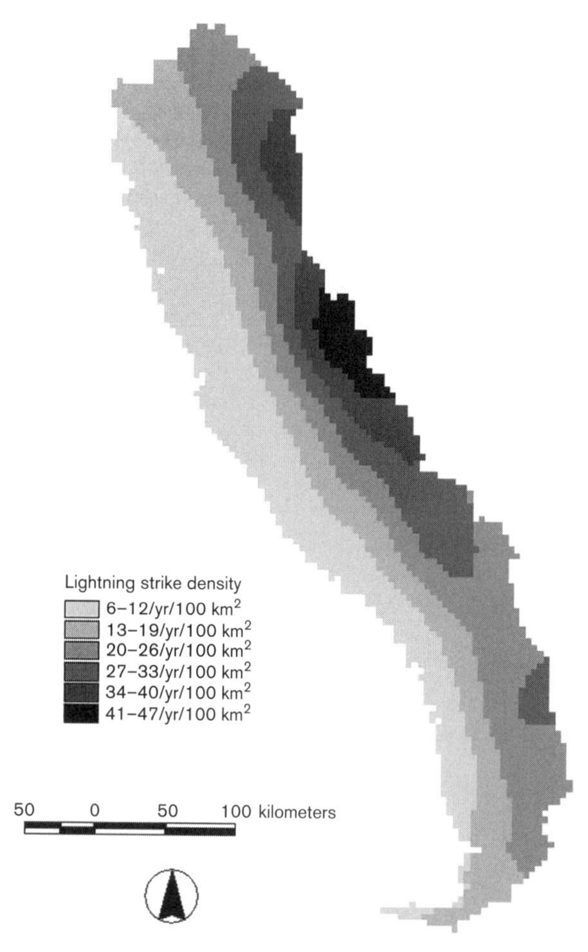

MAP 14.2 Spatial distribution of lightning strikes in the Sierra Nevada bioregion, 1985–2000. The density increases from west to east and reaches a maximum just east of the crest north of Sonora Pass.

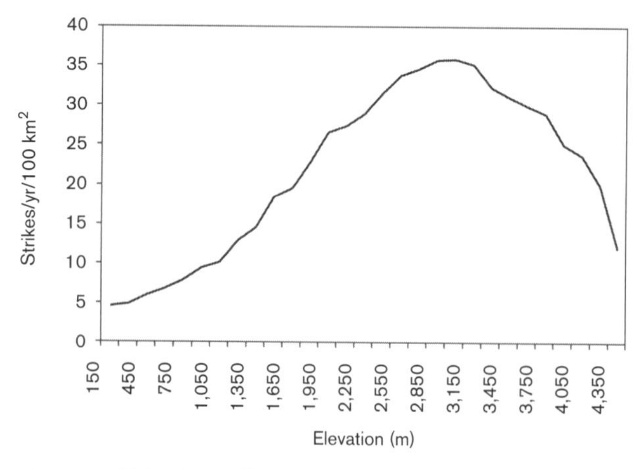

FIGURE 14.3 Lightning strikes by elevation in the Sierra Nevada bioregion, 1985–2000. The density of strikes is greatest at 3,000 m and decreases as elevation increases above that point.

pressure gradients behind the front causing strong winds to blow downslope. The Great Basin High often follows the Pacific High Postfrontal type with air stagnating over the Great Basin. Combined with a surface thermal trough off the California coast, the Great Basin High creates strong pressure gradients and easterly or northeasterly winds across the Sierra Nevada. Although this type is often present during winter months when fires are uncommon, the Great Basin High can produce extreme fire weather during the summer.

During the Subtropical High Aloft type, the belt of westerly winds is displaced northward and a stagnant air pattern effectively blocks advection of moist air from the Gulf of Mexico. High temperature and low relative humidity are associated with this type. The Meridional Ridge with Southwest Flow pattern requires a ridge to the east and a trough to the west, allowing marine air penetration in coastal and inland areas. Above the marine layer in the Sierra Nevada, temperatures are higher and relative humidities are lower as short wave troughs and dry frontal systems pass over the area (Hull et al. 1966).

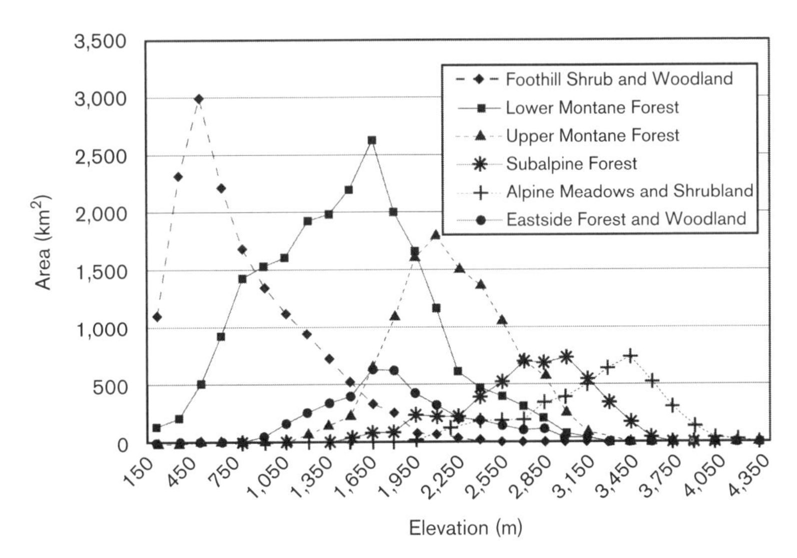

FIGURE 14.4 Area of ecological zones by 500 m elevation bands summed for the entire Sierra Nevada. Each zone is found at progressively higher elevations from the northern Sierra Nevada to the south, hence the broad elevational distributions shown in the figure.

Recent fires such as the Rim fire in 2013 and the King fire in 2014 burned under these conditions.

Ecological Zones

The vegetation of the Sierra Nevada is as variable as its topography and climate. In response to actual evapotranspiration and deficit, the vegetation forms six broad ecological zones that roughly correspond with elevation (Stephenson 1998). These zones include (1) the foothill shrubland and woodland zone, (2) the lower montane forest zone, (3) the upper montane forest zone, (4) the subalpine forest zone, (5) the alpine meadow and shrubland zone, and (6) the eastside forest and woodland zone. These zones are arranged in elevation belts from the Central Valley up to the Sierra Nevada crest and back down to the Great Basin (Fig. 14.4). Because of the effects of latitude on temperature, the belts increase in elevation from the northern to southern Sierra Nevada.

FOOTHILL SHRUBLAND AND WOODLAND

The foothill shrubland and woodland zone covers about 16,000 km^2 (6,200 mi^2) from the lowest foothills at 142 m (466 ft) to occasional stands at 1,500 m (5,000 ft). The primary vegetation types in this zone are foothill pine-interior live oak (*Pinus sabiniana-Quercus wislizeni*) woodlands, mixed hardwood woodlands and chaparral shrublands. Blue oak (*Q. douglasii*) woodlands and annual grasslands occur at lower elevations and are treated in chapter 15.

LOWER MONTANE FOREST

The lower montane forest is the most prevalent zone in the Sierra Nevada bioregion, occupying about 22,000 km^2 (8,500 mi^2) primarily on the west side of the range above the foothill zone. Major vegetation types include California black oak (*Quercus kelloggii*), ponderosa pine, white fir (*Abies concolor*) mixed conifer, Douglas-fir (*Pseudotsuga menziesii* var. *menziesii*) mixed conifer, and mixed evergreen forests. Interspersed within the forests are chaparral stands, riparian forests, and meadows.

UPPER MONTANE FOREST

This ecological zone covers about 11,500 km^2 (4,400 mi^2) and extends from as low as 750 m (2,500 ft) in the northernmost Sierra Nevada to as high as 3,450 m (11,500 ft) in the southern Sierra Nevada. The upper montane forest is most widespread above 1,950 m (6,500 ft), the elevation at which most winter precipitation transitions from rain to snow. Forests within this zone include extensive stands of California red fir accompanied by variable densities of western white pine (*Pinus monticola*) and lodgepole pine (*P. contorta* subsp. *murrayana*). Woodlands with Jeffrey pine (*P. jeffreyi*) and Sierra juniper (*Juniperus grandis*) occupy exposed ridges and warm dry slopes, and meadows and quaking aspen (*Populus tremuloides*) stands occur in moist areas.

SUBALPINE FOREST

The subalpine forest zone ranges from a low of 1,650 m (5,500 ft) in the northern Sierra Nevada to a high of 3,450 m (11,500 ft) in the southern Sierra Nevada. The subalpine zone comprises about 5,000 km^2 (1,930 mi^2) and consists of forests and woodlands of lodgepole pine, and mountain hemlock (*Tsuga mertensiana*), whitebark pine, limber pine (*P. flexilis*), and foxtail pine (*P. balfouriana* subsp. *austrina*), with numerous large meadow complexes.

ALPINE MEADOW AND SHRUBLAND

Sitting astride the crest of the Sierra Nevada is the 4,400-km^2 (1,700-mi^2) alpine meadow and shrubland ecological zone. The zone extends from as low as 2,000 m (7,000 ft) in the northern Sierra Nevada to as high as 4,421 m (14,505 ft) in the southern Sierra Nevada. Willow (*Salix* spp.) shrublands and alpine fell fields containing grasses, sedges, and herbs are the dominant vegetation types.

EASTSIDE FOREST AND WOODLAND

On the eastern side of the Sierra Nevada, forest and woodlands cover a total of about 3,900 km^2 (1,500 mi^2). Single-leaf pinyon pine (*Pinus monophylla*) and western juniper (*Juniperus occidentalis*) woodlands are extensive, as are woodlands and forests dominated by Jeffrey pine and ponderosa pine, often with some component of white fir where moisture is sufficiently available. Shrublands dominated by sagebrush (*Artemisia tridentata*), bitterbrush (*Purshia tridentata*), and other associated species are also widespread. The zone ranges in elevation from a low of 1,050 m (3,500 ft) in the northern Sierra Nevada up to a high of 2,850 m (9,500 ft) in the south.

Overview of Historical Fire Occurrence

Fire has been an ecological force in the Sierra Nevada for millennia. A Mediterranean climate, flammable fuels, abundant ignition sources, and hot, dry summers combine to produce conditions conducive for an active fire role. Fire activity has varied over time, with different patterns characterizing the Early, Middle, and Late Holocene Epochs.

Prehistoric Period

The earliest charcoal evidence of fire in the Sierra Nevada is found in lake sediments between 11,000 and 17,000 years old (Smith and Anderson 1992, Brunelle and Anderson 2003). These records are from glacially scoured lakes and are no older than the time of deglaciation after the Last Glacial Maximum. Core samples from meadow sediments also record fire back to the beginning of the Holocene (Anderson and Smith 1997). Fire scar studies are constrained by the durability of wood, but in the Sierra Nevada the large size and old age of many tree species allow records as far back as 3,000 years before present. The first fire scar studies in the western Sierra Nevada were conducted by Show and Kotok (1924). Since then, numerous fire scar studies have been published (Wagener 1961a, Kilgore and Taylor 1979, Taylor 2000, 2004, Stephens and Collins 2004, Collins and Stephens 2007, Moody et al 2006, Vaillant and Stephens 2009, Scholl and Taylor 2010, Van de Water and North 2010). Additional studies have developed fire histories for the eastern Sierra Nevada including the Lake Tahoe Basin (Gill and Taylor 2009, North et al. 2009, Taylor et al. 2013).

Pollen and macrofossils in lake and meadow cores and fire scars show broadly similar trends in climate, vegetation, and fire in the Sierra Nevada since the beginning of the Holocene Epoch. The Early Holocene (11,700 BP to 8,200 BP) was a period of postglacial warming, with general conditions somewhat cooler and moister than today. At the beginning of the epoch, elevations that currently support lower montane forests were dominated by sagebrush and grass species, with minor presence of pines and juniper (Woolfenden 1996). By the middle of the Early Holocene, however, conifer forests had established themselves in most of these areas (Minnich 2007, Hallett and Anderson 2010). Most paleo records suggest that fires were common, but generally less frequent than during the remainder of the Holocene.

During the Middle Holocene (8,200 BP to 4,200 BP) climate became much warmer and drier, with the warmest and driest conditions of the Holocene occurring between 6,500 BP and 6,000 BP. Fire frequency increased markedly during the Middle Holocene, although some places saw reduced frequencies because they were too dry to support much plant cover (Woolfenden 1996, Beaty and Taylor 2009). The pollen data show that forests of pine and fir were replaced by oak, sagebrush, and juniper in many areas, and forest structure was likely very open, with abundant understory shrubs. Conifers invaded former moist areas of meadow, and desert plant and animal taxa migrated upslope (Anderson 1990, Minnich 2007).

The Late Holocene (4,200 BP to present) has been characterized by general cooling, with some warmer periods. Precipitation increased, and small glaciers began to form again in the Sierra Nevada. According to Millar and Woolfenden (1999), the spatial and compositional outlines of modern Sierra Nevada ecosystems developed by the beginning of the Late Holocene. As temperatures cooled, available moisture rose, fir and incense cedar (*Calocedrus decurrens*) abundance began to increase, and giant sequoias (*Sequoiadendron giganteum*) colonized their current groves. At high elevations, firs and hemlock supplanted pine and chaparral (Anderson 1990). White fir increased in abundance, and oaks and sagebrush declined since the end of the Middle Holocene. Fire frequency dropped from its mid-Holocene high, allowing the development of denser forest types in moist locations.

The last 1,000 years of the Holocene have been marked by short-term changes in temperature and precipitation that have had marked impacts on Sierra Nevada ecosystems (Woolfenden 1996, Millar and Woolfenden 1999, Minnich 2007). Between about 900 and 1100 AD and 1200 to 1350 AD, two long "Medieval Droughts" led to very low levels in lakes and streams and increased fire frequencies. This was followed by a shift to cooler temperatures known as the "Little Ice Age" that lasted from about 1400 to 1880; the period between 1650 and 1850 was the coolest since the Early Holocene (Stine 1974). Glaciers expanded in the Sierra Nevada, tree line dropped, and fire frequencies moderated.

Van de Water and Safford (2011) reviewed all of the available fire scar literature and provided a comprehensive summary of pre-Euro-American settlement fire return intervals for fire regime types as they existed in California before major Euro-American settlement (i.e., before 1850). Their data for the mean, median, mean minimum, and mean maximum fire return intervals for Sierra Nevada vegetation types and characteristic species are included in Table 14.1. Median fire return intervals ranged from 7 years for yellow pine forest and woodland to 132 years for subalpine forest.

Although lightning ignitions would have been present for millennia prior to charcoal appearing in lake sediments 17,000 years ago, ignitions by Native Americans probably did not occur until about 9,000 years ago (Hull and Moratto 1999). Their use of fire was extensive and had specific cultural purposes (Anderson and Rosenthal 2015). It is currently not possible to determine whether charcoal deposits or fire scars were caused by lightning fires or by fires ignited by Native Americans. However, Anderson and Carpenter (1991) attributed a decline in pine pollen and an increase in oak pollen coupled with an increase in charcoal in sediments in Yosemite Valley to expanding populations of Native Americans 650 years ago. Similarly, Anderson and Smith (1997) could not rule out burning by Native Americans as the cause of the change in fire regimes beginning 4,500 years ago. It is reasonable to assume that their contribution to ignitions was significant but varied over the spectrum of inhabited landscapes (Vale 2002), grading from high levels of fire use in the foothill chaparral and woodland belt to little or no use in subalpine forests.

TABLE 14.1
Mean, median, mean minimum, and mean maximum historical fire return intervals for major Sierra Nevada vegetation types (from Van de Water and Safford 2011). Van de Water and Safford summarized fire history records for the entire state of California. Local and regional deviations from the listed means and medians are common, but values will be found between the mean minima and maxima.

Ecological zone	Vegetation type[a] (characteristic species)	Mean	Median	Mean min	Mean max
Foothill shrub and woodland	Chaparral and serotinous conifers (chamise, manzanita, California lilac, knobcone pine)	55	59	30	90
	Oak woodland (blue oak, interior live oak, foothill pine)	12	12	5	45
Lower montane forest	Mixed evergreen forest (canyon live oak, interior live oak, black oak, tan oak, madrone, California bay, Douglas-fir)	29	13	15	80
	Yellow pine forest and woodland (ponderosa pine, sugar pine, black oak)	11	7	5	40
	Dry mixed conifer forest (ponderosa pine, sugar pine, incense cedar, white fir, black oak)	11	9	5	50
	Moist mixed conifer forest (white fir, D ouglas-fir, incense cedar, ponderosa pine, sugar pine, lodgepole pine, giant sequoia)	16	12	5	80
	Montane chaparral (huckleberry oak, California lilac, manzanita, bitter cherry, chinquapin)	27	24	15	50
Upper montane forest	Red fir forest (red fir, western white pine, lodgepole pine)	40	33	15	130
	Western white pine forest	50	42	15	370
	Lodgepole pine woodland	37	36	15	290
	Aspen forest	19	20	10	90
	California juniper	83	77	5	335
Subalpine forest	Subalpine forest (mountain hemlock, lodgepole pine, whitebark pine, limber pine, foxtail pine, western white pine)	133	132	100	420
Eastside woodland and forest	Yellow pine forest and woodland (Jeffrey pine, ponderosa pine)	11	7	5	40
	Pinyon-juniper woodland (single-leaf pinyon pine, juniper)	151	94	50	250
	Big sagebrush (big sagebrush, bitter brush, rabbit brush)	35	41	15	85

[a] Some types occur in more than one vegetation zone, and most of them are listed only once in the zone in which they are most prevalent.

Historic Period

The arrival of Euro-Americans in the Sierra Nevada affected fire regimes in several ways. Native Americans were often driven from their homeland, and diseases brought from Europe decimated their populations. As a result, use of fire by Native Americans was greatly reduced. Settlers further modified fire regimes by introducing cattle and sheep to the Sierra Nevada, setting fires in attempts to improve the range, and suppressing fires to protect timber. Extensive fires occurred as a result of loggers who burned slash, prospectors who burned large areas to enhance the discovery of mineral outcrops, and shepherds who burned to increase forage and ease travel (Leiberg 1902, Vankat and Major 1978, Safford and Stevens 2017).

The effect of Euro-Americans on fire in the Sierra Nevada was to severely reduce its frequency. Evidence of the changed fire regimes is found in charcoal deposits and fire scars. The meadow sediments examined by Anderson and Smith (1997) showed a drop in charcoal particles during the most recent century, which they attributed to fire suppression. Beaty and Taylor (2009) documented a similar pattern in lake sediments from near Lake Tahoe. Essentially all fire scar studies in the Sierra Nevada document the near complete elimination of fire by the beginning of the twentieth century (Wagener 1961a, Kilgore and Taylor 1979, Swetnam 1993, Stephens and Collins 2004, Taylor 2004, North et al. 2005).

Of all the activities affecting fire regimes, the exclusion of fire by suppression forces has had the greatest effect. Beginning in the late 1890s, the US Army attempted to extinguish all fires within the national parks in the Sierra Nevada (van Wagtendonk 1991b). When the Forest Service was established in 1905, it developed both a theoretical basis for systematic fire protection and considerable expertise to execute that theory on national forests (Show and Kotok 1923). This expertise was expanded to the fledgling National Park Service when it was established in 1916. Fire control remained the dominant management practice throughout the Sierra Nevada until the late 1960s.

Fire exclusion resulted in an increase in accumulated surface debris and density of shrubs and understory trees. Although the number of fires and the total area burned decreased between 1908 and 1968, the proportion of the yearly area burned by the largest fire each year increased (McKelvey and Busse 1996). Suppression forces were able to extinguish most fires while they were small but, during extreme weather conditions, they were unable to control the large ones.

Current Period

The National Park Service changed its fire policy in 1968 to allow the use of prescribed fires deliberately set by managers, and to allow fires of natural origin to burn under prescribed conditions (van Wagtendonk 1991b). The Forest Service followed suit in 1974, changing from a policy of fire control to one of fire management (DeBruin 1974). As a result, fire was reintroduced to national parks in the Sierra Nevada through programs of prescribed burning and managed wildfires (Kilgore and Briggs 1972, van Wagtendonk 2007). Similarly, the Stanislaus, Sierra, Sequoia, and Inyo national forests have managed some wilderness wildfires for resource benefit since 2003 (Meyer 2015).

For much of the Sierra Nevada, however, routine fire suppression is still the rule. For lower montane and foothill forests, where most fires occurred before fire exclusion, vegetation structure and fire regimes have been greatly altered. Fire frequencies in these forest types have shifted from frequent to very infrequent (Safford and Van de Water 2014). In contrast forests in the Sierra de San Pedro Mártir, Mexico have experienced fire through much of the twentieth century and are therefore important reference sites (see sidebar 14.1). Steel et al. (2015) estimated that 74% of mixed-conifer forests in California had not experienced a fire in the last century. When fire does occur, there are much greter proportions of high severity relative to the historical period (Miller et al. 2009, Mallek et al. 2013, Lydersen et al. 2014, Harris and Taylor 2015, Safford and Stevens 2017). Some headway is being made in wilderness areas and areas where prescribed fire can be applied safely and effectively, but the overall deficit in fire is huge and widening (North et al. 2012).

Widespread harvesting, primarily targeting overstory trees, beginning in the early twentieth century also contributed to contemporary fire patterns. Large accumulations of slash and vigorous tree establishment and growth resulted in stands that became overstocked with much greater horizontal and vertical fuel continuity than existed prior to harvesting and fire exclusion (Knapp et al. 2013). Consequently many forests now have so much fuel that most fires are likely to move into the canopy, crowning out and killing many overstory trees (Stephens and Moghaddas 2005, Ritchie et al. 2007, Valliant et al. 2009). Under these conditions, prescribed fire without antecedent thinning is difficult to use unless fuel moistures are high, and even then multiple applications may be needed to reduce fuels sufficiently to restore surface burning conditions.

Major Ecological Zones

The six ecological zones of the Sierra Nevada are comprised of different vegetation types and species. Each species has different adaptations to fire and varies in its relation to fire. Similarly, the fire regimes and plant community interactions of the zones vary.

Foothill Shrubland and Woodland

The foothill shrubland and woodland zone is bounded below by the valley grasslands and blue oak woodlands, and above the lower montane zone, which is conifer dominated. The terrain is moderately steep with deep canyons. Metasedimentary, metavolcanic, and granitic rocks form the substrate, and soils are generally thin and well drained. The climate is subhumid with hot, dry summers and cool, moist winters. Lightning is relatively infrequent averaging only 8.25 strikes yr^{-1} 100^{-1} km^{-2} (75.9 strikes yr^{-1} 100^{-1} mi^{-2}).

The vegetation is a mix of large areas of chaparral and live oak woodland with patchy foothill or ponderosa pines (Fig. 14.5); annual grasslands are also common. Except on very unproductive soils, chaparral forms dense continuous stands of vegetation and fuels. Chamise (*Adenostoma fasciculatum*), manzanita (*Arctostaphylos* spp.), and California-lilac (*Ceanothus* spp.) dominate the chaparral. Interior live oaks and canyon live oaks (*Quercus chrysolepis*) are extensive on steep slopes of large canyons, and can be accompanied by Pacific madrone (*Arbutus menziesii*), big-leaf maple (*Acer macrophyllum*), California buckeye (*Aesculus californica*), and Douglas-fir on north-facing slopes or where moisture and soil conditions permit. This forest type is commonly referred to as mixed evergreen. Tall deciduous shrubs or forests dominate riparian areas with dense vertical layering and a cooler microclimate.

FIRE RESPONSES OF IMPORTANT SPECIES

Many foothill species of the Sierra Nevada have fire responses and characteristics that are similar to those of the interior South Coast zone described in chapter 17. Chaparral includes several sprouting shrub species and some that require heat or other fire cues for seed germination. As in southern California, many herb species are strongly benefitted by fire. The two live oaks are vigorous sprouters. The most prevalent conifers are fire resistant, such as ponderosa pine, or have serotinous cones, such as foothill pine and knobcone pine (*Pinus attenuata*). Establishment, survival, and abundance of many species are enhanced by fire. The fire responses for knobcone pine, ponderosa pine, and chamise are covered in more detail in the North Coast (chapter 10), Northeast Plateaus (chapter 13), and South Coast (chapter 17) chapters, respectively. Table 14.2 lists the fire responses of the important species in the foothill zone.

Most chaparral shrubs sprout following fire. These include chamise, birch-leaf mountain-mahogany (*Cercocarpus betuloides* var. *betuloides*), yerba santa (*Eriodictyon californicum*), California coffeeberry (*Frangula californica*), and toyon (*Heteromeles arbutifolia*). Nonsprouting shrubs can be locally dominant, with seeds that are heat resistant and have fire-enhanced germination—such as whiteleaf manzanita (*Arctostaphylos viscida*), chaparral whitethorn (*Ceanothus leucodermis*), and buckbrush (*C. cuneatus* var. *cuneatus*). Some species like chamise, yerba santa, greenleaf manzanita (*A. patula*),

SIDEBAR 14.1 FORESTS IN THE SIERRA DE SAN PEDRO MÁRTIR, BAJA CALIFORNIA, MEXICO

S.L. Stephens and H.D. Safford

FIGURE 14.1.1 Jeffrey pine-mixed conifer forest in the Sierra de San Pedro Mártir, Baja California, Mexico.

Fire suppression and harvesting have impacted the majority of California's ponderosa pine, mixed conifer, and Jeffrey pine forests. These changes have made it difficult to derive reference conditions for management, conservation, and restoration, particularly at landscape scales. One Jeffrey pine-mixed-conifer forest exists where fire suppression did not begin until the 1970s and harvesting was confined to an area less than 10 ha (25 ac); the Sierra de San Pedro Mártir (SSPM) in Baja California, Mexico (Minnich and Franco 1998, Fry et al. 2014) (Fig. 14.1.1). This region includes approximately 20,000 ha (50,000 ac) of Jeffrey pine and mixed conifer forests that are floristically similar to forests of the eastern Sierra Nevada (Dunbar-Irwin and Safford 2016) and southern California mountains (Minnich et al. 1995). Climatic analysis of the SSPM and 17 meteorological stations in the range of Jeffrey pine determined that the SSPM clearly belongs to the general class of Jeffrey pine-dominated yellow pine-mixed-conifer (YPMC) forests in California (Dunbar-Irwin and Safford 2016).

Recent research measured forest structure, fuels, and vegetation and ground cover in the SSPM and in multiple National Forests along the eastern slope of the Sierra Nevada (Dunbar-Irwin and Safford 2016). Live tree density was nearly twice as high in the eastern Sierra Nevada as in SSPM, and dead tree density was 2.6 times higher. Basal area was about 30% higher in the eastern Sierra Nevada even though average tree size was larger in SSPM. Fuel loads and coarse woody debris were very similar between the two sites, and fine fuels (1-h fuels) were actually higher in SSPM. Logging and fire suppression have resulted in denser YPMC forests dominated by smaller trees in California, but Dunbar-Irwin and Safford's (2016) results suggest that fire suppression in SSPM over the last 40 years has increased surface fuel loads. Nonetheless, the Baja California forests still retain an overstory structure created and maintained by centuries of frequent fire.

Fire severity trends in the SSPM and the Sierra Nevada are very different over the last several decades

(continued)

(continued)

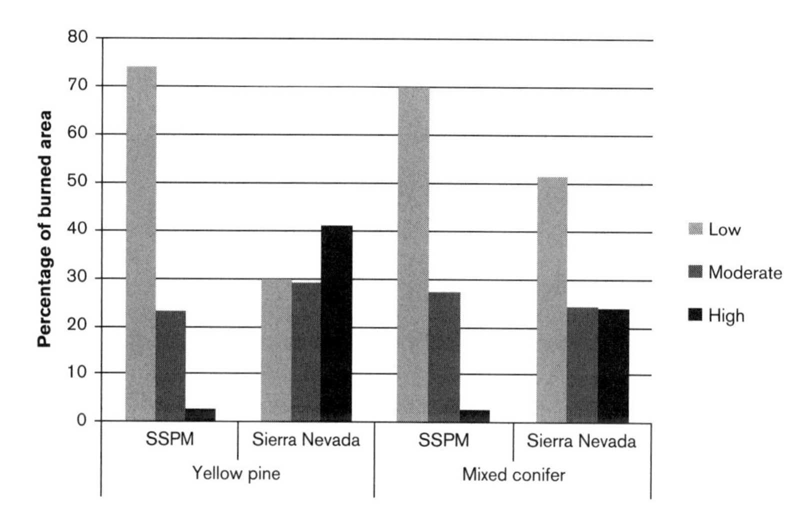

FIGURE 14.1.2 Fire severity distribution in the Sierra de San Pedro Mártir, Baja California, Mexico, versus distribution from the Sierra Nevada, California, USA. Data from 1984 to 2010. The measured variable is % basal area mortality, categorized as low (<25% mortality), moderate (25–90% mortality), and high (>90% mortality). Data from Rivera et al. (2015).

(Rivera-Huerta et al. 2015). LANDSAT data were used to identify 32 fires that burned 26,529 ha (66,555 ac) in the SSPM from 1984 to 2010, of this, 1,993 ha (4,925 ac) burned in YPMC forests in 17 fires. No temporal trends in forest burned area or in the proportion of high severity fire were found in the SSPM, but the mean size of high-severity patches within fires is rising. In the SSPM, the overall proportion of fire area burned at high severity averaged 3% (average high-severity patch = 2.9 ha [7.1 ac], median = 0.6 ha [1.5 ac], largest high-severity patch = 11.3 ha [27.9 ac]) in both yellow pine and mixed-conifer forests (Fig. 14.1.2), similar to severity values found in much of the Sierra Nevada historically (Stephens et al. 2015, Safford and Stevens 2017). In the SSPM, there was no correlation between burned area and proportion of high-severity fire; Rivera-Huerta et al (2015) interpreted this to mean that differences in fuels in the SSPM were more important to fire behavior than weather condition. This is in stark contrast to similar forests in California, which are experiencing fires of sizes and severities that fall far outside the historical range of variation (Stephens et al. 2017).

Resilience to drought followed by fire has also been demonstrated in SSPM forests (Stephens et al. 2008, 2018). Fire effects were moderate especially considering that this 2003 SSPM wildfire occurred at the end of a severe, multiyear (1999–2003) drought. Shrub consumption was an important factor in tree mortality and the dominance of Jeffrey pine increased after fire. The SSPM wildfire enhanced or maintained a patchy forest structure, in contrast to large high-severity fires in California that produce homogeneous structures in large high-severity patches. The cumulative impact of 4 years of severe drought followed by wildfire only killed 20% of the trees (Stephens et al. 2008) in the SSPM demonstrating that YPMC can be very resilient to stressors if properly restored and managed.

and deer brush (*C. integerrimus*) are facultative, and regenerate after fire both by sprouting and by fire-enhanced germination. Many chaparral species produce seed at an early age that can remain viable in the soil for many decades. Buckbrush and deer brush produce seeds at around 4 to 7 years of age. Growing in dominantly single species patches, such species resist burning until decadent or foliar moistures are extremely low. Many years' worth of seeds are often produced before fire returns, enhancing postfire dominance.

Depending on nutrient and water availability, chaparral stands can recover biomass very rapidly after fire. This is especially true where resprouting species dominate the stand. As chaparral stands age, overall biomass levels off and the proportion of dead biomass increases (Oechel and Reid 1983). For

FIGURE 14.5 Foothill shrub and woodland. Foothill pine and interior live oak are dominant overstory species in this stand with nonnative grasses and species of manzanita and California-lilac in the understory. Grazing, thin soils, and periodic fire help keep the understory relative clear.

TABLE 14.2
Fire response types for important species in the foothill shrub and woodland ecological zone

	Type of fire response			
Lifeform	Sprouting	Seeding	Individual	Species
Conifer	None	None	Resistant, killed	Ponderosa pine
	None	Fire-stimulated (seed release)	Resistant, killed	Foothill pine, knobcone pine
Hardwood	Fire-stimulated	None	Top-killed or branch killed	Blue oak, interior live oak, canyon live oak
Shrub	Fire-stimulated	None or unknown	Top-killed	Poison oak, flannelbush, coyote bush, birch-leaf mountain-mahogany, yerba santa, California coffeeberry, Christmas berry
	Fire-stimulated	Fire-stimulated	Top-killed	Chamise, redbud
	None	Fire-stimulated	Killed	Whiteleaf manzanita, chaparral whitethorn, buckbrush
	None	None	Killed	
Forb	Fire-stimulated	None	Top-killed	Soap plant, death camas, mariposa lilies
	None	None		
Grass	Fire-stimulated	None	Top-killed	Deergrass
	None	None	Killed	Cheat grass

example, by the time chamise stands reach about 9 years in age, the combination of dead branches and live resinous foliage make them extremely flammable.

Chaparral supports numerous species of geophytes, or bulb bearing plants that show increased flowering and growth following fire. Common examples are soap plant (*Chlorogalum pomeridianum*), death camas (*Toxicoscordion* spp.), and mariposa lilies (*Calochortus* spp.) (Tyler and Borchert 2007). Most plant species in the postfire flora are annuals, some of which are postfire endemics with fire-enhanced germination. The annuals are responding to the sudden availability of light and nutrients and can recruit either from the soil seed bank or from offsite sources. Nonnative annual species invariably increase after fire, but they tend to disappear as shrubs recover dominance.

Interior and canyon live oaks sprout both from root and canopy crowns following fire. Canyon live oak bark resists low-intensity fires (Paysen and Narog 1993), whereas the relatively thin bark of interior live oak results in top-kill with all but lowest intensity fires (Plumb 1980). Both species can also sprout new branches from epicormic buds on the stem.

Foothill pines are sensitive to fire and the oldest individuals in a chaparral stand are a minimum indication of the time since last burn. Pitch running down the bole is common and increases crown torching (Lawrence 1966). Foothill pines develop cones and seeds at an early age. Their large cones are partially serotinous and the annual accumulation of cones results in a large seed crop available when fire arrives, ready to germinate on open mineral soil. Foothill pine is relatively drought tolerant and grows well in rocky, thin soils, including serpentine soils. In such low productivity conditions fuels are not heavy and fire frequency and intensity may be reduced, enhancing foothill pine survival. Foothill pine seeds are large and wingless, and long distance dispersal of seeds is dependent on rodents and birds.

FIRE REGIME–PLANT COMMUNITY INTERACTIONS

Fire regimes in the foothill zone vary with topography and vegetation (Appendix 1). Fire season is long, beginning in early summer, and lasting through late fall. In the lower areas of the zone, oak grassland savannah areas burned frequently and with low to moderate intensity as described in the Central Valley chapter (chapter 15). Areas dominated by chaparral burned less frequently because lightning is infrequent, especially in the fall when fuel moistures become critically low. On a statewide basis, pre-Euro-American settlement fire return intervals in foothill chaparral probably averaged between 30 years and 90 years (Van de Water and Safford 2011).

Persistence of obligate-seeding shrubs on the landscape can be potentially threatened by long fire return intervals (Zedler 1995). For example, California-lilac species are relatively short-lived (40 to 50 years is common), and fire-free intervals much longer than this could result in both the plants and the majority of the seed pool dying before the return of fire necessary for seed germination. However, other species have seeds that remain viable for very long periods, even after the plants have died (Knapp et al. 2012). As Zedler (1995) noted though, this threat may be more theoretical than real. Keeley et al. (2005) tested this hypothesis in the southern Sierra Nevada and found that even 150-year old chaparral stands in which most of the buckbrush component had died regenerated after fire with densities of over 14,000 seedlings ha^{-1} (5,566 ac^{-1}). A more real threat is the risk of very frequent fire (<15 years) outpacing the ability of either seeding or sprouting species to recover enough seeds or viable buds to remain in the landscape (Zedler 1995).

In Yosemite, Thode et al. (2011) found that the greatest proportion of area burned in the foothill chaparral type was of high severity. Live oaks also burned with high severity, conifer patches with moderate severity, and oak woodlands with low severity. Islands of trees in chaparral landscapes can survive fire by occupying rocky or moist areas such that flames are avoided or attenuated, but longer periods without fire might lead to invasion by chaparral and subsequent stand loss under a more severe fire regime (Lombardo et al. 2009). Where severe fires occur at the upper end of the foothill shrubland zone, the boundary between the shrublands and the lower montane forest may shift. Reestablishment of conifers in such areas may take decades to centuries, and frequent recurring high-severity fires may perpetuate the shrub species.

Frequent fire in foothill grasslands, in part from burning by Native Americans, reduced encroachment by chaparral and expanded the area supporting important herbaceous species that were used for many cultural purposes (Anderson and Rosenthal 2015). The mosaic of vegetation in the foothills is very dynamic. In areas that have soils that are sufficiently fertile and stable to support woody biomass, grasslands would have required frequent fire to persist on the landscape. Such areas dominated by chaparral could be reduced to grassland by frequent fire, or they might succeed to woodland and then forest with the very long absence of fire. On the other hand, oak or pine woodlands that experienced a long fire-free interval might see invasion by chaparral shrubs, with subsequent fire being severe enough to extirpate the oaks (Cocking et al. 2014).

Ponderosa pine occurred at much lower elevations in the Sierra Nevada when Euro-Americans arrived, and the uphill movement of its lower distribution has been primarily driven by logging and changes in land-use (Thorne et al. 2008). Ponderosa pine remains in the foothills in limited patches on more mesic north-facing slopes. Natural reestablishment of ponderosa pine in the foothills is limited by the reduction in fires, which maintained canopy openings and provided mineral soil for successful recruitment and also prevented invasion by chaparral. Warming temperatures, which increase the summer water deficit, could also limit reproduction (Lutz et al. 2010). The increase in large, high-severity fires over the last few decades has also reduced conifer cover in some areas and greatly increased the distance to living sources of seed. In the foothills to the west of Yosemite National Park, recurrent, large and overlapping fires with large stand-replacing patches have resulted in establishment of vast shrubfields and annual grasslands. The 5-year drought from 2012 through 2016 has resulted in almost complete elimination of ponderosa pine in the southern Sierra Nevada foothills (Stephens et al. 2018).

Foothill pine stands respond to the fire regimes of the surrounding chaparral and live oak stands, surviving those of low severity and succumbing to moderate- to high-severity fires. Partial serotiny allows reestablishment after stand-replacing fires. Woody and duff fuel loads are among the lowest of any Sierra Nevada conifer and do not contribute significantly to fire spread and intensity (van Wagtendonk et al. 1998a). Although relatively uncommon, patches of knobcone pine exist in the Sierra Nevada foothills surrounded by chaparral. Like foothill pine, knobcone pine shows a notable tolerance for low productivity soils, which lowers fuel accumulation rates and reduces the probability of fire before stands have amassed sufficient seed to regenerate after fire (Safford and Harrison 2004). Knobcone is more completely serotinous

FIGURE 14.6 Lower montane forest. This open stand of ponderosa pine, incense cedar, and sugar pine with mountain misery in the understory burned in 1978 and in 1996.

than foothill pine and grows in dense mostly single-aged stands that promote high-intensity fires. Current practices of fire exclusion may reduce the persistence of some knobcone pine patches, but counteracting this is the recent increase in large severe fires in the Sierra Nevada.

Lower Montane Forest

The lower montane forest ecological zone is the first continuous zone of conifers as one ascends the Sierra Nevada. The foothills are below with the upper montane forest above. The relatively gentle western slope consists of an alternating series of often flat-topped ridges and river canyons. Metavolcanic, metasedimentary, and granitic rocks form the majority of the geologic substrates and soils are relatively deep and well drained. Summers are hot and dry, and winters are cold and wet. Lightning is moderately frequent, averaging 15.6 strikes yr^{-1} 100^{-1} km^{-2} (40.3 strikes yr^{-1} 100^{-1} mi^{-2}).

Vegetation and fire within the lower montane zone vary with elevation, landscape position, and latitude. At the lowest elevations, California black oak and ponderosa pine dominated large areas before Euro-American settlement, but many stands have been logged or developed, and others have succeeded to mixed-conifer forest dominated by Douglas-fir and incense cedar due to selective logging and fire exclusion (Dolanc et al. 2014b). Intermixed with these forests are various-sized patches of chaparral and canyon live oak. Manzanita and California-lilac species dominate chaparral, whereas canyon live oak is extensive on steep slopes of large canyons. As elevation increases, moist sites and north slopes are dominated by Douglas-fir at lower elevations and white fir at higher elevations. Incense cedar and sugar pine (*Pinus lambertiana*) are found throughout. Fig. 14.6 shows a stand of ponderosa pines, incense cedars, and sugar pines with an understory of mountain misery (*Chamaebatia foliolosa*). Giant sequoia groves are concentrated in several river basins in the central and southern Sierra Nevada, occupying sites where soils are moist and deep. At the highest elevations, at the boundary with upper montane forests, white fir often becomes dominant on all aspects except where soils are shallow or very rocky, where pine or shrub communities often dominate.

Throughout the zone, riparian plant communities characterized by deciduous trees, shrubs, herbs, and grasses occur with varied proportions of conifers. White alder (*Alnus rhombifolia*), mountain alder (*A. incana*), or black cottonwood (*Populus trichocarpa*) dominate larger streams or wetter sites. Big-leaf maple and mountain dogwood (*Cornus nuttallii*) occur along smaller or intermittent streams or where soil moisture is high, especially in the northern Sierra Nevada. Small patches of quaking aspen (*P. tremuloides*) occur in the higher elevation mixed-conifer forests but are more prevalent in the upper montane zone and on the eastside of the range (Potter 2006). Meadows and seeps are mostly small and scattered throughout.

Partly due to increasing precipitation, Douglas-fir becomes important in the north. Mixed-evergreen forests comprised of tanoak (*Notholithocarpus densiflorus* var. *densiflorus*), Pacific madrone, and other montane hardwoods and conifers (especially Douglas-fir and ponderosa pine) occupy large areas in the western Yuba and Feather River basins where precipitation exceeds 152 cm (60 in) annually.

FIRE RESPONSES OF IMPORTANT SPECIES

The majority of lower montane species have characteristics resulting in resistance to fire and often have favorable responses to fire. Sprouting hardwood trees, shrubs, vines, herbs, and grasses are common and mostly fire-enhanced; conifers have at least some fire resistant characteristics.

Giant sequoia, ponderosa pine, sugar pine, Douglas-fir, and white fir have thick bark when mature (Table 14.3). The trees vary in their level of resistance to low- and moderate-intensity fires (Safford and Stevens 2017). Ponderosa pine has a thicker bark as a young tree and is more resistant to fire than the other lower montane conifers. As ponderosa pine grows older, it also self-prunes lower branches, which raises the crown above typical surface flame heights. In addition, ponderosa pine's large, protected buds provide further fire resistance. Ponderosa pine leaf litter is among the most flammable of all

TABLE 14.3
Fire response types for important species in the lower montane ecological zone

Lifeform	Type of fire response: Sprouting	Seeding	Individual	Species
Conifer	None	None	Resistant, killed	Ponderosa pine, Douglas-fir, white fir, sugar pine, incense cedar
	None	Fire-stimulated (seed release)	Resistant, killed except sprouts when young	Giant sequoia
	None	None	Low resistance, killed	Pacific yew
Hardwood	Fire-stimulated	None	Top-killed	Black oak, tanoak, canyon live oak, big-leaf maple, Pacific madrone, white alder
Shrub	Fire-stimulated	None	Top-killed	Mountain misery, greenleaf manzanita, poison oak, hazelnut, , willow
	Fire-stimulated	Fire-stimulated	Top-killed	Deer brush, mountain whitethorn
	None	Fire-stimulated	Killed	Whiteleaf manzanita
Forb	Fire-stimulated	None	Top-killed	Penstemon, many lilies, iris, starflower, trail plant, sanicle
				Lady's slipper orchid
Grass	Fire-stimulated	None	Top-killed	Red fescue, melic, sedges
	None	None	Killed	Cheat grass

North American conifers (van Wagtendonk et al. 1998a), which may represent an adaptation to frequent fires that kill less fire-tolerant competitors. Rapid growth of giant sequoia seedlings produces early fire resistance. Douglas-fir, white fir, and incense cedar have thick bark when mature, but are killed by fire when young because of thin bark, low, flammable crowns, and small, unprotected buds. Sugar pine is intermediate in fire resistance with thick bark and high crowns but potentially more susceptible to cambial or root damage from heat (Haase and Sackett 1998).

Giant sequoias have serotinous cones that are exposed by heat and show increased seedling density with higher intensity fire (Kilgore and Biswell 1971). Giant sequoia and white fir are the only major Sierra Nevada conifers that epicormically (none sprout basally) sprout, but this response is apparently limited to younger trees (Weatherspoon 1986, Hanson and North 2006). More information on the responses of giant sequoias to fire is provided in sidebar 14.2. Pacific yew (*Taxus brevifolia*) and California nutmeg (*Torreya californica*) are uncommon, relict conifers that have thin bark, and they sprout prolifically. They have survived in the fire-prone landscape by their restricted habitats in wet, mostly riparian areas and can apparently survive low-intensity fire, as evidenced by observed fire scars and sprouting (Fites-Kaufman 1997).

The montane hardwoods, including tanoak, Pacific madrone, California black oak, canyon live oak, California bay (*Umbellularia californica*), mountain dogwood, big-leaf maple, white alder, and black cottonwood, all sprout from basal burls or root crowns following fire. Sprouting can be vigorous with up to 100 sprouts produced on individual California black oak stumps (McDonald 1981). Epicormic sprouting from the stem following low-intensity fire was observed in California black oak, tanoak, and mountain dogwood (Kauffman and Martin 1990). California black oak is the only oak species that develops bark sufficiently thick to resist low- to moderate-intensity fire in larger trees (Plumb 1980). Like ponderosa pine, black oak leaf litter is highly flammable (Engber and Varner 2012) but decomposes rapidly. Frequent fires in oak litter kill competitors and maintain open canopy conditions. Riparian hardwoods all sprout following fire. Native Americans burned riparian areas to enhance shoot production of big-leaf maple and hazelnut (*Corylus cornuta*) shrubs (Anderson 1999).

Most shrub species resprout prolifically after fire and some also have heat-stimulated seeds (Kauffman and Martin 1990) (Table 14.3). Sprouters include mountain misery, deer brush, greenleaf manzanita, bush chinquapin (*Chrysolepis sempervirens*), mountain whitethorn (*Ceanothus cordulatus*), and riparian shrubs like hazelnut, thimbleberry (*Rubus parviflorus*), and mountain alder. The burning season can affect sprouting response. Bush chinquapin, Sierra gooseberry (*Ribes roezlii*), deer brush, greenleaf manzanita, and thimbleberry all showed greater sprouting following early spring burns than fall or late spring burns (Kauffman and Martin 1990). Overall, mountain whitethorn showed the greatest postfire sprouting after higher intensity fall burns. Sprouting occurs from burls, root crowns, and rhizomes.

SIDEBAR 14.2 GIANT SEQUOIAS AND FIRE

One is in no danger of being hemmed in by sequoia fires, because they never run fast, the speeding winds flowing only across the treetops, leaving the deeps below calm, like the bottom of the sea. Furthermore, there is no generally distributed fire food in sequoia forests on which fires can move rapidly. Fire can only creep on the dead leaves and burrs, because they are solidly packed. —MUIR (1878)

Probably better than any other species, giant sequoia exemplifies a truly fire-adapted species. Not only does it have thick bark that protects it from periodic surface fires, but also its cones are opened by heat and its regeneration is dependent on exposed mineral soil, such as occurs after a moderately severe fire. Biswell (1961) was one of the first scientists to explore the relationships between giant sequoias and fire. He reported fire scar dates in the Mariposa Grove in Yosemite National Park from as early as 450 AD with periods between fire scars averaging 18 years. He also looked at the number of lightning fires in 93 km^{-2} (36 mi^{-2}) areas surrounding sequoia groves and found that during the years from 1950 through 1959, 36 fires had been suppressed in the Mariposa Grove and 39 in the Tuolumne Grove. These data along with observations of dense thickets of white firs and incense cedars and large increases in forest floor debris led him to conclude the groves should be managed with fire as part of the environment.

Hartesveldt (1964) conducted the first detailed scientific study of giant sequoias and fire in the Mariposa Grove and concluded that the greatest threat to the survival of the big trees was catastrophic fire burning through accumulated surface and understory fuels as a result of decades of fire exclusion. His recommendation was to reintroduce fire to the giant sequoia ecosystem through the use of prescribed burning.

Subsequently, Hartesveldt and Harvey (1967) and Harvey et al. (1980) studied factors associated with giant sequoia reproduction in the Redwood Mountain Grove of Kings Canyon National Park. Using experimental fires and mechanical manipulations, they measured seedling survival and growth and investigated the role of vertebrate animals and arthropods in giant sequoia reproduction. Seedlings established on areas burned the hottest survived at a higher rate than those on other soils. Fire did not greatly affect vertebrate populations, and only one species had a significant effect on sequoia reproduction. The Douglas squirrel (*Tamiasciurus douglasii*) feeds on the scales of 2-year to 5-year-old giant sequoia cones and cuts and caches literally thousands of cones each year. This greatly aids the distribution of cones and, subsequently, seedlings because the squirrels could not relocate most cached cones. Although over 150 arthropods were found to be associated with giant sequoias only two significantly affected regeneration. The gelechiid moth (*Ghelechia* spp.) feeds on one-year old cones, while the small long-horned beetle (*Phymatodes nitidus*) mines the main axis of cones older than five years, which causes them to dry and drop their seeds.

Based on these findings, the National Park Service began a program of prescribed burning and research in giant sequoia groves in Yosemite, Sequoia, and Kings Canyon national parks (Kilgore 1972). Detailed information on fires and minerals (St. John and Rundel 1976), fuel accumulation (Parsons 1978), and fire history (Kilgore and Taylor 1979) added to the knowledge about the role of fire in these forests.

Burning in sequoia groves was not without controversy, however. Charred bark from a prescribed burn in Sequoia National Park prompted an investigation and a report on the burning programs in the groves (Cotton and McBride 1987). As a result, additional research was conducted to refine the scientific basis for the programs (Parsons 1994). Fire history studies extended the fire scar record back to 1125 B.C. with an average interval between fires from 2 to 30 m years (Swetnam 1993). Pollen and charcoal in sediments cores taken in the groves indicated that giant sequoias became more prevalent about 5,000 years ago and that fires occurred throughout the record (Anderson 1994, Anderson and Smith 1997).

Studies on the effects of fire on fungi and insect relationships with giant sequoias led Piirto (1994) to conclude that fire does influence the types and population levels of numerous organisms but that their interactions are not well understood. Other studies looked at the role of fire severity in establishing and maintaining giant sequoia groves. Of particular interest was the finding that patchy, intense fires existed in presettlement times and that these fires were important determinants of grove structure and composition (Stephenson et al. 1991). Contributing to these intense fires in giant sequoia groves are the heaviest woody fuel loads found for any Sierra Nevada conifer species (van Wagtendonk et al. 1998b).

All the research to date indicates that fires have always played an important role in giant sequoia ecology and that the survival of the species depends on the continued presence of fire. Management programs must recognize this fact and must be designed to include fire in as natural a role as practicable. Restoration targets must include process goals as well as structural goals based on sound science (Stephenson 1999). Only through such a program can we ensure the survival of this magnificent fire species.

Shrubs sprouting from deeply buried rhizomes, such as mountain misery, can readily dominate sites with frequent and intense fire. Mountain misery occupies large areas, 4–40 ha (10–100 ac) through extensive networks of rhizomes protected from heat over 20 cm (8 in) below the soil surface. With highly flammable foliage containing volatile oils, mountain misery promotes burning. Rundel et al. (1981) found that regrowth was stimulated by spring and fall burns but that summer burns inhibited resprouting for at least two years. Further enhancing its competitive advantage, mountain misery is able to fix nitrogen from nodules that develop after burning (Heisey et al. 1980).

Heat-stimulated seed of deer brush, buckbrush, or mountain whitethorn can produce extensive dense seedling patches. Seeds produced by these species can persist in the soil for decades or centuries (Knapp et al. 2012). The fire-stimulated sprouting and seed germination responses of deer brush make it particularly successful in rapidly colonizing burned sites. Deer brush germination with wet seed can be greater than from dry heat (Kauffman and Martin 1990). This could explain its greater prevalence, especially after fires, on moister portions of the landscape, such as north and east aspects or lower slopes. Deer brush gains height rapidly but can be limited by deer browsing (Kilgore and Biswell 1971). In the Sierra Nevada, deer brush and mountain whitethorn can dominate postfire habitats in lower montane forest that burns at high severity, reaching heights of 2–4 m (6–12 ft) in 5 years. As shade-tolerant trees succeed in growing through the overlying shrub canopy, sometimes many decades after fire, these shrub species can continue to persist for some years in the understory in a decadent, highly flammable state (Oakley et al. 2006).

The postfire herbaceous flora of lower montane forests supports few species whose germination is enhanced by fire (Keeley et al. 2003). The general lack of a fire-following endemic flora is the result of the long-term absence of widespread high-severity fires in these forests. Certainly many species respond positively to the increase in understory light, moisture availability, and nutrients that fire brings (Collins et al. 2007, Wayman and North 2007, Abella and Springer 2015). However, very severe fire in these forests reduces understory species diversity at scales larger than the plot due to strong homogenization of the flora (Stevens et al. 2015). Numerous perennial plants with underground rhizomes, corymbs, or stolons have been observed to increase in abundance following fire. These include Pacific starflower (*Trientalis latifolia*), trail plant (*Adenocaulon bicolor*), western blue flag (*Iris missouriensis*), Bolander's bedstraw (*Galium bolanderi*), bear-grass (*Xerophyllum tenax*), sanicles (*Sanicula* spp.), many-stemmed sedge (*Carex multicaulis*), Ross' sedge (*C. rossii*), needle grass (*Stipa* spp.), oniongrass (*Melica bulbosa*), and red fescue (*Festuca rubra*). Other plants exhibit sprouting or enhanced flowering following fire, such as Mariposa lilies and penstemons (*Penstemon* spp.). Severe fire also increases the presence of nonnative species in the postfire flora.

FIRE REGIME–PLANT COMMUNITY INTERACTIONS

Fire regime attributes for major vegetation types of the lower montane ecological zone are shown in Appendix 1. Most fires occur between mid-summer and early fall. The fire season is longer in the southern portion of the Sierra Nevada because of drier conditions, and the proportion of fire scars in the growing season is greater (Skinner 2002). Some fires have always occurred in the spring and early summer and occasionally in the winter, but the prevalence of such "out of season" fires is increasing as climates warm.

Before Euro-American settlement, fire was generally frequent in the lower montane zone, with stand-scale fire return intervals averaging 10 to 20 years, and ranging from 5 to 80 years (Van de Water and Safford 2011). At the landscape scale, fire rotations averaged 22 to 31 years in ponderosa pine and mixed-conifer forest, ranging from 11 to 70 years (Mallek et al. 2013). There was noticeable variation in fire pattern with latitude and elevation. Drier areas with longer fire seasons experienced the most frequent and regular fires. These areas are most prevalent in the southern and central Sierra Nevada and throughout the range on south aspects, ridges, and lower elevations. These areas tend to be dominated or codominated by ponderosa pine and California black oak. Throughout the zone, relatively cooler and wetter sites have had somewhat less frequent fire and are more likely to have a presence or dominance of Douglas-fir and white fir. Forests around smaller, headwater streams have similar fire regimes to adjacent uplands, whereas larger streams with cooler, more mesic microclimate have longer fire return intervals and higher fuel loading (Van de Water and North 2010, 2011). Fire severity is also variable throughout the lower montane forest zone. In a summary of 20 years of fire in Yosemite National Park, Thode et al. (2011) found mostly low to moderate severity in ponderosa pine stands, mostly low severity in white fir and California black oak stands, and high severity in riparian and chaparral stands. Under severe fire weather conditions, high-severity patches in white fir did reach 60 ha (148 ac) in size (Collins and Stephens 2010).

Frequent fire creates forest structural diversity at stand and landscape scales associated with several ecosystem processes. The within-stand structure has been characterized as containing three main conditions: individual trees, clumps of trees, and openings or gaps (Lydersen et al. 2013). Stand-level average canopy cover under frequent-fire conditions is typically low (20–45%) compared to modern fire-excluded conditions (typically 55–85%) (Collins et al. 2011). However, within a stand, individual, clump and opening conditions produce heterogeneity such that canopy closure (North and Stine 2012) is highly variable, providing a scattering of areas for plants and animals associated with dense forest conditions. Several studies suggest this fine-scale heterogeneity affects ecosystem conditions and functions, producing a wide range of microclimates (Rambo and North 2009, Ma et al. 2010), a diversity of understory plants (Wayman and North 2007) and soil invertebrates (Marra and Edmonds 2005), variation in soil respiration (Concilio et al. 2005), and limits on insect and pathogen spread (Maloney et al. 2008). The spatial variability also makes these stand conditions more resistant to crown fire because groups of trees are separated by low fuel gaps reducing crown fire spread potential (Agee and Skinner 2005, Stephens and Moghaddas 2005).

At a landscape level, fire often burned with greater intensity associated with changes in topography and species composition (Taylor and Skinner 2003, Lydersen and North 2012). Topography has a direct effect on fire intensity on steeper and drier slopes and an indirect effect on mesic north-facing slopes and canyon bottoms where fuel loads are higher because of greater forest productivity (Lydersen and North 2012, Kane et al. 2014). Diverse and variable species in both the tree and shrub layers result in variable fuel and fire patterns. For example, ponderosa pine fuels are both more energy rich and more loosely packed than those of white fir, allowing the pine fuels to burn more readily (van Wagtendonk et al. 1998a). In addition, ponderosa pine has one of the highest annual deposition rates of both litter and small branches (van Wagtendonk and Moore 2010).

High levels of variation in slope, aspect, elevation, and weather, as well as topographically controlled diurnal changes in fire behavior, overlap with variable fuel patterns to create fine-scale patterns of variation in forest density, structure, and understory vegetation. With fire exclusion, density and uniformity in structure and composition have increased (Lydersen et al. 2013, Safford and Stevens 2017). Across many sites in the mid-elevations of the Sierra Nevada, white fir and incense cedar have increased, shifting composition away from ponderosa pine and sugar pine and creating more uniformly dense forests (Vankat and Major 1978, Parsons and deBenedetti 1979, North et al. 2007, Dolanc et al. 2014a, 2014b). Douglas-fir responds similarly in the northern Sierra Nevada (Fites-Kaufman 1997). At lower elevations bordering the foothills, these shade-tolerant species are less common and ponderosa pine has decreased in density (Parsons and deBenedetti 1979, Fites-Kaufman 1997). Similarly, at higher elevations, white fir dominates but with increased uniformity and density attributed to fewer fires (Parsons and deBenedetti 1979).

Historically, open or more variable forest structure occurred as a result of more frequent fire. Not only did fire favor different species, it also affected forest structure by thinning the young trees leaving a patchier or more open forest, and selectively retaining larger, more fire resistant trees (van Wagtendonk 1985). Exactly how forest and understory vegetation conditions varied across historical landscapes is unknown, but a close reading of early observers' writings (Safford and Stevens 2017) suggests a very heterogeneous landscape dominated by mostly open canopied forests with scattered dense patches in certain landscape positions (Sudworth 1900, Leiberg 1902). These observations are supported by historical photographs (Gruell 2001) and analyses of historical forest inventories conducted in both the central and southern Sierra Nevada (Collins et al. 2015, Stephens et al. 2015). The original inventories were conducted systematically in 1911 across large landscapes greater than 10,000 ha (>25,000 ac), allowing for robust characterization of historical forest conditions in the lower montane zone. Areas with moderate- to high-density forests historically had greater proportions of white fir and Douglas-fir and tended to be associated with higher elevations and topographic settings with greater moisture availability (Collins et al. 2015, Stephens et al. 2015). In addition to heterogeneous forest overstory conditions, understory vegetation was highly variable, including stands with high mountain misery cover (50–80%) and moderate cover of taller shrubs (25–50%).

Montane chaparral patches also appear to have been a distinct feature of historical landscapes that contributed to overall heterogeneity. These patches likely occurred at relatively low proportions (<10%) and small patch sizes (<20 ha [<50 ac]) across landscapes with intact fire regimes (Show and Kotok 1923, Collins and Stephens 2010). In some areas of upper montane forest there has been a loss of chaparral due to tree invasion in the absence of fire (Nagel and Taylor 2005). This is consistent with observations from Show and Kotok (1924) that frequent fire facilitated shrub persistence or expansion, which was one of several justifications for early fire exclusion policies. At the same time, the increasing size and severity of wildfires in the lower montane zone is leading to the origin, maintenance, and expansion of very large areas of montane chaparral that may never succeed to forest given current trends (Safford and Stevens 2017).

Questions remain concerning the intensity and severity of presettlement fires in lower montane forests, but almost all evidence points to fires that were typified by low and moderate severity, with higher severity occurring in areas of heavier fuels or under extreme weather conditions (Collins et al. 2015, Stephens et al. 2015, Safford and Stevens 2017). Steel et al. (2015) showed that time since last fire was positively related to fire severity in lower montane forests, substantiating the general maxim that fuel accumulations in these forests due to a century of fire exclusion is changing modern fire behavior. Various sources of evidence suggest that high-severity patches in lower montane fires were generally small (Sudworth 1900, Stephenson et al. 1991, Collins et al. 2015, Stephens et al. 2015, Safford and Stevens 2017). There is certainly evidence of historical large high-severity fires in the Sierra Nevada, but such fires were the exception rather than the rule. Characteristics of modern fires burning >20,200 ha (>50,000 ac) at a time with >30% high severity and high-severity patches of >400 ha (>1,000 ac) appear to have no analogue in the Late Holocene, at least over the period of the tree-ring record.

There is a lack of historical information on the size or distribution of high-severity fires in the lower montane zone. It is likely that they occurred infrequently and were related to drought cycles, which would create larger areas of highly flammable vegetation. The northern Sierra Nevada experiences higher average annual rainfall than the southern Sierra Nevada and supports denser forest and a higher component of relatively fire-intolerant species. As a result historic fire return intervals in the north tend to be somewhat longer than in the south, and fires probably burned at higher severity (Fites-Kaufman 1997). There is evidence from historical inventories that indicates high proportions of shrub-dominated vegetation within and adjacent to major canyons. These shrubfields could be attributed to centuries of recurring fires perpetuating the shrubs or forests that were converted to shrubs by high-severity fires that occurred after Euro-American settlement (Collins et al. 2015, Airey Lauvaux et al. 2016).

Sierra Nevada lower montane forests are some of the most productive fire-prone forests in the western United States. This results in increased stand densities and reduced decomposition rates that produce high fuel accumulations (Kilgore 1973, Vankat and Major 1978, Agee et al. 2000). This increases the tendency for high-intensity and high-severity fire through increased fuels and increased susceptibility of dense smaller vegetation, especially when fire intervals are increased. Various studies have documented large increases over the last three decades in fire size and severity in the lower montane zone of the Sierra Nevada (Miller et al. 2009, Miller and Safford 2012, Mallek et al. 2013). This trend has not been seen in Yosemite National Park, however, where extensive prescribed burning and managed wildfire programs begun in the 1970s have reduced fuels (van Wagtendonk and Lutz 2007).

Upper Montane Forest

The upper montane forest is located above the lower montane forest and occurs on both sides of the crest of the Sierra Nevada. On the west side elevations are generally lower than on the east, with the differences greater in the south than in the north (Potter 1998). The terrain is relatively moderate on the west side but drops precipitously on the east. The geology underlying this zone is primarily volcanic in the north and granitic in the south. Soils are weakly developed and are

FIGURE 14.7 Upper montane forest. This stand is characterized by large red fir, western white pine, and Jeffrey pine in the overstory with an understory of prostrate and erect manzanita and California-lilac species. Fire is infrequent but can burn extensive areas.

typically medium to coarse textured and often lack a clay-rich subsoil (Potter 1998). The large expanses of exposed rock and shallow soils in the southern Sierra Nevada provide a discontinuous landscape that inhibits fire spread.

The climate of the upper montane forest is characterized by cool summers and cold winters. The transition from the lower montane zone to the upper montane zone is found approximately at the elevation of maximum overall precipitation, where the percentage of precipitation falling as snow rises above 50%, the average freezing-level occurs in winter storms, and the winter snowpack is at its deepest (Major 1988, Barbour et al. 1991, Safford and Van de Water 2014). The upper montane forest zone receives as many lightning strikes as might be expected by chance (van Wagtendonk 1991a). The average number of lightning strikes that occurred in the zone between 1985 and 2000 was 29.3 strikes yr^{-1} 100^{-1} km^{-2} (75.9 strikes yr^{-1} 100^{-1} mi^{-2}) (van Wagtendonk and Cayan 2008).

The vegetation of the upper montane forest is characterized by the presence of California red fir (Potter 1998). Fig. 14.7 shows a stand of California red fir and western white pine with a sparse understory of montane chaparral. Besides thee two species, other species include western white pine, quaking aspen, Sierra juniper, Jeffrey pine, and tufted hair grass (*Deschampsia cespitosa*). Interspersed in the forests are meadows and stands of montane chaparral.

FIRE RESPONSES OF IMPORTANT SPECIES

Many upper montane species have fire resistant characteristics and respond favorably to fire (Table 14.4). Shrubs and hardwood trees typically sprout, whereas herbs and grasses either reseed or regrow quickly after fire. Conifers are protected from fire heat by thick bark layers.

The conifers in the upper montane ecological zone vary in their resistance to fire. California red fir has thin bark when it is young, making it susceptible to fire. As California red fir matures, its bark becomes thicker and it is able to survive most fires (Kilgore 1971). Like ponderosa pine, Jeffrey pine has thicker bark when young than its competitors, which—along with self pruning of lower branches and highly flammable litter—gives it an advantage in resisting frequent fire. Western white pine and Sierra juniper are more susceptible to fire at a young age than California red fir or Jeffrey pine. The percent crown scorch that a species can sustain is also variable. Up to 50% of the buds of a Jeffrey pine can be killed and still survive (Wagener 1961b). The other upper montane conifers can sustain only 30–40% scorch (Kilgore 1971).

Quaking aspen is the primary hardwood species in the upper montane forest and occurs in small stands where moisture is available. It is a vigorous and a profuse sprouter after fire (Brown and DeByle 1987). It becomes increasingly resistant to fire as its diameter increases beyond 15 cm (6 in).

Bush chinquapin, mountain whitethorn, and huckleberry oak (*Quercus vaccinifolia*) form extensive stands in the open and underneath conifers. They are all sprouters and are top-killed by fire (Conard et al. 1985). Mountain whitethorn is also a relatively prolific seeder after fire. Pinemat manzanita (*Arctostaphylos nevadensis* subsp. *nevadensis*) and greenleaf manzanita are usually killed by intense heat, but Sierra Nevada populations of greenleaf mazanita grow lignotunber and can resprout after fire. Both species are able to reestablish by seed the first year after fire. Although these nonsprouting manzanitas are killed by intense heat, they are able to reestablish by seed the first year after fire.

Woolly mule's ears (*Wyethia mollis*) apparently resprouts after fire (Mueggler and Blaisdell 1951). The density of mule's ears has been noted to increase after fire (Young and Evans 1978). Tufted hair grass is one of many grass and sedge species common in wet meadows. Although it burns infrequently, tufted hair grass generally survives all but the most intense fires and sprouts from the root crown, as do most sedges.

FIRE REGIME-PLANT COMMUNITY INTERACTIONS

Although the upper montane forest receives a proportionally higher number of lightning strikes on a per area basis than the lower montane forest, fewer fires result (van Wagtendonk 1994). Lightning is often accompanied with rain, and the

TABLE 14.4
Fire response types for important species in the upper montane forest ecological zone

	Type of fire response			
Lifeform	Sprouting	Seeding	Individual	Species
Conifer	None	None	Resistant, killed	Red fir, Jeffrey pine, western white pine, Sierra juniper
Hardwood	Fire-stimulated	None	Resistant, top-killed	Quaking aspen
Shrub	Fire-stimulated	Abundant seed production	Top-killed	Bush chinquapin, mountain whitethorn, huckleberry oak
	None	Fire-stimulated	Killed	Whiteleaf manzanita, pinemat manzanita
Forb	Fire-stimulated	None	Top-killed	Western mule's ears
	None	None	Top-killed	Corn lily
Grass	Fire-stimulated	Off-site	Top-killed	Tufted hair grass
	Tillers	Off-site	Top-killed	Western needlegrass

compact fuelbeds are not easily ignited. Those fires that do occur are usually of low intensity and spread slowly through the landscape except under extreme weather conditions. Natural fuel breaks such as rock outcrops and moist meadows limit the development of extensive fires (Kilgore 1971).

California red fir fuelbeds are among the heaviest and most compact found for conifers in the Sierra Nevada. Whereas duff weight was just above average, woody fuel weight is surpassed only by giant sequoia (van Wagtendonk et al. 1998a). Relatively high annual deposition of large woody fuels contributes to woody fuel weight (van Wagtendonk and Moore 2010). The bulk density of California red fir duff fuels is above average and the fuelbed bulk density, including woody and litter fuels, is only exceeded by limber pine. Such dense fuels ignite and carry fire only under extremely dry and windy conditions.

Fire regimes tend to be more variable in frequency and severity than those in the lower montane forest (Appendix 1) (Skinner and Chang 1996). Van de Water and Safford (2011) reported mean, median, mean minimum, and mean maximum fire return intervals for red fir of 40, 33, 15, and 130 years, respectively. The mean presettlement fire rotation for Sierra Nevada red fir was 61 years, ranging from 25 years to 76 years (Mallek et al. 2013). The wide range of values may be due to variability in red fir forest conditions and locations. Studies at lower elevations, where the tree species composition suggests drier site conditions, have generally found shorter historical fire return intervals and age structures that suggest frequent pulses of regeneration. In contrast, higher elevation and more mesic site studies often document a fire regime of mixed severities with distinct recruitment pulses following fire events (Taylor 2004, Scholl and Taylor 2006). One study documented a strong linear relationship between fire return interval and elevation, possibly driven by snowpack and its effect on fuel moistures (Bekker and Taylor 2001). Montane chaparral stands in Yosemite National Park burned with high severity, whereas red fir, Jeffrey pine-shrub, Jeffrey pine-western white pine, and aspen stands all burned with low to moderate severity (Thode et al. 2011).

At the higher elevations in the upper montane zone, fire has an important role in the successional relationship between California red fir and lodgepole pine (Kilgore 1971, Taylor 2000). Where lodgepole pine occurs under a California red fir canopy, without fire it is eventually succeeded by California red fir. Pitcher (1987) concluded that fire was necessary for creating openings where young California red fir trees could get established. In areas where crown fires have burned through California red fir forests, montane chaparral species such as mountain whitethorn and bush chinquapin become established. Within a few years, however, California red fir and Jeffery pine begin to overtop the chaparral.

Fires in Jeffrey pine, Sierra juniper, and western white pine stands are usually moderate in intensity, burning through litter and duff or, if present, through huckleberry oak or greenleaf manzanita. Older trees survive these fires, although occasionally an intense fire may produce enough heat to kill an individual tree (Wagener 1961b). Fuelbed bulk density and woody fuels weights are comparable for the three species, but Jeffrey pine has three times as much litter and twice as much duff (van Wagtendonk et al. 1998a). Jeffrey pine also accumulates more fuel annually than either western white pine or Sierra juniper (van Wagtendonk and Moore 2010). As a result, surface fires tend to be more intense in Jeffrey pine stands. Jeffrey pine will be replaced by huckleberry oak and greenleaf manzanita if fires of high severity occur frequently, or by California red fir if the period between fires is sufficiently long (Bock and Bock 1977).

Although quaking aspen stands in the Sierra Nevada usually burn only if a fire from adjacent vegetation occurs at a time when the stands are flammable, the decline of quaking aspen stands has been attributed to the absence of natural fire regimes (Lorentzen 2004). Quaking aspen stands burn in late summer when the herbaceous plants underneath the trees have dried sufficiently to carry fire. Because quaking aspen is a vigorous sprouter, it is able to recolonize burns immediately at the expense of nonsprouting conifers. Recent research has found that aspen seedling establishment is more common than has been assumed, suggesting the species may be able to thrive even with changing climate and fire regime conditions (Krasnow and Stephens 2015). Similarly, meadows consisting primarily of tufted hair grass burn if fires in adjacent forests occur during the late summer. Occasional fires reduce encroachment into the meadows by conifers (deBenedetti and Parsons 1979).

Subalpine Forest

The subalpine forest lies between the upper montane forest and the alpine meadows and shrublands. Extensive stands of subalpine forest occur on the west side of the Sierra Nevada and a thin band exists on the east side of the range. Like the upper montane zone below, the terrain is moderate on the west side and steep on the east. Volcanic rocks are prevalent in the north and granitic rocks occur throughout the zone. Soils are poorly developed.

The climate of the subalpine forest is characterized by cool summers and extremely cold winters. Whereas growth and other ecological processes are more moisture limited in lower elevation forests, growth in the subalpine zone is largely energy limited driven by low temperatures and a short growing season. Other than occasional summer thundershowers, precipitation falls as snow. The snow-free period is short, from mid-June to late October, and the frost-free period is much shorter. Lightning is pervasive in the subalpine forest with many more lightning strikes than might be expected by chance (van Wagtendonk 1991a). Between 1985 and 2000, the average number of strikes was 33.6 strikes yr^{-1} 100^{-1} k^{-2} (87.1 strikes yr^{-1} 100^{-1} mi^{-2}) (van Wagtendonk and Cayan 2008).

In the central and southern Sierra Nevada, the subalpine forest is dominated by lodge pole pine. As tree line is approached, lodgepolepine is replaced by mountain hemlock and whitebark pine (Fig. 14.8). In the northern Sierra Nevada, the forest is dominated by mountain hemlock, western white pine, and whitebark pine. Lodgepole pine also occurs in the north, but there it is more of a riparian or lacustrine fringe species. On the east side of the Sierra Nevada, limber pine occurs with whitebark pine, and in Sequoia National Park, foxtail pine is found at tree line. Extensive meadows of sagebrush sedge (*Carex filifolia* var. *erostrata*) are mixed within the forest.

FIRE RESPONSES OF IMPORTANT SPECIES

Subalpine trees are easily killed by fire at a young age but increase their resistance as they grow older (Table 14.5). Unlike the Rocky Mountain lodgepole pine (*Pinus contorta* var. *latifolia*), the cones of the Sierra Nevada lodgepole are not fully serotinous (some of the cones will hold onto seeds for a few years [Lotan 1975]), and seeding from off-site survivors is often necessary to regenerate a stand lost to high severity fire. Parker (1986) concluded that fire was not necessary for the perpetuation of lodgepole pine, but fire-induced openings supplemented those created by tree-falls. When surface fires occur in Sierra Nevada lodgepole pine forests, individual trees may be killed (deBenedetti and Parsons 1984), while others survive.

FIGURE 14.8 Subalpine forest. Lodgepole pine forms extensive stands in this zone. Fire is infrequent but when it occurs it burns from log to log or creeps through the sparse understory vegetation and litter.

The combination of thin bark, flammable foliage, low hanging branches, and growth in dense groups make mountain hemlocks susceptible to fire (Fischer and Bradley 1987). As the trees mature, the bark thickens giving them some protection. Whitebark pine survives fires because trees grow in rocky or sandy habitats and are scattered in areas of patchy fuels (Keane and Arno 2001). Clark's nutcrackers (*Nucifraga columbiana*) facilitate postfire seedling establishment (Tomback 1986). Bark thickness is moderate and mature trees usually survive low and sometimes moderate-intensity surface fires, whereas smaller trees do not. Limber pines also have moderately thin bark, and young trees often do not survive surface fires (Keeley and Zedler 1998). Terminal buds are protected from the heat associated with crown scorch by the tight clusters of needles around them. Foxtail pine occurs where fuels to carry fires are practically nonexistent (Parsons 1981). The charred remains of trees struck by lightning are evidence that periodic fires do occur, although they seldom spread over large areas.

FIRE REGIME–PLANT COMMUNITY INTERACTIONS

Fire regime attributes for the subalpine zone are listed in Appendix 1. Although lightning strikes are plentiful in the subalpine forest zone, ignitions are infrequent. Between 1930 and 1993, lightning caused only 341 fires in the zone in Yosemite National Park (van Wagtendonk 1994). Those fires

TABLE 14.5
Fire response types for important species in the subalpine forest ecological zone

	Type of fire response			
Lifeform	Sprouting	Seeding	Individual	Species
Conifer	None	Fire-enhanced	Killed	Lodgepole pine
	None	None	Resistant, killed	Mountain hemlock
	None	None	Resistant, killed	Whitebark pine, limber pine, foxtail pine
Grass	Fire-stimulated	None	Top-killed	Brewer's reedgrass

burned only 2,448 ha (5,953 ac) primarily in the lodgepole forest. During the period between 1972 and 1993 when lightning fires were allowed to burn under prescribed condition, only six fires in lodgepole pine grew larger than 123 ha (300 ac).

Lodgepole pine fuelbeds are relatively shallow and compact (van Wagtendonk et al. 1998a). Often herbaceous plants occur in the understory precluding fire spread except under extreme dry conditions. Annual deposition of litter and woody fuels are about average among common Sierra Nevada conifers (van Wagtendonk and Moore 2010). When fires do occur, encroaching California red firs and mountain hemlocks are replaced by the more prolific seeding lodgepole pines. In areas where lodgepole pines have invaded meadows, fires will often kill the trees and halt or reverse the invasion (deBenedetti and Parsons 1984). Stand-replacing fires are rare, but when they do occur, lodgepole pines become reestablished from the released seeds.

Data from fires that have burned in the managed wildfire zone in Yosemite suggest a fire rotation of 579 years (van Wagtendonk 1995). Caprio (2008), however, reported that prior to 1860, widespread fires were recorded in 1751, 1815, and 1846 in lodgepole pine stands in Sequoia National Park. Mallek et al. (2013) found that presettlement fire rotations in Sierra Nevada subalpine forest ranged from 75 years to 721 years, with an average of 394 years. Two fire-scar studies in the Sierra Nevada found that lodgepole had a mean fire return interval of 19 years to 39 years on the eastside (North et al. 2009) and 31 years to 98 years in Sequoia National Park (Caprio 2008). Both studies suggested that unlike many Rocky Mountain populations, Sierra Nevada lodgepole pine forests have a fire regime characterized by a mix of fire severities and that stand regeneration may not be largely dependent on serotinous seed dispersal. In Yosemite, lodgepole pine burned predominantly at low severity, although some high severity occurred (Thode et al. 2011). Severity of white bark pine and mountain hemlock fires was barely detectable.

Little information exists for the role of fire in mountain hemlock forests in the Sierra Nevada. In Montana, however, fires in the cool wet mountain hemlock forests generally occur as infrequent, severe stand-replacing crown fires (Fischer and Bradley 1987). Fire return intervals are estimated to be between 400 and 800 years (Habeck 1985). During the 28-year period prior to 1972, no fires burned in hemlock forests in the managed wildfire zone of Yosemite National Park (van Wagtendonk et al. 2002). Litter and duff fuels of mountain hemlocks were some of the deepest, heaviest, and most compact of any Sierra Nevada conifer, indicating long periods between fires (van Wagtendonk et al. 1998a). For mature stands, mountain hemlock had the highest annual deposition of small diameter twigs among eleven common Sierra Nevada conifers (van Wagtendonk and Moore 2010). Mountain hemlock is replaced by lodgepole pine in areas where both are present before a fire. Seeding from adjacent areas is possible but can take several years to be successful.

Fire seldom burns in the pine stands that occur at tree line. There have been only 25 lightning fires in whitebark pine during a years period in Yosemite (van Wagtendonk 1994). Only four of these fires grew larger than 0.1 ha (0.25 ac), and they burned a total of 4 ha (9 ac). Based on the area burned in the type, van Wagtendonk (1995) calculated a fire rotation of over 27,000 years. Although no records exist showing fires in limber pine stands in the Sierra Nevada, it is reasonable to assume equally long fire return intervals for that species. Scattered pockets of fuel beneath both whitebark pine and limber pine attest to the long period between fires. Limber pine recorded the heaviest litter and duff load of any Sierra Nevada conifer (van Wagtendonk et al. 1998a). On the other hand, foxtail pine had hardly any fuel beneath it. Keifer (1991) found only occasional evidence of past fires in foxtail stands. She noted sporadic recruitment in stands that did not appear to be related to fire and suggested that the thick bark on the mature trees protected them from low-intensity fires.

Little is known about fire in subalpine meadows. These meadows are sometimes ignited when adjacent forests are burning. Short hair reedgrass (*Calamagrostis breweri*) can become reestablished after fire from seeds and rhizomes. Meadow edges are maintained by fire as invading lodgepole pines are killed (deBenedetti and Parsons 1984, Vale 1987).

Alpine Meadow and Shrubland

The alpine meadow and shrubland zone consists of fell fields and riparian willows. The short growing season produces little biomass and fuels are sparse. Lighting strikes occur regularly in the alpine zone but result in few fires (van Wagtendonk and Cayan 2008). Weather, coincident with lightning, is usually not conducive for fire ignition or spread. Fires are so infrequent that they probably did not play a role in the evolutionary development of the plants that occur in the alpine zone. The 70-year record of lightning fires in Yosemite includes only eight fires, burning a total of 12 ha (28 ac), primarily in a single fire (van Wagtendonk 1994).

Eastside Forest and Woodland

The width of the eastside montane zone of the Sierra Nevada varies from 30 to 40 km (18 mi) in the northern and central Sierra Nevada to as little as 1 km (0.6 mi) or in the far south. The area to the north and east of Lake Tahoe basin comprises large expanses of eastside forest and woodland vegetation, as does the area around Mammoth Lakes. Lightning is common in the eastside zone with 28.9 strikes yr^{-1} 100^{-1} k^{-2} (74.8 strikes yr^{-1} 100^{-1} mi^{-2}) for the period between 1985 and 2000 (van Wagtendonk and Cayan 2008). Proportionally more lightning strikes occur in the central part of the zone than in any other zone in the Sierra Nevada.

The vegetation of the eastside of the Sierra Nevada is often transitional between upper montane and lower elevation Great Basin species. A variable, but often coarse-scale mosaic of open woodlands, forests, shrublands, or grasslands, is characteristic. The most prevalent tree-dominated types include Jeffrey pine or mixed Jeffrey and ponderosa pine woodlands, mixed white fir and pine forests, and quaking aspen groves (Fig. 14.9). In some locations, particularly in the central and southern portions, pinyon pine occurs. Douglas-fir occur in small amounts in the northern Sierra Nevada and California black oak is scattered throughout the entire range. Shrublands can be extensive and variable, ranging from typical Great Basin species of sagebrush and bitterbrush to chaparral comprised of snow brush (*Ceanothus velutinus*), greenleaf manzanita, bearbrush (*Garrya fremontii*), and bush chinquapin. Curl-leaf mountain-mahogany (*Cercocarpus ledifolius*) occurs in patches on rocky and dry sites. Riparian and wetland areas occur throughout, and meadows can be extensive. Quaking aspen, black cottonwood, and various willow species dominate the overstory of riparian communities of larger streams. Lodgepole pine is also common in riparian areas or localized areas with cold air drainage.

Because of similarities with the Southern Cascades (chapter 12) and the Northeastern Plateaus (chapter 13) bioregions, the focus of this chapter is on the Jeffrey pine woodlands, mixed Jeffrey pine-white fir forests, and montane chaparral. Additional information on communities dominated by desert species can be found in the Southeastern Deserts chapter (chapter 18).

FIRE RESPONSES OF IMPORTANT SPECIES

Species in this zone tend to be a mixture of those with fire resistant or enhanced characteristics and those that are fire inhibited (Table 14.6). Jeffrey pine has thick, fire resistant bark, and large well-protected buds, self-pruning that often results in high crowns, and highly flammable foliage. Pinyon pine is not very fire resistant, with crowns low to the ground, relatively thin bark, and a tendency to have pitchy bark, making it flammable. In this zone, pinyon pine often occupies rocky sites with sparse vegetation and fuels that decrease the likelihood of frequent fire. After severe fires, pine stands can be replaced by sprouting hardwood species. For example, in the northeastern Sierra Nevada, California black oak woodlands now occur where there were Jeffrey pine forests before the 1987 Clark Fire (Carl Skinner, Shasta lake, California, USA, Pers. comm.).

Shrub species vary from those that have enhanced sprouting or seed germination following fire to those that have little fire resistance. Greenleaf manzanita, bearbrush, bush chinquapin, and snow brush all sprout from basal burls following fire. Where branches are pressed against the soil from snow, layering results in sprouting; however, these sprouts can be more susceptible to fire mortality. Snow brush also has enhanced germination from fire.

FIGURE 14.9. Eastside forest and woodland. This stand of Jeffrey pine and red fir was recently thinned.

FIRE REGIME–PLANT COMMUNITY INTERACTIONS

Fire regimes vary with both vegetation type and landscape location (Appendix 1). Several fire history studies have been conducted in the eastern montane zone. In an area east of the crest near Yosemite, Stephens (2001) found median fire return intervals of 9 years for Jeffrey pine and 24 years for adjacent upper montane forest consisting of California red fir, lodgepole pine, and western white pine. Taylor's (2004) work on the east shore of Lake Tahoe showed a mean fire return interval of 11.4 years for presettlement mixed Jeffrey pine and white fir stands. Sampling 10 different eastside Jeffrey pine forests, North et al. (2009) found fire return intervals that ranged from 9 years to 36 years and suggested that some of the variability was related to how isolated or connected stands were to larger forested areas. As recent, severe fires have burned on the lower slopes of the eastside forests, the boundary between forests and sagebrush has retreated upslope. On the eastern slope of the northern Sierra Nevada, Gill and Taylor (2009) determined that the grand mean and grand median composite fire return intervals for all sites were 12.1 years and 10.4 years, respectively.

The most frequent fires and lowest intensity fires occur in the lower elevation, open pine-dominated areas of this zone with responses similar to that described in the Northeastern Plateaus (chapter 13). On less productive or more southern portions, Jeffrey pine woodlands likely had a fire regime

TABLE 14.6
Fire response types for important species in the eastside forest and woodland ecological zone

Lifeform	Type of fire response: Sprouting	Seeding	Individual	Species
Conifer	None	None	Resistant, killed	Jeffrey pine, ponderosa pine
	None	None	Low resistance, killed	Pinyon pine
Hardwood	Fire-stimulated	None	Top-killed	Quaking aspen, black cottonwood, willow
Shrub	Fire-stimulated	None	Top-killed	Bush chinquapin, greenleaf manzanita, huckleberry oak, Fremont silk tassel, snowberry, willow, bitterbrush*
	Fire-stimulated	Fire-stimulated	Top-killed	Tobacco brush
	None	None	killed	Sagebrush, bitterbrush*
Herb	Fire-stimulated	None	Top-killed	Woolly mule's ears
	None	None		
Grass	Fire-stimulated	None	Top-killed	Sedges
			killed	Cheat grass

* Bitterbrush has a variable sprouting response to fire.

similar to those described for upper montane Jeffrey pine woodlands, with a range of fire return intervals from 5 years to 47 years (Taylor 2004). White fir forests occur in a mosaic with chaparral on the more mesic sites on north slopes and at higher elevations. The fire regimes included a greater variety of severities, due, in part, to less consistent fire intervals and patterns. The fire season was primarily from summer through fall, with longer seasons at lowest elevations in open pine forests.

The fire regime for the white fir-chaparral type apparently included some high-severity fires in the past (Russell et al. 1998), although the importance of settlement activities on contributing to these types of fires is unclear. Regeneration of white fir is continuous (Bock et al.1978, Conard and Radosevich 1982) until a crown fire occurs. Subsequently, portions of the forest are converted to chaparral dominated by sprouting greenleaf manzanita and both sprouting and heat-stimulated germination of snow brush (Conard and Radosevich 1982). The duration of this fire-generated chaparral can last for over 50 years (Russell et al. 1998). Under current projected climate change, these conversions could become permanent (Airey Lauvaux et al. 2016). Numerous sites in locations conducive to high-severity fire may have supported chaparral stands in the past, but, in the absence of fire over the last century, have succeeded to conifer stands, often dominated by white fir, which can survive for decades in the dense shade of the chaparral understory (Nagel and Taylor 2005). The relative amounts of pine and white fir regeneration are affected by fire. Pine regeneration can increase from 25% in forests with no fire to greater than 93% in forests with fire (Bock and Bock 1969). Fire can also serve to control regeneration by limiting the density of white fire recruitment (Bock et al. 1976), but white fir can also regenerate well under the shade of chaparral (Conard and Radosevich 1982).

Management Issues

Private property owners, land managers, and the public in the Sierra Nevada face many issues as a result of changed fire regimes and population growth. Primary among the issues is the accumulation of fuels both on the ground and in tree canopies. Dealing with these fuels has become more complicated by increased urbanization, at risk species, air quality and high densities of dead trees from drought and beetle infestations.

Urbanization

The population of the Sierra Nevada more than doubled between 1970 and 1990 (Duane 1996). Much of this growth has occurred in the foothills of the Sierra Nevada. In particular, the central Sierra Nevada contains one of the largest areas of intermixed urban and wildlands in California. This creates changes in fire patterns and restricts restoration and fuels reduction activities. The relatively high productivity chaparral in the foothills means that the maintenance of fuel reduction areas needs to be more frequent and are more costly. There are two contrasting fire management conditions in the montane and eastern portions of the Sierra Nevada. One is where communities are adjacent to and mixed with wildlands, and the second is where vast areas are undeveloped, often bordering higher elevation wilderness. The former creates conditions where intensive and frequent fuels reduction treatments around communities are important. Property owners demand that fire suppression forces protect their homes first, thus diverting them from protecting resources. In contrast, more remote forests are well suited for managed wildfire, a program that restores naturally occurring fires in a less intensive and expensive means.

Fire and Fuels Management

Each new catastrophic fire increases the clamor to do something about fuels. Homeowners expect fire and land management agencies to act, yet are often unwilling to accept some of the responsibility themselves. The most immediate problem exists around developments and other areas of high societal values. Mechanical removal of understory trees followed by some sort of surface fuel treatment—prescribed burning is ideal from the standpoint of reducing fuels and restoring ecological process—is the most likely to succeed in these areas. Where houses have encroached into shrublands, removal of shrubs up to 30 m (100 ft) may be necessary. Less compelling are treatments in remote areas where there is less development and access is difficult. Prescribed burning and the use of naturally occurring fires are more appropriate in areas beyond the urban wildland interface.

The call to thin forests to prevent catastrophic fires has confused the issue. Only in rare occasions can a fire move independently through the crowns of trees without a surface fire to feed it. Thinning forests without treating surface fuels is likely to increase surface fuels and thus increase severity (Vaillant et al. 2009, McIver et al. 2013). A combination of treatments including thinning from below and prescribed fire will probably be most productive.

Climate Change

Under most climate change projections, fire is projected to increase in frequency, size, and severity (Westerling 2011). These fire regime attributes are already increasing in many places (Miller et al 2009, Miller and Safford 2012, chapter 26), and fire is likely to influence changes in forest cover types. Some site type changes from repeated high-severity fire are already occurring (Stephens et al. 2013). Long-term climate change coupled with decades of fire exclusion will worsen the problem. Young et al. (2017) found that forest mortality in California during the first four years of the drought that began in 2012 increased disproportionately in response to increases in climatic water deficit and stand basal area.

Although current efforts to reduce fuel loads and wildfire severity have been marginal at best, climate change is likely to exacerbate this situation (chapter 26). Greater use of prescribed fire and managed wildfire may be a more effective way to significantly increase the pace and scale of fuels reduction and mitigate the impacts of climate change. For example, van Mantgem et al. (2016) found that in Yosemite, Sequoia, and Kings Canyon national parks, common conifers in plots burned by prescribed fires or managed wildfires had significantly reduced drought-related mortality than in unburned plots.

Species at Risk

All species living in the Sierra Nevada evolved with fire and its effects on habitat vegetation and prey populations. Several at-risk animal species, including the Pacific fisher (*Martes pennanti pacifica*), American marten (*M. americana*), northern goshawk (*Accipiter gentilis*), and California spotted owl (*Strix occidentalis occidentalis*), are associated with habitat characterized by older, dense forest stands with high canopy cover. As a result of fire exclusion, these conditions often have high fuel loads that produce high-severity fire killing most of the large, overstory trees. A challenge for Sierra Nevada forest managers is to provide this habitat while reducing landscape-level fuels sufficiently to reduce wildfire severity (North et al. 2010). Current and projected future trends in climate are likely to accelerate forest change due to fire, especially the loss of older forest and the expansion of early seral habitats (McKenzie et al. 2004, Scheller et al. 2011, Safford and Stevens 2017).

Research has identified the amount, size, and location of dense, high-canopy cover habitat in historical and modern forests with active fire regimes. These conditions are often associated with wetter, more productive sites that burned at lower frequencies, as well as lower severities under moderate weather conditions, but the probability of high severity fire was increased in such stands under severe fire weather conditions (Lydersen and North 2012, Collins et al. 2015, Kane et al. 2015). One study of California spotted owl and Pacific fisher in the southern Sierra Nevada found significantly higher than expected use of these areas given their proportion of the landscape (Underwood et al. 2010). There is also evidence that although these species may rest, den, and nest in these conditions, they often forage in variable forest conditions that include less dense, open canopy stands (Irwin et al. 2007, Roberts et al. 2008, Truex and Zielinski 2013). The forest heterogeneity created by low- to moderate-severity burns is associated with greater evenness in small mammal communities, possibly providing more consistent prey base abundances for higher predators such as the California spotted owl and Pacific fisher (Roberts et al. 2015).

The question becomes how to restore natural fire regimes and forest structures without adversely affecting at risk species and their habitats. To do nothing threatens to make the already tenuous situation worse, predisposing the species and habitats to destruction by catastrophic fire. These species evolved with fire and the answer must include fire (Jones et al. 2016).

Air Quality

Air quality is one of the biggest impediments to conducting prescribed burns and managing wildfires in the Sierra Nevada. Lower to mid-elevation forests inevitably burn and produce smoke. When a wildfire escapes suppression under extreme weather conditions, it often produces heavy concentrations of smoke lasting weeks that winds may direct into rural communities or urban cities. With prescribed burning, local residents can be forewarned and the area burned when wind direction will disperse smoke away from population centers (Rappold et al. 2014). Although air quality regulations exempt wildfire emissions, prescribed and managed wildfires are considered anthropogenic and only permitted when their emissions will not exceed regulatory limits (Engel 2013). Current air regulations and acceptable pollutant concentrations are based on a concept of pristine air quality that does not consider pre-Euro-American settlement levels of burning (chapter 23). Another problem with current regulations is that burn projects are evaluated based on burn area rather than on smoke emissions and do not consider actual impacts to human health (Long et al. 2018).

Model projections suggest smoke emissions are likely to double by the end of the century, but a greatly expanded program of prescribed burning could mitigate much of that effect (Hurteau et al. 2014). Society is faced with deciding to accept periodic episodes of low concentrations of smoke from managed fires or heavy, long duration doses from wildfires. Either reduced emission restrictions for wildland management activities or exemptions for Federal agencies from air pollution control district regulations will be necessary if fire is to be allowed to again play its natural role in the Sierra Nevada.

Research Needs

Skinner and Chang (1996) developed a comprehensive list of research needs during the Sierra Nevada Ecosystem Project. They identified topics in three general areas: (1) spatial and temporal dynamics of fire, (2) presettlement forest conditions, and (3) effects of fire on ecosystem processes. Much has been accomplished on these topics and new issues have arisen. These include forest heterogeneity, smoke dynamics, and post-burn forest restoration.

Forest Heterogeneity

Fire exclusion tends to make forests more homogeneous as trees fill in forest openings and create high, uniform canopy cover. When these forests burn, particularly during high-intensity wildfire, they often perpetuate that homogeneity by creating large areas of dead trees and subsequently uniform areas of high brush cover. A challenge has been to reestablish heterogeneous forests conditions that were historically created by frequent fire and that may have been self-reinforcing. Although there is a general sense of how forest and fuel conditions may have varied with topography and water availability, specific management prescriptions are still vague for creating variable conditions that could be self-perpetuating under an active fire regime. Tree spatial patterns such as clumps and openings likely varied with topography, as did fuels loads in different size classes. Research is needed to identify general target levels for these conditions that would help managers set objectives at site, stand, and landscape scales.

Smoke Dynamics

Current models of the total amount, concentration, and dispersal of smoke produced by wildland fires are fairly general. Many unknowns still limit prediction accuracy such as the amount of fuel consumed, wind patterns at different elevations affecting dispersal, local diurnal weather patterns, and how these affect concentrations of different smoke pollutants such as particulate matter of and ozone. Furthermore, a network of established pollutant sensors is needed to measure where smoke actually accumulates. The concern with wildland fire smoke is human exposure and yet current assessments are based on crude estimates of total production rather than concentration, exposure, and duration.

Post-Burn Forest Restoration

Many wildfires escape suppression and burn with high severity during extreme weather events in fuels-loaded forests. These fires often produce large patches (>250 ha [>1,000 ac]) with near 100% tree mortality. Historically this was a rare occurrence and may now produce relatively novel conditions for vegetation succession, wildlife habitat, and microclimate. Research is needed on how forest regeneration, carbon cycling, and water quality and quantity respond to these conditions. In the past, high-severity areas have been replanted at high density (>80 trees ha^{-1} [>200 trees ac^{-1}]) often with regularly spaced pines. Many of these plantations have burned at high severity in subsequent wildfires. There is also concern that such densities may be too high given drying and warming climate conditions and that the regular spacing does not mimic forest patterns produced by frequent fire. The size and severity of modern wildfires raise many questions about how best to manage these post-burn landscapes and help set them on a trajectory toward recovery. How is the forest to fare if nothing is done in large areas of high severity—will it be able to recover in a reasonable time or will the vegetation likely shift to nonforest vegetation? If the latter is the case, then would planting with other treatments help to mitigate?

Summary

John Muir named the Sierra Nevada the Range of Light; a better name might have been the Range of Fire. Fires have been a part of the Sierra Nevada for millennia and will continue to be so in the future. In this chapter we surveyed the factors that make fire an important process in the ecological zones of the range and the interactions fire has with the vegetation in each zone. The success of our management of the Sierra Nevada is contingent upon our ability and willingness to maintain fire as an integral part of these ecosystems. To not do so is to consign ourselves to failure: fire in the Sierra Nevada is inevitable, and the only real questions are when and under what conditions an area will burn.

References

Abella, S.R., and J.D. Springer. 2015. Effects of tree cutting and fire on understory vegetation in mixed conifer forests. Forest Ecology and Management 335: 281–299.

Agee, J.K., B. Bahro, M.A. Finney, P.N. Omi, D.B. Sapsis, C.N. Skinner, J.W. van Wagtendonk, and C.P. Weatherspoon. 2000. The use of fuel breaks in landscape fire management. Forest Ecology and Management 127: 55–66.

Agee, J.K., and C.N. Skinner. 2005. Basic principles of forest fuel reduction treatments. Forest Ecology and Management 211: 83–96.

Airey Lauvaux, C., C.N. Skinner, and A.H. Taylor. 2016. High severity fire and mixed conifer forest-chaparral dynamics in the southern Cascade Range, USA. Forest Ecology and Management 363: 74–85.

Anderson, R.S. 1990. Holocene forest development and paleoclimates within central Sierra Nevada, California. Journal of Ecology 78: 470–489.

———. 1994. Paleohistory of a giant sequoia grove: the record from Log Meadow, Sequoia National Park. Pages 49–55 in: P.S. Aune, technical coordinator. Proceedings symposium on giant sequoias: their place in ecosystem and society. USDA Forest Service General Technical Report PSW-GTR-151. Pacific Southwest Research Station, Albany, California, USA.

Anderson, M.K. 1999. The fire, pruning, and coppice management of temperate ecosystems for basketry material by California Indian tribes. Human Ecology 27(1): 79–113.

Anderson, R.S., and S.L. Carpenter. 1991. Vegetation changes in Yosemite Valley, Yosemite National Park, California, during the protohistoric period. Madroño 38: 1–13.

Anderson, M.K., and J. Rosenthal. 2015. An ethnobiological approach to reconstructing indigenous fire regimes in the foothill chaparral of the western Sierra Nevada. Journal of Ethnobiology 35: 4–36.

Anderson, R.S., and S.J. Smith. 1997. The sedimentary record of fire in montane meadows, Sierra Nevada, California, USA: a preliminary assessment. Pages 313–327 in: J.S. Clark, H. Cachier, J.G. Goldammer, and B.J. Stocks, editors. Sediment Records of Biomass

Burning and Global Change. NATO ASI Series 51. Springer, Berlin, Germany.

Barbour, M.G., N.H. Berg, T.G.F. Kittel, and M.E. Kunz. 1991. Snowpack and the distribution of a major vegetation ecotone in the Sierra Nevada of California. Journal of Biogeography 18: 141–149.

Beaty, R.M., and A.H. Taylor. 2009. A 14,000 year sedimentary charcoal record of fire from the northern Sierra Nevada, Lake Tahoe Basin, California, USA. The Holocene 19: 347–358.

Bekker, M.F., and A.H. Taylor. 2001. Gradient analysis of fire regimes in montane forests of the southern Cascade Range, Thousand Lakes Wilderness, California, USA. Plant Ecology 155: 15–28.

Biswell, H.H. 1961. The big trees and fire. National Parks Magazine 35: 11–14.

Bock, C.E., and J.H. Bock. 1977. Patterns of post-fire succession on the Donner Ridge burn, Sierra Nevada. Pages 464–469 in: H.A. Mooney and C.E. Conrad, technical coordinators. Proceedings Symposium on the Environmental Consequences of Fire and Fuel Management in Mediterranean Ecosystems. USDA Forest Service General Technical Report WO-3. Pacific Southwest Research Station, Washington, D.C., USA.

Bock, J.H., and C.E. Bock. 1969. Natural reforestation in the northern Sierra Nevada-Donner Ridge burn. Proceedings Tall Timbers Fire Ecology Conference 9: 119–126.

Bock, J.H., C.E. Bock, and V.M. Hawthorne. 1976. Further studies of natural reforestation in the Donner Ridge burn. Proceedings Tall Timbers Fire Ecology Conference 14: 195–200.

Bock, J.H., M. Raphael, and C.E. Bock. 1978. A comparison of planting and natural succession after a forest fire in the northern Sierra Nevada. Journal of Applied Ecology 15: 597–602.

Brown, J.K., and N.V. DeByle. 1987. Fire damage, mortality, and suckering in aspen. Canadian Journal of Forest Research 17: 1100–1109.

Brunelle, A., and R.S. Anderson. 2003. Sedimentary charcoal as an indicator of late-Holocene drought in the Sierra Nevada, California, and its relevance to the future. The Holocene 13: 21–28.

Caprio, A.C. 2008. Reconstructing fire history of lodgepole pine on Chagoopa Plateau, Sequoia National Park, California. Pages 255–262 in: M.G. Narog, technical coordinator. Proceedings of the 2002 Fire Conference: Managing Fire and Fuels in the Remaining Wildlands and Open Spaces of the Southwestern United States. USDA Forest Service General Technical Report PSW-GTR-189. Pacific Southwest Research Station, Albany, California, USA.

Cocking, M.I., J.M. Varner, and E.E. Knapp. 2014. Long-term effects of fire severity on oak-conifer dynamics in the southern Cascades. Ecological Applications 24: 94–107.

Collins, B.M., R.G. Everett, and S.L. Stephens. 2011. Impacts of fire exclusion and managed fire on forest structure in an old growth Sierra Nevada mixed-conifer forest. Ecosphere 2(4): art51.

Collins, B.M., J.M. Lydersen, R.G. Everett, D.L. Fry, and S.L. Stephens. 2015. Novel characterization of landscape-level variability in historical vegetation structure. Ecological Applications 25: 1167–1174.

Collins, B.M., J.J. Moghaddas, and S.L. Stephens. 2007. Initial changes in forest structure and understory plant communities following fuel reduction activities in a Sierra Nevada mixed conifer forest. Forest Ecology and Management 239: 102–111.

Collins, B.M., and S.L. Stephens. 2007. Fire scarring patterns in Sierra Nevada wilderness areas burned by multiple wildland fire use fires. Fire Ecology 3(2): 53–67.

———. 2010. Stand-replacing patches within a 'mixed severity' fire regime: quantitative characterization using recent fires in a long-established natural fire area. Landscape Ecology 25: 927–939.

Conard, S.G., A.E. Jaramillo, K. Cromack, and S. Rose. 1985. The role of the genus *Ceanothus* in western forest ecosystems. USDA Forest Service General Technical Report PNW-GTR-182. Pacific Northwest Forest and Range Station, Portland, Oregon, USA.

Conard, S.G., and S.R. Radosevich. 1982. Post-fire succession in white fir (*Abies concolor*) vegetation of the northern Sierra Nevada. Madroño 29: 42–56.

Concilio, A., S. Ma, Q. Li, J. LeMoine, J. Chen, M.P. North, D. Moorhead, and R. Jensen. 2005. Soil respiration response to prescribed burning and thinning in mixed conifer and hardwood forests. Canadian Journal of Forest Research 35: 1581–1591.

Cotton, L., and J.R. McBride, 1987. Visual impacts of prescribed burning on mixed conifer and giant sequoia forests. Pages 32–37 in: J.B. Davis, and R.E. Martin, technical coordinators. Proceedings of the symposium on wildland fire 2000. USDA Forest Service General Technical Report PSW-101. Pacific Southwest Research Station, Albany, California, USA.

deBenedetti, S.H., and D.J. Parsons. 1979. Natural fire in subalpine meadows: a case description from the Sierra Nevada. Journal of Forestry 77: 477–479.

———. 1984. Post-fire succession in a Sierran subalpine meadow. American Midland Naturalist 111: 118–125.

DeBruin, H.W. 1974. From fire control to fire management: a major policy change in the Forest Service. Proceedings Tall Timbers Fire Ecology Conference 14: 11–17.

Dettinger, M.D., F.M. Ralph, T. Das, P.J. Neiman, and D.R. Cayan. 2011. Atmospheric rivers, floods and the water resources of California. Water 3: 445–478.

Dolanc, C.R., H.D. Safford, S.Z. Dobrowksi, and J.H. Thorne. 2014a. Twentieth century shifts in abundance and composition of vegetation types of the Sierra Nevada, CA, USA. Applied Vegetation Science 17: 442–455.

Dolanc, C.R., H.D. Safford, J.H. Thorne, and S.Z. Dobrowski. 2014b. Changing forest structure across the landscape of the Sierra Nevada, CA, USA, since the 1930s. Ecosphere 5(8): art101.

Duane, T.P. 1996. Human settlement, 1850-2040. Pages 235–360 in: D.C. Erman, general editor. Sierra Nevada Ecosystem Project: Final Report to Congress, Volume II. Wildland Resources Center Report 37. University of California, Davis, California, USA.

Dunbar-Irwin, M., and H.D. Safford. 2016. Climatic and structural comparison of yellow pine and mixed-conifer forests in northern Baja California (México) and the eastern Sierra Nevada (California, USA). Forest Ecology and Management 363: 252–266.

Engber, E.A., and J.M. Varner, III. 2012. Patterns of flammability of the California oaks: the role of leaf traits. Canadian Journal of Forest Research 42: 1965–1975.

Engel, K.H. 2013. Perverse incentives: the case of wildfire smoke regulation. Ecology Law Quarterly 40: 623–672.

Fischer, W.C., and A.F. Bradley. 1987. Fire ecology of western Montana forest habitat types. USDA Forest Service General Technical Report INT-GTR-223. Intermountain Forest and Range Experiment Station, Ogden, Utah, USA.

Fites-Kaufman, J. 1997. Historic landscape pattern and process: fire, vegetation, and environment interactions in the northern Sierra Nevada. Dissertation, University of Washington, Seattle, Washington, USA.

Fry, D.L., S.L. Stephens, B.M. Collins, M.P. North, E. Franco-Vizcaino, and S.J. Gill. 2014. Contrasting spatial patterns in active-fire and fire-suppressed Mediterranean climate old-growth, mixed conifer forests. PLOS ONE, doi:10.1371/journal.pone.0088985.

Gill, L., and H.A. Taylor. 2009. Top-down and bottom-up controls on fire regimes along an elevational gradient on the east slope of the Sierra Nevada, California, USA. Fire Ecology 5(3): 57–75.

Gruell, G.E. 2001. Fire in Sierra Nevada Forests: A Photographic Interpretation of Ecological Change since 1849. Mountain Press, Missoula, Montana, USA.

Haase, S.M., and S.S. Sackett. 1998. Effects of prescribed fire in giant sequoia-mixed conifer stands in Sequoia and Kings Canyon National Parks. Proceedings Tall Timbers Fire Ecology Conference 20: 236–243.

Habeck, J.R. 1985. Impact of fire suppression on forest succession and fuel accumulations in long-fire-interval wilderness habitat types. Pages 110–118 in: J.E. Lotan, B.M. Kilgore, W.C. Fischer, and R.W. Mutch, technical coordinators. Proceedings Symposium and Workshop on Wilderness Fire. USDA Forest Service General Technical Report INT-GTR-182. Intermountain Forest and Range Experiment Station, Ogden, Utah, USA.

Hahm, W.J., C.S. Riebe, C.E. Lukens, and S. Araki. 2014. Bedrock composition regulates mountain ecosystems and landscape evolution. PNAS 111: 3338–3343.

Hallett, D.J., and R.S. Anderson. 2010. Paleofire reconstruction for high-elevation forests in the Sierra Nevada, California, with

implications for wildfire synchrony and climate variability in the late Holocene. Quaternary Research 73: 180–190.
Hanson, C., and M. North. 2006. Post-fire epicormic branching in Sierra Nevada *Abies concolor* (white fir). International Journal of Wildland Fire 15: 31–35.
Harris, L., and A.H. Taylor. 2015. Topography, fuels, and fire exclusion drive fire severity of the Rim Fire in an old-growth mixed-conifer forest, Yosemite National Park, USA. Ecosystems 18: 1192–1208.
Hartesveldt, R.J. 1964. Fire ecology of the giant sequoias: controlled fire may be one solution to survival of the species. Natural History Magazine 73: 12–19.
Hartesveldt, R.J., and H.T. Harvey. 1967. The fire ecology of sequoia regeneration. Proceedings Tall Timbers Fire Ecology Conference 7: 65–77.
Harvey, H.T., H.S. Shellhammer, and R.E. Stecker. 1980. Giant sequoia ecology. National Park Service Science Monograph 12. Washington, D.C., USA.
Heisey, R.M., C.C. Delwiche, R.A. Virginia, A.F. Wrona, and B.A. Bryan. 1980. A new nitrogen-fixing non-legume: *Chamaebatia foliolosa* (Rosaceae). American Journal of Botany 67: 429–431.
Hill, M. 1975. Geology of the Sierra Nevada. University of California Press, Berkeley, California, USA.
Huber, N.K. 1987. The geologic story of Yosemite National Park. U.S. Geological Survey Bulletin 1595. Reston, Virginia, USA.
Hull, K.L., and M.J. Moratto. 1999. Archeological synthesis and research design, Yosemite National Park, California. Yosemite Research Center Publications in Anthropology 21. Yosemite National Park, El Portal, California, USA.
Hull, M.K., C.A. O'Dell, and M.K. Schroeder. 1966. Critical fire weather patterns—their frequency and levels of fire danger. USDA Forest Service Pacific Southwest Forest and Range Experiment Station, Berkeley, California, USA.
Hurteau, M.D., A.L. Westerling, C. Wiedinmyer, and B.P. Bryant. 2014. Projected effects of climate and development on California wildfire emissions through 2100. Environmental Science and Technology 48: 2298–2304.
Irwin, L.L., L.A. Clark, D.C. Rock, and S.L. Rock. 2007. Modeling foraging habitat of California spotted owls. Journal of Wildlife Management 71: 1183–1191.
Jones, G.M., R.J. Gutiérrez, D.J. Tempel, S.A. Whitmore, W.J. Berigan, and M.Z. Peery. 2016. Megafires: an emerging threat to old-forest species. Frontiers in Ecology and the Environment 14: 300–306.
Kane, V.R., J.A. Lutz., C.A. Cansler, N.A. Povak, D.J. Chruchill, D.F. Smith, J.T. Kane, and M.P. North. 2015. Water balance and topography predict fire and forest structure patterns. Forest Ecology and Management 338: 1–13.
Kane, V., M.P. North, J. Lutz, D. Churchill, S.L. Roberts, D.F. Smith, R.J. McGaughey, J. Kane, and M.L. Brooks. 2014. Assessing fire effects on forest spatial structure using a fusion of Landsat and airborne LiDAR data in Yosemite National Park. Remote Sensing of Environment 151: 89–101.
Kauffman, J.B., and R.E. Martin. 1990. Sprouting shrub response to different seasons and fuel consumption levels of prescribed fire in Sierra Nevada mixed conifer ecosystems. Forest Science 36: 748–764.
Keane, R.E., and S.F. Arno. 2001. Restoration concepts and techniques. Pages 367–400 in: D. Tomback, S.F. Arno, and R.E. Keane, editors. Whitebark Pine Communities: Ecology and Restoration. Island Press, Washington, D.C., USA.
Keeley, J.E., D. Lubin, and C.J. Fotheringham. 2003. Fire and grazing impacts on plant diversity and alien plant invasions in the southern Sierra Nevada. Ecological Applications 13: 1355–1374.
Keeley, J.E., A. Pfaff, and H.D. Safford. 2005. Fire suppression impacts on postfire recovery of Sierra Nevada chaparral shrublands. International Journal of Wildland Fire 14: 255–265.
Keeley, J.E., and P.H. Zedler. 1998. Evolution of life histories in Pinus. Pages 219–250 in: D.M. Richardson, editor. Ecology and Biogeography of *Pinus*. Cambridge University Press, Boston, Massachusetts, USA.
Keifer, M.B. 1991. Age structure and fire disturbance in southern Sierra Nevada subalpine forests. Thesis, University of Arizona, Tucson, Arizona, USA.
Kilgore, B.M. 1971. The role of fire in managing red fir forests. Transactions North American Wildlife and Natural Resources Conference 36: 405–416.
———. 1972. Fire's role in a sequoia forest. Naturalist 23: 26–37.
———. 1973. The ecological role of fire in Sierran conifer forests: its application to national park management. Quaternary Research 3: 496–513.
Kilgore, B.M., and H.H. Biswell. 1971. Seedling germination after prescribed fire. California Agriculture 25: 163–169.
Kilgore, B.M., and G.M. Briggs. 1972 Restoring fire to high elevation forests in California. Journal of Forestry 70: 266–271.
Kilgore, B.M., and D. Taylor 1979. Fire history of a sequoia mixed-conifer forest. Ecology 60: 129–142.
Knapp, E.E., C.N. Skinner, M.P. North, and B.L. Estes. 2013. Long-term overstory and understory change following logging and fire exclusion in a Sierra Nevada mixed-conifer forest. Forest Ecology and Management 310: 903–914.
Knapp, E.E., C.P. Weatherspoon, and C.N. Skinner. 2012. Shrub seed banks in mixed conifer forests of northern California and the role of fire in regulating abundance. Fire Ecology 8(1): 32–48.
Konrad, S.K., and D.H. Clark. 1998. Evidence for an early Neoglacial glacier advance from rock glaciers and lake sediments in the Sierra Nevada, California, USA. Arctic and Alpine Research 30: 272–284.
Krasnow, K.D. and S.L. Stephens. 2015. Evolving paradigms of aspen ecology and management: impacts of stand condition and fire severity on vegetation dynamics. Ecosphere 6(1): art12.
Lawrence, G.E. 1966. Ecology of vertebrate animals in relation to chaparral fire in the Sierra Nevada foothills. Ecology 47: 278–291.
Leiberg, J.B. 1902. Forest conditions in the northern Sierra Nevada, California. USGS Professional Paper 8. Washington, D.C., USA.
Lombardo, K.J., T.W. Swetnam, C.H. Baisan, and M.I. Borchert. 2009. Using bigcone Douglas-fir fire scars and tree rings to reconstruct interior chaparral fire history. Fire Ecology 5(3): 32–53.
Lorentzen, E. 2004. Aspen delineation project. Bureau of Land Management, California State Office Resource Note 72. Sacramento, California, USA.
Long, J.W., M.K. Anderson, L. Quinn-Davidson; R.W. Goode, F.K. Lake, and C.N. Skinner. 2016. Restoring California black oak ecosystems to promote tribal values and wildlife. USDA Forest Service General Technical Report PSW GTR-252. Pacific Southwest Research Station, Albany, California, USA.
Long, J.W., L.W. Tarnay, and M.P. North. 2018. Aligning smoke management with ecological and public health goals. Journal of Forestry 116: 76–86.
Lotan, J.E. 1975. Cone serotiny—fire relationships in lodgepole pine. Proceedings Tall Timbers Fire Ecology Conference 14: 267–278.
Lutz, J.A., J.W. van Wagtendonk, and J.F. Franklin. 2010. Climatic water deficit, tree species ranges, and climate change in Yosemite National Park, USA. Journal of Biogeography 37: 936–950.
Lutz, J.A., J.W. van Wagtendonk, A.E. Thode, J.D. Miller, and J.F. Franklin. 2009. Climate, lightning ignitions, and fire severity in Yosemite National Park, California, USA. International Journal of Wildland Fire 18: 765–774.
Lydersen, J., and M. North. 2012. Topographic variation in active-fire forest structure under current climate conditions. Ecosystems 15: 1134–1146.
Lydersen, J., M.P. North, and B. Collins. 2014. Severity of an uncharacteristically large wildfire, the Rim Fire, in forests with relatively restored frequent fire regimes. Forest Ecology and Management 328: 326–334.
Lydersen, J.M., M.P. North, E.E. Knapp, and B.M. Collins. 2013. Quantifying spatial patterns of tree groups and gaps in mixed-conifer forests: reference conditions and long-term changes following fire suppression and logging. Forest Ecology and Management 304: 370–382.
Ma, S., A. Concilio, B. Oakley, M.P. North, and J. Chen. 2010. Spatial variability in microclimate in a mixed-conifer forest before and after thinning and burning treatments. Forest Ecology and Management 259: 904–915.
Major, J. 1988. California climate in relation to elevation. Pages 11–74 in: M.G. Barbour and J. Major, editors. Terrestrial Vegetation of California. Wiley-Interscience, New York, New York, USA.

Mallek, C.R., H.D. Safford, J.H. Viers, and J.D. Miller. 2013. Modern departures in fire severity and area vary by forest type, Sierra Nevada and southern Cascades, California, USA. Ecosphere 4(12): art153.
Maloney, P., T. Smith, C. Jensen, J. Innes, D. Rizzo, and M. North. 2008. Initial tree mortality, and insect and pathogen response to fire and thinning restoration treatments in an old growth, mixed-conifer forest of the Sierra Nevada, California. Canadian Journal of Forest Research 38: 3011–3020.
Marra, J., and R. Edmonds. 2005. Soil arthropod responses to different patch types in a mixed conifer forest of the Sierra Nevada. Forest Science 51: 255–265.
McDonald, P.M. 1981. Adaptations of woody shrubs. Pages 21–29 in: S.D. Hobbs and O.T. Helgerson, editors. Reforestation of Skeletal Soils: Proceedings of a Workshop. Oregon State University, Forest Research Laboratory, Corvallis, Oregon, USA.
McIver, J.D., S.L. Stephens, J.K. Agee, J. Barbour, R.E.J. Boerner, C. Edminster, K.L. Erickson, K.L. Farris, C.J. Fettig, C. Fiedler, S.M. Haase, S. Hart, J. Keeley, E.E. Knapp, J. Lehmkuhl, J. Moghaddas, W. Otrosina, K.W. Outcalt, D. Schwilk, C.N. Skinner, T.A. Waldrop, C.P. Weatherspoon, D. Yaussy, A. Youngblood, and S. Zack. 2013. Ecological effects of alternative fuel reduction treatments: highlights of the U.S. Fire and Fire Surrogate study (FFS). International Journal of Wildland Fire 22: 63–82.
McKelvey, K.S., and K.L. Busse. 1996. Twentieth century fire patterns on Forest Service lands. Pages 1119–1138 in: D.C. Erman, general editor. Sierra Nevada Ecosystem Project: Final Report to Congress, Volume II. Wildland Resources Center Report 37. University of California, Davis, California, USA.
McKenzie, D., Z. Gedalof, D.L. Peterson, and P. Mote. 2004. Climatic change, wildfire, and conservation. Conservation Biology 18: 890–902.
Meyer, M. 2015. Forest fire severity patterns of resource objective wildfires in the southern Sierra Nevada. Journal of Forestry 113: 49–56.
Miles, S.R., and C.B. Goudy, compilers. 1997. Ecological subregions of California. USDA Forest Service Technical Publication R5-EM-TP-005. Pacific Southwest Region, San Francisco, California, USA.
Millar, C.I., and W.B. Woolfenden. 1999. Sierra Nevada forests: where did they come from? Where are they going? What does it mean? Transactions of the North American Wildlife and Natural Resources Conference 64: 206–236.
Miller, J.D., and H.D. Safford. 2012. Trends in wildfire severity 1984-2010 in the Sierra Nevada, Modoc Plateau and southern Cascades, California, USA. Fire Ecology 8(3): 41–57.
Miller, J.D., H.D. Safford, M. Crimmins, and A.E. Thode. 2009. Quantitative evidence for increasing forest fire severity in the Sierra Nevada and southern Cascade Mountains, California and Nevada, USA. Ecosystems 12: 16–32.
Minnich, R.A. 2007. Climate, paleoclimate, and paleovegetation. Pages 43–70 in: M.G. Barbour, T. Keeler-Wolf, and A.A. Schoenher, editors. Terrestrial Vegetation of California. University of California Press, Berkeley, California, USA.
Minnich, R.A., M.G. Barbour, J.H. Burk, and R.F. Fernau. 1995. Sixty years of change in California conifer forests of the San Bernardino Mountains. Conservation Biology 9: 902–914.
Minnich, R.A., and E. Franco. 1998. Land of chamise and pines: historical accounts and current status of northern Baja California's vegetation. University of California Publications in Botany 80: 1–166.
Moody, T.J., J. Fites-Kaufman, and S.L. Stephens. 2006. Fire history and climate influences from forests in the northern Sierra Nevada, USA. Fire Ecology 2(1): 115–141.
Mueggler, W.F., and J.P. Blaisdell. 1951. Replacing wyethia with desirable forage species. Journal of Range Management 4: 143–150.
Muir, J. 1878. The new sequoia forests. Harper's New Monthly Magazine 57: 813–827.
———. 1895. The Mountains of California. The Century Company, New York, New York, USA.
Nagel, T.A., and A.H. Taylor. 2005. Fire and persistence of montane chaparral in mixed conifer forest landscapes in the northern Sierra Nevada, Lake Tahoe Basin, California, USA. Journal of the Torrey Botanical Society 132: 442–457.
North, M.P., B.M. Collins, and S.L. Stephens. 2012. Using fire to increase the scale, benefits, and future maintenance of fuels treatments. Journal of Forestry 110: 392–401.
North, M.P., M. Hurteau, R. Fiegener, and M. Barbour. 2005. Influence of fire and El Niño on tree recruitment varies by species in Sierran mixed conifer. Forest Science 51: 187–197.
North, M.P., J. Innes, and H. Zald. 2007. Comparison of thinning and prescribed fire restoration treatments to Sierran mixed-conifer historic conditions. Canadian Journal of Forest Research 37: 331–342.
North, M.P., and P. Stine. 2012. Clarifying concepts. Pages 149–164 in: M.P. North, editor. Managing Sierra Nevada Forests. USDA Forest Service General Technical Report PSW-GTR-237. Pacific Southwest Research Station, Albany, California, USA.
North, M.P., P. Stine, W. Zielinski, K. O'Hara, and S.L. Stephens. 2010. Harnessing fire for wildlife. The Wildlife Professional 4: 30–33.
North, M.P., K.M. van de Water, S.L. Stephens, and B.M. Collins. 2009. Climate, rain shadow, and human-use influences on fire regimes in the eastern Sierra Nevada, California, USA. Fire Ecology 5(3): 17–31.
Oakley, B., M.P. North, and J. Franklin. 2006. Facilitative and competitive effects of a N-fixing shrub on white fir saplings. Forest Ecology and Management 233: 100–107.
Oechel, W.C., and C.D. Reid. 1983. Photosynthesis and biomass of chaparral shrubs along a fire induced age gradient in southern California. Bulletin de la Société Botanique de France. Actualités Botaniques 131: 399–409.
Parker, A.J. 1986. Persistence of lodgepole pine forests in the central Sierra Nevada. Ecology 67: 1560–1567.
Parsons, D.J. 1978. Fire and fuel accumulation in a giant sequoia forest. Journal of Forestry 76: 104–105.
———. 1981. The role of fire management in maintaining natural ecosystems. Pages 469–488 in: H.A. Mooney, T.M. Bonnicksen, N.L. Christensen, J.E. Lotan, and W.A. Reiners, technical coordinators. Proceedings Conference on Fire Regimes and Ecosystem Properties. USDA Forest Service General Technical Report WO-26. 594p. Pacific Southwest Research Station, Washington, D.C., USA.
———. 1994. Objects or ecosystems: giant sequoia management in national parks. Pages 109–115 in: P.S. Aune, technical coordinator. Proceedings Symposium on Giant Sequoias: Their Place in Ecosystem and Society. USDA Forest Service General Technical Report PSW-GTR-151. Pacific Southwest Research Station, Albany, California, USA.
Parsons, D.J., and S.H. deBenedetti. 1979. Impact of fire suppression on a mixed-conifer forest. Forest Ecology and Management 2: 21–33.
Paysen, T.E., and M.G. Narog. 1993. Tree mortality 6 years after burning a thinned *Quercus chrysolepis* stand. Canadian Journal of Forest Research 23: 2236–2241.
Piirto, D.D. 1994. Giant sequoia insect, disease, and ecosystem interactions. Pages 82–89 in: P.S. Aune, technical coordinator. Proceedings Symposium on Giant Sequoias: Their Place in Ecosystem and Society. USDA Forest Service General Technical Report PSW-GTR-151. Pacific Southwest Research Station, Albany, California, USA.
Pitcher, D.C. 1987. Fire history and age structure of red fir forests of Sequoia National Park, California. Canadian Journal of Forest Research 17: 582–587.
Plumb, T.R. 1980. Response of oaks to fire. Pages 202–215 in: T.R. Plumb, technical coordinator. Proceedings of Symposium on Ecological Management and Utilization of California Oaks. USDA Forest Service, PSW-GTR-44. Pacific Southwest Research Station, Albany, California, USA.
Potter, D.A. 1998. Forested communities of the upper montane in the central and southern Sierra Nevada. USDA Forest Service General Technical Report PSW-GTR-169. Pacific Southwest Research Station, Albany, California, USA.
———. 2006. Riparian plant community classification: west slope, central and southern Sierra Nevada, California. USDA Forest Service Technical Publication R5-TP-022. Pacific Southwest Region, Vallejo, California, USA.
Rambo, T., and M. North. 2009. Canopy microclimate response to pattern and density of thinning in a Sierra Nevada forest. Forest Ecology and Management 257: 435–442.
Rappold, A.G., N.L. Fann, J. Crooks, J. Huang, W.E. Cascio, R.B. Devlin, and D. Diaz-Sanchez. 2014. Forecast-based interventions can reduce the health and economic burden of wildfires. Environmental Science and Technology 48: 10571–10579.

Ritchie, M., C.N. Skinner, and T. Hamilton. 2007. Probability of tree survival after wildfire in an interior pine forest of northern California: effects of thinning and prescribed fire. Forest Ecology and Management 247: 200–208.
Rivera-Huerta, H., H.D. Safford, and J.D. Miller. 2015. Patterns and trends in burned area and fire severity from 1984 to 2010 in the Sierra de San Pedro Mártir, Baja California, Mexico. Fire Ecology 12: 52–72.
Roberts, S.L., D.A. Kelt, J.W. van Wagtendonk, A.K. Miles, and M.D. Meyer. 2015. Effects of fire on small mammal communities in frequent-fire forests in California. Journal of Mammalogy 96: 107–119.
Roberts, S.L., J.W. van Wagtendonk, A.K. Miles, D.A. Kelt, and J.A. Lutz. 2008. Modeling the effects of fire severity and spatial complexity on small mammals in Yosemite National Park, California. Fire Ecology 4(2): 83–104.
Rundel, P.W., G.A. Baker, and D.J. Parsons. 1981. Productivity and nutritional response of *Chamaebatia foliolosa* (*Rosaceae*) to seasonal burning. Pages 191–196 in: N.S. Margaris and H.A. Mooney, editors. Components of Productivity of Mediterranean-Climate Regions. Dr. W. Junk, The Hague, The Netherlands.
Russell, W.H., J.R. McBride, and R. Rowntree. 1998. Revegetation after four stand-replacing fires in the Lake Tahoe Basin. Madroño 45: 40–46.
Safford, H.D., and S.P. Harrison. 2004. Fire effects on plant diversity in serpentine versus sandstone chaparral. Ecology 85: 539–548.
Safford, H.D., and J.T. Stevens. 2017. Natural Range of Variation (NRV) for yellow pine and mixed conifer forests in the Sierra Nevada, southern Cascades, and Modoc and Inyo National Forests, California, USA. USDA Forest Service General Technical Report PSW-GTR-256. Pacific Southwest Research Station, Albany, California, USA.
Safford, H.D., and K.M. Van de Water. 2014. Using Fire Return Interval Departure (FRID) analysis to map spatial and temporal changes in fire frequency on National Forest lands in California. USDA Forest Service Research Paper PSW-RP-266. Pacific Southwest Research Station, Albany, California, USA.
Scheller, R.M., W.D. Spencer, H. Rustigian-Romsos, A.D. Syphard, B.C. Ward, and J.R. Strittholt. 2011. Using stochastic simulation to evaluate competing risks of wildfires and fuels management on an isolated forest carnivore. Landscape Ecology 26: 1491–1504.
Scholl, A.E., and A.H. Taylor. 2006. Regeneration patterns in old-growth red fir-western white pine forests in the northern Sierra Nevada, Lake Tahoe, USA. Forest Ecology and Management 235: 143–154.
———. 2010. Fire regimes, forest change, and self-organization in an old-growth mixed-conifer forest, Yosemite National Park, USA. Ecological Applications 20: 362–380.
Schweickert, R.A. 1981. Tectonic evolution of the Sierra Nevada range. Pages 87–131 in: W.G. Ernst, editor. The Geotectonic Development of California, Rubey Vol. 1. Prentice-Hall, Englewood Cliffs, New Jersey, USA.
Show, S.B., and E.I. Kotok. 1923. Forest fires in California, 1911–1920. U.S. Department of Agriculture Circular 243. Washington, D.C., USA.
———. 1924. The role of fire in the California pine forests. U.S. Department of Agriculture Bulletin 1294. Washington, D.C., USA.
Skinner, C.N. 2002. Influence of fire on dead woody material in forests of California and southwestern Oregon. Pages 445–454 in: W.F. Laudenslayer, P.J. Shea, B.E. Valentine, C.P. Weatherspoon, and T.E. Lisle, technical coordinators. Proceedings of the Symposium on the Ecology and Management of Dead Wood in Western Forests. USDA Forest Service General Technical Report PSW-GTR-181. Pacific Southwest Research Station, Berkeley, California, USA.
Skinner, C.N. and C. Chang. 1996. Fire regimes, past and present. Pages 1041–1069 in: D.C. Erman, general editor. Sierra Nevada Ecosystem Project: Final Report to Congress, Volume II. Wildland Resources Center Report 37. University of California, Davis, California, USA.
Smith, S.J., and R.S. Anderson. 1992. Late Wisconsin paleoecologic record from Swamp Lake, Yosemite National Park, California. Quaternary Research 38: 91–102.
St. John, T.V., and P.W. Rundel. 1976. The role of fire as a mineralizing agent in a Sierran coniferous forest. Oecologia. 25: 35–45.
Steel, Z.L., H.D. Safford, and J.H. Viers. 2015. The fire frequency-severity relationship and the legacy of fire suppression in California forests. Ecosphere 6(1): art8.
Stephens, S.L. 2001. Fire history of adjacent Jeffrey pine and upper montane forests in the eastern Sierra Nevada. International Journal of Wildland Fire 10: 161–167.
Stephens, S.L., J.K. Agee, P.Z. Fulé, M.P. North, W.H. Romme, T.W. Swetnam, and M.G. Turner. 2013. Managing forests and fire in changing climates. Science 342: 41–42.
Stephens, S.L., and B.M. Collins. 2004. Fire regimes of mixed conifer forests in the north-central Sierra Nevada at multiple spatial scales. Northwest Science 78: 12–23.
Stephens, S.L., B.M. Collins, C.J. Fettig, M.A. Finney, C. Hoffman, E.E. Knapp, M.P. North, H. Safford, and R. Wayman. 2018. Drought, tree mortality, and wildfire in forests adapted to frequent fire. BioScience.
Stephens S.L., D.L. Fry, and E. Franco-Vizcano. 2008. Wildfire and forests in Northwestern Mexico: the United States wishes it had similar fire problems. Ecology and Society 13(2): 10.
Stephens, S.L., J.M. Lydersen, B.M. Collins, D.L. Fry, and M.D. Meyer. 2015. Historical and current landscape-scale ponderosa pine and mixed conifer forest structure in the Southern Sierra Nevada. Ecosphere 6(5): art79.
Stephens, S.L., and J.J. Moghaddas. 2005. Experimental fuel treatment impacts on forest structure, potential fire behavior, and predicted tree mortality in a California mixed conifer forest. Forest Ecology and Management 215: 21–36.
Stephenson, N.L. 1998. Actual evapotranspiration and deficit: biologically meaningful correlates of vegetation distribution across spatial scales. Journal of Biogeography 25: 855–870.
———. 1999. Reference conditions for giant sequoia forest restoration: structure, process and precision. Ecological Applications 9: 1253–1265.
Stephenson, N.L., D.J. Parsons, and T.W. Swetnam. 1991. Restoring natural fire to the sequoia-mixed conifer forest: should intense fire play a role? Proceedings Tall Timbers Fire Ecology Conference 17: 321–337.
Stevens, J. T., B. M. Collins, J. D. Miller, M. P. North, and S. L. Stephens. 2017. Changing spatial patterns of stand-replacing fire in California conifer forests. Forest Ecology and Management 406: 28–36.
Stevens, J.T., H.D. Safford, S.P. Harrison, and A.M. Latimer. 2015. Disturbance regimes predict diversity and composition of forest understory plant communities. Journal of Ecology 103: 1253–1263.
Stine, S. 1974. Extreme and persistent drought in California and Patagonia during mediaeval time. Nature 369: 546–549.
Sudworth, G.B. 1900. Stanislaus and Lake Tahoe Forest Reserves, California, and adjacent territory. Annual Report of the U.S. Geological Survey, Part 5: 505-61. Washington, D.C., USA.
Swetnam, T.W. 1993. Fire history and climate change in giant sequoia groves. Science 262: 885–889.
Taylor, A.H. 2000. Fire regimes and forest changes in mid and upper montane forests of the southern Cascades, Lassen Volcanic National Park, California, U.S.A. Journal of Biogeography 27: 87–104.
———. 2004. Identifying forest reference conditions on early cut-over lands, Lake Tahoe basin, USA. Ecological Applications 14: 1903–1920.
Taylor, A.H. and C.N. Skinner. 2003. Spatial and temporal patterns of historic fire regimes and forest structure as a reference for restoration of fire in the Klamath Mountains. Ecological Applications 13: 704–719.
Taylor, A.H., A.M. Vandervlugt, R.S. Maxwell, R.M. Beaty, C. Airey, and C.N. Skinner 2013. Changes in forest structure, fuels and potential fire behaviour since 1873 in the Lake Tahoe Basin, USA. Applied Vegetation Science 17: 17–31.
Thode, A.E., J.W. van Wagtendonk, J.D. Miller, and J.F. Quinn. 2011. Quantifying the fire regime distributions for fire severity in Yosemite National Park, California, USA. International Journal of Wildland Fire 20: 223–239.
Thorne, J.H., B.J. Morgan, and J.A. Kennedy. 2008. Vegetation change over sixty years in the central Sierra Nevada, California, USA. Madroño 55: 223–237.
Tomback, D.F. 1986. Post-fire regeneration of krummholz whitebark pine: a consequence of nutcracker seed caching. Madroño 33: 100–110.

Truex, R.L. and W.J. Zielinski. 2013. Short-term effects of fuel treatments on fisher habitat in the Sierra Nevada, California. Forest Ecology and Management 293: 85–91.
Tyler, C.M., and M.I. Borchert. 2007. Chaparral geophytes: fire and flowers. Fremontia 35(4): 22–24.
Underwood, E.C., J.H. Viers, J.F. Quinn, and M. North. 2010. Using topography to meet wildlife and fuels treatment objectives in fire-suppressed landscapes. Journal of Environmental Management 46: 809–819.
Vaillant, N.M., J. Fites-Kaufman, A.L. Reiner, E.K. Noonan-Wright, and S.N. Dailey. 2009. Effect of fuel treatments on fuels and potential fire behavior in California, USA, national forests. Fire Ecology, 5(2): 14–29.
Vaillant, N.M., J.A. Fites-Kaufman, and S.L. Stephens. 2009. Effectiveness of prescribed fire as a fuel treatment in Californian coniferous forests. International Journal of Wildland Fire 18: 165–175.
Vaillant, N.M., and S.L. Stephens. 2009. Fire history of a lower elevation Jeffrey pine-mixed conifer forest in the eastern Sierra Nevada, California, USA. Fire Ecology 5(3): 4–19.
Vale, T.R. 1987. Vegetation change and park purposes in the high elevations of Yosemite National Park. Annals of the Association of American Geographers 77: 1–18.
———. 2002. Fire, Native Peoples, and the Natural Landscape. Island Press, Washington, D.C., USA.
Van de Water, K., and M. North. 2010. Fire history of coniferous riparian forests in the Sierra Nevada. Forest Ecology and Management 260: 384–395.
———. 2011. Stand structure, fuel loads, and fire behavior in riparian and upland forests, Sierra Nevada Mountains, USA; a comparison of current and reconstructed conditions. Forest Ecology and Management 262: 215–228.
Van de Water, K.M., and H.D. Safford. 2011. A summary of fire frequency estimates for California vegetation before Euro-American settlement. Fire Ecology 7(3): 26–58.
Vankat, J.L., and J. Major. 1978. Vegetation changes in Sequoia National Park, California. Journal of Biogeography 5: 377–402.
van Mantgem, P.J., A.C. Caprio, N.L. Stephenson, and A.J. Das. 2016. Does prescribed fire promote resistance to drought in low elevation forests of the Sierra Nevada, California, USA? Fire Ecology 12(1): 13–25.
van Wagtendonk, J.W. 1985. Fire suppression effects on fuels and succession in short-fire-return interval wilderness ecosystems. Pages 119–126 in: J.E. Lotan, B.M. Kilgore, W.C. Fischer, and R.W. Mutch, technical coordinators. Proceedings Symposium and Workshop on Wilderness Fire. USDA Forest Service General Technical Report INT-GTR-182. Intermountain Forest and Range Experiment Station, Ogden Utah, USA.
———. 1991a. Spatial analysis of lightning strikes in Yosemite National Park. Proceedings Conference on Fire and Forest Meteorology 11: 605–611.
———. 1991b. The evolution of National Park Service fire policy. Fire Management Notes 52(4): 10–15.
———. 1994. Spatial patterns of lightning strikes and fires in Yosemite National Park. Proceedings Conference on Fire and Forest Meteorology 12: 223–231.
———. 1995. Large fires in wilderness areas. Pages 113–116 in: J.K. Brown, R.W. Mutch, C.W. Spoon, and R.H. Wakimoto, technical coordinators. Proceedings Symposium on Fire in Wilderness and Park Management. USDA Forest Service General Technical Report INT-GTR 320. Intermountain Forest and Range Experiment Station, Ogden, Utah, USA.
———. 2007. The history and evolution of wildland fire use. Fire Ecology 3(2): 3–17.
van Wagtendonk, J.W., J.M. Benedict, and W.M. Sydoriak. 1998a. Fuel bed characteristics of Sierra Nevada conifers. Western Journal of Applied Forestry 13: 73–84.
van Wagtendonk, J.W., and D.R. Cayan. 2008. Temporal and spatial distribution of lightning strikes in California in relationship to large-scale weather patterns. Fire Ecology 8(4): 34–56.
van Wagtendonk, J.W., and J.A. Lutz. 2007. Fire regime attributes of wildland fires in Yosemite National Park. Fire Ecology 3(2): 34–52.
van Wagtendonk, J.W., and P.E. Moore. 2010. Fuel deposition rates of montane and subalpine conifers in the central Sierra Nevada, California, USA. Forest Ecology and Management 259: 2122–2132.
van Wagtendonk, K.A., and D.F. Smith. 2016. Prioritizing lightning ignitions in Yosemite National Park with a biogeophysical and sociopolitically informed decision tool. Pages 129–135 in: S. Weber, editor. Engagement, Education, and Expectations—The Future of Parks and Protected Areas: Proceedings of the 2015 George Wright Society Conference on Parks, Protected Areas, and Cultural Sites. George Wright Society, Hancock, Michigan, USA.
van Wagtendonk, J.W., W.M. Sydoriak, and J.M. Benedict. 1998b. Heat content variation of Sierra Nevada conifers. International Journal of Wildland Fire 8: 147–158.
van Wagtendonk, J.W., K.A. van Wagtendonk, J.B. Meyer, and K.J. Paintner. 2002. The use of geographic information for fire management planning in Yosemite National Park. The George Wright Forum 19(1): 19–39.
Wagener, W.W. 1961a. Past fire incidence in Sierra Nevada forests. Journal of Forestry 59: 739–748.
———. 1961b. Guidelines for estimating the survival of fire-damaged trees in California. USDA Forest Service Miscellaneous Paper 60. Pacific Southwest Forest and range Experiment Station, Berkeley, California, USA.
Wayman, R.B., and M. North. 2007. Initial response of a mixed-conifer understory plant community to burning and thinning restoration treatments. Forest Ecology and Management 239: 32–44.
Weatherspoon, C.P. 1986. Silvics of giant sequoia. Pages 4–10 in: C.P. Weatherspooon, Y.R. Iwamoto, and D. Piirto, technical coordinators. Proceedings of the Workshop on Management of Giant Sequoia, Reedly, California. USDA Forest Service General Technical Report PSW-GTR-9. Pacific Southwest Research Station, Albany, California, USA.
Westerling, A., B. Bryant, H. Preisler, T. Holmes, H. Hidalgo, T. Das, and S. Shrestha. 2011. Climate change and growth scenarios for California wildfire. Climatic Change 109: 445–463.
Woolfenden, W.B. 1996. Quaternary vegetation history. Status of the Sierra Nevada. Pages 47–70 in: D.C. Erman, general editor. Sierra Nevada Ecosystem Project: Final Report to Congress, Volume II. Wildland Resources Center Report 37. University of California, Davis, California, USA.
Young, J.A., and R.A. Evans. 1978. Population dynamics after wildfires in sagebrush grasslands. Journal of Range Management 31: 283–289.
Young, D.J.N., J.T. Stevens, J.M. Earles, J. Moore, A. Ellis, A.L. Jirka, and A.M. Latimer. 2017. Long-term climate and competition explain forest mortality patterns under extreme drought. Ecology Letters 20: 78–86.
Zedler, P.H. 1995. Fire frequency in southern California shrublands: biological effects and management options. Pages 101–112 in: J.E. Keeley and T. Scott, editors. Brushfires in California Wildlands: Ecology and Resource Management. International Association of Wildland Fire, Fairfield, Washington, USA.

CHAPTER FIFTEEN

Central Valley Bioregion

ROBIN WILLS

When I first saw this central garden, the most extensive and regular of all the bee-pastures of the state, it seemed all one sheet of plant gold, hazy and vanishing in the distance, distinct as a new map along the foothills at my feet.

JOHN MUIR

Description of Bioregion

The Central Valley creates one of the most important defining physical features of the California landscape. Lying between the Sierra Nevada and the Coast Ranges, this massive valley is nearly 800 km (500 mi) long and up to 120 km (75 mi) across (Map 15.1). Over 15% of the state's total area is included within its boundaries. Although still one of the most spectacular elements of California's geography, this bioregion is among the state's most highly impacted. A long history of significant alteration has resulted in a landscape with little resemblance to its pre-Euro-American settlement vegetation (Huenneke 1989, Williams 2002). The Great Valley has witnessed the most significant extent of land conversion to agriculture of any single California bioregion, affecting 82% of its total land area (USDA NRCS 2012).

The valley once supported a diverse array of prairies, oak (*Quercus* spp.) savannas, semiarid grasslands, freshwater marshes, and riparian woodlands. From its earliest periods of human occupation until today, the Central Valley has supported large and diverse populations of people, plants, and animals. These assemblages thrived in the context of complex vegetation patterns and dynamic ecological processes.

Physical Geography

The Central Valley is most appropriately described as two valleys lying end to end. Each is drained by a major river, which provides the valleys with their respective names. The Sacramento Valley lies north of San Francisco Bay and is drained by the Sacramento River. South of the Bay lies the San Joaquin Valley and its associated San Joaquin River. These two rivers find their confluence at the Sacramento-San Joaquin Delta. The bay-delta formed a massive 1,900 km^2 (734 mi^2) wetland that, today, is one of California's most important agricultural centers. Complex irrigation projects, moving water throughout the valley, have converted the arid southern San Joaquin into one of the world's most productive agricultural sites.

Much of the Central Valley occurs at an elevation approximating sea level. At both its northern and southern ends, elevations rise to near 120 m (400 ft). In the delta, elevations below sea level are recorded. Consequently, river systems in this portion of the Valley are still tidally influenced.

Defining the bioregion's boundary is complex. Ecotones between valley and foothill vegetation types occur in an irregular pattern. Elevation indicators are also confusing. Vegetation types are associated with a series of terraces, which vary on the east and western sides of the Valley (Fig. 15.1). One useful boundary description defines the Central Valley where alluvial soils grade into bedrock features and the landscape becomes dominated by foothill woodland (Schoenherr 1992).

The character of the Valley has been changed most significantly through alteration in the pattern, frequency, and magnitude of flood events. As a result of intensive land-use, fundamental changes have occurred in the valley's hydrology. Flooding was possibly the most important natural process affecting the valley's vegetation. Today, flood control structures dominate nearly every river draining the Sierra Nevada. Structures control the extent, magnitude, and intensity of flood events and in most cases have confined flow to a narrow portion of the original flood plain. The volume of water available to valley ecosystems has also been greatly reduced. For example, flows through the Tuolumne River below the dams during an average water year are from 10% to 20% of what the flows would be without the flood control structures (McBain and Trush 2000). Three massive lakes formally occurred in the southern San Joaquin Valley. Tule, Buena Vista, and Kern lakes have all been drained. The shores of these lakes once supported expansive freshwater marsh systems, providing critical habitat to an impressive array of resident and migratory species (RHJV 2004). These lake shores also concentrated a large portion of the valley's early human occupants (Anderson 2005).

Geologists describe the Central Valley as a large trough of mud. Indeed, the basin has been a repository of sediments during the last 145 million years. Deep sedimentary soils characterize this landscape, with a seemly endless input of alluvial material from the surrounding mountains. Sand and gravel deposits over 9,000 m (30,000 ft) deep cover parent material, which angles westward from the Sierra Nevada. Following that alignment, the deepest sediment layers occur

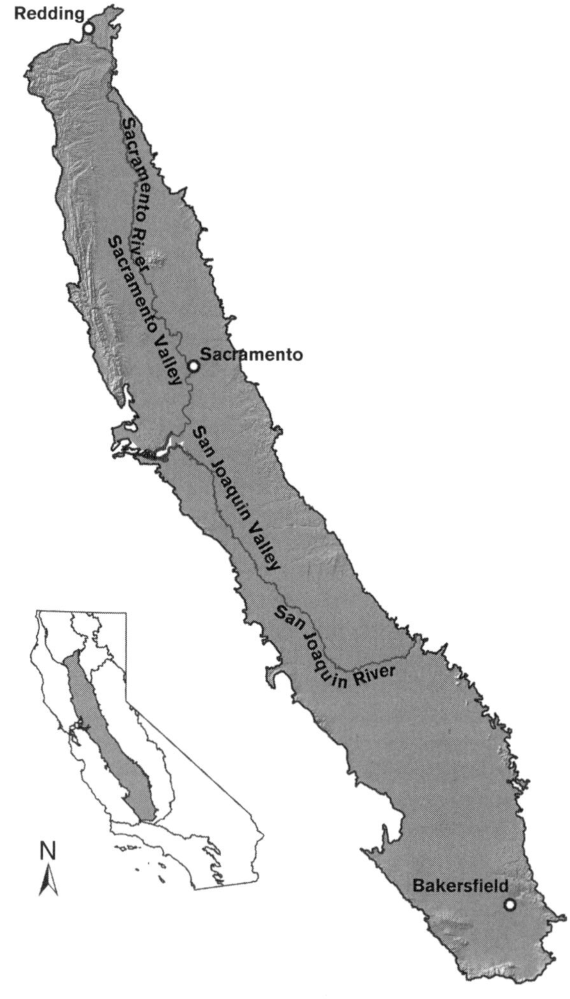

MAP 15.1 The Central Valley bioregion covers over 15% of the area in California and includes both the Sacramento and San Joaquin river valleys.

near the eastern edge of the coast range. Here too, a thrust fault zone parallels the western edge of the Valley. The underlying basement rocks are a continuation of the landform that makes up the Sierra Nevada. The long, slow processes that resulted in the Valley's formation have left much of the underlying rock materials relatively unbroken.

Climatic Patterns

Climate in the Central Valley varies in a north to south gradient. Temperature and rainfall patterns promote herbaceous vegetation types throughout the valley landscape, with moderately wet winters and hot dry summers. Precipitation is generally limited to winter months, although summer rainfall can occur in the northern Sacramento Valley. The north Sacramento Valley receives a moderate amount of annual precipitation 9.8–15.4 cm yr^{-1} (25–39 in yr^{-1}), whereas the San Joaquin Valley can experience rainfall levels similar to desert climates 2.4–3.1 cm yr^{-1} (6–8 in yr^{-1}).

Relative humidity can also vary tremendously across the Sacramento and San Joaquin valleys. During the warm season, humidity is characteristically low. Under the influence of the north winds, humidity readings may drop to below 10%. Surrounding the Bay-Delta, a strong inflow of marine air during the summer creates a transition zone between the high humidity of the coast and the low readings of the valley. Winter humidities are usually moderate to high across the Central Valley.

The weather in the Central Valley is influenced by its location relative to the surrounding bioregions (Map 15.2). Cold-air draining from the surrounding mountains collects in the deep basin of the Central Valley. As temperatures drop at higher elevations, downslope flows pour cold air into the valley, filling it to the crest of the coast range. This dense pool of cold air can persist for days. The resulting temperature inversion produces dense, ground hugging fog, known locally as tule fog, during the winter. In summer and early fall, inversion layers often burn off quickly, and photochemical smog can become a significant problem.

FIRE CLIMATE VARIABLES

The topography of the Central Valley strongly influences important fire climate variables. The Coast Range and Sierra Nevada provide a boundary of high mountains surrounding the valley. This limits offshore air flow and the development of föehn style winds (Mitchell 1969). Prevailing summer winds are southwestern and tend to increase in speed during daylight hours. Synoptic-scale winds do occur and become more common in the fall (Schroeder and Buck 1970). These winds, resulting from the combination of a Great Basin high-pressure cell and Pacific Coast low-pressure trough, come from the north and may be of considerable strength.

Lightning occurs at an extremely low density in the valley with only 3.9 strikes yr^{-1} 100^{-1} km^{-2} (35.8 strikes yr^{-1} 100^{-1} mi^{-2}) (van Wagtendonk and Cayan 2008). Thunderstorms may occur in Central Valley at any time of the year and there appears to be no definite season. When they do occur, these storms are generally light with few positive strikes. Although it is true that lightning strikes on the valley floor may have contributed little to presettlement fire regimes, numerous fires from the surrounding mountains may have eventually spread downslope into these lower elevation plant communities. Storms generating lighting in surrounding bioregions are generally limited to late summer months.

Weather Systems

Central Valley storms develop as areas of low pressure in the north Pacific. Prevailing northwesterly winds track these storm cells down the California coast. The position of the Pacific high will determine how far south the storms will travel. Spinning counterclockwise, these systems continue to pull moisture from the ocean while dumping rainfall on the land. A gradient of increasing land temperatures and air pressure results in significantly more rainfall occurring in northern than southern California. As summer approaches and strong high-pressure systems develop over southern California, storms are deflected toward the Pacific Northwest.

Ecological Zones

The Central Valley has distinctly arranged ecological zones. Variations in topography, soils, water availability, and climate

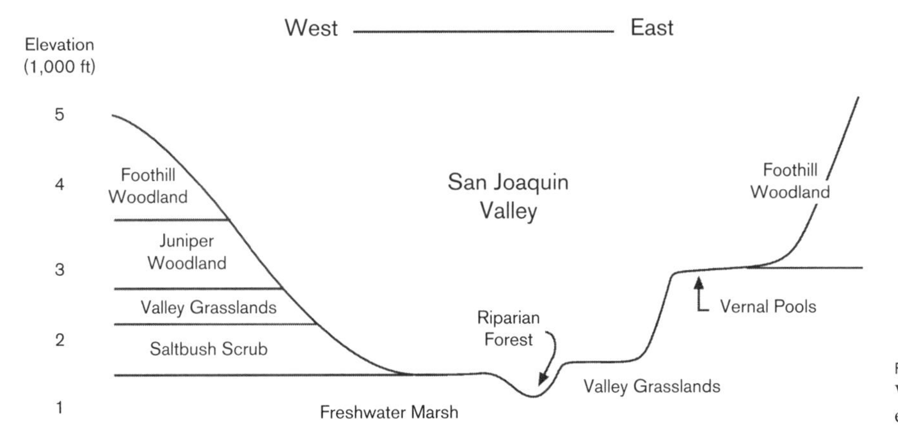

FIGURE 15.1 Vegetation in the Central Valley is neatly organized along elevation and directional gradients.

MAP 15.2 The relationship of the Central Valley to surrounding bioregions has strongly affected weather and possibly the pattern of fire occurrence (map credit Zeke Lunder).

define the pattern of plant occurrence. With numerous mountain draining river systems entering the Valley from both the Sierra Nevada and the Coast Range, vegetation is often partitioned by riparian communities. Although vegetation has certainly changed in response to human disturbance and annual variations in precipitation, it seems reasonable to assume that no major shifts in distribution of plant formations occurred in the Central Valley during the last 6,000 years up to the arrival of Euro-Americans (Huntley and Webb 1988).

An analysis of prehistory evidence indicates a valley dominated by vegetation significantly different than that seen by early Euro-Americans (Stromberg and Griffin 1996). Present-day grasslands are now largely devoid of native perennial species and are instead dominated by nonnative grasses. This pattern exists regardless of some aspects of past land-use history such as grazing.

Humans have lived in California grasslands since the Late Pleistocene Period (Anderson 2005). Native Americans undoubtedly burned grasslands to maintain certain types of plant and animal species, but the extent, periodicity, and full ecological implications of these activities are largely unknown (Erlandson 1994, Keeley 2002, Anderson 2005).

It is likely that much of contemporary California grasslands have been created from other types of vegetation (Huenneke 1989). The current grasslands in the Tulare Lake basin were likely saltbush (*Atriplex* spp.) scrub, oak woodlands, and chaparral, whereas the original grasslands are now converted to intensive agriculture (Adams 1985, Davis 1990). Type conversions from shrublands and other vegetation increased the extent of upland soils under grassland (McBride and Heady 1968, Minnich and Dezzani 1998, Cione et al. 2002). Recent studies suggest that grasses may have been dominant only in coastal grasslands and along riparian corridors (Evett and Bartolome 2013) and that grasses were present in the other grasslands but the dominant species may have been native annual forbs (Schiffman 2007, Minnich 2008).

Still the associated faunal record for the same period is dominated by large grazing herbivores like bison, horses, mammoth, and ground sloth. The presence of these species has been interpreted as evidence for large grasslands dominating this portion of the valley (Fenenga 1991). Fenenga (1994) has suggested some portion of both these models may hold true. The Tulare Lake Basin was likely a mosaic of plant communities, including shrublands with a significant cover of perennial grasses.

The pattern of vegetation does differ between the two valleys. Throughout its history, the Sacramento Valley has maintained some connection to coastal influences via the San Francisco Bay-Delta. The smaller spatial scale, higher latitude, and moderating marine influence in the Sacramento Valley have resulted in more plant communities dominated by woody species. Grassland remnants also support the hypothesis that perennial grasses may have been more prevalent in the Sacramento Valley (Barry 1972).

With its large expanse and strong continental influence, the San Joaquin Valley represents a more xeric landscape. Grasslands in this portion of the valley were likely to have been dominated by annual vegetation and to have exhibited less structural diversity. Saltbush scrub communities were once common in the southern San Joaquin. Juniper woodlands were also a significant component of the western valley foothills (Bauer 1930).

Although tremendous amounts of diversity can be described within these generalized vegetation communities, four primary ecological zones dominated the presettlement Central Valley. Freshwater marsh, valley grasslands, foothill woodlands, and riparian forests were arranged in a mosaic most strongly influenced by moisture and elevation gradients. An array of grassland and vernal pool assemblages was thought to be the most abundant valley vegetation, covering more than 9 million ha (22 million ac). Freshwater marshes, fed by a combination of winter precipitation and spring runoff, may have occupied nearly 1.6 million ha (4 million ac). Long linear riparian woodlands patterned the length of the Valley, covering more than 400,000 ha (1 million ac) in the early 1800s (Katibah 1984). In all cases, these once expansive vegetation types now exist in less than 10% of their former range (Smith 1977).

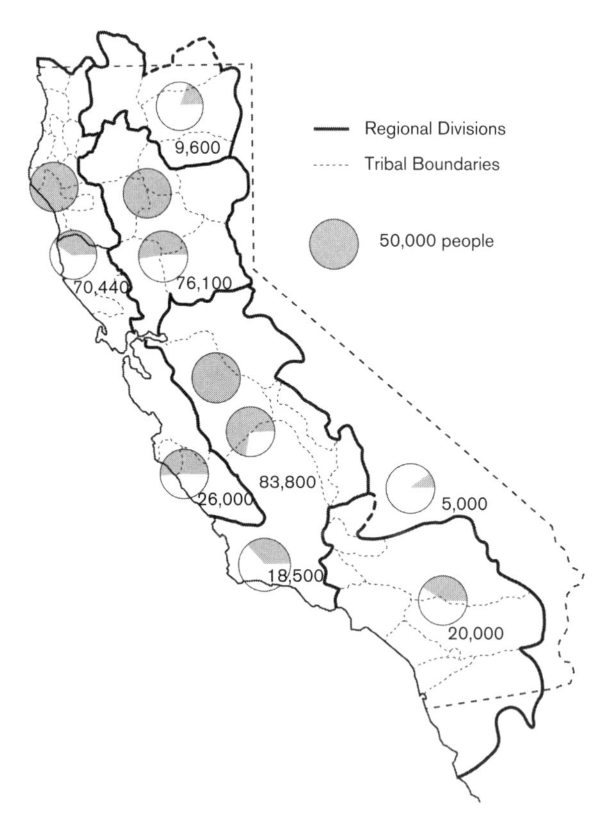

MAP 15.3 The Central Valley supported some of the highest known densities of Native Americans west of the Mississippi River (map credit Cook 1978).

Overview of Historic Fire occurrence

PREHISTORIC

Unlike other bioregions in California, little is known of the pre-Euro-American settlement pattern of fire within the Central Valley. Without long-lived trees, fire occurrence for the several hundred years prior to European contact was not captured in the tree-ring record. Other methods, like sediment core analysis, for documenting the abundance of fire in a given landscape are complicated by the significant input of sediments from the surrounding mountains.

Fire is commonly associated with grass-dominated ecosystems throughout the world (Noss 1994, Pyne 1995, Whelan 1995). The combined presence of abundant ignition sources and receptive fuels results in a high likelihood of frequent fire events. The extent of these events in time and space is currently difficult to determine for the Central Valley. Still, some evidence suggests a rich history of fire throughout the Central Valley. Research supports the idea that reoccurring fire events have been common in parts of California as early as the Mid Miocene (Keeley et al. 2012).

Several studies have documented past fire occurrence in forest communities immediately adjacent to valley ecosystems. Fry and Stephens (2006) examined Klamath Mountain sites with connectivity to the north Sacramento Valley. They found significant fire occurrence that then became less frequent but more extensive after about 1850. Skinner et al. (2009) analyzed the potential effect of land-use practices and climate on changes in fire regimes of the coast range. On dry sites of the eastern coast range, they documented a significant drop in fire occurrence in the twentieth century. These changes in fire regimes align with mid-1800s influx of settlers and intensive cattle grazing to the Sacramento Valley.

The high density of Native Americans occupying the Central Valley lends support to the idea that frequent burning of grasslands and savannah vegetation occurred (Map 15.3). While there are numerous records of burning by Native Americans, there is minimal agreement on the effect of these fires on vegetation (Wickstrom 1987, Anderson and Moratto 1996, Keeley 2002, Stewart et al. 2002). The influence they had on local fire regimes was affected by the background pattern of lightning occurrence. The exceptionally low density of lightning strikes in the Central Valley may have increased the relative impact of burning by Native Americans on vegetation (Keeley and Safford 2016). The importance of Native American fires would have increased in topographically complex areas or where natural ignitions were infrequent (Frost 1998). Burning in the Central Valley was certainly complicated by the continuous nature of fine fuels. Native Americans may have shifted the seasonality of fire, taking advantage of valley wetlands to control fire spread. Much evidence has supported the idea that Native American burning patterns relate strongly to the location of settlements and food production areas (Clark et al. 1996). In many cases, fire was used to protect assets from the threat of future wildfire (Williams 1998).

Population densities may help to describe a pattern of presettlement fire. Before Hispanic contact, Cook (1974) estimated populations of 76,100 and 83,800 for the Sacramento and San Joaquin valleys, respectively. This intense occupation of nonagricultural, hunter-gathers clearly impacted vegetation at both fine and coarse scales. There is some evidence to support Native Californians burning more often in the fall than spring in an effort to promote food plants and limit the impact to deer browse (Lewis 1973).

No evidence currently exists documenting a pattern of repeated fire events in the presettlement Central Valley. What is obvious is a set of conditions creating a high probability of fire events. Weather conditions supporting fire spread, abundant dry fuels and sources of ignition all align for some communities during some years. Stephens et al. (2007) reconstructed presettlement fire occurrence for California, estimating approximately 1.8 million ha (4.5 million ac) burning annually, including a significant contribution from grass and shrublands.

The exact description of past fire regimes within the plant communities of the Central Valley will never be known. Still, most ecologists would agree that a history of fire occurrence helped shape the pattern and structure of plant communities in the Valley. The flat plains, large open fire compartments, available fuel, and high densities of Native Americans indicate a potential fire frequency of 1 to 3 years for the Central Valley (Frost 1998).

HISTORIC

Writings from early visitors to the Central Valley provide evidence of fire use by Native Americans (Roquefeuil 1823, Gordon 1977, Litke 1989). Spanish expeditions describe a well-populated valley with closely spaced, small villages, and occasional large settlements. Father Munoz mentions burning by Indians during his visit of 1806 (Cook 1960). Jepson (1923) interpreted the density of oak stems to be a direct result of annual fire treatments by Native Americans. Later expeditions and small military raids to the interior of California occurred between the years 1820 and 1840 (Cook 1962). Brief reports and letters from this period document frequent use of fire, by both Native Americans and explorers, to facilitate warfare and capture of individuals.

The history of fire in the Central Valley is further complicated by the very early influence of nonnative plant species and land-use. Several early accounts document the presence of nonnatives plants in the Central Valley prior to the introduction of extensive cattle or sheep grazing (Cook 1960). Filaree (*Erodium* spp.), mustard (*Brassica* spp.), and wild oats (*Avena* spp.) were well established in the San Joaquin by the early part of the nineteenth century (Torrey 1859, Webster 1981). The presence of these nonnatives may have changed both the frequency and timing of fire events. Invasion by annual grasses, in particular, has substantially changed fuel characteristics of grassland communities. These thatch forming grasses have likely lengthened the season of burning, increased probabilities of ignitions, and created more continuous fuelbeds than that of presettlement grassland vegetation.

By the 1840s, ranching began to have a significant influence on plant communities in the Central Valley (Webster 1981). Horses were known to occur in the valley as early as 1807, with large numbers being documented after 1830 (Bryant 1848, Fremont 1848, Cook 1960). Although the impact of overgrazing may not have occurred prior to the 1860 droughts, the presence of large numbers of livestock likely influenced elements of historic fire regimes. During the mid-1800s the alignment of low rainfall years and intensive grazing pressure likely reduced fuels and fire occurrence. No direct evidence documents the use of fire by early settlers for rangeland improvement or the clearing of agricultural lands, although it is likely that fire was used for these purposes through the early part of the twentieth century.

CURRENT

Unlike other bioregions in California, the Central Valley has seen less significant effects from changes in fuels or climate. Human-caused ignitions result in frequent fires, but fragmentation and land conversion limits fire size in the contemporary landscape. Much continues to change in the Great Central Valley. From 2005 to 2010, agriculture in the bioregion fluctuated following changes in water allocation as well as persistent drought conditions. Grasslands and shrublands continued to decline, while developed lands increased.

Few fires extend beyond initial suppression actions due to the fuel types and their restricted spatial arrangement. The vast majority of the fires are contained at less than 4 ha (10 ac), and the largest was just over 6,000 ha (15,000 ac). This pattern of fire events has changed little over the last 50 years.

Major Ecological Zones

Foothill Woodland

The Central Valley is characteristically framed by a rim of foothill oak woodlands defining the transition from modern valley grasslands to hillslope vegetation types (Fig. 15.2). This community also grows on the floor of the Valley and in some locations may occur at relatively high elevations. Still, this collection of vegetation types is very strongly associated with the hill terrain ascending from the Central Valley. Foothill woodlands cover several millions hectares and nearly encircle the Central Valley. Though these vegetation types are characterized by the presence of either blue oak (*Quercus douglasii*) or foothill pine (*Pinus sabiniana*), soil depth, aspect, and elevation strongly influence the composition and structure of the tree-dominated communities. Where soils are deep, well-developed valley oak (*Q. lobata*) woodlands are common. In addition to the development of hillslope oak woodlands, ribbons of riparian forest often partition this ecological zone. Blue oak is clearly the dominant tree in the zone and a critical focus for describing fire regimes.

Open savannahs most common in the low foothills rimming the Valley. Blue oak savannahs appear to be strongly correlated with exposed south facing slopes and shallow soils. These communities generally have few additional tree and shrub species present. An occasional California buckeye (*Aesculus californica*) or interior live oak (*Q. wislizeni*) may appear in the canopy. Shrubs in the understory are sparse, with buckbrush (*Ceanothus cuneatus*) and whiteleaf manzanita (*Arctostaphylos viscida*) the most common.

Overstory trees are scattered in a sea of grasses and forbs that has been dramatically altered by a nonnative plants. With the introduction of domestic livestock grazing and nonnative species during the Spanish colonization, herbaceous cover changed from perennial to annual and from native to nonnative (Holmes 1990). In some locations, as much as 95% of the herbaceous layer biomass is made up of nonnative species (Gerlach et al. 1998). Though their percentage of plant biomass may be small, native herbaceous species contribute significantly to species diversity. The presence of Mediterranean annual grasses has significantly increased moisture stress on oak seedlings and may negatively impact recruitment (Adams et al. 1992, Gordon and Rice 1993). Prominent among nonnative forbs is yellow star-thistle (*Centaurea solstitalis*). Introduced during the mid-1800s, it remained limited in its distribution until the 1940s. Since the 1960s, this plant

FIGURE 15.2 Foothill Woodlands support a diverse and structurally complex assemblage of plants. Both California black oak and blue oak dominate the overstory of this mid-elevation foothill site.

has expanded its range at an explosive rate. The result is a significant change in species composition of both grassland and woodland ecosystems and a possible altering of fire regimes (DiTomaso et al. 2006).

Though the percentage of plant biomass may be small, native herbaceous species are abundant in these communities and contribute significantly to species diversity. Within the range of blue oak woodlands, the assemblage of herbaceous species varies considerably. This may be particularly true following fire events (Marty 2002). Borchert et al. (1991) found the compositional pattern of selected herbs to be strongly correlated with crown cover, slope, solar insolation, and elevation. Soil and moisture gradients have also been noted as important factors influencing the arrangement of herbaceous plants in the blue oak understory. Native perennials grasses, purple needle grass (*Stipa pulchra*), California melic (*Melica californica*), squirrel tail (*Elymus elymoides*), and western wildrye (*E. glaucus*) may be common. A rich collection of forbs may be found amongst blue oaks. San Bernardino larkspur (*Delphinium parryi*), dotseed plantain (*Plantago erecta*), johnny-jump-up (*Viola pedunculata*), and rusty popcornflower (*Plagiobothrys nothofulvus*) all add to a diverse display of spring wildflowers in blue oak grasslands.

Further upslope from the valley floor, blue oaks occur in a more continuous overstory. Higher stem densities and the addition of other tree species create a woodland stand structure. Holzman and Allen-Diaz (1991) suggest that overstory cover of this community has generally increased in the contemporary landscape. Additional rainfall or cooler north aspects result in a denser and more structurally complex community. Additional tree species commonly include foothill pine in the north and central portions of the valley. California juniper (*Juniperus californica*) may also be present throughout the southern valley foothills.

Both the density and diversity of shrub species increase significantly in blue oak woodlands. Many shrubs not found in lower elevation savannahs may now be found spilling downslope from adjacent chaparral communities. Birch-leaf mountain-mahogany (*Cercocarpus betuloides* var. *betuloides*), California coffee berry (*Frangula californica*), Western redbud (*Cercis occidentalis*), and western poison oak (*Toxicodendron diversilobum*) can occur in relatively dense patches scattered throughout the understory. The composition of herbaceous species may closely resemble that occurring in savannahs, though their arrangement may be much more discontinuous.

The arrangement of vegetation communities differs on the east and west sides of the valley. In the Sierra Nevada foothills, blue oak savannahs can become established as low as 150 m (500 ft) in elevation. This band of blue oak-dominated vegetation is often sandwiched between vernal pool grassland in the valley and mixed chaparral types above 1,500 m (5,000 ft). On the west side of the valley, blue oaks tend to occur at higher elevations, following juniper woodlands and mixed shrublands above 1,000 m (3,500 ft).

On deep soils, valley oak may form extensive, monotypic stands of large trees with very little understory. Though valley oak is most often the dominant species, coast live oak (*Q. agrifolia*) or interior live oak stems may be present. These stands are most common on recent alluvial terraces of large valleys, but can also appear on hillslope terrain (Griffin 1977). Valley oak woodlands are characteristically heterogeneous with varied structure (Conard et al. 1977). Smaller trees tend to occur at higher densities in a matrix of more open savannah-like structure.

FIRE ECOLOGY OF IMPORTANT SPECIES

A number of studies have shown no positive effect of fire on blue oak seedling establishment or survival (Allen-Diaz and Bartolome 1992, Swiecki and Bernhardt 2002, Fry 2002). Other studies have indicated that fire may not be a limiting factor in seedling recruitment, as many seedlings are vigorously resprouting (Table 15.1) (Kauffman and Martin 1987, Lathrop and Osborne 1991, Haggerty 1994). It has also been suggested that alteration of presettlement fire regimes has contributed to weak seedling recruitment (McClaran and Bartolome 1989, Pavlik et al. 1991). Mature trees seem to persist following fire events and may experience an increase in crown vigor (Mooney 1977, Fry 2002). Indeed, factors other than fire may be more likely to limit seedlings from recruiting into the sapling stage of development. Predicting postfire response of blue oak is

TABLE 15.1
Fire response of selected foothill woodland plant species

	Type of fire response			
Lifeform	Sprouting	Seeding	Individual	Species
Hardwood	Fire-stimulated	Fire neutral	Top-killed or branch killed	Blue oak, valley oak, coast live oak
Conifer	None	Fire-stimulated	Killed	Gray pine
Shrub	Fire-stimulated	Fire neutral	Top-killed	Mountain-mahogany
Herb	Fire-stimulated	Fire-stimulated	Top-killed	Redbud
Grass	Fire-stimulated	Fire neutral	Top-killed	Oniongrass

difficult, as phenological variation, unique to individual fire events, may explain the majority of variation.

Information reported in the literature has often characterized oaks seedlings as ephemeral and short-lived (Biswell 1956, White 1966). Contrary to previous assumptions, Allen-Diaz and Bartolome (1992) found mortality of blue oak seedlings was not well correlated with age or size, and that both fire and sheep grazing appeared to have no effect on the seedling recruitment. Their study supports the idea that blue oaks may not require fire, but have developed a strategy to survive in the presence of frequent fires. Seedlings are always present in the understory, providing advance regeneration for that time when the canopy is removed and release occurs.

Mature oaks suffer very little mortality following low- or moderate-intensity fire events. However, higher intensity events can result in severe damage and significant mortality of mature trees (Plumb 1980). Significant variability can be seen in postfire response of various oak species. Survival may be strongly correlated with sprouting ability and bark thickness (Plumb and Gomez 1983). Rundel (1980) identified suites of traits in many California oak species which allow survival in areas where fires are frequent.

FIRE REGIME–PLANT COMMUNITY INTERACTIONS

Unlike some other bioregions, little evidence of past fire regimes exists for contemporary foothill woodlands (Appendix 1). Although some long-lived tree species are present in these communities, none readily retain fire scars. Using a variety of sources Van de Water and Safford (2011) reconstructed median fire return intervals of 12 years in oak woodlands under presettlement fire regimes. Additional evidence of past fire patterns based on the contemporary arrangement of vegetation is confounded by the strong influence of edaphic factors. Also, because of the dramatic changes in fuel type, load, and arrangement, recent fire events may differ dramatically from those of the past.

Studies using fire scars to reconstruct past patterns of fire and its relationship to age class distribution have indicated that repeated fire events played an important role in maintaining the structure of blue oak woodlands (McClaran and Bartolome 1989, Mensing 1992, Haggerty 1994). Though information remains limited, presettlement fire regimes likely supported frequent, low-severity events. The correlation of fire events and oak regeneration may result from increased growth rates of sprouts as well as reduced competition from herbaceous vegetation (McClaran and Bartolome 1987, Mensing 1990). These same fire effects have resulted in noted improvement of structural elements in blue oak savannahs. Haggerty (1994) found savannah-like characteristics to be, at least temporarily, enhanced within a wildfire perimeter outside Sequoia and Kings Canyon National Parks. Although some small diameter trees were top-killed, little change occurred in relative dominance or basal area of oaks.

Swiecki and Bernhardt (2002) described blue oak seedling growth and survivorship following a 1996 wildfire in the northern San Joaquin Valley. Their observations indicate that fire damage has hindered the advancement of small saplings into the over story. They suggest a 1.5–2 m (4.9–6.6 ft) height or about a 4–5.5 cm (1.6–2.2 in) diameter are critical thresholds for initial complete top kill. In another study at Pinnacles National Monument, Swiecki and Bernhardt (1998) indicate some potential negative impacts from frequent fire events. Plots with fire-free intervals of less than five years were least likely to contain sapling size trees. On a similar site that had been prescribed burned, Tietje et al. (2001) found small blue oak to be very low, with seedlings sprouting very vigorously.

Although actual mortality of blue oaks may be infrequent following fire events, there does not seem to be a demonstrated stimulatory effect on regrowth or regeneration (Bartolome et al. 2002). Most studies seem to indicate that blue oak grows fastest when protected from both grazing pressure and burning. This is not a clear indication that fire is incompatible with successful oak recruitment, but that the timing and frequency of fire events are critical (Bartolome et al. 2002, Swiecki and Bernhardt 2002).

Even with concerns about fire impacts, the model for blue oak regeneration supports a role for reoccurring fire in foothill woodlands (Allen-Diaz and Bartolome 1992, Swiecki and Bernhardt 1998). It is conceivable that fire was a significant agent of change, occasionally killing overstory trees and promoting the transition of seedlings to the sapling stage. Still, blue oak recruitment may not be limited by fire patterns (White 1966), and the effects of browsing may have a more significant effect on recruitment (Vankat and Major 1978). The possibility that stand-replacing fire events, initiating the single age cohorts dominating many stands, seems unlikely as fuels necessary to support this type of fire spread rarely occur within these sites.

FIGURE 15.3 Valley Grasslands represent a rich collection of grasses and forbs that vary tremendously depending on specific site factors and land-use history. Meadow barley and seep monkeyflower are dominant on this mesic grassland site.

Valley Grasslands

Much discussion has occurred regarding the original composition of Central Valley grasslands (Fig. 15.3). Many ecologists have contributed to the perception of plant communities being dominated by native perennial grasses (Clements 1934, Clements and Shelford 1939, Beetle 1947, Munz and Keck 1949, Clark 1956, Benson 1957, Burcham 1957, Heady 1977, Hull and Muller 1977, Küchler 1977). Clements and Shelford (1939) described the community as a diverse assemblage of perennial grasses including purple needle grass, western wild-rye, Nevada blue grass (*Poa secunda*) and deergrass (*Muhlenbergia rigens*). The arrangement of grassland communities in the Central Valley was likely organized by soil types, levels of available soil moisture, and annual precipitation. In wet soils of the central San Joaquin Valley, extensive stands of beardless wild rye (*Elymus triticoides*) were thought to dominate (Barry 2003). Similar sites in the southern San Joaquin, with more alkaline soils, likely supported sacaton-dominated grasslands (*Sporobolus airoides*) or, on drier flats, desert saltgrass (*Distichlis spicata*). Barry (2003) has promoted the idea of a gradient on dry valley sites from needle grass prairie communities to xeric bunchgrass steppe assemblages.

It has been commonly held that native perennial grasses were widespread in the Central Valley (Clements 1934, Heady 1977). However, others have made the case that native forbs were once much more dominant, especially in drier parts of the Valley (Hamilton et al. 1999). Although questions remain concerning the composition of early Central Valley grasslands, it is well documented that by the mid-1800s much of the valley was clearly dominated by nonnatives (Thurber 1880). Some researchers believe the cover of perennial grasses decreased dramatically with the onset of intensive grazing by cattle and sheep (Barry 1972, Heady et al. 1991, Barbour et al. 1993, Noss 1994). It was also assumed that periods of intensive drought during the 1860s and extensive plowing for early dry land farming facilitated the decline of perennial grasses. Although variation in productivity and species composition among within years is heavily influenced by prevailing weather, long-term change in annual grassland and oak woodland has been most strongly influenced by continuing waves of invasion (DiTomaso et al. 2007). This alignment of factors resulted in conditions that promoted introduced European annuals over the native grassland flora (Jackson 1985, Barbour et al. 1993, Knapp and Rice 1994). Cover in coastal and valley grasslands commonly exceeds 90% for nonnative annual grass and forbs (Jackson et al. 2007).

Although not dependent, many of the valley's bunchgrasses appear well adapted to persist in the presence of repeated burning or grazing (Langstroth 1991, Menke 1992). The use of frequent fire by California tribes is well documented, and suggests type conversion may have begun well before the influence of Euro-American settlers (Timbrook et al. 1982, Anderson and Moratto 1996).

The composition of these herbaceous communities likely varied across the valley landscape under presettlement conditions. The arrangement of presettlement grasslands in the Central Valley was likely organized by soil types, levels of available soil moisture, and annual precipitation. The likely spatial variation overlaid with the few, scattered, and generally limited reports of presettlement conditions in the Central Valley often generate contradicting images of the full sweep of communities that were likely present. For example, a number of researchers have questioned the dominance of bunchgrasses in the valley grasslands (Cooper 1922, Biswell 1956, Twisselmann 1967, Wester 1981, Schiffman 1994, Hamilton 1997, Holstein 2001). Although almost certainly present in the presettlement Central Valley, the pattern and organization of native perennial bunch grasses may have differed significantly from common perceptions. Utilizing a combination of survey data and historical accounts, Holstein (2001) has argued that topography and precipitation affected community composition and pattern, with much of the dry southern San Joaquin being dominated by annual species. Documentation from early explorers supports this idea (Williamson 1853).

Barry (2003) has promoted the idea of a gradient on dry valley sites from needle grass prairie communities to xeric bunchgrass steppe assemblages. Still other investigators have suggested a Central Valley dominated by shrublands (Cooper 1922, Bauer 1930, Wells 1964, Naveh 1967, Keeley 1989). Some of these authors indicate that the ultimate loss of shrubs and type conversation to annual vegetation was a result of repeated burning and grazing (Huenneke 1989, Keeley and Fotheringham

2001, Keeley 2002). Father Munoz traveled throughout the southern Central Valley during 1806 (Cook 1960), including what appear to be foothill portions of the west valley. He notes a "great scarcity of grass" as well as areas of alkaline soils. His descriptions of riparian vegetation identify large willow thickets with some good pasture but also his journals note a lack of firewood in this portion of the valley. Father Munoz's descriptions of vegetation appear similar to those of other explorers.

Much of the San Joaquin seems more likely to have been occupied by wetland vegetation or ephemeral annual vegetation on drier sites (Webster 1981, Hamilton 1997, Meyer and Schiffman 1999). In wet soils of the central San Joaquin Valley, extensive stands of beardless wild rye were thought to dominate (Barry 2003). Similar sites in the southern San Joaquin, with more alkaline soils, likely supported sacaton (*S. airoides*) or, on drier flats, desert saltgrass. On extremely dry or alkaline sites, xerophytic shrubs seem more likely to have occurred than bunchgrass prairies (Twisselmann 1967, Menke 1989). Evidence indicates the importance of perennials on more mesic sites in the north valley (Bartolome and Gemmill 1981, Gerlach et al. 1998).

Journals from colonial expeditions to the Central Valley also provide some insights on presettlement vegetation patterns. Father Zalvidea, in July of 1806, described large areas of the southern San Joaquin as characterized by sparse vegetation with no grass (Cook 1960). The only well-developed vegetation identified during this expedition are all associated with riparian systems. Included in these journals are descriptions of large tule stands on the shores of Lake Buena Vista and impressive cottonwood (*Populus* spp.) forest associated with the San Joaquin River. The latter contained an understory of "good pasture."

The stability of current annual vegetation in the Central Valley unique compared with other Mediterranean grasslands (Blumler 1984, Jackson 1985). Very little invasion of perennials seems to occur within existing grassland sites (Bartolome and Gemmill 1981) This resistance of annual grasslands may support the hypothesis that the vegetation on the valley floor was composed primarily of annual vegetation (Piemeisel and Lawson 1937, Hoover 1970, Schiffman 1994). Frenkel (1970) proposed a vegetation type, comprised largely of annual grasses, dominating lower elevation sites. Other evidence, including a number of historical accounts, seems to indicate a plant community dominated by annual forbs (Muir 1894, Roundtree 1936).

Questions regarding the exact makeup of grassland communities prior to the nineteenth century will remain unanswered. What does seem clear is that stands dominated by perennials did occur in some valley locations, with growing importance in the Sacramento Valley, Coast Range, and Sierra Nevada foothills. The dominance of perennials throughout the valley appears to have determined by variations in soil and available moisture. The central and southern portions of the valley were more likely to support annual vegetation on upland sites. Evidence would seem to indicate that forbs were more common than grasses (Hamilton 1998). Additionally, in the southern San Joaquin, many sites presumed to have been grasslands perhaps were occupied by a desert scrub plant assemblage.

FIRE ECOLOGY OF IMPORTANT SPECIES

Given the history of ecological thought in Central Valley grasslands, it may not be surprising that much of the fire effects literature has focused on perennial grasses. Physiology supporting fire response is common among all guilds of valley grassland species (Table 15.2). Many native bunchgrasses are known to respond favorably to fires, with vigorous resprouting and flower production after fire (Keeley 1981, Axelrod 1985, Young and Miller 1985, Glenn-Lewis et al. 1990, Fehmi and Bartolome 2003). In particular, the vigorous response of purple needle grass to fire events has been documented in a number of studies (Bartolome 1979, Ahmed 1983, Menke 1989, Dyer et al. 1996, Wills 2001). Purple needle grass commonly will begin to sprout just days after late spring fire events (Fig. 15.4). Bartolome and Gemmill (1981) noted the success of seedling establishment following fire or grazing. Menke (1992) observed an overall benefit to periodic burning of needle grass-dominated communities and recommended treatment every three to four years. Rates of regrowth for Idaho fescue (*Festuca idahoensis*) following fire were shown to increase with no significant mortality recorded for a variety of fire severities (Defosse and Robberecht 1996).

Marty et al. (2005) found both burning and grazing affected bunchgrass populations; mortality was 10% higher in burned versus unburned plots but was not significantly different among grazing treatments. Still resulting seedling density was 100% higher in burned versus unburned plots two years after the burn. The interaction of grazing and burning had no significant impacts on the purple needle grass populations except on the diameters of adult bunchgrasses which were highest in the lightly grazed, unburned treatments. Overall, there appeared to be few interactive effects of grazing and burning but the separate treatments did affect bunchgrass growth, reproduction, and mortality, and these effects were modulated by the ubiquitous effects of climatic fluctuations. It has been suggested that the persistence of needle grass on Central Valley sites is a result of strong adaptation to frequent fire and heavy grazing (Sampson 1944, Jones and Love 1945, Bartolome and Gemmill 1981, Ahmed 1983).

In some cases, fire can negatively impact native grasses. For example, if water is limited, fire may reduce the productivity of perennial grasses during the first postfire growing season (Blaisdell 1953, Daubenmire 1968, Wright 1974, Robberecht and Defosse 1995). In contrast to much of what had previously appeared in the literature, Marty (2002) found burn treatments to negatively affect stands of purple needle grass. Across a variety of grazing treatments bunchgrass mortality was 10% higher in burned versus unburned plots. Burning treatments did elicit a positive seedling response, with densities 100% higher in burned versus unburned plots two years after the burn. Still, during the period of this research seedling densities did not attain pre-burn levels.

Annual forbs represent a critical component of valley grassland communities; however, little is known of the specific tolerance individuals species may have to burning. However, some research in California grasslands has shown that certain native species increase in the postfire environment (Parsons and Stohlgren 1989, Meyer and Schiffman 1999). For example, on a grassland site at Beale Air Force Base, Marty (2002) found rose leptosiphon (*Leptosiphon rosacea*) absent from all plots prior to treatment. Following burning and grazing rose leptosiphon was found only in plots which had experienced fire.

Parsons and Stohlgren (1989) documented increases in the biomass of several native forbs following prescribed fires during a variety of seasons. Following three consecutive fall prescribed fires, valet tassels (*Castilleja attenuatus*) experienced a significant increase in biomass. Following the same number

TABLE 15.2
Fire response of selected valley grassland plant species

Lifeform	Type of fire response			Species
	Sprouting	Seeding	Individual	
Herb	Fire-stimulated	Fire neutral	Top-killed	Brodiaea, navarretia
Grass	Fire-stimulated	Fire neutral	Top-killed	Purple needle grass, creeping wildrye, alkaline sacaton
	None	Fire-stimulated	Killed	One-sided bluegrass

FIGURE 15.4 Purple needle grass sprouts vigorously three days following a spring prescribed fire.

of spring burns, small-head clover (*Trifolium microcephalum*) also increased in treated plots. Chilean bird's-foot trefoil (*Acmispon wrangelianus*) also increased in biomass following both fall and spring burns.

On a foothill site, York (1997) found eight native annuals and one perennial to respond positively to a single fire event. Dwarf brodiaea (*Brodiaea terrestris* subsp. *terrestris*), common stickyseed (*Blennosperma nanum*), California goldfields (*Lasthenia californica*), Marigold navarretia (*Navarretia tagetina*), and butter-and-eggs (*Triphysaria eriantha*) all seemed to demonstrate some adaptation to growing season fires. Pygmyweed (*Crassula connata*) and water chickweed (*Montia fontana*) seemed to respond well to the removal of thatch.

Many of the nonnative species invading California's oak woodland and grasslands are not well adapted to fire (especially spring and summer), and some prescribed fire treatments have proven effective at controlling Scotch broom (*Cytisus scoparius*), yellow star-thistle, and various annual grasses (Bossard et al. 2000, DiTomaso et al. 2006). Much attention has been given to fire effects on nonnative species currently occupying much of the valley's former grassland habitat. Although certain nonnative species positively responded to the burn treatments, some effective fire prescriptions have been developed. Yellow star-thistle was significantly less abundant in burned versus unburned plots of several studies (DiTomaso et al. 1999). DiTomaso et al. (1999) found that several years of repeated burning was necessary to sufficiently control yellow star-thistle. The effect of fire on this and other nonnative species is covered in greater detail in chapter 25.

FIRE REGIME–PLANT COMMUNITY INTERACTIONS

Although accurate estimates of presettlement fire return intervals do not exist for the Central Valley, investigations from other grasslands systems may help to interpret fire regimes (Appendix 1). Greenlee and Langeheim (1990) estimated that the fire frequency in grasslands of the Monterey Bay area were between 1 and 15 years prior to 1880. The timing of fire events is less clear, though it is likely that some burning by Native Americans took place as soon as fuels were receptive. Biswell (1989) stated fires were set by Native Americans as soon as patches of vegetation dried sufficiently to sustain fire spread. Summer and fall fires continued following both natural (lightning strikes) and anthropogenic causes (Vogl 1974, Biswell 1989).

D'Antonio et al. (2002) analyzed existing data on the effects of fire and grazing on contemporary California grasslands and concluded there has been much variation in community response. They found that precipitation has generally been more important than the type of burning treatment in influencing the pattern of perennial grasses and forbs. Native forbs benefited most from burning, but this same response was commonly seen in nonnative forbs. They also found little long-term effect on the abundance of native grasses demonstrated in the literature. With the exception of purple needle grass and California oat grass (*Danthonia californica*), there was a lack of data on native perennial grass species. Although prescribed fire treatments may temporarily increase the cover of natives and suppress some nonnative annual grasses, the specific approach to successful fire application is not well defined and may have significant spatial and temporal variability (Reiner 2007).

Investigating the impacts of grazing and burning on spatial patterns of purple needle grass, Fehmi and Bartolome (2003) recommend repeated burning and grazing exclusion as the most effective management regime for mixed, coastal grasslands. On sites where perennial grasses occupy significant cover, varied burning regimes appear to be effective conservation strategies (Wills 1999). Still, caution should be exercised when applying these results to Central Valley systems, as historic and contemporary conditions appear to differ significantly.

Questions addressing the effect of timing on grassland diversity have drawn little attention. Meyer and Schiffman (1999) compared the effects of different seasons of burning and mulch reduction annual grassland plant communities at the Carrizo Plain. Their results document a significant increased native species cover and diversity following late spring and fall burns while winter burning showed minimal effect on native cover or diversity. Other studies have also documented changes in diversity patterns following fire events. York (1997) demonstrated a significant change in grassland species composition and diversity following a single fire event. Marty (2002) noted increases in species richness across all functional groups. Similar groups of native forbs increased in biomass and cover following spring fires in southern California grasslands (Wills 2001, Klinger and Messer 2001). Numerous authors have noted that fire can reduce competition from annual grasses (Hervey 1949, Zavon 1982, Ahmed 1983, Keeley 1990, George et al. 1992). Hopkinson et al. (1999) noted timing and intensity as significant variables affecting the response of nonnative annual grasses.

Riparian Forest

Riparian forests in California reached their highest level of development in the Central Valley (Fig. 15.5). The Sacramento, the San Joaquin, and their tributaries all supported extensive gallery forests. Growing on raised terraces slightly above the stream channel, these forests varied in their size, structure, and composition. An impressive collection of winter-deciduous trees crowded the dense canopy of this community. Western Sycamore (*Platanus racemosa*), box elder (*Acer negundo*), Fremont Cottonwood (*Populus fremontii*), black willow (*Salix gooddingii*), and valley oak all occurred in complex multistory stands.

Estimates suggest about 400,000 ha (1 million ac) of riparian forest occupied the valley in the mid-nineteenth century. Although riparian forest certainly occurred in the southern San Joaquin, historical accounts seem to indicate a larger portion of the Sacramento Valley being occupied by this vegetation type (von Kotzebue 1830). William Dane Phelps traveled up the Sacramento in 1841 and noted tall trees along the high banks (Busch 1983). Williamson (1853), in a survey of the southern San Joaquin to identify potential railroad routes to the Pacific, stated that his views of the valley were obscured by large riparian forests. Very little riparian forest is left today, and less still has maintained a semblance of presettlement structure.

FIGURE 15.5 Rich galley forest once followed river corridors throughout the Central Valley. Levees follow most major rivers in the valley today, limiting both flooding and riparian forest development.

FIRE ECOLOGY OF IMPORTANT SPECIES

It seems clear that plants of the Central Valley riparian forest are not dependent upon fire to successfully regenerate. Still, fire has influenced the structure and composition of riparian vegetation (Bendix 1994, Ellis 2001). At the watershed scale, spatial patterns of native riparian richness were driven by herbaceous species, whereas woody species were largely cosmopolitan across the nearly 38,000 km^2 (14,672 mi^2) study area (Viers et al. 2012). At the floodplain scale riparian floras reflected species richness and dissimilarity patterns related to hydrological and disturbance-driven successional sequences (Vaghti and Greco 2007). Although the dominant tree and shrub species may not be specifically adapted to fire events, most species are physiologically capable of surviving a variety of other disturbance events (Table 15.3) (Ellis 2001).

TABLE 15.3
Fire response of selected riparian woodland plant species

Lifeform	Type of fire response: Sprouting	Seeding	Individual	Species
Hardwood	Fire-stimulated	Fire-stimulated	Survive or top-killed	Arroyo willow, California sycamore
	Fire neutral	Fire-inhibited	Survive or top-killed	Black cottonwood
Shrub	Fire-stimulated	Fire-stimulated	Top-killed	Coyote brush, Mexican elderberry
	Fire-stimulated	Fire neutral	Top-killed	Mule fat
Grass	Fire-stimulated	Fire neutral	Top-killed	Deergrass

In more coastal riparian vegetation, Davis et al. (1989) found sprouting to be the dominant means of recovery for tree species. Sycamore showed rapid rates of recovery, whereas other less vigorous sprouters like alder were very slow to respond. Although easily killed by moderate- or high-severity fire, cottonwood has also demonstrated the ability to persist following fires (Ellis 1999). Studies on other southwestern riparian forest document variable responses and inefficient recovery of native trees (Busch and Smith 1993). Cottonwood was eliminated from burn sites on the lower Colorado River, while on these same sites willow did persist (Busch 1995). Size may contribute significantly to sprouting ability, with smaller diameter trees responding more vigorously than larger ones (Blaisdell and Mueggler 1956). Halofsky and Hibbs (2009) found abundant postfire regeneration in riparian areas in Oregon, and the self-replacing of hardwood and conifer-dominated communities indicate high resilience of these fire-adapted plant communities.

Severity appears to significantly affect response of both trees and shrubs. Even moderate-severity fire events may result in complete overstory mortality and limited response of sprouting trees (Busch 1995, Ellis 2001). The number of fires may also significantly affect community structure and composition (Busch 1995). Compressed fire return intervals may facilitate the invasion of weedy species.

FIRE REGIME–PLANT COMMUNITY INTERACTIONS

Very little documentation of fire effects on riparian communities is available for the Central Valley. More research has been initiated in riparian systems of the Southwest. Several investigators have suggested fire to be a novel event in riparian plant communities (Dobyns 1981, Bahre 1985, Kirby et al. 1988, Busch 1995). For 3 years after prescribed fire treatment, Arkle and Pilliod (2010) found that the extent and severity of riparian vegetation burned was substantially lower after prescribed fire compared to nearby wildfires. Additional research on prescribed fire effects indicates an increase in abundance and diversity of both herpetofauna and small mammals following fall burns in two Central Valley riparian systems (Hankins 2009).

In spite of the lack of actual fire observations in riparian woodlands of the Central Valley, it is possible to predict fire regime components for these systems (Appendix 1). Evidence supports claims that in riparian woodland vegetation, fires are increasing in both frequency and severity (Busch 1995, Busch and Smith 1995, Stuever et al. 1997). The presence of more flammable, nonnative vegetation may be driving shifts in fire regimes (Ohmart and Anderson 1982, Busch and Smith 1993). More fundamental changes in salinity level, available nutrients, and plant–water relations may be transitioning gallery forests to a riparian scrubland (Busch and Smith 1993). The lack of flooding in these systems may also contribute to excessive fuel accumulation and more severe fire events (Ellis 2001).

Assuming a repeated pattern of fire in upland vegetation types, riparian forests of the Central Valley must also have experienced periodic fires. During some seasons, conditions could have made riparian systems effective barriers to fire spread. These systems may have been significant boundaries to fire events, hence defining an important component of valley fire regimes. The wide spread distribution of riparian forests partitioned the valley into discrete blocks of upland vegetation. It is not unreasonable to assume Native Americans made use of these barriers to control fire spread and ultimately fire size.

It is also fair to assume that during periods of prolonged drought, riparian forests did burn readily, increasing fire size and altering vegetation patterns. Contemporary examples of valley riparian forest often exhibit high fuel loads in a continuous vertical and horizontal arrangement. Fire events on these sites can be of extremely high severity, with all overstory trees being killed (Stuever 1997). Fire behavior in some recent events has been strongly influenced by nonnative species, most notably giant reed (*Arundo donax*) and tamarisk (*Tamarix* spp.) (Busch 1995, Coffman 2007). The past pattern of fire within these native vegetation types will remain unknown, though contemporary evidence would indicate an important role for riparian forest in the fire regimes of the Central Valley.

Freshwater Marsh

Our contemporary image of the Central Valley has been so strongly affected by changes wrought by land-use that it has become difficult to imagine its presettlement splendor. Possibly the single most striking feature of the precontact Central Valley landscape was the scale and scope of its massive fresh-

FIGURE 15.6 Freshwater marshes once covered significant portions of the valley, creating unique plant communities and important wildlife habitat.

water wetlands (Fig. 15.6). This element of the valley has been changed so fundamentally that it may be impossible to fully define its former extent. European explorers noted a valley hydrology quite different from today's, with large expanses of open water and marshes (Williamson 1853, Morgan 1960, Cronise 1868). Missing from the contemporary landscape are the interconnected Tulare, Buena Vista, and Kern Lakes that formed the largest wetland in California. These large, former lakes of the Central Valley combined with an intricate network of sloughs and marshes wherever water settled in low portions of the landscape. This incredible wetland system was estimated to contain over 3,400 km (2,100 mi) of shoreline prior to active dewatering (Schoenherr 1992).

In all cases marsh vegetation is arranged in distinct units based on water depth and ecological requirements. It appears that near shore portions of the valley's wetlands were dominated by various reed-like plants. A variety of cattail species (*Typha* spp.), southern bulrush (*Schoenoplectus californicus*), and slender-beaked sedge (*Carex athrostachya*) may all be abundant in these locations (Sawyer and Keeler-Wolf 1995). Following the gradient to higher, better drained soils, willows are also an important component of freshwater marsh habitats. Black willow (*Salix gooddingii*), red willow (*S. laevigata*), and pacific willow (*S. lasiandra*) may all be present. Although the tree form of these species is present, more individuals will branch close the ground and resemble large shrubs. On drier, more alkaline sites, mule fat (*Baccharis salicifolia*) may replace willows.

FIRE ECOLOGY OF IMPORTANT SPECIES

Several studies have demonstrated significant change following fires in wetland ecosystems (Johnson and Knapp 1993, Gabrey et al. 1999, Clark and Wilson 2001). Following prescribed fires, Johnson and Knapp (1993) noted pronounced increases in soil pH, organic matter, and available nutrients in rush (*Juncus* spp.) and cord grass (*Spartina* spp.) marsh vegetation. Changes occurring in soils may contribute to significantly greater biomass production, inflorescence density, and plant height in annually burned cord grass wetlands (Johnson and Knapp 1993). Smith and Kadlec (1985) noted no difference in production of broad-leaved cattail (*Typha latifolia*), slender bulrush (*Schoenoplectus heterochaetus*), and salt marsh bulrush (*Bolboschoenus maritimus* subsp. *paludosus*) on burned sites versus controls. Increased protein in postfire vegetation may have resulted in preferential grazing by wetland vertebrates (Smith and Kadlec 1985). Specific response of Central Valley marsh vegetation is still in need of further investigation, though some adaptations to fire seem clear (Table 15.4).

FIRE REGIME–PLANT COMMUNITY INTERACTIONS

No spatially explicit records of fire patterns exist for presettlement marsh habitats of the Central Valley. Still, historical descriptions place native Californians and active burning in locations of abundant freshwater marsh habitat. Much research in other locations has documented a history of naturally occurring fires in wetland habitats (Viosa 1931, Loveless 1959, Cohen 1974, Wade et al. 1980, Wilbur and Christensen 1983, Izlar 1984). Based on these studies, fire regime components in freshwater marshes of the Central Valley can be postulated (Appendix 1).

The effect these long, linear, mesic environments had on fire spread is unknown. Certainly, during portions of the year the combinations of riparian forest and freshwater marsh provided an effective barrier to fire spread. This extremely wet environment split the valley in half and may have partitioned much of the east side (Fig. 15.7). During years of above average rain fall, these features likely held abundant water all summer.

Freshwater marsh vegetation may have also contributed significantly to enlarging fire events. Particularly in the San Joaquin, much of the uplands may have produced low volumes of discontinuous fuels. Marsh habitats are known to be extremely productive (Bakker 1972) and would have contributed tremendous amounts of fuel to the valley's fire environment. Although species like cattails may not be available for burning during significant portions of the year, these fuels do burn readily with great intensity, under a particular alignment of factors. The linear arrangement of these vegetation types may have contributed to extensive north-south fire spread, during periods of low rainfall and dry conditions.

Management Issues

The Central Valley is arguably the most highly altered bioregion in California. Its long history of occupation and intensive use has resulted in a contemporary landscape largely

TABLE 15.4
Fire response of selected freshwater marsh plant species

Lifeform	Type of fire response: Sprouting	Seeding	Individual	Species
Grass	Fire-stimulated	Fire-stimulated	Top-killed	Cord grass
	Fire inhibited	Fire inhibited	Top-killed	Saltgrass
	Fire neutral	Fire neutral	Top-killed	Bulrush
	Fire-stimulated	Fire neutral	Top-killed	Cattail

FIGURE 15.7 Wetlands and small tributaries partitioned significant parts of the Central Valley landscape and may have formed effective barriers to fire spread.

devoid of natural vegetation. This conversion has been occurring over long periods of time. Land-use stretches deep into the valley's past, though it is modern agriculture that remains the single most significant agent of change. With few exceptions, the conservation landscape here has been defined long ago. Lands fall into clear patterns of protected or intensively used. There are very few areas in which remnant, unprotected plant communities still exist.

Given the condition of the valley's open space, restoration is the obvious priority for management action. Restoring Central Valley plant communities has been a high priority for conservation in this portion of the state. Yet, with so few enact remnants left, reference conditions have been difficult to define. Much work is still needed to identify vegetation targets at multiple scales.

Agreement on restoration goals is only helpful when supported by adequate research, technologies, and resources. The hard work of many has improved our ability to establish particular elements of valley plant communities. Still a number of important guilds of species remain unavailable and untested in restoration settings. The process of restoration, particularly in grasslands and riparian forest, is further complicated by the presence of invasive nonnative species, the control of which is absolutely critical to management success.

Recent efforts have demonstrated how commonly used management tools affect grassland composition and how various environmental factors interact with these disturbances. Late-spring (June–July) burning has become an often used tool for increasing native species richness and cover in grasslands or preparing sites for more active restoration. Prescriptive grazing has been used to similar effect. On sites of various conditions, moderate levels of grazing do not seem to have an overwhelming negative or positive effect on species composition.

Even these modest goals for prescribed fire may see limited success. Throughout much of the twentieth century agricultural burning was a common practice in the Central Valley. The fall burning of rice straw stubble, in particular, resulted in notoriously poor air quality. Non-attainment in many of the valley's air sheds has reduced acceptable burn windows to a small number of days each year. Burn permitting is competitive, with both agricultural and wildland units attempting to implement during the same periods. The unlikely resolution to the air quality issue will continue to minimize fire applications in the Central Valley.

References

Adams, D.P. 1985. Quaternary pollen records from California. Pages 125–140 in: V.M. Bryant and R.C. Holloway, editors. Pollen Records of Late-Quaternary North American Sediments. American

Association of Stratigraphic Palynologists Foundation, Dallas, Texas, USA.
Adams, T.E., P.B. Sands, W.H. Weitkamp, and N.K. McDougald. 1992. Oak seedlings establishment on California rangelands. Journal of Range Management 45(1): 93–98.
Ahmed, E.O. 1983. Fire ecology of *Stipa pulchra* in California annual grassland. Dissertation, University of California, Davis, California, USA.
Allen-Diaz, B.H., and J.W. Bartolome. 1992. Survival of *Quercus douglasii* seedlings under the influence of fire and grazing. Madroño 39: 47–53.
Anderson, M.K. 2005. Tending the Wild: Native American Knowledge and the Management of California's Natural Resources. University of California Press, Berkeley, California, USA.
Anderson, M.K., and M.J. Moratto. 1996. Native American land-use practices and ecological impacts. Pages 187–206 in: D.C. Erman, general editor. Sierra Nevada Ecosystem Project: Final Report to Congress, Volume II. Wildland Resources Center Report 37. University of California, Davis, California, USA.
Arkle, R.S., and D.S. Pilliod. 2010. Prescribed fires as ecological surrogates for wildfires: a stream and riparian perspective. Forest Ecology and Management 259: 893–903.
Axelrod, D.I. 1985. Rise of the grassland biome, central North America. Botanical Review 62: 1165–1183.
Bahre, C.J. 1985. Wildfire in southeastern Arizona between 1859 and 1890. Desert Plants 7: 190–194.
Bakker, E. 1972. An Island Called California. University of California Press, Berkeley, California, USA.
Barbour, M., B. Pavlik, F. Drysdale, and S. Lindstrom. 1993. California's Changing Landscape. California Native Plant Society, Sacramento, California, USA.
Barry, W.J. 1972. California Prairie Ecosystems (Vol. 1): The Central Valley Prairie. State of California, Resources Agency, Department of Parks and Recreation, Sacramento, California, USA.
———. 2003. California primeval grasslands and management in the State Park System. Grasslands 13: 4–8.
Bartolome, J.W. 1979. Germination and seedling establishment in California annual grassland. Journal of Ecology 67: 273–281.
Bartolome, J.W., and B. Gemmill. 1981. The ecological status of *Stipa pulchra* (Poaceae) in California. Madroño 28: 172–184.
Bartolome, J.W., M.P. McClaran, B.H. Allen-Diaz, J. Dunne, L.D. Ford, R.B. Standiford, N.K. McDougald, and L.C. Forero 2002. Effects of fire and browsing on regeneration of blue oak. Pages 281–286 in: R.B. Standiford, D. McCreary, and K.L. Purcell, technical coordinators. Proceedings of the Fifth Symposium on Oak Woodlands: Oaks in California's Changing Landscape. USDA Forest Service General Technical Report PSW-GTR-184. Pacific Southwest Research Station, Albany, California, USA.
Bauer, H.L. 1930. Vegetation of the Tehachapi Mountains, California. Ecology 11: 263–280.
Beetle, A.A. 1947. Distribution of the native grasses of California. Hilgardia 17: 309–357.
Bendix, J. 1994. Among-site variation in riparian vegetation in the Southern California Transverse Ranges. American Midland Naturalist 132: 136–151.
Benson, L. 1957. Plant Classification. D.C. Heath, Boston, Massachusetts, USA.
Biswell, H.H. 1956. Ecology of California grasslands. Journal of Range Management 9: 19–24.
———. 1989. Prescribed Burning in California Wildlands Vegetation Management. University of California Press, Berkeley, California, USA.
Blaisdell, J.P. 1953. Ecological effects of planned burning of sagebrush-grass range on the upper Snake River plains. USDA Technical Bulletin 1075. Washington, D.C., USA.
Blaisdell, J.P., and W.F. Mueggler. 1956. Sprouting of bitterbush (*Purshia tridentate*) following burning or top removal. Ecology 37: 365–370.
Blumler, M.A. 1984. Climate and the annual habitat. Thesis, University of California, Berkeley, California, USA.
Borchert, M.I., F.W. Davis, and B.H. Allen-Diaz. 1991. Environmental relationships of herbs in blue oak (*Quercus douglasii*) woodlands of central coastal California. Madroño 38: 249–266.
Bossard, C., J. Randall, and M. Hoshovsky, editors. 2000. Wildland weeds of California. California Native Plant Society, Sacramento, California, USA.
Bryant, E. 1848. What I Saw in California. Appleton, New York, New York, USA.
Burcham, L.T. 1957. California Rangeland. California Department of Natural Resources, Division of Forestry, Sacramento, California, USA.
Busch, B.D., editor. 1983. Alta California 1840-1842: The Journal and Observations of William Dane Phelps. Arthur H. Clark Co., Glendale, California, USA.
Busch, D.E. 1995. Effects of fire on southwestern riparian plant community structure. Southwestern Naturalist 40: 259–267.
Busch, D.E., and S.D. Smith. 1993. Effects of fire on water and salinity relations of riparian woody taxa. Oecologia 94: 186–194.
———. 1995. Mechanisms associated with decline of woody species in riparian ecosystems of the southwestern U.S. Ecological Monographs 65: 347–370.
Cione, N.K., P.E. Padget, and E.B. Allen. 2002. Restoration of a native shrubland impacted by exotic grasses, frequent fire, and nitrogen deposition in southern California. Restoration Ecology 10(2): 376–384.
Clark, A.K. 1956. The impact of exotic invasion on the remaining New World mid-altitude grasslands. Pages 737–762 in: W.L. Thomas, editor. Man's Role in Changing the Face of the Earth. University of Chicago Press, Chicago, Illinois, USA.
Clark, J.S., T. Hussey, and P.D. Royall. 1996. Presettlement analogs for Quaternary fire regimes in eastern Northern America. Journal of Paleolimnology 16: 79–96.
Clark, D.L., and M.V. Wilson. 2001. Fire, mowing, and hand-removal of woody species in restoring a native wetland prairie in the Willamette Valley of Oregon. Wetlands 21: 135–144.
Clements, F.E. 1934. The relict method in dynamic ecology. Journal of Ecology 22: 39–68.
Clements, F.E., and V.E. Shelford. 1939. Bioecology. Wiley, London, UK.
Coffman, G.C. 2007. Factors influencing invasion of giant reed (*Arundo donax*) in riparian ecosystems of Mediterranean-type climates. Dissertation, University of California, Los Angeles, California, USA.
Cohen, A.D. 1974. Evidence of fires in the ancient everglades and coastal swamps of southern Florida. Pages 213–218 in: P.J. Gleason, editor. Environments of South Florida Past and Present. Miami Geological Society Memoir 2. Miami, Florida, USA.
Conard, S.G., R.L. MacDonald, and R.F. Holland. 1977. Riparian vegetation and flora of the Sacramento Valley. Pages 47–55 in: A. Sands, editor. Riparian Forest in California: Their Ecology and Conservation. Institute of Ecology Publication 15. University of California, Davis, California, USA.
Cronise, T.F. 1868. The Natural Wealth of California. H.H. Bancroft, San Francisco, USA.
Cook, S.F. 1962. Colonial expeditions to the interior of California: Central Valley, 1820-1840. University of California Anthropological Records 20(5): 151–214.
———. 1960. Colonial Expeditions to the interior of California: Central Valley, 1800-1820. University of California Anthropological Records 16(6): 239–292.
———. 1974. The Esselen: Territory, Villages, and Populations. Quarterly of the Monterey County Archaeological Society 3(2). Carmel, California, USA.
———. 1978. Historical demography. Pages 91–98 in: R.F. Heizer, editor. Handbook of North American Indians, Vol. 8, California. Smithsonian Institute, Washington, D.C., USA.
Cooper, W.S. 1922. The broad-sclerophyll vegetation of California. Carnegie Institute of Washington Publication 319. Washington, D.C., USA.
D'Antonio, C.D., S. Bainbridge, C. Kennedy, J.W. Bartolome, and S. Reynolds. 2002. Ecology and restoration of California grasslands with special emphasis on the influence of fire and grazing on native grassland species. Report to the David and Lucille Packard Foundation. University of California, Berkeley, California, USA.
Daubenmire, R.F. 1968. Ecology of fire in grasslands. Pages 209–267 in: J.B. Cragg, editor. Advances in Ecological Research. Academic Press, New York, New York, USA.
Davis, O.K. 1990. Preliminary report of the pollen analysis of Tulare Lake. Newsletter of the Tulare Lake Archaeological Research Group 3(8): 2–4.

Davis, F.W., E.A. Keller, A. Parikh, and J. Florsheim. 1989. Recovery of the chaparral riparian zone after fire. Pages 194–203 in: D. Abell, technical coordinator. Proceedings of the California Riparian Systems Conference: Protection, Management and Restoration for the 1990s. USDA Forest Service General Technical Repot PSW-GTR-110. Pacific Southwest Forest and Range Experiment Station, Berkeley, California, USA.

Defosse, G.E., and R. Robberecht. 1996. Effects of competition on the post fire recovery of 2 bunchgrass species. Journal of Range Management 49: 137–142.

DiTomaso, J.M., S.F. Enloe, and M.J. Pitcairn. 2007. Exotic plant management in California annual grasslands. Pages 281–296 in: M.R. Stromberg, J.D. Corbin, and C.M. D'Antonio, editors. California Grasslands. University of California, Berkeley, California, USA.

DiTomaso, J.M., G.B. Kyser, S.B. Orloff, and S.F. Enloe. 1999. Integrated strategies offer site-specific control of yellow starthistle. California Agricultural 54(6): 30–36.

DiTomaso, J.M., G.B. Kyser, and M.J. Pitcairn 2006. Yellow starthistle management guide. Cal-IPC Publications 2006-3. California Invasive Plant Council, Berkeley, California, USA.

Dobyns, H.F. 1981. From Fire to Flood: Historic Human Destruction of Sonoran Desert Riverine Oases. Ballena Press, Socorro New Mexico, USA.

Dyer, A., H. Fossum, and J.W. Menke 1996. Emergence and survival of *Nassella pulchra* in a California grassland. Madroño 43: 316–366.

Ellis, L.M. 1999. Floods and fire along the Rio Grande: the role of disturbance in the riparian forest. Dissertation, University of New Mexico, Albuquerque, New Mexico, USA.

———. 2001. Short-term response of woody plants to fire in a Rio Grande riparian forest, Central New Mexico, USA. Biological Conservation 97: 159–170.

Erlandson, J.M. 1994. Early Hunter-Gatherers of the California Coast. Plenum, New York, New York, USA.

Evett, R.R., and J.W. Bartolome. 2013. Phytolith evidence for the extent and nature of prehistoric Californian grasslands. The Holocene 23: 1644–1649.

Fehmi, J.S., and J.W. Bartolome. 2003. Impacts of livestock and burning on the spatial patterns of grass *Nassella pulchra* (Poaceae). Madroño 50: 8–14.

Fenenga, G.L. 1991. A preliminary analysis of faunal remains from early sites in the Tulare Lake Basin. Pages 11–22 in: W.J. Wallace and F.A. Riddell, editors. Contributions to Tulare Lake Archaeology I: Background to a Study of Tulare Lake's Archaeological Past. Tulare Lake Archaeological Research Group, Redondo Beach, California, USA.

———. 1994. Alternative interpretations of Late Pleistocene paleoecology in the Tulare Lake Basin, San Joaquin Valley, California. Kern County Archaeological Society Journal 5: 105–117.

Fremont, J.C. 1848. Geographical memoir upon upper California. 30th Congress, 1st Session, Senate Miscellaneous Document 148. Washington, D.C., USA.

Frenkel, R.E. 1970. Ruderal Vegetation along Some California Roadsides. University of California Press, Berkeley, California, USA.

Frost, C.C. 1998. Presettlement fire frequency regimes of the United States: a first approximation. Proceedings Tall Timbers Fire Ecology Conference 20: 70–81.

Fry, D.L. 2002. Effects of a prescribed fire on oak woodland stand structure. Pages 235–242 in: R.B. Standiford, D. McCreary, and K.L. Purcell, technical coordinators. Proceedings of the Fifth Symposium on Oak Woodlands: Oaks in California's Changing Landscape. USDA Forest Service General Technical Report PSW-GTR-184. Pacific Southwest Research Station, Albany, California, USA.

Fry, D.L., and S.L. Stephens. 2006. Influence of humans and climate on the fire history of a ponderosa pine-mixed conifer forest in the southeastern Klamath Mountains, California. Forest Ecology and Management. 223: 428–438.

Gabrey, S.W., A.D. Afton, and B.C. Wilson. 1999. Effects of winter burning and structural marsh management on vegetation and winter bird abundance in the Gulf Coast Chenier Plain, USA. Wetlands 19: 594–606.

George, M.R., J.R. Brown, and W.J. Clawson. 1992. Application of nonequilibrium ecology to management of Mediterranean grasslands. Journal of Range Management 45: 436–440.

Gerlach, J., A. Dyer, and K. Rice. 1998. Grassland and foothill woodland ecosystems of the central valley. Fremontia 26(4): 39–43.

Glenn-Lewis, D.C., L.A. Johnson, T.W. Jurik, A. Akey, M. Loeschke, and T. Rosburg. 1990. Fire in central North American grasslands: vegetative reproduction, seed germination, and seedling establishment. Pages 28–45 in: S.L. Collins and L.L. Wallace, editors. Fire in North American Tallgrass Prairies. University of Oklahoma Press, Norman, Oklahoma, USA.

Gordon, B.I. 1977. Monterey Bay Area: Natural History and Cultural Imprints. Boxwood, Pacific Grove, California, USA.

Gordon, D.R., and K.J. Rice. 1993. Competitive effects of grassland annuals on soil water and blue oak (*Quercus douglasii*) seedlings. Ecology 74: 68–82.

Greenlee, J.M., and J.H. Langeheim. 1990. Historic fire regimes and their relation to vegetation patterns in the Monterey Bay Area of California. American Midland Naturalist 124: 239–253.

Griffin, J.R. 1977. Oak woodland. Pages 383–415 in: M.G. Barbour and J. Major, editors. Terrestrial Vegetation of California. John Wiley and Sons, New York, New York, USA.

Haggerty, P.K. 1994. Damage and recovery in southern Sierra Nevada foothill oak woodland after a severe ground fire. Madroño 41: 185–198.

Halofsky, J.E., and D.E. Hibbs. 2009. Controls on early post-fire woody plant colonization in riparian areas. Forest Ecology and Management 258: 1350–1358.

Hamilton, J.G. 1997. Changing perceptions of pre-European grasslands in California. Madroño 44: 311–333.

Hamilton, J.G., C. Holzapfel, and B.E. Mahall. 1999. Coexistence and interference between a native perennial grass and non-native annual grasses in California. Oecologia 121: 518.

Hankins, D. 2009. The effects of indigenous prescribed fire on herpetofauna and small mammals in two Central Valley California riparian ecosystems. California Geographer 49: 31–50.

Heady, H.F. 1977. Valley grassland. Pages 491–514 in: M.G. Barbour and J. Major, editors. Terrestrial Vegetation of California. John Wiley and Sons, New York, New York, USA.

Heady, H.F., J.W. Bartolome, M.D. Pitt, G.D. Savelle, and M.C. Stroud. 1991. California prairie. Pages 313–335 in: R.T. Coupland, editor. Natural Grasslands. Series: Ecosystems of the World. Vol. 8A. Introduction and Western Hemisphere. Elsevier, New York, New York, USA.

Hervey, D.F. 1949. Reaction of a California annual-plant community to fire. Journal of Range Management 2: 116–121.

Holmes, T.H. 1990. Botanical trends in northern California oak-woodland. Rangelands 12: 3–7.

Holstein, G. 2001. Pre-agricultural grassland in central California. Madroño 48: 253–264.

Holzman, B.A., and B.H. Allen-Diaz. 1991. Vegetation change in blue oak woodlands in California. Pages 189–193 in: R.B. Standiford, technical coordinator. Proceedings of Symposium on Oak Woodlands and Hardwood Rangeland Management. USDA Forest Service General Technical Report PSW-GTR-126. Pacific Southwest Research Station, Berkeley, California, USA.

Hoover, R.F. 1970. The Vascular Plants of San Luis Obispo County, California. University of California Press, Berkeley, California, USA.

Hopkinson, P.J., S. Fehmi, and J.W. Bartolome. 1999. Summer burns reduce cover but not spread of barbed goatgrass in California grassland. Ecological Restoration 17: 168–169.

Huenneke, L.F. 1989. Distribution and regional patterns of California grasslands. Pages 1–12 in: L.F. Huenneke and H.A. Mooney, editors. Grassland Structure and Function: California Annual Grassland. Kluwer Academic, Dordrecht, The Netherlands.

Hull, J.C., and C.H. Muller. 1977. The potential for dominance by *Stipa pulchra* in a California grassland. The American Midland Naturalist 97: 147–175.

Huntley, B., and T. Webb, III. 1988. Vegetation History. Kluwer Academic Publishers, Dordrecht, The Netherlands.

Izlar, R.L. 1984. Some comments on fire and climate in the Okefenokee swamp-marsh complex. Pages 70–85 in: A.D. Cohen, D.J. Casagrande, M.J. Andrejko, and G.R. Best, editors. The Okefenokee Swamp: Its Natural History, Geology and Geochemistry. Wetlands Surveys, Los Alamos, New Mexico, USA.

Jackson, L.E. 1985. Ecological origins of California's Mediterranean grasses. Journal of Biogeography 12: 349–361.

Jackson, L.E., M. Potthoff, K.L. Steenwerth, A.T. O'Geen, M.R. Stromberg, and K.M. Scow. 2007. Soil biology and carbon sequestration in grasslands. Pages 107–118 in: M.R. Stromberg, J.D. Corbin, and C. D'Antonio, editors. California grasslands:

Ecology and Management. University of California Press, Berkeley, California, USA.
Jepson, W.L. 1923. The Trees of California. 2nd ed. Sather Gate, Berkeley, California, USA.
Johnson, S.R., and A.K. Knapp. 1993. The effect of fire on gas exchange and aboveground biomass production in annually vs biennially burned *Spartina pectinata* wetlands. Wetlands 13: 299–303.
Jones, B.J. and R.M. Love. 1945. Improving California ranges. University of California Agricultural Experiment Station Circular 129: 1–48. Berkeley, California, USA.
Katibah, E.F. 1984. A brief history of riparian forests in the Central Valley of California. Pages 23–29 in: R.E. Warner and K.M. Hendrix, editors. California Riparian Systems: Ecology, Conservation, and Productive Management. University of California Press, Berkeley, California, USA.
Keeley, J.E. 1981. Reproductive cycles and fire regimes. Pages 231–277 in: H.A. Mooney, T.M. Bonnicksen, N.L. Christensen, J.E. Lotan, and W.A. Reiners, editors. Proceedings of the Conference Fire Regimes and Ecosystems Properties. USDA Forest Service General Technical Report WO-26. Washington, D.C., USA.
———. 1989. The California valley grassland. Pages 3–23 in: A.A. Schoener, editor. Endangered plant communities of southern California. Southern California Botanists Special Publication 3. Southern California Botanists, Claremont, California, USA.
———. 1990. The California valley grassland. Pages 2–23 in: A.A. Schoenherr, editor. Endangered Plant Communities of Southern California. Southern California Botanists Special Publication 3. California State University, Fullerton, California, USA.
———. 2002. Native American impacts on fire regimes of the California coast ranges. Journal of Biogeography 29: 303–320.
Keeley, J.E., W.J. Bond, R.A. Bradstock, J.G. Pausas, and P.W. Rundel 2012. Fire in Mediterranean Ecosystems: Ecology, Evolution and Management. Cambridge University Press, Cambridge, UK.
Keeley, J.E., and C.J. Fotheringham. 2001. The historic fire regime in southern California shrublands. Conservation Biology 15: 1536–1548.
Keeley, J.E. and H.D. Safford. 2016. Fire as an ecosystem process. Pages 27–45 in: H. Mooney and E. Zavaleta, editors. Ecosystems of California. University of California Press, Oakland, California, USA.
Kirby, R.E., S.J. Lewis, and T.N. Sexton. 1988. Fire in North American wetland ecosystems and fire-wildfire relations: an annotated bibliography. U.S. Fish and Wildlife Service Biological Report 88 (1). Washington, D.C., USA.
Klinger, R.C., and I. Messer. 2001. The interaction of prescribed burning and site characteristics on the diversity and composition of a grassland community on Santa Cruz Island, California. Pages 66–80 in: K.E.M. Galley and T.P. Wilson, editors. Proceedings of the Invasive Species Workshop: The Role of Fire in the Control and Spread of Invasive Species. Fire Conference 2000: The First National Congress on Fire Ecology, Prevention, and Management. Tall Timbers Research Station, Tallahassee, Florida, USA.
Knapp, E.E., and K.J. Rice. 1994. Starting from seed: genetic issues in using native grasses for restoration. Restoration and Management Notes 12: 40–45.
Küchler, A.W. 1977. The map of the natural vegetation of California. Pages 909–938 in: M.G. Barbour and J. Major, editors. Terrestrial Vegetation of California. Wiley-Interscience, New York, New York, USA.
Langstroth, R. 1991. Fire and grazing ecology of *Stipa pulchra* grassland: a field study at Jepson Prairie, California. Thesis, University of California, Davis, California, USA.
Lathrop, E.W., and C.D. Osborne, 1991. Influence of fire on oak seedlings and saplings in southern oak woodland on the Santa Rosa Plateau Preserve, Riverside California. Pages 366–370 in: R.B. Standiford, technical coordinator. Proceedings of Symposium on Oak Woodlands and Hardwood Rangeland Management. USDA Forest Service General Technical Report PSW-GTR-126. Pacific Southwest Research Station, Berkeley, California, USA.
Lewis, H.T. 1973. Patterns of Indian burning in California: ecology and ethnohistory. Anthropological Papers 1: 1–101. Ballena Press, Ramona, California, USA.
Litke, F.P. 1989. From the diary of Fedor P. Lutke [Litke] during his circumnavigation aboard the Sloop Kamchatka, 1817-1819: observations on California. Pages 257–285 in: E.A.P. Crownhart-Vaugh, B. Dmytryshyn, and T. Vaughan, editors. The Russian American Colonies: Three Centuries of Russian Eastward Expansion 1798-1867. Oregon Historical Society Press, Portland, Oregon, USA.
Loveless, C.M. 1959. A study of vegetation in the Florida Everglades. Ecology 40: 1–9.
Marty, J.T. 2002. Managing and restoring California annual grassland species: an experimental field study. Dissertation, University of California, Davis, California, USA.
Marty, J.T., S.K. Collinge, and K.J. Rice. 2005. Responses of a remnant California native bunchgrass population to grazing, burning and climatic variation. Plant Ecology 181: 101–112.
McBain and Trush. 2000. Habitat Restoration Plan for the Lower Tuolumne River Corridor. Report prepared for The Tuolumne River Technical Advisory Committee. McBain and Trush, Arcata, California, USA.
McBride, J.R., and H.F. Heady. 1968. Invasion of grassland by *Baccharis pilularis* DC. Journal of Range Management. 21: 106–108.
McClaran, M.P., and J.W. Bartolome. 1987. Factors associated with oak regeneration in California. Pages 86–91 in: T.R. Plumb and N.H. Pillsbury, technical coordinators. Proceedings of the Symposium on Multiple-Use Management of California's Hardwood Resources. USDA Forest Service General Technical Report PSW-GTR-100. Pacific Southwest Research Station, Berkeley, California, USA.
———. 1989. Fire related recruitment in stagnant *Quercus douglasii* populations. Canadian Journal of Forest Research 19: 580–585.
Menke, J.W. 1989. Management limits on productivity. Pages 173–199 in: L.F. Heunneke and H.A. Mooney, editors. Grassland Structure and Function: California Annual Grassland. Kluwer Academic Publishers, Dordrecht, The Netherlands.
———. 1992. Grazing and fire management for native perennial grass restoration in California grasslands. Fremontia 20(2): 22–25.
Mensing, S.A. 1990. Blue oak regeneration in the Tehachapi Mountains. Fremontia 18(3): 38–41.
———. 1992. The impact of European settlement on blue oak (*Quercus douglasii*) regeneration and recruitment in the Tehachapi Mountains, California. Madroño 39: 36–46.
Meyer, M.D., and P.M. Schiffman. 1999. Fire season and mulch reduction in California annual grassland: a comparison of restoration strategies. Madroño 46: 25–37.
Minnich, R.A. 2008. California's fading wildflowers. University of California Press, Berkeley, California, USA.
Minnich, R.A., and R.J. Dezzani. 2002. Historical decline of coastal sage scrub in the Riverside-Perris Plain, California. Western Birds 29: 366–391.
Mitchell, V.L. 1969. The regionalization of climate in montane areas. Dissertation, University of Wisconsin, Madison, Wisconsin, USA.
Mooney, H.A., editor. 1977. Convergent Evolution in Chile and California Mediterranean Climate Ecosystems. Dowden, Hutchinson and Ross, Stroudsberg, Pennsylvania, USA.
Morgan, D.L. 1960. California as I saw it. Incidents of Travel by Land and Water by William M'Collum. The Talisman Press, Los Gatos, California, USA.
Muir, J. 1894. The Mountains of California. The Century Co., New York, New York, USA.
Munz, P.A., and D.D. Keck. 1949. California plant communities. Aliso 2: 87–105.
Naveh, Z. 1967. Mediterranean ecosystems and vegetation types in California and Israel. Ecology 48: 445–459.
Noss, R.F. 1994. Managing rangelands. Pages 220–262 in: R.F. Noss and A. Cooperrider, editors. Saving Nature's Legacy. Island Press, Washington, D.C., USA.
Ohmart, R.D., and B.W. Anderson. 1982. North American desert riparian ecosystems. Pages 433–479 in: G.G. Bender, editor. Reference Handbook on the Deserts of North America. Greenwood Press, Westport, Connecticut, USA.
Parsons, D.J., and T.J. Stohlgren 1989. Effects of varying fire regimes on annual grasslands in the southern Sierra Nevada of California. Madroño 36: 154–168.
Pavlik, B.M., P.C. Muick, S. Johnson, and M. Popper. 1991. Oaks of California. Cachuma Press, Los Olivos, California, USA.
Piemeisel, R.L., and F.R. Lawson. 1937. Types of Vegetation in the San Joaquin Valley of California and Their Relation to the Beet Leafhopper. U.S. Department of Agriculture, Washington, D.C., USA.
Plumb, T.R. 1980. Response of oaks to fire. Pages 205–215 in: T.R. Plumb, technical coordinator. Proceedings on the Symposium on

the Ecology, Management and Utilization of California Oaks. USDA Forest Service PSW-GTR-44. Pacific Southwest Forest and Range Experiment Station, Berkeley, California, USA.

Plumb, T.R., and A.P. Gomez. 1983. Five southern California oaks: identification and post fire management. USDA Forest Service General Technical Report PSW-GTR-71. Pacific Southwest Forest and Range Experimental Station, Berkeley, California, USA.

Pyne, S.J. 1995. World Fire. Henry Holt and Company, New York, New York, USA.

Reiner, R.J. 2007. Fire in California grasslands. Pages 207–217 in: M.R. Stromberg, J.D. Corbin, and C.M. D'Antonio, editors. California Grasslands: Ecology and Management. University of California Press, Berkeley, California, USA.

RHJV (Riparian Habitat Joint Venture). 2004. Version 2.0. The Riparian Bird Conservation Plan: A Strategy for Reversing the Decline of Riparian Associated Birds in California. California Partners in Flight, Stinson Beach, California, USA.

Robberecht. R., and G.E. Defosse. 1995. The relative sensitivity two bunchgrass species to fire. International Journal of Wildland Fire. 5: 127–134.

Roquefeuil, M.C. 1823. A Voyage Round the World between the Years 1816-1819. Phillips, London, UK.

Roundtree, L. 1936. Hardy Californians. McMillan, New York, New York, USA.

Rundel, P.R. 1980. Adaptations of Mediterranean-climate oaks to environmental stress. Pages 43–54 in: T.R. Plumb, technical coordinator. Proceedings on the Symposium on the Ecology, Management and Utilization of California Oaks. USDA Forest Service PSW-GTR-44. Pacific Southwest Forest and Range Experiment Station, Berkeley, California, USA.

Sampson, A.W. 1944. Plant succession on burned chaparral lands in northern California. California Agricultural Experiment Station Bulletin 685: 1–144. Berkeley, California, USA.

Sawyer, J.O., and T. Keeler-Wolf. 1995. A Manual of California Vegetation. California Native Plant Society, Sacramento, California, USA.

Schiffman, P.M. 1994. Promotion of exotic weed establishment by endangered giant kangaroo rats (*Dipodomys ingens*) in a California grassland. Biodiversity and Conservation 3: 524–537.

———. 2007. Ecology of native animals in California grasslands. Pages 180–190 in: M.R. Stromberg, J.D. Corbin, and C.M. D'Antonio, editors. California Grasslands: Ecology and Management. University of California Press, Berkeley, California, USA.

Schoenherr, A.A. 1992. A Natural History of California. University of California Press, Berkeley, California, USA.

Schroeder, M.J. and C.C. Buck. 1970. Fire weather: a guide for application of meteorological information to forest fire control operations. USDA Forest Service Agricultural Handbook 360. Washington, D.C., USA.

Skinner, C.N., C.S. Abbott, D.L. Fry, S.L. Stephens, A.H. Taylor, and V. Trouet. 2009. Human and climatic influences on fire occurrence in California's North Coast Range, USA. Fire Ecology 5(3): 76–99.

Smith, F.E. 1977. A short review of the status of riparian forest in California. Pages 1–2 in: A. Sands, editor. Riparian Forest in California; Their Ecology and Conservation. Institute of Ecology Publication 15. University of California, Davis, California, USA.

Smith, L.M., and J.A. Kadlec. 1985. Comparisons of prescribed burning and cutting of Utah marsh plants. Great Basin Naturalist 45: 462–466.

Stephens, S.L., R.E. Martin, and N.E. Clinton. 2007. Prehistoric fire area and emissions from California's forests, woodlands, shrublands and grasslands. Forest Ecology and Management 251: 205–216.

Stewart, O.C., H.T. Lewis, and K.A. Anderson. 2002. Forgotten fires: Native Americans and the transient wilderness. University of Oklahoma Press, Norman, Oklahoma, USA.

Stromberg, M.R., and J.R. Griffin. 1996. Long-term patterns in coastal California grasslands in relation to cultivation, gophers, and grazing. Ecological Applications 6: 1189–1211.

Stuever, M.C. 1997. Fire induced mortality of Rio Grande cottonwood. Thesis, University of New Mexico, Albuquerque, New Mexico, USA.

Swiecki, T.J., and E.A. Bernhardt. 1998. Understanding blue oak regeneration. Fremontia 26(1): 19–26.

———. 2002. Effects of fire on naturally occurring blue oak (*Quercus douglasii*) saplings. Pages 251–259 in: R.B. Standiford, D. McCreary, and K.L. Purcell, technical coordinators. Proceedings of the Fifth Symposium on Oak Woodlands: Oaks in California's Changing Landscape. USDA Forest Service General Technical Report PSW-GTR-184. Pacific Southwest Research Station, Albany, California, USA

Thurber, G. 1880. Gramineae. Pages 253–328 in: S. Watson, editor. Geological Survey of California: Botany of California, Vol. II. John Wilson and Son, University Press, Cambridge, Massachusetts, USA.

Tietje, W.D., J.K. Vreeland, and W.H. Weitkamp. 2001. Live oak saplings survive prescribed fire and sprout. California Agriculture 55(2): 18–22.

Timbrook, J., J.R. Johnson, and D.D. Earle. 1982. Vegetation burning by the Chumash. Journal of California and Great Basin Anthropology 4: 163–186.

Torrey, J. 1859. Botany of the boundary. In: W.H. Emory, editor. Report on the United States and Mexican Boundary Survey. Vol. 2, Pt. 1 24th Congress. 1st Session, Exec. Doc. 135. Nicholson, Washington, D.C., USA.

Twisselman, E.C. 1967. A Flora of Kern County. University of San Francisco Press, San Francisco, California, USA.

USDA, Natural Resources Conservation Service. 2012. Natural Resources Inventory Summary Report, August 2012. USDA, Natural Resources Conservation Service, Washington, DC, USA, and Center for Survey Statistics and Methodology, Iowa State University, Ames, Iowa, USA. http://www.nrcs.usda.gov/technical/nri/12summary.

Vaghti, M.G., and S.E. Greco. 2007. Riparian vegetation of the Great Valley. Pages 425–455 in: M.G. Barbour, T. Keeler-Wolf, and A.A. Schoenherr, editors. Terrestrial Vegetation of California. 3rd ed. University of California Press, Berkeley, California, USA.

Van de Water, K.M., and H.D. Safford. 2011. A summary of fire frequency estimates for California vegetation before Euro-American settlement. Fire Ecology 7(3): 26–58.

Vankat, J.L., and J. Major. 1978. Vegetation changes in Sequoia National Park, California. Journal of Biogeography 5: 377–402.

van Wagtendonk, J.W., and D.R. Cayan. 2008. Temporal and spatial distribution of lightning strikes in California in relation to large-scale weather patterns. Fire Ecology 41(1): 34–56.

Viers, J.H., A. Fremier, R. Hutchinson, J. Quinn, J. Thorne, and M. Vaghti. 2012. Multiscale patterns of riparian plant diversity and implications for restoration. Restoration Ecology 20: 160–169.

Viosa, P., Jr. 1931. Spontaneous combustion in the marshes of southern Louisiana. Ecology 12: 439–442.

Vogl, R. 1974. Effect of fire on grasslands. Pages 139–194 in: T.T. Kozlowski and C.E. Ahlgren, editors. Fire and Ecosystems. Academic Press, New York, New York, USA.

von Kotzebue, O. 1830. A New Voyage Round the World in the Years 1823–1826. De Capo Press, New York, New York, USA.

Wade, D., J. Ewel, and R. Hosfstetter. 1980. Fire in south Florida ecosystems. USDA Forest Service General Technical Report SE-GTR-17. Southeastern Forest Experiment Station, Asheville, North Carolina, USA.

Webster, L. 1981. Composition of native grasslands in the San Joaquin Valley, California. Madroño 28: 231–241.

Wells, P.V. 1964. Antibiosis as a factor in vegetation patterns. Science 144 (3620): 889.

Whelan, R.J. 1995. The Ecology of Fire. Cambridge University Press, Cambridge, UK.

White, K.L. 1966. Structure and composition of foothill woodland in central coastal California. Ecology 47: 229–237.

Wickstrom, C.K.R. 1987. Issues concerning Native American use of fire: a literature review. Yosemite Research Center Publications in Anthropology 6. Yosemite National Park, El Portal, California, USA.

Wilbur, R.B., and N.L. Christensen. 1983. Effects of fire on nutrient availability in a North Carolina coastal plain pocosin. American Midland Naturalist 110: 54–61.

Williams, G.W. 1998. References on the American Indian use of fire in ecosystems. Manuscript and bibliography. USDA Forest Service, Washington, D.C., USA.

———. 2002. Aboriginal use of fire: Were there Any "natural" plant communities? Pages 179–214 in: C.E. Kay and R.T. Simmons, editors. Wilderness and Political Ecology: Aboriginal Land

Management–Myths and Reality. University of Utah Press, Logan, Utah, USA.
Williamson, R.S. 1853. Reports of exploration and surveys to ascertain the most practical and economical route for the railroad from the Mississippi River to the Pacific Ocean. Vol. 5. Beverley Tucker, Washington, D.C., USA.
Wills, R. 1999. Effects of varying fire regimes in a California native grassland. Ecological Restoration 19: 292–309.
———. 2001. Effects of varying fire regimes in a California native grasslands. Ecological Restoration 19: 109.
Wright, H.A. 1974. Effect of fire on southern mixed prairie grass. Journal of Range Management 27: 417–419.
York, D. 1997. A fire ecology study of the Sierra Nevada foothill basaltic mesa grassland. Madroño 44: 374–383.
Young, J.A., and R.F. Miller. 1985. Response of *Sitanion hystrix* (Nutt.) J.G. to prescribed burning. American Midland Naturalist 113: 182–187.
Zavon, J.A. 1982. Grazing and fire effect on annual grassland composition and sheep diet selectivity. Thesis, University of California, Davis, California, USA.

CHAPTER SIXTEEN

Central Coast Bioregion

MARK I. BORCHERT AND FRANK W. DAVIS

Description of Bioregion

The Central Coast bioregion includes the Central California Coast and Central California Coast Ranges Sections (Map 16.1) (Miles and Goudey 1997) in the California Coastal Chaparral Forest and Shrub Province of the Mediterranean Division of the Humid Temperate Domain (Bailey 1995). The bioregion extends from Napa County south to northern Santa Barbara County, altogether covering 38,830 km^2 (14,992 mi^2) (Map 16.1). The eastern boundary of the bioregion adjoins the western edge of the San Joaquin Valley. Familiar coastal landmarks include San Francisco Bay, Monterey Bay, Big Sur, and Morro Bay. Notable interior landmarks include Mt. Diablo, San Benito Mountain, and the Carrizo Plain.

Physical Geography

The topography of the region consists of rugged, northwest-to-southeast trending ranges including the Santa Cruz Mountains, the Santa Lucia Ranges, San Rafael Mountains, Diablo Range, Gabilan Range, and Temblor Range. Expansive intervening valleys include the Santa Clara, Salinas, and Santa Maria River valleys. Elevations range from sea level to over 1,800 m (5,906 ft). Half the area in the Central California Coast Section (261A in Map 16.1) is above 160 m elevation (525 ft), versus 488 m (1,600 ft) in the Central California Coast Ranges (M262A in Map 16.1).

Geology exerts a strong control on landforms, soils, and vegetation of the region (Wells 1962, Griffin 1975). Lithology of the region is dominated by folded and faulted Cenozoic marine and nonmarine sediments, except for the northern Santa Lucia Range and northern Gabilan Range, which are composed of Mesozoic granitic and Triassic metamorphic rocks. Marine sediments are predominantly inter-bedded sandstones and shales.

In the Central California Coast Section, rugged terrain, complex geology, local topo-climatic variability, anthropogenic disturbance, and fire history result in complex local vegetation mosaics (Wells 1962). In general, upland natural vegetation changes from coastal prairies and coastal sage scrub below 300 m (984 ft), to chaparral-dominated slopes around 1,200 m (3,937 ft) to montane hardwood and mixed hardwood forests at the higher elevations. Conifer forests prevail above 1,500 m (4,921 ft). Annual grasslands, oak woodlands, and chaparral dominate interior valleys and foothills to the east of the coastal ridges in the Central California Coast Ranges.

Azonal grasslands, shrublands, and woodlands in the bioregion are associated with scattered serpentine outcrops of the Mesozoic Franciscan Complex, a mélange of metamorphosed sedimentary and volcanic rocks. These outcrops are especially widespread in the south coastal Santa Lucia Range. In the interior Diablo Range, San Benito Mountain is basically a highly altered ultrabasic plug with large patches of highly mineralized serpentine (Griffin 1975). Along the coast, stabilized Pleistocene sand dunes support distinctive maritime chaparral in the Santa Maria and Salinas River valleys, east of Pismo and Morro Bays and along coastal stretches as far north as Sonoma County (Van Dyke and Holl 2001).

Climatic Patterns

The regional climate is strongly Mediterranean with cool wet winters and warm dry summers. Over 80% of seasonal rain falls between November and March, primarily due to occluded fronts arriving from the west-northwest. Precipitation generally decreases from north to south and from the coast inland but topography exerts a strong influence on climate with the highest rainfall in the coastal mountains and lowest rainfall in rain shadows along the eastern edge of the region (Map 16.2). The highest rainfall generally is associated with El Niño years and lower rainfall with La Niña years (Cayan et al. 2001).

Seasonal temperatures vary markedly with latitude, elevation, and distance from the coast (Map 16.3) (Thornton et al. 1997, Johnstone and Dawson 2010, Vasey et al. 2012, Vasey et al. 2014). At coastal stations, mean monthly temperatures at sea level range from 10–13°C (50–55.4°F) in the winter months to 16–18°C (60.8–64.4°F) in the summer, with highest temperatures in August through October. Inland temperature regimes are considerably more continental (Vasey et al. 2014).

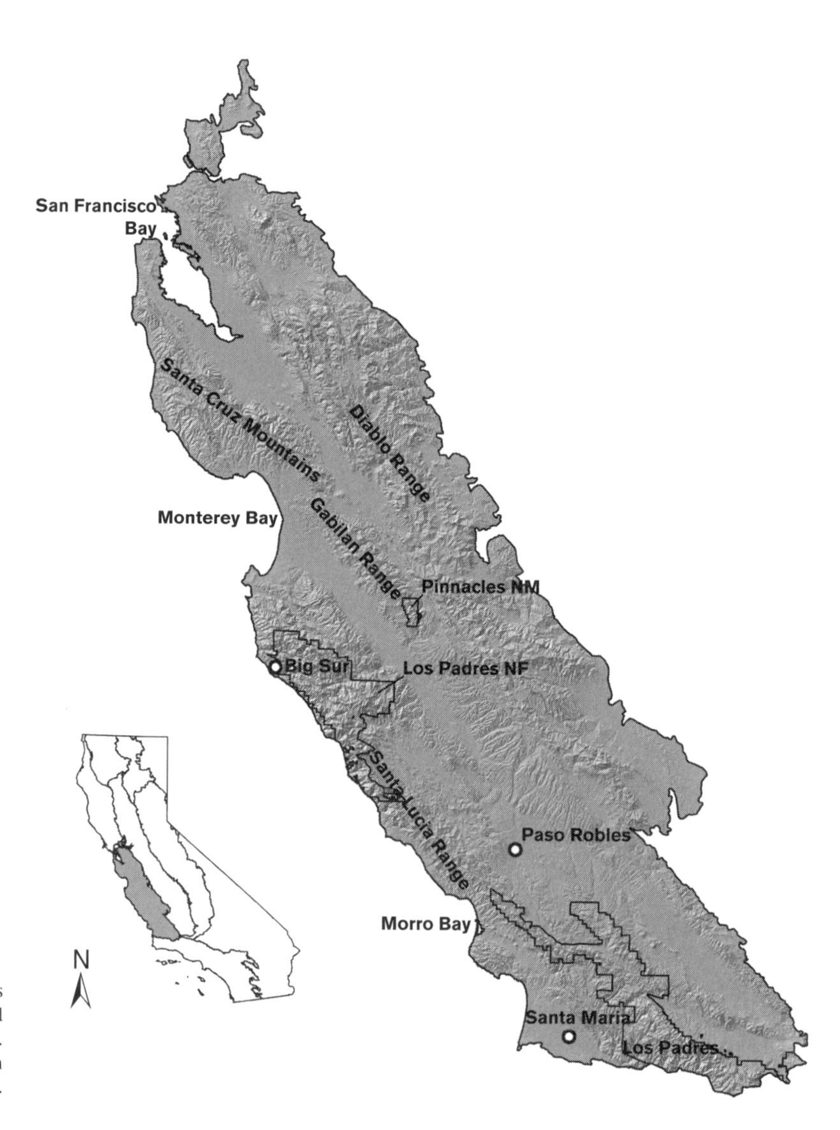

MAP 16.1 The Central Coast bioregion extends from Napa south to Santa Barbara counties and west to include the coast mountain ranges. Sections 261A and M262A are taken from Miles and Goudy (1997).

During much of the rainless season from May to October, a persistent marine layer develops along the coast of much of the bioregion. Cold, upwelling water from the California current interacts with the warm atmosphere creating low-level stratus clouds that form below 400–500 m (1,312–1,640 ft) in the summertime. Within these clouds, a fog layer frequently develops between 200 and 400 m (656–1,312 ft). Below 200 m (656 ft), the stratus clouds provide shade (overcast) rather than fog (Fischer et al. 2009). Coastal ranges prevent this marine layer from reaching more than a few km inland, except where it penetrates large river valleys, sometimes up to 60 km (37.3 mi) (Sawaske and Freyberg 2014). Coastal stratus clouds play a critical role in reducing ecosystem water loss and drought stress and therefore have a profound effect on the vegetation. The layer reduces plant water demand by (1) absorbing and reflecting large amounts of solar radiation, (2) lowering air temperatures, (3) increasing humidity, and (4) directly inputting moisture as mist and/or fog drip (Fischer et al. 2009, Sawaske and Freyberg 2014).

Weather Systems

In southern coastal California hot, windy Santa Ana events greatly increase the risk of large wildfires. However, in the Central Coast bioregion the relative location of the high-pressure center over Utah and Nevada, as well as the northwest-southeast axis of the Central Coast Ranges, limits the development of strong Santa Ana winds (Hughes and Hall 2010), although Santa Ana events become increasingly prevalent in the Santa Ynez Mountains at the southern end of the bioregion (Moritz 1997).

Summer convective storms and accompanying lightning activity are uncommon in the Central Coast due to a strong atmospheric inversion and cool coastal marine layer. In fact, except for the northern California coast, the Central Coast bioregion has the lowest incidence of lightning strikes of any region in the State. Lightning network data collected since 1985 indicate an average of only 2.99 strikes yr^{-1} 100^{-1} km^{-2} in the Central Coast region compared to 27.3 strikes yr^{-1} 100^{-1} km^{-2} for the Sonoran Desert Region of California or 19.6 strikes yr^{-1} 100^{-1} km^{-2} for the Sierra Nevada Region (van Wagtendonk and Cayan 2008). Although lightning-caused wildfires are rare, in the right combination of weather and fuel moisture conditions, fires can become quite large, as exemplified by the 72,500 ha (179,150 ac) Marble Cone fire of 1977 and in 2008 by the 65,918 ha (162,818 ac) Basin Complex Fires in the northern Santa Lucia Mountains.

Chaparral live fuel moisture peaks in April and declines steadily to minimum levels in September and October

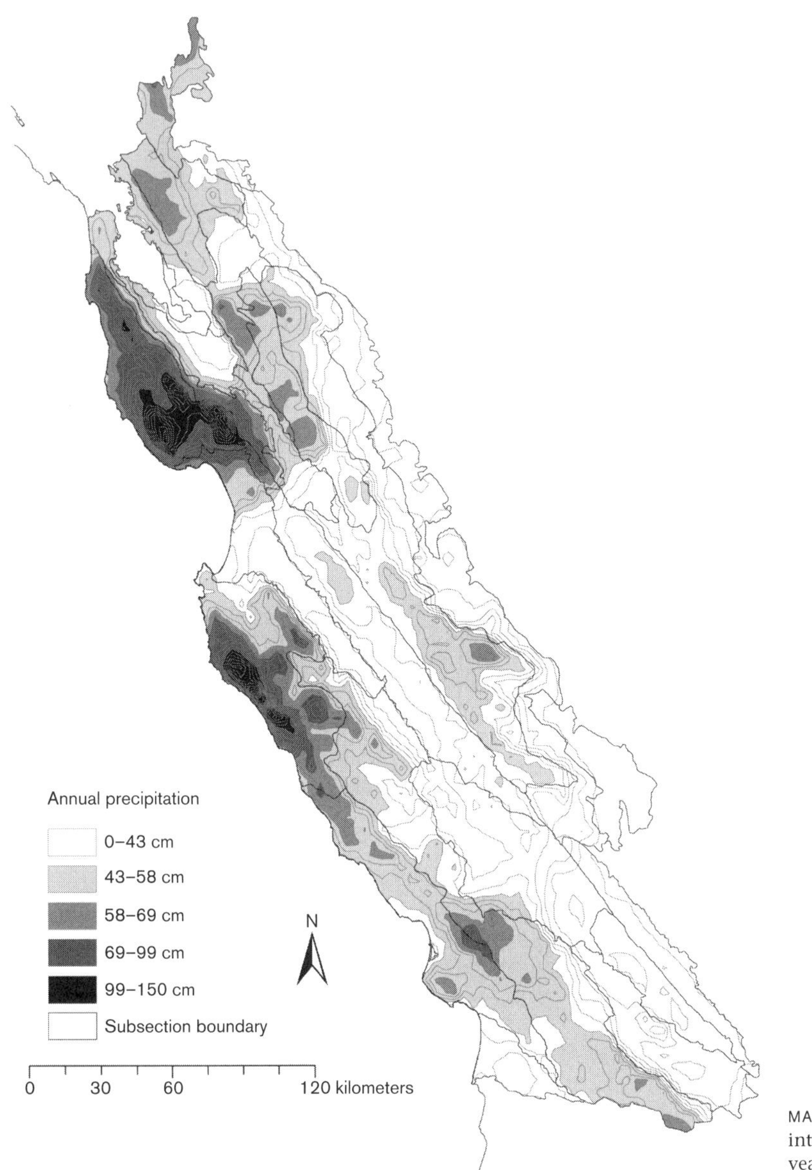

MAP 16.2 Isohyets of mean annual precipitation interpolated from weather station data for the years 1961–1990.

(Fig. 16.1). This fuel moisture pattern is typical of stations throughout the region, although there is high local variation as well as marked interannual variability associated with late winter and spring precipitation (Davis and Michaelsen 1995, Peterson et al. 2011). For example, between 1976 and 1999, the lowest summer fuel moisture levels occurred during the multiyear drought from 1984 to 1988 and 2012 to 2014. In contrast, the highest summer fuel moisture levels followed strong El Niño years (1982–1983, 1994–1995, and 1997–1998).

Long-term rainfall data from station records and tree rings indicate 2- to 7-year wet-dry cycles in coastal California between San Francisco and San Diego, at least for the past 400 to 600 years (Michaelsen et al. 1987, Haston and Michaelsen 1994, 1997). On the other hand, there is no clear relation between annual precipitation and the El Niño Southern Oscillation (ENSO) in the long-term record. The Central Coast often shows a precipitation pattern opposite that of southern California (south of Point Conception): wet years in the south often coincide with dry years in the north, and vice versa (Haston and Michaelsen 1997).

There is evidence of 20- to 50-year fluctuations in Central Coast rainfall, and even longer term precipitation patterns, including a generally wetter climate during the sixteenth and seventeenth centuries and relatively drier climate during the eighteenth and nineteenth centuries (Haston and Michaelsen 1994). The magnitude of climate variability has fluctuated considerably over 50- to 150-year periods and this variability has been increasing in past 30 to 40 years (Haston et al. 1988, Haston and Michaelsen 1997).

Human Geography

Since the early Holocene, native peoples have occupied the Central Coast bioregion at relatively high densities, especially along the immediate coastal plains, foothills, and valleys, where densities may have averaged one to three persons km^{-2} (2.6–7.8 persons mi^{-2}) (Beals and Hester 1974). Spanish settlement began in earnest in the last quarter of the eighteenth century with the construction of missions and the slow but

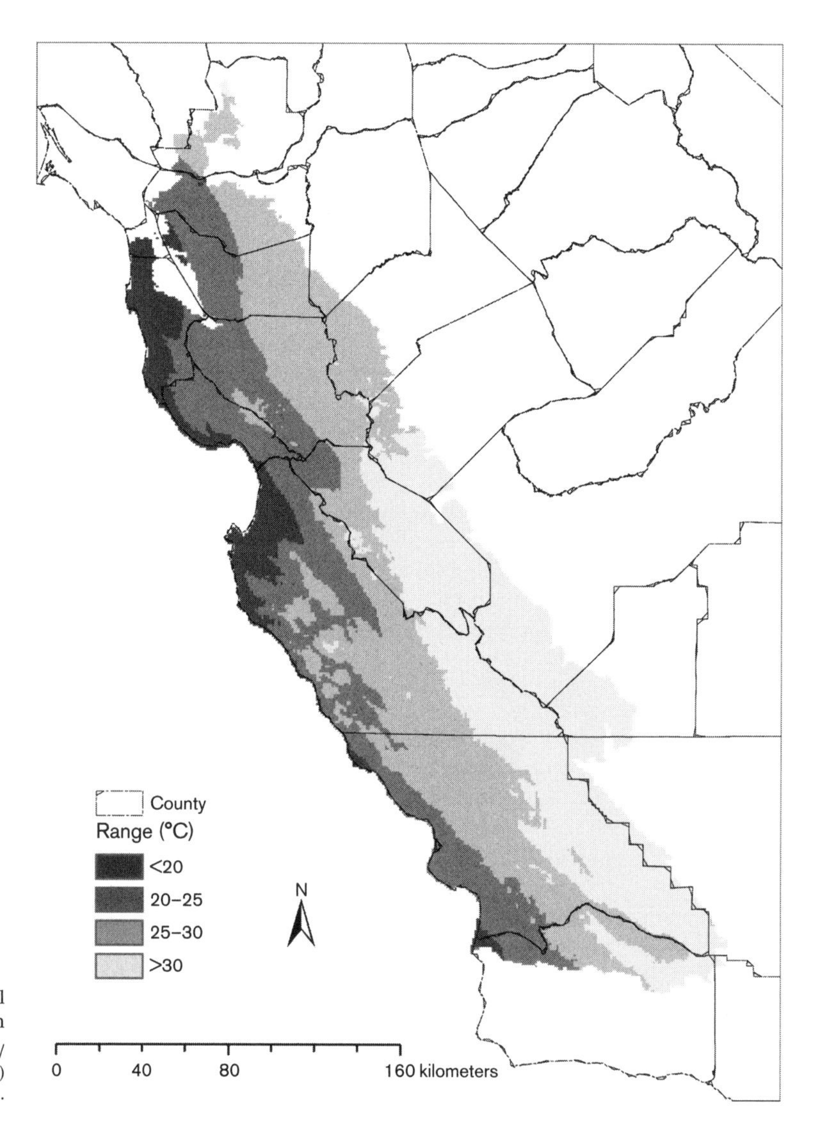

MAP 16.3. Intervals of mean annual temperature for the bioregion

SOURCE: DAYMET US Data Center, http://https://daymet.ornl.gov/. See Thornton et al. (1997) for details of the method.

steady influx of new settlers during the Mexican era from 1836 to 1850. The population climbed gradually through the early twentieth century and rapid growth did not commence in the region until the 1940s, especially in San Francisco Bay area counties (Fig. 16.2).

Most of the population increase has occurred near the coast, whereas the interior of the region remains rural. With the exception of the San Francisco Bay area and smaller urban centers such as Santa Cruz, Monterey, Salinas, Paso Robles, San Luis Obispo, and Santa Maria, the region remains sparsely settled. Eighty-seven percent of the region has a housing density of less than one house per 8 ha (20 ac). Late twentieth and early twenty-first century land-use change including the pronounced expansion of urban centers and extensive vineyard development in grasslands and oak savannas (Sleeter et al. 2011, Cameron et al. 2014). With the exception of large wilderness areas on federal lands, a dense network of public and private roads accesses rural lands.

Seventy-eight percent of the region is privately owned. Los Padres National Forest is the largest public landowner with 983,300 ha (2,429,720 ac), including large wilderness areas in the northern and southern Santa Lucia Ranges. Other large tracts of public land include Fort Hunter-Liggett, comprising 66,800 ha (165,066 ac) that adjoins Los Padres National Forest in southern Monterey County, Fort Ord (11,237 ha [27,767 ac]) on Monterey Bay, and Pinnacles National Monument (5,396 ha [13,333 ac]) in San Benito County.

Overview of Historical Fire Occurrence

Prehistoric Period

It is now widely accepted that Native Americans used fire extensively to manage vegetation in central coastal California. Native Americans occupied the entire coast at densities averaging one to three persons km^{-2} (7.8 persons mi^{-2}) (Keeley 2002b). The Ohlone (Castanoans) inhabited an area from San Francisco to Point Sur and regularly burned coastal vegetation to stimulate the seed production of preferred species (Lewis and Bean 1973, Gordon 1979, Cuthrell et al. 2012).

The Esselen Indians occupied a comparatively small area from Point Sur to Big Creek and inland to the Salinas River. South and east of the Esselens, the Salinans reached San Carpoforo Creek. Unfortunately, we know little about fire use by either group (Henson and Usner 1993). South of the Salinans,

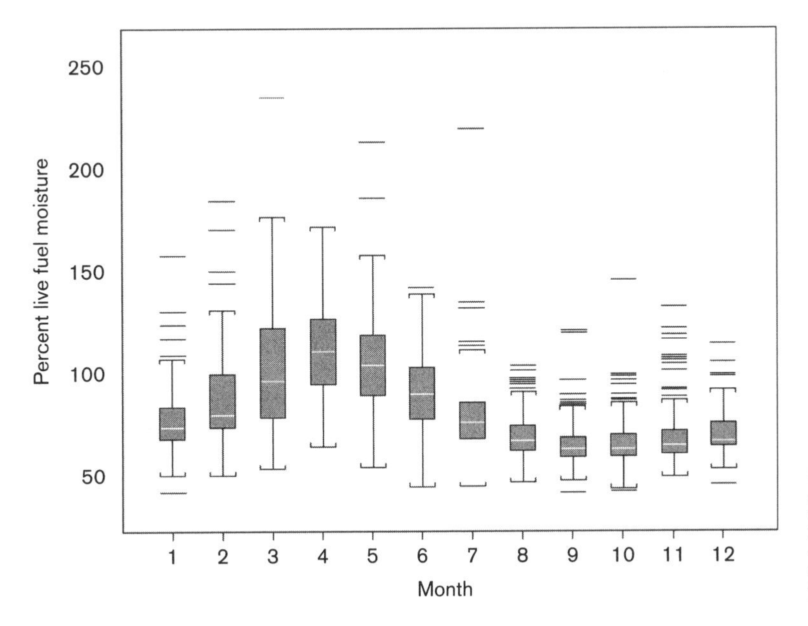

FIGURE 16.1 Boxplots of monthly green fuel moisture for chamise vegetation sampled from Monterey County to Santa Barbara County (seven locations) in Los Padres National Forest. Data are from 1976 to 1999 and are courtesy of Los Padres National Forest, Supervisor's Office.

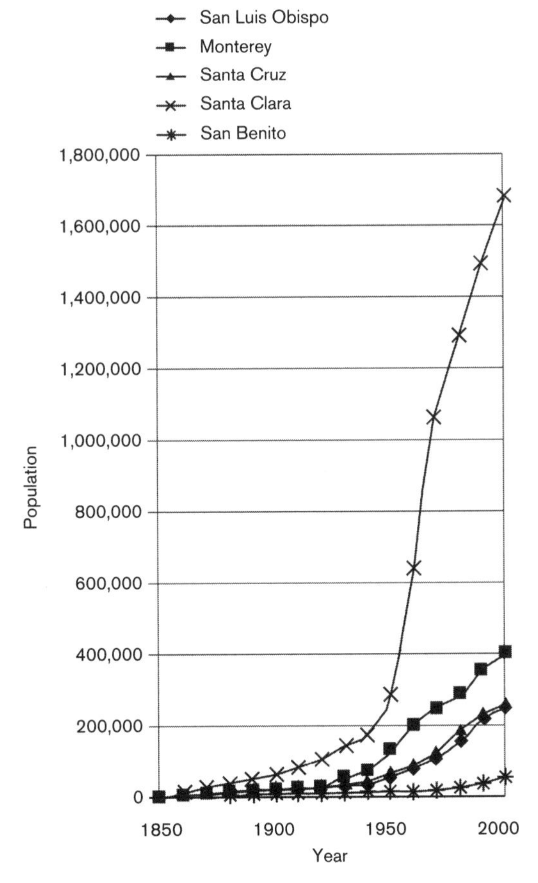

FIGURE 16.2 Population trends for five selected counties in the Central Coast bioregion based on data from the California Department of Finance Demographic Research Unit "Historical Census Populations of California State, Counties, Cities, Places, and Towns, 1850–2000."

the Chumash territory stretched from the Santa Maria River to the Santa Clara River on the coast and east to the upper Cuyama Valley (Keeley 2002b). Using diaries and journals of early explorers and clerics, Timbrook et al. (1982) documented that the Chumash, like the Ohlone to the north, regularly employed burning to encourage the growth of bulbs, green shoots and seeds of herbs like chia (*Salvia columbariae*) and Brewer's redmaids (*Calandrinia breweri*).

Further south, burning likely occurred not only in coastal prairie and oak woodlands, but also in chaparral and coastal sage scrub (Keeley 2002b). Frequent burning likely converted many areas of coastal sage scrub and chaparral to grasslands, although there were certainly large, rugged, remote areas that probably were little affected by Native American burning (Keeley 2002b). Similarly, there is little doubt that Native Americans augmented vegetation burning in the interior ranges just as they did along the coast. Greenlee and Moldenke (1982) posit that the Castanoans were burning grasslands and oak woodlands annually or semiannually. Chaparral and foothill pine (*Pinus sabiniana*) woodland were likely thinned or reduced in extent by the high frequency of deliberate or inadvertent fires in the region (Keeley 2002b).

The longest fire history records for the Santa Cruz Mountains have been reconstructed pollen and charcoal data in sediment cores. Cowart and Byrne (2013) analyzed a 3,000-year-old core from Skylark Pond near Point Año Nuevo. At Skylark Pond, fire was apparently not important between 1240 BC and 1400 AD. However, from 1400 to 1750 AD the charcoal content of the core increased suggesting regular burning by the Quiroste tribe. Anderson et al. (2013) examined a 6,200-year-old core from Glenmire in Point Reyes National Seashore. Between 436 and 1486 BC, they found that surface fires were frequent, perhaps set by semisedentary native populations. From 436 BC to 494 AD there was a sharp increase in charcoal deposition, which they attributed to the proliferation of sedentary native communities. This increase in purposeful burning likely changed regional vegetation to more inflammable types.

Analysis of charcoal particles in varved sediments from the Santa Barbara Basin from AD 1425 to 1985 furnishes the most detailed, long-term fire history for the southern Santa Lucia Ranges and western Transverse Ranges to the south (Byrne et al. 1977, Mensing et al. 1999). Mensing et al. (1999) recorded 23 fires >20,000 ha (<49,400 ac) in a 560-year record. The average interval between fires was 24 years (SD ± 18.4, $n = 22$) with a range of 5 to 75 years. Remarkably, the mean interval between large fires changed little from prehistoric to modern periods. Mensing et al. (1999) compared a time series analyses of tree-ring data from bigcone Douglas-fir (*Pseudotsuga macrocarpa*) (Haston and Michaelsen 1994) to the varve record and found large fires to be most common in the early years of multiyear droughts and at the end of wet periods.

The fire history depicted in the varve sediments probably only applies to the southern Santa Lucia Mountains from Santa Barbara north to San Luis Obispo, or possibly to Morro Bay. Still, large fires periodically burned along the coast of the northern Santa Lucia Mountains, as evidenced by the Marble Cone Fire (72,500 ha [179,075 ac]) in 1977, the Kirk Complex (35,100 ha [86,697 ac]) in 1999, and the 2008 Basin Complex (65,918 ha [162,818 ac]), all caused by lightning. Even before fire suppression began around 1910, a 20,000 ha (49,400 ac) human-caused fire burned in 1903, and a 60,000-ha (148,200 ac) fire occurred in 1906 (Henson and Usner 1993).

Fires >20,000 ha (>49,400 ac) probably have a long history in this region. Prehistoric mudflows in the Big Sur River (Jackson 1977) suggest that the average interval between large fires may have been longer than the interval gleaned from varve cores in the Santa Barbara region. The two most recent mudflow events both coincided with large fires in watersheds of the Big Sur River drainage. Assuming mudflows have followed all large fires, the mean interval between fires over the period 1370 AD to 1972 AD is estimated as 75 years (SD ± 19.7, $n = 8$). However, fire recurrence estimates from varve sediments and mudflows are not directly comparable since varve sediments represent a much larger area than watersheds of the Big Sur River.

Historic Period

In the Santa Cruz Mountains, the historic period begins around 1750. In the late 1790s, Spanish Franciscans arrived followed by Mexican nationals and finally by Euro-Americans in the mid-nineteenth century. The arrival of these emigrants brought lasting changes in land-use and fire regimes over much of the subregion. Spanish prohibitions on burning in 1739, the movement of Native Americans to missions and their depopulation, and cattle grazing during the mission era of the late eighteenth and early nineteenth centuries all significantly reduced fire use by the Castanoans (Anderson et al. 2013). However, after Euro-American settlement, burning increased once again. Fires became especially frequent between 1800 and 1930 when logging was prevalent in coast redwood forests (Brown et al. 1999, Stephens and Fry 2005, Anderson et al. 2013, Cowart and Byrne 2013, Jones 2014).

Similar trends in fire occurrence have been documented in the southern and interior portions of the Coast Ranges. Burning by aborigines declined with the advent of the Mission Period in the last quarter of the eighteenth century (Greenlee and Langenheim 1990). By the time Mexico ceded California to the United States in 1850, regular burning of coastal prairies and oak woodlands by Native Americans had ceased but chaparral burning probably expanded both to increase rangeland area and to facilitate travel. Fires still occurred but they were more likely accidental or lightning-caused rather than deliberate.

Current Period

The Developed Plains, Valleys, and Terraces (see subregion analysis below) are heavily urbanized. However, in the counties of San Francisco's East Bay region, grassland and shrubland vegetation is still abundant. Since 1945, the number of ignitions has increased along with an increasing population until ~1980 when ignitions plateaued (Keeley 2005). Over this 60-year period (1945–2005) fire size has decreased dramatically so that the largest percentage of fires is now <4 ha. Despite the overall increase in ignitions and reduction in fire size, the total area burned has shown little directional change in the 60-year period (Keeley 2005).

Since 1930, fire suppression has successfully controlled most wildfires in the Santa Cruz Mountains subregion but a reduction in ignitions also has been an important factor reducing the incidence of fire. Fire suppression has reduced fire frequency in all major vegetation types, but most dramatically in coastal prairie, where Greenlee and Langenheim (1990) estimate a modern mean interval of >1,000 years. Their estimated mean fire intervals for chaparral/coastal sage scrub, oak woodland, and mixed evergreen forest are on the order of 150 years to 250 years. Recent burning has been most prevalent in coast redwood forests, but even in that type the mean fire interval is estimated to be 100 years to 150 years.

Although separated by less than 75 km (46.6 mi), the northern Santa Lucia Ranges and Santa Cruz Mountains provide a dramatic contrast in modern fire histories. Fire is much more widespread in the Santa Lucia Ranges, where roughly one-quarter of the region has burned at least once since 1950 (Oneal et al. 2006). Furthermore, fires that have occurred in recent times in the Santa Lucia Ranges are far larger than those in the Santa Cruz Mountains (Map 16. 4). The largest modern fires in the Santa Cruz Mountains, including the Pine Mountain Fire an 6,400 ha (15,808 ac) in 1948, the 1320-ha (3,260 ac) Lincoln Hill Fire in 1962, and the 5,314 ha (13,125 ac) Lexington Fire of 1985 (which mainly burned east of the Santa Cruz Mountains in the Leeward Hills), are an order of magnitude smaller than the largest fires in the northern Santa Lucia Ranges just south of Monterey Bay.

Relatively detailed records of twentieth-century fires on Los Padres National Forest have been analyzed by Davis and Michaelsen (1995), Mensing et al. (1999), and Moritz (1997, 2003). A large fraction of area burned in the Santa Lucia Ranges since 1900 can be attributed to a few very large fires. Most of these large fires have been human-ignited (except the Marble Cone, Kirk Complex and Basin Complex fires) and many of the large fires at the southern end of the region have spread under severe weather conditions of high temperature and winds (Davis and Michaelsen 1995, Moritz 1997, 2003).

Based on analysis of Los Padres fire history data, Moritz (1997, 2003) concluded that fire hazard in the Santa Lucia Ranges is not significantly related to fuel age but is controlled instead by extreme weather events. A combination of rugged terrain and poor access into remote wilderness areas has limited the ability of firefighting agencies to control fire spread in these weather conditions, and, despite improved suppression, there does not seem to be a temporal trend in large fire frequency in this region (Moritz 1997), although the introduction of fixed-wing aircraft and helicopters after 1950 has proved effective in reducing fire spread under more moderate conditions (Moritz 1997).

In the Inner Coast Ranges, maps of fires that burned between 1950 and 1998 were compiled by the Forest Service and California Department of Forestry and Fire Protection (CDF&FP) (Map 16. 4). Only fires >120 ha (>300 ac) are mapped on private lands. We also obtained fire history records from Pinnacles National Monument, which provide a more complete and accurate history of fires from the interior Diablo and Gabilan Ranges. In comparing the two datasets we found the

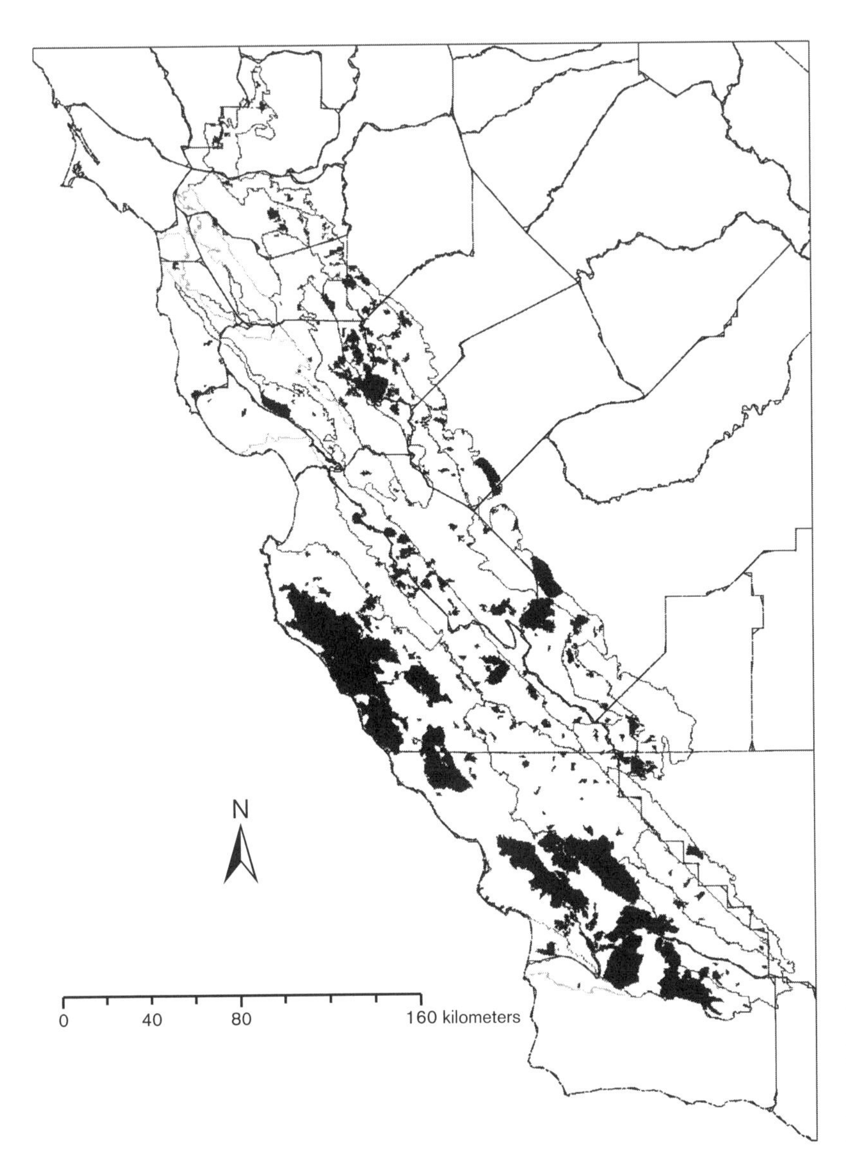

MAP 16.4 Areas >120 ha (300 ac) that have burned at least once since 1950. The burn areas (shaded black) are superimposed on the four subregions displayed in Map 16.5. Fire perimeter data are from the California Department of Forestry and Fire Protection (http://frap.cdf.ca.gov/projects/fire_data/fire_perimeters/index.asp).

CDF&FP data to be incomplete for lands outside of the national forests. Nevertheless, these data provide a good general picture of fire frequency and size across the region. Based on these records, at least 40% of the region has burned at least once since 1950, with fires concentrated in shrublands and mixed evergreen forests of the northern Santa Lucia Ranges.

Although uncommon, lightning fires occur with greater frequency in the Interior Ranges than in the Santa Lucia Ranges or Santa Cruz Mountains. Between 1930 and 1979, fire history data record 142 lightning-caused fires out of a total of 3,086 fires (4.6%) in the Gabilan and Diablo Ranges (Greenlee and Moldenke 1982). Eighty-six percent of these lightning fires occurred in grasslands or oak woodlands and, with one exception, all lightning fires started between May and October. Nearly half of the fires burned in September. Humans started 95% of all recorded fires from 1930 to 1979 (Greenlee and Moldenke 1982). Some of this is due to the widespread use of controlled burning for rangeland improvement. Sixteen percent of the fires recorded during this period were characterized as deliberate "brush burning." In the period 1951 to 1978, 22,814 ha (56,350 ac) were deliberately burned in San Benito and western Fresno Counties alone.

At least 15% of the Interior Ranges burned at least once between 1950 and 1998 in fires >120 ha (296 ac) (Map 16.4), compared to 40% of the Santa Lucia Ranges and 3% of the Santa Cruz Mountains. Large fires in the Interior Ranges are most common in southern San Benito County (Hepsedam Peak and San Benito Mountain) and in the Gabilan Ranges north and east of Pinnacles National Monument. Based on fire scar data and maps of fire perimeters, Greenlee and Moldenke (1982) concluded that fire suppression efforts have reduced fire frequency in the region from every 10 to 30 years prior to 1930 to a current recurrence interval of 25 to 35 years, depending on vegetation, topography, and exposure.

Major Ecological Zones

Subregion Analysis

To help discriminate systematic geographical variation in environmental conditions and associated fire regimes in the Central Coast bioregion, we subjected the subsection data (Table 16.1) to Principal Components Analysis (PCA) using

TABLE 16.1

Twelve variables for the 23 subsections of the Central Coast bioregion used in the Principal Components Analysis. Subsection names and codes are taken from Miles and Goudey (1997)

The analysis produced four distinct subregions shown in Map 16.4. Climate statistics are based on 1-km² grids (Thornton et al. 1997). Percentages in different land-use and land cover types are based on 100-m resolution, multisource land cover data (California Wildlife Habitat Relationship System) obtained from the California Department of Forestry and Fire Protection (CDF&FP). "Burned" is the percentage of the subsection that burned at least once since 1950 using fire history data obtained from CDF&FP. Relief is derived from 100-m resolution digital elevation data using the mean and standard deviation of elevation (m) values taken from a 3 × 3 moving window passed over the subsection

Subsection code	Subsection name	Area (sq. km)	Annual ppt (1)	Ann. temp. tange (2)	August max temp. (3)	Chaparral (4)	Crop (5)	Grassland (6)	Blue oak woodland (7)	Redwood–Douglas-Fir (8)	Montane hardwood–mixed hardwood conifer (9)	Urban (10)	% Burned (11)	Relief (12)
261Ad	East Bay Terraces and Alluvium	505	21.08	21.54	26.74	0.00	1.38	1.49	0.02	0.00	0.00	94.51	0.18	2.48
261Ai	San Francisco Peninsula	195	26.67	16.91	23.03	3.92	0.23	1.34	0.03	0.00	0.00	91.92	3.78	7.73
261Ab	Bay Flats	272	19.52	21.90	26.91	0.00	1.10	1.30	0.00	0.00	0.00	80.63	0.00	0.79
261Ae	Santa Clara Valley	1,180	19.79	23.43	27.66	0.05	27.10	13.67	0.00	0.00	0.00	57.70	0.69	3.08
261Ah	Watsonville Plain–Salinas Valley	1,622	20.64	20.87	25.39	3.24	55.92	10.54	0.78	1.14	0.00	15.88	0.79	5.15
261Al	Santa Maria Valley	620	21.41	20.98	25.65	7.08	45.91	10.83	4.14	0.00	0.33	19.16	6.31	3.72
261Af	Santa Cruz Mountains	1,995	38.42	20.25	23.50	21.07	3.02	7.16	0.44	36.04	0.65	8.88	2.60	24.72
261Aj	North Coastal Santa Lucia Range	2,522	37.41	22.61	23.63	42.98	0.19	9.98	0.12	2.38	10.83	2.77	51.59	34.50
M262Ae	Interior Santa Lucia Range	4,974	23.92	27.50	28.62	44.15	4.57	19.26	2.14	0.00	1.54	0.91	40.79	20.62
261Ak	South Coastal Santa Lucia Range	2,257	30.23	23.49	26.22	22.77	4.69	31.45	2.11	0.17	6.67	4.29	27.87	21.11
261Aa	Suisun Hills and Valleys	889	23.68	26.63	30.52	0.15	7.32	47.86	0.64	0.00	1.01	33.97	6.84	10.28
261Ac	East Bay Hills–Mount Diablo	1,190	21.56	23.87	28.04	11.82	0.15	24.76	0.12	0.38	1.37	35.08	4.25	19.36
M262Ad	Eastern Hills	3,567	15.45	31.74	33.09	3.83	1.00	83.07	2.97	0.00	0.00	2.23	16.58	16.43
M262Aa	Fremont–Livermore Hills and Valleys	9,85	18.12	26.14	29.22	1.11	5.58	51.59	2.14	0.00	0.85	21.27	2.52	16.55
261Ag	Leeward Hills	653	28.55	23.61	26.18	16.44	2.99	14.77	0.17	1.99	0.30	25.12	10.41	23.44
M262Ac	Diablo Range	4,739	19.58	32.51	31.91	19.74	0.77	33.42	25.98	0.00	0.55	0.35	14.16	24.48
M262Ab	Western Diablo Range	1,352	19.33	28.66	29.15	8.95	0.78	32.72	10.98	0.00	3.90	0.34	7.98	29.11
M262Af	Gabilan Range	2,424	20.15	29.10	30.05	20.49	1.34	49.01	0.55	0.00	0.51	0.19	14.77	17.29
M262Ag	Kettleman Hills and Valleys	1,055	11.09	31.83	34.40	0.00	38.04	52.88	0.06	0.00	0.00	6.22	3.46	4.02
M262Ah	Paso Robles Hills and Valleys	2,574	21.75	29.82	31.13	2.81	7.01	73.65	0.75	0.00	0.07	2.55	7.51	8.36
M262Ak	Temblor Range	939	15.03	32.31	32.52	2.19	0.08	72.97	0.04	0.00	0.00	0.28	3.72	17.81
M262Ai	Carrizo Plain	857	13.96	31.27	31.87	0.11	0.01	88.27	0.04	0.00	0.00	0.01	4.26	4.35
M262Aj	Caliente Range–Cuyama Valley	1,462	15.72	30.39	30.79	21.82	7.37	44.78	0.82	0.00	0.38	1.22	13.62	13.37

the correlation matrix. Although the modern fire history differs considerably from the historic regimes (before 1910), we included modern fire history in the analysis because it is correlated with environmental factors such as vegetation, climate, topography, population density, and land-use.

The analysis revealed four distinct geographical and environmental clusters (Map 16.5) which we refer to as: (1) Developed plains, valleys, and terraces, (2) the Santa Cruz Mountains, (3) the Santa Lucia Ranges, and (4) the Interior Coast Ranges. We discuss fire history, plant communities, and modern fire regimes in the four ecological subregions and provide a more detailed analysis of the fire ecology and plant community—fire regime interactions for selected tree species and plant communities.

The developed plains, valleys, and terraces subregion (Map 16.5, square 1) consists of flat areas dominated by urban and agricultural land-use. There are three discrete areas in this subregion: the Santa Maria Valley, the Watsonville Plain and Salinas Valley, and the San Francisco Bay peninsula, bay flats, and East Bay terraces. In the East Bay of San Francisco, ignition frequency is high, fire size is small, and the area burned has remained largely unchanged between 1945 and 2002 (Keeley 2005). Overall, 16% of the Central Coast region has been converted to urban and agricultural uses, and by 2050, the conversion is expected to increase to 26% (Riordan and Rundel 2009). Grassland cover has decreased and tree and shrub cover has increased in remaining open spaces over the past 75 years (Russell and McBride 2003).

Of the remaining subregions, the Santa Cruz Mountains subregion (Map 16.4, square 2) is distinctive in combining high rainfall (Map 16.2), high topographic relief, extensive forests of Douglas-fir (*Pseudotsuga menziesii* var. *menziesii*), and coast redwoods (*Sequoia sempervirens*) but low fire occurrence.

The southern coastal, northern coastal, and interior Santa Lucia Ranges together form the Santa Lucia Ranges subregion (Map 16.4, square 3) which is characterized by extreme ruggedness, moderate rainfall and continentality, extensive shrublands, montane hardwood forests and mixed hardwood-conifer forests, and a high occurrence of wildfires.

The Interior Coast Ranges subregion (Map 16.4, square 4) is characterized by moderate relief, low rainfall, high summer temperatures, extensive grasslands and oak woodlands, and an intermediate wildfire frequency. Two subsections, the East Bay Hills-Mount Diablo (261Ac in Table 16.1) and the Leeward Hills (261Ag) are intermediate in character between the Santa Cruz Mountains and the Interior Coast Ranges but rather than create a separate subregion, we combined them with the Interior Coast Ranges subregion.

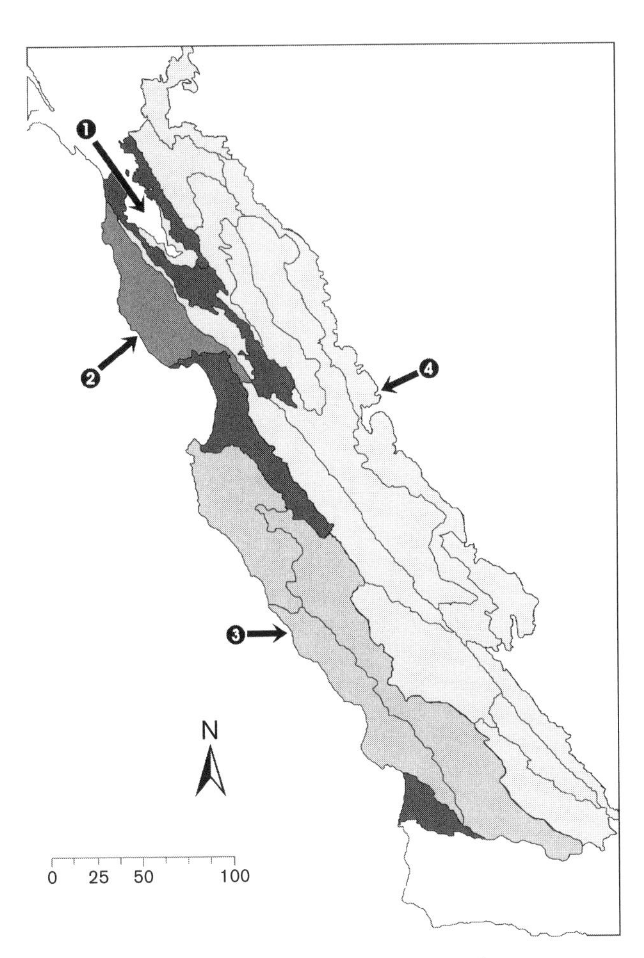

MAP 16.5 Subregion map based on climate, topography, vegetation, post-1950 fire history, and land-use (Table 16.1). Subregions are: (1) developed plains, valleys, and terraces; (2) the Santa Cruz Mountains; (3) the Santa Lucia Ranges; and (4) the Interior Coast Ranges.

Santa Cruz Mountains Subregion

Roughly 12% of the Santa Cruz Mountains has been converted to urban and agricultural uses. Major ecological zones include: (1) coastal prairie and coastal sage scrub, (2) coast redwood–Douglas-fir forest, and coast redwood–mixed evergreen forest, (3) chaparral, and (4) oak forests and woodlands.

The coastal prairie and coastal scrub zone is most prevalent below 300 m (984 ft) elevation between Point Año Nuevo and Pillar Point (Fig. 16.3). Characteristic species of these types include coyote brush (*Baccharis pilularis*), seaside woolly sunflower (*Eriophyllum staechadifolium*), bracken fern (*Pteridium aquilinum* var. *pubescens*), tufted hair grass (*Deschampsia cespitosa*), and California oat grass (*Danthonia californica*).

Coast redwood–Douglas-fir forests and coast redwood–mixed evergreen forests cover many slopes of the Santa Cruz Mountains above 300 m (984 ft). Coast redwood is particularly common in more mesic sites on the western slopes. Mixed evergreen forests of tanoak (*Notholithocarpus densiflora* var. *densiflora*), coast live oak (*Quercus agrifolia*), Pacific madrone (*Arbutus menziesii*), and California bay (*Umbellularia californica*) tend to occur on drier sites in topo-mosaics with coast redwood–Douglas-fir forests, and are more widespread in the more interior portions of the Santa Cruz Mountains.

Chaparral and oak woodlands occupy xeric sites but also form an extensive zone below 600 m (1,968 ft) along the eastern interior edge of the Santa Cruz Mountains. Characteristic species include: chamise (*Adenostoma fasciculatum*), buckbrush (*Ceanothus cuneatus* var. *cuneatus*), coast live oak, and valley oak (*Q. lobata*).

Chapter 10 (North Coast Bioregion) describes the fire ecology, fire regimes, and community dynamics for coast redwood forests in northwest California. The degree of fire-dependence in this community type is best viewed as a continuum and that more southern and drier coast redwoods are more fire-prone (Oneal et al. 2006) than those to the north. Because coast redwoods and other widespread types such as chaparral and mixed evergreen forests are treated in

FIGURE 16.3 Grassland, chaparral, and coastal scrub mosaic in the foothill and lower montane zones of the southern Santa Lucia Ranges subregion. View is looking northwest from Cuesta Grade toward Morro Bay, San Luis Obispo County (photo by Christopher Cogan).

detail in chapters 10 and 17 (South Coast Bioregion), we limit our discussion to two distinctive types: Bishop pine (*Pinus muricata*) and Monterey pine (*P. radiata*) forests.

BISHOP PINE ECOLOGICAL ZONE

Fire Ecology The Central Coast bioregion harbors eight closed-cone conifers, four pines, and four cypresses. All three members of the pine subsection Oocarpae are represented: Bishop pine, knobcone pine (*Pinus attenuata*), and Monterey pine. Coulter pine (*P. coulteri*) is the only closed-cone member of the Sabinianae subsection that includes Torrey pine (*P. torreyana*) and foothill pine. Cypress species are Monterey cypress (*Hesperocyparis macrocarpa*), Gowen cypress (*H. goveniana* subsp. *goveniana*), Santa Cruz cypress (*H. abramsiana*), and Sargent cypress (*H. sargentii*). Seven of the closed-cone conifers are patchily distributed in the bioregion near the coast and in the coastal mountains. Only Coulter pine extends inland, i.e., into the Diablo Range and San Benito Mountains in the north and into the La Panza Range in the south.

Bishop pine is nonsprouting, moderately serotinous conifer (Keeley and Zedler 1998) (Appendix 1), that has distinct northern and southern populations. Bishop pine does not require fires to free seeds from the cones; cones also open on hot days and seedlings sometimes establish in the understories of mature stands.

In 1995, a high-intensity wildfire burned 423 ha (1,045 ac) of the mature Bishop pine forest at Point Reyes. Seedling establishment in the first year postfire was high throughout the burned pine forest with densities averaging 249,750 trees ha^{-1} (617,145 trees ac^{-1}) (Harvey and Holzman 2014). Although seedlings predictably established in the prefire forest, they also expanded into surrounding coastal sage scrub, and to a lesser extent into grasslands, well beyond the prefire forest boundary. In fact, Bishop pine forest increased 359 ha (887 ac) (85%) beyond its prefire size (Forrestel et al. 2011). Although much of the areal increase was due to high winds blowing small seeds into coastal sage scrub and grasslands, isolated burned trees outside the contiguous forest were an surprising source of recruits: as many as 1,800 trees ha^{-1} (4,448 ac^{-1}) recruited in stands over 300 m (984 ft) from the nearest prefire tree (Harvey et al. 2011).

Fire Regime–Plant Community Interactions Serotiny in Bishop pine increases from northern to southern populations (Duffield 1951). Compared to the southern populations, the northern populations burn relatively infrequently and the average fire size tends to be much smaller (Greenlee and Langenheim 1990). Even within the northern population, Millar (1986) describes an uncharacteristically high degree of serotiny in five Bishop pine populations growing inland of coastal populations. It would be interesting to know if other fire-related traits in Bishop pine such as bark thickness, self-pruning, foliage flammability, age of cone production, and seed dispersal also vary systematically with the north to south change in fire regime.

MONTEREY PINE ECOLOGICAL ZONE

Of the three members of the Oocarpae, Monterey pine is the most restricted in its distribution and is the least variable genetically (Millar et al. 1988). Like Bishop pine, Monterey pine is a moderately serotinous nonsprouter, and seed release is highest following fires but some cones open every year, providing continuous seed input for inter-fire regeneration (McDonald and Laacke 1990) (Appendix 1). It is short-lived (80 years to 120 years) and has thick basal bark. In unburned stands seedling establishment is highest where shrub cover and duff depth are low but canopy cover is high (O'Brien et al. 2007).

Fire Regime–Plant Community Interactions There are no long-term fire ecology studies of Monterey pine within its native distribution. In 1994, White (1999) revisited 38 pine stands on the Monterey Peninsula that he had first sampled in 1965 and 1966. He concluded that, while seedling establishment is most abundant following fire, long unburned pine stands do not convert to coast live oak forests, the most frequent understory tree, but instead continue to be dominated by Monterey pine (Fig. 16.4).

A fire history study of Monterey pine in Santa Cruz County covering a period from 1893 to 1976 revealed that forests

FIGURE 16.4 Monterey pine forest understory near Cambria dominated by blackberry and poison oak. This site has not burned for many decades.

mostly burned in low- and mixed-severity fires. Mean fire return interval ranged from 11 to 20 years (Stephens et al. 2004). Many Monterey pine forests now are threatened by "senescent risk" (Zedler 1995), i.e., ageing trees are losing their aerial seed banks (closed cones) before fire opens them. In addition to trees dying of old age, pitch canker (*Fusarium circinatum*) outbreaks have killed trees or caused extensive crown dieback (Wikler et al. 2003).

Santa Lucia Ranges Subregion

Widespread low-elevation coastal alliances in this subregion include California annual grassland, coyote brush, California sagebrush (*Artemisia californica*), blue blossom, and other as yet undescribed alliances in the Diablan, Franciscan, and Lucian coastal sage scrub associations. Localized coastal vegetation types of special interest include closed-cone pine forests such as Bishop pine forests, knobcone pine forests, and Monterey pine forests along the Monterey and Big Sur Coast, as well as maritime chaparral in which chamise codominates with coast live oak and a number of endemic ceanothus and manzanita species.

Maritime chaparral is associated with sandy substrates on level or rolling terrain within 10–20 km (6–12 mi) of the coast. Stands of northern and central maritime chaparral are scattered along the coast from northern Santa Barbara County north to Sonoma County. These areas experience a strong maritime climate characterized by frequent summer fog and low annual temperatures (Vasey et al. 2014). Maritime chaparral supports many rare and endemic plants (Fischer et al. 2009) and has received a fair amount of scientific study, because it has been heavily reduced and fragmented by coastal residential development and military operations (Lambrinos 2000, Van Dyke and Holl 2001).

Foothill and lower montane vegetation includes a diverse variety of localized types that generally can be characterized as a mosaic of coastal scrub, chaparral, grassland, and foothill woodlands. Widespread scrub types include black sage (*Salvia mellifera*), purple sage (*S. leucophylla*), and California buckwheat (*Eriogonum fasciculatum*). Widespread chaparral alliances include chamise, buckbrush, and scrub oak (*Quercus berberidifolia*). Blue oak and coast live oak are the most widespread woodland alliances. Sargent cypress forests are one of the many distinctive mid-elevation plant communities associated with infertile ultramafic soils.

Coast live oak occurs in many coastal and foothill settings and occurs as closed forests (typically >60% crown cover), open woodlands (40–60%), and savannas (10% crown closure with an herbaceous understory), and mixed with coastal sage scrub, riparian forests, and chaparral (Griffin 1988, Allen et al. 1991, Peinado et al. 1997, Sawyer and Keeler-Wolf 2009). Coast live oak forests often occupy mesic north-facing slopes, shaded canyons, or border riparian zones, while coast live oak woodlands and savannas grow in a broad range of settings at low-to-mid elevations.

Mixed evergreen forests (oaks, Pacific madrone, tanoak, California bay, and big-leaf maple [*Acer macrophyllum*]) are abundant in both the Santa Cruz Mountains and in the northern Santa Lucia Range. Hardwood-conifer forests (ponderosa pine [*Pinus ponderosa*], Coulter pine, sugar pine [*P. lambertiana*], canyon live oak [*Q. chrysolepis*], and tanoak) are widespread in the north coastal Santa Lucia Ranges at the highest elevations (Griffin 1975, Davis et al. 2010). Common single-species alliances include canyon live oak, California bay, tanoak, and Coulter pine.

Endemic bristlecone fir (*Abies bracteata*) forests are patchily distributed at mid-to-high elevations, mainly in the watersheds of the Big Sur, Little Sur, and Upper Carmel Rivers in Monterey County (Talley 1974). High-elevation coastal forests include ponderosa pine and sugar pine alliances. In the southern Santa Lucia Mountains mixed evergreen forests, conifer-hardwood forests and montane conifer forests all but disappear or become highly localized. Here, Coulter pine/montane chaparral is the most abundant conifer type, especially in the La Panza Range (Borchert et al. 2004).

From the perspective of regional fire ecology, chaparral and mixed evergreen forest are the most widespread fire-prone vegetation types in the Santa Lucia Range subregion. Chapter 17 discusses the fire ecology of chaparral in South Coast bioregion and applies to chaparral in the Central Coast as well, although there are likely systematic differences in fire regimes due to climate, weather, and human population densities. In the following sections we limit our discussion to coast live oak forests and woodlands, maritime chaparral, Coulter pine woodlands, knobcone pine forests, and Sargent cypress forests.

FIGURE 16.5 Patches of coast live oak forest in a matrix of annual grassland on northern footslopes of the Purisima Hills, Los Alamos Valley, Santa Barbara County.

COAST LIVE OAK FOREST AND WOODLAND ECOLOGICAL ZONE

Fire Ecology Coast live oak is one of the most fire-resistant oaks in California. Only the more geographically restricted Engelmann oak (*Quercus engelmannii*) is perhaps more fire-tolerant (Lathrop and Osborne 1991). Coast live oak seedlings and saplings survive low-intensity surface fires, which burn in the matrix of annual grasslands (Snow 1980), although seedling mortality increases during high-intensity surface fires and crown fires (Fig. 16.5). Adult trees exhibit a number of fire adaptations, including a dense outer bark, a thick inner bark with high insulating capacity, and an ability to resprout from the base and crown (epicormic sprouts) following severe wildfires (Plumb 1980). Adult survival rates exceeding 95% have been documented following severe wildfire, and canopy volume may return to prefire levels within 5 to 10 years (Plumb 1980, Dagit 2002). Mortality rates are higher in late-season fires and for oaks growing in chaparral or coastal sage scrub where fire intensity is more extreme (Wells 1962, Plumb and Gomez 1983, Davis et al. 1988).

Fire Regime–Plant Community Interactions Pollen records from the Santa Barbara Channel and Zaca Lake in northern Santa Barbara County indicate that coast live oak was relatively stable for many centuries prior to European settlement but has increased in the last quarter of the nineteenth century (Mensing 1998, Dingemans et al. 2014).

Coast live oak is widespread in grasslands and oak savannas of the region but appears to be declining in these settings due to tree removal and low recruitment of tree-sized individuals caused by drought, and herbivory by rodents, deer, cattle, and insects (Plumb and Hannah 1991, Callaway and Davis 1998, Parikh and Gale 1998, Dunning et al. 2003).

In coast live oak forests the litter layer is often deep, and perennials such as western poison oak (*Toxicodendron diversilobum*), toyon (*Heteromeles arbutifolia*), and bracken fern form a discontinuous herb and shrub understory (Campbell 1980, Allen et al. 1991). For much of the year, fuel moisture of the shrubs and litter is high and conditions are not conducive to surface-fire ignition and spread. However, forests become increasingly susceptible to fire during the long rainless months of summer and fall.

Callaway and Davis (1993) documented a shifting mosaic of four vegetation types associated with coast live oak in Gaviota State Park and on neighboring ranchlands between 1947 and 1989 in the Santa Ynez Mountains. Coast live oak is shade tolerant (Callaway 1992) and recruits into both chaparral and coastal sage scrub on many substrates as well as into more mesic settings such as north-facing slopes and areas bordering riparian areas (Wells 1962, Callaway and Davis 1993, 1998, Parikh and Gale 1998, Bendix and Cowell 2010). Nevertheless, high-intensity fires in shrublands probably kill most seedlings and saplings, thereby reversing any increase in oak cover that takes place during fire-free periods (Wells 1962, Callaway and Davis 1993, Van Dyke and Holl 2001). It appears that on many sites the presence of oak woodland versus chaparral or coastal sage scrub depends on whether sufficient time has elapsed between fires for oaks to establish and grow large enough to endure high-intensity fires (Appendix 1).

MARITIME CHAPARRAL ECOLOGICAL ZONE

Fire Ecology Maritime chaparral is distributed from Mendocino County south to northern Baja but has its highest species diversity in central coastal California. It occupies the immediate coast and coastal uplands where summers are relatively cool and moist compared to the hotter, drier inland areas (Inner Coast Ranges) (Vasey et al. 2014). Compared to inland chaparral, maritime chaparral has nearly twice the number of shrub species, many of which are endemic. Moreover, it has a much higher ratio of obligate seeders to facultative seeders than inland chaparral which is dominated by facultative seeders. Vasey et al. (2012, 2014) attribute the high shrub endemism of maritime chaparral to increased availability of water in the summer, increased annual rainfall and a reduced vulnerability to freezing (Coale et al. 2011).

Maritime chaparral usually is dominated by chamise in combination with several locally endemic *Ceanothus* and *Arctostaphylos* species. In the Central Coast California Bioregion, endemic shrub species include obligate-seeders such as Santa Barbara ceanothus (*Ceanothus impressus*), Lompoc ceanothus (*C. cuneatus* var. *fascicularis*), La Purisima manzanita (*Arctostaphylos purissima*), Hooker's manzanita (*A. hookeri* subsp. *hookeri*), sandmat manzanita (*A. pumila*), Pajaro manzanita (*A. pajaroensis*), and Morro manzanita (*A. morroensis*), and the resprouting Sand Mesa manzanita (*A. rudis*) (Griffin 1978a,

FIGURE 16.6 Maritime chaparral on Burton Mesa, northern Santa Barbara, 2 years after burning. Peak rush-rose (*Crocanthemum scoparium*) is a conspicuous member of the diverse postfire community.

Davis et al. 1988). Subshrub and herb layer diversity can be high, especially for the first 5 years following fire (Fig. 16.6).

Many rare and endemic shrubs in maritime chaparral are fire-dependent. For example, Odion and Tyler (2002) observed high levels of fire-induced mortality in the soil seed bank of the endangered Morro manzanita and concluded that the species may require considerably longer than 40 years between burns to establish a seed bank that is adequate to replace adults killed by fire.

Fire Regime–Plant Community Interactions Lightning is rare along the coast and it is probably safe to assume that the fire regime of maritime chaparral has been anthropogenic for millennia, especially given the prehistoric densities of Native Americans in coastal areas supporting maritime chaparral (e.g., lower Santa Ynez River Valley, Pismo Bay, Morro Bay, and Monterey Bay [Keeley 2002b]). Greenlee and Langenheim (1990) estimated prehistoric fire return intervals near Monterey Bay to be on the order of 10 years to 100 years. Today human-caused ignitions are frequent in maritime chaparral but wildfires are quickly suppressed at roads and fuel breaks. As a result, fires now rarely exceed 100 ha (247 ac) (Davis et al. 1988, Van Dyke and Holl 2001, Odion and Tyler 2002). In maritime chaparral to the east of Vandenberg Air Force Base near Lompoc, Davis et al. (1988) documented only 27 fires larger than 1 ha (2.4 ac) between 1938 and 1986 that occurred in 10–15% of the maritime chaparral area under investigation.

As in other chaparral communities, many maritime chaparral species are dependent on or promoted by fire (Appendix 1). Postfire regeneration occurs both by sprouting and recruitment from buried seeds. Exogenous seed sources do not play an important role in maritime chaparral succession (Davis et al. 1988, Odion and Davis 2000). Chronosequence studies suggest that postfire succession is largely a function of prefire floristic composition, differential seedling survival, and species differences in longevity.

Vegetation in the first several years after fire vegetation is a diverse combination of annuals and short-lived perennials recruited from the seed bank and resprouting geophytes and woody plants (Davis et al. 1988). Unlike other chaparral communities, the flush of postfire annuals appears to be driven in part by removal of the shrub canopy rather than fire-induced germination of refractory seeds (Tyler 1996). Herb layer biomass and diversity drop rapidly with increasing closure of the shrub canopy 5 to 10 years after burning. After 20 to 40 years, short-lived shrubs senesce and the community is increasingly dominated by long-lived chamise, manzanitas, and coast live oak.

Shrub dieback during the fire-free period creates conspicuous gaps in the chaparral canopy where herbaceous species can grow and augment the soil seed bank. In the next fire these gaps experience reduced soil heating and associated seed mortality and thus become microsites for seedling recruitment by both herbaceous and woody species (Davis et al. 1989, Odion and Davis 2000, Odion and Tyler 2002).

Several researchers of maritime chaparral argue that, in the absence of fire, maritime chaparral would ultimately be replaced by coastal oak or pine forests (Cooper 1922, McBride and Stone 1976, Griffin 1978a). Davis et al. (1988) reported a significant positive correlation between oak canopy cover and time since burning but noted that the increase in oaks varied widely depending on distance from the coast, soil characteristics, and fire severity. In the coastal sandhills of northern Monterey County, Van Dyke and Holl (2001) found that in the long absence of fire, remnants of Prunedale maritime chaparral had undergone significant changes in species composition and stand structure. They posited that in the continued absence of fire, coast live oak would gradually replace the long-lived obligate seeder Pajaro manzanita, eventually converting maritime chaparral to coast live oak woodland.

COULTER PINE ECOLOGICAL ZONE

Fire Ecology Coulter pine is the most widely distributed closed-cone species in the Central Coast bioregion. Its range is more or less linear and extends from northern Diablo Range in Contra Costa County south along the coastal Santa Lucia Range to Figueroa Mountain. Inland populations are more scattered and trail down the Diablo Range (Ledig 2000).

FIGURE 16.7 Aerial view of annual grassland and oak forest topo-mosaics of the Interior Coast Ranges, northeast of Salinas in San Benito County.

In the southern Coast Ranges, Coulter pine exhibits considerable cone-habit variation that appears to be directly related to fire regime (Borchert 1985). Over much of its range, Coulter pine grows as an overstory tree in a matrix of dense montane chaparral (Borchert et al. 2004). In this setting crown fires are inevitable because fire carries from the shrub layer into the crowns of the pines that self-prune poorly.

In an environment of repeated stand-replacing fires Coulter pine tends to be highly serotinous. Heat from the burning understory chaparral and the porous canopy of long needles breaks the resinous seal of the cones and seeds fall *en masse* into the ash bed. Although winged, the heavy seeds do not disperse far from the tree except perhaps in strong winds (Borchert et al. 2003, Johnson et al. 2003). Once on the ground, rodents and birds quickly harvest seeds and bury them in caches of 1 to 15 seeds per cache. Most seedlings emerge from unrecovered caches. Seedling mortality is relatively low, especially when compared to early seedling mortality of other serotinous pines (Borchert et al. 2003).

At about age 10, saplings begin to produce cones but, because they are heavy, they require the support of the tree bole and only appear on the ends of branches after the limbs become sufficiently stout. As the tree grows, closed cones accumulate creating a persistent aerial seed bank. Seeds encased in the closed cones receive a high degree of protection, as evidenced by a seed viability of 95% in 25-year-old cones (Borchert 1985).

Fire Regime–Plant Community Interactions Coulter pine is not confined to chaparral in its distribution. On the Central Coast it frequently associates with coast live oak (Campbell 1980, Borchert et al. 2002) and occasionally with valley oak and other hardwoods. In these forests Coulter pine is not serotinous, or only moderately so. Where sites are productive and trees grow large, the continuous shrub understory is absent or poorly developed. Thus, surface fires are more common than crown fires and adult Coulter pine mortality is low. Seedlings establish after fires from cones of the current year or they establish and grow to a fire-tolerant size in the interval between fires (Borchert 1985) (Appendix 1).

KNOBCONE PINE ECOLOGICAL ZONE

Fire Ecology Knobcone pine is a medium-sized, relatively short-lived conifer that frequently grows in dense stands. Because trees self-prune poorly, they are easily killed in chaparral fires and depend on fire for the release of cone-stored seeds onto the burned soil surface (Appendix 1).

Despite its widespread distribution in California, there are remarkably few postfire studies of knobcone pine. Keeley et al. (1999) studied the regeneration of this species in the central Santa Lucia Mountains after a fire in 1994. They examined the regeneration of knobcone stands that re-burned 8 years after an earlier fire. Seedling recruitment following the 1985 fire was abundant and because knobcone pine produces cones at an early age (2 years), a partial aerial seed bank already had developed before the second fire. Although seedling recruitment after the second fire was low (one to two seedlings m^{-2} [11–22 ft^{-2}) and patchily distributed compared to recruitment after the 1985 fire, local extinction was averted by the presence of the relatively small number of new cones.

Fire Regime–Plant Community Interactions Knobcone pine, like Coulter pine and Sargent cypress, often grows with highly inflammable vegetation like chaparral. As a result, knobcone pine stands usually burn in stand-replacing fires (Fry et al. 2012) and at frequencies matching the surrounding vegetation. Nevertheless, like Bishop pine and Monterey pine, knobcone pine also produces uneven-aged stands after fire. Occasionally, stands burn in surface fires which generate enough heat to open cones but not enough heat to kill the overstory trees. Where surface fires occur, seedlings grow beneath older trees resulting in multiaged forests (Appendix 1).

SARGENT CYPRESS ECOLOGICAL ZONE

Within the region Sargent cypress forms an archipelago of small stands that extend from the northern Santa Lucia Range to the southern part of the bioregion above San Luis Obispo.

Among the four cypress (*Hesperocyparis*) species in the region, Sargent cypress is the only species that has been studied after fire (Appendix 1). Sargent cypress is a fire-dependent, obligate seeding species that releases prodigious numbers of small, wingless seeds after crown fires.

Fire Regime–Plant Community Interactions After a wildfire swept the Cuesta Ridge Botanical Area in 1994, Ne'eman et al. (1999) reconstructed prefire stand characteristics (adult density, cone and seed densities, age, etc.) using the skeletal remains of trees in even-aged stands that ranged from 20 years to 95 years. The number of cones per tree increased rapidly after 80 years as tree densities thinned from 0.8 m^{-2} (2.6 ft^{-2}) in young-aged stands to 0.4 m^{-2} (1.3 ft^{-2}) in the oldest ones.

Seedling densities ranged from 6.3 to 81.7 m^{-2} (20.7–268.0 ft^{-2}) but seedling density was negatively correlated with tree density. The highest seedling densities occurred in stands <60 years rather than in the oldest stands with the highest number of cones per tree. They attributed low seedling densities in the oldest stands to either reduced seed viability with age or to higher intensity fires. Indeed, for some cypress species seed viability decreases rapidly with age (De Magistris et al. 2001). They concluded that fires burning at intervals as short as 20 years posed little risk (immaturity risk) to the regeneration of the species at this site, presumably because 20-year-old stands had an adequate cone bank. Nevertheless, the fire that burned these Sargent cypress forests also re-burned the knobcone pine forests described above (Keeley et al. 1999) just a few kilometers away. Had the Sargent cypress stands burned after 8 years, as some of the knobcone pine forests did, much of the cypress forest may have been lost.

Interior Coast Ranges Subregion

The interior valleys and foothills are dominated by alliances such as California annual grassland, California buckwheat, blue oak, and blue oak-foothill pine. The latter two alliances are especially widespread in the Diablo Range. Remnant valley oak woodlands are present in larger stream valleys on deeper loamy soils across the region. Mid-to-high elevations support blue oak woodlands, chamise chaparral, and other chaparral alliances such as buckbrush and Eastwood manzanita (*Arctostaphylos glandulosa*). The only montane forests in the inland region are the unusually open stands of Jeffrey pine, incense cedar (*Calocedrus decurrens*), and Coulter pine that inhabit the serpentine areas of San Benito Mountain.

The fire ecology and fire regimes of the widespread vegetation types in this zone such as blue oak woodland, annual grassland, and chaparral are covered in chapter 15 (Central Valley Bioregion).

Management Issues and Climate Change

Vegetation Management

Sound vegetation management using fire relies on a thorough understanding of the fire ecology of the species or plant communities. Thus, it is somewhat surprising that the Central California Coast, which has one of the richest arrays of plant communities in the State, has so few fire ecology studies, especially in vegetation types that cover much of the bioregion. For example, four associations of coastal sage scrub are represented in the bioregion and although Venturan, Lucian, Diablan, and Franciscan cover more than 2,500 km^2 (965 mi^2) (Davis et al. 1998), we know of only one fire ecology study in Franciscan coastal sage scrub by Ford (1991). Similarly, chaparral makes up 20% (7,765 km^2 [2,998 mi^2]) of the region but there are only two post-burn studies of prescribed fires in nonmaritime chaparral, one in Pinnacles National Monument (Florence 1985) and the other in the Mount Hamilton Range (Dunne et al. 1991). By comparison, the number of fire ecology studies of chaparral in the South Coast Bioregion (chapter 17) number more than 100. Finally, this bioregion is one of the major repositories of mixed evergreen forests (1,625 km^2 [627 mi^2]) in the State but, except for limited post-Marble Cone observations by Griffin (1978b), there are no formal fire ecology studies in this highly variable type (but see Davis et al. 2010).

In sharp contrast to many of the common alliances in the bioregion, rare alliances have received considerable research attention. This is perhaps not surprising since a number of rare types like maritime chaparral and Monterey pine forests are at risk from lack of fire, urbanization, and other land conversions (Cylinder 1997, Hillyard 1997, Lambrinos 2000, Stephens et al. 2004).

In this chapter we have highlighted some similarities but also some major differences in fire regimes among the geographic subregions and major vegetation types of the Central California Coast Bioregion. This heterogeneity and pressing need for more ecological research notwithstanding, we would be remiss if we did not reiterate four management issues that face fire and natural resource managers: climate change, fire and exotic species, management of fire-dependent species, and fire management at the wildland-urban interface.

Fire and Nonnative Species

Coastal areas are particularly vulnerable to nonnative invasive plants (Seabloom et al. 2006, Lambert et al. 2010). Coastal and foothill vegetation types of the region are now extensively invaded by nonnative plant species and this trend is likely to continue, especially given the rapid increases in human populations and development in parts of the bioregion (Seabloom et al. 2006).

Deliberate use of fire to convert shrublands and closed woodlands to grasslands has promoted invasion of nonnative plants (Keeley 2001). In the past, the spread of nonnatives into shrublands was undoubtedly promoted by postfire seeding although this practice has become far less prevalent in recent years. There is now increasing interest by managers in using fire to control nonnative plant species, despite the mixed success of efforts to date (D'Antonio 1993, Keeley 2001, Keeley et al. 2011).

Recent widespread mortality of tanoak, black oak, and coast live oak at the northern end of the bioregion has been linked to the nonnative sudden oak death syndrome pathogen (*Phytophthora ramorum*) (Rizzo et al. 2002). The disease now extends over at least 300 km of the Central Coast bioregion. In heavily infested areas of Marin County, mortality of tanoak and coast live oak has been as high as 18–50% and 15–20%, respectively (Kelly and Meentemeyer 2002). The widespread tanoak death had profound effects on coast redwood mortality after the 2008 Basin Fire consumed large swaths of the Big Sur (Kuljian and Varner 2010, Metz et al. 2013).

Management of Fire-Dependent Species

As noted in the sections on maritime chaparral and closed-cone conifers, the distribution and abundance of many rare and endemic species in the region is closely tied to fire regime. Fire management for many of these species has become increasingly difficult due to their close proximity to residential areas and restrictions on prescribed burning. In some areas managers have resorted to mechanically clearing fuel breaks, setting prescribed burns in relatively cool and damp spring or early winter conditions (Keeley 2006), and/or shortening the time between burns to prevent excessive fuel buildup. Such management practices can have unintended negative impacts on native species and plant communities (Merriam et al. 2006, Keeley et al. 2011). Mechanical clearing can promote nonnative species and native vegetation may be slow to recover (Stylinski and Allen 1999). Moreover, burning outside of the normal fire season and high frequency of burning favors species like sprouters and can operate strongly against obligate seeding species (Keeley 2002a).

Wildland-Urban Interface

By 2050, 36,000 km^2 (13,900 mi^2) of agricultural land and 10,000 km^2 (3,861 mi^2) of pastureland will experience housing growth in California and much of this growth will be concentrated on lands with very high wildfire severity (Mann et al. 2014). Counties in the bioregion expected to experience the highest housing growth include Alameda, Contra Costa, Santa Clara, and Santa Barbara (Mann et al. 2014).

Although many of the cities are located in agricultural areas (i.e., the Developed Plains, Valleys, and Terraces subregion) and are immune from wildland fires, others, like San Luis Obispo (which has been threatened twice in the last 34 years by chaparral fires), are vulnerable to fires in nearby wildland areas. In 1991, the deadly Tunnel Fire killed 25 people and destroyed 3,810 dwelling in the Oakland Hills. A combination of drought-dry vegetation, high temperatures, low humidity, steep topography, and Santa Ana-strength winds that forced the fire downslope created a firestorm that defied control for several days (Ewell 1995). This historically unprecedented urban fire provides a vivid, worst-case example of fire management problems that other areas in the bioregion will face as California's population continues to grow exponentially and the populace pushes into the wildland-urban interface.

Climate Change

Between the historical and modern periods, California has experienced an annual mean temperature increase of ~1.68°C (3.02°F), primarily driven by a marked increase in annual minimum temperatures (Rapacciuolo et al. 2014). Near the coast, twentieth-century maximum temperature increases were highest for spring (March–May) and fall (September–October) and relatively unchanged during the summer (Cordero et al. 2011). In the inner Coast Ranges, maximum temperatures increased at the steepest rate during summer months (Cordero et al. 2011).

Downscaled global climate models all project accelerated warming over the twenty-first century depending on assumptions about increases in greenhouse gas concentrations. On average, downscaled models using "business as usual" emission scenarios project warming for the Central Coast of around 2°C (3.6°F) by 2060 to 2069 and 4–5°C (7.2–9.0°F) by the end of the century (Pierce et al. 2013, Maloney et al. 2014). Temperatures are projected to increase during all seasons, with greatest warming in summer and fall, especially further from the coast (Pierce et al. 2013). Model experiments from phase 5 of the Coupled Model Intercomparison Project tend to project winter precipitation increasing by 0.5–1.5 mm d-1 (0.02–0.06 in d-1) by the end of the century (Neelin et al. 2013).

These changes will likely lead to greatly increased likelihood of wildfires across much of the region (Lenihan et al. 2008, Westerling and Bryant 2008). As the temperature increases, large wildfire occurrence, burned area, and the length of the fire season are projected to increase, particularly in the northern Santa Lucia Ranges (Lenihan et al. 2008, Westerling et al. 2011, Bryant and Westerling 2014). However, changes in fire regime will depend on seasonal timing and the interactions between temperature and precipitation (e.g., Keeley and Syphard 2015) as well as scenarios of land-use change and fire management (Bryant and Westerling 2014).

References

Allen, B.H., B.A. Holzman, and R.R. Evertt. 1991. A classification system for California hardwood rangelands. Hilgardia 59: 1–45.

Anderson, R.S., A. Ejarque, P.M. Brown, and D.J. Hallett. 2013. Holocene and historical vegetation change and fire history on the north-central coast of California, USA. The Holocene 23: 1797–1810.

Bailey, R.G. 1995. Description of the ecoregions of the United States. USDA Forest Service Miscellaneous Publication 1391. Washington, D.C., USA.

Beals, R., and J. Hester. 1974. California Indians I, Vol. 1. Garland Press, New York, New York, USA.

Begley, E. 2010. Growth rates in a southern California population of knobcone pine (*Pinus attenuata*). Ecological Restoration 28: 10–12.

Bendix, J., and C.M. Cowell. 2010. Impacts of wildfire on the composition and structure of riparian forests in southern California. Ecosystems 13: 99–107.

Borchert, M.I. 1985. Serotiny and cone-habit variation in populations of *Pinus coulteri* (Pinaceae) in the southern Coast Ranges of California. Madroño 32: 29–48.

Borchert, M.I., M. Johnson, D. Schreiner, and S.B. Vander Wall. 2003. Early postfire seed dispersal, seedling establishment and seedling mortality of *Pinus coulteri* (D. Don) in central coastal California, USA. Plant Ecology 168: 207–220.

Borchert, M.I., A. Lopez, C. Bauer, and T. Knowd. 2004. Field guide to coastal sage scrub and chaparral series of Los Padres National Forest. USDA Forest Service Technical Publication R5-TP-019. Pacific Southwest Region, Vallejo, California, USA.

Borchert, M.I., D. Schreiner, T. Knowd, and T. Plumb. 2002. Predicting postfire survival in Coulter pine (*Pinus coulteri*) and gray pine (*Pinus sabiniana*) after wildfire in central California. Western Journal of Applied Forestry 17: 134–138.

Brown, P.M., M.W. Kaye, and D. Buckley. 1999. Fire history in Douglas-fir and coast redwood forests at Point Reyes National Seashore, California. Northwest Science 73: 205–216.

Bryant, B.P., and A.L. Westerling. 2014. Scenarios for future wildfire risk in California: links between changing demography, land use, climate, and wildfire. Environmetrics 25: 454–471.

Byrne, R., J. Michaelsen, and A. Soutar. 1977. Fossil charcoal as a measure of wildfire frequency in Southern California: a preliminary analysis. Pages 361–367 in: H.A. Mooney and C.E. Conrad, technical coordinators. Proceedings of the Symposium on the Environmental Consequences of Fire and Fuel Management in Mediterranean Ecosystems. USDA Forest Service General Technical Report WO-3. Washington, D.C., USA.

Callaway, R.M. 1992. Morphological and physiological-responses of 3 California oak species to shade. International Journal of Plant Sciences 153: 434–441.
Callaway, R.M., and F.W. Davis. 1993. Vegetation dynamics, fire, and the physical-environment in coastal central California. Ecology 74: 1567–1578.
———. 1998. Recruitment of *Quercus agrifolia* in central California: the importance of shrub-dominated patches. Journal of Vegetation Science 9: 647–656.
Cameron, D.R., J. Marty, and R.F. Holland. 2014. Whither the rangeland?: protection and conversion in California's rangeland ecosystems. PLoS One 9(8): e103168.
Campbell, B. 1980. Some mixed hardwood forest communities of the coastal ranges of southern California. Phytocoenologia 8: 297–320.
Cayan, D.R., M.D. Dettinger, S.A. Kammerdiener, J.M. Caprio, and D.H. Peterson. 2001. Changes in the onset of spring in the western United States. Bulletin of the American Meteorological Society 82: 399–415.
Coale, T.H., A.J. Deveny, and L.R. Fox. 2011. Growth, fire history, and browsing recorded in wood rings of shrubs in a mild temperate climate. Ecology 92: 1020–1026.
Cooper, W.S. 1922. The broad-sclerophyll vegetation of California; an ecological study of the chaparral and its related communities. Carnegie Institute of Washington Publications 319. Washington, D.C., USA.
Cordero, E.C., W. Kessomkiat, J.T. Abatzoglou, and S.A. Mauget. 2011. The identification of distinct patterns in California temperature trends. Climatic Change 108: 357–382.
Cowart, A., and R. Byrne. 2013. A paleolimnological record of late Holocene vegetation change from the Central California coast. California Archeology 5: 337–352.
Cuthrell, R.Q., C. Striplen, M. Hylkema, and K.G. Lightfoot. 2012. Pages 153–172 in: T.L. Jones and K.E. Perr, editors. A Land of Fire: Anthropogenic Burning on the Central Coast of California. Contemporary Issues in California Archaeology. Left Coast Press, San Francisco, California, USA.
Cylinder, P. 1997. Monterey pine forest conservation strategy. Fremontia 25(2): 21–26.
Dagit, R. 2002. Post-fire monitoring of coast live oaks (*Quercus agrifolia*) burned in the 1993 Old Topanga Fire. Pages 243–249 in: R.B. Standiford, D. McCreary, and K.L. Purcell, editors. Proceedings of the 5th Symposium on Oak Woodlands: Oaks in California's Changing Landscape. USDA Forest Service General Technical Report PSW-GTR-184. Pacific Southwest Research Station, Albany, California, USA.
D'Antonio, C.M. 1993. Mechanisms controlling invasion of coastal plant communities by the alien succulent *Carpobrotus edulis*. Ecology 74: 83–95.
Davis, F.W., M.I. Borchert, R.K. Meentemeyer, A. Flint, and D.M. Rizzo. 2010. Pre-impact forest composition and ongoing tree mortality associated with sudden oak death in the Big Sur region, California. Forest Ecology and Management 259: 2342–2354.
Davis, F.W., M.I. Borchert, and D. Odion. 1989. Establishment of microscale vegetation pattern in maritime chaparral after fire. Vegetatio 84: 53–67.
Davis, F.W., D. Hickson, and D.C. Odion. 1988. Composition of maritime chaparral related to fire history and soil, Burton Mesa, California. Madroño 35: 169–195.
Davis, F.W., and J. Michaelsen. 1995. Sensitivity of fire regime in chaparral ecosystems to climate change. Pages 435–456 in: J.M. Moreno and W.C. Oechel, editors. Global Change and Mediterranean-Type Ecosystems. Springer, New York, New York, USA.
Davis, F.W., D.M. Stoms, A.D. Hollander, K.A. Thomas, P.A. Stine, D. Odion, M.I. Borchert, J.H. Thorne, M.V. Gray, R.E. Walker, K. Warner, and J. Graae. 1998. The California gap analysis project–final report. University of California, Santa Barbara, California, USA.
De Magistris, A.A., P.N. Hashimoto, S.L. Masoni, and A. Chiesa. 2001. Germination of serotinous cone seeds in *Cupressus* ssp. Israel Journal of Plant Science 49: 253–258.
Dingemans, T.D., S.A. Mensing, S.J. Feakins, M.E. Kirby, and S.R.H. Zimmerman. 2014. 3000 years of environmental change at Zaca Lake, California, USA. Frontiers in Ecology and Evolution 2: 34.
Duffield, J.W. 1951. Interrelationships of the California closed-cone pines with special reference to *Pinus muricata* D. Don. Dissertation, University of California, Berkeley, California, USA.
Dunne, J., A. Dennis, J.W. Bartolome, and R.H. Barrett. 1991. Chaparral response to a prescribed fire in the Mount Hamilton, Santa Clara County, California. Madroño 38: 21–29.
Dunning, C.E., R.A. Redak, and T.D. Paine. 2003. Preference and performance of a generalist insect herbivore on *Quercus agrifolia* and *Quercus engelmannii* seedlings from a southern California oak woodland. Forest Ecology and Management 174: 593–603.
Ewell, P.L. 1995. The Oakland-Berkeley Hills fire of 1991. Pages 7–10 in: D.R. Weise and R.E. Martin, editors. The Biswell Symposium: Fire Issues and Solutions in Urban Interface and Wildland Ecosystems. USDA Forest Service General Technical Report PSW-GTR-158. Pacific Southwest Research Station, Albany, California, USA.
Fischer, D.T., C.J. Still, and A.P. Williams. 2009. Significance of summer fog and overcast for drought stress and ecological functioning of coastal California endemic plant species. Journal of Biogeography 36: 783–799.
Florence, M.A. 1985. Successional trends in plant species composition following fall, winter and spring prescribed burns of chamise chaparral in the central Coast Range of California. Thesis, California State University, Sacramento, California, USA.
Ford, L.D. 1991. Post-fire dynamics of northern coastal scrub, Monterey County, California. Dissertation, University of California, Berkeley, California, USA.
Forrestel, A.B., M.A. Moritz, and S.L. Stephens. 2011. Landscape-scale vegetation change following fire in Point Reyes, California USA. Fire Ecology 7(2): 114–128.
Fry, D.L., J. Dawson, and S.L. Stephens. 2012. Age and structure of mature knobcone pine forests in the northern California Coast Range, U.S.A. Fire Ecology 8(1): 49–62.
Gordon, B.L. 1979. Monterey Bay Area: Natural History and Cultural Imprints. Boxwood Press, Pacific Grove, California, USA.
Greenlee, J.M., and J.H. Langenheim. 1990. Historic fire regimes and their relation to vegetation patterns in the Monterey Bay Area of California. American Midland Naturalist 124: 239–253.
Greenlee, J.M., and A. Moldenke. 1982. History of wooldand fires in the Gabilan Mountains region of central coastal California. Unpublished Report. National Park Service, Pinnacles National Monument, Paicines, California, USA.
Griffin, J.R. 1975. Plants of the highest Santa Lucia and Diablo Range Peaks, California. USDA Forest Service Research Paper PSW-RP-110. Pacific Southwest Forest and Range Experiment Station, Albany, California, USA.
———. 1978a. Maritime chaparral and endemic shrubs of the Monterey Bay region, California. Madroño 25(2): 65–81.
———. 1978b. The Marble-Cone fire ten months later. Fremontia 6(2): 8–14.
———. 1988. Oak woodland. Pages 383–416 in: M.G. Barbour and J. Major, editors. Terrestrial Vegetation of California. John Wiley and Sons, New York, New York, USA.
Harvey, B.J., and B.A. Holzman. 2014. Divergent successional pathways of stand development following fire in a California closed-cone pine forest. Journal of Vegetation Science 25: 88–99.
Harvey, B.J., B.A. Holzman, and J.D. Davis. 2011. Spatial variability in stand structure and density-dependent mortality in newly established post-fire stands of a California closed-cone pine. Forest Ecology and Management 262: 2042–2051.
Haston, L., F.W. Davis, and J. Michaelsen. 1988. Climate response functions for bigcone spruce: a Mediterranean climate conifer. Physical Geography 9: 81–97.
Haston, L., and J. Michaelsen. 1994. Long-term central coastal California precipitation variability and relationships to El-Nino-Southern Oscillation. Journal of Climate 7: 1373–1387.
———. 1997. Spatial and temporal variability of southern California precipitation over the last 400 yr and relationships to atmospheric circulation patterns. Journal of Climate 10: 1836–1852.
Henson, P., and D.J. Usner. 1993. The Natural History of Big Sur. University of California Press, Berkeley, California, USA.
Hillyard, D. 1997. Challenges in conserving Monterey pine forest. Fremontia 25(2): 16–20.
Hughes, M., and A. Hall. 2010. Local and synoptic mechanisms causing southern California's Santa Ana Winds. Climate Dynamics 34: 847–857.
Jackson, L.E., Jr. 1977. Dating and recurrence frequency of prehistoric mudflows near Big Sur, Monterey County, California. U.S. Geological Survey Journal of Research 5(1): 17–32.

Johnson, J., S.B. Vander Wall, and M.I. Borchert. 2003. A comparative analysis of seed and cone characteristics and seed-dispersal strategies of three pines in the subsection Sabinianae. Plant Ecology 168: 69–84.
Johnstone, J.A., and T.E. Dawson. 2010. Climatic context and ecological implications of summer fog decline in the coast redwood region. PNAS 107: 4533–4538.
Jones, G. 2014. Coast redwood fire history and land use in the Santa Cruz Mountains, California. Thesis, San Jose State University, San Jose, California, USA.
Keeley, J.E. 2001. Fire and invasive species in Mediterranean-climate ecosystems of California. Invasive species workshop: the role of fire in the control and spread of invasive species. Miscellaneous Publication 11. Tall Timbers Research Station, Tallahassee, Florida, USA.
———. 2002a. Fire management of California shrubland landscapes. Environmental Management 29: 395–408.
———. 2002b. Native American impacts on fire regimes of the California coastal ranges. Journal of Biogeography 29: 303–320.
———. 2005. Fire history of the San Francisco East Bay region and implications for landscape patterns. International Journal of Wildland Fire 14: 285–296.
———. 2006. Fire management impacts on invasive plants in the western United States. Conservation Biology 20: 375–384.
Keeley, J.E., J. Franklin, and C. D'Antonio. 2011. Fire and invasive plants on California landscapes. Pages 193–221 in: D.M. McKenzie, C.A. Miller, and D.A. Falk, editors. The Landscape Ecology of Fire. Springer, New York, New York, USA.
Keeley, J.E., G. Ne'eman, and C.J. Fotheringham. 1999. Immaturity risk in a fire-dependent pine. Journal of Mediterranean Ecology 1: 41–48.
Keeley, J.E., and A.D. Syphard. 2015. Different fire–climate relationships on forested and non-forested landscapes in the Sierra Nevada ecoregion. International Journal of Wildland Fire 24: 27–36.
Keeley, J.E., and P.H. Zedler. 1998. Evolution of life histories in *Pinus*. Pages 219–249 in: D.M. Richardson, editor. Ecology and Biogeography of *Pinus*. Cambridge University Press, Cambridge, UK.
Kelly, M., and R.K. Meentemeyer. 2002. Landscape dynamics of the spread of sudden oak death. Photogrammetric Engineering and Remote Sensing 68: 1001–1009.
Kuljian, H., and J.M. Varner. 2010. The effects of sudden oak death on foliar moisture content and crown fire potential in tanoak. Forest Ecology and Management 259: 2103–2110.
Lambert, A.M., C.M. D'Antonio, and T.L. Dudley. 2010. Invasive species and fire in California ecosystems. Fremontia 38(2): 29–36.
Lambrinos, J.G. 2000. The impact of the invasive alien grass *Cortaderia jubata* (Lemoine) Stapf on an endangered Mediterranean-type shrubland in California. Diversity and Distributions 6: 217–231.
Lathrop, E.W., and C.D. Osborne. 1991. Influence of fire on oak seedlings and saplings in southern oak woodland on the Santa Rosa Plateau Preserve, Riverside County, California. Pages 366–370 in: R.B. Standiford, editor. Proceedings of the Symposium on Oak Woodlands and Hardwood Rangeland Management. USDA Forest Service General Technical Report PSW-GTR-126. Pacific Southwest Research Station, Berkeley, California, USA.
Ledig, F.T. 2000. Founder effects and the genetic structure of Coulter pine. The Journal of Heredity 91: 307–315.
Lenihan, J.M., D. Bachelet, R.P. Neilson, and R. Drapek. 2008. Response of vegetation distribution, ecosystem productivity, and fire to climate change scenarios for California. Climatic Change 87: 215–230.
Lewis, H.T., and L.J. Bean. 1973. Patterns of Indian burning in California: ecology and ethnohistory. Ballena Press Anthropological Papers No. 1. Socorro, New Mexico, USA.
Maloney, E.D., S.J. Camargo, E. Chang, B. Colle, R. Fu, K.L. Geil, Q. Hu, X. Jiang, N. Johnson, K.B. Karnauskas, J. Kinter, B. Kirtman, S. Kumar, B. Langenbrunner, K.J. Lombardo, L.N. Long, A. Mariotti, J.E. Meyerson, K.C. Mo, J.D. Neelin, Z. Pan, R. Seager, Y. Serra, A. Seth, J. Sheffield, Stroeve, J. Thibeault, S.-P. Xie, C. Wang, B. Wyman, and M. Zhao. 2014. North American climate in CMIP5 experiments: Part III: assessment of twenty-first-century projections. Journal of Climate 27: 2230–2270.
Mann, M.L., P. Berck, M.A. Moritz, E. Batllori, J.G. Baldwin, C.K. Gately, and D.R. Cameron. 2014. Modeling residential development in California from 2000 to 2050: integrating wildfire risk, wildland and agricultural encroachment. Land Use Policy 41: 438–452.
McBride, J.R., and E.C. Stone. 1976. Plant succession on sand dunes of Monterey Peninsula, California. American Midland Naturalist 96: 118–132.
McDonald, P.M., and R.J. Laacke. 1990. *Pinus radiata* D. Don. Pages 443–441 in: R.M. Burns and B.H. Honkala, technical coordinators. Silvics of North America: Conifers. USDA Forest Service Handbook No. 654. Washington, D.C., USA.
Mensing, S.A. 1998. 560 years of vegetation change in the region of Santa Barbara, California. Madroño 45: 1–11.
Mensing, S.A., J. Michaelsen, and R. Byrne. 1999. A 560-year record of Santa Ana fires reconstructed from charcoal deposited in the Santa Barbara basin, California. Quaternary Research 51: 295–305.
Merriam, K.E., J.E. Keeley, and J.L. Beyers. 2006. Fuel breaks affect nonnative species abundance in Californian plant communities. Ecological Applications 16: 515–527.
Metz, M.R., J.M. Varner, K.M. Frangioso, R.K. Meentemeyer, and D.M. Rizzo. 2013. Unexpected redwood mortality from synergies between wildfire and an emerging infectious disease. Ecology 94: 2152–2159.
Michaelsen, J., L. Haston, and F.W. Davis. 1987. 400 years of central California precipitation variability reconstructed from tree-rings. Water Resources Bulletin 23: 809–817.
Miles, S.R., and C.B. Goudey. 1997. Ecological subregions of California: section and subsection descriptions. USDA Forest Service Technical Publication R5-EM-TP-005. Vallejo, California, USA.
Millar, C.I. 1986. Bishop pine (*Pinus muricata*) of inland Marin County, California. Madroño 33: 123–129.
Millar, C.I., S.H. Strauss, M.T. Conkle, and R.D. Westfall. 1988. Allozyme differentiation and biosystematics of the Californian closed-cone pines (*Pinus* subsect. Oocarpae). Systematic Botany 13: 351–370.
Moritz, M.A. 1997. Analyzing extreme disturbance events: fire in Los Padres National Forest. Ecological Applications 7: 1252–1262.
———. 2003. Spatiotemporal analysis of controls on shrubland fire regimes: age dependency and fire hazard. Ecology 84: 351–361.
Neelin, J.D., B. Langenbrunner, J.E. Meyerson, A. Hall, and N. Berg. 2013. California winter precipitation change under global warming in the Coupled Model Intercomparison Project 5 ensemble. Journal of Climate 26: 6238–6256.
Ne'eman, G., C.J. Fortheringham, and J.E. Keeley. 1999. Patch to landscape patterns in post fire recruitment of a serotinous conifer. Plant Ecology 145: 235–242.
O'Brien, M.J., K.L. O'Hara, N. Erbilgin, and D.L. Wood. 2007. Overstory and shrub effects on natural regeneration processes in native *Pinus radiata* stands. Forest Ecology and Management 240: 178–185.
Odion, D.C., and F.W. Davis. 2000. Fire, soil heating, and the formation of vegetation patterns in chaparral. Ecological Monographs 70: 149–169.
Odion, D.C., and C.M. Tyler. 2002. Are long fire-free periods necessary to maintain the endangered, fire-recruiting shrub, *Arctostaphylos morroensis* (Ericaceae)? Conservation Ecology 6: 4.
Oneal, C.B., J.D. Stuart, S.S. Steinberg, and L. Fox, III. 2006. Geographic analysis of natural fire rotation in the California redwood forest during the suppression era. Fire Ecology 2(1): 73–99.
Parikh, A., and N. Gale. 1998. Coast live oak revegetation on the central coast of California. Madroño 45: 301–309.
Peinado, M., J.L. Aguirre, and J. Delgadillo. 1997. Phytosociological, bioclimatic and biogeographical classification of woody climax communities of western North America. Journal of Vegetation Science 8: 505–528.
Peterson, S.H., M.A. Moritz, M.E. Morais, P.E. Dennison, and J.M. Carlson. 2011. Modelling long-term fire regimes of southern California shrublands. International Journal of Wildland Fire 20: 1–16.
Pierce, D.W., T. Das, D.R. Cayan, E.P. Maurer, N.L. Miller, Y. Bao, M. Kanamitsu, K. Yoshimura, M.A. Snyder, L.C. Sloan, G. Franco, and M. Tyree. 2013. Probabilistic estimates of future changes in California temperature and precipitation using statistical and dynamical downscaling. Climate Dynamics 40: 839–856.

Plumb, T.R. 1980. Response of oaks to fire. Pages 202–215 in: T.R. Plumb, technical editor. Proceedings of the Symposium on the Ecology, Management, and Utilization of California Oaks. USDA Forest Service PSW-GTR-44. Pacific Southwest Forest and Range Experiment Station, Berkeley, California, USA.
Plumb, T.R., and A.P. Gomez. 1983. Five southern California oaks: identification and postfire management. USDA Forest Service General Technical Report PSW-GTR-71. Pacific Southwest Forest and Range Experiment Station, Berkeley, California, USA.
Plumb, T.R., and B. Hannah. 1991. Artificial regeneration of blue and coast live oaks in the Central Coast. Pages 74–80 in: R.B. Standiford, editor. Proceedings of the Symposium on Oak Woodlands and Hardwood Rangeland Management. USDA Forest Service General Technical Report PSW-GTR-126. Pacific Southwest Research Station, Berkeley, California, USA.
Rapacciuolo, G., S.P. Maher, A.C. Schneider, T.T. Hammond, M.D. Jabis, R.E. Walsh, K.J. Iknayan, G.K. Walden, M.F. Oldfather, D.D. Ackerly, and S.R. Beissinger. 2014. Beyond a warming fingerprint: individualistic biogeographic responses to heterogeneous climate change in California. Global Change Biology 20: 2841–2855.
Riordan, E.C., and P.W. Rundel. 2009. Modelling the distribution of a threatened habitat: the California sage scrub. Journal of Biogeography 36: 2176–2188.
Rizzo, D.M., M. Garbelotto, J.M. Davidson, G.W. Slaughter, and S.T. Koike. 2002. *Phytophthora ramorum* as the cause of extensive mortality of *Quercus* spp. and *Lithocarpus densiflorus* in California. Plant Disease 86: 205–214.
Russell, W.H., and J.R. McBride. 2003. Landscape scale vegetation-type conversion and fire hazard in the San Francisco Bay Area open spaces. Landscape and Urban Planning 64: 201–208.
Sawaske, S.R., and D.L. Freyberg. 2014. Fog, fog drip, and streamflow in the Santa Cruz Mountains of California Coast Range. Ecohydrology 8: 695–713.
Sawyer, J.O., and T. Keeler-Wolf. 2009. A manual of California Vegetation. California Native Plant Society, Sacramento, California, USA.
Seabloom, E.W., J.W. Williams, D. Slayback, D.M. Stoms, J.H. Viers, and A.P. Dobson. 2006. Human impacts, plant invasion, and imperiled plant species in California. Ecological Applications 16: 1338–1350.
Sleeter, B.M., T.S. Wilson, C.E. Soulard, and J. Liu. 2011. Estimation of late twentieth century land-cover change in California. Environmental Monitoring and Assessment 173: 251–266.
Snow, G.E. 1980. The fire resistance of Engelmann and coast live oak seedlings. Pages 62–66 in: T.R. Plumb, technical editor. Proceedings of the Symposium on the Ecology, Management, and Utilization of California Oaks. USDA Forest Service PSW-GTR-44. Pacific Southwest Forest and Range Experiment Station, Berkeley, California, USA.
Stephens, S.L., and D.L. Fry. 2005. Fire history in coast redwood stands in the northeastern Santa Cruz Mountains, California. Fire Ecology 1(1): 2–19.
Stephens, S.L., D.D. Piirto, and D.F. Caramagno. 2004. Fire regimes and resultant forest structure in the native Año Nuevo Monterey pine (*Pinus radiata*) forest, California. American Midland Naturalist 152: 25–36.
Stylinski, C.D., and E.B. Allen. 1999. Lack of native species recovery following severe exotic disturbance in southern Californian shrublands. Journal of Applied Ecology 36: 544–554.
Talley, S.N. 1974. The ecology of Santa Lucia fir (*Abies bracteata*), a narrow endemic of California. Dissertation, Duke University, Durham, North Carolina, USA.
Thornton, P.E., S.W. Running, and M.A. White. 1997. Generating surfaces of daily meteorology variables over large regions of complex terrain. Journal of Hydrology 190: 214–251.
Timbrook, J., J.R. Johnson, and D.D. Earle. 1982. Vegetation burning by the Chumash. Journal of California and Great Basin Anthropology 4: 163–186.
Tyler, C.M. 1996. Relative importance of factors contributing to postfire seedling establishment in maritime chaparral. Ecology 77: 2182–2195.
Van Dyke, E., and K.D. Holl. 2001. Maritime chaparral community transition in the absence of fire. Madroño 48: 221–229.
van Wagtendonk, J.W., and D.R. Cayan. 2008. Temporal and spatial distribution of lightning strikes in California in relation to large-scale weather patterns. Fire Ecology 4(1): 34–56.
Vasey, M.C., M.E. Loik, and V.T. Parker. 2012. Influence of summer marine fog and low cloud stratus on water relations of evergreen woody shrub (*Arctostaphylos*: Ericaceae) in the chaparral of central California. Oecologia 170: 325–337.
Vasey, M.C., V.T. Parker, K.D. Holl, M.E. Loik, and S. Hiatt. 2014. Maritime climate influence on chaparral composition and diversity in the coast range of central California. Ecology and Evolution 4: 3662–3674.
Wells, P.V. 1962. Vegetation in relation to geological substratum and fire in San Luis Obispo Quadrangle, California. Ecological Monographs 32: 79–103.
Westerling, A.L., and B.P. Bryant. 2008. Climate change and wildfire in California. Climatic Change 87: 231–249.
Westerling, A.L., B.P. Bryant, H.K. Preisler, T.P. Holmes, H.G. Hidalgo, T. Das, and S.R. Shrestha. 2011. Climate change and growth scenarios for California wildfire. Climatic Change 109: 445–463.
White, K.L. 1999. Revisiting native *Pinus radiata* forests after twenty-nine years. Madroño 46: 80–87.
Wikler, K., A.J. Storer, W. Newman, T.R. Gordon, and D.L. Wood. 2003. The dynamics of an introduced pathogen in a native Monterey pine (*Pinus radiata*) forest. Forest Ecology and Management 179: 209–221.
Zedler, P.H. 1995. Fire frequency in southern California shrublands: biological effects and management options. Pages 101–112 in: J.E. Keeley and T. Scott, editors. Brushfires in California Wildlands: Ecology and Resource Management. International Association of Wildland Fire, Fairfield, Washington, USA.

CHAPTER SEVENTEEN

South Coast Bioregion

JON E. KEELEY AND ALEXANDRA D. SYPHARD

There is fire on the mountain and lightning in the air . . . "Fire on the mountain".

MARSHAL TUCKER BAND 1975

Description of Bioregion

Physical Geography

The South Coast bioregion includes the Southern California Coast and the Mountains and Valleys sections as defined by Miles and Goudey (1997). The region is bordered on the north by the Transverse Ranges and on the south by the US border with Mexico, on the east by the Peninsular Ranges and on the west by the Pacific Ocean (Map. 17.1). The Transverse and northern Peninsular ranges bound the Los Angeles Basin, an extensive floodplain comprising millions of years of alluvial outwash deposited by the Los Angeles, San Gabriel, and Santa Ana rivers. Minor mountain ranges occur within the Basin, including the Palos Verde Peninsula on the west, Puente Hills on the east, and San Joaquin Hills on the south. The Orange and San Diego coastlines comprise a series of Pleistocene marine terraces and an uplifted peneplain now incised by east-west running rivers including the Santa Margarita, San Luis Rey, and San Diego rivers (Schoenherr 1992).

The Transverse Ranges are one of the few east-west ranges in western North America and are composed of five mountain systems separated by broad alluvial valleys (Norris and Webb 1976). The Channel Islands are geologically part of this formation. On the western end, the Transverse Ranges fork into a northern extension, the Santa Ynez Mountains, and a southern extension, the Santa Monica Mountains (Map. 17.1). Both extend nearly to the ocean leaving a very narrow coastal plain, and both are relatively low elevation (1,430 m [4,691 ft] and 949 m [3,114 ft]) mountains largely lacking significant coniferous forests. The San Gabriel Mountains (3,074 m [9,997 ft]) are a largely granitic fault block, with the Castaic Mountains (1,765 m [5,791 ft]) extending to the northwest. These mountains have experienced considerable uplift in recent geological time, contributing to the rather steep and highly dissected front that faces the Los Angeles Basin. The easternmost San Bernardino Mountains (3,508 m [11,509]) are also a rapidly rising granitic and metamorphic fault block, but, unlike the San Gabriel Mountains, they have conifer-dominated mid-elevation plateaus between 1,700 and 2,300 m (5,600–7,500 ft), which includes Arrowhead and Big Bear lakes.

The Peninsular Ranges are a north-south trending series of fault blocks including the San Jacinto Mountains in the north, with a maximum elevation of 3,300 m (10,800 ft), and to the southeast the drier Santa Rosa mountains with several peaks over 1,800 m (5,900 ft). To the west is another series of lower elevation ranges separated by broad valleys that include the Santa Ana, Agua Tibia, and Laguna Mountains, mostly below 1,800 m (5,900 ft).

The South Coast bioregion includes numerous peaks over 3,000 m (13,000 ft); however, most of the region is below 500 m (1,600 ft) (Davis et al. 1995). The extraordinary range of elevations represented in this region translates into a high diversity of vegetation types and fire regimes. The high ranges separated by broad valleys dissected by riparian corridors all play important roles in determining the extent of fire spread. The widespread distribution of substrates such as gabbro has possibly reduced fire frequency by limiting the rate of biomass productivity.

Despite comprising only 8% of the land area of the state, this region contains 56% of the total human population (Davis et al. 1995). This pattern of development has placed immense pressure on natural resources and has created a fire management problem of extraordinary proportions.

Climate

This is a region of Mediterranean-climate, consisting of cool, wet winters and hot, dry summers. Average rainfall varies spatially from 200 to 1,200 mm (7.9–47.2 in) annually, two-thirds of which falls November to April in storms of several days duration, and dry spells of a month or more may occur during the wet season (Major 1977). The distribution of winter rainfall is strongly controlled by the orographic effect, and precipitation increases with elevation. Above 2,100 m (6,900 ft) a significant portion of winter precipitation comes as snow. Significant summer precipitation is rare near the coast but occasional Mexican monsoons from the Gulf of California affect the interior mountains (Douglas et al. 1993).

Mean winter temperatures near the coast are greater than 10°C (50°F) and decrease toward the interior and with

MAP 17.1 South Coast bioregion extends south from the Transverse ranges to Mexico and from the coast east to the Peninsular Ranges.

elevation. Coastal sites may occasionally experience below freezing temperatures with lethal effects on some native species (Collett 1992). In the interior foothills and valleys summer temperatures often exceed 40°C (104°F), but are more moderate at higher elevations and toward the coast. Late spring and early summer marine air delays the onset of the fire season in the coastal mountains (Coffin 1959).

Lightning accompanies both winter and summer storms, and the proportion varies markedly between the coastal and the interior ranges (Fig. 17.1). In the interior ranges lightning strikes peak in August, a pattern typical in other interior parts of California and the Southwest (van Wagtendonk and Cayan 2008); however, in the coastal ranges lightning strikes peak in August (Fig. 17.1B). The coastal ranges also differ in that 35% of all strikes occur in winter and spring, whereas in the interior these two seasons represent less than 12% (Fig. 17.1A, B). Due to high fuel moisture, winter and spring lightning strikes play essentially no role as ignition sources for wildfires, thus a substantially smaller proportion of lightning strikes are effective at igniting fires in these coastal ranges.

Regionally, lightning strike density is substantially greater in the interior mountain ranges, which receive several times more strikes than at comparable elevations in the coastal ranges (Fig. 17.2; Wilcoxin's signed rank test, $P < 0.001$, $n = 10$). Broad comparisons of lightning strike density reveal roughly 25 to 40 lightning strikes yr^{-1} 100^{-1} km^{-2} for interior ranges, and this is 5 to 10 times higher than coastal valleys (Fig. 17.2). Based on the number of fires ignited by lightning in these two regions (Keeley 1982) it appears that only 3–5% of all lightning strikes ignite a fire. In the interior ranges lightning-ignited fires are predictable most years, whereas in the coastal ranges they are sporadic. For example in the Santa Monica Mountains lightning-ignited fires occurred rarely in fire records between 1919 and 1980 (Radtke et al. 1982). Since 1981 there have been two lightning-ignited fires, one in 1982 and one in 1998 (Santa Monica Mountains National Recreation Area fire records, supplied by Marti Witter, NPS, 2015). Thus, in the coastal ranges lightning is a potential but relatively uncommon source of fire ignitions. In the interior ranges they are annual events, with the vast majority of lightning-ignited fires in forests and fewer than 20% in shrublands (Keeley 1982). Humans are responsible for the majority of all fires in the state, but even more so in southern California where they are responsible for over 86% on Forest Service lands and more than 98% on the lower elevation Cal Fire protected lands (Keeley and Safford 2016).

A factor of local importance in southern California is the Santa Ana wind. This foehn wind results from a high-pressure cell in the Great Basin coupled with a low pressure trough off the coast of southern California, which drives dry air toward the coast. This pressure gradient is steep and thus foehn winds may exceed 100 km hr^{-1} (60 mi h^{-1}) and bring high temperatures and low humidity. Santa Ana winds create the most severe fire weather conditions in the country because several days or weeks of these winds occur every autumn at the time natural fuels are at their driest (Schroeder et al. 1964). The winds have a distinct diurnal pattern with offshore surface winds building up overnight and peaking in the early morning, but as the ground heats up, convection currents typically hold these winds aloft through the afternoon (Keeley et al. 2009). Under Santa Ana wind conditions fire spread is rapid, sometimes covering 30,000 ha (74,100 ac) in a single day through fuels of any age class (Phillips 1971, Moritz et al. 2010).

The overwhelming importance of Santa Ana winds is illustrated by the relation between burning patterns and climate. Throughout the western United States there is a strong relation between antecedent drought and fire activity (Westerling et al. 2002). However, in southern California during the twentieth century, there is a surprisingly weak relation between antecedent drought and area burned (Keeley 2004a), but length of drought does play a role in extending the fire

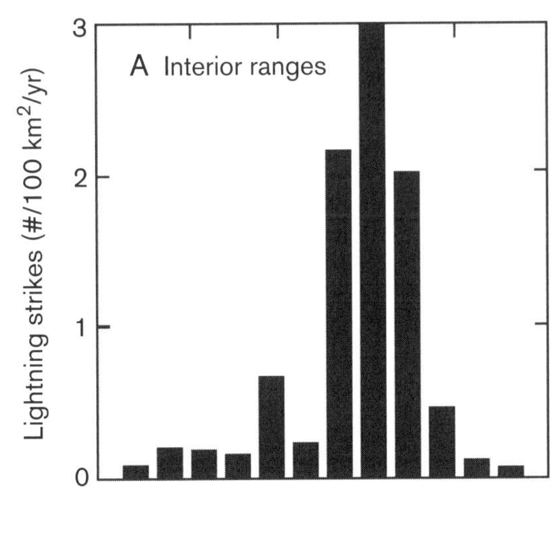

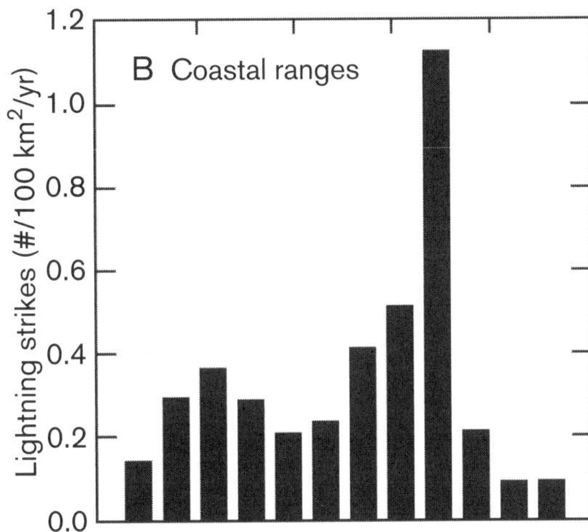

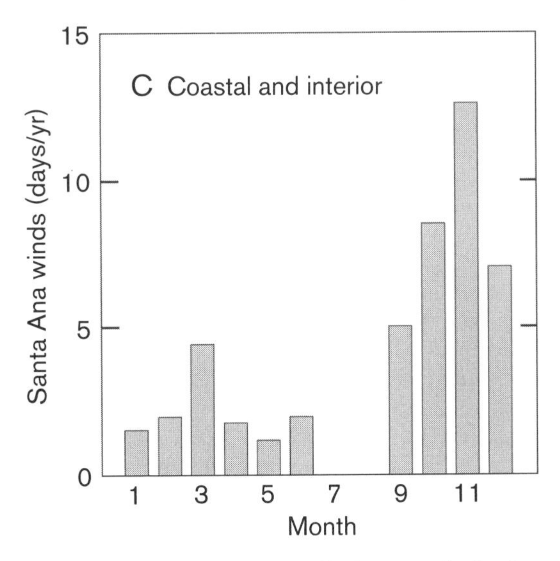

FIGURE 17.1 Monthly distribution of lightning strike density in the (A) interior and (B) coastal zones of southern California as defined by Miles and Goudey 1997 (data from van Wagtendonk and Cayan 2008) and (C) Santa Ana wind days for Los Angeles County (data from Weide 1968), based on (i) 3-mb decrease in mean sea level pressure from Santa Monica to Palmdale, (ii) ≥10°C (50°F) decrease in temperature from Santa Monica to Palmdale, (iii) northerly wind speeds of 16 km h^{-1} (10 mi h^{-1}) in the San Fernando Valley and 50 km h^{-1} (31 mi h^{-1}) in the Riverside San Bernardino area, and (iv) relative humidity ≤30%.

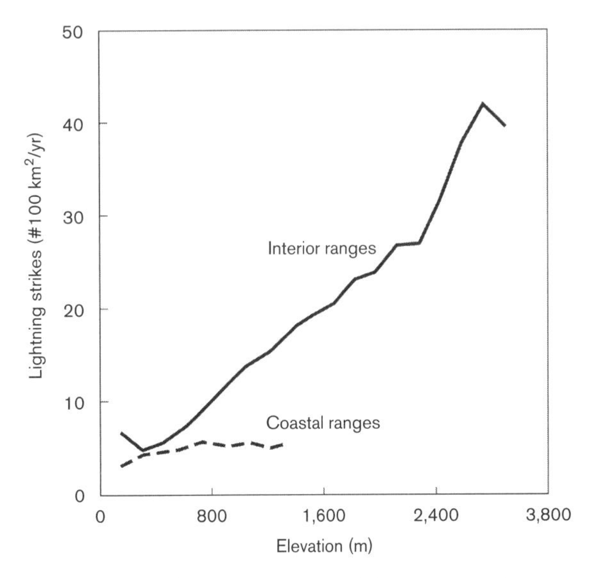

FIGURE 17.2 Elevational distribution of lightning strike density for the interior and coastal ranges. Due to limited area in the highest elevations these were combined into the last elevational category plotted (data from van Wagtendonk and Cayan 2008).

season (Dennison et al. 2008). Anomalous drought events of year or more are, however, associated with extreme fire events in this region (Keeley and Zedler 2009). It is hypothesized that the primary reason these events produce extreme fires is tied to vegetation dieback, which potentially increases the likelihood of spotting fires that challenge fire control and ultimately contribute to greater fire size (Keeley and Zedler 2009).

Although the weather conditions that produce lightning do not overlap with conditions that produce foehn winds, the seasons do overlap and Santa Ana winds may occur within days or weeks of a lightning-ignited fire (Fig. 17.1C). While the vast majority of contemporary burning in the region is due to human-ignited fires during Santa Ana wind conditions, it is hypothesized that under natural conditions occasionally lightning-ignited fires persisted into the autumn foehn wind season, and these Santa Ana wind-driven fires accounted for the majority of area burned (Keeley and Fotheringham 2001a).

One of the difficulties in sorting out the role of climate in driving fire regimes is that ignitions by human play a major role in this bioregion and complicate the interpretation (Syphard and Keeley 2015). For example, focusing on just the decadal average Palmer Drought Severity Index for Riverside County, it is apparent that it matches closely the area burned during the twentieth century (Fig. 17.3). However, human population growth parallels these changes in area burned, and increased ignitions are likely a major contributor to the late twentieth century increase in burning in that county.

Ecological Zones

The South Coast bioregion is a complex mosaic of grassland, shrubland, forest, and woodland (Fig. 17.4) that forms a relatively fine-grained landscape relative to most wildfires, which usually burn large enough areas to encompass a diversity of vegetation types and associations. Thus, fire regimes vary on a rather coarse scale, and within a vegetation type there is

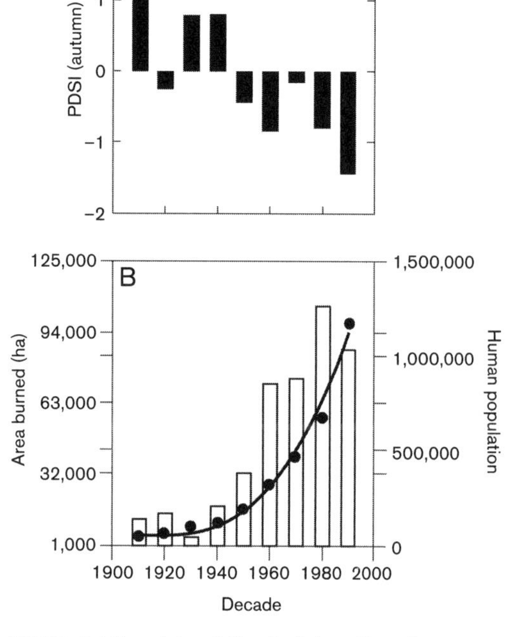

FIGURE 17.3 Decadal variation in Palmer Drought Severity Index, area burned and human population for Riverside County (data from Keeley et al. 1999).

limited association of fire regimes with specific plant associations. In this respect there are two broad ecological zones, the coastal valleys and foothill zone and the montane zone. Within each of these zones fire response is best examined at the scale of broadly defined vegetation alliances (Sawyer et al. 2009).

COASTAL VALLEY AND FOOTHILL ZONE

In the coastal and foothill zone the pattern of grasslands, sage scrub, chaparral, and woodlands is driven largely by a combination of soil moisture, fire, and other disturbances. At the arid end of the gradient herbaceous and semi-woody subshrub associations predominate with woodlands and chaparral at the moister end of the gradient. However, there is an important interaction between these drivers, such that under xeric conditions far less fire or disturbance is required to displace the dominant woody association with herbaceous vegetation (Keeley 2002a, Keeley et al. 2005a).

Southern California grasslands include small, highly fragmented patches of native grassland and more widespread, anthropogenically created nonnative-dominated annual grasslands. The native grasslands are dominated by perennial bunchgrasses, purple needle grass (*Stipa pulchra*), Nevada blue grass (*Poa secunda*), and June grass (*Koeleria macrantha*), and are distributed on fine-grained soils in localized patches within a mosaic of shrublands and woodlands (Huenneke 1989, Keeley 1993). Native grasslands also include a rich diversity of annual forbs but very few native annual grasses. Significant native grasslands still persist in valleys and shallow slopes on clay soils, which, due to the water-holding capacity of these fine-grained soils, are stressful substrates for competing shrubs.

Most of the grasslands in the region lack any native grasses and are dominated by nonnative annual grasses and forbs (D'Antonio et al. 2007). They bear little resemblance, in both ecology and distribution, to native grasslands. Many annual grasslands were derived from fire-induced type conversion of shrublands and subsequent invasion by nonnative grasses and forbs (see sidebar 17.1). They exhibit no strong relationship with soil type or precipitation and are the result of anthropogenic disturbance (Cooper 1922, Wells 1962, Keeley 1990b, 1993, 2002a). Although these generalizations are broadly applicable across the region, on a more limited scale there are some annual grasslands that do appear to be derived from native grassland due to intensive livestock grazing (Oberbauer 1978) or soil tillage (Kellogg and Kellogg 1990, Stromberg and Griffin 1996).

Semi-deciduous sage scrub dominates lower elevations along the coast and in interior valleys, and in these two subregions sage scrub forms very different species assemblages (Davis et al. 1994). Thus, it is not surprising that there are significant differences in fire response between coastal and interior sage scrub associations. This appears to be driven partly by floristic differences (Kirkpatrick and Hutchinson 1977) and partly by the impact of the more arid conditions in interior valleys. The fact that it forms very different associations in coastal versus interior sites makes the term Coastal Sage Scrub rather confusing and some authors have preferred to just use the term Sage Scrub (Keeley et al. 2005a, Rundel 2007) and perhaps a better term is California Sage Scrub (CSS; D. Sandquist, University of Utah, Salt Lake City, Utah, USA, pers. comm. 2009).

At lower elevations the taller stature evergreen chaparral replaces sage scrub on north-facing exposures or on deeper soils (Fig. 17.5), and with increasing elevation completely displaces it throughout much of the foothills and lower montane. Chaparral comprises a rich array of different floristic associations (Sawyer et al. 2009), some of which, such as chamise (*Adenostoma fasciculatum*) and black sage (*Salvia mellifera*), have such broad distributions throughout the region that they exhibit more than a single fire regime. Although fire response between dominants of different associations varies, different associations are often subject to similar fire regimes because of the fine-grain distribution of associations due to small-scale topographic variation. For example, monotypic stands of the facultative seeder chamise on south-facing slopes are often juxtaposed with diverse north-facing slopes of obligate resprouters such as scrub oak (*Quercus berberidifolia*), California coffeeberry (*Frangula californica*), and hollyleafed cherry (*Prunus ilicifolia* subsp. *ilicifolia*). They can also sometimes be separated by ridge tops dominated by the obligate seeder California-lilac species (e.g., hoaryleaf ceanothus [*Ceanothus crassifolius*], Mojave ceanothus [*C. vestitus*], and buckbrush [*C. cuneatus* var. *cuneatus*]). Most fires burn broad portions of the landscape, and thus these three associations typically burn in the same high-intensity fire. Although fire severity may vary in these different associations, the impact of fires on community resilience may be as much a function of pre- and postfire climate as it is on floristic differences (Fig. 17.5).

Within chaparral dominated landscapes there are isolated patches of serotinous cypress species Tecate cypress (*Hesperocyparis forbesii*) or Cuyamaca cypress (*H. stephensonii*). They typically occur above 700 m (2,300 ft) in widely disjunct populations, perhaps as relicts of a once wider distribution. They range from 330 to 1,700 m (1,083–5,577 ft), primarily on north-facing slopes, and all populations appear to date back to the last fire (Armstrong 1966, Zedler 1977). These small stature cypress form even-aged monotypic stands from less

SIDEBAR 17.1 TYPE CONVERSION

Type conversion is the change in vegetation type due to unnatural disturbances and is usually applied to situations where there has been a fundamental change in functional types, such as from shrublands to grasslands. In chaparral, type conversion is often precipitated by short-interval fires that specifically target nonresprouting or obligate seeding shrubs such as certain species of California-lilac (*Ceanothus*) (Zedler et al. 1983). Such stands become more open, and often are subsequently invaded by nonnative herbaceous species. Fire return intervals shorter than six years have been shown to be highly detrimental to the persistence of nonresprouting chaparral species (Jacobson et al. 2004, Lippitt et al. 2013); in fact, multiple fires within a six-year interval have even reduced resprouting species, further opening the chaparral environment (Haidinger and Keeley 1993).

Grasses alter the fire regime and create a positive feedback process (Keeley et al. 2012) and continued short-interval fires may further degrade from a chaparral–grassland mosaic to annual grasslands of mostly alien grasses and forbs. The speed of the type conversion process can be increased dramatically by numerous variables such as drought, cool season fires, livestock grazing, herbicides, and mechanical clearance activities (Bentley 1967, Keeley 2006a). It was hypothesized that this process began with the late Pleistocene entry of people into California (Cooper 1922, Wells 1962) and resulted in substantial loss of woody vegetation and an increase in more productive seed bearing annual plants (Keeley 2002a). During the nineteenth century Mexicans and, later, Americans increased type conversion for the purpose of enhancing rangeland for livestock, and in the twentieth century this practice was promoted by State programs (Keeley and Fotheringham 2003).

Documentation of these alien invasions has shown that the major driver is short-interval fires (Jacobson et al. 2004, Keeley and Brennan 2012). An important question is the extent of this contemporary threat. Talluto and Suding (2008) found that over a 76-year period 49% of the sage scrub shrublands in one southern California county had been replaced by annual grasses and a substantial amount of this could be attributed to fire frequency. Fire could play an even greater role in this region since this study only included large fire events as reported in the Fire Resource Assessment Program (FRAP) database, which includes only about 5% of all fires (J. Keeley, US Geological Survey, Three Rivers, California, USA, unpublished data) but does include the largest fires in the state.

Despite what would seem to be a convincing case for type conversion, Meng et al. (2014) raised some skepticism about the ability of repeat fire to effect type conversion. They noted that early in the twentieth century range managers could not effectively displace chaparral by prescribe fire alone and often relied on herbicides and mechanical destruction for converting chaparral to rangeland. Based on this Meng et al. (2014) posited that type conversion of chaparral to grassland from short fire intervals is a rare phenomenon. However, as pointed out by Keeley and Brennan (2012), managers only utilize fire under narrow prescription conditions, which are generally not capable of carrying repeat fires at short fire return intervals; in contrast, wildfires burn outside prescription, often with 100 km h^{-1} (about 62 mi h^{-1}) wind gusts and relative humidity less than 5% and under these conditions short-interval fires effecting type conversion are a very real phenomenon. Nitrogen deposition has been implicated in effecting type conversion and considering the extraordinary regional variation in rates of deposition (Padgett and Allen 1999), it could play very different roles throughout region. On coastal sites nitrogen interacted with drought to slow postfire shrub recovery but by itself was not obviously a driver of type conversion (Kimball et al. 2014).

than 1 ha (2.47 ac) to more than 100 ha (247 ac) in size usually on fine clay soils derived from gabbro or other low productivity substrates (McMillan 1956, Zedler 1995a).

Several pines are closely associated with chaparral shrublands. In the eastern Transverse and northern Peninsular ranges are pockets of knobcone pine (*Pinus attenuata*) (Pequegnat 1951, Vogl 1973, Minnich 1980a), a species more common further north in the coastal ranges. Coulter pine (*P. coulteri*) is widely distributed above 1,200 m (3,900 ft) throughout the Transverse and Peninsular ranges, occurring in both chaparral and mixed-conifer forests (Griffin and Critchfield 1972). Torrey pine (*P. torreyana*) is restricted to a few coastal populations in San Diego County, with one disjunct population on Santa Rosa Island.

FIGURE 17.4 Landscape mosaic of southern oak woodland, grassland, arid chaparral on south-facing exposures (slopes on right side of photo), and mesic chaparral on north-facing slopes (photo by J. Keeley, location near Black Mountain north of Ramona, San Diego County).

FIGURE 17.5 Mosaic of coastal sage scrub surrounding patch of larger stature chaparral in the Santa Ana Mountains, Orange County at approximately 700 m (2,300 ft) (photo by J. Keeley).

Bigcone Douglas-fir (*Pseudotsuga macrocarpa*) occurs in low to high density populations between 400 and 1,800 m (1,300–5,900 ft), although small populations also occur at lower and higher elevations (Gause 1966). Approximately two-thirds of this forest type (19,000 ha [47,000 ac]) are distributed in the San Gabriel Mountains (Stephenson and Calcarone 1999). They are most abundant on north-facing slopes, particularly on interior sites (Bolton 1966, Minnich 1978). This tree was heavily exploited for timber during the early settlement period because it was the lowest elevation conifer in the region (Minnich 1988). This and increased fire frequency accompanying settlement in the Los Angeles Basin are considered factors accounting for the upward movement of its elevational boundary as well as its contemporary restriction to mostly localized populations (Gause 1966). Today bigcone Douglas-fir is commonly confined to rather steep ravines in association with canyon live oak (*Q. chrysolepis*) within large areas of chaparral and oak woodland (McDonald and Litterell 1976).

Riparian woodlands and shrublands today occupy a relatively small percentage of the bioregion; it is estimated that 90% of them have been lost (Faber et al. 1989). Historically the greatest extent of riparian habitat was the Los Angeles coastal plain, much of which was a vast seasonal wetland (Gumprecht 1999). Riparian woodlands have the potential for affecting landscape patterns of fire spread. The dominant woody plants willow, alder, poplar, elderberry, maple, baccharis, ash, camphorweed, poison oak, and sycamore (*Salix*, *Alnus*, *Populus*, *Sambucus*, *Acer*, *Baccharis*, *Fraxinus*, *Pluchea*, *Toxicodendron*, and *Platanus*, respectively) are nearly all are winter-deciduous (Rundel and Sturmer 1998). In foothill and coastal environments this contrasts with a landscape otherwise dominated by either summer deciduous or evergreen lifeforms.

Woodlands dominate undisturbed valleys and north-facing slopes, most commonly including the evergreen coast live oak (*Q. agrifolia*), sometimes in association with the much rarer evergreen Engelmann oak (*Q. engelmannii*). Typically these woodlands are relatively open and in many instances are best considered oak savannas. Today the understory consists mostly of nonnative grasses and forbs, and these are the primary fuels that carry fire. Prior to human entry into California, the oak understory is thought to have been comprised of sage scrub subshrubs (Wells 1962), and examples of this association are still readily observed today along Hwy 101, north of Gaviota Pass in Santa Barbara County. This plant community has long been desired for human development, and thus extensive portions of woodlands have been lost. Recruitment within existing stands appears to be good for some oak species but not others (Tyler et al. 2006). Interior live oaks (*Q. wislizeni*) often form closed canopy forests on mesic sites, often mixing with broadleaf chaparral shrubs. Valley oak (*Q. lobata*) woodlands occur in several valleys of the northwestern Transverse Ranges. These represent a southern extension of such woodlands and are covered in more detail in the Central Coastal bioregion (chapter 16).

One largely endemic association is the California black walnut-dominated woodland, widely distributed in the foothills of the Los Angeles Basin (Quinn 1990). This winter-deciduous tree often forms associations with arborescent forms of evergreen chaparral shrubs such as toyon (*Heteromeles arbutifolia*), hollyleafed cherry, and hollyleaf redberry (*Rhamnus ilicifolia*) (Keeley 1990a).

MONTANE ZONE

In the lower montane zone, winter-deciduous California black oak (*Quercus kelloggii*) woodlands often dominate in association with conifer forests between the lower elevation evergreen sclerophyllous chaparral and the higher elevation evergreen conifers. This winter-deciduous woodland is more widespread in the other interior ranges of the region.

Montane forest types considered here all exhibit substantial ecological overlap with Sierra Nevada forests Ponderosa (*Pinus ponderosa*) and Jeffrey (*P. jeffreyi*) segregate out according to specific climatic parameters, with the latter being predominant on drier and/or colder sites (Thorne 1977, Stephenson 1990). At their lower margins they are associated with Coulter pine and at higher elevations with white fir (*Abies concolor*). On more mesic slopes white fir and incense cedar (*Calocedrus decurrens*) dominate with a smaller percentage of sugar pine (*P. lambertiana*) and Jeffrey pine. Lodgepole pine (*P. contorta* var. *murrayana*) forests have limited distribution in the highest elevations of the eastern Transverse and northern interior Peninsular ranges. Unlike the lodgepole populations of the northern Rocky Mountains, these southern California pines are not fire-dependent serotinous species, and in this respect are most similar to the Sierra Nevada lodgepole pine forests.

Historical Fire Occurrence

Prehistorical Era

Chaparral genera, including *Arctostaphylos*, *Ceanothus*, *Cercocarpus*, *Frangula*, *Fremontodendron*, *Heteromeles*, *Prunus*, *Quercus*, *Rhamnus*, and *Rhus* (manzanita, California-lilac, coffeeberry, fremontia, chaparral holly, plum, oak, redberry, sumac, respectively), were present in the south coast region since at least mid-late Miocene, 5 million to 15 million years ago (Axelrod 1939). Middle Miocene deposits in the Santa Monica Mountains contain charcoal and partially burned chaparral fragments (Weide 1968), and there is evidence that fire has been an important ecosystem process for tens of millions of years in these shrublands (Keeley et al. 2012). Throughout the Pliocene and Pleistocene epochs the region was dominated by closed-cone cypress and pines and included Monterey cypress (*Hesperocyparis macrocarpa*), Bishop pine (*Pinus muricata*), and Monterey pine (*P. radiata*), species that today are displaced 560 km (348 mi) northward (Axelrod 1938, Axelrod and Govean 1996, but cf. Millar 1999). The closed-cone character indicates fire has been a predictable feature of this landscape for at least two million years, and this is supported by fossil charcoal records (Johnson 1977; but cf. Axelrod 1980 for a contrary opinion). The facts that these species typically burn in stand-replacing crown fires are moderately long-lived obligate seeders sensitive to short fire return intervals (Keeley and Zedler 1998), and growth rates were lower than today due to 40% less CO_2 during much of the Pleistocene (Ward et al. 2005) suggest the hypothesis that fires were infrequent, on the order of once or twice a century.

The present distribution of vegetation types is a product of Holocene climatic variability and coincides with the first entry of humans into the region. The concomitant emergence of contemporary climate, vegetation distribution, and humans presents a challenge to accurately discern natural and anthropogenic fire regimes. However, although humans were present more than 10,000 years BP, a widespread presence significant enough to have an impact on the local ecology is undoubtedly more recent, perhaps only the last few thousand years or less (Jones 1992). Throughout the first half of the Holocene, lightning was likely the dominant or only ignition source in most parts of the region. It has been hypothesized that the coastal region burned when the occasional August or September lightning-ignited fire persisted until picked up and carried by the autumn foehn winds (Keeley and Fotheringham 2003, Keeley and Zedler 2009). At the other extreme, montane coniferous forests in the interior subjected to frequent lightning ignitions would have burned on much shorter intervals, controlled largely by the rate of fuel accumulation (Minnich 1988).

By middle Holocene a great diversity of Indian groups had settled in the region, and fire was one management practice they all held in common. Numerous uses for burning have been reported but the one that likely had the greatest impact was burning shrublands for type conversion to herb-dominated associations (Timbrook et al. 1982, Keeley 2002a). The clearest statement of this management practice is the 1792 report by Spanish explorer Jose Longinos Martinez who wrote, ". . . in all of New California from Fronteras northward the gentiles have the custom of burning the brush" (Simpson 1938). It has been said that Native American presence was widespread and touched every part of this region (Anderson et al. 1998). The archeological record in San Diego County alone documents over 11,000 widely distributed Indian sites (Christenson 1990). Two-thirds of the settlements were in the coastal valleys and foothills in what currently is chaparral or sage scrub vegetation.

The sudden emergence of charcoal deposits and replacement of woody elements by herbaceous taxa around 5,000 year BP in coastal southern California (Davis 1992) is thought to be a reflection of increased burning by Native Americans (Anderson et al. 2013). Higher fire frequency is also a likely contributing factor to the demise of closed-coned cypress along the San Diego coast 1,800 years BP (Anderson and Byrd 1998), because it is not tied to climatic patterns that might account for the extirpation of this species in the area. Cessation of burning by Indians following settlement is thought to have been a major contributor to the apparent nineteenth century lowering of the elevational distribution of oak woodlands (Byrne et at. 1991).

Historical Era

Much of our historical information is from written records that often contain misinformation and exhibit various biases.

TABLE 17.1
Historical fires over 50,000 ha (123,553 ac) recorded from the south coast region

Year	Fire	Cause	County	Month	Area		Damage	
					Hectares	Acres	Lives	Structures
1889	Santiago Cyn	Campfire	Orange	Sept	125,000	308,900	0	0
1932	Matilija	Campfire	Santa Barbara	Sept	8,900	220,200	0	0
1970	Laguna	Pwr ln	San Diego	Sept	70,500	174,200	5	382
1985	Wheeler #2	Arson	Ventura	July	49,700	122,800	0	26
2003	Cedar	Flares	San Diego	Oct	109,500	270,600	15	2,820
2006	Day	Debris burning	Ventura	Sept	65,500	161,900	0	11
2007	Zaca	Equipment	Santa Barbara	July	97,300	240,400	0	1
2007	Witch	Pwr ln	San Diego	Oct	80,200	198,200	2	1,650
2009	Station	Arson	Los Angeles	Sept	65,000	160,600	2	209
2017	Thomas	Pwr ln and campfire	Ventura and Santa Barbara	Dec	114,078	281,193	1	1,063

Early reports on fire cause are suspect when it comes to evaluating lightning ignitions. For example, Sargent (1884) reported for 1881 that 138,000 ha (340,000 ac) burned in California forests but none were attributed to lightning, whereas hunters, campers, and Indians were thought to be the primary sources of ignition. This report is highly suspect because today we know lightning accounts for a substantial portion of all wildfires in the state.

The first Spanish settlement in California was the Mission San Diego de Alcala, founded in 1769. In addition to initiating a relentless wave of settlers, these first Europeans dramatically changed the landscape by introducing numerous nonnative grasses and forbs (Mack 1989). These species spread rapidly, perhaps facilitated by the highly modified landscape resulting from a long history of frequent burning by Indians (Keeley 2002a, 2004c) and further promoted by the Mexican vaqueros habit of expanding grazing lands by "burning off the brush" (Kinney 1887).

American settlement brought increased competition for grazing land and consequently the extent of chaparral conversion (Kinney 1900, Plummer 1911). In rural San Diego County a typical late nineteenth century land-use pattern involved homesteads of 65 ha (160 ac) centered in small grassy potreros or dry meadows that were generally too small to provide a sound economic basis for subsistence (Lee and Bonnicksen 1978). Chaparral dominated the slopes surrounding these potreros (e.g., Fig. 17.5), and ranchers routinely burned the shrublands to supplement meadow grazing with additional forage and to allow easier movement from one canyon to another.

In remote areas sheep herders routinely used fire to enhance mountain grazing lands, and miners used fire to open the brush and facilitate exploration. Surveys done in the three southern California forest reserves in the late nineteenth century provide some of the best quantitative data on prefire-suppression burning patterns. For the chaparral zone, which comprised 50% (San Bernardino Mountains) to 90% (San Gabriel Mountains) of the new reserves, USGS biologist J.B. Leiberg presented estimates of burning in annual reports filed in the late 1890s. Based on the area of chaparral in each reserve and his estimate of how much chaparral had burned in the previous 4 years to 9 years, we calculate that he observed rates of burning, expressed as percentage of chaparral landscape burned per decade, were 2.8%, 3.3%, and 5.4% for the San Jacinto (Leiberg 1899a, 1900c), San Bernardino (Leiberg 1899b, 1900b), and San Gabriel (Leiberg 1899c, 1900a) reserves, respectively. Even this limited extent of burning cannot be entirely ascribed to natural sources. For example Leiberg (1899c) stated "It is also noteworthy that the worst-burnt areas in the three reserves examined are to be found in the San Gabriel Reserve in the region of the most extensive mining operations." Although Leiberg's estimates likely did not capture all of the burning on these landscapes, they do indicate that natural fire rotation in the lower elevation chaparral zone was on the scale of once or twice a century, and certainly not on the scale of every few decades as was apparently the case in higher elevation conifer forests.

These historical records suggest a presuppression model of burning in chaparral landscapes of many modest-size summer lightning-ignited fires that burned a relatively small portion of the landscape, punctuated one or two times a century by massive autumn Santa Ana wind-driven fires (Keeley and Fotheringham 2001a). This is supported by the historical analysis of fires in the San Gabriel Mountains during a 5-year period in the late 1890s where it was found that a small portion of the range burned in small fires (Minnich 1987). Longer time periods of a decade or two are required to encounter massive fire events responsible for burning much of this landscape, such as happened in that range during 1919 (Keeley and Zedler 2009). This fire regime of small lightning-ignited fires interrupted by extreme fire events is supported by the historical marine charcoal record of infrequent large Santa Ana fires over the past 500 years (Mensing et al. 1999) and by tree-ring studies in the region (Lombardo et al. 2009). Also, based on the lightning frequency in coastal regions, fire

SIDEBAR 17.2 COMPARISON OF FIRE REGIMES BETWEEN SOUTHERN CALIFORNIA AND NORTHERN BAJA CALIFORNIA

Following on ideas first published by Dodge (1975), Minnich (1983) used Landsat remote imagery to contrast burning in shrublands on both sides of the border between 1972 and 1980. He concluded that during this 9-year period fires were larger north of the border; however, critics have found there is no statistically significant difference between these regions (Strauss et al. 1989, Keeley and Fotheringham 2001a). The primary motivation for concluding fires was larger north of the border was based on the inclusion of two fires reported in written historical records north of the border (Minnich 1983). Because written records were not available south of the border, it has been argued that this is a biased comparison (Keeley and Fotheringham 2001a). In order to compensate for the lack of written records south of the border, Minnich (1989, 1995, 1998, Minnich and Dezzani 1991, Minnich and Chou 1997) estimated fire size for an 80-year period south of the border using historical aerial photographs, and concluded that during this period large fires were absent from the northern Baja California landscape. However, the long gaps between images and lack of appropriate controls make these studies open to interpretation (Keeley and Fotheringham 2001a). The conclusion from Minnich's studies is that the smaller fires south of the border are reflective of the natural southern California fire regime and larger fires north of the border are a modern artifact of fire suppression and illustrate the need for doing landscape scale prescription burning in southern California.

This conclusion has been criticized because historical records show that prior to fire suppression in southern California, large fires were always part of this landscape (Barrett 1935, Keeley et al. 1999, Mensing et al. 1999, Weise et al. 2002, Lombardo et al. 2009). In 1878 one of the largest fires in Los Angeles County history burned 24,000 ha, and in 1889 the Santiago Canyon Fire burned over 100,000 ha in Orange and surrounding counties (Barrett 1935, Lee and Bonnicksen 1978, Keeley and Zedler 2009). In San Diego County it was reported in 1885 that "At least one third of the land covered with brush, grass and oak timber in the southern part of this county has been burnt off by settlers within the past eighteen months, doing a great deal of damage, not only as regards pasturage, timber, and bees, but also decreasing the reservoirs of water, which the absence of brush will effect . . .," and in 1889 a huge 37,000 ha fire burned coastal shrublands in the county (Barrett 1935).

Additionally, it has been proposed that the present fire regime in Baja California is not likely representative of the natural regime, and that it is largely the result of contemporary land-use practices (Keeley and Fotheringham 2001a, 2001b). These latter authors contended that the patterns observed north and south of the border should not be the basis for southern California fire management policy (see also Keeley et al. 2012).

frequencies of once or twice a century would require big fire events, otherwise fire rotations would be thousands of years long and this seems unlikely (Keeley and Zedler 2009). Such a fire regime is also supported by life history characteristics of many dominant woody species that are resilient to long fire-free intervals and sensitive to short intervals on the scale of once a decade (Keeley 1986, Zedler 1995a).

Although remote regions were experiencing limited burning, rapidly developing rural areas reported some rather massive wildfires. In 1878 one of the largest fires in Los Angeles County history burned 24,100 ha (60,000 acres) and in 1889 well over 100,000 ha (247,000 acres) burned in a single fire in Orange County (Barrett 1935, Lee and Bonnicksen 1978, Keeley and Zedler 2009). Historical studies show that the region has experienced nine fires greater than 50,000 ha (123,500 ac) (Table 17.1). The largest of these was during a Santa Ana wind event in September 1889. Of particular interest is the fact that no one died and no structures were destroyed in that fire. Since then large fire events have occurred at sporadic intervals with increasing destruction. It is clear that the occurrence of large fire events has not changed, but human population growth has put increasing numbers of people at risk.

It has been proposed that a view of the natural southern California fire regime can be gained by examination of burning patterns south of the US border, where it is presumed humans have had less impact on the natural fire regime. These studies are discussed in sidebar 17.2 on fire regimes in southern California and northern Baja California.

Current Era

The nineteenth-century pattern of large fires has continued through the twentieth century, and Moritz (1997)

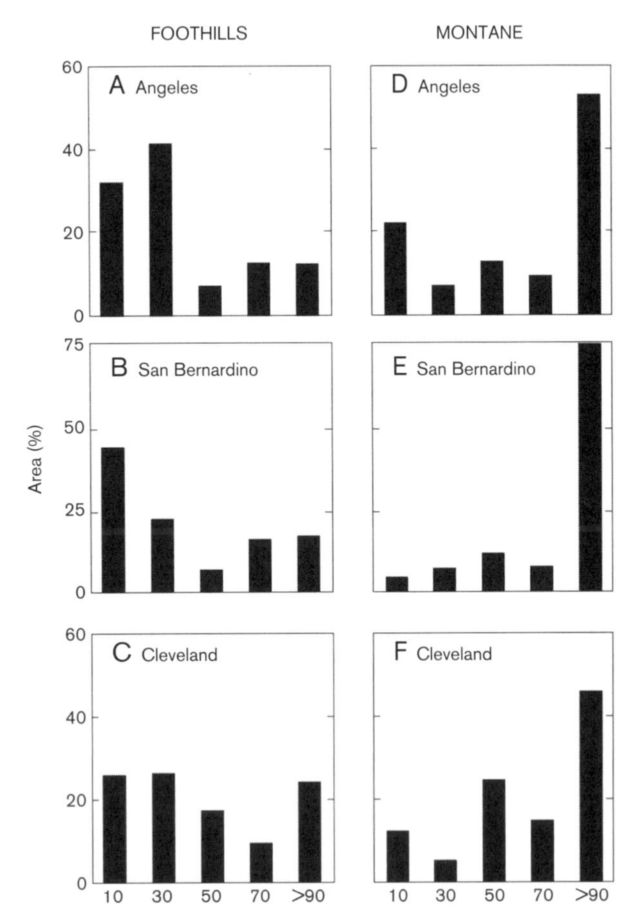

FIGURE 17.6 Time since last fire for foothill (A, B, C) and montane (D, E, F) environments in the Angeles, San Bernardino, and Cleveland USFS national forests. The Angeles includes most of the San Gabriel Mountains and surrounding foothills, the San Bernardino includes the San Bernardino and San Jacinto mountains, and the Cleveland includes largely the Santa Ana and Laguna mountains (data from Stephenson and Calcarone 1999).

demonstrated that the probabilities for large fires in the Central Coast bioregion have not changed due to fire suppression. Recent decades have experienced an increasing frequency of smaller fires (Moritz 1997, Keeley et al. 1999). Throughout the century there has been an increase in fire frequency that parallels human population growth (Keeley and Fotheringham 2003). Much of the population growth has been in coastal valleys and foothills, where natural fire frequencies were lowest. In the latter half of the twentieth century fire rotations shortened and average fire size decreased in all southern California counties (Conard and Weise 1998, Keeley et al. 1999, Weise et al. 2002). Human impacts on montane environments have been slower to develop, although air pollution has caused substantial changes in forest structure (McBride and Laven 1999).

Although the region has long been susceptible to massive fire events (Table 17.1), in recent years the frequency of such fires has greatly increased. In southern California over half of the really massive fire events have occurred since 2002. Factors considered responsible are drought events over the past several decades coupled with increased population and a trend toward increasing ignitions in more interior sites where Santa Ana winds carry the fire further toward the coast (Keeley and Zedler 2009).

Major Ecological Zones

Burning patterns in the lower elevation coastal valley and foothill zone show that, despite vigorous fire suppression efforts throughout the twentieth century, over half of the landscape has burned within the past several decades (Fig. 17.6A, B, C). This in fact is an underestimate of the extent of twentieth-century burning because these data are for just national forest lands (Stephenson and Calcarone 1999), and much of the foothill landscape outside the forests has experienced higher rates of burning (Keeley 1982). Indeed, the extent of burning in lower elevation coastal southern California led Safford and Van de Water (2014) to conclude that this region comprises a totally different fire regime from much of the more northern part of the state. Fire regime attributes for the vegetation types in the South Coast bioregion are listed in Appendix 1.

Widespread burning should not be interpreted to suggest that fire suppression activities have had no impact. Considering the lower foothills and coastal zone where humans account for over 98% of all wildfires (Keeley and Safford 2016), it seems reasonable that the primary impact of suppression has been to maintain these landscapes closer to their natural fire regime than if fires were not suppressed (Keeley et al. 1999). Within this zone it appears that roughly 10–25% of the landscape has never had a recorded fire (Table 17.2). Based on estimates of burning made by Leiberg, and discussed in the historical section, it would appear that this proportion of unburned landscape is probably not far outside the natural range of variability for the region.

Burning patterns for the montane zone (Fig. 17.6D, E, F) suggest that vigorous fire suppression efforts throughout the twentieth century have had a substantial impact in excluding fires from this landscape (Stephenson and Calcarone 1999). Roughly 50–75% of these forests have never had a recorded fire (Table 17.2), which is likely far outside the natural range of variability for these conifer forests. Effective fire exclusion in this zone contrasts markedly with the foothill zone where suppression has not been very effective at excluding fire (Fig. 17.6A, B, C).

Grasslands

A substantial proportion of native bunchgrass rootstocks survive grassland fires and resprout as moisture becomes available in autumn (Table 17.3). These grasses flower and disperse seeds in the first postfire spring, and seedling recruitment is commonly very abundant in the second and the next few postfire years (Keeley 1990b), although recruitment may only be successful in years of high precipitation (Hamilton et al. 1999). These patterns are the same for a rich diversity of native herbaceous perennial forbs as well, including species of mariposa lily (*Calochortus*), blue dicks (*Dichlostemma*), sanicle (*Sanicula*), and lomatium (*Lomatium*) among others. Although annual species (both native and nonnative) lack much long-term seed storage, annual recruitment and seed production ensure a ready seed bank at the time of fire.

Grasslands are resilient to a range of fire frequencies (Reiner 2007). Native grassland species are mostly perennials and are adapted to persisting through the dry season with dormant rhizomes, bulbs, or seeds. In the absence of fire for extended periods of time, natural grassland patches persist because of the greater competitive ability of perennial grasses on fine-grained soils. These substrates retain moisture in the upper

TABLE 17.2

Percentage of area burned in foothills (below 900 m [2,300 ft]) and montane (above 1,500 m [4,900 ft]) portions of the three southern California national forests (from Stephenson and Calcarone 1999)

	Percentage of area burned							
	Foothills				Montane			
	NUMBER OF RECORDED FIRES				NUMBER OF RECORDED FIRES			
National Forest	0	1	2	>3	0	1	2	>3
Angeles	11	34	33	22	52	31	14	3
San Bernardino	16	38	25	21	73	22	4	1
Cleveland	24	37	24	16	45	47	8	0

TABLE 17.3

Fire response types for important species in grasslands in the South Coast bioregion

	Type of fire response			
Lifeform	Resprouting	Seeding	Individual	Species
Herbaceous perennial	Yes as a normal phenological response, not fire-stimulated	No dormant seed bank. Seeds are produced by first year resprouts generate seedlings in subsequent postfire years	Aboveground portions of plants usually dead at time of fire	*Calochortus* spp. *Dichelostemma capitatum* *Lomatium* spp. *Stipa pulchra* *Sanicula* spp.
Annual	None	Yes, largely from current year seed bank	Plants dead at the time of fire	*Avena* spp. *Bromus* spp. *Brassica* spp.

soil profiles and reduce percolation to deeper soil layers, which is lethal for shrub taproots that depend upon it for surviving summer drought. Annual grassland persistence in the absence of fire is more complicated and a function of plant community interactions.

FIRE REGIME–PLANT COMMUNITY INTERACTIONS

Historically, fire return intervals were likely heavily influenced by proximity to Native American settlements (Keeley 2002a), being frequent in those areas, but on remote sites, where lightning was the primary ignition source, more on the order of one to several times a century. Today grassland fires are almost entirely ignited by humans, and fire frequency is very high, particularly at the interface of urban areas. Historically, it is to be expected that most fires burned between summer and fall, but wherever human ignitions have been or are a factor, the fire season is extended to include late spring and early winter, and year round during droughts.

Burning beginning with Native Americans and continued by Euro-American settlers has greatly expanded grassland distribution at the expense of sage scrub, chaparral, and woodland (Wells 1962, Keeley 1990b, 2002a, Hamilton 1997). Initially these "grasslands" were likely dominated by native forbs, as the state lacks aggressive colonizing native annual grasses. It is to be expected that where these type conversions occurred adjacent to native grasslands that bunchgrasses such as purple needle grass and nodding needle grass (*Stipa cernua*) would have invaded, although the long and extreme summer drought in this region might have made persistence on well-drained coarse substrates precarious. One native bunchgrass, foothill needle grass (*S. lepida*), which is a common component of sage scrub (Keeley and Keeley 1984), likely persisted in these anthropogenic disturbance-maintained grasslands. However, based on the available flora of native colonizing species, prior to European entry into California, these "grasslands" probably were dominated by annual forbs, many of which were important food resources for Native Americans (Timbrook et al. 1982, Keeley 2002a).

During the first half of the nineteenth-century California grasslands, both bunchgrass-dominated and annual forb-dominated grasslands were rapidly and thoroughly invaded by nonnative grasses and forbs brought by the European colonizers (Mack 1989). Overgrazing and drought during the mid- to late-nineteenth century are commonly given as

TABLE 17.4
Fire response types for important species in sage scrub in the South Coast bioregion

Lifeform	Type of fire response: Resprouting	Seeding	Individual	Species
Shrub	Fire-stimulated	Fire-stimulated	Top-killed	*Malosma laurina*
Shrub	Fire-stimulated	None	Top-killed	*Rhus integrifolia*
Subshrub	Fire-stimulated	Polymorphic seed bank, a portion is dormant and fire-stimulated	Top-killed or killed	*Artemisia californica* *Eriogonum cinereum* *Eriogonum fasciculatum* *Mimulus aurantiacus* *Salvia leucophylla* *Salvia mellifera*
Subshrub	Fire-stimulated	No dormant seed bank. Seeds are produced by first year resprouts generate seedlings in subsequent postfire years	Top-killed	*Encelia californica* *Hazardia squarrosa*
Suffrutescent	None	Fire-stimulated	Killed or present only as dormant seeds	*Acmispon glaber*

causal factors behind our present nonnative-dominated grasslands. However, some species, such as wild oats (*Avena fatua*) and black mustard (*Brassica nigra*), appear to have dominated rapidly in the absence of overgrazing (Heady 1977). The primary factors driving the rapid nonnative invasion of grasslands over much of the landscape were: (1) Holocene climate warming, which favored annuals over perennials, (2) long coevolution of European annuals with human disturbance, which selected for aggressive colonizing ability (Kimball and Schiffman 2003), and (3) the disequilibrium in native ecosystems created by high frequency of burning by Native Americans (Keeley 2002a), which favored rapid invasion.

Annual grasslands thrive on frequent fires due to copious seed production and high seed survival under low-intensity fires. These nonnative-dominated grasslands originated from anthropogenic disturbance and are not restricted by substrate, being found on soils that support most other vegetation types (Wells 1962). In the absence of fire or other disturbances their persistence is a function of recolonization ability of woody associations (Wells 1962, DeSimone and Zedler 1999, 2001). Chaparral species colonize open grassland sites poorly, because species most tolerant of open xeric sites have weak dispersal and species with high dispersal establish poorly on open sites (Keeley 1998b). In the absence of fire or disturbance, sage scrub species recolonize much more readily than chaparral (McBride 1974, Freudenberger et al. 1987, Callaway and Davis 1993).

California Sage Scrub

On sites near the coast the most common subshrubs in this community are obligate resprouters after fire (Table 17.4), including California brittlebush (*Encelia californica*), saw-toothed goldenbush (*Hazardia squarrosa*), and nodding wild buckwheat (*Eriogonum cinereum*), although in the long absence of fire these species are capable of continued canopy regeneration through basal sprouting and layering (Malanson and Westman 1985). Many subshrubs are facultative seeders and regenerate after fire from resprouts and dormant seed banks, although on many sites resprouting is the dominant means for recapturing postfire sites (Malanson and O'Leary 1982). The most common facultative seeders include monkey flower (*Mimulus aurantiacus*), California sagebrush (*Artemisia californica*), purple sage (*Salvia leucophylla*), and black sage. Seeds produced by the first year resprouts are largely nondormant and produce a massive flush of seedlings in the second postfire year (Keeley and Keeley 1984, Keeley et al. 2006). Several of these species are known to produce polymorphic seed banks with a portion of the seeds having deep dormancy broken by smoke or other combustion products (Keeley and Fotheringham 2000). Deerweed (*Acmispon glaber*) is the only obligate seeding woody species typical of sage scrub, and it exhibits massive seedling recruitment in the first postfire year from dormant seed banks. In this, and all other Fabaceae, chemical cues such as smoke play no role in stimulating germination, rather heat shock triggers germination in the first postfire year. As a consequence germination may be triggered in the absence of fire on open substrates due to solar heating of the soil. For example, deerweed continues to recruit at low levels in subsequent years, particularly following high rainfall (Keeley et al. 2006). This species is not long-lived, perhaps a decade or two, and occasionally is subject to mass die off as was observed on many 5-year postfire sites in spring following the extremely wet El Niño winter of 1997–1998 (J. Keeley, personal observations).

Two larger stature evergreen shrubs, laurel sumac (*Malosma laurina*) and lemonade berry (*Rhus integrifolia*), are widely dispersed throughout sage scrub. They are both vigorous resprouters but have very different seedling recruitment dynamics. The former species often has substantial seedling recruitment following fire (Keeley et al. 2006), whereas the latter species typically recruits under fire-free conditions (Lloret and Zedler 1991).

Much of the postfire woody flora in southern California is rather promiscuously distributed in both chaparral and sage scrub (Keeley et al. 2005a). Some species such as deerweed, golden yarrow (*Eriophyllum confertiflorum*), California buckwheat (*E. fasciculatum*), bush-mallow (*Malacothamnus fasciculatus*), laurel sumac, and black sage are equally common in both vegetation types. Coastal sage communities also have a very rich annual and herbaceous perennial flora, many of which are ephemeral postfire followers. The floristic overlap between chaparral and sage scrub in both the annual and herbaceous floras is even more pronounced than with woody species, thus they are discussed in the chaparral section. On the more mesic slopes rhizomatous grasses including giant wild-rye (*Elymus condensatus*) and species of bent grass (*Agrostis* spp.) often dominate, and due to their vigorous resprouting are resilient to high fire frequencies.

On interior sites there exists a more arid version of sage scrub that lacks some of the most vigorous resprouting species, and the facultative seeders on these sites tend to behave more like obligate seeders, as fire-caused mortality often can be 100% on interior sites. Resprouting success appears to be more closely tied to plant age than to fire severity. In a study of several thousand burned shrubs it was found that for both California buckwheat and California sagebrush mortality was not related to height of the burned skeleton, suggesting fire severity was not a factor determining resprouting (Keeley 1998a, Keeley 2006b). Rather stem diameter (an indicator of plant age) was the primary determinant of resprouting success: as stems increased in diameter their probability of resprouting declined, a phenomenon shared by the northern California coastal scrub dominant coyotebrush (*Baccharis pilularis*) (Hobbs and Mooney 1985). Greater resprouting in young plants appears to derive from the loss of functional adventitious buds due to wood production, and may be the result of the evolution of secondary wood production in taxa derived from herbaceous perennial ancestors (Keeley 2006b). Other interior shrubs regenerating heavily from seed are black sage, white sage (*S. apiana*), deerweed, and chaparral mallow; and all except brittlebush (*E. farinosa*) are shared with coastal sage scrub associations. Most of these species have light wind dispersed seed, so if populations are extirpated from a site they readily disperse in from nearby source populations (Wells 1962, DeSimone and Zedler 1999, 2001).

Interior sage scrub communities have a similar ephemeral postfire flora of annual and herbaceous perennial species that overlap greatly with coastal sage scrub and chaparral associations (Keeley et al. 2005a).

FIRE REGIME–PLANT COMMUNITY INTERACTIONS

Historically fires burned primarily summer to winter but today the bulk of this landscape burns in the fall. As with grasslands, ignitions by humans have increased both the frequency of fires (Wells et al. 2004) and length of the fire season; and during severe droughts fires can burn year around. Size of fires is extremely variable and can be very large, on the order of thousands of hectares (thousands of acres), but modal fire size has decreased (Keeley et al. 1999), due to increasingly effective fire suppression, as well as habitat fragmentation, despite increased human-caused ignitions.

Lightning-ignited fires are rare in the low-elevation coastal sites; however, under autumn Santa Ana wind conditions fires would have readily burned in from adjoining chaparral and woodlands. Contemporary burning patterns in the western end of the Transverse Ranges results in much of this vegetation burning at roughly 5 year intervals, and nearly all of this type burns before 20 years of age (McBride and Jacobs 1980, Keeley and Fotheringham 2003). It is unlikely that lightning-ignited fires ever burned at such a short fire return interval. Although this vegetation is reasonably resilient to high fire frequencies, the current levels are near the lower threshold of tolerance due to time required to develop an adequate seed bank, and many such sites are experiencing accelerated loss of natives and type conversion to nonnatives (Keeley 2001, 2004c).

Vogl (1977) hypothesized that sage scrub was an artifact of fire suppression and livestock grazing and that its current distribution was greatly expanded due to human disturbance. However, empirical studies demonstrate sage scrub species colonize grassland sites when grazing pressure is relieved (McBride 1974, Hobbs 1983, Freudenberger et al. 1987, Callaway and Davis 1993). These patterns are more consistent with the model that sage scrub loses ground to annual grassland under frequent fire and grazing, although it is capable of recolonizing when their frequency declines. Sage scrub also expands into chaparral sites when fire frequencies exceed the minimum regeneration time needed for chaparral species (Cooper 1922, Wells 1962, Freudenberger et al. 1987, Callaway and Davis 1993).

Increased burning in interior valleys in recent decades, as illustrated for Riverside County (Fig. 17.3), is certainly a major stress for these shrubs, most of which regenerate from seed. Historical studies show that there has been a substantial type conversion of sage scrub to nonnative grasslands during the twentieth century (Minnich and Dezzani 1998). It has been hypothesized that in addition to increased fire frequency, nitrogen deposition in this region of high air pollution has contributed to this shift from sage scrub to nonnative grasslands (Allen et al. 2000). It is unclear to what extent this affects postfire invasion, because soil nitrate levels normally increase by an order of magnitude in the first spring after fire (Christensen 1973). Consistent with the role of air pollution driving the invasion process is the much greater presence of nonnatives during the first 5 postfire years in interior sage scrub than in coastal sage scrub (Keeley et al. 2005c). However, there is also a very strong negative relationship between shrub cover and nonnative presence in both the coastal and interior sage scrub associations. Certainly a contributing factor to the greater invasibility of interior sage scrub is the much slower shrub canopy recovery rate resulting from more limited resprouting and greater reliance on seedling recruitment (Fig. 17.7).

Chaparral

The primary differences in chaparral fire response are best understood by contrasting responses on arid, usually south-facing slopes and ridges, with mesic, north-facing exposures. As a general rule, species that recruit seedlings after fire (Table 17.5) tend to occupy the more xeric sites, i.e., low elevations and south-facing exposures (Keeley 1986, Meentemeyer and Moody 2002, Pausas and Keeley 2014). Resprouting species that do not recruit after fire, i.e., obligate resprouters, are usually on the more mesic slopes (Keeley 1998b, Keeley et al. 2012). These are extremes of a gradient, and each comprises a collection of floristically different associations not readily distinguished from one another by fire ecology or fire regime. Distribution of arid and mesic chaparral types is commonly

TABLE 17.5
Fire response types for important species in chaparral in the South Coast bioregion

Lifeform	Type of fire response			Species
	Resprouting	Seeding	Individual	
Tree	None	Serotinous cones	Killed	*Hesperocyparis forbesii* *Hesperocyparis arizonica* *Pinus attenuata*
Tree	None	Variable, low-level serotiny. Often dependent on parent tree survival	Killed or survive	*Pinus coulteri* *Pinus torreyana*
Subshrub or suffrutescent	None	Fire-stimulated	Largely present only as dormant seed banks	*Eriogonum fasciculatum* *Acmispon glaber* *Salvia*
Shrub	None	Fire-stimulated	Killed	*Ceanothus tomentosa* *Ceanothus greggii* *Arctostaphylos glauca*
Shrub	Fire-stimulated	Fire-stimulated	Top-killed or killed	*Adenostoma fasciculatum* *Arctostaphylos glandulosa* *Ceanothus leucodermis* *Ceanothus spinosus* *Eriodictyon californium* *Fremontodendron californica*
Shrub	Fire-stimulated	None	Top-killed or killed	*Cercocarpus betuloides* *Garrya* spp. *Heteromeles arbutifolia* *Prunus ilicifolia* *Quercus berberidifolia* *Frangula californica* *Frangula crocea*
Herbaceous perennials	Fire-stimulated in some, in others resprouting is normal phenological stage	None	Aboveground portions of plants dead at time of fire	*Acourtia microcephala* *Calochortus* spp. *Chlorogalum* spp. *Delphinium* spp. *Dichelostemma capitatum* *Lomatium* spp. *Marah macrocarpus* *Melica imperfecta* *Zigadenus fremontii*
Annuals	None	None	Present only as a dormant seed bank	*Antirrhinum* spp. *Calandrinia ciliata* *Calyptridium monandrum* *Camissonia* spp. *Chaenactis artemisiifolia* *Cryptantha* spp. *Eucrypta chrysanthemifolia* *Emmenanthe penduliflora* *Gilia* spp. *Lotus* spp. *Lupinus* spp. *Mentzelia micrantha* *Papaver californicum* *Phacelia* spp. *Silene multinervia* Salvia columbariae

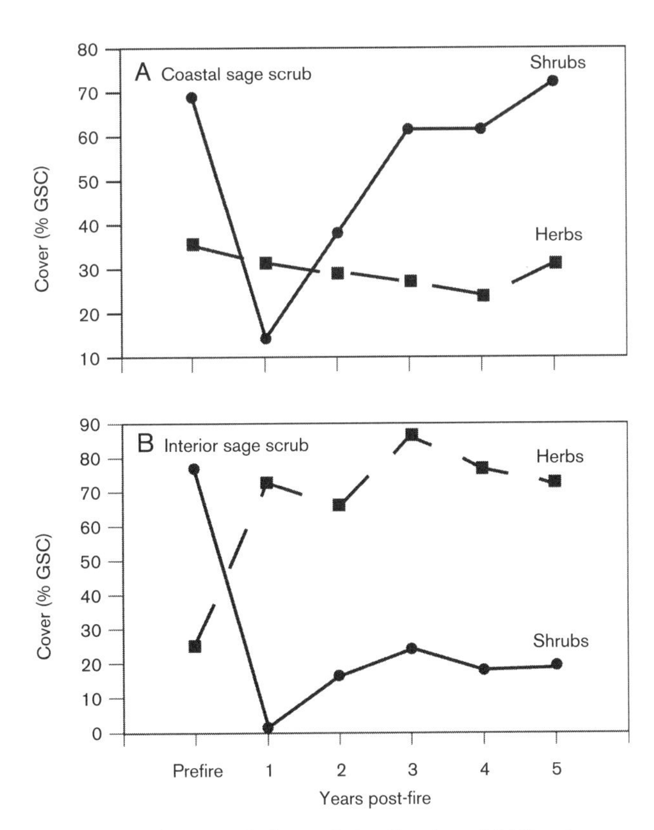

FIGURE 17.7 Foliar cover before and after fire at coastal (A) and interior (B) sage scrub sites. (Redrawn from O'Leary and Westman 1988.)

determined by aspect, and thus landscapes usually comprise a mixture of different vegetation and postfire ecologies because of complex topography. This fine-grained distribution of different associations results in a great deal of diversity of fire behaviors and postfire responses arising out of large fires that cover extensive portions of the landscape.

ARID SITES

Chamise is the nearly ubiquitous dominant on most arid chaparral sites. With respect to fire response, this needle-leaved shrub is a seeder and often exhibits massive seedling recruitment in the first postfire year from a dormant seed bank (Keeley 2000). There is substantial site-to-site variation in the ratio of seedlings to resprouts, on some sites regenerating almost entirely from seedling recruitment and other sites from resprouts (Keeley et al. 2006). Fire intensity, season of burn, and level of precipitation in the first postfire winter appear to be the primary factors determining these patterns (Moreno and Oechel 1994, Keeley and Fotheringham 2003, Keeley et al. 2005a). Other facultative seeders include Eastwood's manzanita (*Arctostaphylos glandulosa*), chaparral whitethorn (*Ceanothus leucodermis*), California yerba santa (*Eriodictyon californicum*), and flannelbush (*Fremontodendron californicum*), although these latter broadleaved species often occupy less arid sites.

Also at the arid end of the gradient are species that lack any ability to resprout after fire, and their persistence is entirely a function of successful seedling recruitment from a dormant seed bank. These shrubs are termed obligately seeding (Wells 1969) or obligate seeder species (Keeley 1977). Chaparral and three of the other Mediterranean-climate shrublands are unique among woody shrublands of the world in having a large proportion of species that lacks the ability to resprout following fire or other disturbance (Keeley 1986). The obligate seeding species of manzanita and California-lilac recruit heavily in the first postfire year from dormant seed banks, and recruitment is typically nil until the next fire on the site (Keeley et al. 2006). The bulk of the seed bank is dispersed locally (Zedler 1995b) and with deep dormancy, thus these species disperse more in time than in space. They are very sensitive to repeat fires and may be extirpated from a site if fires occur too frequently (Keeley 2000). Some coastal species appear to require more than 40 years without fire in order to establish seed banks sufficient to maintain populations (Odion and Tyler 2002).

Demography of obligate seeders is variable. Bush poppy (*Dendromecon rigida*) is a short-lived shrub that has massive recruitment in the first postfire year (Bullock 1989). Shrubs die within a decade after fire and seed banks remain dormant until the next fire. Obligate seeding California-lilac varies markedly in longevity (Zedler 1995b). Some species, e.g., woolyleaf ceanothus (*C. tomentosa*), appear to be relatively short-lived, on the order of three to five decades, whereas others persist far longer, e.g., Mojave ceanothus. Mortality appears to be driven by competition and is increased on more mesic slopes (Keeley 1992a). As stands age there is a shifting balance in the competitive relations resulting in a successional replacement of obligate seeders by species more competitive under long fire-free conditions. Even where populations experience complete mortality, they are highly resilient to long fire-free periods as soil-stored seed banks survive hundreds of years (Zavitokovski and Newton 1968, Franklin et al. 2001). Obligate seeding manzanita generally have smaller seed banks, take longer to establish seed banks, recruit fewer seedlings, exhibit greater early seedling survivorship, and perhaps greater adult longevity than California-lilac (Keeley 1977, Odion and Tyler 2002). All postfire-seeding species exhibit structural and physiological characteristics selected to tolerate the severe conditions on open sites during the long summer-autumn drought (Davis et al. 1998, Keeley 1998b).

This community comprises a rich diversity of herbaceous species, the bulk of which form an ephemeral postfire successional flora. Most of this flora is composed of annuals that arise from dormant seed banks, are endemic to burned sites, and do not persist more than 1 or 2 years (Keeley et al. 2005a, 2005b). Herbaceous perennials are almost all obligate resprouters, arising from dormant bulbs, corms, or rhizomes (Keeley 2000), although a few are short-lived and are present in postfire environments only as seedlings. Postfire resprouts flower and disperse nondormant seeds that readily recruit in the subsequent postfire years. Once the canopy closes these species may remain entirely dormant for many years (Epling and Lewis 1952) or continue to produce foliage but not flower (Tyler and Borchert 2002).

Dormant seed banks are triggered to germinate by either heat shock or chemicals from smoke or charred wood (Keeley and Fotheringham 2000). In general, species are stimulated by one or the other of these cues; however, there are a few species where both appear to play a role; e.g., chamise is equally stimulated by either smoke or heat (J. Keeley, US Geological Survey, Three Rivers, California, USA, unpublished data). There is a very strong phylogenetic component to germination response, with certain families responding only to heat and others only to chemicals. For example, all California-lilac species are stimulated by heat, and smoke plays no role in their germination. This pattern is typical for other fire-stimulate Rhamnaceae, e.g., phylica (*Phylica* spp.) in South African fynbos (Keeley and Bond 1997). Other common families in chaparral and sage scrub with heat-stimulated germination include the Fabaceae, Convolvulaceae, Malvaceae, and Sterculiaceae. However, the majority of species with seedling recruitment in the postfire flora of both chaparral and sage scrub are not stimulated by heat, but rather by smoke or other chemicals produced during combustion (Keeley and Fotheringham 2000). This behavior is common in woody and herbaceous species in the Asteraceae, Boraginaceae, Brassicaceae, Caryophyllaceae, Hydrophyllaceae, Lamiaceae, Onagraceae, Papaveraceae, Polemoniaceae, Scrophulariaceae, and Solanaceae (Keeley 1991, Keeley and Fotheringham 2000).

One factor that plays an important role in determining postfire patterns is fire intensity. Davis et al. (1989) demonstrated that microscale patterns in postfire regeneration resulted from spatial variation in seed banks and soil temperature. Fire intensity impacts are complex because seedling recruitment of some species is greatly inhibited by high fire intensity, whereas for others it is enhanced (Keeley et al. 2005a, 2006).

MESIC SITES

Broadleaved evergreen shrubs dominate these sites and include scrub oak, California coffeeberry, spiny redberry (*Rhamnus crocea*), silk tassel bush (*Garrya* spp.), hollyleafed cherry, and toyon. Throughout much of the region scrub oak is dominant but in the interior ranges, interior live oak dominates. The proportion of shrubs that are killed by fire and fail to resprout is generally very low in most of these species, and they seldom recruit seedlings except in the long absence of fire (Keeley 1992a, 1992b). These obligate resprouting species avoid the stressful summer drought by deep roots that gain access to water in deep rock and soil layers; however, this option is unavailable to seedlings and thus recruitment success is dependent upon establishment in favorable mesic microsites (Keeley 1998b). All of these shrubs have animal dispersed fruits and successful establishment is dependent upon finding safe sites with shade and highly organic soils, usually in the understory of chaparral or woodlands. Seedlings that recruit in the understory often remain suppressed for decades; however, they do resprout after fire, and it may be that fires are required for successful emergence into the canopy (Keeley 1992a).

There are a number of facultative seeders such as greenbark ceanothus (*Ceanothus spinosus*), chaparral whitethorn, and Eastwood manzanita that often form mixed chaparral on either arid or mesic sites, and their response to fire is similar to chamise, often recruiting seedlings and resprouting. Another species that seems to have a wide range of tolerances, occurring on both xeric and mesic sites, is birch-leaf mountain-mahogany (*Cercocarpus betuloides* var. *betuloides*), a postfire obligate resprouter. However, unlike other obligate resprouters, it does not recruit in the understory of mature chaparral, rather it appears to be dependent upon openings created by other types of disturbance (Keeley 1992a).

Mesic sites contribute to higher fuel moisture conditions that may reduce fire intensity and fire severity under moderate but not Santa Ana conditions. However, these mesic sites lead to higher primary production and under severe fire weather they experience much higher fire intensity (Keeley and Fotheringham 2003). When these productive sites burn in low-intensity fires, much of the larger woody fuels is not consumed, thus putting the community on a trajectory for potentially more intense fires during the next fire.

FIRE REGIME–PLANT COMMUNITY INTERACTIONS

With rare exceptions, chaparral always burns as active crown fires. Natural (before Euro-American settlement) frequencies in southern California were on the order of one to several times a century (Van de Water and Safford 2011), but modern fire frequencies are overwhelmingly influenced by human-caused ignitions. For example, many areas in the lower southern California foothills that once supported shrublands have burned 5 to 10 times in the last century, transforming these sites to annual grass and forb lands dominated by nonnative species. Spatial complexity is commonly low to moderate, and is a function of topography, vegetation mosaic, antecedent climate, and weather. Fire intensity is generally high, but variable dependent upon fuels and weather.

Today fire frequency is highest in the summer, but the bulk of the landscape burns in the fall under Santa Ana conditions (Keeley and Fotheringham 2003). Historically, fire frequency also would have peaked in the summer due to the timing of lightning ignitions. These summer fires spread slowly and often reached sizes of only 100–1,000 ha (250–2,500 ac) after months of burning (Minnich 1987). Fire size would have increased markedly if lightning-ignited fires persisted until the autumn Santa Ana winds season, when, within a day, these fires could have expanded to 100,000 ha (250,000 acres), and with typical Santa Ana wind episodes lasting several days to a week, would have generated fires on the order of 10 million ha (25 million ac) (Keeley et al. 2004). Today modal fire size class has greatly decreased due to habitat fragmentation coupled with effective fire suppression, despite high number of ignitions by humans; however, the bulk of the landscape still burns in large fires (Strauss et al. 1989) at rotations of 30 years to 40 years (Keeley et al. 1999).

One exception to this rule is the chaparral on the Channel Islands. The ecosystems on these islands are broadly similar to those on the mainland and share a substantial number of plant species with the mainland. The presence of fire-dependent shrubs chamise and species of California-lilac and manzanita (Minnich 1980b) and closed-cone pines (Linhart et al. 1967) are evidence that fires have historically been a predictable feature of this landscape. However, since settlement by Euro-Americans in the nineteenth century fires have been relatively rare. During the period 1830–1986, only 73 fires were recorded in written records, 65% less than 10 ha (2.5 ac) and only 7% greater than 1,000 ha (247 ac) (Carroll et al. 1993). These numbers provide crude estimates of fire rotation periods on the order of hundreds of years. Also, during this period only three lightning-ignited fires were recorded; however, this is likely below the historical range of variability because of greatly reduced fuels due to intense grazing pressure during the period of record.

An alternative hypothesis has been expressed that large fires in southern California are an artifact of twentieth century fire suppression policy (Philpot 1974, Minnich 1983, 1998, Minnich and Chou 1997). These authors contend that fire exclusion has resulted in an unnatural coarse mosaic of old chaparral stands on the southern California landscape, which is directly responsible for increasing fire size. The hypothesis that fire

SIDEBAR 17.3 CHAPARRAL FUELS AND FIRE

Several lines of evidence cast doubt on the strict control of fires by fuel age in these crown fire regimes. For example, large fires (>5,000 ha) in the westernmost portion of the Transverse Range are fueled primarily in shrublands that are less than 20 years of age (Keeley et al. 1999). In addition, across all of Los Angeles County there is no statistically significant change in probability of burning after approximately 20 to 25 years (Schoenberg et al. 2003). A thorough analysis of the role of fuel age by Moritz et al. (2004) included 10 chaparral landscapes from Baja California to Monterey, a span of 500 km, with each data set representing tens of thousands of hectares and hundreds of fires over a period 85 years in most cases. For 9 of the 10 landscapes they observed a near constant probability of burning with age. Hazard functions calculated with the Weibull function indicated that the rate of fire hazard did not change over time, ranging from 2–4% in year 20 and 4–7% in year 60. Collectively, these data refute the hypothesis that large fires are determined by a buildup of dead fuels, and several lines of evidence suggest the primary determinant of fire size is the coincidence of ignitions and Santa Ana winds (Davis and Michaelsen 1995, Conard and Weise 1998, Keeley and Fotheringham 2001a, 2001b, Moritz 2003, Keeley and Zedler 2009).

This does not mean that fuel age has no effect on chaparral fires. In general it appears that fuel age is primarily a controlling factor under moderate weather conditions (e.g., the Zaca Fire of 2007, Keeley et al. 2009), and relatively unimportant under severe weather conditions accompanying Santa Ana winds (Keeley et al. 2004). However, Zedler and Seiger (2000) contend that it is unlikely that fuel age alone has ever maintained a landscape pattern of small fires. Their modeling studies demonstrate that it would take only a single large Santa Ana fire to set the landscape to the same age class, and thus, if subsequent fires were controlled by fuel age, the landscape would be forever doomed to burn in large fires.

suppression has excluded fire from brushlands in southern California has been tested and shown to be unsupported (Moritz 1997, Conard and Weise 1998, Keeley et al. 1999, Weise et al. 2002). Indeed, during the twentieth century much of this landscape received a higher frequency of burning (30 year to 40 year rotations) than would be expected under natural conditions (Keeley and Fotheringham 2003). The role of fuel age in controlling fires is further developed in sidebar 17.3.

Extensive chaparral dieback has the potential for causing a rapid change in fire hazard (Riggan et al. 1994). Dieback commonly hits large patches of a single species, typically species of California-lilac in the subgenus *Cerastes*. This was once thought to be driven by fungal pathogens, but experimental studies demonstrate conclusively that it is a result of drought (Davis et al. 2002). These are obligate seeding shrubs whose shallow root systems subject these plants to excessive water stress during extended droughts. Such dieback has been implicated in producing anomalously large wildfires (Keeley and Zedler 2009).

Nonnative annuals frequently invade postfire chaparral sites and may persist for several years until shrub canopies return (Keeley et al. 2005c). Invasion is largely a function of proximity of postfire seed sources, and this is largely determined by prefire stand age. Mature chaparral stands generate sufficient fuels to produce fire intensities that kill seed banks of nonnative grasses. However, when fires occur too frequently, shrub canopies fail to close and a substantial nonnative grass flora persists (Zedler et al. 1983). These surface fuels readily ignite and increase the chances of another fire, and because the fuels are a mixture of shrubs and grasses, they generate lower soil temperatures (Odion and Davis 2000) and greater nonnative seed bank survival (Keeley 2001, 2004c).

Closed-Cone Cypress

Fire regime characteristics match that of chaparral. Both Tecate cypress and Cuyamaca cypress have characteristics shared by closed-cone (i.e., serotinous) pines, specifically, high stand density, lack of self-pruning and thin bark, and relatively long fire return intervals; characteristics typical of stand-replacing crown fire regimes (Keeley and Zedler 1998). These obligate seeding trees disperse seeds shortly after fire from aerial seed banks and recruit heavily in the first postfire spring (Table 17.5). They form monotypic even-aged stands that may require many decades of fire-free conditions in order to develop a seed bank sufficient to withstand a repeat fire (Zedler 1977). Studies of seed production show that populations less than 30 years of age are extremely vulnerable to extinction, and seed production continues up to a century and perhaps beyond (Zedler 1995b). Canopy fuels typically produce extremely high fire intensities and sterilize the soil seed bank of competing plants. Cypress seeds are protected by cones and once the cones are opened by heat, they generally disperse beneath the parent skeleton, and thus the seedlings are released from competition compared to microsites just outside the canopy shadow (Ne'eman et al. 1999).

Historical fire-frequencies were undoubtedly very long, perhaps once a century. Contemporary fire frequencies are much higher (Zedler 1995a), and stands may be extirpated, though recolonization from metapopulations that survive in ravines or other fire-free refugia are likely an important means of long-term persistence (Zedler et al. 1984). Fire frequencies greater than once or twice a century appear to be the primary threat to the persistence of these cypress populations (Zedler 1977, Reveal 1978).

Low-Elevation Pines

These pines are typically distributed in patches within a mosaic of chaparral, and thus their fire regime characteristics match that of chaparral. Knobcone pine is similar to the closed-cone cypress species in its deeply serotinous cones that initiate dense monotypic stands following crown fires (Keeley and Zedler 1998, chapter 10).

Coulter pine spans a range of habitats from chaparral to forests. Fire regimes vary from stand-replacing to mixed severity (stand-thinning) fire regimes, largely dependent upon associated vegetation (Dodge 1975, Minnich 1977). When associated with chaparral, cones are serotinous and recruitment is synchronized to the immediate postfire environment, whereas on forested sites cones are not serotinous and recruitment may occur between fires (Table 17.5), and consequently stands are uneven-aged (Borchert 1985, Borchert et al. 2002). Torrey pine is associated with chaparral and naturally exposed to stand-replacing fires. Although most recruitment is tied to fire, stands comprise different age cohorts that recruited after different fire events (Wells and Getis 1999).

Coulter and Torrey pines share characteristics that are tied to the stand-replacing fire regime characteristic of chaparral shrublands. They typically do not self-prune lower branches, enhancing their flammability and increasing the probability of them as well as their neighbors not surviving fires (Keeley and Zedler 1998, Schwilk and Ackerly 2001, Schwilk and Kerr 2002, Schwilk 2003). Some populations of Coulter pine in chaparral stands are strongly serotinous (Borchert 1985), but Torrey pine does not require heat for opening cones, although some seeds persist in the cones for years after seed maturity (McMaster and Zedler 1981). This delayed seed dispersal is thought by McMaster and Zedler (1981) to result from selection in an environment where large crown fires occur at long and unpredictable intervals, often longer than one generation. Such relaxed serotiny is also found in other closed-cone pines when distributed in coastal environments where natural lightning-ignited fires are rare (e.g., Millar 1986). These coastal pines also have a very subtle form of serotiny in that seeds dispersed in the year of maturity typically disperse months later than most pine species, thus dispersing seeds in winter, following the autumn fire season (Keeley and Zedler 1998).

Both of these pines have large seeds with limited wing development and exhibit far less capacity for wind dispersal than montane pines such as ponderosa pine (Johnson et al. 2002). Localized dispersal in these chaparral-associated pines has likely been selected for because large stand-replacing fires generate suitable habitat close to the now dead parent tree. In contrast, low-severity surface fire or mixed severity fire regimes typical of ponderosa pine forests present fewer safe sites for recruitment, thus selection for the ability to disperse into gaps produced by the patchiness of mixed severity fires (see yellow pine forest section).

FIGURE 17.8 Bigcone Douglas-fir tree in the early stages of resprouting from epicormic buds following wildfire in the San Gabriel Mountains (photo by J. Keeley).

Bigcone Douglas-fir

Bigcone Douglas-fir is capable of resprouting after fire, typically reprouts, from epicormic buds present throughout the length of the bole and branches (Fig. 17.8). Trees less than 10 cm (2.5 in) dbh fail to resprout (Bolton and Vogl 1969); however, Minnich (1980a) reported that sprouting success appeared to be dependent primarily on fire severity rather than tree size (Table 17.6). In this regard, pines provide a useful study in that in this large genus the few species that resprout exist in stand-replacing fire regimes conducive to vegetative regeneration (Keeley and Zedler 1998).

Seedling recruitment occurs sporadically during fire-free periods often under chaparral or oak woodland (Bolton and Vogl 1969, McDonald and Litterell 1976). Seedlings and saplings are very sensitive to burning, and thus successful establishment requires extended fire-free periods of at least two decades (Minnich 1980a). Unless recruitment occurs in a very

TABLE 17.6
Fire response types for important species in bigcone Douglas-fir forests in the South Coast bioregion

Lifeform	Type of fire response: Resprouting	Seeding	Individual	Species
Tree	Fire-stimulated, only epicormic	None	Top-killed or survive	*Pseudotsuga macrocarpa*
Tree	Fire-stimulated basal or epicormic	None	Top-killed or survive	*Quercus chrysolepis*

sheltered environment, a further hiatus in burning for many more decades is apparently required for successful establishment of mature trees (Bolton and Vogl 1969). Seedling recruitment is uncommon on south-facing exposures, and it seems likely that the more mesic conditions on north-facing slopes, coupled with longer fire-free intervals due to higher fuel moisture, are factors contributing to greater seedling recruitment on those slopes.

Aspects of this species' life history seem anomalous. Adult trees survive and resprout after fires only if fires are of low to moderate intensity. The habitat of this tree includes chaparral and oak woodlands, and low-intensity fires in these fuels are most likely if they occur frequently. However, seedling and sapling recruitment is eliminated under frequent fires of any intensity, and successful emergence into a canopy level tree may require 50 years to 100 years of fire-free conditions. In short, frequent fires favor adult persistence and infrequent fires favor population expansion.

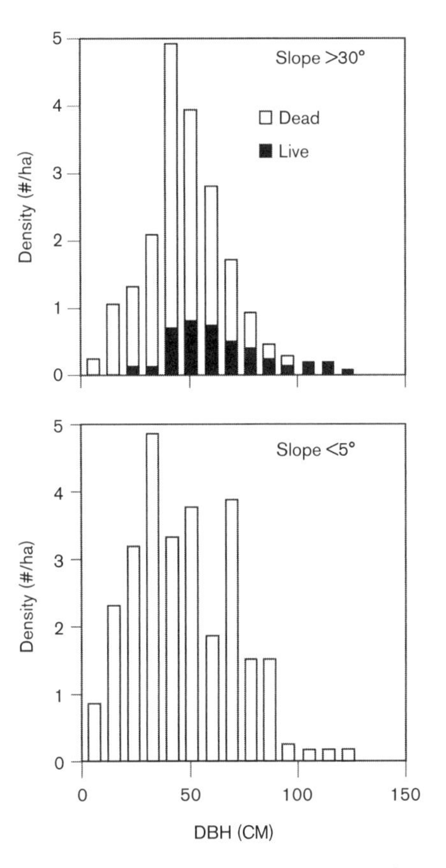

FIGURE 17.9 Live and dead bigcone Douglas-fir trees 13 years after a wildfire in the San Gabriel Mountains on sites of different incline. Approximate elevations were, site A, 1,700 m (5,600 ft) and site B, 1,500 m (4,900 ft). Sites burned in the Sage Fire of 1979 and were sampled in autumn of 1992 (data from Keeley, Crandell, and Calderon, unpublished data).

FIRE REGIME–PLANT COMMUNITY INTERACTIONS

Fire regimes in bigcone Douglas-fir forests vary spatially and temporally. Historically, higher elevation groves associated with conifer forests likely experienced fires on decadal time scales, but lower elevations more on the scale of once or twice a century. Fires can burn as surface fires, but more typically as crown fires (Steel et al. 2015), and thus complexity is high, being heavily influenced by topography and associated vegetation.

Bigcone Douglas-fir forests are typically restricted to ravines and steep slopes, comprising groves of a few trees to thousands of trees. There is the widespread belief that this species has become more restricted with Euro-American and Mexican settlement due to increased fire frequency or more severe fires during severe fire weather conditions (Leiberg 1900b, Horton 1960, Gause 1966). Groves that occur amidst chaparral are particularly vulnerable to high-severity fire that results in substantial mortality, whereas forests in more sheltered ravines and steep slopes are often less prone to destruction (Minnich 1988). In the San Bernardino Mountains between 1938 and 1975, Minnich (1980a) reported 25% mortality of this species and that survivorship increased with increasing slope inclination, from 37% on slopes less than 20° to greater than 90% on slopes exceeding 40°. In a study of bigcone Douglas-fir mortality on two different aspects following a lightning-ignited wildfire in the San Gabriel Mountains (J. Keeley, unpublished data) it was found that populations on relatively level terrain suffered 100% mortality, whereas a population on a steep north-facing slope had substantial survivorship, but only of large trees (Fig. 17.9A, B).

Bolton and Vogl (1969) observed that bigcone Douglas-fir could invade chaparral on gentler slopes only in the absence of fire, usually during a series of wet years following a severe fire that had destroyed the existing vegetation. These invasions were most often terminated with fire, unless the trees had grown large enough in the fire-free period to become fire resistant. Dense chaparral on shallow slopes is generally free of bigcone Douglas-fir because of frequent high-severity fires (e.g., Fig. 17.9B).

TABLE 17.7
Fire response types for important species in riparian woodland-shrubland in the South Coast bioregion

Lifeform	Type of fire response: Resprouting	Seeding	Individual	Species
Tree	Fire-stimulated	None	Top-killed or survive	*Alnus rhombifolia* *Platanus racemosa* *Populus fremontii* *Salix* spp.
Shrub	Fire-stimulated	None	Top-killed or survive	*Baccharis salicifolia* *Rubus* spp. *Toxicodendron diversifolia*

Minnich (1988) suggested that infrequent high-severity fires in these forests are an artifact of fire exclusion, and under natural conditions low-intensity fires would burn at frequent intervals under more moderate summer weather conditions. As a consequence fire intensity would be less and tree mortality would be limited. Whereas this model would ensure survival of existing bigcone Douglas-fir, frequent fires greatly restrict population expansion, because recruitment of seedlings and saplings is dependent upon long fire-free periods.

Lombardo et al. (2009) used the tree-ring record to reconstruct past fire occurrence in bigcone Douglas-fir stands in southern California. They determined that large extensive fires were apparent before and after the turn of the twentieth century, and that smaller fires were more prevalent during the nineteenth century. A shift to more severe fires occurred after the late nineteenth century in the stands. They suggested that land management practices, human-caused ignitions, or climatic variation all could have played a role in modern fire regimes. Although Lombardo et al. (2009) assumed the fire regime was being determined by the surrounding chaparral landscape, Steel et al. (2015) felt that a sizeable portion of their study site was originally dominated by grasslands, oak woodlands, or mixed evergreen forest.

Long-term dynamics of this chaparral conifer are best explained by a metapopulation model involving fire-free refugia. This model is based on the premise that frequent fires will be conducive to maintaining static adult populations, but a dynamic population capable of expansion would require extended fire-free conditions. During extreme fire events, such as illustrated in Fig. 17.9B, populations would retreat to steep fire-resistant slopes that act as refugia (Fig. 17.9A). From these refugia the species disperses back into more hazardous habitats during extended fire-free conditions. Subsequent fires on these sites will eliminate all of the younger trees and the extent to which mature trees survive will be dependent on fuels and severity of the fire weather. Thus, understanding the dynamics of bigcone Douglas-fir requires an understanding of the long-term role of metapopulations and not the short-term effects of fires on individual populations. This model is based on a similar idea proposed by Zedler (1981) who highlighted the importance of refugia to the long-term persistence of closed-cone cypress, particularly in the face of increased anthropogenic fires. The general dynamic of population contraction and expansion under altering fire regimes also is a good explanation for foothill pine (*Pinus sabiniana*) population dynamics (Schwilk and Keeley 2006).

Riparian Woodland and Shrublands

Response to fire is uniform, as nearly all are vigorous resprouters, either from rootcrowns or rhizomes (Davis et al. 1989). None of the riparian species are postfire seeders because all have transient seed banks; indeed, most have extraordinarily short-lived seeds measured on the order of days or weeks (Table 17.7). A number of factors account for the lack of selection for postfire seedling recruitment: (1) the unpredictability of fire penetrating riparian woodlands, (2) the ready water source favors rapid and dense vegetative regrowth that results in intense competition for seedings, and (3) the severe surface scour resulting from annual winter flooding, which carries away any soil stored seed banks. Differences in species response to fire are largely unexplored, although these are to be expected based on species differences in distribution along disturbance gradients (Harris 1987, Bendix 1994).

FIRE REGIME–PLANT COMMUNITY INTERACTIONS

Riparian zones maintain communities that, relative to upland species, have higher fuel moistures during severe fire weather. As a consequence they potentially pose an important barrier to fire spread, and thus play a role in determining landscape patterns of burning (Dwire and Kauffman 2003). The deciduous habit of dominant species, coupled with increased water availability during winter to summer, greatly limits the fire season, except during severe droughts. Although most fires burn as crown fires with high intensity, fire severity is generally low due to the predominance of vigorous sprouting species.

Riparian areas are of particular ecological concern because of their disproportionate share of biodiversity and propensity to nonnative invasion (Rundel and Sturmer 1998). A major threat to southern California riparian communities is the highly aggressive nonnative giant reed (*Arundo donax*), which has displaced substantial portions of the native vegetation in riparian ecosystems (Rieger and Kreager 1989). It has invaded thousands of hectares (ac) of riparian habitat, including all major drainages from Ventura to San Diego counties (Bell 1997). It is extremely flammable and can change green riparian corridors that are barriers to fire spread into highly combustible fire conduits. Giant reed is a very vigorous resprouter capable of dominating riparian sites far more quickly than the native species.

Oak and Walnut Woodlands

The dominant oaks in this association have thick bark and are resistant to frequent fires. Based on bark thickness and mortality following fires (Table 17.8), it appears that Engelmann oak is more fire resistant than coast live oak (Lawson 1993). Where understory fuels have accumulated, fires may carry into the canopy, producing lethal crown fires, because these older trees seldom resprout from the base when top-killed. They have long juvenile periods and thus frequent fires may inhibit successful establishment (Plumb 1980, Lawson 1993). Seedlings and saplings often resprout from the base after fire, but if fires are frequent enough they remain suppressed for many decades. They apparently require an extended fire-free period to emerge to a size capable of withstanding fires, a pattern typical of recruitment in many savanna trees (e.g., Bond et al. 2001). For these southern California oaks, Lawson (1993) reported a negative relationship between fire frequency and both density of saplings and density of mature trees.

In southern California black walnut woodlands, all of the dominant species are vigorous basal resprouters after fire. There is no information on how burning during the growing versus dormant season affects resprouting of this winter-deciduous tree. Seedling recruitment is limited following fires, but in the long absence of fire, walnut seedling recruitment is abundant, and successful establishment generates an uneven age structure that appears to form a stable age distribution (Keeley 1990a). Also in the absence of fire, mesic-type chaparral shrubs persist as "gap phase" species capable of seedling recruitment into openings in the walnut canopy, as well as gaps in other woodland associations (Campbell 1980).

FIRE REGIME–PLANT COMMUNITY INTERACTIONS

Fire regimes in oak and walnut woodlands are variable depending upon moisture status, woodland density, and location in the landscape. Fires occur on the order of one to several times a century. On mesic sites higher fuel moisture acts in concert with the closed canopy, which reduces drying of surface fuels to reduce fire occurrence. On these sites, fire complexity can be high with patches of crown fire and surface fire interspersed with unburned patches. At the xeric end of the gradient more open woodlands today support surface fuels of annual grasses, which increase fire frequency, but reduce fire intensity and complexity. The rapid growth rates of understory subshrubs and herbs make it likely that historically the primary limitation to fires was ignitions. Thus, it is to be expected that fire return intervals varied along a gradient from the coast to the interior in conjunction with the distribution of summer lightning storms. Due to the mosaic distribution of woodlands, fires typically burn in to them from adjacent shrublands.

With the mid-Holocene shift of Native American diets to emphasize acorns and other seeds (Erlandson and Glassow 1997), settlements centered around oak woodlands increased fire frequency to maintain herbaceous understories and reduce pathogens (Anderson 1993). Wells (1962) hypothesized that prior to extensive Native American burning, oak woodlands in coastal California supported subshrub understories, which would have contributed to a greater incidence of high-severity fire. He contended that throughout the region Native American burning converted oak woodland sage scrub associations to oak savannas (Cooper 1922, Wells 1962), although remnants of this woodland sage scrub association are still evident in the central coast. Historical studies of oak woodlands indicate that twentieth century fire regimes have had no discernable impact on woodland cover in San Diego County (Scheidlinger and Zedler 1980).

California Black Oak Woodlands

At higher elevations California black oak woodland sometimes interfaces between the crown fire chaparral ecosystem below and the surface fire yellow pine forests above. Although little specific information is available, this community likely had a mixed fire regime such that surface fires burning down slope from pine forests were low-intensity surface fires and those burning upslope from chaparral were high-intensity crown fires. Considering the elevational distribution of lightning fires it seems likely that fires more often burned in from higher elevation conifer forests. Thick bark would have contributed to surviving surface fires, although these trees would succumb to high-intensity crown fires. Fire-free conditions in adjacent yellow pine forests are said to promote the invasion by California black oak, but frequent fire—especially fire severe enough to kill conifers—and logging of conifers promote persistence of oaks as they resprout whereas conifers generally do not.

Yellow Pine Forests

The dominant ponderosa and Jeffrey pines are adapted to a mix of low-severity surface fires and high-intensity crown fires (Table 17.9). Thick bark and self-pruning of dead branches ensure survival of the larger trees under low-intensity surface fires (Keeley and Zedler 1998), but a mix of high and low fire severity patches is requisite for reproduction. Seedling recruitment requires gaps produced by localized high-severity fire, which exposes mineral soil by removing surface duff, and opens the canopy to sunlight. Critically important is the size of these gaps, because recruitment is dependent on survival of parent seed trees in adjacent patches of forest that are either unburned or burned by low-severity surface fire. Patches of high-severity crown fires are also critical because these openings accumulate fuels at a slower rate and inhibit subsequent fires from burning in and destroying recruitment cohorts (Keeley and Stephenson 2000).

Understory species have received limited attention in these forests, although many of the patterns observed in Sierra Nevada forests following fires of different severities (Stephens et al. 2015) likely hold here as well. The dominant species are typically greenleaf manzanita (*Arctostaphylos patula*) and deerbrush (*Ceanothus integerrimus*), which often germinate *en masse* after fire from soil seed banks. Recruitment, however, is not restricted to burned sites, and these species often recruit into gaps caused by other disturbances. Herbaceous species are mostly perennials that resprout after fire, and it is to be expected that seedling recruitment would follow in subsequent years.

TABLE 17.8
Fire response types for important species in oak and walnut woodlands in the South Coast bioregion

	Type of fire response			
Lifeform	Resprouting	Seeding	Individual	Species
Tree	Fire-stimulated, basal when young, epicormic in mature trees	None	Top-killed or survive	*Quercus agrifolia* *Quercus engelmannii*
Tree	Fire-stimulated	None	Top-killed	*Juglans californica*

TABLE 17.9
Fire response types for important species in montane coniferous forests in the South Coast bioregion and includes yellow pine forests on drier sites and white fir-dominated mixed-conifer forests on more mesic sites

	Type of fire response			
Lifeform	Resprouting	Seeding	Individual	Species
Tree	None	Recruitment from seeds dispersed by parent trees that survive fire, in first year or later	Survive or killed	*Abies concolor* *Calocedrus decurrens* *Pinus jeffreyi* *Pinus lambertiana* *Pinus ponderosa*
Shrub	None	Fire-stimulated	Killed	*Arctostaphylos patula* var. *platyphylla*
Shrub	Fire-stimulated	Fire-stimulated	Top-killed	*Ceanothus cordulatus* *Ceanothus integerrimus*
Shrub	Fire-stimulated	None	Top-killed	*Chrysolepis sempervirens* *Prunus emarginata* *Ribes* spp. *Symphoricarpos parishii*
Herbaceous perennial	Yes as a normal phenological response, not fire-stimulated	None	Top-killed	*Arabis* spp. *Castilleja* spp. *Elymus elymoides* *Phacelia imbricata* *Silene lemmonii* *Viola purpurea*
Annual	None	A few fire-stimulated, mostly transient seed banks that escape high-intensity fire	Killed or dead at time of fire	*Gayophytum heterozygum* *Lotus* spp. *Mentzelia* spp. *Mimulus* spp.

FIRE REGIME–PLANT COMMUNITY INTERACTIONS

Fire history studies in the San Bernardino Mountains show that the presettlement fire return intervals were slightly longer than for Sierra Nevada forests, with estimates of 10 years to 14 years for ponderosa pine forests and longer intervals for the drier Jeffrey pine forests (McBride and Laven 1976, McBride and Jacobs 1980). Historically, fires were primarily from summer lightning storms and were predominately surface fires with patches of passive crown fire (Minnich 1974). Due to the short fire season and vertical gap between surface and crown fuels, which leads to a preponderance of low-intensity surface fires, humans have been very effective at suppressing fires, and as a consequence the contemporary fire return interval is much longer and far outside the historical range of variability. As with other parts of the western United States, during the twentieth century much of this landscape has escaped fire entirely (Stephenson and Calcarone 1999).

Evidence of the impacts of management practices on forest structure in the eastern Transverse Ranges is presented by Minnich et al. (1995). They utilized Vegetation Type Map (VTM) plot data from 1932 as a baseline for measuring historical changes in these forests by comparing with contemporary samples. Their study showed increased in-growth of white fir and incense cedar into yellow pine forests over the 60 years of study, reflecting the impact of fire exclusion. This generalization primarily applies to ponderosa forests as Jeffrey pine forests have had far less white fir in-growth under long fire-free periods (Minnich et al. 1995). These ponderosa pine forests present a greater fire hazard for crown fires than would be expected under natural fire regimes, though it is worth noting that stand densities of 200–300 ha^{-1} (500–700 ac^{-1}) are substantially lower than commonly observed in Sierra Nevada forests (Rundel et al. 1977).

Other historical changes in mixed-conifer forest structure reported by Minnich et al. (1995) include 79% mortality of all ponderosa pine greater than 30 cm (12 in) dbh during the 60 years between the 1930s VTM plots and their samples, attributed to ozone pollution. However, other studies have not reported twentieth-century mortality levels anywhere near these values (McBride and Laven 1999), and some of this estimated mortality may be tied to errors associated with use of VTM plots (Bouldin 1999, Keeley 2004b). A general decline in large conifer trees outside of logged areas is apparent across California, including southern California, and McIntyre et al. (2015) tied these losses largely to increasing climatic water deficit. High levels of ponderosa pine mortality have certainly been observed in the early years of the twenty-first century, and are attributed to severe drought and associated beetle damage (USDA Forest Service 2004).

Although the success of fire suppression activities is certainly a factor in driving these structural changes, past logging practices have perhaps created a bigger problem. Data from Minnich et al. (1995) show that stand density was 51% higher on logged, as opposed to unlogged sites. This is important in light of the fact that over a 30-year period beginning in 1950, roughly 350 million board feet were removed from the rather limited conifer forests of the Transverse Ranges (McKelvey and Johnston 1992). In-growth from this cutting has greatly added to the contemporary fire hazard in these forests.

The high fire hazard in these forests has contributed to some significant crown fire damage. For example, crown fires with tree mortality of 90% or more in ponderosa and Jeffrey pine forests have been reported for several twentieth- and twenty-first-century fires in the Peninsular Ranges (Dodge 1975, Halsey 2004), and similar fires have occurred in the eastern Transverse Ranges (Minnich 1988). However, it is unclear how far outside the natural range of variation such fires are because historical records show large high-intensity fires did occur in these forests during the nineteenth century (Minnich 1978, p.134). Also, evidence of former conifer forests now occupied by chaparral indicates high-intensity crown fires prior to the twentieth century (Minnich 1999). These historical large fires occurred primarily after Euro-American settlement.

Mixed-Conifer Forests

These are white fir-dominated forests with incense cedar, sugar pine, ponderosa pine, and Jeffrey pine as codominants. These forests are more mesic than yellow pine forests and occur either at higher elevations or lower on north-facing slopes. Productivity, and thus fuel loads, are typically higher and largely comprise needles and branches, with relatively little of the fuel load from herbaceous fuels as in the drier ponderosa type. Jeffrey pine replaces ponderosa pine on drier slopes, in the interior, and at higher elevation.

Both white fir and incense cedar can persist and recruit seedlings and saplings in the absence of fire. This characteristic allows them to expand into drier yellow pine forests when fires are excluded. Under a mixed severity fire regime both white fir and incense cedar are capable of taking advantage of fire-induced gaps and recruiting into them. These species have thick fire resistant bark and are tolerant of surface fires, although less tolerant than associated yellow pines. Although both white fir and incense cedar can recruit after fire, unlike the yellow pines, recruitment is not fire-dependent. Sugar pine is very long-lived, capable of surviving low-intensity surface fires, and recruiting both after and between fires.

Associated shrubs that are typically restricted to permanent or temporary gaps in the forest canopy include bush chinquapin (*Chrysolepis sempervirens*), huckleberry oak (*Quercus vaccinifolia*), bitter cherry (*Prunus emarginata*), and mountain whitethorn (*Ceanothus cordulatus*). All of these are vigorous resprouters after fires or other disturbances, and mountain whitethorn recruits seedlings after fire.

FIRE REGIME–PLANT COMMUNITY INTERACTIONS

Fire return intervals historically were frequent, 10 years to 30 years, but since the turn of the century fires have been excluded from much of the higher elevation mixed-conifer landscape (McBride and Laven 1976, Everett 2003). Historical studies of fire return intervals in mixed-conifer forests of the Sierra Nevada show that north-facing slopes have had longer fire return intervals, presumably because high fuel moisture levels inhibit ignition more often than on south facing slopes (Tony Caprio, Sequoia National Park, Three Rivers, California, USA, unpublished data). The importance of fuel moisture is suggested by the observation that in the eastern Transverse Ranges severe fires consistently burn a greater percentage of the landscape on south-facing than on north-facing slopes (Minnich 1999).

Savage (1994, 1997) examined patterns of mortality in forest dieback in the San Jacinto Mountains and recorded that the bulk of the mortality occurred in the older age classes of ponderosa and Jeffrey pines, with relatively few trees older than 60 years surviving. Roughly half of the white fir over 40 years died, whereas sugar pine and incense cedar suffered little or no mortality. Periodic severe drought followed by bark beetle (*Dendroctonus* spp.) attack seems to be the primary factors affecting pine survival. These mortality events may be a natural part of this system and would contribute sufficient fuels to produce high-intensity crown fires resulting in a burning mosaic of gaps, suggesting that mixed fire regimes of low and high severity are likely within the range of natural variation.

Whereas fire suppression appears to have excluded natural fires over much of the mixed-conifer landscape (Fig. 17.6D, E, F), it is not clear how much structural change in these forests can be attributed to this fire management policy. In particular, a substantial portion of Forest Service lands have been repeatedly logged over in the last century and some young plantations have such dense fuels resulting from silvicultural practices more than from fire suppression. The natural fire regime in southern California mixed-conifer forests included substantial surface fire burning, but high-severity crown fires occurred, although their contribution to overall burned area was probably lower than today. Other historical fire regime parameters like size and complexity are still poorly understood.

Lodgepole Pine Forests

In this region lodgepole pine forests are at the highest end of the elevational gradient for forests exposed to fire on any regular basis. Lodgepole pine has thin bark and is not tolerant of even moderately intense surface fires. Recruitment is not tied to fire and stands tend to be uneven-aged (Keeley and Zedler 1998).

FIRE REGIME–PLANT COMMUNITY INTERACTIONS

These forests are characterized by a short growing season, low fuel accumulation, infrequent droughts, and lower lightning fire incidence than characterize lower elevation forest types (Sheppard and Lassoie 1998). As is common in other high-elevation pines, lodgepole pine is quite long-lived (>600 years). Periodic seedling recruitment is restricted to gaps created by wind-thrown trees or to meadows experiencing a drop in water table, or seedlings may initiate primary succession on granitic outcrops (Rundel 1975). Lodgepole pine forests are not replaced by more shade-tolerant species, perhaps due to the shallow granitic soils (Rundel et al. 1977). Where this species comes together with more stress-tolerant taxa, e.g., Jeffrey pine, forests appear to result from a chance coexistence, which derives from infrequent, species-specific pulses of seedling recruitment in response to differing sets of environmental conditions.

These high-elevation forests receive a frequent bombardment of lightning strikes but ignitions are limited by fuels and fuel moisture. Low productivity in these high-elevation forests and the lack of continuity of fuels are important limitations to fire spread. Fires in this type generally are low-severity surface fires that usually are rather localized and spread slowly, often being contained by lack of fuel continuity. Minnich (1988) reviewed a number of historical accounts that report crown fires on the scale of hundreds to thousands of hectares. It appears that on a frequency of perhaps once a century the surface fire regime may be punctuated by larger high-intensity fires that cover thousands of hectares (Sheppard and Lassoie 1998). The impact of these high-intensity fire events on forest structure is apparently highly variable. Sheppard and Lassoie (1998) report that such fire events reduce the smaller size classes but have little impact on the larger trees.

Management Issues

Fire regimes vary markedly between the crown fire regimes of the foothill zone and the mixed surface and crown fire regimes of the montane zone, and fire suppression policy has played a very different role in each (Safford and Van de Water 2014).

Montane environments have experienced an unnaturally low level of fires this past century, and there is reason to believe fire exclusion has resulted from both a fire suppression policy and the loss of herbaceous surface fuels due to livestock grazing (Minnich et al. 1995). Fire hazard is currently higher than was likely the case historically and is largely the result of increased surface fuels and increased in-growth of saplings (Everett 2003, USDA Forest Service 2004). Although higher fuel loads are often blamed on fire exclusion, an equally important factor is past logging practices that promoted recruitment of young trees induced by the removal of overstory canopy trees (Minnich 1980a, Everett 2003). Severe drought during 2002 and 2003 resulted in substantial pine mortality in southern California forests, potentially exacerbating the fire hazard situation (USDA Forest Service 2004). The bulk of mortality occurred in ponderosa pine, which is at the southernmost limits of its distribution on the West Coast, and it is unclear how much of this is natural mortality to be expected at the limits of a species' range or is the result of increased competition for water due to increased forest density during the twentieth century (Keeley et al. 2004, McIntyre et al. 2015).

In October 2003, the San Diego County Cedar Fire burned 113,000 ha (280,000 ac), making it the largest fire recorded for the twentieth century in California (Table 17.1). Although most of the area burned was shrubland, the fire also burned through 5,700 ha (14,000 ac) of fire-suppressed mixed-conifer forest at very high severity. There was significant conifer mortality and low subsequent seedling establishment and forest recovery throughout the burned area (Franklin et al. 2006, Franklin and Bergman 2011). This loss has prompted a large-scale, controversial reforestation project on state park lands in which the naturally regenerating shrublands are being removed to plant trees.

Many yellow pine and mixed-conifer forests may benefit from some form of fuel manipulation to reduce the fire hazard for human safety and diminish negative ecosystem impacts (Safford et al. 2012). Prescription burning is a viable option in parts of these ranges but not widely applicable in areas highly fragmented by private in-holdings. In addition, air quality constraints greatly limit the window of opportunity for prescription burning. Mechanical thinning is perhaps the only viable option in many of these forests; however, it is costly, particularly if only the smaller trees most responsible for fire hazard are removed, and impossible to implement on much of the land base due to slope steepness, road limitations, and areas such as wilderness (North et al. 2015). Thinning projects can pay for themselves, but usually only if larger trees are taken. This creates a dilemma because removing large trees promotes further in-growth and exacerbates the fire hazard problem, requiring future mechanical thinning at shorter intervals. It is unknown if this required reentry interval is long enough to allow sufficient growth to regenerate larger, commercially acceptable trees needed to fund these operations. Equally important is the finding that mechanical treatments cannot serve as a surrogate for fire when it comes to many important ecological parameters (McIver et al. 2012).

In the lower coastal valley and foothill zone, fire suppression has not reduced the natural fire frequency (Table 17.2). Today there are many more fires than historically, but fire suppression policy has kept many of these in check so that, although fire rotations are shorter, most stands burn within the range of natural variability, albeit at the lower end. However, in landscapes surrounding urban areas, fires have been of sufficient frequency to effect widespread type conversion to nonnative grasslands (Keeley and Fotheringham 2003).

During severe fire weather a small percentage of fires escape containment and account for the bulk of the landscape that burns in this region (Minnich and Chou 1997, Moritz 1997). The limited lightning ignitions in this zone, coupled with historical accounts of relatively little area burned by these summer fires (Minnich 1987), suggests they were not responsible for most of the area burned (Keeley and Zedler 2009). However, the hypothesized occurrence of infrequent lightning-ignited fires persisting until the Santa Ana wind season (Fig. 17.1) could account for burning huge fragments of the landscape in a matter of days (Keeley and Fotheringham 2003).

Maintaining this natural fire regime creates a major management dilemma because chaparral and sage scrub share this landscape with people in an ever-expanding urban sprawl. On these landscapes, fire is as much a social and political problem as it is a fire management problem (Rodrigue 1993, Pincetl et al. 2008), particularly because the high number of homes destroyed by fires continues to increase in the region. The primary driver of the large fires most catastrophic to humans is the coincidence of fire ignitions and Santa Ana winds (Davis and Michaelsen 1995, Moritz 1997, 2003, Conard and Weise 1998, Moritz et al. 2004). Because humans are responsible for more than 95% of fire ignitions (Syphard et al. 2007), the high rate of population growth and rapidly expanding urban footprint (Syphard et al. 2012) have increased the potential for this coincidence to occur.

Under severe fire weather conditions, shrubland fuel age has shown to be ineffective as a barrier to fire spread (Keeley and Fotheringham 2001b, 2003, Keeley et al. 2004). Thus, landscape-scale prescription burning to create a mosaic of age classes in chaparral and sage scrub will not prevent the spread of Santa Ana wind driven fires (Dunn 1989, Keeley 2002b, Keeley and Fotheringham 2003, Keeley and Zedler 2009). Illustrative of this are the massive wildfires that burned over 300,000 ha (740,000 ac) in the last week of October 2003 (Halsey 2004). Within the perimeters of these large fires were substantial areas that had burned by either prescription burns or wildfires within the previous 10 years (Keeley et al. 2004, 2009). Fires either burned through or skipped over or around these younger age classes.

Research conducted at a coarser time and space scale has shown that, over a 29-year time span, antecedent fire in the form of prescribed fire or wildfire has had no effect on reducing the amount of subsequent fire, even after controlling for weather (Price et al. 2012). This does not suggest that prescription burning has no place in chaparral management, only that it needs to be used strategically with a clear understanding of its limitations. During severe fire weather, young fuels do reduce fire intensity and thus may provide defensible space for fighting fires. This argues for the use of prescription burning and other fuel manipulations in strategic sites that could be safely accessed during extreme fire events.

Because of the challenges in conducting prescribed burns near communities, mechanical fuel treatments in the form of linear fuel breaks and more extensive areal treatments created through mastication, cutting, or chipping have largely replaced prescribed fire in interface areas; and in fact, they have also increased across the landscape (Brennan and Keeley 2015). These treatments are premised on the notion that fuel reduction inhibits fire spread; thus, their limitations are similar to those of prescribed fire. Recent research on the long-term, broad-scale role of mechanical fuel breaks across the southern California region has shown that these treatments do not stop fires on their own. Rather, their most effective role has been to provide safe firefighter access in strategic locations (Syphard et al. 2011a, 2011b). Although these fuel treatments provide clear strategic benefits in managing fires, their effectiveness is diminished during weather-driven fire events with large or multiple fires; when fire fighter resources are scarce; and access is limited (Syphard et al. 2011a, 2011b).

Broader application of fuel manipulations may be warranted for managing fires that occur under mild weather conditions and are not driven by strong winds. In these fires, the flames are nearly vertical and fuels ahead of the fire front are not preheated by hot gases (Morvan and Dupuy 2004). Under these conditions, age mosaics of young chaparral fuels can provide barriers to fire spread as suggested by anecdotal observations (Philpot 1974). Age mosaics also may control burning patterns south of the US-Mexican border where wildfires are generally fanned by mild onshore breezes, and seldom under Santa Ana conditions (Minnich 1989, 1998, Minnich and Everett 2002). In southern California, fires that ignite under mild weather are seldom responsible for major property damage or loss of lives (Halsey 2004). Thus, the critical question about the widespread use of prescription burning to inhibit the spread of these fires is whether or not it is cost effective, and further research on this question is needed. This is particularly important when considering the high cost of burning in crown fire ecosystems (Conard and Weise 1998).

Costs of prescription burning also need to be evaluated in terms of impacts on resources. The lower elevation chaparral and sage scrub in this region are already stressed with unnaturally high fire frequencies that threaten to replace much of this landscape with nonnative annual plants (Keeley 2001, 2004c). The primary concern with prescription burning is that it makes these stands vulnerable to type conversion by reburning due to the high frequency of anthropogenic fires (see sidebar 17.1). Other suggested benefits of prescription burning include the effects of mosaic age classes on reducing postfire flooding and mudflows (Loomis et al. 2003); however, winter precipitation is a far greater factor, and high frequency fires increase the chances of fire being followed by a winter of high rainfall (Keeley et al. 2004). This analysis suggests that the appropriate fuel management strategy for southern California chaparral is the strategic placement of fuel manipulation projects. For much of southern California this represents a major paradigm shift from the widely accepted model of landscape scale rotational burning.

It is important to recognize that there is much intraregional variation in fire regimes, evident in different levels of fire hazard (Moritz 2003, Moritz et al. 2004) and different fire seasons (Keeley and Fotheringham 2003) throughout central and southern California chaparral. Although fire suppression has had little impact on chaparral stands in southern California, there have been significant impacts in other parts of the state. For example, marked fire exclusion is documented for chaparral in the southern Sierra Nevada (Keeley et al. 2005d, Keeley and Safford 2016), which suggests a need for different approaches to fire management in different regions.

A problem that is likely to capture more and more attention in the future is the impact of past rehabilitation problems. One illustration is the widespread use of black mustard (*Brassica nigra*) and shortpod mustard (*Hirschfeldia incana*) for postfire rehabilitation projects in the 1940s and 1950s. These species were abandoned long ago because of their aggressive invasive tendencies, particularly into economically important citrus orchards. However, we still see the ghost of past seedlings because these species produce dormant seed banks that are fire-stimulated and thus, after fire, many shrublands are rapidly invaded from within. Most of the data to date do not support use of these or any other nonnative species for postfire management because they are not particularly effective and they have serious negative ecosystem impacts (Conard et al. 1995).

Another area in serious need of consideration is the impact of fuel breaks on invasive species (Keeley 2004a, Merriam et al. 2006). By removing the native shrub and tree cover these sites act as sources for nonnative species to further spread into wildlands. The fact that fuel breaks typically form long corridors makes them ideal mechanisms for transporting nonnative species into remote wildlands. When sites are burned, the nonnative seed banks in shrubland vegetation are generally killed by the high fire intensities, but are not killed by lower fire intensities in the fuel breaks, and thus provide a ready seed source for invasion of burns. Colocating fuelbreaks with roadways may be one way to minimize the nonnative invasion of wildland areas.

In recent years, increased attention is being focused on the need to consider a wider range of management options and alternatives to minimize fire risk to communities and to protect resources. Although fire protection via suppression or fuels management has traditionally been the responsibility of state and local agencies, many management alternatives shift some of the responsibility to other interests. Studies show that homeowner actions or actions by policy-makers and planners may not only be more effective at reducing fire risk to communities than traditional fire management (Penman et al. 2014), but they may also provide the greatest protection of resources (Syphard et al. 2013).

Most homes are destroyed by fires because they are ignited by embers flying ahead of a fire, not by the actual flames of the fire itself (Maranghides and Mell 2009). Therefore, the flammability of a structure's building materials or the vegetation immediately adjacent can be a significant determinant of that structure's survival (Quarles et al. 2010). Therefore, homeowners can play a role in their own fire safety by preventing easy ignition of whatever an ember might land on during a fire. In some cases, a structure's building materials can be designed to be more resistant to ember attacks, and homeowners may choose to retrofit existing homes with improved roof materials, windows, or attic vents. County fire codes are also being updated that regulate building standards to improve fire safety for new construction.

Although retrofitting homes may be challenging or expensive for the homeowner, creating areas of defensible space in the landscaping around a home may be more feasible; and in hazardous areas of California, this is required by law. California homeowners are required to create 30 m (100 ft) of defensible space according to a specific set of guidelines dictating how to minimize vegetation (Cal Fire 2006). An empirical evaluation of defensible space distances and their role in home survival has shown that the most effective distance ranged from 5 to 20 m (16–58 ft); and anything beyond 30 m (100 ft) provided no added benefit (Syphard et al. 2014). Furthermore, defensible space did not require excessive reduction in vegetation to be effective; reduction in cover up to 40% immediately adjacent to the structure provided the maximum benefit. These results are important not only in reducing the burden on homeowners to excessively manipulate their landscaping, but they also have serious implications for resources, as the cumulative effects of large defensible space projects can result in significant habitat loss (Keeley et al. 2012).

In the same study of defensible space, Syphard et al. (2014) found that the arrangement of housing development was even more significant than defensible space in explaining whether a home was destroyed by a fire. Other studies have also suggested that land use planning decisions that influence the location and arrangement of new development could alter community risk to fire across the landscape (Syphard et al. 2012, 2013, Alexandre et al. 2016). This is not only because some areas of a landscape are inherently more fire-prone than others (e.g., Syphard et al. 2008), but is also related to the juxtaposition of humans and wildland vegetation. In general, exurban areas have both the highest fire ignition densities and highest number of homes destroyed because they have both people to start fires and sufficient wildland areas to facilitate fire spread (Syphard et al. 2007, 2012). The reason that land-use planning could be so effective is that it targets the source of fire risk, which is the exposure of homes to fires. In addition, the development patterns that result in the lowest exposure are also the ones that produce a smaller and less fragmented housing footprint on the landscape (Syphard et al. 2012), and thus might be less detrimental to resources.

What about planning for the future? Anticipated climate changes under the most plausible emissions scenarios, temperatures in Southern California are projected to rise by 2.5°C to 4.4°C (4.5–8.1°F) by 2100, and precipitation will remain similar to today or drop, with winter precipitation not likely to change much, but precipitation outside of winter is projected to drop by 11–44% (Cayan et al. 2008). That scenario may not be borne out in chaparral shrublands since contemporary catastrophic fires are not closely linked to fuel level, although extreme droughts are implicated with large fires (Keeley and Zedler 2009). Predicted global warming is likely to have more marked impacts on fire hazard in montane forests where historical burning is associated with spring and summer temperatures (Keeley and Syphard 2015). However, predictions are complicated, and the modeling efforts thus far fall short of including all of the important ecosystem processes. For example, because the primary fuels in conifer forests are dead material on the surface, a change in decomposition due to the expected higher temperatures and lower moisture could affect fuel loads. On the other hand, the greater mineralization and release of nutrients could, through increased productivity, have the opposite impact. Regardless of how climate affects fuels, based on current patterns of burning it appears that throughout this region the primary threat to future fire regimes is more tied to future patterns of human demography than to climate.

References

Alexandre, P.M., S.I. Stewart, M.H. Mockrin, N.S. Keuler, A.D. Syphard, A. Bar Massada, M.K. Clayton, and V.C. Radeloff. 2016. The relative impacts of vegetation, topography and spatial arrangement on building loss to wildfires in case studies of California and Colorado. Landscape Ecology 31: 415–430.

Allen, E.B., S.A. Eliason, V.J. Marquez, G.P. Schultz, N.K. Storms, C.D. Stylinski, T.A. Zink, and M.F. Allen. 2000. What are the limits to restoration of coastal sage scrub in southern California? Pages 253–262 in: J.E. Keeley, M.B. Keeley, and C.J. Fotheringham, editors. 2nd Interface between Ecology and Land Development in California. U.S. Geological Survey Open File Report OFR-00-62. Reston, Virginia, USA.

Anderson, K. 1993. Native Californians as ancient and contemporary cultivators. Pages 151–174 in: T.C. Blackburn and K. Anderson, editors. Before the Wilderness: Environmental Management by Native Californians. Ballena Press, Menlo Park, California, USA.
Anderson, M.K., M.G. Barbour, and V. Whitworth. 1998. A world of balance and plenty. Pages 12–47 in: R.A. Gutierrez and R.J. Orsi, editors. Contested Eden: California before the Gold Rush. University of California Press, Los Angeles, California, USA.
Anderson, R.S., and B.F. Byrd. 1998. Late-Holocene vegetation changes from the Las Flores Creek coastal lowlands, San Diego County, California. Madroño 45: 171–182.
Anderson, R.S., A. Ejarque, P.M. Brown, and D.J. Hallett. 2013. Holocene and historical vegetation change and fire history on the north-central coast of California, USA. Holocene 23: 1797–1810.
Armstrong, W.P. 1966. Ecological and taxonomic relationships of *Cupressus* in southern California. Thesis, California State University, Los Angeles, California, USA.
Axelrod, D.I. 1938. A Pliocene flora from the Mount Eden Beds, southern California. Pages 125–183 in: Miocene and Pliocene Floras of Western North America. Carnegie Institution of Washington, Washington, D.C., USA.
———. 1939. A Miocene flora from the western border of the Mohave Desert. Carnegie Institute of Washington Publication 516: 125–183.
———. 1980. History of the maritime closed-cone pines, Alta and Baja California. University of California Publications in Geological Sciences 120: 1–143.
Axelrod, D.I., and F. Govean. 1996. An early Pleistocene closed-cone pine forest at Costa Mesa, Southern California. International Journal of Plant Science 157: 323–329.
Barrett, L.A. 1935. A Record of Forest and Field Fires in California from the Days of the Early Explorers to the Creation of the Forest Reserves. USDA Forest Service, San Francisco, California, USA.
Bell, G.P. 1997. Ecology and management of *Arundo donax*, and approaches to riparian habitat restoration in southern California. Pages 103–133 in: J.H. Brock, M. Wade, P. Pysek, and D. Green, editors. Plant Invasions: Studies from North America and Europe. Backhuys Publishers, Leiden, The Netherlands.
Bendix, J. 1994. Among-site variation in riparian vegetation of the southern California transverse ranges. American Midland Naturalist 132: 136–151.
Bentley, J.R. 1967. Conversion of chaparral to grassland: techniques used in California. Washington D.C. USDA Forest Service, Agricultural Handbook 328. Washington, D.C., USA.
Bolton, R.B., Jr. 1966. Ecological requirements of big-cone spruce (*Pseudotsuga macrocarpa*) in the Santa Ana Mountains. Thesis, California State University, Los Angeles, California, USA.
Bolton, R.B., and R.J. Vogl. 1969. Ecological requirements of *Pseudotsuga macrocarpa* (Vasey) Mayr. in the Santa Ana Mountains. Journal of Forestry 69: 112–119.
Bond, W.J., K.A. Smythe, and D.A. Balfour. 2001. *Acacia* species turnover in space and time in an African savanna. Journal of Biogeography 28: 117–128.
Borchert, M.I. 1985. Serotiny and cone-habit variation in populations of *Pinus coulteri* (Pinaceae) in the southern Coast Ranges of California. Madroño 32: 29–48.
Borchert, M.I., D. Schreiner, T. Knowd, and T. Plumb. 2002. Predicting postfire survival in Coulter pine (*Pinus coulteri*) and gray pine (*Pinus sabiniana*) after wildfire in Central California. Western Journal of Applied Forestry 17: 134–138.
Bouldin, J.R. 1999. Twentieth-century changes in forests of the Sierra Nevada, California. Dissertation, University of California, Davis, California, USA.
Brennan, T.J., and J.E. Keeley. 2015. Effect of mastication and other mechanical treatments on fuel structure in chaparral. International Journal of Wildland Fire 24: 949–963.
Bullock, S.H. 1989. Life history and seed dispersal of the short-lived chaparral shrub *Dendromecon rigida* (Papaveraceae). American Journal of Botany 76: 1506–1517.
Byrne, R., E. Edlund, and S.A. Mensing. 1991. Holocene changes in the distribution and abundance of oaks in California. Pages 182–188 in: R.B. Standiford, editor. Proceedings of the Symposium on Oak Woodlands and Hardwood Rangeland Management. USDA Forest Service, Pacific Southwest Research Station, Albany, California, USA.
Cal Fire. 2006. General Guidelines for Creating Defensible Space. California Department of Forestry and Fire Protection, Sacramento, California, USA.
Callaway, R.M., and F.W. Davis. 1993. Vegetation dynamics, fire, and the physical environment in coastal central California. Ecology 74: 1567–1578.
Carroll, M.C., L.L. Laughrin, and A.C. Bromfield. 1993. Fire on the California Islands: does it play a role in chaparral and closed cone pine forest habitats? Pages 73–88 in: F.G. Hochberg, editor. Third California Islands Symposium: Recent Advances in Research on the California Islands. Santa Barbara Museum of Natural History, Santa Barbara, California, USA.
Campbell, B.M. 1980. Some mixed hardwood forest communities of the coastal range of southern California. Phytocoenologia 8: 297–320.
Cayan, D.R., E.P. Maurer, M.D. Dettinger, M. Tyree, and K. Hayhoe. 2008. Climate change scenarios for the California region. Climatic Change 87: 21–42.
Christensen, N.L. 1973. Fire and the nitrogen cycle in California chaparral. Science 181: 66–68.
Christenson, L.E. 1990. The late prehistoric Yuman people of San Diego County, California: their settlement and subsistence system. Dissertation, Arizona State University, Tempe, Arizona, USA.
Coffin, H. 1959. Effect of marine air on the fire climate in the mountains of southern California. USDA Forest Service Technical Paper 39. Pacific Southwest Forest and Range Experiment Station, Berkeley, California, USA.
Collett, R. 1992. Frost report: towns and temperatures. Pacific Horticulture 53: 7–9.
Conard, S.G., J.L. Beyers, and P.M. Wohlgemuth. 1995. Impacts of postfire grass seeding on chaparral systems—what do we know and where do we go from here? Pages 149–161 in: J.E. Keeley and T. Scott, editors. Wildfires in California Brushlands: Ecology and Resource Management. International Association of Wildland Fire, Fairfield, Washington, USA.
Conard, S.G., and D.R. Weise. 1998. Management of fire regime, fuels, and fire effects in southern California chaparral: lessons from the past and thoughts for the future. Proceedings Tall Timbers Ecology Conference 20: 342–350.
Cooper, W.S. 1922. The broad-sclerophyll vegetation of California: an ecological study of the chaparral and its related communities. Publication No. 319. Carnegie Institution of Washington, Washington, D.C., USA.
D'Antonio, C.M., C. Malmstrom, S.A. Reynolds, and J. Gerlach. 2007. Ecology of invasive non-native species in California grassland. Pages 67–83 in: M.R. Stromberg, J.D. Corbin, and C.M.D. Antonio, editors. California Grasslands: Ecology and Management. University of California Press, Los Angeles, California, USA.
Davis, O.K. 1992. Rapid climatic change in coastal southern California inferred from pollen analysis of San Joaquin Marsh. Quaternary Research 37: 89–100.
Davis, S.D., F.W. Ewers, J.S. Sperry, K.A. Portwood, M.C. Crocker, and G.C. Adams. 2002. Shoot dieback during prolonged drought in Ceanothus (Rhamnaceae) chaparral of California: a possible case of hydraulic failure. American Journal of Botany 89: 820–828.
Davis, F.W., E.A. Keller, A. Parikh, and J. Florsheim. 1989. Recovery of the chaparral riparian zone after wildfire. Pages 194–203 in: D.L. Abell, editor. Proceedings of the California Riparian Systems Conference: Protection, Management, and Restoration for the 1990s. USDA Forest Service General Technical Report PSW-GTR-110. Pacific Southwest Forest and Range Experiment Station, Albany, California, USA.
Davis, S.D., K.J. Kolb, and K.P. Barton. 1998. Ecophysiological processes and demographic patterns in the structuring of California chaparral. Pages 297–310 in: P.W. Rundel, G. Montenegro, and F.M. Jaksic, editors. Landscape Disturbance and Biodiversity in Mediterranean-Type Ecosystems. Springer-Verlag, New York, New York, USA.
Davis, F.W., and J. Michaelsen. 1995. Sensitivity of fire regime in chaparral ecosystems to climate change. Pages 435–456 in: J.M. Moreno and W.C. Oechel, editors. Global Change and Mediterranean-Type Ecosystems. Springer-Verlag, New York, New York, USA.
Davis, F.W., P.A. Stine, and D.M. Stoms. 1994. Distribution and conservation status of coastal sage scrub in southwestern California. Journal of Vegetation Science 5: 743–756.

Davis, F.W., P.A. Stine, D.M. Stoms, M.I. Borchert, and A.D. Hollander. 1995. Gap analysis of the actual vegetation of California 1. The southwestern region. Madroño 42: 40–78.
Dennison, P.E., M.A. Moritz, and R.S. Taylor. 2008. Evaluating predictive models of critical live fuel moisture in the Santa Monica Mountains, California. International Journal of Wildland Fire 17: 18–27.
DeSimone, S.A., and P.H. Zedler. 1999. Shrub seedling recruitment in unburned Californian coastal sage scrub and adjacent grassland. Ecology 80: 2018–2032.
———. 2001. Do shrub colonizers of southern Californian grassland fit generalities for other woody colonizers? Ecological Applications 11: 1101–1111.
Dodge, J.M. 1975. Vegetational changes associated with land use and fire history in San Diego County. Dissertation, University of California, Riverside, California, USA.
Douglas, M.W., R.A. Maddox, and K. Howard. 1993. The Mexican monsoon. Journal of Climate 6: 1665–1677.
Dunn, A.T. 1989. The effects of prescribed burning on fire hazard in the chaparral: toward a new conceptual synthesis. Pages 23–29 in: N.H. Berg, editor. Proceedings of the Symposium on Fire and Watershed Management. USDA Forest Service General Technical Report PSW-GTR-109. Pacific Southwest Forest and Range Experiment Station, Albany, California, USA.
Dwire, K.A., and J.B. Kauffman. 2003. Fire and riparian ecosystems in landscapes of the western USA. Forest Ecology and Management 178: 61–74.
Epling, C., and H. Lewis. 1952. Increase of the adaptive range of the genus *Delphinium*. Evolution 6: 253–267.
Erlandson, J.M. and M.A. Glassow, editors. 1997. Archaeology of the California coast during the middle Holocene (Perspectives in California Archaeology), Volume 4. Institute of Archaeology, University of California, Los Angeles, California, USA.
Everett, R.G. 2003. Grid-based fire-scar dendrochronology and vegetation sampling in the mixed-confer forests of the San Bernardino and San Jacinto Mountains of southern California. Dissertation, University of California, Riverside, California, USA.
Faber, P.M., E.A. Keller, A. Sands, and B.M. Massey. 1989. The ecology of riparian habitats of the southern California coastal region: a community profile. Biological Report 85 (7.27). Fish and Wildlife Service, Sacramento, California, USA.
Franklin, J., and E. Bergman. 2011. Patterns of pine regeneration following a large, severe wildfire in the mountains of southern California. Canadian Journal of Forest Research 41: 810–821.
Franklin, J., L.A. Spears-Lebrun, D.H. Deutschman, and K. Marsden. 2006. Impact of a high-intensity fire on mixed evergreen and mixed conifer forests in the Peninsular Ranges of southern California, USA. Forest Ecology and Management 235: 18–29.
Franklin, J., A.D. Syphard, D.J. Mladenoff, H.S. He, D.K. Simons, R.P. Martin, D. Deutschman, and J.F. O'Leary. 2001. Simulating the effects of different fire regimes on plant functional groups in southern California. Ecological Modelling 142: 261–283.
Freudenberger, D.O., B.E. Fish, and J.E. Keeley. 1987. Distribution and stability of grasslands in the Los Angeles Basin. Bulletin of the Southern California Academy of Sciences 86: 13–26.
Gause, G.W. 1966. Silvical characteristics of bigcone Douglas-fir. USDA Forest Service Research Paper PSW-RP-39. Pacific Southwest Forest and Range Experiment Station, Berkeley, California, USA.
Griffin, J.R., and W.B. Critchfield. 1972. The distribution of forest trees in California. USDA Forest Service Research Paper PSW-RP-82. Pacific Southwest Forest and Range Experiment Station, Berkeley, California, USA.
Gumprecht, B. 1999. The Los Angeles River. Johns Hopkins University Press, Baltimore, Maryland, USA.
Haidinger, T.L., and J.E. Keeley. 1993. Role of high fire frequency in destruction of mixed chaparral. Madroño 40: 141–147.
Halsey, R.W. 2004. Fire, Chaparral and Survival in Southern California. Sunbelt Publications, El Cajon, California, USA.
Hamilton, J.G. 1997. Changing perceptions of pre-European grasslands in California. Madroño 44: 311–333.
Hamilton, J.G., C. Holzapfel, and B.E. Mahall. 1999. Coexistence and interference between a native perennial grass and non-native annual grasses in California. Oecologia 121: 518–526.
Harris, R.R. 1987. Occurrence of vegetation on geomorphic surfaces in the active floodplain of a California alluvial stream. American Midland Naturalist 118: 393–405.
Heady, H.F. 1977. Valley grassland. Pages 491–514 in: M.G. Barbour and J. Major, editors. Terrestrial Vegetation of California. John Wiley and Sons, New York, New York, USA.
Hobbs, E.R. 1983. Factors controlling the form and location of the boundary between coastal sage scrub and grassland in southern California. Dissertation, University of California, Los Angeles, California, USA.
Hobbs, R.J., and H.A. Mooney. 1985. Vegetative regrowth following cutting in the shrub *Baccharis pilularis* ssp. *consanguinea* (DC) C.B. Wolf. American Journal of Botany 72: 514–519.
Horton, J.S. 1960. Vegetation types of the San Bernardino Mountains. USDA Forest Service Technical Paper PSW-44. Pacific Southwest Forest and Range Experiment Station, Berkeley, California, USA.
Huenneke, L.F. 1989. Distribution and regional patterns of Californian grasslands. Pages 1–12 in: L.F. Huenneke and H.A. Mooney, editors. Grassland Structure and Function: California Annual Grasslands. Kluwer Academic Publishers, Dordrecht, The Netherlands.
Jacobson, A.L., S.D. Davis, and S.L. Babritius. 2004. Fire frequency impacts non-sprouting chaparral shrubs in the Santa Monica Mountains of southern California. In: M. Arianoutsou and V.P. Papanastasis, editors. Ecology, Conservation and Management of Mediterranean Climate Ecosystems. Millpress, Rotterdam, The Netherlands.
Johnson, D.L. 1977. The California ice-age refugium and the Rancholabrean extinction problem. Quaternary Research 8: 149–153.
Johnson, M., S.B. Vander Wall, and M. Borchert. 2002. A comparative analysis of seed and cone characteristics and seed-dispersal strategies of three pines in the subsection *Sabinianae*. Plant Ecology 168: 69–84.
Jones, T.L. 1992. Settlement trends along the California coast. Pages 1–37 in: T.L. Jones, editor. Essays on the Prehistory of Maritime California. Center for Archaeological Research, Davis, California, USA.
Kimball, S., and P. Schiffman. 2003. Differing effects of cattle grazing on native and alien plants. Conservation Biology 17: 1681–1693.
Keeley, J.E. 1977. Seed production, seed populations in soil, and seedling production after fire for two congeneric pairs of sprouting and non-sprouting chaparral shrubs. Ecology 58: 820–829.
———. 1982. Distribution of lightning and man-caused wildfires in California. Pages 431–437 in: C.E. Conrad and W.C. Oechel, editors. Proceedings of the Symposium on Dynamics and Management of Mediterranean-Type Ecosystems. USDA Forest Service General Technical Report PSW-GTR-58. Pacific Southwest Forest and Range Experiment Station, Berkeley, California, USA.
———. 1986. Resilience of Mediterranean shrub communities to fire. Pages 95–112 in: B. Dell, A.J.M. Hopkins, and B.B. Lamont, editors. Resilience in Mediterranean-Type Ecosystems. Dr. W. Junk Publisher, Dordrecht, The Netherlands.
———. 1990a. Demographic structure of California black walnut (*Juglans californica*; Juglandaceae) woodlands in southern California. Madroño 37: 237–248.
———. 1990b. The California valley grassland. Pages 2–23 in: A.A. Schoenherr, editor. Endangered Plant Communities of Southern California. Southern California Botanists, Fullerton, California, USA.
———. 1991. Seed germination and life history syndromes in the California chaparral. Botanical Review 57: 81–116.
———. 1992a. Demographic structure of California chaparral in the long-term absence of fire. Journal of Vegetation Science 3: 79–90.
———. 1992b. Recruitment of seedlings and vegetative sprouts in unburned chaparral. Ecology 73: 1194–1208.
———. 1993. Native grassland restoration: the initial stage-assessing suitable sites. Pages 277–281 in: J.E. Keeley, editor. Interface between Ecology and Land Development in California. Southern California Academy of Sciences, Los Angeles, California, USA.
———. 1998a. Postfire ecosystem recovery and management: the October 1993 large fire episode in California. Pages 69–90 in:

J.M. Moreno, editor. Large Forest Fires. Backhuys Publishers, Leiden, The Netherlands.
———. 1998b. Coupling demography, physiology and evolution in chaparral shrubs. Pages 257–264 in: P.W. Rundel, G. Montenegro, and F.M. Jaksic, editors. Landscape Diversity and Biodiversity in Mediterranean-Type Ecosystems. Springer-Verlag, New York, New York, USA.
———. 2000. Chaparral. Pages 203–253 in: M.G. Barbour and W.D. Billings, editors. North American Terrestrial Vegetation. Cambridge University Press, Cambridge, UK.
———. 2001. Fire and invasive species in Mediterranean-climate ecosystems of California. Pages 81–94 in: K.E.M. Galley and T.P. Wilson, editors. Proceedings of the Invasive Species Workshop: The Role of Fire in the Control and Spread of Invasive Species. Miscellaneous Publication No. 11. Tall Timbers Research Station, Tallahassee, Florida, USA.
———. 2002a. Native American impacts on fire regimes of the California coastal ranges. Journal of Biogeography 29: 303–320.
———. 2002b. Fire management of California shrubland landscapes. Environmental Management 29: 395–408.
———. 2004a. Impact of antecedent climate on fire regimes in southern California. International Journal of Wildland Fire 13: 173–182.
———. 2004b. VTM plots as evidence of historical change: goldmine or landmine? Madroño 51: 372–378.
———. 2004c. Invasive plants and fire management in California Mediterranean-climate ecosystems. In: M. Arianoutsou and V.P. Papanastasis, editors. Ecology, Conservation and Management of Mediterranean Climate Ecosystems. Millpress, Rotterdam, The Netherlands.
———. 2006a. Fire management impacts on invasive plant species in the western United States. Conservation Biology 20: 375–384.
———. 2006b. Fire severity and plant age in postfire resprouting of woody plants in sage scrub and chaparral. Madroño 53: 373–379.
Keeley, J.E., and W.J. Bond. 1997. Convergent seed germination in South African fynbos and Californian chaparral. Plant Ecology 133: 153–167.
Keeley, J.E., W.J. Bond, R.A. Bradstock, J.G. Pausas, and W. Rundel. 2012. Fire in Mediterranean Climate Ecosystems: Ecology, Evolution and Management. Cambridge University Press, Cambridge, UK.
Keeley, J.E., and T.J. Brennan. 2012. Fire driven alien invasion in a fire-adapted ecosystem. Oecologia 169: 1043–1052.
Keeley, J.E., and C.J. Fotheringham. 2000. Role of fire in regeneration from seed. Pages 311–330 in: M. Fenner, editor. Seeds: The Ecology of Regeneration in Plant Communities. 2nd ed. CAB International, Oxon, UK.
———. 2001a. Historic fire regime in Southern California shrublands. Conservation Biology 15: 1536–1548.
———. 2001b. History and management of crown-fire ecosystems: a summary and response. Conservation Biology 15: 1561–1567.
———. 2003. Impact of past, present, and future fire regimes on North American Mediterranean shrublands. Pages 218–262 in: T.T. Veblen, W.L. Baker, G. Montenegro, and T.W. Swetnam, editors. Fire and Climatic Change in Temperate Ecosystems of the Western Americas. Springer, New York, New York, USA.
Keeley, J.E., C.J. Fotheringham, and M. Baer-Keeley. 2005a. Determinants of postfire recovery and succession in Mediterranean-climate shrublands of California. Ecological Applications 15: 1515–1534.
———. 2005b. Factors affecting plant diversity during postfire recovery and succession of Mediterranean-climate shrublands in California, USA. Diversity and Distributions 11: 535–537.
———. 2006. Demographic patterns of postfire regeneration in Mediterranean-climate shrublands of California. Ecological Monographs 76: 235–255.
Keeley, J.E., C.J. Fotheringham, and M. Morais. 1999. Reexamining fire suppression impacts on brushland fire regimes. Science 284: 1829–1832.
Keeley, J.E., C.J. Fotheringham, and M.A. Moritz. 2004. Lessons from the October 2003 wildfires in southern California. Journal of Forestry 102: 26–31.
Keeley, J.E., and S.C. Keeley. 1984. Postfire recovery of California coastal sage scrub. American Midland Naturalist 111: 105–117.
Keeley, J.E., M.B. Keeley and C.J. Fotheringham. 2005c. Alien plant dynamics following fire in Mediterranean-climate shrublands of California. Ecological Applications 15: 2109–2125.
Keeley, J.E., A.H. Pfaff, and H.D. Safford. 2005d. Fire suppression impacts on postfire recovery of Sierra Nevada chaparral shrublands. International Journal of Wildland Fire 14: 255–265.
Keeley, J.E., and H.D. Safford. 2016. Fire as an ecosystem process. Chapter 3 in: H. Mooney and E. Zavaleta, editors. Ecosystems of California. University of California Press, Berkeley, California, USA.
Keeley, J.E., H. Safford, C.J. Fotheringham, J. Franklin, and M. Moritz. 2009. The 2007 southern California wildfires: lessons in complexity. Journal of Forestry 107: 287–296.
Keeley, J.E., and N.L. Stephenson. 2000. Restoring natural fire regimes in the Sierra Nevada in an era of global change. Pages 255–265 in: D.N. Cole, S.F. McCool, and J. O'Loughlin, editors. Wilderness Science in a Time of Change Conference. USDA Forest Service RMRS-P-15-VOL-5. Rocky Mountain Research Station, Fort Collins, Colorado, USA.
Keeley, J.E., and A.D. Syphard. 2015. Different fire-climate relationships on forested and non-forested landscapes in the Sierra Nevada ecoregion. International Journal of Wildland Fire 24: 27–36.
Keeley, J.E., and P.H. Zedler. 1998. Evolution of life histories in *Pinus*. Pages 219–250 in: D.M. Richardson, editor. Ecology and Biogeography of *Pinus*. Cambridge University Press, Cambridge, UK.
———. 2009. Large, high intensity fire events in southern California shrublands: debunking the fine-grained age-patch model. Ecological Applications 19: 69–94.
Kellogg, E.M., and J.L. Kellogg. 1990. A study of the distribution and pattern of perennial grassland on the Camp Pendleton Marine Corps Base. Contract No. M00681-88-P03161. U.S. Marine Corps, Camp Pendleton, California, USA.
Kimball, S., M.L. Goulden, K.N. Suding, and S. Parker. 2014. Altered water and nitrogen input shifts succession in a Southern California coastal sage community. Ecological Applications 24: 1390–1404.
Kinney, A. 1887. Report on the forests of the counties of Los Angeles, San Bernardino, and San Diego, California. First Biennial Report. California State Board of Forestry, Sacramento, California, USA.
———. 1900. Forest and Water. Post Publishing Company, Los Angeles, California, USA.
Kirkpatrick, J.B., and C.F. Hutchinson. 1977. The community composition of California coastal sage scrub. Vegetatio 35: 21–33.
Lawson, D.M. 1993. The effect of fire on stand structure of mixed *Quercus agrifolia* and *Quercus engelmannii* woodlands. Thesis, San Diego State University, San Diego, California, USA.
Lee, R.G., and T.M. Bonnicksen. 1978. Brushland watershed fire management policy in southern California: biosocial considerations. Contribution No. 172. California Water Resources Center, University of California, Davis, California, USA.
Leiberg, J.B. 1899a. San Jacinto Forest Reserve. U.S. Geological Survey, Annual Report 19: 351–357. Washington, D.C., USA.
———. 1899b. San Bernardino Forest Reserve. U.S. Geological Survey, Annual Report 19: 359–365. Washington, D.C., USA.
———. 1899c. San Gabriel Forest Reserve. U.S. Geological Survey, Annual Report 19: 367–371. Washington, D.C., USA.
———. 1900a. San Gabriel Forest Reserve. U.S. Geological Survey, Annual Report 20: 411–428. Washington, D.C., USA.
———. 1900b. San Bernardino Forest Reserve. U.S. Geological Survey, Annual Report 20: 429–454. Washington, D.C., USA.
———. 1900c. San Jacinto Forest Reserve. U.S. Geological Survey, Annual Report 20: 455–478. Washington, D.C., USA.
Linhart, Y.B., B. Burr, and M.T. Conklin. 1967. The closed-cone pines of the northern Channel Islands. Pages 151–177 in: R.N. Philbrick, editor. Proceedings of the Symposium on the Biology of the California Islands. Santa Barbara Botanic Garden, Santa Barbara, California, USA.
Lippitt, C.L., D. Stow, J.F. O'Leary, and J. Franklin. 2013. Influence of short-interval fire occurrence on post-fire recovery of fire-prone shrublands in California, USA. International Journal of Wildland Fire 22: 184–193.
Lloret, F., and P.H. Zedler. 1991. Recruitment of *Rhus integrifolia* in chaparral. Journal of Vegetation Science 2: 217–230.
Lombardo, K.J., T.W. Swetnam, C.H. Baisan, and M.I. Borchert. 2009. Using bigcone Douglas-fir fire scars and tree rings to reconstruct interior chaparral fire history. Fire Ecology 5(3): 35–56.

Loomis, J., P. Wohlgemuth, A. Gonzale-Caban, and D. English. 2003. Economic benefits of reducing fire-related sediment in southwestern fire-prone ecosystems. Water Resources Research 39(No 9, WES 3): 1–8.
Mack, R.N. 1989. Temperate grasslands vulnerable to plant invasions: characteristics and consequences. Pages 155–179 in: J.A. Drake, H.A. Mooney, F. di Castri, R.H. Groves, F.J. Kruger, M. Rejmanek, and M. Williamson, editors. Biological Invasions: A Global Perspective. John Wiley and Sons, New York, New York, USA.
Major, J. 1977. California climate in relation to vegetation. Pages 11–74 in: M.G. Barbour and J. Major, editors. Terrestrial Vegetation of California. John Wiley and Sons, New York, New York, USA.
Malanson, G.P., and J.F. O'Leary. 1982. Post-fire regeneration strategies of California coastal sage shrubs. Oecologia 53: 355–358.
Malanson, G.P., and W.E. Westman. 1985. Postfire succession in Californian coastal sage scrub: the role of continual basal sprouting. American Midland Naturalist 113: 309–318.
Maranghides, A., and W. Mell. 2011. A case study of a community affected by the Witch and Guejito Fires. Fire Technology 47: 379–420.
McBride, J.R. 1974. Plant succession in the Berkeley Hills, California. Madroño 22: 317–380.
McBride, J.R., and D.F. Jacobs. 1980. Land use and fire history in the mountains of southern California. Pages 85–88 in: M.A. Stokes and J.H. Dieterich, editors. Proceedings of the Fire History Workshop. USDA Forest Service General Technical Report RM-GTR-81. Rocky Mountain Forest and Range Experiment Station, Fort Collins, Colorado, USA.
McBride, J.R., and R.D. Laven. 1976. Scars as an indicator of fire frequency in the San Bernardino Mountains, California. Journal of Forestry 74: 439–442.
———. 1999. Impact of oxidant air pollutants on forest succession in the mixed conifer forests of the San Bernardino Mountains. Pages 338–352 in: P.R. Miller and J.R. McBride, editors. Oxidant Air Pollution Impacts in the Montane Forests of Southern California: A Case Study of the San Bernardino Mountains. Springer, New York, New York, USA.
McDonald, P.M., and E.E. Litterell. 1976. The bigcone Douglas-fir–canyon live oak community in southern California. Madroño 23: 310–320.
McIntyre P.J., J.H. Thorne, C.R. Dolanc, A.L. Flint, L.E. Flint, M. Kell, and D.D. Ackerly. 2015. Twentieth-century shifts in forest structure in California: denser forests, smaller trees, and increased dominance of oaks. Proceedings of the National Academy of Sciences 112: 1458–1463.
McIver, J.D., S.L. Stephens, J.K. Agee, J. Barbour, R.E.J. Boerner, C.B. Edminster, K.L. Erickson, K.L., Farris, C.J. Fettig, C.E. Fielder, S. Haase, S.C. Hart, J.E. Keeley, E.E. Knapp, J.F. Lehmkuhl, J.J. Moghaddas, W. Otrosina, K.W. Outcalt, D.W. Schwilk, C.N. Skinner, T.A. Waldrop, C.P. Weatherspoon, D.A. Yaussy, A. Youngblood, and S. Zack. 2012. Ecological effects of alternative fuel-reduction treatments: highlights of the National Fire and Fire Surrogate study (FFS). International Journal of Wildland Fire 22: 63–82.
McKelvey, K.S., and J.D. Johnston. 1992. Historical perspectives on forests of the Sierra Nevada and the Transverse Ranges of Southern California: forest conditions at the turn of the century. Pages 225–246 in: J. Verner, K.S. McKelvey, B.R. Noon, R.J. Gutierrez, G.I. Gould, and T.W. Beck, editors. The California Spotted Owl: A Technical Assessment of Its Current Status. USDA Forest Service General Technical Report PSW-GTR-133. Pacific Southwest Research Station, Albany, California, USA.
McMaster, G.S., and P.H. Zedler. 1981. Delayed seed dispersal in *Pinus torreyana* (Torrey pine). Oecologia 51: 62–66.
McMillan, C. 1956. Edaphic restriction of *Cupressus* and *Pinus* in the coast ranges of central California. Ecological Monographs 26: 177–212.
Meentemeyer, R.K., and A. Moody. 2002. Distribution of plant history types in California chaparral: the role of topographically-determined drought severity. Journal of Vegetation Science 13: 67–78.
Meng, R., P.E. Dennison, C.M. D'Antonio, and M.A. Moritz. 2014. Remote sensing analysis of vegetation recovery following short-interval fires in southern California shrublands. PLoS One 9(10): e110637.
Mensing, S.A., J. Michaelsen, and R. Byrne. 1999. A 560-year record of Santa Ana fires reconstructed from charcoal deposited in the Santa Barbara Basin, California. Quaternary Research 51: 295–305.
Merriam, K.E., K.E. Keeley, and J.L. Beyers, 2006. Fuel breaks affect nonnative species abundance in Californian plant communities. Ecological Applications 16: 515–527.
Miles, S.R., and C.B. Goudey. 1997. Ecological subregions of California. USDA Forest Service Technical Publication R5-EM-TP-005. Pacific Southwest Region, Vallejo, California, USA.
Millar, C.I. 1986. Bishop pine (*Pinus muricata*) of inland Marin County, California. Madroño 33: 123–129.
Minnich, R.A. 1974. The impact of fire suppression on southern California conifer forests: a case study of the Big Bear fire, November 13 to 16, 1970. Pages 45–57 in: M. Rosenthal, editor. Symposium on Living with the Chaparral, Proceedings. Sierra Club, San Francisco, California, USA.
———. 1977. The geography of fire and big cone Douglas-fir, Coulter pine and western conifer forest in the east Transverse Ranges, southern California. Pages 343–350 in: H.A. Mooney and C.E. Conrad, editors. Proceedings of the Symposium on Environmental Consequences of Fire and Fuel Management in Mediterranean Ecosystems. USDA Forest Service General Technical Report WO-3. Washington, D.C., USA.
———. 1978. The geography of fire and conifer forests in the eastern Transverse Ranges, California. Dissertation, University of California, Los Angeles, California, USA.
———. 1980a. Wildfire and the geographic relationships between canyon live oak, Coulter pine, and bigcone Douglas-fir forests. Pages 55–61 in: T.R. Plumb, editor. Proceedings of the Symposium on Ecology, Management and Utilization of California Oaks. USDA Forest Service General Technical Report PSW-GTR-44. Pacific Southwest Forest and Range Experiment Station, Berkeley, California, USA.
———. 1980b. Vegetation of Santa Cruz and Santa Catalina Islands. Pages 123–139 in: D.M. Power, editor. The California Islands: Proceedings of a Multidisciplinary Symposium. Santa Barbara Museum of Natural History, Santa Barbara, California, USA.
———. 1983. Fire mosaics in southern California and northern Baja California. Science 219: 1287–1294.
———. 1987. Fire behavior in southern California chaparral before fire control: the Mount Wilson burns at the turn of the century. Annals of the Association of American Geographers 77: 599–618.
———. 1988. The biogeography of fire in the San Bernardino Mountains of California: a historical study. University of California Publications in Botany 28: 1–120.
———. 1989. Chaparral fire history in San Diego County and adjacent northern Baja California: an evaluation of natural fire regimes and the effects of suppression management. Pages 37–47 in: S.C. Keeley, editor. The California Chaparral: Paradigms Reexamined. Natural History Museum of Los Angeles County, Los Angeles, California, USA.
———. 1995. Fuel-driven fire regimes of the California chaparral. Pages 21–27 in: J.E. Keeley and T. Scott, editors. Brushfires in California wildlands: ecology and resource management. International Association of Wildland Fire, Fairfield, Washington, USA.
———. 1998. Landscapes, land-use and fire policy: where do large fires come from? Pages 133–158 in: J.M. Moreno, editor. Large Forest Fires. Backhuys Publishers, Leiden, The Netherlands.
———. 1999. Vegetation, fire regimes, and forest dynamics. Pages 44–83 in: P.R. Miller and J.R. McBride, editors. Oxidant Air Pollution Impacts in the Montane Forests of Southern California: A Case Study of the San Bernardino Mountains. Springer, New York, New York, USA.
Minnich, R.A., M.G. Barbour, J.H. Burk, and R.F. Fernau. 1995. Sixty years of change in Californian conifer forests of the San Bernardino Mountains. Conservation Biology 9: 902–914.
Minnich, R.A., and Y.H. Chou. 1997. Wildland fire patch dynamics in the chaparral of southern California and northern Baja California. International Journal of Wildland Fire 7: 221–248.
Minnich, R.A., and R.J. Dezzani. 1991. Suppression, fire behavior, and fire magnitudes in Californian chaparral at the urban/wildland interface. Pages 67–83 in: J.J. DeVries, editor. California Watersheds at the Urban Interface, Proceedings of the Third

Biennial Watershed Conference. University of California, Davis, California, USA.
———. 1998. Historical decline of coastal sage scrub in the Riverside-Perris Plain, California. Western Birds 29: 366–391.
Minnich, R.A., and R.G. Everett. 2002. What unmanaged fire regimes in Baja California tell us about presuppression fire in California Mediterranean ecosystems. Pages 325–338 in: N.G. Sugihara, M.E. Morales, and T.J. Morales, editors. Proceedings of the Symposium: Fire in California Ecosystems: Integrating Ecology, Prevention and Management. Association for Fire Ecology Miscellaneous Publication No. 1. Sacramento, California, USA.
Moreno, J.M., and W.C. Oechel. 1994. Fire intensity as a determinant factor of postfire plant recovery in southern California chaparral. Pages 26–45 in: J.M. Moreno and W.C. Oechel, editors. The Role of Fire in Mediterranean-Type Ecosystems. Springer-Verlag, New York, New York, USA.
Moritz, M.A. 1997. Analyzing extreme disturbance events: fire in the Los Padres National Forest. Ecological Applications 7: 1252–1262.
———. 2003. Spatiotemporal analysis of controls on shrubland fire regimes: age dependency and fire hazard. Ecology 84: 351–361.
Moritz, M.A., J.E. Keeley, E.A. Johnson, and A.A. Schaffner. 2004. Testing a basic assumption of shrubland fire management: does the hazard of burning increase with the age of fuels? Frontiers in Ecology and the Environment 2: 67–72.
Moritz, M.A., T.J. Moody, M.A. Krawchuk, M. Huges, and A. Hall. 2010. Spatial variation in extreme winds predicts large wildfire locations in chaparral ecosystems. Geophysical Research Letters 37: L04801.
Morvan, D. and J.L. Dupuy. 2004. Modeling the propagation of a wildfire through a Mediterranean shrub using a multiphase formulation. Combustion and Flame 138: 199–210.
Ne'eman, G., C.J. Fotheringham, and J.E. Keeley. 1999. Patch to landscape patterns in post fire recruitment of a serotinous conifer. Plant Ecology 145: 235–242.
Norris, R.M., and R.W. Webb. 1976. Geology of California. John Wiley and Sons, New York, New York, USA.
North, M., A. Brough, J. Long, B.M. Collins, P. Bowden, D. Yasuda, J.D. Miller, and N.G. Sugihara. 2015. Constraints on mechanized treatment significantly limit mechanical fuels reduction extent in the Sierra Nevada. Journal of Forestry 113: 40–48.
Oberbauer, A.T. 1978. Distribution dynamics of San Diego County grasslands. Thesis, San Diego State University, San Diego, California, USA.
Odion, D.C., and F.W. Davis. 2000. Fire, soil heating, and the formation of vegetation patterns in chaparral. Ecological Monographs 70: 149–169.
Odion, D., and C. Tyler. 2002. Are long fire-free periods needed to maintain the endangered, fire-recruiting shrub *Arctostaphylos morroensis* (Ericaceae)? Conservation Ecology 6: 4.
O'Leary, J.F., and W.E. Westman. 1988. Regional disturbance effects on herb succession patterns in coastal sage scrub. Journal of Biogeography 15: 775–786.
Padgett, P.E., and E.B. Allen. 1999. Differential responses to nitrogen fertilization in native shrubs and exotic annuals common to Mediterranean coastal sage scrub of California. Plant Ecology 144: 93–101.
Pausas, J.G., and J.E. Keeley. 2014. Evolutionary ecology of resprouting and seeding in fire-prone ecosystems. New Phytologist 204: 55–65.
Penman, T.D., L. Collins, A.D. Syphard, J.E. Keeley, and R.A. Bradstock. 2014. Influence of fuels, weather and the built environment on the exposure of property to wildfire. PLoS One 9(10): doi:10.1371.0111414
Pequegnat, W.E. 1951. The biota of the Santa Ana Mountains. Journal of Entomology and Zoology 42: 1–84.
Phillips, C.B. 1971. California Aflame! September 22–October 4, 1970. State of California, Department of Conservation, Division of Forestry, Sacramento, California, USA.
Philpot, C.W. 1974. The changing role of fire on chaparral lands. Pages 131–150 in: M. Rosenthal, editor. Symposium on Living with the Chaparral, Proceedings. Sierra Club, San Francisco, California, USA.
Pincetl, S., P.W. Rundel, J.C. DeBlasio, D. Silver, T. Scott, J.E. Keeley, and R. Halsey. 2008. It's the land use, not the fuels: fire and land development in southern California. Real Estate Review 37(1): 25–42.
Plumb, T.R. 1980. Response of oaks to fire. Pages 202–215 in T.R. Plumb, editor. Proceedings of the Symposium on Ecology, Management and Utilization of California Oaks. USDA Forest Service General Technical Report PSW-GTR-44. Pacific Southwest Forest and Range Experiment Station, Berkeley, California, USA.
Plummer, F.G. 1911. Chaparral studies in the dwarf forests, of elfin-wood, of southern California. USDA Forest Service Bulletin 85. Washington, D.C., USA.
Price, O.F., R.A. Bradstock, J.E. Keeley, and A.D. Syphard. 2012. The impact of antecedent fire area on burned area in southern California coastal ecosystems. Journal of Environmental Management 113: 301–307.
Quarles, S.L., Y. Valachovic, G.M. Nakamura, G.A. Nader, and M.J. DeLasaux. 2010. Home survival in wildfire-prone areas: building materials and design considerations. Agriculture and Natural Resources Publication 8393. University of California, Berkeley, California, USA.
Quinn, R.D. 1990. The status of walnut forests and woodlands (*Juglans californica*) in southern California. Pages 42–54 in: A.A. Schoenherr, editor. Endangered Plant Communities of Southern California. Southern California Botanists, Fullerton, California, USA.
Radtke, K.W.-H., A.M. Arndt, and R.H. Wakimoto. 1982. Fire history of the Santa Monica Mountains. Pages 438–443 in: C.E. Conrad and W.C. Oechel, editors. Proceedings of the Symposium on Dynamics and Management of Mediterranean-Type Ecosystems. USDA Forest Service General Technical Report PSW-GTR-58. Pacific Southwest Forest and Range Experiment Station, Berkeley, California, USA.
Reiner, R.J. 2007. Fire in California grasslands. Pages 207–217 in: M.R. Stromberg, J.D. Corbin, and C.M. D'Antonio, editors. California Grasslands: Ecology and Management. University of California Press, Los Angeles, California, USA.
Reveal, J.L. 1978. A report on the autecology and status of Cuyamaca cypress (*Cupressus arizonica* var. *stephensonii*). USDA Forest Service Unpublished Report, Order no. 1191-PSW-77. Cleveland National Forest California, San Diego, California, USA.
Rieger, J.P., and D.A. Kreager. 1989. Giant reed (*Arundo donax*): a climax community of the riparian zone. Pages 222–225 in: D.L. Abell, editor. Proceedings of the California Riparian Systems Conference: Protection, Management, and Restoration for the 1990s. USDA Forest Service General Technical Report PSW-GTR-110. Pacific Southwest Forest and Range Experiment Station, Berkeley, California, USA.
Riggan, P.J., S.E. Franklin, J.A. Brass, and F.E. Brooks. 1994. Perspectives on fire management in Mediterranean ecosystems of southern California. Pages 140–162 in: J.M. Moreno and W.C. Oechel, editors. The Role of Fire in Mediterranean-Type Ecosystems. Springer-Verlag, New York, New York, USA.
Rodrigue, C.M. 1993. Home with a view: chaparral fire hazard and the social geographies of risk and vulnerability. California Geographer 33: 29–42.
Rundel, P.W. 1975. Primary succession on granite outcrops in the montane southern Sierra Nevada. Madroño 23: 209–219.
———. 2007. Sage scrub. Pages 208–228 in: M.G. Barbour, T. Keeler-Wolf, and A.A. Schoenherr, editors. Terrestrial Vegetation of California. 3rd ed. University of California Press, Los Angeles, California, USA.
Rundel, P.W., D.J. Parsons, and D.T. Gordon. 1977. Montane and subalpine vegetation of the Sierra Nevada and Cascade Ranges. Pages 559–599 in: M.G. Barbour and J. Major, editors. Terrestrial Vegetation of California. John Wiley and Sons, New York, New York, USA.
Rundel, P.W., and S.B. Sturmer. 1998. Native plant diversity in riparian communities of the Santa Monica Mountains, California. Madroño 45: 93–100.
Safford, H.D., J.T. Stevens, K. Merriam, M.D. Meyer, and A.M. Latimer. 2012. Fuel treatment effectiveness in California yellow pine and mixed conifer forests. Forest Ecology and Management 274: 17–28.
Safford, H.D., and K.M. Van de Water. 2014. Using fire return interval departure (FRID) analysis to map spatial and temporal changes in fire frequency on National Forest lands in California. USDA Forest Service Research Report PSW-RP-266. Pacific Southwest Research Station, Albany, California, USA.

Sargent, C.S. 1884. Report on the forests of North America (exclusive of Mexico). Report on the Productions of Agriculture as Returned at the Tenth Census, Volume 9. U.S. Government Printing Office, Washington, D.C., USA.
Savage, M. 1994. Anthropogenic and natural disturbance and patterns of mortality in a mixed conifer forest in California. Canadian Journal of Forest Research 24: 1149–1159.
———. 1997. The role of anthropogenic influences in a mixed-conifer forest mortality episode. Journal of Vegetation Science 8: 95–104.
Sawyer, J.O., T. Keeler-Wolf, and J.M. Evens. 2009. A Manual of California Vegetation. 2nd ed. California Native Plant Society, Sacramento, California, USA.
Scheidlinger, C.R., and P.H. Zedler. 1980. Change in vegetation cover of oak stands in southern San Diego County: 1928-1970. Pages 81–85 in: T.R. Plumb, editor. Proceedings of the Symposium on the Ecology, Management, and Utilization of California Oaks. USDA Forest Service General Technical Report PSW-GTR-44. Pacific Southwest Forest and Range Experiment Station, Berkeley, California, USA.
Schoenberg, F.P., R. Peng, Z. Huang, and P. Rundel. 2003. Detection of non-linearities in the dependence of burn area on fuel age and climatic variables. International Journal of Wildland Fire 12: 1–6.
Schoenherr, A.A. 1992. A Natural History of California. University of California Press, Los Angeles, California, USA.
Schroeder, M.J., M. Glovinsky, V. Hendricks, F. Hood, M. Hull, H. Jacobsen, R. Kirkpatrick, D. Krueger, L. Mallory, A. Oertel, R. Reese, L. Sergius, and C. Syverson. 1964. Synoptic weather types associated with critical fire weather. U.S. Department of Commerce, National Bureau of Standards, Institute for Applied Technology AD 449-630. Washington, D.C., USA.
Schwilk, D.W. 2003. Flammability is a niche construction trait: canopy architecture affects fire intensity. American Naturalist 162: 725–733.
Schwilk, D.W., and D.D. Ackerly. 2001. Flammability and serotiny as strategies: correlated evolution in pines. Oikos 94: 326–336.
Schwilk, D.W., and J.E. Keeley. 2006. The role of fire refugia in the distribution of *Pinus sabiniana* (Pinaceae) in the southern Sierra Nevada. Madroño 53: 364–372.
Schwilk, D.W., and B. Kerr. 2002. Genetic niche-hiking: an alternative explanation for the evolution of flammability. Oikos 99: 431–442.
Sheppard, P.R., and J.P. Lassoie. 1998. Fire regime of the lodgepole pine forest of Mt. San Jacinto, California. Madroño 45: 47–56.
Simpson, L.B. 1938. California in 1792. The Expedition of Jose Longinos Martinez, The Huntington Library, San Marino, California, USA.
Steel, Z.L., H.D. Safford, and J.H. Viers. 2015. The fire frequency-severity relationship and the legacy of fire suppression in California forests. Ecosphere 6(1): art8.
Stephens, S.L., J.M. Lydersen, B.M. Collins, D.L. Fry, and M.D. Meyer. 2015. Historical and current landscape-scale ponderosa pine and mixed conifer forest structure in the southern Sierra Nevada. Ecosphere 6(5): art79.
Stephenson, N.L. 1990. Climatic control of vegetation distribution: the role of the water balance. American Naturalist 135: 649–670.
Stephenson, J.R., and G.M. Calcarone. 1999. Southern California mountains and foothills assessment. USDA Forest Service General Technical Report PSW-GTR-172. Pacific Southwest Research Station, Albany, California, USA.
Strauss, D., L. Dednar, and R. Mees. 1989. Do one percent of forest fires cause ninety-nine percent of the damage? Forest Science 35: 319–328.
Stromberg, M.R., and J.R. Griffin. 1996. Long-term patterns in coastal California grasslands in relation to cultivation, gophers, and grazing. Ecological Applications 6: 1189–1211.
Syphard, A.D., T.J. Brennan, and J.E. Keeley. 2014. The role of defensible space for residential structure protection during wildfires. International Journal of Wildland Fire 23: 1165–1175.
Syphard, A.D., and J.E. Keeley. 2015. Location, timing and extent of wildfire vary by cause of ignition. International Journal of Wildland Fire 24: 37–47.
Syphard, A.D., J.E. Keeley, and T.J. Brennan. 2011a. Factors affecting fuel break effectiveness in the control of large fires on the Los Padres National Forest, California. International Journal of Wildland Fire 20: 764–775.
———. 2011b. Comparing the role of fuel breaks across southern California national forests. Forest Ecology and Management 261: 2038–2048.
Syphard, A.D., J.E. Keeley, A.B. Massada, T.J. Brennan, and V.C. Radeloff. 2012. Housing arrangement and location determine the likelihood of housing loss due to wildfire. PLoS One 7(3): 1–13.
Syphard, A.D., A.B. Massada, V. Butsic, and J.E. Keeley. 2013. Land use planning and wildfire: development policies influence future probability of housing loss. PLoS One 8(8): L04801.
Syphard, A.D., V.C. Radeloff, J.E. Keeley, T.J. Hawbaker, M.K. Clayton, S.I. Stewart, and R.B. Hammer. 2007. Human influence on California fire regimes. Ecological Applications 17: 1388–1402.
Syphard, A.D., V.C. Radeloff, N.S. Keuler, R.S. Taylor, T.J. Hawbaker, S.I. Stewart, and M.K. Clayton. 2008. Predicting spatial patterns of fire on a southern California landscape. International Journal of Wildland Fire 17: 602–613.
Talluto, M.V. and K.N. Suding. 2008. Historical change in coastal sage scrub in southern California, USA in relation to fire frequency and air pollution. Landscape Ecology 23: 803–815.
Thorne, R.F. 1977. Montane and subalpine forests of the Transverse and Peninsular Ranges. Pages 537–557 in: M.G. Barbour and J. Major, editors. Terrestrial Vegetation of California. John Wiley and Sons, New York, New York, USA.
Timbrook, J., J.R. Johnson, and D.D. Earle. 1982. Vegetation burning by the Chumash. Journal of California and Great Basin Anthropology 4: 163–186.
Tyler, C., and M. Borchert. 2002. Reproduction and growth of the chaparral geophyte, *Zigadenus fremontii* (Liliaceae), in relation to fire. Plant Ecology 165: 11–20.
Tyler, C.M., B. Kuhn, and F.W. Davis. 2006. Demography and recruitment limitations of three oak species in California. Quarterly Review of Biology 81: 127–152.
USDA Forest Service. 2004. Draft environmental impact statement for revised land management plans. Angeles National Forest, Cleveland National Forest, Los Padres National Forest, San Bernardino National Forest. USDA Forest Service R5-MB-05. Pacific Southwest Region, Vallejo, California, USA.
Van de Water, K.M., and H.D. Safford. 2011. A summary of fire frequency estimates for California vegetation before Euro-American settlement. Fire Ecology 7(3): 26–58.
van Wagtendonk, J.W., and D.R. Cayan. 2008. Temporal and spatial distribution of lightning strikes in California in relationship to large-scale weather patterns. Fire Ecology 4(1): 34–56.
Vogl, R.J. 1973. Ecology of knobcone pine in the Santa Ana Mountains, California. Ecological Monographs 43: 125–143.
———. 1977. Fire frequency and site degradation. Pages 151–162 in: H.A. Mooney and C.E. Conrad, editors. Proceedings of the Symposium on Environmental Consequences of Fire and Fuel Management in Mediterranean Ecosystems. USDA Forest Service General Technical Report WO-3. Washington, D.C., USA.
Ward, J.K., J.M. Harris, T.E. Cerling, A. Wiedenhoeft, M.J. Lott, M.-D. Dearing, J.B. Coltrain, and J.R. Ehleringer. 2005. Carbon starvation in glacial trees recovered from the La Brea tar pits, southern California. Proceedings of the National Academy of Sciences 102: 690–694.
Weide, D.L. 1968. The geography of fire in the Santa Monica Mountains. Thesis, California State University, Los Angeles, California, USA.
Weise, D.R., J.C. Regelbrugge, T.E. Paysen, and S.G. Conard. 2002. Fire occurrence on southern California national forests—has it changed recently? Pages 389–391 in: N.G. Sugihara, M.E. Morales, and T.J. Morales, editors. Proceedings of the Symposium: Fire in California Ecosystems: Integrating Ecology, Prevention and Management. Miscellaneous Publication No. 1. Association for Fire Ecology, Sacramento, California, USA.
Wells, P.V. 1962. Vegetation in relation to geological substratum and fire in the San Luis Obispo quadrangle, California. Ecological Monographs 32: 79–103.
———. 1969. The relation between mode of reproduction and extent of speciation in woody genera of the California chaparral. Evolution 23: 264–267.

Wells, M.L., and A. Getis. 1999. The spatial characteristics of stand structure in *Pinus torreyana*. Plant Ecology 143: 153–170.

Wells, M.L., J.F. O'Leary, J. Franklin, J. Michaelsen, and D.E. McKinsey. 2004. Variations in a regional fire regime related to vegetation type in San Diego County, California (USA). Landscape Ecology 19: 139–152.

Westerling, A.L., A. Gershunov, D.R. Cayan, and T.P. Barnett. 2002. Long lead statistical forecasts of area burned in western U.S. wildfires by ecosystem province. International Journal of Wildland Fire 11: 257–266.

Zavitokovski, J., and M. Newton. 1968. Ecological importance of snowbrush, *Ceanothus velutinus*, in the Oregon Cascades. Ecology 49: 1134–1145.

Zedler, P.H. 1977. Life history attributes of plants and the fire cycle: a case study in chaparral dominated by *Cupressus forbesii*. Pages 451–458 in: H.A. Mooney and C.E. Conrad, editors. Proceedings of the Symposium on Environmental Consequences of Fire and Fuel Management in Mediterranean Ecosystems. USDA Forest Service General Technical Report WO-3. Washington, D.C., USA.

———. 1981. Vegetation change in chaparral and desert communities in San Diego County, California. Pages 406–430 in: D.C. West, H.H. Shugart, and D. Botkin, editors. Forest Succession: Concepts and Applications. Springer-Verlag, New York, New York, USA.

———. 1995a. Fire frequency in southern California shrublands: biological effects and management options. Pages 101–112 in: J.E. Keeley and T. Scott, editors. Wildfires in California Brushlands: Ecology and Resource Management. International Association of Wildland Fire, Fairfield, Washington, USA.

———. 1995b. Plant life history and dynamic specialization in the chaparral/coastal sage shrub flora in southern California. Pages 89–115 in: M.T.K. Arroyo, P.H. Zedler, and M.D. Fox, editors. Ecology and Biogeography of Mediterranean Ecosystems in Chile, California and Australia. Springer-Verlag, New York, New York, USA.

Zedler, P.H., C.R. Gautier, and P. Jacks. 1984. Edaphic restriction of *Cupressus forbesii* (Tecate Cypress) in southern California. U.S.A.—A Hypothesis. Pages 237–243 in: N.S. Margaris, M. Arianoustou-Farragitaki, and W.C. Oechel, editors. Being Alive on Land: Tasks for Vegetation Science. Dr. W. Junk Publishers, The Hague, The Netherlands.

Zedler, P.H., C.R. Gautier, and G.S. McMaster. 1983. Vegetation change in response to extreme events: the effect of a short interval between fires in California chaparral and coastal scrub. Ecology 64: 809–818.

Zedler, P.H., and L.A. Seiger. 2000. Age mosaics and fire size in chaparral: a simulation study. Pages 9–18 in: J.E. Keeley, M. Baer-Keeley, and C.J. Fotheringham, editors. 2nd Interface between Ecology and Land Development in California. U.S. Geological Survey Open File Report OFR-00-62. Reston, Virginia, USA.

CHAPTER EIGHTEEN

Southeastern Deserts Bioregion

MATTHEW L. BROOKS, RICHARD A. MINNICH,
AND JOHN R. MATCHETT

> Because of the inescapably close correlation between prevalence of fire and amount of fuel, deserts are characteristically less affected by fire than are most ecosystems . . . however, even though fire frequency and severity may be relatively low in any rating scale, their effect on the ecosystem may be extreme.
>
> HUMPHREY (1974)

Description of Bioregion

Physical Geography

The Southeastern Deserts bioregion (desert bioregion) occupies the southeastern 27% of California (11,028,300 ha, 110,283 km^2, or 27,251,610 ac) (Miles and Goudy 1997) (Map 18.1). The desert bioregion is within the basin and range geomorphic province of western North America, and includes two ecoregional provinces comprised of five ecological sections. The American Semi-Desert and Desert Province (warm deserts) includes the Mojave Desert, Sonoran Desert, and Colorado Desert sections in the southern 83% of the desert bioregion. The Intermountain Semi-Desert Province (cold deserts) includes the Southeastern Great Basin and Mono sections in the northern 17% of the desert bioregion.

The desert bioregion is characterized by isolated mountain ranges with steep slopes separated by broad basins containing alluvial fans, lava flows, dunes, and playas. Elevations range from −86 m (−282 ft) below sea level in Death Valley to 4,342 m (14,246 ft) above sea level in the White Mountains. Soil taxa range widely from hyperthermic or thermic, aridic Aridisols and Entisols in the Colorado, Sonoran, and Mojave Desert sections, to thermic, mesic, frigid, or cryic, aridic, xeric, or aquic Alfisols, Aridisols, Entisols, Inceptisols, Mollisols, and Vertisols in the Mono and Southeastern Great Basin sections (Miles and Goudy 1997). This wide range in geomorphology and soil conditions translates into variable vegetation and fuel types, which include shrublands, grasslands, woodlands, and forests.

Climate

Although frontal cyclones of the jet stream pass through the region during winter (November through April), most of the desert bioregion is arid due to rain shadows of the Sierra Nevada, and the Transverse and Peninsular ranges (chapter 2). Precipitation increases locally with orographic lift in desert ranges, particularly those that rise above 2,000 m (6,096 ft). From July to early September, the region experiences 10 to 25 days of afternoon thunderstorms from the North American monsoon originating in the Gulf of California and Mexico. Thunderstorm cells tend to concentrate over high terrain, especially the eastern escarpments of the Sierra Nevada, Transverse, and Peninsular ranges, the mountains of the eastern Mojave Desert, and the high basin and range terrain between the White Mountains and Death Valley. Average annual precipitation on valley floors ranges from 10 to 20 cm (3.9–7.9 in) in the Mojave Desert and Southeastern Great Basin, to 7 to 10 cm (2.8–3.9 in) in the Colorado and Sonoran deserts. The percentage of annual precipitation falling during summer (May through October) ranges from approximately 20% in the southeastern Great Basin to 40% at the Colorado River in the Sonoran Desert.

Interannual variation in precipitation and fuels is relatively high compared to other California bioregions, resulting in highly variable frequency and extent of fires among years. High precipitation produces fine fuels from annuals and subshrubs that promote fire spread, especially in the warm deserts and low elevations where fuels are otherwise sparse. Prolonged periods of low precipitation cause shrub and tree mortality, which reduces woody fuel moisture and can promote fire spread in the cold deserts where woody fuel cover is relatively high.

The desert bioregion has high seasonal variation in temperature due to its isolation from the stabilizing influence of the Pacific Ocean. Average January temperatures on valley floors range from −3°C to 0°C (27–32°F) in the Southeastern Great Basin to 11°C to 13°C (52–55°F) in the Sonoran and Colorado deserts. July average temperatures on valley floors range from 18°C to 20°C (64–68 °F) in the Southeastern Great Basin to 30°C to 35°C (86–95°F) in the Sonoran and Colorado deserts. Temperature decrease with altitude resulting in decreased evapotranspiration, which coupled with increasing precipitation results in a corresponding increase in woody biomass of ecosystems. Light snowpacks 10–15 cm (3.9–5.9 in) deep can develop in winter, but typically disappear by spring above 2,000 m (6,562 ft), although deeper snow of 100 cm (39.4 in) can persist into the spring in subalpine forests higher than 3,000 m (9,842 ft).

Relative humidity in the afternoon during the summer when most fires occur is very low throughout the desert

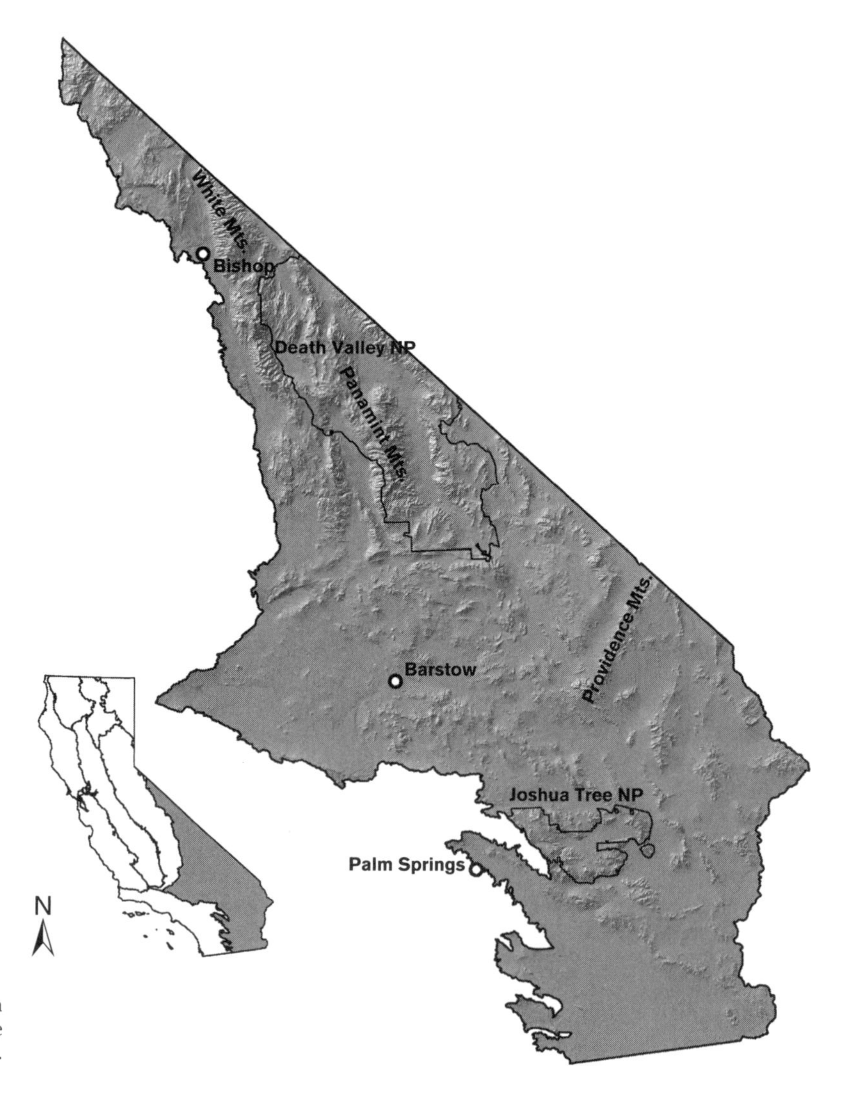

MAP 18.1 The Southeastern Deserts bioregion includes parts of the Great Basin Desert, Mojave Desert, Sonoran Desert, and Colorado Desert.

bioregion. Average relative humidity in July ranges from 20% to 30% in the northeastern Mojave Deserts and from 10% to 20% in the Mojave, Sonoran, and Colorado deserts. The lowest humidity of the year (frequently less than 10%) typically occurs in late June in phase with highest temperatures and highest solar zenith angles, just before the arrival of the North American monsoon.

Lightning frequency is higher in the desert than in any other California bioregion (van Wagtendonk and Cayan 2008). Lightning strikes yr^{-1} 100^{-1} km^{-2} averaged 27 (SD = 16) from 1985 through 2001, ranging from 32 strikes in the Mono to 12 strikes in the Colorado Desert sections (Brooks and Minnich 2006; Table 16.2, van Wagtendonk and Cayan 2008). The bioregions with the next most-frequent lightning strikes were the Northeastern Plateaus (22 strikes yr^{-1} 100^{-1} km^{-2}) and Sierra Nevada (20 strikes yr^{-1} 100^{-1} km^{-2}) bioregions. Most lightning in the desert bioregion occurred from July to September (78%), resulting from the North American monsoon that developed in the Colorado, Sonoran, and eastern Mojave deserts, and from summer storms that developed in the Sierra Nevada mountains and drifted into the southeastern Great Basin and Mono sections. Lightning also occurred primarily during daylight hours, with 81% between 0600 and 1800.

Major Ecological Zones

From a fire regime perspective, much of the variation in the desert bioregion relates to patterns of vegetation (fuels), topography, and lightning occurrence, all which vary locally with elevation. Accordingly, we consider elevation to be the primary determinant of ecological zones in the desert bioregion (Fig. 18.1). The ecological zones described below are listed in order of increasing elevation, except for the desert riparian zone which is not fixed within a particular elevational range. The elevational ecotones between each zone shift downward in elevation as one proceeds from the warm deserts in the south to the cold deserts in the north.

The desert bioregion contains 43 LANDFIRE vegetation biophysical setting types representing burnable wildland areas (www.landfire.gov, Rollins 2009), which were grouped into 13 general vegetation types for the purposes of his chapter (Brooks and Matchett 2016). These burnable wildland areas comprised the majority of this bioregion (82%; 9,025,152 ha [22,301,636 ac]) and were used as the basis for all analyses. Other nonburnable (e.g., open water/barren rock, sand, or clay) and nonwildland (e.g., developed urban/agriculture/

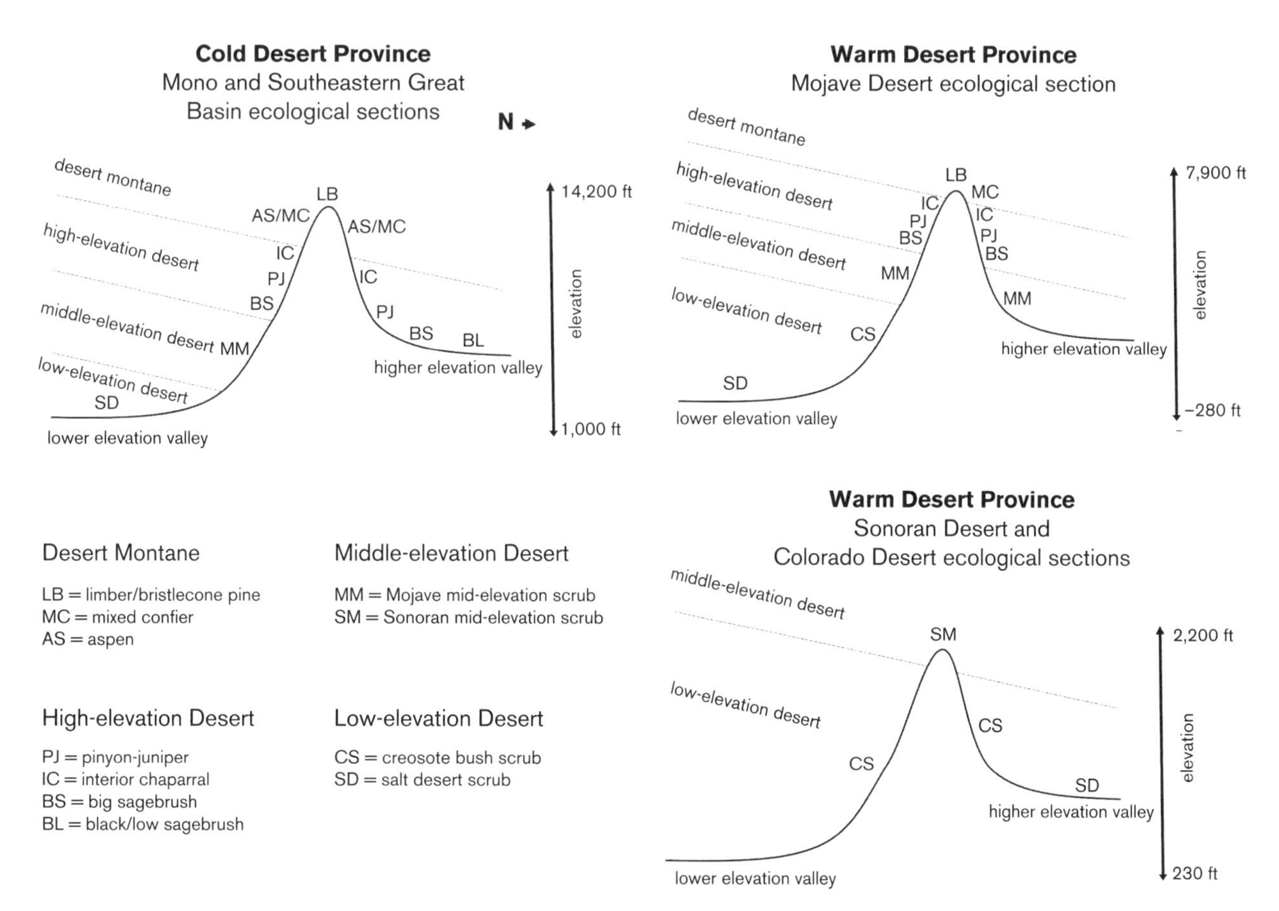

FIGURE 18.1 Elevational characteristics of the desert montane, high-elevation desert, middle-elevation desert, and low-elevation desert zones of the Southeastern Deserts bioregion. The Desert Riparian zone is not represented in this figure because it transcends elevational zones.

roads) areas comprised a smaller percentage of this bioregion (18%; 2,003,148 ha [4,949,887 ac]), and these areas were omitted from analyses.

The dominant vegetation types across the desert bioregion are creosote bush scrub (42%), salt desert scrub (20%), Mojave mid-elevation scrub (17%), and riparian (6%) (Table 18.1). The proportions of dominant types vary greatly across this vast bioregion, especially between the warm and cold deserts. Understanding how vegetation varies is critical for interpreting how fire regimes and fire effects vary. To facilitate this understanding, the 13 general vegetation types were assigned to each of four elevation-based ecological zones plus one riparian zone according to their constituent plant associations (Brooks and Matchett 2016). The resulting five ecological zones were then intersected with the boundaries of the five ecological sections in the desert bioregion to create a map (Map 18.2) and associated statistics for burnable wildland area (Table 18.1). A diagram was also created illustrating the relative elevational positions of each ecological zone and vegetation type along a latitudinal gradient from the warm to cold deserts (Fig. 18.1).

The low-elevation desert zone characterizes most of the burnable wildland area in the desert bioregion (5,579,096 ha [13,786,246 ac], 62% of total) (Table 18.1). It occurs at the lowest elevations within each of the five ecological sections (Fig. 18.1) and is the predominant zone in the Mojave and Sonoran ecological sections (Map 18.2).

The middle-elevation desert zone comprises the second most burnable wildland area in the desert bioregion (2,105,171 ha [5,201,991 ac], 23% of total area) (Table 18.1). It occurs at the middle elevations within the Mono, Southeastern Great Basin, and Mojave ecological sections and at the highest elevations within the Sonoran and Colorado ecological sections (Fig. 18.1) and is the predominant zone in the Southeastern Great Basin (Table 18.1) (Map 18.2).

The high-elevation desert zone comprises the third most burnable wildland area in the desert bioregion (738,945 ha [1,825,973 ac], 8% of total area) (Table 18.1). It is the dominant ecological zone in the Mono ecological section (Table 18.1) (Map 18.2) where it occurs between the low-elevation and desert montane zones (Fig. 18.1). In the Southeastern Great Basin and Mojave ecological sections, it occurs at the tops of most mountain ranges and below the desert montane zone in the highest ranges (Map 18.2). It also occurs along the margins of the Sierra Nevada and the Transverse and Peninsular mountain ranges.

The desert montane zone is limited in total area, occupying the least amount of area in the desert bioregion (85,430 ha [211,102 ac], 1% of total area) (Table 18.1). It occurs at the tops of some mountains in the Mojave and Southeastern Great Basin ecological sections (Map 18.2), but primarily in the Mono ecological section, where it occupies 11% of the total burnable wildland area (Table 18.1).

The desert riparian zone comprises the fourth largest area in the desert bioregion (516,512 ha [1,276,328 ac], 6% of total

TABLE 18.1
Burnable wildland area within each ecological section and constituent ecological zones and dominant vegetation types in the Southeastern Deserts bioregion

Ecological section[a]	Area (ha)[b]	Percentage of bioregion	Ecological zones[c] (percentage of ecological section)					Dominant vegetation types[d] (percentage of ecological section)
			Low	Middle	High	Montane	Riparian	
Mono	723,283	8	16	3	67	11	3	Big sagebrush (42), salt desert scrub (16), black/low sagebrush (14), pinyon-juniper (11), mixed conifer (8)
SE Great Basin	1,078,597	12	36	42	15	<1	7	Mojave mid-elevation scrub (42), creosote bush scrub (24), salt desert scrub (12), pinyon-juniper (7), riparian (7)
Mojave	5,828,432	65	74	19	2	<1	6	Creosote bush scrub (52), salt desert scrub (22), Mojave mid-elevation scrub (18), riparian (6)
Sonoran	1,011,885	11	62	31	<1	0	7	Creosote bush scrub (41), Sonoran mid-elevation scrub (30), salt desert scrub (21), riparian (7)
Colorado	382,955	4	42	52	<1	0	6	Sonoran mid-elevation scrub (50), salt desert scrub (27), creosote bush scrub (15), riparian (6)
Ecological sections combined	9,025,151	100	62	23	8	1	6	Creosote bush scrub (42), salt desert scrub (20), Mojave mid-elevation scrub (17), riparian (6)

NOTES:
a. Miles and Goudy (1997).
b. Area only includes that within the five ecological zones and does not include the category of "omitted" as defined in Brook and Matchett (2016).
c. Low-elevation desert, middle-elevation desert, high-elevation desert, desert montane, desert riparian (see detailed descriptions in the text).
d. Vegetation types comprising ≥5% of the ecological section.

area) (Table 18.1). It is equally represented at 6–7% of total burnable wildland area among all ecological sections except the Mono, where it only represents 3%.

Overview of Historic Fire Occurrence

The predominant view of fire in the desert bioregion has traditionally been focused on the Mojave ecological section and the low- and middle-elevation zones, which encompass most of the bioregion (Table 18.1). The paradigm has been that fires were historically infrequent, all fires are detrimental because native species are generally not adapted to them, and significant management responses are often required to mitigate their negative effects on natural resources. Although this view of fire is plausible across most of the desert bioregion, some areas have differing fuel and fire history characteristics that lead to alternative implications for fire effects and management. Understanding how fire history has varied across the entire desert bioregion and among the ecological zones is essential for evaluating the causes of recent fire patterns and trends and placing them within the appropriate evolutionary and land management context.

Prehistoric Period

Fire occurrence during the prehistoric period is inferred primarily from Holocene climate and vegetation data derived from packrat middens, Sierra Nevada glaciers, shoreline elevations of prehistoric lakes, and oxygen stable isotope records. Direct evidence of prehistoric fire is very limited in desert bioregion, largely because the usual tools for reconstructing fire histories, such as analyzing trees for fire scars or coring sediments in swamps or lakes for charcoal deposits, are not readily applied where forests and lakes are scarce. Anthropological descriptions of how indigenous humans utilized fire in desert regions are also used to infer their potential effects on fire occurrence.

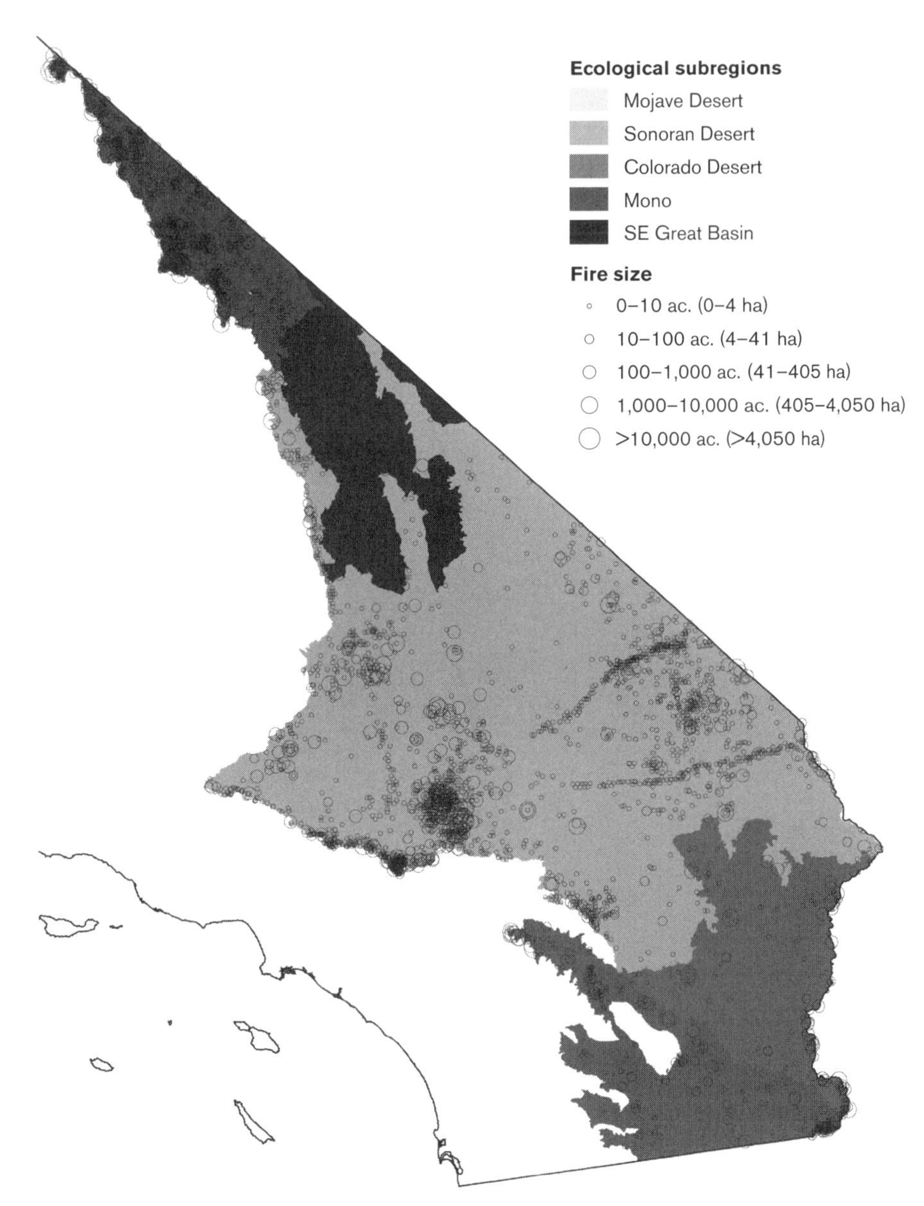

MAP 18.2 Ecological sections and zones within the Southeastern Deserts bioregion.

CLIMATE EFFECTS

The deserts of North America began to develop during the early Pleistocene (2 million years ago), and possibly as early as the Miocene (5 million years ago), when uplifting mountain ranges established rain shadows reducing precipitation from storms moving eastward from the Pacific Ocean into the western United States (Axelrod 1995, Minnich 2007). During the glacial cycles of the Pleistocene, climate fluctuated between hotter/drier and colder/wetter periods, during which the ecotones between ecological zones shifted upward and downward in elevation within the desert bioregion. As the ecological zones and their constituent vegetation types shifted, so too did their characteristic fire regimes.

The most recent major ecotonal shift was upward in elevation as conditions became hotter/drier at the transition from the Pleistocene to the Holocene approximately 12,000 years ago (Van Devender 1977, Van Devender and Spaulding 1979, Koehler et al. 2005, Minnich 2007). Perhaps the most dramatic example is from the central Mojave Desert, which is currently one of the most arid regions in North America. Prior to the dawn of the Holocene, landscapes above 1,000 m (3,280 ft) in that region were characterized by high-elevation zone vegetation types dominated by singleleaf pinyon pine (*Pinus monophylla*), Utah juniper (*Juniperus osteosperma*), Mexican cliffrose (*Purshia mexicana*), curl-leaf mountain-mahogany (*Cercocarpus ledifolius* var. *ledifolius*), and desert almond (*Prunus fasciculata*) (Koehler et al. 2005). Currently, these elevations are characterized by low- and middle-elevation zone vegetation types, dominated by creosote bush (*Larrea tridentata*) and blackbush (*Coleogyne ramosissima*), two species that did not become abundant in the central Mojave Desert until approximately 4,500 and 3,000 years ago, respectively (Koehler et al. 2005, Fig. 2). Similarly, at the beginning of the Holocene at the regional ecotone between the Great Basin and Mojave Desert, the lower limit of pinyon-juniper woodlands was approximately 1,000 m (3,280 ft), which is 850 m (2,789 ft) lower than its contemporary lower limit (Spaulding 1990). Nearby in the Sheep Range of the eastern Mojave Desert, mesic juniper woodlands were replaced by xeric juniper woodlands 9,500 years ago at middle elevations, which were then replaced by blackbush 7,500 years ago and remain blackbush today. At higher elevations, bristlecone pine forests from the last glacial period were replaced by xeric juniper woodlands approximately 9,500 years ago and still persist.

Although conditions are currently hotter/drier than at the beginning of the Holocene, there have also been less dramatic variations in climate that occurred during the course of this epoch (Koehler et al. 2005). The two notable periods when the climate temporarily shifted back toward colder/wetter conditions were the Neoglacial (5,000 to 4,000 years ago) and the Little Ice Age (600 to 300 years ago) (Minnich 2007). These two time periods were also associated with increased fire occurrence at a site in the western Mojave Desert based on evidence from a seismic trench study conducted on the Garlock fault (McGill and Rockwell 1998). Charate (burned wood fragments) was collected and dated using radiometric carbon methods along a depositional profile, and although this was specifically done to document seismic events, it also documented fire occurrence within a local (250 ha [618 ac]) watershed. Charate fragments were found mostly in sediment deposited during the Neoglacial and Little Ice Age, indicating that fire occurrence was concentrated during these periods. Although these results do not provide evidence of discrete fires or fire return intervals, they do indicate that charate flux (average rates of burning per unit area over time) was positively associated with 100- to 1,000-year precipitation peaks during the Holocene.

The past vegetation conditions associated with increased charate flux at the McGill and Rockwell (1998) study site can be attributed to two likely factors. One factor is that colder/wetter conditions led to increased volume and cover of perennial plants and dominance of species characteristic of the middle-elevation zone which currently occurs within the upper reaches of this watershed. These vegetation shifts would have increased fuelbed continuity and biomass and the potential for fire (Brooks and Matchett 2006, Brooks et al. 2015). These types of vegetation shifts have been observed over much shorter multi-decadal time scales during the 1900s, so it's highly plausible that they occurred during past multi-century time scales. For example, a doubling of perennial biomass and cover and shifts in species composition warranting reclassification of plant community types were associated with increased precipitation during the middle to late 1900s in the northern Mojave Desert (Webb et al. 2003). Similar changes have been observed when comparing matched photos from a dry early-1900 period, with a wetter late-1900 period, throughout the Mojave Desert (K. Berry, US Geological Survey, Box Springs, California, unpublished data).

Another likely factor contributing to increase charate flux is the association of long-term periods of higher precipitation with increased frequency of high-precipitation years (Tagestad et al. 2016) which lead to high production of annual plants that fill interspaces between perennial plants, thereby promoting fire spread. Although nonnative annual grasses are well-known for their biomass pulses that promote fire following high-precipitation years in modern desert landscapes (Brooks and Pyke 2001, Balch et al. 2013, Brooks et al. 2016), native herbaceous species can also produce large amounts of ephemeral fuels that promote fire. For example, extremely high winter precipitation during 2004–2005 produced 907–1,814 kg ha^{-1} (809–1,618 lb ac^{-1}) of native annual biomass that helped promote fires during that summer in the western Mojave Desert (Minnich 2008). Specific low- to middle-elevation zone fires that have been facilitated by native herbaceous fuels include the Landers Fire (32,375 ha [80,000 acre]) fueled largely by desert fiddleneck (*Amsinckia tessellata*) in the western Mojave Desert (R. Minnich, University of California, Riverside, USA, personal observation), a fire in Joshua Tree National Park (1,619 ha [4,000 ac]) fueled by threeawn (*Aristida* spp.) and grama (*Boutelua* spp.), and the King Valley Fire (11,700 ha [28,911 ac]) facilitated by desert Indianwheat (*Plantago ovata*) in the Sonoran Desert of southwestern Arizona (Esque et al. 2013).

HUMAN EFFECTS

Indigenous humans have used fire since they arrived in California 12,000 years ago (chapter 19). Fire was used for hunting game, clearing vegetation, managing pest species, and facilitating the growth of vegetation possessing desirable properties. Examples include the use of fire in the cold deserts as part of strategies for hunting deer and managing pinyon pine nut resources. In the warm deserts, fire was used to help convert Colorado river floodplains to farming, promote water yield, manage palm oases, and promote growth of food and fiber plants at springs and seeps (e.g., mesquite [*Prosopis* spp.] and willow [*Salix* spp.]). With increased aridity, decreased productivity, and decreased human presence, especially since the Pleistocene-Holocene transition, the spatial extent of anthropogenic fire undoubtedly declined across the desert bioregion, although perhaps less so in riparian areas and oases.

Historic Period

The historic period began in the mid-1800s with the first Euro-American settlers who produced photographs and written accounts of fire. The written information is focused primarily on large fire years that attracted enough attention to warrant writing a report or newspaper article. Historical photographic data are spatiotemporally discontinuous and their use for documenting fires is mostly opportunistic, because photographs were typically created for other purposes. Thus, although there is some direct evidence of historical fire, the record is incomplete.

CLIMATE EFFECTS

The first direct accounting of extensive historic fire in the desert bioregion estimated that 20% of the 161,875 ha (400,000 ac) of blackbush that occurred in the eastern Mojave Desert of southern Nevada and eastern California burned during the late 1930s and early 1940s (Croft 1950). Analyses of historic aerial photos from Joshua Tree National Park also indicate that there were periodic 121 ha (300 ac) to 607 ha (1,500 ac) acre fires in middle- to high-elevations zones prior to 1942 at the nexus between the Mojave, Colorado, and Sonoran deserts (Minnich 2003). These fires occurred following a multi-decadal period of high rainfall that began in 1905 and was punctuated by very high rainfall years from 1938 to 1941, the last year of which was one of the highest precipitation years on record, eclipsed only by precipitation totals during 1983 and 2005 (Hereford et al. 2006, Tagestad et al. 2016, National Climate Data Center—www.ncdc.noaa.gov). Land management agency reports (Croft 1950, Dimock 1960, Holmgren 1960) and archived oblique photographs from southern Nevada (Bureau of Land Management, Caliente, Nevada, USA) clearly indicate that the nonnative annual grass red brome (*Bromus madritensis* subsp. *rubens*) was a major landscape feature that likely facilitated fire spread during this period (Brooks et al. 2007, Brooks et al. 2013a) (for a visual example see Brooks et al. 2016, Fig. 2.6). It appears that multiple years of above-average precipitation promoting perennial plant growth and fuel accumulation, capped by the highest rainfall years producing abundant fine fuels, were the cause for these early-century fires.

During the mid-century drought from 1942 to 1975 there were relatively few fires documented in the Mojave Desert (Minnich 2003, Brooks et al. 2007, McKinley et al. in press). At Joshua Tree National Park only three small fires were documented during this time period, all of which occurred in Joshua tree (*Yucca brevifolia*) woodlands of the middle-elevation zone (Minnich 2003). This has been attributed to lower fuel loads and fuelbed continuity and less prevalent monsoonal conditions (with its high winds and lightning) during the middle decades of the 1900s (Tagestad et al. 2016, McKinley et al. in press).

HUMAN EFFECTS

When Euro-Americans began occupying the desert bioregion in the mid-1800s they brought with them livestock, which are known to reduce perennial plant cover, especially cover of perennial grasses (Brooks and Berry 2006). These effects on vegetation may have led to reduced flame lengths and fire spread rates, at least in the cold desert regions (Diamond et al. 2009). At the same time that fuel loads were reduced due to livestock grazing, ignition rates by indigenous humans likely declined as their traditional practices of burning were curtailed by settlers and the US military (e.g., in the high-elevation zone; Biondi et al. 2011). However, settlers also set their own fires in attempts to convert shrublands into grasslands and increase forage production for livestock, especially in the middle- and high-elevation zones of the Southeastern Great Basin and Mojave Desert sections (Croft 1950, Holmgren 1960, Minnich 2003, Brooks et al. 2007). Thus, the net effect of settlement was a reduction in fire use by indigenous human and an increase in fire use by Euro-Americans.

Current Period

The current period began in the 1970s as the mid-century drought ended, perennial vegetation biomass began to slowly increase, and annual biomass increased tremendously in response to years of high precipitation. It was these fine fuels that were associated with one of the first large fires of the current period, a 2,428 ha (6,000 ac) fire during 1978 in the middle- and high-elevation zones at Joshua Tree National Park (Minnich 2003). This fire was fueled largely by old stands of pinyon-juniper woodlands, shrubs, and perennial grasses, but fire spread was facilitated by nonnative annual grasses red brome and cheat grass (*Bromus tectorum*), especially where fire passed through previously burned areas where cover of these grasses was very high according to National Park Service fire reports. Thus, it did not take very long after the mid-century drought for vegetation to produce fuels sufficient to carry fire.

SUMMARY OF AGENCY FIRE RECORDS

Land management agency records that contain fire point occurrence information have been used to describe current fire patterns in the desert bioregion. These studies have alternatively focused on the warm deserts 1980 to 1995 (Brooks and Esque 2002), the Mojave Desert 1980 to 2004 (Brooks and Matchett 2006), National Park Service units 1992 to 2011 (Hegeman et al. 2014), and the entire desert bioregion 1980 to 2001 (Brooks and Minnich 2006; Table 16.2). Although these point occurrence databases include fires of all sizes, they underrepresent nonfederal lands and can contain a high degree of error (up to 30% for Department of the Interior lands, Brown et al. 2002).

The current fire history data summarized below and in Table 18.2 were derived using Landsat satellite imagery from the Monitoring Trends in Burn Severity (MTBS) program to precisely document fire area (area within fire perimeters) for fires ≥405 ha (1,000 ac) between 1984 and 2013 (www.mtbs.gov; accessed June 30, 2015). Although these large MTBS fires only represent 1–2% of the total number of fires in a given area, they comprised 93% of the total fire area in a recent study from the Mojave Desert (McKinley et al. in press) and approximately 95% of fire area across the western United States (Eidenshink et al. 2007). In addition, the slight underestimate (~5%) of actual fire area due the focus on fires ≥405 ha (1,000 ac) is roughly offset by a similar overestimation of actual burned area within MTBS perimeters due to the presence of unburned inclusions as reported for large expanses of the cold deserts in western North America (Kolden and Weisberg 2007, Brooks et al. 2015).

Fire area encompassed 156,598 ha (386,962 ac) of the 9,025,151 ha (22,301,634 ac) burnable wildland area in the desert bioregion from 1984 to 2013 (Table 18.2). Burning at a rate of 5,220 ha yr^{-1} (12,899 ac yr^{-1}), the estimated time to burn an area equal to the entire bioregion (the fire rotation) was 1,729 years. In contrast, more than twice as much fire area (404,761 ha [1,000,186 ac]) burned during a 36-year period (1972 to 2007) within the warm/cold desert ecotone in Clark and Lincoln counties of southern Nevada (McKinley et al. in press, Table 2), and that region was only half the size (53%, 4,439,476 ha [10,970,184 ac]) of the California desert bioregion. The southern Nevada region burned at a rate of 13,492 ha yr^{-1} (33,339 ac yr^{-1}) resulting in a fire rotation estimate of 430 years which is much lower than the 1,964 and 2,359 year rotations in the ecologically similar Mojave and Southeastern Great Basin sections of California (Table 18.2). This difference is largely attributed to higher precipitation in Nevada, including a significant amount associated with the lightning and high winds of the North American Monsoon (Tagestad et al. 2016, McKinley et al. in press).

Among the ecological sections in the desert bioregion, the majority of the fire area occurred within the Mojave (57% of total fire area) and Mono (30%) sections, with much less in the Southeastern Great Basin (9%), and the least amount in the Colorado (2%) and Sonoran (2%) sections (Table 18.2). These fire area results were generally similar to those derived using fire point occurrence data from 1980 to 2001 (Brooks and Minnich 2006; calculated from Table 16.2). Most of the fire area occurred in the middle- and high-elevation zones (each 38% of total fire area), an intermediate amount in the

TABLE 18.2
Fire area and fire rotation in each ecological section and ecological zone of the Southeastern Deserts bioregion, 1984–2013

Ecological section[a]	Ecological zones[b]					Ecological zones combined
	Low	Middle	High	Montane	Riparian	
Mono						
Total Fire Area (ha)	2,538	231	38,319	6,373	553	48,014
Mean Annual Fire Area (ha)	85	8*	1,277	212	18	1,600
Fire Rotation (years)	1,354	2,990	379	382	1,125	452
SE Great Basin						
Total Fire Area (ha)	189	10,216	1,695	6	1,610	13,716
Mean Annual Fire Area (ha)	6	341	57	<1	54	457
Fire Rotation (years)	61,918	1,315	2,818	17,507	1,467	2,359
Mojave						
Total Fire Area (ha)	15,859	48,921	19,252	61	4,821	89,013
Mean Annual Fire Area (ha)	529	1,631**	642	314	164	2,967
Fire Rotation (years)	8,111	687	149	314	1,975	1,964
Sonoran						
Total Fire Area (ha)	742	78	0	0	2,033	2,854
Mean Annual Fire Area (ha)	25	3	0	0	68	95
Fire Rotation (years)	25,341	120,154	NA	NA	1,051	10,636
Colorado						
Total Fire Area (ha)	2,416	88	2	0	495	3,000
Mean Annual Fire Area (ha)	81	3	<1	0	16	100
Fire Rotation (years)	1,997	68,075	7,386	NA[c]	1,333	3,829
Ecological Sections Combined						
Total Fire Area (ha)	21,744	59,535	59,263	6,440	9,611	156,598
Mean Annual Fire Area (ha)	725	1,984	1,976	215	320	5,220
Fire Rotation (years)	7,697	1,061	374	398	1,612	1,729

NOTES:
a. Miles and Goudy (1997).
b. Low-elevation desert, middle-elevation desert, high-elevation desert, desert montane, desert riparian (see detailed descriptions in the text).
c. NA = not applicable because there was not enough fire area.
* $p \leq 0.05$ one-tailed test for increasing trend in annual fire area.
** $p \leq 0.10$ one-tailed test for increasing trend in annual fire area.

low-elevation zone (14%), and the least amount in the desert riparian (6%) and desert montane (4%) zones (Table 18.2).

Fires in the warm deserts of California are clustered in regional hot spots where they are more frequent and burn more proportional area than desert-wide averages (Brooks and Esque 2002). These areas all occur in the Mojave ecological section, with one hot spot at the ecotone with the Colorado section in the vicinity of Joshua Tree National Park (Brooks and Esque 2002, Fig. 3). Detailed analyses of fire occurrence 1972 to 2007 in the Mojave Desert of southern Nevada displayed similar regional clustering of fires (McKinley et al. in press). These results indicate that although fire occurrence across large parts of the warm deserts may be relatively low, they can be much higher and pose significant land management challenges in localized areas.

The majority of fire area in the Mojave section of California occurred in the middle-elevation zone (48,912 ha [593,480 ac], 55%) (Table 18.2), the same zone that contained most of the fire area in another study encompassing the entire Mojave Desert 1980 to 2004 (240,173 ha; 88%) (Brooks and Matchett 2006). In the middle elevations of the Mojave Desert there was also evidence of a significant increase in annual fire area

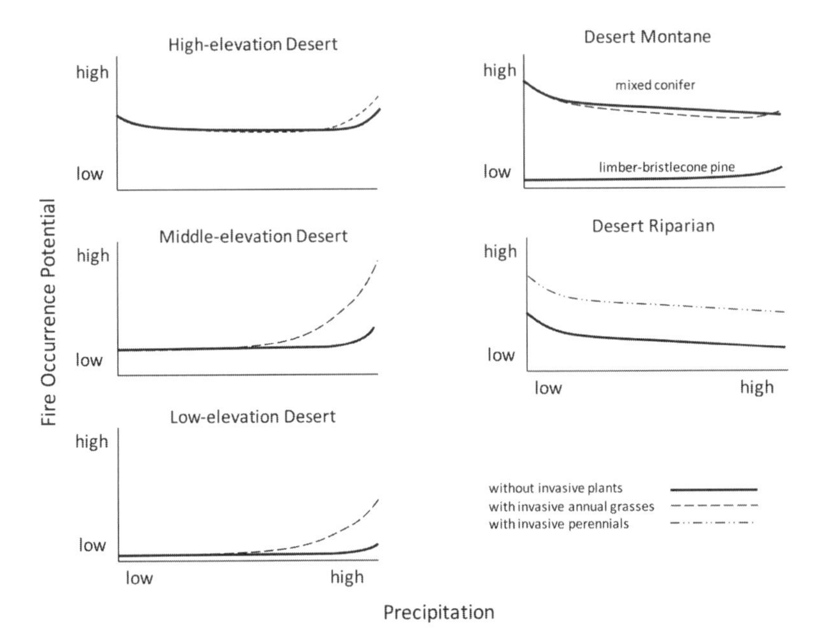

FIGURE 18.2 Relationships between precipitation and fire-occurrence potential in the ecological zones of the Southeastern Deserts bioregion.

($p \leq 0.10$) (Table 18.2). Another analysis of point occurrence fire data in the warm deserts found that annual fire frequency increased significantly from 1980 to 1995 ($r^2 = 0.27$) (Brooks and Esque 2002), but the increase was only significant in the low- and middle-elevation zones below 1,280 m (4,200 ft) ($r^2 = 0.32$, 1980 to 2001). The increase in fire frequency was associated with increased numbers of fires caused by humans, because the number of lightning-caused fires remained constant (Brooks and Esque 2002). Although most human fires were small and started along roadsides, the less frequent large fires typically occurred in remote areas far from major roads, and were started by lightning.

Fires in the cold deserts were mostly within the Mono section (48,014 ha [118,645 ac], 78% of cold desert fire area) and specifically within the high-elevation zone (38,319 ha [94,688 ac], 80% of section fire area) (Table 18.2). In contrast, fire area was much less within the Southeastern Great Basin section (13,716 ha [33,892 ac], 22% of cold desert fire area), where it mostly occurred within the middle-elevation zone (10,216 ha [25,244 ac], 75% of section fire area).

The fire rotation estimate of 374 years in the high-elevation zone was the lowest of all ecological zones (Table 18.2), reflecting the relatively high fire-occurrence potential of this area (Fig. 18.2). This rotation value falls within the general range of historical fire rotations in Wyoming big sagebrush (*Artemisia tridentata* subsp. *wyomingensis*) (171–342 yr) and mountain big sagebrush (*A. tridentata* subsp. *vaseyana*) (137–217 yr) landscapes of the western United States (Bukowski and Baker 2013). It is similar to estimates of current fire rotations in big sagebrush (150 yr) and black/low sagebrush (435 yr) from the southern Great Basin (Brooks et al. 2015). Fire rotation of 398 years was also relatively low in the desert montane zone (Table 18.2). Comparison with other published values are problematic because this zone includes both mixed conifer (69% of area) and limber (*Pinus flexilis*)/bristlecone pine (*P. longaeva*) (20% of area), two vegetation types with very high and low fire-occurrence potentials, respectively (Fig. 18.2).

Fire rotation was intermediate for the middle-elevation zone (1,061 yr) (Table 18.2). The Mojave mid-elevation scrub that predominates in this zone (73%) is dominated by blackbush, which is thought to take many hundreds of years to recover after burning (Bowns 1973, Webb et al. 1988). This zone may be the most susceptible to type conversion via the grass/fire cycle (sidebar 18.1). Fire-occurrence potential can increase dramatically following years of high precipitation and ensuing high fuel loads of nonnative annual grasses (Fig. 18.2). The high fire rotation value we report suggests that most of middle-elevation zone burned close to its historical rate at the scale of the desert bioregion, although it clearly burned at higher rates within localized areas (Brooks and Esque 2002, Brooks et al. 2013a, McKinley et al. in press). Fire rotation was also intermediate for the desert riparian zone (1,612 yr) but this value may be skewed high due to the propensity for riparian fires to be smaller than the 405 ha (1,000 ac) lower threshold for fires in the databases used to generate fire statistics in Table 18.2. Fire rotation was longest in the low-elevation zone (7,697 yr) (Table 18.2), on par with historical estimates (Brooks and Minnich 2006; Table 16.2). This zone is susceptible to increased fire area following years of high precipitation and annual grass fuel production (Fig. 18.2).

THE EXTREME FIRE YEARS OF 2005 AND 2006

The summer of 2005 experienced 385,357 ha (952,238 ac) of total fire area, an amount equal to 132% of the cumulative fire area during the previous 25 years in that region (Brooks and Matchett 2006). The majority occurred in southern Nevada, where 249,219 ha (615,833 ac) burned, which was 51.63 SD units above the long-term (1972–2007) annual mean of fire area for that portion of the Mojave Desert (McKinley et al. in press). The 2006 fire season was also notable in southern Nevada, when 50,534 ha (124,871 ac) burned, which was 9.82 SD above the long-term annual fire area mean. These fires occurred toward the end of a multi-decadal period of high precipitation that began in 1976, and immediately following two winters of very high precipitation in 2004 and 2005 that

SIDEBAR 18.1 NONNATIVE ANNUAL GRASSES AND ALTERED DESERT FIRE REGIMES

Matthew L. Brooks

Nonnative annual grasses have become dominant components of many ecosystems in western North America during the twentieth century. These invasions have negatively affected native plant species through competition for limited soil nutrients and water, and by altering ecosystem properties such as fuel characteristics and fire regimes. The positive feedback between nonnative grasses and fire, or the grass/fire cycle (D'Antonio and Vitousek 1992), is the most clearly understood and well-documented example of the many ways that plant invasions can alter fire regimes (Brooks et al. 2004, Brooks 2008).

Mediterranean split-grass (*Schismus arabicus* and *S. barbatus*) is the dominant nonnative annual grass where soil moisture is lowest and temperature is highest at low elevations of the warm desert province (Brooks 2000, Brooks and Chambers 2011). Red brome (*Bromus madritensis* susp. *rubens*) dominates where soil moisture and temperature are moderate at middle elevations of the warm desert and low elevations of the cold desert province (Brooks et al. 2016). Cheat grass (*B. tectorum*) dominates where soil moisture is highest and temperature is lowest at the high elevations of the warm desert and middle to high elevations of the cold desert province. All of these species, plus ripgut brome (*B. diandrus*), can be codominants in desert riparian areas. These annual grass species alter desert fire regimes by increasing fuelbed continuity and flammability which can lead to increased frequency and size of fires, further promoting the dominance of these species (Brooks 2008, Brooks et al. 2016). Grass/fire cycles represent ecosystem shifts to alternative stable states that likely persist unless fuels, climate, or ignition patterns significantly change (Brooks et al. 2004, Brooks 2008).

Increased fire frequencies beyond their historical range and variation can have dramatic and far-reaching ecological effects. For example, invasion of cheat grass has shortened fire return intervals from a range of 30 to 100 years down to only 5 years in the Great Basin of the cold desert province (Whisenant 1990). This new fire regime promotes the dominance of cheat grass over native species, resulting in conversion of large expanses of high-diversity, native sagebrush shrubland and steppe to low-diversity, nonnative annual grassland (Brooks and Pyke 2001, Balch et al. 2013, Chambers et al. 2014, Brooks et al. 2016). These habitat conversions have negatively affected animals that require sagebrush steppe for forage and cover such as the greater sage-grouse (*Centrocercus urophasianus*) and Cassin sage sparrow (*Amphispiza belli*), and prey species such as black-tailed jackrabbits (*Lepus* spp.) and the Paiute ground squirrel (*Spermophilus mollis*), which are important for golden eagles (*Aquila chrysaetos*) and prairie falcons (*Falco mexicanus*) (Knick and Rotenberry 1995, Knick et al. 2003, Earnst and Holmes 2012). Although similar large-scale higher-order effects have been less documented in the warm desert province, grass/fire cycles have clearly degraded habitat in some areas by reducing the abundance of native plants that provide forage and thermal cover for the desert tortoise (*Gopherus agassizii*) (Brooks and Esque 2002; sidebar 18.3).

Nonnative annual grasses such as cheat grass persist in the cold desert province because precipitation is typically sufficient to support reproduction during any given year (Brooks and Chambers 2011, Brooks et al. 2016). Although the fuelbeds they create may only significantly affect fire behavior following years of high precipitation, their populations typically persist even during years of low precipitation. As a result, nonnative grasses and the altered fire regimes they cause are now relatively permanent features in many parts of the cold desert province.

The hot desert province receives less annual precipitation than the cold desert province, increasing the chances of population reductions of nonnative annual grasses such as cheat grass and red brome because soil moisture and temperature conditions are at the edge of the biophysical ranges for these species (Brooks et al. 2016). Precipitation events as small as 5 mm (2 in) can stimulate their germination, and when there is scant subsequent precipitation, the plants often die before reproducing (Minnich 2008, M. Brooks, personal observation), potentially depleting their soil seed bank. This is probably why red brome was not detected at two low-elevation sites immediately after the late 1980s drought and at one low-elevation site after the late 1990s drought (Minnich 2008). However, red brome remained detectable at many higher-elevation sites following these same drought periods, in some cases surviving to reproduce in more mesic microsites such as beneath the north side of shrub canopies and in rock crevices (M. Brooks, personal observation). It should also be noted that even when nonnative annual grasses represent <1% vegetative cover and may be overlooked, they can still be a dominant component of the soil seed bank (3,259 m^{-2} [35,080 ft^{-2}]) (Brooks et al. 2013b), making it difficult to detect and evaluate their population trends.

Although populations of nonnative annual grasses can be depleted by drought, they can recover to ecologically significant numbers relatively quickly. For example, red brome was a dominant feature of the Mojave Desert during the multi-decadal period of high precipitation in the early 1900s, but then declined significantly in abundance during the mid-century drought (Holmgren 1960, Brooks et al. 2007, Brooks et al. 2016). It only took one year between the last year of the drought (1975) and the first year of higher precipitation (1976) for red brome density and biomass to rebound 700% and 150%, respectively, at a Mojave Desert/Great Basin ecotone in southern Nevada (Hunter 1991). By 1988 the increase above 1975 levels reached 15,646% for density and 1,596% for biomass. During this time interval, density and biomass of native annuals decreased (Hunter 1991), whereas the frequency and size of fires across the Mojave Desert increased (Brooks and Esque 2002). In addition, the shorter 1987–1991 drought was followed by a significant fire year in 1993 (Brooks 2012), and the spread of most of these fires was facilitated by substantial fine fuelbeds of red brome and cheat grass (M. Brooks, personal observation). Similarly, the 1999–2002 drought was followed by the two largest fire years ever recorded in the Mojave Desert in 2005, and to a lesser degree 2006 (Brooks et al. 2013b). Although the aboveground dominance of nonnative annual grasses clearly varies more in response to precipitation in the warm deserts than in the cold deserts, and their chances of becoming locally extirpated are greater there, they still persist regionally through droughts and recover rapidly following droughts.

were each > 1 SD above the long-term precipitation average across the entire Mojave Desert.

Although much of the desert bioregion that burned during 2005 was fueled by native annuals in the western Mojave Desert (Minnich 2008), the vast majority of fire area occurred in the eastern Mojave Desert and was fueled mostly by nonnative annual grasses (Bauer et al. 2011). The Hackberry Complex burned 28,697 ha (70,912 ac) over seven days during late June, 2005. Spread rates during this fire reached 35 mi hr^{-1} (56 km hr^{-1}) within middle- and high-elevation zones (C. Heard, Mojave National Preserve, Barstow, California, USA, pers. comm.). The southern Nevada Complex burned 299,078 ha (739,037 ac) over 19 days in late June and early July, 2005. Accounts by agency fire mangers indicate that these fires exhibited extreme fire behavior not previously observed in the Mojave Desert, and they attributed this largely to continuous cover of red brome up to 2 ft (0.61 m) tall in some areas (Bauer et al. 2011, Appendix 1.1).

Lessons Learned from Fire History Information

Fire occurrence has been, and still is, primarily fuel-limited in the desert bioregion. Rates of burning increase following multi-decadal periods of high precipitation when perennial vegetation cover expands. Fire potential can increase even further when one or more years of high precipitation triggers high annual plant growth that can fill interspaces between perennials (Brooks and Matchett 2006, Van Linn et al. 2013, Gray et al. 2014, Hegeman et al. 2014, Rao et al. 2015, Tagestad et al. 2016). The potential is highest when high-precipitation years occur toward the end of a multi-decadal period of high precipitation. Examples include the years 1938 through 1941 and 2005 through 2006. Prediction reliability increases substantially when fire-occurrence models couple winter precipitation with monsoonal summer precipitation (and its associated high winds and frequent lighting strikes) (Tagestad et al. 2016). Nonnative annual grasses are the primary fine fuel component following years of high precipitation, although native annuals can also produce fuels sufficient to carry fire (sidebar 18.1).

Most of the recent fire area in the warm deserts has occurred just to the east of California in the Mojave Desert of southern Nevada, southwest Utah, and northwest Arizona (Tagestad et al. 2016, McKinley et al. in press). It appears that lower combined winter and summer precipitation in the California desert bioregion is the primary reason for lower fire occurrence there during the twentieth century. The conditions immediately east of California may be an analog of more mesic past conditions in the California desert bioregion, possibly as recently as a few hundred years ago during the Little Ice Age (see Prehistoric Period section above). The frequency of years with deviations ≥1 SD above the long-term precipitation mean are predicted to increase during the next century in the Mojave Desert (Tagestad et al. 2016, Fig. 8). If the current positive precipitation/fire relationships continue, then the future potential for fire may increase during the twenty-first century in the desert bioregion.

Major Ecological Zones

In this section we describe the fire ecology of the predominant plant species in each ecological zone. We also discuss patterns of postfire succession, and interactions among plant communities, fire behavior, and fire regimes. More details on the fire ecology of a wider range of desert species can be found on the Fire Effects Information System web site (www.fs.fed.us/database/feis) and in Brown and Smith (2000).

FIGURE 18.3 The low-elevation desert shrubland ecological zone depicted by creosote bush scrub.

Low-Elevation Desert Zone

Vegetation is dominated by shrublands and includes creosote bush scrub (67% of the zone) and salt desert scrub (33%) derived from six biophysical setting types (Brooks and Matchett 2016). Surface fuel loads and continuity are typically low, hindering the spread of fire (Fig. 18.3).

FIRE RESPONSES OF IMPORTANT SPECIES

Most shrub species in the low-elevation zone do not survive after being completely consumed by fire (Humphrey 1974, Wright and Bailey 1982), but fires in this zone are often patchy and low intensity, with plants frequently surviving in unburned islands. Low fire temperatures in interspaces and high temperatures beneath woody shrubs likely result in relatively higher seed bank mortality for annual plants that frequent beneath-shrub than interspace micro-habitats (Brooks 2002). Perennial species that evolved to resprout after floods often do so after burning, especially after low-intensity fires. Such species include desert-willow (*Chilopsis linearis* subsp. *arcuata*), catclaw (*Senegalia greggii*), smoke tree (*Psorothamnus spinosus*), and white bursage (*Ambrosia dumosa*) (Brooks and Minnich 2006; Table 16.3). Cheesebush can have almost 100% survival rates even after being totally consumed by fire (Brooks and Minnich 2006; Table 16.4). Yucca species such as Joshua tree and Mojave yucca (*Yucca schidigera*) often survive burning, but Joshua trees typically die within the first few years after fire due to drought and herbivory stress (DeFalco et al. 2010). Cacti are usually only scorched during fires, as flames propagate through their spines but their stems do not ignite due to their high moisture content. High levels of scorching typically lead to death from uncontrolled desiccation that occurs postfire, whereas partially scorched cacti can survive from rooting of fallen unburned stem fragments.

The most frequently encountered and dominant shrub in this zone, creosote bush, can have 25–80% survival rates eight years postfire when it is only scorched (1–10% biomass loss), and 0–12% survival rates by year eight when it is consumed by fire (11–100% biomass loss) in the Mojave Desert (Brooks and Minnich 2006; Table 16.4). Individuals with slight to moderate scorching had 30–40% survival in the Sonoran Desert (Dalton 1962). Resprouting potential is inversely correlated with fire intensity and duration (White 1968). Large variability in survival rates by creosote bush appears to be associated with its variable physiognomy and variable fuel loads beneath their canopies, which translate into variable fire intensity and vertical continuity from surface to canopy fuels. Individuals with canopies in the shape of inverted cones tend to occur in water-limited environments (De Soyza et al. 1997), resulting in relatively low fuel loads in their lower canopies and relatively low probability of total consumption by fire. In contrast, individuals with hemispherical canopies that extend to the ground tend to occur in less water-limited environments and have higher fuel loads in their lower canopies. This results in higher probability of total consumption by fire. Resprouting rates in creosote bush appear to vary throughout its geographical range, and are more prevalent at ecotones with vegetation types that support more frequent burning. For example, moderate (O'Leary and Minnich 1981, Brown 1984) to high (Brown 1984) rates of postfire resprouting were reported at the ecotone of the western Colorado Desert with the chaparral landscape in the Peninsular Ranges.

FIRE REGIME–PLANT COMMUNITY INTERACTIONS

This is the zone to which Humphrey (1974, p.337) referred when he stated that in desert shrublands ". . . fires are a rarity, and the few fires that do occur cause little apparent damage to the various aspects of the ecosystem " Fuels are discontinuous and characterized by sparse 8–15% cover of woody shrubs, and the large interspaces between shrubs are mostly devoid of vegetation, inhibiting fire spread (Fig. 18.3). A summary of fire regimes of the United States (Schmidt et al. 2002) assumed that Küchler's "barren vegetation type" (Küchler 1964), which covers most of the low-elevation zone, is mostly devoid of vegetation and therefore fireproof. However, 21,744 ha (53,730 ac) (14%) of the total area burned between 1984 and 2013 occurred within the low-elevation zone (Table 18.2). Thus, fires do occur in the low-elevation zone, but at lower frequencies than in the other zones of the desert bioregion (Fig. 18.2).

Fire behavior and fire regimes in this zone are affected primarily by the ephemeral production of fine fuels from annual plants. Years of high winter and spring precipitation can increase continuity of fine fuels by stimulating the growth of annual plants that fill interspaces and allow fire to spread between perennial plants (Brown and Minnich 1986, Rogers and Vint 1987, Schmid and Rogers 1988, Brooks 1999). Among native annuals, those that produce the most persistent interspace fuels include sixweeks grass (*Festuca octoflora*) and small fescue (*F. microstachys*), and the large forbs desert fiddleneck, tansy mustard (*Descurainia pinnata*), and lacy phacelia (*Phacelia tanacetifolia*), compared with a whole suite of smaller native forbs (119 species, Brooks 1999). Other native forbs that have been observed fueling desert fire include chia (*Salvia columbariae*), California suncup (*Eubolus californica*), and desert pincushion (*Chaenactis steveoides*) (Minnich 2008). During rare successive years of high precipitation, native annuals build up fine fuel loads sufficient to carry fire across the interspaces between larger perennial plants thus increasing the potential for fire (Fig. 18.2). Low-elevation fires carried by native annual fuel loads typically only burn dead annual plants and finely textured subshrubs, leaving many of the larger woody shrubs such as creosote bush unburned. Thus, the historic fire regime was likely characterized by relatively small, patchy, low-severity surface fire and a truncated long fire return interval synchronous with long-term climatic cycles (Appendix 1).

The invasion of nonnative annual grasses into the desert bioregion introduced new fuel conditions. Although senesced native annuals can persist to some degree as fuels for over a year (Minnich 2008), nonnative species such as red brome, common Mediterranean grass (*Schismus barbatus*), and Arabian schismus (*S. arabicus*) provide more persistent and less patchy fine fuelbeds, breaking down more slowly and persisting longer into subsequent years (Brooks 1999). These new fuel conditions have the potential to increase the size and shorten intervals between fires (sidebar 18.1), especially following years of high precipitation (Fig. 18.2). These fire regime changes have occurred over a small fraction of the low-elevation zone, and fire regimes over the vast majority of this zone are still within the historic range of variation. Mediterranean grass is the most widespread and abundant nonnative annual grass in the low-elevation zone, although red brome may predominate under large shrubs or in the less arid areas. Mediterranean grass has fueled fires as large as 41 ha (100 ac) (Bureau of Land Management DI-1202 records), and interspace fuel loads of as little as 112 kg ha^{-1} (100 lb ac^{-1}) are sufficient to carry fire (Brooks 1999). Because these fires promoted by Mediterranean grass tend to burn with low intensity, soil heating is negligible and most woody shrubs are left unburned.

The recent spread of Sahara mustard (*Brassica tournefortii*) in the low-elevation zone poses a new fire threat. During years of high precipitation this nonnative annual can exceed 1 m (3.3 ft) in height with a rosette of basal leaves 1 m (3.3 ft) across, and even moderately sized plants can produce as many as 16,000 seeds (Trader et al. 2006). Plants can remain rooted and upright through the summer, and when they finally do break off they blow like a tumbleweed and lodge in shrubs or fences, accumulating piles of fuels similar to Russian thistle (*Salsola tragus*). The combination of senesced Sahara mustard with red brome in the understory helped fuel a 20.2 ha (50 ac) fire in creosote bush scrub in the northeastern Mojave Desert (M. Brooks, personal observation). It affected fire behavior both by augmenting interspace fuel loads and spread rates and by increasing flame lengths where its biomass was highest in the vicinity of individual creosote bushes. During the first years after this fire Sahara mustard and red brome dominated the site while creosote bush failed to show signs of recovery. Thus, Sahara mustard has the potential to increase fire spread rates and intensity, especially when coupled with other fine fuels.

Middle-Elevation Desert Zone

Vegetation is dominated by shrublands and includes Mojave mid-elevation scrub (73% of the zone) and Sonoran mid-elevation scrub (27%) derived from four biophysical settings (Brooks and Matchett 2016). Both of these general vegetation types include the upper elevations of creosote bush scrub and contain higher perennial plant cover than do the lower elevations of this vegetation type (Rowlands 1980). These types also include Joshua tree woodland in the Mojave Desert and Southeastern Great Basin sections, and Shadscale (*Atriplex confertifolia*) scrub and blackbush in the Mono, Southeastern Great Basin, and Mojave Desert sections. Perennial grasses such as sand rice grass (*Stipa hymenoides*), desert needle grass (*S. speciosa*), and big galleta (*Hilaria rigida*) can codominate with major shrub species creating shrub-steppe communities. Surface fuel characteristics are variable, but loads and continuity can be relatively high compared with the low-elevation zone, facilitating the spread of fire (Fig. 18.4).

FIRE RESPONSES OF IMPORTANT SPECIES

Higher fuel loads and more continuous fuelbeds in the middle-elevation zone result in higher-intensity fires and more top-killing, although also more postfire resprouting, than in the low-elevation zone. Perennial grasses such as desert needle grass, galleta (*Hilaria jamesii*), and desert ricegrass readily resprout after burning (Brooks and Minnich 2006; Table 16.3). Spiny menodora (*Menodora spinescens*) and Mormon tea (*Ephedra* spp.) often survive fire because their foliage does not easily burn and they readily resprout. Some shrub species such as blackbush and winterfat (*Krascheninnikovia lanata*) rarely survive burning.

Blackbush is one of the most flammable native shrubs in the desert bioregion due to its high proportion of fine fuels and optimal packing ratio. As a result, once individuals ignite they tend to completely combust and are killed. Historic time-series photographic records from the Grand Canyon (Webb et al. 1988) and southern Nevada (M. Brooks, unpublished data) support the common view that blackbush stands require centuries to recover (Bowns 1973, Webb et al. 1988). However, other time-series photographs from Joshua Tree National Park (Minnich 2003) and southern Nevada (M. Brooks, unpublished data) suggest that blackbush stands might be able to recover their pre-burn cover within 50 to 75 years. This differing evidence is likely due to variation in the degree to which blackbush individuals are consumed when burned, and potentially their local adaptations for resprouting. For example, individuals that were only partially burned during an experimental fire at a middle-elevation site in the Mono ecological section survived or resprouted from the root crown within the first few postfire years (Bates 1984). A revisit of that same site found these same partially burned

FIGURE 18.4 The middle-elevation desert zone depicted by blackbrush.

individuals alive 3 decades later (M. Brooks, personal observation, 2013). Resprouting rates appear to vary across its wide geographic range, which extends from the Colorado Plateau and southern Great Basin on through the Mojave Desert.

Leaf succulents in the *Yucca* genus such as Joshua tree, Mojave yucca, banana yucca (*Yucca baccata* var. *baccata*), and chaparral yucca (*Hesperoyucca whipplei*) are typically scorched as flames propagate through the shag of dead leaves along their trunks. The relatively small size and dense packing ratio of dead Joshua tree leaves compared with dead Mojave or banana yucca leaves increase the frequency at which they are completely burned and may explain why Joshua trees are more frequently killed by fire. All four yucca species readily resprout after fire, but resprouts are often eaten by herbivores (DeFalco et al. 2010). Postfire recruitment of new Joshua trees is infrequent, and likely occurs during years of high precipitation. No seedlings or saplings were observed in burned areas less than 10 years old, and fewer than 10 individuals ha^{-1} (3 individuals ac^{-1}) were present on burned areas more than 40 years old in Joshua Tree National Park (Minnich 2003). Joshua tree populations along the extreme western edge of the desert bioregion near the Sierra Nevada and Transverse Ranges often resprout and survive more readily after fire than those further east (M. Brooks, personal observation). A cycle of relatively frequent fire and resprouting can result in short, dense clusters of Joshua tree clones, such as those found near Walker Pass, in the western end of the Antelope Valley, and in pinyon-juniper woodlands at ecotones with the Transverse Ranges such as Cajon Pass. High resprouting rates of Joshua trees in these areas may have evolved in local ecotypes that became adapted to shorter fire return intervals along the western desert ecotones than in other parts of the desert bioregion.

FIRE ECOLOGY AND FIRE REGIME–PLANT COMMUNITY INTERACTIONS

The most continuous native upland fuels in the desert bioregion occur at the upper elevations of this zone, especially in areas dominated by blackbush (Fig. 18.4). Nonnative annual grasses have contributed to increased fire frequencies since the 1970s (Brooks and Esque 2002), although the native perennial vegetation in this zone can at times be sufficient alone to carry fire during extreme fire weather conditions (Humphrey 1974).

At the lower elevations within this zone, where creosote bush is codominant with a wide range of other shrubs and perennial grasses, fire spread is largely dependent on high production of fine fuels filling interspaces during years of high precipitation (Brown and Minnich 1986, Rogers and Vint 1987, Schmid and Rogers 1988, Brooks 1999). At higher elevations within this zone, where blackbush is often the dominant species, fire spread is not as dependent on the infilling of shrub interspaces during years of high precipitation, with lower interannual variability in fire occurrence. The historic fire regime was likely characterized by relatively moderate- to large-sized, patchy to complete, moderate-severity, surface-to-crown fire and a long fire return interval (Appendix 1).

Blackbush fires remove cover of woody shrubs, which is soon replaced by similar cover of herbaceous perennials and annual plants (Brooks and Matchett 2003). Nonnative species such as red brome, cheat grass, and red-stemmed filaree typically increase in cover after fire if precipitation is sufficient for their germination and growth. Although perennial plant cover may approach unburned conditions within the first four postfire decades, the species composition typically does not (Abella 2009, Engel and Abella 2011).

Red brome is the dominant nonnative grass in the middle-elevation zone (Brooks et al. 2016). Brome grasses produce higher fuel loads and fuel depths than does Mediterranean grass, and accordingly produce longer flame lengths that carry fire into the crowns of large woody shrubs more readily producing more-intense fires (Brooks 1999). They are also primarily responsible for increased fire-occurrence potential following years of high precipitation in this zone (Fig. 18.2). Brome grasses are less susceptible to periodic population declines at middle- compared to low-elevation sites, because conditions more frequently remain within its biophysical range of temperature and precipitation tolerance (Brooks et al. 2016). Their cover increases in continuity and amount after fire (Brooks 2012), thereby promoting further fire.

FIGURE 18.5 The high-elevation desert zone depicted by pinyon-juniper woodland.

High-Elevation Desert Zone

Vegetation is dominated by shrublands with some woodlands and includes big sagebrush (*Artemisia tridentata*) (49% of the zone), pinyon-juniper (24%), black sagebrush/low sagebrush (*A. nova*/*A. arbuscula*) (18%), and interior chaparral (9%). Although it occupies a small proportion of the desert bioregion (8%, Table 18.1), this zone is ecologically diverse, representing 17 biophysical setting types (Brooks and Matchett 2016). Big sagebrush and pinyon-juniper woodland occurs in the Mono, Southeastern Great Basin, and Mojave Desert sections. Among the pinyon-juniper vegetation types, the Utah juniper–singleleaf pinyon pine association is the most widespread, occurring in the Mono, Southeastern Great Basin, and eastern Mojave Desert ecological sections of California (Minnich and Everett 2001). The California juniper (*Juniperus californica*)–singleleaf pinyon pine association occurs along the desert slopes of the Transverse Ranges at the edge of the Mojave Desert section. Singleleaf pinyon pine occurs with Sierra Juniper (*J. grandis*) and canyon live oak (*Quercus chrysolepis*) in the northeastern San Bernardino Mountains (Wangler and Minnich 1996). Black sagebrush and low sagebrush occur mostly within the high-elevation zone of Mono ecological section, although the former also extends southward into the mountain ranges of the Mojave Desert section. Interior chaparral has limited distribution in this ecological zone, largely on the middle slopes of the Transverse Ranges and the Peninsular Ranges at the western fringes of the desert bioregion. Surface fuel loads and continuity are highest where sagebrush scrub and chaparral dominate, facilitating the spread of fire. In closed pinyon-juniper woodlands, surface fuels are replaced by very high loads of crown fuels from trees and fires only occur under extreme fire weather conditions and are typically very intense (Fig. 18.5).

FIRE RESPONSES OF IMPORTANT SPECIES

Relatively high fuel loads result in high fire intensity, but plant mortality rates vary widely among species. Big sagebrush, Wyoming sagebrush, and mountain sagebrush are typically killed by fire, but can reestablish rapidly from wind-dispersed seeds. Cliffrose is typically killed by fire, whereas its close relative, antelope bitterbrush (*Purshia tridentata*), can resprout (Brooks and Minnich 2006; Table 16.3). Most interior chaparral species resprout, including Muller's oak (*Quercus cornelius-mulleri*), shrub live oak (*Q. turbinella*), birch-leaf mountain-mahogany (*Cercocarpus betuloides* var. *betuloides*), Eastwood's manzanita (*Arctostaphylos glandulosa*), canyon live oak, and beargrass (*Nolina* spp.). The obligate seeding manzanitas and Mojave (*Ceanothus vestitus*) reestablish from seed banks after fire.

Singleleaf pinyon pine, Colorado pinyon (*Pinus edulis*), Utah juniper, and California juniper are relatively thin-barked and have low crown bases that act as ladder fuels and facilitate consumption of entire canopies in active stand-replacement fires. They are also killed by passive surface fires that damage the cambium. However, these woodlands reestablish at time scales of a century. Juniper typically reestablishes from seed sooner than pinyon pine. Initial establishment of singleleaf pinyon pine appears to be delayed 20 to 30 years by sun scald and/or freeze-thaw soil heaving until the establishment of the shrub layer and young juniper trees, which act as nurse plants (Wangler and Minnich 1996). The establishment of a pinyon pine canopy after about 75 years eventually reduces freeze-thaw processes, setting off a chain reaction of spatially random recruitment. Pinyon pines reach complete canopy closure after 100 to 150 years, which is accompanied by a decline in the surface vegetation, due apparently to shrub senescence caused by shading from the overstory.

FIRE REGIME–PLANT COMMUNITY INTERACTIONS

Fuel continuity is similar to that of the middle-elevation zone, but high-elevation fuels are woodier making them more difficult to ignite. High plant cover and prevalence of steep slopes facilitate the spread of fire. Large fires are among the most intense encountered in the desert bioregion due to high woody biomass of juniper and pinyon pine, as well as a subcanopy of sagebrush, bitterbrush, cliffrose, and shrub live oak.

Fires can occur almost any year in sagebrush steppe, although the probability increases slightly when precipitation and woody fuel moisture are very low, or precipitation and fine fuel loads (mostly cheat grass and native forbs) are very high (Fig. 18.2). Fire potential is also increased during periods of high winds and low relative humidity, regardless of fine fuel conditions. Fires are patchy to complete, moderate-severity passive to active fires, depending on the continuity of the woody shrub fuels. Fire spread in pinyon-juniper woodlands typically coincides with low live fuel moisture, low relative humidity, and high winds. A surface to passive crown fire regime is characteristic at the interface between sagebrush steppe and pinyon-juniper woodland. Fires spread through woody and herbaceous surface fuels and occasionally torch woodland fuels, especially younger trees. The historic fire regime was likely characterized by relatively large, patchy to complete, moderate-severity surface-to-crown fires, and long fire return intervals (Appendix 1).

Sagebrush stands generally require 30 to 100 years to recover following fire (Whisenant 1990). High abundance of nonnative annual grasses may lead to shorter fire return intervals that do not allow sufficient regeneration times for native sagebrush species to persist. These conditions also may lead to large homogeneous fires that hinder seed dispersal of native perennial species back into burned areas and result in intense competition between invaders and native seedlings for available resources. Where cheat grass has shortened fire return intervals, especially in the lower-elevation Wyoming big sagebrush communities, sagebrush steppe has been converted to nonnative annual grassland (sidebar 18.1). Such type conversion is less common in higher-elevation mountain sagebrush communities because the native shrubs and perennial grasses are more resilient to fire (Brooks and Chambers 2011).

A century or more of fire suppression coupled with removal of fine fuels by livestock grazing has allowed pinyon-juniper woodlands to encroach on sagebrush steppe across much of the western United States (Miller and Tausch 2001). This has been documented in the northeast bioregion of California (Schaefer et al. 2003) and in the Mono section of desert bioregion (A. Halford, Bureau of Land Management, Bishop, California, USA, pers. comm.). Less woodland encroachment has occurred in the more arid warm deserts due to low primary productivity rates. Resampling of 1929 to 1934 California Vegetation Type Map (VTM) survey plots revealed no significant changes in woodland densities at the western edge of the warm deserts (Wangler and Minnich 1996). Pinyon-juniper woodlands adjacent to the Transverse Ranges have experienced long periods between stand-replacement fires both before and after fire suppression began (fire rotation ~450 years; Wangler and Minnich 1996). Exceptionally large fires can break out in pinyon-juniper during episodes of high precipitation and abundant herbaceous cover. Such long fire intervals permit recolonization of pinyon-juniper, afforded by long-range seed dispersal by corvids and mammals (Vander Wall and Balda 1977).

Fires in pinyon-juniper woodlands are least frequent in open stands at lower elevations and most frequent in dense forests at higher elevations, in response to increasing productivity and fuel accumulation gradients with increasing elevation and precipitation. The upper-elevation ecotones between pinyon-juniper woodlands and mixed-conifer forest are typically very narrow, due to truncated gradients related to fire behavior and stem mortality (Minnich 1988). The thin bark of pinyon pine prevents their survival in the frequent surface fire regime more typical of mixed-conifer forests. Postfire surface fuels appear to lack sufficient biomass to support high-frequency fires, and as canopy closure occurs in pinyon and juniper woodlands, surface fuel loads and continuity are further reduced. The arid limit of Jeffrey pine (*Pinus jeffreyi*) with pinyon pine is constrained by rare episodes of extreme drought. During the drought of 2002, the driest year in instrumental records since 1849, whole forests perished on south facing slopes while associated pinyon pine woodlands survived intact along the western desert ecotones (Minnich et al. 2016). Thus, a historic discontinuity in fire return intervals probably existed along the ecotones between mixed-conifer forests and pinyon woodlands in which frequent understory surface fire at high elevations shift to less frequent stand-replacement crown fire at lower elevations in response to differences in stand structure, fire behavior, and tree survivorship (Minnich 1988).

Desert Montane Zone

There are four general vegetation types in this ecological zone (Brooks and Matchett 2016). Mixed-conifer pine forests occur primarily in the Mono section and bristlecone/limber pine forests occur in the Mono and Southeastern Great Basin sections (Rowlands 1980). Stands of quaking aspen occur almost exclusively in the Mono section, and small areas of sparse-herbaceous alpine fell-fields occur above timberline, primarily in the White Mountains. Small white fir forest enclaves also occur on north-facing slopes in the New York, Clark, and Kingston mountains of the Mojave Desert section. Surface fuel loads and continuity can be very high, facilitating fire spread, although vertical continuity of ladder fuels and horizontal continuity of fuels from trees are often insufficient to carry fire from surface shrubs and grasses into the crowns of trees (Fig. 18.6).

FIRE RESPONSES OF IMPORTANT SPECIES

Bristlecone pine and limber pine are thin-barked trees with low resillience to fire. These tree species can survive only low-intensity surface fire or lightning strikes that do not girdle cambial tissue (Johnson 2001, Fryer 2004). The presence of exposed burned heartwood in millennial-aged, bark-striped bristlecone pine trees is likely a consequence of single tree fires directly struck by lightning, as reported for lodgepole-limber pine forest on Mt. San Jacinto in southern California (cf. Sheppard and Lassoie 1998). These lighting strike fires do not typically kill trees, and likely do not result in fires spreading beyond these trees. Sparse surface fuels result in small fires and low fire intensities with limited effects on small woody shrubs and herbaceous species.

Ponderosa pine (*Pinus ponderosa*) is one of the most fire-adapted conifer species in North America. Its adaptations include open crowns, self-pruning branches, thick bark, thick bud scales, tight needle bunches protecting meristems, high foliar moisture, and a deep rooting habit (Howard 2003). Jeffrey pine is similarly resistant to fire (Miller 2000). Widely spaced older trees display higher fire survival rates than more densely packed and younger trees. White fir (*Abies concolor*) is a thin-barked tree species with branches and foliage from top to bottom that make it highly vulnerable to mortality from fire. Mountain-mahogany (*Cercocarpus* spp.), Gambel oak (*Quercus gambelii*), manzanita (*Arctostaphylos* spp.), and snowberry (*Symphoricarpos albus* var. *laevigatus*) are all elements of

FIGURE 18.6 The desert montane woodland and forest ecological zone depicted by a bristlecone pine forest.

the interior chaparral vegetation type often found in the understory of mixed-conifer sites. All species except mountain mahogany (a nonsprouter) are extremely tolerant of fire (Brooks et al. 2007).

FIRE REGIME–PLANT COMMUNITY INTERACTION

Fuels in bristlecone/limber pine woodlands are very discontinuous, and ephemeral production by annuals during years of high precipitation adds limited biomass to the fuelbed, due to shallow soils, low temperatures, and a short growing season. Surface fires are rare, and most spread through the crowns of bristlecone and limber pines only during extreme fire weather conditions. These fires are small (≤1 ha [≤2.5 ac]). Analysis of Google earth imagery records from the White Mountains revealed only one mass snag forest (prehistoric burn) of 25 ha (62 ac) at latitude 37.56 and longitude 118.20. The historic fire regime in bristlecone/limber pine is characterized by truncated small, patchy, variable severity, passive crown fire and a truncated long fire return interval (Appendix 1).

Fuels in mixed-conifer woodlands and forests are characterized by widely spaced woody vegetation, and small amounts of fine fuels that are relatively densely packed hindering fire spread. Steep, north-facing canyons can have more abundant woody fuels. Long fire intervals allow the accumulation of understory and ladder fuels which generally result in high-severity crown fires that can threaten the persistence of isolated mixed-conifer stands in the desert bioregion. Stand-replacement events are often followed by initial dominance by understory chaparral or sagebrush species which may persist if conditions are no longer conducive to pine establishment. The historic fire regime is characterized by truncated small, patchy, variable severity, passive crown fire and a moderate fire return interval (Appendix 1).

Desert Riparian Zone

The zone includes diverse vegetation types that do not fit into any single elevational range and are not well discerned among its six constituent LANDFIRE biophysical setting types (Brooks and Matchett 2016). Vegetation is generally classified into oases and riparian woodlands, shrublands, grasslands, and marshes (Rowlands 1980). Riparian woodlands occur primarily along the Colorado and Mojave rivers adjacent to low-elevation shrublands in the warm deserts. Other examples can be found in the Amargosa Gorge, Whitewater River, Andreas Canyon, and Palm Canyon. In the cold deserts, riparian woodlands occur along the Owens and Walker rivers and the many creeks along the east slope of the Sierra Nevada. Oasis woodlands occur in isolated stands such as the Palm Canyon, Thousand Palms, and Twentynine palms oases, as well as a string of oases along the eastern escarpment of the Peninsular Ranges of the Colorado Desert section. Surface fuels in the desert riparian zone can rival those of the high-elevation zone, and when coupled with the prevalence of humans who are drawn to these areas, the frequency of fire can be high compared to other ecological zones of the desert bioregion (Fig. 18.7).

FIRE RESPONSES OF IMPORTANT SPECIES

Periodic fire coupled and flooding in this zone has resulted in riparian plant species that are generally resilient to fire. Woodland dominants such as Fremont cottonwood (*Populus fremontii* subsp. *fremontii*), honey mesquite (*Prosopis glandulosa* var. *torreyana*), and willows typically resprout after being top-killed (Brooks and Minnich 2006; Table 16.3). Resprouting individuals and seedlings are susceptible to mortality during recurrent fires. Oasis species such as California fan palm (*Washingtonia filifera*) benefit from frequent, low-severity fire, which reduces competition for water from other plants growing in their understories, and allows new palm seedlings to become established.

The degree to which fire may have affected the evolution of plant species in the vicinity of springs is related to the fire regime of the surrounding vegetation type. Current species assemblages may also have been affected by historical fire used by Native Americans in desert riparian areas. In addition, spring sites dominated by nonnative tamarisk (*Tamarix* spp.) can increase fire hazard unless they are actively managed (sidebar 18.2).

FIGURE 18.7 The desert riparian woodland and oasis ecological zone depicted by riparian shrubland and woodland.

FIRE REGIME–PLANT COMMUNITY INTERACTIONS

Fuel characteristics and fire behavior are extremely variable, due to the wide range of vegetation types that characterize the desert riparian zone. Fuels are typically continuous and fuel loads high, but fuel moisture content is also often high. Fires may not spread except under extreme fire weather conditions. The historic fire regime is characterized by small- to moderate-sized, complete, high-severity passive-to-active crown fire and a short-to-moderate fire return interval (Appendix 1). Fires typically establish *in situ* in oases because surrounding desert plant assemblages are typically low-elevation zone types that have characteristically low fire spread potential.

In riparian woodlands, the nonnatives tamarisk and less frequently giant reed (*Arundo donax*) create ladder fuels that allow fire to spread from surface fuels of willow, saltbush (*Atriplex* spp.), sedge (*Carex* spp.), bulrush (*Schoenoplectus* spp.), and arrow-weed (*Pluchea sericea*) into the crowns of overstory Fremont cottonwood trees, top-killing them. After fire these nonnatives quickly recover and surpass prefire dominance, which results in larger, more intense and more frequent fires than occurred historically (sidebar 18.2) (Dwire and Kauffman 2003). Although some native riparian species may not survive increased fire frequency and intensity (e.g., cottonwood trees), others are clonal and are capable of surviving fire and resprouting (e.g., willows). The altered fire regime accompanied by vigorous postfire resprouting and seedling establishment of nonnatives creates an intense competitive environment that can suppress native species (sidebar 18.2). Thus, the net effects of these dual stressors on natives can be significant even though they may be generally tolerant of fire and other disturbances. For example, in palm oases, postfire dominance of tamarisk can eliminate the potential benefits of fire to Washington fan palms, and actually increase threats to their persistence.

Management Issues

Annual Plant Fuels Management

The fuel condition of greatest concern in the desert bioregion is when the nonnative annual grasses red brome, cheat grass, and Mediterranean grass accentuate herbaceous fuel loads and continuity during years of high precipitation (sidebar 18.1). Although native herbaceous species can also facilitate fire, the fuels they create are not as persistent or flammable (Brooks 1999), plus they are a component of the natural fire regime so they are not typically considered a fundamental fire management problem. New grass invaders such as crimson fountaingrass (*Pennisetum setaceum*), buffel grass (*P. ciliare*), and nonnative mustards such as Sahara mustard may pose additional fire hazards in the future. For example, in the Sonoran Desert, buffel grass invasion coupled with frequent fire has converted desert scrub to nonnative grassland in Mexico (Búrquez et al. 1996), created fuels sufficient to carry fire in Arizona, and recently appeared in southeastern California. Land managers who once lamented the damage caused by fires fueled by red brome in southern Arizona are even more concerned now about the potential effects of buffel grass (S. Rutman, Organ Pipe Cactus National Monument, Ajo, Arizona, USA, pers. comm.). Thus, fine fuels management should be closely tied to invasive plant management from the perspective of both managing invasive plant fuels that are currently present and preventing the establishment of new invasive plants that may change fuel structure and potentially cause even greater fire management challenges in the future.

Managed fire may provide some limited control of nonnative annual grasses under specific circumstances (DiTomaso et al. 2006). For example, burning before cheat grass seeds disperse to the ground can reduce seed densities 400–1,000% during the following spring in the cold deserts (Pechanec and Hull 1945), because seeds suspended aboveground within inflorescences are more susceptible to lethal temperatures than seeds located on or beneath the soil surface (Rasmussen 1994). Regardless of what happens during the first few postfire years, cheat grass and red brome typically return to or exceed prefire dominance during subsequent years in the Mojave (Beatley 1966, Hunter 1991), Sonoran (Brown 1984), and Great Basin deserts (Callison et al. 1985, Rasmussen 1994). These results indicate that fire can be used to temporarily reduce the dominance of nonnative annual grasses, but these grasses recover to or exceed their previous abundances after only a few postfire years.

SIDEBAR 18.2 SALTCEDAR INVASIONS CAN CHANGE RIPARIAN FIRE REGIMES

Tom T. Dudley and Matthew L. Brooks

Saltcedar, also known as tamarisk (*Tamarix* spp.), was introduced to North America in the early 1800s by European colonists as a horticultural plant, and by the early 1900s was widely used to provide windbreaks and erosion control along railways and other erosion-prone sites. Its ability to tolerate periodic drought and harsh soil conditions helped ensure its persistence where other species failed. It was recognized as an invader of desert water courses around the 1920s, and, with the advent of water control and diversion projects, took advantage of the altered conditions to expand its range during the middle and latter part of the century (Robinson 1965).

Saltcedar is deciduous and produces a fine-structured, water-repellent litter layer that is highly flammable in late summer and fall. Because stand densities can be very high, and litter is slow to decompose, a nearly continuous layer of surface fuels can develop that facilitates the spread of fire (Busch and Smith 1992, Drus et al. 2013). The standing trees are flammable when green and actively growing, when the deciduous foliage is senescing, and during the winter when dried foliage remains lodged in its branches. Fire can thus be carried from surface fuels up into the canopies of mature saltcedar as well as nearby native riparian trees (Drus 2013). These fuel characteristics can create a frequent, high-intensity, crown fire regime where an infrequent, low- to moderate-intensity, surface fire regime previously existed. After burning, saltcedar resprouts readily and benefits from nutrients released by fire, whereas native riparian plants such as cottonwoods and willows do not resprout as vigorously (Ellis 2001). Recurrent high-intensity fire may lead to monoculture stands of saltcedar, turning watercourses from barriers of fire movement into pathways for fire spread (Lambert et al. 2010).

As stands of saltcedar increase in density and cover, native cottonwood and willow trees decrease. In some cases, this is coincident with changing environmental conditions that do not favor native species (e.g., decreased water tables caused by water diversion projects; Everitt 1998), but in other cases, it is clear that saltcedar is responsible for the decline in native trees, directly through competition and indirectly through altered fire regimes (Busch and Smith 1995, Hultine and Dudley 2013). Because it provides lower quantity and quality of shade, forage, and insect prey species, wildlife generally avoid large stands of saltcedar in preference for native stands (Shafroth et al. 2005). This includes numerous threatened and declining riparian birds which find better nesting and feeding resources on native trees. In addition, saltcedar can have higher transpiration rates than native trees, potentially reducing water tables (Sala et al. 1996).

Mechanical and chemical methods are typically used to manage saltcedar; however, they can be very expensive ($300–$6,000/ha; Shafroth et al. 2005), and recovery of native species to replace them can be very slow (Ostoja et al. 2014). After more than a decade of prerelease testing, a foliage-feeding beetle from Eurasia, the northern tamarisk beetle (*Diorhabda carinulata*), was released as a biological control agent against saltcedar (Dudley et al. 2000, Bean et al. 2013). From introduction sites in Nevada, Utah, Colorado, and other western states this beetle has spread rapidly and defoliated several thousand hectares of saltcedar (Pattison et al. 2011). The physiological stress experienced by defoliated saltcedar leads to reduced nutrient storage, lowered live fuel moisture, and an increase in the amount of dead wood and foliage (Hudgeons et al. 2007, Drus et al. 2013). This defoliation stress is similar to that caused by herbicides, and can increase saltcedar mortality rates when followed by fire, especially high-intensity summer fire (Drus et al. 2014). In the short term, biocontrol may increase the chance of extreme fire behavior, but in the long run the conversion of saltcedar stands back toward native riparian woodlands will likely reduce fire hazards.

Herbicides have been used to control nonnative annual grasses and to reduce fine fuel loads, but collateral effects on nontarget plants may occur. Atrazine (Aatrex®) and Sulfometuron methyl (Oust®) can reduce biomass of brome grasses, but their negative effects on native plants can persist for at least 8 years (Evans and Young 1977, Hunter et al. 1978, Pellant et al. 1999). Fluazifop-p-butyl (Fusilade®) can be used to control annual and perennial grasses, and at low doses can be used to more selectively kill Mediterranean grass (Steers and Allen 2010). Glyphosate (Roundup®) is widely used to control invasive nonnatives and is one of the least expensive herbicides available. It can control cheat grass (Blackshaw 1991, Beck et al. 1995) and red brome (Larry Jensen, Helena Chemical Company, Las Vegas, Nevada, USA, pers. comm.) with minimal collateral effects on native perennials, although native annuals are typically affected.

Following the extensive Mojave Desert fires of 2005 and 2006, land managers considered using herbicides to manage nonnative annual grass fuels along the margins of dirt and paved roads to create fuelbreaks to impede the spread of subsequent fire. The idea was that corridors of managed fuels could be used to facilitate fire suppression efforts and potentially reduce the frequency of fire starts from vehicles travelling along roads. These actions were proposed primarily in the northern and eastern Mojave Desert where nonnative annual grasses were very abundant at the time (37% cover, Klinger et al. 2011a; Table 6.8). However, even if herbicides are temporarily effective at reducing dominance of nonnative annual grasses, they would need to be applied at least every high-precipitation year as a part of a regular maintenance program to maintain fuelbreaks. For these reasons, they are not typically used to manage nonnative annual fuels at landscape scales in desert regions.

Livestock grazing has been mentioned as a possible tool for managing fine fuels in the desert bioregion (Minnich 2003, Brooks et al. 2007). It may temporarily reduce fine fuel loads and be effective for managing fuels in specific areas such as the wildland-urban interface. Although fuel consumption by livestock grazing has been shown to reduce flame lengths and fire spread rates in the Great Basin (Diamond et al. 2009), it may be effective only in early successional vegetation stands dominated by herbaceous species (Launchbaugh et al. 2008). Intensive livestock grazing can be associated with reduced cover of red brome (Brooks et al. 2006), although the intensity of grazing required to significantly alter fire behavior may actually facilitate the long-term dominance of nonnative grasses (Brooks et al. 2007). If livestock grazing is used to manage fine fuels, then it should be applied with attention to the net response of all dominant plant species, both in the short-term based on the intensity and phenologic stage during which they are grazed, and in the long-term based on the life history characteristics and interrelationship among species.

Scientists and land managers alike are increasingly realizing that climate is the ultimate determinant of nonnative grass dominance in the desert bioregion, and there are no management actions that can overcome its effects at the landscape scale. Aboveground biomass and seed bank density of these grasses can vary more between years of contrasting precipitation than between burned and unburned areas (Brooks et al. 2013b). This calls in question the justification for emergency management of these species in postfire landscapes when a similar level of urgency is not placed upon managing them during years of high precipitation. Clearly, the efficacy of any landscape-level control effort for nonnative annual grasses must consider the effects of climate.

Woody Perennial Plant Fuels Management

Where native perennial cover is sufficient to carry fire without additional fuels from nonnative annual grasses, coarse woody fuels and fine perennial grass fuels may be a concern for fire managers. These types of fuels can be found in the middle-elevation, high-elevation, desert montane, and desert riparian zones. Specific vegetation types include blackbush, sagebrush, and interior chaparral shrublands, pinyon-juniper woodlands and riparian woodlands, and mixed-conifer forests. Although infrequent, intense, stand-replacing fires are a natural part of the fire regime for most of these vegetation types, such fires are not desirable when they occur near human habitations or where they may damage cultural resources such as historic buildings or prehistoric sites. Once these fires start, they often burn intensely and require indirect firefighting tactics to suppress, which complicates efforts to protect specific areas.

Managed fire has been used to shift the composition of plant communities and alter fuel characteristics to facilitate future fire suppression efforts in middle elevations of the desert bioregion. In the early 1900s, settlers used fire in the middle-elevation zone to promote the growth of livestock forage (Croft 1950, Holmgren 1960, Minnich 2003, Brooks et al. 2007). However, these actions often resulted in a dramatic increase in the dominance of red brome which posed a greater fire hazard than the shrublands they replaced (Holmgren 1960). The dominance of red brome on those landscapes has proved persistent 63 years postfire (Brooks et al. 2016, Fig. 2.6). Fire was also used in 1988 (Nolina Prescribed Fire), 1989 (Nolina II Prescribed Fire), and 1993 (Nolina III Prescribed Fire) to evaluate its efficacy in reducing blackbush fuels and creating a managed fuel zone between Joshua Tree National Park and the adjacent town of Yucca Valley (T. Patterson, Joshua Tree National Park, Twentynine Palms, California, USA, pers. comm.). Although most woody fuels were dramatically reduced following these fires, by spring 1998 following a winter of high precipitation, cheat grass and other herbaceous fuels were observed in amounts capable of easily carrying fire (M. Brooks, personal observation). Three years later in 2001 cheat grass remained one of the two most dominant annual species in on these burns, even though two of three previous winters experienced precipitation at less than half of the long-term average (Brooks and Matchett 2003). The persistent dominance of cheat grass following these prescribed fires was a primary reason that fire has not been used since as a fuels management tool for middle elevations at Joshua Tree National Park.

Managed fire clearly has a place in controlling fuels and providing important ecosystem functions in the high-elevation, desert montane, and desert riparian zones where native vegetation is most resilient to fire. It is typically the preferred management tool in mixed-conifer woodlands and forests, often following understory fuel thinning to minimize the chances of crown fire. Nonnative annual grasses are much less likely to become postfire management problems and alter fire regimes at the higher elevations, although they can create problems, along with tamarisk, in riparian areas (sidebar 18.2).

Mechanical control methods such as chaining, cabling, and brush beating, sometimes coupled with the herbicide Tebuthiuron (Spike®), have been used to reduce blackbush (Bowns 1973) and sagebrush (Monsen et al. 2004) cover. Mechanical thinning of understory fuels in mature mixed-conifer stands is also sometimes necessary prior to the reintroduction of low- to moderate-intensity surface fires. This same approach of understory thinning followed by low- to moderate-intensity fire can be used in areas exhibiting pinyon and juniper encroachment. If fuelbeds are already conducive to low- to moderate-intensity fire, then fire alone may be considered a preferred alternative. Livestock grazing can also reduce cover of perennial vegetation in the Mojave Desert (Brooks and Berry 2006). For example, grazing of shrubs by goats has been used to reduce dry stem biomass and promote the regrowth of more succulent live shoots (Provenza et al. 1983) which can reduce flammability.

It should be noted that fuelbreaks, thinnings, or any other type of fuels reduction projects can also have negative effects, such as facilitating spread and dominance of invasive plants (Merriam et al. 2006, Brooks et al. 2007) or creating other unintended fire management problems. Accordingly, both the economic cost and ecological effects associated with the creation and maintenance of managed fuel zones should always be weighed against their efficacy in slowing or stopping fires, the additional costs and efficacy of suppression efforts where fuelbreaks are not present, and the ecological effects of increased burned areas where fires attain larger size due to the absence of managed fuel zones.

Fire Suppression

Fire suppression is arguably the most effective way to protect the majority of landscapes in the warm deserts, and the low- and middle-elevation zones of the cold deserts, from the potential negative effects of fire. This requires aggressive fire suppression efforts that may include aerial water and retardant drops and off-road travel by suppression equipment (e.g., engines, bulldozers). The use of these tactics should not be taken lightly, because phosphate-based retardants may promote dominance by nonnative annual plants (Brooks and Lusk 2008) and off-road travel, especially the use of bulldozers, can damage both natural and cultural resources. There is specific concern about the effect of these activities on the federally threatened desert tortoise (*Gopherus agassizii*) (sidebar 18.3), although guidelines are available to minimize these impacts (Duck et al. 1997).

Fire suppression in desert wilderness areas became a significant issue after The California Desert Protection Act of 1994 (Public Law 103-433) applied the wilderness designation to many areas within the desert bioregion. Wilderness areas in deserts often encompass mountain ranges where locally high fuel loads and steep slopes facilitate the spread of fire. Fire suppression options are generally more limited in these areas by the constraints outlined in wilderness management plans, and often the primary tactic is to wait for fire to spread downslope and attempt to stop it along preexisting roads. This can result in large portions of desert mountain ranges burning during a single event. The critical question is, do activities associated with aggressive firefighting (e.g., construction of hand or bulldozer control lines, fire retardant drops) cause greater ecological damage than large-scale or recurrent fire occurring where fires were historically small and infrequent? Fire suppression can be a high priority where fire frequency has been recently high in regional hot spots (Brooks and Esque 2002) and nonnative grass/fire cycles have become locally established (sidebar 18.1), where local populations of nonnative plants may be poised to expand their range and landscape dominance following fire (especially in the middle-elevation zone), or where there are other management reasons to exclude fire (e.g., the desert tortoise, sidebar 18.3).

Managed wildland fire, rather than full fire suppression, may alternatively be the preferred fire management tactic in the desert montane zone and some parts of the high-elevation zone in the cold deserts. Fire spread potential is minimal in the alpine and bristlecone pine ecosystem type, but if these conditions change in the future (e.g., due to climate change or plant invasions) then active fire suppression may be necessary. Periodic low- to moderate-intensity fire is a desirable natural ecosystem process in the mixed-conifer pine forests and interior chaparral shrublands, and to a lesser degree in the pinyon and juniper and some sagebrush ecosystem types. Fire suppression in these vegetation types can be limited to what is required to protect the wildland-urban interface and/or to limit the spread of fire into excessively large fires that could threaten the persistence of relatively small isolated vegetation stands.

Postfire Management

Burned Area Emergency Rehabilitation teams have developed postfire stabilization/rehabilitation plans after large fires in the desert bioregion. These plans have sometimes prescribed postfire seeding of fast growing plants to reduce soil erosion and compete with and reduce the cover of nonnative grasses associated with the grass/fire cycle (sidebar 18.1). Although postfire seeding has been done extensively for years in the cold deserts, its use in the warm deserts has been much less frequent.

Numerous studies have been published on the effects of postfire seeding from the cold deserts. These studies indicate that establishment success of seeding projects depends on precipitation (Wirth and Pyke 2009) and that very high seeding rates may be required in areas at the lower end of the precipitation range (Thompson et al. 2006). The highest vegetation cover is generally achieved when nonnatives are seeded, specifically the perennial grass desert wheat-grass (*Agropyron cristatum* subsp. *pectinatum*) and/or the shrub forage kochia (*Kochia prostrata*). These studies also suggest that successful seeding can lead to lower invasive plant abundance (Evans and Young 1978, Thompson et al. 2006, Wirth and Pyke 2009), although poorly implemented seeding efforts may actually increase invasive plant abundance (Ratzlaff and Anderson 1995). Seedings with high initial plant establishment are more likely to reduce nonnative annual abundance than those with low initial establishment, and time since seeding increases the chances of detecting an effect (Pyke et al. 2013). In general, establishment of seeded species and associated management objectives are more likely to be met at high-elevation or high-precipitation sites in the cold deserts (Knutson et al. 2014).

Only a few studies have evaluated postfire seeding in the warm deserts. One aerial seeding study following the 2005 Mojave Desert fires suggests that establishment may only occur in the high-elevation compared to the middle-elevation zone (Klinger et al. 2011b). That study also indicates that perennial seedlings may only appear in measureable numbers where both precipitation is high and density of nonnative annual plants is low (Klinger et al. 2011a). Another aerial seeding study evaluated evidence of establishment 2 years to 16 years after postfire aerial seedings spanning low- to high-elevation zones (Brooks and Klinger 2011). Only half of the species included in the seed mixes were present when sampled, and occurrence rates and cover of the seeded species were not higher in seeded than unseeded areas. A hand-broadcast seeding study suggested that mechanical soil-pitting in the low-elevation zone of the Mojave Desert can improve establishment rates of perennial seeds, but not annual seeds (Scoles-Sciulla et al. 2011). Thus, postfire seeding alone may not be the correct tool to control nonnative

SIDEBAR 18.3 FIRE EFFECTS ON THE DESERT TORTOISE

Todd C. Esque and Matthew L. Brooks

Changing fire regimes threaten 12 of the 40 major tortoise species worldwide (Swingland and Klemens 1989). Only general habitat destruction is listed as a threat for more species (23 of 40 species). In general, tortoises are poorly adapted to fire because they evolved in arid or semiarid habitats where fire was historically rare. The Mojave Desert tortoise (*Gopherus agassizii*) is a Federally Threatened species listed partly because of threats posed by fire.

Fires can kill desert tortoises, especially in the spring and early summer when tortoises are most active aboveground throughout their range (Esque et al. 2002). Years of high precipitation lead to high production of annual plants that is required for desert tortoise reproduction, but it also contributes to fire occurrence, especially at the low- and middle-elevation zones within the desert tortoise range. Thus, years when tortoise growth and reproduction are expected to be greatest can be coincident with years of increased fire occurrence. Although mortality from individual fires is generally considered insignificant for wildlife populations compared with the habitat changes that can follow, loss of a few individuals may be catastrophic for local populations of species that are already in decline (Esque et al. 2003).

Fires can affect desert tortoises indirectly by changing habitat structure and plant species composition. Loss of cover vegetation that provides protection from the sun and predators, loss of native forage plants, and increases in nonnative annual grasses are specific examples of potential negative effects of fire (Brooks and Esque 2002, Esque and Schwalbe 2002, Drake et al. 2015). Individual fires which are often patchy, leaving unburned islands of native vegetation, may have relatively small indirect effects on adult desert tortoises which can readily move to those habitat islands. The combination of increased dominance by nonnative grasses and loss of perennial vegetation cover could have more severe consequences for juvenile desert tortoises that have relatively small home ranges (Drake et al. 2015). Recurrent fires pose an even greater threat, as they often burn through previously unburned islands of vegetation and can produce broad landscapes devoid of shrub cover and dominated by nonnative annual grasses (Brooks 2012, Brooks et al. 2016). These conditions occur within a number of regional hot spots in the desert bioregion (Brooks and Esque 2002).

When fighting fires within desert tortoise habitat, land managers follow guidelines developed to reduce the chance of killing desert tortoises. Options include leaving habitat islands unburned, checking under tires before moving vehicles, and walking ahead of vehicles when they are required to travel off-road (Duck et al. 1997). Results of firefighting in desert tortoise habitats have shown that the benefits of fighting fires far outweigh the potential for damaging tortoises and their habitats as long as appropriate guidelines are followed. Postfire restoration of tortoise habitat can be challenging, but recent advances in plant genomics and seed bank ecology have been used to identify seed transfer zones and build expertise in plant materials development to increased restoration success (DeFalco and Esque 2014, Shryock et al. 2015).

annual plant populations, but hand broadcast with soil pitting may be effective for establishing perennials.

Fire Management Planning

One of the biggest challenges in fire management planning is determining desired future conditions to use as management goals. In cases where historic fire regimes can be reconstructed (e.g., mixed-conifer forest), the natural range and variation of historic fire regime characteristics may be a realistic and appropriate target. Management goals may be elusive where historic fire regimes cannot be easily reconstructed, such as across most of the desert bioregion. If plant invasions have shifted fuel characteristics outside of their natural range of historic variation, then restoration of historic fire regimes may be impossible (Brooks et al. 2004). Although it appears that fire regimes, and at least woody fuel conditions, across much of the desert bioregion may be within their historic range of variation, it is difficult to quantify the impact that nonnative plant invasions have had, aside from recognizing that fire regimes have been altered dramatically in some regional hot spots (Brooks and Esque 2002). Further complicating this process are the potential future effect of shifting climate patterns (Tagestad et al. 2016), and potentially increasing levels of atmospheric CO_2 (Mayeaux et al. 1994) and nitrogen deposition (Brooks 2003, Rao et al. 2015), on fuel conditions and fire regimes. All of these potential variables need to be considered when determining fire management goals in the desert bioregion.

References

Abella, S.R. 2009. Post-fire plant recovery in the Mojave and Sonoran Deserts of North America. Journal of Arid Environments 73: 699–707.

Axelrod, D.I. 1995. Outline history of California vegetation. Pages 139–193 in: M.G. Barbour and J. Major, editors. Terrestrial Vegetation of California. Special Publication Number 9. California Native Plant Society, Sacramento, California, USA.

Balch, J.K., B.A. Bradley, C.M. D'Antonio, and J. Gomez-Dans. 2013. Introduced annual grass increases regional fire activity across the arid western USA (1980–2009). Global Change Biology 19: 173–183.

Bates, P.A. 1984. The role and use of fire in blackbrush (*Coleogyne ramosissima* Torr.) communities in California. Dissertation, University of California, Davis, California, USA.

Bauer, K.L., M.L. Brooks, L.A. DeFalco, L. Derasary, K. Drake, N. Frakes, D. Gentilcore, R. Klinger, J.R. Matchett, R.A. McKinley, K. Prentice, and S.J. Scoles-Sciulla. 2011. Southern Nevada Complex emergency stabilization and rehabilitation final report. U.S. Department of the Interior, Bureau of Land Management, Ely, Nevada, USA.

Bean, D., T. Dudley, and K. Hultine. 2013. Bring on the beetles! The biology of tamarisk biocontrol. Pages 377–403 in: A. Sher and M.G. Quigley, editors. *Tamarix*: A Case Study of Ecological Change in the American West. Oxford University Press, New York, New York, USA.

Beatley, J.C. 1966. Ecological status of introduced brome grasses (*Bromus* spp.) in desert vegetation of southern Nevada. Ecology 47: 548–554.

Beck, G.K., J.R. Sebastian, and P.L. Chapman. 1995. Jointed goatgrass (*Aegilops cylindrica*) and downy brome (*Bromus tectorum*) control in perennial grasses. Weed Technology 9: 255–259.

Biondi, F., L.P. Jamieson, S. Strachan, and J. Sibold. 2011. Dendroecological testing of the pyroclimatic hypothesis in the central Great Basin, Nevada, USA. Ecosphere 21: 1–20.

Blackshaw, R.E. 1991. Control of downy brome (*Bromus tectorum*) in conservation fallow systems. Weed Technology 5: 557–562.

Bowns, J.E. 1973. An autecological study of blackbrush (*Coleogyne ramosissima* Torr.) in southwestern Utah. Dissertation, Utah State University, Logan, Utah, USA.

Brooks, M.L. 1999. Alien annual grasses and fire in the Mojave Desert. Madroño 46: 13–19.

———. 2000. *Schismus arabicus* Nees, *Schismus barbatus* (L.) Thell. Pages 287–291 in: C. Bossard, M. Hoshovsky, and J. Randall, editors. Invasive Plants of California's Wildlands. University of California Press, Berkeley, California, USA.

———. 2002. Peak fire temperatures and effects on annual plants in the Mojave Desert. Ecological Applications 12: 1088–1102.

———. 2003. Effects of increased soil nitrogen on the dominance of alien annual plants in the Mojave Desert. Journal of Applied Ecology 40: 344–353.

———. 2008. Plant invasions and fire regimes. Pages 33–46 in: K. Zouhar, J. Kapler Smith, S. Sutherland, and M.L. Brooks, editors. Wildland Fire in Ecosystems: Fire and Nonnative Invasive Plants. USDA Forest Service General Technical Report RMRS-GTR-42-VOL 6. Rocky Mountain Research Station, Fort Collins, Colorado, USA.

———. 2012. Effects of high fire frequency in creosote bush scrub vegetation of the Mojave Desert. International Journal of Wildland Fire 21: 61–68.

Brooks, M.L. and K.H. Berry. 2006. Dominance and environmental correlates of alien annual plants in the Mojave Desert. Journal of Arid Environments 67: 100–124.

Brooks, M.L., C.S. Brown, J.C. Chambers, C.M. D'Antonio, J.E. Keeley, and J. Belnap. 2016. Exotic annual *Bromus* invasions: comparisons among species and ecoregions in the western United States. Pages 11–60 in: M. Germino, J.C. Chambers, and C.S. Brown, editors. Exotic Brome-Grasses in Arid and Semi-Arid Ecosystems of the Western US: Causes, Consequences, and Management Implications. Springer International Publishing, Cham, Switzerland.

Brooks, M.L., and J.C. Chambers. 2011. Resistance to invasion and resilience to fire in desert shrublands of North America. Rangeland Ecology and Management 64: 431–429.

Brooks, M.L., J.C. Chambers, and R.A. McKinley. 2013a. Fire history, effects, and management in southern Nevada. Pages 75–96 in: J.C. Chambers, M.L. Brooks, B.K. Pendleton, and C.B. Raish, editors. The Southern Nevada Agency Partnership Science and Research Synthesis: Science to Support Land Management in Southern Nevada. USDA Forest Service General Technical Report RMRS-GTR-303. Rocky Mountain Research Station, Fort Collins, Colorado, USA.

Brooks, M.L., C.M. D'Antonio, D.M. Richardson, J.B. Grace, J.E. Keeley, J.M. DiTomaso, R.J. Hobbs, M. Pellant, and D. Pyke. 2004. Effects of invasive alien plants on fire regimes. BioScience 54: 677–688.

Brooks, M.L., and T.C. Esque. 2002. Alien annual plants and wildfire in desert tortoise habitat: status, ecological effects, and management. Chelonian Conservation and Biology 4: 330–340.

Brooks, M.L., T.C. Esque, and T. Duck. 2007. Creosotebush, blackbrush, and interior chaparral shrublands. Pages 97–110 in: S. Hood and M. Miller, editors. Fire Ecology and Management of the Major Ecosystems of Southern Utah. USDA Forest Service General Technical Report RMRS-GTR-202. Rocky Mountain Research Station, Fort Collins, Colorado, USA.

Brooks, M. L., and R. Klinger. 2011. Establishment of plants from postfire aerial seeding treatments implemented 1993 to 2007 in the Eastern Mojave Desert. U.S. Geological Survey, Cooperator Publication. Yosemite Filed Station, El Portal, California, USA.

Brooks, M.L., and M. Lusk. 2008. Fire Management and Invasive Plants: A Handbook. U.S. Fish and Wildlife Service, Arlington, Virginia, USA.

Brooks, M.L., and J.R. Matchett. 2003. Plant community patterns in unburned and burned blackbrush (*Coleogyne ramosissima*) shrublands in the Mojave Desert. Western North American Naturalist 63: 283–298.

———. 2006. Spatial and temporal patterns of wildfires in the Mojave Desert, 1980–2004. Journal of Arid Environments 67: 148–164.

———. 2016. Fire Patterns among ecological zones in the California Desert, 1984–2013. U.S. Geological Survey Data Release. Reston, Virginia, USA. https://doi.org/10.5066/F75B00MP.

Brooks, M.L., J.R. Matchett, and K. Berry. 2006. Alien and native plant cover and diversity near livestock watering sites in a desert ecosystem. Journal of Arid Environments 67: 100–124.

Brooks, M.L., J.R. Matchett, D.J. Shinneman, and P.S. Coates. 2015. Fire patterns in the range of greater sage-grouse, 1984–2013—implications for conservation and management. U.S. Geological Survey Open-File Report 2015-1167. Reston, Virginia, USA.

Brooks, M.L., and R.A. Minnich. 2006. Southeastern deserts bioregion. Pages 391–414 in: N.G. Sugihara, J.W. van Wagtendonk, J. Fites-Kaufman, K.E. Shaffer, and A.E. Thode, editors. Fire in California Ecosystems. University of California Press, Berkeley, California, USA.

Brooks, M.L., S.M. Ostoja, and R.C. Klinger. 2013b. Fire effects on seed banks and vegetation in the Eastern Mojave Desert: implications for post-fire management. Final Report for Project #06-1-2-02 of the Joint Fire Science Program. U.S. Geological Survey, El Portal, California, USA.

Brooks, M.L., and D. Pyke. 2001. Invasive plants and fire in the deserts of North America. Pages 1–14 in: K. Galley and T. Wilson, editors. Proceedings of the Invasive Species Workshop: The Role of Fire in the Control and Spread of Invasive Species. Fire Conference 2000: The First National Congress on Fire, Ecology, Prevention and Management. Miscellaneous Publications No. 11. Tall Timbers Research Station, Tallahassee, Florida, USA.

Brown, D.E. 1984. Fire and changes in creosote bush scrub on the western Colorado Desert, California. Thesis, University of California, Riverside, California, USA.

Brown, D.E., and R.A. Minnich. 1986. Fire and creosote bush scrub of the western Sonoran Desert, California. American Midland Naturalist 116: 411–422.
Brown, T.J., B.L. Hall, C.R. Mohrle, and H.J. Reinbold. 2002. Coarse assessment of federal wildland fire occurrence data. CEFA Report 02-04. Desert Research Institute, Division of Atmospheric Sciences, Reno, Nevada, USA.
Brown, J.K., and J.K. Smith, editors. 2000. Wildland fire in ecosystems: effects of fire on flora. USDA Forest Service General Technical Report RMRS-GTR-42-VOL-2. Rocky Mountain Research Station, Fort Collins, Colorado, USA.
Bukowski, B.E., and W.L. Baker. 2013. Historical fire regimes, reconstructed from land-survey data, led to complexity and fluctuation in sagebrush landscapes: Ecological Applications 23: 546–564.
Búrquez, A.M., A.Y. Martinez, M. Miller, K. Rojas, M.A. Quintana, and D. Yetman. 1996. Mexican grasslands and the changing arid lands of Mexico: an overview and a case study in north-western Mexico. Pages 21–32 in: B. Tellman, D.M. Finch, E. Edminster, and R. Hamre, editors. The Future of Arid Grasslands: Identifying Issues, Seeking Solutions. USDA Forest Service Proceedings RMRS-P-3. Rocky Mountain Research Station, Fort Collins, Colorado, USA.
Busch, D.E., and S.D. Smith. 1992. Fire in a riparian shrub community: postburn water relations in the *Tamarix-Salix* association along the lower Colorado River. Pages 52–55 in: W.P. Clary, M.E. Durant, D. Bedunah, and C.L. Wambolt, compilers. Proceedings, Ecology and Management of Riparian Shrub Communities. USDA Forest Service General Technical Report INT-GTR-28. Intermountain Forest and Range Experiment Station, Ogden, Utah, USA.
———. 1995. Mechanisms associated with decline of woody species in riparian ecosystems of the southwestern U.S. Ecological Monographs 65: 347–370.
Callison, J., J.D. Brotherson, and J.E. Bowns. 1985. The effects of fire on the blackbrush (*Coleogyne ramosissima*) community of southwestern Utah. Journal of Range Management 38: 535–538.
Chambers, J.C., B.A. Bradley, C.S. Brown, C. D'Antonio, M.J. Germino, J.B. Grace, S.P. Hardegree, R.F. Miller, and D.A. Pyke. 2014. Resilience to stress and disturbance, and resistance to *Bromus tectorum* L. invasion in cold desert shrublands of western North America. Ecosystems 17: 360–375.
Croft, A.R. 1950. Inspection of black brush burn, May 12, 1950. Memorandum to the unpublished report, Bureau of Land Management, State Supervisor for Nevada. Unpublished Report on File. U.S. Department of the Interior, Bureau of Land Management, Caliente, Nevada, USA.
Dalton, P.D. 1962. Ecology of the creosotebush *Larrea tridentata* (D.C.) Cov. Dissertation, University of Arizona, Tucson, Arizona, USA.
D'Antonio, C.M., and P.M. Vitousek. 1992. Biological invasions by exotic grasses, the grass/fire cycle, and global change. Annual Review of Ecology and Systematics 3: 63–87.
DeFalco, L.A., and T.C. Esque. 2014. Soil seed banks: preserving native biodiversity and repairing damaged desert shrublands. Fremontia 42(2): 20–23.
DeFalco, L.A., T.C. Esque, S.J. Scoles-Sciulla, and J. Rodgers. 2010. Desert wildfire and severe drought diminish survivorship of the long-lived Joshua tree (*Yucca brevifolia*: Agavaceae). American Journal of Botany 97: 243–250.
De Soyza, A.G., W.G. Whitford, E. Martinez-Meza, and J.W. Van Zee. 1997. Variation in creosotebush (*Larrea tridentata*) canopy morphology in relation to habitat, soil fertility, and associated annual plant communities. American Midland Naturalist 137: 13–26.
Diamond, J.M., C.A. Call, and N. Devoe. 2009. Effects of targeted cattle grazing on fire behavior of cheatgrass-dominated rangeland in the northern Great Basin, USA. International Journal of Wildland Fire 18: 944–950.
Dimock, D.E. 1960. Report on blackbrush burn observations, April 18–20, 1960. Memorandum to the Bureau of Land Management. State Supervisor for Nevada, Carson City, Nevada, USA.
DiTomaso, J.M., M.L. Brooks, E.B. Allen, R.A. Minnich, P.M. Rice, and G.B. Kyser. 2006. Control of invasive weeds with prescribed burning. Weed Technology 20: 535–548.
Drake, K.K., T.C. Esque, K.E. Nussear, L.A. DeFalco, S.J. Scoles-Sciulla, A.T. Modlin, and P.A. Medica. 2015. Desert tortoise use of burned habitat in the eastern Mojave Desert. Journal of Wildlife Management 79: 618–629.
Drus, G.M. 2013. Fire ecology of *Tamarix*. Pages 240–255 in: A. Sher and M.G. Quigley, editors. *Tamarix*: A Case Study of Ecological Change in the American West. Oxford University Press, New York, New York, USA.
Drus, G.M, T.L. Dudley, M.L. Brooks, and J.R. Matchett. 2013. The effect of leaf-beetle herbivory on the fire behaviour of tamarisk (*Tamarix ramosissima* Lebed.). International Journal of Wildland Fire. 22: 446–458.
Drus, G.M., T.L. Dudley, C.M. D'Antonio, T.J. Even, M.L. Brooks, and J.R. Matchett. 2014. Synergistic interactions between leaf beetle herbivory and fire enhance tamarisk (*Tamarix* spp.) mortality. Biological Control 77: 29–40.
Duck, T.A., T.C. Esque, and T.J. Hughes. 1997. Fighting wildfires in desert tortoise habitat, considerations for land managers. Pages 7–13 in: J.M. Greenlee, editor. Fire Effects on Rare and Endangered Species and Habitats. International Association for Wildland Fire, Fairfield, Washington, USA.
Dudley, T.L., C.J. DeLoach, J.E. Lovich, and R.I. Carruthers. 2000. Saltcedar invasion of western riparian areas: impacts and new prospects for control. Pages 345–381 in: R.E. McCabe and S.E. Loos, editors. Transactions 65th North American Wildlife Management. Institute, Washington, D.C., USA.
Dwire, K.A., and J.B. Kauffman. 2003. Fire and riparian ecosystems in landscapes of the western USA. Forest Ecology and Management 178: 61–74.
Earnst, S.L., and A.L. Holmes. 2012. Bird-habitat relationships in interior Columbia Basin shrubsteppe. Condor 11: 15–29.
Eidenshink, J., B. Schwind, K. Brewer, Z. Zhu, B. Quayle, and S. Howard. 2007. A project for monitoring trends in burn severity. Fire Ecology 3(1): 3–21.
Ellis, L.M. 2001. Short-term response of woody plants to fire in a Rio Grande riparian forest. Biological Conservation 97: 159–170.
Engel, C.E., and S.R. Abella. 2011. Vegetation recovery in a desert landscape after wildfires: influences of community type, time since fire and contingency effects. Journal of Applied Ecology 48: 1401–1410.
Esque, T.C., A.M. Búrquez, C.R. Schwalbe, T.R. VanDevender, M.J.M. Nijhuis, and P. Anning. 2002. Fire ecology of the Sonoran desert tortoise. Pages 312–333 in: T.R. Van Devender, editor. The Sonoran Desert Tortoise: Natural History, Biology, and Conservation. Arizona-Sonora Desert Museum and the University of Arizona Press, Tucson, Arizona, USA.
Esque, T.C., and C.R. Schwalbe. 2002. Alien annual plants and their relationships to fire and biotic change in Sonoran desertscrub. Pages 165–194 in: B. Tellman, editor. Invasive Exotic Species in the Sonoran Region. Arizona-Sonora Desert Museum and the University of Arizona Press, Tucson, Arizona, USA.
Esque, T.C., C.R. Schwalbe, L.A. DeFalco, R.B. Duncan, and T.J. Hughes. 2003. Effects of wildfire on desert tortoise (*Gopherus agassizii*) and other small vertebrates. The Southwestern Naturalist 48: 103–111.
Esque, T.C., R.H. Webb, C.S.A. Wallace, C. van Riper, III, C. McCreedy, and L. Smythe. 2013. Desert fires fueled by native annual forbs: effects of fire on communities of plants and birds in the lower Sonoran Desert of Arizona. The Southwestern Naturalist 58: 223–233.
Evans, R.A., and J.A. Young. 1978. Effectiveness of rehabilitation practices following wildfire in a degraded big sagebrush-downy brome community. Journal of Range Management 31: 185–188.
———. 1977. Weed control-revegetation systems for big sagebrush-downy brome rangelands. Journal of Range Management 30: 331–336.
Everitt, B.L. 1998. Chronology of the spread of tamarisk in the central Rio Grande. Wetlands 18: 658–668.
Fryer, J.L. 2004. *Pinus longaeva*. In: Fire Effects Information System. USDA Forest Service, Rocky Mountain Research Station, Fire Sciences Laboratory, producer. http://www.fs.fed.us/database/feis/. Accessed December 28, 2011.
Gray, M.E., B.G. Dickson, and L.J. Zachmann. 2014. Modeling and mapping dynamic variability in large fire probability in the lower Sonoran Desert of south-western Arizona. International Journal of Wildland Fire 23: 1108–1118.
Hegeman, E.E., B.G. Dickson, and L.J. Zachmann. 2014. Probabilistic models of fire occurrence across National Park Service units

within the Mojave Desert Network, USA. Landscape Ecology 29: 1587–1600.
Hereford, R., R.H. Webb, and C.I. Longpre. 2006. Precipitation history and ecosystem response to multidecadal precipitation variability in the Mojave Desert region, 1893–2001. Journal of Arid Environments 67: 13–34.
Holmgren, R.C. 1960. Inspection tour of old blackbrush burns in BLM District N-5, southern Nevada. Unpublished Report. USDA Forest Service, Intermountain Forest and Range Experiment Station, Reno Research Center, Reno, Nevada, USA.
Howard, J.L. 2003. *Pinus ponderosa* var. *scopulorum*. In: Fire Effects Information System. USDA Forest Service, Rocky Mountain Research Station, Fire Sciences Laboratory, producer. http://www.fs.fed.us/database/feis/. Accessed December 28, 2014.
Hudgeons, J.L., A.E. Knutson, K.M. Heinz, C.J. DeLoach, T.L. Dudley, R.R. Pattison, and J.R. Kiniry. 2007. Defoliation by introduced *Diorhabda elongata* leaf beetles (Coleoptera: Chrysomelidae) reduces carbohydrate reserves and regrowth of *Tamarix* (Tamaricaceae). Biological Control 43: 213–221.
Hultine, K., and T.L. Dudley. 2013. *Tamarix* from organism to landscape. Pages 149–167 in: A. Sher and M. Quigley, editors. *Tamarix*: A Case Study of Ecological Change in the American West. Oxford University Press, New York, New York, USA.
Humphrey, R.R. 1974. Fire in deserts and desert grassland of North America. Pages 365–401 in: T.T. Kozlowski and C.E. Ahlgren, editors. Fire and Ecosystems. Academic Press, New York, New York, USA.
Hunter, R. 1991. *Bromus* invasions on the Nevada Test Site: present status of *B. rubens* and *B. tectorum* with notes on their relationship to disturbance and altitude. Great Basin Naturalist 51: 176–182.
Hunter, R., A. Wallace, and E. Romney. 1978. Persistent atrazine toxicity in Mojave Desert shrub communities. Journal of Range Management 31: 199–203.
Johnson, K.A. 2001. *Pinus flexilis*. In: Fire Effects Information System. USDA Forest Service, Rocky Mountain Research Station, Fire Sciences Laboratory, producer. http://www.fs.fed.us/database/feis/. Accessed December 28, 2014.
Klinger, R., M.L. Brooks, N. Frakes, J.R. Matchett, and R. McKinley. 2011a. Vegetation trends following the 2005 Southern Nevada Complex Fire. Pages 118–194 in: K.L. Bauer, M.L. Brooks, L.A. DeFalco, L. Derasary, K. Drake, N. Frakes, D. Gentilcore, R. Klinger, J.R. Matchett, R.A. McKinley, K. Prentice, and S.J. Scoles-Sciulla, editors. Southern Nevada Complex Emergency Stabilization and Rehabilitation Final Report. U.S. Department of the Interior, Bureau of Land Management, Ely, Nevada, USA.
Klinger, R., M.L. Brooks, N. Frakes, J.R. Matchett, R. McKinley, and K. Prentice, K. 2011b. Establishment of aerial seeding treatments in blackbrush and pinyon-juniper sites following the 2005 Southern Nevada Complex. Pages 92–117 in: K.L. Bauer, M.L. Brooks, L.A. DeFalco, L. Derasary, K. Drake, N. Frakes, D. Gentilcore, R. Klinger, J.R. Matchett, R.A. McKinley, K. Prentice, and S.J. Scoles-Sciulla, editors. Southern Nevada Complex Emergency Stabilization and Rehabilitation Final Report. U.S. Department of the Interior, Bureau of Land Management, Ely, Nevada, USA.
Knick, S.T., D.S. Dobkin, J.T. Rotenberry, M.A. Schroeder, W.M. Vander Hagen, and C. Van Riper, III. 2003. Teetering on the edge or too late? Conservation and research issues for avifauna of sagebrush habitats. The Condor 105: 611–634.
Knick, S.T., and J.T. Rotenberry. 1995. Landscape characteristics of fragmented shrubsteppe habitats and breeding passerine birds. Conservation Biology 9: 1059–1071.
Knutson, K., D. Pyke, T.A. Wirth, R.S. Arkle, D. Pilliod, M.L. Brooks, J.C. Chambers, and J.B. Grace. 2014. Long-term effects of seeding after wildfire on vegetation in Great Basin shrubland ecosystems. Journal of Applied Ecology 51: 1414–1424.
Koehler, P.A., R.S. Anderson, and W.G. Spaulding. 2005. Development of vegetation in the central Mojave Desert of California during the late Quaternary. Paleogeography, Paleoclimatology, Paleoecology 215: 297–311.
Kolden, C.A., and P.J. Weisberg. 2007. Assessing accuracy of manually-mapped wildfire perimeters in topographically dissected areas. Fire Ecology 3(1): 22–31.
Küchler, A.W. 1964. Potential natural vegetation of the conterminous United States. Special Publication 36. American Geographical Society, New York, New York, USA.
Lambert, A.M., C. D'Antonio, and T.L. Dudley. 2010. Invasive species and fire in California ecosystems. Fremontia 38(2): 29–36.
Launchbaugh, K., B. Brammer, M.L. Brooks, S.C. Bunting, P. Clark, J. Davidson, M. Fleming, R. Kay, M. Pellant, D. Pyke, and B. Wylie. 2008. Interactions among livestock grazing, vegetation type, and fire behavior in the Murphy Wildland Fire Complex in Idaho and Nevada, July 2007. U.S. Geological Survey Open-File Report 2008-1214. Reston, Virginia, USA.
Mayeaux, H.S., H.B. Johnson, and H.W. Polley. 1994. Potential interactions between global change and Intermountain annual grasslands. Pages 95–110 in: S.B. Monsen and S.G. Kitchen, editors. Proceedings of Ecology and Management of Annual Rangelands. USDA Forest Service General Technical Report INT-GTR-313. Intermountain Research Station, Ogden, Utah, USA.
McGill, S., and T. Rockwell. 1998. Ages of late Holocene earthquakes on the central Garlock fault near El Paso Peaks, California. Journal of Geophysical Research 103: 7265–7279.
McKinley, R.A., M.L. Brooks, and R.C. Klinger. Fire history of Clark and Lincoln Counties in Southern Nevada. U.S. Geological Survey Open File Report. Sioux Falls, South Dakota, USA. (In press.)
Merriam, K.E., J.E. Keeley, and J.L. Beyers. 2006. Fuel breaks affect nonnative species abundance in Californian plant communities. Ecological Applications 16: 515–527.
Miles, S.R., and C.B. Goudy. 1997. Ecological subregions of California: section and subsection descriptions. USDA Forest Service Technical Publication R5-EM-TP-005. Pacific Southwest Region, Vallejo, California, USA.
Miller, M. 2000. Fire autecology. Pages 9–34 in: J.K. Brown and J.K. Smith, editors. Wildland Fire in Ecosystems: Effects of Fire on Flora. USDA Forest Service General Technical Report RMRS-GTR-42-VOL-2. Rocky Mountain Research Station, Ogden, Utah, USA.
Miller, R.F., and R.J. Tausch. 2001. The role of fire in juniper and pinyon woodlands: as descriptive analysis. Pages 15–30 in: K. Galley and T. Wilson, editors. Proceedings of the Invasive Species Workshop: The Role of Fire in the Control and Spread of Invasive Species. Fire Conference 2000: The First National Congress on Fire, Ecology, Prevention and Management. Miscellaneous Publications 11. Tall Timbers Research Station, Tallahassee, Florida, USA.
Minnich, R.A. 1988. The biogeography of fire in the San Bernardino Mountains of California. University of California Publications in Geography 28: 1–121.
———. 2003. Fire and dynamics of temperate desert woodlands in Joshua Tree National Park. Report Submitted to the National Park Service, Joshua Tree National Park. Joshua National Park, Twentynine Palms, California, USA.
———. 2007. Climate, paleoclimate, and paleovegetation. Pages 43–70 in: M.G. Barbour, T. Keeler-Wolf, and A.A. Schoenherr, editors. Terrestrial Vegetation of California. University of California Press, Berkeley, California, USA.
———. 2008. California's Fading Wildflowers: Lost Legacy and Biological Invasions. University of California Press, Berkeley, California, USA.
Minnich, R.A., and R.G. Everett. 2001. Conifer tree distributions in southern California. Madroño 48: 177–197.
Minnich, R.A., B.R. Goforth, and T.D. Paine. 2016. Follow the water: extreme drought and the conifer forest pandemic of 2002–2003 along the California borderland. Pages 859–890 in: T.D. Paine and F. Lieutier, editors. Insects and Diseases of Mediterranean Forest Systems. Springer International Publishing, Cham, Switzerland.
Monsen, S., R.R. Stevens, and N. Shaw. 2004. Restoring western ranges and wildlands. USDA Forest Service General Technical Report RMRS-GTR-136. Rocky Mountain Research Station, Fort Collins, Colorado, USA.
O'Leary, J.F., and R.A. Minnich. 1981. Postfire recovery of creosote bush scrub vegetation in the western Colorado Desert. Madroño 28: 61–66.
Ostoja, S.M., M.L. Brooks, T. Dudley, and S.R. Lee. 2014. Short-term vegetation response following mechanical control of saltcedar (*Tamarix* spp.) on the Virgin River, USA. Invasive Plant Science and Management 7: 310–319.
Pattison, R.R., C.M. D'Antonio, T.L. Dudley, K.K. Allander, and B. Rice. 2011. Early impacts of biological control on canopy cover and water use of the invasive saltcedar tree (*Tamarix* spp.) in western Nevada, USA. Oecologia 165: 605–616.

Pyke, D.A., T.A. Wirth, and J.L. Beyers. 2013. Does seeding after wildfires in rangelands reduce erosion or invasive species? Restoration Ecology 21: 415–421.
Pechanec, J.C., and A.C. Hull. 1945. Spring forage lost through cheatgrass fires. National Wool Grower 35: 13.
Pellant, M., J. Kaltenecker, and S. Jirik. 1999. Use of OUST herbicide to control cheatgrass in the northern Great Basin. Pages 322–326 in: S.B. Monson and R. Stevens. Proceedings: Ecology and Management of Pinyon-Juniper Communities within the Interior West. USDA Forest Service Proceedings RMRS-P-9. Rocky Mountain Research Station, Ft Collins, Colorado, USA.
Provenza, F.D., J.E. Bowns, P.J. Urness, J.C. Malechek, and J.E. Butcher. 1983. Biological manipulation of blackbrush by goat browsing. Journal of Range Management 36: 518–529.
Rao, L.E., J.R. Matchett, M.L. Brooks, R.F. Johnson, R.A. Minnich, and E.B. Allen. 2015. Relationships between annual plant productivity, nitrogen deposition and fire size in low elevation California desert scrub. International Journal of Wildland Fire 24: 48–58.
Rasmussen, G.A. 1994. Prescribed burning considerations in sagebrush annual grassland communities. Pages 69–70 in: S.B. Monsen and S.G. Kitchen, editors. Proceedings—Ecology and Management of Annual Rangelands. USDA Forest Service General Technical Report INT-GTR-313. Intermountain Research Station, Ogden, Utah, USA.
Ratzlaff, T.D., and J.A. Anderson. 1995. Vegetal recovery following wildfire in seeded and unseeded sagebrush steppe. Journal of Range Management 48: 386–391.
Robinson, T.W. 1965. Introduction, spread and areal extent of saltcedar (*Tamarix*) in the western states. US Geological Survey Professional Paper 491-A. Reston, Virginia, USA.
Rogers, G.F., and M.K. Vint. 1987. Winter precipitation and fire in the Sonoran Desert. Journal of Arid Environments 13: 47–52.
Rollins, M.G. 2009. LANDFIRE—a nationally consistent vegetation, wildland fire, and fuel assessment. International Journal of Wildland Fire 18: 235–249.
Rowlands, P.G. 1980. The vegetational attributes of the California Desert Conservation Area. Pages 135–183 in: J. Lattin, editor. The California Desert: An Introduction to Its Resources and Man's Impact. California Native Plant Society Special Publication 5. Sacramento, California, USA.
Sala, A., S.D. Smith, and D.A. Devitt. 1996. Water use by *Tamarix ramosissima* and associated phreatophytes in a Mojave Desert floodplain. Ecological Applications 6: 888–898.
Schaefer, R.J., D.J. Thayer, and T.S. Burton. 2003. Forty-one years of vegetation change on permanent transects in northeastern California: implications for wildlife. California Fish and Game 89: 55–71.
Schmid, M.K., and G.F. Rogers. 1988. Trends in fire occurrence in the Arizona upland subdivision of the Sonoran Desert, 1955 to 1983. The Southwestern Naturalist 33: 437–444.
Schmidt, K.M., J.P. Menakis, C.C. Hardy, W.J. Hann, and D.L. Bunell. 2002. Development of coarse-scale spatial data for wildland fire and fuel management. USDA Forest Service General Technical Report, RMRS-GTR-87. Rocky Mountain Research Station, Fort Collins, Colorado, USA.
Scoles-Sciulla, S.J., K.L. Bauer, K.K. Drake, and L.A. DeFalco. 2011. Effectiveness of post-fire seeding in desert tortoise critical habitat following the 2005 Southern Nevada Fire Complex. Pages 43–76 in: L. Derasaray, N. Frakes, D. Gentilcore, T. Lenard, and K. Prentice, compilers. Southern Nevada Complex Emergency Stabilization and Rehabilitation Final Report. U.S. Department of the Interior, Bureau of Land Management, Administrative Report. Ely, Nevada, USA.
Shafroth, P.B., J.R. Cleverly, T.L. Dudley, J.P. Taylor, C. van Riper, III, E.P. Weeks, and J.N. Stuart. 2005. Control of *Tamarix* in the western United States: implications for water salvage, wildlife use, and riparian restoration. Environmental Management 35: 231–246.
Sheppard, P.R. and J.P. Lassoie. 1998. Fire regime of the lodgepole pine forest of Mt. San Jacinto, California. Madroño 45: 47–56.
Shryock, D.F., C.A. Havrilla, T.C. Esque, N.A. Custer, T. Nakazato, L.A. DeFalco, and T.E. Wood. 2015. Landscape genomics of *Sphaeralcea ambigua* in the Mojave Desert: a multivariate, spatially-explicit approach to guide its use in ecological restoration. Conservation Genetics 16: 1303–1317.
Spaulding, W.G. 1990. Vegetational and climatic development of the Mojave Desert: the last glacial maximum to the present. Pages 166–199 in: J.L. Betancourt, T.R. Van Devender, and P.S. Martin, editors. Packrat Middens: The Last 40,000 Years of Biotic Change. The University of Arizona Press, Tucson, Arizona, USA.
Steers, R.J., and E.B. Allen. 2010. Post-fire control of invasive plants promotes native succession in a burned desert shrubland. Restoration Ecology 18: 334–343.
Swingland, I.R., and M.W. Klemens. 1989. The conservation biology of tortoises. Occasional Papers of the IUCN Species Survival Commission (SSC), No. 5. International Union for Conservation of Nature and Natural Resources, Gland, Switzerland.
Tagestad, J., M.L. Brooks, V. Cullinan, J. Downs, and R. McKinley. 2016. Precipitation regime classification for the Mojave Desert: implications for fire occurrence. Journal of Arid Environments 124: 388–397.
Thompson, T.W., B.A. Roundy, E.D. McArthur, B.D. Jessop, B. Waldron, and J.N. Davis. 2006. Fire rehabilitation using native and introduced species: a landscape trial. Rangeland Ecology and Management 59: 237–248.
Trader, M.R., M.L. Brooks, and J.V. Draper. 2006. Seed production by the non-native *Brassica tournefortii* (Sahara mustard) along desert roadsides. Madroño 53: 313–320.
Vander Wall, S.B., and R.P. Balda. 1977. Co-adaptation of the Clark's nutcracker and pinyon pine for efficient seed harvest and dispersal. Ecological Monographs 47: 89–111.
Van Devender, T.R. 1977. Holocene woodlands in the Southwestern deserts. Science 198: 189–192.
Van Devender, T.R., and W.G. Spaulding. 1979. Development of vegetation and climate in the southwestern United States. Science 204: 701–710.
Van Linn, P.F., K.E. Nussear, T.C. Esque, L.A. DeFalco, R.D. Inman, and S.R. Abella. 2013. Estimating wildfire risk on a Mojave Desert landscape using remote sensing and field sampling. International Journal of Wildland Fire 22: 770–779.
van Wagtendonk, J.W., and D. Cayan. 2008. Temporal and spatial distribution of lightning strikes in California in relationship to large-scale weather patterns. Fire Ecology 4(1): 34–46.
Wangler, M., and R.A. Minnich. 1996. Fire and succession in pinyon-juniper woodlands of the San Bernardino Mountains. Madroño 43: 493–514.
Webb, R.H., M.B. Murov, T.C. Esque, D.E. Boyer, L.A. DeFalco, D.F. Haines, D. Oldershaw, S.J. Scoles, K.A. Thomas, J.B. Blainey, and P.A. Medica. 2003. Perennial vegetation data from permanent plots on the Nevada Test Site, Nye County, Nevada. U.S. Geological Survey Open-File Report 2003-336. Reston, Virginia, USA.
Webb, R.H., J.W. Steiger, and E.B. Newman. 1988. The response of vegetation to disturbance in Death Valley National Monument, California. U.S. Geological Survey Bulletin 1793. Reston, Virginia, USA.
Whisenant, S.G. 1990. Changing fire frequencies on Idaho's Snake River Plains: ecological and management implications. Pages 4–7 in: E.D. McArthur, E.D. Romney, E.M. Smith, and S.D. Tueller, editors. Proceedings—Symposium on Cheatgrass Invasion, Shrub Die-Off, and Other Aspects of Shrub Biology and Management. USDA Forest Service General Technical Report INT-GTR-276. Intermountain Research Station, Ogden, Utah, USA.
White, L.D. 1968. Factors affecting the susceptibility of creosotebush (*Larrea tridentata* [D.C.] Cov.) to burning. Dissertation, University of Arizona, Tucson, Arizona, USA.
Wirth, T.A., and D.A. Pyke. 2009. Final report for emergency stabilization and rehabilitation treatment monitoring of the Keeney Pass, Cow Hollow, Double Mountain, and Farewell Bend fires. U.S. Geological Survey Open-File Report 2009-1152. Reston, Virginia, USA.
Wright, H.E., and A.W. Bailey. 1982. Fire Ecology, United States and Canada. Wiley, New York, New York, USA.

PART THREE

FIRE MANAGEMENT ISSUES IN CALIFORNIA'S ECOSYSTEMS

In the first two parts of the book, we have presented details about fire as an ecological process and the role fire plays in the nine bioregions of California. It should be obvious that fire has been a dynamic force in these ecosystems and will continue to be so in the future. How then, must we reconcile this ecological fact with need for human society to coexist with fire? There are several issues Californians must face if they are to continue live in a fire-prone landscape. In Part Three, we address these issues.

Chapter 19 takes a look at Native American use of fire and discusses how this important cultural activity has influenced ecosystems and how it can continue in light of conflicting land management objectives. Since European settlement, our population has expanded, and fire policies have evolved as our understanding of fire has become more sophisticated. Chapter 20 chronicles this evolution and asks how policies must change in the future. One of the greatest challenges faced by Californians is the accumulation of fuels in areas within and surrounding communities. Dealing with this challenge is the subject of chapter 21. If fire is to be reintroduced as an ecological process and a management tool, conflicting societal goals must be reconciled. Chapter 22 discusses the impacts of fire on watersheds and aquatic resources, chapter 21 addresses smoke and air quality concerns, chapter 24 deals with the problem of invasive plants, and chapter 25 deals with the conflict between fire and at-risk species. The specter of climate change as it affects fire is explored in chapter 26. As human settlements have encroached on former wildlands, the social dynamics of fire have become increasingly important. These dynamics are discussed in chapter 27. In chapter 28, we summarize the concepts developed in the book and present challenge Californians to accept the fact that they live in fire-prone ecosystems.

CHAPTER NINETEEN

The Use of Fire by Native Americans in California

M. KAT ANDERSON

> Speech, tools, and fire are the tripod of culture and have been so, we think, from the beginning.
>
> SAUER (1981)

The use of fire is an important dimension of human evolution. Its advent enabled our species to move around the world, occupy higher latitudes and elevations, thrive in cold environments, and, perhaps most importantly, cook food and thereby extract calories from it with much greater efficiency (Wrangham 2009). The use of fire by humans may be more than 400,000 years old (Weiner et al. 1998). It is about this time in the archaeological record that proper hearths consisting of rings of stones, burned bones, and other clear evidence of fire used for cooking become common throughout Europe (McCrone 2000). Fire appears to have been used at this time to drive and hunt wildlife as well (Boyd 1999).

If learning how to use fire figured prominently in human evolution, then gaining the knowledge needed to influence vegetation patterns with fire was surely one of the most important achievements of the human species; it shifted our status from that of foragers to managers of nature (Lewis and Anderson 2002). Because of the power over natural resources and productivity that it gives humans, the use of fire as a land management tool has been virtually universal across human cultures since well before the beginning of settled agriculture. Thus, the first humans to occupy California over 12,000 years ago most likely brought with them knowledge of how to burn the landscape to favor wildlife and increase hunting success (Rosenthal and Fitzgerald 2012). Certainly by the time Europeans first came into contact with them, the indigenous people of California had developed to a high art the practice of burning vegetation to achieve a variety of cultural objectives. When Spanish explorer Juan Rodríguez Cabrillo anchored in San Pedro Bay in October of 1542, it was the chaparral fires that gave him the signal that the coast was occupied by humans (Kelsey 1986, p.143) (Map 19.1). A succession of explorers, missionaries, and settlers coming to California thereafter would continually note in their journals the "smoky air" from these fires in every corner of the state—in the coastal redwood (*Sequoia sempervirens*) forests, the tule (naked-stemmed bulrush [*Schoenoplectus* spp.]) marshes of the Delta, the southern oak (*Quercus* spp.) woodlands, the mixed-conifer forests, and the northern hazelnut (*Corylus* spp.) flats (Thompson [1916] 1991, Sutter 1939, Timbrook et al. 1993).

The success of indigenous economies depended on setting fires. A large proportion of most tribes' food supplies depended directly or indirectly on management of plants and habitats with fire. In many areas of California, setting fires was also integral to the maintenance of cordage and basketry production systems—two essential cultural-use categories that required enormous quantities of high-quality plant material to satisfy human needs. Only widespread, careful, and effective fire management could have supplied the phenomenal quantities of food and raw materials required to support the large numbers of people—estimated to have been about 310,000 (Cook 1971)—that lived in prehistoric California. Given the large populations and its needs, estimates of the area that was burned annually by California's earliest humans are impressive. Martin and Sapsis (1992) estimated that between 2.3 and 5 million ha (5.7–12.4 million ac) of California burned annually under both lightning-caused and indigenous peoples' fires. A more recent synthesis by fire ecologists yields the similar estimate of area burned annually, which equates to 6–16% of California per year (excluding the southern deserts) (Stephens et al. 2007).

California Indians used a number of technologies including digging sticks, seed beaters, knocking sticks, knives, stone axes, and deer antlers that may appear "primitive" and unable to affect vast areas. But once Indians developed fire-making technologies—which probably occurred even before their entry into California—they had at their disposal a powerful tool that could alter whole landscapes. By rapidly rotating a slender wooden shaft in a hole in a stationary board to create intense heat by friction (drilling) or by striking two stones together to create sparks (percussion), Indians could make fire when needed (Driver and Massey 1957). Most tribes could also transport fire and start it without the aid of a fire drill by utilizing a slow match or torch, which consisted of a tightly packed flammable material that would smolder at one end for a considerable period of time (Dixon 1905, Barrett 1907). With these technologies, California Indians could start fires in nearly any kind of vegetation when weather conditions were advantageous.

Former indigenous burning patterns are a significant part of the historical ecology of many environments in California—a

MAP 19.1 The territories associated with California Indian language groups. Names in bold represent a language family of two or more languages and multiple dialects.

fact that makes this topic relevant to ecologists, conservation biologists, and land managers interested in restoring various areas to their pre-European settlement condition.

Native American Uses of Fire

Native Americans' uses of fire pervaded their everyday lives. Many of the food items they ate and many of the cultural objects they used and wore owed their existence, at least in part, to fire. They used fire to keep the country open, provide forage for wildlife, drive and capture animals, fell trees, manage pests and diseases, encourage the growth of plant material that could be used to manufacture cultural items—particularly baskets—and enhance the growth of plants, plant parts, and fungi used for food and medicine.

Keeping the Country Open

Burning areas to "decrease the brush" was a nearly ubiquitous practice in California. It facilitated hunting, made travel easier, increased visibility, and enhanced safety—thick underbrush could hide enemies and harbor dangerous animals such as rattlesnakes.

Writing about the Indians of northern California, de Massey (1926) reported: "The Indians, particularly in the spring and autumn, set the stubble in the pastures on fire to destroy the insects and reptiles, and to make hunting easier." The former openness of the country is substantiated by the fact that numerous tribes ran down deer, which would have been impossible in a brushy landscape.

Keeping the country open also had the advantage of reducing the chances of severe fire which, in a tribal territory, would not only have meant immediate loss of life, but would have spelled disaster for the long-term well-being of a village. If, for example, the kind of stand-replacing fire witnessed in modern times had occurred, it would have eliminated thousands of hectares of important tree food resources. Although these trees would eventually have been replaced through seed or vegetative means, it would not have been within the lifetime of the inhabitants or their children (Anderson 1993). Thus, it was not in tribes' best interest to allow catastrophic fires to occur. During a study conducted on indigenous burning in the Yosemite, Sequoia, and Kings Canyon regions, Native American elders commonly stressed that Indian-set fires "did not hurt the big trees" (Anderson 1993). Burning to keep the brush down provided the environmental context within which more localized burning could then be conducted for specific cultural purposes.

Managing and Hunting Wildlife

An important reason for setting fires was to increase forage for wildlife. Anthropologist Harold Driver (1939) recorded that the Wiyot burned every two or three years to increase feed for deer. José Joaquin Moraga, a chronicler of the second Anza expedition, jotted down in 1776 that "The heathen [probably the Ohlone] had burned many patches [southeast of the Mission of San Francisco], which doubtless would produce an abundance of pasturage."

Today, Native American elders from different tribes substantiate the importance of burning for wildlife. Sierra Miwok elder Bill Franklin learned about burning from his father and grandfather: "They said the Indians used to burn in the fall—October and November. They set the fires from the bottom of the slope to decrease the snowpack, get rid of the debris so there's no fire danger and they burned in the hunting areas so there was more food for the deer. They burned every year and in the same areas."

In addition to using fire to increase the food supply for valued animals, Native Californians employed it in an abundant number of ingenious ways to lure, capture, or drive wildlife. The Tubatulabal waved torches under trees where quail were roosting at night; as the birds flew down they were easily clubbed (Voegelin 1938). Ishi (Yahi) told of hunters using fire to kill black bears (*Ursus americanus*). A number of men would surround an animal, building a circle of fire about him. They then would discharge arrows at him, attempting to shoot him in the mouth. If the bear charged an Indian, he defended himself with a firebrand (Heizer and Kroeber 1979). The California golden beaver (*Castor canadensis* subsp. *subauratus*) was hunted by the Sierra Miwok by first burning off the tule around its pond, thus exposing the entrances to the animal's house, and creating bare ground in which to dig out the beaver (Barrett and Gifford 1933). Many tribes drove deer using fire, including the Pit River and the Owens Valley Paiute (Steward 1935, Olmstead and Stewart 1978). The use of fire to capture red-legged grasshoppers (*Melanoplus femurrubrum*) was a widespread phenomenon in California. It was a tool of the Yuki, Pomo, Pit River, and many other tribes (Foster 1944, Merriam 1955, Olmstead and Stewart 1978). Both grasslands and meadows were burned to retrieve grasshoppers, including the grassland understory within mixed-conifer forests.

Creating Plant Material for the Manufacture of Baskets and other Cultural Items

Cultural items made from plant material constituted the bulk of the material culture of all California Indian tribes. These fire-dependent cultural items included weapons, armor, cordage, household utensils, tools, fire drills, structures, baskets, traps and snares, fishing gear, fish weirs, musical instruments, clothing, ceremonial regalia, games, and boats. California Indians burned two general types of plants to create materials for cultural items: in one group were shrubs and trees, and, in the other group, perennial grasses. Each group responds in somewhat different ways to fire and produces materials with different characteristics and usages.

Although older wood made significant contributions to tribes' material cultures—particularly in the form of firewood and the support members of structures—it was not utilized in nearly as diverse ways as the young material produced by plants after burning.

After many shrubs and trees are burned, the young shoots they produce—termed "sprouts" or "suckers" by horticulturists and "epicormic" or "adventitious" shoots by plant morphologists—have vigorous growth characterized by both an upsurge of vertical development and retardation of lateral branching. In general, sprouts have a suite of characteristics that make them suitable for use in making cultural items: they are flexible, straight, long, unvarying in diameter, and free of lateral branches, buds, and blemishes. These shoots also tend to exhibit juvenile characteristics; flowering is generally absent until the shoots have reached a certain stage of maturity. These specialized growth forms do not occur readily in nature in the absence of perturbations (e.g., flooding, fire, herbivory); they are adaptive traits that enhance the plants' survival and regeneration in environments with frequent disturbance (Keeley and Zedler 1978, Philpot 1980, Kauffman and Martin 1990).

Of all the cultural items made from postfire sprouts, baskets were simultaneously the most important and the most demanding of raw materials with very specific properties. Baskets were the single most ubiquitous and essential possessions of a family (Anderson 2005a); each woman utilized workbaskets several times every day for activities as varied as seed harvesting, transport, food storage, and cooking. Because baskets often required precise construction for both aesthetics and functionality, they could be woven only from materials that met exacting criteria (Fig. 19.1) (Bates 1984, Anderson 1993, 1999, Mathewson 1998). To a large extent, plants produced satisfactory materials for baskets only when managed with fire. Pruning could also yield suitable sprouts (and was also used extensively to manage basketry materials), but it was more labor-intensive and did not have the auxiliary benefits of burning. In the absence of management, shrubs and trees used for basketry materials are largely composed of old, brittle, and crooked branch growth that is useless for basketry and sometimes harbors insects and diseases besides (Anderson 1999).

Among the shrub and tree genera used most widely for basketry materials and managed with burning were redbud (*Cercis*), California lilac (*Ceanothus*), hazelnut (*Corylus*), sumac or sourberry (*Rhus*), and willow (*Salix*) (Merrill 1923, Potts 1977, Farmer 1993, Anderson 1999). These genera have widespread distributions, exhibit suitable characteristics for basketry and

FIGURE 19.1 Eliza Coon, a Pomo woman, weaving a basket. Note the long straight branches protruding from the basket, signifying that the shrub from which the branches were obtained was pruned or burned prior to harvest (photo courtesy of the Smithsonian Institution, National Anthropological Archives, #47,749-D; photo taken by H. W. Henshaw, circa 1892–1893).

FIGURE 19.2 Contrasting plant architectures of managed versus wild sourberry (*Rhus trilobata*) (read from bottom to top). Weavers select branches with strict parameters: they must be flexible, straight, long, and with no lateral branching. Weavers burn or prune shrubs to create this young growth. Wild growth, which is several to many years old, on the other hand, exhibits many short lateral branches that are crooked and unsuitable for weaving.

other cultural items, and readily regenerate after repeated burning or pruning. Less widely used genera included dogwood (*Cornus*), maple (*Acer*), oak (*Quercus*), and plum (*Prunus*). Large quantities of young shoots from these plants were needed for basketry; one cradleboard, for example, required 500 to 675 sourberry sticks from at least six separate patches that had been burned or pruned prior to being harvested (Figs. 19.2 and 19.3).

The perennial bunchgrass most extensively used as a source of materials for baskets in California was deer grass (*Muhlenbergia rigens*), a large native bunchgrass occurring below 2,150 m (7,000 ft) along streams and in chaparral, oak woodland, and other plant community types (Peterson 1993). It was used as basketry material by tribes whose territories covered more than half of California. The part of the plant used for basketry was the flower stalk, which was essential in making the foundation of many kinds of coiled baskets (Merrill 1923). Like woody plants, deer grass was traditionally managed with fire, but the burning served a different function. Fire did not cause the plant to produce material with particular characteristics, as was the case for woody plants; rather, fire was necessary for maintaining populations, ensuring the vigor of individual plants, and maximizing the production of flower stalks. Fire had these effects because it decreased detritus, reduced competition, and recycled nutrients (Shipek 1989, Anderson 1996) and because it helped create and maintain the openings in the canopy conducive to the sunlight requirements of deer grass (Lathrop and Martin 1982).

Because of the importance of deer grass for basketry and the large quantity of flower stalks needed for a single basket—a Western Mono cooking basket would take about 3,750 flower stalks to complete, a quantity that required harvest from at least three dozen large bunchgrass plants—it must have existed in great abundance in former times (Beetle

FIGURE 19.3 Wahnomkot (Wukchumni Yokuts) displaying coils of split basketry materials that come from hundreds of branches of young shrubs and rhizomes from sedges (courtesy of the Yosemite Museum, Yosemite National Park; photograph by Frank F. Latta).

1947). Although deer grass is still gathered today by weavers of various tribes, it is more and more difficult to find in large colonies; plants are usually found in small, scattered populations of less than a dozen plants along roads, streams, and in meadows. And frequently the grasses contain very few stalks. This decline in deer grass populations may be due to the cessation of indigenous management with fire (Anderson 1996).

Bear-grass (*Xerophyllum tenax*) was another important source of basketry material that requires periodic burning. The young leaves of this plant were gathered by the Wailaki, Karuk, Tolowa, Yurok, and other tribes in northern California for the making of baskets (Clarke Memorial Museum 1985, Turnbaugh and Turnbaugh 1986) and are still highly valued by contemporary basket weavers (Heffner 1984). Because they are more pliable, stronger, and thinner than older leaves, the new green leaves produced after a burn makes the best basketry material. Burning of areas to enhance bear-grass has been recorded among the Karuk, Yurok, and Chilula in northwestern California (Gifford 1939, Kroeber 1939, Gibbs 1851, Clarke Memorial Museum 1985, p.51).

Enhancing Food Production

Many food plants with edible parts—bulbs, leaves, fruits, and seeds—occurred in open woods, meadows, prairies, or grasslands in California and required systematic burning to keep their populations healthy and abundant, and to keep surrounding vegetation from encroaching. These food plants included the oak trees of various species, which produced the acorns that were the staff of life for Native American cultures. Many tribes in California used fire as a vegetation management tool to ensure continual yields of high-quality acorns (Fig. 19.4) (Schenck and Gifford 1952, McCarthy 1993).

Also important for food were the seeds of many native grasses and diverse wildflowers, and tribes assured abundant harvests of these foods by burning the areas in which they grew at the appropriate times. Burning of areas supporting seed-producing food plants was carried out to facilitate harvesting, stimulate seed production, protect the perennial stock, replenish the annual stock, recycle nutrients, and remove detritus to allow for new growth. Burning meadows and open hillsides for these purposes has been documented for indigenous groups in many parts of California (Driver and Massey 1957, Anderson 2005a). The Paiute, for example, burned the brush in the hills near their winter villages and then broadcast seeds of blazing star (*Mentzelia* spp.) and pigweed (*Chenopodium* spp.) (Steward 1938).

Another food source was the underground swollen stems, bulbs, corms, and tubers of the various geophytic species referred to as "Indian potatoes," and these, too, were managed with fire. Their habitats, as well as specific plant populations, were burned to reduce plant competition, facilitate gathering, recycle nutrients, and increase the size and number of bulbs and tubers (Baxley 1865, Anderson 1993) (Fig. 19.5). There is solid archaeological evidence for the use of bulbs, corms, and tubers for food in California beginning over 9,000 years ago (Eric Wohlgemuth, archaeologist, pers. comm. 2015).

Fire was also used to enhance the production of greens, the edible leaves and stems of various herbaceous plants, which were attractive for their vitamins and minerals. Many plant species that produce edible greens required regular burning to maintain their quality and quantity. Clover (*Trifolium* spp.) patches, for example, were burned by the Wukchumni Yokuts, North Fork Mono, and Pomo (Peri et al. 1982, Anderson 1993). Aginsky (1943) records the "burning of herbage for better wild crops" among the Valley Yokuts, Chukchansi Yokuts, Western Mono, and Southern, Central, and Northern Miwok. The Maidu burned areas to encourage the growth of bulbs and greens (Duncan 1964).

Fruits were gathered in substantial quantities and often dried and stored for winter use. Fire was used as a management tool to maintain or increase the fruit production of native shrubs such as manzanita (*Arctostaphylos* spp.), elderberry (*Sambucus* spp.), western choke-cherry (*Prunus virginiana* var. *demissa*), strawberry (*Fragaria* spp.), blackberry (*Rubus ursinus*), California wild grape (*Vitis californica*), and currants (*Ribes* spp.). The Pit River, for example, burned fields and forests to stimulate the growth of seed and berry plants (Garth 1953). Peri et al. (1982) reported that the Pomo people burned manzanita shrubs and that their berries provided food. The Karuk burned huckleberry areas to enhance shrub growth and productivity (Harrington 1932). The Maidu, Foothill Yokuts, Western Mono, and Miwok tribes burned shrubs in order to thin dense canopies, reduce insect activity, and increase fruit production (Jewell 1971, Anderson 1993).

Fire also played a role in managing the substantial food source represented by various species of fungi. Tribes gathered many kinds of edible fungi in grasslands, shrublands, and forests, cutting the aboveground fruiting bodies with a

FIGURE 19.4 Mrs. Freddie, a Hupa woman, pouring water from a basket cup into acorn meal being leached in a hollow in the sand. To her right is an acorn-collecting basket. Setting fires under various kinds of oaks ensured a continual supply of nonwormy, disease-free acorns (photo courtesy of the Phoebe A. Hearst Museum of Anthropology and the Regents of the University of California: photographed by Pliny E. Goddart, 1902 Catalog No. #15-3329).

FIGURE 19.5 Alferetta and Grapevine Tom (both Pit River) digging *búlidum'* (*Lomatium californicum*) near Black Tom Bar (courtesy of the Santa Barbara Museum of Natural History, circa 1931–1932). The tubers were probably used medicinally and ceremonially. It was a common practice in many parts of California to dig the many different kinds of bulbs and tubers with hardwood digging sticks, replant propagules, and burn over areas to increase the numbers, densities, and size of these wild plants' subterranean organs.

stone knife and leaving undisturbed the "fine threads" (mycelium) under the ground (Anderson and Lake 2013). The mushrooms were dried in large quantities and prepared by baking, boiling, or roasting. Some tribes set fires to foster the growth of mushrooms; species known to benefit from fire include coccora (*Amanita calyptroderma*), sweet tooth (*Hydnum repandum*), fried chicken mushroom (*Lyophyllum decastes*), black morel (*Morchella elata*), woodland cup (*Peziza sylvestris*), coral fungus (*Ramaria violaceibrunnea*), and ponderosa mushroom (*Tricholoma magnivelare*). Traditional burning made mushrooms more plentiful and increased their size, likely by stimulating mycelial growth and releasing mineral nutrients, and cleared away thick duff that would block mushroom emergence. Today Native Mono elders comment that with years of fire suppression in the mixed-conifer forests of the Sierra Nevada, the "duff is too thick and it needs to have a fire come through." According to a Yurok woman interviewed by Lake (2007, p.649), "a lot of the underbrush" in the mixed evergreen forests of northwestern California "should be taken out" by "light burns, controlled burns" in order to "let the mushrooms grow."

Fire to Combat Insects and Diseases

A wide range of insects and diseases were in direct competition with Indians for plants important for their foods, technologies, and medicines in aboriginal California. Native people used fire as one method for controlling these pests and pathogens. Without the effective biological-control effects of fire, pathogenic and insect agents—capable of rendering plant parts completely useless for weaving or consumption—would likely have been a more significant cause of malnutrition and starvation, and indigenous fiber technologies might never have reached the level of sophistication they achieved.

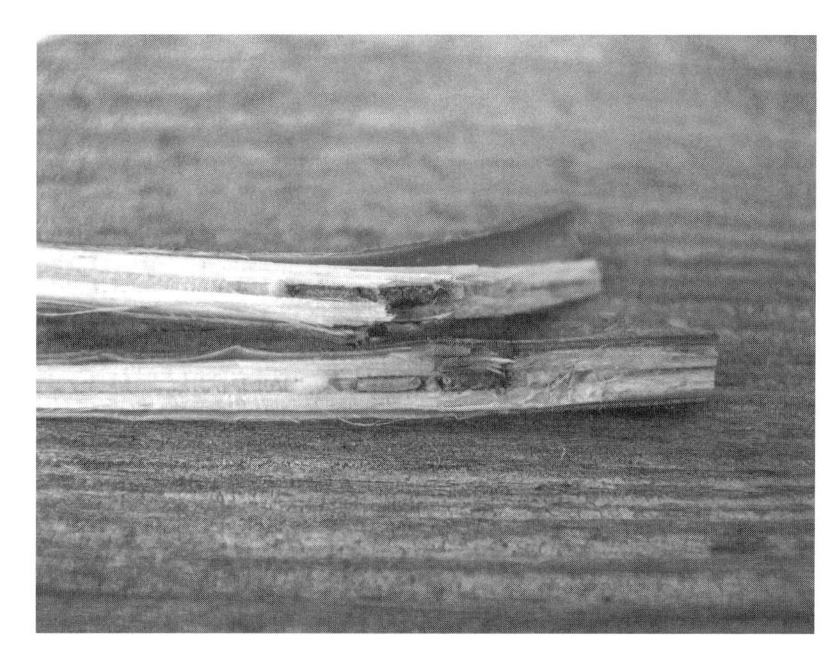

FIGURE 19.6 Basketry willow (*Salix exigua* var. *exigua*) with Agromyzidae (Diptera) larval chambers, probably *Hexomyza simplicoides* (photo by Frank K. Lake, Forest Service, in 2005; insect identified by Stephen Heydon, Senior Museum Scientist, Bohart Museum, UC Davis).

John Hudson interviewed Pomo basket weavers in the early 1900s who explained that the young branches of narrow-leafed willows (*Salix exigua* var. *exigua*) important for their craft contained a parasite "which destroys the product" (Hudson, n.d.). This may have been sawflies (*Euura* spp.), which are also a problem for Southern Sierra Miwok weavers in the Yosemite region and Karuk and Yurok weavers that use willows in northwestern California (Fig. 19.6) (Anderson, unpublished field notes 2001, Lake 2007). Pathogenic fungus can also cause gross deformities in plants important for basketry. Black knot (*Apiosporina morbosa*), for example, infects limbs of western chokecherry (*Prunus virginiana* var. *demissa*), an important basketry material to the Sierra Miwok, Maidu, and other tribes, forming cankerous swellings and dieback. Showy milkweed (*Asclepias speciosa*), an important plant for cordage and medicine, is susceptible to leaf spot (*Passalora* spp.) that blackens the leaves, pods, and stems and can sweep through whole stands (David Rizzo, Professor of Plant Pathology, UCD, pers. comm. 2013). Leaf blight (*Alternaria alternata*) attacks yerba santa (*Eriodictyon californicum*), a medicinal plant with widespread tribal use, causing lesions and appearing as a brown mold (Sinclair et al. 1987, Michael Davis, pers. comm. 2014).

Insects and diseases presented a formidable threat to food crops as well as to basketry materials and medicinals. For example, the achenes of mule's ears (*Wyethia* spp.) are subject to attack by a host of different insects, such as fruit flies in the genera *Neotephritis* and *Trupanea* that prey on developing ovules and maturing seeds, causing significant reductions in total seed yields in wild populations (Johnson 2008). Similarly, filbert weevils (*Curculio* spp.) and filbertworms (*Cydia latiferreana*) can damage a high percentage of the acorns produced by an oak tree (Fig. 19.7) (Swiecki and Bernhardt 2006).

FIGURE 19.7 Acorn of blue oak (*Quercus douglasii*) with insect damage of filbert weevil (*Curculio pardus*) in the larval stage. Different tribes burned under black, blue, and tanoak trees to help rid areas of this pest (courtesy of and photographed by Tedmund Swiecki, Principal/Plant Pathologist, Phytosphere Research).

Because diseases can infect whole stands or patches of plants through wind dispersal, wildlife dissemination, or other means, and insects can fly from one plant to the next, control efforts focused on individual plants are only minimally effective (Hardison 1976). Furthermore, insects and pathogens reproduce prolifically and pathogens can spend long periods of inactivity as dormant propagules, making them difficult to eradicate and a menace to economic security (Strange and Gullino 2010, Schumann and D'Arcy 2012). Indigenous people realized that the application of fire to patches and stands of plants covering small to large areas was the most effective tool for biological control of these organisms.

Dry Creek and Cloverdale Pomo weavers informed David Peri et al. (1982, p.117) that one of the main reasons for burning was to eliminate "parasites occurring on trees and shrubs." Florence Shipek's (1977, p.118) Luiseño consultants told her that "regular burning destroyed insect pests and parasites, such as dodder, which damaged food crops" and she recorded burning by the Luiseño to eliminate insect pests and parasites that damaged seed crops. In the fall of the year, the Yurok burned patches of oak, California hazels (*Corylus cornuta* var. *californica*), and California huckleberries (*Vaccinium ovatum*) to eliminate fungus and insects and improve the crop of nuts and berries (Warburton and Endert 1966).

Possible Ecological Impacts of Indigenous Burning

Although indigenous use of fire was carried out to realize specific cultural objectives, as discussed above—reinvigorating a particular patch of bear-grass to increase its production of basketry materials, for example, or capturing a deer to eat on a feast day—these uses of fire all had ecological consequences. Whether such consequences were the intended result of the burning or not, indigenous use of fire, in aggregate and over long periods of time, had impacts on the ecosystems, vegetation, and landscapes of California. The requirements for substantial amounts of raw materials with precise, fire-shaped qualities and the food needs of the large numbers of people who lived in many areas of the state point to extensive and regular use of fire over broad areas. Further, Native Californians' ability to produce very specific results through burning strongly suggests that they understood very well the reproductive responses of plants to fire, as well as its ecological effects at different levels of biological organization (Blackburn and Anderson 1993, Anderson 2005a). Combining these observations with the well-documented fact that indigenous people burned specifically to alter the character of vegetation, it becomes apparent that at the time Europeans first came on the scene, the landscapes, vegetation, and ecosystems of much of California had been significantly altered from their unpeopled, pristine conditions by California Indian tribes' use of fire.

The slow match or torch gave Native Americans the technological capability to burn both small patches and extensive tracts of vegetation in a systematic fashion. The existence of vegetation types—such as grasslands—that occur as continuous fuelbeds meant that fires could conceivably burn uninterrupted for miles. Fire was used for type conversions of areas for villages and, in southeastern California, for conversion of riparian habitat and floodplains for farming. Burning and hand weeding of young conifers or hardwoods were used in tandem to keep trees from encroaching on meadows or prairies. Galen Clark, guardian of the Yosemite grant for many years, observed burning and weeding among the Southern Sierra Miwok/Mono Lake Paiute in Yosemite Valley (Clark 1894). During the period of European settlement, fire was so commonly used by Native Americans as a habitat management tool that it threatened the agricultural, ranching, lumbering, and gold mining plans of the new settlers, causing the white authorities to draw up edicts, agreements, and proclamations prohibiting burning by American Indians.

Indigenous burning had discernable effects at every level of biological organization. It affected individual organisms, populations of plants and animals, the structure and composition of ecological communities, and the overall makeup of the landscape. And because it was practiced over many millennia, indigenous burning was likely to have influenced organisms at the genetic level, thus playing a role in the evolution of the flora.

Individual Organisms

Specific shrubs and trees were manipulated through spot burning, weeding, pruning, and knocking to enhance production, improve the quality of desired plant parts, and shape plant architecture (Fig. 19.8). Tribes in different parts of California purposefully pruned individual shrubs and trees repeatedly or piled brush onto individual shrubs and set them on fire to induce the sprouting of young shoots for arrows, looped stirring sticks, musical instruments, traps, baskets, regalia, cages, and many other items. These practices tended to keep plants in a physiologically young state, which may have prolonged their life spans.

Plant Populations

The tending of plant populations by burning—and by other techniques that often accompanied burning, such as sowing, tilling, and weeding—changed the distribution of the populations in space, affecting species' densities and abundances. Burning and sowing seeds of wildflowers such as gray mule's ears (*Wyethia helenoides*), farewell-to-spring (*Clarkia* spp.), and blazing star (Hudson 1901, Steward 1938) probably promoted high concentrations of these favored species in an area, and encouraged them to grow in a clumped or aggregated patterns. Over time, Native Americans assert, these techniques expanded the size of the gathering tracts of certain species. Selection for these desirable species probably led to the reduction of other less desirable species that grew in association with them.

The effect of encouraging populations of plant species at particular gathering sites was a high degree of "patchiness," with many medium-scale areas devoted to a single species. Many journals of early settlers describe landscapes made up of wildflower patches, each of a different color (Purdy 1976, Mayfield 1993). Patches of basketry grasses also were encouraged, such as deer grass colonies in ponderosa pine forests, chaparral, and blue oak woodlands. These patches were burned to increase flower stalk production for basketry, decrease dead material, and expand the tract.

Plant Communities

Indigenous burning practices were likely to have changed the physiognomy, structure, and composition of many communi-

The grade of basketry materials

- Flexibility
- Straightness
- Anthocyanins present
- Bark blemishes absent
- Long length
- Even diameter
- Lateral branching absent

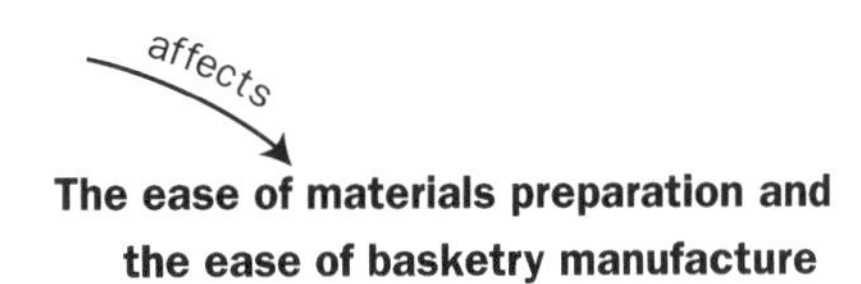

The ease of materials preparation and the ease of basketry manufacture

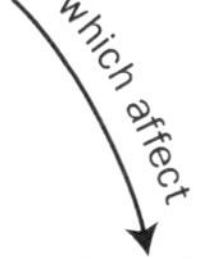

The functionality of the basket

- Long lasting
- Strong
- Watertight
- Holds shape
- Greater variety of shapes possible
- Ease of manipulating small particles
- Aesthetically pleasing

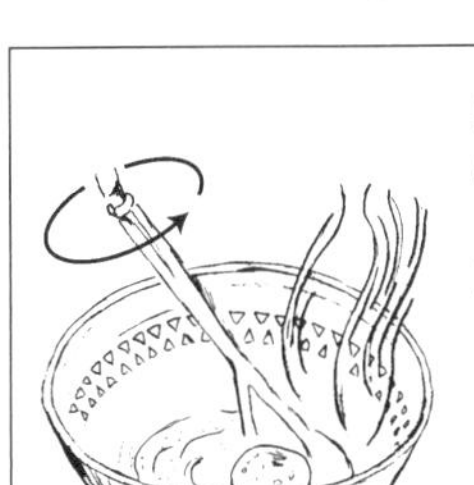

Example—cooking acorn mush:

A stirring implement, which holds a heated pumice stone, is rapidly circled in uncooked acorn meal and water in a watertight basket until the mush is cooked.

FIGURE 19.8 There are strong links between quality and quantity of plant material, ease of manufacture, and functionality of finished product. Frequently, native plants were not abundant enough or of the proper grade in their wild state, necessitating fire management.

ties. Burning, along with pruning, weeding, sowing, and replanting of bulblets, encouraged a higher level of species diversity than would have existed otherwise because it increased spatial heterogeneity, regularly reintroduced "intermediate" disturbance, and offered more efficient cycling of nutrients. Under indigenous burning regimes, hardwood and softwood forests, for example, had wider spacing between trees and greater proportions of large, mature trees compared to their unmanaged counterparts. In these communities, increased insolation on the forest floor and larger areas of exposed soil heightened the seed germination rates of herbaceous plants, and probably led to an increase in plant species diversity on an area basis (Fig. 19.9). Furthermore, Indian burning created and maintained larger areas of transition zones or ecotones than would have existed without anthropogenic fire, expanding the abundance of edge-effect niches and thus the populations of organisms that require them. Similarly, Indian burning tended to produce a mosaic of areas at different stages of succession within plant communities because Indian economies depended on foods and raw materials produced in the greatest quantities and in the best condition by plants and patches at different stages of growth and maturity (Anderson and Rosenthal 2015).

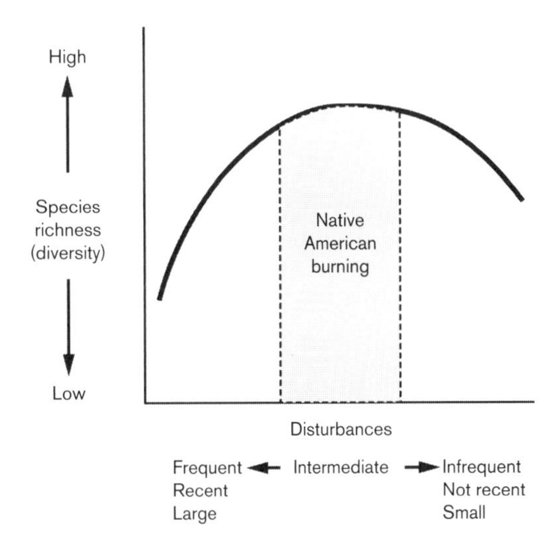

FIGURE 19.9 Intermediate disturbance hypothesis, after Connell 1978. It can be postulated that the temporal and spatial scales of Native American burning and other indigenous disturbances most closely fit the "intermediate" zone (gray region).

The plant communities exhibiting the greatest effects from indigenous management were those subjected to the most frequent burning. Typically these were communities that harbored a diversity of different resources, each of which required a different regime of fire-based management. The black oak–ponderosa pine forests in the Sierra Nevada of California are a good example; they were managed by the Western Mono, Sierra Miwok/Mono Lake Paiute, Foothill Yokuts, and other tribes for at least eight purposes: increasing woodland cup and black morel production, facilitating acorn collection, increasing rapid elongation of epicormic branches on oaks for the manufacture of items, reducing the incidence of the insect pests that inhabit acorns (filbert weevils and filbertworms) (Fig. 19.7), curtailing diseases that attack the trees with smoke from ground fires, promoting useful understory grasses and forbs, promoting a vegetative structure that increased acorn production, and eliminating brush to inhibit catastrophic fires (Anderson 1993). In resource-rich communities such as these, an anthropogenically created plant community structure was both the indirect consequence and the intended goal of indigenous management with fire.

Native American groups recognized that some plant community types covering small land surface areas, such as ponds, marshes, meadows, and prairies, harbored extremely useful and varied plant and animal life and therefore merited special attention in the form of hand clearing and careful burning. Burning maintained and may have expanded some of these special plant community types and subtypes by arresting the process of succession (in the case of dry meadows surrounded by forest, for example) and aiding in the cycling of nutrients held in dead biomass (such as in patches

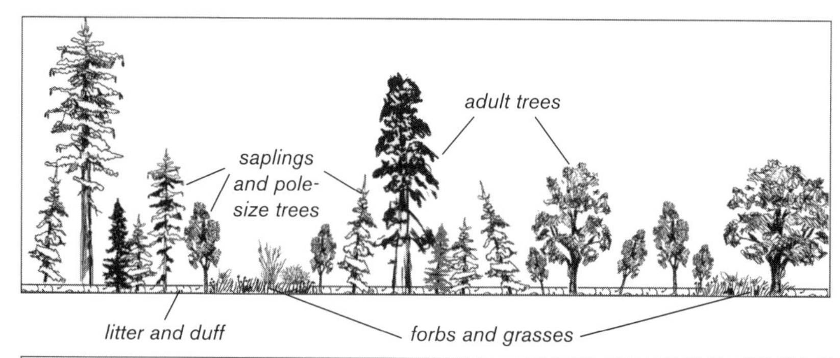

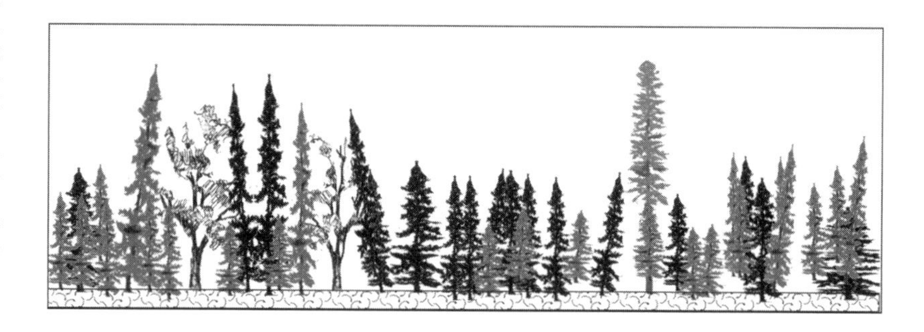

FIGURE 19.10 Mixed-conifer forest types created by three local fire regimes. At top, medium-density forest with fires due to lightning only (every 15 years); medium biodiversity, with medium rate of nutrient cycling and medium depth of litter and duff. In the middle, park-like, pine-dominated forest with fires due to Native American burning (every 2 to 5 years) and lightning (every 15 years); greatest biodiversity, with highest rate of nutrient cycling and thinnest layer of litter and duff. At bottom, dense, fir-dominated forest due to fire suppression management or rocky area not susceptible to burning by lightning or Native Americans; least biodiversity, with lowest rate of nutrient cycling and thickest layer of litter and duff (adapted from Anderson and Barbour 2003).

of tules and cattails [*Typha* spp.] in marshlands) (Anderson 1993, Anderson and Moratto 1996).

The removal of burning by Indians has contributed to the loss of certain forests that are defined by the dominance of large, culturally significant trees such as sugar pine (*Pinus lambertiana*), California black oak (*Quercus kelloggii*), and tanoak (*Notholithocarpus densiflorus* var. *densiflorus*), which harbor tremendous vertebrate and invertebrate biodiversity. These forests include black oak–ponderosa pine-sugar pine forest in the Sierra Nevada, Douglas-fir (*Pseudotsuga menziesii* var. *menziesii*)–tanoak forest in the North Coast Ranges, and black oak in mixed-conifer forests of the Western Klamath Mountains (Fig. 19.10) (Bowcutt 2015, Long et al. 2015). Grassland and meadow communities once favored and kept open through burning are also shrinking in number and size. These include prairies in coastal redwood forests, coastal prairies, valley grasslands, montane meadows, thousands of smaller stringer meadows in our mountain ranges, and grassy-forb understories of open woodlands and forests.

The Landscape

Landscapes can be viewed as mosaics of ecosystems, generated by physical and ecological processes (Pickett 1976). By influencing some natural processes and altering the constituent ecosystems, Indian management with fire had effects at this broadest scale of biological organization. In particular, it probably maximized landscape heterogeneity in many areas of the state (Fig. 19.11). Greater spatial heterogeneity in the landscape resulted from the maintenance and expansion of certain valued community types such as meadows, the intentional maximization of heterogeneity within forest and woodland communities with broad extent, and the variability deriving from the uneven pattern of indigenous occupation on the land.

The distribution of some plant communities over the California landscape today, along with their composition and extent, may be in part a product of Indian burning. As the climate changed, causing woody species to invade coastal habitats and the incidence of lightning to decline, Indians continued to keep some habitats open with fire in a kind of holding pattern. In fact, entire habitats may be dependent upon this continued human intervention for their survival; these include coastal prairies, open woodlands and forests, and early successional Labrador tea (*Rhododendron columbianum*) wetlands along the northern California coast (Blackburn and Anderson 1993, Guerrant et al. 1998).

Indigenous use of fire also had effects on the landscape's physical processes. Native people assert, for example, that regular burning promoted an abundance of water in springs and creeks (Duncan 1964, James Rust, Southern Sierra Miwok, pers. comm. 1989). This phenomenon may have been the result of fire reducing the total leaf surface area of plant com-

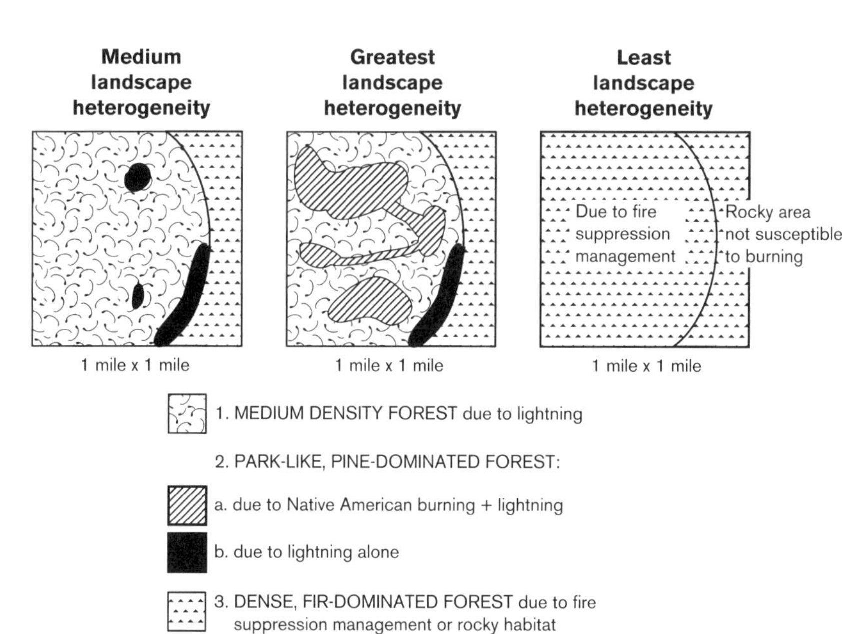

FIGURE 19.11 Hypothetical mosaic as seen from above of mixed-conifer forest types created by three different regional fire regimes. Types 1, 2, and 3 are those shown in Figure 19.10. It can be postulated that lightning and Native American burning resulted in the greatest landscape heterogeneity.

munities, which would have reduced transpiration and the uptake of soil moisture. In the Sierra Nevada, burning at mid-elevations was for the express purpose of shaping the nature of the snowpack. By removing shrub and duff layers and reducing foliage interception of snow, burning promoted a more tightly assembled, denser snowpack that melted off more slowly in the spring, reducing flooding and causing ephemeral creeks and streams to run longer in the summer (Jewell 1971).

Extent and Degree of Fire-Mediated Human Influence

Given California's diverse habitat types and wide variations in their likelihood of carrying a fire, along with the spatial unevenness of indigenous occupation, the influence of indigenous burning across the landscape was far from uniform. While fire was applied regularly and intensively in many areas, others probably experienced very infrequent Indian-set fires. Among the places in the latter category were the subalpine forests, the driest desert regions of southern California, and the alkali flats and serpentine balds with widely spaced plants, all of which do not burn readily. Also largely excluded from Indian burning were some remote mountainous areas not frequently visited by Native Americans and certain areas in many tribes' territories that were considered off limits to burning.

The unevenness of indigenous application of fire on the California landscape can be viewed as a continuum from very little or no Native American influence to fully human-created ecosystems. The serpentine barrens and subalpine areas of various parts of California would qualify as uninfluenced wilderness at one end point, and the agricultural fields of the Mojave, the coastal prairies of the northwest coastal tribes of California, and the desert fan palm oases of the Cahuilla in southern California would be among the heavily fire-influenced communities at the other end point. Other communities and vegetation types fall somewhere in between these two extremes, each reflecting some degree of indigenous influence with fire. The most heavily influenced landscapes make up perhaps 20% of California.

In the regions, communities, and vegetation types in which fire was undoubtedly a significant factor in shaping the vegetation, there is still some debate about the relative importance of human-set versus lightning-ignited fires. The incidence of lightning increases substantially from the coast up to the higher elevations of the Sierra Nevada (van Wagtendonk and Cayan 2008), leading most fire ecologists to conclude that prehistoric fire regimes in Sierra Nevada montane forests, as indicated by fire scar records, are more a function of natural ignition and vegetation inflammability than of Indian burning. Additionally, if fire ecologists can show that fire regimes are intimately linked to wet and dry cycles, then they are apt to conclude that fires are correlated with climate changes and therefore are natural, not of Indian origin (Swetnam 1993).

During the past few decades, however, researchers have been increasingly able to tease apart the history of human- and lightning-caused fires by using interdisciplinary studies incorporating archaeological and ethnographic data; new techniques such as charcoal analysis of soils, phytolith analysis, and pyro-dendrochronological studies; paleoecological data from pollen analyses; and data from the automated lightning detection systems now in place all over the state. This new research is providing strong correlations between human activity, fire, and vegetation in such places as Yosemite Valley, the southern Sierra Nevada, and the western Klamath Mountains (Crawford et al. 2015, Klimaszewski-Patterson et al. 2015). For example, in Yosemite Valley, a dramatic increase in oak pollen and a decline in pine pollen after Miwok settlement 500 years ago corresponded with increased fire and anthropogenic activity. Lightning fires are relatively infrequent in the valley and thus a direct cause-and-effect relationship was inferred between human-set fires and vegetation change (Anderson and Carpenter 1991). Ecologists Bill Kuhn

FIGURE 19.12 Michael Bonillas, Amah Mutsun, conducting a cultural burn in deer grass (*Muhlenbergia rigens*) habitat in Pinnacles National Park to recycle nutrients, stimulate new growth, and increase flower stalks for basketry. This is a collaboration between the Amah Mutsun, Pinnacles National Park, and UC Santa Cruz (photograph by Rick Flores, U.C. Santa Cruz Arboretum, 2011).

and Brent Johnson (2008) support this conclusion, arguing that Native Americans changed the fire regime in Yosemite Valley substantially, instituting a regime of frequent fires that created this iconic landscape.

Research in the past 25 years has also found that in zones with very low frequency of lightning strikes, such as along California's coast, whole ecosystems exist that are clearly fire-dependent. These coastal ecosystems, which include bishop pine (*Pinus muricata*) forests, Oregon oak (*Quercus garryana*) woodlands, coastal redwood forests on drier sites, and early successional Labrador tea wetlands, show evidence of relatively frequent fire in prehistory, most of which must have been set by humans because lightning would have been insufficient. What has happened to these coastal ecosystems since the removal of indigenous cultures and their fire management is further evidence that they were managed and maintained by Indian burning: they are all in decline, and whole habitat types with their unique suites of species are disappearing (Guerrant et al. 1998, Stuart and Stephens 2006). Further, ecologists are recognizing that a general decline in biodiversity in coastal areas is likely linked to the absence of indigenous management with fire.

Martin and Sapsis (1992) offer a way of transcending the debate over whether the prehistoric fire regime in California was driven by human or natural causes. They believe that Indian-set and lightning fires together resulted in a high level of *pyrodiversity*, which had the beneficial effect of creating greater biodiversity. This fire-dependent biodiversity was reflected in more diverse habitat types, more diverse physiognomies in woodlands and forests, more mosaics of resources at different stages of succession, and greater representation of species in forests that are "keystones" both ecologically and culturally (Long et al. 2015).

Because indigenous people had the technological capability and the economic motivation to expand the season of burn and to light fires in hard-to-reach places with varying topographies and soils whenever and wherever conditions permitted, they were able to push beyond the spatial and temporal constraints of lightning ignitions. They did this to varying extents in different parts of the state, but nowhere did they rely on lightning alone to start the fires their economies required. By intervening to change the frequency, season, intensity, and pattern of burning, they effectively took over control of the fire regime in certain areas.

Implications for Management Today

The legacy of Indian burning has much meaning for environmental management today. Most fundamentally, perhaps, the recognition that many of the "wild" landscapes and plant communities that we value for their biodiversity, ecological services, resources, and aesthetic values were in fact shaped and maintained by indigenous burning provides important perspective on both the role of fire in environmental management and on management itself. It validates management as an activity, reveals the wrong-headedness of fire suppression, puts into question the leave-it-alone approach to wilderness and wilderness-quality wildlands, and brings to the forefront of management policy the concept that in many ecosystems indigenous burning existed as *part* of ecological processes.

Indian burning also lends itself to emulation by non-Indian land managers. Many of the cultural goals that Native Americans set for burning various habitats in the past are consistent with the goals of contemporary ecologists and forest and range managers of our public and private lands, land trusts, fire safe councils, and communities today. Native people, like non-Indians, valued beauty in the natural world, great plant and wildlife diversity, stewardship of resources that provide for human needs, low levels of insect and disease pests, and the safety of human communities. These goals can be accomplished today by judicious application of burning that mimics what Native people did.

FIGURE 19.13 Skip Lowry, Yurok, Maidu, and Pit-River, conducting a cultural burn on the Yurok Reservation to rejuvenate a hazelnut (*Corylus cornuta* subsp. *californica*) patch for future basketry material and foods. According to Skip, these cultural burns are essential to restoring the ecological health of the land and responsibly bringing back the traditional ways of the ancestors. This is a collaboration between the Yurok, Cultural Fire Management Council, the Nature Conservancy's Fire Learning Network, and Terra Fuego Resource Foundation (photograph by Elizabeth Azzuz, Yurok, CFMC Secretary, March 2015).

Redwood National Park's prescribed burning program in the Bald Hills honors Native American burning practices by simulating their techniques. It has increased feed for Roosevelt elk, encouraged the oaks, kept the prairies open, and fostered the sun-loving plants in the prairies just as the Hupa, Karuk, and Yurok once did with their fire management. In Whiskeytown National Recreation Area, where black oak woodlands were traditionally managed with fire by Wintu cultural groups, prescribed burning and silvicultural treatments favoring black oak and open diverse understories are restoring these landscapes by preventing encroachment by Douglas-fir. Significantly, this program seeks restoration specifically for the sake of "tribal values" (Long et al. 2015).

Fire-based management informed by knowledge of prehistoric practices is also being carried out by Native people themselves (Figs. 19.12 and 19.13). Some tribal elders and indigenous resource managers still retain detailed knowledge of how, why, and when to apply fire to the land. With a grant from the USDA NRCS, Lois Conner Bohna (Mono/Chukchansi) removes brush and burns under California black oak trees on leased land near the town of O'Neal to bathe the trees with smoke to curtail mistletoe and kill insects that overwinter in the duff, as her ancestors had done for eons (see Sidebar 19.1). Members of the Amah Mutsun Land Trust, a tribally owned trust, are bringing back onto their traditional lands the practices of seed beating, tillage of sedge beds for basketry, and burning of deer grass and other plants in partnership with Pinnacles National Park, UC Santa Cruz, and the Midpeninsula Open Space District in the San Francisco Bay Area. In Northern California, the Orleans/Somes Bar Fire Safe Council and private non-Indian land owners are teaming up with the Karuk to treat tanoak stands with fire to get better acorn crops and burn hazelnut flats to generate better basketry material. The US Forest Service is working with the Karuk and local communities around Happy Camp to restore controlled burns to high country ridge systems to create landscape-level fuel breaks. The US Fish and Wildlife Service Recovery Plan for the Western lily (*Lilium occidentale*), an important food plant to the Karuk, recognizes that "burning by Native Americans may have been a significant factor" in maintaining its habitat in early successional bogs and recommends research on the use of fire as a possible management approach for recovering endangered populations.

As valuable as fire can be as a management tool, it must be used with great care on today's landscapes. Since frequent, patchy, low-severity fires have not swept the land since indigenous people ceased to be its stewards a century or more ago, the wildfire cycle is out of kilter in many areas, increasing the threat of large and destructive fires like the 2014 Rim Fire, the largest in Sierra Nevada history. In today's overstocked forests, woodlands, and shrublands, fire may exhibit very different behavior than it would have in prehistory, thus creating vastly different ecological effects. Many of the nonnative weeds advancing into various habitats thrive with burning, so reintroducing fire where they have established may result in the exact opposite of what managers intend. At the same time, the arrival of exotic diseases and insects such as sudden oak death and the golden spotted oak borer—which have indiscriminately killed millions of tanoaks and coast live oaks, drastically changing the composition of some of our forests—introduces new wildcards for land managers and restorationists to consider. Climate change brings further complexities, putting forests under greater stress, drying out fuels earlier in the season, and increasing the frequency of weather patterns conducive to severe fires. Under these circumstances, the reintroduction of indigenous-type fire management may offer important benefits as long as it is pursued as part of management and restoration plans that take into account the new variables and changed environmental context. Indigenous-type burning may have as its most important role the stemming of the overall decline in diversity that we are seeing at all levels of biological organization. When guided by traditional ecological knowledge, prescribed burning of the sort applied by Native Americans can be essential in promoting species recovery and restoring declining habitats.

SIDEBAR 19.1 THE IMPORTANCE OF GOOD FIRES FOR BLACK OAK USED BY NATIVE AMERICANS

Jonathan W. Long and M. Kat Anderson

Native Americans have traditionally used fire in tending groves of California black oak to produce large amounts of high-quality acorns, a substantial food source that has remained important since ancient times. Many Native American families have long depended on being able to gather hundreds of pounds of black oak acorns, which are often preferred over any other acorn except tanoak. Beyond the tremendous utilitarian value of acorns and wood products from black oak, Native Americans cultural traditions celebrate black oak through first acorn ceremonies and emphasize the importance of the tree for sustaining valued wildlife species.

For millennia, Native Americans have set low-intensity fires (Figs. 19.1.1 and 19.1.2) that consumed litter, duff, damaged acorns, and dead branches; kept risk of severe fires low; controlled acorn pests such as the filbert weevil and filbert worm; and promoted desired understory plants and other conditions favorable for gathering to sustain people (Anderson 2005a). In addition, burning stimulated formation of sprouts used for basketry, tools, clothing, and other items. Oral histories indicate that tribal burning in black oak forests managed for acorn production may have been conducted as frequently as surface fuels and weather conditions allowed for a contiguous understory burn (Anderson 2005a), with fire return intervals below the typically reported median of 7 years to 9 years for the ponderosa pine and dry mixed-conifer forests that commonly have black oak trees (Van de Water and Safford 2011). Over time, tending practices are likely to have promoted desired conditions for high acorn production that include widely spaced oaks with broad, rounded crowns, large girth, and low branches. In contrast, wildfire-dominated systems appear to promote narrower trees without low branches (Fig. 19.1.3).

FIGURE 19.1.1 During a cultural burn under California black oaks (*Quercus kelloggii*) to prepare the grounds at an Indian Mission site for the traditional bear dance (photos by Danny Manning).

FIGURE 19.1.2 After a cultural burn under black oaks to prepare the grounds at an Indian Mission site for the traditional bear dance (photos by Danny Manning).

California black oaks are vulnerable to damage from fire, as McDonald (1969, p.15) described fire as "black oak's worst enemy"; however, large black oaks appear to tolerate fire and indeed thrive under the frequent low-intensity burning practiced by Native Americans and often emulated by Euro-American settlers and land managers. Frequent fires provide shade-intolerant black oak with openings and resources needed to flourish within forests otherwise dominated by conifers. Douglas-fir, ponderosa pine, incense cedar, and white fir tend to grow underneath black oaks (Fig. 19.1.4) and eventually overtop them (Fig. 19.1.5), which reduces the amount of light available to the oaks (Cocking et al. 2012). As California forests have become more overgrown without frequent fires, oaks often appear spindly, less vigorous, and produce fewer

FIGURE 19.1.3 A tall, narrow black oak maintained by wildfire in the Beaver Creek Pinery (photo by Carl Skinner).

FIGURE 19.1.4 A large black oak encroached by conifers on the Sierra National Forest (photo by Jonathan Long).

FIGURE 19.1.5 An overtopped black oak in Yosemite National Park (photo by Jonathan Long).

acorns, except where roadsides and other openings allow them to grow fuller crowns. The absence of fire encourages accumulations of needles, branches, small trees, and other fuels that increase the likelihood that oaks will be damaged when fire returns.

California black oaks have a distinctive capacity to resprout; in the wake of wildfires that kill tree overstories, this adaptation gives them a competitive edge over conifer trees and even allows them to quickly overtop understory shrubs. When fires have returned to forests under modern fuel and weather conditions, they have often been at intensities that have killed the trunks of mature black oaks (Fig. 19.1.6). Furthermore, fire can also be so intense that it completely kills black oak trees. Even with resprouting, recovery of mature oaks capable of producing acorns in large quantities may be set back by 80 years or more (McDonald 1969). Consequently, severe fires constitute a loss of important ecosystem services, even though black oaks, as a species and as a dominant forest type, may even increase as wildfires, temperatures, and drought take a toll on competing conifer trees.

That black oak both needs fire and is threatened by it suggests a Goldilocks effect in which fuels and structural conditions need to fall within particular margins for fire to favor large trees. By keeping fuel loads low, Native American tending would have perpetuated such outcomes. Studies of historical fire patterns based upon fire scars indicate that forests with black oak experienced frequent fires, and that Native burning likely increased that frequency in many areas. Restoring frequent, low-severity fire, particularly in formerly harvested groves and other areas that are accessible to gatherers, would help promote and conserve large black oaks that are not only important to Native Americans, but also to many birds, mammals, and other wildlife that depend on acorns and large cavities. Black oak also provides valuable functional diversity as a deciduous, drought-tolerant tree capable of resprouting within forests dominated by evergreen conifers. Consequently, mature black oak is an important focus for restoring heterogeneity and reestablishing the critical role of frequent fire in California forests.

FIGURE 19.1.6 A dead trunk of black oak (with curving black fire scar) within a stand of conifers within a high-severity patch of the Rim Fire (photo by Jonathan Long).

References

Aginsky, B.W. 1943. Culture element distributions, XXIV, Central Sierra. University of California Anthropological Records 8(4): 393–468.

Anderson, M.K. 1993. Indian fire-based management in the sequoia-mixed conifer forests of the central and southern Sierra Nevada. Final Report to the Yosemite Research Center, Yosemite National Park. United States Department of Interior, National Park Service, Western Region, Cooperative Agreement Order Number 8027–002. San Francisco, California, USA.

———. 1996. The ethnobotany of deergrass, *Muhlenbergia rigens* (Poaceae): its uses and management by California Indian tribes. Economic Botany 50(4): 409–422.

———. 1999. The fire, pruning, and coppice management of temperate ecosystems for basketry material by California Indian tribes. Human Ecology 27(1): 79–113.

———. 2005a. Tending the Wild: Native American Knowledge and the Management of California's Natural Resources. University of California Press, Berkeley, California, USA.

———. 2005b. Pre-agricultural plant gathering and management. Pages 1055–1060 in: R.M. Goodman, editor. Encyclopedia of Plant and Crop Science. Marcel Dekker, New York, New York, USA.

Anderson, M.K., and M.G. Barbour. 2003. Simulated indigenous management: a new model for ecological restoration in national parks by Anderson and Barbour published in Ecological Restoration 21: 269–277.

Anderson, R.S., and S.L. Carpenter. 1991. Vegetation change in Yosemite Valley, Yosemite National Park, California, during the protohistoric period. Madroño 38: 1–13.

Anderson, M.K., and F.K. Lake. 2013. California Indian ethnomycology and associated forest management. Journal of Ethnobiology 33(1): 33–85.

Anderson, M.K., and M.J. Moratto. 1996. Native American landuse practices and ecological impacts. Pages 187–206 in: D.C. Erman, general editor. Sierra Nevada Ecosystem Project: Final Report to Congress, Volume II. Wildland Resources Center Report 37. University of California, Davis, California, USA.

Anderson, M.K., and J. Rosenthal. 2015. An ethnobiological approach to reconstructing indigenous fire regimes in the foothill chaparral of the western Sierra Nevada. Journal of Ethnobiology 35(1): 4–36.

Barrett, S.A. 1907. The material culture of the Klamath Lake and Modoc Indians of northeastern California and southern Oregon. University of California Publications in American Archaeology and Ethnology 6(4): 239–260.

Barrett, S.A., and E.W. Gifford. 1933. Miwok Material Culture: Indian Life of the Yosemite Region. Yosemite Association, Yosemite National Park, California, USA.

Bates, C.D. 1984. Traditional Miwok basketry. American Indian Basketry and Other Native Arts 4(13): 3–18.

Baxley, W.H. 1865. What I Saw on the West Coast of South and North America and at the Hawaiian Islands. D. Appleton and Company, New York, New York, USA.

Beetle, A.A. 1947. Distribution of the native grasses of California. Hilgardia 17(9): 309–357.

Blackburn, T.C., and K. Anderson. 1993. Introduction: managing the domesticated environment. Pages 15–25 in: T.C. Blackburn and K. Anderson, editors. Before the Wilderness: Native Californians as Environmental Managers. Ballena Press, Menlo Park, California, USA.

Bowcutt, F. 2015. The Tanoak Tree: An Environmental History of a Pacific Coast Hardwood. University of Washington Press, Seattle, Washington, USA.

Boyd, R.T., editor. 1999. Indians, Fire, and the Land. Oregon State University Press, Corvallis, Oregon, USA.

Clark, G. 1894. Letter to the board of commissioners of the Yosemite Valley and Mariposa Big Tree Grove. August 30. Contained in the Yosemite Research Library. Yosemite National Park, California, USA.

Clarke Memorial Museum. 1985. The Hover Collection of Karuk Baskets. Clarke Memorial Museum, Eureka, California, USA.

Cocking, M.I., J.M. Varner, and R.L. Sherriff. 2012. California black oak responses to fire severity and native conifer encroachment in the Klamath Mountains. Forest Ecology and Management 270: 25–34.

Connell, J.H. 1978. Diversity in tropical rainforests and coral reefs. Science 199: 1302–1310.

Cook, S.F. 1971. The aboriginal population of Upper California. Pages 66–72 in: R.F. Heizer and M.A. Whipple, editors. The California Indians: A Source Book. University of California Press, Berkeley, California, USA.

Crawford, J.N., S.A. Mensing, F.K. Lake, and S.R.H. Zimmerman. 2015. Late Holocene fire and vegetation reconstruction from the western Klamath Mountains, California, USA: a multi-disciplinary approach for examining potential human land-use impacts. The Holocene 25: 1341–1357.

de Massey, E. 1926. A Frenchman in the goldrush, Part V. Translated by M.E. Wilbur. California Historical Society, San Francisco, California, USA.

Dixon, R.B. 1905. The northern Maidu. The Huntington California Expedition. Bulletin of the American Museum of Natural History 17(3): 119–346.

Driver, H.E. 1939. Culture element distributions X: northwest California. University of California Anthropological Records 1(6): 297–433.

Driver, H.E., and W.C. Massey. 1957. Comparative studies of North American Indians. Transactions of the American Philosophical Society 47(2): 165–456.

Duncan, J.W., III. 1964. Maidu ethnobotany. Thesis, Sacramento State University, Sacramento, California, USA.

Farmer, J.F. 1993. Preserving Diegueno basket weaving. Pages 141–147 in: C.L. Moser, editor. Native American Basketry of Southern California. Riverside Museum Press, Riverside, California, USA.

Foster, G.M. 1944. A summary of Yuki culture. University of California Anthropological Records 5(3): 155–244.

Garth, T.R. 1953. Atsugewi ethnography. UCPAR 14(2): 129–212.

Gibbs, G. 1851. Journal of the expedition of Colonel Redick through northwestern California, 1851. Pages 99–177 in: H.R. Schoolcraft, editor. Archives of Aboriginal Knowledge. Containing all the original paper laid before Congress respecting the history, antiquities, language, ethnology, pictography, rites, superstitions, and mythology of the Indian tribes of the United States. Vol. III. J.B. Lippincott & Co., Philadelphia, Pennsylvania, USA.

Gifford, E.W. 1939. Karok field notes, Part 1. Ethnological Document No. 174 in Department and Museum of Anthropology, University of California; Manuscript in University Archives. Bancroft Library, Berkeley, California, USA.

Guerrant, E.O., Jr., S.T. Schultz, and D.K. Imper. 1998. Recovery plan for the endangered western lily (*Lilium occidentale*). Prepared for Region 1 U.S. Fish and Wildlife Service, Portland, Oregon, USA.

Hardison, J.R. 1976. Fire and flame for plant disease control. Annual Review of Phytopathology 14: 355–379.

Harrington, J.P. 1932. Tobacco among the Karuk Indians of California. Bureau of American Ethnology Bulletin 94. Washington, D.C., USA.

Heffner, K. 1984. Following the smoke: contemporary plant procurement by the Indians of northwest California. Unpublished Document. Six Rivers National Forest, Eureka, California, USA.

Heizer, R.F., and T. Kroeber. 1979. Ishi the Last Yahi. University of California Press, Berkeley, California, USA.

Hudson, J.W. n.d. Vocabulary for basketry. G.H.N. Acc. No. 21,170. Collection of Grace Hudson Museum, Ukiah, California, USA.

Hudson, J.W. 1901. Field notebook. G.H.M. Acc. No. 20,004. Collection of Grace Hudson Museum and Sun House, Ukiah, California, USA.

Jewell, D. 1971. Letter to R. Riegelhuth, Sequoia and Kings Canyon National Parks. On file, Research Office, Sequoia and Kings Canyon National Parks, Three Rivers, California, USA.

Johnson, R.L. 2008. Impacts of habitat alteration and predispersal seed predation on the reproductive success of Great Basin forbs. Dissertation, Brigham Young University, Provo, Utah, USA.

Kauffman, J.B., and R.E. Martin. 1990. Sprouting shrub response to different seasons and fuel consumption levels of prescribed fire in Sierra Nevada mixed conifer ecosystems. Forest Science 36: 748–764.

Keeley, J.E., and P.H. Zedler. 1978. Reproduction of chaparral shrubs after fire: a comparison of the sprouting and seeding strategies. American Midland Naturalist 99: 142–161.

Kelsey, H. 1986. Juan Rodríguez Cabrillo. Huntington Library, San Marino, California, USA.

Klimaszewski-Patterson, A., S. Mensing, and L. Gassaway. 2015. Potential non-climate forest structure change in the southern

Sierra Nevada range using paleoenvironmental and archaeological proxy data. Pacific Climate Workshop 2015. March 8–11, 2015. Pacific Grove, California, USA.
Kroeber, A.L. 1939. Unpublished field notes on the Yurok. University Archives, Bancroft Library (quoted from Stewart 2002). University of California, Berkeley, California, USA.
Kuhn, B., and B. Johnson. 2008. Status and trends of black oak (*Quercus kelloggii*) populations and recruitment in Yosemite Valley (a.k.a. preserving Yosemite's oaks). Final Report to the Yosemite Fund. San Francisco, California, USA.
Lake, F.K. 2007. Traditional ecological knowledge to develop and maintain fire regimes in northwestern California, Klamath-Siskiyou bioregion: management and restoration of culturally significant habitats. Dissertation, Oregon State University, Corvallis, Oregon, USA.
Lathrop, E., and B. Martin. (1982) Fire ecology of deergrass (*Muhlenbergia rigens*) in Cuyamaca Rancho State Park, California. Crossosoma 8(5): 1–4, 9–10.
Lewis, H.T., and M.K. Anderson. 2002. Introduction. Pages 3–16 in: O.C. Stewart, H.T. Lewis, and M.K. Anderson, editors. Forgotten Fires: Native Americans and the Transient Wilderness. University of Oklahoma Press, Norman, Oklahoma, USA.
Long, J.W., L.N. Quinn-Davidson, R.W. Goode, F.K. Lake, and C.N. Skinner. 2015. Restoring California black oak to support tribal values and wildlife. Pages 113–122 in: R.B. Standiford, editor. Proceedings of the 7th California Oak Symposium. USDA Forest Service General Technical Report PSW-GTR-251. Pacific Southwest Research Station, Albany, California, USA.
Martin, R.E., and D.B. Sapsis. 1992. Fires as agents of biodiversity—pyrodiversity promotes biodiversity. Pages 150–157 in: H.M. Kerner, editor. Proceedings of the Symposium on Biodiversity of Northwestern California. Wildland Resources Center Report 29. Division of Agriculture and Natural Resources. University of California, Berkeley, California, USA.
Mathewson, M. 1998. The living web: contemporary expressions of Californian Indian basketry. Dissertation, University of California, Berkeley, California, USA.
Mayfield, T.J. 1993. Indian summer: traditional life among the Choinumne Indians of California's San Joaquin Valley. Heyday Books, Berkeley, California, USA.
McCarthy, H. 1993. Managing oaks and the acorn crop. Pages 213–228 in: T.C. Blackburn and K. Anderson, editors. Before the Wilderness: Native Californians as Environmental Managers. Ballena Press, Menlo Park, California, USA.
McCrone, J. 2000. Fired up. New Scientist 166(2239): 30–34.
McDonald, P.M. 1969. Silvical characteristics of California black oak (*Quercus kelloggii* Newb.). USDA Forest Service Research Paper PSW-RP-53. Pacific Southwest Forest and Range Experiment Station, Berkeley, California, USA.
Merriam, C.H. 1955. Studies of California Indians. University of California Press, Berkeley, California, USA.
Merrill, R.E. 1923. Plants used in basketry by the California Indians. University of California Publications in American Archaeology and Ethnology, Berkeley 20(13): 215–242.
Olmstead, D.L., and O.C. Stewart. 1978. Achumawi. Pages 225–235 in: R.F. Heizer, editor. Handbook of North American Indians. Vol. 8. California. Smithsonian Institution, Washington, D.C., USA.
Peri, D.W., S.M. Patterson, and J.L. Goodrich. 1982. Ethnobotanical mitigation Warm Springs Dam—Lake Sonoma California. In: E. Hill and R.N. Lerner, editors. Report of Elgar Hill, Environmental Analysis and Planning. Penngrove, California, USA.
Peterson, P.M. 1993. *Muhlenbergia*: muhly. Pages 1272–1274 in: J.C. Hickman, editor. The Jepson Manual: Higher Plants of California. University of California Press, Berkeley, California, USA.
Philpot, C.W. 1980. Vegetative features as determinants of fire frequency and intensity. Pages 202–215 in: H.A. Mooney and C.E. Conrad, editors. Proceedings of the Symposium on the Environmental Consequences of Fire and Fuel Management in Mediterranean Ecosystems. USDA Forest Service General Technical Report WO-3. Washington, D.C., USA.
Pickett, S.T.A. 1976. Succession: an evolutionary interpretation. The American Naturalist 110(971): 107–119.
Potts, M. 1977. The Northern Maidu. Naturegraph, Happy Camp, California, USA.
Purdy, C. 1976. My life and my times. Privately published by E.E. Humphrey and M.E. Humphrey (no location listed).
Rosenthal, J.S., and R.T. Fitzgerald. 2012. The paleo-archaic transition in Western California. Pages 67–103 in: C.B. Bousman and B.J. Vierra, editors. From the Pleistocene to the Holocene: Human Organization and Cultural Transformations in Prehistoric North America. Texas A&M University Press, College Station, Texas, USA.
Sauer, C.O. 1981. The agency of man on the earth. Pages 330–363 in: Selected Essays 1963–1975. University of California Press, Berkeley, California, USA.
Schenck, S.M., and E.W. Gifford. 1952. Karok ethnobotany. Anthropological Records 13(6): 377–392.
Schumann, G.L., and C.J. D'Arcy. 2012. Hungry Planet: Stories of Plant Diseases. APS Press, St. Paul, Minnesota, USA.
Shipek, F.C. 1977. A strategy for change: the Luiseño of southern California. Dissertation, University of Hawaii, Manoa, Hawaii, USA.
———. 1989. An example of intensive plant husbandry: the Kumeyaay of southern California. Pages 159–170 in: D.R. Harris and G.C. Hillman, editors. Foraging and Farming: The Evolution of Plant Exploitation. Unwin-Hyman Publishers, London, UK.
Sinclair, W.A., H.H. Lyon, and W.T. Johnson. 1987. Diseases of Trees and Shrubs. Cornell University Press, Ithaca, New York, USA.
Stephens, S.L., R.E. Martin, and N.E. Clinton. 2007. Prehistoric fire area and emissions from California's forests, woodlands, shrublands, and grasslands. Forest Ecology and Management 251: 205–216.
Steward, J.H. 1935. Indian Tribes of Sequoia National Park Region. U.S. Department of Interior, National Park Service, Berkeley, California, USA.
———. 1938. Basin-Plateau aboriginal sociopolitical groups. Smithsonian Institution Bureau of American Ethnology Bulletin 120. United States Government Printing Office, Washington, D.C., USA.
Strange, R.N. and M.L. Gullino, editors. 2010. The role of plant pathology in food safety and food security: plant pathology in the 21st century. Contributions to the 9th International Congress. Springer, Dordrecht, The Netherlands.
Stuart, J., and S.L. Stephens. 2006. North coast bioregion. Pages 147–169 in: N.G. Sugihara, J.W. van Wagtendonk, K.E. Shaffer, J. Fites-Kaufman, and A.E. Thode, editors. Fire in California's Ecosystems. University of California Press, Berkeley, California, USA.
Sutter, J.A. 1939. New Helvetia Diary: A Record of Events Kept by John A. Sutter and His Clerks at New Helvetia, California, from September 9, 1845, to May 25, 1848. Grabhorn Press and the Society of California Pioneers, San Francisco, California, USA.
Swetnam, T.W. 1993. Fire history and climate change in giant sequoia groves. Science 262: 885–889.
Swiecki, T.J. and E.A. Bernhardt. 2006. A field guide to insects and diseases of California oaks. USDA Forest Service General Technical Report PSW-GTR-197. Pacific Southwest Research Station, Albany, California, USA.
Thompson, L. [1916] 1991. To the American Indian: Reminiscences of a Yurok Woman. Heyday Press, Berkeley, California, USA.
Timbrook, J., J.R. Johnson, and D.D. Earle. 1993. Vegetation burning by the Chumash. Pages 117–149 in: T.C. Blackburn and K. Anderson, editors. Before the Wilderness: Environmental Management by Native Californians. Ballena Press, Menlo Park, California, USA.
Turnbaugh, S.P., and W.A. Turnbaugh. 1986. Indian Baskets. Schiffer Publishing, West Chester, Pennsylvania, USA.
Van de Water, K.M., and H.D. Safford. 2011. A summary of fire frequency estimates for California vegetation before Euro-American settlement. Fire Ecology 7(3): 26–58.
van Wagtendonk, J.W., and D.R. Cayan. 2008. Temporal and spatial distribution of lightning strikes in California in relationship to large-scale weather patterns. Fire Ecology 8(4): 34–56.
Voegelin, E.W. 1938. Tubatulabal ethnography. University of California Anthropological Records 2(1): 1–84.
Warburton, A.D., and J.F. Endert. 1966. Indian lore of the North California Coast. Pacific Pueblo Press, Santa Clara, California, USA.
Weiner, S., Q. Xu, P. Goldberg, J. Liu, and O. Bar-Yosef. 1998. Evidence for the use of fire at Zhoukoudian, China. Science 281(5374): 251–253.
Wrangham, R. 2009. Catching Fire: How Cooking Made us Human. Basic Books, New York, New York, USA.

CHAPTER TWENTY

Fire Management and Policy since European Settlement

SCOTT L. STEPHENS AND NEIL G. SUGIHARA

Everywhere, and from the earliest times, humans have altered the natural fire regimes they have encountered.

PYNE ET AL. (1996)

I request permission [to burn] from the State Forester and the USFS DuBois. Both refused. We proceeded to burn anyway, and Chief Forester Graves came out from Washington and DuBois and many others with cameras and notebooks to get damaging evidence. They stayed several days and followed the burning, with comment by Graves that the work was excellent. DuBois apologized to me for panning me in the newspapers previously.

CLINTON WALKER (1938), Red River Lumber Co., Southern Cascades, CA

Since European explorers first touched the shores of California, their activities, shaped by their needs and values, have altered the state's fire regimes. Fire regime changes have resulted, directly and indirectly, from a variety of human activities. At times these influences have been unintentional consequences of other land management activities; while in other instances they have been well planned and even codified.

Formal fire policy since Euro-American settlement is a response to society's and institutions' views of fire. Our understanding of the historical relations between fire and society is greatly enhanced if we review the setting in which that society existed. It is common for us to blame our current fire situation on the shortcomings and lack of perspective of past land managers, but this is rarely the case. The needs and values of society were the driving force of the past policies, and those needs and values have changed and will continue to change.

European Exploration Era

The first significant impacts on fire regimes that the European civilization brought to California actually predate the arrival of large-scale permanent settlers by over a century. The impacts were the introduction of the human diseases that decimated the populations of indigenous peoples and the introduction of plants from other parts of the world. Although both of these impacts involved the expansion of the historic ranges of biological organisms, they had very different mechanisms for influencing fire regimes. Both of these actions were inadvertent but were to have enormous impacts on California ecosystems that continue to the present.

Removal of Native American Fire Use

Manipulation of fire by Native Americans had many important impacts on the character and geographic distribution of California's fire regimes, and greatly modified fire as an ecological process (Stephens et al. 2007, Anderson and Rosenthal 2015). The removal of the Native Americans and their fire use had variable effects on California's ecosystems. Although there was little or no change to ecosystems where fire was very rare or where fire regimes were not altered by the activities of the Native Americans, there was often a profound change to ecosystems in the areas where they actively managed with fire, including many oak (*Quercus* spp.) woodlands, montane meadows, grasslands, wetlands, and coniferous forests (Anderson 2005). These ecosystems now supported a different burning pattern, replacing the specific pattern of ignitions by Native Americans and lightning with a new combination of burning by the new settlers and lightning (Taylor et al. 2016). Coastal areas experience little lightning and fire regimes in these areas were dominated by anthropogenic ignitions (Keeley 2002, Stephens and Fry 2005). Removal of anthropogenic fire from these ecosystems has brought changes in species composition, both by encroachment of invasive species and by conversion to other vegetation types, and increased fire hazards.

Introduction of Invasive Nonnative Plant Species

The introduction of nonnative invasive plant species began when the explorers visited the California coast during the 1500s and 1600s. The establishment of the Jesuit Missions starting in the late 1700s with their livestock expanded the ranges of nonnative grasses and forbs (Menke et al. 1996). Cattle ranching in California began on the coast, and by 1823 livestock grazing was an established activity at all 21 coastal missions. At their peak, the missions may have had more than 400,000 cattle grazing one-sixth of California's land area. In 1860 the US Census reported nearly a million beef cattle (not including open range cattle), just over a million sheep, and 170,000 horses in California. As range quality declined, sheep ranching gained in favor with a peak of 5.7 million animals in 1880 (Barbour et al. 1993). Among the newly introduced plants were several species that were adapted to the rangelands in the Mediterranean climates and outcompeted the native species.

Many of California's grasslands and woodlands are currently dominated by invasive nonnative species (Menke et al. 1996), and changes in species composition have influenced fire regimes. For example, fuel from introduced annual grasslands cures earlier, recovers faster, and has greater continuity than the native vegetation that previously occupied the sites. This extends the fire season into the earlier spring, shortens fire return intervals, and increases fire size. Nonnative species also colonize some ecosystems when openings are created by mechanical methods or high-severity fire. Chapter 24 includes an in-depth treatment of invasive species and fire in California ecosystems.

Early European Settlement Era

California Missions–Alta California

During the time period from 1769 to 1823 a string of Spanish missions were established in California. Private Spanish land grants to establish pueblos or towns were introduced in California starting in 1775. Each grantee was required to build a storehouse and to stock his holdings with at least 2,000 head of cattle. By 1790, there were 19 private rancheros in California. During the Spanish and Mexican periods, over 800 large grants of land were given to immigrants who settled in California. The effect of these grants was to expand the area of California being utilized by domestic livestock.

Although no large-scale direct manipulations of the fire regimes were made during this time period, the forced movement of Native Americans to the missions and introduction of nonnative grasses resulted in drastic alterations in the role of fire in California's ecosystems. The removal of widespread and often focused burning by Native Americans was accomplished indirectly by relocating the people and decimating the population by introducing diseases. This was an especially important change to fire regimes where focused, long-term, burning by Native American maintained vegetation in conditions that would not persist without them. As detailed in chapter 19 these ecosystems were often manipulated using fire to maintain game animal habitat and to increase the reliability of food crops, cordage materials, and medicines (Anderson 2005). This abrupt halt in the use of fire initiated ecosystem change in many areas of California.

The Gold Rush and Early Statehood

The miners and early settlers had several direct and indirect impacts on the fire regimes during this era. Early settlers used fire to clear land for improved grazing and to facilitate the search for gold. Sheepherders often set fires on the way out of the mountains in the fall to improve forage for the next year (Biswell 1989). The developing railroad system served as a new source of ignitions along the expanding railroad corridors. Widespread logging occurred in many areas near major mining centers to support both the mines and the population needed to work them. Much of the Lake Tahoe basin and areas of the western Sierra Nevada were logged to support the mines (Laudenslayer and Darr 1990). Meanwhile, fire suppression efforts were confined to the immediate protection of structures.

Fire Control Era

Federal fire policy was formally started with establishment of large-scale forest reserves during the late 1800s and early 1900s. In 1891 Congress authorized President Harrison to establish forest reserves, later to be known as national forests (Pinchot 1907, Stephens and Ruth 2005). The Forest Service was established as a separate agency in 1905 with Gifford Pinchot as its first chief. Under his direction, a national forest fire policy was initiated and the agency began systematic fire suppression including the development of an infrastructure of fire control facilities, equipment, fire stations, lookouts, and trails. The forest reserves were created partly because Congress believed the nation's forests were being destroyed by fire and reckless cutting (Pinchot 1907). Pinchot declared that one of the objectives of the national forests was to make sure that "timber was not burnt up."

A policy of fire suppression was not universally supported during this period. One of the most vocal groups that argued for the use of fire in forest management was a group of private foresters from the southern Cascades (Clar 1959, Pyne 1982). In the late 19th century 1880s they promoted the concept of "light burning" modeled after earlier Native American uses of fire. The main objective of this burning was to reduce fuel loads and associated damage when the inevitable wildfire occurred. Federal managers disagreed with this policy because of the damage to small trees and problems with fire escapes. Most foresters at this time believed that western forests were under-stocked and elimination of fire would ultimately produce higher yields of timber (Show and Kotok 1924).

Specific information on the early debate regarding fire management is contained in a letter dated January 21, 1938, when Clinton Walker wrote to the Red River Lumber Company Board, a company that once owned nearly 344,000 ha (850,000 ac) of timberland in northern California:

> "I do not see how anyone can help but come to the conclusion that the elimination of the customary summer fires which the forests have grown up with is the direct cause of the various pests which have assumed such damaging and alarming proportion since the advent of fire protection. The present stand of timber is very old, the larger Sugar and Ponderosa being five hundred [years] or perhaps more of age. That being the case, it seems certain that the forests have passed through every known cycle of climatic conditions, wet years, dry years, dry cycles, windstorm, etc. Even now with all the "Prevent Fires" agitation, 75 to 80% of the forest fires are started by lightning. So that even without the incendiary fires started or alleged started by Indians, lightning has kept the forests on fire from summer to summer whenever there was sufficient forest liter to support a running fire. The forest thrived on these fires, some damage was done, but the general condition of the forests when the white man first came into

California was very excellent. When we first began cruising and Father [T. B. Walker] began buying, the fire risk was considered as being very small and the beetle infection was not even given thought. Then came the foresters from Yale University and put the tourniquet on the forests."

Walker concluded in 1938:

> "With preliminary preparatory work, the use of bulldozers to crush down the thickets under the big trees, power tractors, pulled burners and a small crew of men to patrol the fires, I think they could again handle the light burning successfully. . . . I should like to see an earnest and competent attempt made to burn some area. It seems little short of a crime to sit by idly and allow the pests to destroy our magnificent forests."

Walker's observations turned out to be correct regarding the impacts of fire exclusion. It is interesting that some private forest managers and owners recognized these issues early in the 1900s, but unfortunately their views were not accepted by most people.

The Great Idaho Fire of 1910 was pivotal in the development of early fire policy (Pyne 2001, 2015). In this wildfire 78 firefighters were killed and over a million ha (2.5 million ac) of national forest lands were burned. It was time to "do something" about the wildfire threat and the 1910 fires instigated the creation of a national system of wildland fire protection. This fire also emphasized the need and importance of the new Forest Service, as before this fire some many US Senators and Representatives were working to dissolve it and move its lands to private interests. The Great Idaho Fire changed this dynamic and kept these vast lands under federal ownership.

Henry Graves, the second Chief of the Forest Service, was against the "light burning" practice declaring "the first measure necessary for the successful practice of forestry is protection from fire" (Graves 1910), although Graves did consider using fire in forest management. The earliest comprehensive federal fire control policy was written shortly after Graves was appointed Chief (DuBois 1914). William Greeley, the third Forest Service Chief, took over the agency in 1920 and continued the strong endorsement of fire suppression stating "The conviction burned into me that fire prevention is the number one job of American foresters" (Greeley 1951). During Greeley's 9-year tenure fire suppression was paramount in federal and private forest management.

A scientific study was initiated in California on the merits of fire suppression versus light underburning and its conclusions supported a strong fire suppression policy (Show and Kotok 1924). The philosophy that nature could be dominated and controlled contributed to these early policies. Passage of the federal Clarke-McNary Act in 1924 tied federal appropriations to the state first adopting fire suppression and this law effectively created a national fire suppression policy.

A new type of fire began in this era, one that burned at the interface of homes and wildlands. The Berkeley fire of September 17, 1923, is such a fire in the new and rapidly growing urban area that has become all too familiar to Californians in recent decades. Biswell (1989) describes the progression of the fire: "A strong hot, dry northeast wind quickly drove a fire through the grasslands and eucalyptus groves along the crest of the ridge above Berkeley. Firebrands from the eucalyptus allowed the fire to spread out of the wildlands onto the shingle rooftop of the first house at 2:20 pm. Within 40 minutes the fire spread throughout a one-half square mile area. Over a period of just two hours, 625 houses and other buildings were destroyed. At about 4:30 pm the cool moist coastal breeze took over and the firefighters extinguished the blaze." Today California has enacted some state and local legislation to improve land use planning and home construction in the urban interface but substantial improvements still elude the state.

In 1935 federal forest fire policy was updated to incorporate the "10 AM" policy, aimed at increasing suppression efficiency. This policy directed that all fires should be controlled in the first burning period or by 10 AM the following morning. To accomplish this objective a large labor force and improved access to wildlands were necessary. The newly created Civilian Conservation Corps (CCC) provided thousands of workers to assist in this effort (Anderson et al. 1941, Pyne 1982). Efforts were made to increase the effectiveness of fire suppression by developing better access to further reduce response times, mapping vegetation and fuel hazards, and keeping detailed records of any large fires that occurred (van Wagtendonk 1991).

The first national education campaign designed to influence public behavior regarding forest fires began when the Forest Service created the Cooperative Forest Fire Prevention Program in 1942 (USDA 1995a). This program encouraged citizens nationwide to make a personal effort to prevent forest fires. The campaign was modified three years later to produce the national "Smokey Bear" campaign that is still in existence. Smokey Bear has been one of the most successful public education campaigns in the United States. Indeed Smokey still visits thousands of primary schools each year and he has a web site and twitter account (http://www.smokeybear.com/).

World War II had a lasting influence on fire suppression. During the war, fire suppression efforts were minimal due to the war effort. However, after the war there was a new much more intensive fire suppression effort that included the widespread use of the tools that were developed and refined in the war. Aerial retardant drops, helitack crews, bulldozers, and smokejumpers became the new tools of choice and this new fire fighting force was very effective in continuing the policy of full fire suppression (USDA 1960, van Wagtendonk 1991). Tactics developed during wartime were shifted to civilian crews to improve fire suppression.

Fire Management Era

Beginning of Fire Use

Using fire to manage California private rangelands was common in the early 1900s through the 1960s (Biswell 1989, McClaran and Bartolome 1989, Stephens 1997). Private ranchers pooled resources and burned rangelands to increase forage for livestock. Fire use was tolerated by the state agencies that had oversight authority on private lands, but the burning was done by private citizens (Biswell 1989). Grassland and shrubland burning was accepted during this period but forest burning was not.

In 1962 the Secretary of the Interior requested a report on the elk management situation in Yellowstone National Park, and it identified fire suppression as a policy that was adversely impacting wildlife habitats and threatening forests (Leopold et al. 1963). In 1968 the Leopold Committee report was incorporated into Department of Interior policy and for the first time since 1916, the National Park Service viewed fire as a natural process rather than a menace (van Wagtendonk 1991). Shortly after the report was commissioned the first western federal prescribed fire occurred in California at Sequoia-Kings Canyon National Parks (1968) and two years later in Yosemite National Park (USDI 1968, van Wagtendonk 1991); earlier prescribed fires occurred in US National Park lands in the southeast US. This was the beginning of the use of fire on federal forested lands in the western United States. Creation of the national Wilderness System in 1964 also advanced the idea of wildland fire use in remote forested areas (Pyne 1982).

In 1968 the first experimental managed wildfire program in Sequoia and Kings Canyon National Parks was created (USDI 1968, Kilgore 1974, van Wagtendonk 2007). Lightning-ignited fires were allowed to burn and monitored to achieve resource benefits. This program became possible because of earlier research on the fire ecology of mixed-conifer forest in the Sierra Nevada (Biswell 1961, Hartesveldt and Harvey 1967) and because of the recent change in National Park Service fire policy. The era of wildland fire use in the National Park Service had begun; the long era of total suppression had ended (Kilgore 1974).

Prescribed fire in the California State Park system was initiated in 1972 at Montaña de Oro State Park on the Central Coast and then Calaveras Big Trees State Park in the Sierra Nevada (Biswell 1989). The program was initially the target of high levels of political and academic criticism, but persistent efforts to evaluate the biological impacts of fire and fire exclusion have supported the need for prescribed fire to maintain ecosystems.

Forest fire policy in the Forest Service changed from fire control to fire management in 1974. Henry DeBruin, Director of Fire and Aviation Management for the Forest Service, stated "we are determined to save the best of the past as we change a basic concept from fire is bad to fire is good and bad" (DeBruin 1974). This was a major shift for the Forest Service, but fire suppression was still to dominate for the coming decades. Some Forest Service wilderness areas such as the Selway-Bitterroot Wilderness in Idaho and Montana, and the Gila Wilderness in New Mexico began a program of prescribed natural fire but areas with similar management philosophies were few in number (Stephens and Ruth 2005), but this is beginning to change (Meyer 2015).

In 1978 the National Park Service further refined its fire management policy to describe the conditions under which fire could be used and specified that any managed wildfire would be suppressed if it posed a threat to human life, cultural resources, physical facilities, threatened or endangered species, if it threatened to escape from predetermined zones, or exceeded prescription (van Wagtendonk 1991). In 1988 a series of very large fires in the Greater Yellowstone Area again brought federal fire policy under review. While fire policy was generally reaffirmed, greater emphasis was placed on the development of fire management plans (van Wagtendonk 1991).

Cal Fire has used their Vegetation Management Program to encourage partnerships among private land owners to reduce fire hazards, primarily in the urban-wildland interface. Although this program has been successful in a relatively small number of communities, Cal Fire still emphasizes fire suppression. Certainly the complexity of Cal Fire not owning the majority of the lands where it has fire responsibility makes it difficult to initiate fuels management programs, diverse land ownership has resulted in many small landowners which also make it difficult to manage fuels. Cal Fire is preparing a Programmatic Environmental Impact Report (PEIR) that will serve as the overarching CEQA-compliant process for all vegetation treatment projects on State Responsible lands in the state of California as administered through Cal Fire. The Vegetation Treatment Program will be administered and implemented through contracts between Cal Fire and private landowners, other agencies, and other organizations. The overall goal is to create more partnerships with land owners to implement fuel reduction treatments in appropriate vegetation types.

One area where the state and counties have tried to improve fire management has been by passing and enforcing laws that mandate the use of combustion resistant construction materials and designs and defensible space for every structure in the wildland-urban interface. Presently the counties have the majority of the jurisdiction in this area, but in most cases, they have been unwilling or incapable of providing this essential oversight. Losses of life and property in the wildland-urban interface will only be reduced when both private homeowners and adjacent wildland managers take steps to reduce fire hazards and risks. Unless both sides of the interface (private homeowners and adjacent wildland managers) take steps to reduce their vulnerability to wildfire, high losses will continue in this expanding area of California. Australia has been more successful in engaging its citizens in working to solve the problems of fire in the urban interface (Gill and Stephens 2009, Stephens et al. 2009, Gibbens et al. 2012, Gill et al. 2013, Sneeuwjagt et al. 2013).

In 1989 Harold Biswell asked, "Is fire management on a collision course with disaster?" and answered " Perhaps, because wildfires continue to become more intense and destructive of resources, and expenses in fire control are increasing at an astronomical rate" (Biswell 1989). Today, the practice of managing ecosystems to modify future fire behavior has become an important land management activity (Stephens et al. 2012). The wildland-urban interface has become one of the focal points for application of fuel management in California but much more needs to be done (North et al. 2012, 2015). While society as a whole continues to value the suppression of wildfire and we are still impacting ecosystems by reducing the impact of fire as an ecosystem process, there is an increasing recognition among land managers that the long-term exclusion of fire has changed fuel dynamics (especially in forests that once burned frequently with low-moderate-intensity fire regimes) and is changing the pattern of uncontrollable wildfires (Stephens et al. 2016, 2018). This challenge has become even more difficult with warming climates (Stephens et al. 2013).

Overview of Some Key Historic Fires

The human interaction with fire during the period since Euro-Americans first settled in California has primarily been a struggle to reduce negative impacts of wildfire on society. Throughout this period, there has been effort to remove fire from California's ecosystems. Even now that fire is widely recognized as an important ecological process, efforts to restore and manage fire are dwarfed by suppression of wildfires. As new technologies develop and resources become available, they are continuously used to control wildfire. However, despite our intensive efforts over decades to remove fire as a threat to society and natural resources, large-scale fires continue to occur at an alarming rate (Westerling and Bryant 2008). Although we have often altered its pattern, we have not eliminated fire from California's ecosystems, nor have we eliminated fire as a threat to human life and property. In fact, we now have larger, more severe, and more costly fires than ever. Why?

Although fire suppression is very effective over most of California, there are a few situations in which very large, damaging, and deadly fires still occur despite the intensive effort and firefighting resources applied to suppressing them. These are settings and conditions in which fire intensity is extremely high over large areas. It is a relatively rare, but predictable set of circumstances that leads to conditions that can produce the largest, deadliest, and most destructive fires in California.

Several important patterns are evident in these fires. The largest, most destructive, and deadliest fires have occurred in a relatively few definable settings under a small number of weather conditions and continue to occur during those conditions in the same ecosystems.

Large Fire Settings

There are three settings described here that support large fires currently or in the recent past. We summarize information on the 22 largest, most damaging and deadliest fires in California from 1923 to 2017 (Tables 20.1–20.3, and Figs. 20.1 and 20.2). Figures 20.1 and 20.2 chart the occurrence of these fires by decade and by month, respectively.

SETTING 1: CHAPARRAL AND WOODLAND FIRES

Large chaparral, deciduous oak woodland, live oak woodland, and grassland fires occur in the South Coast, Central Coast, and North Coast Bioregions, and in the Sierra Nevada foothills in two distinct scenarios.

Long Duration Chaparral Fires During the annual extended periods of hot dry summer weather, fires can burn for several weeks or even months in very steep inaccessible terrain with primarily chaparral vegetation. The 2016 Soberanes, 2009 Station, 2008 Basin Complex, 2007 Zaca, 1999 Kirk, and 1977 Marble Cone Fires in the Central Coast and portions of the McNally fire in the Southern Sierra Nevada are examples.

Wind-Driven Fires The second type of very large chaparral and woodland fire occurs mostly in chaparral, live and deciduous oak woodlands, and grasslands during the late summer, fall, and early winter during short periods of extreme fire weather characterized by dry and hot foehn winds, locally known as Santa Ana, Sundowner, Diablo, or North Winds. Since there is virtually no lightning activity during this part of the year, these fires historically may have started with holdover fires burning in from higher altitude vegetation types or deliberate attempts to reduce shrubland by Native American burning, but are now almost universally human-caused. Multiple fires can occur during intense extreme fire weather events in what have become known as *fire sieges*. The October 2017 fires, which were mostly in Sonoma, Mendocino, and Napa counties (including the Tubbs, Nuns, Atlas, Redwood Valley, and many other fires) of the North Coast Bioregion, were wind-driven fires causing fatalities and structure loss during a fairly short duration foehn wind event. The fires spread rapidly in the oak woodlands and grasslands, generating large amounts of heat in the chaparral and conifer forests, eventually burning nearly 100,000 ha (245,000 ac), 8,400 structures, and killing at least 43 people, making it the deadliest and most destructive fire siege in California history. Earlier fires in 2003 included the Cedar, Old, and Simi, and a 2007 wind event that produced the Witch, Harris, and Slide fires. These fires are responsible for very high structure losses, including the 1991 Tunnel Fire (2,900 structures lost), 2003 Cedar (2,820), 2007 Witch (1,650), and Old (1,003) Fires, 1990 Paint Fire (641), and 1923 Berkeley Fire (584). Wind-driven chaparral and woodland fires in the South Coast, Central Coast, and North Coast bioregions account for nearly all of California's deadliest fires. The lone exception is the 2008 Iron Alps Complex where several firefighters were killed in a helicopter crash. The management challenge is landscape fuel treatments have limited impact on foehn wind driven fires in chaparral (Moritz et al. 2004) but in some case local changes in fuels can be used to enhance suppression operations (T. Porter, pers. comm. 2013).

SETTING 2: FOREST FIRES

Very large fires also occur in the Klamath Mountain, Southern Cascade, and Sierra Nevada bioregions. These fires can be human-caused, or started by numerous lighting strikes over a large area, overwhelming fire suppression forces.

Multiple Lighting Strike Fire Complexes The Klamath Theater Complex (2008), Stanislaus Complex Fires (1987), Biscuit (2002), Big Bar Complex (1999), and other June 2008 fires are examples of lightning-ignited multiple fire complexes. While most of these fires have burned in uninhabited landscapes, they have the potential to cause a great deal of structure loss when they occur in landscapes with numerous developments mixed into the wildlands. These fires also produce high amounts of smoke for long periods.

Human-Caused Large Forest Fires Recently some very large human-caused fires have burned in the Sierra Nevada including the Rim (2013), McNally (2002), Moonlight (2008), and King Fires (2014). These fires are burning very large areas and high-severity patches created within these fires are a cause for concern regarding forest sustainability (Collins and Roller 2013, Welch et al. 2016). The 1992 Fountain, 2007 Angora, 1988 49er, 1992 Old Gulch, and 2015 Valley fires are examples of large human-caused forest fires that also burned numerous structures.

SETTING 3: LARGE VALLEY FIRES

Prior to the widespread development of agriculture, road networks, cities, and fire control, it is likely that the largest fires occurred in the central valley and surrounding foothills. The continuous surface fuels, long fire season, and fuelbeds that can regenerate in a single year, light surface fire could potentially burn much of the valley each year (Skinner et al. 2009). Since these fires are easily suppressed, the fuels have changed and the fuel continuity is fragmented, large fires are now uncommon in these ecosystems.

Patterns of Large, Destructive, and Deadly Fires

The largest fires are clearly occurring at an increasing rate. The 22 fires burning the largest area in California from 1923 to 2017 (Table 20.1) are illustrated by decade in Fig. 20.1 and

TABLE 20.1

The 22 largest fires by acres burned in California from 1923 to 2017

Data from CAL FIRE (http://www.fire.ca.gov/communications/downloads/fact_sheets/20LACRES.pdf)

Fire name	Date	Bioregion	Hectares	Acres
Thomas	December 2017	South Coast	114,078	281,893
Cedar	October 2003	South Coast	110,579	273,246
Rush	August 2012	Northeastern Plateau	110,038	271,911
Rim	August 2013	Sierra Nevada	104,134	257,314
Zaca	July 2007	South Coast and Central Coast	97,210	240,207
Matilija	September 1932	South Coast	89,031	220,000
Witch	October 2007	South Coast	80,125	197,990
Klamath Theater Complex	June 2008	Klamath Mountains	77,717	192,038
Marble Cone	July 1977	Central Coast	71,980	177,866
Laguna	September 1970	South Coast	70,992	175,425
Basin Complex	June 2008	Central Coast	65,892	162,818
Day	September 2006	South Coast	65,845	162,702
Station	August 2009	South Coast	64,977	160,557
McNally	July 2002	Sierra Nevada	60,985	150,696
Stanislaus Complex	August 1987	Sierra Nevada	59,076	145,980
Big Bar Complex	August 1999	Klamath Mountains	57,040	140,948
Happy Camp Complex	August 2014	Klamath Mountains	54,251	134,056
Campbell Complex	August 1990	North Coast	50,947	125,892
Rough	July 2015	Sierra Nevada	48,186	119,069
Wheeler	July 1985	South Coast	47,753	118,000
Simi	October 2003	South Coast	43,790	108,204
King	September 2014	Sierra Nevada	39,545	97,717

by month in Fig. 20.2. In the 95 years of record, all but one (95%) of the largest fires occurred in the last 47 years (after 1970), and most (68%) were in the years from 2001 to 2017. The 16 years since 2001 have produced the five largest fires, and more than twice as many of the largest fires than the first 78 years of the recorded period. Most of the largest fires occurred in steep inaccessible terrain. The fires were well distributed during the four-month period from July to October with two starting in June and burning well into the summer. Increasing fuel continuity and loads from successful fire exclusion, climate change, and increasing population have contributed to the sudden increase in large fires since 1970, especially in forested ecosystems.

The 22 fires burning the most structures in California since 1923 were responsible for the destruction of a total of 25,304 structures (Table 20.2) and are illustrated by decade in Fig. 20.1 and by month in Fig. 20.2. While there were a variety of specific causes, all of the ignitions except one were human-caused or under investigation. Nineteen of the 22 fires occurred since 1990 including those causing the 12 highest structure losses, and all except two have occurred since 1970. Most (82%) of the 22 largest structure loss fires occurred during the months of September, October, and November. There is clearly an increasing rate of occurrence of the state's largest and most destructive fires (Stephens et al. 2009). This is despite increasing efforts, effectiveness, and expenditures for fire suppression and improving house design and fuel clearances. All of the fires occurred during extreme fire weather conditions and all were actively suppressed.

The 22 wildland fires with four or more direct fatalities are listed in Table 20.3 and illustrated by decade in Fig. 20.1 and month in Fig. 20.2. All of the fires were in the South Coast, North Coast, or Central Coast Bioregions except for one. All of the fires burned between July and November in some combination of chaparral, grass, coastal scrub, and oak woodland settings. The one dissimilar fatality fire was the

TABLE 20.2

The 22 most damaging fires by structures destroyed in California from 1923 to 2017

Data from CAL FIRE (http://www.fire.ca.gov/communications/downloads/fact_sheets/Top20_Damaging.pdf)

Fire name	Date	Bioregion	Structures lost	Hectares	Acres
Tubbs	October 2017	North Coast	5,300	14,895	36,807
Tunnel	October 1991	Central Coast	2,900	647	1,600
Cedar	October 2003	South Coast	2,820	110,579	273,246
Valley	September 2015	North Coast	1,955	30,783	76,067
Witch	October 2007	South Coast	1,650	80,125	197,990
Nuns	October 2017	North Coast	1,355	4,042	9,985
Old	October 2003	South Coast	1,003	36,940	91,281
Thomas	December 2017	South Coast	1,063	114,078	281,893
Jones	October 1999	Central Valley	954	10,603	26,200
Butte	September 2015	Sierra Nevada	921	28,679	70,868
Paint	June 1990	South Coast	641	1,983	4,900
Fountain	August 1992	Southern Cascades	636	25,884	63,960
Sayre	November 2008	South Coast	604	4558	11,262
City of Berkeley	September 1923	Central Coast	584	53	130
Harris	October 2007	South Coast	548	36,601	90,440
Redwood Valley	October 2017	North Coast	543	14,780	36,523
Bel Air	November 1961	South Coast	484	2,465	6,090
Laguna	October 1993	South Coast	441	5,842	14,437
Atlas	October 2017	North Coast	481	20,891	51,624
Erskine	June 2016	Sierra Nevada	386	18,892	46,684
Laguna	September 1970	South Coast	382	70,992	175,425
Cascade	October 2017	Sierra Nevada	365	4,042	9,989

Iron Alps Complex (2008) where a helicopter crash killed 10 firefighters.

Current Fire Policies, Initiatives, and Direction

In recent years, societal concerns for managing natural resources have shifted to include the role of fire as a dynamic and predictable part of wildland ecosystems (Calkin et al. 2011, Thompson and Calkin 2011). This has served to thrust fire management into the forefront of wildland management. It is now widely recognized that fire plays an important role in the functioning of most of California's natural ecosystems (Stephens et al. 2007). It is also recognized that if we value ecosystem components such as plant and animal species and their habitats, water, and air, then fire must be managed rather than eliminated from most ecosystems. The focus of fire policy and management has shifted away from the overall goal of removing fire toward the much more complex goal of managing fire (Calkin et al. 2011, Thompson et al. 2013).

Federal Fire Policy Reviews

In response to 14 firefighter fatalities at the South Canyon Fire in Colorado in 1994, the growing recognition that fire problems are caused by fuel accumulation, the Bureau of Land Management, Forest Service, National Park Service, US Fish and Wildlife Service, Bureau of Indian Affairs, and the National Biological Service released a joint Federal Wildland Fire Management Policy and Program Review in 1995 (Philpot et al. 1995, USDA 1995b). The key findings of this policy and program review are:

TABLE 20.3
The 22 deadliest California wildfires
Fires from 1923 to 2017 with four or more fatalities are included
Data from CAL FIRE (http://calfire.ca.gov/communications/downloads/fact_sheets/Top20_Deadliest.pdf)

					Area		
Fire name	Date	Bioregion	Deaths	Structures lost	Hectares	Acres	Cause*
Griffith Park	October 1933	South Coast	29	0	19	47	H
Tunnel	October 1991	Central Coast	25	2,843	647	1,600	H
Tubbs	October 2017	North Coast	22	5,300	14,895	36,807	?
Cedar	October 2003	South Coast	15	2,820	110,579	273,246	H
Rattlesnake	July 1953	North Coast	15	0	542	1,340	H
Loop	November 1966	South Coast	12	0	821	2,028	H
Inaja	November 1956	South Coast	11	0	17,768	43,904	H
Hauser Creek	October 1943	South Coast	11	0	5,320	13,145	H
Iron Alps Complex	August 2008	Klamath Mts.	10	10	42,839	105,855	L
Redwood Valley	October 2017	North Coast	8	543	14,780	36,523	?
Harris	October 2007	South Coast	8	548	36,601	90,440	H
Canyon	August 1968	South Coast	8	0	8,983	22,197	H
Atlas	October 2017	North Coast	6	481	20,664	51,064	?
Old	October 2003	South Coast	6	1,003	36,940	91,281	H
Decker	August 1959	South Coast	6	1	577	1,425	H
Hacienda	September 1955	South Coast	6	0	465	1,150	H
Esperanza	October 2006	South Coast	5	54	16,269	40,200	H
Laguna	September 1970	South Coast	5	382	70,992	175,425	H
Cascade	October 2017	Sierra Nevada	4	365	4,042	9,989	?
Valley	September 2015	North Coast	4	1,955	30,783	76,067	H
Panorama	November 1980	South Coast	4	325	9,551	23,600	H
Clampitt	September 1970	South Coast	4	86	42,579	105,212	H

* Causes are human (H), lightning (L), or under investigation (?).

1. Reaffirm that the protection of human life as the first priority in wildland fire management.
2. Property and natural and cultural resources jointly became the second priority.
3. Wildland fire, as a critical natural process, must be reintroduced into the ecosystem. This will be accomplished across agency boundaries and will be based upon the best available science.
4. Treatment of hazardous fuel buildups, particularly in the wildland-urban interface, and approved Fire Management Plans, are needed.

The review concluded "agencies and the public must change their expectations that all wildfires can be controlled or suppressed." For the first time the federal land management agencies jointly took responsibility for managing fire as a natural process in America's wildlands.

The 1995 fire policy was reviewed and updated in the aftermath of the 2000 Cerro Grande Fire. This incident started with an escaped prescribed fire that eventually burned 235 structures in and around Los Alamos, New Mexico, and threatened the Los Alamos National Laboratory in May of 2000 (Douglas et al. 2001). The review supported the 1995 fire policy and fine-tuned its implementation. The key findings of the review were:

1. The 1995 policy is generally sound and continues to provide a solid foundation for wildland fire management activities and for natural resources management activities of the federal government.
2. As a result of fire exclusion, the condition of fire-adapted ecosystems continues to deteriorate; the fire hazard situation in these areas is worse than previously understood.

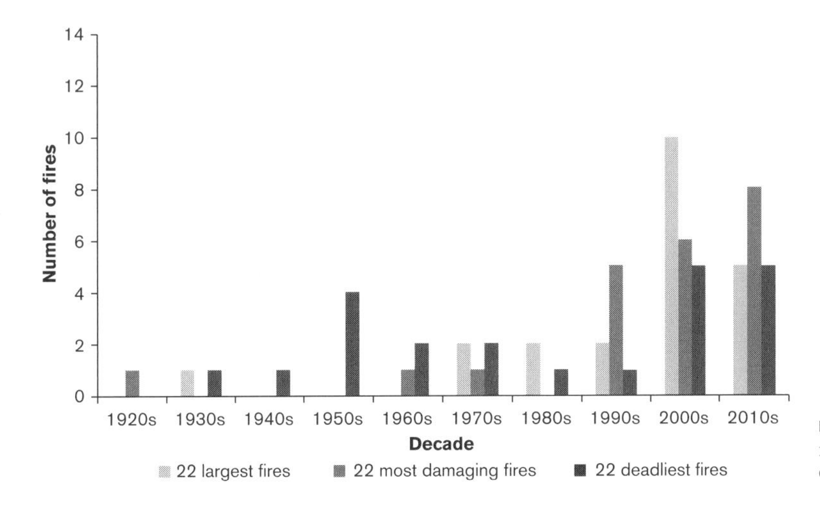

FIGURE 20.1 Distribution of the 22 largest, most damaging, and deadliest fires occurring in California by decade between 1923 and 2017.

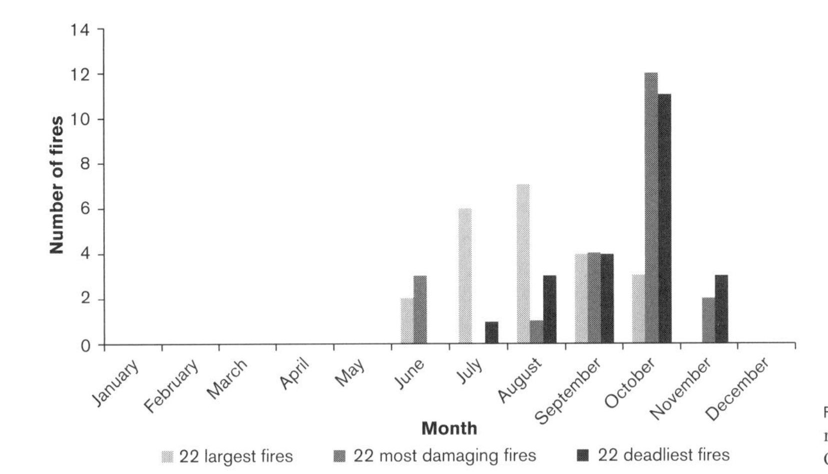

FIGURE 20.2 Distribution of the 22 largest, most damaging, and deadliest fires ocurring in California by month between 1923 and 2017.

3. The fire hazard situation in the wildland-urban interface is more complex and extensive than understood in 1995.

National Fire Plan

The National Fire Plan, established in "A Report to the President in Response to the Wildfires of 2000" (USDA–USDI 2000), was implemented using the "Collaborative Approach for Reducing Wildfire Risks to Communities and the Environment: Ten-Year Comprehensive Strategy" (WGA 2001). The Plan and the Strategy state: "unless hazardous fuel is reduced, the number of severe wildland fires and the costs associated with suppressing them will continue to increase." Implementation of the National Fire Plan is designed to be a long-term, multibillion-dollar effort (GAO 2003). The Ten-Year Comprehensive Strategy recognizes that key decisions in setting priorities for restoration, fire, and fuel management should be made collaboratively at local levels. As such, the Strategy requires an ongoing process whereby local, Tribal, State, and Federal land management, scientific, and regulatory agencies exchange the required technical information to assist the decision-making process.

Federal Policy Implementation Guidelines

A significant interpretation of the 2001 federal fire policy was produced in 2009 as a set of guiding principles for implementation that included:

1. Firefighter and public safety is the first priority in every fire management activity.
2. The role of wildland fire as an essential ecological process and natural change agent will be incorporated into the planning process.
3. Fire management plans, programs, and activities support land and resource management plans and their implementation.
4. Sound risk management is a foundation for all fire management activities.
5. Fire management programs and activities are economically viable, based upon values to be protected, costs, and land and resource management objectives.
6. Fire management plans and activities are based upon the best available science.
7. Fire management plans and activities incorporate public health and environmental quality considerations.

8. Federal, state, tribal, local, interagency, and international coordination and cooperation are essential. Increasing costs and smaller work forces require that public agencies pool their human resources to successfully deal with the ever-increasing and more complex fire management tasks.
9. Standardization of policies and procedures among federal wildland fire management agencies is an ongoing objective.

These guidelines stated that a wildland fire may be concurrently managed for one or more objectives and objectives can change as the fire spreads across the landscape. These guiding principles give more flexibility in fire management operations but still focus on the critical goal of restoring fire as a critical ecosystem process in federal lands.

Federal Land Assistance, Management, and Enhancement Act

In the Federal Land Assistance, Management, and Enhancement Act of 2009 (FLAME Act), Congress mandated the development of a national cohesive wildland fire management strategies across all lands in the United States. The culmination of the cohesive strategy effort is the National Strategy (2014) and the National Action Plan (2014). The Cohesive Strategy Vision (fhttp://www.forestsandrangelands.gov/strategy/thestrategy.shtml) of the next century is "To safely and effectively extinguish fire, when needed; use fire where allowable; manage our natural resources; and as a Nation, live with wildland fire." It sets these basic priority levels:

1. Safe and effective response to wildfires is the highest priority of the National Strategy, and includes enhancing wildfire response preparedness with an emphasis on both structural protection and wildfire prevention to maximize the effectiveness of initial response.
2. The second priority is vegetation and fuels management, and is perhaps the most challenging issue. General guidance in this area includes designing and prioritizing fuel treatments; strategically placing fuel treatments (Finney et al. 2007); increasing use of wildland fire for meeting resource objectives; and continuing and expanding the use of all methods to improve the resiliency of our forests and rangelands.
3. The third priority involves engaging homeowners and communities in taking proactive action prior to wildfires.
4. The fourth priority includes emphasizing programs and activities, tailored to meet identified local needs, which seek to prevent human-caused ignitions.

Continued Fire Challenges in California

In October of 2003 a series of ten fires burned over 303,514 ha (750,000 ac) of chaparral, woodlands, and forest in and around the cities of the greater San Diego and Los Angeles areas. Over 3,400 houses were burned and 22 lives were lost over a one-week period. In October of 2017 another series of fires burned nearly 100,000 ha (245,000 ac) north of the San Francisco Bay Area mostly in Sonoma, Mendocino, and Napa Counties. Over 8,400 homes were destroyed and at least 43 people killed. These are dramatic demonstrations of several points that define our current place in the history of fire ecology, management, and policy:

1. Very high-intensity wildland fires continue to operate on large landscapes.
2. Fire management can be effective in modifying fire behavior in certain ecosystems. However, in ecosystems characterized by very high-intensity or very rapidly moving fire, such as chaparral and grasslands management of wildlands as natural ecosystems is not likely to reduce the threat to adjacent urban areas. In such cases it is likely that we must rely more on the design of structures and the development of buffers between the wildlands and urban areas.
3. We must continue to improve our understanding of the nature of fire in ecosystems, how that serves as a threat to society, and how to coexist with fire in California's wildlands (Moritz et al. 2015).

Fires in a Changing Climate

Global climate will further complicate fire management in California (Westerling et al. 2006, Fried et al. 2008, Stephens et al. 2013). Climate change may lead to differences in plant distributions (Bachelet et al. 2001), lightning frequency (Romps et al. 2015), and the length of fire season, which could increase ignitions and further exacerbating wildfire effects. The great challenge in the next one to three decades is to produce resilient ecosystems that can incorporate drought, tree-killing insects, and fire while still conserving key conservation values and ecosystem services. This will entail novel approaches to wildland fire policy and management. Without such innovation ecosystems of California will continue to be degraded from inappropriate fire regimes and human communities will also negatively impacted. The impacts of climate change should accelerate the pace and scale of restoration, especially in ecosystems that once burned frequently with low-moderate-intensity fire regimes (North et al. 2015) and will probably require some new ideas in fire and forest policy (Table 20.4).

Conclusion

The challenge today is to develop fire policies, management actions, public support, expertise, and budgets that recognize the need for both fire suppression and the management of fire as an ecological process and hazard reduction tool (Calkin et al. 2015, Stephens et al. 2016). Wildland fire is clearly impacting human life and property at an increasing rate and the trend shows no sign of changing. Fire will continue as an important agent of change in many western ecosystems but we must strive to produce conditions where fire can become a positive force in most of California. This is a challenge of magnitude and complexity that is unprecedented in the history of wildland resource management. The stakes are extremely high and the future of both the relationship of wildlands to society and the integrity of natural ecosystems is at risk.

TABLE 20.4

Summary of policy recommendations (ranked most important –1 to least important –5) that could improves US federal forest and fire management (from Stephens et al. 2016)

Priority	Recommendation
	Procedural and legal
1	Make forest resilience a stand-alone, top land management priority and connect it to managing long-term for endangered species
2	Ensure fire suppression funding does not impede restoration efforts
3	Switch default environmental review rules for managed wildfire
4	Provide for equal balance in decision making between long-and short-term impacts
	Planning and implementation
1	Mandate evaluation of opportunities for ecologically beneficial fire in land management planning
2	Expand capacity and utilization of interagency modules dedicated to increasing beneficial fire
3	Develop specific performance measures for forest resilience
4	Increase educational requirements for wildfire and forest management personnel to included fire ecology and ecosystem dynamics
5	Analyze long-term impacts of continued suppression
	Outreach and collaboration
1	Engage in collaborative planning to ensure Firewise development in the WUI, including appropriate fire suppression cost sharing
2	Increased public–private collaborations for landscape-level fuel management (including expanding ecologically beneficial fire)
3	Increase education and outreach to non-federal resource agencies on inevitability of fire

References

Anderson, M.K. 2005. Tending the Wild: Native American Knowledge and the Management of California's Natural Resources. University of California Press, Berkeley, California, USA.

Anderson, R.E., B.L. Rasmussen, and V.V. Church. 1941. Adapting advanced principles of organization and fire-line construction to CCC suppression crews. Fire Control Notes 5(3): 123–128.

Anderson, M.K., and J. Rosenthal. 2015. An ethnobiological approach to reconstructing indigenous fire regimes in the foothill chaparral of the western Sierra Nevada. Journal of Ethnobiology 35(1): 4–36.

Bachelet, D., R.P. Neilson, J.M. Lenihan, and R.J. Drapek. 2001. Climate change effects on vegetation distribution and carbon budget in the United States. Ecosystems 4: 164–185.

Barbour, M., B. Pavlik, F. Drysdale, and S. Lindstrom. 1993. California's changing landscapes. California Native Plant Society, Sacramento, California, USA.

Biswell, H.H. 1961. The big trees and fire. National Parks and Conservation Magazine 35: 11–14.

———. 1989. Prescribed Burning in California Wildland Vegetation Management. University of California Press, Berkeley, California, USA.

Calkin, D.C., M.A. Finney, A.A. Ager, M.P. Thompson, and K.M. Gebert. 2011. Progress towards and barriers to implementation of a risk framework for US federal wildland fire policy and decision making. Forest Policy and Economics 13: 378–389.

Calkin, D.E., M.P. Thompson, and M.A. Finney. 2015. Negative consequences of positive feedbacks in US wildfire management. Forest Ecosystems 2(9): 1–10.

Clar, C.R. 1959. California government and forestry. Division of Forestry, State of California, Sacramento, California, USA.

Collins, B.M., and G.B. Roller. 2013. Early forest dynamics in stand-replacing fire patches in the northern Sierra Nevada, California, USA. Landscape Ecology 28: 1801–1813.

DeBruin, H.W. 1974. From fire control to fire management: a major policy change in the Forest Service. Proceedings of the Tall Timbers Fire Ecology Conference 14: 11–17.

Douglas, J., T.J. Mills, D. Artly, D. Ashe, A. Bartuska, R.L. Black, S. Coloff, J. Cruz, M. Edrington, J. Edwardson, R.T. Gale, S.W. Goodman, L. Hamilton, R. Landis, B. Powell, S. Robinson, R.J. Schuster, P.K. Stahlschmidt, J. Stires, and J. van Wagtendonk. 2001. Review and update of the 1995 federal wildland fire management policy. U.S. Department of the Interior and U.S. Department of Agriculture, Washington, D.C., USA.

DuBois, C. 1914. Systematic fire protection in the California forests. USDA Forest Service, Washington, D.C., USA.

Finney, M.A., R.C. Seli, C.W. McHugh, A.A. Ager, B. Bahro, and J.K. Agee. 2007. Simulation of long-term landscape-level fuel treatment effects on large wildfires. International Journal of Wildland Fire 16: 712–727.

Fried, J.S., J.K. Gilles, W.J. Riley, T.J. Moody, C.S. de Blas, K. Hayhoe, M.A. Moritz, S.L. Stephens, and M. Torn. 2008. Predicting the effect of climate change on wildfire behavior and initial attack success Climatic Change 87(Suppl. 1): S251–S264.

GAO. 2003. Forest service fuels reduction. US General Accounting Office Report GAO-03-689R. Washington, D.C., USA.

Gibbons, P., L. van Bomme, A.M. Gill, G.J. Cary, D.A. Driscoll, R.A. Bradstock, E. Knight, M.A. Moritz, S.L. Stephens, and D.B. Lindenmayer. 2012. Land management practices associated with house loss in wildfires. PLoS One 7(1): e29212.

Gill, A.M., and S.L. Stephens. 2009. Scientific and social challenges for the management of fire-prone wildland-urban interfaces. Environmental Research Letters 4(3): 0304014.
Gill, A.M., S.L. Stephens, and G.J. Carry. 2013. The world-wide "wildfire" problem. Ecological Applications 23: 438–454.
Graves, H.S. 1910. Protection of forests from fire. US Forest Service Bulletin 82. Washington, D.C., USA.
Greeley, W.B. 1951. Forests and Men. Doubleday Publishing, Garden City, New York, USA.
Hartesveldt, R.J., and H.T. Harvey. 1967. The fire ecology of sequoia regeneration. Proceedings Tall Timbers Fire Ecology Conference 6: 65–77.
Keeley, J.E. 2002. Native American impacts on fire regimes of the California coastal ranges. Journal of Biogeography 29: 303–320.
Kilgore, B.M. 1974. Fire management in national parks: an overview. Proceedings Tall Timbers Fire Ecology Conference 14: 45–57.
Laudenslayer, W.F., and H.H. Darr. 1990. Historical effects of logging on forests of the Cascade and Sierra Nevada Ranges of California. Transaction of the Western Section of the Wildlife Society 26: 12–23.
Leopold, S.A., S.A. Cain, C.A. Cottam, I.N. Gabrielson, and T.L. Kimball. 1963. Wildlife management in the national parks. American Forestry 69: 32–35, 61–63.
McClaran, M.P., and J.W. Bartolome. 1989. Fire-related recruitment in stagnant *Quercus douglasii* populations. Canadian Journal of Forest Research 19: 580–585.
Menke, J.W., C. Davis, and P. Beesley. 1996. Rangeland assessment. Pages 901–172 in: D.C. Erman, general editor. Sierra Nevada Ecosystem Project, Final Report to Congress, Volume III, Assessments, Commissioned Reports, and Background Information. Wildland Research Center, University of California, Davis, California, USA.
Meyer, M.D. 2015. Forest fire severity patterns of resource objective wildfires in the Southern Sierra Nevada. Journal of Forestry 113: 49–56.
Moritz, M.A., E. Batllori, R.A. Bradstock, A.M. Gill, J. Handmer, P.F. Hessburg, J. Leonard, S. McCaffrey, D. Odion, T. Schoennagel, and A.D. Syphard. 2015. Learning to coexist with wildfire. Nature 515: 58–66.
Moritz, M.A., J.E. Keeley, E.A. Johnson, and A.A. Schaffner. 2004. Testing a basic assumption of shrubland fire management: how important is fuel age? Frontiers in Ecology and the Environment 2: 67–72.
North, M.P., B.M. Collins, and S.L. Stephens. 2012. Using fire to increase the scale, benefits and future maintenance of fuels treatments. Journal of Forestry 110: 392–401.
North, M.P., S.L. Stephens, B.M. Collins, J.K. Agee, G. Aplet, J.F. Franklin, and P.Z. Fulé. 2015. Reform forest fire management: agency incentives undermine policy effectiveness. Science 18: 1280–1281.
Philpot, C.W., C. Schechter, A. Bartuska, K. Beartusk, D. Bosworth, S. Coloff, J. Douglas, M. Edrington, R.T. Gale, M.J. Lavin, L.K. Rosenkrance, R. Streeter, and J. van Wagtendonk. 1995. Federal wildland fire management policy & program review. U.S. Department of the Interior and U.S. Department of Agriculture, Washington, D.C., USA.
Pinchot, G. 1907. The use of the National Forests. USDA Forest Service. Washington, D.C., USA.
Pyne, S.J. 1982. Fire in America: A Cultural History of Wildland and Rural Fire. Princeton University Press, Princeton, New Jersey, USA.
———. 2001. Year of the Fires: The Story of the Great Fires of 1910. Viking Publishing, New York, New York, USA.
———. 2015. Between Two Fires: A History of Contemporary America. The University of Arizona Press, Tucson, Arizona, USA.
Pyne, S.J., P.L. Andrews, and R.L. Laven. 1996. Introduction to Wildland Fire. 2nd ed. John Wiley and Sons, New York, New York, USA.
Romps, D.M., J.T. Seeley, D. Vollaro, and J. Molinari. 2015. Projected lightning strikes in the United States due to global climate warming. Science 346: 851–854.
Show, S.B., and E.I. Kotok. 1924. The role of fire in the California pine forests. USDA Bulletin 1294. Washington, D.C., USA.
Sneeuwjagt, R.J., T.S. Kline, and S.L. Stephens. 2013. Opportunities for improved fire use and management in California: lessons from Western Australia. Fire Ecology 9(2): 14–25.
Skinner, C.N., C.S. Abbott, D.L. Fry, S.L. Stephens, A.H. Taylor, and V. Trouet. 2009. Human and climatic influences on fire occurrence in California's North Coast Range, USA. Fire Ecology 5(3): 76–99.
Stephens, S.L. 1997. Fire history of a mixed oak-pine forest in the foothills of the Sierra Nevada, El Dorado County, California. Pages 191–198 in: N.H. Pilsbury, J. Verner, and W.D. Tietje, technical coordinators. Symposium on Oak Woodlands: Ecology, Management, and Urban Interface Issues. USDA Forest Service General Technical Report PSW GTR-160. Pacific Southwest Research Station, Albany, California, USA.
Stephens, S.L., M. Adams, J. Hadmer, F. Kearns, B. Leicester, J. Leonard, and M.A Moritz. 2009. Urban-wildland fires: how California and other regions of the US can learn from Australia. Environmental Research Letters 4(1): 014010.
Stephens, S.L., J.K. Agee, P.Z. Fulé, M.P. North, W.H. Romme, T.W. Swetnam, and M.G. Turner. 2013. Managing forests and fire in changing climates. Science 342: 41–42.
Stephens, S.L., B.M. Collins, E. Biber, and P.Z. Fulé. 2016. U.S. federal fire and forest policy: emphasizing resilience in dry forests. Ecosphere 7(11): e01584.
Stephens, S.L., B.M. Collins, C.J. Fettig, M.A. Finney, C.H. Hoffman, E.E. Knapp, M.P. North, H. Safford, and R.B. Wayman. 2018. Drought, tree mortality and wildfire in forests adapted to frequent fire. Bioscience 68: 77–88.
Stephens, S.L., and D.L. Fry. 2005. Fire history in coast redwood stands in the Northeastern Santa Cruz Mountains, California. Fire Ecology 1(1): 2–19.
Stephens, S.L., R.E. Martin, and N.E. Clinton. 2007. Prehistoric fire area and emissions from California's forests, woodlands, shrublands and grasslands. Forest Ecology and Management 251: 205–216.
Stephens, S.L., J.D. McIver, R.E.J. Boerner, C.J. Fettig, J.B. Fontaine, B.R. Hartsough, P. Kennedy, and D.W. Schwilk. 2012. Effects of forest fuel reduction treatments in the United States. BioScience 62: 549–560.
Stephens, S.L., and L.W. Ruth. 2005. Federal forest fire policy in the United States. Ecological Applications 15: 532–542.
Taylor, A.H., V. Trouet, C.N. Skinner, and S.L. Stephens. 2016. Socio-ecological transitions trigger fire regime shifts and modulate fire-climate interactions in the Sierra Nevada, USA 1600-2015 CE. Proceedings of the National Academy of Sciences 113(48): 13684–13689.
Thompson, M.P., and D.E. Calkin, 2011. Uncertainty and risk in wildland fire management: a review. Journal of Environmental Management 92: 1895–1909.
Thompson, M.P., D.E. Calkin, M.A. Finney, K.M. Gebert, and M.S. Hand. 2013. A risk-based approach to wildland fire budgetary planning. Forest Science 59: 63–77.
USDA. 1960. Air attack on forest fires. US Forest Service Agricultural Information Bulletin 229. Washington, D.C., USA.
———. 1995a. Smokey Bear: the first fifty years. USDA Forest Service Publication FS-551. Washington, D.C., USA.
———. 1995b. Course to the future: positioning fire and aviation management. Department of Fire and Aviation Management. Washington, D.C., USA.
USDA–USDI. 2000. A Report to the President in Response to the Wildfires of 2000. www.fireplan.gov\president.cfm. Accessed May 22, 2003.
USDI. 1968. Compilation of the fire administrative policies for the national parks and monuments of scientific significance. National Park Service. Washington, D.C., USA.
van Wagtendonk, J.W. 1991. The evolution of national park fire policy. Fire Management Notes 52(4): 10–15.
———. 2007. The history and evolution of wildland fire use. Fire Ecology 3(2): 3–17.
Welch, K.R., H.D. Safford, and T.P. Young. 2016. Predicting conifer establishment post wildfire in mixed conifer forests of the North American Mediterranean-climate zone. Ecosphere 7(12):e01609. 10.1002/ecs2.1609.
Westerling, A.L., and B.P. Bryant. 2008. Climate change and wildfire in California. Climatic Change 87(Suppl. 1): S231–S249.
Westerling, A.L., H.G. Hidalgo, D.R. Cayan, and T.W. Swetnam, T.W. 2006. Warming and earlier spring increase western U.S. forest wildfire activity. Science 313: 940–943.
WGA. 2001. A Collaborative Approach for Reducing Wildland Fire Risk to Communities and the Environment: 10-Year Comprehensive Strategy. Western Governors' Association. www.westgov.org /wga/initiatives/fire/final_fire_rpt.pdf. Accessed May 25, 2003.

CHAPTER TWENTY-ONE

Fire and Fuel Management

SCOTT L. STEPHENS, SUSAN J. HUSARI, H. TOM NICHOLS, NEIL G. SUGIHARA, AND BRANDON M. COLLINS

Even though fire is itself an inexorable force of nature, we need not view its worst effects as inevitable.

ARNO AND ALLISON-BUNNELL (2002)

That maximum protection or fire exclusion inevitably increases hazard by the encouragement of undergrowth is, of course, true, but such added hazard in no way vitiates the reasons for protection.

SHOW AND KOTOK (1924)

Introduction

The complex set of tasks we now characterize as fire management evolved from the single-minded pursuit of fire control. Fuel management, which involves manipulating fuel loads and arrangement in order to modify subsequent fire behavior and effects, is a great example of this evolution. In the last 25 years fuel management has come to play a leading role in managing ecosystems and natural resources (Biswell 1989, Omi 2015). During this period scientists and managers have improved their shared understanding of the importance of natural processes in ecosystem function (North et al. 2009, North 2012a). Attempts to exclude fires merely delay, alter, and intensify subsequent fires in most ecosystems (Biswell 1989). This is especially true as climate change leads to longer and drier fire seasons in many areas of California and the western United States (Westerling et al. 2006, Trouet et al. 2010).

Fuel Management Objectives

Common goals of fuel treatments include the reduction of potential fire intensity and rate of spread, reducing the severity of fire effects, or producing ecosystems that are more resilient to fire (Agee and Skinner 2005, Finney et al. 2007, Fulé et al. 2012). Achieving these goals will improve resiliency in those fire regimes anticipated to occur in the future with changing climates (Millar et al. 2007, Stephens et al. 2010). Developing restoration programs at appropriate spatial scales to modify fire behavior and effects is critical to the conservation of many forests and woodlands in California (North et al. 2012).

The land management objectives for fuel treatment projects are diverse and often complex (Collins et al. 2010). Reestablishing or restoring historical fire regimes can be, but are not always the objective of fuel management. It is also important to note that while fuel management is important for a wide range of ecosystems, it is not appropriate or effective everywhere (Keeley 2002).

Fuel Management Basics

Fuel is accumulated live and dead plant biomass. Chapter 4 discusses how the characteristics of fuel influence its potential to burn. Fuel characteristics that are typically manipulated are:

Fuel quantity and size–The overall amount of fuel in the ecosystem is an important factor determining the character and impact of fires. The metric used to describe the quantity of fuel is oven dry weight per unit area (T ha^{-1} [ton ac^{-1}]). This dry weight is often subdivided into size classes. The size classes are related to the time the fuel takes to reach equilibrium with moisture in the air. Small fuels, such as pine needles, respond to changes in relative humidity more rapidly than large, dense fuel, such as logs. Reduction of fine fuel is a primary focus of many fuel management projects.

Surface fuel–Surface fuel is composed of small shrubs, grasses, seedlings/saplings and plant debris lying on the surface of the ground. Continuous surface fuel is necessary for fire to spread across landscapes unless there is high wind or low humidity (<30%) to support spot fire propagation. Surface fuel continuity can be interrupted to achieve fuel management goals.

Crown fuel–The fine branches and foliage of the trees and large shrubs (greater than approximately 2 m [6.5 ft] in height) make up crown fuel. Continuous crown fuel is required for fire to spread through the crowns of trees (also known as independent crown fire). Localized crown fires may also occur in discontinuous stands of trees if supported by surface fire, known as passive crown fire. Wind speed and, to a lesser extent, foliar moisture are important to propagation of crown fires. Crown fire risk reduction may be accomplished through removing surface, ladder, and crown fuels in the order of their influence on potential fire behavior in frequent fire-adapted forest (Agee and Skinner 2005, Stephens et al. 2009a).

Ladder fuel—Intermediate-sized trees or shrubs provide a fuel continuity that can allow a surface fire to "climb" into the crown fuel. Fuel treatments can remove shrubs, small trees, or the lower branches of larger trees to reduce ladder fuel (Kramer et al. 2014).

Types of Fuel Treatments

Fuel treatments take on a wide assortment of forms but can generally be divided into two categories, fire treatments and mechanical treatments. Fire treatments are the application, use or management of wildfire or prescribed fire to modify fuel. Mechanical treatments rely on a variety of methods to modify or remove fuel.

The National Study of Fire and Fire Surrogates (FFS) has three research sites in California (Goosenest Demonstration Forest, Blodgett Research Forest, Sequoia National Park); these sites are part of a larger network of 13 sites across the United States. This and other studies (e.g., Teakettle Fire and Forest Health Experiment, North et al. 2007) have demonstrated that both prescribed fire and its mechanical surrogates were generally successful in meeting short-term fuel-reduction objectives such that treated stands are more resilient to high-intensity wildfire (North et al. 2007, Safford et al. 2009, Stephens et al. 2009a, Vaillant et al. 2009, Fulé et al. 2012). In addition, most available evidence suggests that these objectives are typically accomplished with few unintended consequences, because most ecosystem components (e.g., vegetation, soils, small mammals, song birds, bark beetles, carbon sequestration) exhibit only very subtle effects or no measurable effects at all (Amacher et al. 2008, Hurteau and North 2009, Kobziar et al. 2009, Ma et al. 2010, Stephens et al. 2012a). However, fuel treatments effects on wildlife species with large home-ranges need further research (Stephens et al. 2014a) and innovation (Stephens et al. 2016a). Although mechanical treatments do not serve as complete surrogates for fire (Stephenson 1999, Wayman and North 2007) and in some case they increase nonnative understory plants (Collins et al. 2007), their application can help mitigate wildfire costs and liability in some areas (Schwilk et al. 2009). Results from this research should give forest managers assurance that treatments designed to reduce fire hazards and restore forests can be implemented with few negative consequences (Meyer et al. 2007, Moghaddas and Stephens 2007, 2008, North et al. 2007, Wayman and North 2007, Stephens et al. 2009b, 2012a, Stark et al. 2013, Collins et al. 2014, Hurteau et al. 2014).

Fire Treatments

Fire treatments may include prescribed fires purposely ignited to achieve established objectives, or naturally caused fires allowed to burn in designated locations under specific range of conditions (Collins et al. 2009, Collins and Stephens 2010). Both types of fire treatments maintain the presence of fire as an ecological process, but prescriptions may or may not be designed to mimic the historic influences of fire.

PRESCRIBED FIRE

The uses and purposes of prescribed fire in California are widely varied. A general differentiation can be made between restoration burns, in which the current ecological condition is modified, often due to suppression of natural ignitions or occurrence of human-caused ignitions, and maintenance burns, in which existing conditions are maintained within a specified range. Other goals may include the reduction of hazardous amounts of dead and down fuel, the stimulation of fire-dependent species, improvement of range condition, or the creation of wildlife habitat (Fig. 21.1).

Prescriptions for burning consider the variables that influence fire behavior, the ecological role of fire, and the ability to control the fire to minimize the potential for escapes. Site considerations include slope, aspect, topographic position, and role of fire in the project area. Prescribed conditions at the time of burning include the season, weather, fuel conditions, and the availability of qualified personnel. Methods for ignition can greatly influence the intensity and severity of prescribed fires. Ignition patterns are used to modify and control fireline intensity and fire severity (Martin et al. 1989). Whether for restoration or maintenance purposes, the establishment of measurable objectives, and monitoring methods to measure them, are critical.

MANAGING WILDFIRE FOR MULTIPLE OBJECTIVES

The concept of allowing lightning fires to burn (old names: prescribed natural fires; wildland fire use, wildfires used for resource benefits) originated in 1968 in Sequoia and Kings Canyon National Parks, followed by other agencies and other units (Kilgore 1974). Much like maintenance prescribed fires, areas in which managed wildfires occur are generally considered to generate outcomes that achieve multiple objectives, which may include resource management objectives and ecological purposes (Fig. 21.2). The original justification for such areas is that they were sufficiently remote to have been unaffected by management activities (van Wagtendonk 2007) (see sidebar 21.1 for a contrast of how fire is managed on national parks and national forests in California).

Mechanical Treatments

Many kinds of vegetation management remove, rearrange, or modify biomass (such as tree thinning). Mechanical fuel treatments are differentiated in that they must also include the objective of modifying potential fire behavior and effects. Removal of both live and dead woody fuel can utilize equipment such as feller bunchers, skidders, and grapplers. Thinning, crushing, chipping, shredding, chopping, and other mechanical methods of changing the fuel characteristics are commonly used. Mechanical treatments are more precise than prescribed fire. Smoke impacts and damage from scorching are avoided when mechanical methods are used instead of fire, but mechanical treatments are not complete surrogates for fire (Schwilk et al. 2009). In some cases the fuel can be removed from the area and used to produce wood products or to generate electricity (Hartsough et al. 2008).

Mechanical treatments are limited to areas with moderate topography and where legal designations permit such activities (e.g., non-wilderness). Over 60% of Forest Service lands in the

FIGURE 21.1 Prescribed fire in Sierra Nevada mixed conifer forests at the University of California Blodgett Forest Research Station in October, 2002 (photo by S. Stephens).

FIGURE 21.2 High severity patch 3.5 ha (8.6 ac) created by the 1994 Horizon Fire in red fir-Jeffrey pine forests in the Illilouette Creek basin, Yosemite National Park. Mount Star King in the background (photo taken in 2010 by S. Stephens).

SIDEBAR 21.1 CONTEMPORARY FIRE PATTERNS ON PARK SERVICE VERSUS FOREST SERVICE LANDS

Brandon M. Collins and Andrea E. Thode

Our ability to understand the different fire regime attributes in the contemporary period is easier compared to the historical period (see chapter 6). Fire severity mapping has allowed for explicit characterization of fire patterns, but the record of data only extends back to the mid 1980s, giving a temporal scale of only decades. That said, recent studies have looked at contemporary fire patterns in reference sites in California, primarily in Yosemite National Park (Collins et al. 2009, Lutz et al. 2011, Thode et al. 2011, van Wagtendonk et al. 2012). The term reference sites is used to describe forests which have never been harvested for timber and have minimally altered or restored fire regimes (Collins et al. 2016). These areas have been under management for 35 to 45 years that has sought to restore more natural fire regimes using managed lightning ignitions and prescribed fire. These fire restoration efforts have resulted in contemporary fire patterns that more closely resemble historical fires than most other forested areas, where fire suppression and exclusion tend to be the dominant fire management objective. When looking at trends of high-severity fire across landscapes, increases in the area of high-severity fire per year have been evident (Dillon et al. 2011, Miller and Safford 2012), and in the Sierra Nevada, the proportion and patch sizes of high-severity fire have also increased in recent decades (Miller et al. 2009). In contrast, a natural fire area in Yosemite National Park saw no changes in the proportions of area burned by severity class through time (Collins et al. 2009). With more fire activity, these national park and wilderness areas tend to have more reburns, or areas that have burned more than once. Several studies suggest that severely burned areas are more likely to reburn at high severity (van Wagtendonk et al. 2012) and that the time since the previous fire increases the severity of reburns (Collins et al. 2009, van Wagtendonk et al. 2012).

A comparison of contemporary fire patterns (extent and severity) between lands managed by the Forest Service and National Park Service in the Sierra Nevada revealed a significant distinction in fire severity patterns between the two agencies (Miller et al. 2012). Across the forest types that were analyzed, Miller et al. (2012) demonstrated that the proportion of high-severity fire and high-severity patch size were smaller for National Park Service fires than for Forest Service fires. In addition, their results showed that overall fire extent was less on National Park Service lands. The authors point out that although in recent years the Forest Service has begun to manage more wildfires for resource benefit, a policy of full suppression was in effect on most fires that occurred during their study period. In contrast, the National Park Service areas that they analyzed (all within Yosemite National Park) have a policy of only suppressing lightning-ignited fires when they occur outside their fire use zone or fall out of prescription, which resulted in most fires being managed for resource benefit. Miller et al. (2012) suggested that by allowing most lightning fires to burn relatively unimpeded, Yosemite has been able to achieve fire patterns that are closer to what may have occurred historically. This is not the case for the Forest Service fires they analyzed, which tended to burn under more extreme fire weather conditions, as these are the conditions under which fires generally escape initial fire suppression efforts (Finney et al. 2011). The authors noted that the National Park Service and Forest Service lands included in the study have different land management histories, particularly with respect to timber harvesting, which have resulted in different contemporary stand structures (Collins et al. 2017a), and could contribute to differential fire patterns. The National Park Service and Forest Service lands in the study also had very different landscape contexts, with Forest Service lands exhibiting a wider range of topographic and geomorphic configurations that would ultimately affect fire behavior. However, we should also note that many of the Forest Service fires analyzed were in the Cascade Range and Modoc Plateau, where gentler, less complex landscapes more easily facilitate large fires and major fire runs compared with the rocky, interrupted landscapes of Yosemite National Park that were the focus of the Miller et al. (2012) study.

When assessing contemporary fire patterns, human influences from different land management histories (fire suppression/exclusion, timber harvesting, grazing, etc.), combined with more recent external stressors (e.g., climate change, invasive species), can greatly affect the patterns and trends we see. Contemporary analyses in reference sites can give some insight into these changes but many of these sites are not perfect reference sites. Assessing contemporary fire regimes is easier with current data collection and new technology but understanding the drivers of change with the underlying confounding factors of different human and climate influences can prove to be difficult.

southern Sierra Nevada are unsuitable for mechanical treatments under current guidelines (North et al. 2015). Mechanical treatments can also be the most costly (Hartsough et al. 2008), and treatment rates can be modest in comparison to fire treatments. The application of mechanical methods on a scale matching the fuel problem in California is dependent on continued partnerships with industry to find uses for the material and cost-effective methods of removing it (Hartsough et al. 2008). Other fuel reduction methods may include grazing to remove fine fuel.

FOREST THINNING

Thinning is used as a treatment to modify the fuel structure in forests which have become denser due to fire exclusion (Collins et al. 2011b, Knapp et al. 2013). Thinning projects that reduce ladder fuel can be effective at moderating crown fire behavior as long as they are coupled to the reduction of surface fuels which are key in reducing crown fire behavior (Agee and Skinner 2005, Stephens et al. 2009a, Fulé et al. 2012).

Thinning can remove trees to create specified stand densities, patterns, distributions, and species compositions (O'Hara 1998). Thinning is an effective fuel management method if it reduces the likelihood that a surface fire will transition into a crown fire by breaking up vertical and horizontal fuel continuity. The thinning specifications, by density and by diameter classes of trees, are important characteristics of thinning prescriptions. Considerable progress has been made in developing guidelines to implement fuel reduction goals through thinning projects (Peterson et al. 2004, Fulé et al. 2012). Thinning for fuel management objectives normally retains the largest trees because of their importance to ecosystems (Lutz et al. 2012) and higher resistance to fire-induced mortality (Agee and Skinner 2005).

California's legislature has recognized the need for new statutes that encourage private timber companies and smaller landowners to reduce fire hazard on their properties through the use of thinning. These statutes provide incentives and exemptions that reduce the cost of planning for harvests that reduce fire hazard. Landowners may harvest timber under an exemption to create 46 m (150 ft) of fire safe clearance around permitted structures. The extent to which private property owners take advantage of these rules and incentives is dependent on the cost of transporting logs to the mill for processing and the economics of the timber market.

MASTICATION

Mastication is the mechanical grinding, crushing, shredding, chipping, and chopping of fuel to reduce fireline intensity and the rate of fire spread. Mastication and some other mechanical modifications of the fuel are used to reduce potential fire behavior by reducing fuelbed depth and thereby increasing the packing ratio (Kane et al. 2009). An ever-increasing selection of mechanical equipment is available to accomplish these tasks.

Mastication and other mechanical methods can effectively accomplish the modification of potential fire behavior with a great deal of precision (Kobziar et al. 2009). Mastication, like other mechanical treatments, is most commonly applied in the restoration rather than in the maintenance of ecosystems. Mastication commonly leaves elevated surface fuel conditions which can reduce the initial effectiveness of the treatment but decomposition can reduce these activity fuels in some cases if sufficient time (>7 to 10 years) elapses (Stephens et al. 2012b).

GRAZING AND VEGETATION MANAGEMENT PROGRAMS

Prior to the arrival of domestic livestock, native grazers undoubtedly had a great influence on herbaceous fuel. Cattle grazing or browsing by goats for the specific purpose of reducing fuel is applied on a limited scale, mostly on the wildland-urban interface in shrublands or grasslands. Its use is growing as a maintenance tool on fuel breaks and other linear fuel reduction projects. Goats are confined by fences and browse fuel breaks in the East Bay Regional Parks in the San Francisco Bay Area. The removal of fine fuel by domestic animals shortens the fire season and reduces fire potential. However, grazing can negatively impact some ecosystems, especially riparian areas or other mesic areas, if not carefully designed and monitored.

The California Board of Forestry is preparing a Programmatic Environmental Impact Report (PEIR) that will serve as the overarching CEQA-compliant process for all vegetation treatment projects (other than those that produce commercial timber) on State Responsibility Areas (SRAs) in California. The Vegetation Treatment Program will be administered and implemented through contracts between Cal Fire and private landowners, other agencies, and other organizations. The overall goal is to create more partnerships with land owners to implement fuel reduction treatments in appropriate vegetation types.

Fuel Management Phases: Restoration and Maintenance

Management of areas with altered fire regimes will often require both an initial restoration phase and a long-term maintenance phase. The importance of analyzing the costs and the frequency of restoration and maintenance has been highlighted by efforts by Federal agencies to define condition class and fire regime on a spatial basis. These concepts are designed to assist fire managers and the public in setting priorities for fuel management based on the frequency and severity of fire under pre-Euro-American conditions (fire regime) and departure from these regimes that has occurred during the fire suppression era (condition class) (Schmidt et al. 2002). The growing availability of spatial data describing fuel characteristics allows managers to set priorities for fuel treatments, and to quantify the extent of the fire hazard problem, at a variety of scales (Collins et al. 2010, Moghaddas et al. 2010).

The *restoration phase* can be designed to reestablish the historic or desired fuel structure and composition. During this phase, the techniques that are used are not necessarily the ones that occurred historically or the ones that will be prescribed for a long-term program. Mechanical treatments such as thinning of overly dense forest stands and other mechanical treatments are important tools (Stephens et al. 2009a). These treatments must maintain the desired focal characteristics of the landscape while accelerating the progress toward desired conditions (Miller and Urban 2000, Collins et al. 2011b, Van de Water and North 2011). Restoration can effectively and efficiently set up the landscape for fire to operate as an ecosystem process, enabling continuance during the maintenance phase (North et al. 2012). Prescribed fires and managed wildfire can also be used in this phase to manipulate fuel conditions before the maintenance phase.

The *maintenance phase* is the long-term application of mechanical methods, prescribed fire, or managed wildfire to the landscape. The maintenance phase can be accomplished once the restoration phase is completed. If the landscape is already in a condition that can support the use of prescribed fire or managed wildfire then restoration is not necessary. Maintenance phase treatments can be characterized by greater variability, which can be accomplished by fire applications under a range of fuel moisture and weather conditions. In some cases, application of herbicides, scraping, chopping, mastication, or other methods are used to maintain mechanical fuel treatments. As more areas are restored, the cost of completing maintenance on previously treated areas multiplies. A recent idea is to maintain such areas with lightning-ignited wildfires which are allowed to burn under prescribed conditions. This would allow managers to concentrate the use of prescribed fire and mechanical treatments on more areas needing restoration (North et al. 2012). New restoration work done at the expense of maintenance work can result in the loss of the original investment and redevelopment of wildfire hazard.

The dynamic nature of forest ecosystems imposes an important temporal consideration on fuels management planning. As time since treatment increases, vegetation growth will contribute to fuel pools and rebuild fuel continuity (Agee and Skinner 2005, Stephens et al. 2012b). Empirical studies from wildfires (Collins et al. 2009, Martinson and Omi 2013) and studies based on modeled fire (Collins et al. 2011a, Stephens et al. 2012b) suggest that treatments can be expected to reduce fire behavior for 10 to 20 years. Obviously a number of factors contribute to this longevity: type and intensity of treatment, site productivity, forest type, etc., but this only emphasizes the continued need for long-term fuels/restoration programs in fire-adapted forests. Fire in riparian areas is another important ecological and management concern in California (sidebar 21.2).

Choosing Management Methods on Complex Landscapes

We know how to design and evaluate the effects of individual fuel treatments (Stephens et al. 2009a, 2012a). It is far more challenging to design and evaluate treatments at a landscape scale (Collins et al. 2010, Moghaddas et al. 2010, Stephens et al. 2014a). This requires the application of large numbers of treatments over entire watersheds and the development of measures to evaluate the interaction of wildfires with these treatments over time. Essential questions include how to arrange fuel treatments, how often to maintain them, how much of the landscape should be treated, and how these treatments interact with critical wildlife habitat (Tempel et al. 2015) and riparian areas (Van de Water and North 2010) in order to mitigate unwanted wildfire behavior and effects. The answers will require both complex tools and trade-offs (Collins et al. 2010, Stephens et al. 2014a, 2016b). Methods have been created to place fuel treatments at the landscape scale with the goal of reducing landscape fire behavior and effects (Finney 2004).

Recently an evaluation of the system of shaded fuel breaks installed in the northern Sierra Nevada has been published (Stephens et al. 2014a). To enhance forest resilience, a coordinated network of shaded fuel breaks was installed which reduced the potential for hazardous fire, despite constraints for wildlife protection that limited the extent and intensity of treatments. Small mammal and songbird communities were largely unaffected by this landscape strategy, but the number of territories occupied by California spotted owl (*Strix occidentalis occidentalis*) declined. It is possible that these effects on owls could have been mitigated by increasing the spatial heterogeneity of fuel treatments and by using more prescribed fire or managed wildfire to better mimic historic vegetation patterns and processes (North et al. 2009). Research in Yosemite National Park suggests that California spotted owls are not adversely affected by low- to moderate-severity fire (Roberts et al. 2011) and that management of this species should be compatible with producing forests with high resiliency. Recent research has determined that high canopy cover forests (>70% canopy cover) most commonly associated with California spotted owl nesting habitat are being lost to wildfire at a high rate which could result in the development of new owl conservation strategy (Stephens et al. 2016a).

The social and political aspects of fuel management, while always factors in land management, are of even greater consequence today (McCaffery et al. 2008). It is of increasing importance for land managers to articulate and quantify the nature of the hazardous fuel issue and the rationale behind the selection of tools and programs to mitigate it, as well as demonstrating that such treatments are effective in modifying wildfire behavior and effects. Recent research has demonstrated that the majority of the public is in support of fuels and forest restoration projects (McCaffrey and Olsen 2012).

Air quality regulators commonly want land managers to consider the use of mechanical fuel reduction methods before the use of prescribed fire. Recent research has determined that mechanical treatments cannot access the majority of Forest Service lands in the central and southern Sierra Nevada (North et al. 2015) because of their remote and rugged nature. Managed wildfire is the most likely option for meeting restoration goals in these areas, which has been shown to generally meet restoration objectives when used in the southern and central Sierra Nevada (Collins et al. 2009, Collins and Stephens 2010, van Wagtendonk et al. 2012, Meyer 2015). State and federal efforts to implement active prescribed fire and managed wildfire programs to reduce the occurrence of large, long duration, and severe wildfire events and their emissions are being limited due to smoke regulations. State air regulatory agencies could therefore consider wildfire emissions which will be generated from untreated fuels when determining compliance with air quality standards (Stephens et al. 2016b).

History of Fuel Management—The evolution of a Fuel Emergency

As detailed in chapter 19, Native Americans used fire to modify ecosystems for several thousand years prior to the arrival of European settlers (Stephens et al. 2007). Some Native American fire practices were continued in modified forms by early Euro-American settlers. Ironically, the first widespread modification of the Native American fuel patterns came with the introduction of large numbers of domestic livestock, which severely reduced most of the surface fuel (grasses and herbs) and damaged ecosystems over large areas during the late 1800s (Fig. 21.3).

SIDEBAR 21.2 FIRE AND CALIFORNIA'S RIPARIAN ECOSYSTEMS

Amy G. Merrill and Tadashi J. Moody

Although the area of stream-influence, referred to as the riparian zone, occupies a small portion of California's landscapes, this network of relatively high moisture and high productivity helps maintain clear, cool streams and provides important habitat and landscape connectivity for aquatic, wetland, and terrestrial wildlife (Rundel and Sturmer 1998, Naiman et al. 2005). Vegetation structure and moisture characteristics, as well as landscape position, combine in various ways to affect fire frequency and fire behavior in the riparian zone. Dense vegetation with well-developed vertical structure and plentiful down wood and debris that is typical of many riparian zones can translate to high fuel loads that can support intense fire. In contrast, microclimate factors, such as dense shade, high moisture compared to uplands, locally dampened winds due to canopy complexity, and the fact that riparian areas are inherently located in local topographic minima, can reduce fire intensity and rate of spread (Dwire and Kauffman 2003).

Although riparian corridors are by definition in local low areas, they can wind down steep narrow valleys or across wide open plains. This variation in topographic context can importantly affect fire behavior as well: steep-sided valleys or canyons can create wind tunnels that increase fire intensity, whereas the local riparian tree canopy can baffle winds blowing across wide valleys to create local quiet areas that can lessen fire intensity and spread (Taylor and Skinner 1998, Dwire and Kauffman 2003). Over the past decade, observations of fire behavior in different kinds of riparian environments have led to a growing appreciation of the complexity of fire-riparian interactions and the need to better understand how human impacts and management actions can affect fire behavior in these areas. Several recent reviews of existing scientific literature on wildfire behavior in riparian zones highlight the need for additional research and reporting in this area (e.g., Bisson et al. 2003, Dwire and Kauffman 2003, Reeves et al. 2006, Pettit and Naiman 2007, Fryer 2015).

Riparian Fire Ecology

Recent studies reveal that, unlike the previous assumption that riparian areas burned less often and less intensely than their surrounding uplands, western forest riparian zones historically burned with frequencies ranging from tens to thousands of years, with both more and less burning than surrounding uplands (Benda et al. 1998, Dwire and Kauffman 2003, Van de Water and North 2011). Factors driving this variation have not been clearly articulated. Several authors report differences in fire return interval based upon stream order, with the lower order stream corridors having fire return intervals more similar to surrounding uplands than corridors along higher order streams (Agee 1998, Skinner 2002). Some have noted that fire frequencies in drier environments may be similar between riparian and upland areas (e.g., Dwire and Kauffman 2003, Van de Water and North 2010), while others report similar average fire intervals but higher variation in riparian areas (Skinner 2003).

Van de Water and North (2011) measured current fuel loads, tree diameters, heights, and height to live crown in 36 paired riparian and upland plots in the Sierra Nevada, reconstructed historic conditions, and then modeled fire frequency and intensity under both current and reconstructed historic conditions. The authors report that fire exclusion altered plant species composition and increased fuel loads in many montane riparian corridors of the Sierra Nevada compared to reconstructed historical conditions (Van de Water and North 2011). As a result, many riparian forests are more prone to fire than uplands, a reversal from reconstructed historical conditions in many areas. Van de Water and North (2011) also report that modeled fire severity in riparian areas of the Sierra Nevada is greater under current than reconstructed historical conditions.

Many site-specific and fire-specific factors affect fire severity. Wind speeds may be lower than in surrounding uplands, which can lower fire intensities in riparian areas, although channeling of winds in steeper riparian areas could occur. Variation in fuel moisture, vegetation composition, and stream order may result in complex patterns of fire frequency, severity, and behavior within and among riparian zones. Thus, depending on conditions during a given fire, some riparian areas are likely to act as barriers to spread while others might burn more readily and act as 'fuses' (Taylor and Skinner 2003, Pettit and Naiman 2007, North 2012a, 2012b).

River flow and the associated flood regime (frequency, timing, and severity) are all important factors affecting composition and the process of recovery from fire in riparian systems. In the American West, where stream flow changes by orders of magnitude in any given year, riparian vegetation is adapted to flooding through both resistance and recovery from disturbance. Adaptive traits, such as thick bark,

(continued)

(continued)

sprouting, and/or water- and wind-dispersed seeds also make it possible for many native riparian plant species to recover quickly following flood or fire (Pettit and Naiman 2007). Associated aquatic stream communities in Mediterranean landscapes were found to recover more quickly than stream communities in non-Mediterranean landscapes because of rapid recovery of riparian vegetation in Mediterranean areas (Verkaik et al. 2013). The timing and lateral extent of postfire flooding in the riparian zone can affect the rate of recovery and composition of riparian forests following fire as much or more than fire severity (Russell and McBride 2001, Reardon et al. 2008).

Fuel and Fire Management in Riparian Environments

Historical logging practices did not discriminate between upland and riparian forests; however, the current management policy of riparian environments in the Sierra Nevada and elsewhere—often called Stream Environment Zones (SEZs) in the federal system—greatly restricts fuel management within these zones. This raises questions about the degree to which fire suppression has altered natural fire frequencies and severities in riparian zones. Wildfire observations and other anecdotal evidence suggest that "unmanaged" or unaltered riparian zones might, under certain conditions, exhibit rapid rates of fire spread (North 2012b). Deciding whether or not to leave riparian zones in their current state requires consideration of possible trade-offs among competing risks and values (e.g., water quality, habitat, and fire hazard), and acceptance of potential compromise. To address this issue, it is also necessary to have some understanding of the prehistoric conditions within riparian zones under consideration, an assessment of the fire behavior and expected effects if left unaltered, and a clear definition of desired future outcomes and related inherent trade-offs.

Given the lack of knowledge on reference conditions for these areas and the heterogeneity of fire behaviors expected in riparian zones, it is still unclear what constitutes "uncharacteristically severe" fire in riparian areas. Although the Angora Fire in the Lake Tahoe Basin was catastrophic in human terms, was the high-severity riparian burning that occurred during extreme weather conditions outside the natural range of variability (North 2012b)? Due to higher biomass productivity in riparian zones, some sections will be capable of carrying higher severity fires during periods of drought and dieback or during episodes of extreme fire weather (Agee 1998). Accommodation of a range of natural disturbance severities, both due to fire and other physical processes, is actually necessary to maintain riparian habitats and biodiversity (Bisson et al. 2003). Some degree of high-severity burning is therefore to be expected in riparian zones, similar to mixed severity fire regimes of higher elevation coniferous forests.

In terms of fire regime restoration alone, riparian zones have been classified as relatively low priorities for fuel treatments. Where concerns over fire hazard are considered of paramount importance, riparian areas are still viewed as sensitive to mechanical fuel reduction techniques, and prescribed fire is seen as the most appropriate tool (Weatherspoon and Skinner 1996, Brown et al. 2004). Prescribed fire has successfully been employed in riparian areas with minimal short-term effects on several biotic and abiotic characteristics (e.g., Beche et al. 2005), and channel water quality is also relatively unaffected by prescribed fire (e.g., Stephens et al. 2004).

There is some evidence for shifts in species composition in riparian areas due to fire suppression, such as conifer encroachment (Kobziar and McBride 2006, Van de Water and North 2011). The lack of fire and the relatively high site productivity in and near riparian areas has resulted in the production of many large trees in the past 100 years. In one study, high-intensity prescribed fire was applied to reduce fuel loads and increase the light for deciduous plants near streams, but it was not successful in reducing tree density in mixed-conifer forests in the Sierra Nevada (Beche et al. 2005). Thus, where trees have become large enough to be very difficult to kill by prescribed fire, mechanical methods might be required to limit or reverse conifer tree encroachment. For riparian areas below dams where river flow regulation and diversion has also reduced channel meander and scouring effects that reset the succession of riparian forests, mechanical management of large trees could be the best option for restoring what is understood as historical tree size distribution. If mechanical methods are deemed necessary for habitat restoration in riparian zones, approaches would need to limit soil disturbance, compaction, and erosion.

Due to relatively high moisture levels, medium to high order riparian zones in Sierra Nevada forests are not expected to act as "fuses" that carry fire across portions of the landscape where efforts to limit fire spread would otherwise be successful (Weatherspoon and Skinner 1996). However, until more studies increase the clarity of how we understand and model fire in riparian zones, management decisions intended to balance short-term risks to riparian functionality against risks of fire hazard in riparian areas will ultimately be made in the face of uncertainty and competing values (Bisson et al. 2003).

FIGURE 21.3 Calaveras County near Gardner. Band of 1900 sheep grazing in pine forest. Bare surface. (Photo taken by George Sudworth in 1899, photo in the UC Berkeley Bioscience and Natural Resources Library.)

A fuel condition emergency has been developing over the century. Since 2002 five states—California, Arizona, New Mexico, Washington, and Colorado—have experienced their largest wildfires fires ever recorded. The annual cost of suppression and rehabilitation are exceeding 2 billion dollars in most years today. These catastrophic situations were foreseen decades ago by the pioneers in prescribed fire use in California. As noted by Carle (2002), government advocates of fire protection overrode ranchers, loggers, and other practitioners of "light burning." Interestingly, many of the arguments used against the practice of light burning are still used today as they were in 1924 (Show and Kotok 1924): fire damages young trees, is expensive, may escape control, and is difficult to adapt to variations in fuel and topography. Fire suppression was viewed as more straightforward and practical. The difference today is the understanding we now have about the effects on both ecosystems and fire behavior which result from the accumulation of understory vegetation and fuel in the absence of fire, primarily in forests and woodlands that once burned frequently (Stephens et al. 2013).

The research and teaching of Dr. Harold H. Biswell remains the cornerstone of the use of fire in California (Biswell 1989). Therefore, it is useful to examine in some detail the techniques that Dr. Biswell and others used to support the transition of policy from fire control to fire management. The translation of Dr. Biswell's techniques from small demonstration burns to large prescribed fire units to landscape or drainage-sized prescribed fire projects has continued to be a challenge. The teachings of Dr. Biswell, however, contain the solution to this transition: patience and public education.

Dr. Biswell's lesson of patience has two elements. One is the more obvious technique of conducting prescribed fires slowly and carefully, not exceeding the holding capacity of the personnel present, as well as not exceeding the capacity of the ecosystem to absorb the effects of fire and to avoid undesirable amounts of damage. The other is bringing along the public, agency administrators, politicians, and cooperators slowly enough that their comfort level with the use of prescribed fire is not exceeded. It is often said that it's taken a century of fire suppression to cause the fuel condition of ecosystems we see today, and it might take a generation or more of prescribed fire and managed wildfire to restore these ecosystems.

The difficult task of the modern fire manager is to increase the size and magnitude of fuel management programs so that the fuel can be reduced and ecosystems restored at a

SIDEBAR 21.3 FOREST LANDSCAPE-SCALE FUEL TREATMENT DESIGN

Brandon M. Collins and Scott L. Stephens

Large wildfire occurrence is increasing for most ecoregions in the western United States, which is particularly evident for the Sierra Nevada, southern Cascade Range, and Klamath Mountains (Dennison et al. 2014). These increases emphasize the pressing need to implement fire mitigation efforts at a scale that noticeably reduce large fire spread and intensity. However, implementing fuels treatment across an entire landscape may not be consistent with desired conditions or may not be operationally feasible (because of such issues as funding, access, and land designations) (Collins et al. 2010, North et al. 2015). In response, fire scientists and fire managers have conceptually developed and are refining methods for the strategic placement of treatments across landscapes (Weatherspoon and Skinner 1996, Finney 2004, Bahro et al. 2007, Finney et al. 2007). The basic idea is that an informed deployment of treatment areas (i.e., a deployment that covers only part of the landscape) can modify fire behavior and effects for the entire landscape. Owing to the complexity of modeling fire and fuels treatment across landscapes (e.g., data acquisition, data processing, model execution, etc.), fuels treatment project design is often based on local knowledge of both the project area and past fire patterns. Recent studies in the Sierra Nevada and southern Cascade Range suggest that these types of landscape-level fuels treatment projects (where treatment arrangement is based more on local knowledge and basic fire behavior modeling rather than intensive modeling associated with an optimization approach) can be quite effective at reducing potential fire behavior at the landscape scale (Schmidt et al. 2008, Moghaddas et al. 2010, Collins et al. 2011a, 2013).

Although only a few studies have explicitly modeled effectiveness of strategic landscape fuels treatments using different proportions of treated area, there are some common findings. First, noticeable reductions in modeled fire size, flame length, and spread rate across the landscape relative to untreated scenarios occurred with 10% of the landscape treated, but the 20% treatment level appears to have the most consistent reductions in modeled fire size and behavior across multiple landscapes (Ager et al. 2007, Finney et al. 2007, Schmidt et al. 2008, Ager et al. 2010). Second, increasing the proportion of area treated generally results in further reductions in fire size and behavior; however, the rate of reduction diminishes more rapidly when more than 20% of the landscape is treated (Ager et al. 2007, Finney et al. 2007). Third, random placement of treatments requires substantially greater proportions of the landscape to be treated compared to optimized or regular treatment placement (Finney et al. 2007, Schmidt et al. 2008); however, Finney et al. (2007) noted that the relative improvement of optimized treatment placement breaks down when larger proportions of the landscape (about 40–50%) are excluded from treatment because of land management constraints that

significant rate while also building public support (North et al. 2009, McCaffrey and Olsen 2012). Many land management agencies conduct prescribed fires, but at a rate and size that is trivial when compared to recent wildfire and the overall scale of wildland fire risk (North et al. 2012). It is the advancement of the size of prescribed fire and managed wildfire programs from small burns to ecologically significant landscape burns that is needed today (Boisramé et al. 2017a,b) (sidebar 21.3).

Fire policy itself has become more complex, and in particular the measurement and mitigation of risk associated with prescribed fire operations. Although very few prescribed fires escape and cause damage to structures or property, the few that do cause such damage receive much media attention and result in even more risk mitigation policies, and managers become even more risk adverse and cautious. Escaped prescribed fires in California have had a profound impact on interagency cooperation in the implementation of prescribed fire programs. In 1999, the BLM Lowden Ranch Fire escaped and burned through part of the town of Lewiston, destroying 23 homes. This escape served to increase public anxiety about the use of prescribed fire adjacent to towns and homes and increased the interest of communities in northern California in finding mechanical alternatives to prescribed fire in the wildland-urban interface. The Lowden Ranch escape also had significant impacts on fire management professionals. In the recent years, managers and prescribed burn bosses have become more risk averse and concerned about both personal liability and the risk associated with performance of their jobs (Stephens and Ruth 2005).

Invasive plants, which may take advantage of prescribed fire or mechanical activities to become established (Keeley limit treatment activities. It should be emphasized that this is not to preclude treating more than 20% of a landscape to achieve forest restoration, resilience, adoption, or other resource objectives. These studies suggest that when beginning to deal with fire hazard in a landscape, the initial objective would be to strategically reduce fire hazard on between 10% and 20% of the area to effectively limit the ability of uncharacteristically high-intensity fire to easily move across the area. This would buy time to allow needed restoration activities to progress in the greater landscape.

In designing landscape-level fuels treatment or restoration projects, there are often conflicts between reducing potential fire behavior and protecting/conserving other resources (Collins et al. 2010). One common conflict is habitat for wildlife species of concern (e.g., California spotted owl [*Strix occidentalis occidentalis*] and Pacific fisher [*Martes pennanti*]). Often these species prefer multistoried stands or closed canopies for nesting or denning habitat (Weatherspoon et al. 1992, Seamans and Gutierrez 2007, Tempel et al. 2014). Although it has been argued that fire suppression and past harvesting practices have created much of the habitat that is being called "desirable" for many of these species (Spies et al. 2006, Lyderson et al. 2018), the species-specific approach toward managing forests continues to prevail (Stephens and Ruth 2005, White et al. 2013). This approach limits the timing and intensity of fuels treatments. As a consequence, the ability to modify potential fire behavior, particularly fast-moving, high-intensity fire, in forests with prolonged fire exclusion is restricted but recent work has proposed some new ideas to conserve the California spotted owl (Stephens et al. 2016a). Furthermore, regulations on forest management within and around nesting centers or natal dens, and riparian buffer zones affect the size and placement of fuels treatments across landscapes. Therefore, there is limited opportunity to apply "optimal" placement of fuels treatments to maximize the reduction in spread of intense fire across the landscape. Additionally, these protected areas are often highly productive and contain large amounts of live and dead fuel. Thus, these areas may be prone to exacerbated fire behavior, creating effects not only within these protected areas (Healey et al. 2008), but also carrying into adjacent stands (Jones et al. 2016).

With climate change already extending the length of the dry season (Westerling et al. 2006) and increasing incidence of high fire weather conditions (Collins 2014) we can expect greater occurrence of large fires in California forests (Westerling et al. 2011a). Current conditions of the vast majority of California forests that once burned under a frequent, low-moderate-intensity fire regime (i.e., ponderosa pine [*Pinus ponderosa*], Jeffrey pine [*P. jeffreyi*], mixed conifer, xeric Douglas-fir [*Pseudotsuga menziesii* var. *menziesii*]) places them at high risk to uncharacteristically severe fire, especially with regard to the creation of large stand-replacement patches where >95% of the trees have been killed. The only way to reduce this risk is to implement coordinated fuels treatments at the landscape scale, which at the current rate of implementation is inadequate (North et al. 2012). Once areas have been strategically treated moving some of these landscapes (e.g., more remote watersheds) to a managed wildfire maintenance regime may be appropriate (North et al. 2012). One thing is certain, the time has come to develop and implement landscape strategies to reduce the vulnerabilities of California's forested landscapes to uncharacteristically severe fire and the impacts of drought and insects.

et al. 2003, Collins et al. 2007, Schwilk et al. 2009), are a growing concern (see chapter 24). This problem is particularly severe in the deserts and chaparral of California and is discussed in chapters 17 and 18. The effects of prescribed fire on sensitive species are often poorly understood; managers are often reluctant to allow prescribed fires which may have deleterious effects on sensitive species or their habitats (Stephens et al. 2014a, 2016a) (see chapter 25).

One of the most encouraging developments regarding fuels management is the cooperative education and planning efforts of fire managers and the public. Cal Fire, in trying to develop community involvement at the local level through the State's California Fire Plan, encourages the development of Fire Safe Councils throughout the state. These councils have sprung up in both rural and suburban communities. The primary objective of the councils is to involve citizens in creating defensible space around their homes and in working together to design protection strategies for their communities. These groups are increasingly interested, as they should be, in influencing fuel management priorities on areas adjacent to their communities and are influencing the design of projects and the expenditures of funds by federal, State, and local government managers.

These are encouraging trends in a State that clearly needs a force to galvanize interagency cooperation and cohesion in development and execution of fuel management strategies. The continued implementation of fuel management at a significant scale in California requires the involvement of the public in issues such as smoke management, the occurrence of wildfires due to large-scale forest mortality from drought and bark beetles (Stephens et al. 2018), and the protection of wildlife and its habitat from damage by wildfires, as well as cooperation between a variety of agencies and regulators. The ability of land managers to educate and engage the public to gain their support for prescribed fire and managed wildfire programs is directly linked to the survival and increasing scale of these programs (North et al. 2015).

Managing Fuel in Twenty-first–Century California

Recent fire history in California is characterized by what have been come to be called "fire sieges" (California Department of Forestry and Fire Protection and USDA Forest Service 2004) or mega-fires (Williams 2013). These are periods when large fires overwhelm the considerable fire suppression capability of federal, State, and local governments. These events are costly and large enough to capture the attention of media, government, and the people of California, the United States, and sometimes the world (Stephens et al. 2014b). Such fires directly impact the cities and towns in the paths of the fires and can inundate adjacent communities and far-away cities with smoke for months. Homes, businesses, and lives are lost. Closed highways and public areas such as national parks and forests, electrical power disruptions, and large-scale evacuations impact the economies of large areas of the State during these events.

Mega-fires

Recent fire behavior in some wildfires in California, such as exhibited in the Moonlight Fire in 2007, Rim Fire in 2013, and King Fire in 2014, has been classified as extreme and the term mega-fire is appropriately applied to them (Williams 2013). Mega-fires are produced by three factors—climate change, fire exclusion, and antecedent disturbance, collectively referred to as the "mega-fire triangle" (Stephens et al. 2014b). Climate change influences on large fire occurrence may be manifested in greater frequency of high- to extreme-fire weather conditions, which have been observed in the northern Sierra Nevada since the mid-1990s (Collins 2014). Some characteristics of mega-fires may emulate historical fire regimes and can therefore sustain healthy fire-prone ecosystems, but other attributes decrease ecosystem resiliency such as patches of tree mortality outside of desired ranges (Collins and Roller 2013). Although extreme, the Rim Fire was influenced by a fuel mosaic as it entered Yosemite National Park which was created over many years by prescribed fires and managed wildfires (Miller et al. 2012) and contributed to a reduction in the severity of the Rim Fire as it burned into the park (Lydersen et al. 2014).

Crown-fire-adapted ecosystems such as southern California chaparral and Rocky Mountain lodgepole pine (*Pinus contorta* subsp. *latifolia*) are likely at higher risk of frequent mega-fires as a result of climate change, as compared with other ecosystems once subject to frequent, less severe fires (Westerling et al. 2011a, Stephens et al. 2014b). In surface-fire-adapted ecosystems, management actions can be taken today to reduce the negative consequences of subsequent mega-fires (i.e., minimize high-severity patch size) and simultaneously achieve restoration objectives (Stephens et al. 2012a); in crown-fire-adapted ecosystems, however, such action (fuel treatments) is largely beyond the scope of restoration objectives (Stephens et al. 2013). Frequent high-severity burning will disrupt the ability of crown fire-adapted ecosystems to regenerate since seeds require sufficient time between fires to mature and vegetative resprouting can be exhausted by repeated fires. Increases in mega-fire abundance in these ecosystems may severely reduce resilience because thresholds could be crossed that change ecosystem states (e.g. forest to shrublands or shrublands to grasslands) over extensive areas (Westerling et al. 2011b).

Each of the fire sieges or mega-fires has had a profound impact on the fuel management program in the State, stimulating fact-finding, reviews, and investigations. Reports, commissions, and local government bodies uniformly recommend that fuel management be expanded to protect communities, improve the efficiency of fire suppression, and improve the resiliency of forest ecosystems. There is a surge of public and governmental support for fuel management programs in the wake of these events. This support translates into increases in funding for all types of fuel treatment, at least on the short term. In every case, there has been a marked increase in treated areas subsequent to these events, but a failure to carry through on the recommendations over the longer term (LHC 2018) and gather vital landscape level data to adapt and improve programs in the future.

Fires in the Wildland-Urban Interface

It has been said that the wildland-urban interface is a defining fire management issue of the twenty-first century. Pyne (1982), however, noted that interface issues have been a part of fire management as long as fires have burned from wildlands into communities. The 1991 Oakland Hills Fire, for example, had a precedent in the 1923 Berkeley Fire. What is new is the number of homes and communities which have been, and continue to be, constructed in the interface, greatly increasing the number of people and amount of property at risk (Gill and Stephens 2009, Stephens et al. 2009c, Gibbons et al. 2012, Moritz et al. 2015). Many of these homes are at risk not so much because of the buildup of fuel, but because homes and towns are constructed in vegetation types which naturally burn with high intensity and rapid spread such as chaparral. Comparisons of structure density in the footprint of the Laguna Fire of 1970, an area burned again in the Cedar Fire of 2003, showed that the number of structures has increased fivefold (Husari et al. 2004); some areas burned in the 2017 North Bay fires also previously burned in the 1960s and 1980s. Fire management in the interface is clearly one of the major issues for California in the twenty-first century.

Continued development in the interface area is reorienting fire resources and funding for fuel management programs. Allocating firefighting resources to protect homes and communities, rather than to suppress or manage the fire itself, is one reason for increasing costs and size of wildfires (Gude et al. 2013). Increasing loss of homes in the interface spurred Congress to allocate more fuel management funds to treat more areas at risk. Congress had accepted the argument made by land managers in 2000 that an expanded fuels management program working in concert with suppression resources is an important part of the solution to the escalating costs and effects of wildfires.

However, as suppression costs continued to rise even with the allocation of more fuels funding, the Office of Management and Budget (OMB) became skeptical that fuels management reduces wildfire expenditures. It provided direction to give priority to the use of federal fuels funds to projects near the wildland-urban interface, and fuels funding was also reduced. As fuels funding has been reduced and redirected, in part to pay for the rising suppression costs, there is uncertainty within federal fire agencies concerning the availability of funds and staff to execute fuels projects in areas other than the wildland-urban interface, making it very difficult to move forward on this important problem (Calkin et al. 2014, Stephens et al. 2016b). Beyond implementing more extensive fuel treatments, land-use planning for future developments in the interface will have to incorporate measures to reduce exposure to wildfires (Syphard et al. 2013).

Post-Wildfire Management

The increased incidence of mega-fires in California is forcing a greater attention toward postfire management. The most problematic aspects of these fires for forest management are the overall proportions of high-severity effects and the large, contiguous patches of high-severity fire (Miller et al. 2009, Miller and Safford 2012). Recently a new metric has been developed to better evaluate the ecological impacts of high severity patches (Collins et al. 2017b, Stevens et al. 2017). Large patches of high-severity fire in these forest types present challenges for reestablishing forests mainly due to the lack of direct mechanisms for seed persistence or long-distance dispersal (Barton 2002, Goforth and Minnich 2008, Keeley 2012, Crotteau et al. 2013, Welch et al. 2016). In the absence of management intervention, shrubs will often occupy these patches at high density (Collins and Roller 2013) creating the additional concern of a continuous fuelbed of live woody vegetation interspersed with heavy coarse woody loads (as fire-killed trees fall). This condition can also lead to repeat high-severity effects when areas are reburned with a potential conversion from forest to shurbland (Coppoletta et al. 2016).

Harvesting of fire-killed trees (salvage) is often proposed in areas burned at high severity. Management objectives for salvage-harvesting include: recovering economic value of timber, increasing personnel safety for reforestation efforts, and reducing large woody surface fuel accumulation (Ritchie et al. 2013). While salvage-harvesting generally achieves these objectives, it has very few short-term (<10 yrs) ecological benefits (McGinnis et al. 2010). In particular, there can be increased fine woody fuel loads (Donato et al. 2009, Peterson et at. 2015) and negative impacts on habitat associated with the removal of fire-killed trees (Swanson et al. 2011). Over the long-term (>20 yrs) the trade-offs of salvage-harvesting versus leaving fire-killed stands unaltered may change. A salvage-harvested and reforested area may return to mature conifer forest more quickly than an unaltered burned area, as well as have substantially reduced coarse woody fuel loads (Peterson et al. 2015). Some have argued there presently is too much emphasis on managing large areas of dead trees versus pro-actively managing live forests to reduce their vulnerability to wide-spread mortality (Stephens et al. 2018).

High-severity fire does, however, facilitate development of alternate vegetation types (e.g., montane chaparral or California black oak forests) which may be underrepresented in some contemporary landscapes due to fire suppression (Nagel and Taylor 2005, Cocking et al. 2012, 2014). Spatial scale is a critical consideration when balancing these trade-offs. For example, if patches of stand-replacing fire are large (e.g., >40 ha [>100 ac]) and left unaltered, the potential for colonization by montane chaparral across the entire patch is high (Conard and Radosevich 1982), resulting in a homogenization of landscape vegetation rather than an increase in vegetation diversity.

The Risk of Fuel Management Projects

Changes in the federal policy emphasizing the importance of the restoration of the natural role of fire have led to a greater use of prescribed and managed wildfire, at least in certain areas of the country. This, in turn, has led to the potential for escaped fires or for smoke episodes from larger or multiple burns. It is important to recognize that fuel management, particularly prescribed fire and managed wildfire, has inherent risk to both natural resources and communities, although the risk can be analyzed, mitigated, and reduced with better identification of the nature of the risk.

Fire fighter and public safety continue to be the highest priority during all fires. Similarly, health and visibility impacts from smoke are also a source of liability. Planning, modeling, and mitigation of the volume and extent of smoke are needed to anticipate and manage smoke impacts. As with other aspects of prescribed or managed wildfire, the use of monitoring equipment to show that health standards have, or have not, been violated is a basic part of the program. It should be noted that emissions from prescribed or managed fire were found to be a small component total particulate matter of the San Joaquin airshed adjacent to the southern Sierra Nevada, one of the poorest air quality airsheds in California (Cisneros et al. 2014, Preisler et al. 2015). The influence of prescribed and managed wildfire in carbon cycles is the subject of current discussion over allocation of California's funding for greenhouse gas reduction programs (Gonzalez et al. 2015). Over time a transition from a high-severity fire regime created by years of fire suppression to the restoration of a low-to moderate-severity surface fire regime is needed for forests adapted to frequent fire. It is important to note that this does not mean less fire, but often more frequent, lower-severity fire and appropriate mechanical treatments (Stephens et al. 2012a, North et al. 2012).

The Cost of Not Doing Fuel Treatments

Since the 1940s, many wildfire experts have warned of the effects of allowing fuel to accumulate far beyond natural levels. Harold Weaver, Harold Biswell, Roy Komarek, Ed Komarek, Herbert Stoddart, Bruce Kilgore, and Bob Martin, all stressed the need for the restoration of fire and for the need to manage fuel loads to mitigate the occurrence of high-intensity, destructive wildfires. Their predictions have proven to have been all too accurate and even more so as human-caused climate change lengthens fire seasons. The cost of not doing fuel management projects, therefore, is increasing damage to cultural and natural resources, higher suppression costs, degraded air quality, reduced water quality loss of revenue to communities, increased potential for tree mortality from drought and bark beetles, and, of paramount importance, increased risk to firefighter and public safety.

Conclusion

Fuel management programs in California do not suffer from a lack of public interest. However, they consistently have lacked a clear focus and scale that could carry across the multiple jurisdictions, varied fire regimes, and political and demographic landscapes of California in a way that could truly influence large fires and sustain biodiversity and ecosystem health (LCH

2018). Every fire siege or mega-fire, expensive fire season, or escaped prescribed fire generates new fire and fuels management policies and initiatives before previous decisions have been fully tested or evaluated. Recent federal fire policy and implementation guidance are more flexible and innovative (see the Fire Policy, chapter 20) but more can be done in this area (Stephens et al. 2016b).

Priority setting for fuel management will always be a difficult task in California. The sheer number of vegetation types and fire regimes described in the preceding chapters illustrates the difficulty of sorting out fuel management techniques appropriate to each of these assemblages of fire adapted plants and animals. The selection of fuel management techniques is further complicated by how the land is used and how many people live nearby. Is it wilderness or is it private land? Is it allowable or practical to use mechanical treatment methods to reduce accumulated fuel? These factors also influence how fuel treatments are distributed in the landscape. With all the barriers to implementation, it is clear that we should use every tool available to us, including, and especially, decision support and risk analysis tools, which help us in distributing fuel treatments in the most efficient way, and monitoring, to help us understand what works and what does not. Fire managers should not overlook the use of the various forms of social media to educate, inform, and discuss their program with the many stakeholders affected by it.

It is also important to note that as wildfires worsen, there is a tendency by fire and land managers to increasingly concentrate all fire resources on the suppression of current wildfires at the expense of planning and executing fuels projects which ironically will reduce the risk of future wildfires. The most difficult decision of all may be to proceed with the use of mechanical treatments, prescribed fire, and managed wildfire to reduce wildfire hazard in one part of the country, while unwanted wildfires are happening in other parts of the country. The alternative is that we lose much of what we have worked so hard to describe in this book, in all its beauty, variety, and complexity.

References

Agee, J.K. 1998. The landscape ecology of western forest fire regimes. Northwest Science 72: 24–34.

Agee, J.K., and C.N. Skinner. 2005. Basic principles of fuel reduction treatments. Forest Ecology and Management 211: 83–96.

Ager, A.A., M.A. Finney, B.K. Kerns, and H. Maffei. 2007. Modeling wildfire risk to northern spotted owl (*Strix occidentalis caurina*) habitat in Central Oregon, USA. Forest Ecology and Management 246: 45–56.

Ager, A.A., N.M. Vaillant, and M.A. Finney. 2010. A comparison of landscape fuel treatment strategies to mitigate wildland fire risk in the urban interface and preserve old forest structure. Forest Ecology and Management 259: 1556–1570.

Amacher, A.J., R.H. Barrett, J.J. Moghaddas, and S.L. Stephens. 2008. Preliminary effects of fire and mechanical fuel treatments on the abundance of small mammals in the mixed-conifer forest of the Sierra Nevada. Forest Ecology and Management 255: 3193–3202.

Arno, S.F., and S. Allison-Bunnell. 2002. Flames in Our Forest–Disaster or Renewal? Island Press, Washington, D.C., USA.

Bahro, B., K.H. Barber, J.W. Sherlock, and D.A. Yasuda. 2007. Stewardship and fireshed assessment: a process for designing a landscape fuel treatment strategy. Pages 41–54 in: R.F. Powers, technical editor. Restoring Fire-Adapted Ecosystems: 2005 National Silviculture Workshop. USDA Forest Service General Technical Report PSW-GTR-203. Pacific Southwest Research Station, Albany, California, USA.

Barton, A.M. 2002. Intense wildfire in southeastern Arizona: transformation of a Madrean oak-pine forest to oak woodland. Forest Ecology and Management 165: 205–212.

Beche, L.A., S.L. Stephens, and V. Resh. 2005. Effects of prescribed fire on a Sierra Nevada (California, USA) stream and its riparian zone. Forest Ecology and Management 218: 37–59.

Benda, L., D. Miller, T. Dunne, J. Agee, and G. Reeves. 1998. Dynamic landscape systems. Pages 261–288 in: R.B. Naiman and R. Bilby, editors. River Ecology and Management: Lessons from the Pacific Coastal Ecoregion. Springer, New York, New York, USA.

Bisson, P.A., B.E. Rieman, C. Luce, P.F. Hessburg, D.C. Lee, J.L. Kershner, G.H. Reeves, and R.E. Gresswell. 2003. Fire and aquatic ecosystems of the western USA: current knowledge and key questions. Forest Ecology and Management 178: 213–229.

Biswell, H.H. 1989. Prescribed Burning in California Wildland Vegetation Management. University of California Press, Berkeley, California, USA.

Boisramé, G.,S. Thompson, B. Collins, and S.L. Stephens. 2017a. Managed wildfire effects on forest resilience and water in the Sierra Nevada. Ecosystems 20: 717–732.

Boisramé, G., S.E. Thompson, M. Kelly, J. Cavalli, K.M. Wilkin, and S.L. Stephens. 2017b. Vegetation change during 40 years of repeated managed wildfires in the Sierra Nevada, California. Forest Ecology and Management 402: 241–252.

Brown, T.J., B.L. Hall, and A.L. Westerling. 2004. The impact of twenty-first century climate change on wildland fire danger in the western United States: an applications perspective. Climatic Change 62: 365–388.

California Department of Forestry and Fire Protection and USDA Forest Service. 2004. The story. California fire siege 2003. California Department of Forestry and Fire Protection and USDA Forest Service, Sacramento, California, USA.

Calkin, D.E., J.D. Cohen, M.A. Finney, and M.P. Thompson. 2014. How risk management can prevent future wildfire disasters in the wildland-urban interface. Proceedings of the National Academy of Sciences 111: 746–751.

Carle, D. 2002. Burning Questions—America's Fight with Nature's Fire. Praeger Publishers, Westport, Connecticut, USA.

Cisneros, R., D. Schweizer, H. Preisler, D. Bennett, G. Shaw, and A. Bytnerowicz. 2014. Spatial and seasonal patterns of particulate matter less than 2.5 microns in the Sierra Nevada Mountains, California. Atmospheric Pollution Research 5: 581–590.

Cocking, M.I., J.M. Varner, and E.E. Knapp. 2014. Long-term effects of fire severity on oak–conifer dynamics in the southern Cascades. Ecological Applications 24: 94–107.

Cocking, M.I., J.M. Varner, and R.L. Sherriff. 2012. California black oak responses to fire severity and native conifer encroachment in the Klamath Mountains. Forest Ecology and Management 270: 25–34.

Collins, B.M. 2014. Fire weather and large fire potential in the northern Sierra Nevada. Agricultural and Forest Meteorology 189–190: 30–35.

Collins, B.M., A.J. Das, J.J. Battles, D.L. Fry, K.D. Krasnow, and S.L. Stephens. 2014. Beyond reducing fire hazard: fuel treatment impacts on overstory tree survival. Ecological Applications 24: 1879–1886.

Collins, B.M., R.G. Everett, and S.L. Stephens. 2011b. Impacts of fire exclusion and recent managed fire on forest structure in old growth Sierra Nevada mixed-conifer forests. Ecosphere 2(4): art51.

Collins, B.M., D.L. Fry, J.M. Lydersen, R. Everett, and S.L. Stephens. 2017a. Impacts of different land management histories on forest change. Ecological Applications 27: 2475–2486.

Collins, B.M., H.A. Kramer, K. Menning, C. Dillingham, D. Saah, P.A. Stine, and S.L. Stephens. 2013. Modeling hazardous fire potential within a completed fuel treatment network in the northern Sierra Nevada. Forest Ecology and Management 310: 156–166.

Collins, B.M., J.M. Lydersen, D.L. Fry, K. Wilkin, T. Moody, and S.L. Stephens. 2016. Variability in vegetation and surface fuels across mixed-conifer-dominated landscapes with over 40 years of natural fire. Forest Ecology and Management 381: 74–83.

Collins, B.M., J.D. Miller, A.E. Thode, M. Kelly, J.W. van Wagtendonk, and S.L. Stephens. 2009. Interactions among wildland fires in a long-established Sierra Nevada natural fire area. Ecosystems 12: 114–128.

Collins, B.M., Moghaddas, J.J., and Stephens, S.L. 2007. Initial changes in forest structure and understory plant community

following fuel reduction activities in a Sierra Nevada mixed conifer forest. Forest Ecology and Management 239: 102–111.
Collins, B.M., and G.B. Roller. 2013. Early forest dynamics in stand-replacing fire patches in the northern Sierra Nevada, California, USA. Landscape Ecology 28: 1801–1813.
Collins, B.M. and S.L. Stephens. 2007. Managing natural fires in Sierra Nevada wilderness areas. Frontiers in Ecology and the Environment 5: 523–528.
———. 2010. Stand-replacing patches within a mixed severity fire regime: quantitative characterization using recent fires in a long-established natural fire area. Landscape Ecology 25: 927–939.
Collins, B.M., S.L. Stephens, J.J. Moghaddas, and J.J. Battles. 2010. Challenges and approaches in planning fuel treatments across fire-excluded forested landscapes. Journal of Forestry 108: 24–31.
Collins, B.M., S.L. Stephens, G.B. Roller, and J.J. Battles. 2011a. Simulating fire and forest dynamics for a landscape fuel treatment project in the Sierra Nevada. Forest Science 57: 77–88.
Collins, B.M., J.T. Stevens, J.D. Miller, S.L. Stephens, P.M. Brown, and M.P. North. 2017b. Alternative characterization of forest fire regimes: incorporating spatial patterns. Landscape Ecology 32: 1543–1552.
Conard, S.G., and S.R. Radosevich. 1982. Post-fire succession in white fir (*Abies concolor*) vegetation of the northern Sierra Nevada. Madroño 29: 42–56.
Coppoletta, M., K.E. Merriam, and B.M. Collins. 2016. Post-fire vegetation and fuel development influences fire severity patterns in reburns. Ecological Applications 26: 686–699.
Crotteau, J.S., J.M. Varner, and M.W. Ritchie. 2013. Post-fire regeneration across a fire severity gradient in the southern Cascades. Forest Ecology and Management 287: 103–112.
Dennison, P.E., S.C. Brewer, J.D. Arnold, and M.A. Moritz. 2014. Large wildfire trends in the western United States, 1984-2011. Geophysical Research Letters 41: 2928–2933.
Dillon, G.K., Z.A. Holden, P. Morgan, M.A. Crimmins, E.K. Heyerdahl, and C.H. Luce. 2011. Both topography and climate affected forest and woodland burn severity in two regions of the western US, 1984 to 2006. Ecosphere 2(12): art130.
Donato, D.C., J.B. Fontaine, J.L. Campbell, W.D. Robinson, J.B. Kauffman, and B.E. Law. 2009. Conifer regeneration in stand-replacement portions of a large mixed-severity wildfire in the Klamath-Siskiyou Mountains. Canadian Journal of Forest Research 39: 823–838.
Dwire, K.A., and J.B. Kauffman. 2003. Fire and riparian ecosystems in landscapes of the western USA. Forest Ecology and Management 178: 61–74.
Finney, M.A. 2004. Landscape fire simulation and fuel treatment optimization. Pages 117–131 in: J.L. Hayes, A.A. Ager, and J.R. Barbour, editors. Methods for Integrated Modeling of Landscape Change. USDA Forest Service General Technical Report PNW-GTR-610. Pacific Northwest Research Station, Portland, Oregon, USA.
Finney, M.A., C.W. McHugh, I.C. Grenfell, K.L. Riley, and K.C. Short. 2011. A simulation of probabilistic wildfire risk components for the continental United States. Stochastic Environmental Research and Risk Assessment 25: 973–1000.
Finney, M.A., R.C. Seli, C.W. McHugh, A.A. Ager, B. Bahro, and J.K. Agee. 2007. Simulation of long-term landscape-level fuel treatment effects on large wildfires. International Journal of Wildland Fire 16: 712–727.
Fryer, J.L. 2015. Fire regimes of montane riparian communities in California and southwestern Oregon. In: Fire Effects Information System. http://www.fs.fed.us/database/feis/fire_regimes/CA_montane_riparian/all.html. Accessed July 5, 2015.
Fulé, P.Z., J.E. Crouse, J.P. Roccaforte, and E.L. Kalies. 2012. Do thinning and/or burning treatments in western USA ponderosa or Jeffrey pine-dominated forests help restore natural fire behavior? Forest Ecology and Management 269: 68–81.
Gibbons, P., L. van Bomme, A.M. Gill, G.J. Cary, D.A. Driscoll, R.A. Bradstock, E. Knight, M.A. Moritz, S.L. Stephens, and D.B. Lindenmayer. 2012. Land management practices associated with house loss in wildfires. PLoS One 7(1): e29212.
Gill, A.M., and S.L. Stephens. 2009. Scientific and social challenges for the management of fire-prone wildland-urban interfaces. Environmental Research Letters 4: 034014.
Goforth, B.R., and R.A. Minnich. 2008. Densification, stand-replacement wildfire, and extirpation of mixed conifer forest in Cuyamaca Rancho State Park, southern California. Forest Ecology and Management 256: 36–45.
Gonzalez, P., J.J. Battles, B.M. Collins, T. Robards, and D. Saah. 2015. Aboveground live carbon stock changes of California wildland ecosystems, 2001–2010. Forest Ecology and Management 348: 68–77.
Gude, P.H., K. Jones, R. Rasker, and M.C. Greenwood. 2013. Evidence for the effect of homes on wildfire suppression costs. International Journal of Wildland Fire 22: 537–548.
Hartsough, B.R., S. Abrams, R.J. Barbour, E.S. Drews, J.D. McIver, J.J. Moghaddas, D.W. Schwilk, and S.L. Stephens. 2008. The economics of alternative fuel reduction treatments in western United States dry forests: financial and policy implications from the National Fire and Fire Surrogate Study. Forest Economics and Policy 10: 344–354.
Healey, S.P., W.B. Cohen, T.A. Spies, M. Moeur, D. Pflugmacher, M.G. Whitley, and M. Lefsky. 2008. The relative impact of harvest and fire upon landscape-level dynamics of older forests: lessons from the Northwest Forest Plan. Ecosystems 11: 1106–1119.
Hurteau, M., and M. North. 2009. Fuel treatment effects on tree-based forest carbon storage and emissions under modeled wildfire scenarios. Frontiers in Ecology and the Environment 7: 409–414.
Hurteau, M., T. Robards, D. Stevens, D. Saah, M. North, and G. Koch. 2014. Modeling climate and fuel reduction impacts on forest carbon stocks in Sierran mixed-conifer forest. Forest Ecology and Management 315: 30–42.
Husari, S.J., D. Brown, N. Cleaver, G. Glotfelty, D. Golder, R. Green, T. Hatcher, K. Hawk, N. Hustedt, P. Kidder, J. Millar, M, Sandeman, and T. Walsh. 2004. The 2003 San Diego County fire siege fire safety review. Unpublished Report on File at the Cleveland National Forest. San Diego, California, USA.
Jones, G.M., R.J. Gutiérrez, D.J. Tempel, S.A. Whitmore, W.J. Berigan, and M.Z. Peery. 2016. Megafires: an emerging threat to old-forest species. Frontiers in Ecology and the Environment 14: 300–306.
Kane, J.M., J.M. Varner, and E.E. Knapp. 2009. Novel fuelbed characteristics associated with mechanical mastication treatments in northern California and south-western Oregon, USA. International Journal of Wildland Fire 18: 686–697.
Keeley, J.E. 2002. Fire management of California shrublands. Environmental Management 29: 395–408.
———. 2012. Ecology and evolution of pine life histories. Annals of Forest Science 69: 445–453.
Keeley, J.E., D. Lubin, and C.J. Fotheringham. 2003. Fire and grazing impacts on plant diversity and alien plant invasions in the southern Sierra Nevada. Ecological Applications 13: 1355–1374.
Kilgore, B.M. 1974. Fire management in national parks: an overview. Proceedings Tall Timbers Fire Ecology Conference 14: 45–57.
Knapp, E.E., C.N. Skinner, M.P. North, and B.L. Estes. 2013. Long-term overstory and understory change following logging and fire exclusion in a Sierra Nevada mixed-conifer forest. Forest Ecology and Management 310: 903–914.
Kobziar, L.N., and J.R. McBride. 2006. Wildfire burn patterns and riparian vegetation response along two northern Sierra Nevada streams. Forest Ecology and Management 222: 254–265.
Kobziar, L.N., S.L. Stephens, and J.R. McBride. 2009. The efficacy of fuels reduction treatments in a Sierra Nevada pine plantation. International Journal of Wildland Fire 18: 791–801.
Kramer, H.A., B.M. Collins, M. Kelly, and S.L. Stephens. 2014. Quantifying ladder fuels: a new approach using LiDAR. Forests 5: 1432–1453.
LHC 2018. Little Hoover Commission—Fire on the Mountain: Rethinking Forest Management in the Sierra Nevada. Report #242, Sacramento, California, USA. http://www.lhc.ca.gov/report/fire-mountain-rethinking-forest-management-sierra-nevada.
Lutz, J.A., C.H. Key, C.A. Kolden, J.T. Kane, and J.W. van Wagtendonk. 2011. Fire frequency, area burned, and severity: a quantitative approach to defining a normal fire year. Fire Ecology 7(2): 51–65.
Lutz, J.A., A.J. Larson, M.E. Swanson, and J.A. Freund. 2012. Ecological importance of large diameter trees in a temperate mixed-conifer forest. PLoS One 7: e36131.
Lydersen, J.M., and B.M. Collins. 2018. Change in vegetation patterns over a large forested landscape based on historical and contemporary aerial photography. Ecosystems (in press).

Lydersen, J., M. North, and B. Collins. 2014. Severity of an uncharacteristically large wildfire, the Rim Fire, in forests with relatively restored frequent fire regimes. Forest Ecology and Management 328: 326–334.

Ma, S.Y., A. Concilio, B. Oakley, M. North, and J.Q. Chen. 2010. Spatial variability in microclimate in a mixed-conifer forest before and after thinning and burning treatments. Forest Ecology and Management 259: 904–915.

Martin, R.E., J.B. Kauffman, and J.D. Landsberg. 1989. Use of prescribed fire to reduce wildfire potential. Pages 11–16 in: N.H. Berg, technical coordinator. Proceedings of the Symposium on Fire and Watershed Management. USDA Forest Service General Technical Report PSW-GTR-109. Pacific Southwest Forest and Range Experiment Station, Berkeley, California, USA.

Martinson, E.J., and P.N. Omi. 2013. Fuel treatments and fire severity: a meta-analysis. USDA Forest Service Research Paper RMRS-RP-103WWW. Rocky Mountain Research Station, Fort Collins, Colorado, USA.

McGinnis, T.W., J.E. Keeley, S.L. Stephens, and G.B. Roller. 2010. Fuel buildup and potential fire behavior after stand-replacing fires, logging fire-killed trees and herbicide shrub removal in Sierra Nevada forests. Forest Ecology and Management 260: 22–35.

McCaffery, S., J.J. Moghaddas, and S.L. Stephens. 2008. Survey results from fire and fire surrogate fuel treatments in a Sierran mixed conifer forest, California, USA. International Journal of Wildland Fire 17: 224–233.

McCaffrey, S.M., and C.C. Olsen. 2012. Research perspectives on the public and fire management: a synthesis of current social science on eight essential questions. USDA Forest Service General Technical Report NRS-GTR-104. Northern Research Station, Newtown Square, Pennsylvania, USA.

Meyer, M.D. 2015. Forest fire severity patterns of resource objective wildfires in the southern Sierra Nevada. Journal of Forestry 113: 49–56.

Meyer, M., D. Kelt, and M. North. 2007. Effects of burning and thinning on lodgepole chipmunks (*Neotamias speciosus*) in the Sierra Nevada, California. Northwestern Naturalist 88: 61–72.

Millar, C.I., N.L. Stephenson, and S.L. Stephens. 2007. Climate change and forests of the future: managing in the face of uncertainty. Ecological Applications 17: 2145–2151.

Miller, C., and D. Urban. 2000. Modeling the effects of fire management alternatives on Sierra Nevada mixed conifer forests. Ecological Applications 10: 85–94.

Miller, J.D., B.M. Collins, J.A. Lutz, S.L. Stephens, J.W. van Wagtendonk, and D.A. Yasuda. 2012. Differences in wildfires among ecoregions and land management agencies in the Sierra Nevada region, California, USA. Ecosphere 3(9): art80.

Miller, J.D. and H.D. Safford. 2012. Trends in wildfire severity: 1984-2010 in the Sierra Nevada, Modoc Plateau, and southern Cascades, California, USA. Fire Ecology 8(3): 41–57.

Miller, J.D., H.D. Safford, M.A. Crimmins, and A.E. Thode. 2009. Quantitative evidence for increasing forest fire severity in the Sierra Nevada and southern Cascade Mountains, California and Nevada, USA. Ecosystems 12: 16–32.

Moghaddas, J.J., B.M. Collins, K. Menning, E.E.Y. Moghaddas, and S.L. Stephens. 2010. Fuel treatment effects on modeled landscape-level fire behavior in the northern Sierra Nevada. Canadian Journal of Forest Research 40: 1751–1765.

Moghaddas, E.E.Y., and S.L. Stephens. 2007. Thinning, burning, and thin-burn fuel treatment effects on soil properties in a Sierra Nevada mixed-conifer forest. Forest Ecology and Management 250: 156–166.

———. 2008. Mechanized fuel treatment effects on soil compaction in Sierra Nevada mixed-conifer stands. Forest Ecology and Management 255: 3098–3106.

Moritz, M.A., E. Batllori, R.A. Bradstock, A.M. Gill, J. Handmer, P.F. Hessburg, J. Leonard, S. McCaffrey, D. Odion, T. Schoennagel, and A.D. Syphard. 2015. Learning to coexist with wildfire. Nature 515: 58–66.

Naiman, R.J., N. Décamps, and M.E. McClain. 2005. Riparia: ecology, conservation, and management of streamside communities. Elsevier, New York, New York, USA.

Nagel, T.A., and A.H. Taylor. 2005. Fire and persistence of montane chaparral in mixed conifer forest landscapes in the northern Sierra Nevada, Lake Tahoe Basin, California, USA. Journal of the Torrey Botanical Society 132: 442–457.

North, M., editor. 2012a. Managing Sierra Nevada forests. USDA Forest Service General Technical Report PSW-GTR-273. Pacific Southwest Research Station, Albany, California, USA.

———. 2012b. Riparian zones pose severe wildfire threat. California Forests 16: 10–11.

North, M., A. Brough, J. Long, B. Collins, P. Bowden, D. Yasuda, J. Miller, and N. Suighara. 2015. Constraints on mechanized treatment significantly limit mechanical fuels reduction extent in the Sierra Nevada. Journal of Forestry 113: 40–48.

North, M.P., B.M. Collins, and S.L. Stephens. 2012. Using fire to increase the scale, benefits and future maintenance of fuels treatments. Journal of Forestry 110: 392–401.

North, M., J. Innes, and H. Zald. 2007. Comparison of thinning and prescribed fire restoration treatments to Sierran mixed-conifer historic conditions. Canadian Journal of Forest Research 37: 331–342.

North, M., P. Stine, K. O'Hara, W. Zielinski, and S.L. Stephens. 2009. An ecosystem management strategy for Sierran mixed-conifer forests. USDA Forest Service General Technical Report PSW-GTR-220. Pacific Southwest Research Station, Albany, California, USA.

O'Hara, K. 1998. Silviculture for structural diversity: a new look at multiaged systems. Journal of Forestry 96: 4–10.

Omi, P.N. 2015. Theory and practice of wildland fuels management. Current Forestry Reports 1: 100–117.

Peterson, D.W., E.K. Dodson, and R.J. Harrod. 2015. Post-fire logging reduces surface woody fuels up to four decades following wildfire. Forest Ecology and Management 338: 84–91.

Peterson, D.L., M.C. Johnson, J.K. Agee, T.B. Jain, D. McKenzie, and E.D. Reinhardt. 2004. Fuel planning: science synthesis and integration-forest structure and fire hazard. USDA Forest Service General Technical Report PNW-GTR-628. Pacific Northwest Research Station, Portland, Oregon, USA.

Pettit, N.E., and R.J. Naiman. 2007. Fire in the riparian zone: characteristics and ecological consequences. Ecosystems 10: 673–687.

Preisler, H.K., D. Schweizer, R. Cisneros, T. Procter, M. Ruminski, and L. Tarnay. 2015. A statistical model for determining impact of wildland fires on particulate matter (PM2.5) in Central California aided by satellite imagery of smoke. Environmental Pollution 205: 340–349.

Pyne, S.J. 1982. Fire in America: A Cultural History of Wildland and Rural Fire. Princeton University Press, Princeton, New Jersey, USA.

Reardon, J.R., K.C. Ryan, L.F. DeBano, and D.G. Neary. 2008. Wetlands and riparian systems. Pages 149–170 in: D.G. Neary, K.C. Ryan, and L.F. DeBano, editors. Wildland Fire in Ecosystems: Effects of Fire on Soil and Water. USDA Forest Service General Technical Report RMRS-GTR-42-VOL-4. Rocky Mountain Research Station, Fort Collins, Colorado, USA.

Reeves, G.H., P.A. Bisson, B.F. Rieman, and L.E. Benda. 2006. Postfire logging in riparian areas. Conservation Biology 20: 994–1004.

Ritchie, M.W., E.E. Knapp, and C.N. Skinner. 2013. Snag longevity and surface fuel accumulation following post-fire logging in a ponderosa pine dominated forest. Forest Ecology and Management 287: 113–122.

Roberts, S.L., J.W. van Wagtendonk, A.K. Miles, and D.A. Kelt. 2011. Effects of fire on spotted owl site occupancy in a late-successional forest. Biological Conservation 144: 610–619.

Rundel, P.W., and S.B. Sturmer. 1998. Native plant diversity in riparian communities of the Santa Monica Mountains, California. Madroño 45: 93–100.

Russell, W.H., and J.R. McBride. 2001. The relative importance of fire and watercourse proximity in determining stand composition in mixed conifer riparian forests. Forest Ecology and Management 150: 259–265.

Safford, H.D., D.A. Schmidt, and C.H. Carlson. 2009. Effects of fuel treatments on fire severity in an area of wildland–urban interface, Angora Fire, Lake Tahoe Basin, California. Forest Ecology and Management 258: 773–787.

Schmidt, K.M., J.P. Menakis, and C.C. Hardy. 2002. Development of coarse scale spatial data for wildland fire and fuel management. USDA Forest Service General Technical Report RMRS-GTR-87. Rocky Mountain Research Station, Fort Collins, Colorado, USA.

Schmidt, D.A., A.H. Taylor, and C.N. Skinner. 2008. The influence of fuels treatment and landscape arrangement on simulated fire behavior, Southern Cascade range, California. Forest Ecology and Management 255: 3170–3184.

Schwilk, D.W., J.E. Keeley, E.E. Knapp, J. McIver, J.D. Bailey, C.J. Fettig, C.E. Fiedler, R.J. Harrod, J.J. Moghaddas, K.W. Outcalt, C.N. Skinner, S.L. Stephens, T.A. Waldrop, D.A. Yaussy, and A. Youngblood. 2009. The National Fire and Fire Surrogate study: effects of fuel reduction methods on forest vegetation structure and fuels. Ecological Applications 19: 285–304.

Seamans, M.E., and R.J. Gutierrez. 2007. Habitat selection in a changing environment: the relationship between habitat alteration and Spotted Owl territory occupancy and breeding dispersal. The Condor 109: 566–576.

Show, S.B., and E.I. Kotok. 1924. The role of fire in the California pine forests. USDA Bulletin number 1294. U.S. Department of Agriculture, Washington, D.C., USA.

Skinner, C.N. 2002. Fire history in riparian reserves of the Klamath Mountains. Pages 164–169 in: N.G. Sugihara, M. Morales, and T. Morales, editors. Fire in California Ecosystems: Integrating Ecology, Prevention and Management: Proceedings of the Symposium. Association for Fire Ecology Publication Miscellaneous 1. Davis, California, USA.

———. 2003. A tree-ring based fire history of riparian reserves in the Klamath mountains. Pages 116–119 in: P.M. Faber, editor. California Riparian Systems: Processes and Floodplain Management, Ecology, and Restoration. Pickleweed Press, Mill Valley, California, USA.

Spies, T.A., M.A. Hemstrom, A. Youngblood, and S. Hummel. 2006. Conserving old-growth forest diversity in disturbance-prone landscapes. Conservation Biology 20: 351–362.

Stark, D.L., D.L. Wood, A.J. Storer, and S.L. Stephens. 2013. Prescribed fire and mechanical thinning effects on bark beetle caused tree mortality in a mid-elevation Sierran mixed-conifer forest. Forest Ecology and Management 306: 61–67.

Stephens, S.L., M. Adams, J. Hadmer, F. Kearns, B. Leicester, J. Leonard, and M. Moritz. 2009c. Urban-wildland fires: how California and other regions of the US can learn from Australia. Environmental Research Letters 4: 014010.

Stephens, S.L., J.K. Agee, P.Z. Fulé, M.P. North, W.H. Romme, T.W. Swetnam, and M.G. Turner, 2013. Managing forests and fire in changing climates. Science 342: 41–42.

Stephens, S.L., S.W. Bigelow, R.D. Burnett, B.M. Collins, C.V. Gallagher, J. Keane, D.A. Kelt, M.P. North, S.L. Roberts, P.A. Stine, and D.H. Van Vuren. 2014a. California spotted owl, songbird, and small mammal responses to landscape fuel treatments. BioScience 64: 893–906.

Stephens, S.L., N. Burrows, A. Buyantuyev, R.W. Gray, R.E. Keane, R. Kubian, S. Liu, F. Seijo, L. Shu, K.G. Tolhurst, and J.W. van Wagtendonk. 2014b. Temperate and boreal forest mega-fires: characteristics and challenges. Frontiers in Ecology and the Environment 12: 115–122.

Stephens, S.L., B.M. Collins, E. Biber, and P.Z. Fulé. 2016b. U.S. federal fire and forest policy: emphasizing resilience in dry forests. Ecosphere 7(11): e01584.

Stephens, S.L., B.M. Collins, C.J. Fettig, M.A. Finney, C.H. Hoffman, E.E. Knapp, M.P. North, H. Safford, and R.B. Wayman. 2018. Drought, tree mortality and wildfire in forests adapted to frequent fire. Bioscience 68: 77–88.

Stephens, S.L., B.M. Collins, and G. Roller. 2012b. Fuel treatment longevity in a Sierra Nevada mixed conifer forest. Forest Ecology and Management 285: 204–212.

Stephens, S.L., R.E. Martin, and N.E. Clinton. 2007. Prehistoric fire area and emissions from California's forests, woodlands, shrublands and grasslands. Forest Ecology and Management 251: 205–216.

Stephens, S.L., J.D. McIver, R.E.J. Boerner, C.J. Fettig, J.B. Fontaine, B.R. Hartsough, P. Kennedy, and D.W. Schwilk. 2012a. Effects of forest fuel reduction treatments in the United States. BioScience 62: 549–560.

Stephens, S.L., T. Meixner, M. Poth, B. McGurk, and D. Payne. 2004. Prescribed fire, soils, and stream water chemistry in a watershed in the Lake Tahoe Basin. International Journal of Wildland Fire 13: 27–35.

Stephens, S.L., C.I. Millar, and B.M. Collins. 2010. Operational approaches to managing forests of the future in Mediterranean regions within a context of changing climates. Environmental Research Letters 5: 024003.

Stephens, S.L.,J.D. Miller, B.M., Collins, M.P. North, J.J. Keane, and S.L. Roberts. 2016a. Wildfire impacts on California Spotted Owl nesting habitat in the Sierra Nevada. Ecosphere 7: e01478.

Stephens, S.L., J.J. Moghaddas, C. Ediminster, C.E. Fiedler, S. Hasse, M. Harrington, J.E. Keeley, J.D. McIver, K. Metlen, C.N. Skinner, and A. Youngblood. 2009a. Fire treatment effects on vegetation structure, fuels, and potential fire severity in western U.S. forests. Ecological Applications 19: 305–320.

Stephens, S.L., J.J. Moghaddas, B. Hartsough, E. Moghaddas, and N.E. Clinton. 2009b. Fuel treatment effects on stand level carbon pools, treatment related emissions, and fire risk in a Sierran mixed conifer forest. Canadian Journal of Forest Research 39: 1538–1547.

Stephens, S.L., and L.W. Ruth. 2005. Federal forest fire policy in the United States. Ecological Applications 15: 532–542.

Stephenson, N.L. 1999. Reference conditions for giant sequoia forest restoration: structure, process, and precision. Ecological Applications 9: 1253–1265.

Stevens, J.T., B.M. Collins, J.D. Miller, M.P. North, and S.L. Stephens. 2017. Changing spatial patterns of stand-replacing fire in California conifer forests. Forest Ecology and Management 406: 28–36.

Swanson, M.E., J.F. Franklin, R.L. Beschta, C.M. Crisafulli, D.A. Dellasala, R.L. Hutto, D.B. Lindenmayer, and F.J. Swanson. 2011. The forgotten stage of forest succession: early-successional ecosystems on forest sites. Frontiers of Ecology and the Environment 9: 117–125.

Syphard, A.D., A.B. Massada, V. Butsic, and J.E. Keeley. 2013. Land use planning and wildfire: development policies influence future probability of housing loss. PLoS One 8(8): e71708.

Taylor, A.H., and C.N. Skinner. 1998. Fire history and landscape dynamics in a late-successional reserve, Klamath Mountains, California, USA. Forest Ecology and Management 111: 285–301.

———. 2003. Spatial patterns and controls on historical fire regimes and forest structure in the Klamath Mountains. Ecological Applications 13: 704–719.

Tempel, D.J., R.J. Gutierrez, J.J. Battles, D.L. Fry, Y. Su, Q. Guo, M.J. Reetz, S.A. Whitmore, G.M. Jones, B.M. Collins, S.L. Stephens, M. Kelly, W.J. Berigan, and M.Z. Perry. 2015. Evaluating short- and long-term impacts of fuels treatments and simulated wildfire on an old-forest species. Ecosphere 6(12): art26.

Tempel, D.J., R.J. Gutierrez, S.A. Whitmore, M.J. Reetz, R.E. Stoelting, W.J. Berigan, M.E. Seamans, and M.Z. Peery. 2014. Effects of forest management on California Spotted Owls: implications for reducing wildfire risk in fire-prone forests. Ecological Applications 24: 2089–2106.

Thode, A.E., J.W. van Wagtendonk, J.D. Miller, and J.F. Quinn. 2011. Quantifying the fire regime distributions for severity in Yosemite National Park, California, USA. International Journal of Wildland Fire 20: 223–239.

Trouet, V., A.H. Taylor, E.R. Wahl, C.N. Skinner, and S.L. Stephens. 2010. Fire-climate interactions in the American West since 1400 CE. Geophysical Research Letters 37(4): L04702.

Vaillant, N.M., J. Fites-Kaufman, and S.L. Stephens. 2009. Effectiveness of prescribed fire as a fuel treatment in Californian coniferous forests. International Journal of Wildland Fire 18: 165–175.

Van de Water, K., and M. North. 2010. Fire history of coniferous riparian forests in the Sierra Nevada. Forest Ecology and Management 260: 383–395.

———. 2011. Stand structure, fuel loads, and fire behavior in riparian and upland forests, Sierra Nevada Mountains, USA; a comparison of current and reconstructed conditions. Forest Ecology and Management 262: 215–228.

van Wagtendonk, J.W. 2007. The history and evolution of wildland fire use. Fire Ecology 3(2): 3–17.

van Wagtendonk, J.W., K.A. van Wagtendonk, and A.E. Thode. 2012. Factors associated with the severity of intersecting fires in Yosemite National Park, California, USA. Fire Ecology 8(1): 11–31.

Verkaik, I., M. Rieradevall, S.D. Cooper, J.M. Melack, T.L. Dudley, and N. Prat. 2013. Fire as a disturbance in Mediterranean climate streams. Hydrobiologia 719: 353–382.

Wayman, R. and M. North. 2007. Initial response of a mixed-conifer understory plant community to burning and thinning restoration treatments. Forest Ecology and Management 239: 32–44.
Weatherspoon, C.P., S.J. Husari, and J.W. van Wagtendonk. 1992. Fire and fuels management in relation to owl habitat in forests of the Sierra Nevada and Southern California. Pages 247–260 in: J. Verner, K.S. McKelvey, B.R. Noon, R.J. Gutierrez, G.I. Gould, Jr., and T.W. Beck, editors. The California spotted owl: a technical assessment of its current status. USDA Forest Service General Technical Report PSW-GTR-133. Pacific Southwest Research Station, Albany, California, USA.
Weatherspoon, C.P., and C.N. Skinner. 1996. Landscape-level strategies for forest fuel management. Pages 1471–1492 in: D.C. Erman, general editor. Sierra Nevada Ecosystem Project: Final Report to Congress, Volume II. Wildland Resources Center Report 37. University of California, Davis, California, USA.
Welch, K.R., H.D. Safford, and T.P. Young. 2016. Predicting conifer establishment post wildfire in mixed conifer forests of the North American Mediterranean-climate zone. Ecosphere 7(12): e01609, doi:10.1002/ecs2.1609.
Westerling, A., B. Bryant, H. Preisler, T. Holmes, H. Hidalgo, T. Das, and S. Shrestha. 2011a. Climate change and growth scenarios for California wildfire. Climatic Change 109: 445–463.
Westerling, A.L., H.G. Hidalgo, D.R. Cayan, and T.W. Swetnam. 2006. Warming and earlier spring increase western US forest wildfire activity. Science 313: 940–943.
Westerling, A.L., M.G. Turner, E.H. Smithwick, W.H. Romme, and M.G. Ryan. 2011b. Continued warming could transform Greater Yellowstone fire regimes by mid-21st Century. Proceedings of the Notational Academy of Sciences USA 108: 13165–13170.
Williams, J. 2013. Exploring the onset of high-impact mega-fires through a forest land management prism. Forest Ecology and Management 294: 4–10.
White, A.M., E.F. Zipkin, P.N. Manley, and M.D. Schlesinger. 2013. Conservation of avian diversity in the Sierra Nevada: moving beyond a single-species management focus. PLoS One 8(5): e63088.

CHAPTER TWENTY-TWO

Fire, Watershed Resources, and Aquatic Ecosystems

JAN L. BEYERS, ANDREA E. THODE, JEFFREY L. KERSHNER, KEN B. ROBY, AND LYNN M. DECKER

> When we study the individual parts or try to understand the system through discrete quantities, we get lost. Deep inside the details, we cannot see the whole. Yet to understand and work with the system, we need to be able to observe it as a system, in its wholeness.
>
> WHEATLEY (1999)

Introduction

Fire is a major process within many California watersheds, causing fluctuations of water, nutrients, and sediment that are transmitted through aquatic ecosystems. However, many historic fire regimes and other watershed processes have shifted and changed with anthropogenic disturbances. In addition, as a result of urban expansion more homes and communities are located along streams and in floodplains. These changes cause social and ecological issues to arise concerning fire as a functional ecological process within watersheds. This chapter examines the current perspective of fires in watersheds and the social and ecological reactions to changes in the historic fire regimes. It then discusses challenges surrounding the restoration of fire regimes within watersheds, and ends with the concept of integrated watershed restoration.

The Effects of Fire on Watershed Processes and Functions

The effects of fire on watershed processes in California are as diverse as the landscape itself and involve a variety of ecosystems and physical processes. Watersheds can include a mosaic of intermittent, ephemeral, and perennial streams of various sizes. All are shaped by the dynamics of water and sediment, which ultimately determine the size and function of streams and rivers. This balance of water and sediment is influenced by physical and biological perturbations, including fire, at a variety of scales.

The relationship of water and sediment in watersheds is shaped by a number of factors including climate, topography, geology, soils, vegetation, and the frequency and magnitude of ecological processes and human disturbances. Ecological processes such as fire play an important role in determining the timing and magnitude of sediment, nutrient, and water delivery to streams. Timing and magnitude are influenced by aspects of the fire regime including fire size, type, seasonality, return interval, intensity, severity, and spatial complexity.

The influence of fire on watershed erosion is manifested in many different ways (chapter 7). Surface erosion occurs where canopy cover, surface litter, and duff are removed by wildfire, allowing direct rain drop impact and overland flow, with the subsequent displacement of soil. Eroded products of sediment mobilization are transported down slope into streams. Debris flows can ensue when postfire storm events mobilize sediment and dead woody material that has accumulated in channels (see chapter 7). Large erosional events such as landslips may be initiated several years after fire where slopes are steep, roots of fire-killed vegetation have decayed, and prolonged heavy rains occur (Rice and Foggin 1971, Ziemer 1981).

Once sediment reaches the channels, it moves downstream, alternately transported and deposited through the stream network. This sequence of transport and deposition is influenced by the timing and amount of runoff and sediment, local stream gradient, and the hydraulic complexity of the channels. Sediment is delivered to channels in pulses for the first few years postfire and is reworked and resorted for a number of years, until eventually sediment loads return to prefire levels (Fig. 22.1) (Minshall et al. 1989). Responses to fire of other components of stream ecosystems follow a similar trajectory (Pettit and Naiman 2007). The direction, duration, and magnitude of watershed ecosystem response are governed by the intensity, severity, and spatial complexity of the fire event. Uniformly high-intensity wildfires, resulting in severe effects that cover large landscapes, will generate a more dramatic response than low-intensity fires that remove little ground cover and are patchy in their effects.

Aquatic Habitats

The impacts of wildfires on streams are generally viewed as "pulses" (Detenbeck et al. 1992) that may be initially severe but are generally short-lived. In some cases these pulses may actually renew and rejuvenate stream habitats, for example by redistributing sediment during postfire flood flows (Benda et al. 2003) and bringing in new supplies of gravel (for spawning areas) and woody debris (Burton 2005). Postfire recovery of aquatic ecosystems is influenced by the integrity of the surrounding landscape and the proximity of high-quality habitats as sources of new colonists for stream organisms that

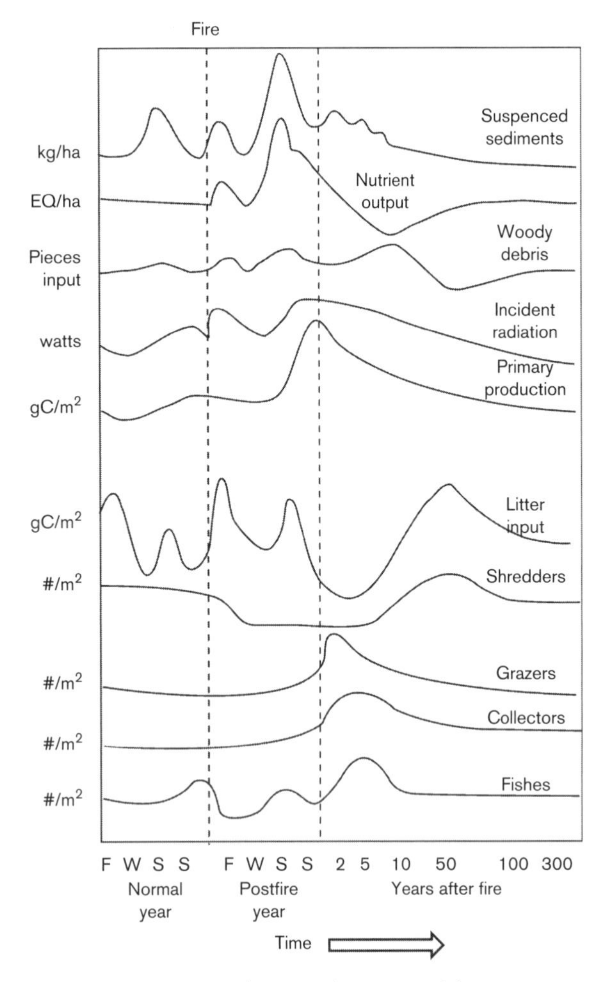

FIGURE 22.1 A conceptual, temporal sequence of the recovery of components of stream ecosystems in response to fire (Minshall et al. 1989). Short-term inputs of sediment and nutrients may occur in the first few years after a fire. Most other major fire-related disturbances to streams, such as debris flows or flooding, also typically occur within a few years after fire.

have been locally extirpated. Recovery of the stream/riparian interface generally parallels the surrounding vegetation recovery (Minshall et al. 1989).

Chronic or "press" events may truncate or retard the recovery of watershed processes and aquatic systems. Many parts of California present highly altered landscapes, where historic fire regimes have been modified (Stephens et al. 2007, and see earlier chapters) and watersheds influenced by past and present land management activities that include water development, mining, road building, urbanization, fire suppression, timber harvest, and recreation. Aquatic ecosystems have been substantially altered from conditions that existed prior to Euro-American settlement. Stream and watershed recovery after fire may be inhibited or truncated by these disturbances (Dunham et al. 2003). For example, postfire sediment delivery will decrease as vegetation regrowth creates ground cover and roots promote soil cohesion. However, postfire vegetative removal (e.g., salvage logging) and/or road building may stimulate continued inputs of sediment into streams (Kattleman 1996). This additional sediment may delay the recovery of aquatic organisms.

Major ecological effects occur when local extirpations of fish and other organisms are caused by fire effects, such as stream heating, temporary changes in stream chemistry, or postfire debris torrents (Spencer et al. 2003, Rinne and Jacoby 2005). In well-connected stream networks, individuals from surrounding areas can recolonize recovering streams once living conditions are again favorable (Burton 2005). Migration barriers from poorly placed road culverts, irrigation diversions, stream channelization, and dams may limit the movement of organisms through stream networks. This impact can be particularly severe where small population fragments exist in headwaters susceptible to wildfire and downstream habitats have been heavily altered.

The range of responses of postfire aquatic ecosystems will vary with the intensity, severity, and temporal length of the fire event (Bisson et al. 2003). In some cases, wildfire may benefit aquatic ecosystems by restoring ecological processes altered by past management practices. In others, the combination of wildfire and anthropogenic influences may compound effects to severely alter aquatic ecosystems for much longer time periods. Ultimately, the speed and trajectory of postfire recovery will be influenced by the amount and location of anthropogenic disturbance.

Whereas aquatic ecosystem recovery after fire is a long-term process, there are often immediate public concerns over the short-term effects of fire on water quality, particularly where the water is used for human consumption. A variety of laws and policies are in place that guide the management of water quality within California (see sidebar 22.1, Water Quality Laws and Policy). Where there are concerns for downstream water quality or potential threats to life or property, federal agencies may undertake postfire stabilization treatments to augment the recovery process.

Postfire Watershed Stabilization, Rehabilitation, and Restoration

Land management agencies have a long history of undertaking postfire stabilization and rehabilitation activities in watersheds. The Forest Service has implemented treatments since the 1930s, with a formal program initiated in the 1970s; the Bureau of Land Management (BLM) has had programs in place at least since the 1960s. The intention of these programs is to minimize risks to life, property, and natural or cultural resources related to fire effects, including flood runoff, erosion, off-site sedimentation, and hydrologic damage (Robichaud et al. 2000). In addition, other nonemergency postfire activities may be planned and implemented. These often include capturing economic value from burned trees (salvage logging) and actions intended to promote longer-term restoration of ecosystem products and services (tree planting).

The definitions of emergency stabilization, rehabilitation, and restoration are often confused. Emergency stabilization actions are taken during the first year after a fire to mitigate postfire effects on physical ecosystem processes, such as erosion and runoff. Longer-term rehabilitation and restoration activities are more focused on the biotic part of the ecosystem, such as recovery of habitat for rare species, preventing spread of invasive nonnative species, and reestablishment of timber stands (Robichaud et al. 2009). There is some overlap in goals—efforts may be undertaken the first year after fire to

SIDEBAR 22.1 FEDERAL AND STATE WATER QUALITY LAWS AND POLICIES

Andrea E. Thode

When discussing prescribed fire and mechanical fuel treatments in California watersheds, there are several water quality laws and policies that can come into play. The main water quality issue that stems from fuels treatments is sedimentation, which is addressed through nonpoint source pollution laws. However, herbicide use to eliminate or suppress understory growth also presents water quality issues and could entail regulation as point source pollution. It is critical to understand the general structure and relationships between federal and state laws to understand how these laws and policies can affect fuels treatments. A summary of the main points of water quality law and policy in California follows.

Three principal laws govern water quality in California: the federal Clean Water Act of 1972 (33 USC 1251-1387); the state's Porter-Cologne Water Quality Control Act (California Water Code 13000-14958); and the federal Coastal Zone Management Act (16 USC 1451 et seq.).

Water quality protection is the responsibility of the State Water Resources Control Board (State Board) and the regional water quality control boards. The state is split into nine regions based on watersheds. Within each region, the regional boards establish regional water quality plans and work under the policies of the State Board (Ruffolo 1999). Under the Clean Water Act and the Coastal Zone Management Act, California is required to adopt water quality standards and to submit those standards for approval to the US Environmental Protection Agency (EPA) and to the National Oceanic and Atmospheric Administration (NOAA).

The Clean Water Act

The Clean Water Act establishes nationwide minimum standards for water quality control in waters subject to federal jurisdiction (i.e., "navigable" waters and tributaries to navigable waters); it addresses water quality pollution from two distinct sources—point sources and nonpoint sources. *Point sources* include direct discharge from factories, sewers, and other sources that have direct contact with surface waters. Section 402 of the Clean Water Act and regulations set by the EPA (33 USC 1342, 40 CFR 122, et seq.) require that these point sources are regulated by permits or requirements that implement the National Pollutant Discharge Elimination System (NPDES). *Nonpoint sources* are those associated with activities such as agriculture and forestry practices, where many activities are spread across large areas and water quality is affected by runoff from the activity areas, rather than from a discrete location. Specific sections of the Clean Water Act provide guidance on the assessment and treatment of nonpoint source pollution, and these are applicable to wildfire and fuel treatment activities. However, the Clean Water Act does not require nonpoint sources to be regulated under the NPDES; instead, nonpoint sources are regulated more indirectly through section 303d.

Section 303d of the Clean Water Act provides a structure for states to assess water quality (Houck 2002). States must evaluate all water bodies and identify those that fail to meet water quality standards despite the application of effluent limits for point sources. Total Maximum Daily Loads (TMDLs) of pollutants must be established for these waters (Ruffolo 1999, Houck 2002). TMDLs are defined as the maximum amount of a pollutant that can occur daily in a waterbody to meet water quality standards. In practice, the term "daily" has been expanded to include various times such as weekly, monthly or annually. This part of the act was dormant until the 1990s when environmental and citizen groups began to utilize its power in court (Houck 2002), and refinement of the total daily load process is ongoing. Current interpretation of this section and the implications of TMDLs to nonpoint source polluters and regulators are in flux.

Section 319, added to the Clean Water Act in 1987, established the nonpoint source grant program that requires states to develop nonpoint source management plans (CWA section 319, 33 USC 1329). These plans generally depend on nonregulatory approaches to control nonpoint source pollution and include voluntary efforts, incentives, education, and training to help implement "Best Management Practices" (BMPs) (Ruffolo 1999). BMPs are guidelines designed to offset or mitigate land-use practices with potential to affect water quality from nonpoint sources. These practices have been developed by state and federal agencies for a variety of land-use activities including timber harvest, road construction and maintenance, fire suppression, fuel management, and many others.

The Porter-Cologne Water Quality Control Act

Section 303d of the Clean Water Act requires states to set TMDLs and section 319 requires states to develop management plans that depend on nonreg-

(continued)

(continued)

ulatory actions. However, the actual regulatory authority over nonpoint source pollution lies within California's Porter-Cologne Act (Ruffolo 1999). This act does not distinguish between point and nonpoint sources, but instead focuses on "discharges" of waste to water bodies and regulates the quality of those discharges. Regulatory power over nonpoint sources comes through the state's "waste discharge requirement" permitting program. Waste discharge requirement permits must be obtained by any person or entity that intends to discharge waste into California's waters.

The Coastal Zone Management Act

The 1972 federal Coastal Zone Management Act set a framework for management, protection, development, and beneficial use of the coastal zone (SWRCB and CCC 2000). However, this act did not specifically deal with water quality until the 1990 Coastal Zone Act Reauthorization Amendments. The reauthorization required that California's Water Resources Control Board and the California Coastal Commission jointly develop a coastal nonpoint source pollution control program (SWRCB and CCC 2000). This was done with the realization that nonpoint source pollution was a major polluter of coastal resources as well as the state's fresh water resources.

California's Nonpoint Source Programs—The 1998 and 2000 Plans

In response to section 319 of the Clean Water Act, California developed the 1988 *Nonpoint Source Management Plan* (1988 Plan), which applies three tiers, or levels, of regulatory involvement (Ruffolo 1999). The first tier relies on voluntary implementation of BMPs, and for the most part California has used this level of regulation. The second tier is regulatory-based encouragement of BMPs where the regional water quality control boards have discretion over the waiver of waste discharge requirements. Regional boards may enter into agreements with other agencies capable of enforcing BMPs. These agreements have been initiated with the Forest Service, the California Board of Forestry and the California Department of Forestry and Fire Protection to control discharges of waste and pollutants associated with silvicultural activities (SWRCB and CCC 2000). The third tier contemplates the issuance and enforcement of waste discharge requirements prescribing BMPs and effluent limits for proposed or existing nonpoint source discharges (SWRCB 1988, Ruffolo 1999).

The first significant revision of the 1988 plan occurred in 1998 with the *Plan for California's Nonpoint Source Pollution Control Program* (SWRCB and CCC 2000). Because of the 1990 changes in the Coastal Zone Management Act, the 1998 plan was written in conjunction with the California Coastal Commission. This revision adds 61 management measures as goals for six nonpoint source pollution categories. Two pollution categories—forestry and urban areas—include management measures that relate to fire suppression, prescribed fire, and fuel treatments. In 2004 the State Board developed the *Policy for the Implementation and Enforcement of the Nonpoint Source Pollution Control Program*, which explains how the 1998 plan will be implemented and enforced. In 2015 the State Board issued the *California Nonpoint Source Program Implementation Plan 2014–2020* to describe activities planned by the state and regional water quality control boards; this document updates the 1998 plan.

protect or stabilize habitat for a rare species, for example, or to locate and eliminate incipient nonnative plant invasions—and similar methods may be employed.

Postfire Stabilization

From the perspective of ecosystems, fires are not an emergency, especially when fire effects are within the range of the historic fire regime for the vegetation type. However, there is a social desire to respond to large wildfires, particularly when they are in proximity to communities and municipal water resources. Where wildfires exhibit low spatial complexity, high intensity, and/or high severity, they often result in increased erosion and overland flow of water from burned watersheds (see chapter 7). The effects are initially expressed during the first rainy seasons after fire, resulting in flooding, debris flows, and sedimentation. The magnitude of these effects will depend on how much of the watershed was burned at high severity and the intensity of postfire rain events. Life and property may be at risk when developments are located downstream. Federal land management agencies are required to assess burned areas and determine if threats to resources or human health and safety exist. Assessment teams—called Burned Area Emergency Response (BAER) teams by the Forest Service and National Park Service, and Emergency Stabilization and Rehabilitation (ESR) teams by the BLM—are convened to evaluate: (1) threats to life and property; (2) potential for loss of control of water; (3) likely reduction in water quality; and (4) potential for loss of soil productivity in the burned area (U.S. GAO 2003). Threats to sensitive natural and cultural resources are also assessed. If an emergency is found to exist, assessment teams prescribe short-term stabilization treatments designed to alleviate the immediate threats. Treatments are undertaken when analyses show that the planned actions are likely to significantly reduce risk of adverse impacts in a cost-effective manner. The US Fish and Wildlife Service and Bureau of Indian Affairs have similar programs. State agencies may do a postfire assessment across private lands, and there are programs through the Natural Resources Conservation Service (NRCS) that can aid land owners in identifying and applying treatments.

A report (Robichaud et al. 2000) and a U.S. General Accounting Office (GAO) audit (U.S. GAO 2003) evaluated and questioned the efficacy and cost of many of these treatments. Expenditures for postfire rehabilitation have escalated over the decades—reported BAER treatment costs for the Forest Service, for example, rose from $71 million in the 1990s to $310 million for the 2000s (Robichaud et al. 2014). In response to the GAO report, efforts to consistently monitor the effectiveness of Forest Service and Department of the Interior BAER and ESR treatments increased, and new research has been conducted on the effectiveness and ecological impacts of various stabilization methods (reviewed by Foltz et al. 2008, Peppin et al. 2010, Robichaud et al. 2010, Pyke et al. 2013).

In California, postfire grass seeding—formerly routinely applied after almost every fire (Barro and Conard 1987)—is now seldom done based on the results of this research (Wohlgemuth et al. 2009). Where increased ground cover is deemed necessary to reduce erosion after high-severity fire because critical values are at risk, BAER teams are now more likely to prescribe some kind of mulch (Wohlgemuth et al. 2011, Hubbert et al. 2012, Robichaud et al. 2013a, 2013b, 2013c, 2013d). The shift from aerial application of inexpensive nonnative grass seed to heavier mulches that are complex to deliver accounts for some of the increase in postfire stabilization treatment costs (Robichaud et al. 2014). Treatments are targeted to areas of greatest threat to values at risk. On many fires, increased runoff from roads and damage due to undersized culverts is identified as a major threat, such that road treatments have become a large part of postfire stabilization expense (Furniss et al. 1998, Foltz and Robichaud 2013, Robichaud et al. 2014). Grass seeding—and indeed other postfire stabilization treatments—can have unknown or unintended effects on ecosystem recovery, the extent of which is still understudied (Beyers 2004, Robichaud et al. 2009, Peppin et al. 2010, Berryman et al. 2014).

Postfire Rehabilitation and Restoration

Most vegetation communities eventually recover from fire without human intervention. This is especially the case when the fire falls within the historic fire regime for a vegetation community. Most fires in chaparral, fires in grasslands, low- to moderate-severity fires in ponderosa pine (*Pinus ponderosa*) or mixed-conifer forests, and high-severity fire in closed-cone conifer stands are good examples. However, there are other cases where rehabilitation and restoration actions are deemed useful or necessary to accelerate ecosystem recovery. These include landscapes where fire effects have been severe, large, and homogenous (outside the historic range of variation for the fire regime), where fire occurred soon after a previous fire and an adequate seed bank has not formed, or where undesirable nonnative species may grow back more quickly than the native flora.

In sagebrush (*Artemisia tridentata* subspp.) rangelands, native or well-adapted nonnative bunch grasses, native forbs, and shrubs are often seeded after fire in an attempt to reduce invasion by cheat grass (*Bromus tectorum*), other nonnative grasses, or noxious weeds (e.g., thistles [*Cirsium* spp. or *Centaurea* spp.]) (Pyke et al. 2013). These invasive species can spread rapidly into burned areas. Seeding with native species to prevent nonnative species invasion can be both a short-term stabilization and longer-term rehabilitation treatment. Similarly, direct control of noxious weeds using hand or chemical methods can be implemented the first year after a wildfire as emergency stabilization as well as continue under different funding mechanisms for vegetation rehabilitation (e.g., up to 3 years of Burned Area Rehabilitation [BAR] funding at Department of the Interior agencies).

Restoration is generally a longer-term process than rehabilitation. Most restoration treatments, such as tree or shrub planting, are not carried out until after emergency stabilization treatments have been implemented, and in the case of tree planting may not occur until after any salvage logging has occurred. Treatments that control competing vegetation around planted seedlings, such as grazing, herbicide application, or prescribed fire, may continue for many years.

Many burned areas do not need emergency rehabilitation or vegetation restoration treatments. If the characteristics of the fire are within the historic range of variation for the fire regime and human or natural resource values are not at risk, a burned area can be left to recover on its own.

Ecological and Social Values

It is apparent that ecological and social values may suggest conflicting actions when it comes to postfire rehabilitation and restoration. As discussed earlier, fire has historically been an integral part of ecosystem processes, influencing fluctuations in water and sediment yield. However, there are now homes and entire communities along streams, in floodplains, and adjacent to fire-prone ecosystems in much of California. The propensity for burned watersheds to move large amounts of water and sediment into people's yards, houses, and businesses creates a social demand for postfire stabilization treatment from both safety and economic perspectives.

Postfire restoration activities, such as tree planting, are used to bring burned landscapes back to prefire conditions faster than would occur naturally. The goals and objectives are not necessarily ecologically based as much as socially based: they focus on "quick recovery," which is a human concept rather than an ecological one. Recent very large fires in the Sierra Nevada (e.g., 2013 Rim fire, 2014 King fire) with extensive patches of severely burned conifer forest have stimulated renewed debate about the need for and ecological consequences of salvage logging and subsequent replanting (Long et al. 2014a). Social benefits cited for salvage logging of dead trees include creating local jobs, generating revenue that may be used for replanting, and storing carbon in forest products rather than letting it return to the atmosphere through decomposition. Industrial forest lands are quickly harvested and replanted after stand-replacing fire. There can be high public support for salvage logging on public lands in communities near large fires, but public attitudes generally may depend on the perceived ecological impacts of the practice (summarized in Long et al. 2014b).

Unharvested high-severity burn patches provide habitat for early seral stage plants and animals, including nitrogen-fixing shrubs and woodpeckers drawn to feed on bark beetles that quickly invade dead and dying trees. Predators find good hunting in rodent populations that increase in abundance during early postfire recovery (see chapters 8 and 9). Watershed conditions are cited as a concern on both sides of postfire logging debates. Negative effects of salvage logging can include soil compaction, disruption of understory vegetation recovery and tree seedling recruitment, loss of snags and large downed wood recruitment (both upland and in streams), and increased sedimentation into aquatic habitats from skid trails and logging roads (usually short term if best management practices are applied) (Karr et al. 2004). Watershed benefits of salvage logging may include reduction in surface fuel loads and subsequent soil burn severity in future fires, especially in stands that were overly dense due to past fire suppression (Reeves et al. 2006, Peterson et al. 2009, Ritchie et al. 2013). However, Donato et al. (2006) found that salvage logging resulted in elevated surface fuels levels compared to unlogged areas on the Biscuit fire in Oregon.

Restoring Fire as a Process in Watersheds

Watersheds are dynamic, and fire is part of that dynamism. Ideally, reintroducing fire as an ecological process and restoring fire regimes would be an integral part of watershed restoration. Socially, this would mean less expenditure on fire suppression and rehabilitation, and less loss of property and lives in the long run. Ecologically it would mean the restoration of sediment and runoff regimes that more closely resembled those that occurred prior to European settlement. Numerous obstacles to restoring historic fire regimes exist; some of the most important, and possible approaches to overcome them, are discussed below.

Existing Watershed and Habitat Condition

As discussed earlier, aquatic and riparian conditions in many California watersheds are in an altered or degraded state due to past human activities (Kattleman 1996, Hunsaker et al. 2014). In headwater streams, disturbances include road construction and logging. These activities are also present in larger stream systems, along with historic and current impacts from mining activities, grazing, recreation use, and urban development. Farther downstream, dams, diversions, and flood control structures have radically altered flow regimes and channel conditions. In many locations, the combination of dams and road crossings has disrupted the longitudinal connection of channel systems, affecting fish movement (Moyle et al. 1996). Aquatic communities have been further stressed by the widespread introduction of nonnative species. In this context of anthropogenic change, proposals for fuel management and fire treatments may be viewed by some as just adding to existing disturbance (Rieman and Clayton 1997).

Treatment in Riparian Areas

Riparian areas are defined by the presence of vegetation dependent on year-round water in the root zone (e.g., willows [*Salix* spp.], western sycamore [*Platanus racemosa*], alders [*Alnus* spp.] or similar plants) or as a certain distance from the bankfull border of a perennial or seasonal stream. The importance of riparian areas to the condition and function of streams and other aquatic systems is well documented (Swanson et al. 1982). Many management activities are restricted or excluded from riparian areas, including timber harvest, livestock access, mechanical fuel treatments, and prescribed fire (USDA Forest Service 2004).

Relatively little research has evaluated the natural role of fire in riparian system function or the effects of fire management activities in riparian zones (Pettit and Naiman 2007, Stone et al. 2010). However, a few studies in formerly frequent-fire forest types support the idea that riparian areas previously experienced considerable fire. Skinner (2002, 2003) found that fire return intervals in the Klamath Mountains varied with stream type and topographic location. Stream sides higher in the watershed, with steeper channels and vegetation more like the surrounding slopes, had shorter return intervals compared to low gradient channels with riparian vegetation. Van de Water and North (2010) documented historic fire return intervals for riparian areas and their surrounding upland forests in the northern Sierra Nevada. In most sites they found similar fire frequency and seasonality, generally burning in the late summer to early fall dormant period.

It is being recognized that streamside forests have undergone the same stand densification from fire exclusion as upland forests and may thus be at equally great risk of uncharacteristic fire effects should wildfire occur (Meyer et al. 2012). Van de Water and North (2011) modeled fire behavior under current and reconstructed riparian forest conditions for sites in the northern Sierra Nevada. Both riparian and upland modern stands had greater basal area, stand density, crown bulk density, and fuel load than reconstructed stands, and predicted flame lengths, probability of torching, and fire-caused tree mortality were higher as well. This suggests that untreated riparian forests are more susceptible to high-intensity fire than in the presuppression era.

Like uplands, riparian areas may benefit from prescribed fire. Huntzinger (2003) found two to three times as many butterfly species in riparian areas that had been prescribed burned than in unburned controls. Beche et al. (2005) found that a small (only 14% of the watershed), low- to moderate-severity prescribed fire in the Sierra Nevada resulted in few short-term effects and reduced fuel loads in the riparian zone up to 80%. However, prescribed fire may not mimic all of the ecological effects of wildfire on riparian or in-stream environments. More of the riparian forest vegetation burned in wildfires studied by Arkle and Pilliod (2010) than in nearby prescribed fires, and other ecosystem attributes varied more after wildfire than prescribed fire (e.g., water temperature, fish occupancy, large woody debris).

Mechanical fuel reduction treatments have seldom been used in riparian areas (Dwire et al. 2005) but may be necessary to reduce fuels before prescribed fire can be applied. A survey of western US Forest Service district-level fire management officers found that fuel treatments in riparian areas (mechanical or prescribed fire) were seldom used and generally covered only a small area, but they were most likely to be conducted at the wildland-urban interface, where fire intensity reduction is most critical (Stone et al. 2010).

Treatment Scale: Site versus Landscape Level

Unless fuel treatments are frequently applied over a broad area, they have low potential for influencing fire behavior at a landscape scale. Treatments have a high probability of altering fire severity at the site scale, and they are likely to reduce some of the impacts to soils and hydrology that are described in chapter 7. However, the utility of such treatments to affect high-severity fire across a watershed and at a landscape scale is unclear.

Existing fuel conditions and air quality concerns seriously constrain the use of prescribed fire, often necessitating mechanical fuel manipulation. Strategic placement of mechanical fuel reduction treatments will be necessary to effectively modify wildfire behavior and severity in forested landscapes. Various modeling approaches can be used to project the effectiveness of different fuel modification types and their placement on the landscape in moderating fire behavior (Schmidt et al. 2008, Collins et al. 2011b).

However, there are constraints on applying mechanical fuel treatments over large areas as well. These include land designations such as wilderness, special status species habitat requirements, economic considerations (timber or small diameter product value relative to extraction cost), and terrain too steep for mechanical operations (Collins et al. 2010). For example, North et al. (2015) estimated that only 20% of subwatersheds on Forest Service land in the Sierra Nevada have enough unconstrained land to effectively contain or suppress wildfire after mechanical treatment alone. They suggest that mechanical treatments could function as anchor points from which to strategically expand prescribed or managed fire into the greater landscape.

Integrating Fire with other Watershed Resources

The placement and design of mechanical fuel treatments affect more than just future fire behavior and severity. Efforts have focused on describing fuel treatment strategies that produce vegetation structure more similar to historic stand conditions than simple efforts to reduce canopy continuity or tree density (North et al. 2009, North 2012). Spatially explicit historic forest stand datasets portray vegetation communities that not only have fewer stems and less basal area per unit area than modern forests, but also a more clumped distribution of trees with much larger openings than found today (e.g., Collins et al. 2011a, Knapp et al. 2013). Shrubs and herbaceous vegetation were more abundant in the gaps as well. Such conditions probably created more habitat heterogeneity for wildlife, a hypothesis being tested with stand- and landscape-level research in the Sierra Nevada (Knapp et al. 2012, Stephens et al. 2014). Tree density and spatial distribution can be varied by aspect as well when applying treatments on a landscape scale (North et al. 2009), benefitting wildlife diversity and abundance.

Restoring watershed ecological processes and functions may require more than just reestablishing desired aspects of historic vegetation structure and fire regimes. Possible actions include removing roads from riparian areas to prevent chronic sediment delivery and reconstructing roads and road crossings to reduce their hydrologic impacts (Luce et al. 2001, Madej et al. 2012). There may be opportunities to reconnect fragmented aquatic habitats by renovating culverts and road crossings (Rieman et al. 2010). However, some barriers are valuable in that they prevent nonnative fish from invading headwater refugia for now-rare native species (Fausch et al. 2009). Without efforts to eradicate or reduce populations of nonnative species, removing some barriers could have negative biological consequences.

As noted in chapter 7, wildfire often increases streamflow for several years due to reduced transpiration by the burned plant community. Prescribed fire and mechanical fuel reduction can have similar effects. Biswell (1989) described several examples of spring flow increases after prescribed burning in chaparral rangelands. Robles et al. (2014) projected an average 20% increase in runoff from thinned ponderosa pine forests in central Arizona using a modeling approach based on results from small watershed studies. The increase in runoff was short-lived (about six years) once all area treatments were complete, but they suggest that maintenance prescribed burning could help extend the runoff gains. Increases in runoff from thinned or fire-treated forests in California could help augment streamflows projected to decrease or change in timing under future climate change (Bales et al. 2011). In the Sierra Nevada, there has been an observed shift to earlier runoff peaks with a change in precipitation from snow to rain at middle elevations and earlier snowmelt resulting from warmer temperatures (summarized in Hunsaker et al. 2014). This trend is expected to increase, along with increased length of the freeze-free season and fire season, increased storm severity, decreased snowpack, and increased drought (Jardine and Long 2014). Water stored in snowpack provides 35% of the water used for irrigation and domestic purposes in California (Hunsaker et al. 2014). Watershed treatments that restore forest stands to less dense, more fire-resilient condition could also increase water available for both ecosystem and human use.

Reestablishment of fire as a watershed process requires an integrated landscape approach at large (river basin) scales, with fire, wildlife, terrestrial, and aquatic specialists working together to develop priorities for treatment. Short-term negative effects from fire management activities could be balanced and surpassed by short-term improvement in sediment regime and long-term benefits to all these processes. Monitoring and evaluation of integrated, landscape-level projects is critical. Given the long time frames necessary to assess effectiveness of treatments, it will be necessary to supplement short-term monitoring (e.g., treatment implementation, evaluation of effects on soils, riparian vegetation response) with modeling of long-term processes (e.g., fire behavior, sediment yield). Schemes for effectiveness monitoring at both the site (USDA Forest Service 1992) and landscape scale (Kershner et al. 2004) have been implemented and might serve as models for other monitoring efforts.

There are places where restored fire regimes will not be possible or desired, such as some wildland-urban interfaces (Jensen and McPherson 2008). Developments in fire-prone areas will complicate the implementation of desired aspects of historic fire regimes. Use of fuel modification zones around human communities that reduce the chance of wildfire impinging on homes may ultimately increase social comfort with using fire for ecosystem restoration further into nearby wildlands (Long et al. 2014b). Modified vegetation close to communities may also provide habitat heterogeneity for wildlife.

Summary

Fire is a natural process within most California watersheds and must be considered when managing for riparian, aquatic, and water resources. Sedimentation, mass-wasting, and flooding are also natural processes that work within watersheds. Issues concerning fire and watershed resources often mix social and ecological values. Communities, homes, and municipal water supplies are located at the base of large watersheds, and increased water and sediment flows from large wildfires may threaten these investments. In addition, many aquatic species are endangered or extremely limited in their habitat and populations due to a variety of anthropogenic activities. Fires out of their natural regime can and do cause serious damage to these ecological assets. Such concerns have focused fire-related watershed management on postfire stabilization and rehabilitation actions. Reintroducing fire as a natural process at the watershed scale would help move toward or restore natural conditions of vegetation structure, water flow, and other ecological functions, but many challenges lie ahead, including the effects of climate change. Interdisciplinary planning and collaboration will be needed to implement fire and fuel treatments that fully integrate with watershed restoration goals.

References

Arkle, R.S., and D.S. Pilliod. 2010. Prescribed fires as ecological surrogates for wildfires: a stream and riparian perspective. Forest Ecology and Management 259: 893–903.

Bales, R.C., J.J. Battles, Y. Chen, M.H. Conklin, E. Holst, K.L. O'Hara, P. Saksa, and W. Stewart. 2011. Forests and water in the Sierra Nevada: Sierra Nevada Watershed Ecosystem Enhancement Project. Sierra Nevada Research Institute Report 11.1. University of California, Merced, California, USA.

Barro, S.C., and S.G. Conard. 1987. Use of ryegrass seeding as an emergency revegetation measure in chaparral ecosystems. USDA Forest Service General Technical Report PSW-GTR-102. Pacific Southwest Forest and Range Experiment Station, Berkeley, California, USA.

Beche, L.A., S.L. Stephens, and V.H. Resh. 2005. Effects of a prescribed fire on a Sierra Nevada (California, USA) stream and its riparian zone. Forest Ecology and Management 218: 37–59.

Benda, L., D. Miller, P. Bigelow, and K. Andras. 2003. Effects of post-wildfire erosion on channel environments, Boise River, Idaho. Forest Ecology and Management 178: 105–119.

Berryman, E.M., P. Morgan, P.R. Robichaud, and D. Page-Dumroese. 2014. Post-fire erosion control mulches alter belowground processes and nitrate reductase activity of a perennial forb, heartleaf arnica (*Arnica cordifolia*). USDA Forest Service Research Note RMRS-RN-69. Rocky Mountain Research Station, Fort Collins, Colorado, USA.

Beyers, J.L. 2004. Postfire seeding for erosion control: effectiveness and impacts on native plant communities. Conservation Biology 18: 947–956.

Bisson, P.A., B.R. Rieman, C. Luce, P.F. Hessburg, D.C. Lee, J.L. Kershner, G.H. Reeves, and R.E. Gresswell. 2003. Fire and aquatic ecosystems: current knowledge and key questions. Forest Ecology and Management 178: 213–229.

Biswell, H.H. 1989. Prescribed Burning in California Wildlands Vegetation Management. University of California Press, Berkeley, California, USA.

Burton, T.A. 2005. Fish and stream habitat risks from uncharacteristic wildfire: observations from 17 years of fire-related disturbances on the Boise National Forest, Idaho. Forest Ecology and Management 211: 140–149.

Collins, B.M., R.G. Everett, and S.L. Stephens. 2011a. Impacts of fire exclusion and recent managed fire on forest structure in old growth Sierra Nevada mixed-conifer forests. Ecosphere 2(4): art51.

Collins, B.M., S.L. Stephens, J.J. Moghaddas, and J. Battles. 2010. Challenges and approaches in planning fuel treatments across fire-excluded forested landscapes. Journal of Forestry 108: 24–31.

Collins, B.M., S.L. Stephens, G.B. Roller, and J.J. Battles. 2011b. Simulating fire and forest dynamics for a landscape fuel treatment project in the Sierra Nevada. Forest Science 57: 77–88.

Detenbeck, N.E., P.W. DeVore, G.J. Niemi, and A. Lima. 1992. Recovery of temperate-stream fish communities from disturbance: a review of case studies and synthesis of theory. Environmental Management 16: 33–53.

Donato, D.C., J.B. Fontaine, J.L. Campbell, W.D. Robinson, J.B. Kauffman, and B.E. Law. 2006. Post-wildfire logging hinders regeneration and increases fire risk. Science 311: 352.

Dunham, J.B., M.K. Young, R.E. Gresswell, and B.E. Rieman. 2003. Effects of fire on fish populations: landscape perspectives on persistence of native fishes and nonnative fish invasions. Forest Ecology and Management 178: 183–196.

Dwire, K.A., C.C. Rhoades, and M.K. Young. 2005. Chapter 10. Potential effects of fuel management activities on riparian areas. Pages 175–205 in: W.J. Elliot, I.S. Miller, and L. Audin, editors. Cumulative watershed effects of fuel management in the western United States. USDA Forest Service General Technical Report RMRS-GTR-231. Rocky Mountain Research Station, Fort Collins, Colorado, USA.

Fausch, K.D., B.E. Rieman, J.B. Dunham, M.K. Young, and D.P. Peterson. 2009. Invasion versus isolation: trade-offs in managing native salmonids with barriers to upstream movement. Conservation Biology 23: 859–870.

Foltz, R.B., and P.R. Robichaud. 2013. Effectiveness of post-fire Burned Area Emergency Response (BAER) road treatments: results from three wildfires. USDA Forest Service General Technical Report RMRS-GTR-313. Rocky Mountain Research Station, Fort Collins, Colorado, USA.

Foltz, R.B., P.R. Robichaud, and R. Hakjun. 2008. A synthesis of postfire road treatments for BAER teams: methods, treatment effectiveness, and decisionmaking tools for rehabilitation. USDA Forest Service General Technical Report RMRS-GTR-228. Rocky Mountain Research Station, Fort Collins, Colorado, USA.

Furniss, M.J., T.S. Ledwith, M.A. Love, B.C. McFadin, and S.A. Flanagan. 1998. Response of road-stream crossings to large flood events in Washington, Oregon, and northern California. USDA Forest Service Technology and Development Program Report 9877 1806-SDTDC. San Dimas Technology Development Center, San Dimas, California, USA.

Houck, O.A. 2002. Clean Water Act TMDL Program: Law, Policy, and Implementation. Environmental Law Institute, Washington, D.C., USA.

Hubbert, K.R., P.M. Wohlgemuth, and J.L. Beyers. 2012. Effects of hydromulch on post-fire erosion and plant recovery in chaparral shrublands of southern California. International Journal of Wildland Fire 21: 155–167.

Hunsaker, C.T., J.W. Long, and D.B. Herbst. 2014. Watershed and stream ecosystems. Pages 265–322 in: J.W. Long, L. Quinn-Davidson, and C. Skinner, editors. Science Synthesis to Support Socioecological Resilience in the Sierra Nevada and Southern Cascade Range. USDA Forest Service General Technical Report PSW-GTR-247. Pacific Southwest Research Station, Albany, California, USA.

Huntzinger, M. 2003. Effects of fire management practices on butterfly diversity in the forested western United States. Biological Conservation 113: 1–12.

Jardine, A., and J. Long. 2014. Synopsis of climate change. Pages 71–81 in: J.W. Long, L. Quinn-Davidson, and C. Skinner, editors. Science Synthesis to Support Socioecological Resilience in the Sierra Nevada and Southern Cascade Range. USDA Forest Service General Technical Report PSW-GTR-247. Pacific Southwest Research Station, Albany, California, USA.

Jensen, S.E., and G.R. McPherson. 2008. Living with Fire. University of California Press, Berkeley, California, USA.
Karr, J.R., J.J. Rhodes, G.W. Minshall, F.R. Hauer, R.L. Beschta, C.A. Frissell, and D.A. Perry. 2004. The effects of postfire salvage logging on aquatic ecosystems in the American West. BioScience 54: 1029–1033.
Kattleman, R. 1996. Hydrology and water resources. Pages 855–920 in: D.C. Erman, general editor. Sierra Nevada Ecosystem Project: Final Report to Congress, Volume II. Wildland Resources Center Report 37. University of California, Davis, California, USA.
Kershner, J.L., E.K. Archer, M. Coles-Ritchie, E.R. Cowley, R.C. Henderson, K. Kratz, C.M. Quimby, D.L. Turner, L.C. Ulmer, and M.R. Vinson. 2004. Guide to effective monitoring of aquatic and riparian resources. USDA Forest Service General Technical Report RMRS-GTR-121. Rocky Mountain Research Station, Fort Collins, Colorado, USA.
Knapp, E., M. North, M. Benech, and B. Estes. 2012. The variable-density thinning study at Stanislaus-Tuolumne Experimental Forest. Pages 127–139 in: M. North, editor. Managing Sierra Nevada Forests. USDA Forest Service General Technical Report PSW-GTR-237. Pacific Southwest Research Station, Albany, California, USA.
Knapp, E.E., C.N. Skinner, M.P. North, and B.L. Estes. 2013. Long-term overstory and understory change following logging and fire exclusion in a Sierra Nevada mixed-conifer forest. Forest Ecology and Management 310: 903–914.
Long, J.W., C. Skinner, S. Charney, K. Hubbert, L. Quinn-Davidson, and M. Meyer. 2014a. Post-wildfire management. Pages 187–220 in: J.W. Long, L. Quinn-Davidson, and C. Skinner, editors. Science Synthesis to Support Socioecological Resilience in the Sierra Nevada and Southern Cascade Range. USDA Forest Service General Technical Report PSW-GTR-247. Pacific Southwest Research Station, Albany, California, USA.
Long, J.W., C. Skinner, M. North, C.T. Hunsaker, and L. Quinn-Davidson. 2014b. Integrative approaches: promoting socioecological resilience. Pages 17–54 in: J.W. Long, L. Quinn-Davidson, and C. Skinner, editors. Science Synthesis to Support Socioecological Resilience in the Sierra Nevada and Southern Cascade Range. USDA Forest Service General Technical Report PSW-GTR-247. Pacific Southwest Research Station, Albany, California, USA.
Luce, C.H., B.E. Rieman, J.B. Dunham, J.L. Clayton, J.G. King, and T.A. Black. 2001. Incorporating aquatic ecology into decisions on prioritization of road decommissioning. Water Resources Impact 3(3): 8–14.
Madej, M.A., G. Bundros, and R. Klein. 2012. Assessing effects of changing land use practices on sediment loads in Panther Creek, north coastal California. Pages 101–110 in: R.B. Standiford, T.J. Weller, D.D. Piirto, and J.D. Stuart, technical coordinators. Proceedings of Coast Redwood Forests in a Changing California: A Symposium for Scientists and Managers. USDA Forest Service General Technical Report PSW-GTR-238. Pacific Southwest Research Station, Albany, California, USA.
Meyer, K.E., K.A. Dwire, P.A. Champ, S.E. Ryan, G.M. Riegel, and T.A. Burton. 2012. Burning questions for managers: fuel management practices in riparian areas. Fire Management Today 72(2): 16–22.
Minshall, G.W., J.T. Brock, and J.D. Varley. 1989. Wildfires and Yellowstone's stream ecosystems. BioScience 39: 707–715.
Moyle, P.B., R.M. Yoshiyama, and R.A. Knapp. 1996. Status of fish and fisheries. Pages 953–973 in: D.C. Erman, general editor. Sierra Nevada Ecosystem Project: Final Report to Congress, Volume II. Wildland Resources Center Report 37. University of California, Davis, California, USA.
North, M., editor. 2012. Managing Sierra Nevada forests. USDA Forest Service General Technical Report PSW-GTR-237. Pacific Southwest Research Station, Albany, California, USA.
North, M., A. Broughton, J. Long, B. Collins, P. Bowden, D. Yasuda, J. Miller, and N. Sugihara. 2015. Constraints on mechanized fuel treatment significantly limit mechanical fuel reduction extent in the Sierra Nevada. Journal of Forestry 113: 40–48.
North, M., P. Stine, K. O'Hara, W. Zielinski, and S. Stephens. 2009. An ecosystem management strategy for Sierran mixed-conifer forests. USDA Forest Service General Technical Report PSW-GTR-220. Pacific Southwest Research Station, Albany, California, USA.
Peppin, D., P.Z. Fulé, C. Hull Sieg, J.L. Beyers, and M.E. Hunter. 2010. Post-wildfire seeding in forests of the western United States: an evidence-based review. Forest Ecology and Management 260: 573–586.
Peterson, D.L., J.K. Agee, G.H. Aplet, D.P. Dykstra, R.T. Graham, J.F. Lehmkuhl, D.S. Pilliod, D.F. Potts, R.F. Powers, and J.D. Stuart. 2009. Effects of timber harvest following wildfire in western North America. USDA Forest Service General Technical Report PNW-GTR-776. Pacific Northwest Research Station, Portland, Oregon, USA.
Pettit, N.E., and R.J. Naiman. 2007. Fire in the riparian zone: characteristics and ecological consequences. Ecosystems 10: 673–687.
Pyke, D.A., T.A. Wirth, and J.L. Beyers. 2013. Does seeding after wildfires in rangelands reduce erosion or invasive species? Restoration Ecology 21: 415–421.
Reeves, G.H., P.A. Bisson, B.E. Rieman, and L.E. Benda. 2006. Postfire logging in riparian areas. Conservation Biology 20: 994–1004.
Rice, R.M., and G.T. Foggin, III. 1971. Effect of high intensity storms on soil slippage on mountainous watersheds in southern California. Water Resources Research 7: 1485–1496.
Rieman, B., and J. Clayton. 1997. Wildfire and native fish: issues of forest health and conservation of sensitive species. Fisheries 22(11): 6–15.
Rieman, B.E., P.F. Hessburg, C. Luce, and M.R. Dare. 2010. Wildfire and management of forests and native fishes: conflict or opportunity for convergent solutions? BioScience 60: 460–468.
Rinne, J.N., and G.R. Jacoby. 2005. Aquatic biota. Pages 136–148 in: D.G. Neary, K.C. Ryan, and L.F. DeBano, editors. Wildland Fire in Ecosystems: Effects of Fire on Soils and Water. USDA Forest Service General Technical Report RMRS-GTR-42-vol.4. Rocky Mountain Research Station, Fort Collins, Colorado, USA.
Ritchie, M.W., E.E. Knapp, and C.N. Skinner. 2013. Snag longevity and surface fuel accumulation following post-fire logging in a ponderosa pine dominated forest. Forest Ecology and Management 287: 113–122.
Robichaud, P.R., L.E. Ashmun, R.B. Foltz, C.G. Showers, J.S. Groenier, J. Kesler, C. DeLeo, and M. Moore. 2013a. Production and aerial application of wood shreds as a post-fire hillslope erosion mitigation treatment. USDA Forest Service General Technical Report RMRS-GTR-307. Rocky Mountain Research Station, Fort Collins, Colorado, USA.
Robichaud, P.R., L.E. Ashmun, and B.D. Sims. 2010. Post-fire treatment effectiveness for slope stabilization. USDA Forest Service General Technical Report RMRS-GTR-240. Rocky Mountain Research Station, Fort Collins, Colorado, USA.
Robichaud, P.R., J.L. Beyers, and D.G. Neary. 2000. Evaluating the effectiveness of postfire rehabilitation treatments. USDA Forest Service General Technical Report RMRS-GTR-63. Rocky Mountain Research Station, Fort Collins, Colorado, USA.
Robichaud, P.R., P. Jordan, S.A. Lewis, L.E. Ashmun, S.A. Covert, and R.E. Brown. 2013b. Evaluating the effectiveness of wood shred and agricultural straw mulches as a treatment to reduce post-wildfire hillslope erosion in southern British Columbia, Canada. Geomorphology 197: 21–33.
Robichaud, P.R., S.A. Lewis, R.E. Brown, and L.E. Ashmun. 2009. Emergency post-fire rehabilitation treatment effects on burned area ecology and long-term restoration. Fire Ecology 5: 115–128.
Robichaud, P.R., S.A. Lewis, J.W. Wagenbrenner, L.E. Ashmun, and R.E. Brown. 2013c. Post-fire mulching for runoff and erosion mitigation. Part 1: effectiveness at reducing hillslope erosion rates. Catena 105: 75–92.
Robichaud, P.R., H. Rhee, and S.A. Lewis. 2014. A synthesis of post-fire Burned Area Reports from 1972 to 2009 for western US Forest Service lands: trends in wildfire characteristics and post-fire stabilization treatments and expenditures. International Journal of Wildland Fire 23: 929–944.
Robichaud, P.R., J.W. Wagenbrenner, S.A. Lewis, L.E. Ashmun, R.E. Brown, and P.M. Wohlgemuth. 2013d. Post-fire mulching for runoff and erosion mitigation. Part 2: effectiveness in reducing runoff and sediment yields from small catchments. Catena 105: 93–111.
Robles, M.D., R.M. Marshall, F. O'Donnell, E.B. Smith, J.A. Haney, and D.F. Gori. 2014. Effects of climate variability and accelerated

forest thinning on watershed-scale runoff in southwestern USA ponderosa pine forests. PLoS One 9(10): e111092.
Ruffolo, J. 1999. TMDLs: the revolution in water quality regulation. California Research Bureau CRB-99-005. Sacramento, California, USA.
Schmidt, D.A., A.H. Taylor, and C.N. Skinner. 2008. The influence of fuels treatment and landscape arrangement on simulated fire behavior, Southern Cascade range, California. Forest Ecology and Management 255: 3170–3184.
Skinner, C.N. 2002. Fire history in riparian reserves of the Klamath Mountains. Pages 164–169 in: N.G. Sugihara, M. Morales, and T. Morales, editors. Fire in California Ecosystems: Integrating Ecology, Prevention and Management. Association for Fire Ecology Miscellaneous Publication 1. Davis, California, USA.
———. 2003. A tree-ring based fire history of riparian reserves in the Klamath Mountains. Pages 116–119 in: P.M. Faber, editor. California Riparian Systems: Processes and Floodplains Management, Ecology, and Restoration. Riparian Habitat and Floodplains Conference Proceedings, March 12–15, 2001. Riparian Habitat Joint Venture, Sacramento, California, USA.
Spencer, C.N., F.R. Hauer, and K.O. Gabel. 2003. Wildfire effects on stream food webs and nutrient dynamics in Glacier National Park, USA. Forest Ecology and Management 178: 141–153.
Stephens, S.L., S.W. Bigelow, R.D. Burnett, B.M. Collins, C.V. Gallagher, J. Keane, D.A. Kelt, M.P. North, S.L. Roberts, P.A. Stine, and D.H. Van Vuren. 2014. California spotted owl, songbird, and small mammal responses to landscape fuel treatments. BioScience 64: 893–906.
Stephens, S.L., R.E. Martin, and N.E. Clinton. 2007. Prehistoric fire area and emissions from California's forests, woodlands, shrublands and grasslands. Forest Ecology and Management 251: 205–216.
Stone, K.R., D.S. Pilliod, K.A. Dwire, R.C. Rhodes, S.P. Wollrab, and M.K. Young. 2010. Fuel reduction management practices in riparian areas of the western USA. Environmental Management 46: 91–100.
Swanson, F.J., S.V. Gregory, J.R. Sedell, and A.G. Campbell. 1982. Land-water interactions: the riparian zone. Pages 267–291 in: R.L. Edmonds, editor. Analysis of Coniferous Forest Ecosystems in the Western United States. Hutchinson Ross, Stroudsburg, Pennsylvania, USA.
SWRCB. 1988. Nonpoint Source Management Plan. State Water Resources Control Board, Division of Water Quality, Sacramento, California, USA.
SWRCB and CCC. 2000. Volume I: Nonpoint Source Program Strategy and Implementation plan, 1998-2013 (PROSIP). In: Plan for California's Nonpoint Source Pollution Control Program. State Water Resources Control Board and the California Coastal Commission, Sacramento, California, USA.
USDA Forest Service. 1992. Investigating water quality in the Pacific Southwest Region. Best management practices evaluation program: a user's guide. USDA Forest Service, Pacific Southwest Region. San Francisco, California, USA.
———. 2004. Sierra Nevada forest plan amendment: final supplemental environmental impact statement. USDA Forest Service, Pacific Southwest Region. Vallejo, California, USA.
U.S. GAO. 2003. Wildland fires: better information needed on effectiveness of emergency stabilization and rehabilitation treatments. GAO-03-430. U.S. General Accounting Office, Washington, D.C., USA.
Van de Water, K., and M. North. 2010. Fire history of coniferous riparian forests in the Sierra Nevada. Forest Ecology and Management 260: 384–395.
———. 2011. Stand structure, fuel loads, and fire behavior in riparian and upland forests, Sierra Nevada Mountains, USA; a comparison of current and reconstructed conditions. Forest Ecology and Management 262: 215–228.
Wheatley, M.J. 1999. Leadership and the New Science: Discovering Order in a Chaotic World. 2nd ed. Berrett-Koehler Publishers, San Francisco, California, USA.
Wohlgemuth, P.M., J.L. Beyers, and K.R. Hubbert. 2009. Rehabilitation strategies after fire: the California, USA experience. Pages 511–535 in: A. Cerdà and P.R. Robichaud, editors. Fire Effects on Soils and Restoration Strategies. Science Publishers, Enfield, New Hampshire, USA.
Wohlgemuth, P.M., J.L. Beyers, and P.R. Robichaud. 2011. The effectiveness of aerial hydromulch as an erosion control treatment in burned chaparral watersheds. Pages 162–167 in: N. Medley, G. Patterson, and M.J. Parker, editors. Observing, Studying, and Managing for Change—Proceedings of the Fourth Interagency Conference on Research in the Watersheds. Scientific Investigations Report 2011-5169. U.S. Geological Survey, Reston, Virginia, USA.
Ziemer, R.R. 1981. Roots and the stability of forested slopes. Pages 343–361 in: T.R.H. Davies and A.J. Pearce, editors. Proceedings of the International Symposium on Erosion and Sediment Transport in Pacific Rim Steeplands. International Association of Hydrology Science Publication No. 132. Christchurch, New Zealand.

CHAPTER TWENTY-THREE

Fire, Air Quality, and Greenhouse Gases

SURAJ AHUJA AND TRENT PROCTER

There is no fire without some smoke.

PROVERBS, HEYWOOD (1546)

Introduction

Fire is an important part of California's ecosystems, but it also produces combustion byproducts that are not only potentially harmful to human health and welfare, but impair visibility and increase the concentrations of greenhouse gases (GHGs). These trap heat and are warming our planet. Fire also increases atmospheric ozone (O_3) concentrations and contributes to nitrogen (N) deposition. On the other hand, health and survival of ecosystems—such as mixed-conifer forests with giant sequoia (*Sequoiadendron giganteum*) and ponderosa pine (*Pinus ponderosa*), or chaparral with various California-lilac species (*Ceanothus* spp.)—depend on reoccurrence of fires (Hardy and Leenhouts 2001, Krupa 2009). This sometimes puts land management and resource agencies at odds with air quality regulatory agencies and the public. A challenge in managing wildland fire is balancing public interest and regulatory objectives while still sustaining ecological integrity. Minimizing adverse effects of smoke on human health, welfare, and climate change, while maximizing the effectiveness of using wildland fire, is becoming—and needs to continue as—an integrated and collaborative activity.

Awareness of air quality regulations, smoke production, transport, and effects from planned prescribed fires and unplanned wildfires will enable land managers to refine existing smoke management strategies. Additionally, they can develop better smoke management plans and programs in the future. Land managers need to ensure that using fire for any given planned area is the most effective alternative for achieving their land management objectives. At the same time, federal, state, local, and tribal air resource regulators must ensure that air quality rules, regulations, and policies are equitable (Engel-Cox et al. 2013). The sociopolitical nature of the California nuisance law discourages a "risk taking" regulatory atmosphere toward planned or managed fire. Public support plays an important role in achieving the objectives of planned and managed fire.

Vegetation has the capacity to both sequester and emit carbon dioxide (CO_2), a significant greenhouse gas, which contributes to climate change. Pollutants also can be transported thousands of miles away from the source and add complexity for attaining National Ambient Air Quality Standards (NAAQS), and reaching Reasonable Progress Goals (RPG) required under regional haze rules. Fuel treatments are needed to restore wildlands to a positive carbon fire resistant regime that produces low-intensity fire with minimum emissions, and sequesters more carbon than it emits. California is such a large and ecologically diverse state that equitable fuel treatment and air quality solutions for the variations in landscapes, vegetation types, public sensitivity, and air quality exposure are extremely complex.

Composition of Smoke

Smoke from wildland fires is a complex mixture of chemical compounds produced from combustion of fuel. Carbon dioxide and water are the two products of complete combustion and generally make up over 90% of the total emissions by weight. In the incomplete combustion that occurs under wildland conditions, smoke is composed of carbon monoxide, particulate matter, hydrocarbons, other organic compounds, nitrogen oxides, trace minerals, and several thousand other compounds (Ryan and McMahon 1976, Peterson and Ward 1991, Akagi et al. 2011).

Particulates from wildland fire smoke have been a major concern. These can be coarse with a diameter of up to 10 µm (PM_{10}), fine with a diameter of up to 2.5 µm ($PM_{2.5}$), or ultrafine with a diameter smaller than 0.1 µm (Boman et al. 2006). Particles can also be formed through physical and/or chemical transformations. These particles can be elemental carbon or organic carbon. Inorganic or elemental carbon, known as Black Carbon, is a product of incomplete combustion. Studies indicate that 90% of smoke particles emitted during wildland burning are particles less than 10 µm in size (PM_{10}), and about 90% of these are less than 2.5 µm ($PM_{2.5}$) (Hardy et al. 2001).

Other pollutants of concern include carbon monoxide, nitrogen oxides, and hydrocarbons. Carbon monoxide levels are highest during the smoldering stages of a fire. Nitrogen oxides are produced primarily from oxidation of the nitrogen contained in wildland fuels. Most fuel contains less than 1% nitrogen, of which about 20% is converted to nitrogen oxides

SIDEBAR 23.1 SMOKE EMISSIONS CALCULATION PROCESS

Knowledge of the source, including burned acreages, burn period, fuel characteristics and conditions, fuel consumed, and emissions factors for a particular fuel are needed to derive emissions released into the atmosphere (Ottmar et al. 2001). The largest error and uncertainties for emission estimation arise from fuel characteristics like fuel load and amount of fuel consumed. Total emission for a pollutant can be calculated by the following equation:

$$E_a = \Sigma^j_i A_i \times FL_i \times \%C_i \times EF_{ia} \times 1/2{,}000$$

where:

E_a = Emissions produced (tons) for pollutant type "*a*"

A_i = Number of acres consumed for *i*th vegetation type

FL_i = Fuel load per acre for *i*th type veg. (tons)

$\%C_i$ = % of fuel consumed for the *i*th type

EF_{ia} = Emissions factor for pollutant type "*a*" (lbs/ton) of fuel type *i* consumed

1/2,000 = Conversion factor from lbs to tons.

Emission factors (pounds of emissions per ton of fuel consumed) have been identified and described by Ward (1990), Ward and Hardy (1991), and Akagi et al. (2011). Table 23.3 shows pounds of emissions produced from one ton of fuel consumed, by vegetation type and combustion phase (Hardy et al. 1996).

Fuel load is highly variable and estimation is one of the parameters that can introduce a major error in estimation of emissions. Changes in burn intensity can result in large differences in the amount of fuel consumed, emissions to the atmosphere, and the capacity of the atmosphere to recover carbon after a fire. Postfire shift to a different vegetation type in response to changing climate, unusually high burn severities, or other factors also has the potential to affect the amount and rate of carbon storage on the landscape.

when burned (Hardy et al. 2001). Other unhealthy air toxics such as acrolein, benzene, and formaldehyde are present in smoke, but in much lower concentrations than particulate matter and carbon monoxide. These air toxics have been known to be carcinogenic (Kane and Alarie 1977, Boman et al. 2006).

The process of burning itself involves converting organic carbon to CO_2 (a major GHG). Other greenhouse gas emissions such as CH_4, and N_2O, while emitted in small amounts, can have much stronger climate forcing coefficients per unit mass. Black carbon is also emitted during wildfires. The process for calculating smoke emissions is provided in sidebar 23.1.

Smoke and Fire Effects

Smoke and other emissions from forest fires affect human health, environment (particularly visibility), and greenhouse gases.

Human Health

The level and duration of exposure, age, and susceptibility of the individual (and other factors) play significant roles in determining whether someone will experience smoke-related problems. The effects of smoke range from eye and respiratory tract irritation to more serious disorders including reduced lung function, bronchitis, exacerbation of asthma, and premature death.

Particulate matter is the principal pollutant of concern to human health from wildland fire smoke for short-term exposures, which are typically experienced by firefighters and the public. Studies have found that fine particles (alone or with other pollutants) are linked to increased mortality and aggravation of preexisting respiratory and cardiovascular diseases, particularly among sensitive populations (children and elderly). Additionally, particles are respiratory irritants, and laboratory studies show that high concentrations of particulate matter can cause persistent coughing, phlegm, wheezing, and difficulty breathing. These can also affect healthy people, causing respiratory symptoms, transient reductions in lung function, and pulmonary inflammation. Particulate matter can challenge the body's immune system and make it more difficult to remove inhaled foreign materials, such as pollen and bacteria, from the lungs.

Carbon monoxide enters the bloodstream through the lungs, reducing oxygen delivery to the body's organs and tissues. Individuals experiencing health effects (e.g., chest pain and cardiac arrhythmias) from lower levels of carbon monoxide include those with cardiovascular disease. At higher levels, carbon monoxide exposure causes headaches, dizziness, visual impairment, reduced work capacity, and reduced manual dexterity, even in otherwise healthy individuals. At even higher concentrations (seldom associated solely with wildfire), carbon monoxide can be deadly.

Formaldehyde and acrolein are two of the principal chemicals that add to the cumulative irritant properties of smoke, even though the concentrations of these chemicals individually may be below levels of public health concern. People exposed to toxic air pollutants at sufficient concentrations and durations may have slightly increased risks of cancer or other serious health problems.

The National Institute for Occupational Safety and Health (NIOSH) has recommended short-term exposure limits for CO, CO_2, benzene, acrolein, and formaldehyde to be 200 ppm, 30,000 ppm, 1 ppm, 0.3 ppm, and 0.1 ppm, respectively. However, in general, the long-term risk from short-term smoke exposure is quite low (Reinhardt 2000, Reinhardt et al. 2000).

Environment

Besides its impacts on human health, air pollution has long been recognized as having adverse effects on the environment. Westerling et al. (2006) found that wildfire activity in the United States has increased significantly since the mid-1980s with greater frequency, longer wildfire seasons, and longer wildfire duration. Fire is known to have major impacts on the carbon (C) and nitrogen (N) cycles of semiarid forests, including those in the Sierra Nevada. Because of its low volatilization temperature most N is also volatilized from the materials that burn, in contrast to elements such as calcium (Ca) that are left behind in ash. Thus, fire always causes an immediate, short-term (1 to 3 years) loss of ecosystem C and N capital. Hydrocarbons and nitrogen oxides from large wildland fires contribute to increased ozone formation (Burley et al. 2016) under certain conditions. Ozone is not emitted directly, but is formed from nitrogen oxides (NO_x) and Volatile Organic Compounds (VOCs) in the presence of sunlight. Rising background ozone concentrations increase the potential for alterations to the structure and function of ecosystems (Miller et al. 1989, Plymale et al. 2003, Procter et al. 2003). In addition to damaging agricultural and timber yields, it also affects global climate.

Sulfur and nitrogen released during wildfire can affect land, air, and water resources. Wildlands are experiencing nitrogen deposition from aerosols generated from auto exhaust, industrial facilities, wildland fires, agricultural treatment, and oil and gas production. The effects of too much nitrogen can include altered plant and lichen communities, enhanced growth of invasive species, eutrophication and acidification of lands and waters, and habitat deterioration for native species (Fenn et al. 2011). Acidification of soil and water from N and S deposition is widespread in the east. Wildland fires also contribute to particulate mercury (Hg), well known for its effect on aquatic life (Finley et al. 2009). Its effects on fish in inland and coastal waters have resulted in fish consumption advisories in all states.

Particulate matter from smoke has a size range near the wavelength of visible light (0.4–0.7 μm) that scatters light and reduces visibility. Reduced visibility is a significant impact of drifting fire smoke. Smoke has led to marine, aircraft, and auto accidents with fatalities. Table 23.1 shows the PM_{10} health protective and visibility range values for different concentrations of PM_{10} (Sharkey 1997).

TABLE 23.1
The health-protective and visibility range values for particulate matter concentrations

Concentration of PM_{10} (μg/m^3)	Health-protective value	Visibility range (miles)
0–38	Good	11 and up
39–88	Moderate	6–10
89–138	Unhealthy for sensitive groups	3–5 (mild smoke conditions)
139–350	Unhealthy	1½–2¾ (moderate smoke conditions)
351–526	Very unhealthy	1–1¼ (heavy smoke conditions)
>526	Hazardous	1 or less (extremely heavy smoke conditions)

NOTE: μg/m^3 = microgram per cubic meter.

In wildfire smoke, most particles are less than one micrometer, so the values obtained by measuring either PM_{10} or $PM_{2.5}$ are virtually interchangeable. Therefore, the different particle levels can be measured using either PM_{10} or $PM_{2.5}$ monitors.

SOURCE: http://www.arb.ca.gov/carpa/toolkit/data-Revised July 2008, with 2008 AQI values.

Greenhouse Gases (GHGs)

All planned and unplanned fires release organic carbon as CO_2 (a major greenhouse gas), thereby causing a loss of ecosystem carbon capital and an increase of the atmospheric carbon. Other major GHGs, besides CO_2 emitted, are CH_4, N_2O along with black carbon, which is a non-GHG (affecting snow melt) and impacts climate change. Thus, the recent trend toward more frequent and severe forest fires may produce a net loss of C as GHGs to the atmosphere and contribute to positive feedback in global warming.

California's Air Environment

California is the nation's most populous state (39.4 million in December, 2016) and the third largest in terms of land area (403,466 km^2 or 155,779 mi^2). It integrates 58 counties, nearly 500 cities and towns, and 35 air districts. California's growing population, along with weather conditions and terrain, favor a buildup of pollutants. More than one-third of its area is in nonattainment for $PM_{2.5}$ and ozone national ambient air quality standards.

Managing smoke from fires in California requires knowledge of airflow, pollution sources and patterns, and an understanding of the state's regulatory framework, population patterns, meteorology, and physical features related to fire emissions production and transport prediction. For example, in the Central Valley, horizontal air movement is restricted by mountain ranges: the coastal mountains to the west, the Tehachapi Mountains to the south, and the Sierra Nevada to the east. In the spring and summer when the marine layer is shallow, westerly winds enter through low coastal gaps (primarily the Carquinez Straits) and flow down the valley toward the southeast. Daytime wind speeds increase as the valley heats up and are

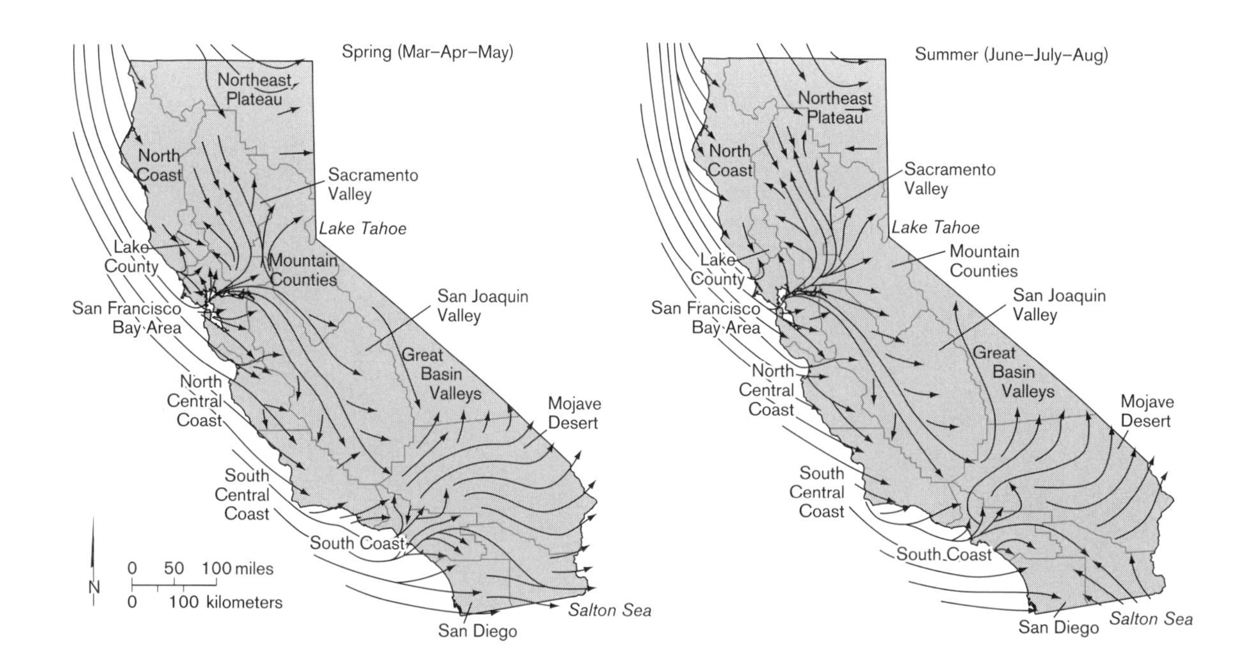

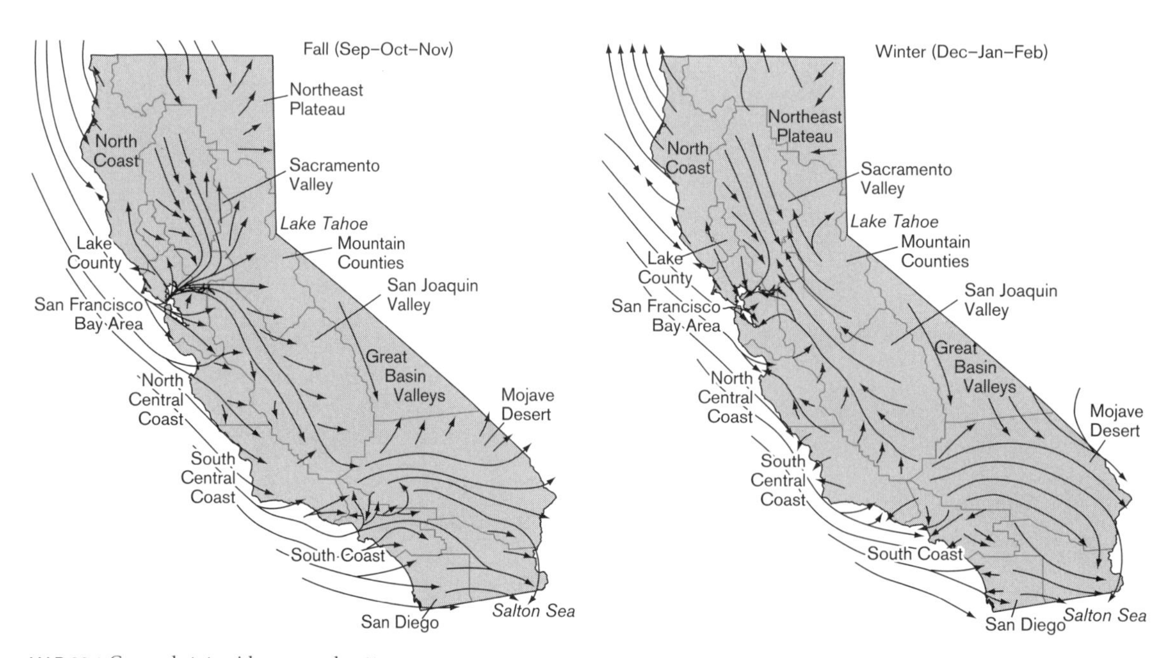

MAP 23.1 General statewide seasonal patterns

SOURCE: California Air Resources Board Aerometric Data Division, April 1984.

strongest in the afternoon. During storm-free periods in the fall and winter, flow is more variable with light wind speeds resulting in less air movement (Map 23.1).

To estimate the sources and quantities of pollution, CARB, in cooperation with APCDs, AQMDs, and industry, maintains an inventory of emission sources. These sources are divided into four categories: *Stationary* source emissions are based on estimates made by the APCDs. *Area-wide* emissions are estimated by CARB. Prescribed fire and managed wildfire come under the area-wide category. *Mobile* source emissions are regulated and estimated by CARB. *Natural* sources are estimated by CARB staff and air districts. Wildfires are classified under the natural sources category.

For California, the estimated daily emissions (by major source category) for Total Organic Gases (TOG), Reactive Organic Gases (ROG), carbon monoxide (CO), nitrogen oxides (NOx), sulfur oxides (SOx), and particulate matter less than 10 µm (PM_{10}) are summarized for 2013 in Table 23.2 (CARB 2014). The numbers are daily averages based on annual estimates. Table 23.2 shows that wildland fires contribute a significant portion of total emissions. Because wildfires are event-based and emissions are not emitted every day, wildfire emissions are highest during the actual event days. Table 23.3 shows emission factors for PM, CO, CO_2, CH_4, and NMHC for planned (prescribed) and unplanned (wildfires).

TABLE 23.2

Statewide emission inventory 2012

Source	Emissions (tons/day, annual average)						
	TOG	ROG	CO	NOx	SOx	PM_{10}	$PM_{2.5}$
Stationary sources	2952.37	384.08	259.68	283.58	51.84	122.80	61.70
Fuel combustion	134.93	26.18	223.79	210.73	28.35	25.57	22.30
Waste disposal	1984.70	37.94	4.46	4.11	1.25	1.52	0.65
Cleaning and surface coatings	216.26	139.31	0.52	0.22	0.11	2.67	2.57
Petroleum production	559.11	131.49	6.91	4.48	4.63	1.83	1.59
Industrial processes	57.38	49.16	24	64.03	17.49	91.21	34.58
Area-wide sources	1998.36	608.80	970.65	75.08	6.11	1213.23	271.01
Solvent evaporation	405.79	353.57	–	–	–	0.03	0.03
Miscellaneous processes	1592.57	255.23	970.65	75.08	6.11	1213.19	270.97
Residential fuel combustion	131.03	57.41	404.29	60.20	2.51	57.39	55.44
Fires	0.98	0.69	9.47	0.23	–	1.19	1.12
Managed burning and disposal	76.19	45.59	556.17	14.63	3.60	57.27	50.48
Mobile sources	819.47	745.74	6141.96	1746.97	47.14	123.83	85.31
On road motor vehicles	441.20	403.50	3937.59	1023.52	5.38	77.56	43.07
Other mobile sources	378.27	342.24	2204.36	723.45	41.75	46.27	42.25
Grand total for statewide	5770.20	1738.62	7372.29	2105.63	105.08	1459.86	418.01
Natural sources (includes biogenic wildfires and volcanic)	2857.40	2424.42	4847.42	56.35	29.89	461.46	391.13
Wildfires	595.51	339.32	4847.42	56.35	29.89	461.46	391.13
Grand total for statewide	20971.72						

NOTE: Does not include biogenic sources. These summaries do not include emissions from windblown dust exposed Lake beds from Owens and Mono Lakes. TOG = total organic gases, ROG = reactive organic gases, CO = carbon monoxide, NOx = nitrogen oxides, SOx = sulfur oxides, PM_{10} = particles less than 10 microns in size.

The inventory also provides data by air basin and county, which helps local air regulatory and land management agencies find background-level emissions for management purposes. Inyo County has the highest PM_{10} emission levels, primarily due to windblown dust from the Owens and Mono lakes. Los Angeles and south San Diego lead in nitrogen oxide emissions that result from onsite and mobile sources.

Airborne pollutants result—in large part—from human activities, and urban growth generally has a negative impact on air quality. Despite substantial increases in population, California is fortunate that emission controls imposed by the state have resulted in significant air quality improvements. Ozone levels have improved in California over the last two decades. Nearly 63% of Californians live in areas that meet the current federal standard for ozone, compared with only 24% in 1990 (CARB 2014). Highest ozone concentrations occur in the anterior portion of the South Coast air basin where the peak 1-hour indicator is more than twice the state standard, indicating mobile and industrial sources being the biggest contributors to the problem. Ozone concentrations are lower near the coast and in rural areas. This can be explained in part by the characteristics of ozone, including pollutant reactivity, transport, and deposition. Based on the current ozone concentrations and the Environmental Protection Agency's (EPA) revised O_3 standards, substantial additional emission control measures will be needed to attain the standard. From 1992 through 2011, statewide maximum 8-hour ozone values decreased 3%. There has been a consistent decline in number of days above the national 8-hour standard. Lake Tahoe and North Coast went from 2 days in 1995 to 0 days in 2005; South Coast was the highest, with 150 days in 1995 to 102 days in 2010. The Salton Sea went from 130 days to 63 days during the same period. Ambient annual average $PM_{2.5}$ values in nondesert areas also showed a 32%

TABLE 23.3

Pounds of emissions per ton of fuel consumed by combustion phase

Fuel/fire configuration	Combustion phase	Emission factors						
		PM	PM_{10}	$PM_{2.5}$	CO	CO_2	CH_4	NMHC
Broadcast burned slash								
Douglas-fir/hemlock	Flaming	24.7	16.6	14.9	14.3	3,383	4.6	4.2
	Smoldering	35	27.6	26.1	463	2,804	15.2	8.4
	Average	29.6	23.1	21.83	312	3,082	11	7.2
Hardwoods	Flaming	23	14	12.2	92	3,389	4.4	5.2
	Smoldering	38	25.9	23.4	366	2,851	19.6	14
	Average	37.4	25	22.4	256	3,072	13.2	10.8
Ponderosa pine/ lodgepole pine	Flaming	18.8	11.5	10	89	3,401	3	3.6
	Smoldering	48.6	36.7	34.2	285	2.971	14.6	9.6
	Average	39.6	25	22	178	3,202	8.2	6.4
Mixed conifer	Flaming	22	11.7	9.6	53	3,458	3	3.2
	Smoldering	33.6	25.3	23.6	273	3,023	17.6	13.2
	Average	29	20.5	18.8	201	3,165	12.8	9.8
Juniper	Flaming	29.1	15.3	13.9	82	3,401	3.9	5.5
	Smoldering	35.1	25.8	23.8	250	3,050	20.5	15.5
	Average	28.3	20.4	18.7	163	3,231	12	10.4
Pile and burn slash								
Tractor piled	Flaming	11.4	7.4	6.6	44	3,492	2.4	2.2
	Smoldering	25	15.9	14	232	3,124	17.8	12.2
	Average	20.4	12.4	10.8	15.3	3,271	11.4	8
Crane piled	Flaming	22.6	13.6	11.8	101	3,349	9.4	8.2
	Smoldering	44.2	33.2	31	232	3,022	30	20.2
	Average	36.4	25.6	23.4	185	3,143	21.7	15.2
Average piles	Average	28.4	19	17.1	169	3,207	16.6	11.6
Broadcast burned brush								
Sagebrush	Flaming	45	31.8	29.1	155	3,197	7.4	6.8
	Smoldering	45.3	29.6	26.4	212	3,118	12.4	14.5
	Average	45.3	29.9	26.7	206	3,126	11.9	13.7
Chaparral	Flaming	31.6	16.5	13.5	119	3,326	3.4	17.2
	Smoldering	40	24.7	21.6	197	3,144	9	30.6
	Average	34.1	20.1	17.3	154	3,257	5.7	19.6
Wildfires (in forests)	Average		30	27				

NOTE: Adapted from Hardy et al. (2001). Fire Averages are weighted averages based on measured carbon flux. PM_{10} values are derived from known size class distribution of particulates using PM and $PM_{2.5}$.

decrease from 1999 to 2011. This air quality improvement occurred over the same time-period that the state's population increased 21%, and the average daily VMT increased 41%. While the air quality improvements are impressive, additional emission controls will be needed to offset future growth.

It is now well documented that the ecological effects of fuel treatments are beneficial and can reduce potential fire behavior at the stand level (Fulé et al. 2012, Stephens et al. 2012). Air regulatory agencies are concerned about allowing burning without application of Emission Reduction Techniques (ERTs) and consideration of alternatives to burning in nonattainment areas. Land managers are generally concerned about the unbalanced use of other treatments that do not replicate the ecological role that fire plays in the environment. They are also concerned that any constraints placed by regulators on fuel reduction projects will reduce their ability to achieve a positive outcome on large, more intense wildfires. There is a need to develop coordination among land managers and regulators to agree on strategies that can help reduce smoke impact, but still achieve the use of fire that is desired. In order to develop management strategies, it is important for land managers to understand air quality laws, rules, and regulations, and for regulators to know the role of fire in ecosystem maintenance.

Air Regulatory Framework

Air quality is managed through federal, state, and local laws and regulations. The EPA has the primary federal role of ensuring compliance with the requirements of the Clean Air Act. The EPA issues national air quality regulations, approves and oversees State and Tribe Implementation Plans (SIPs/TIPs), and conducts major enforcement actions. If a proposed or active SIP is deemed inadequate or unacceptable, the EPA can take over enforcing all or parts of the Clean Air Act requirements for that state or tribe through implementation of a Federal Implementation Plan (FIP).

The Clean Air Act (CAA) recognizes that states should take the lead in carrying out its provisions because appropriate and effective design of pollution control programs requires an understanding of local industries, geography, transportation, meteorology, urban and industrial development patterns, and priorities. States have direct responsibility for meeting requirements of the Federal Clean Air Act and corresponding federal regulations. It also allows the states/tribes and local agencies to establish laws, rules, and regulations that are stricter than federal standards. CARB and local APCDs/AQMDs have the primary responsibility of carrying out the development and execution of SIPs, which must provide for the attainment and maintenance of air quality standards.

California Clean Air Act

CARB administers the California Clean Air Act of 1988. CARB added several requirements concerning plans and control measures to attain and maintain the state ambient air quality standards. One such requirement is for the board to establish designation criteria and to designate areas of the state as *attainment*, *nonattainment*, or *unclassified* for any state standards (see Table 23.4 for national and state standards). Besides standards for criteria pollutants, California has also established ambient air quality standards for sulfate, hydrogen sulfide, vinyl chloride, and visibility reducing particles.

As authorized by Division 26 of the California Health and Safety Code, CARB is directly responsible for regulating emissions from mobile sources. However, authority to regulate stationary sources has been delegated to APCDs/AQMDs at the county and regional levels. The state still has oversight authority to monitor the performance of district programs. It can also assume authority to conduct district functions if the district fails to meet certain responsibilities.

California Air Basins and APCDs/AQMDs

California contains a wide variety of climates, physical features, and emission sources adding complexity to tasks leading to improvement in air quality. To better manage the air quality problems, California is divided into 15 air basins as shown in Map 23.2.

An air basin has similar meteorological and geographical conditions and thus is supposed to be similar in air quality. Most of the boundary lines follow along political county lines even though air moves freely from basin to basin. As a result, pollutants such as ozone and PM_{10} can be transported across air basin boundaries. Thus, interbasin/interregional transport adds to the complexity of managing air quality addressed by the state. California is a geographically diverse state with many unique basins with common sources of air pollution and air quality, which is why it is divided into 35 APCDs/AQMDs.

The Federal Clean Air Act

The original CAA was passed in 1963. This act was followed by the CAA Amendments in 1970, 1977, and 1990 (Public Law 1963, 1967, 1970, 1977, 1990). Only the important sections that apply to forest activities are summarized below.

National Ambient Air Quality Standards (NAAQS)

The amendments of 1970, Section 109, required the EPA to develop primary NAAQS to protect human health and secondary standards to protect welfare. To protect human health, the EPA established NAAQS for six pollutants referred to as *Criteria Pollutants*: Particulates (PM_{10}, $PM_{2.5}$), sulfur dioxide, nitrogen dioxide, ozone, carbon monoxide, and lead. The federal standards are shown along with California Standards in Table 23.4. If the federal standards are violated in an area, that area is designated as "nonattainment" for that pollutant, and the state must develop an SIP for bringing that area back into "attainment."

Ecological thresholds for air pollution, such as critical loads for nitrogen and sulfur deposition, are not included in the regulatory process for emissions control. Ecological thresholds provide a scientifically sound approach to protecting and restoring ecosystems, and a tool for natural resource management and policy. Fenn et al. (2010) found nitrogen compound deposition exceeding Critical Loads that have already impacted lichen and diatom species distribution and abundance (Map 23.3).

TABLE 23.4
National and California ambient air quality standards

Pollutant	Averaging time	Primary standards: Federal	Primary standards: State
PM_{10}	Annual		20 µg/m³
	24 hours	150 µg/m³	50 µg/m³
$PM_{2.5}$	Annual	12 µg/m³	12 µg/m³
	24 hour	35 µg/m³	None
NO_2	Annual	0.053 ppm (100 µg/m³)	0.030 ppm (57 µg/m³)
	1 hour	100ppb (188 µg/m³)	0.18 ppm (339 µg/m³)
CO	8 hours	9 ppm (10 mg/m³)	9ppm (10 mg/m³)
	1 hour	35 ppm (40 mg/m³)	20 ppm (23 mg/m³)
SO_2	Annual	0.030ppm	None
	24 hours	0.14 ppm	0.04 ppm (105 µg/m³)
	3 hours	None	None
	1 hour	75 ppb (196 µg/m³)	0.25 ppm (655 µg/m³)
O_3	1 hour	None	0.09 ppm (180 µg/m³)
	8 hour	0.075 ppm	0.07ppm (137 µg/m³)
Pb	Calendar quarter	1.5 µg/m³	None
	30-day average	None	1.5 µg/m³

NOTE: Annual standards are never to be exceeded. Other standards are not to be exceeded more than once a year.

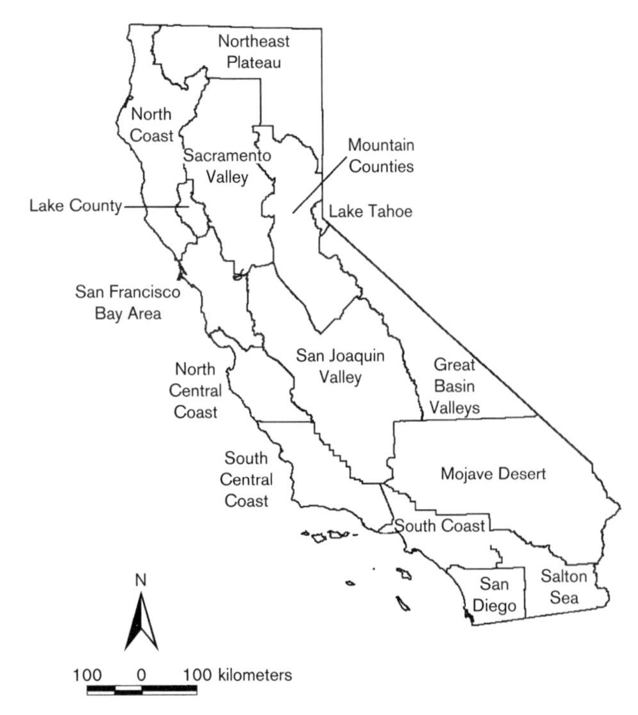

MAP 23.2 California air basins.

Prevention of Significant Deterioration (PSD) and Regional Haze

To prevent further degradation of air quality in typically clean and remote areas, congress established the Prevention of Significant Deterioration (PSD) Program. Under the CAA Amendments of 1977, Congress declared "the remedying of any existing impairment of visibility in 156 class 1 public parks and wilderness areas which impairment results from manmade air pollution." California has 29 such class I areas: 20 wildernesses and 9 national parks (Map 23.4). The PSD regulations authorized federal land managers to review permit applications for construction of industrial facilities within 100 miles of class I areas to maintain and improve visibility in these areas. Visibility impairment occurs as a result of the scattering and absorption of light by particles and gases in the atmosphere. However, over the years and in many parts of the United States, fine particles have significantly reduced the range that people can see. Without the effects of pollution, a natural visual range is approximately 225 km (140 mi) in the west while the current range is 97–145 km (60–90 mi).

The Grand Canyon Visibility Transport Commission (GCVTC) was established under the 1990 amendment. The Commission recognized that fire plays a significant role in visibility degradation, so haze regulations were developed in

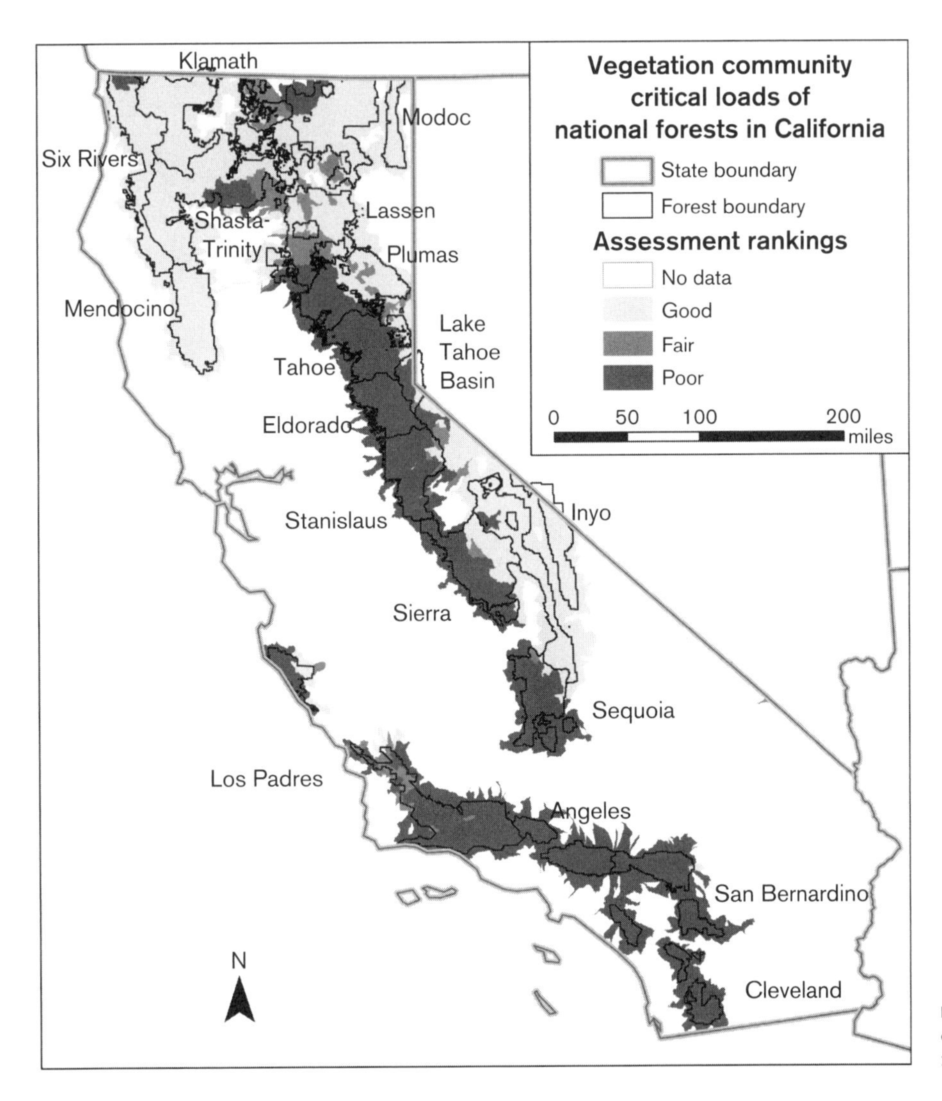

MAP 23.3 Critical loads of nitrogen deposition in California national forests.

response to Recommendations for Improving Western Vistas, in the report provided to the EPA (GCVTC 1996a, 1996b, 1996c). The Commission recommended the implementation of programs to minimize emissions and visibility impacts from planned fires. The EPA issued final regional haze regulations on July 1, 1999 (EPA 1999) to improve visibility, or visual air quality, in 156 class I areas across the country. The rule required all states to develop a visibility SIP to bring all class I areas to the natural background level by 2064. Based on the GCVTC recommendations, 16 western states joined together as the Western Regional Air Partnership (WRAP) to develop tools and guidance necessary for the states to prepare a visibility SIP.

On February 27, 2015, the EPA Region 9 approved the California Regional Haze Plan 2014 Progress Report as a revision to the California visibility, or Regional Haze (RH), SIP. CARB had demonstrated in the report that the existing California RH SIP was sufficient to enable the state to meet all established visibility goals for 20% best visibility days (known as *Reasonable Progress Goals* or *RPGs*) in 2018. California demonstrated in its progress report a detailed analysis of the impact of documented wildfire events on specific class I areas where the state was unable to meet the goal of 20% worst visibility days. The EPA does not expect states to include wildfire in addressing the requirement, though it does affirm its importance. The EPA stated that this type of information is critical to understanding the effect of wildfire on visibility.

Federal Conformity Regulation

Under the 1990 amendment, the EPA issued regulations for nonattainment areas on November 30, 1993, known as "Conformity Rules" (Maps 23.5A, B, C). The Clean Air Act, section 176(C) (1), states that "No department, agency, or instrumentality of the federal Government shall engage, support in any way or provide financial assistance for, license or permit, or approve any activity which does not conform to an implementation plan." The EPA, on April 5, 2010, issued revised General Conformity regulations stating that federal actions conform to the appropriate SIP/TIP or Federal Implementation Plans (FIP) for attaining clean air. Besides ensuring that the federal actions are in conformance with the plans, the regulations encourage consultations between federal agencies, states, tribes, and local air pollution control agencies before and after environmental review.

Under the revised rules, the EPA has included a presumption of conformity for prescribed fires that are conducted in compliance with a Smoke Management Program (SMP). The purpose of an SMP is to mitigate nuisance smoke and public

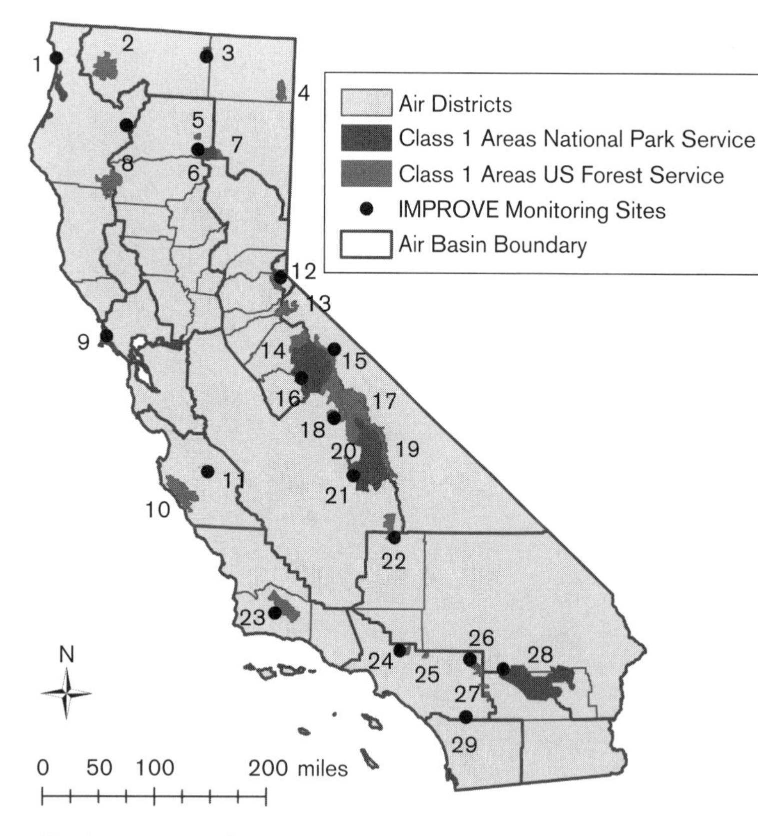

1. Redwood National Park
2. Marble Mountain Wilderness Area
3. Lava Bed Wilderness Area
4. South Warner Wilderness Area
5. Thousand Lakes Wilderness Area
6. Lassen Volcanic National Park
7. Caribou Wilderness Area
8. Yolla Bolly Middle Eel Wilderness Area*
9. Point Reyes Wilderness Area
10. Ventana Wilderness Area
11. Pinnacles Wilderness Area
12. Desolation Wilderness Area
13. Mokelumne Wilderness Area
14. Emigrant Wilderness Area
15. Hoover Wilderness Area
16. Yosemite National Park
17. Ansel Adams Wilderness Area
18. Kaiser Wilderness Area
19. John Muir Wilderness Area
20. Kings Canyon National Park
21. Sequoia National Park
22. Dome Lands Wilderness Area*
23. San Rafael Wilderness Area
24. San Gabriel Wilderness Area
25. Cucamonga Wilderness Area
26. San Gorgonio Wilderness Area*
27. San Jacinto Wilderness Area
28. Joshua Tree Wilderness Area
29. Agua Tibia Wilderness Area

*Portions of these Class 1 Areas are managed by the federal Bureau of Land Management

MAP 23.4 Class I areas in California including national parks larger than 6,000 acres, national wilderness areas greater than 5,000 acres, and national wildlife refuges in existence as of August 17, 1977.

safety hazards, prevent NAAQS violations, protect public health, and address visibility impacts in class I areas. The SMPs establish procedures and requirements for minimizing impacts (page 17,264 of the conformity rule dated April 5, 2010). Burns conducted in accordance with revised Title 17 in California would qualify for a presumption of conformity because Title 17 is an approved SMP. In the absence of an SMP, the EPA encourages states and federal agencies to work together to develop and finalize SMPs, or to conduct burns in accordance with SMPs as actions that are "Presumed to Conform." Because no conformity rules have been proposed by the state, conformity applies to federal standards only.

Policy and Tool Development

EPA's Interim Air Quality Policy

The GCVTC had recognized the impact of wildland fires on visibility in 1996. On the other hand, federal, tribal, and state wildland owners/managers were planning to increase the use of prescribed fires to achieve resource benefits in the wildlands. Therefore, to meet these two opposing objectives, the EPA in 1998 prepared an Interim Air Quality Policy for Wildland and Prescribed Fire. It required states to develop an SMP to manage smoke from prescribed burns and wildland fire use for resource benefit. The policy statement integrates two public policy goals: (1) to allow fire to function, as much as possible, in its natural role in maintaining healthy wildland ecosystems; and (2) to protect public health and welfare by mitigating the impacts of air pollutant emissions on air quality and visibility.

2009 Fire Policy Implementation Guidelines

At present, prescribed fire is conducted at much lower levels than is needed to maintain healthy ecosystems. FLMs have arguably underutilized the opportunity for "unplanned fires for resource benefits" policy. Daily decisions continue to favor suppression, given the high risk to public safety, property, and health. The FLMs and air regulatory agencies needed to examine and balance the long-term public interest and short-term impacts of both air quality and fuels risk. Therefore in 2009, federal land management agencies developed fire policy implementation guidelines. The policy revised the terminology from "wildfire use," "prescribed fire," and "wildfire" to "planned" or "unplanned" only. Unplanned fire can have objectives for suppression and/or resource benefit, which is not limited to one objective but can be for multiple objectives. It can be interchanged from suppression and resource benefits, for the same event during the life of the fire. This process can lead to achieving similar results of large planned burns.

MAP 23.5A. 2015 attainment and nonattainment areas for PM_{10} National Ambient Air Quality Standards.

SOURCE: CARB 2015.

California's SMP, Communication Protocol, and Prescribed Fire Information Reporting System (PFIRS)

California's agricultural burning guidelines (Title 17) were established in 1971. These were in response to statewide legislation which recognized the need to reduce the harmful health effects caused by smoke from unrestrained, open burning of vegetative material on public and private lands. CARB revised Title 17 guidelines in 2000 in response to the EPA's Interim Air Quality Policy in order to elevate the importance of smoke management planning, collaboration, and consultation between land management agencies and air agencies. The revision led to an EPA-approved SMP in California. CARB and the APCDs/AQMDs are responsible for developing and administering the SMPs; CARB has oversight-authority and assists districts through daily forecasts to make daily burn/no-burn decisions. The SMPs were based on the following two objectives: (1) to accommodate large increases in prescribed burning (including wildfires managed for resource benefit) without unacceptable decreases in air quality; and (2) to define a workable management plan for California's smoke management by working with all affected parties, local air districts, land management agencies, private industry, and the public. These amendments place primary emphasis on smoke management through improved planning, collaboration, and consultation between burners, including federal and state land management agencies, and air agencies. The amendments contain new basic provisions such as: requirements for a "burn authorization system," requirements for a "smoke management plan," requirements for "post-burn evaluations and monitoring" for large burns, and provisions for the use of a "marginal" burn day.

Federal land management agencies sometimes manage unplanned (naturally ignited fires) to achieve resource benefits. Planning for unplanned fires is obviously limited, but the agencies require fire management plans to be included in land-use plans for an area before a naturally ignited fire can be managed for resource benefits (natural ignitions in areas without fire management plans are "unwanted" or "wildfires"). Therefore, a communication protocol/guide was developed between burn

MAP 23.5B. 2015 attainment and nonattainment areas for PM_{25} National Ambient Air Quality Standards.

SOURCE: CARB 2015.

and regulatory agencies to coordinate unplanned fires managed for resource benefits (https://www.arb.ca.gov/smp/smp.htm).

CARB, with the cooperation of other agencies, has developed a centralized Prescribed Fire Information Reporting System (PFIRS), which is operational. This is assisting burners, local regulators, and the public in air quality smoke management planning and decision-making. Use of PFIRS is voluntary, but to be fully effective all burn agencies, agricultural owners, and regulators must utilize it. The burners enter data into PFIRS and then submit it to the APCDs for a permit as required under the state SMP. The objective of the PFIRS is to ensure that agricultural and prescribed burning can be tracked, authorized, and takes place only on forecasted burn days. This proposed strategy would not reduce emissions from agricultural and prescribed burning, but would ensure that burning does not inadvertently exceed either the state or federal standards. Monitoring of smoke (Ahuja 2002, Cisneros et al. 2014) will help strategize future planning.

CANSAC and Blue Sky Modeling Framework

The California and Nevada Smoke and Air Committee (CANSAC) is an affiliation of land management and air quality agencies that collaboratively fund operational high resolution meteorological modeling. CANSAC products are used to enhance the BlueSky smoke plume location and $PM_{2.5}$ concentrations. One aspect of collaborative work is exemplified by the use of integrated, linked models to assess a variety of fuel, emissions, and weather conditions to predict smoke intensity, dispersion, and duration on impacted landscapes. BlueSky is one such modeling framework: BlueSky modularly links a variety of independent models of fire information, fuel load, fire consumption, fire emissions, and smoke dispersion (http://www.airfire.org/bluesky).

Few real-time Smoke Prediction Systems have been developed to help land management agencies accomplish fuel treatment without the risk of violating NAAQS (the air quality

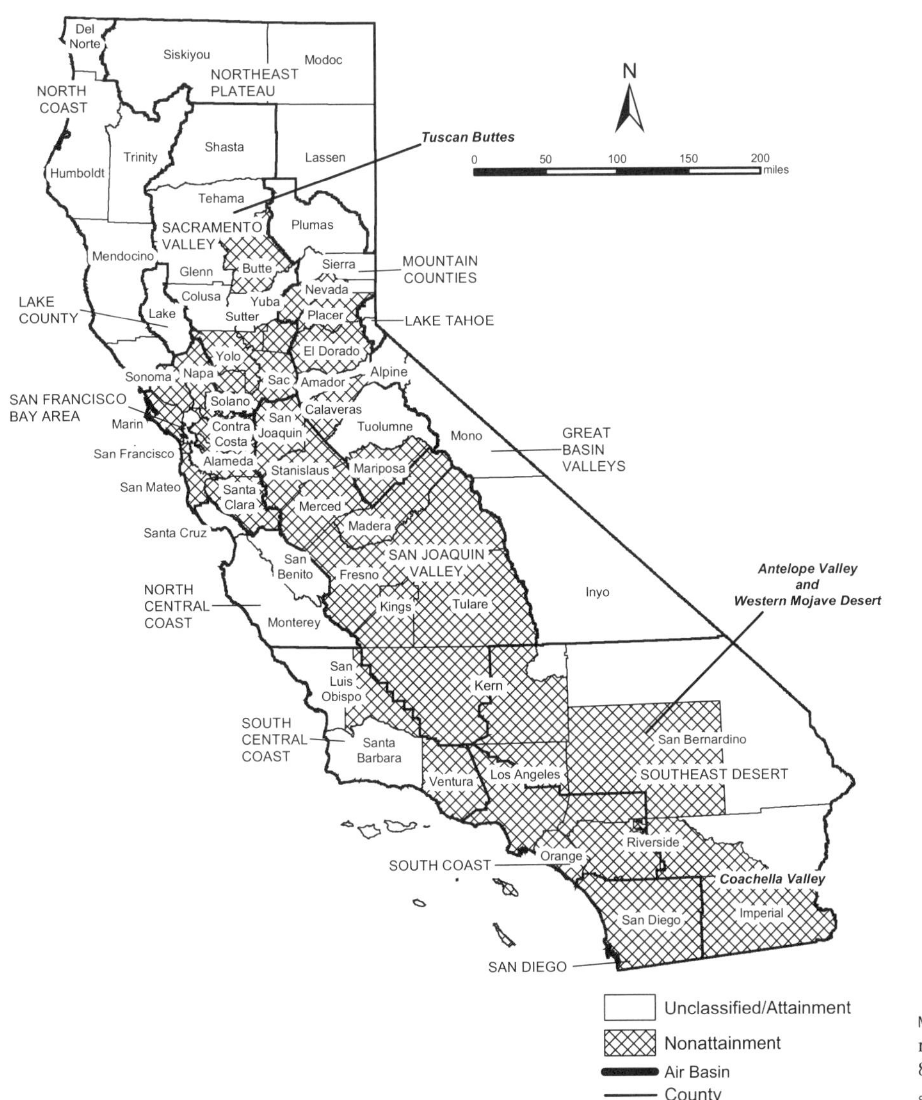

MAP 23.5C. 2015 attainment and nonattainment areas for ozone 8-hour federal standards

SOURCE: CARB 2015.

standards established to protect public health). The late Sue Ferguson was an avalanche and fire weather forecaster and a founder of Atmosphere and Fire Interactions Research and Engineering (AirFire). Her vision led to the development of the BlueSky Modeling Framework as a way to incorporate a series of models into a workflow that can predict the impact of smoke, based on either manually entered emissions in the BlueSky playground, or based on automatically detected fires and assumed emissions using the SMARTFIRE system.

Meteorological data reports are generated by the Weather Research and Forecasting (WRF) model. The wildfire data are drawn from ICS-209 reports (the daily ICS-209 report contains current area burned, and ignition location for wildfire typically greater than 100 acres), or detected by satellite through NASA's Moderate Resolution Image Spectroradiometer (MODIS). Those data are interpreted as emissions, for input into an automated system that predicts smoke dispersion. Fuel load is obtained from one of the two 1 km critical load grid national maps: the National Fire Danger Rating System (NFDRS) and the Fuel Characteristics Classification System (FCCS). Fire growth is entered to generate fire line concentrations for the period to be predicted. BlueSky assumes "next day" fire growth equals today's fuel consumption, and emissions are modeled using the Consume/Emission production model (EMP). Emission models supply emission factors for CO, CO_2, $PM_{2.5}$, and PM_{10} to consumed biomass resulting in the amount of emissions produced. Dispersion and Transport models (CALPUFF & HYSPLIT) are then utilized to track and predict smoke plumes and concentrations (primarily $PM_{2.5}$) at the needed sensitive sites.

GHGs, AB 32, and Cap-and-Trade

Greenhouse Gases (GHGs)

According to the Intergovernment Panel on Climate Change (IPCC 2013, 2014), total anthropogenic GHG emissions have continued to increase from 1970 to 2010 with lager decadal increases toward the end of this period. Total GHG emissions were the highest in human history from 2000 to 2010 and reached 49 $GtCO_2eq$ per year in 2010. Globally, about 76% of emissions come from CO_2, along with 16%, 6% and 2% from CH4, N2O, and others, respectively. Globally, economic and population growth continued to be the most important drivers of increases in CO_2 emissions from fossil fuel combustion. The contribution of population growth between 2000 and 2010 remained roughly identical to the previous three decades, while the contribution of economic growth has risen sharply.

GHGs are considered as having a role in warming the atmosphere through the ability to absorb thermal radiation, which helps keep the surface of the planet warm. Major GHGs in the atmosphere are CO_2, CH_4, N_2O, and fluorinated gases. N_2 and O_2 are the dominant gases in the atmosphere but these gases do not play any role in absorbing radiant energy (Abatzoglou et al. 2014). The heat-absorbing power of a gas depends upon the efficiency in absorbing heat energy, concentration, distribution, and its lifetime in the atmosphere. Water vapor also has the ability to absorb the radiant energy, but its lifetime is only a few days (usually less than 10 days). The heat-absorbing efficiency is determined by a unit called "Global Warming Potential" (GWP), in units of CO2e. It is the ratio of energy trapped for a given mass of gas. The GWP for CO_2 is considered as 1. Methane has a GWP of 25, while N_2O and halocarbon have GWPs of 300 and 5000, respectively. Emission components other than CO_2 such as CH_4, N_2O, black carbon, and other aerosols—while emitted in small quantities—can have much stronger climate-forcing coefficients per unit mass (IPCC 2007).

Wildfires in the United States contribute approximately 213 million metric tons (234 million tons) of CO_2eq to the atmosphere, which is equivalent to emissions from 60 coal-fired power plants or 44 million automobiles. According to Wiedinmyer and Hurteau (2010), studies show that climate change will lengthen the fire season, which can create more fires and the release of more GHGs. Increases in fire occurrence, caused by global warming and increased stress from drought and insects, may have significant effects on growth, regeneration, long-term distribution, and abundance of forest species, as well as short-term carbon sequestration. The effects of stress-complexes will be magnified given a warming climate (McKenzie et al. 2009). Emission components other than CO_2 such as CH_4, N_2O, and other aerosols, while emitted in small quantities, can have much stronger climate forcing coefficients per unit mass (IPCC 2007). In the absence of natural events or human disturbances, all forests and grassland systems would be carbon sinks, as they were during the Paleozoic and Mesozoic times when much of today's fossil fuels were sequestered. As forests grow, annual carbon sequestration reaches the maximum rate at 30–100 years (or more) and then the rate begins to decline. Even mature forests continue to store carbon, although the rate of sequestration declines (Conard and Solomon 2009).

Mobile emissions are one of the biggest sources of CO_2 emissions. One gallon (3.78 liters) of gasoline releases 2.5 kg (5.5 lbs) of CO_2 into the atmosphere. The EPA (2013, 2016), estimates that CH_4 has increased 158% since the industrial revolution because of human activity. Fossil fuel and biomass combustion also produce black carbon and other aerosols, and when deposited on snow-packed mountains, can increase solar radiation absorption resulting in more and rapid melting of snow. Black carbon has a very short life and some scientists have suggested that to control global warming, Short-Lived-Pollutants (SLPs) should be managed first. The SLPs are black carbon (BC), CH_4 and some HFCs.

According to the IPCC (2014), ambient levels of CO_2 have increased from 280 ppm in the mid-1850s to almost 400 ppm now (measured at the Mauna Loa observatory in Hawaii). Mitigation scenarios, in which it is likely that the temperature change caused by anthropogenic GHG emissions can be kept to less than 2°C relative to preindustrial levels, are characterized by atmospheric concentrations in 2100 of about 450 ppm CO_2eq.

AB 32—The California Global Warming Solutions Act of 2006

AB 32, passed in 2006, requires California to reduce its GHG emissions to 1990 levels by 2020—a reduction of approximately 15% below emissions expected under a "business as usual" scenario (CARB 2014) (http://www.arb.ca.gov/cc/ab32/ab32.htm). It includes the major GHGs and groups of GHGs that are being emitted into the atmosphere. These gases include: CO_2, CH_4, N_2O, Hydrofluorocarbons (HFCs), Perfluorocarbons (PFCs), Sulfur hexafluoride (SF_6), and Nitrogen trifluoride. Initially, NF_3 was not listed in AB 32, but was subsequently added to the list via legislation.

Pursuant to AB 32, ARB must adopt regulations to achieve the maximum technologically feasible and cost-effective GHG emission reductions. It requires ARB to develop a Scoping Plan which lays out California's strategy for meeting the goals. In December, 2008, the Board approved the initial Scoping Plan, which included a suite of measures to sharply cut GHG emissions. Reductions in GHG emissions will come from virtually all sectors of the economy and will be accomplished from a combination of policies, planning, direct regulations, market approaches, incentives, and voluntary efforts. These efforts target GHG emission reductions from cars and trucks, electricity production, fuels, and other sources (CARB 2014).

ARB annually updates a statewide GHG inventory, which includes estimates of GHGs emitted into the atmosphere by human activities in California. The California Inventory published in 2016 for the years 2000 to 2014 shows California's gross emissions of GHGs decreased by 5.2%—from 465.9 million metric tons (MMT) of CO_2e in 2000 to 441.5 MMT in 2014, with a maximum of 492.7 MMT in 2004. During the same period, California's population grew by 11% and VMT increased by 19%. The US emissions of GHGs in 2014 were 6870.5 MMT of CO_2e, having increased by 7.4% from 1990 to 2014.

Cap-and-Trade

According to Hansen et al. (2008), a higher price on GHG emissions and payment for carbon sequestrations are surely needed to make the drawdown of airborne CO_2 a reality. A 50 ppm drawdown via agricultural and forestry practices seems plausible, but a strategy to increase carbon sequestration while decreasing the output of carbon from the sources is needed to achieve this goal. This strategy could be Cap-and-Trade.

According to the EPA, Cap-and-Trade is an environmental policy tool that delivers results with a mandatory cap on emissions, while providing sources the flexibility to determine how they comply. The Cap-and-Trade program (http://www.arb.ca.gov/cc/capandtrade.htm) passed by CARB in October, 2011, is a central element of AB 32 and covers major sources of GHG emissions in the state, such as refineries, power plants, industrial facilities, and transportation fuels. The regulation includes an enforceable cap on GHG emissions that will decrease over time to 1990 baselines by 2020. ARB will distribute allowances, which are tradable permits, equal to the emission allowed under the cap. This is a 15% reduction compared to what the emissions would be in 2020 without any programs in place (the so-called "business-as-usual" level). CARB offset credits are GHG emission reductions or sequestered carbon which meet regulatory criteria, and may be used by an entity to meet up to 8% of its triennial compliance obligation under the Cap-and-Trade program. Each ARB offset credit is equal to 1 metric ton of carbon dioxide equivalent ($MTCO_2e$), and can only be generated through implementation of an offset project for which CARB has adopted a compliance offset protocol.

One of the Compliance Offset Protocol is *forestry management*. This protocol provides requirements and methods for quantifying the net climate benefits of activities that sequester carbon on forestland. It provides offset project eligibility rules, methods to calculate an offset project's net effects on GHG emissions, procedures for assessing the risk that carbon sequestered by a project may be reversed (i.e., released back to the atmosphere), and approaches for long-term project monitoring and reporting.

Strategies and Mitigation Measures for Emission Reduction and their Impacts

The strategies described below are for land managers and/or regulators, and can reduce impacts (or can be modified to reduce impacts) of fire emissions on air quality.

Land Management Strategies

For land managers the purpose and effect of prescribed fire and wildfire, managed for resource benefit, are to mimic the natural fire cycle. The purpose of planned fire is to "approximate the natural vegetative disturbance of periodic fire occurrence" and to "maintain fire dependent ecosystems and restore those outside their natural balance." This is done by mimicking the effect of natural recurring fire regimes (prescribed fire can be considered "natural") (WRAP 2005). Hubbard (2014) stated that 26–33 million ha (65–82 million ac), or 42% of national forest lands, are in need of fuel and forest health treatments, whereas only about 80,000 ha (2,000,000 ac) of land are treated annually.

One of the most effective tools to reduce the incidence and severity of unplanned wildfires is prescribed burning that reduces the buildup of vegetation, which had accumulated from years of wildfire suppression, and maintains the forest in a healthy ecological condition. At present, the number of prescribed fire areas treated fall far short of the numbers needed to restore a natural ecosystem and reduce impacts from wildfires (North et al. 2012). Regulatory requirements, nuisance rules, risk avoidance, and policy are some of the factors that contribute to the under-provision of prescribed fire (Collins et al. 2010). The exclusion of wildfire emissions, but inclusion of prescribed fire (anthropogenic), in controlling emissions perpetuate the difficulties that land management agencies and air regulatory agencies have in mediating conflicting legislation and missions (Engel-Cox et al. 2013). The conflict overrides the long-term public interest in reduced wildfire risk, healthier forests, and air quality.

Planned prescribed fires, or the appropriate management of natural ignitions, significantly limit the precursor emissions from larger uncharacteristic events that contribute to ozone and particulate matter (PM) pollution (Cisneros et al. 2012, Cisneros et al. 2014, Schweizer and Cisneros 2014). In contrast to large, high-severity fire, the authors conclude that, "the more extensive air quality impacts documented with large high-intensity fire may be averted by embracing the use of fire to prevent unwanted high-intensity burns."

Air and Land Managers Working Group

In 2003, representatives from the ARB Executive Office began meeting with leadership from the Forest Service, the Bureau of Land Management, the National Park Service, CAL FIRE, and the California Air Pollution Control Officers Association (CAPCOA) to discuss issues of mutual concern regarding smoke management. Through this initial contact, the Air and Land Managers (ALMs) group was formed. ALM meets twice per year and provides a forum for decision-makers to coordinate policy decisions and maintain consistency in federal, state, and local issues associated with smoke management in California. This has led to the development of PFIRS and the support for CANSAC. Support from an interagency effort will continue to be critical for the maintenance of PFIRS and CANSAC, in order to move forward in resolving smoke air quality issues.

Smoke Forecasting and Messaging

In recent years, land management agencies, air regulatory agencies, and public health agencies have developed a process of smoke forecasting and messaging, informed by comprehensive ground-level monitoring, coupled with modeling during unplanned fire events. The Forest Service (FS), National Park Service (NPS), CARB, and APCDs have been instrumental in developing a coordinated smoke emergency response plan during large fire events. Because fire behavior, fuels, and suppression-tactical decisions are critical to successful smoke forecasting, the position of Air Resource Advisors (ARAs) has been added to Incident Command Teams on most of the larger fires in California. An ARA is tasked with analyzing fire data, meteorology, $PM_{2.5}$ monitoring, and modeling data in an effort to forecast smoke concentrations and transport patterns in daily reports. Those forecasts are then shared in daily conference calls with air quality and public health agencies, at which time messaging is coordinated to clearly inform the public of any health risks. The forecasts are also very valuable to the incident command team in evaluating aircraft operations, road conditions, firefighter exposure, and firefighter camp and staging locations. This well-coordinated unplanned fire strategy may be able to transition over to reducing public exposure in proposed large-scale and planned landscape burns, and gain public support for large planned burns.

Emission Reduction and Distribution

According to Hardy et al. (2002) there are two general strategies to managing wildland fire smoke: (1) emission reduction and (2) emission redistribution. Emission reduction techniques vary widely in their applicability and effectiveness by vegetation type, objective, county, or region, and whether fuel is natural or activity-generated (Ottmar et al. 2001). All pollutants, except NOx and CO, are negatively correlated with combustion efficiency, so actions that reduce one pollutant result in the reduction of all except NOx and CO. NOx and CO can increase if the emission reduction technique improves combustion efficiency. Optimal use of reduction techniques can reduce emissions by approximately 20–25%, assuming all other factors (vegetation types, area, etc.) were held constant and land management goals were still met. Strategies that can effectively reduce emissions and applied are: reducing the area burned, reducing fuel load, reducing fuel production, reducing fuel consumption, scheduling burning before new fuel appears, and increasing combustion efficiency (for details see Hardy et al. 2001).

Emission Redistribution Strategy

Emissions can be spatially and temporally redistributed by burning during periods of good atmospheric dispersion (*dilution*), and when prevailing winds will transport smoke away from sensitive areas (*avoidance*) so that air quality standards are not violated. Total emissions are not necessarily reduced under this strategy, though impacts to the receptor sites are. Strategies that are commonly applied to achieve the redistribution/avoidance of emissions are: burning when dispersion is good, allocation of the air basins, and burning more frequently.

Strategies for Reduction of GHGs

An existing approach to removing carbon from the atmosphere is to grow plants that sequester CO_2 in their biomass. Methods for sequestering CO_2 through afforestation have already been accepted as tradable "carbon offsets" under the Kyoto Protocol, and CARB (2011) through its Cap-and-Trade policy. Through sustainable management and protection, forests can play a positive and significant role to help address global climate change. The Forest Offset Protocol is designed to address the forest sector's unique capacity to sequester, store, and emit GHGs and to facilitate the positive role that forests can play to address climate change. Public forests are excluded from participating in the Forest Offset Protocol, but biochar production from forest slash and application to soil can bring similar results of carbon offsets.

Plants remove CO_2 from the atmosphere through the process of photosynthesis and emit CO_2, CH_4, and N_2O during combustion and CH_4 during their decay. The accumulated surface fuel, in excess of ecological need, can lead to catastrophic wildfires. Before human intervention, frequent natural ignitions effectively thinned the understory. These lower intensity fires could renew vitality in fire-dependent ecosystems and provide the opportunity for germination and reproduction, which many fire-evolved plant species require. This allows trees to become larger, increasing fire resistance. Human intervention has led to the accumulation of hazardous fuels which must be removed in order to restore the forest to a healthy natural condition. Excess biomass can be burned under favorable conditions, removed for cogeneration, or to manufacture biochar as described below.

Slash and Biochar

Forest activities like thinning timber result in a large amount of slash left on the forest surface, or piled and later treated as a planned burn. An alternative to burning is conversion to biochar, which can retain carbon in the soil without decomposition for centuries (Hurteau and North 2010). Biochar (short for "bio-charcoal") is plant-derived carbon obtained through the process known as Pyrolysis. Biomass is heated in the absence of oxygen which produces biochar, liquids, and gases. Biochar comes in the form of small, black, light, and highly porous fragments. The relative yield of products from pyrolysis varies with temperature. Temperatures of 400°C to 500°C (752°F–932°F) produces more char, while temperatures above 700°C (1,292°F) favors the yield of liquid and gas fuel components. High temperature pyrolysis, also known as gasification, produces primarily syngas.

Biochar is a lower-risk strategy compared to other sequestration options. Stored carbon can be released in forest fires, by converting no-tillage lands to tillage, or by leaks from geological carbon storage. Once biochar is incorporated into soil, it is difficult to imagine any incident or change that would cause a sudden loss of stored carbon. Plant biomass decomposes in a relatively short period of time, whereas biochar is orders-of-magnitude more stable. Half of the carbon that cycle annually through plants can be taken out of its natural cycle and sequestered in a much slower biochar cycle.

Biochar has been shown to improve the structure and fertility of soils, thereby improving biomass production. As stated previously, biochar applied to soil can store large quantities of GHGs in the soil for centuries. The presence of biochar in the soil can improve water quality, increase soil fertility, raise soil productivity, improve water holding capacity, and reduce wildfire (by reduction of hazardous fuel used to manufacture biochar).

Carbon Balance/Accounting

Depending on how forests are managed or impacted by natural events, they can be a net source of emissions (resulting in a decrease to the reservoir, or an increase of CO_2 to the reservoir). In other words, forests may have a net negative or net positive impact on CO_2 sequestration. Carbon is also stored in the soils that support forest systems, as well as the understory plants and litter on the forest floor, and wood products that are harvested from forests. Global Carbon (C) stored in vegetation and the top 1 m (3.3 ft) of soil are about 2,500 Gt, with 81% of this in soils and the balance in aboveground vegetation (Bolin and Sukuman, 2000). This terrestrial carbon storage is slightly over three times the amount of carbon in the atmosphere (760 Gt [836 Gtn]). Land managers could consider developing land-use plans on the basis of carbon balance accounting by Hydrologic Unit Code 6 (HUC6) watersheds.

The burning and natural decomposition of biomass adds large amounts of CO_2 to the atmosphere. Any mitigation measure that results in reduced fuel combustion would also reduce short-term GHG release. Any measure that leads to production of greater biomass will result in greater short-term carbon sequestration. The return to historic fire cycles maximizes long-term reduction of GHG, release and increased carbon sequestration and residence time. Reducing overstocking to restore degraded ecosystems, encouraging species for increased tolerance to environmental stressors, implementing ERTs, applying nonburning alternatives, and harvesting timber for wood products can result in a net positive carbon reservoir.

Conclusion and Recommendations

The following conclusions and recommendations are presented for consideration:

1. Large areas of California forests are outside of their natural balance and in need of restoration. A widespread increase in the use of fire for ecological benefit may provide the resiliency needed in forests and could mitigate other negative impacts on human health, visibility, GHGs, and forest ecosystems. Regulatory agencies are often hesitant to allow large planned burns due to the risk associated with noncompliance of the NAAQS, attainment status, and nuisance regulations. Regulatory legislation and operations are oriented toward short-term outcomes, but managing fire only for the short term often results in long-term negative effects. A coordinated decision-making process between burn and regulatory agencies could provide opportunities to maximize use of fire for resource benefit, and minimize fire suppression. The EPA intends to revise the Interim Air Quality Policy issued in 1998. It may provide an opportunity for the EPA to help minimize the existing conflicting rules and policies that hamper the utilization of "unplanned fires for resource benefit" and "large planned burns."
2. Over the last two decades CO_2 levels in the atmosphere have increased to over 400 ppm, resulting in impacts from recent climate-related extremes such as heat waves, droughts, floods, cyclones, and wildfires, which reveal significant vulnerability and exposure of some ecosystems to current climate variability. Forests can sequester and contribute to atmospheric reduction of CO_2. In fact, the goal of the CARB's forestry-management-compliance-offset-protocol (from a forest offset project) is to ensure that the net GHG reductions and GHG removal enhancements are accounted for in a complete, consistent, transparent, accurate, and conservative manner for issuing ARB offset credits. Forests can burn and release CO_2 into the atmosphere; therefore, managers of fire-adapted ecosystems should emphasize a science-based use of fire. Wherever appropriate, alternatives to burning, such as biomass for bioenergy production and/or biochar production could be considered. Biochar application can improve soil and maintain sequestered CO_2 for centuries. Unfortunately, CARB has not yet recognized biochar as an offset protocol though its production and utilization can meet the criteria as an offset under the Cap-and-Trade program. Although public forests are not allowed to participate in Forest Offset Protocols, they can still consider a proactive biochar strategy.
3. There is a need to quantify complex processes that are used to predict how wildfires will affect terrestrial carbon stocks. Burn severity differences can result in varying amounts of fuel consumed, and varying postfire recovery rates. Land management agencies and research scientists can consider collaboratively developing a Carbon Balance Accounting process for forest planning that is based on watershed HUC units.
4. CARB is on track to meet all established regional haze visibility goals, except the goal of "20% worst visibility days" for class I areas that are impacted by unplanned wildfires. Fires ignited by lightning and Native Americans have been a component of most California ecosystems for thousands of years (Stephens et al. 2007). This prehistoric burning suggests a significant amount of haze can be attributed to natural background conditions during summer and fall in California. The EPA needs to revise the "20%-worst-visibility-days" goal for 2064 to properly account for prehistoric fire conditions.
5. Although BlueSky and other predictive tools are currently in use, the Forest Service, public health organizations, air regulators, scientists, and conservation organizations seek significantly expanded outreach, education, and communication systems to better notify and protect public health—especially those most at risk from short-duration smoke impacts.
6. Smoke from wildfires can move long distances, both at upper and lower levels in the atmosphere, and may affect air quality in areas thousands of kilometers/miles away. This can affect atmospheric profiles of $PM_{2.5}$, CO, CO_2 domes, N_2O, and other aerosols. Fire contribution to ozone and nitrogen deposition needs more research. Models predicting these, and their quantitative impacts to biological processes, need refinement for the proper utilization of fire in ecological management.

References

Abatzoglou, J.T., F. Joseph, P. Doughman, and S. Nespor. 2014. A primer on global climate change science. Pages 15–52 in: J.F.C. Dimento and P. Doughman, editors. Climate Change. The MIT Press, Cambridge, Massachusetts, USA.

Ahuja, S. 2002. Monitoring fire emissions to notify the public mobile monitoring. Pages 45–49 in: N.G. Sugihara, M. Morales, and T. Morales, editors. Proceedings of the Symposium: Fire in California Ecosystems: Integrating Ecology, Prevention, and Management. Association for Fire Ecology Miscellaneous Publication 1. Davis, California, USA.

Akagi, S., R. Yokelson, C. Siedinmyer, M. Alvarado, J. Reid, T. Karl, J. Crounce, and P. Wennberg. 2011. Emission factors for open and domestic biomass burning for use in atmospheric models. Atmospheric Chemistry and Physics 11: 4039–4072.

Bolin, B., and R. Sukuman. 2000. Global perspective. Pages 23–51 in: R.T. Watson, I.R. Noble, B. Bolin, N.H. Ravindrath, D.J. Verardo, and D.J. Dokken, editors. Land Use, Land Use Change, and Forestry, Special Report of the IPCC on Climate Change. Cambridge University Press, Cambridge, UK.

Boman, C., B. Forsberg, and T. Sandström. 2006. Shedding new light on wood smoke: a risk factor for respiratory health. European Respiratory Journal 27: 446–447.

Burley, J.D., A. Bytnerowicz, M. Buhler, B. Zielinska, D. Schweizer, R. Cisneros, S. Schilling, J.C. Varela, M. McDaniel, M. Horn, and

D. Dulen. 2016. Air quality at Devils Postpile National Monument, Sierra Nevada Mountains, California, USA. Aerosol and Air Quality Research 16: 2315–2332.
CARB. 2011. Cap-and-Trade Program October 2011. http://www.arb.ca.gov/cc/capandtrade.htm.
———. 2014. California Regional Haze Plan 2014 Progress Report. http://www.arb.ca.gov/planning/reghaze/progress/carhpr2014.pdf
———. 2015. Area Designations Maps / State and National. https://www.arb.ca.gov/desig/adm/adm.htm.
Cisneros, R., D. Schweizer, H. Preisler, D.H. Bennett, G. Shaw, and A. Bytnerowicz. 2014. Spatial and seasonal patterns of particulate matter less than 2.5 microns in the Sierra Nevada Mountains, California. Atmospheric Pollution Research 5: 581–590.
Cisneros, R., D. Schweizer, S. Zhong, K. Hammond, M.A. Perez, Q. Gou, S. Traina, A. Bytnerowicz, and D.H. Bennett. 2012. Analyzing the effects of the 2002 McNally fire on air quality in the San Joaquin Valley and southern Sierra Nevada, California. International Journal of Wildland Fire 21: 1065–1075.
Collins, B.M., S.L. Stephens, J.M. Moghaddas, and J. Battles. 2010. Challenges and approaches in planning fuel treatments across fire-excluded forested landscapes. Journal of Forestry 108: 24–31.
Conard, S.G., and A.M. Solomon. 2009. Effects of wildland fire on regional and global carbon stocks in a changing environment. Pages 109–138 in: A. Bytnerowicz, M. Arbaugh, A. Riebau, and C. Anderson, editors. Wildland Fires and Air Pollution. Development in Environmental Science 8. Elsevier, Amsterdam, The Netherlands.
Engel-Cox, J., N.T.K. Oanh, A. van Donkelaar, R.V. Martin, and E. Zell. 2013. Toward the next generation of air quality monitoring: particulate matter. Atmospheric Environment 80: 584–590.
———. 1999. 40 CFR Part 51: Regional Haze Regulations; Final Rule. Vol. 64. No. 126. July 1, 1999. Washington, D.C., USA.
———. 2013. Inventory of U.S. greenhouse gas Emissions and sinks: 1990–2011. EPA 430-R-13-001. Environmental Protection Agency, Washington, D.C., USA.
———. 2016. Inventory of U.S. greenhouse gas Emissions and sinks: 1990–2014. EPA 430-R-16-002. Environmental Protection Agency, Washington, D.C., USA.
Fenn, M.E., E.B. Allen, S.B. Weiss, S. Jovan, L.H. Geiser, G.S. Tonnesen, R.F. Johnson, L.E. Rao, B.S. Gimeno, F. Yuan, T. Meixner, and A. Bytnerowicz. 2010. Nitrogen critical loads and management alternatives for N-impacted ecosystems in California. Journal of Environmental Management 91: 2404–2423.
Fenn, M.E., K.F. Lambert, T.F. Blett, D.A. Burns, L.H. Pardo, G.L. Lovett, R.A. Haeuber, D.C. Evers, C.T. Driscoll, and D.S. Jefferies. 2011. Setting Limits: using air pollution thresholds to protect and restore U.S. ecosystems. Issues in Ecology Report 14. Ecological Society of America, Washington, D.C., USA.
Finley, B.D., P.C. Swartzendruber, and D.A. Jaffe. 2009. Particulate mercury emissions in regional fire plumes observed at the Mount Bachelor Observatory, 43. Atmospheric Environment 43: 6074–6082.
Fulé, P.Z., J.F. Crouse, J.P. Roccaforte, and E.L. Kalies. 2012. Do thinning and/or burning treatments in western USA ponderosa or Jeffrey pine-dominated forests help restore natural fire behavior? Forest Ecology and Management 269: 68–81.
GCVTC (Grand Canyon Visibility Transport Commission). 1996a. Alternative Assessment Committee Report. Western Governors' Association, Denver, Colorado, USA.
———. 1996b. Recommendations for Improving Western Vistas. Western Governors' Association, Denver, Colorado, USA.
———. 1996c. Report of the Grand Canyon Visibility Transport Commission to the United States Environmental Protection Agency. Western Governors' Association, Denver, Colorado, USA.
Hansen, J., M. Sato, P. Kharecha, D. Beerling, R. Berner, V. Masson-Delmotte, M. Pagani, M. Raymo, D.L. Royer, and J.C. Zachos, 2008: Target atmospheric CO_2: where should humanity aim? Open Atmospheric Science Journal 2: 217–231.
Hardy, C.C., S.G. Conard, J.C. Regelbrugge, and D.T. Teesdale. 1996. Smoke emissions from prescribed burning of southern California chaparral. USDA Forest Service Research Paper PNW-RP-486. Pacific Northwest Research Station, Portland, Oregon, USA.
Hardy, C.C., S. Hermann, and R. Mutch. 2002. The wildland fire imperative. Pages 11–19 in: C.C. Hardy, R.D. Ottmar, J.L. Peterson, J.E. Core, and B. Leenhouts, editors. Smoke Management Guide for Prescribed and Wildland Fire. 2001 ed. PMS 420-2. National Wildfire Coordinating Group, Boise, Idaho, USA.
Hardy, C.C., and B. Leenhouts. 2001. Why do we need a smoke management guide? In: C.C. Hardy, R.D. Ottmar, J.L. Peterson, J.E. Core, and B. Leenhouts, editors. Smoke Management Guide for Prescribed and Wildland Fire. 2001 ed. PMS 420-2. National Wildfire Coordinating Group, Boise, Idaho, USA.
Hardy, C.C., R.D. Ottmar, J.L. Peterson, J.E. Core, and B. Leenhouts, editors. 2001. Smoke management guide for prescribed and wildland fire. 2001 ed. PMS 420-2. National Wildfire Coordinating Group, Boise, Idaho, USA.
Hubbard, J. 2014. Statement of James Hubbard, Deputy Chief of State and Private Forestry, U.S. Forest Service, U.S. Department of Agriculture, to the U.S. Senate Committee on Indian Affairs, May 14, 2014. USDA, Forest Service, Washington, D.C. USA.
Hurteau, M.D., and North, M. 2010. Carbon recovery rates following different wildfire risk mitigation treatments. Forest Ecology and Management, 260: 930–937.
IPCC (Intergovernmental Panel for Climate Change). 2007. 4th Assessment Report—Climate Change 2007 (AR4). World Meteorological Organization, Geneva, Switzerland.
———. 2013. 5th Assessment Report—Climate Change 2013 (AR5). World Meteorological Organization, Geneva, Switzerland.
———. 2014. Climate Change 2014: Synthesis Report. In: Core Writing Team, R.K. Pachauri, and L.A. Meyer, editors. Contribution of Working Groups I, II and III to the Fifth Assessment Report of the Intergovernmental Panel on Climate Change. IPCC, Geneva, Switzerland.
Kane, L.E., and Y. Alarie. 1977. Sensory irritation to formaldehyde and acrolein during single and repeated exposures in mice. American Industrial Hygiene Association Journal 38: 509–522.
Krupa, S. 2009. Preface. Pages xliii-xliv in: A. Bytnerowicz, M. Arbaugh, A. Riebau, and C. Anderson. Wildland Fires and Air Pollution. Development in Environmental Science 8. Elsevier, Amsterdam, The Netherlands.
McKenzie, G., D.L. Peterson, and J.J. Littell. 2009. Global warming and stress complexes in forests of Western North America. Pages 319–337 in: A. Bytnerowicz, M. Arbaugh, A. Riebau, and C. Anderson, editors. Wildland Fires and Air Pollution. Development in Environmental Science 8. Elsevier, Amsterdam, The Netherlands.
Miller, P., J. McBride, S. Schilling, and A. Gomez. 1989. Trends of ozone damage to conifer forests between 1974 and 1988 in the San Bernardino Mountains of southern California. Pages 309–323 in: Proceedings Air Pollution Effects on Western Forests. 32nd Annual Meeting. Air Waste Management. Association, Pittsburg, Pennsylvania, USA.
North, M.P., B.M. Collins, and S.L. Stephens. 2012. Using fire to increase the scale, benefits and future maintenance of fuels treatments. Journal of Forestry 110: 392–401.
Ottmar, R.D., J.L. Peterson, B. Leenhouts, and J.E. Core. 2001 Smoke management techniques to reduce or redistribute emissions. Pages 141–162 in: C.C. Hardy, R.D. Ottmar, J.L. Peterson, J.E. Core, and B. Leenhouts, editors. Smoke Management Guide for Prescribed and Wildland Fire. 2001 ed. PMS 420-2. National Wildfire Coordinating group, Boise, Idaho, USA.
Ottmar, R.D., R.E. Vihnanek, H.S. Miranda, M.N. Sato, and S.M.A. Andrade. 2001. Stereo photo series for quantifying cerrado fuels in central Brazil—Volume I. USDA Forest Service General Techical Report PNW-GTR-519. Pacific Northwest Research Station. Portland, Oregon, USA.
Peterson, J.L., and D. Ward. 1991. An inventory of particulate matter and air toxic emissions from prescribed fires in the United States for 1989. Final Report. U.S. EPA Office of Air Quality Programs and Standards. Environmental Protection Agency, Washington, D.C., USA.
Plymale, E., M. Arbaugh, T. Procter, S. Ahuja, G. Smith, and P. Temple. 2003. Towards an air pollution effects monitoring system for the Sierra Nevada. Pages 285–298 in: A. Bytnerowicz, R. Alonso, and M. Arbaugh, editors. Ozone Air Pollution in the Sierra Nevada—Distribution and Effects on Forests. Development in Environmental Science 2. Elsevier, Amsterdam, The Netherlands.
Procter, T., S. Ahuja, and M. McCorison. 2003. Managing air pollution effected forests in the Sierra Nevada. Pages 359–370 in: A. Bytnerowicz, R. Alonso, and M. Arbaugh, editors. Ozone Air Pollution in the Sierra Nevada—Distribution and Effects on

Forests. Development in Environmental Science 2. Elsevier, Amsterdam, The Netherlands.
Public Law. 1963. Public Law 88-206. Clean Air Act of 1963. Act of December 17, 1963, 77 Stat. 392.
———. 1967. Public Law 90-148. Air Quality Act of 1967. Act of November 1, 1967. 42 U.S.C. 7401. 81 Stat. 485, 501.
———. 1970. Public Law 91-604. Clean Air Act Amendments of 1970. Act of December 31, 1970. 42 USC 1857h-7 et seq.
———. 1977. Public Law 95-95. Clean Air Act as Amended August 1977. 42 U.S.C. s/s 1857 et seq.
———. 1990. Public Law 101-549. Clean Air Act as Amended. November 15, 1990. 104 Stat. 2399.
Reinhardt, T.E. 2000. Effects of Smoke on Wildland Firefighters. URS/Radian International, Seattle, Washington, USA.
Reinhardt, T.E., R.D. Ottmar, and A.J.S. Hanneman. 2000. Smoke exposure among firefighters at prescribed burns in the Pacific Northwest. USDA Forest Service Research Paper PNW-RP-526. Pacific Northwest Research Station, Portland, Oregon, USA.
Ryan, P.W., and C.K. McMahon. 1976. Some chemical characteristics of emissions from forest fires. Pages 2–3 in: Proceedings of the 69th Annual Meeting of the Air Pollution Control Association, Portland, Oregon. Air Pollution Control Association Paper 76. Pittsburgh, Pennsylvania, USA.
Schweizer, D., and R. Cisneros. 2014. Wildland fire management and air quality in the southern Sierra Nevada: using the Lion Fire as a case study with a multi-year perspective on $PM_{2.5}$ impacts and fire policy. Journal of Environmental Management 144: 265–278.
Sharkey, B., editor. 1997. Health hazards of smoke: recommendations of the April 1997 consensus conference. USDA Forest Service Technical Report 9751-2836-MTDC. Missoula Technology and Development Center, Missoula, Montana, USA.
Stephens, S.L., R.E. Martin, and N.E. Clinton. 2007. Prehistoric fire area and emissions from California's forests, woodlands, shrublands, and grasslands. Forest Ecology and Management 251: 205–216.
Stephens, S.L., J.D. McIver, R.E.J. Boerner, C.J. Fettig, J.B. Fontaine, B.R. Hartsough, P. Kennedy, and D.W. Schwilk. 2012. Effects of forest fuel reduction treatments in the United States. BioScience 62: 549–560.
Ward, D.E. 1990. Airborne monitoring and smoke characterization of prescribed fires on forest lands in western Washington and Oregon. USDA Forest Service General Technical Report PNW-GTR-251. Pacific Northwest Research Station, Portland, Oregon, USA.
Ward, D.E., and C.C. Hardy. 1991. Smoke emissions from wildland fires. Environmental International 17: 117–134.
Westerling, A.L., H.G. Hidalgo, D.R. Cayan, and T.W. Swetnam. 2006. Warming and earlier spring increase western US forest wildfire activity. Science 313: 940–943.
Wiedinmyer, C., and M.D. Hurteau. 2010. Prescribed fire as a Means of reducing forest carbon emissions in the western United States. Environmental Science and Technology 44: 1926–1932.
WRAP (Western Regional Air Partnership). 2005. Policy for categorizing fire emissions. Prepared by Natural Background Task Team of Fire Emission Joint Forum and Approved by Consensus: Western Regional Air Partnership. November 15, 2001.

CHAPTER TWENTY-FOUR

Fire and Invasive Plants

ROBERT C. KLINGER, MATTHEW L. BROOKS, AND JOHN M. RANDALL

> Just as there is honor among thieves, so there is solidarity and co-operation among plant and animal pests. Where one pest is stopped by natural barriers, another arrives to breach the same wall by a new approach. In the end every region and every resource get their quota of uninvited ecological guests.
>
> ALDO LEOPOLD (1949)

Introduction

One of the most significant issues in conservation is the effect of invasive plant species on natural communities and ecosystems. Conservation scientists generally agree that only habitat destruction poses a greater threat to native biological diversity and the integrity of ecosystems (Wilcove et al. 1998, Mack et al. 2000). Invasive species, which are defined as nonnative species that spread into areas and cause ecological change and/or economic damage (Rejmanek 1995, Richardson et al. 2000), can have impacts on populations of individual species as well as entire communities (Vitousek et al. 1996, Lonsdale 1999). Over the last several decades their impact on ecosystem processes, including fire regimes, has been increasingly well documented (Mack and D'Antonio 1998, Mack et al. 2001, Zouhar et al. 2008). Although these impacts are highly variable (Parker et al. 1999, Jauni et al. 2015), there is little argument that invasive species have long-term implications for management of many (or even most) natural areas. In a world where biological diversity is considered imperiled, the transformation of ecological interactions and ecosystem properties by invasive species poses both an immediate and a chronic threat to preservation, management, and restoration of parks, refuges, and reserves (Harty 1986, Gordon 1998).

Interest regarding the inter-relationship between invasive plant species and fire is not a recent phenomenon (Pickford 1932, Hervey 1949), but in the early 1990s the publication rate of studies on fire and invasive species began to increase exponentially (Fig. 24.1), as did the geographic scope, method of study, and emphasis of research (see Table 1 in D'Antonio [2000] for an overview). Over the last 25 years many reviews on invasive species and fire have been written, ranging from global (D'Antonio and Vitousek 1992, Mack and D'Antonio 1998, D'Antonio 2000, D'Antonio et al. 2000, Brooks et al. 2004, Brooks 2008, Klinger 2011) to more regional patterns (Galley and Wilson 2001, Keeley 2006, Klinger et al. 2008, Zouhar et al. 2008, Keeley et al. 2011). Two significant recent studies have used meta-analysis to: 1) compare several measures of response by invasive plants to wildfire and prescribed fire relative to those of native species (Alba et al. 2015) and 2) evaluate the degree to which diversity and abundance of nonnative plants varied among different disturbance types and fire (Jauni et al. 2015). These overviews as well as many individual studies have effectively described the existence of broad patterns in the interaction between fire and invasive species, but also that there are many variations to these general patterns depending on regional and local conditions

The management of fire and invasive plants creates unique challenges. Fire exclusion has resulted in extensive changes in many ecosystems in North America, thus the restoration of fire is often regarded as a critical component for managing natural communities where it was a naturally occurring process (Biswell 1999). However, many invasive species are known to exploit areas burned by fires (Hobbs and Huenneke 1992, Alba et al. 2015). By exposing soil and reducing competition, fire opens the way for invasion by new species or expansion of those already resident in the community. There are few if any areas in North America without nonnative species, thus invasions into burn sites and, oftentimes, alterations to fire regimes have been reported in many regions of the continent (Zouhar et al. 2008).

Plant invasions, especially by annual grasses (e.g., bromes [*Bromus* spp.] and wild oats [*Avena* spp.]) have especially affected fire regimes in regions with Mediterranean climates (Klinger 2011, Brooks et al. 2016). This is problematic in terms of conservation, because these regions contain some of the highest recorded levels of biodiversity outside of the tropics (Wilson and Peter 1988, Cowling et al. 1996). They also tend to have relatively high human population densities, and human activities both transport nonnative species and disturb vegetation and soils, making systems more vulnerable to invasion. Other North American ecosystems where alteration of fire regimes by invasive species has been particularly severe include the cold and warm deserts (Whisenant 1990, Brooks and Pyke 2001, Brooks et al. 2016) and Midwest grasslands (Grace et al. 2001).

Nevertheless, fire can be an effective tool for managing some species of invasive plants (Nuzzo 1991, Lonsdale and Miller 1993, DiTomaso et al. 1999, Myers et al. 2001). Compared to other cultural and mechanical types of management (e.g., herbicides, grazing, plowing, and cutting), fire is a

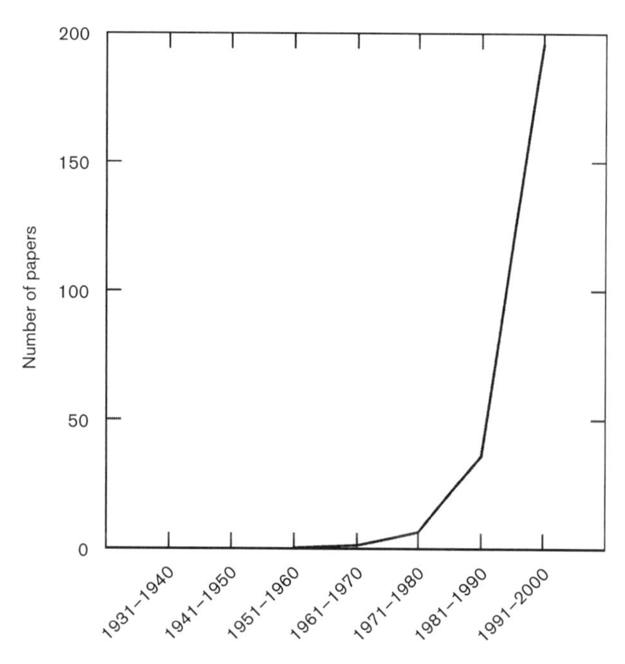

FIGURE 24.1 The number of studies published on fire and invasive species.

relatively inexpensive method that can be applied in a short period of time. Moreover, the timing, frequency, and extent of burning can be managed to a certain extent. But responses of plants to fire are complex, species-specific, and can vary both within a site and between sites. This complexity and variability are reflected in the outcomes of many efforts to control invasive species with fire (Keeley 2006, Klinger et al. 2008).

In this chapter we examine the relationship between fire and invasive species in California from three different perspectives: 1) the general interrelationships between fire and invasive plants, 2) specific examples from within or near California, and 3) the use of fire as a management tool to control nonnative species. We conclude by synthesizing these three perspectives in the context of agents of global change and more general ecological processes.

Inter-relationships between Fire and Invasive Plants

General Patterns and Processes

Much of the research on fire and invasive species has focused on how fire facilitates plant invasions, and how invasive species in turn alter fire regimes (Whisenant 1990, D'Antonio and Vitousek 1992, D'Antonio 2000, D'Antonio et al. 2000, Brooks et al. 2004, Brooks 2008, Brooks and Chambers 2011). Many invasive species often exploit areas of disturbance (Hobbs and Huenneke 1992, Jauni et al. 2015). When these areas burn or are adjacent to burned areas, conditions favorable to the nonnatives lead to their spread and/or increase in abundance in the burned areas (D'Antonio 2000). As invasive species increases in abundance, plant species composition and ecosystem processes change, which in turn alters fuel characteristics and, ultimately, fire regimes (Holmes and Cowling 1997, Brooks et al. 2004).

Probably the most widespread mechanism of alteration of fire regimes by invasive species is through the "grass-fire cycle" (D'Antonio and Vitousek 1992). Invasive grass species become established in an area dominated by woody vegetation, either as a result of disturbance or, in some instances, where there has been a long history of fire suppression (Grace et al. 2001). As the invasive grasses increase in abundance, a continuous layer of highly combustible fine fuel develops. The fine fuel has higher probability of ignition than the woody vegetation, resulting in increased fire starts, rates of fire spread, and fire frequency. Then, as a result of shortened fire return intervals, areas that once were shrublands or forest dominated by native species are converted to grasslands dominated by nonnatives (Fig. 24.2).

The grass-fire cycle has been reported in many parts of the world, including Hawaii, South Africa, Australia, and the western United States (D'Antonio 2000, Brooks et al. 2004). In the western United States, the most notable examples are invasion of cheat grass (*Bromus tectorum*) in the Great Basin (Mack 1981), red brome (*Bromus madritensis* subsp. *rubens*) and Mediterranean split grass (*Schismus* spp.) in the Mojave desert (Brooks 1999), and rye grass (*Festuca perennis*) in chaparral and coastal scrub areas of southern California (Zedler et al. 1983, Haidinger and Keeley 1993). Brooks et al. (2004) presented a more general model of the invasive plant-fire regime cycle (later refined by Brooks 2008) which is similar to the phases of the invasion cycle (Lodge et al. 2006) but provides a broader conceptual understanding of how the colonization, establishment, and spread of any nonnative plant species might interact with fire regimes to create an altered ecosystem state.

Examples of invasive species altering fire regimes by reducing fire frequency are far less common than examples of those that increase fire frequencies, but alteration of ecosystem properties can be as severe. In the southeastern United States, Chinese tallow (*Triadica sebifera*; synonym *Sapium sebiferum*) has invaded extensive areas of coastal prairie, marsh and swamp along the Gulf coastal plain, forming dense thickets that suppress herbaceous species growth that would otherwise produce continuous surface fuel (Grace et al. 2001). The lowered levels of surface fuel results in reduced fire frequency, which promotes Chinese tallow dominance and leads to the establishment of nearly monotypic stands (Bruce et al. 1995).

Whereas fire can promote invasion or dominance of nonnative plants in many ecosystems, its occurrence alone is not necessarily sufficient for predicting what levels of dominance they might reach in postfire communities. There must be a combination of suitable environmental conditions, spatial structure among communities, and species demographic traits (survival, recruitment, and dispersal) for nonnative plants to reach high levels of abundance after a fire. The invasive plant-fire regime model proposed by Brooks et al. (2004, 2008) takes into consideration suitable conditions in terms of areas with a history of high fire frequency as well as demographic traits in terms of dispersal by species from unburned areas. But it does not explicitly address two important spatial issues: populations of nonnative plants that already occur in an area that burns, and how burned patches relate spatially to adjacent communities, whether they are burned or not. In situations where nonnative plants are uncommon or do not occur within a burned area, sufficient propagule pressure from adjacent unburned communities is required for colonization, establishment, and spread (Brooks et al. 2004). However, recruitment from populations already established within a burned area frequently occurs (Keeley et al. 2005), as can a combination of dispersal and high recruitment from

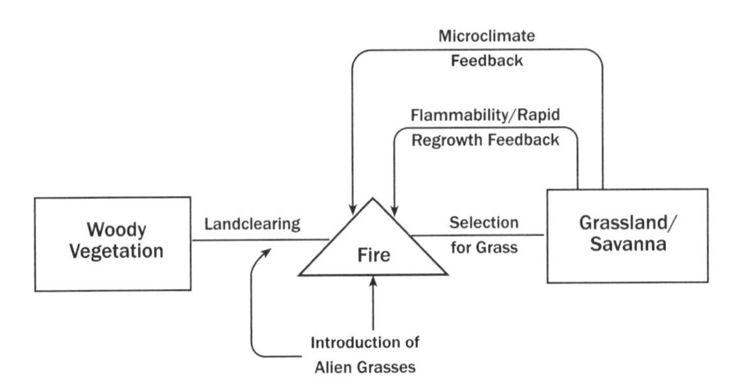

FIGURE 24.2 Schematic representation of the grass-fire cycle (redrawn from D'Antonio and Vitousek 1992).

already established populations. Moreover, propagule pressure into burned areas can be enhanced when the burned areas occur in a mosaic with unburned communities where there are patches of abundant nonnative species (Moody and Mack 1988, Keeley et al. 2003).

An important concept developed over the last 15 years that attempts to explain spatial dynamics among communities is the *metacommunity*, defined as a set of communities linked by dispersal (Leibold et al. 2004). The metacommunity is a general model of interactions among communities (Mouquet and Loreau 2002), thus it provides a useful framework for understanding how plant assemblages dominated by nonnatives can be created and/or maintained following a fire (Miller et al. 2010). For example, prior to burning there could be a network of communities dominated by native species, but patches within the network might have populations of one or more abundant nonnative plants. From a metacommunity perspective, this could be an example of species sorting, where species with similar resource requirements form local assemblages along one or more environmental gradients. Following a fire, nonnatives could colonize burned areas from unburned patches where their abundance was high (Brooks et al. 2004), a paradigm in the metacommunity concept known as mass effects. Habitat conditions in early postfire years may not allow colonizers to initially achieve positive population growth rates, but dispersal from the unburned patches could subsidize their populations and allow them to persist. As conditions improve over time, recruitment becomes positive and these populations eventually become self-sustaining.

Dispersal of nonnative species also occurs among patches of suitable habitat within the landscape mosaic of burned areas, an important process in the metacommunity framework termed *patch dynamics* (Pickett and White 1985, Leibold et al. 2004). Populations or assemblages of nonnative species that persist after burning often outcompete natives (Coffman et al. 2010), but in other instances competitive interactions can potentially be altered by stochastic processes. Two particularly important processes are *priority effects* and *storage effects*. Priority effects occur as a result of a species arriving at (or persisting in) a site prior to other species, and then impacting the community by reducing resources or altering abiotic conditions (Körner et al. 2008, Kardol et al. 2013). Storage effects occur when a species exploits favorable environmental conditions and "stores" the gains, either demographically (e.g., propagules, seed bank) or in performance (e.g., increased leaf area or root mass), that help it survive during periods of time or in patches where resources are low or of poor quality (Chesson and Huntly 1988, Angert et al. 2007, Angert et al. 2009).

The spatial dynamics of invasive plants are important from a postfire management perspective. In situations where propagule pressure into burned areas is expected to be high, early detection programs could potentially reduce the number of nonnative populations that become established (Brooks and Klinger 2009). However, if there is high recruitment from populations already established within a burned area, then broader control efforts might be appropriate (DiTomaso et al. 2006). Despite its clear ecological and management relevance, there have been few studies explicitly linking the metacommunity concept with development and maintenance of postfire plant communities, especially in regards to invasive plants (Miller et al. 2010, Tucker and Cadotte 2013). Despite this, there is sufficient evidence that metacommunity dynamics may often be very significant. In the following sections, we describe specific patterns in the relationship between nonnative plants and fire in the major vegetation formations and highlight the potentially important metacommunity dynamics that might result before and after fire.

The Fire and Invasive Plant Situation in California

Although the spread of species into new areas is a common and natural occurrence, the rate of introduction, escape from cultivation, and subsequent spread of nonnative plants in California increased tremendously with the influx of European and American settlers since the late 1700s (Fig. 24.3). In just over 200 years, 1,360 intentionally and unintentionally introduced plant species have established populations outside cultivation in the state (Rejmanek and Randall 1994, Randall et al. 1998, Hrusa et al. 2002). The ranges of many of these species have not increased substantially beyond their points of introduction, or if their ranges did increase they never reached levels of abundance to have significant, detectable effects on natural systems. But other species did flourish, and as early as the late 1800s major alterations to some plant communities were being noted (Mooney et al. 1989). It was not uncommon to consider some of these community-level changes to be beneficial, especially in terms of improved range forage for livestock and of erosion control. However, it was also clear that nonnative species could have undesirable effects on communities (Fig. 24.3), and over the past several decades the magnitude of the threats that biotic invasions pose to California's wildlands have become quite apparent to scientists and land managers (Jackson 1985, Rejmanek et al. 1991, Randall 1996, Randall et al. 1998).

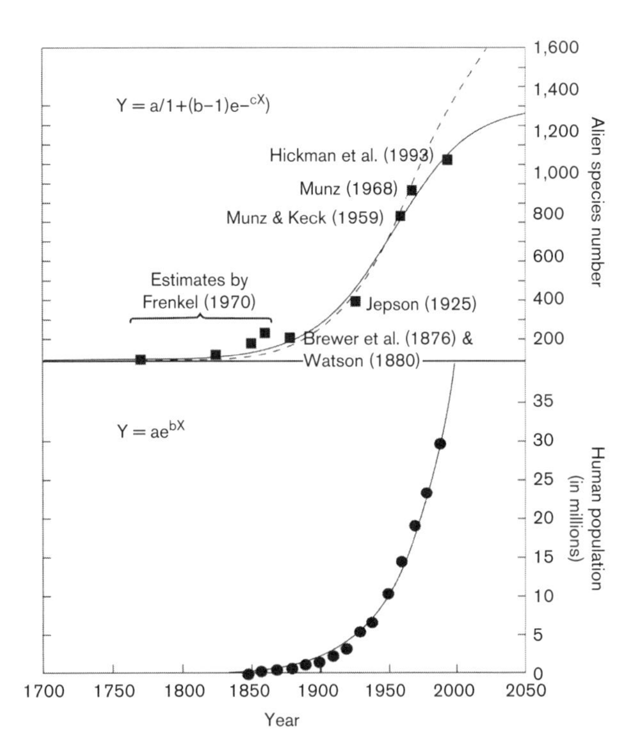

FIGURE 24.3 The increase in invasive plant species recorded in California (redrawn from Randall et al. 1998). The dashed line is from data analyzed by Randall et al. (1998) and the solid line from an earlier analysis by Frenkel (1970). Note the higher number of species considered invasive in 1998 relative to the prediction of the 1970 analysis.

GRASSLANDS

California's grasslands have been particularly susceptible to invasions by nonnative species (Bossard et al. 2000, D'Antonio et al. 2006). The native California prairie has been converted to grasslands dominated by nonnative annual grasses and forbs (Heady 1988), but it is important to note that this was not due solely to fire. Most likely it was a result of a number of interacting factors, including climate change, human disturbance, the introduction of large numbers of nonnative grazing mammals by Euro-American settlers, seed transport on the hides and in the feed of the grazers, and fire (Jackson 1985, Mack 1989, Hamilton 1997, Brooks et al. 2016). Many studies have shown that nonnative annual grasses exert strong negative competitive effects on native herbaceous species (Bartolome and Gemmil 1981, Menke 1992, Dyer and Rice 1997, Brown and Rice 2000), but several studies also indicate that, over time, native perennial grasses can have strong competitive effects on nonnative annual grasses (Seabloom et al. 2003, Corbin and D'Antonio 2004, Lulow et al. 2007). This suggests an interaction between demographic traits of nonnative annual grasses and priority effects have likely played an important role in development and maintenance of most contemporary California grasslands. Although fire was a natural ecosystem process in California grasslands, it was also most likely a key factor that helped nonnative grasses reach levels of abundance where they exerted priority effects on native species attempting to become reestablished in burned areas (Seabloom et al. 2003, Lulow et al. 2007).

Precisely how the shift in grasslands to dominance by nonnative species has affected fire regimes is unknown (Heady 1988, also see chapter 15). The size of fires in these grasslands has likely decreased as a result of fire suppression and reduced fuel continuity caused by anthropogenic features arising from agriculture and urban developments. And while seasonality and frequency of fire play a role in the relative abundance of particular nonnative species in grasslands (Foin and Hektner 1986, Meyer and Schiffman 1999), fire alone typically does not change dominance of these systems by nonnatives (Parsons and Stohlgren 1989, Klinger and Messer 2001).

Although many of the nonnative species in California grasslands are highly invasive (Heady 1988), this has not resulted in any single nonnative species having a particularly strong individual effect on fire behavior. For example, yellow star thistle (*Centaurea soltitialis*) is one of the most invasive species in California grasslands, but it has had relatively little effect on fire regimes. Fire behavior in California's grasslands is generally determined by the assemblage of herbaceous species, and there is little dispute that nonnative annual grasses comprise the most important fuel type in the system.

MEDITERRANEAN SHRUBLANDS

In contrast with California's grasslands, fire has certainly played a central role in the conversion of native-dominated shrublands into grasslands dominated by nonnative species (Keeley 2001). Type conversion of shrublands to grasslands is a major concern in southern California, especially in chaparral and coastal scrub ecosystems (Keeley 2001). These ecosystems tend to have a continuous horizontal and vertical arrangement of woody vegetation and far fewer nonnative species than surrounding grasslands (Keeley 2001). Two interrelated mechanisms have led to widespread conversion of chaparral and coastal scrub communities to grasslands in southern California: 1) an increase in fire frequency, especially ones started by humans after long periods of fire suppression and the subsequent buildup of fuel and 2) greater fragmentation of shrub patches as a result of urbanization.

Many shrub communities in California are adapted to a fire regime with a return interval of 20–50 years (Biswell 1999). If fire return intervals remain within this range, invasive nonnative plants can become established in the burned area, but it is difficult for them to persist. This is because most nonnative species that invade burned areas are herbaceous and not shade-tolerant. As the canopy closes in, the nonnative species that have become established are shaded out.

However, when return intervals decrease to just 1–15 years, the more frequent fires kill many shrub seedlings and saplings, and may stunt or even kill resprouting adults. The end result is that the shrub canopy cannot reestablish, and invasive nonnative herbaceous species (mainly annual grasses and forbs) become the community dominants. Once dense stands of grass develop, it becomes extremely difficult for woody and herbaceous native species to become established and regenerate (Eliason and Allen 1997). When a stand of chaparral or coastal scrub is converted to grassland, not only is the woody community lost but the community that replaces it is comprised primarily of nonnative species. Once this happens, nonnatives can alter soil properties enough to further reduce the chance of establishment of native species (Dickens and Allen 2014).

Although its magnitude varies among different regions of California, fragmentation of chaparral and coastal scrub ecosystems has generally been substantial (O'Leary 1995, Syphard et al. 2009). In many parts of California this has created a heterogeneous landscape comprised of a mix of shrubland, grassland, and woodland communities. Seed banks in this matrix often have high densities of invasive plants (Cox and Allen 2008), resulting in many sources of colonizing species. Mass effects in the first several years after burning can be very high (Keeley et al. 2005), a process that is especially important in high severity burns where invasive plants typically do not survive (Foin and Hektner 1986). The heterogeneous landscape likely results in significant patch dynamics that increases the extent of the propagule pressure.

There appears to be a negative relationship between the distance of a burned area from areas infested with nonnative species and rates of invasion into the burn (Giessow and Zedler 1996). This is important in areas fragmented by urban development, especially for smaller burns with a high edge to area ratio. Fragmented stands are often in close proximity to degraded plant communities with a high proportion of invasive species. When these fragments burn invasion rates into them from surrounding degraded vegetation can be very high (Allen 1998, Minnich and Dezzani 1998).

Fragmentation of plant communities has been exacerbated by management programs in which propagule density of nonnative herbaceous species is artificially increased above that of native species that regenerate naturally in the burned areas. The most common way for this to occur was seeding nonnative grasses and forbs across burned sites to prevent erosion in the first months and years following a fire (Beyers et al. 1998). These areas act as propagule sources for nonnative species, particularly where shrublands are already highly fragmented (Zedler 1995, Allen 1998). Moreover, many seeded areas continue to be dominated by nonnatives. Different studies have shown the practice is usually of little benefit and it is uncommon for it to now occur in chaparral and coastal sage scrub communities (Conard et al. 1995, Beyers et al. 1998, Peppin et al. 2011, Pyke et al. 2013) (chapter 22).

Fire management and suppression activities (e.g., fuel break construction, heavy equipment and vehicle use) can also promote invasions into burned shrublands. These activities create disturbances that promote establishment and spread of nonnatives (Stylinski and Allen 1999). They also create corridors or act as vectors of transport that invasive species can exploit (Giessow 1996, Merriam et al. 2006). These activities are essential components of fire management programs, but their role in spreading invasive plants and leading to undesirable outcomes is now recognized by fire management agencies. Thus, it is more commonplace now for agencies to plan and implement measures to reduce the degree to which these unintended invasions occur (Cal-IPC 2012).

DESERT SHRUBLANDS AND WOODLANDS

The invasion of cheat grass into the Great Basin Desert of eastern Oregon, Idaho, and Nevada and its effect on fire regimes has been well studied (see chapter 18, sidebar 18.1). Although relatively few of those studies have been conducted in the Great Basin regions of eastern California, there is little doubt that similar alteration in vegetation structure and fire regimes has resulted there. The storage and priority effects of cheat grass are now so strong that fire has only a relatively small and very short lived effect on its seedbank (Ellsworth and Kauffman 2013).

In contrast to the changes in fire regimes in the Great Basin resulting predominantly from cheat grass, changes in the Mojave Desert have resulted from invasion by red brome and, to a lesser extent, Mediterranean split-grass and cheat grass (chapter 18, sidebar 18.1). These species were introduced into the Mojave Desert through livestock grazing, military activities, or off-road vehicle use (Brooks and Pyke 2001) and have spread throughout the region. These invasive annual grasses produce fuel that is more persistent and continuous than fuel produced by native annuals (Brooks 1999), facilitating the spread of fire where fire was previously infrequent (Brooks and Pyke 2001, Brooks et al. 2016, Fig. 24.4).

Some field and greenhouse experiments have shown red brome has strong competitive effects on several native perennial species (DeFalco et al. 2003, 2007), but other studies have shown it may not exert as consistently strong effect across a broad spectrum of natives (Abella et al. 2011). Moreover, timing of germination (i.e., winter vs. spring) appears to be important in the magnitude of red brome's competitive advantage effect (DeFalco et al. 2007). The relation between variability in red brome's competitive ability and seasonality suggest that storage effects and, potentially, priority effects may be important mechanisms promoting its establishment and eventual dominance of some postfire communities in the Mojave. However, plant resources in arid systems are patchily distributed, so it may be that storage and priority effects are structured along resource gradients and result in strong patterns of species sorting. Red brome uses water more effectively early in the growing season than native perennial species (DeFalco et al. 2003), but it does not necessarily reduce native species ability to take up water (DeFalco et al. 2007). This implies that, following fire in areas of the Mojave that have not burned in many decades, red brome is most likely to reach high abundance in areas with low to moderate moisture levels. Indirect evidence indicates this may be occurring. Postfire succession in the Mojave Desert is highly variable in areas that have only burned once in the last 30 years (Engel and Abella 2011), and red brome's aboveground cover and seed bank density can be lower in burned than unburned areas in some of these cases (Abella 2009). Thus, while there is evidence that the amount of area burning in the Mojave is increasing and a grass-fire cycle may be playing a role (Brooks and Esque 2002, Brooks 2012) (chapter 18), these changes may be in specific regions (such as the mid elevation ecological zone) and may be occurring at particular times, such as the growing season following a particularly wet winter (Brooks and Matchett 2006).

Postfire seeding is used in desert ecosystems as in Mediterranean climate shrublands. It has been found to be of marginal or little benefit for promoting establishment of native species or controlling erosion, but in contrast with Mediterranean shrublands it is still used on a relatively widespread basis in sagebrush shrublands (Pyke et al. 2013, Knutson et al. 2014).

CONIFEROUS FORESTS

Fire regimes in montane conifer forest ecosystems have been drastically altered by fire suppression practices, but impacts from invasive species have not been as great in these systems

FIGURE 24.4 Burned/invaded creosotebush scrub (A) compared with unburned/uninvaded creosotebush scrub (B) in the Opal Mountain region of the central Mojave Desert in California. Nonnative annual grasses have rapidly invaded and now dominate the burned area.

as others and have had relatively little effects on their fire regimes (Keeley 2001, Klinger et al. 2006, Brooks et al. 2016). Most invasive species in montane conifer forests are herbaceous (Klinger et al. 2006), although there are concerns that some woody nonnative species present potentially severe threats as well. These include scotch broom (*Cytisus scoparius*), tamarisk (*Tamarix* spp.), Russian olive (*Elaeagnus angustifolia*), and tree-of-heaven (*Ailanthus altissima*) (Schwartz et al. 1996).

An important aspect of the relatively low rates of invasion into montane conifer forests is that there is a negative relationship between elevation and nonnative species richness (Schwartz et al. 1996, Randall et al. 1998, Keeley 2001, Klinger et al. 2006). Factors that may reduce the likelihood of establishment by invasive species in higher elevation plant communities include less activity by humans, relatively intact shrub and tree canopies, and relatively harsh climates and other physical conditions. It is telling that sites in montane conifer forests where invasive species have become established and are relatively abundant are often those that had previously been disturbed by human activities, such as livestock grazing, logging, mining, trails, and historic dwelling sites (Harrod and Reichard 2001).

Invasive species are more abundant in conifer forests where the canopy is broken (Keeley 2001, Keeley et al. 2003), which is probably because reduced light availability in stands with intact canopies prevents or minimizes establishment (Klinger et al. 2006). Elevation per se probably has little to do with the decreasing incidence of invasive species at higher elevations, but instead is correlated with physical factors that do affect establishment and growth of plants (reduced moisture, low temperatures, short growing seasons, etc.). Rejmánek (1989) pointed out that as a general rule more extreme environments, such as high-elevation conifer forests, may be less susceptible to invasion.

Historically there was concern that alteration of fire regimes in montane conifer forests could occur, primarily because of increasing invasion by annual species such as cheat grass and disturbance resulting from land use and fire suppression activities (Harrod and Reichard 2001). A study from ponderosa pine (*Pinus ponderosa*) forests in Arizona indicated that nonnative invasive species were changing postburn succession patterns (Crawford et al. 2001), but data from California were limited (Randall and Rejmanek 1993, Bossard and Rejmanek 1994, Schoennagel and Waller 1999). Over the last 10–15 years, there has been increasing evidence indicating that the level of invasive plant dominance in burned conifer stands depends on fire frequency and severity (Keeley 2001, Keeley et al. 2003, Franklin et al. 2006, Franklin 2010, Kaczynski et al. 2011). In Sequoia and Kings Canyon National Parks, Keeley et al. (2003) found that frequency and cover values of invasive nonnative species were low in virtually all burned conifer sites. Levels of invasion were greater in areas with higher severity fires, which was likely due to less competition from native species. In Yosemite National Park, Kaczynski et al. (2011) found that, in addition to fire severity, landscape position affected invasive occurrence. Nonnative species were detected on 73% of the high-severity lowland sites and on 50% of the moderate-severity lowland sites.

Based on an analysis of two different data sets from Yosemite National Park, Klinger et al. (2006) found that species richness and cover of nonnatives were relatively low in conifer forests and did not differ significantly between burned and unburned sites (Fig. 24.5). Klinger et al. (2006) also found that after an initial surge in the first 10 years after burning there was a negative relationship between richness and cover of nonnative species and time since burn (Fig. 24.6). This was consistent with the Keeley et al. (2003) study and indicated that even though nonnative plants invaded burned conifer forests they were shaded out as the canopy closed. This was because fire frequencies were low enough that the canopy had time to reestablish itself (Keeley et al. 2003). In cases where this does not happen then there can be rapid and heavy invasion of annual grass, which appears to be occurring in conifer forests in the Peninsular ranges (Franklin et al. 2006, Franklin 2010).

RIPARIAN ECOSYSTEMS

California's riparian systems have been heavily invaded by nonnative species. In terms of extent of distribution and impacts to native species and ecosystem properties the two taxa of greatest concern are giant reed (*Arundo donax*) and tamarisk. Both have invaded riparian areas throughout much of California, although giant reed tends to be more common and have greater impacts in coastal parts of the state (Bell 1997), while tamarisk has had greater impacts in the southeastern desert region (Ostoja et al. 2014) (sidebar 18.2). Both species are also present in the Central Valley and inner Coast Ranges, where they are known to have altered species composition as well as flooding regimes and other aspects of riparian

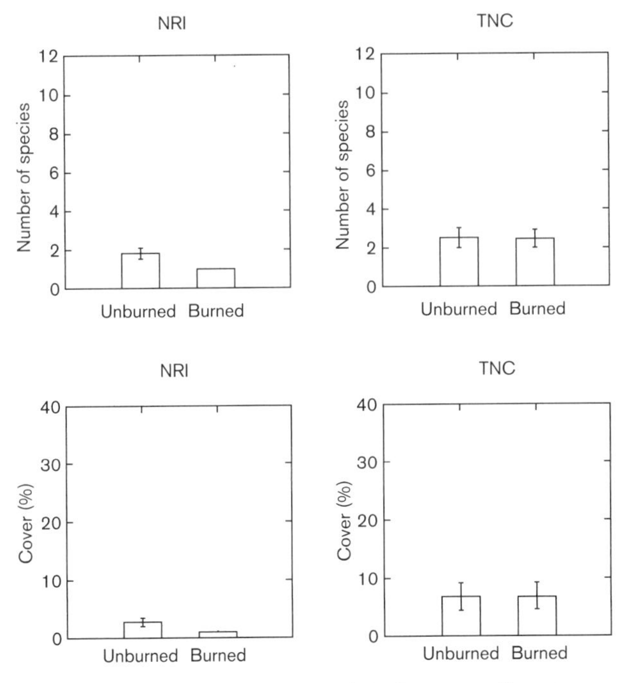

FIGURE 24.5 Mean species richness and absolute cover of invasive alien species in burned and unburned plots in Yosemite National Park. The data are from 356 0.1 ha (0.247 ac) plots sampled by the National Park Service in 1991–1993 (NRI) and 236 plots ranging in size from 0.05 ha to 0.1 ha (0.124 ac to 0.247 ac) sampled by The Nature Conservancy in 1998–1999 (TNC).

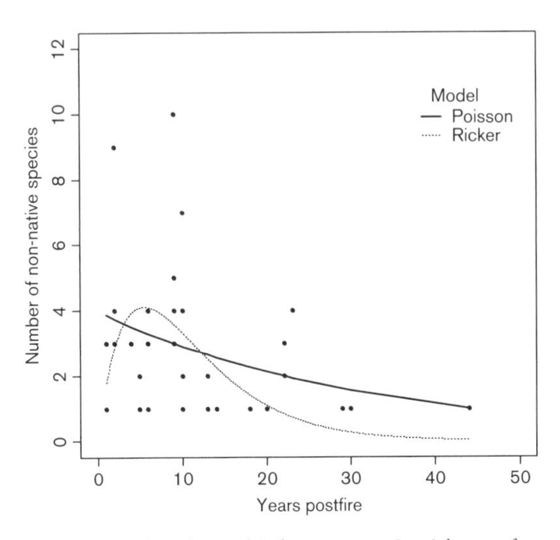

FIGURE 24.6 The relationship between species richness of nonnative plants and the number of years since burning in Yosemite National Park, California. The Poisson model is representative of low to moderate severity post-fire conditions where there is a surge of recruitment and/or colonization and establishment of non-native species in the initial years after burning, with a general linear decline afterwards. The Ricker model is representative of high severity conditions where there is low recruitment and establishment in the initial years after burning, followed by a surge and then a monotonic decline in later years. Data are from Klinger et al. (2006).

hydrology. However, both have also had some effect on fire regimes. Bell (1997) reported fires occurred more frequently and were more intense in areas where giant reed dominates riparian communities, and Coffman et al. (2010) described the strong response of giant reed to wildfire and how it promoted fire spread into adjacent shrublands. Busch (1995) concluded that riparian forests along the Colorado River in Arizona have been drastically changed because tamarisk resprouts vigorously after a burn, while most native trees (e.g., Fremont cottonwood, *Populus fremontii* subsp. *fremontii*) do not.

Fire and the Management of Invasive Species

Fire is often recommended as a way of managing invasive species (Randall et al. 1998, Bossard et al. 2000, DiTomaso and Johnson 2006, Pyke et al. 2010), and there are a number of examples where it has been used successfully for this purpose (Rice and Kapler Smith 2008). But there are also examples where the use of fire alone is not effective at decreasing the abundance of target invasive species or guilds, or where there is an increase in the abundance of other invasive species (see sidebar 24.1: case study from Santa Cruz Island). Germination of African lovegrass (*Eragrostis lehmanniana*) was enhanced following a prescribed burn in Arizona (Ruyle et al. 1988). Prescribed fire increased biomass and seed production of Dalmation toadflax (*Linaria dalmatica*) after a single spring burn in Montana (Jacobs and Shelley 2003). In the prairies of the mid-western United States the abundance of native herbaceous species often increases following fires, but so does the abundance of nonnative grasses and forbs (Howe 1995).

Such inconsistent patterns underscore the importance of the interactions between fire regimes and the natural history of different species. Invasive species show the same range of adaptations and responses to fire as do native species (Rowe 1981). The details of the fire regime in the system they invade may differ from those in their natural range, but the species general adaptations to fire may still allow them to persist or even thrive (Pyke et al. 2010). Consequently, it is reasonable to expect there to be a tremendous range of responses by individual species to burning (Fig. 24.7). These complex relations make broad management prescriptions for multiple species difficult to devise (Alba et al. 2015).

Fire and the Control of Invasive Herbaceous Species

Fire has been used to try and control or manage invasive species in California for over 50 years, with the most extensive use being in grasslands (Table 24.1). Nonnative annual grasses and forbs have flourished in California for at least 150 years (Mack 1989). These nonnatives suppress regeneration by native species through competitive interactions (Dyer and Rice 1997, Eliason and Allen 1997), primarily by development of a dense layer of "thatch" that reduces light and increases moisture (Heady 1988). Many nonnative annual grass and forb species can exploit moisture and do better under conditions of reduced light than native species (Evans and Young 1970, Dyer and Rice 1997). Therefore, prescribed burning has been recommended as a way of reducing competition intensity in grasslands by increasing the amount of light available to native species for germination and development of seeds and seedlings, and also a way of directly reducing the abundance of nonnative grasses and forbs (Dix 1960, Heady 1972, Menke 1992).

SIDEBAR 24.1 THE DILEMMA OF CONTROLLING INVASIVE SPECIES IN COMMUNITIES WITH MANY INVADERS: A CASE STUDY FROM SANTA CRUZ ISLAND

There are few ecosystems in California that have been impacted by just a single nonnative species. Most ecosystems have been invaded by multiple species, which presents many challenges for developing, implementing, and evaluating control programs. Although this issue has been receiving increasing attention recently (Kuebbing et al. 2013), it has been a concern for many years (Zavaleta et al. 2001, Klinger 2007). Organizations that manage parks, forests, refuges, and preserves have to carefully evaluate which invasive species pose the greatest threats, what the biological traits of the species are that make them both invasive and pose a threat, and how the species can potentially be controlled or, in some instances, eradicated. But perhaps the most critical aspect of controlling invasive species is what the outcomes of a control or eradication program are. Eradication and control programs are usually very expensive and long-term projects, and expectations about their outcomes can be quite high. But while expectations of success for these programs can be high, a number of different factors can affect their outcomes. Evaluating the success of a control program requires not just focusing on the target species, but analyzing how other species are affected as well. Some of the complexities involved in conducting and evaluating alien species control programs using fire can be illustrated from several studies conducted on Santa Cruz Island in the 1990s.

Like many islands throughout the world, the California Channel Islands off the coast of southern California have been subject to myriad impacts from accidentally and intentionally introduced nonnative plant and animal species. These impacts have been well described in a series of symposiums dealing with the biology and management of the islands, and include outright destruction of vegetation communities, dominance of some vegetation communities by nonnative species (e.g., grasslands, the understory of oak woodlands), species extinctions, altered hydrology, and extreme erosion. In addition to the impacts from nonnative species, other human activities have played a major role in the transformation of the islands' ecosystems. One of these activities has been fire suppression. None of the islands had any extensive burns from the 1850s up through the 1990s. Although the historic fire regimes on the islands are not well understood, many of the species on the islands show the same range of adaptations to fire that their counterparts on the mainland do (Carroll et al. 1993, Ostoja and Klinger 2000). So, presumably, fire once played an important role on the islands. Santa Cruz Island is the largest and most diverse of the Channel Islands, and is held in joint ownership and managed as an ecological preserve by The Nature Conservancy (TNC) and the National Park Service (NPS). There are 42 plants on Santa Cruz that are endemic to the Channel Islands, eight of which occur only on Santa Cruz itself (Junak et al. 1995). However, 25% of the flora on the island is nonnative (Junak et al. 1995). Grasslands are the most extensive vegetation community on the island, but this is likely a result of type conversion from coastal scrub and chaparral communities from overgrazing by feral sheep. Nonnative grasses and forbs can comprise more than 90% of the cover of grasslands and the understory of oak woodlands. The existing stands of chaparral, Bishop pine (*Pinus muricata*) forest, and coastal scrub communities are comprised predominantly of native species, but they have been severely fragmented from overgrazing and are surrounded by grasslands. Recruitment of some shrub and tree species was extremely low as a result of either pathogens or high seed and seedling mortality from feral pig rooting.

Because conditions on the island had been so drastically altered, an evaluation of the contemporary role of fire needed to be done as the first step in a more comprehensive fire management plan. A central question in this evaluation was how fire affected the abundance of nonnative species. Fire could potentially decrease abundance of some nonnative species, but it might also lead to an increase in others. A series of exploratory experiments was undertaken by TNC and NPS in the 1990s to try to answer two questions: (1) could prescribed burns be used to control the spread of fennel (*Foeniculum vulgare*) on the island? and (2) would prescribed burns in grassland reduce the species richness and cover of nonnative species and increase that of native species?

Fennel is considered one of the most invasive plants in California (CALEPPC 1999). Following the removal of cattle in 1988 and the end of a drought in 1991, the distribution and density of fennel on Santa Cruz increased dramatically. A series of three experiments from 1991 to 2001 was conducted to determine an effective method for controlling fennel, and to evaluate the effects on groups of species other than fennel as a result of the control methods (Brenton and Klinger 1994, 2002). One of these experiments tested whether fire alone or in combination with an herbicide (Triclopyr®) reduced fennel cover (Ogden and Rejmánek 2005). Fire alone was ineffective at reducing fennel cover and was no more effective at reducing fennel cover when combined with herbicide spraying than spraying alone was. However, in areas where fennel cover was reduced alien grasses and forbs dominated the community. Native species richness and cover increased in the burned and sprayed plots, but they comprised less than 10% of the mean cover and less than 15% of the species (Ogden and Rejmánek 2005).

Two experiments that differed greatly in scale were conducted in grasslands on the island from 1993 to 2000. Although the experiments each had their own specific objectives, a common objective was to determine whether fire alone was effective at reducing abundance of nonnative species and increasing that of natives (Klinger and Messer 2001). A large-scale experiment (1993–1998) involved burning three different areas (270–490 ha [667–1,211 ac]) once in the fall in separate years (1993–1995) and then monitoring vegetation response in each area for three years post-burn. The response of alien species to the burns was generally consistent; there was an initial decrease in cover of annual grasses but an increase in that of annual forbs. However, the response of native species in each area varied. In one area there was an initial increase in native forbs, in another area there was no change, and in the third area they decreased. Of most interest was that the response of both alien and native species varied by topography (primarily aspect) as well as rainfall patterns (Klinger and Messer 2001). Burning had a greater effect on species composition than topography in the first year post-burn, but it declined in importance each year after.

The second grassland burning experiment (1996–2000) was much smaller in scale (burn plots were 50 m^2 [538 ft^2]), but rather than using just a single fall burn, both season (spring and fall) and frequency (0 to 3 burns in consecutive years) of burning were manipulated. The experimental plots were arranged along an aspect gradient, so that the interaction between burn season, frequency, and topography could be analyzed. There was strong variation in species composition from year to year, but this was due mainly to variation in interannual rainfall patterns. There were some differences in species composition as a result of the fire treatments within any given year, but aspect explained much more of this variation than did burning. As in the larger-scale grassland burn study, nonnative grasses and forbs dominated both species composition and cover in all years.

These three studies underscore the challenges of using fire to try to control nonnative species in heavily invaded systems. Fire alone was not effective at controlling fennel, and did not substantially reduce fennel cover more when used in conjunction with spraying than did spraying alone. The best justification for using fire with herbicide spraying was that this was the only treatment where native species showed a significant increase in species richness and cover. Nevertheless, reduction in fennel cover was compensated for primarily by nonnative annual grasses and forbs. In both grassland studies, factors other than burning had a more pronounced effect on variation in species composition than burning did; fire effects were transient and varied from year to year or site to site, and nonnative species continued to dominate the communities. In all three studies, responses to fire were species specific, so that reduction in cover of one species was compensated for by an increase in cover by another.

Are the outcomes from these studies unique to Santa Cruz Island? It appears that they are not, and may actually be quite representative of broader patterns throughout the state. Two meta-analyses, one focused on studies conducted in California (D'Antonio et al. 2006) and another primarily on studies in the United States and Australia (Alba et al. 2015) found that fire generally has little long-term effect on nonnative species. The meta-analysis for studies done in California found that nonnative annual grasses are often reduced the first year after burning but by the second or third year after a burn their cover has returned to pre-burn levels. The reduction in cover of annual grasses is compensated by an increase in nonnative forbs, although this may vary depending on whether a site is burned repeatedly and/or grazed. Native plant species, especially forbs, increase in the first year after burning in ungrazed sites and in the second year after burning in grazed sites. Season of burning had no significant effect on postfire response for native or nonnative taxa. Precipitation did influence response of different taxa, although the magnitude of this effect and how different guilds responded to precipitation were highly variable.

Although these outcomes do not necessarily argue against using fire as a management tool for nonnative species, they do highlight the need to set appropriate management expectations and goals and to recognize the complexity inherent in the systems being managed. They also show that development of control methods that are part of a more comprehensive restoration program will require a long-term investment in research before full-scale implementation is initiated over large areas. Even then, results are likely to vary from year to year at any given site because of factors that are beyond the control of the managing organization.

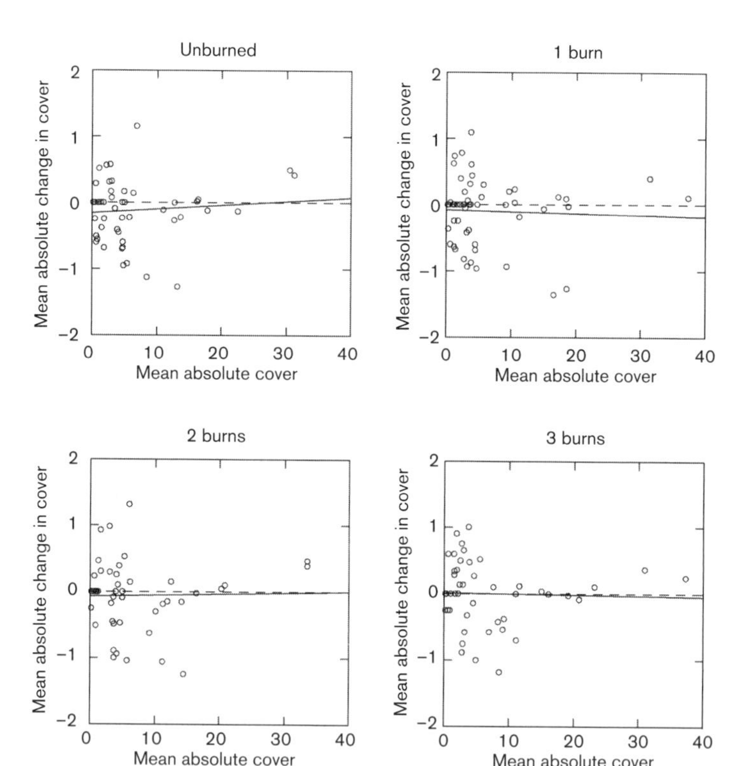

FIGURE 24.7 The relative change in abundance of nonnative herbaceous species (individual points) in control (unburned) and treatment (burn frequency) conditions in grasslands on Santa Cruz Island, California, 1996–2000. Each point represents the mean annual change in cover over a five year period for a given species. Note the flat regression lines, indicating that regardless of how frequently an area was burned the number of species that increased was about the same as those that decreased. These results indicate the highly individualistic response of species, and that management prescriptions for collectively trying to control guilds of species (e.g., nonnative annual grasses) have a low likelihood of succeeding.

Studies on managing invasive species with fire in California's grasslands have varied broadly in their spatial scale, site characteristics, site history (e.g., land use patterns), and climate conditions (Table 24.2). Their objectives have also been quite varied, and have included range improvement (Hervey 1949), increasing diversity of native species (Dyer and Rice 1999, Hatch et al. 1999), reduction of populations of particular nonnative species (Pollak and Kan 1998, DiTomaso et al. 1999, Hopkinson et al. 1999), and community-level decreases in abundance of nonnative species and increased abundance of natives (Parsons and Stohlgren 1989, Meyer and Schiffman 1999, Wills 2000, Klinger and Messer 2001, Marty 2002).

Despite the varying objectives, some generalities can be drawn from the studies. First, timing and frequency of burning are critical factors for reducing the abundance of individual nonnative species. DiTomaso et al. (2001) found that barbed goat grass (*Aegilops triuncialis*) could be controlled with two late spring or early summer burns in consecutive years, while Hopkinson et al. (1999) reported that single summer burns were effective at reducing the cover but not the spread of goat grass. Pollak and Kan (1998) reported that medusa head (*Elymus caput-medusae*) was virtually eliminated the first growing season after a spring burn in a vernal pool/grassland system. DiTomaso et al. (1999) reported a 91% reduction in cover of yellow starthistle following three consecutive summer burns.

Second, study duration is critical to interpreting postburn outcomes. For example, despite their initial encouraging results (DiTomaso et al. 1999), Kyser and DiTomaso (2002) found that within four years of the cessation of burning to control yellow starthistle the seedbank, seedling density, and cover of yellow starthistle rose rapidly. In studies focused on community-level changes, a short term (≤2 years) decrease in cover (or biomass) of annual grasses and short-term increase (1–2 years) in annual forbs usually occurs beginning in the first, and sometimes the second, growing season after the fire (but see Marty 2002). However, nonnative species often continue to dominate burned sites, and within 2–3 years burn effects are largely gone (Parsons and Stohlgren 1989, Klinger and Messer 2001), a pattern that has been recognized for many decades (Hervey 1949, Heady 1972). The mean duration of studies on fire as means of controlling invasive species in California has been approximately two years (Table 24.2), so conclusions based on studies of such short duration are likely misleading.

Third, variation in the spatial extent of studies likely has an important influence on interpreting postfire responses by invasive plants. When the extent increases environmental variability and factors interacting with fire can have as, or even more, important effects on both nonnative and native species than fire alone. For example, Kyser et al. (2008) found that the success of controlling medusa head with fire varied substantially across different geographic regions of the state. Harrison et al. (2003) reported species richness of nonnative plants in a burned grassland community in the Central Coast Range to be greater on more productive soils. Klinger and Messer (2001) found that species composition in burned grasslands on Santa Cruz Island was determined largely by an interaction between rainfall, topography (aspect and elevation), and fire, and that native species were suppressed during years of high rainfall, apparently because of increased cover of nonnative grass and forbs in wetter years. Hatch et al. (1999) also found that fire effects were modified by topography, and that some native species showed a positive response to burning while others did not. Pollak and Kan (1998) reported different responses by both native and nonnative species depending on variation in topography.

Fourth, the most commonly manipulated component of fire regimes has been burn season (Table 24.2). It has often been assumed spring is the most effective season to conduct prescribed burns in grasslands because there would be much higher mortality of invasive species seed crops (Menke 1992, Dyer 2002). Indeed, there is evidence spring burns can lead to

TABLE 24.1

Examples of studies on the use of fire to control invasive species and increase abundance or diversity of native species in grasslands in California

Target is whether the goal of burning was to alter abundance of a particular species or produce community-level changes in species composition

	Target	Native species response	Alien species response
Hervey 1949	Community	None reported	Decrease in annual grass and increase in annual forbs
Furbush 1953	*Elymus capat-medusae*	None reported	*E. capat-medusae* nearly eliminated, but other annual herbaceous species dominated postfire
Parsons and Stohlgren 1989	Community	Short-term increase in forbs; no change in grass biomass	Short-term decrease in grass; increase in forb biomass
Ahmed 1983	*Stipa pulchra*	Results varied among response variables, but generally positive response	Burning reduced cover and biomass of alien grass, primarily *Bromus hordeaceus*
Dyer et al. 1996	*Stipa pulchra*	Not measured	
Dyer and Rice 1997	*Stipa pulchra*	Short-term increase in size; depended on reduced competition	Not measured
Pollak and Kan 1998	*Elymus capat-medusae* Community	"Early" annual forbs increased	*E. capat-medusae* reduced; annual forbs increased and dominated cover
DiTomaso et al. 1999	*Centaurea soltitialis*	Short-term increase in forbs and grass in some areas	No change in cover of grasses and forbs
Hatch et al. 1999	*Danthonia californica Stipa lepida Stipa pulchra*	No change in *Danthonia* or *S. lepida*, *S. pulchra* response varied unpredictably with topographic position	Not measured

greater reduction of nonnative species and greater increases of native species than fall burns (DiTomaso et al. 1999, Meyer and Schiffman 1999, Wills 2000). Other studies though that have used fall burns have shown significant decreases in cover of nonnative species and significant increases in species richness and cover of native species, albeit over short-term periods (Parsons and Stohlgren 1989, Klinger and Messer 2001). Moreover, very few studies have manipulated both season and frequency of burn (e.g., Parsons and Stohlgren 1989), and none have manipulated fire return interval. Other aspects of fire regimes that are difficult or virtually impossible to manipulate but can be measured (e.g., fire line intensity, rate of spread) have rarely been quantified.

The only study where burning led to an unequivocal and relatively long-term positive response by native species has been on the Santa Rosa Plateau in eastern riverside county (Wills 2000). Following a long-term program of spring burning, the percent cover of native species (bunchgrasses and forbs) was roughly equal or even greater than that of nonnative species. What is significant about this study is that the initial (pretreatment) cover of native species on the Santa Rosa Plateau was far greater than in other studies. Similarly, Harrison et al. (2003) reported that the dominant groups of species prior to burning were the ones that tended to be most abundant after a burn, a phenomena they called the ". . .rich getting richer and the poor getting poorer."

Collectively, the implications of these findings is that fire will likely be most effective as a means of maintaining areas that already have a significant component of native species. In other areas where native species are not particularly abundant, fire will probably have to be integrated with other techniques (e.g., seeding, planting, etc.) if the goal is enhancement of native species diversity. Aside from Wills' (2000) study, there is little evidence grasslands in California can be restored to significant levels of native species cover simply by burning (D'Antonio et al. 2006). These results are not surprising because most invasive nonnative species in California grasslands: 1) evolved in a fire-prone environment with a Mediterranean climate, 2) are annuals and can respond rapidly to fire, and 3), what is likely the most important factor, nonnative herbaceous species dominate the above-ground cover of most grasslands in California. This last factor likely means they dominate the seed bank as well as above ground cover (Major and Pyott 1966, Maranon and Bartolome 1989), and suggests that storage effects and priority effects are likely to tilt in favor of nonnative species more so than natives.

Fire and the Control of Invasive Woody Species

Management of woody invasive species in California with fire has been used less frequently than with herbaceous species,

TABLE 24.2

Location, duration, and components of fire regimes in selected studies on the use of fire to control invasive species and increase abundance or diversity of native species in grasslands in California

Duration is the length of the study, Season is the temporal divisions of the year (characterized by weather and daylight) that burns were conducted, and Frequency is how often plots were burned, with interval being the amount of time between repeated burns

Reference	Location	Duration	Season	Frequency
Hervey 1949	Berkeley foothills (Alameda County)	1 year	Summer	1
Furbush 1953	Coast Range (Mendocino County)	2 years	Summer	1
Parsons and Stohlgren 1989	Sierra foothills (Sequoia National Park)	4 years	Spring and fall	1–3 years with 1 year interval
Dyer et al. 1996	Central Valley grassland (Solano County)	2 years	Late spring	1
Dyer and Rice 1997	Central Valley grassland (Solano County)	4 years	Late spring	1
Pollak and Kan 1998	Central Valley grassland (Solano County)	1 year	Late spring	1
DiTomaso et al. 1999	Coast Range (Sonoma County)	3 years	Late spring	3 with 1 year interval
Hatch et al. 1999	Coastal grassland (San Mateo County)	1 year	Fall	1
Hopkinson et al. 1999	Coast Range (Contra Costa County)	3 years (effectively 2)	Summer	2 with 3 years interval
Meyer and Schiffman 1999	Carrizo Plain (San Luis Obispo County)	1 year	Late spring, fall, winter	1
DiTomaso et al. 2001	Coast Range (Mendocino County)	2 years	Late spring	1
Klinger and Messer 2001	Santa Cruz Island (Santa Barbara County)	5 years	Fall	1
Marty 2002	Sierra foothills (Yuba County)	2 years	Summer	1
Alexander and D'Antonio 2003	Coastal grassland (Marin County)	1 year	Not specified	1–3 years (interval not specified)

primarily because most invasive plants in California are herbaceous (Randall et al. 1998). Accordingly, the effect of fire on most woody invasive species in the state is largely unstudied. Burning does not seem to be effective at controlling tamarisk (Busch 1995, sidebar 18.2), but some studies indicate burning can reduce density and/or the seedbank of Scotch broom and French broom (*Genista monspessulana*) (Roja and Popenoe 1998, Odion and Hausenback 2002, Alexander and D'Antonio 2003). Using fire to control brooms is complicated by nonnative annual grasses rapidly becoming established in the burned areas (Swezey and Odion 1997).

As with herbaceous species, the timing and frequency of burning are important factors for controlling woody invasive species as well as enhancing habitat for native species (DiTomaso and Johnson 2006, Pyke et al. 2013). However, one particular type of prescribed burn that is effective at controlling spread or abundance of woody invasive species may not increase richness or abundance of native species. Alexander and D'Antonio (2003) reported that multiple burns done in consecutive years did not contribute to further reduction of the seed bank of French broom than a single burn. However, the seed bank of native forbs increased and that of nonnative grasses decreased when burns were done in three consecutive years (Alexander and D'Antonio 2003).

Many woody invasive species either have biological traits or occur in areas that make them inappropriate candidates for burning programs. A particularly good example of this is Tasmanian blue gum (*Eucalyptus globules*), which is rated as one of the most invasive species in California (Cal-IPC 2006). While there is conservation justification for controlling Tasmanian blue gum, the species has certain biological traits that complicate management options (Boyd 1997). Tasmanian blue gum evolved in fire-prone environments in Australia, and the bark on adult trees is extremely thick and resists burning in all but the hottest fires. The leaves have a high content of volatile oils, and fire intensity in the stands is extremely intense (Sapsis et al. 1995, Cole 2002). Seeds of Tasmanian blue gum have a tough coat that resists burning, and annual seed production is prolific. Although Tasmanian blue gum seeds don't require fire to germinate, germination rates after even intense burns are very high (R. Klinger personal observation). In addition, Tasmanian blue gum frequently occurs near urban areas. If fire were intense enough to ignite adults in mature stands, the extreme fire behavior would make control tenuous and present a severe hazard to human life and property.

Summary: Fire and the Control of Invasive Species

The complex and often unpredictable outcomes of trying to control invasive species with fire do not imply that using fire for this purpose is inappropriate. On the contrary, fire is often a useful tool for managing invasive species while they are still only locally abundant and before they begin to spread, dominate, and alter ecological systems (Brooks and Klinger 2009). However, in most instances resource managers are faced with managing sites that have already been invaded. In some cases, carefully timed fires will be effective at reducing abundance of certain nonnative species. But the outcomes from burning will become less predictable in sites that have been invaded by many nonnative species (Kuebbing et al. 2013), especially when the management goal is to produce wholesale shifts in species composition in an entire community (see sidebar 24.1). A number of factors will modify burn effects and variation in outcomes between sites and between years within a site are likely (Corbin et al. 2004).

Selecting where, when, and how fire is used is critical to the success of an invasive species management program. In many instances, the chances of achieving management goals may be increased by integrating burning with other control practices (Gillespie and Allen 2004, DiTomaso et al. 2006, Pyke et al. 2013). In fact, an investment in a focused and comprehensive research and monitoring program will, more times than not, prevent investment in expensive, ineffective, and/or possibly even counterproductive management programs (DiTomaso and Johnson 2006).

Synthesis and Conclusions: Fire and Invasive Species in an Era of Global Change

Ultimately, the reason to prevent and control biological invasions is to maintain ecosystem properties that humans find desirable. Because invasive species can rapidly alter these properties, control (or eradication) is really only a step in a long-term management process that focuses on maintaining or restoring conditions beneficial to native species (Randall et al. 1997). This means that controlling invasive species is not an end in itself, but must be evaluated relative to its contribution to maintaining or helping to restore desired ecosystem properties. However, this approach will be complicated because so many of our desired future conditions will be moving targets. Climate shifts, nitrogen deposition, invasive species, and other forces are interacting with human demography, economic practices, and transportation systems so that some ecosystems are being transformed to such a degree that their structure, species composition, and functional processes have no analogs in recent times (Suding et al. 2004). This presents great management challenges (Seastedt et al. 2008), and fire is likely to be among the most significant of these (Dennison et al. 2014, Rocca et al. 2014).

A range of forces other than climate can change fire regimes in short periods of time and result in alternative states, many of which can be dominated by invasive plants species (Pausus and Keeley 2014). Nevertheless, climatic change is generally considered to be the most likely force that, directly or indirectly, will bring about major changes in fire regimes in the coming century (Rocca et al. 2014). In the context of invasive plants, fire, and climate, a great deal of focus has been put on how warming might improve environmental conditions for some nonnative species and result in extensions of their range and/or increased abundance (Peterson 2003, Bradley 2009, Dukes et al. 2011, Concilio et al. 2013). This is especially so for *transformer* species (Richardson et al. 2000) such as annual grasses, which could exacerbate already existing problems or lead to new ones in ecosystems that currently have low impacts from nonnative plants (Hellman et al. 2008, Abatzoglou and Kolden 2011, Taylor et al. 2014).

Changes in range and/or abundance of animal species could also lead to changes in distribution and/or abundance of nonnative plant species. Dispersal of animal species into new areas can lead to plant-animal interactions (e.g., seed predation and seed dispersal) that increase rates of establishment of invasive plants into burned areas (Rost et al. 2012). Alternatively, herbivory can release existing populations of nonnative species from competition and result in an increase in their abundance (Johnson and Cushman 2007). Not only could these processes lead to changes in the structure and composition of species assemblages at higher trophic levels (Ostoja and Schupp 2009, Freeman et al. 2014), but also facilitate shifts in fire regimes because of altered fuel loads and ignition probabilities.

Global change, though, does not necessarily mean that invasive plants will be the main force transforming ecosystems. Rather, they often will be utilizing favorable conditions brought about by other processes (MacDougall and Turkington 2005). For instance, humans have altered atmospheric and geochemical processes to such a degree that some factors, such as nitrogen deposition (Fenn et al. 2003), are thought to be facilitating invasions and bringing about changes in fire regimes in the southern coastal scrub and western Mojave Desert regions (Rao and Allen 2009). However, other lines of evidence indicate fire is necessary for invasive plants to establish even in areas where deposition rates of nitrogen are high (Keeley et al. 2005, Keeley et al. 2011). It appears then that nitrogen deposition is not allowing invasive plants to transform fire regimes. Rather, nitrogen might be facilitating increased invasion rates in areas that have already burned.

How ecologists, land managers, and natural resource management agencies react to biological invasions is conditioned to a great extent on human and institutional values and perceptions (Coates 2006). Biological invasions have effects on ecosystem properties, of which altered fire regimes are only one. As we have seen throughout this text, fire can be used as a management tool for many different reasons. It is not necessarily applied to solely achieve ecological or conservation goals and the potential proliferation of invasive species may be a secondary concern in some, or even many, fire management programs. Even in fire management programs where controlling invasive species is important, conflicts with other management programs may exist. Examples include management of threatened or endangered species (Germano et al. 2001) or bio-control programs (Briese 1996, Fellows and Newton 1999).

It is difficult to predict what our attitudes towards invasive species will be in the coming century. They are likely to be modified as we learn more about how they interact within natural systems, some of which may in some cases be desirable or at least benign (Pec and Carlton 2014, Roy et al. 2014). Nevertheless, the transformation that many systems have undergone are largely irreversible (Yelenik and D'Antonio 2013) and it must be accepted that those invasive species now established as a dominant component of an ecosystem will be difficult to control or, even less likely, eradicate. Setting targets for restoration and management programs such as returning a system to a state similar to that prior to European settlement will, in most cases, likely not be successful. Historical information is useful, but primarily as a measure of what the range of variability in an ecosystem was and not as a specific target; the changes that are now occurring globally require that we look forward much more than we look back.

References

Abatzoglou, J.T., and C.A. Kolden. 2011. Climate change in western US deserts: potential for increased wildfire and invasive annual grasses. Rangeland Ecology and Management 64: 471–478.

Abella, S.R. 2009. Post-fire plant recovery in the Mojave and Sonoran deserts of western North America. Journal of Arid Environments 73: 699–703.

Abella, S.R., D.J. Craig, L.P. Chiquoine, K.A. Prengaman, S.M. Schmid, and T.M. Embrey. 2011. Relationships of native desert plants with red brome (*Bromus rubens*): toward identifying invasion-reducing species. Invasive Plant Science and Management 4: 115–124.

Ahmed, E.O. 1983. Fire Ecology of *Stipa pulchra* in California Annual Grassland. Dissertation, University of California, Davis, California, USA.

Alba, C., H. Skalova, K.F. McGregor, C.M. D'Antonio, and P. Pysek. 2015. Native and exotic plant species respond differently to wildfire and prescribed fire as revealed by metaanalysis. Journal of Vegetation Science 26: 122–129.

Alexander, J.M., and C.M. D'Antonio. 2003. Seed bank dynamics of French Broom in Coastal California grasslands: effects of stand age and prescribed burning on control and restoration. Restoration Ecology 11: 185–197.

Allen, E.B. 1998. Restoring habitats to prevent exotics. Pages 41–44 in: M. Kelly, E. Wagner, and P. Warner, editors. Proceedings California Exotic Pest Plant Council Symposium Volume 4. CALEPPC, Ontario, California, USA.

Angert, A.L., T.E. Huxman, G.A. Barron-Gafford, K.L. Gerst, and D.L. Venable. 2007. Linking growth strategies to long-term population dynamics in a guild of desert annuals. Journal of Ecology 95: 321–331.

Angert, A.L., T.E. Huxman, P.L. Chesson, and D.L. Venable. 2009. Functional tradeoffs determine species coexistence via the storage effect. Proceedings of the National Academy of Science 106: 11641–11645.

Bartolome, J.W., and B. Gemmil. 1981. The ecological status of *Stipa pulchra* (Poaceae) in California. Modroño 28: 172–184.

Bell, G.P. 1997. Ecology and management of *Arundo donax* and approaches to riparian habitat restoration in southern California. Pages 103–113 in: J.H. Brock, editor. Plant Invasions: Studies from North America and Europe. Blackhuys Publishers, Leiden, The Netherlands.

Beyers, J.L., C.D. Wakeman, P.M. Wohlgemuth, and S.G. Conard. 1998. Effects of postfire grass seeding on native vegetation in southern California chaparral. Proceedings of the Annual Forest Vegetation Management Conference 19: 52–64.

Biswell, H.H. 1999. Prescribed burning in california wildlands vegetation management. 2nd ed. University of California Press, Berkeley, California, USA.

Bossard, C., J.M. Randall, and M.C. Hoshovsky. 2000. Invasive Plants of California's Wildlands. University of California Press, Berkeley, California, USA.

Bossard, C., and M. Rejmanek. 1994. Herbivory, growth, seed production, and resprouting of an exotic invasive shrub *Cytisus scoparius*. Biological Conservation 67: 193–200.

Boyd, D. 1997. Eucalyptus removal on Angel Island. Pages 73–75 in: M. Kelly, E. Wagner, and P. Warner, editors. Proceedings California Exotic Pest Plant Council Symposium Volume 3, Concord, California, USA.

Bradley, B.A. 2009. Regional analysis of the impacts of climate change on cheatgrass invasion shows potential risk and opportunity. Global Change Biology 15: 196–208.

Brenton, B., and R.C. Klinger. 1994. Modeling the expansion and control of fennel *Foeniculum vulgare* on the Channel Islands. Pages 497–504 in: W.L. Halvorson and G.J. Maender, editors. The Fourth California Islands Symposium: Update on the Status of Resources. Santa Barbara Museum of Natural History, Santa Barbara, California, USA.

———. 2002. Factors influencing the control of fennel *Foeniculum vulgare* using Triclopyr on Santa Cruz Island. Natural Areas Journal 22: 135–147.

Brewer, W.H., H.S. Watson, and A. Gray. 1876. Polypetalae. In: W.H. Brewer, S. Watson, and A. Gray. Geological Survey of California. Botany, Volume I. First Edition. Welch, Beiglow and Company, Bosyton, Massachusetts, USA.

Briese, D.T. 1996. Biological control of weeds and fire management in protected natural areas: are they compatible strategies? Biological Conservation 77: 135–141.

Brooks, M.L. 1999. Alien annual grasses and fire in the Mojave desert. Madroño 46: 13–19.

———. 2008. Plant invasions and fire regimes. Pages 33–46 in: K. Zouhar, J. Kapler Smith, S. Sutherland, and M.L. Brooks, editors. Wildland Fire in Ecosystems: Fire and Nonnative Invasive Plants. USDA Forest Service General Technical Report RMRS-GTR-42-VOL-6. Rocky Mountain Research Station, Fort Collins, Colorado, USA.

———. 2012. Effects of high fire frequency in creosote bush scrub vegetation of the Mojave Desert. International Journal of Wildland Fire 21: 61–68.

Brooks, M.L., C.S. Brown, J.C. Chambers, C.M. D'Antonio, J.E. Keeley, and J. Belnap. 2016. Exotic annual Bromus invasions: comparisons among species and ecoregions in the western United States. Pages 11–60 in: M.J. Germino, J.C. Chambers, and C.S. Brown, editors. Exotic Brome Grasses in Arid and Semi-arid Ecosystems of the Western US: Causes, Consequences, and Management Implications. Springer International Publishing, Cham, Switzerland.

Brooks, M.L., and J.C. Chambers. 2011. Resistance to invasion and resilience to fire in desert shrublands of North America. Rangeland Ecology and Management 64: 431–429.

Brooks, M.L., C.M. D'Antonio, D.M. Richardson, J.B. Grace, J.E. Keeley, J.M. DiTomaso, R.J. Hobbs, M. Pellant, and D.A. Pyke. 2004. Effects of invasive alien plants on fire regimes. BioScience 54: 677–688.

Brooks, M.L., and T.C. Esque. 2002. Alien plants and fire in desert tortoise (*Gopherus agassizii*) habitat of the Mojave and Colorado Deserts. Chelonian Conservation and Biology 4: 330–340.

Brooks, M.L., and R.C. Klinger. 2009. Practical considerations for early detection monitoring of plant invasions. Pages 9–33 in: Inderjit, editor. Management of Non-Native Invasive Plant Species. Springer, New York, New York, USA.

Brooks, M.L., and J.R. Matchett. 2006. Spatial and temporal patterns of wildfires in the Mojave Desert, 1980-2004. Journal of Arid Environments 67: 148–164.

Brooks, M.L., and D.A. Pyke. 2001. Invasive plants and fire in the deserts of North America. Pages 1–14 in: K.E.M. Galley and T.P. Wilson, editors. Proceedings of the Invasive Species Workshop: The Role of Fire in the Control and Spread of Invasive Species. Tall Timbers Research Station, Miscellaneous Publication 11. Tallahassee, Florida, USA.

Brown, C.S., and K.J. Rice. 2000. The mark of zorro: effects of the exotic annual grass *Vulpia myuros* on California native perennial grasses. Restoration Ecology 8: 10–17.

Bruce, K.A., G.N. Cameron, and P.A. Harcombe. 1995. Initiation of a new woodland type on the Texas Coastal Prairie by the Chinese tallow tree (*Sapium sebiferum* (L.) Roxb.). Bulletin of the Torrey Botanical Club 122: 215–225.

Busch, D.E. 1995. Effects of fire on southwestern riparian plant community structure. Southwestern Naturalist 40: 259–267.

CALEPPC. 1999. The CalEPPC List: Exotic Pest Plants of Greatest Ecological Concern California, October, 1999. California Exotic Pest Plant Council (CALEPPC), San Juan Capistrano, California, USA.

Cal-IPC. 2012. Preventing the spread of invasive plants: best management practices for land managers, 3rd edition. Cal-IPC Publication 2012-03. California Invasive Plant Council, Berkeley, California, USA.

———. 2006. California Invasive Plant Inventory; February 2006. California Exotic Pest Plant Council (CALEPPC), Sacramento, California, USA.

Carroll, M.C., L.L. Laughrin, and A. Bromfield. 1993. Fire on the California islands: does it play a role in chaparral and closed cone pine forest habitats. Pages 73–88 in: F.G. Hochberg, editor. The Third California Islands Symposium: Recent Advances in Research on the California Islands. Santa Barbara Museum of Natural History, Santa Barbara, California, USA.
Chesson, P.L., and N. Huntly. 1988. Community consequences of life-history traits in a variable environment. Annales Zoologici Fennici 25: 5–16.
Coates, P. 2006. American Perceptions of Immigrant and Invasive Species: Strangers on the Land. University of California Press, Berkeley, California, USA.
Coffman, G.C., R.F. Ambrose, and P.W. Rundel. 2010. Wildfire promotes dominance of invasive giant reed (*Arundo donax*) in riparian ecosystems. Biological Invasions 12: 2723–2734.
Cole, D. 2002. California's urbanizing wildlands and the "Fire of the Future". Pages 231–238 in: N.G. Sugihara, M. Morales, and T. Morales, editors. Proceedings of the Symposium: Fire in California Ecosystems: Integrating Ecology, Prevention, and Management. Association for Fire Ecology Miscellaneaous Publication 1. Davis, California, USA.
Conard, S.G., J.L. Beyers, and P.M. Wohlgemuth. 1995. Impacts of postfire grass seeding on chaparral systems C what do we know and where do we go from here? Pages 149–161 in: J.E. Keeley and T. Scott, editors. Wildfires in California Brushlands: Ecology and Resource Management. International Association of Wildland Fire, Fairfield, Washington, USA.
Concilio, A.L., M.E. Loik, and J. Belnap. 2013. Global change effects on *Bromus tectorum* L. (Poaceae) at its high-elevation range margin. Global Change Biology 19: 161–172.
Corbin, J.D., and C.M. D'Antonio. 2004. Competition between native perennial and exotic annual grasses: implications for an historical invasion. Ecology 85: 1273–1283.
Corbin, J.C., C.M. D'Antonio, and S.J. Bainbridge. 2004. Tipping the balance in the restoration of native plants: experimental approaches to changing the exotic: native ratio in California grasslands. Pages 154–179 in: M. Gordon and S. Bartol, editors. Experimental Approaches to Conservation Biology. University of California Press, Los Angeles, California, USA.
Cowling, R.M., I.A.W. MacDonald, and M.T. Simmons. 1996. The Cape Peninsula, South Africa: physiographical, biological and historical background to an extraordinary hot-spot of biodiversity. Biodiversity and Conservation 5: 527–550.
Cox, R.D., and E.B. Allen. 2008. Composition of soil seed banks in southern California coastal sage scrub and adjacent exotic grassland. Plant Ecology 198: 37–46.
Crawford, J.A., C.H.A. Wahren, S. Kyle, and W.H. Moir. 2001. Responses of exotic plant species to fires in *Pinus ponderosa* forests in northern Arizona. Journal of Vegetation Science 12: 261–268.
D'Antonio, C.M. 2000. Fire, plant invasions, and global changes. Pages 65–93 in: H.A. Mooney and R.J. Hobbs, editors. Invasive Species in a Changing World. Island Press, Washington, D.C., USA.
D'Antonio, C.M., S.J. Bainbridge, C. Kennedy, J.W. Bartolome, and S. Reynolds. 2006. Ecology and restoration of California grasslands with special emphasis on the influence of fire and grazing on native grassland species. A Report to the David and Lucille Packard Foundation. University of California, Santa Barbara, California, USA.
D'Antonio, C.M., J.T. Tunison, and R.K. Loh. 2000. Variation in the impact of exotic grasses on native plant composition in relation to fire across an elevation gradient in Hawaii. Austral Ecology 25: 507–522.
D'Antonio, C.M., and P.M. Vitousek. 1992. Biological invasions by exotic grasses, the grass/fire cycle and global change. Annual Review of Ecology and Systematics 23: 63–87.
DeFalco, L.A., D.R. Bryla, V. Smith-Longozo, and R.S. Nowak. 2003. Are Mojave Desert annual species equal? Resource acquisition and allocation for the invasive grass *Bromus madritensis* subsp. *rubens* (Poaceae) and two native species. American Journal of Botany 90: 1045–1053.
DeFalco, L.A., G.C.J. Fernandez, and R.S. Nowak. 2007. Variation in the establishment of a non-native annual grass influences competitive interactions with Mojave Desert perennials. Biological Invasions 9: 293–307.
Dennison, P.E., S.C. Brewer, J.D. Arnold, and M.A. Moritz. 2014. Large wildfire trends in the western United States, 1984–2011. Geophysical Research Letters 41: 2928–2933.
Dickens, S.J.M., and E.B. Allen. 2014. Exotic plant invasion alters chaparral ecosystem resistance and resilience pre- and post-wildfire. Biological Invasions 16: 1119–1130.
DiTomaso, J.M., K.L. Hiese, G.B. Kyser, A.M. Merenlender, and R.J. Keiffer. 2001. Carefully timed burning can control barb goatgrass. California Agriculture (November–December 2001) 55: 47–53
DiTomaso, J.M., and D.W. Johnson, editors. 2006. The use of fire as a tool for controlling invasive plants. Cal-IPC Publication 2006-01. California Invasive Plant Council, Berkeley, California, USA.
DiTomaso, J.M., G.B. Kyser, and M.S. Hastings. 1999. Prescribed burning for control of yellow starthistle (*Centaurea soltitialis*) and enhanced native plant diversity. Weed Science 47: 233–242.
DiTomaso, J.M., G.B. Kyser, J.R. Miller, S. Garcia, R.F. Smith, G. Nader, J.M. Connor, and S.B. Orloff. 2006. Integrating prescribed burning and clopyralid for the management of yellow starthistle (*Centaurea solstitialis*). Weed Science 54: 757–767.
Dix, R.L. 1960. The effects of burning on the mulch structure and species composition of grasslands in western North Dakota. Ecology 41: 49–56.
Dukes, J.S., N.R. Chiariello, S.R. Loarie, and C.B. Field. 2011. Strong response of an invasive plant species (*Centaurea solstitialis* L.) to global environmental changes. Ecological Applications 21: 1887–1894.
Dyer, A.R. 2002. Burning and grazing management in a California grassland: effect on bunchgrass seed viability. Restoration Ecology 10: 107–111.
Dyer, A.R., H.C. Fossum, and J.W. Menke. 1996. Emergence and survival of *Nassella pulchra* in a California grassland. Madroño 43: 316–333.
Dyer, A.R., and K.J. Rice. 1997. Intraspecific and diffuse competition: the response of *Nassella pulchra* in a California grassland. Ecological Applications 7: 484–492.
———. 1999. Effects of competition on resource availability and growth of a California bunchgrass. Ecology 80: 2697–2710.
Eliason, S.A., and E.B. Allen. 1997. Exotic grass competition in suppressing native shrubland reestablishment. Restoration Ecology 5: 245–255.
Ellsworth, L.M., and J.B. Kauffman. 2013. Seedbank responses to spring and fall prescribed fire in mountain big sagebrush ecosystems of differing ecological condition at Lava Beds National Monument, California. Journal of Arid Environments 96: 1–8.
Engel, E.C., and S.R. Abella. 2011. Vegetation recovery in a desert landscape after wildfires: influences of community type, time since fire and contingency effects. Journal of Applied Ecology 48: 1401–1410.
Evans, R.A., and J.A. Young. 1970. Plant litter and establishment of alien annual weed species in rangeland communities. Weed Science 18: 697–703.
Fellows, D.P., and W.E. Newton. 1999. Prescribed fire effects on biological control of leafy spurge. Journal of Range Management 52: 489–493.
Fenn, M.E., R. Haeuber, G.S. Tonnesen, J.S. Baron, S. Grossman-Clarke, D. Hope, D.A. Jaffe, S. Copeland, L. Geiser, H.M. Rueth, and J.O. Sickman. 2003. Nitrogen emissions, deposition, and monitoring in the western United States. BioScience 53: 391–403.
Foin, T.C., and M.M. Hektner. 1986. Secondary succession and the fate of native species in a California coastal prairie. Madroño 33: 189–206.
Franklin, J. 2010. Vegetation dynamics and exotic plant invasion following high severity crown fire in a southern California conifer forest. Plant Ecology 207: 281–295.
Franklin, J., L.A. Spears-Lebrun, D.H. Deutschman, and K. Marsden. 2006. Impact of a high-intensity fire on mixed evergreen and mixed conifer forests in the Peninsular Ranges of southern California, USA. Forest Ecology and Management 235: 18–29.
Freeman, E.D., T.R. Sharp, R.T. Larsen, R.N. Knight, S.J. Slater, and B.R. McMillan. 2014. Negative effects of an exotic grass invasion on small-mammal communities. PLoS One 9(9): 1–7.
Frenkel, R.E. 1970. Ruderal Vegetation Along Some California Roadsides. University of California Press, Berkeley, California, USA.
Furbush, P. 1953. Control of medusa-head on California ranges. Journal of Forestry 51: 118–121.

Galley, K.E.M., and T.P. Wilson, editors. 2001. Proceedings of the invasive species workshop: the role of fire in the control and spread of invasive species. Tall Timbers Research Station, Miscellaneous Publications No. 11. Tallahassee, Florida, USA.
Germano, D.J., G.B. Rathbun, and L.R. Saslaw. 2001. Managing exotic grasses and conserving declining species. Wildlife Society Bulletin 29: 551–559.
Giessow, J. 1996. The effects of fire frequency and distance to firebreak on the distribution and abundance of exotic species in coastal sage scrub. American Journal of Botany 83: 158.
Giessow, J., and P. Zedler. 1996. The effects of fire frequency and firebreaks on the abundance and species richness of exotic plant species in coastal sage scrub. Pages 86–94 in: J. Lovich, J. Randall, and M. Kelly, editors. Proceedings California Exotic Pest Plant Council Symposium Volume 2. California Exotic Pest Council, San Juan Capistrano, California, USA.
Gillespie, I.G., and E.B. Allen. 2004. Fire and competition in a southern California grassland: impacts on the rare forb *Erodium macrophyllum*. Journal of Applied Ecology 41: 643–652.
Gordon, D.R. 1998. Effects of invasive, nonindigenous plant species on ecosystem processes: lessons from Florida. Ecological Applications 8: 975–989.
Grace, J.B., M.D. Smith, S.L. Grace, S.L. Collins, and T.J. Stohlgren. 2001. Interactions between fire and invasive plants in temperate grasslands of North America. Pages 40–65 in: K.E.M. Galley and T.P. Wilson, editors. Proceedings of the Invasive Species Workshop: The Role of Fire in the Control and Spread of Invasive Species. Tall Timbers Research Station, Miscellaneous Publication 11. Tallahassee, Florida, USA.
Haidinger, T.L., and J.E. Keeley. 1993. Role of high fire frequency in destruction of mixed chaparral. Madroño 40: 141–147.
Hamilton, J.G. 1997. Changing perceptions of pre-European grasslands in California. Madroño 44: 311–333.
Harrison, S., B.D. Inouye, and H.D. Safford. 2003. Ecological heterogeneity in the effects of grazing and fire on grassland diversity. Conservation Biology 17: 837–845.
Harrod, R.J., and S. Reichard. 2001. Fire and invasive species within the temperate and boreal coniferous forests of North America. Pages 95–101 in: K.E.M. Galley and T.P. Wilson, editors. Proceedings of the Invasive Species Workshop: The Role of Fire in the Control and Spread of Invasive Species. Tall Timbers Research Station, Miscellaneous Publication 11. Tallahassee, Florida, USA.
Harty, F.M. 1986. Exotics and their ecological ramifications. Natural Areas Journal 6: 20–26.
Hatch, D.A., J.W. Bartolome, J.S. Fehmi, and D.S. Hillyard. 1999. Effects of burning and grazing on a coastal California grassland. Restoration Ecology 7: 376–381.
Heady, H.F. 1972. Burning and the grasslands in California. Proceedings Tall Timbers Fire Ecology Conference 12: 97–107.
———. 1988. Valley grassland. Pages 491–514 in: M.G. Barbour and J. Major, editors. Terrestrial Vegetation of California. John Wiley and Sons, New York, New York, USA.
Hellman, J.J., J.E. Byers, B.G. Bierwagen, and J.S. Dukes. 2008. Five potential consequences of climate change for invasive species. Conservation Biology 22: 534–543.
Hervey, D.F. 1949. Reaction of a California annual-plant community to fire. Journal of Range Management 2: 116–121.
Hickman, J.C., editor. 1993. The Jepson Manual: Higher Plants of California. University of California Press, Berkeley, California, USA.
Hobbs, R.J., and L.F. Huenneke. 1992. Disturbance, diversity, and invasion: implications for conservation. Conservation Biology 6: 324–337.
Holmes, P.M., and R.M. Cowling. 1997. Diversity, composition and guild structure relationships between soil-stored seed banks and mature vegetation in alien plant-invaded South African fynbos shrublands. Plant Ecology 133: 107–122.
Hopkinson, P., J.S. Fehmi, and J.W. Bartolome. 1999. Summer burns reduce cover, but not spread, of barbed goatgrass in California grassland. Ecological Restoration 17: 168–169.
Howe, H.F. 1995. Succession and fire season in experimental prairie plantings. Ecology 76: 1917–1925.
Hrusa, F., B. Ertter, A. Sanders, G. Leppig, and E. Dean. 2002. Catalogue of non-native vascular plants occurring spontaneously in California beyond those addressed in the Jepson Manual—Part I. Madroño 49: 61–98.
Jackson, L.E. 1985. Ecological origins of California's Mediterranean grasses. Journal of Biogeography 12: 349–361.
Jacobs, J.S., and R.L. Shelley. 2003. Prescribed fire effects on dalmation toadflax. Journal of Range Management 56: 193–197.
Jauni, M., S. Gripenberg, and S. Ramula. 2015. Non-native plant species benefit from disturbance: a meta-analysis. Oikos 124: 122–129.
Jepson, W.L. 1925. Manual of the Flowering Plants of California. University of California Press, Berkeley, California, USA.
Johnson, B.E., and J.H. Cushman. 2007. Influence of a large herbivore reintroduction on plant invasions and community composition in a California grassland. Conservation Biology 21: 515–526.
Junak, S., T. Ayers, R. Scott, D. Wilken, and D. Young. 1995. A Flora of Santa Cruz Island. Santa Barbara Botanic Garden, Santa Barbara, California, USA.
Kaczynski, K.M., S.W. Beatty, J.W. van Wagtendonk, and K.N. Marshall. 2011. Burn severity and non-native species in Yosemite National Park, California, USA. Fire Ecology 7(2): 66–80.
Kardol, P., L. Souza, and A.T. Classen. 2013. Resource availability mediates the importance of priority effects in plant community assembly and ecosystem function. Oikos 122: 84–94.
Keeley, J.E. 2001. Fire and invasive species in Mediterranean-climate ecosystems of California. Pages 81–94 in: K.E.M. Galley and T.P. Wilson, editors. Proceedings of the Invasive Species Workshop: The Role of Fire in the Control and Spread of Invasive Species. Tall Timbers Research Station, Miscellaneous Publication 11. Tallahassee, Florida, USA.
———. 2006. Fire management impacts on invasive plants in the western United States. Conservation Biology 20: 375–384.
Keeley, J.E., M. Baer-Keeley, and C.J. Fotheringham. 2005. Alien plant dynamics following fire in Mediterranean-climate Califronia shrublands. Ecological Applications 15: 2109–2125.
Keeley, J.E., J. Franklin, and C.M. D'Antonio. 2011. Fire and invasive plants on California landscapes. Pages 193–221 in: D.L. McKenzie, C. Miller, and D.A. Falk, editors. The Landscape Ecology of Fire. Springer, Dordrecht, The Netherlands.
Keeley, J.E., D. Lubin, and C.J. Fotheringham. 2003. Fire and grazing impacts on plant diversity and alien plant invasions in the southern Sierra Nevada. Ecological Applications 13: 1355–1374.
Klinger, R.C. 2007. Ecosystem engineers and the complex dynamics of non-native species management on California's Channel Islands. Pages 343–366 in: K. Cuddington, J. Byers, W. Wilson, and A. Hastings, editors. Ecosystem Engineers: Plants to Protists. Academic Press—Elsevier, San Diego, California, USA.
———. 2011. Fire regimes. Pages 223–228 in: D. Simberloff and M. Rejmanek, editors. Encyclopedia of Biological Invasions. University of California Press, Berkeley, California, USA.
Klinger, R.C., and I. Messer. 2001. The interaction of prescribed burning and site characteristics on the diversity and composition of a grassland community on Santa Cruz Island, California. Pages 66–80 in: K.E.M. Galley and T.P. Wilson, editors. Proceedings of the Invasive Species Workshop: The Role of Fire in the Control and Spread of Invasive Species. Tall Timbers Research Station, Miscellaneous Publication 11. Tallahassee, Florida, USA.
Klinger, R.C., E.C. Underwood, and P.E. Moore. 2006. The role of environmental gradients in non-native plant invasions into burnt areas of Yosemite National Park, California. Diversity and Distributions 12: 139–156.
Klinger, R.C., R.D. Wills, and M.L. Brooks. 2008. Fire and nonnative invasive plants in the Southwest Coastal Bioregion. Pages 175–196 in: K. Zouhar, J. Kapler Smith, S. Sutherland, and M.L. Brooks, editors. Wildland Fire in Ecosystems: Fire and Nonnative Invasive Plants. USDA Forest Service General Technical Report RMRS-GTR-42-VOL-6. Rocky Mountain Research Station, Fort Collins, Colorado, USA.
Knutson, K.C., D.A. Pyke, T.A. Wirth, R.S. Arkle, D.S. Pilliod, M.L. Brooks, J.C. Chambers, and J.B. Grace. 2014. Long-term effects of seeding after wildfire on vegetation in Great Basin shrubland ecosystems. Journal of Applied Ecology 51: 1414–1424.
Körner, C., J. Stöcklin, L. Reuther-Thiébaud, and S. Pelaez-Riedl. 2008. Small differences in arrival time influence composition and productivity of plant communities. New Phytologist 177: 698–705.
Kuebbing, S.E., M.A. Nunez, and D. Simberloff. 2013. Current mismatch between research and conservation efforts: the need to study co-occurring invasive plant species. Biological Conservation 160: 121–129.

Kyser, G.B., and J.M. DiTomaso. 2002. Instability in a grassland community after the control of yellow starthistle (*Centaurea solstitialis*) with prescribed burning. Weed Science 50: 648–657.

Kyser, G.B., M.P. Doran, N.K. McDougald, S.B. Orloff, R.N. Vargas, R.G. Wilson, and J.M. DiTomaso. 2008. Site characteristics determine the success of prescribed burning for medusahead (*Taeniatherum caput-medusae*) control. Invasive Plant Science and Management 1: 376–384.

Leibold, M.A., M. Holyoak, N. Mouquet, P. Amarasekare, J.M. Chase, M.F. Hoopes, R.D. Holt, J.B. Shurin, R. Law, D. Tilman, M. Loreau, and A. Gonzalez. 2004. The metacommunity concept: a framework for multi-scale community ecology. Ecology Letters 7: 601–613.

Lodge, D.M., S. Williams, H.J. MacIsaac, K.R. Hayes, B. Leung, S. Reichard, R.N. Mack, P.B. Moyle, M. Smith, D.A. Andow, J.T. Carlton, and A. McMichael. 2006. Biological invasions: recommendations for U.S. policy and management. Ecological Applications 16: 2035–2054.

Lonsdale, W.M. 1999. Global patterns of plant invasions and the concept of invasibility. Ecology 80: 1522–1536.

Lonsdale, W.M., and I.L. Miller. 1993. Fire as a management tool for a tropical woody weed: *Mimosa pigra* in North Australia. Journal of Environmental Management 33: 7–87.

Lulow, M.E., T.P. Young, J.L. Wirka, and J.H. Anderson. 2007. Variation in the initial success of seeded native bunchgrasses in the rangeland foothills of Yolo County, California. Ecological Restoration 25: 20–28.

MacDougall, A.S., and R. Turkington. 2005. Are invasive species the drivers or passengers of change in degraded ecosystems? Ecology 86: 42–55.

Mack, R.N. 1981. Invasion of *Bromus tectorum* L. into western North America: an ecological chronicle. Agro-Ecosystems 7: 145–165.

———. 1989. Temperate grasslands vulnerable to invasions: characteristics and consequences. Pages 155–179 in: J.A. Drake, H.A. Mooney, F. DiCastri, R.H. Groves, F.J. Kruger, M. Rejmanek, and M. Williamson, editors. Biological Invasions: A Global Perspective. John Wiley and Sons, New York, New York, USA.

Mack, M.C., and C.M. D'Antonio. 1998. Impacts of biological invasions on distrubance regimes. Trends in Ecology and Evolution 13: 195–198.

Mack, M.C., C.M. D'Antonio, and R.E. Ley. 2001. Alteration of ecosystem nitrogen dynamics by exotic plants: a case study of C4 grasses in Hawaii. Ecological Applications 11: 1323–1335.

Mack, R.N., D. Simberloff, W.M. Lonsdale, H. Evans, M. Clout, and F.A. Bazzaz. 2000. Biotic invasions: causes, epidemiology, global consequences, and control. Ecological Applications 10: 689–710.

Major, J., and W.T. Pyott. 1966. Buried, viable seeds in two California bunchgrass sites and their bearing on the definition of a flora. Vegetatio 13: 253–282.

Maranon, T., and J.W. Bartolome. 1989. Seed and seedling populations in two contrasted communities: open grassland and oak (*Quercus agrifolia*) understory in California. Acta Oecologia 10: 147–158.

Marty, J.T. 2002. Spatially-dependent effects of fire and grazing in a California annual grassland plant community (Chapter 3). Dissertation, University of California, Davis, California, USA.

Menke, J.W. 1992. Grazing and fire management for native perennial grass restoration in California grasslands. Fremontia 20: 22–25.

Merriam, K.E., J.E. Keeley, and J.L. Beyers. 2006. Fuel breaks affect nonnative species abundance in Californian plant communities. Ecological Applications 16: 515–527.

Meyer, M.D., and P.M. Schiffman. 1999. Fire season and mulch reduction in a California grassland: a comparison of restoration strategies. Madroño 46: 25–37.

Miller, T.J., P.F. Quintana-Ascencio, S. Maliakal-Witt, and E.S. Menges. 2010. Metacommunity dynamics over 16 years in a pyrogenic shrubland. Conservation Biology 26: 357–366.

Minnich, R.R., and R.J. Dezzani. 1998. Historical decline of coastal sage scrub in the Riverside-Perris Plain. Western Birds 29: 366–391.

Moody, M.E., and R.N. Mack. 1988. Controlling the spread of plant invasions: the importance of nascent foci. Journal of Applied Ecology 25: 1009–1021.

Mooney, H.A., S.P. Hamburg, and J.A. Drake. 1989. The invasions of plants and animals into California. Pages 250–272 in: J.A. Drake, H.A. Mooney, F. DiCastri, R.H. Groves, F.J. Kruger, M. Rejmanek, and M. Williamson, editors. Biological Invasions: A Global Perspective. John Wiley & Sons, New York, New York, USA.

Mouquet, N., and M. Loreau. 2002. Coexistence in metacommunities: the regional similarity hypothesis. American Naturalist 159: 420–426.

Munz, P.A. 1968. Supplement to a California Flora. University of California Press, Berkeley, California. USA.

Munz, P.A., and D.D. Keck. 1959. A California Flora. University of California Press, Berkeley, California. USA.

Myers, R.L., H.A. Belles, and J.R. Snyder. 2001. Prescribed fire in the management of *Melaleuca quinquenervia* in subtropical Florida. Pages 132–140 in: K.E.M. Galley and T.P. Wilson, editors. Proceedings of the Invasive Species Workshop: The Role of Fire in the Control and Spread of Invasive Species. Tall Timbers Research Station, Miscellaneous Publication 11. Tallahassee, Florida, USA.

Nuzzo, V.A. 1991. Experimental control of garlic mustard (*Alliaria petiolata* [Bieb.] Cavara & Grande) in northern Illinois using fire, herbicide, and cutting. Natural Areas Journal 11: 158–167.

Odion, D.C., and K. Hausenback. 2002. Response of French Broom to fire. Pages 296–307 in: N.G. Sugihara, M. Morales, and T. Morales, editors. Proceedings of the Symposium: Fire in California Ecosystems: Integrating Ecology, Prevention, and Management. Association for Fire Ecology Miscellaneaous Publication 1. Davis, California, USA.

Ogden, J.A.E., and M. Rejmánek. 2005. Recovery of native plant communities after the control of a dominant invasive plant species, *Foeniculum vulgare:* Implications for management. Biological Conservation 125: 427–439.

O'Leary, J.F. 1995. Coastal sage scrub: threats and current status. Fremontia 23: 27–31.

Ostoja, S.M., M.L. Brooks, T. Dudley, and S.R. Lee. 2014. Short-term vegetation response following mechanical control of saltcedar (*Tamarix* spp.) on the Virgin River, Nevada, USA. Invasive Plant Science and Management 7: 310–319.

Ostoja, S.M., and R.C. Klinger. 2000. The relationship of Bishop pine *Pinus muricata* morphology to serotiny on Santa Cruz Island, California. Pages 167–171 in: D.R. Browne, K.L. Mitchell, and H.W. Chaney, editors. Proceedings of the Fifth California Islands Symposium. Santa Barbara Museum of Natural History, Santa Barbara, California, USA.

Ostoja, S.M., and E.W. Schupp. 2009. Conversion of sagebrush shrublands to exotic annual grasslands negatively impacts small mammal communities. Diversity and Distributions 15: 863–870.

Parker, I.M., D. Simberloff, W.M. Lonsdale, K. Goodell, M. Wonham, P.M. Kareiva, M.H. Williamson, B. Von Holle, P.B. Moyle, J.E. Byers, and L. Goldwasser. 1999. Impact: toward a framework for understanding the ecological effects of invaders. Biological Invasions 1: 3–19.

Parsons, D.J., and T.J. Stohlgren. 1989. Effects of varying fire regimes on annual grasslands in the southern Sierra Nevada of California. Madroño 36: 154–168.

Pausus, J.G., and J.E. Keeley. 2014. Abrupt climate-independent fire regime changes. Ecosystems 17: 1109–1120.

Pec, G.J., and G.C. Carlton. 2014. Positive effects of non-native grasses on the growth of a native annual in a Southern California ecosystem. PLoS One 9(11): e112437.

Peppin, D.L., P.Z. Fule, C.H. Sieg, J.L. Beyers, M.E. Hunter, and P.R. Robichaud. 2011. Recent trends in post-wildfire seeding in western US forests: costs and seed mixes. International Journal of Wildland Fire 20: 702–708.

Peterson, A.T. 2003. Predicting the geography of species invasions via ecological niche modeling. Quarterly Review of Biology 78: 419–433.

Pickett, S.T.A., and P.S. White, editors. 1985. The Ecology of Natural Disturbance and Patch Dynamics. Academic Press, New York, New York, USA.

Pickford, G.D. 1932. The influence of continued heavy grazing and promiscuous burning on spring-fall ranges in Utah. Ecology 13: 159–171.

Pollak, O., and T. Kan. 1998. The use of prescribed fire to control invasive exotic weeds at Jepson Prairie Preserve. Pages 241–249 in: C.W. Witham, E.T. Bauder, D. Belk, W.R.F., Jr., and R. Onduff, editors. Ecology, Conservation, and Management of Vernal Pool Ecosystems. California Native Plant Society, Sacramento, California, USA.

Pyke, D.A., M.L. Brooks, and C.M. D'Antonio. 2010. Fire as a restoration tool: a decision framework for predicting the control or enhancement of plants using fire. Restoration Ecology 18: 274–284.

Pyke, D.A., T.A. Wirth, and J.L. Beyers. 2013. Does seeding after wildfires in rangelands reduce erosion or invasive species? Restoration Ecology 21: 415–421.

Randall, J.M. 1996. Weed control for the preservation of biological diversity. Weed Technology 10: 370–383.

Randall, J.M., R.R. Lewis, and D.B. Jensen. 1997. Ecological restoration. Pages 205–219 in: D. Simberloff, D.C. Schmitz, and T.C. Brown, editors. Strangers in Paradise. Island Press, Washington, D.C., USA.

Randall, J.M., and M. Rejmanek. 1993. Interference of bull thistle (*Cirsium vulgare*) with growth of ponderosa pine (*Pinus ponderosa*) seedlings in a forest plantation. Canadian Journal of Forest Research 23: 1507–1513.

Randall, J.M., M. Rejmanek, and J.C. Hunter. 1998. Characteristics of the exotic flora of California. Fremontia 26(4): 3–12.

Rao, L.E., and E.B. Allen. 2009. Combined effects of precipitation and nitrogen deposition on native and invasive winter annual production in California deserts. Oecologia 162: 1035–1046.

Rejmánek, M. 1989. Invasibility of plant communities. Pages 369–388 in: J.A. Drake, H.A. Mooney, F. DiCastri, R.H. Groves, F.J. Kruger, M. Rejmanek, and M. Williamson, editors. Biological Invasions: A Global Perspective. John Wiley and Sons, New York, New York, USA.

———. 1995. What makes a species invasive? Pages 3–13 in: P. Pysek, K. Prach, M. Rejmanek, and P.M. Wade, editors. Plant Invasions. SPB Academic Publishing, The Hague, The Netherlands.

Rejmanek, M., and J. Randall. 1994. Invasive alien plants in California: 1993 summary and comparison with other areas in North America. Madroño 41: 161–177.

Rejmanek, M., C.D. Thomsen, and I.D. Peters. 1991. Invasive vascular plants of California. Pages 81–101 in: R.H. Groves and F. DiCastri, editors. Biogeography of Mediterranean Invasions. Cambridge University Press, New York, New York, USA.

Rice, P.M., and J. Kapler Smith. 2008. Use of fire to manage populations of nonnative invasive plants. Pages 47–60 in: K. Zouhar, J. Kapler Smith, S. Sutherland, and M.L. Brooks, editors. Wildland Fire in Ecosystems: Fire and Nonnative Invasive Plants. USDA Forest Service General Technical Report RMRS-GTR-42-VOL-6. Rocky Mountain Research Station, Fort Collins, Colorado, USA.

Richardson, D.M., P. Pysek, M. Rejmanek, M.G. Barbour, F.D. Pannetta, and C.J. West. 2000. Naturalization and invasion of alien plants: concepts and definitions. Diversity and Distributions 6: 93–107.

Rocca, M.E., C.F. Miniat, and R.J. Mitchell. 2014. Introduction to the regional assessments: climate change, wildfire, and forest ecosystem services in the USA. Forest Ecology and Management 327: 265–268.

Roja, D., and J. Popenoe. 1998. Fire effects on first-year scotch broom in Redwood National and State Parks. Pages 59–60 in: M. Kelly, E. Wagner, and P. Warner, editors. Proceedings California Exotic Pest Plant Council Symposium Volume 4. Ontario, California, USA.

Rost, J., P. Pons, and J.M. Bas. 2012. Seed dispersal by carnivorous mammals into burnt forests: an opportunity for non-indigenous and cultivated plant species. Basic and Applied Ecology 13: 623–630.

Rowe, J.S. 1981. Concepts of fire effects on plant individuals and species. Pages 135–154 in: R.W. Wein and D.A. Maclean, editors. The Role of Fire in Northern Circumpolar Ecosystems. John Wiley and Sons, New York, New York, USA.

Roy, B.A., K. Hudson, M. Visser, and B.R. Johnson. 2014. Grassland fires may favor native over introduced plants by reducing pathogen loads. Ecology 95: 1897–1906.

Ruyle, G.B., B.A. Roundy, and J.R. Cox. 1988. Effects of burning on germinability of Lehmann lovegrass. Journal of Range Management 41: 404–406.

Sapsis, D.B., D.V. Pearman, and R.E. Martin. 1995. Progression of the Oakland/Berkeley Hills "Tunnel Fire". Pages 187–189 in: The Biswell Symposium: Fire Issues and Solutions in Urban Interface and Wildland Ecosystems. USDA Forest Service General Technical Report PSW-GTR-158. Pacific Southwest Research Station, Albany, California, USA.

Schoennagel, T.L., and D.M. Waller. 1999. Understory responses to fire and artificial seeding in an eastern Cascades Abies grandis forest, U.S.A. Canadian Journal of Forest Research 29: 1393–1401.

Schwartz, M.W., D.J. Porter, J.M. Randall, and K.E. Lyons. 1996. Impact of nonindigenous plants. Pages 1203–1218 in: D.C. Erman, general editor. Sierra Nevada Ecosystem Project: Final Report to Congress, Volume II. Wildland Resources Center Report 37. University of California, Davis, California, USA.

Seabloom, E.W., W.S. Harpole, O.J. Reichman, and D. Tilman. 2003. Invasion, competitive dominance, and resource use by exotic and native California grassland species. Proceedings of the National Academy of Sciences of the United States of America 100: 13384–13389.

Seastedt, T.B., R.J. Hobbs, and K.N. Suding. 2008. Management of novel ecosystems: are novel approaches required? Frontiers in Ecology and the Environment 6: 547–553.

Stylinski, C.D., and E.B. Allen. 1999. Lack of native species recovery following severe exotic disturbance in southern Californian shrublands. Journal of Applied Ecology 36: 544–554.

Suding, K.N., K.L. Gross, and G.R. Houseman. 2004. Alternative states and positive feedback in restoration ecology. Trends in Ecology and Evolution 19: 46–53.

Swezey, M., and D.C. Odion. 1997. Fire on the mountain: a land managers manifesto for broom control. Pages 76–81 in: M. Kelly, E. Wagner, and P. Warner, editors. Proceedings California Exotic Pest Plant Council Symposium Volume 3, Concord, California, USA.

Syphard, A.D., V.C. Radeloff, Hawbaker, T.J., and S.I. Stewart. 2009. Conservation threats due to human-caused increases in fire frequency in Mediterranean-climate ecosystems. Conservation Biology 23: 758–769.

Taylor, K., T. Brummer, L.J. Rew, M. Lavin, and B.D. Maxwell. 2014. *Bromus tectorum* response to fire varies with climate conditions. Ecosystems 17: 960–973.

Tucker, C.M., and M.W. Cadotte. 2013. Fire variability, as well as frequency, can explain coexistence between seeder and resprouter life histories. Journal of Applied Ecology 50: 594–602.

Vitousek, P.M., C.M. D'Antonio, L.L. Loope, and R. Westbrooks. 1996. Biological invasions as global environmental change. American Scientist 84: 468–478.

Watson, S. 1880. Gamopetalae. In: W.H. Brewer, S. Watson, and A. Gray. Geological Survey of California. Botany, Volume I, Second Edition. Little, Brown, and Company, Boston, Massachusetts, USA.

Whisenant, S.G. 1990. Postfire population dynamics of *Bromus japonicus*. American Midland Naturalist 123: 301–308.

Wilcove, D.S., D. Rothstein, J. Dubow, A. Phillips, and E. Losos. 1998. Quantifying threats to imperiled species in the United States. BioScience 48: 607–615.

Wills, R.D. 2000. Effective fire planning for California native grasslands. Pages 75–78 in: J.E. Keeley, M.B. Keeley, and C.J. Fotheringham, editors. 2nd Interface between Ecology and Land Development in California. U.S. Geological Survey Open File Report OFR-00-62. Reston, Virginia, USA.

Wilson, E.O., and F.M. Peter, editors. 1988. Biodiversity. National Academy Press, Washington, D.C., USA.

Yelenik, S.G., and C.M. D'Antonio. 2013. Self-reinforcing impacts of plant invasions change over time. Nature 503: 517–520.

Zavaleta, E.S., R.J. Hobbs, and H.A. Mooney. 2001. Viewing invasive species removal in a whole-ecosystem context. Trends in Ecology and Evolution 16: 454–459.

Zedler, P. 1995. Fire frequency in southern California shrublands: biological effects and management options. Pages 101–112 in: J.E. Keeley and T. Scott, editors. Brushfires in California: Ecology and Resource Management. International Association of Wildland Fire, Fairfield, Washington, USA.

Zedler, P.H., C.R. Gautier, and G.S. McMaster. 1983. Vegetation change in response to extreme events: the effect of a short interval between fires in California chaparral and coastal scrub. Ecology 64: 809–818.

Zouhar, K., J. Kapler Smith, S. Sutherland, and M.L. Brooks, editors. 2008. Wildland fire in ecosystems: fire and nonnative invasive plants. USDA Forest Service General Technical Report RMRS-GTR-42-VOL-6. Rocky Mountain Research Station, Fort Collins, Colorado, USA.

CHAPTER TWENTY-FIVE

Fire and At-Risk Species

KEVIN E. SHAFFER AND SHAULA J. HEDWALL

> Nothing is more priceless and more worthy of preservation than the rich array of animal life with which our country has been blessed.
>
> PRESIDENT RICHARD NIXON, Signing the Federal Endangered Species Act, December 28, 1973

> Is fire management on a collision course with disaster? Perhaps, because wildfires continue to become more intense and destructive of resources, and expenses in fire control are increasing at an astronomical rate.
>
> BISWELL (1989)

Introduction

A Summary of the Conflict

Previous chapters have described the essential, ecological function of fire and society's struggle to control and manage it. As wildfire size and severity have increased due to over a century of suppression efforts, fuel accumulation, and extended drought; the environmental, economic, and social costs of fire have also increased (Gill et al. 2013, North et al. 2015). These mounting costs and the loss of human life and property have resulted in the potential for greater conflicts associated with fire management and the associated management of plant and animal species that may also be impacted by fire and suppression actions. As described in Chapter 20, modern-era management of fire commenced at the turn of the nineteenth century, and even the issue of wildland fire burning at the urban interface has been part of the California landscape since the 1920s. The middle of the twentieth century not only represented a growing attention to fire management and the use of fire, but also society's growing attention to the importance of a healthy and functioning environment that continued to support native animals and plants. In 1973, the federal Endangered Species Act (ESA) was signed into law. The primary purpose of the ESA is the conservation of endangered and threatened species and the ecosystems upon which they depend. In 1986, the State of California passed its own law to protect its native plants and animals, the California Endangered Species Act (CESA).

In addition to the ESA and CESA, several other federal and state laws were passed from the 1970s through the 1990s that were designed to protect native plant and animals (Table 25.1).

As California's human population has increased and urban development has pushed into the State's wildlands, homeowners' and entire communities' experience and familiarity with wildfire has become more frequent and more disconcerting. Growing human population and expanding urban areas not only have brought more people into conflict with wildfire but has also affected how wildlands are managed. For example as described in Chapter 21, the ability to use prescribed fire or allow wildfires to burn has become exceedingly difficult due to the proximity of fire-adapted systems to human dwellings, private property, and infrastructure. Population growth into wildlands has made management of vegetation more complex. Agencies responsible for conserving wilderness and natural resources (California Department of Fish and Wildlife [CDFW], U.S. Fish and Wildlife Service [USFWS], California Department of Parks and Recreation CDPR], National Park Service [NPS], Forest Service [FS], Bureau of Land Management [BLM]) and agencies responsible for addressing vegetation fuel on public lands (Cal Fire, NPS, FS, BLM, USFWS, Bureau of Indian Affairs [BIA], local fire districts) increasingly spend more time and money addressing fuel management. Frequently, agency policies come into conflict when trying to manage fire and fuel while protecting ecological values and at-risk species. California citizens, though wanting to protect the State's native biota, have become increasingly concerned with the impacts of landscape level high-severity fires to lives, property, and finances.

Laws protecting at-risk species were enacted when both the California and federal governments recognized the ever-increasing threats to many species and the ecosystems in which they occur. Rectifying the causes of endangerment or the threat of future endangerment of extinction are the essential goals of both ESA and CESA. The primary intents of the other laws and policies are conservation and protection of at-risk species before they become threatened or endangered. Fire and fuel management regulations, policies, and activities sometimes conflict with protecting at-risk species because loss and alteration of habitat and human activities often create a need for increased protection (Box 25.1). Fire suppression,

TABLE 25.1
Laws and conventions applicable in California to protect animals, plants, and habitat, year enacted, and regional scope of authority

Protective act	Year enacted (last amended)	Governing authority and responsibility
Migratory Bird Treaty Act	1918, as amended	Canada, United States, Mexico, Japan, and Russia
Bald and Golden Eagle Protection Act	1940, as amended	National
Convention on International Trade in Endangered Species of Wild Fauna and Flora (CITES)	1973, as amended	International
Endangered Species Act (ESA)	1973, as amended	National
Native Plant Protection Act (NPPA)	1977	State
California Endangered Species Act (CESA)	1986 (2003)	State
California Fish and Game Commission-Board of Forestry and Fire Protection Joint Policy on Pre-, During, and PostFire Activities	1993	State
Neotropical Migratory Bird Conservation Act (NMBCA)	2000	Western hemisphere

prescribed burning, vegetation reduction, postfire restoration, air quality control, and protection of homes and lives rarely fully consider or integrate conservation or protection of at-risk species. The reality that each of these laws and activities do not consider adequately the other may seem understandable, but this lack of integration, or perhaps adjustment, may be the basic reason for the conflicts that have occurred over the last three decades in California. To date, this conflict has pitted species protection against protection of humans, homes, and other natural resource values.

With summer and early autumn come news reports of wildfires raging somewhere in California. In southern California, fire season is now considered to be year-round (Jin et al. 2015). As the length of the fire season has increased, the perceived and real conflicts between species protection and fire management have also escalated. Though this situation is not unique to California, it may be more evident due to the following factors:

- Hundreds species are either state or federally listed. As of October 2015, 151 animal and 405 plant species are listed under CESA and/or ESA (CDFW 2015a, 2015b);
- Large areas burn annually. The average number of acres burned per year from 1972 to 2010 was 100,837 ha (249,173 ac) (Cal Fire 2011);
- Large human population. Approximately 38,715,000 people inhabit California (as of January 1, 2015—California Department of Finance). The presence of human residences within and adjacent to wildlands has resulted in increased human-caused fire starts (increased fire frequency), prioritization of structure protection over fire ecology, and added pressures to wildlife habitat; and,
- Economic costs due to wildfire. The annual average dollar damage from 1972 to 2010 was $97,266,559.00 (Cal Fire 2011).

The struggle begins before fire season. Activities to reduce the dry and dead vegetation that feeds wildfires are scrutinized, in part because of their possible or actual effects to native plants and animals and habitat. Residents and communities clear, chip, or crush brush, burn off grass, or thin trees. Agencies and land conservancies apply broader mechanical treatment or conduct prescribed burns, and develop long-term land management plans, which include fire and fuel management. In many circumstances, at-risk species are affected by these actions. Addressing species and habitat needs take planning and time and more resources. Simply put, to-date in California fire and fuel management and at-risk species conservation and protection have more often been in conflict than in accord with one another because of the lack of planning, time, and resources dedicated to the conflicts. There is no data to show that the enforcement of ESA and CESA has interfered with fire management activities or affected parties participating in fuel or fire management activities. It has been the perceived constraints that ESA and CESA placed on fire and fuel management activities that form the basis for the conflict in California. Nonetheless, the conflict is real and has set various organizations and agencies at odds with one another.

What Is an At-Risk Species?

'At risk' refers to a species' legal or land management status; it is not a biological trait. The terms candidate, rare, threatened, endangered, sensitive, and special concern are included under 'at-risk'. The phrase refers to native plants and animals identified through a variety of formal means, by one or more organizations or agencies, to be deserving of special protection and/or land management attention. These species are typically: (1) formally identified and protected by law; (2) given special management protection by agency policy; (3) specially

BOX 25.1. FIRE AND FUEL MANAGEMENT ACTIVITIES THAT POTENTIALLY AFFECT AT-RISK SPECIES

Air quality management:

- Restricted season for prescribed burning

Invasive plant species control:

- Timing of prescribed burning
- Mechanical treatment
- Herbicide application

Vegetation management to lower risk of fire:

- Chemical application to vegetation
- Mechanical removal or crushing of vegetation
- Road construction
- Prescribed burning
- Use of domestic livestock to reduce vegetation

Fire suppression:

- Mechanical removal or clearance of vegetation
- Use of heavy machinery (i.e., compaction of soil, noise)
- Road construction
- Back-burning and burning out of vegetation
- Water drafting
- Use of aircraft
- Application of fire-retardant or water drops
- Establishing incident command camps and additional locales for fire-fighting equipment and staff

Post-wildfire rehabilitation and watershed restoration:

- Application of seed
- Additional removal or clearance of vegetation
- Fire incident mop-up
- Road construction
- Engineering technologies for erosion control

monitored because of vulnerable status or circumstances. At-risk species include those with legal protection, such as those species listed under CESA or ESA (i.e., northern spotted owl [*Strix occidentalis caurina*], marbled murrelet [*Brachyramphus marmoratus*]). Legal protection also is afforded to some animals under the California's Forest Practice Rules. The California Environmental Quality Act (CEQA) defines 'sensitive' to

TABLE 25.2

Legal protection and special consideration afforded California native plant and animals under various federal and State laws

Federal protection opportunities	California protection opportunities
Species listing and critical habitat designation, ESA	Species listing, CESA
Interagency Cooperation (Section 7 consultation), ESA	Incidental take permits, CESA
Habitat conservation planning, ESA	Recovery planning, CESA
Recovery planning, ESA	Natural community conservation planning, NCCP
Evaluating and mitigating significant impacts to species, NEPA	Special sensitive species listing and protection, CFPR
Evaluation, planning, and avoidance, NMBCA	Evaluating and mitigation significant impacts to species, CEQA

include species determined to meet the criteria for listing under CESA and needing investigation and consideration for impacts that may be significant during the development of environmental impact reports. Results of such investigations must be made available to the public. Lastly, some agencies give special attention to species identified as 'sensitive' (e.g., FS and BLM) or 'of special concern' (e.g., CDFW) due to their rarity or sensitivity. These designations do not usually include regulatory or legal protection but do allow agencies to develop policies and programs for conserving species before listing becomes necessary (Table 25.2).

Chapters 20 and 21 discussed how land management agencies and the California public have been keenly interested and involved in managing fire and fuels for decades. This chapter focuses on the fundamental issues that surround efforts to protect and conserve at-risk species while managing fire, fuels, and vegetation. The chapter will briefly highlight the essential aspects of species protection laws and policies that are relevant to fire and fuel management. It describes how fuel treatments can impact at-risk species and how protection of at-risk species can influence fuel reduction. Finally, the chapter puts forth ideas and perspective on how protecting native species and their habitat and managing wildfire could and should be better integrated.

Protecting At-Risk Species

Agency and public attention is focused on threatened and endangered species. These are the most obvious at-risk species. Protection under ESA and CESA is not identical in all aspects, but both have specific elements that are relevant in fire and fuel management (Table 25.2). The most obvious element is the listing of species as threatened or endangered. Such a designation means that many fuel management activities must consider whether potentially harming or killing listed species could occur as part of the action. Such consideration has been seen as restricting or preventing fuel management activities. Under ESA, critical habitat is also designated, and these areas require special considerations for management. Agencies may pursue permits to allow for incidental take to their otherwise lawful, fuel management activities. Non-federal entities dealing with federally listed species, may also pursue Habitat Conservation Plan (which includes an incidental take statement), where all land management activities are evaluated and planned for in relation to species needs and the permitted actions. Both ESA and CESA conduct recovery planning for listed species, and this planning and implementation process allows both the consideration of (1) the impacts that might occur from fuel and fire management activities, (2) recommendations for conducting these actions in a listed species habitat, and (3) discussion of what benefit such activities may provide for the species and its habitat.

Several other conventions and laws addressing protection of plants and animals (Table 25.1) afford less direct protection or address a much smaller group of at-risk animals and plants. Nine species of birds are listed as *sensitive* under Forest Practice Rules of the California Forest Practices Act and are given special rules for forestry practices, including vegetation management and broadcast burning. Native birds migrating through or to the state are protected by the Neotropical Migratory Bird Treaty Act, but the federal courts have found that federal agencies are exempt to protections provided by the act when conducting fire and fuel management activities Hence, protection only applies to private, local, and state agencies and private, regional, and state property. Two additional laws, the National Environmental Protection Act (NEPA) and CEQA, address and attempt to minimize effects to species listed under ESA and CESA, as well as species identified as sensitive by the land management agency. CEQA also affords consideration to species that meet the criteria of CESA. But for both CEQA and CESA, the consideration of species protection is not absolute and often fuel and fire management activities are deemed less than significant or even exempt from consideration. For example, under CEQA, there are classes of projects that are considered not to have a significant effect on the environment, and they are declared to be categorically exempt from the requirement for the preparation of environmental documents. One example of this is Categorical Exclusion 18.36.060 Class 4—Minor alterations to land(J), which allows fuel management activities within 30 feet of structures to reduce the volume of flammable vegetation, provided that the activities will not result in the taking of endangered, rare, or threatened plant or animal species or significant erosion and sedimentation of surface waters. This exemption also applies to fuel management activities within 100 feet of a structure if the public agency having fire protection responsibility for the area has determined that 100 feet of fuel clearance is required due to extra hazardous fire conditions.

The Interface between At-Risk Species and Fire and Fuel Management

Questions arise when addressing the interface between managing fire and fuel and protecting at-risk species. What means of reducing fuel are available that do not impact species or at least minimize the effects? How do we balance potential short-term adverse effects to species and habitats with the potential for long-term benefits? How does management of invasive species nonnative affect native species? How can fires best be suppressed, burned areas restored and species be either minimally affected or actually benefited by such actions? How can land management actions regarding fire be undertaken without contributing to a future California or federal listing? And possibly most important of all, what is the role of fire in restoring, enhancing, or simply maintaining the ecological integrity of a landscape so that the native animals and plants dependent on that landscape may be viable into the future? Walter et al. (2005) warn that assumptions about fire's positive or necessary role in sustaining sensitive vegetation communities and species do not always bear out desired results. These are but a few of the questions that local, State, and federal agencies, land owners and conservancies, fire departments and districts, and conservation and environmental groups are asking and trying to answer.

As with the immediate and longer-term effects that wildfire has on plants and animals (Chapters 8 and 9), management actions also can have immediate and longer-term effects on at-risk species. Fuel reduction activities, prescribed fire, and post-fire seeding historically have been the primary issues of concern, but other issues, such as fire suppression tactics, use of chemicals and livestock grazing to reduce fuel, and use of heavy machinery after fires to remove burned vegetation are either potential issues or have some example where they have posed a threat to at-risk species. As the size and extent of high severity fires become more common in California, the concerns of these fire effects on wildlife and habitat increases. Many at-risk species have population characteristics such as small population size, a short or long-term decline in size and distribution, and geographic isolation. Notwithstanding habitat resilience and wildlife being a necessary ecological element, addressing high fire severity as a management strategy for at-risk animal and plant species may be a key action is sustaining these species. Sidebar 25.1 presents an example of managing an endangered anadromous fish in a fire-prone stream environment.

There is a range of issues related to fuel and fire management and species protection, from conducting prescribed burns (e.g., as a tool for recovering at-risk species versus reducing fuel load) to addressing potential effects from fire suppression tactics (e.g., blading or constructing fire line) and postfire restoration (e.g., impacts to native flora from grasses seeded to reduce erosion). Table 25.3 lists some of the general conflicts that have arisen between fuel and fire management and species conservation. These activities span the entire range of fire and fuel management: fuel reduction and prescribed burning, fire suppression, and postfire rehabilitation. Planning and implementation of activities can be altered to protect and conserve species and habitat, but the necessary changes are often not seen as being conducive to the primary goals of the fire or fuel management objectives. In essence, protection of at-risk species and their habitats is often identified as a hindrance, even a barrier, to planning and implementing fuel and fire management activities because of the perceived belief that at-risk species protection cannot be done in conjunction with these management actions.

Fire scientists and ecologists have become more involved in fire and fuel management, providing technical advice and recommendations on fire incidents and postfire restoration and recovery teams, aiding in development of fuel reduction and prescribed fire plans, studying the effects of fire and fire

SIDEBAR 25.1 WILDFIRE SEVERITY, ANADROMOUS AT-RISK FISH, AND FOOD WEBS IN A CALIFORNIA COASTAL STREAM

Michael P. Beakes and Sean A. Hayes

In the fire-prone landscapes of California's Mediterranean climate ecosystems wildfire plays an important role in structuring stream temperatures and productivity (Verkaik et al. 2013). The magnitude of stream ecosystem change due to wildfire is generally dependent on burn severity, where more severe wildfires typically result in a greater degree of steam temperature warming and higher likelihood of secondary disturbances such as debris flows (Gresswell 1999, Dunham et al. 2007, Verkaik et al. 2013). In recent decades the frequency and severity of wildfires in California has significantly increased (Westerling et al. 2006, Westerling and Bryant 2008) leading to concern regarding how at-risk species will respond to wildfire-related alterations of stream ecosystems (Wenger et al. 2011).

Small or isolated populations of at-risk species and populations near the edge of their species' distribution may be disproportionately affected by wildfires (Isaak et al. 2010, Wenger et al. 2011). In California, the southernmost population of EESA and CESA coho Salmon (*Oncorhynchu kisutch*, listed as endangered) resides in Scott Creek, a central California coastal stream. Scott Creek also contains ESA-listed steelhead trout (*Oncorhynchu mykiss*) listed as threatened) and is a rain-dominated watershed that drains 78 km^2 of the Santa Cruz Mountains into the Pacific Ocean. From August 12 to 23, 2009, the Lockheed wildfire burned approximately 41% (32 km^2 [12.4 mi^2]) of the Scott Creek watershed prompting a rapid response from local academic and government organizations to investigate the wildfire's impact on listed salmonids. This investigation was aided by an unprecedented amount of prefire data derived from seven years of watershed-wide research on salmon biology. As well, the spatial burn pattern of the Lockheed wildfire included burned and unburned tributaries that provided an ideal "natural" experiment. Much of the Lockheed fire research was conducted during the summer when California coastal streams have relatively low flow, less invertebrate prey available to drift-feeding fishes, seasonally high water temperatures, and are generally considered stressful for at-risk salmonids (Grantham et al. 2012, Sogard et al. 2012, Sloat et al. 2013).

The effects of the Lockheed wildfire on Scott Creek scaled with the fire's severity. The burn intensity of the Lockheed wildfire was relatively moderate over much of the salmonid-bearing tributaries of Scott Creek. However, more intensely burned stream reaches had elevated light levels that were correlated with increased stream temperatures. Mean daily stream temperatures in the most severely burned stream pools were 0.6°C (1.1°F) warmer on average one year after the wildfire (Beakes et al. 2014). As a result of increased stream temperatures, larger fish in the burned region of the watershed experienced a 4.0% increase in energy costs and 6.04 kJ (5.72 Btu) of additional energetic expense over 48 summer days. To mitigate for this increase in energetic demand individual fish would have to consume an additional ~260 g (0.57 lb) (dry mass) of prey (Cummins and Wuycheck 1971, Beakes et al. 2014), lose energy reserves, or seek less energetically costly habitat (Hayes et al. 2008, Sogard et al. 2009, Grantham et al. 2012). Previous studies have shown that warm summer water temperatures can drive changes in the abundance and distribution of salmonids when they are bioenergetically stressed (Sestrich et al. 2011, Sloat et al. 2013) and the Lockheed wildfire investigation at Scott Creek illustrates how wildfire can exacerbate bioenergetically stressful conditions for at-risk salmonids by elevating stream temperatures (Beakes et al. 2014).

Although stream temperatures increased resulting in potential energetic costs to fish, the effects of the Lockheed wildfire may also result in increased stream productivity. Along with increased light, concentrations of some nutrients (e.g., nitrate, NO_3^- µM) were elevated sixfold in a burned region relative to an unburned reference region of Scott Creek during the winter. By burning streamside vegetation, wildfires can increase nutrient availability (Hauer and Spencer 1998, Wan et al. 2001) and light (Dunham et al. 2007) stimulating primary productivity (Gresswell 1999, Verkaik et al. 2013). In some cases, increased primary productivity from wildfires can lead to elevated benthic invertebrate production (Malison and Baxter 2010). Although stream temperatures in Scott Creek have warmed on average in the burned regions of the watershed, elevated stream productivity may increase prey resources and provide compensation for added energetic costs of wildfire to fish over longer time scales.

There were minimal short-term effects of the Lockheed wildfire on the prey base for salmonids in Scott Creek. In general, spatiotemporal variation in terrestrial and aquatic invertebrate prey available to salmonids was predominantly driven by differences among seasons and years rather than differences that could be attributed to the wildfire. The highest postfire rates of terrestrial invertebrate flux in the

(continued)

(continued)

burned region of Scott Creek occurred in the spring and fall at rates similar in timing and magnitude to unburned regions of Scott Creek and to other unburned watersheds on the California coast (Rundio and Lindley 2008). Postfire aquatic invertebrate abundance and biomass were similar between the burned and unburned regions of Scott Creek and increased in the summer and fall relative to the winter and spring, which is typical of stream ecosystems controlled by Mediterranean climates (McElravy et al. 1989, Power et al. 2008, Power et al. 2013). As well, comprehensive analysis of pre- and postfire LiDAR data revealed no evidence of large-scale erosion or debris flows resulting from the Lockheed wildfire that could otherwise have resulted in dramatic changes to the aquatic invertebrate assemblage in Scott Creek. Indeed, it is not surprising that there were no significant differences in the invertebrate prey base for salmonids between the burned and unburned regions of Scott Creek considering the moderate nature of the Lockheed wildfire (Minshall 2003, Malison and Baxter 2010, Jackson et al. 2012).

In contrast to the Lockheed wildfire, previous research has shown that high severity wildfires can profoundly affect stream ecosystems. For example, severely burned watersheds can experience stream temperature increases of up to 10°C (18°F) (Albin 1979) resulting in potentially lethal conditions for thermally-sensitive fishes such as salmon and trout. Severe wildfires can also dramatically reduce prey resources to at-risk fishes. Key resource subsidies such as terrestrial arthropod inputs to streams (Kawaguchi and Nakano 2001) can be reduced by twofold in severely burned watersheds compared to unburned watersheds (Jackson et al. 2012). Severe wildfires can also induce secondary disturbances such as catastrophic debris flows (Dunham et al. 2007), flooding, and sedimentation from denuded landscapes that can profoundly affect aquatic communities (Earl and Blinn 2003, Vieira et al. 2004). Thus, it is imperative to take preventative measures (e.g., fuel load reduction) that may mitigate the intensity of wildfires in fire-prone landscapes and thereby reduce the potential for adverse effects of wildfire on at-risk species.

Results from the Lockheed wildfire investigation corroborate the conclusion that the effects of wildfire on stream ecosystems are governed by fire severity. In response to moderate-severity fires such as the Lockheed wildfire, stream ecosystems that are governed by Mediterranean climates such as Scott Creek appear to be resistant to wildfire-associated change (Verkaik et al. 2013). Over longer time scales, elevated primary productivity in Scott Creek due to increased light and nutrient availability in the burned portions of the watershed may result in net neutral or beneficial effects of fire to at-risk fishes. Thus, management of at-risk species such as listed anadromous salmonids should incorporate management actions aimed at preventing severe wildfires where at-risk species exist and consider the possibility that at-risk species and stream ecosystems may ultimately benefit from low-to-moderate severity wildfires.

surrogates on biotic communities and species, and describing historical and current fire regimes in various ecosystems. The first conference dedicated to fire effects and rare and endangered species and habitats took place in Coeur d' Arlene, Idaho (Greenlee 1997) because of the growing knowledge about the co-evolution of fire-adapted ecosystems and at-risk species and the growing understanding that we can implement prescribed fire regimes varied enough to meet conservation needs, in spite of the real and perceived obstacles (Myers 1998). Integration of local, state, and federal agency activities in fuel management, prescribed fire, and wildland fire response have greatly improved considerations for at-risk species in fire and fuel management. There is still work to be done though as wildland fire suppression and postfire assessment and rehabilitation following the 2007 wildfires in southern California demonstrated that wildlife issues and wildlife specialists were not completely integrated into agency responses (Shaffer et al. 2008).

There are many considerations for protecting at-risk species and their habitat when planning and implementing fire and fuel management activities (Box 25.2), and such considerations can result in restrictions and conditions on land management activities. Prescribed burning in the spring may be desirable because of safety and air quality issues, but may pose a threat to migratory birds nesting, ground-nesting birds, sprouting or flowering plants, or even nursing mammals. Using particular types of machinery or treating particular stands or quantities of vegetation might be advantageous for reducing fuel before fire season, during a wildfire, or even as postfire treatment but may threaten unique or rare at-risk plant populations, animal habitat, or disturb critical animal behavior (this relates to pre-season treatment specifically). However, including conservation measures for at-risk species can become part of the fuels and fire program if these measures are known up-front and agreed to during the planning process. Natural resource management is always complicated

TABLE 25.3
Potential conflicts between mechanical fuel reduction and management of at-risk species

Consideration or technique	Potential conflict
Timing (seasonal)	Feasibility to do project and attaining desired results versus impacting crucial aspect of species life cycle
Level of complexity and detail	Affordability, time required to conduct treatment, level of personnel needed versus planning for the special needs and variety of species involved
Scope (spatial)	Attaining a lower risk by treating a small area that has minimal impacts to plant and animal species versus treating an area large enough to all allow fire to be reintroduced but that may result in adverse effects to a species
Re-entry for further treatment	The need for additional or continual treatment versus repeated affects to species
Techniques: - fuel buffer - fuel break - shaded fuel break	The treatment needed to significantly reduce fire risk (e.g., increase vegetative crown height, alteration of the vegetative community, removal of vegetation) versus managing for specific plant and animal habitat needs (e.g., cover, shelter, migration corridors)

by the need and desire to have multiple resource objectives and involves balancing of considerations for many different objectives.

BOX 25.2. IMPORTANT CONSIDERATIONS FOR AT-RISK SPECIES WHEN CONDUCTING FIRE AND FUELS MANAGEMENT ACTIVITIES

Air quality management:

- What is the effect of limiting the burn window for prescribed fire

Planning:

- Are incidental take permits required

Invasive plant species control:

- What are the immediate and longer-term effects to native species and habitat from
  - Prescribed burning
  - Herbicide application
  - Mechanical treatment

Fuel reduction and prescribed burning:

- Will species or habitat be directly impacted
- Will species habitat be impacted longer-term
- What is the pre-treatment condition

Fire suppression:

- Are there tactical options that would avoid or minimize impact to species or habitat
- Can unburned areas be left as biological refugia

Post-fire rehabilitation:

- What are the effects of post-fire clean-up
- What are the effects of seeding on native plant communities and herbivorous animals
- What are the effects of check dams, silt fences and other technologies on riparian and aquatic ecosystems
- Will monitoring occur to verify and validate both effectiveness of efforts and effects to species and habitat

Protection of life and property has been universally accepted as paramount, so planning and implementation of tactics is always done within the context of judgment and decisions of the agency(ies) responsible for suppressing the wildfire.

Vegetation Management to Lower Risk of Fire

There are four general categories of techniques available to reduce vegetation and fire risk: chemical, biological, mechanical, and prescribed burning. Each category has various applications and each application has various degrees of effectiveness and potential impact to at-risk species and their habitats.

Chemical and Biological Treatments

Chemical and biological treatments are rarely used. For all intents and purposes, chemical treatments are not used by public agencies and rarely used by private landowners to control or reduce fuel load and do not currently represent a threat to at-risk species. If chemical treatments were to be used by agencies in the future, there clearly would be a need to evaluate the impacts to at-risk species, especially plants and amphibious and aquatic animals. Biological treatment is also a rarely used strategy in California. Biological treatment is the use of domestic livestock (i.e., cattle, sheep, goats) to reduce fuel by strategically or broadly allowing herds to graze and browse grasses and shrubs. There is the potential that the domestic livestock to reduce fuels would adversely impact plants (e.g., trampling, consumption) and animals (e.g., disturbance in animal behavior, short-term displacement, trampling of nests) and habitat (e.g., encroachment of invasive plant species, destruction of native seedbank, alteration of solar radiation penetration). Perhaps the best known example of the interaction between a biological treatment and at-risk species is at Laguna Beach. Goats have been used in the

Laguna Beach area since the Laguna Fire of 1993. There has been the concern that the goats' consumption of surrounding vegetation, disturbance of the soil, and the resulting increased solar radiation penetration has reduced the vigor and viability of some populations of Laguna Beach dudleya (*Dudleya stolonifera*), a plant listed as threatened under both ESA and CESA. Due to a lack of monitoring, the effect of the goats on both Laguna Beach dudleya and reducing the risk or impact of future fires is unclear.

Whereas chemical and biological vegetation treatments are relatively minor to or non-existent issues currently, the use of mechanical treatments and prescribed fire are significant issues. Both types of treatments have the potential to both reduce the negative wildfire effects and contribute to the conservation of at-risk species.

Mechanical Treatments

Mechanical treatments represent some of the most common and desired treatments currently used to reduce vegetation fuel due to its economic viability, ready availability of operators, variety of tools and techniques, the social acceptance, and relative low risk to field personnel and local residents. Effects to at-risk species and their habitats occur from the immediate physical changes to the vegetation community and soil and longer-term ecological changes that mechanical treatments cause. Alteration and removal of vegetation equates to alteration of plant and animal habitat. Mechanical treatment upturns and can compact soil, affecting plants and surface and subterranean animals. Removing overstory trees and shrubs alters arboreal animal habitat, allowing increased solar radiation to reach understory and herbaceous plants, and exposing understory and surface animal cover and shelter. The use of mechanical equipment and presence of field personnel can disturb animal behavior, and many techniques disturb soil and vegetation in such a way as to promote invasive plant species to expand or colonize an area (also see Chapter 22). In many regions of California, especially in the urban intermix, managing vegetation for the purpose of protecting human life and property has immediate and longer-term, ecological consequences for native species.

Mechanical treatments can also include or result in the protection of at-risk species and their habitats. Many scientists and conservationists agree that the application of fire is necessary to recover or stabilize many at-risk species; however, the current fuel loads do not allow for the application of fire due to the risk and potential for undesired fire behavior. Therefore, mechanical treatment may be a preferred prerequisite to reducing fuel loads in many places of California so that fire could eventually fulfill its ecological role again. The role of fire in promoting at-risk species is not absolute (Walter et al. 2005), and the merit and function of reducing vegetation along with prescribed fire may be a valuable strategy (Sneeuwjagt et al. 2013).

Mechanical treatments can create conditions where fire would both be safe for humans and within an appropriate fire regime to benefit biota and natural communities. There are many aspects of mechanical treatments that can be modified to provide for at-risk species protection. The timing of the treatment, the techniques chosen, or the degree of vegetation alteration are all examples of how mechanical treatments could be modified to meet fuels objectives and manage for at-risk species. Planning and implementing mechanical treatment that addresses at-risk species may initially require additional planning time, knowledge, and detail and the involvement of more agencies and experts and more resources, time, and, often, staffing, in the field. Windows of operation may be shorter so as to avoid crucial, biological phenomena (e.g., flowering, seeding seed, migration, nesting, rearing) and types of appropriate activities might be fewer. For example, the use of larger machinery might not be preferred where there are plant populations that would be impacted by compaction or disturbance of soil or where the use of machinery would disturb critical animal behavior or activities. In addition, sometimes additional actions, such as marking off of sensitive areas within a treatment unit, even individual plants or stands of shrubs and trees, might have to occur prior to the primary activities commencing. Therefore, although these modifications and involvement of additional staff could be considered to be disincentives to integrating at-risk species protection with mechanical treatments, these conservation measures may also allow for mechanical treatments to occur in areas that were previously not considered due to the presence of at-risk species.

Interestingly, implementation of ESA and CESA has been indicted as both halting fuel treatment projects and even contributing to wildfires by disallowing fuel reduction. This allegation was a prominent issue following the wildland fire storms in 1993 that saw 21 fires burn approximately 80,000 ha (197,000 ac) in six counties in southern California, and up until 2003, represented the largest fires to impact any region of California. Though no example of such conflicts between fire and fuel management and species protection was ever confirmed, agencies decided to pursue agreements to (1) establish cooperative efforts to integrate species needs with fire and fuel management, (2) demonstrate that ESA and CESA protection would be secondary to fire safety regulations, and (3) provide fire management agencies and organizations the flexibility necessary to treat fuel loading without the concern of being in conflict with either law (Box 25.3). Since 1993, and even with these agreements, there still exists tension and the lack of clarity related to developing mutual benefits between fuel reduction and species protection.

Prescribed Burning

Many of the issues related to managing for at-risk species and mechanical fuel treatment are pertinent to prescribed fire: the seasonal timing and spatial scope to achieve prescribed fire goals, the necessity of including additional personnel, funding, and knowledge. But the use of fire also presents unique problems and offers unique opportunities different from all other fuel or fire management techniques. Fundamentally, use of fire restores fire back to natural communities; at-risk species have existed in fire-adapted ecosystems, and the long-term viability of these species is tied to ecological processes such as the fire regime. Wildfire's ability to perform its ecological functions has become rare, and as wildfires become more of a threat to human populations, prescribed fire becomes the only feasible alternative to fill the role of vital role of wildfire. On the other hand, the (1) threat of prescribed burning to human life and property, (2) cost for staffing and equipment to control burns, and (3) public's opposition to prescribe fire's smoke and resulting poorer air quality (see Chapter 23), have prevented the broad application of pre-

BOX 25.3. INTERAGENCY AGREEMENTS AND POLICY DIRECTION INVOLVING CESA AND ESA AND FIRE AND FUEL MANAGEMENT PRACTICES

1994 BOF and California Fish and Game Commission adopt a joint policy for wildland fire in an attempt to coordinate fire and fuels management and wildlife protection and conservation between Cal Fire and CDFW.

1995 CDFW adopts policy and procedures to address activities regarding pre-, during-, and post-fire, especially in conjunction with Cal Fire. This includes a CDFW policy of conducting postfire seeding.

1995 Cal Fire convenes two multi-agency working groups to review and make recommendations for revision of its Vegetation Management Program (VMP) and Emergency Watershed Protection Program. One result of this effort was the revision of Calfire's VMP programmatic environmental impact report.

1995 Cal Fire signs a Memorandum of Understanding (MOU) exempting land owners and agencies conducting activities to promote fire safety around homes and structures from prosecution for incidental take of listed species under CESA if those activities were conducted under the conditions of fire safety regulations (California Code, §4260-4263). The MOU expired September 1999 and was not renewed.

1996 Cal Fire, CDFW, and USFWS enter into MOUs in Riverside and San Diego counties to address the incidental take of species under both ESA and CESA resulting from fuels management activities for the purposes of protecting life and property.

2002 USFWS and NOAA issued a memorandum with guidance providing alternative approaches for streamlining section 7 consultation under the ESA for hazardous fuels treatment projects.

scribed fire for any purposes, let alone conservation of at-risk species.

Studies have demonstrated the benefits of prescribed fire to native animals and plants. For most of the twentieth century, the focus was on the benefits to wildlife game species and the vegetation communities they inhabit (Biswell 1957, 1969). There has been an increasing appreciation of the potential recovery role of prescribed fire for California's native, at-risk biota. Since the middle of the 1980s, prescribed fire has been proposed or actually used to manage and benefit at-risk species, such as the California spotted owl (*Strix occidentalis occidentalis*) (Weatherspoon et al. 1992), Stephen's kangaroo rat (*Dipodomys stephensi*) (Price and Tayler 1993), butterfly diversity (Huntzinger 2003), at-risk native plants (USFWS 2002a), and Alameda whipsnake (*Masticophis lateralis euryxanthus*) and pallid manzanita (*Arctostaphylos pallida*) (USFWS 2002b). But logistical, safety, and human health (smoke) conflicts have prevented prescribed fire from becoming a major tool for conservation. The needs and priorities for both the public and agencies conducting burns have prevented the majority of prescribed fires from being carried out during periods of greatest ecologic value (e.g., late summer and autumn) and allowed prescribed fire during times when fire is either of less ecological value or of ecological impact (i.e., spring burning). It has been postulated that in a state with so many fire-adapted or fire-dependent natural communities and species, prescribed fire may be a mandatory tool for the conservation and recovery of at-risk species (Biswell 1989). It also is likely that in a state with so many people and so much urban-wildland interface, prescribed fire either will have a selective, minor role or Californians will have to make a wholesale re-examination of prescribed burning in the future.

The use of prescribed fire is neither a universal solution nor is it easy to undertake in California. Walter et al. (2005) point out that unique vegetation communities (e.g., Bishop pine [*Pinus muricata*], over-age chaparral, chaparral-mixed forest) in southern California could be at greater risk or not enhanced by fire. Using fire to benefit at-risk species in California (e.g., California spotted owl, northern flying squirrel [*Glaucomys sabrinus*], desert bighorn sheep [*Ovis canadensis nelsoni*]) has been shown to have complex effects on habitat and food availability, diversity, and distribution (Meyer et al. 2008, Roberts et al. 2011, Holl et al. 2012). Sneeuwjagt et al. (2013) compared the prescribed fire programs in Western Australia to California, finding that California struggles in comparison to Western Australia to institute and maintain a successful program. The authors list recommendations to develop a more successful program in California. Those changes included: train and maintain larger crews available to land managers to conduct burns; plan for and be allowed to stage burns in complex fuel conditions; improve state, federal, and local collaboration to approximate the single agency responsibility that exists in western Australia; develop and implement an extensive public and education program; improve the relationship with air quality agencies and remove disincentives regarding prescribed fire emissions. As California's human population growth and climate change both contribute to large, severe wildfire, the use of prescribed fire may become a vital tool in reducing the potential damage of severe wildfire incidents and creating socially acceptable fire effects beneficial to at risk species and their habitats.

The Effects of Fire Suppression Activities on At-Risk Species

Fire Suppression

Fire suppression activities represent an interesting issue and threat to at-risk species and habitat. On one hand, there is the cumulative effect of fire suppression activities over approximately the last 80 years across much of California, particularly in northwest California and the Sierra Nevada where suppression is the main contributor of increased forest fuels and fire risk (Safford and Van de Water 2014). On the other hand, there are the specific effects of fire suppression activities for an individual fire. The success of fire suppression in excluding fire from many vegetation communities is now considered to have been one of the major causes for critical, deleterious shifts in those vegetation communities. The loss of the normal range of fire regime attributes and the subsequent (1) loss of more natural fire behavior and (2) substitution of altered fire behavior has resulted in negative effects to biotic communities. In fact, recent endeavors to alter wildfire management have cited the adverse impacts of successful fire suppression to ecological health, wildlife, at-risk species, and habitat as being a reason for modifying fire and fuels management (CBOF 1999; USDA/USDI 2001, USA 2002).

Absence of fire or major shifts in fire regime (e.g., from frequent to infrequent or vice versa) have had a critical effect on fire-adapted or dependent at-risk plant species, and there are examples of these changes being identified as a cause of decline (e.g., Baker's manzanita [*Arctostaphylos bakeri* subsp. *bakeri*], USFWS 1998a). Plant survival and population viability are linked both to seed bank viability during the time between fires and the effects that subsequent fires have both on plants and seed (see Chapter 8). The absence of fire also has profound effects on vegetation structure and composition and thus has consequences to at-risk animal species. In many California bioregions these changes to vegetation communities due to a lack of fire and the resulting consequences have been the topic of discussion between conservationists, land managers, and the general public with regard to many at-risk species (Table 25.4).

TABLE 25.4
Bioregion examples of at-risk species affected negatively by the exclusion of fire

Bioregion	Example of at-risk species
Sierra Nevada	California spotted owl, *Strix occidentalis occidentalis*
Central Valley	Colusa grass, *Neostapfia colusana*
Central Coast	Monterey pine, *Pinus radiata*
South Coast	California gnatcatcher, *Polio californica*
Southeast Deserts	Bighorn sheep, *Ovis canadensi nelsonis*

Unlike other fuel and fire management, wildfire suppression is immediate and often emergency in nature. Hence, the ability to alter suppression activities in lieu of at-risk species is less, and the actual impacts to at-risk species tend to be more acceptable because human life and property are also at risk. In recent years, agencies responsible for suppressing fire have appreciated the needs of at-risk species and habitats to the point of having biologists and ecologists as part of the fire incident command team to act as technical resource advisors (i.e., READs), giving fire fighters and fire incident commanders options for addressing the fire and minimizing or avoiding impacts to at-risk species and their habitats. Some of the more important suppression activities that have been examined include the use of heavy machinery to clear away vegetation, burning out unburned areas after a fire front passes, drafting water, applying fire surfactant or retardant to vegetation, and the placement of incident base camp and other fire-fighting bases. Table 25.5 lists some of potential impacts and solutions of fire suppression activities on at-risk species. In some cases, fire suppression activities have contributed to the declined status of an at-risk species (e.g., western lily [*Lilium occidentale*], USFWS 1998b), while in other cases, guidelines and recommendations have been developed specifically for the fire activities in an attempt to promote conservation of an at-risk species (e.g., California red-legged frog [*Rana aurora draytonii*], USFWS 2002a).

Post-Wildfire Restoration and Recovery

Post-wildfire restoration and recovery (postfire recovery) has similarities to both vegetation management and fire suppression, with some activities being considered emergency in nature (e.g., application of wood straw on severely burned slopes) and others considered to be longer-term management (e.g., tree planting). Several activities conducted during postfire recovery can impact at-risk species, including fire mop-up, removal of additional vegetation, compaction of soil (i.e., when using heavy machinery), use of engineered structures (e.g., check dams), and application of seed to burned areas. The potential for immediate impacts to at-risk species may be relatively low, and there are few cases where there has been concern of such impacts. One exception is Laguna Beach. As with the use of sheep for vegetation control (see above), Laguna Beach dudleya was impacted by postfire seeding (hydro and not aerial) following the Laguna Fire in 1993. There may be many reasons for the lack of documented impacts, including that (1) restoration or recovery areas are areas that have already been burned or affected by suppression activities, and thus species were already affected or had left the area or (2) there usually is not sufficient time for planning and technical input from scientists to avoid or minimize effects to habitat and species (Shaffer et al. 2008).

Some postfire recovery activities have been seen as posing a real longer-term threat to vegetation communities, species, and habitat (Peppin et al. 2010). Such activities all fall within actions taken to minimize or control delivery of sediment off and downstream of burned slopes. Hay bales, log dams, check dams, mulch, netting, and seeding all are activities undertaken to trap or hold sediment back (e.g., dams) or stabilize the slopes themselves (e.g., netting, aerial seeding). Most of these activities are localized both in their application and effect, though a check dam does affect all down-stream aquatic and riparian habitats as long as the dam in kept in place. The major exception is seeding. This postfire rehabilitation practice is considered to have had the greatest impact due to the negative effects of seeding on native vegetation recovery and persistence (Peppin et al. 2010, Stella et al. 2010).

For most of the twentieth century, seeding using nonnative annual grasses was seen as the most viable and cost-efficient means of stabilizing large areas immediately following fires. Research and postfire monitoring in the 1980s and, especially, in 1990s and into the 2000s has questioned the effectiveness of seeding but also recognized the threats to biodiversity and the native vegetation communities (Conard et al. 1995, Keeler-Wolf 1995, O'Leary 1995). Impacts to native biota from early twentieth century seeding or from postfire rehabilitation since enactment of ESA and CESA are difficult to identify due lack of recognition of the issue and not enough time having elapsed, respectively. To date, attention has been focused on the effects to the native plant species that define the local plant community and not to at-risk plant or animal species. Nevertheless, in the mid-1990s, CDFW and the California Native Plant Society adopted a policy and statement, respectively, highly critical of the use of nonnative grasses

TABLE 25.5
Potential issues and solutions for fire suppression impacts to at-risk species

Fire suppression activity	Potential impact	Potential solutions
Use of heavy machinery	a. Soil: impacting future vegetation and increasing sedimentation b. Vegetation: destruction of subterranean, surface, and arboreal species habitat	Evaluate the options for: other locations – amount of area that needs to be cleared (e.g., the blade width of a fuel break) – fuel buffers vs. fuel breaks – use of existing or natural fuel breaks
Burning out	Critical postfire, biologic refugia for wildlife	Setting aside unburned refugia
Water drafting	Movement of invasive and/or native aquatic species, fatality of native aquatic species, modification of aquatic habitat	Alternate drafting sites, using screens to avoid collecting or moving aquatic animals
Application of fire-surfactant or retardant	Native plants, impacts to aquatic areas	Avoid application to at-risk plant populations and near aquatic habitats
Incident command camps and other fire-fighting bases	Modification and/or destruction of vegetation and soil, noise disturbance	Locate incident command centers or other fire bases away from sensitive habitats (i.e., previously disturbed areas)

and encouraging the use of native grasses or natural reseeding wherever possible. It is to be seen if and how fire management agencies will alter the nearly century-long practice of the use of nonnative grass seed for emergency rehabilitation when down-stream values are considered at risk. And it is to be seen what research and monitoring might tell of effects to at-risk species on the slopes or down-stream of seeding activities. In Arizona and across the western United States, research conducted in has found that postfire seeding using native or nonnative seeds is likely ineffective (Stella et al. 2010, Peppin et al. 2010).

Potential Solutions and the Future

One of the most interesting trends in recent years has been a seemingly ubiquitous, government agency policy perspective that reducing vegetation fuel loading is necessary and beneficial to protecting, enhancing, or restoring ecosystem health and at-risk species. In California, this viewpoint, and subsequent collaboration, funding, and land management planning, can be tied both to the development of the California Fire Plan (CBOF 1999) and National Fire Plan (USDA/USDI 2001). Several critical documents affecting significant wildland regions, if not the entire State, have been developed (e.g., Sierra Nevada Forest Plan [USDAFS 2003]), regionally (e.g., Association of Western Governor's 10-Year Comprehensive Strategy [WGA 2001]), or nationally (National Fire Plan). All of these efforts to plan and implement substantial fuel reduction in shrub, woodland, and forest communities are now in their second decade of development and implementation. Each endeavor has the potential to improve or degrade the natural environment, or do both. It is to be seen if such comprehensive plans and activities will alter fire regimes to benefit natural communities, and if at-risk species will benefit from these alterations. It is also unknown what impacts will occur to at-risk species during activities to alter fuel load, and ultimately, fire behavior, and if any at-risk species might not endure to ever benefit from longer-term changes. The ongoing and future effects of climate change on management as well as to existing and future biotic communities are unknown. And lastly, it is unknown if there will be the longer-term commitment by agencies and the public essential to achieving changes to the landscape that benefit natural communities, let alone the commitment necessary to conserve at-risk species. Notwithstanding these uncertainties, at-risk species protection will depend, at least in part, on the success of these comprehensive plans to manage both wildlands and fire.

Success in achieving both the goals of fuel and fire management and conservation of species and habitat already has a history in California. And with (1) a growing knowledge of the fire ecology of the bioregions of California and many of its native flora and (2) the ever-increasing capabilities and variety fire and fuel management techniques, the potential to meet both goals should be greater than ever before. Combined with the many land management plans and collaborations mentioned above, protecting California's at-risk species, even with an ever-growing urban-wildland intermix, may be more of possibility in the future than at any time in the State's past.

Achieving Both Goals

Fuel management can be done in a way that (1) minimizes impacts, (2) is neutral, or (3) is beneficial to biotic communities. Often, this requires a combination of mechanical treatment and prescribed fire and the longer-term planning and management of the area in question. Since the 1980s, such planning and implementation has been undertaken in several regions in California (Table 25.6). The efforts have protected life and property, maintained ecological function, and promoted the recovery of at-risk species. In places like the Santa Rosa Plateau, the Channel Islands, and the Jepson

TABLE 25.6
Examples in California of integrating fire and fuels management and conservation and protection of at-risk species and habitat

Integrated land management	Management activities
Santa Rosa Plateau Ecological Reserve, southern Riverside County	Mechanical treatment of vegetation, prescribed burning
Western Riverside County Multiple Species Habitat Conservation Plan	Mechanical treatment of vegetation
Channel Island National Park	Mechanical treatment of vegetation, prescribed burning, control of invasive plant species
U.C. Davis Jepson Prairie Reserve, Solano County	Mechanical treatment of vegetation, prescribed burning, control of invasive plant species
Pine Hills Preserve, El Dorado County	Mechanical treatment of vegetation, prescribed burning

Prairie, successful conservation of at-risk species had to be integrated with the management of fire, fuel, and invasive plant species.

As mentioned above, comprehensive planning and implementation of land management activities represent a practical solution to accomplishing protection and conservation of at-risk species while allowing for effective fuel and fire management. Many plans cover large areas with multiple, at-risk species, and involve many agencies and landowners. To protect at-risk species and habitats, management plans must address the temporal and spatial characteristics of vegetation communities and fire regimes in regard to the fire ecology of at-risk species and the short-term and longer-term consequences that any activity might have on the at-risk populations and habitats. Though this may seem to be a difficult mission to accomplish, the several examples in California that have already occurred demonstrate that such planning and implementation is possible.

Other land management options can initially come from either the perspective of species protection or fire and fuel management and then integrate the needs of the other. From the perspective of fuel and fire management, the California Fire and National Fire plans provide for integrating vegetation communities, species, and habitat needs into fuel and fire management. From the perspective of at-risk species, California's Natural Community Conservation Planning program, recovery plans under CESA and ESA, and Habitat Conservation Plans under ESA allow for reciprocal planning. Under the California Fire Plan, Cal fire has successfully conducted fuel treatment and prescribed burning projects that have addressed the conservation needs of at-risk species in the southern California (e.g., rodents, birds, butterflies, and plants), the western foothills of the Sierra Nevada (e.g., plants), and several regions' grassland and vernal pool ecosystems (e.g., crustaceans and plants). Yosemite National Park and Calaveras Big Trees State Park have managed vegetation and conducted prescribed burning for several years benefiting the forest ecosystem and their native species while addressing fire and fuel management goals (Biswell 1989). Such management does the most to prevent native species from becoming at-risk. Even such ecosystem-based management should investigate and address specific fire ecology needs of already at-risk species; conditions and activities that benefit the majority of native species might not benefit particular, at-risk species.

> **BOX 25.4. EXAMPLES OF SMALL-SCALE RECOVERY EFFORTS FOR AT-RISK SPECIES THAT HAVE UTILIZED VEGETATION AND FIRE MANAGEMENT ACTIVITIES**
>
> Large-flowered fiddleneck, *Amsinckia grandiflora*
> Morro manzanita, *Arctostaphylos morroensis*
> Ione manzanita, *Arctostaphylos myrtifolia*
> Santa Cruz tarplant, *Holocarpha macrodenia*

Not all efforts integrating species protection and fuel management need be large-scale. Many smaller, local efforts have been attempted across the State. Often, these activities focus on a single, at-risk species, and to-date, often address an at-risk plant (Box 25.4, USFWS 1997b). It is not explicitly known why this is the case. There is both an agency and public sensitivity to incidental take of animal species and a more obvious, ecological relationship between fire and plant population viability. These factors may be the foundation for a relative lack of application of fire and fuel management to animal recovery. The programs that have addressed rodent and bird species in South Coast bioregion and the potential, extensive fire and fuel program in the East San Francisco Bay region may signal the acknowledgment that at-risk animal recovery can benefit from fuel and fire management activities also.

As mentioned above, the use of prescribed fire may provide the last, best opportunity for at-risk species and habitat where fire has been absent or where, because of human development, is not likely to be allowed to return naturally. Prescribed burning has been identified as being a tool for recovery for a variety of at-risk species across many regions of California, including the San Francisco Bay area, Central Valley grasslands, the central Sierra Nevada foothills, south-

TABLE 25.7
Examples of at-risk species recovery where prescribed fire has been recommended, researched, or used as a recovery action

Region to be treated with prescribed fire	Targeted habitat for at-risk species
San Francisco Bay	More than 12 serpentine-soil plants and dependent insects; East Bay chaparral species, including Alameda whipsnake
Central Sierra Nevada foothills	6 rare plants
Channel Islands	More than 12 rare plants
Vernal pools (from Butte to San Diego County)	More than 30 plants and crustaceans, including several grass species and fairy shrimp (*Branchinecta* ssp.)
Southern California shrub and chaparral	Species of coastal and interior chaparral and coastal sage brush communities, including California gnatcatcher and cactus wren

ern California chaparral and shrublands, and the Channel Islands (Table 25.7). One effort that may provide a model for the future has been evolving in the foothills east of the San Francisco Bay region. Agencies (i.e., USFWS, CDFW, CDPR), water districts, and landowners in counties east of San Francisco Bay worked to develop a broad, large-scale management plan that would include fuel load reduction and prescribed burning to protect suburban communities and watersheds, promote recovery of two federally listed species, the Alameda whipsnake and pallid manzanita, and protect an additional four sensitive species (USFWS 1997a). Planning efforts like this represent the integration of complex land management needs across a vast area of multiple land ownerships.

There are numerous examples of across California's nine bioregions demonstrating that fire and fuel management activities, whether vegetation or invasive species control and reduction, prescribed burning, fire suppression, or postfire rehabilitation and restoration can be integrated with conserving and protecting at-risk species and their habitats and ecological processes needed to promote species recovery. Such integration has (1) been done on small and large scales, (2) addressed single- and multiple species conservation, and (3) occurred in isolated, wild areas as well as the urban intermix. And possibly as important, has (1) not resulted in loss of property or life and (2) has not caused insurmountable or excessive financial burden to either the private or public sectors of California. Human communities cannot afford the risk that landscape scale, high-intensity fire presents. Aquatic and terrestrial species at risk, and the ecological systems they depend on, cannot be sustained or recovered without proximate and longer-term ecological functioning provided by fire. Clearly, the viability of both human communities and at-risk species depend on integrating fire and fuel management with at-risk species conservation and protection wherever both activities overlap.

References

Albin, D.P. 1979. Fire and stream ecology in some Yellowstone Lake tributaries. California Fish and Game 65: 216–238.

Beakes, M.P., J.W. Moore, S.A. Hayes, and S.M. Sogard. 2014. Wildfire and the effects of shifting stream temperature on salmonids. Ecosphere 5: art. 63.

Biswell, H.H. 1957. The use of fire in California chaparral for game habitat improvement. Pages 151–155 in: Proceedings, Society of American Foresters, Syracuse, New York, New York, USA.

———. 1969. Prescribed burning for wildlife in California brushlands. Transactions of the 34th North American Wildlife and Natural Resources Conference, Wildlife Management Institute, Washington, D.C., USA.

———. 1989. Prescribed fire in California wildlands vegetation management. University of California Press, Berkeley, California, USA.

California Board of Forestry and Fire Protection [CBOF]. 1999. California Fire Plan. California Department of Forestry and Fire Protection, Sacramento, California, USA.

California Department of Finance. 2015. New population report. Sacramento, California, USA. http://www.dof.ca.gov/research/demographic/reports/estimates/e-1/documents/E-1_2015Press Release.pdf

California Department Fish and Wildlife [CDFW]. 2015a. State and federally listed endangered and threatened animals of California. Sacramento, California, USA.

———. 2015b. State and federally listed endangered, threatened, and rare plants of California. Sacramento, California, USA.

California Department of Forestry and Fire Protection [Cal Fire]. 2011. Cal Fire jurisdiction fires, acres, and dollar damage: 1933-2010. Sacramento, California, USA.

Conard, S.G., J.L. Beyers, and P.M. Wohlgemuth. 1995. Impacts of postfire grass seeding on chaparral systems—what do we know and where do we go from here? Pages 52–64 in: J.E. Keeley and T. Scott, editors. Brushfires in California wildlands: Ecology and Resource Management. International Association for Wildland Fire, Fairfield, Washington, USA.

Cummins, K.W., and J.C. Wuycheck. 1971. Caloric equivalents for investigations in ecological energetics. International Association of Theoretical and Applied Limnology 18: 1–158.

Dunham, J.B., A.E. Rosenberger, C.H. Luce, and B.E. Rieman. 2007. Influences of wildfire and channel reorganization on spatial and temporal variation in stream temperature and the distribution of fish and amphibians. Ecosystems 10: 335–346.

Earl, S.R., and D.W. Blinn. 2003. Effects of wildfire ash on water chemistry and biota in South-Western U.S.A. streams. Freshwater Biology 48: 1015–1030.

Gill, A.M., S.L. Stephens, and G.J. Cary. 2013. The worldwide "wildfire" problem. Ecological Applications 23: 438–454.

Grantham, T.E., D.A. Newburn, M.A. McCarthy, and A.M. Merenlender. 2012. The role of streamflow and land use in limiting oversummer survival of juvenile steelhead in California streams. Transactions of the American Fisheries Society 141: 585–598.

Greenlee, J.M., editor. 1997. Proceedings: First conference on fire effects on rare endangered species and habitats. International Association for Wildland Fire, Fairfield, Washington, USA.

Gresswell, R.E. 1999. Fire and aquatic ecosystems in forested biomes of North America. Transactions of the American Fisheries Society 128: 193–221.

Hauer, F.R., and C.N. Spencer. 1998. 'Phosphorus and nitrogen dynamics in streams associated with wildfire: a study of immediate and long-term effects. International Journal of Wildland Fire 84: 183–198.
Hayes, S.A., M.H. Bond, C.V. Hanson, E.V. Freund, J.J. Smith, E.C. Anderson, A.J. Ammann, and R.B. MacFarlane. 2008. Steelhead growth in a small central California watershed: upstream and estuarine rearing patterns. Transactions of the American Fisheries Society 137: 114–128.
Holl, S.A., V.C. Bleich, B.W. Callenberger, and B. Bahro. 2012. Simulated effects of two fire regimes on bighorn sheep: the San Gabriel Mountains, California USA. Fire Ecology 8(3): 88–103.
Huntzinger, P.M. 2003. Effects of fire management practices on butterfly diversity in the forested western United States. Biological Conservation 113: 1–12.
Isaak, D.J., C.H. Luce, B.E. Rieman, D.E. Nagel, E.E. Peterson, S.P. Horan, and G.L. Chandler. 2010. Effects of climate change and wildfire on stream temperatures and salmonid thermal habitat in a mountain river network. Ecological Applications 20: 1350–1371.
Jackson, B.K., S.M.P. Sullivan, and R.L. Malison. 2012. Wildfire severity mediates fluxes of plant material and terrestrial invertebrates to mountain streams. Forest Ecology and Management 278: 27–34.
Jin, Y., M.L. Goulden, N. Faivre, S. Veraverbeke, F. Sun, A. Hall, M.S. Hand, S. Hook, and J.T. Randerson. 2015. Identification of two distinct fire regimes in Southern California: implications for economic impact and future change. Environmental Research Letters 10(9).
Kawaguchi, Y., and S. Nakano. 2001. Contribution of terrestrial invertebrates to the annual resource budget for salmonids in forest and grassland reaches of a headwater stream. Freshwater Biology 46: 303–316.
Keeler-Wolf, T. 1995. Post-fire emergency seeding and conservation in southern California shrublands. Pages 127–140 in: J.E. Keeley and T. Scott, editors. Brushfires in California Wildlands: Ecology and Resource Management. International Association for Wildland Fire, Fairfield, Washington, USA.
Malison, R.L., and C.V. Baxter. 2010. Effects of wildfire of varying severity on benthic stream insect assemblages and emergence. Journal of the North American Benthological Society 29: 1324–1338.
Meyer, M.D., M.P. North, and S.L. Roberts. 2008. Truffle abundance in recently prescribed burned and unburned forests in Yosemite National Park: implications for mycophagous mammals. Fire Ecology 4(2): 105–114.
McElravy, E.P., G.A. Lamberti, and V.H. Resh. 1989. Year-to-year variation in the aquatic macroinvertebrate fauna of a northern California stream. Journal of the North American Benthological Society 8: 51–63.
Minshall, G.W. 2003. Responses of stream benthic macroinvertebrates to fire. Forest Ecology and Management 178: 155–161.
Myers, R.K., K.J. Hofeldt, and D.H. Van Lear. 1998. Constraints to using fire after Hurricane Hugo to restore fire-adapted ecosystems in South Carolina. Pages 167–172 in T.L. Pruden and L.A. Brennan, editors. Fire in ecosystem management: shifting the paradigm from suppression to prescription. Tall Timbers Fire Ecology Conference Proceedings No. 20. Tall Timbers Research Station, Tallahassee, Florida, USA.
North, M.P., S.L. Stephens, B.M. Collins, J.K. Agee, G. Aplet, J.F. Franklin, and P.Z. Fule. 2015. Reform forest fire management: agency incentives undermine policy effectiveness. Science 349: 1280–1281.
O'Leary, J.F. 1995. Potential impacts of emergency seeding on cover and diversity pattern of California's shrubland communities. Pages 141–148 in: J.E. Keeley and T. Scott, editors. Brushfires in California Wildlands: Ecology and Resource management. International Association for Wildland Fire, Fairfield, Washington, USA.
Peppin, D.L., P.Z. Fule, C.H. Sieg, J.L. Beyers, and M.E. Hunt. 2010. Post-fire seeding in forests of the west: trends, costs, effectiveness, and use of native seeds. Final Report to the Joint Fire Sciences Program Project ID 08-2-1-11. Bureau of Land Management, Boise, Idaho, USA.
Power, M.E., J.R. Holomuzki, and R.L. Lowe. 2013. Food webs in Mediterranean rivers. Hydrobiologia 719: 119–136.
Power, M.E., M.S. Parker, and W.E. Dietrich. 2008. Seasonal reassembly of a river food web: floods, droughts, and impacts of fish. Ecological Monographs 78: 263–282.
Price, M.V. and K.E. Tayler. 1993. The potential value of fire for managing Stephens' kangaroo rate habitat at Lake Perris State Recreational Area. Final Report. Lake Perris Recreational Area Stephens' kangaroo rat habitat restoration. California Department of Parks and Recreation. Sacramento, California, USA.
Roberts, S.L., J.W. van Wagtendonk, A.K. Miles, and D.A. Kelt. 2011. Effects of fire on spotted owl site occupancy in a late-successional forest. Biological Conservation 144: 610–619.
Rundio, D.E., and S.T. Lindley. 2008. Seasonal patterns of terrestrial and aquatic prey abundance and use by *Oncorhynchus mykiss* in a California Coastal Basin with a Mediterranean climate. Transactions of the American Fisheries Society 137: 467–480.
Safford, H.D., and K.M. Van de Water. 2014. Using fire return interval departure (FRID) analysis to map spatial and temporal changes in fire frequency on national forest lands in California. USDA Forest Service Research Paper PSW-RP-266. Pacific Southwest Research Station ,Albany, California, USA.
Sestrich, C.M., T.E. McMahon, and M.K. Young. 2011. Influence of fire on native and nonnative salmonid populations and habitat in a western Montana basin. Transactions of the American Fisheries Society 140: 136–146.
Shaffer, K.E., T. Stewart, D. Blankenship, and K. Schmoker. 2008. Consequences of 2007 fire response and post fire restoration to wildlife habitat in southern California. Pacific Coast Fire Conference: Changing Fire Regimes, Goals and Ecosystems, December 1–4, 2008, San Diego, California. Unpublished manuscript. California Department of Fish and Wildlife, Sacramento, California, USA.
Sloat, M.R., A.-M.K. Osterback, and P. Magnan. 2013. Maximum stream temperature and the occurrence, abundance, and behavior of steelhead trout (*Oncorhynchus mykiss*) in a southern California stream. Canadian Journal of Fisheries and Aquatic Sciences 70: 64–73.
Sneeuwjagt, R.J., T.S. Kline, and S.L. Stephens. 2013. Opportunities for improved fire use and management in California: lessons from Western Australia. Fire Ecology 9(2): 14–25.
Sogard, S.M., J.E. Merz, W.H. Satterthwaite, M.P. Beakes, D.R. Swank, E.M. Collins, R.G. Titus, and M. Mangel. 2012. Contrasts in habitat characteristics and life history patterns of *Oncorhynchus mykiss* in California's Central Coast and Central Valley. Transactions of the American Fisheries Society 141: 747–760.
Sogard, S.M., T.H. Williams, and H. Fish. 2009. Seasonal patterns of abundance, growth, and site fidelity of juvenile steelhead in a small coastal California stream. Transactions of the American Fisheries Society 138: 549–563.
Stella, K.A., C.H. Sieg, and P.Z. Fulé. 2010. Minimal effectiveness of native and non-native seeding following three high severity wildfires. International Journal of Wildland Fire 19: 746–758.
Stralberg, D. 2000. Landscape-level urbanization effects on chaparral birds: a Santa Monica Mountains case study. Pages 125–136 in: J.E. Keeley, M. Baer-Keeley, and C.J. Fotheringham, editors. 2nd Interface between Ecology and Land Development in California. U.S. Geological Survey Open-File Report OFR-00-62. Reston, Virginia, USA.
United States Fish and Wildlife Service [USFWS]. 1997a. Recovery plan for large-flowered fiddleneck, *Amsinckia grandiflora*. Region 1, U.S. Fish and Wildlife Service, Portland, Oregon, USA.
———. 1997b. Final Rule: Endangered and Threatened Wildlife and Plants; Determination of Endangered Status for the Callippe Silverspot Butterfly and the Behren's Silverspot Butterfly and Threatened Status for the Alameda Whipsnake. Federal Register 62(234): 64306–64320.
———. 1998a. Recovery plan for serpentine soil species of the San Francisco Bay area. Region 1, U.S. Fish and Wildlife Service, Portland, Oregon, USA.
———. 1998b. Final recovery plan for the endangered western lily (*Lilium occidentale*). Region 1, U.S. Fish and Wildlife Service, Portland, Oregon, USA.
———. 2002a. Recovery plan for the California red-legged frog (*Rana aurora draytonii*). Region 1, U.S. Fish and Wildlife Service, Portland, Oregon, USA.

———. 2002b. Draft recovery plan for chaparral and scrub community species east of San Francisco Bay, California. Region 1, U.S. Fish and Wildlife Service, Portland, Oregon, USA.

United States Department of Agriculture Forest Service [USDAFS]. 2003. Draft Supplemental Environmental Impact Statement. USDA Forest Service R5-MB-019. , Pacific Southwest Region, Vallejo, California, USA.

United States Departments of Agriculture and Interior [USDA/USDI]. 2001. The National Fire Plan: A Report to the President In Response to the Wildfires of 2000 September 8, 2000. Managing the Impact of Wildfires on Communities and the Environment. Washington, D.C., USA.

United State of America, Office of the President. 2002. Healthy Forests: *An Initiative for Wildfire Prevention and Stronger Communities*. Washington, D.C., USA.

Verkaik, I., M. Rieradevall, S.D. Cooper, J.M. Melack, T.L. Dudley, and N. Prat. 2013. Fire as a disturbance in mediterranean climate streams. Hydrobiologia 719: 353–382.

Vieira, N.K.M., W.H. Clements, L.S. Guevara, and B.F. Jacobs. 2004. Resistance and resilience of stream insect communities to repeated hydrologic disturbances after a wildfire. Freshwater Biology 49: 1243–1259.

Walter, H.S., T. Brennan, and C. Albrecht. 2005. Fire management in some California ecosystems. Pages 257–260 in: B.E. Kus and J.L. Beyers, technical coordinators. Planning for biodiversity: bringing research and management together. USDA Forest Service General Technical Report PSW-GRT-195. Pacific Southwest Research Station, Albany, California, USA.

Wan, S., D. Hui, and Y. Luo. 2001. Fire effects on nitrogen pools and dynamics in terrestrial ecosystems: a meta-analysis. Ecological Applications 11: 1349–1365.

Wenger, S.J., D.J. Isaak, C.H. Luce, H.M. Neville, K.D. Fausch, J.B. Dunham, D.C. Dauwalter, M.K. Young, M.M. Elsner, B.E. Rieman, A.F. Hamlet, and J.E. Williams. 2011. Flow regime, temperature, and biotic interactions drive differential declines of trout species under climate change. Proceedings of the National Academy of Sciences 108: 14175–14180.

Westerling, A.L., and B.P. Bryant. 2008. Climate change and wildfire in California. Climatic Change 87: 231–249.

Westerling, A.L., H.G. Hidalgo, D.R. Cayan, and T.W. Swetnam. 2006. Warming and earlier spring increase western U.S. forest wildfire activity. Science 313: 940–943.

Western Governors' Association [WGA]. 2001. A Collaborative Approach for Reducing Wildland Fire Risks to Communities and the Environment 10-Year Comprehensive Strategy. Western Governors' Association, Washington, D.C., USA.

Weatherspoon, C.P., S.J. Husari, and J.W. van Wagtendonk. 1992. Fire and fuels management in relation to owl habitat in forests of the Sierra Nevada and southern California. Pages 247–260 in: J. Verner, K.S. McKelvey, B.R. Noon, R.J. Gutierrez, G.I. Gould, and T.W. Beck, technical coordinators. The California Spotted Owl: A Technical Assessment of Its Current Status. USDA Forest Service General Technical Report PSW-GTR-133. Pacific Southwest Research Station, Albany, California, USA.

CHAPTER TWENTY-SIX

Fire and Climate Change

CHRISTINA M. RESTAINO AND HUGH D. SAFFORD

The only thing that is constant is change.

HERACLITUS

The ecological impacts of climate change present major challenges for resource managers and policy makers in the state of California. State and federal agencies are now requiring that vulnerability to climate change be addressed in land and resource management plans, and major efforts are underway to augment resilience in California ecosystems (USDA 2012, Obama 2013, Cal-NRA 2014). One of the major effects of climate change in California will be its influence on ecological processes like fire (Westerling and Bryant 2008, Safford et al. 2012). Public and private lands in the state are experiencing increasingly large and destructive wildfires, from desert and Great Basin ecosystems, to southern California chaparral and the forested mountains in the north (Miller et al. 2009, Miller and Safford 2012, Barbero et al. 2014). The impacts of climate change on fire in California are accompanied by accelerating population growth in one of the largest and most complex wildland urban interfaces in the nation, and the impact of fire on human assets is rising and is projected to continue to rise as the climate warms (Bryant and Westerling 2014). It seems clear that climate change driven alterations in fire regimes will provoke great changes in California landscapes, and under even the best scenarios the momentum of these effects will not be curbed for many decades.

The influences of climate change on fire are both direct and indirect and have the potential to affect two of the three legs of the fire behavior triangle. Local weather and fuels are both largely controlled by macroclimate and its interactions with topography. It has been argued that climate change will principally modify fire through altering fuel condition, fuel volume, and fuel ignitions (Hessl 2011), suggesting that fire weather will play a less significant role than the fuel itself. Others argue that the role of climate varies depending on whether the system is mostly conditions-limited (e.g., fire weather dependent) or resource-limited (e.g., fuel dependent; Krawchuk and Moritz 2011). Either way, climate change is expected to have a significant impact on both the extreme weather conditions that drive large catastrophic wildfire and the production and drying of fuels that are necessary for fire spread. Climate change is likely to influence fire in three principal ways (Krawchuk and Moritz 2012): (1) direct effects of changes on fire weather conditions such as drought, high temperatures, winds, and their seasonality, (2) an indirect effect on fire through vegetation—that is, by climate altering the structure, abundance, and energetics of biomass to burn, and (3) through changes in ignition potential due to shifting spatial or temporal patterns of lightning and human behavior in response to factors such as climate policy and environmental management.

Our ability to model and infer the potential impacts of climate change on fire in California depends on our understanding of the relationship between fire, fire weather, and fuels. In this chapter, we outline what is known about the relationship between fire and climatic change, summarize modeled and other projections for future fire activity in California, and discuss potential feedbacks that may alter the fire landscapes that we currently manage.

Fire–Climate Change Interactions

All attributes of the fire regime are connected to climatic variability in some way. Indeed, while climate is only one leg of the fire behavior triangle, under some conditions its effect can be so pervasive that it overrides the other components that determine fire behavior. Climate properties that influence fire include temperature, precipitation, humidity, wind speed and direction, and lightning. Such properties vary spatially and temporally across orders of magnitude, ranging, as Gedalof (2011) notes, "from a sunfleck that might dry a few square meters for a minute or two, to a megadrought that might persist throughout a given region for decades or more." Gedalof (2011) provides a succinct summary of how climate variability affects fire, ranging from short-term controls on fine fuel moisture, ignition frequencies, and fire spread rates; to intermediate-scale (annual to interannual) effects on the abundance and continuity of fine fuels and the abundance and moisture content of coarser fuels; to long term (decadal to centennial) influences on the pool of species that can persist in a given location. Interactions between the physical characteristics of these species and more direct influences of climate on fire lead to the distinctive fire regime and vegetation structure that we consider "characteristic" of a place, and changes in these variables and their interactions can have major ecological implications.

The historical record demonstrates how these complex linkages between climate change and fire operate. Tree ring records and sedimentary charcoal make evident that fire frequency and area burned are closely linked with the duration, frequency, and intensity of droughts (Marlon et al. 2012). Historically, the periods of highest fire activity during the Holocene Epoch coincided with periods of drought and/or climatic change, for example during the Altithermal or Xerothermic Period (ca. 7,000–4,000 years ago) and the Medieval Droughts (ca. 900 to about 1350 AD) (Whitlock et al. 2003, Beaty and Taylor 2009). Similarly, fire activity increased markedly at the end of the Younger Dryas stadial (ca. 12,800–11,500 years ago; Berger 1990), when global temperatures are thought to have increased at rates at least as fast as the current ones (Marlon et al. 2009). Changes in fire regimes provoked, and were provoked by changes in vegetation. Ancient pollen assemblages in Sierra Nevada lakes and peaty soils show local and regional shifts from forest to shrubland or grassland and back, or cycling between groups of more moisture loving, fire intolerant species and more xeric, fire-tolerant species. These types of vegetation shifts fed back into important changes in burning conditions, as biomass became more or less flammable, and fuelbeds more or less dense and continuous (Safford and Stevens, 2017).

Current changes in global climates are writing the next chapter in this age-old saga. The ten warmest years in the formal human record system (1880 to present) have all occurred since 1990 (Jones and Palutikof 2006) and 2016 was the warmest year on record (www.noaa.gov). In the high California mountains, earlier spring snowmelt and warmer summer temperatures have increased the length of the growing season as well as the fire season (Westerling et al. 2006), less precipitation is falling as snow (Knowles et al. 2006), and there is an overall decline in snow accumulation at all but the highest elevations (Mote et al. 2005). In northern California forests as a whole, the influence of climate on fire size and annual burned area has increased by two to four times over the last century. Whereas temperature was the primary climate driver of these fire variables in the first half of the twentieth century, today it is the variability in precipitation—and primarily during the fire season, not before—that is the most important factor (Miller et al. 2009, 2012).

Climate Change Predictions for California

Many empirical or statistically based climate projections are available at global and continental scales, but there are fewer statewide or regional projections due to the difficulty of obtaining precisely downscaled climate data. Most global circulation models (GCMs) produce raster outputs on grids of >10,000 km^2 and downscaling introduces errors and uncertainty to the projections. Additionally, there are many different ways to downscale climate data and different downscaling techniques lead to disparate results, even from the same dataset. Precipitation is particularly difficult to downscale given high levels of uncertainty in how the mechanisms driving precipitation may change as a result of complicated feedbacks. For example, early models varied from a 26% increase to an 8% decrease in precipitation per 1°C (2.1°F) temperature increase in California (Gutowski et al. 2000). That said, efforts have been made to provide finer-scale climate projections, and every year there are more—and more trustworthy—data to support decision-making and adaptation efforts in California (Cayan et al. 2008a).

Cayan et al. (2008a, b) summarized climate scenarios and downscaled models for California from two of the models used in the Third and Fourth Intergovernmental Panel on Climate Change (IPCC) Assessments: the Parallel Climate Model (PCM; Meehl et al. 2003) and the National Oceanic Atmospheric Association (NOAA) Geophysical Fluid Dynamics Laboratory (GFDL) model (Stouffer et al. 2006). These models were chosen because they met a series of conditions, including a minimum grid size and daily outputs, realistic simulations of historical climate, realistic representation of temporal and spatial variability in the California climate, and differing levels of sensitivity to greenhouse forcing (GFDL is much more sensitive to climate forcing than PCM and predicts greater warming). The models incorporate two greenhouse gas emission scenarios (B1—doubling of CO_2 emissions followed by decrease to below current by 2100, and A2—tripling of CO_2 emissions) (Cayan et al. 2008b), so there were four separate simulations carried out, ranging from moderate change compared to today (PCM-B1) to severe change from today (GFDL-A2). Outputs include projected precipitation and temperature, which are summarized below and in Table 26.1.

Process-based hydrologic models are an alternative approach that provides projections that are spatially relevant to ecological processes and can reconcile precipitation inconsistencies across climate models. Hydrologic models incorporate runoff, recharge and soil properties, creating a more holistic picture of water availability in ecosystems. Examples include the Basin Characterization Model (BCM; Flint et al. 2013) and the Variable Infiltration Capacity Hydrologic Model (Nijssen et al. 1997, Liang et al. 1994), both of which link climate and hydrologic models and output historical and future climate datasets. The historical component of the dataset allows for robust model validation that is not as easily accomplished with statistical models (Cuddington et al. 2013). Below we summarize projected climate data (Cayan et al. 2008b) and report the results of the BCM process-based hydrologic model for California (Thorne et al. 2015). BCM also uses the GFDL and PCM models used in the IPCC.

Temperature

By 2100, Cayan et al.'s (2008b) simulations project mean annual temperature will rise by 1.5°C to 4.5°C (34.7°F to 40.1°F) in California; three of the four simulations project greater temperature increases in summer than in winter (Table 26.1). These projections fall within the spread of future temperature projections from most other climate modeling exercises (*c.* 2°C to 7°C [4.2°F to 14.8°F]; Dettinger 2005). "Cool" summers will largely be a thing of the past by the second half of the century. The number of extremely hot days (those that fall above the 99.9th percentile of days between June 1 and September 30, using 1961–1990 as the reference period) is projected to rise by 50 to 500 times (to up to 23% of days) by 2100 in northern California, and slightly less dramatically in southern California; oceanic influences will reduce warming near the coast (see Cayan et al. 2008b for details). These results parallel those of most other published climate change predictions for California (e.g., Gutowski et al. 2003, Hayhoe et al. 2004).

TABLE 26.1

Temperature and precipitation changes from GFDL and PCM B1 and A2 simulations for northern and southern California

Mean values are for historical period (1961–1990). Changes between successive 30 year periods are shown in columns for the models and emission scenarios. Units are °C for temperature, mm for precipitation, and % for precipitation changes. Tables are reproduced with permission from Cayan et al. (2008b)

			2005–2034 change				2035–2064 change				2070–2099 change			
	Mean 1961–1990		GFDL		PCM		GFDL		PCM		GFDL		PCM	
	GFDL	PCM	A2	B1	A2	B1	A2	B1	A2	B1	A2	B1	A2	B1
Northern California														
Annual °C	9.3	8	1.5	1.4	0.5	0.5	2.3	2.2	1.3	0.8	4.5	2.7	2.6	1.5
Annual °C (JJA)	21.5	17.9	2.1	1.7	0.9	0.6	3.4	2.6	1.7	1.1	6.4	3.7	3.3	1.6
Annual °C (DJF)	−0.46	0.08	1.4	1.3	0.1	0.7	1.7	2.1	0.9	2.4	3.4	2.3	2.3	1.7
Annual mm/%	1098	750	+0.3	+2	−0.4	+7	−3	−2	−2	+3	−18	−9	−2	0
Summer mm/%	14	14	−29	+28	+28	+44	−67	−13	+35	−18	−68	−43	−30	−4
Winter mm/%	649	386	−1	−5	−5	+13	+6	−0.1	−5	−2	−9	−6	+4	+4
Southern California														
Annual °C	12.2	14.3	1.3	1.3	0.5	0.6	2.3	2.1	1.2	0.8	4.4	2.7	2.5	1.6
Annual °C (JJA)	23.2	23.4	1.7	1.6	0.4	0.5	3.1	2.3	1.3	0.8	5.3	3.2	2.6	1.5
Annual °C (DJF)	2.4	5.4	1.0	1.0	0.2	0.7	1.7	1.6	1.0	0.6	3.3	2.0	2.4	1.6
Annual mm/%	537	342	−6	−2	+7	+18	−2	−11	+7	−2	−26	−22	+8	+7
Summer mm/%	7	5	+49	−13	−7	+6	−60	−50	+35	+33	−44	−63	−11	+2
Winter mm/%	320	187	−0.7	+0.8	+1	+32	+9	−9	+6	−6	−2	−26	+8	−0.8

Precipitation

The simulations run by Cayan et al. (2008b) suggest that California's Mediterranean-type precipitation regime will not change, with most of the precipitation continuing to fall in the months between November and May. Cayan et al.'s (2008b) simulations do not project increased monsoonal influence in California, but drivers of interannual monsoonal variability in North America are poorly understood and difficult to model (Adams and Comrie 1997). Overall precipitation is also not expected to change much, with slight increases or no change under the PCM simulations, and 10% to 20% decreases under GFDL. Earlier summaries of different climate change models and simulations suggested broadly similar patterns, with most simulations showing no change or modest increases in annual precipitation in California by 2100 (Gutowski et al. 2000, Hakkarinen and Smith 2003, Maurer 2007). While results do not suggest significant change in annual precipitation, Cayan et al.'s (2008b) simulations project an increase in the frequency and intensity in extreme precipitation events in northern California, but not much change in southern California (which is already characterized by a preponderance of such events; Dettinger et al. 2011). Interannual to decadal variability in precipitation is projected to continue to be very high (California supports the highest interannual variability in precipitation in the United States [Dettinger et al. 2011]), and Cayan et al.'s (2008b) simulations do not suggest a change in the broad periodicity of the El Niño Southern Oscillation (ENSO), which is an important driver of climate variability in California, especially in the south. Note that, like the North American monsoon, it is very difficult to model climate change impacts on the ENSO system because it is a highly dynamic process influenced by both atmospheric and ocean circulation patterns (Diaz et al. 2001). Changes in ENSO patterns would affect precipitation differently in the northern and southern regions of California as a function of the ENSO dipole (Brown and Comrie 2004).

Snow Accumulation

Although overall precipitation averages may not change, higher temperatures will influence the amount of precipitation that falls as snow and accumulates on the ground. Current trends in increasing snow:rain ratios (Mote et al. 2005) are projected by Cayan et al. (2008b) and Thorne et al. (2015) to continue and accelerate. In California, April 1 snowpack (defined as snow-water equivalent, SWE) has declined by an average of 10 mm (0.4 in) annually, with large declines in the Cascade Range (–33 mm [–1.3 in]) and the Sierra Nevada (–29 mm [–1.1 in]; Fig. 26.1). Cayan et al. (2008b) and Thorne et al. (2015) project further losses in April 1 SWE across northwestern California, the Modoc Plateau and the Sierra Nevada. By 2100, Cayan et al. (2008b) project decreases in SWE of 32% to 79% (compared to today), with the largest losses at elevations <2,000 m (6,561 ft); snow loss in the Sierra Nevada will be greatest in the northern and central part of the range because elevations there are much lower than in the south.

Climatic Water Deficit

In the semiarid landscapes of western North American, water availability is a major driver of ecosystem distribution and condition (Major 1988, Loik et al. 2004). In terms of water availability in ecosystems, it is probably less valuable to consider changes in precipitation or temperature independently and more valuable to consider how the integration of water and energy manifest into climate conditions that are actually experienced by plants. Climatic water deficit (CWD; potential evapotranspiration minus actual evapotranspiration) is a climate variable that represents available water for plant use by incorporating precipitation and temperature into one "biologically relevant" measure (Stephenson 1998). CWD has been used as a metric for understanding water balance, drought stress, and fire vulnerability in ecosystems throughout California and the western United States. Since the period 1951–1980, the average annual CWD has increased (i.e., water is less available) in California by an average of 17 mm, but this change is less than one standard deviation from the historical record, and trends have been geographically variable. For example, much of northwestern California and parts of the Sierra Nevada west slope have experienced increasing water availability (decreasing CWD), due to precipitation increases over the period that have counteracted warming temperatures (Thorne et al. 2015; Fig. 26.2). Future changes under the BCM scenarios are projected to increase CWD by 40 mm to 160 mm (1.6 in to 6.3 in) in most of California, depending on the simulation and the location. The largest projected changes will occur east of the Sierra Nevada-Cascades crest. Many of these changes will depart from the historical record by between 1.5 and 2 standard deviations, implying a very significant decrease in water availability by the end of the century (Fig. 26.2; Thorne et al. 2015).

Projected Climate Change Impacts on California Fire Regimes

The combined effects of changing temperatures and shifts in the timing and magnitude of precipitation will undoubtedly alter California fire regimes. In many montane forests, a century of fire exclusion has resulted in the accumulation of fuels such that uncharacteristic weather conditions can more quickly accelerate a small fire event into a major wildfire. A majority of future fire models agree that changes in climate will directly and indirectly increase the frequency and area burned across most of the western United States, including California (Lenihan et al. 2008, Gedalof 2011, Westerling et al. 2011, Safford et al. 2012). It has been notably more difficult to predict changes in fire severity and intensity because of nonlinear relationships and complex feedbacks between vegetation, climate, and fire (Flannigan et al. 2009, van Mantgem et al. 2013), but most models project increases in fire severity/intensity as well, depending on the fire regime and vegetation type in question (Lenihan et al. 2003b, Flannigan et al. 2013). Undoubtedly, changing fire regimes will alter vegetation composition and structure and the water balance as well (Miller and Urban 1999), ultimately altering fuel conditions. In some cases, decreases in fire activity are projected in certain components of the fire regime towards the end of the next century due to ecosystem transitions caused by originally increased fire activity and/or from altered species composition (i.e., woody vegetation to grass). It is important to remember that most models that project future fire activity are built on empirical relationships between past climate and fire, and then these relationships are applied to future climate scenarios. Empirical models assume stationarity in the fire–climate

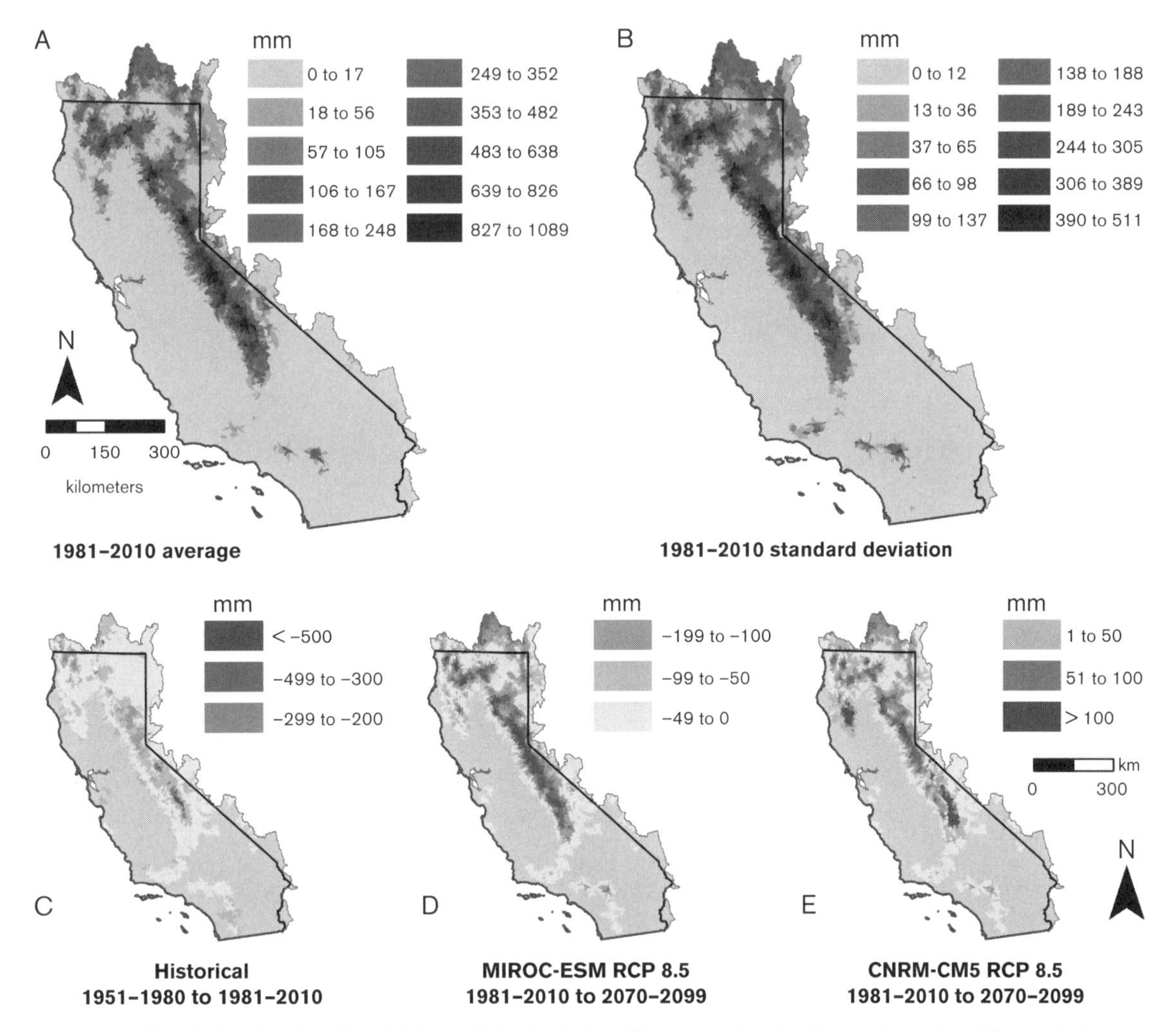

FIGURE 26.1 Historical and projected April 1 Snow Water Equivalent. Figure reproduced with permission from Thorne et al. (2015).

relationship and presume the relationship between a fire regime attribute and a climate variable will remain quantitatively similar into perpetuity. This is not always a valid assumption given that the relationship between climate and fire is specific to vegetation on the landscape that is burning (Littell et al. 2009, Hessl 2011), and future vegetation assemblages may be very different from those today or in the past (Williams and Jackson 2007). Here we summarize this literature, and differentiate projections for forest, shrubland, and grassland ecosystems where the research results permit.

Vegetation Shift Impacts on Fire Frequency and Probability

Miller and Urban (1999) conducted simulations of future fire regimes and species composition across an elevational gradient in the southern Sierra Nevada. Their results highlighted how strongly fire activity depends on not just the abundance but the composition of fuel. In the two lowest elevation sites in the analysis (yellow pine and mixed conifer forests), fire frequencies increased during the first century of simulation, but then declined gradually as woody biomass was consumed and then disappeared altogether from the ecosystem. At the end of their 400 year simulations, woody biomass decreased at the lower elevation (1,800 m [5,905 ft]) site from ~200 mg $ha^{-1}$to 0 mg ha^{-1}, providing minimal fuel for consumption. Under Miller and Urban's (1999) most extreme scenario, forest fuels at this elevation were completely replaced by grassy fuels and fire frequency increased. At the middle elevation (2,200 m [7,218 ft]) site, biomass loss was high but not as extreme as at the 1,800 m (5,905 ft) site, and fire frequencies remained similar to baseline conditions although fire area decreased with the decrease in biomass. The highest elevation (2,600 m [8,530 ft]) site, which is currently in red fir forest but was predicted to transition to a mixed conifer composition, experienced very large increases in fire frequency. Results from Miller and Urban (1999) predict significant transitions in the fire regime that are largely driven by changes in fuel characteristics—lower elevation woody ecosystems will contain more flashy fuels which will increase fire frequency and higher elevation forests will experience more frequent fire as species composition shifts toward more flammable, fire-prone species.

The Changed Climate Fire Modeling System (CCFMS; Fried and Torn 1990) links GCMs to local weather, fire records, population density, fuel type, and slope to simulate area burned and the number of potential escaped fires (their model

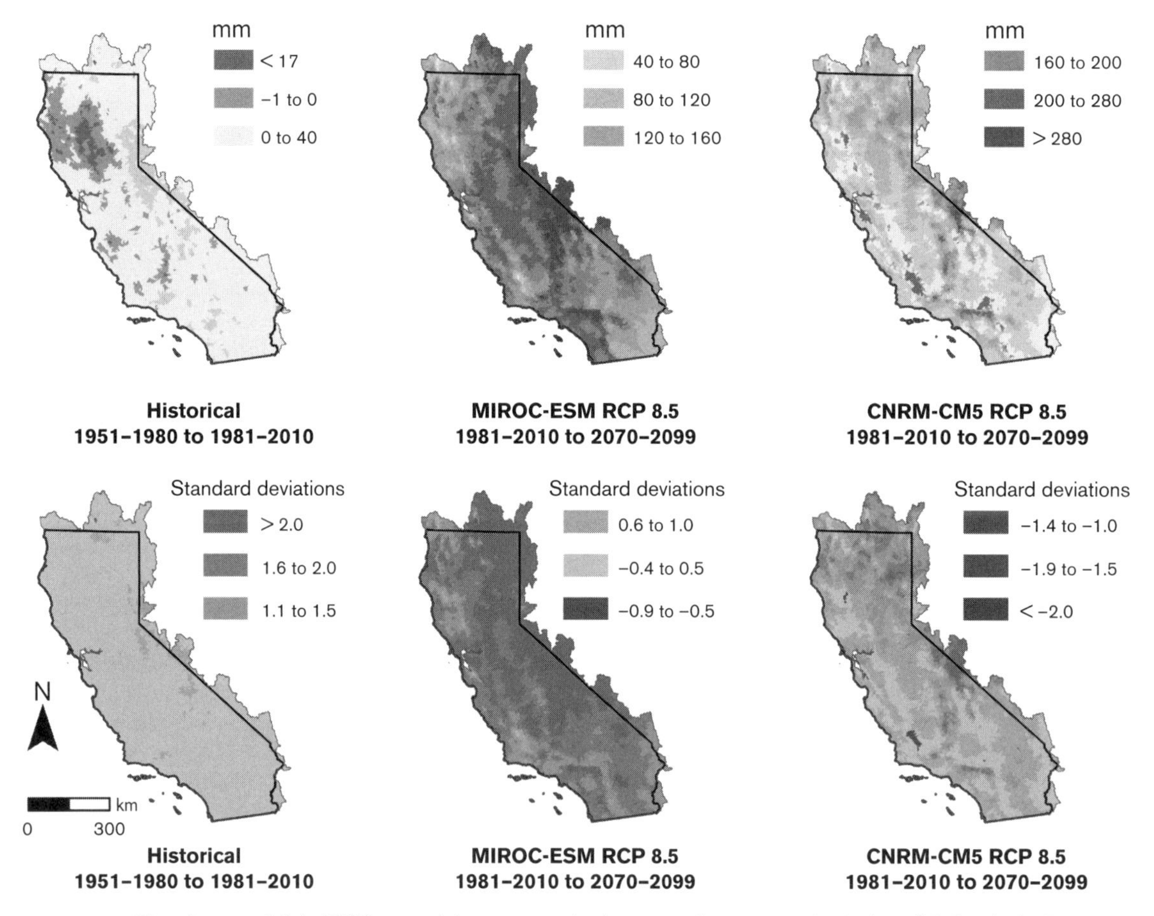

FIGURE 26.2 Climatic water deficit (CWD; potential evapotranspiration—actual evapotranspiration) modeled with the Basin Characterization Model. Historical and future values are displayed as differences between two time periods (top panels). Data is alternatively displayed as standard deviations to better show variability and significant departure from historical conditions (bottom panel). Under the warmer and drier scenario (GFDL) CWD will increase by more than one standard deviation throughout the entire state. Under the warmer and wetter scenario (PCM) CWD will increase by one standard deviation in the eastern regions of the state. Figure reproduced with permission from James Thorne.

includes the effects of fire suppression). Fried et al. (2004) applied the model to northern California under a future climate with twice the atmospheric CO_2 as today. Their projections suggested that climate change will lead to increases in the frequency of weather conditions that are associated with high fire risk and therefore increases in the frequency of fire events. Like Miller and Urban (1999), Fried et al.'s (2004) results underlined the key role that fuel type plays in driving ecosystem response to climate change. Areas vegetated with grass and shrubs were projected to experience many more escaped fires under warming, but the lack of a crown fire module probably led to an underestimate of the number of escapes in forest fuels. Fried et al. (2004) projected that the number of escaped fires would increase by 125% in the Sierra Nevada and 51% in the south Bay Area, but there was no projected change in the North Coast region because of the slower wind speeds and higher humidities produced from the downscaled GCM. Because the CCFMS model does not include any internal feedbacks from fire to vegetation, Fried et al. (2004) underlined that their results represented the minimum expected change.

Krawchuk and Moritz (2012) developed fire projections for California using statistical relationships that relate fire counts and climate variables (e.g., precipitation seasonality, maximum temperature, CWD) to determine the probability of fire under future climate scenarios. In Krawchuk and Moritz's (2012) model, future fire probability varied along two primary axes: (1) the climate scenario used (different GCMs), and (2) the primary limiting factor of the fire regime (fuel-quantity or "burning condition," the latter driven by fuel moisture and climate). Under warmer and drier scenarios, fire probability was not projected to increase in regions where lack of fuel is a major limitation to burning (e.g., deserts of southeastern California and unforested parts of the Modoc Plateau), but the converse was true for the warmer and wetter climate scenarios. In fuel-rich ecosystems, there was agreement between GCMs that future changes in climate will result in increased fire probability. Increases in area burned were projected in northwestern California, the Cascade Ranges and the Sierra Nevada, whereas decreases in area burned were projected for the Mojave and Sonoran Deserts (Krawchuk and Moritz 2012). Mean fire return interval was also modelled and, in general, was reduced across the state suggesting more frequent fire.

Batllori et al. (2013) used maximum entropy modeling to investigate the outcomes of warmer-drier climate scenarios vs. warmer-wetter scenarios with respect to fire occurrence in the world's five Mediterranean-type climate regions. Results for California suggested that under either type of climate scenario, fire probabilities would likely increase over the coming century on 60–80% of the landscape (depending on the scenario and the time period) and decrease on 15–30% of the landscape. Parts of the state that experienced increases or decreases in fire probability changed depending on the climate scenarios however, and Batllori et al. (2013) ascribed this to differences in the way climate and fire interact in fuel- and condition-limited ecosystems.

Burned Area

In Miller and Urban's (1999) model, area burned at their 1,800 m (5,905 ft) and 2,200 m (7,218 ft) sites rose strongly during the first century of their climate change simulations, but then decreased over time as woody biomass was gradually lost. At the lowest site, little woody biomass remained at the end of their simulation and the abundance of grassy fuels led to a large increase in area burned. At the 2,200 m (7,218 ft) site, fire area decreased as biomass was lost over time (Miller and Urban 1999). The red fir forest site at 2,600 m (8,530 ft) experienced very large increases in area burned.

McKenzie et al. (2004) calculated correlations for the twentieth century between mean summer temperature and precipitation and annual burned area for 11 western states, and then used regression to project burned area into the future under two climate scenarios. They found strong relationships between their summer climate variables and fire area for all states but California and Nevada, and concluded that most of the western United States was likely to experience large increases in annual area burned by wildfire in the twenty-first century. However, they conclude that "fire in California and Nevada appears to be relatively insensitive to summer climate, and area burned in these states may not respond strongly to changed climate." This curious statement appears to result from McKenzie et al.'s (2004) combination of southern and northern California into a single dataset. Southern and northern California (with the Tehachapi Mountains being the boundary) each contribute about half of California's total burned area in an average year, but general fire–climate relationships in the two regions are very different (Safford and Van de Water 2014). McKenzie et al.'s (2004) analysis thus buries the relatively strong relationship that exists between fire and summer climate variables in the Sierra Nevada, Klamath Mountains and North Coast Range (Westerling et al. 2006; Miller et al. 2009, 2012; Keeley and Syphard 2015) under the southern California fire–climate relationship, which is essentially independent of summertime temperature or precipitation (Keeley 2004). In summary, changes in summer temperature and precipitation may not have strong effects on southern California fire area, but McKenzie et al.'s (2004) predictions for the western United States in general are certainly valid for most of the assessment area.

The MC1 Dynamic General Vegetation Model is a dynamic vegetation model that simulates physiognomically defined vegetation types; the movement of carbon, nitrogen, and water through ecosystems; and fire. Unlike most other models described here, MC1 incorporates feedbacks between fire and vegetation, and temporal changes in fuel loadings and types can be accounted for. Lenihan et al. (2008) simulated the responses of vegetation distribution, carbon, and fire to climate change in California using MC1. Fire events were determined as a function of moisture content of the largest dead fuels and fire spread thresholds, and the occurrence of fire was limited to extreme events only. Lenihan et al. (2008) found that the average annual area burned across the state was 9–15% higher at the end of the twenty-first century than under current conditions. Greater burned area was strongly driven by the increasing area of grassland promoted by both climatic and fire effects, especially in the central and southern coasts, Modoc Plateau and lower elevation Sierra Nevada. Regional variations in projected fire activity were dependent on vegetation productivity and the distribution of woodlands and grasslands. Lenihan et al. (2008) conclude that increases in burning will have serious consequences for carbon storage as grasslands replace woodlands in many parts of California. In Yosemite National Park, Lutz et al. (2009) used a statistical model to predict that annual burned area would increase by around 20% by 2020–2049 due to projected decreases in snowpack in mid- and high-elevation forests.

Modeling reported by the National Research Council (NRC 2011) projected that, compared to the average of the 1950–2003 period, median annual area burned would increase by over 300% for the northern California mountains with a 1°C (2.1°F) increase in average temperature; increases in over 200% were projected for vegetation of the Central Valley and foothills, and over 70% for the southern California deserts; coastal California south of San Francisco was not modeled. Over time, the report noted that extensive warming and wildfire could ultimately exhaust much of the fuel for fire in some regions, as forests were completely burned (NRC 2011).

Westerling et al. (2011) modeled burned area across California under a range of future climate and development scenarios. They found that, under the most realistic future climate and emissions scenarios and compared to the average of the period 1960–1990, projected area burned by wildfire increased by over 200% by 2085 for most of the forested area of northern California. Middle and higher elevation forests were among the most severely impacted, with some future climate scenarios producing increases in burned area of more than 300%. Naturally vegetated areas in most of central and southern coastal California was projected to see increases of around 100% in annual burned area, except for the southeastern deserts, where burned area remained steady or dropped (Westerling et al. 2011).

Fire Intensity and Severity

The MC1 dynamic vegetation model was used to model vegetation and fire response to two different GCM-based future climate scenarios specific to California (Lenihan et al. 2003a, b). One of the mid-stream outputs of the model is fireline intensity. Under the warmer-wetter climate scenario (Hadley CM2), fireline intensity was projected to increase in grass-dominated systems, to slightly increase in desert and shrubland ecosystems, and to remain steady or slightly increase in most forested ecosystems. Under the warmer-drier future scenario (PCM), simulations projected that increased fire frequency and area burned would remove woody vegetation from coastal areas, the Klamath Mountains and the North Coast Ranges. Ultimately, the reduction or absence of woody material was projected to result in lower fireline intensities.

Flannigan et al. (2000) modeled the seasonal severity rating (SSR)—a measure of the difficulty of fire control—across North America under two GCM scenarios for the year 2060, and found that SSR increased by an average of 10% under both scenarios for California. Flannigan et al. (2013) linked the Canadian Forest Fire Weather Index to three GCMs and predicted the Cumulative Severity Rating (CSR), a fire danger metric based on weather conditions, for the northern and southern hemispheres for the periods 2041–2050 and 2091–2100. They projected that by 2100 severity as measured by CSR would increase by around 10% in California. In Yosemite National Park, the total area burned at high severity in mid- and high-elevation forests was projected by Lutz et al. (2009) to increase 22% between the current (1984–2005) and mid-21st century (2020–2049) periods, due mostly to declines in snowpack.

Van Mantgem et al. (2013) showed that high prefire CWD increases the probability of postfire tree mortality, thus—aside from their well-known effects on fuel moisture—climate warming and increasing growing season drought can enhance fire severity independently of fire intensity. This suggests that future fire severities could be even higher than predicted by fire–climate modeling studies.

Fire Season

Changes in the length and character of the fire season can be attributed to decreases in the snow:rain ratio and the increased incidence of extreme temperatures. The snow:rain ratio has been decreasing across California in the past 75 years (Safford et al. 2012) and combined with warming temperatures, these negative trends in snow amount and storage result in earlier drying of fuels and a lengthening of the fire season. Current trends and projections of future patterns in the snow:rain ratio and snowpack persistence thus portend longer fire seasons (Mote 2006, Mote et al. 2005, Safford et al. 2012, Westerling et al. 2006). Collins (2014) showed that since 1992, 17% to 20% and 8% to 12% of days in the fire season exceeded the 90th and 95th percentile, respectively, in terms of fire weather thresholds. Collins (2014) suggests that extreme weather in a given fire season controls growth of individual fire events and therefore the increased incidence of extreme fire weather (one characteristic of the fire season) is contributing to increases in area burned.

A number of authors have used fire–climate models to project changes in the duration and timing of the fire season as climate warming continues. The Keetch–Byram Drought Index was modelled under a number of different future GCM-based scenarios as a proxy for "wildfire potential" (Liu et al. 2010). Fire season was projected to become a couple of months longer for much of the contiguous United States, including California, by the end of the twenty-first century. Flannigan et al. (2013) projected that fire season length would increase by more than 20 days for all of northern California by 2100. Yue et al. (2013) projected a median increase of more than three weeks in the fire season by the middle of the twenty-first century.

Fire Ignitions

Price and Rind (1994) simulated the distribution and frequency of lightning using downscaled GCMs for California. Results indicate that lightning frequency could increase by as much as 30% globally. Romps et al. (2014) found similar results, based on a simple linear relationship between lightning flash rate and the product of precipitation (per hour) × convective available potential energy (a measure of atmospheric convective instability). Results project a 12% average increase in lightning per 1°C [2.1°F] of temperature rise. By the end of the twenty-first century this could translate into 50% more lightning across much of the United States (Romps et al. 2014). Lightning strike densities (LSDs) are relatively low in California, but areas of high topography (especially in the Sierra Nevada) still see LSDs of 15 to 35 strikes yr^{-1} 100^{-1} km^{-2} (van Wagtendonk and Cayan 2008). The combination of greater lightning incidence, warmer climates, and drier fuels strongly suggests that fire activity will likely rise in most semiarid parts of California that support fuel, even where human ignitions can be reduced.

Fire Effects on Vegetation

Fire is a major driver of vegetation change in both space and time. The effects of fire on vegetation in California will depend greatly on precipitation trends, but Bachelet et al. (2007) note that in either wetter or drier conditions, forest could be notably reduced in much of the western United States in a warmer future. Under drier conditions, enhanced fire frequency could favor drought-tolerant grasses, which would further enhance ecosystem flammability and reduce woody cover. Under wetter conditions, expansion of woody plants might promote more intense fires and high mortality when drought conditions occur, ultimately reducing tree biomass. Bachelet et al. (2007) projected that most of California would see an increase in biomass consumption by fire during the twenty-first century, whether warming was extreme or moderate.

Using the same vegetation dynamics model as Bachelet et al. (2007), Lenihan et al. (2008) simulated the future distribution of terrestrial ecosystems in California under three GCM-based future climate scenarios. Fire drove grassland expansion into former shrublands and woodlands, even under the coolest and wettest future scenario; by 2099, under the warmest and driest scenario, grassland almost completely replaced shrublands on the Sierra Nevada west slope and also expanded greatly in the California Great Basin. Broadleaf woodland and forest replaced large areas of evergreen conifer forest under all three scenarios, with fire playing an important role in the transition, especially in the relatively warmer and drier scenarios (Lenihan et al. 2008). These types of vegetation transitions will have a major impact on fire regimes. For example, grasslands are characterized by flashy fuels that ignite easily and burn rapidly and propagate more grasslands, creating a positive feedback loop (the classic "grass-fire cycle").

Summary of Projected Climate Change Impacts

All the described climate models generally agree that fire frequency and area burned will increase in most California ecosystems. However, spatial variability in fire activity, burned area, and fire severity and intensity will depend on future precipitation patterns and the dynamic relationship among vegetation, fuels, and fire. The effects of climate change will likely vary for fire regimes that are driven principally by fuel quantity versus those driven principally by burning conditions (fuel quality) and climate. Even if increases in fire activity and burned area are in store for ecosystems at both

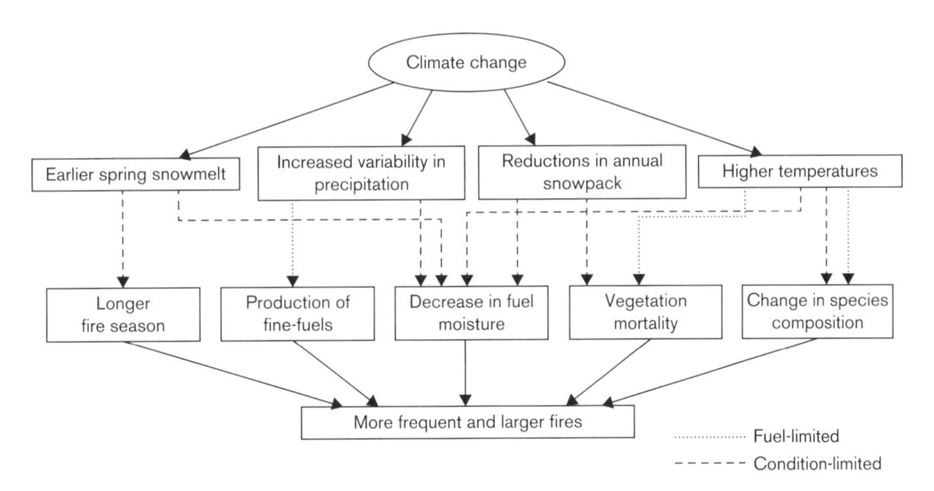

FIGURE 26.3 The effects of climate change will alter temperature and water availability in ecosystems, changing fuel availability and condition, all of which will lead to more frequent and larger fires. Major pathways of change will vary for fuels- versus conditions-limited ecosystems. Snowpack effects will be more significant for conditions-limited systems because most fuels-limited systems experience rain-dominated precipitation. Higher temperatures of course reduce fuel moisture in all systems, but fuel-limited systems are typically dry enough to burn in any given fire season, so the enhancement effect will be more limited than in conditions-limited systems which typically support heavier fuels.

ends of this spectrum, the mechanisms of change will differ (Fig. 26.3). For example, in California, where interannual variability in precipitation is extreme (Dettinger et al. 2011), further increases in precipitation variability could enhance an already prevalent pattern of very wet years catalyzing fine fuel production in fuels-limited regions, and subsequent very dry years leading to large areas burned. Alternatively, low precipitation years that occur simultaneously with projected higher temperatures will drive increased fire activity in conditions-limited systems where there is already abundant fuel (i.e., densely forested landscapes), independently of antecedent years.

Decreases in snowpack or earlier timing of spring snowmelt combine to decrease fuel moisture and increase the length of the fire season, leading to more fire in conditions-limited systems. Higher temperatures and greater water stress may result in vegetation mortality (Allen et al. 2010), creating fine fuels (in the short term) and coarse woody debris (in the medium and long term) for combustion that could promote fire activity or possibly augment the ecological effects of fire in either fuels- or conditions-limited systems. It is likely that increased variability in precipitation will affect the fire regime in nonforested regions of California the most through increases in fuel production, and that higher temperatures will affect all regions in California via fuel moisture reduction. Regional drought stress will lead to increased annual area burned when fuels are not limiting (via fuel moisture reduction) and will result in the addition of fuels (via mortality) in conditions-limited systems (Williams et al. 2013).

Inferences about future fire can also be made from our understanding of the relationship among fire, vegetation, and climate rather than relying on models. Annual area burned is driven largely by climate (Trouet et al. 2009, Keeley and Syphard 2015), especially spring-summer vapor pressure deficit (VPD) because VPD determines atmospheric moisture demand and therefore drives the reduction of fuel moisture (Williams et al. 2014). The frequency of high fire-weather indices is responsible for a lengthening of the dry season resulting in more fires later into the fall (Collins 2014). Antecedent climate also influences fire regimes as drought stress can result in the production of fine fuels that are available for the next year's fire season. Parks et al. (2014) demonstrate that area burned varies with actual evapotranspiration and that greater CWDs lead to more severe fires. Within a given ecosystem, fire severity also increases with fuel quantity (Parks et al. 2014, Steel et al. 2015). Importantly, the strength and sometimes the nature of these relationships can modulate across ecosystem types. The generalization about the differences between fuels-limited (fuel quantity) and conditions-limited (fuel quality and climate) ecosystems (Agee 1993, Mallek et al. 2013, Steel et al. 2015) is highly germane to considerations of the future in California. Overall, these mechanistic relationships between climate and fire are expected to hold into the future where, for example, an increase in spring-summer VPD would lead to increases in annual area burned. Projected extremes in weather and decreases in the snow:rain ratio both portend increases in fire activity under a changing climate. We can expect, therefore, changes in fire regimes commensurate with the magnitude of changes in climate variables.

Understanding how climate change interacts with fire in ecosystems with different fire regimes is fundamental to understanding how undesirable changes might be avoided, for example through fire and fuels management (Fig. 26.3).

A Future with More Fire

Modeling results conclude that more and larger fires are highly probable with a warmer climate in the state of California. Such trends are already obvious. Recent examples of "megafires" (Stephens et al. 2014) in California include the Rim, King, and Rough fires in the Sierra Nevada, but record-breaking fires have also recently occurred in other western states such as Arizona, New Mexico, Oregon, and Washington. In the end, perhaps the key question is "what will our

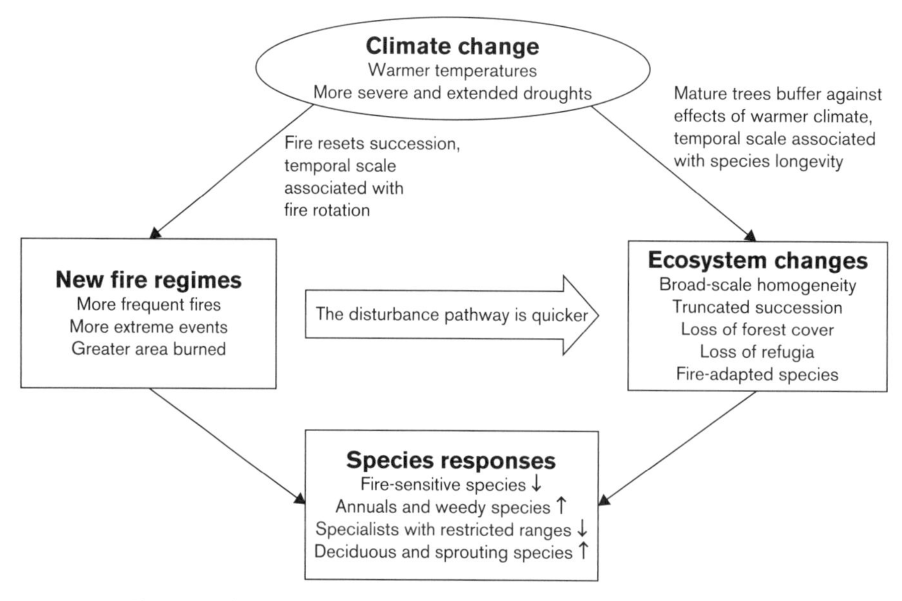

FIGURE 26.4 Changes in fire regimes will be the catalyst for ecosystem change. Fires modify landscapes on shorter temporal scales than climate change alone. Climatic conditions will likely be different after fire than when vegetation established, resulting in changes in species composition. Figure edited and reproduced with permission from Don McKenzie.

management landscapes look and act like if burned area and fire sizes increase as predicted?" It is likely that climate change will have its most dramatic effects on California landscapes through its direct and indirect connection to ecosystem-altering fire (Dale et al. 2001). When fires of novel sizes, or frequencies, or severities (and their combination) occur on landscapes, the likelihood of abrupt vegetation change or local extirpation is increased (Fig. 26.4). On this newly disturbed landscape, conditions—soil, light, nutrients, climate—may be vastly different from when the vegetation established. For example, many shrub-dominated landscapes in southern and eastern California have transitioned to annual grass dominated ecosystems as a result of steep positive trends in fire frequency driven by synergy among climate warming, droughts, ignitions, and invasive species (Keeley and Safford 2016). These trends, which are very difficult to reverse, are likely to accelerate and expand geographically.

In forested regions, areas dominated by long-lived conifers—which have experienced marked swings in climate over the centuries since they established—are being subjected to rapid increases in growing season temperatures and deeper late summer droughts. These conditions are already challenging for mature individuals, but when severe fires kill them, reestablishment of forest species is not a given, especially in the huge patches of stand-replacing fire that are more and more commonplace in California forest fires. This sort of dynamic is already apparent in lower elevation conifer forests in central and southern California and is at the heart of the major conifer forest loss projected by dynamic vegetation models (e.g., Lenihan et al. 2003a, 2008). These models also predict that oak species—which can resprout after fire—will largely benefit from these dynamics, and a trend toward higher hardwood density in lower elevation conifer forests is already apparent in the Sierra Nevada and southern California (Safford et al. 2012).

Increasing frequencies, areas, and severities of fire (which won't necessarily occur in tandem) are likely to alter more than just the overstory vegetation. More burning and more open vegetation stands created by increased burning will be less hospitable to species adapted to moist, cool habitats and more hospitable to drought-tolerant, sun-loving species (Stevens et al. 2015). Many invasive plants, mostly annual grasses but some forbs as well, will find it easier to expand into wildlands. Those that are highly flammable will further transform fire regimes and the ecological landscape. Animal species adapted to dense, old forest stands—spotted owls, fishers, goshawks—will find it increasingly difficult to locate suitable habitat, while animals preferring severely burned landscapes—many woodpeckers and other birds, certain rodents—will benefit (McKenzie et al. 2004, Mallek et al. 2013). As precipitation variability continues to increase, the effects of increased fire on soil erosion and sedimentation in streams are likely to become more pronounced, especially on steep less consolidated bedrock and where woody vegetation has been lost. If current and projected fire trends continue, smoke production will also rise, with the concomitant effects on human health and aesthetics.

Fires do not occur in isolation and are usually accompanied by other processes that may be intrinsic or extrinsic to the ecosystem. Climate change is documented to have impacts on forests through direct heat-related mortality (Bréda et al. 2006, Allen et al. 2010) or water stress (McDowell et al. 2008) and indirectly through the exacerbation of bark beetle outbreaks (Raffa et al. 2008), fungal diseases, and human-caused disturbances such as ozone pollution (Fowler et al. 1999). The co-occurrence of multiple stressors is known as a "stress-complex," where the cumulative impacts of multiple stressors can greatly alter ecosystem processes and patterns (McKenzie et al. 2008). For example, the combination of higher temperatures, longer warm seasons, and ozone can result in pine mortality and epidemic bark beetle outbreaks, resulting in further tree mortality (Fig. 26.5). Fire exclusion in the same forests leads to high stand densities which, coupled with increased fuel accumulation from increased tree mortality, can result in

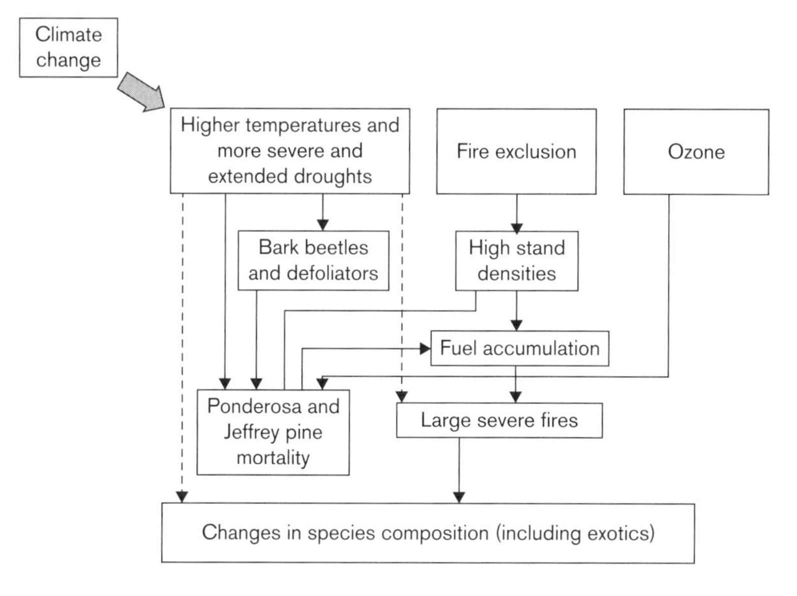

FIGURE 26.5 Example of a stress complex from Sierra Nevada forests. The combination of higher temperatures (imposed from climate change), fire exclusion, and ozone pollution result in multiple feedbacks triggering higher tree mortality and fuel accumulation that leads to larger wildfires. More severe droughts and wildfires can lead to a shift in species composition towards more annuals and fire-adapted species (including nonnatives). Figure reproduced with permission from McKenzie et al. (2008).

larger, more severe fires. All of these perturbations combine to alter species composition, potentially favoring nonnative species that are better adapted to higher fire frequencies or severities. We hypothesize that stress complexes will become more common under a changing climate (and an increasing human population), resulting in an increase in the frequency of ecological processes and an amplification of their effects.

References

Adams, D.K., and A.C. Comrie. 1997. The North American monsoon. Bulletin of the American Meteorological Society 78(10): 2197–2213.

Agee, J.K. 1993. Fire ecology of Pacific Northwest Forests. Island Press, Washington, D.C., USA.

Allen, C.D., A.K. Macalady, H. Chenchouni, D. Bachelet, N. McDowell, M. Vennetier, and N. Cobb. 2010. A global overview of drought and heat-induced tree mortality reveals emerging climate change risks for forests. Forest Ecology and Management 259: 660–684.

Bachelet, D., J.M. Lenihan, and R.P. Neilson. 2007. Wildfires and global climate change. The importance of climate change for future wildfire scenarios in the western United States. Pew Center on Global Climate Change, Arlington, Virginia, USA.

Batllori, E., M.A. Parisien, M.A. Krawchuk, and M.A. Moritz. 2013. Climate change-induced shifts in fire for Mediterranean ecosystems. Global Ecology and Biogeography 22: 1118–1129.

Barbero, R., J.T. Abatzoglou, E.A. Steel, and N.K. Larkin. 2014. Modeling very large-fire occurrences over the continental United States from weather and climate forcing. Environmental Research Letters 9: 124009.

Beaty, R.M. and A.H. Taylor. 2009. A 14 000 year sedimentary charcoal record of fire from the northern Sierra Nevada, Lake Tahoe Basin, California, USA. The Holocene 19(3): 347–58.

Berger, W.H. 1990. The Younger Dryas cold spell–a quest for causes. Global and Planetary Change 3: 219–237.

Bréda, N., R. Huc, A. Granier, and E. Dreyer. 2006. Temperate forest trees and stands under severe drought: a review of ecophysiological responses, adaptation processes and long-term consequences. Annals of Forest Science 63: 625.

Brown, D.P., and A.C. Comrie. 2004. A winter precipitation 'dipole' in the western United States associated with mulitdecadal ENSO variability. Geophysical Research Letters 31: 1–4.

Bryant, B.P., and A.L. Westerling. 2014. Scenarios for future wildfire risk in California: links between changing demography, land use, climate, and wildfire. Environmetrics 25(6): 454–471.

Cal-NRA. 2014. Safeguarding California: reducing climate risk. State of California, Natural Resources Agency, Sacramento, California, USA. http://resources.ca.gov/docs/climate/Final_Safeguarding_CA_Plan_July_31_2014.pdf. Accessed July 22, 2015.

Cayan, D.R., A.L. Luers, G. Franco, M. Hanemann, B. Croes, and E. Vine. 2008a. Overview of the California climate change scenarios project. Climatic Change 87(1): S1–S6.

Cayan, D.R., E.P. Maurer, M.D. Dettinger, M. Tyree, and K. Hayhoe. 2008b. Climate change scenarios for the California region. Climatic Change 87(1): S21–S42.

Collins, B.M. 2014. Fire weather and large fire potential in the northern Sierra Nevada. Agricultural and Forest Meteorology 189: 30–35.

Cuddington, K., M.J. Fortin, L.R. Gerber, A. Hastings, A. Liebhold, M. O'Connor, and C. Ray. 2013. Process-based models are required to manage ecological systems in a changing world. Ecosphere 4(2): art20.

Dale, V.H., L.A. Joyce, S. McNulty, R.P. Neilson, M.P. Ayres, M.D. Flannigan, P.J. Hanson, L.C. Irland, A.E. Lugo, C.J. Peterson, and D. Simberloff. 2001. Climate change and forest disturbances: climate change can affect forests by altering the frequency, intensity, duration, and timing of fire, drought, introduced species, insect and pathogen outbreaks, hurricanes, windstorms, ice storms, or landslides. BioScience 51: 723–734.

Dettinger, M.D. 2005. From climate-change spaghetti to climate-change distributions for 21st century California. San Francisco Estuary and Watershed Science 3(1). http://repositories.cdlib.org/jmie/sfews/vol3/iss1/art4. Accessed July 22, 2015.

Dettinger, M.D., F.M. Ralph, T. Das, P.J. Neiman, and D.R. Cayan. 2011. Atmospheric rivers, floods and the water resources of California. Water 3: 445–478.

Diaz, H.F., M.P. Hoerling, and J.K. Eischeid. 2001. ENSO variability, teleconnections and climate change. International Journal of Climatology 21: 1845–1862.

Flannigan, M., A.S. Cantin, W.J. de Groot, M. Wotton, A. Newbery, and L.M. Gowman. 2013. Global wildland fire season severity in the 21st century. Forest Ecology and Management 294: 54–61.

Flannigan, M.D., B.J. Stocks, and B.M. Wotton. 2000. Climate change and forest fires. Science of the Total Environment 262: 221–229.

Flint, L.E., A.L. Flint, J.H. Thorne, and R. Boynton. 2013. Fine-scale hydrologic modeling for regional landscape applications: model development and performance. Ecological Processes 2: 25.

Fowler, D., Cape, J.N., Coyle, M., Flechard, C., Kuylenstierna, J., Hicks, K., Derwent, D., Johnson, C., and D. Stevenson. 1999. The global exposure of forests to air pollutants Water, Air & Soil Pollution 116: 1–2.

Fried, J.S., and M.S. Torn. 1990. Analyzing localized climate impacts with the Changed Climate Fire Modeling System. Natural Resource Modeling 4(2): 229–252.
Fried, J.S., M.S. Torn, and E. Mills. 2004. The impact of climate change on wildfire severity: a regional forecast for Northern California. Climatic Change 64(1): 169–191.
Gedalof, Z. 2011. Climate and spatial patterns of wildfire in North America. Pages 89–116, in: D. McKenzie, C. Miller, and D.A. Falk, editors. The Landscape Ecology of Fire. Springer, New York, New York, USA.
Gutowski, Jr, W.J., S.G. Decker, R.A. Donavon, Z. Pan, R.W. Arritt, and E.S. Takle. 2003. Temporal-spatial scales of observed and simulated precipitation in central US climate. Journal of Climate 16(22): 3841–3847.
Gutowski, W.J., Z. Pan, C.J. Anderson, R.W. Arritt, F. Otieno, E.S. Takle, J.H. Christensen, and O.B. Christensen. 2000. What RCM data are available for California impacts modeling? California Energy Commission Workshop on Climate Change Scenarios for California, 12–13 June, 2000. California Energy Commission, Sacramento, California, USA.
Hakkarinen, C., and J. Smith. 2003. Appendix I. Climate scenarios for a California Energy Commission study of the potential effects of climate change on California: summary of a June 12–13, 2000, workshop. Global climate change and California: potential implications for ecosystems, health, and the economy. Electric Power Research Institute, Palo Alto, California, USA.
Hayhoe, K., D. Cayan, C.B. Field, P.C. Frumhoff, E.P. Maurer, N.L. Miller, S.C. Moser, S.H. Schneider, K.N. Cahill, E.E. Cleland, L. Dale, R. Drapek, R.M. Hanemann, L.S. Kalstein, J. Leniahn, C.K. Lunch, R.P. Neilson, S.C. Sheridan, and J.H. Verville. 2004. Emissions pathways, climate change, and impacts on California. Proceedings of the National Academy of Sciences 101: 12422.
Hessl, A.E. 2011. Pathways for climate change effects on fire: models, data, and uncertainties. Progress in Physical Geography 35: 393–407.
Jones P., and P. Palutikof. 2006. Global temperature record. Climate research unit, University of East Anglia. http://www.cru.uea.ac.uk/cru/info/warming/. Accessed July 22, 2015.
Keeley, J.E. 2004. Impact of antecedent climate on fire regimes in coastal California. International Journal of Wildland Fire 13(2): 173–182.
Keeley, J.E., and H.D. Safford. 2016. Fire as an ecosystem process. Chapter 3 in: H.A. Mooney and E. Zavaleta, editors. Ecosystems of California. University of California Press, Berkeley, California, USA.
Keeley, J.E., and A. Syphard. 2015. Different fire-climate relationships on forested and no-forested landscapes in the Sierra Nevada ecoregion. International Journal of Wildland Fire 42: 27–36.
Knowles, N., M.D. Dettinger, and D.R. Cayan. 2006. Trends in snowfall versus rainfall in the western United States. Journal of Climate 19: 4545–4559.
Krawchuk, M.A., and M.A. Moritz. 2011. Constraints on global fire activity vary across a resource gradient. Ecology 92: 121–132.
———. 2012. Fire and climate change in California. California Energy Commission. Publication number: CEC-500-2012-026. Sacramento, California, USA.
Lenihan, J.M., D. Bachelet, R.P. Neilson, and R. Drapek. 2008. Response of vegetation distribution, ecosystem productivity, and fire to climate change scenarios for California. Climatic Change 87(1): S215–S230.
Lenihan, J.M., R. Drapek, D. Bachelet, and R.P. Neilson. 2003a. Climate change effects on vegetation distribution, carbon, and fire in California. Ecological Applications 13: 1667–1681.
Lenihan, J.M., R. Drapek, R.P. Neilson, and D. Bachelet. 2003b. Appendix IV. Climate change effects on vegetation distribution, carbon stocks, and fire regimes in California. 60 pp., in: Global climate change and California: potential implications for ecosystems, health, and the economy. Final report to California Energy Commission. Project 500-03-058CF_A04. Electric Power Research Institute, Palo Alto, California, USA.
Liang, X., D.P. Lettenmaier, E.F. Wood, and S.J. Burges. 1994. A simple hydrologically based model of land surface water and energy fluxes for general circulation models. Journal of Geophysical Research 99: 14415–14428.
Littell, J.S., D. McKenzie, D.L. Peterson, and A.L. Westerling. 2009. Climate and wildfire area burned in western US ecoprovinces, 1916-2003. Ecological Applications 19: 1003–1021.
Liu, Y., S. Stanturf, and S. Goodrick. 2010. Trends in wildfire potential in a changing climate. Forest Ecology and Management 259: 685–697.
Loik, M., D. Breshears, W. Lauenroth, and J. Belnap. 2004. A multi-scale perspective of water pulses in 4396 dryland ecosystems: climatology and ecohydrology of the western USA. Oecologia 14: 269–281.
Lutz, J.A., J.W. van Wagtendonk, A.E. Thode, J.D. Miller, and J.F. Franklin. 2009. Climate, lightning ignitions, and fire severity in Yosemite National Park, California, USA. International Journal of Wildland Fire 18: 765–774.
Mallek, C.R., H.D. Safford, J.H. Viers, and J. Miller. 2013. Modern departures in fire severity and area vary by forest type, Sierra Nevada and southern Cascades, California, USA. Ecosphere 4(12): Art153.
Major, J. 1988. California climate in relation to vegetation. Pages 11–74 in M.G. Barbour and J. Major, editors. Terrestrial Vegetation of California. California Native Plant Society, Sacramento, California, USA.
Marlon, J.R., P.J. Bartlein, M.K. Walsh, S.P. Harrison, K.J. Brown, M.E. Edwards, P.E. Higuerea, M.J. Power, R.S. Anderson, C. Briles, A. Brunelle, C. Carcaillet, M. Daniels, F.S. Hu, M. Lavoie, C. Long, T. Minckley, P.J.H. Richard, A.C. Scott, D.S. Shafer, W. Tinner, C.E. Umbanhower, Jr, and C. Whitlock. 2009. Wildfire responses to abrupt climate change in North America. Proceedings of the National Academy of Sciences 106: 2519–2524.
Maurer, E.P. 2007. Uncertainty in hydrologic impacts of climate change in the Sierra Nevada Mountains, California under two emissions scenarios. Climatic Change 82: S309–S325.
McDowell, N., W.T. Pockman, C.D. Allen, D.D. Breshears, N. Cobb, T. Kolb, J. Plaut, J. Sperry, A. West, D.G. Williams, and E. Yepez. 2008. Mechanisms of plant survival and mortality during drought: why do some plants survive while others succumb to drought? New Phytologist 178: 719–739.
McKenzie, D., Z. Gedalof, D.L. Peterson, and P. Mote. 2004. Climatic change, wildfire, and conservation. Conservation Biology 18(4): 890–902.
McKenzie, D., D.L. Peterson, and J. Littell. 2008. Global warming and stress complexes in forests of western North America. Developments in Environmental Science 8: 319–337.
Meehl, G.A., W.M. Washington, T.M.L. Wigley, J.M. Arblaster, and A. Dai. 2003. Solar and greenhouse gas forcing and climate response in the twentieth century. Journal of Climate 16: 426–444.
Miller, C., and D.L. Urban. 1999. Forest pattern, fire, and climatic change in the Sierra Nevada. Ecosystems 2: 76–87.
Miller, J.D., and H.D. Safford. 2012. Trends in wildfire severity: 1984-2010 in the Sierra Nevada, Modoc Plateau, and southern Cascades, California, USA. Fire Ecology 8(3): 41–57.
Miller, J.D., H.D. Safford, M. Crimmins, and A.E. Thode. 2009. Quantitative evidence for increasing forest fire severity in the Sierra Nevada and southern Cascade Mountains, California and Nevada, USA. Ecosystems 12(1): 16–32.
Miller, J.D., C.N. Skinner, H.D. Safford, E.E. Knapp, and C.M. Ramirez. 2012. Trends and causes of severity, size, and number of fires in northwestern California, USA. Ecological Applications 22: 184–203.
Mote, P.W. 2006. Climate-driven variability and trends in mountain snowpack in western North America. Journal of Climate 6209–6220.
Mote, P.W., A.F. Hamlet, M.P. Clark, and D.P. Lettenmaier. 2005. Declining mountain snowpack in western North America. Bulletin of American Meteorological Society 86: 39–49.
National Research Council [NRC]. 2011. Climate Stabilization Targets: Emissions, Concentrations, and Impacts over Decades to Millennia. The National Academies Press, Washington, D.C., USA.
Nijssen, B.N., D.P. Lettenmaier, X. Liang, S.W. Wetzel, and E.F. Wood. 1997. Streamflow simulation for continental-scale river basins. Water Resources Research 33: 711–724.
Obama, B. 2013. Preparing the United States for the impacts of climate change. Executive Order 13653, November 1, 2013. https://www.whitehouse.gov/the-press-office/2013/11/01/executive-order-preparing-united-states-impacts-climate-change. Accessed July 22, 2015.
Parks, S.A., M.A. Parisien, C. Miller, and S.Z. Dobrowski. 2014. Fire activity and severity in the western US vary along proxy gradients representing fuel amount and fuel moisture. PLOS One 9(6): e99699.

Price, C., and D. Rind. 1994. Possible implications of global climate change on global lightning distributions and frequencies. Journal of Geophysical Research 99(D5): 10823–10831.

Raffa, K.F., B.H. Aukema, B.J. Bentz, A.L. Carroll, J.A. Hicke, M.G. Turner, and W.H. Romme. 2008. Cross-scale drivers of natural disturbances prone to anthropogenic amplification: the dynamics of bark beetle eruptions. Bioscience 58: 501–517.

Romps, D.M., J.T. Seeley, D. Vollaro, and J. Molinari. 2014. Projected increase in lightning strikes in the United States due to global warming. Science 346: 851–854.

Safford, H.D., M.P. North, and M.D. Meyer. 2012. Climate change and the relevance of historical forest conditions. Pages 23–46, in: M.P. North, editor. Managing Sierra Nevada Forests. USDA Forest Service General Technical Report PSW-GTR-237. Pacific Southwest Research Station, Albany, California, USA.

Safford, H.D., and J.T. Stevens. 2017. Natural Range of Variation (NRV) for yellow pine and mixed conifer forests in the Sierra Nevada, southern Cascades, and Modoc and Inyo National Forests, California, USA. USDA Forest Service General Technical Report PSW-GTR-256, Pacific Southwest Research Station, Albany, California, USA.

Safford, H.D., and K.M. Van de Water. 2014. Using Fire Return Interval Departure (FRID) analysis to map spatial and temporal changes in fire frequency on National Forest lands in California. USDA Forest Service Research Paper PSW-RP-266, Pacific Southwest Research Station, Albany, California, USA.

Steel, Z.L., H.D. Safford, and J.H. Viers. 2015. The fire frequency-severity relationship and the legacy of fire suppression in California forests. Ecosphere 6(1): art8.

Stephens, S.L., M. Burrows, A. Buyantuyev, R.W. Gray, R.E. Keane, R. Kubian, S. Liu, F. Seijo, L. Shu, K.G. Tolhurst, and J.W. Van Wagtendonk. 2014. Temperate and boreal forest mega-fires: characteristics and challenges. Frontiers in Ecology and the Environment 12(2): 115–122.

Stephenson, N.L. 1998. Actual evapotranspiration and deficit: biologically meaningful correlates of vegetation distribution across spatial scales. Journal of Biogeography 25: 855–870.

Stevens, J.T., H.D. Safford, S.P. Harrison, and A.M. Latimer. 2015. Forest disturbance accelerates thermophilization of understory plant communities. Journal of Ecology 103: 1253–1263.

Stouffer, R.J., A.J. Broccoli, T.L. Delworth, K.W. Dixon, R. Gudgel, I. Held, R. Hemler, T. Knutson, H. Lee, M.D. Shwarzkopf, B. Soden, M.J. Spelman, M. Winton, and F. Zeng. 2006. GFDL's CM2 global coupled climate models–Part 4: Idealized climate response. Journal of Climate 19: 723–740.

Thorne, J.H., R.M. Boynton, L.E. Flint, and A.L. Flint. 2015. The magnitude and spatial patterns of historical and future hydrologic change in California's watersheds. Ecosphere 6(2): 1–30.

Trouet, V., A.H. Taylor, A.M. Carleton, C.N. Skinner. 2009. Interannual variations in fire weather, fire extent, and synoptic-scale circulation patterns in northern California and Oregon. Theoretical and Applied Climatology 95(3–4): 349–360.

USDA. 2012. Planning rule. US Department of Agriculture, Forest Service, Washington, D.C., USA. http://www.fs.usda.gov/detail/planningrule/home/?cid=stelprdb5359471. Accessed July 22, 2015.

Van Mantgem, P.J., J.C.B. Nesmith, M. Keifer, E.E. Knapp, A. Flint, and L. Flint. 2013. Climatic stress increases forest fire severity across the western United States. Ecology Letters 16: 1151–1156.

Van Wagtendonk, J.W., and D.R. Cayan. 2008. Temporal and spatial distribution of lightning strikes in California in relation to large-scale weather patterns. Fire Ecology 4(1): 34–56.

Westerling, A.L., and B.P. Bryant. 2008. Climate change and wildfire in California. Climatic Change 87(1): S231–S249.

Westerling, A.L., B.P. Bryant, H.K. Preisler, T.P. Holmes, H.G. Hidalgo, T. Das, and S.R. Shrestha. 2011. Climate change and growth scenarios for California wildfire. Climatic Change 109(1): S445–S463.

Westerling, A.L., H.G. Hidalgo, D.R. Cayan, and T.W. Swetnam. 2006. Warming and earlier spring increase western US forest wildfire activity. Science 313: 940–943.

Whitlock, C., S.L. Shafer, and J. Marlon. 2003. The role of climate and vegetation change in shaping past and future fire regimes in the northwestern US and the implications for ecosystem management. Forest Ecology and Management 178: 5–21.

Williams, A.P., C.D. Allen, A.K. Macalady, D. Griffin, C.A. Woodhouse, D.M. Meko, T.W. Swetnam, S.A. Rauscher, R. Seager, H.D. Grissino-Mayer, J.S. Dean, E.R. Cook, C. Gangodagamage, M. Cai, and N.G. McDowell. 2013. Temperature as a potent driver of regional forest drought stress and tree mortality. Nature Climate Change 3: 292–297.

Williams, A.P., R. Seager, A.K. Macalady, M. Berkelhammer, M.A. Crimmins, T.W. Swetnam, A.T. Trugman, N. Buenning, D. Noone, N.G. McDowell, N. Hryniw, C.I. Mora, and T. Rahn. 2014. Correlations between components of the water balance and burned area reveal new insights for predicting forest fire area in the southwest United States. International Journal of Wildland Fire 24: 14–26.

Williams, J.W., and S.T. Jackson. 2007. Novel climates, no-analog communities, and ecological surprises. Frontiers in Ecology and the Environment 5: 475–482.

Yue, X., L.J. Mickley, J.A. Logan, J.O. Kaplan. 2013. Ensemble projections of wildfire activity and carbonaceous aerosol concentrations over the western United States in the mid 21st century. Atmospheric Environment 77: 767–780.

CHAPTER TWENTY-SEVEN

Social Dynamics of Wildland Fire in California

SARAH M. MCCAFFREY, GUY L. DUFFNER, AND LYNN M. DECKER

Introduction

A useful question for managers to ask is: why do we care about understanding fire's ecological processes? At a theoretical level, knowledge itself may be the goal. However, at a more practical level, funding and research interest tends to reflect a desire to understand how to manipulate ecological processes to favor one or several *preferred* management outcomes. Although nature is indifferent about whether fire leads to regrowth of existing species or to wholesale change to a different vegetation type, humans usually favor one outcome over another. Which outcomes are viewed as more or less desirable will depend on trade-offs between diverse human beliefs and values. Many disputes over land management practices, including fire, are due to fundamental differences in what various individuals, organizations, and cultures value and different views on how an ecological process might affect those values. A process that may be perceived as destructive by one person or group may be seen as constructive by others depending on the scale and timeframe each is considering and which value each cares most about—whether it is commodity production, recreation, ceremonial purposes, or habitat for a particular animal species. Although scientific knowledge of an ecological process can help inform management decisions, ultimately it is only one of numerous considerations.

Fire is thus not just an ecological process, but also a social process. Just as rainstorms are normal events but become a concern when they cause flooding, fire is a natural biophysical event that merits more human attention when it begins to have a significant impact on something that people value. This dynamic is what, by definition, makes wildfire a natural hazard. Wildfires have received growing attention in recent years because they have had increasing effects on an array of human values. These increased effects have led to growing debate and discussion about how to minimize fire's negative impacts while fostering the positive effects. Whether an effect is deemed positive or negative is obvious in some situations (e.g., loss of human lives and structures) and less obvious in others.

Decades of research on various natural hazards have provided many useful insights for understanding human responses to wildfire (McCaffrey 2004). Fire is different from most other natural hazards (such as tornadoes and earthquakes) in that it plays an integral ecological function in many ecosystems and, to some degree, can be managed by humans. Allowing fire to play its beneficial ecological role while mitigating negative impacts on things people value will involve some level of fire management but it will also require changes in human behavior. Identifying the best way to do this within the context of the current natural and human environment in California is a highly complex endeavor that necessitates understanding the range of values at risk, how people perceive wildfire management, and actions that might be taken to foster more desirable outcomes. Better-informed management decisions thus will require scientific understanding of social processes as well as ecological processes.

In California, a key reason why concern about wildfires has increased in recent years is due to its growing impact on lives and property. Between 2000 and 2014, 31 civilians (CAL FIRE 2015a) and 25 firefighters died in California due to wildfires (USFA 2014), and from 2000 to 2013 a total of approximately 9,500 homes were lost in the state (NIFC 2014). Across the country, an average of about 1,400 homes and 1,300 outbuildings were lost to wildfire each year between 2000 and 2013 (NIFC 2014). Although geographic distribution of these losses varies each year, California often sustains some of the higher losses. In 2013, for example, 1,093 residences, 945 outbuildings, and 97 commercial buildings were lost to wildfire nationally: although Colorado lost the largest number (520) of residences, California had the highest overall state level losses (184 residences, 521 outbuildings, and 10 commercial buildings) (NICC 2013). Although these numbers demonstrate the more obvious negative outcomes, wildfires also can affect a wide range of things people care about including air quality, cultural resources, a valued animal's habitat, or recreational amenities.

This chapter provides a basic overview of several important social considerations related to wildfire. After a brief cultural history of fire management, the chapter will discuss research findings about the range of variables that do or do not shape the beliefs and attitudes of people who live in fire prone areas. It will then provide a brief description of some of the national and state programs that work to foster public mitigation efforts and an overview of the institutional structure of fire management in California.

Brief Cultural History of Fire Management

Fire has always been part of the human landscape. Humans have tended to live where fire could be used to improve living conditions—by hunters to flush out game, by herders to create grasslands, and by farmers to clear fields (Pyne 1997). Fire has shaped human development and, in turn, has been used by humans to shape the surrounding landscape. As such, it is as much a cultural as a natural phenomenon. "Fire and mankind enjoy a symbiosis: most fires are set, directly or indirectly, by man, and even natural fires are tolerated according to human criteria" (Pyne 1997, p.166). Because definitions of proper use and management of fire can vary distinctly between different cultural groups, management decisions often have more to do with cultural than ecological criteria.

Fire suppression in the United States is a relatively recent phenomenon. As is described in chapter 19, Native Americans were active resource managers who had a sophisticated knowledge of fire use. Early Euro-American settlers also used fire for reasons similar to Native Americans, often learning specific practices from local tribes, as well as to convert forest into agricultural land. Occasionally the burning was done in a haphazard manner, but more often than not a strong set of rules and social mores regulated the methods and season for implementing a controlled burn (Pyne 1997). As permanent settlements were established, settlers organized to suppress any fires that threatened private or community resources: activities included creation of firebreaks, fire brigades, and use of prescribed burning to reduce fuel loads (Pyne 1997, p.219).

The official shift toward a policy of suppressing all wildfires began in the late 1800s in step with the development of the Progressive Conservation movement which developed in part out of growing public concern about mismanaged public domain lands and resource scarcity. To address these concerns, it was argued that resource management decisions needed to be taken out of the hands of legislators and local individuals who were thought to be interested more in immediate profit than in long-term resource use. Science and active government involvement was seen as the means of achieving these goals: science provided an objective means by which trained personnel could make rational decisions, and the federal government was, at the time, the only entity able to engage in such a large-scale comprehensive venture. This ultimately led to the withdrawal of large areas of public domain lands from settlement to be placed in forestry reserves that would be managed by the government to maximize their use for present and future generations. In 1905, the United States Forest Service was established to manage these lands.

Professional forestry was seen as the best means to manage the newly reserved lands. The overall approach was imported from Germany where forests had been actively managed for centuries; fire suppression was an integral part of the approach. The emphasis on full control of fire may have been an appropriate professional management method for the German ecological and cultural environment but not necessarily for the American environment. Ecologically, German forestry was developed in less fire-dependent ecosystems, making its emphasis on suppression perhaps less appropriate for US landscapes that are fire dependent. Culturally, professional forestry was formed in a country with a limited land base and a highly structured society where stability and certainty were valued more than innovation. The basic assumptions—scarcity, stability, and certainty—did not apply in the United States, a rather volatile country where resources were abundant and innovation was valued over certainty (Behan 1975). Thus, despite the emphasis on using science, by adopting management practices from a country with such different ecological and socio-political conditions, the Forest Service was importing a cultural approach to forest and fire management as much as than a scientific or an ecological one.

However, the dominant ecological thinking of the time also supported the focus on fire suppression. In 1910, Frederick Clements published arguably the first fire ecology study, *Life History of Lodgepole Pine Forests* and "a milestone in the theoretical scientific justification for fire protection" (Pyne 1997, p.237). Clements' model of succession became the accepted ecological model of vegetation change. In his model, a plant community would move in a linear fashion over time, from an initial pioneer species through intermediary plant communities to culminate in a final climax stage. As an equilibrial state, persistent over time, climax was seen as the best stage and therefore the one to manage for. Any disturbance that drove the system backward in succession was seen as undesirable and fire was such a disturbance. Suppression was thus a means of helping move the succession process toward the climax state and was therefore a scientifically supported practice (Pyne 1997).

The shift to complete suppression did not happen overnight or without controversy. Effective fire control methods and resources needed to be developed and various members of the public brought on board with the policy. Notably, a social tension initially existed between the professional foresters' belief in the need for suppression and early settler beliefs that fire was a useful management tool. A fair amount of agency effort therefore went into educational campaigns against traditional or light burning practices. The ability to suppress fires was given an important boost during the depression with the availability of a large labor pool to fight fires (Nelson 1979, Pyne 1997). During this period, evidence also began to develop that, in certain ecosystems, fire had an important ecological function. Ecologists working in the southeastern longleaf pine forests repeatedly found evidence of the positive benefits of burning only to not publish the information, sometimes due to fear of "administrative reproof" and at other times due to loyalty to the agency that employed many of them, the Forest Service (Schiff 1962).

Full fire suppression was finally approached with the arrival of World War II due to the added impetus created by the increasing value of forest products and fear of Japanese incendiary bombs setting fire to the West Coast (Pyne 1997). Fire was no longer just the enemy of professional foresters but the enemy of the entire nation and advertising agencies (in conjunction with the Wartime Advertising Council) were enlisted to develop fire prevention ad campaigns. Wartime fire prevention posters had slogans such as "Our Carelessness, Their Secret Weapon," "Forest Fires Aid the Enemy," and "Forests are Vital to National Defense." This effort ultimately led to the creation of Smokey Bear in 1945 and the slogan

"Remember Only You Can Prevent Forest Fires" in 1947 (Forest Service 1984).

In the late 1960s, the emphasis of federal policy on fire control began to shift as it became increasingly evident that a technological threshold had been reached and full fire control was not possible, and due to growing evidence that fire was an integral part of many ecosystems (Biswell 1989). By this time ecological thought had begun to shift from Clements' linear model to a more complex, nonlinear view where fire was not inherently a negative process but one that merely led to different, and potentially desirable, ecological states. As fire became seen as an integral part of many ecosystems, agencies began to change their policies from fire control to fire management. The National Park Service incorporated fire management into its policy in 1968 and the Forest Service in 1974, although the 10 AM policy, which said that any reported fire should be suppressed by 10 AM on the following day, remained in place until 1978. The guiding principle was now one of balancing the cost of suppression with the value of the resources protected (Nelson 1979). However, shifting to a more pluralistic fire management approach has not been a straightforward process, as evidenced by the continuing emphasis on suppression. Changing policy alone is not sufficient: cultural perceptions, for both fire managers and the public, of appropriate fire management also need to shift. Ironically, there was now a need to shift the thinking of both practitioners and the public *back* to earlier beliefs that fire is a useful management tool.

In addition, the landscape in which fire use was being proposed had changed dramatically. For much of the twentieth century, as wildfires occurred mostly away from more settled areas and posed little threat to private resources, wildland firefighting and structural firefighting were separate and distinct activities. However, in the six decades after World War II, the US population doubled and the number of housing units tripled: much of this housing growth occurred in what has become known as the wildland urban interface (WUI). Although often used in relation to wildland fires in the western United States, the WUI has no explicit fire component but simply identifies areas where housing meets or intermixes with natural vegetation regardless of how fire prone or fire dependent that vegetation may be. Today roughly one-third of the US population resides in the WUI. In 2010, over 77 million ha (190 million ac) of land in the United States was classified as WUI (about 9.5% of the total US area), encompassing a population of almost 99 million (32.2% of the total population) and almost 44 million homes (33.5% of total homes) (Martinuzzi et al. 2015). Surprising to some is the fact that the majority of WUI land is in the eastern United States, which tends to be more densely developed and have less public land. In 2010, a relatively small amount of California land was classified as WUI (2.73 million ha or 6.6% of the state's area (compared to North Carolina's 5.42 million ha or 39.8% of the state) and a smaller portion of California's houses (32.6%) and population (30.2%) were in the WUI than a number of western and northeastern states such as Montana (64.1% and 62.4%, respectively) and New Hampshire (82.6% and 81.7%). However, in terms of sheer numbers in 2010 California was the state with the largest number of people (11.24 million) and housing units (4.46 million) located in the WUI (Martinuzzi et al. 2015).

More houses in the WUI increases the values at risk and also has created additional challenges for long-term fire and land management in terms of fighting fires, habitat fragmentation, and fire mitigation planning on private property. Over time, the increased number of people living in fire prone WUI areas erased the "buffer" that previously existed between wildland fire and structural fire considerations. This has led to increased tensions between private resources and government wildfire management in a number of areas, particularly in relation to roles and responsibilities for mitigation of the fire risk and fire protection during an event. Increasingly fragmented property ownership also raises challenges around how to have consistent fire management across diverse owners. Reintroducing fire into this more complex landscape is therefore not just a case of reducing the fuels that have built up due to suppression. An effective management strategy will require transcending numerous boundaries, not solely in relation to land ownership but also in terms of differing organizational mandates, management objectives, and values.

Public Perceptions

Engaging in effective fire and land management in this more complex landscape requires an accurate understanding of the attitudes and beliefs about fire management of those who live in fire prone areas, including their support for different fuel management practices on public lands and reasons they do or do not choose to mitigate fire risk on their land. One of the challenges with understanding public perceptions is that without systematically gathered data it is easy to assume that the perspectives one hears most frequently are representative of the population. However, such assumptions may actually reflect a small proportion of the population and result in sampling bias, as individuals tend to be more vocal when they have a strong opinion about something. As social science research on wildfire management was limited prior to 1998, understanding of what the public thought about wildfires was primarily based on such anecdotal evidence. This contributed to a general narrative that the public demands fire suppression and sees all fire as bad. However, increased research funding (with the establishment of the Joint Fire Science Program in 1998 and the National Fire Plan in 2000) has led to development of a substantial body of fire social science research that provides empirical insights into public perceptions and beliefs and calls into question many of the common assumptions and narratives about public response to wildfire.

One common assumption is that there is a need to make individuals more aware of the fire risk: that lack of proactive response amongst private landowners is due to lack of recognition of the fire risk. However, evidence from multiple studies is quite clear that the vast majority of those living in fire prone areas understand the fire risk (McCaffrey and Olsen 2012). For example, when asked in 2004 to rate their risk on a 10-point scale (with 10 = fire is certain), residents around San Bernardino, California, gave an average rating of 9.3. One problem with the belief that the main issue is low risk perception is the underlying assumption that recognition of a risk will automatically lead to action. However, decades of research on risk perception in relation to other natural and technical hazards have shown that risk perception is a complex and subjective concept. Risk is determined by the probability of an event (for a specific timeframe and spatial extent) times the specific consequence being considered. Differences in risk perception thus may simply reflect a difference in the spatial area or the consequence each individual or organization is considering. For example, although the average rating for general fire risk in San Bernardino was 9.3, the rating for risk to the individual's house was 4.3 (McCaffrey 2008). Although the lower rating may suggest that the individuals do not understand the wildfire risk to their home, it is actually a logical reflection of the smaller temporal area and specific consequence under consideration. It may also reflect actions individuals have taken on their property to mitigate their risk. Even if two individuals have similar risk perception assessments, they may still respond differently due to a number of considerations that have been found to influence mitigation decisions (including efficacy of the action in reducing risk, social norms, and resource constraints (e.g., time, money, physical ability) (Toman et al. 2013). Ultimately, barriers to proactive public responses in relation to wildfire management are less often an issue of information deficit (not understanding the risk) than of resource deficits.

An assumption related to the information deficit belief is that a key problem is the new people moving to fire prone areas who do not understand the fire risk because they have had no exposure to fire issues and that this lack of understanding is the major barrier to their being proactive. However, as indicated above, lack of information about the fire risk is rarely the primary barrier to individuals being proactive, including newer residents. Research finds little evidence that those who have lived in their home for less time are less likely to understand the fire risk or undertake mitigation activities; most studies find no significant difference for length of residence, and when a difference has been found it is often the newer residents who are more proactive. Two dynamics may explain why an intuitively reasonable assumption does not hold empirically. First, census data indicate that around 60% of moves in the United States take place within the same county and less than 20% are interstate moves (US Census Bureau 2014). This means that, although a "new" resident may be new to that street or neighborhood, they likely have still had significant exposure to the area's fire issues. Confirmation bias may also be at play. This is a psychological process where individuals who have formed an opinion about a topic (e.g., longer term residents who have had time to develop an opinion about the fire risk in the area) tend to discount any new information that contradicts that view. Thus, residents who are truly new to an area may actually be more receptive to new wildfire information than longer term residents.

Another common narrative is that most of the public thinks all fire is bad. Again, there is little evidence that this is the case. Studies have consistently found that the vast majority of individuals living in fire prone areas have a good, often quite sophisticated, understanding of the beneficial ecological role fire plays in many ecosystems. In focus groups and interviews, individuals also often discussed the need to reintroduce fire, generally with a preference for use of fire in less populated areas. In addition, one of the most consistent findings across studies is that under the right conditions roughly 80% of the public thinks prescribed fire is an acceptable management tool. There is also a clear preference for active forest management—whether via thinning or use of prescribed fire (McCaffrey and Olsen 2012). For example, in a survey of individuals who had taken a field tour of different fuels treatments sites in the Sierra Nevada, the vast majority of respondents felt prescribed fire (89%), use of prescribed fire in conjunction with mechanical treatments (83%), and mechanical treatment (69%), were acceptable practices, whereas barely half (52%) felt taking no action was an acceptable management approach (McCaffrey et al. 2008). Such high levels of support for prescribed burning and recognition of the need to reintroduce fire clearly shows that most individuals have a more nuanced view of fire than simply "all fire is bad."

A key tension point created by the increased intermingling of public and private land affected by fire is the issue of responsibility. A common assumption is that private landowners do not feel responsible for addressing their fire risk—given that government land management agencies have historically had primary responsibility for handling wildland fire. However, there is little evidence this is the case. Comments in focus groups and interviews indicate that, rather than displacing the responsibility onto fire and/or land management organizations, residents see the responsibility as a shared one, where each landowner is responsible for mitigating the risk on their land (McCaffrey and Olsen 2012). Both quantitative and qualitative studies show that a large majority of individuals living in fire prone areas believe that it is their responsibility to mitigate the fire risk on their property. A consistent finding across studies is that at least 2/3 of homeowners surveyed have undertaken some level of vegetation management on their property (McCaffrey and Olsen 2012). For example, in a survey of homeowners in Ventura County, California; Alachua County, Florida; and around Helena, Montana, 58% of respondents indicated that to prepare for wildfire they had done a great deal of vegetation management and 31% indicated they had done some work (McCaffrey and Winter 2011).

Overall, studies have found that key dynamics that influence public response are fairly consistent across the country; key processes in California are similarly important in Florida. Syntheses of fire social science research have found limited evidence that region of the country or socio-demographic variables, such as education or type of residence, consistently account for differences in how individuals respond to various wildfire issues. Instead, studies suggest that more intangible factors such as group membership, worldview, or specific elements related to local context (ecological, historical events, etc.) are more influential in explaining differences in individual attitudes, beliefs, and actions in relation to fire management (McCaffrey and Olsen 2012).

What Does Influence Acceptance of a Practice?

The two variables most consistently associated with a more proactive response are familiarity with a fire management practice and trust in those implementing it. Numerous studies have found, particularly in relation to fuels management treatments on public lands (i.e., prescribed fire and thinning), that higher levels of knowledge or experience with a practice are associated with higher acceptance levels for the practice, and lower levels of concern about potential negative outcomes such as smoke from prescribed fire. Studies have also found a clear relationship between acceptance of practices and trust in those implementing them (Toman et al. 2013). For example, one study found the only factor predictive of acceptance of a fuel treatment was confidence (a form of trust) in those who were implementing the treatment and that the effect could be quite substantial: a one-unit increase in confidence predicted a 6.2 unit increase in acceptance of thinning and 4.6 unit increase in acceptance of prescribed fire in neighborhoods (Toman et al. 2011).

The fact that familiarity and trust are key variables associated with higher acceptance highlights the importance of relationships in facilitating more proactive views. Interactive communication methods (e.g., conversations, field tours) tend to be seen as the most useful and trustworthy information sources. Studies have also found a preference for one-on-one interactions and that personal relationships with agency personnel can have a positive effect on assessments of mitigation activities, whether the activity being assessed is of agency personnel doing a prescribed burn or the decision to implement defensible space measures (McCaffrey and Olsen 2012). Interactive communication also has been shown to be the most effective means of communication for changing behavior as it enhances the ability to ask questions, address specific concerns, clarify misperceptions, and build trust. This conclusion is consistent with findings in the natural hazards and adult learning fields that have found that adults tend to learn better through interactive exchange and that interactive communication is more influential than unidirectional measures in influencing behavior change (Monroe et al. 2006). Research has also shown how various social interactions can help build individual and community capacity to be prepared for wildfire. Social interactions between residents and fire personnel can facilitate not only information exchange but resource sharing. Social interactions amongst residents can help build social networks and a sense of community that, for many individuals, can increase motivation to be more prepared for wildfires. While studies have shown that outreach programs can help increase such social interactions, they have also shown that, as consideration of local context is critical, no single program will be appropriate everywhere (McCaffrey 2015).

Key Policies and Programs Supporting Public Mitigation Efforts

National Level

Two recent Congressional Acts have had a significant influence in fostering a more interactive model of fire management. The Healthy Forest Restoration Act of 2003 (HFRA) provided incentives to implement fuels treatments on federal lands as well as in the WUI. It also placed an emphasis on community planning by prioritizing availability of federal grants for hazardous fuels reduction projects to locations that had established a Community Wildfire Protection Plan (CWPP). The CWPP must be developed collaboratively between fire agencies, local government and area residents. As of 2010 an estimated 317 communities in California had a CWPP in place, in addition to some areas with a countywide CWPP (FRAP 2010).

The second notable federal action stems from the 2009 Federal Land Assistance, Management, and Enhancement Act (FLAME Act), which along with other actions required federal fire agencies to develop a national cohesive wildland fire management strategy in collaboration with state, local, and tribal stakeholders. The resultant National Cohesive Wildland Fire Strategy addresses three key areas: restoration and maintenance of landscapes, wildfire response, and fire-adapted communities. Fire-adapted communities are human communities that understand their fire risk and are taking the full range of actions to mitigate their risk: the more actions taken the more fire adapted the community becomes. The inclusion of fire-adapted communities as a key goal of the Cohesive Strategy has increased the focus of both agency personnel and various local stakeholders in identifying how communities can most effectively learn to live with fire. Developed in three phases the final Cohesive Strategy and accompanying National Action Plan was published in 2014 to help decision-makers weigh the consequences of different management options relative to the three Cohesive Strategy goals.

There are three national level outreach programs, funded in large part by the Forest Service, that support different aspects of a fire-adapted community. Firewise USA is a program administered by the National Fire Protection Association that focuses on providing resources directly to homeowners and their neighbors to help create communities that can live with and limit losses from wildfire. Communities that have met a basic set of standards can apply for designation as an official Firewise Community (NFPA 2014). A second national program (Ready, Set, Go!), administered by the International Association of Fire Chiefs, is designed to improve dialog and interaction between local fire departments and residents in their community (IAFC 2015). Finally, the Fire-Adapted Community Learning Network (administered by The Nature Conservancy and The Watershed Research and Training Center in Northern California) was established in 2013 to connect and support people and communities who are working to live more safely with wildfire. The network helps facilitate the sharing of innovations and best practices in resiliency and fire adaptation—including actions that can be taken before, during, and after wildfires—by exchanging information and supporting communities and groups working together at multiple scales (FACLN 2016).

State Level Efforts

In addition, a number of state level initiatives facilitate development of fire-adapted communities and are relatively unique to California. As of 2016, California is the only state with statewide wildfire related building codes and vegetation management requirements. In 1991, California Public Resources Code 4290 created minimum defensible space related fire safety standards, including access and vegetation management considerations, for all buildings in State Responsibility Areas (see next section). In 2005, California Public Resources Code 4291 established requirements for vegetation management within 100 feet (inside the property line) of all structures and fire-resistant building codes for new structures built on fire prone lands (California State Legislature 2015).

In terms of outreach programs, Fire Safe Councils (FSC) are an important and long-standing initiative within California. FSC began forming in California in the early 1990s in response to an increasing number of wildfires threatening people and property. Essentially, a FSC is an organized group of community members and agency representatives that work to minimize the wildfire risk of their local area. They can be structured very differently depending on the needs of the group and their project location. Some Fire Safe Councils are very small, consisting of a homeowner's association or subdivision, whereas others can cover an entire city, county, or region. FSC help develop and implement communities' priority projects, such as engaging with locals as part of fire-adapted community education campaigns, creating defensible space and implementing Firewise guidelines, and planning

SIDEBAR 27.1 TRADITIONAL USE–CASE STUDY

Northern California is home to a number of tribes including the Hoopa, Yurok, and Karuk. Although these tribal groups have a tradition of using fire as a land management tool they have not been allowed to openly practice this tradition for over a century even though many of the tanoak (*Notholithocarpus densiflorus*) woodlands in Northern California are adapted to frequent, low and mixed severity fires. As part of local efforts to reintroduce fire the Orleans/Somes Bar Fire Safe Council formed in 2001 with the mission of increasing community protection and restoring historic fire regimes on National Forest Service land, lands held by the Karuk tribe, and private lands on the middle section of the Klamath River. Karuk knowledge and experience with using fire as a land management tool influenced the work of the council from the very start.

This work led to the creation of a 2012 memorandum of understanding between the Karuk tribe and the Six Rivers National Forest to allow the reestablishment of the world renewal ceremony on Offield Mountain. Part of the ceremony involves rolling burning logs down the mountain to ignite fires, an activity that is both symbolic of renewal and functional as the fire helps grow fresh medicines, food, and weaving fibers. For the past few years, the tribe, in collaboration with the Forest Service and the Orleans/Somes Bar Fire Safe Council, has been doing smaller prescribed burns and fuel treatments to prepare the area for the ceremony. Bill Tripp, the Eco-Cultural Restoration Specialist for the Karuk Tribe, believes that the project can serve as a demonstration site that can show how communities can gain the skills and knowledge to both use fire to restore fire resilient landscapes and suppress fires when necessary to protect communities and resources—all in an area that links current fire management practices with traditional Native American fire use.

fuels treatments and fire breaks on area landscapes. Although FSC can vary in their individual missions, most serve as important links between communities and the local fire agencies, filling a role in which more structured governmental bodies have historically struggled. On a statewide level, the California Fire Safe Council is an incorporated entity that acts as a grant clearinghouse to distribute nearly $3 million (in 2015), to local FSC for Community Wildfire Protection Plans, fuels mitigation, and wildfire education campaigns. In early 2015, there were approximately 145 recognized Fire Safe Councils in California (CFSC 2015).

Besides federal and state programs that support activities to learn to live with fire, there are numerous similar local and regional efforts throughout the country, such as Nevada's Living with Fire program (See sidebar case studies for additional examples, sidebars 27.1–27.3). It is a complex endeavor for a community to become fire adapted, and no single program is likely to be effective for all activities or for all locations (McCaffrey 2015). Instead, a range and variety of programs facilitates development of activities and messaging that is tailored to the conditions and needs of the local context.

Institutional Structure of Fire Management in California

Finally, an important, albeit often overlooked, consideration in understanding social dynamics is the institutional structure that exists around fire management. In California there are a multiplicity of agencies which have a stake in shaping fire outcomes, each with different missions and goals. This includes state and federal air quality agencies, which can determine the ability to implement a prescribed fire; city and county governments, which generally determine where and how land can be developed; and law enforcement, which is responsible for evacuations during fires. Although too complicated to enumerate the full range of institutional stakeholders in detail, it is illustrative to simply look at organizations with responsibilities for fire response. A number of federal land management agencies have significant fire management responsibilities, including the Forest Service, four agencies within the US Department of Interior (Bureau of Land Management, Fish and Wildlife Service, Bureau of Indian Affairs, and National Park Service), and the Department of Defense. As land management agencies, these entities must work to balance fire safety with ecological fire needs. Of the approximately 42 million ha (104 million ac) in California roughly half is federal land.

In addition to the federal agencies, most states have a fire or forestry agency with state level fire management responsibilities. Although in some states these organizations can be quite small and act primarily in an advisory manner, in California the Department of Forestry and Fire Protection, more commonly referred to as CAL FIRE, plays a significant role in fire management. The agency was established in 1905 as the State Board of Forestry by the California legislature in response to

SIDEBAR 27.2 NORTHERN CALIFORNIA PRESCRIBED FIRE COUNCIL–CASE STUDY

The Northern California Prescribed Fire Council (NCPFC) was formed in 2009 as a forum for land managers, tribes, researchers, and other stakeholders to work together to protect and expand the use of prescribed fire. Since its inception, the council has focused on community building and shared learning, connecting fire practitioners, scientists, and others around their shared interest in increasing use of prescribed fire. The council has also worked to build burning capacity and support for policy changes that would facilitate the use of fire as a fuel reduction and biodiversity conservation tool. Over the last five years, there has been a gradual but perceptible shift in fire management culture in California, which has been attributed, in part, to the increasingly resonant voices of the council (Quinn-Davidson 2016).

The NCPFC is part of a national network of state and regional councils, collectively called the Coalition of Prescribed Fire Councils. Councils were first established in the southeastern United States. Built on the notion that humans and fire both have a place on the landscape, prescribed fire councils have two defining features: 1) They are grassroot organizations that work to promote a culture of positive, solution-oriented fire management that empowers change from the ground up; 2) Their member base is a diverse group of committed practitioners from a wide range of backgrounds, including state and federal agencies, tribes, research institutions, municipal fire departments, environmental groups, ranchers (Quinn-Davidson 2016).

growing public concern over forest health and available timber supply following several large, destructive forest fires and somewhat uninhibited timber harvesting and land clearing practices by homesteaders. The position of State Forester was also created at this time (Thornton 2012). The State Forester and a few office employees based in Sacramento made up the new forestry "department." To assist in fire patrols and emergencies across California, the State Forester was later granted the right to appoint local fire wardens who were initially funded by the counties in which they were located. In 1919, the state began hiring rangers and since then the agency has increased in size and responsibility. In 2014, CAL FIRE employed roughly 4,300 full-time and 2,400 seasonal employees with an annual budget of $1.4 billion (CAL FIRE 2014). With a formal mission to serve and safeguard the people and protect the property and resources of California the organization mainly focuses on fire response, but also manages programs in resource management, environmental restoration, communications, and fire and resource assessment. Through the Office of the State Fire Marshal, they also work on fire prevention and engineering, education, and code enforcement (CAL FIRE 2015b). Through these various programs the agency works with landowners and communities to foster fuels and fire management, environmental restoration, and community preparedness.

In California, primary fire responsibility is divided into three general classifications: Federal Responsibility Areas (FRA), State Responsibility Areas (SRA) and Local Responsibility Areas (LRA). Each federal land management agency has administrative and fire response responsibilities for land under its management. Although all five of the federal land management agencies have fire response obligations, in California the Forest Service and BLM are responsible for the majority of wildland fire protection on FRA lands, an area of around 15 million ha (Artley 2009). Specific local authorities have responsibility for LRA lands which are generally incorporated areas (or lands not classified as either SRA or FRA) (USDA, USDOI, and State of California 2007). Finally, CAL FIRE is primarily responsible for providing fire protection for the SRA. These lands are defined by population density, land use, and land ownership and do not include densely populated areas, incorporated cities, federal government, and agricultural lands. SRA designation is reviewed by the State Board of Forestry every five years and in 2012 consisted of over 12.6 million hectares (FRAP 2012). CAL FIRE provides direct protection and emergency services on nearly 9.6 million hectares of that total and shares the responsibility with other local, state, and federal government agencies on the remainder (FRAP 2012). Additionally, county fire departments provide fire protection on 1.4 million hectares of State Responsibility Area in six "Contract Counties:" Kern, Los Angeles, Marin, Orange, Santa Barbara, and Ventura (CAL FIRE 2012). Conversely, CAL FIRE has agreements to provide some type of emergency services on LRA lands for 150 cities, counties, and districts (CAL FIRE 2014).

Finally, at the local level there are roughly 835 individual county, municipal, and volunteer fire departments in California (USFA 2015). Of these, approximately 29% are volunteer fire departments, 25% are composed entirely of career firefighters, and the remaining departments have a mix of

SIDEBAR 27.3 MOUNTAIN AREA SAFETY TASKFORCE–CASE STUDY

In the aftermath of a nearby wildland fire, the Mountain Area Safety Taskforce (MAST) of San Bernardino and Riverside Counties was formed in 2002 to address potential public safety issues exacerbated by a multiyear drought and a bark beetle infestation that heightened wildfire risk. As the mountain communities above San Bernardino have only three main egress routes, there were significant concerns over the ability to evacuate tens of thousands of residents and visitors during a wildfire while still allowing access by firefighting crews (Newcombe Sr. 2015). San Bernardino County Fire worked with local residents, Firesafe Councils, volunteer organizations, and representatives of a diverse group of agencies and organizations (including the Forest Service, the Natural Resources Conservation Service, CALFIRE, Caltrans, and Southern California Edison) to identify ways to create a safer environment in the San Bernardino Mountains and surrounding communities (Martinez 2015).

MAST developed an extensive outreach campaign that initially focused on raising awareness of the drought-related bark beetle infestation and facilitating the removal of diseased, dying, and dead trees from private and public lands (MAST 2013). Other activities included developing:

- Emergency response and evacuation plans for residents and area visitors.
- A comprehensive local hazard guide and map book for incoming wildland fire resources and management teams.
- Defensible space guidance for homeowners.
- Educational materials describing contributions of fire to healthy forests.
- Disposal methods for removed trees.

Due in part to its massive outreach campaign, MAST was successful in relaying its message to both area residents and visitors. They were directly credited for the effectiveness in evacuating 20,000 to 30,000 people during the 2003 Old Fire (Wilson-Goure et al. 2006). During the 2007 Grass Valley Fire, law enforcement personnel relied heavily on the evacuation and access plans developed as a result of MAST's efforts. They were prepared to utilize area school buses to evacuate attendees of summer camps and mobilize local resources from a prepared list for livestock transport (Newcombe Sr. 2015). Fuels treatments and tree removal along evacuation routes facilitated safe evacuation for thousands of residents. By 2008, over 1.5 million dead or diseased trees were removed from the project scope, over half of which came from private lands (MAST 2013).

The San Bernardino and Riverside County MAST coalition has provided an example of effective collaboration for a number of emergency planning groups including the Forest Area Safety Taskforce (FAST) which was created in 2003 in San Diego County (Martinez 2015). MAST continues to operate, albeit with less funding and to some degree coordination. However, their original work is still supported by many of the local Fire Safe Councils and other involved individuals, agencies and organizations in and along the San Bernardino Mountains.

volunteer and career staff. In general, municipal fire departments are responsible for specific incorporated areas and are staffed with paid full-time firefighters. Volunteer fire departments tend to operate in rural areas and can be supported by a range of resources including local tax assessment, grants, and donations.

Each of the organizations listed above is an independent entity. While, over time, a generally efficient and effective coordinated response for fighting wildfires has developed between the organizations, such coordination before and after fires is more limited, in part because organizations often have different institutional mandates. For example, although agencies with a land management focus may want to use fire on occasion to meet ecological goals, emergency response agencies that focus on human and property safety may see little value or have no authority to engage in such activities. Ultimately, this large number of independent organizations creates a number of challenges for changing fire management outcomes including: developing and implementing any policy change, mitigating risk across land ownership, developing a consistent message, and potential displacement of risk or responsibility. In response to these challenges, a few interagency efforts—for example, the California Wildfire Coordinating Group, the Southern California Association of Forestry and Fire Wardens, and the California-Nevada-Hawaii Forest Fire Council—have developed to try to coordinate the fire management effort. Even so, a disparity still remains in fire management between traditional fire suppression organizations and resource managers.

Conclusion

As wildfires have increasing social impacts, understanding social processes that affect management choices will be as important as understanding the ecological processes associated with fire. The significant loss of life and homes in the 2017 northern and southern California fires further reinforces the importance of better understanding the full range of social dynamics around wildfires in California. A first step is recognizing that a range of human values underlie any land or fire management decision. Nature is quite comfortable with change (e.g., evolution); it is humans that keep trying to restore or maintain ecosystems in a specific state. This chapter provides an initial step in understanding the complexity of social dynamics and range of issues that may need to be considered to restore fire dependent ecosystems and improve future fire management outcomes. Although at one point land management agencies could act as the dominant players in wildland fire management with little conflict, this is no longer the case. Wildfire management is no longer an activity that can be handled solely by land management agencies. The values and needs of all effected stakeholders will need to be taken into account in order to develop ways where we can live with fire in a manner that allows beneficial ecological processes to occur while minimizing undesired outcomes. Traditional narratives that reflect an essentially dichotomous "management versus public" view may no longer be an appropriate one for wildfire management as few of the narratives about the public that underlie the dichotomy hold up when examined empirically. Instead, thinking about fire management from a partnership and collaborative perspective may be more productive. And, in fact, research findings suggest that this is often the way many stakeholders already approach wildfire management. Taking the plurality of views and values into account in fire and land management will be critical for moving forward and developing new and durable approaches and solutions to current and future fire challenges.

References

Artley, D.K. 2009. Wildland Fire Protection and Response in the United States. https://www.iafc.org/files/wild_MissionsProject.pdf. Accessed January 16, 2015 from IAFC.

Behan, R.W. 1975. Forestry and the end of innocence. American Forests 81(16--19): 38–49.

Biswell, H.H. 1989. Prescribed Burning in California Wildlands Vegetation Management. University of California Press, Berkeley, California, USA.

CAL FIRE. 2012. Contract Counties. http://www.fire.ca.gov/fire_protection/fire_protection_coop_efforts_contractcounties.php. Accessed September 28, 2014 from California Department of Forestry and Fire Protection.

———. 2014. CAL FIRE at a Glance. http://calfire.ca.gov/communications/downloads/fact_sheets/Glance.pdf. Accessed June 5, 2015 from CAL FIRE newsroom.

———. 2015a. Incident Information. http://cdfdata.fire.ca.gov/incidents/incidents_statsevents#2000. Accessed April 15, 2015 from California Department of Forestry and Fire Protection.

———. 2015b. CAL FIRE Programs. http://calfire.ca.gov. Accessed June 5, 2015 from CAL FIRE Home.

California State Legislature. 2015. Public Resources Code. http://leginfo.legislature.ca.gov/faces/codes_displayText.xhtml?lawCode=PRC&division=4.&title=&part=2.&chapter=3.&article=. Accessed February 7, 2016 from California Legislative Information.

CFSC. 2015. California Fire Safe Council. http://www.cafiresafecouncil.org. Accessed January 10, 2015 from California Fire Safe Council.

FACLN. 2016. Fire Adapted Communities Learning Network. http://fireadaptednetwork.org/about/. Accessed February 18, 2016.

Forest Service. 1984. Remember, Only—You. . .: 1944–1984, Forty Years of Preventing Forest Fires, Smokey's 40th Birthday. USDA, Forest Service, Washington, DC, USA.

FRAP. 2010. California's Forests and Rangelands: 2010 Assessment. Fire and Resource Assessment Program. California Department of Forestry and Fire Protection, Sacramento, California, USA.

———. 2012. Historical Wildfire Activity Statistics (Redbooks). http://www.fire.ca.gov/downloads/redbooks/2012Redbook/2012_Redbook_SRA_Protected_OtherAgencies.pdf. Accessed September 27, 2014 from California Department of Fire and Forestry Protection.

IAFC. 2015. Ready, Set, Go! http://www.wildlandfirersg.org. Accessed January 15, 2015.

Martinez, T. 2015. San Bernardino County Fire Public Information Officer (G. Duffner, Interviewer).

Martinuzzi, S., S.I. Stewart, D.P. Helmers, M.H. Mockrin, R.B. Hammer, and V.C. Radeloff. 2015. The 2010 Wildland-Urban Interface of the Conterminous United States. USDA Forest Service Resource Map NRS-RMAP-8. Northern Research Station, Newtown Square, Pennsylvania, USA.

MAST. 2013. A Thinner Forest is a Healthy Forest. http://www.calmast.org. Accessed April 9, 2015 from Mountain Area Safety Taskforce.

McCaffrey, S. 2008. Understanding public perspectives of wildfire risk. Pages 11–22 in: W.E. –Martin, C. Raish, and B. Kent, editors. Wildfire Risk, Human Perceptions and Management Implications. Resources for the Future, Washington, DC, USA.

McCaffrey, S., J.J. Moghaddas, and S.L. Stephens. 2008. Different interest group views of fuels treatments: survey results from fire and fire surrogate treatments in a Sierran mixed conifer forest, California, USA. Internationl Journal of Wildland Fire 17: 224.

McCaffrey, S.M. 2004. Thinking of wildfire as a natural hazard. Society and Natural Resources 17: 509–516.

———. 2015. Community wildfire preparedness: a global state-of-the-knowledge summary of social science sesearch. Current Forstry Reports 1(2): 81–90.

McCaffrey, S.M., and C.S. Olsen. 2012. Research Perspectives on the Public and Fire Management : A Synthesis of Current Social Science on Eight Essential Questions. USDA Forest Service General Technical Report NRS-GTR-104. Northern Research Station, Newtown Square, Pennsylvania, USA.

McCaffrey, S.M., and G. Winter. 2011. Understanding homeowner preparation and intended actions when threatened by a wildfire. Pages 88–95 in: S.M. McCaffrey and C.L. Fisher, editors. Proceedings of the Second Conference on the Human Dimensions of Wildland Fire. USDA Forest Service General Technical Report NRS-P-84. Northern Research Station, Newtown Square, Pennsylvania, USA.

Monroe, M.C., L. Pennisi, S. McCaffrey, and D. Mileti. 2006. Social Science to Improve Fuels Management: A Synthesis of Research Related to Communicating with the Public on Fuels Management Efforts. USDA Forest Service General Technical Report NC-GTR-267. North Central Research Station, St. Paul, Minnesota, USA.

Nelson, T.C. 1979. Fire management policy in the national forests: a new era. Journal of Forestry 77: 723–725.

Newcombe Sr., G. 2015. President, Arrowhead Communities Fire Safe Council. (G. Duffner, Interviewer).

NFPA. 2014. Firewise Communites. http://www.firewise.org. Accessed January 15, 2015 from Firewise Communities.

NICC. 2013. Wildland Fire Summary and Statistics Annual Report 2013. http://predictiveservices.nifc.gov/intelligence/2013_Statssumm/annual_report_2013.pdf. Accessed April 14, 2015 from National Interagency Coordination Center.

NIFC. 2014. Structures destroyed by wildfire only 1999-present. External Affairs. National Interagency Fire Center, Boise, Idaho, USA.

Pyne, S.J. 1997. Fire in America: A Cultural History of Wildland and Rural Fire. University of Washington Press, Seattle, Washington, USA.

Quinn-Davidson, L. 2016. Director, Northern California Prescribed Fire Council. Personal communication, February 17, 2016.
Schiff, A.L. 1962. Fire and Water: Scientific Heresy in the Forest Service. Harvard University Press, Cambridge, Massachusetts, USA.
Thornton, M.V. 2012. General History, Part 1. http://www.calfire.ca.gov/about/about_calfire_history.php. Accessed September 25, 2014 from California Department of Forestry and Fire Protection.
Toman, E., M. Stidham, S. McCaffrey, and B. Shindler. 2013. Social Science at the Wildland Urban Interface: A Compendium of Research Results to Create Fire-Adapted Communities. USDA Forest Service General Technical Report NRS-GTR-111. Northern Research Station, Newtown Square, Pensylvania, USA.
Toman, E., M. Stidham, B. Shindler, and S. McCaffrey. 2011. Reducing fuels in the wildland urban interface: Community perceptions of agency fuels treatments. International Journal of Wildland Fire 20: 340–349.
US Census Bureau. 2014. Migration/Geographic Mobility. http://www.census.gov/hhes/migration/data/cps/historical.html. Accessed April 13, 2015 from Census.gov.
USDA, USDI, and State of California. 2007. California Master Cooperative Wildland Fire Management and Stafford Act Response Agreement. http://www.fs.usda.gov/Internet/FSE_DOCUMENTS/stelprdb5380381.pdf. Accessed January 16, 2015.
USFA. 2014. Summary Incident Report. http://apps.usfa.fema.gov/firefighter-fatalities/fatalityData/search?lastName=&deathYear=2000-2014&state.id=13&state=%5Bid%3A13%5D&mnStatus=0&basicSearch=Search&searchAgainAction=index&city=&incidentDataReport=true. Accessed January 15, 2015 from US Fire Administration.
———. 2015. National Fire Department Census Quick Facts. https://apps.usfa.fema.gov/census/summary.cfm. Accessed April 14, 2015 from US Fire Administration.
Wilson-Goure, S., N. Houston, and A. Vann Easton. 2006. Case Studies: Assessment of the State of the Practice and State of the Art in Evacuation Transportation Management. http://ops.fhwa.dot.gov/publications/fhwahop08014/task3_case.pdf. Accessed April 15, 2015 from US Department of Transportation.

CHAPTER TWENTY-EIGHT

The Future of Fire in California's Ecosystems

NEIL G. SUGIHARA, JAN W. VAN WAGTENDONK,
SCOTT L. STEPHENS, KEVIN E. SHAFFER, ANDREA E. THODE,
AND JO ANN FITES-KAUFMAN

> A thing is right when it tends to preserve the integrity, stability, and beauty of the biotic community. It is wrong when it tends otherwise.
>
> LEOPOLD (1952)

The arrival of humans several thousand years ago brought many changes to the already complex pattern and process of fire, which continue to change to this day. To make sense of the complexity of effects, fire ecologists have constructed a foundation of basic concepts and principles that apply to fire and ecosystem interactions. Application of this foundation allows us to understand how fire operates in ecosystems and influences biotic and abiotic components, and gives us tools to project future patterns in a changing world. From these basics we can start to explain and understand how and why changes have occurred and how we can influence the future of fire in California.

"In its natural role, fire is not a disturbance that impacts ecosystems, but rather an ecological process that is as much a part of the environment as precipitation, wind, flooding, soil development, erosion, predation, herbivory, carbon and nutrient cycling, and energy flow." (chapter 5). Fire is an integral part of wildland ecosystems, and ecologists have come to view fire less as an exogenous disturbance and more as an incorporated process. California's diverse climate and topographic patterns have facilitated the development of a rich array of vegetation and habitats where ecological processes including fire, have sculpted the landscapes and plant communities into complex, continuously changing ecosystems. Although during much of the twentieth century, fire was characterized as a retrogressive event that delays progress toward hypothetical, static, climatic climax vegetation, today fire is recognized as a vital, incorporated ecosystem process.

Fire ecology is an emerging and rapidly expanding field of science—and there are still many gaps in our knowledge. Initially, fire ecology research concentrated on chaparral in the South Coast and mixed conifer forests of the Sierra Nevada, and has expanded to cover the entire state. Given the diversity of flora and fire regimes across California, we are far from having a comprehensive body of research on fire ecology. Broader, more comprehensive treatments of statewide fire ecology information are becoming available in this and other publications (Sawyer et al. 2009)

Within the span of a few decades, our understanding of fire in California's ecosystems has grown from a few concepts and opinions into a well-developed, complex, and cohesive discipline. This book covers a wide array of topics that are unified by wildland fire. We have developed a good understanding of how fire operates as an ecological process and its history and current patterns. Predicting the future of fire in California's ecosystems is clearly more speculative and uncertain.

Maintaining Fire Regimes

Pyrodiversity—the variability within fire regimes over long periods of time—promotes biological diversity, is crucial to restoring and maintaining ecosystems (Martin and Sapsis 1992). Pyrodiversity is particularly important where the variation within fire regimes provides much of the fine-scale habitat variability and biodiversity in fire regimes that are characterized by short fire return interval, low-intensity, and low-moderate severity surface fires. Pyrodiversity does promote biodiversity in most California ecosystems, so restoring and maintaining historic levels of pyrodiversity are an important considerations in supporting historic biodiversity.

Over millennia, climate and fire regimes change continuously and biotic communities respond by shifting geographic distribution and range. Modification of fire regimes during post-Euro-American settlement has changed some of the boundaries between ecological zones, allowing white fir (*Abies concolor*) in the Sierra Nevada to expand to lower elevations, and forcing ponderosa pine (*Pinus ponderosa*) to retreat upward, replaced by chaparral and foothill oak woodland. The influences of climate change, fire exclusion and other human activities on post-Euro-American fire and biota are intertwined, and vital considerations in understanding the current ecological role of fire in California.

Human Interactions with Fire

Humans influence many ecological processes and our past and current actions shift naturally occurring fire regimes. Fire's pattern and ecological role have greatly changed. Fire suppression—commonly singled out as the cause for changing fire regimes—is often part, but rarely the entire explanation. Although fire suppression is usually effective, it is far from the only influence changes that fire regimes. Fire suppression is a tool for protecting public safety, natural resource and human property protection, and other quality of life values—ecological impacts are largely unintended and only recently recognized.

Fire suppression is most effective when initial attack puts fires out while they are very small or burning under moderate weather conditions. The effect is to reduce the area burned. Those that do escape suppression usually do so under extreme weather conditions (that is why they were able to escape), shifting a greater proportion of the lower intensity fire to the high fire intensity portion of the range.

Historic fire regimes represent the fire patterns with which current biotic communities developed and persisted, providing a baseline for assessing contemporary changes in fire and ecosystems. Most of California's current fire regimes differ greatly from those prior to the gold rush and early Euro-American settlement era. Stephens et al. (2007) estimate that approximately1,800,000 ha (4,448,000 ac) of fires burned during a typical year prior to the arrival of Euro-American settlers. Only a small fraction of that area burns in typical years now, and even the area burned during our recent "extreme fire years" does not approach the area of historic fire regimes. Today, very few California landscapes still support their previous characteristic fire regimes.

The Challenging Environment Facing Current Fire Management

Fire regime-ecosystem changes that are often attributed to fire exclusion and land management decisions are truly multifaceted. The chapters in this book describe characteristic relations of fire and ecosystems on complex, diverse landscapes, and changing human influences. Differing from historic fire regimes, many of California's current fire regimes can be considered transitional to future ecological condition and fire regimes.

There is an illusion that fire regimes were once stable, static patterns. This is false; fire is a complex ecological process that has been and always will be changing. It is the nature of fire regimes to be dynamic entities, responding to the dynamic patterns of ignition, fuel condition, weather, fire landscapes, and human responses. All of these factors change continuously over a number of time and spatial scales. So why would fire regimes remain static? They don't and won't. Continuing changes to fire regimes in California are not merely probable in the future, they will happen, and knowledge of why and how fire regimes change is key to managing them.

Because there is an intimate relation between vegetation type and fire regime, changes to fire patterns will have resultant effects to vegetation. Understanding these relations is critical to the effective long-term management of ecological systems on large landscapes. We have detailed the fire-vegetation relations for major plant communities, and developed many of the major issues influencing fire pattern changes into the future. However, our perception of how the current and future fire regimes are related to past fire regimes is more complex. To structure the discussion of changing fire regimes, Box 28.1 defines the change of fire regimes from intact, to a finite period where they are altered, and eventually to a permanent converted condition. It is essential that there is an understanding that converted fire regimes will usually no longer support the previous vegetation type and lead to vegetation type conversion. Once the vegetation type conversion is complete, the new "intact" fire regime is characteristic of the area.

Altered Fire Regimes

Most California fire regimes are altered and nearly all native biota and communities have been and are affected by these alterations. Only some long fire return intervals, and some areas of shrublands, with fire return intervals of several decades have remained intact. The most altered fire regimes are typically those that had the shortest characteristic fire return intervals. There are two general types of altered fire regimes—deficient and surplus—both of which lead to fire regime and then ecosystem conversion. In deficient fire regimes, the characteristic fire frequency is reduced, leaving the system deficient of fires, allowing fuel to accumulate and increasing the intensity and severity of the subsequent fires. In surplus fire regimes, the characteristic fire frequency increases leaving the system with a surplus of fires, fuel accumulation is lower, decreasing the intensity and severity of the next fire. Altered fire regimes often facilitate conversion to another vegetation type.

Converted Fire Regimes

For many altered fire regimes, it is not productive to view vegetation and fire regimes as they were, but rather to view them as what they are moving toward. The change is both a fire regime adjustment and vegetation type conversion. Once the vegetative type conversion has been completed, the new fire regime becomes that of the new vegetation type. Most fire regime conversions in the western United States are either, from frequent low intensity/severity fires to infrequent high intensity/severity regimes, or conversion to non-native grasslands with increased fire frequency.

The Future of fire in California's Ecosystems

Management of Fire in California Ecosystems Must Be Based in Ecology and Human Values

The long-term future of fire in California's ecosystems is a social decision. California is the most populous state in the United States, and the challenge of living with fire is ever present. We choose to inhabit fire-prone ecosystems, and need to balance the relative importance of: 1) allowing fire to occur on its own terms, 2) adjusting our communities to fire, and 3) excluding fire and continuing to interfere with the natural range of fire regimes and essential ecological function of fire itself. This is not a simple issue with clear rights and wrongs. There are clear reasons to both exclude fire and to incorporate it in wildland management. How we as a society decide to accommodate—or interfere with—fire will say a great deal about our social ecological sophistication and how much we value our native biota and natural environment.

Globally, humans have been using fire for hundreds of thousands of years to manipulate their environment. The use of fire shifted our status from foragers to cultivators and

BOX 28.1. FIRE REGIME STATUS

1) *Intact fire regime*—The fire regime supports the persistence of the current vegetation type. Examples include most chaparral types and long fire return interval subalpine and alpine forests.

2) *Altered fire regime*—The characteristic fire regime is modified outside the characteristic range of variation. If this modification continues, type conversion of the vegetation can occur. Vegetation and fuel conditions are transitional to another type of fire regime. There are two types of altered fire regimes:
   a. *Fire deficit*—For most altered fire regimes, the FRI is too long to sustain the current vegetation type. Fuel accumulation can change fire intensity, severity, type, spatial complexity, and size. Examples include many of the mixed conifer and eastside pine forests in the Sierra Nevada.
   b. *Fire surplus*—When the FRI is too short to sustain the current vegetation type, increased fire frequency can change fire intensity, severity, type, spatial complexity, and size. Examples include many chaparral types near the Wildland-Urban interface in south coastal California, and many desert shrublands.

3) *Converted fire regime*—Fire regime and fuel structure are converted to another fire regime and its characteristic fuel structure. Type conversion for the vegetation has occurred. There are two types of fire regime conversions:
   a. *Existing*—The fire regime and vegetation have been converted to another type that was already present in the area. Examples are Oregon white oak to Douglas-fir, ponderosa pine forest to mixed conifer forest, and Sagebrush steppe to western juniper.
   b. *New*—The fire regime and vegetation have been converted to another type that was not present in the area. This is typically limited to areas invaded by nonnative invasive plant species. Examples include chamise chaparral to annual grassland, and several types of low elevation desert shrublands to red brome and Mediterranean grass.

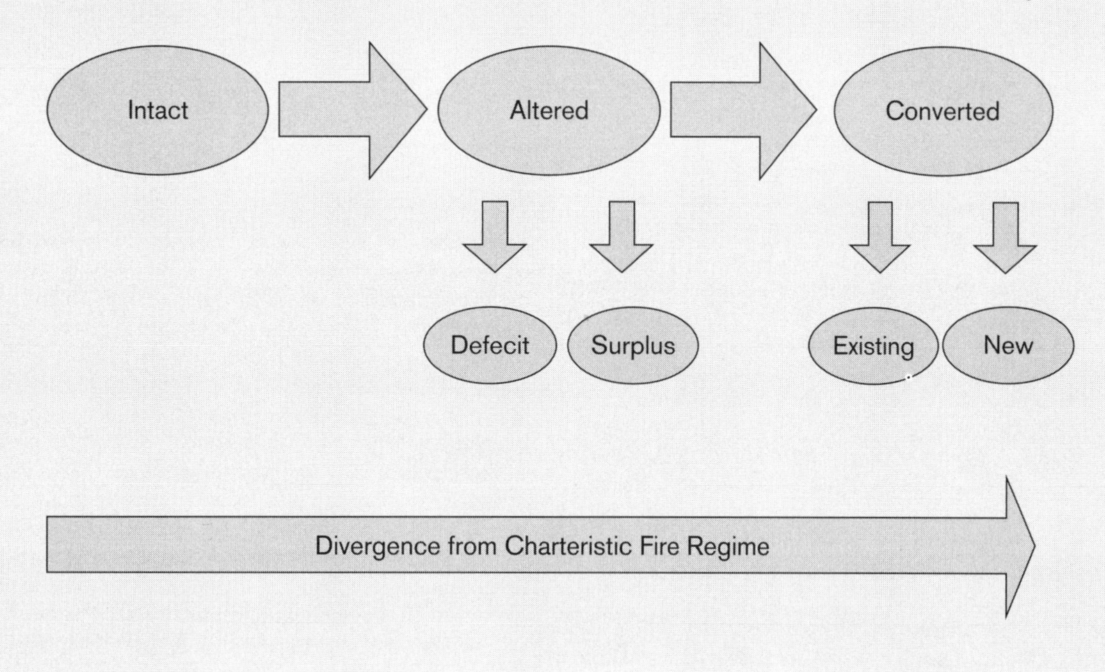

contributed to enabling our species to expand around the world. Fire application to California landscapes is as ancient as the first human occupation about 11,000 years ago. Fire was the most significant, effective, efficient, and widely employed vegetation management tool utilized by California Native Americans, and they conducted purposeful burning to meet specific cultural objectives and maintain specific plant communities. Their influence on California ecosystems has varied across a spectrum from little to none in remote areas to considerable in human-maintained ecosystems.

Since European explorers arrived in 1542, human activities have altered the state's fire regimes. Removal of traditional anthropogenic fire, permanent Euro-American population, and formal fire policy excluding fire all reduced fire occurrence throughout most of California. Starting during the 1960s and 1970s and codified in the 1995 unified federal fire policy, federal fire management recognized altered fire regimes and changed to incorporate a combination of fire suppression and fire management. During the late 1990s and early 2000s, fire and land management have started focusing on managing wildland fuel and ecosystem management treatments. Fire has become recognized as an important ecological process, and fire management is increasingly addressing ecosystem values by focusing on the restoration of characteristic fire regimes.

As is often stated, it's not a matter of if fire will occur but when it will occur.

Fire suppression is necessary, and remains the dominant relation that our society has with wildland fire. Since the Berkeley fire in 1923, the issue of fire in the wildland-urban interface has become one of the most important land management issues facing Californians. Despite intensive efforts and the application of great amounts of technology and money to the effort to exclude wildfires, they continue to occur and have great effects on society and ecosystems. The fires in October 2017 in Sonoma, Mendocino, Napa, and other counties clearly illustrate that this situation continues to worsen, and that without intensive efforts to manage the WUI, we can expect even more deadly and destructive fires in the near future. To moderate the impact of fires burning out of the wildlands into the mixed and urban landscapes, management acknowledgement of both the characteristic fire regimes wildlands and the fuel characteristics of the urban areas is needed.

We value our health, clean air, and clean water, and would like to protect all of our native species and ecosystems. Historically, there was a lot of fire, a lot of smoke, and a lot of fire-accelerated erosion in California. Wildland fire produces smoke and other combustion byproducts that can be harmful to human health and particulate matter that reduces visibility. Fire increases erosion, reduces water quality, and kills vegetation. Although society might not like these effects because they can have detrimental effects on human health and quality of life, they are, to a large extent, natural. Today, we have excluded fire to the point where we have experienced, and are expecting, far less fire impacts to air and water quality than existed at any time during the several millennia before Euro-American-settlement of California.

Today, watersheds and fire regimes are highly altered by human activities. Water development, mining, road building, urbanization, fire suppression, timber harvesting, and recreation are impacting watersheds. Largest erosion events typically follow very large, uniformly high-severity wildfires in steep, erosive landscapes. Fire and its associated pulses of sedimentation, mass wasting, and flooding are natural processes that work within ecosystems and are part of the process that creates and maintains watersheds. However, like air quality management, the focus of watershed management is often to minimize the impacts of prescribed and managed fire because it is considered discretionary. With severe limitations on fire use, the unplanned fires will be more uniformly high in severity and often cause an elevated pulses of watershed instability and tree mortality.

One of the most significant fire and ecosystem changes was the arrival of nonnative, invasive species, with the earliest European contact in the 1500s. In the dynamic cycle between grasses and fire, invasive grass species become established in an area dominated by woody vegetation. As the invasive grasses increase in abundance, a continuous layer of highly combustible fine fuel develops, resulting in increased rates of fire spread and fire frequency. Shrublands and forests composed of native species are converted to grasslands comprised mainly of nonnative species. Invasive species are responsible for converting fire regimes in large areas in Southern California chaparral, the Great Basin, the Central Valley, and the Mojave Desert.

> To keep every cog and wheel is the first precaution of intelligent tinkering. (*Leopold,* 1952)

The Federal Endangered Species Act and the California Endangered Species Act were enacted to protect native plants and animals that are threatened or endangered with extinction.

In California, fire and fuel management and at-risk species conservation and protection have more often been in conflict than in accord. Species protection often has meant fire exclusion; however, a recent research has demonstrated that California spotted owl nesting habitat will not be conserved with a fire exclusion strategy (Stephens et al. 2016). Opportunities for fire management to aid in the protection of at-risk species may provide the best opportunity for these species where the absence of fire has degraded habitat or where fire is not likely to be allowed to return naturally. There are numerous examples across California where fire and fuel management activities, prescribed burning, fire suppression, or post-fire rehabilitation and restoration have been integrated while conserving and protecting at-risk species, their habitats, and ecological processes. Many at-risk species, and the biological communities and ecological systems they depend on, cannot be sustained or recovered without the immediate and long-term ecological functioning provided by fire. Fire as an ecological process is a necessary part of California's ecosystems, and if we really intend to keep all of the parts, fire should be returned to the inventory of California's diverse "cogs" and "wheels."

Where Do We Go from Here?

> I believe that the type of wildfire that occurs is also discretionary and this needs to be better recognized by regulators. The type of wildfire is largely dependent on how the area has been managed. (*Carl Skinner,* personal communication)

Wildfire is both an important ecological process and a powerful, intimidating force that threatens our safety and way of life. Although this book synthesizes and consolidates our understanding of fire, it does not answer the question of what we want our relationship with fire to be. This is fundamentally a social question, and societal wants and needs are as dynamic as fire regimes and ecosystems. One of the basic goals of this book is to articulate the basic relationship between fire and ecosystems, to provide society and additional perspective in developing future fire management options.

The largest fires in California's recorded history are getting larger, more destructive, and are occurring at an increasing rate. The explanations for these trends lie in the nature of fire-ecosystem interactions and the history of our management practices. In terms of human loss, the most destructive fires burn out of the wildlands and into the rapidly expanding urban development. Where fire exclusion caused fire regime alterations or conversions have occurred in wildlands interfacing with our communities, risk to that community is often elevated. Destructive fires cannot be eliminated, but the design of the built environment, including both wildland fuel in strategic locations and creation of fire-resilient structures, buffers and barriers in the interface can moderate the level of damage when fires do occur. Further social engagement of the people living in the interface on possible solutions to this public safety problem is needed.

The Challenge of Future of Fire and Land Management

To effectively manage the emerging and potential issues and trends in managing fire in California's ecosystems, we must look forward with knowledge of history. Anticipation of the future of fire in California ecosystems is greatly enhanced by knowledge of history and ecological events that influenced fire as a process. Past fire regimes do not necessarily reflect the current and future condition of fire in most California ecosystems. The decision to accept ecological change or to restore the historic ecosystems needs to be made based on desired conditions and knowledge of the fire regime that is needed to restore and maintain that system.

No matter how important fire is to ecosystems, we will not universally restore fire to its historical role. Many ecosystems are impacted beyond the point where restoration is feasible, and most fire regimes are altered. Discontinuities exist throughout the natural landscape, preventing fires from achieving their historic patterns. The only way that fire regimes can be fully restored to California is if humans were to value the restoration of historic ecosystems and processes to the exclusion of all other land uses—and that is simply not going to happen. Focused, prioritized management of fire regimes in selected ecosystems and locations is needed.

In wildlands where natural ecosystems and processes are the priority, fire regimes can be incorporated into long-term management plans. Although land management planning includes many social values, prescriptions can incorporate the variability of fire regimes. Narrowly focused prescriptions that apply only parts of the historic fire regime or use mean values for the fire regime attributes do not recognize the variability of historic fire patterns. Without the dynamic nature of natural fire regimes, restored ecosystems are not likely to approximate historic levels and patterns of species distribution and diversity.

A few details are clear when looking into the future of fire and land management. The population of California will continue to grow, the wildland-urban interface will continue to expand, wildlands will be valued as both habitat and open space, and the regulation of fire and other land management activities will continue to increase. We will continue to care about fire management decisions and the short- and long-term effects of those decisions. If we are going to maintain functioning ecosystems, our current pattern of near-universal fire suppression will gradually be replaced by fire management systems that incorporate and apply our rapidly expanding knowledge of fire's role in the ecosystem and how it influences the wildland resources that we value. The understanding of fire and its role in ecosystems is a vital part of making informed, intelligent land management decisions.

> You can't restore fire without fire. *Sue Husari* (personal communication)

The restoration of fire as an ecosystem process is a complex undertaking. Substituting mechanical treatments that can only mimic some aspects of fire will accomplish only portions of fire's role. There is one simple rule that applies to the restoration of fire into ecosystems: to completely restore fire as an ecological process, there is no substitute for fire.

Intentionally and unintentionally, we are affecting fire regimes on all wildlands in which we manage or suppress fire. The management of fire regimes remains among the most important land management activities on most wildlands. Fire exclusion has altered ecosystems on a massive scale, influencing the habitats for thousands of species in hundreds of ecosystems. Whether we implement a detailed fire regime or decide to suppress all fires, we are prescribing a desired fire regime; there is no real "no-action alternative." This decision to impose a fire regime on an ecosystem should be taken seriously.

Managing fuel is an extension of managing fire regimes. Because both surface and crown fires rely on surface fuel to generate fire spread, treating surface fuel is an essential step in effective fuel management programs. Fuel treatments can make fire exclusion more effective by facilitating fire suppression. In other cases, fuel management is the first step in restoring historic fuel conditions for the purposes of restoring historic fire regimes. The intentional manipulation of fuel to achieve desired fire conditions should be the focus of a variety of fire and land management activities.

We will never know everything that we would like to know about fire in California ecosystems, but we do need to use what we know. We can largely determine wildland fire occurrence by suppressing or manage wildland fires, and applying prescribed fire. The key to successful long-term fire management is to manage fire regimes.

Clearly, California ecosystems will not be the same without fire playing its ecological role. If we are to maintain California ecosystems for future generations, it is time to decide if, where, and how we will move forward with restoration of fire and where we will not. And this time, humans will be almost solely responsible for determining future fire regimes. There is no simple answer. The challenge to understand and incorporate fire regimes into land management while providing dynamic ecosystems for habitat, natural resources, and public recreation and safety, is the largest, most complex, and risky endeavor ever attempted in land management, anywhere. To succeed, our commitment must match the task.

Wildland fire is a part of life in California. We have an unprecedented opportunity, a responsibility, an obligation, to decide the direction of fire into the future. Let's use this knowledge to make informed decisions that will help shape the ecosystems of the future.

References

Leopold, A. 1952. A Sand County Almanac and Sketches Here and There. Oxford University Press, New York, New York, USA.

Martin, R.E., and D.B. Sapsis. 1992. Fires as agents of biodiversity: pyrodiversity promotes biodiversity. Pages 150–157 in: R.R. Harris and D.C. Erman, editors. Proceedings of the Symposium on Biodiversity of Northwestern California. Wildland Resources Center, Report 29. University of California, Berkeley, USA.

Sawyer, J.O., T. Keeler-Wolf, and J.M. Evens. 2009. A Manual of California Vegetation, 2nd ed. California Native Plant Society. Sacramento, California, USA.

Stephens, S.L., R.E. Martin, and N.E. Clinton. 2007. Prehistoric fire area and emissions from California's forests, woodlands, shrublands and grasslands. Forest Ecology and Management 251: 205–216.

Stephens, S.L., J.D. Miller, B.M. Collins, M.P. North, J.J. Keane, and S.L. Roberts. 2016. Wildfire impacts on California spotted owl nesting habitat in the Sierra Nevada. Ecosphere 7(11): e01478.

APPENDIX ONE

Fire Regime Attributes for Each Vegetation Type Discussed in The Bioregional Chapters

Fire Regime Terms Used in This Table are Defined in Chapter 5

	Fire regime attribute						
	Temporal		Spatial		Magnitude		
VEGETATION TYPE	SEASONALITY	FIRE RETURN INTERVAL	SIZE	SPATIAL COMPLEXITY	FIRE LINE INTENSITY	FIRE SEVERITY	FIRE TYPE
			North Coast				
N. Coastal scrub and prairie	Summer-fall	Long	Small	Low	Multiple	High	Passive–active crown
N. Coastal chaparral	Summer-fall	Moderate-long	Moderate	Low	High	Low-high	Active crown
N. Coastal pine forest	Summer-fall	Truncated-medium	Medium	Low	High	Very high	Passive–active crown
Sitka spruce forest	Summer-fall	Truncated-long	Small-medium	Moderate	Moderate	Multiple	Passive–active crown
Redwood forest	Summer-fall	Short-long	Small-medium	Moderate	Moderate	Low	Surface
Douglas-fir /tanoak forest	Summer-fall	Short	Medium	Moderate	Multiple	Low-moderate	Surface
Oregon oak woodland	Summer-fall	Truncated-short	Medium	Low	Low	Low	Surface
			Klamath Mountains				
Douglas-fir	Summer-fall	Short	Medium-large	Moderate-high	Low-moderate	Low-moderate	Surface
Canyon live oak	Summer-fall	Short-medium	Medium	Moderate-high	Low-moderate	Low-moderate	Surface
California black oak	Summer-fall	Short	Medium	Low-moderate	Low	Low-moderate	Surface
Oregon oak	Summer-fall	Short	Medium-large	Low-moderate	Low	Low-moderate	Surface
Buckbrush	Summer-fall	Medium-long	Medium	Moderate	High	High	Passive–active crown

Brewer oak	Summer-fall	Short-medium	Small-large	Moderate-high	Low-high	Moderate-high	Surface-passive crown
Whiteleaf manzanita	Summer-fall	Medium-long	Medium-large	Low-high	Low-high	Moderate-high	Surface-active-independent crown
Jeffrey pine	Summer-fall	Short	Small-large	Moderate	Low-moderate	Multiple	Surface
White fir	Summer-fall	Short-medium	Small-large	Moderate-high	Low-moderate	Multiple	Surface-passive crown
Shasta red fir	Late summer-early fall	Short-medium	Medium	Moderate-high	Low-moderate	Multiple	Surface-active crown
Knobcone pine	Late summer-early fall	Truncated-medium	Medium	Low-moderate	Multiple	Moderate-high	Active crown
Port Orford cedar	Late summer-fall	Short-medium	Small-medium	Moderate-high	Low-high	Low-high	Surface-active crown
California pitcher plant	Late summer-fall	Short-medium	Small	Moderate-high	Low-moderate	Low-high	Surface
Greenleaf manzanita	Summer-fall	Medium-long	Small-large	Low-high	Moderate-high	Moderate-high	Active crown
Subalpine	Late summer-fall	Short-long	Small-medium	Moderate-high	Low	Multiple	Surface
			Southern Cascades				
SW foothills-woodlands	Summer-fall	Short	Medium-large	Moderate	Low-moderate	Low-moderate	Surface
SW foothills-conifers	Summer-fall	Short	Medium-large	Moderate	Low-moderate	Low-moderate	Surface
SW foothills-shrubs	Summer-fall	Medium	Medium-large	Multiple	Multiple	Moderate-high	Active crown
Mid-montane confers W-side	Summer-fall	Short	Small-medium	Moderate	Low-moderate	Multiple	Surface
Mid-montane shrubs W-side	Summer-fall	Short	Small-medium	Moderate	Low-moderate	Multiple	Passive crown
Mid-montane conifers E-side	Summer-fall	Short	Small-large	Moderate	Low-moderate	Multiple	Surface
Knobcone pine	Summer-fall	Truncated-medium	Medium	Low-moderate	Multiple	Moderate-high	Active crown
Mid-montane shrubs E-side	Summer-fall	Medium-long	Small-large	Low-high	Low-high	Moderate-high	Active crown
Jeffrey pine/ponderosa pine	Summer-fall	Short	Small-large	Moderate-high	Low-moderate	Low-moderate	Surface
True fir	Late summer-fall	Short-long	Small-medium	Moderate-high	Low-high	Low-high	Surface
Lodgepole pine	Late summer-fall	Medium-long	Small-medium	Low-high	Low-high	Moderate-high	Surface
Upper-montane shrubs	Late summer-fall	Medium-long	Medium	Low-high	Moderate-high	Moderate-high	Active crown

Northeastern Plateaus							
Low sage	Summer-early fall	Truncated-medium	Medium	High	Moderate	Very high	Surface
Basin big sage	Summer-early fall	Truncated medium	Small	High	Moderate	Very high	Multiple
Wyoming sage	Summer-early fall	Truncated medium	Large	Multiple	Moderate	Very high	Multiple
Mountain big sage	Summer-early fall	Truncated medium	Large	Multiple	Moderate	High	Multiple
Bitterbrush	Summer-early fall	Truncated medium	Medium	Multiple	Moderate	Multiple	Multiple
Mountain mahogany	Summer-early fall	Truncated medium	Small	Multiple	High	Very high	Multiple
Western juniper	Late summer-early fall	Truncated long	Small-large	Multiple	High	Very high	Multiple
Quaking aspen	Late summer	Medium	Small	Low	Multiple	High	Surface
Cheat grass	Summer-early fall	Short	Large	Low	Low	Low	Surface
Jeffrey pine	Summer-early fall	Short-medium	Medium	Medium-high	Low	Low	Surface
Ponderosa pine	Summer-early fall	Short	Medium	Low-medium	Low	Low	Surface
Jeffrey pine/ ponderosa pine	Summer-early fall	Short-medium	Medium	Low-high	Low-moderate	Low-moderate	Surface
White fir and incense cedar	Summer-early fall	Medium	Medium	Medium	Multiple	Multiple	Multiple
Sierra Nevada							
Foothill chaparral	Summer-fall	Medium	Large	Low	High	High	Active crown
Oak woodland/ grasslands	Summer-fall	Short	Large	Low	Low	Low	Surface
Foothill conifer patches	Summer-fall	Medium	Small	Low	High	Moderate	Passive crown
Ponderosa pine	Summer-fall	Short	Large	Low	Low	Low-moderate	Surface
California black oak	Summer-fall	Short	Large	Low	Low	Low-moderate	Surface
Douglas-fir /white fir	Summer-fall	Short	Large	Multiple	Low-moderate	Low-moderate	Surface-multiple
Tanoak/mixed evergreen	Summer-fall	Medium	Medium	Multiple	Multiple	Multiple	Multiple
California red fir	Late summer-fall	Medium	Medium	Multiple	Multiple	Low	Multiple
Montane Chaparral	Summer-fall	Medium	Medium	Low	Moderate-High	High	Active crown
Jeffrey/western white pine	Summer-fall	Medium	Truncated small	Low	Low	Low-moderate	Surface
Tufted hairgrass	Late summer-fall	Long	Small	Low	Low	Low	Surface

Lodgepole pine	Late summer-fall	Long	Small	Low	Low	Low	Surface
Mountain hemlock	Late summer-fall	Long	Small	Low	Low	Low	Surface
Whitebark/limber/ foxtail pine	Late summer-fall	Truncated-Long	Small	Low	Low	Low	Surface
E-side Jeffrey/ ponderosa pine	Summer-fall	Short	Small-medium	Multiple	Low	Low	Surface
E-side White fir mixed conifer	Summer-fall	Medium	Medium	Multiple	Multiple	Multiple	Multiple
			Central Valley				
Foothill woodlands	Summer-fall	Short	Small-medium	High	High	High	Surface
Grasslands	Summer-fall	Short	Medium-large	Low	Low	Moderate-high	Surface
Riparian woodlands	Summer-fall	Long	Small-medium	High	Low-high	Low-high	Surface-passive crown
Freshwater marshes	Summer-fall	Short	Small-medium	High	High	High	Surface
			Central Coast				
Monterey pine	Late summer-early fall	Moderate-high	Moderate-large	Low-moderate	Low-high	Moderate-high	Surface-passive crown
Bishop pine	Late spring-late fall	Short-moderate	Small-large	Low-moderate	Low-high	Low-high	Surface, Passive–active crown
Maritime chaparral	Early summer-late fall	Moderate-very long	Moderate-large	Low	Moderate-high	Low-high	Active crown
Coulter pine	Early summer-late fall	Moderate-long	Moderate-large	Low-moderate	Low-moderate	Low-high	Surface, passive–active crown
Knobcone pine	Early summer-late fall	Moderate-long	Large	Low	High	High	Active crown
Coast live oak	Late spring-early summer	Short-moderate	Small-large	Low-moderate	Low-high	Low-high	Surface-Passive crown
Sargent cypress	Spring-summer-fall	Medium truncated-long	Large	Low-moderate	High	Moderate-high	Active crown
			South Coast				
Grasslands	Summer-fall	Moderate-long	Moderate-large	Low	Low	Low	Active crown
Sage scrub	Late summer-fall	Moderate -long	Moderate-large	Low	Moderate	Low-high	Passive–active crown
Chaparral	Summer-fall	Moderate-long	Moderate-large	Low	High	Low-high	Active crown
Big-cone Douglas-fir	Late summer-fall	Moderate-long	Small-moderate	High	Low-high	Low-high	Surface-active–passive crown

Riparian woodland/ shrublands	Fall	Moderate-long	Moderate	Low	Moderate-high	Low	Active crown
Southern oak woodlands	Summer-fall	Moderate-long	Moderate-long	Low-high	Low-high	Low-high	Surface-passive crown
Montane conifers	Summer-fall	Short-moderate	Small-moderate	Moderate-high	Low-high	Low-high	Surface passive crown
			Southeastern Deserts				
Low-elevation desert	Spring-summer-fall	Truncated-long	Small	High	Low	Moderate	Surface passive crown
Middle-elevation desert	Spring-summer-fall	Long	Moderate-large	Multiple	Moderate	Moderate-high	Active crown
High-elevation desert	Spring-early fall	Long	Moderate-large	Low-moderate	Moderate-high	High	Active crown
Desert montane	Summer-early fall	Truncated-long	Truncated-small moderate	Moderate	Multiple	Multiple	Passive–active crown
Desert riparian	Spring-fall	Short-moderate	Small-moderate	Low	High	Multiple	Passive–active crown

INDEX

Note: Page number followed by *f, t, m, b,* and *sb* indicates figure, table, map, box, and sidebar, respectively.

6 m (20 ft) wind, 32

Abatzoglou, J. T., 314, 452, 471
Abbott, C. S., 161, 179, 206
Abella, S. R., 264, 366, 463
Abrams, S., 412, 415
accelerated mass-spectrometry, 71
Acea, M. J., 92, 101
Ackerly, D. D., 104, 111, 314
acres/area burned, 189; between 1910 and 2013 in the northeast California, 221*f*; annual maximum fire size and, 179; fire perimeter locations and trends in, 223*f*; fire severity, 179; fire suppression, 201; by human-caused fires, 179; ignition probability, 221; lightning-caused fires, 172–74; in Modoc National Forest, 222, 222*t*; in North Coast, 154, 155*f*; in southwestern foothills, 202
active-independent crown fire, 68–69
Adam, B. A., 41, 49
Adam, D. P., 221, 281
Adams, D. K., 15
Adams, G. C., 335
Adams, M., 404
Adams, T. E., 283
adaptive management, strategies for, 241
adiabatic compression, 220
adiabatic cooling, 31, 35
adiabatic processes, 27
Adin Mountains, fire history of, 233*t*
advection, 98
aestivation, initiation of, 133
African lovegrass (*Eragrostis lehmanniana*), 465
African Serengeti, 139
Afton, A. D., 291
Agee, J. K., 59, 62, 74, 106, 115, 123, 210, 272, 408, 417
Ager, A. A., 405
age-to-depth models, 71
Aginsky, B. W., 385, 396
agricultural burning: in Central Valley, 292; guidelines for, 449
agricultural burning guidelines, 449
Agrippa, H. C., 87
Aguirre, J. L., 309
Ahlgren, C. E., 94–95
Ahlgren, I. F., 94–95
Ahmed, E. O., 287, 289, 469
Ahuja, S., 450
air and land managers (ALMs), 453
airborne pollutants, 443
air environment, of California, 441–45
Airey, C., 75, 77–78, 83
Airey Lauvaux, C., 75, 265, 271
air–fire interactions: on air quality, 97–98; on pollution transport, 98; smoke effects on ecosystems and, 98–99
air lifting mechanism, 33
air managers working group, 453
air mass circulation, 32
air pollutants, 87, 97
air pollution, 98; ecological thresholds, 445; man-made, 98; urban-generated, 98
air pollution control districts (APCDs), 98
air quality, 33; air basins, 446*f*; attainment and nonattainment areas, 451*m*; Blue Sky modeling framework, 450–51; California and Nevada Smoke and Air Committee (CANSAC), 450–51; critical loads, 97; Environmental Protection Agency (EPA), 97, 448; fire policy implementation guidelines, 448–49; general statewide seasonal patterns, 442*m*; greenhouse gases (GHGs), 451; impact of fire on, 87, 97–98; impact of smoke on, 87, 97; improvements, 445; National and California ambient air quality standards, 446*t*; ozone 8-hour federal standards, 451*m*; PM_{10} standard, 449*m*; PM_{25} standard, 450*m*; prescribed fire information reporting system (PFIRS), 449–50; regulations, awareness of, 439; smoke management program (SMP), 447, 449; standards of, 98
air regulatory framework, 445
air resource advisors (ARAs), 453
Akafuah, N. K., 41, 49
Akagi, S., 439–40
Akey, A., 287
Alameda whipsnake (*Masticophis lateralis euryxanthus*), 485
Alba, C., 459, 465, 467Albin, D. P., 482
Albini, F. A., 27, 43, 46–48, 51, 53
Albrecht, C., 485
Alder, J. R., 178
Aleksoff, K. C., 158
Alexander, H. M., 112
Alexander, J. M., 470
Alexander, M. E., 50–52
Alexandre, P. M., 344
algae growth, 112
Allenbaugh, G. L., 235
Allen, B. H., 309–10
Allen-Diaz, B. H., 284–85
Allendorf, J. W., 134
Allen, E. B., 92, 331
Allen, M. F., 331
Allen, T. F. H., 60
Allison-Bunnell, S., 213, 411
alluvial channels, 96
alluvial valleys, 175
alpine meadow(s), 65, 269
altered fire regime, 519
Alvarado, M., 439–40
Álvarez, G., 134
Amacher, A. J., 412
Amaranthus, M., 188
Amarasekare, P., 461
Ambrose, R. F., 117
American pine martens (*Martes americana*), 125, 272
Ammagarahalli, B., 134
ammonia (NH_3), 92
amphibians, 125, 127, 129, 131; aestivation, 133; effect of fire on habitat of, 138; egg masses, 128; risk due to climate change, 140; tadpoles, 128; toads, 127
anabatic circulations, 34
Anderson, D. R., 141, 187
Anderson, H. E., 43, 45–46
Anderson, H. W., 95, 97
Anderson, M. K., 3, 6, 77, 325
Anderson, R. S., 6, 73–74, 119
Andreae, M. O., 97
Andrews, P. L., 1, 47–48, 51–53
Angeles National Forest, 75
Angert, A. L., 461
animal foraging and predation, fire's influence on, 134

animal habitats zones, 125*f*
animal interactions with fire: animal use of snags, dead wood, and downed wood, 136*b*; aquatic fauna, 127–29, 138–42; arboreal fauna, 126; bird community response to fire, 129; climate change and, 140; direct effects of fire on individuals, 124–29; due to plant community shifts, 137–38; ecosystem feedbacks, 139; ecosystem relations, 139–40; effects from loss of snags and woody debris, 136; fatalities, 124–25; fire and ecosystem interactions, 139–40; fire and fire regimes, 123; fire effects on, 123; fire return interval (FRID), 124; food availability and, 137; foraging and predation, 134; foraging areas, 131; genetic exchange, 130; habitat distribution, 136; heat avoidance, 126*b*; magnitude attributes of, 124; on migration, 130–31; on nutritional value of plant materials, 137; peak abundance, 130; physiology and behavior, 123; plant community composition, 136–38; postfire habitat conditions, 130; postfire rain events, 129; relative mobility and response to fire, 127*b*; on reproduction, 131–34; residence and displacement, impact on, 129–30; rodent community response to fire, 130; snags and downed wood, effects of, 135–36; soil and litter, 135; spatial attributes of, 124; stochastic events and, 129; subterranean fauna, 125; surface-dwelling fauna, 125; temporal attributes of fire, 123–24; terrestrial animals, 134–38; vulnerability assessments of, 140; watershed effects, 128
animal migration: annual movement, 131; changes due to fire regime attributes, 132*t*; fire's influence on, 130–31; postfire movement, 131; ungulate migration, 131; wildlife residence and displacement, 131*t*
animal reproduction: age-classes, 133; effects of fire seasonality on, 133; fire regime attributes and, 133*t*; fire return interval and, 131; fire's influence on, 131–34; life stages, 133; of rodent species, 132*t*; seasonality of fire, effect of, 131; short- and long-term changes, 131; and survival of animal species, 133
Anna's hummingbird (*Calypte anna*), 129
Anning, P., 374
annosus root rot (*Heterobasidion annosum*), 98
annual grassland, 18, 30, 67, 92, 119, 163
antelope bitterbrush (*Purshia tridentata*), 108
Anthony, R. G., 179
Anthropocene, 2
anticyclone, 15
Antonovics, J., 112
Aplet, G., 408
aquatic ecosystem, 87, 138; effects of fire on, 128, 429; heat influx from fire to, 138; postfire recovery of, 429–30; recovery of, 430; stream and watershed recovery, 430; vernal pools, 129; water quality, 96
aquatic fauna, 127–29; decreased oxygen, effect of, 129; effects of wildfire on, 127; indirect effects of fire on, 138–42; life stages of, 127–28, 130; riparian communities, 139; thermal regime, 138; watershed effects on, 128; water temperature, effect of, 129; woody debris, sediment, and water, 138
Araki, S., 250
Arblaster, J. M., 19
arboreal fauna, 126; habitat zone, 127*f*
Arcto-Tertiary Flora, 2
Arguello, L., 162
Arguello, L. A., 64
Arkle, R. S., 290, 373
Armesto, J. J., 60
Armstrong, W. P., 157, 207, 230, 322
Arndt, A. M., 320
Arnold, J. D., 420
Arnold, R. K., 45
Arno, S. F., 74
aromatic oil, 90
Arthur, D., 188
Arthur, W., 103–5
Artly, D., 406
asexual regeneration, supported by recurring fire, 3
Ashe, D., 406
Asian-American settlement, fire regimes in, 6–7
asphyxiation, due to smoke, 134
Association of Western Governor, 487
atmospheric convection, 33
atmospheric instability, 220
atmospheric inversions, 31; marine inversion, 31; overnight radiation cooling inversion, 31; subsidence inversion, 31
atmospheric models, of fire behavior, 52
atmospheric moisture, 28–30; atmospheric stability and, 30; diurnal temperature ranges and, 30; effects on fuels, 30
atmospheric nitrogen deposition, impacts of, 98
atmospheric rivers, 250
atmospheric stability: atmospheric temperature lapse rates, 30–31; definition of, 30; diurnal pattern of, 31; effects on wildland fire behavior, 30; factors determining, 30
atmospheric temperature, 27–28, 30–31; adiabatic processes, 27; daytime surface temperatures, 28; effects on fuels, 30; lapse rates and stability, 30–31; *versus* relative humidity (RH), 28*f*; and solar radiation, 28
Atrazine (Aatrex®), 371
at-risk species: affected negatively by exclusion of fire, 486*t*; in California coastal stream, 481–82; conflicts between mechanical fuel reduction and management of, 483*t*; effects of climate change on, 487; effects of fire suppression activities on, 485–87; fire and fuel management activities affecting, 479*b*; food webs, 481–82; goals of conservation of, 487–89; interagency agreements for protection of, 485*b*; interface with fire and fuel management, 480–83; laws protecting, 477, 479*t*; meaning of, 478–79; post-wildfire restoration and recovery, 486–87; protection of, 477, 479–80; small-scale recovery efforts, 488*b*; solutions for conservation of, 487; vegetation management and, 483–85; wildfire severity and, 481–82
Atzet, T., 185
Austin, A. T., 155
Australia, 106, 402
autoignition (spontaneous ignition), 45
Avers, P. E., 3
Axelrod, D. I., 2
azonal grasslands, 299

Babb, R. D., 123
Babrauskas, V., 39
Babritius, S. L., 323
Bachelet, D., 162, 188, 408
Bachmann, J., 90
Baer-Keeley, M., 157
Bagne, K. E., 125, 134, 140
Bahre, C. J., 290
Bahro, B., 420
Bailey, A. W., 7, 103–5
Bailey, R. G., 2, 3
Bainbridge, S., 288
Baisan, C. H., 72, 74–75
Baja California, Mexico, 17, 19, 21, 23, 257–58*sb*; patchy chaparral burn in, 23*f*
Baker cypress (*Hesperocyparis bakeri*), 206–7, 230
Baker, G. A., 72, 77
Baker, M. B., Jr., 93, 95–97
Baker's manzanita (*Arctostaphylos bakeri* subsp. *bakeri*), 486
Baker, W. L., 75, 78, 110
Bakker, E., 291
Balch, J. K., 6
Balda, R. P., 368
Bald Hills, 162, 165, 393
Baldwin, B. G., 2
Baldwin, J. G., 314
Bales, R. C., 435
Balfour, D. A., 339
Balfour, V. N., 94
Ballantyne, IV, F., 112
Ballard, W. B., 123
Ballouche, A., 71
banana yucca (*Yucca baccata* var. *baccata*), 366
Banks, S. C., 188
Bao, Y., 314
barbed goat grass (*Aegilops triuncialis*), 468
Barber, K. H., 420
Barbour, J., 123, 272
Barbour, M. G., 1–2, 5, 18, 75, 150, 155
Barbour, R. J., 412, 415
Barco, J., 96
bark beetle (*Dendroctonus* spp.), 341; trees defense against, 108
bark flammability, 106
bark-infesting beetles, 134
Bar Massada, A., 344
Barnett, T. P., 19
Barnhardt, S. J., 162
Barrett, J. W., 233
Barrett, L. A., 154
Barrett, R. H., 412
Barrett, S. A., 381, 383
Barron, J. A., 178
Barro, S. C., 433
Barry Point Fire (2012), 233
Barry, W. J., 150, 155, 286, 287
Bartlein, P., 71, 73
Bartlein, P. J., 73–74
Bartolome, J. W., 284–85, 288–89
Barton, A. M., 114
Barton, K. P., 333
Bartuska, A., 405–6
Bar-Yosef, O., 381
basalt plains, 219
basin characterization model (BCM), 494
basketry, of Native Americans, 383
basketry willow (*Salix exigua* var. *exigua*), 387*f*
Bate, L., 123–24
Bates, C. D., 383
Bates, J. D., 226
Bates, P. A., 365
Batllori, E., 314
Battles, J. J., 421, 423, 435
Bauer, C., 312

Bauer, H. L., 282, 286
Bauer, K. L., 363
Baxley, W. H., 385
Baxter, C. V., 481–82
Baxter, W. T., 159–60
beach pine (*P. contorta*), 65
Beakes, M. P., 481
Beall, R. C., 227, 242
Beals, R., 301
Bean, D., 371
Bean, L. J., 77
bearbrush (*Garrya fremontii*), 203
beardless wild rye (*Elymus triticoides*), 286
Bear Fire (1994), 185
bear-grass (*Xerophyllum tenax*), 264, 385
Beartusk, K., 405
Beatley, J. C., 370
Beaty, R. M., 74–75
Beaufait, W. R., 72
Beaver Creek Pinery (BCP), 203, 204*f*; seedlings in understory of, 205*f*
Beche, L. A., 418
Beck, G. K., 372
Beck, J. L., 225
Beesley, P., 400
Beetle, A. A., 384
Behar, J. V., 34
Behave model, of fire behavior, 47, 53
BehavePlus model, of fire behavior, 47–48, 51
Beier, P., 123
Beissinger, S. R., 314
Bekker, M. F., 201–2, 209–11, 267
Bellanger, L. A., 159
Bell, G. P., 338, 464–65
Belnap, J., 374
Benda, L., 417
Bendell, J. F., 125
Bendix, J., 289, 310, 338
Benedict, J. M., 269
Bennett, D., 423
Benson, L., 286
Benson, N. C., 79–80, 82
bent grass (*Agrostis* spp.), 331
Bentley, J. R., 323
Bentz, B. J., 139
Berck, P., 314
Bergeron, Y., 111
Bergman, E., 342
Berigan, W. J., 123–24, 134, 141
Berkeley fire (1923), 401, 403, 422, 520
Bernhardt, E. A., 284–85, 387
Bernoulli effect, 34
Berryman, E. M., 433
Best, D. W., 96
Best Management Practices (BMPs), 431, 434
Betancourt, J. L., 20
Bevins, C. D., 53
Bevis, K., 221
Beyers, J. L., 95
Biber, E., 409
Bicknell, S. H., 155
bigcone Douglas-fir (*Pseudotsuga macrocarpa*), 98, 324, 336–37
Bigelow, S., 53
big galleta (*Hilaria rigida*), 365
Bigg, D. J., 155
Bigham, J. M., 90
bighorn sheep (*Ovis canadensis nelson*), 124
big-leaf maple (*Acer macrophyllum*), 110–11, 150, 160, 175, 200, 256, 309
bigpod ceanothus (*Ceanothus megacarpus*), 111
big sagebrush (*Artemisia tridentata*), 367
Big Valley Mountains, 229–30
Binford, M. W., 71
Binkley, D., 92
bio-charcoal, 454
biomass, 108; consumption by flaming front, 53; pulses, 358
Biondi, F., 359
bioregions, in California landscape, 2–4, 4*m*
biotic community, 130, 140
birch-leaf mountain-mahogany (*Cercocarpus betuloides*), 182, 203, 256, 284, 334, 367
birds, 123, 125–28, 133–36, 138, 140. *See also* specific species
bird species diversity, effect of fire on, 130*t*
Biscuit fire (2002), 172, 434
Bishop pine (*Pinus muricata*), 58, 150, 308, 392, 485; biological and ecological levels of organization for, 118*b*
Bisson, P. A., 128
Biswell, H. H., 137, 139, 142. *See also* Harold Biswell, v
bitterbrush (*Purshia tridentata*), 199, 208, 222, 225, 229, 254
bitter cherry (*Prunus emarginata*), 208
black ash, 88, 90
black bears (*Ursus americanus*), 125
blackberry (*Rubus ursinus*), 385
Blackburn, T. C., 155
blackbush (*Coleogyne ramosissima*), 357, 365
blackbush fires, 366
black carbon (BC), 92, 97, 452
black cottonwood (*Populus trichocarpa*), 237, 261
black knot (*Apiosporina morbosa*), 387
black morel (*Morchella elata*), 386
Black Mountain Experimental Forest, 140
black mustard (*Brassica nigra*), 330, 343
black oak (*Quercus kelloggii*), 253, 325, 390; encroached by conifers, 395*f*; importance of fires, 394–95*sb*; maintained by wildfire, 394*f*; in Sierra Nevada bioregion, 253; in South Coast bioregion, 325; in Southern Cascades bioregion, 200; in Yosemite National Park, 395*f*
Black, R. L., 406
black sage (*Salvia mellifera*), 309, 322
black sagebrush/low sagebrush (*A. nova*/*A. arbuscula*), 367
Blackshaw, R. E., 372
Blacks Mountain Ecological Research Project (BMERP), 211, 214
Blacks Mountain Experimental Forest (BMEF), 196, 211
Blacks Mountain Research Natural Area (BMRNA), 211
black willow (*Salix gooddingii*), 289, 291
Blainey, J. B., 358
Blaisdell, J. P., 224–27
Blakesley, J. A., 141
Blankenship, D., 482, 486
Blank, R. R., 92
blazing star (*Mentzelia* spp.), 385
Bleich, V. C., 133–34, 485
BLM Lowden Ranch Fire, 420
Block, W. M., 123–24
blue bunch wheat grass (*Elymus spicatus*), 226
Blue Canyon, 251
blue dicks (*Dichlostemma*), 328
Blue Fire (2001), 233
blue oak (*Quercus douglasii*), 64, 175, 197, 198*f*, 253, 283, 285, 309, 387*f*; in Central Valley bioregion, 283, 285; in Klamath Mountain bioregion, 175; in North Coast bioregion, 150; in Sierra Nevada bioregion, 253, 256; in Southern Cascades bioregion, 197, 198*f*, 203
Blumler, M. A., 287
Board, D. I., 225
Bock, C. E., 271
Bock, J. H., 271
Boerner, R. E. J., 93, 123, 272
Boersma, O. H., 90
Böhm, M., 34
Boisramé, G., 420
Bolander pine (*Pinus contorta* subsp. *bolanderi*), 150, 157
bolander's bedstraw (*Galium bolanderi*), 264
Bolan Lake, 177
Bolen, W. B., 90
Bolin, B., 454
Bolsinger, C. L., 223
Boltz, M., 225
Boman, C., 439–40
Bond, W. J., 64, 109, 111–12, 339
Bonnicksen, T. M., 326–27
Borchert, M. I., 72, 74, 108–11, 114, 284, 312, 336
Borgias, D., 186
Bork, J. L., 229–30
Bormann, B. T., 179
Bossard, C., 288
Bosworth, D., 405
Bouldin, J. R., 78, 341
boundary-layer air masses, 98
Boutin, M., 108
Bouverie Preserve, 165*f*
Bowcutt, F., 390
Bowden, P., 342
Bowen, W. A., 18
Bowns, J. E., 361, 365, 370, 372–73
Bowyer, T. R., 234
box elder (*Acer negundo*), 289
Boyce, M. S., 110–11
Boyd, C. S., 242
Boyer, D. E., 358
Bozek, M. A., 138
bracken fern (*Pteridium aquilinum* var. *pubescens*), 307
Bradbury, J. P., 221
Bradley, B. A., 6
Bradshaw, R., 71
Bradshaw, S. D., 110
Bradstock, R. A., 108–9, 111–12
Brammer, B., 372
Brand, S., 18
Brass, J. A., 96
Brehme, C. S., 134–35
Brennan, T., 72, 480, 484–85
Brenner, G., 104
Brenton, B., 467
Brewer, K., 80
Brewer oak (*Quercus garryana* var. *breweri*), 175, 182
Brewer, S. C., 420
Brewer spruce (*Picea breweriana*), 115, 175
Brewer's redmaids (*Calandrinia breweri*), 303
Briese, D. T., 471
Briggs, G. M., 256
Briles, C. E., 74
bristlecone fir (*Abies bracteata*), 113
bristlecone pine (*Pinus longaeva*), 65
British Columbia, 11, 15, 138, 161, 195
British thermal units (Btu), 40
Britton, C. M., 227
Brklacich, M., 140
broadleaf lupine (*Lupinus latifolius*), 92
broad-leaved cattail (*Typha latifolia*), 291
Brock, J. T., 129
brome (*Bromus*) genera, 36
Bromfield, A. C., 111
Brooks, M. L., 6, 368–74
Brotherson, J. D., 226, 370
Brough, A., 342

Brown, C. D., 115, 118–19
Brown, C. S., 374
Brown, D., 422
Brown, D. E., 366, 370
Brown, J. A. H., 95–97
Brown, J. K., 42, 53
Brown, K. J., 73–74, 90
Brown, P. M., 74, 181, 187
Brunelle, A., 74
Bruner, A., 227
brush canopy, 125
Bryan, B. A., 264
buckbrush (*Ceanothus cuneatus*), 103, 163, 182, 197, 256, 258, 283, 307
Buck, C. C., 28–31, 33–34, 44–45
buds: before and after sprouting, 106*f*; fire protection, 104
buffel grass (*P. ciliare*), 370
búlidum' (*Lomatium californicum*), 386*f*
bulrush (*Schoenoplectus* spp.), 3, 370
Buma, B., 115, 118–19
Bunnell, D. L., 62
Bunting, S. C., 225–27, 375
buoyancy, 49, 97
Burcham, L. T., 286
Bureau of Land Management (BLM), 15, 430, 477
Burgan, R. E., 41–45, 47–49
Burgy, R. H., 90
Burke, M. P., 96
Burkhardt, J. W., 227–28
Burk, J. H., 75
Burley, L. L., 159
Burn Area Emergency Rehabilitation (BAER), 52, 79, 79*t*, 80, 373
burned area emergency response (BAER), 433
burned area in fires, distribution of, 65
Burned Area Reflectance Classification (BARC) maps, 80
Burnham, K. P., 187
Búrquez, A. M., 370, 374
Burr, B., 334
Burrows, G. E., 104, 108
Burton, T. A.
Burton, T. S., 368
Busbey, C. L., 227
bush chinquapin (*Chrysolepis sempervirens*), 175, 200, 262
Bushey, C. L., 110
bush-mallow (*Malacothamnus fasciculatus*), 331
bush poppy (*Dendromecon rigida*), 333
Businger, J. A., 12
Busse, K. K., 215
Busse, K. L., 256, 276
Busse, M. D., 92, 215, 243
Butcher, J. E., 373
Butsic, V., 79
butter-and-eggs (*Triphysaria eriantha*), 288
butterflies, 133, 187, 488
Butte Valley, 208
Byram, G. M., 39, 40, 46–47, 50
Byrd, B. F., 325
Byrne, R., 6, 72, 325
Bytnerowicz, A., 92, 98, 423

Cabaneiro, A., 92
cacti, 364
Cafferata, P., 138
Cain, S. A., 401
Calaveras Big Trees State Park, 488
Calcarone, G. M., 324, 328–29, 341
Caldwell, B. T., 158
Caldwell, T. G., 92
California air basins, 443; and APCDs/AQMDs, 445
California Air Pollution Control Officers Association (CAPCOA), 453
California and Nevada Smoke and Air Committee (CANSAC), 450
California ash (*Fraxinus dipetala*), 182
California bay (*Umbellularia californica*), 150, 160, 182, 203, 262, 307
California black oak (*Quercus kelloggii*), 112, 150, 160, 175, 182, 197, 200, 229, 390, 395
California brittlebush (*Encelia californica*), 330
California buckeye (*Aesculus californica*), 150, 182, 256, 283
California buckwheat (*Eriogonum fasciculatum*), 309, 331
California Coastal Commission, 432
California coffee berry (*Frangula californica*), 110, 184, 256, 284, 322
California Department of Fish and Wildlife (CDFW), 477
California Department of Forestry and Fire Protection (CDF&FP), 304–5
Department of Forestry and Fire Protection, California, 512
California Department of Parks and Recreation (CDPR), 477
California ecosystems, 521
California Endangered Species Act (CESA), 477
California fan palms (*Washingtonia filifera*), 3, 106, 369
California Fire and National Fire plans, 488
California forest fires, 502
California forest reserves, 326
California Global Warming Solutions Act (2006), 452
California gnatcatcher (*Polioptila californica*), 129
California golden beaver (*Castor canadensis* subsp. *subauratus*), 383
California goldfields (*Lasthenia californica*), 281, 288, 462
California Gold Rush (1849), 5, 77, 400, 518
California hazels (*Corylus cornuta*), 110, 388
California huckleberry (*Vaccinium ovatum*), 150, 388
California including national parks, 448*m*
California Indian tribes, 383
California juniper (*Juniperus californica*), 203, 284, 367
California landscape: bioregions and, 2–4, 4*m*; diversity of, 2; fire regimes of, 4–7; relative size of disturbance area and landscape units, 60*f*
California-lilacs (*Ceanothus* spp.), 68, 110, 138, 150, 184, 200, 256, 261, 323, 383, 439
California melic (*Melica californica*), 284
California national forests, 329*t*; fire return intervals in, 124; nitrogen deposition, critical loads of, 447*m*
California nutmeg (*Torreya californica*), 262
California oak (*Quercus* spp.), 58
California oat grass (*Danthonia californica*), 155, 289, 307
California pitcher plant (*Darlingtonia californica*), 186, 186*f*
California pocket mouse (*Chaetodipus californicus*), 140
California red fir (*Abies magnifica*), 65, 75, 119, 200, 233, 266; fuelbeds, 267
California sagebrush (*Artemisia californica*), 116, 309, 330
California's ecosystems, 518; characteristic of, 62; climatic climax, 58; components of, 57; fire in context of ecological theory, 57–60; fireline intensity of, 66–67; fire regimes of, 60–63, 65; fire return interval, 64–65; fire severity in, 67–68; fire size, 65–66; fire types in, 68–69; human-induced alteration of, 57; patterns of fire occurrence, 57; seasonality in, 63–64; spatial complexity of, 66
California spotted owl (*Strix occidentalis occidentalis*), 124, 272, 416, 485
California suncup (*Eubolus californica*), 364
California Transverse Ranges, 72
California Wildfire Coordinating Group, 514
California wild grape (*Vitis californica*), 385
California yerba santa (*Eriodictyon californicum*), 203, 333
Calkin, D. C., 405
Calkin, D. E., 408
Callas, R., 238
Callaway, R. M., 310
Call, C. A., 359, 372
Callison, J., 370
Calo, D. B., 221
calories, 40
Camargo, S. J., 314
cambial death, 106
cambium necrosis, 106
Cameron, D. R., 314
Campbell, J. L., 179
Campbell Lake, 177
Campbell, R. E., 88, 90, 92–94
campfire analogy, for studying fire behavior, 47
camp fires, 77
canopy cover, 80
Cansler, C. A., 80
canyon live oak (*Quercus chrysolepis*), 160, 175, 182, 256, 367
Cap-and-Trade, 452–53
Caprio, A. C., 82
Carballas, T., 92
carbohydrate storage, 109
carbohydrate-to-water energy ratio, 11, 20; factors that increase, 36
carbon balance/accounting, 454–55
carbon dioxide (CO_2), 92, 97
carbon monoxide (CO), 92, 97
Carcaillet, C., 71, 74
Caribou Wilderness, 211
Carle, D., 419
Carleton, A. M., 174
Carlson, J. M., 301
Carpenter, S. L., 254, 391
Carroll, C., 123
Carroll, M. C., 111
Carruthers, R. I., 371
Carry, G. J., 402
Cascio, W. E., 272
catclaw (*Senegalia greggii*), 364
cat faces, 135
cation exchange capacity (CEC), 93
cattail species (*Typha* spp.), 291
Cavalli, J., 420
Cayan, D. R., 15, 19, 79, 314, 501
CCFMS model, 498
ceanothus (*Ceanothus velutinus*), 222
Central California Coast Section, 299
Central Coast bioregion, 299; annual grassland/oak forest topo-mosaics, aerial view of, 312*f*; Bishop pine (*Pinus muricata*), 308; burned area, 305*m*; climate change, 314; climate, topography, 307*m*; climatic patterns, 299–300; coast live oak forest, patches of, 310, 310*f*; Coulter pine, 311–12; current period, 304–5; description of, 299; ecological zones, 308; fire-dependent species, 314;

fire/nonnative species, 313; grassland, chaparral, and coastal scrub mosaic, 308*f*; historical fire occurrence, overview of, 302; historic period, 304; human geography, 301–2; interior coast ranges subregion, 313; Knobcone pine, 312; management issues/climate change, 313; maritime chaparral, 310–11, 311*f*; mean annual precipitation, isohyets of, 301*m*; mean annual temperature, intervals of, 302*m*; Monterey pine (*P. radiata*), 308–9, 309*f*; monthly green fuel moisture for chamise vegetation, boxplots of, 303*f*; from Napa south to Santa Barbara counties, 300*m*; physical geography, 299; population trends, 303*f*; prehistoric period, 302–4; Santa Cruz Mountains, 307–8; Santa Lucia Range, 309–10; Santa Lucia Ranges Subregion, 309–10; Sargent cypress, 312–13; subregion analysis, 305–6; subsection of, 306*t*; vegetation management, 313; weather systems, 300–301; wildland-urban interface, 314
Central Coast Range, 468
Central Valley bioregion, 279, 280*m*; climatic patterns, 280; current, 283; description of, 279; ecological zones, 280–83; fire climate variables, 280; fire ecology of important species, 284–85, 287–88, 289–91; fire regime-plant community interactions, 285, 288–89, 291; foothill woodlands, 283–84, 284*f*, 285*t*; forest once followed river corridors, 289*f*; freshwater marsh, 290–91; freshwater marshes, 291*f*, 292*t*; historic fire occurrence, overview of, 282; management issues, 291–92; physical geography, 279–80; prehistoric, 282–83; purple needle grass sprouts, 288*f*; relationship of, 281*m*; Riparian forests, 288; riparian woodland plant species, 290*t*; support to, 282*m*; valley grasslands, 286–87, 286*f*, 288*t*; vegetation, 281*f*; weather systems, 280; wetlands and small tributaries, 292*f*
Central Valley riparian forest, 280, 289
Cerling, T. E., 325
Cermak, R. W., 77
Chambers, J. C., 92, 225, 242, 373–74
chamise (*Adenostoma fasciculatum*), 66, 72, 112, 138, 150, 256, 307, 322, 333
chamise communities, 114
Chang, C., 115, 213
Chang, E., 314
Changed Climate Fire Modeling System (CCFMS), 497
channel erosion, 96
channel hydrology, 95
Channel Islands, 74
chaparral ecosystems, 137, 163–64; conifer, 338; fire hazard reduction study, 164*f*; fire regime–plant community interactions, 157; fire regimes, characterizing of, 72; fire responses of important species, 156–57; fuel reduction treatments, 163*f*; landscapes, 72; live fuel moisture, 300; long duration fires, 403; management implications, 164; mastication, 164; prescribed fire, 164; season of treatment, 164
chaparral whitethorn (*Ceanothus leucodermis*), 112, 256, 333
chaparral yucca (*Hesperoyucca whipplei*), 366
Chapman, P. L., 372
Chapman, T. B., 117
charcoal, 40, 92, 103; accumulations, 71; chronology, 73; dating of, 71; deposition, 6; inferred fire events, 73*f*; in Klamath Mountains, 73; sedimentary, 71–74; in Sierra Nevada, 73
Chase, C. H., 51
Chase, E., 188
Chase, J. M., 461
cheat grass (*Bromus tectorum*), 116, 197, 359, 433, 460
Chen, J-Y., 45
Chen, Y., 435
Cheo, P. C., 116
Chesson, P. L., 461
chia (*Salvia columbariae*), 303, 364
Chiesa, A., 313
Childs, H. E., Jr., 125
Chilean bird's-foot trefoil (*Acmispon wrangelianus*), 288
China Mountain, 186
Chinese tallow (*Triadica sebifera*), 460
Chinook salmon (*Oncorhynchus tshawytscha*), 140
Chiquoine, L. P., 463
chlorophyll, 80
Chong, G. W., 6
Choromanska, U., 92
Chou, Y. H., 17, 21–24
Churchill, D. J., 187
Cisneros, R., 423
Cissel, J. H., 187
Clancy, C. G., 96, 123, 129
Clark, D. R., 134–35
Clarke, P. J., 104, 108
Clark, J., 80
Clark, J. S., 282
Clark, L. A., 272
Clark, P., 372
Clark, P. E., 94
Clark, R. G., 227
Clark's nutcrackers (*Nucifraga columbiana*), 268
Clayton, M. K., 344
clean air, 98
Clean Air Act (CAA), 98, 445; Regional Haze Regulations of, 98–99
Cleaver, N., 422
Clements, F. E., 57–59
climate change, 18, 20; and animal response, 140; influence of fire on (*See* fire, climate change caused by); long-term, 118–19; in North Coast bioregion, 162; short-term, 117–18
climate–fire relationships, 159
climates in California, 11–18, 517; in Central Coast bioregion, 299–300; in Central Valley bioregion, 280; evaporation and runoff, 17–18; fire history and variability in, 20–24; interannual and decadal variability in, 18–19; lightning, 15–17; mean 500-mbar contours, 12*f*; North American monsoon and, 14–15; in North Coast bioregion, 149–50; plant phenology and the fire season, 18; precipitation, 13–14; in Sierra Nevada bioregion, 250–53; snowfall, 14; temperature, 12–13
climatic climax, 59; primary, 58; secondary, 58
climatic water deficit, 251
closed-cone conifer ecosystems, 68
Closed Pine Forest sites (CPF), 207
cloud cover, 28
cloud-to-ground lightning strikes, 15, 45
coastal Douglas-fir (*Pseudotsuga menziesii*), 136
coastal marine layer, 11–12
coastal prairie zone, 154–55
Coastal Sage Scrub Communities (CSS), 129
coast fog zone, 177
coast live oak (*Quercus agrifolia*), 284, 307, 310
coast redwoods (*Sequoia sempervirens*), 307
coccora (*Amanita calyptroderma*), 386
Cochran, P. H., 233
Coe, S. J., 125, 134, 140
Coffman, G. C., 117
Cohen, J. D., 41, 43, 45, 49
Cohen, W. B., 421
Coho Salmon (*Oncorhynchus kisutch*), 138
cold-front winds, 33
Cole, K. L., 157, 207, 230
Collaborative Approach for Reducing Wildfire Risks to Communities and the Environment: Ten-Year Comprehensive Strategy, 407
Collins, B., 314, 342, 420
Collins, B. M., 74–75, 77–78, 80, 83, 187, 188, 408–9, 421, 423
Collins, C. D., 112
Collins, S. L., 60
Colman, E. A., 93–95
Colman, J., 52
Coloff, S., 405–6
Colombaroli, D., 74
colonization by plants, 110–11
Colorado pinyon (*Pinus edulis*), 367
Coltrain, J. B., 325
Colver, C. G., 96
combustion: chemical equation for, 39; chemistry of, 39; fire triangle, 39, 40*f*; flaming phase of, 45; fuels for, 41–44, 41*f*; gaseous phase of, 40; heat of, 39; heat transfer, 40–41; hydration phase of, 40; moist threshold of, 23; oxidation processes, 39; phases of, 39–40, 41*f*; of plant matter, 71; preheating phase of, 40; smoldering phase of, 40, 41*f*
common stickyseed (*Blennosperma nanum*), 288
Community Wildfire Protection Plan (CWPP), 511
Composite Burn Index (CBI), 80
Comrack, L. A., 140
Comrie, A. C., 15
Conard, S. G., 95
conduction, process of, 41
cone wildfire burning, 214
Conformity Rules, 447
conifer forests, 5, 18, 67–68, 76, 106, 161, 171, 175, 199, 206, 209, 254; in Sierra Nevada, 112; in Upper montane zone, 200; of western North America, 51
Conklin, M. H., 435
Conklin, M. T., 334
Conrad, C. E., 90, 93–96
consume/emission production model (EMP), 451
convection, process of, 41
converted fire regime, 519
Cook, A. R., 164
Cook, S. F., 6
Cooperative Forest Fire Prevention Program (1942), 401
Cooper, A. W., 75, 77
Cooper, S. D., 418
Cooperrider, A., 123
Cope, A. B., 211
Coppoletta, M., 187–88
coral fungus (*Ramaria violaceibrunnea*), 386
Cordero, E. C., 314
cord grass (*Spartina* spp.), 291
Cordova, C., 71
Core, J., 98
cork cambium, 106

Cornelius, D. R., 226
Cottam, C. A., 401
Cotton, L., 263
cottonwood (*Populus* spp.), 110
Coulter pine (*Pinus coulteri*), 98, 308, 323; mortality of, 312; spans, 336
Countryman, C. M.
Cowart, A., 303–4
Cowell, C. M., 310
Cowling, R. M., 459–60
Cox, C. B., 113
coyotebrush (*Baccharis pilularis*), 150, 307, 331
Craig, D. J., 463
Crain, P. K., 140
Cramer, O. P., 34–35
Crane, M. F., 186, 229
Crank Fire (1987), 233
Crater Creek watershed, 186
Crawford, H. S., 123, 126
Crawford, J. C., 242
Crawford, J. N., 162
Creasy, R. M., 80
creosote bush (*Larrea tridentata*), 357
Crimmins, M. A., 80
crimson fountaingrass (*Pennisetum setaceum*), 370
Critchfield, W. B., 235
critical mass flux, 45
Crocker, M. C., 335
Croft, A. R., 358–59, 372
Cromack, Jr. K., 88
Cronise, T. F., 291
Crooks, J., 272
Crookston, N. L., 238
Crotteau, J. S., 206, 423
Crounce, J., 439–40
crown fire, 124; active, 49, 50*f*; active-independent, 68–69; bulk densities, 50, 50*f*; Crowning Index of, 51; hazard indices of, 50; ignition and torching, 165; impact on plant population, 115; independent, 49–50, 50*f*; initiation models of, 44; passive, 49, 50*f*; passive-active, 68; plume-dominated, 50; removal of sunlight-blocking canopy, 103; spread of, 49–51; stages of, 49; surface-passive, 68; Torching Index of, 51; Van Wagner's models of, 50
crown-fire-adapted ecosystems, 422
crown fraction burned, 51
Crowning Index, 51
crowns and foliage, fire protection of, 104–5
Cruz, J., 406
Cruz, M. G., 51, 52
Cub Creek Research Natural Area (CCRNA), 202; fire frequency in, 206; topographic distribution of fire severity patterns in, 203*f*
Cullinan, V., 358–59, 363, 375
Cummins, K. W., 481
cumulative severity rating (CSR), 500
cumulonimbus cloud, 33
Cunniffe, N. J., 164
Cunningham, S. C., 123
curl-leaf mountain-mahogany (*Cercocarpus ledifolius*), 175, 200, 223, 226, 235, 270, 357
currants (*Ribes* spp.), 385
Cushon, G. H., 44
Custer, N. A., 374
Cuthrell, R. Q., 302
Cuyamaca cypress (*Hesperocyparis stephensonii*), 313, 322
Cuyamaca Peak, 14
Cylinder, P., 313
Czuhai, E., 123, 126

Dagit, R., 310
Dahlberg, B., 92
Dailey, S. N., 254, 272
Dale, V. H., 59–60
Dalke, P. D., 225
dalmation toadflax (*Linaria dalmatica*), 465
Daniels, M., 74
D'Antonio, C. M., 8, 36, 115, 374, 459, 465, 467
Das, T., 314, 501
datasets, of forest inventories, 77
dating, of charcoal/macrofossils, 71
Daubenmire, R. F., 58, 287
Davidson, J., 372
Davies, K. W., 226
Davis, C., 400
Davis, F. W., 2, 72, 114, 284
Davis, K. P., 52
Davis, S. D., 163, 323, 333, 335
Dawson, J., 74–75, 156
dead surface fuels: fire behavior, 43; moisture timelag, 43
deadwood cycling, 135
DeAngelis, D. L., 60
Dean, W. E., 221
Dearing, M.-D., 325
death camas (*Toxicoscordion* spp.), 108, 260
DeBano, L. F., 88–90, 92–94, 96–97, 106, 418
deBenedetti, S. H., 265, 268–69
de Blas, C. S., 408
debris flows, postfire, 96
Decker, M., 128
decomposition, process of, 39
decrease the brush, 382
Deeming, J. E., 43, 45, 47
deerbrush (*Ceanothus integerrimus*), 175, 182, 258, 339; heat-stimulated seed of, 264
Deer Creek watershed, 203
deer firing (kupi't), 222
deer grass (*Muhlenbergia rigens*), 286, 384, 392*f*
deerweed (*Acmispon glaber*), 330
DeFalco, L. A., 358, 363, 374
Dekker, L. W., 90
DeLasaux, M. J., 344
Delgadillo, J., 309
DelGiudice, G. D., 137
Dell, J. D., 105
Del Norte Coast Redwoods State Park, 159
DeLoach, C. J., 371
Del Tredici, P., 137
DeLuca, T. H., 92
Delwiche, C. C., 264
Dennison, P. E., 72, 301, 420
Derasary, L., 363
desert almond (*Prunus fasciculata*), 357
Desert Floristic Province, 2
desert needle grass (*Stipa speciosa*), 365
desert pincushion (*Chaenactis steveoides*), 364
desert saltgrass (*Distichlis spicata*), 286
desert scrub, 3, 36, 370
desert tortoise (*Gopherus agassizii*), 129
desert-willow (*Chilopsis linearis* subsp. *arcuata*), 364
desert zone, characteristics of, 355
Despain, D. G., 73
"destructive" fire, 77
Dettinger, M. D., 19, 501
de Valpine, P., 72
Devil's Garden area, 229
Devlin, R. B., 272
Devoe, N., 359, 372
dew point, 28
Dezzani, R. J., 19, 23
Diamond, J. M., 359, 372
Diamond Mountains, 229
Diaz, H. F., 19
Diaz-Sanchez, D., 272
Dickson, B. G., 363
Dicus, C. A., 163
Dieterich, J. H., 75, 90, 93–96
Dietrich, W. E., 482
differenced Normalized Burn Ratio (dNBR), 80
DiGaudio, R. T., 140
Dilts, T. E., 174
Dingemans, T. D., 310
Di Orio, A. P., 238
diseases, human, 255
dispersal of seeds, 110–11, 117; by animals, 111
distributions curves, of fire regimes: for closed-cone conifer ecosystem, 63*f*; fireline intensity, 67*f*; fire return interval, 63*f*; severity, 67*f*; size, 65*f*; spatial complexity, 66*f*
DiTomaso, J. M., 286
Dixon, K. W., 110
Dobson, A. P., 313
Doehring, D. O., 96
Doerr, S. H., 90
dogwood (*Cornus* spp.), 110, 384
Donaghy Cannon, M., 179
Donato, D. C., 115, 118–19
Donato, D. E., 179
dormant seed banks, 333
dotseed plantain (*Plantago erecta*), 284
Doughman, P., 452
Douglas-fir (*Pseudotsuga macrocarpa*), 265, 304
Douglas-fir (*Pseudotsuga menziesii*), 52, 64, 72, 104, 150, 157, 253, 307, 390
Douglas-fir-tanoak (*Notholithocarpus densiflorus*), 150
Douglas-fir-tanoak forests, 153*f*; fire history studies of, 161; fire regime–plant community interactions, 161; fire responses of important species, 160–61
Douglas-fir tussock moth (*Orgyia pseudotsugata*), 232
Douglas iris (*Iris douglasiana*), 160
Douglas, J., 405–6
Douglas squirrel (*Tamiasciurus douglasii*), 263
Drake, K., 363
Drapek, R., 162, 188
Drapek, R. J., 408
Drews, E. S., 412, 415
drought, 329; impact on burning rates, 20; predictor of, 20
Drysdale, F., 1
dry season, of California, 34–36; autumn, 35–36; summer, 34–35
Duane, T. P., 271
Dubow, J., 459
Dudley, T., 371
Dudley, T. L., 371, 418
duff and litter soil layers, 45, 53, 88, 386, 391, 394, 429
Dunne, J., 285, 417
Dunne, T., 285
Dunn, P. H., 88, 90, 92–94
Dupuy, J. L., 343
Durham, J. B., 134
dusky-footed woodrats (*Neotoma fuscipes*), 130
dwarf birch (*Betula nana*), 237
dwarf brodiaea (*Brodiaea terrestris* subsp. *terrestris*), 288
Dwire, K. A., 109

Dyer, A., 283, 287
Dyrness, C. T., 158

Eagle Lake, 221
Earle, D. D., 286
Earl, S. R., 482
Earnst, S. L., 362
Earth Resources Observation and Science (EROS), 80
earthworms, 125
East San Francisco Bay region, 488
Eastwood manzanita (*Arctostaphylos glandulosa*), 313, 333, 367
Eby, L. A., 123, 129
ecological–fire–climate relationships, 226
ecoregions, of California, 5*m*
ecosystem management, 1, 519; fire regimes during era of, 7
ecosystems, in California, 517; altered fire regimes, 518; converted fire regimes, 518; ecology and human values, 518–20; fire, 518; fire management, 518; fire regime status, 519*b*; future challenges, of fire/land management, 520–21; human interactions with fire, 517–18; maintaining fire regimes, 517; theory of, 59; wildfire, 520
ecotones, 279
Ediminster, C., 51
Edinger, J. G., 34
Edlund, E., 325
Edminster, C., 272
Edminster, C. B., 123
Edmonds, R., 264
Edrington, M., 405–6
Edwards, M. E., 74
Edwardson, J., 406
Egan, D., 7
Ehle, D., 75
Ehleringer, J. R., 325
Eidenshink, J., 80
Ejarque, A., 74
elderberry (*Sambucus* spp.), 385
Eldorado County, 113
Elford, R. C., 149
Eliason, S. A., 331
Elliot, W. J., 95
El Niño/Southern Oscillation (ENSO), 18, 19, 174, 195, 301, 496
Elsner, J. B., 19
Ely, L. A., 96
Embrey, T. M., 463
emergency stabilization, definitions of, 430
emission redistribution strategy, 454
emission reduction techniques (ERTs), 445, 454
endemic bristlecone fir (*Abies bracteata*), 309
Endert, J. F., 388
endosperm, 111
energy dissipation, concept of, 59
energy flow rates, concept of, 50, 59
energy units, 40
Engbeck, J. H., Jr., 158
Engber, E. A., 64
Engel-Cox, J., 439, 453
Engelmann oak (*Quercus engelmannii*), 310, 324
Engelmann spruce (*P. engelmannii*), 175
England, M. H., 19
English, J. D., 41, 49, 343
engraver beetle (*Scolytus ventralis*), 232
Enloe, S. F., 286
Enright, N. J., 104, 108, 111, 115
Environmental Protection Agency (EPA), 97–98, 443; Smoke Management Program (SMP), 447, 449
episodic tree regeneration, 77
Epling, C., 333
Erbilgin, N., 308
Erickson, K. L., 123, 272
Erlandson, J. M., 281, 339
Erman, D. C., 213
Escoto-Rodríguez, M., 15, 20
Esque, T. C., 358, 374
Estes, B. L., 172–73
Euro-American settlements, 264, 416; fire regimes in, 6–7, 75, 77, 82, 155
European settlement, fire management/policy, 399; California missions-Alta California, 400; challenges, in California, 408; chaparral/woodland fires, 403; climate changing, 408; current fire policies, 405; damaging fires in California from 1923 to 2017, 405*t*; deadliest California wildfires, 406*t*, 407*f*; European exploration era, 399; Federal fire policy, implementation guidelines, 407–8; Federal fire policy, reviews, 405–7; Federal Land Assistance, Management, and Enhancement Act of 2009 (FLAME Act), 408; fire control era, 400–401; fire use, beginning of, 401–2; forest fires, 403; Gold Rush/early statehood, 400; invasive nonnative plant species, 399–400; large/destructive/deadly fires patterns, 403–5; large fire settings, 403; largest fires in California from 1923 to 2017, 404*t*; large valley fires, 403; National Fire Plan, 407; Native American fire use, removal of, 399; policy recommendations, 409*t*; some key historic fires, overview of, 402–3
evapotranspiration, 11, 94, 103, 150, 177
Evens, J. M., 62, 69
Everett, R. G., 75, 77–78
evergreen plants, 90
Everham, E. H., 59, 60
Everitt, B. L., 371
Eversman, S., 112
Evertt, R. R., 155–56, 162, 309, 310
Ewers, F. W., 335

Faber, P. M., 324
facultative resprouters, 108
Fahnestock, G. R., 158
Fairbanks, D. H. K., 213–14
Falk, D. A., 75
Fann, N. L., 272
farewell-to-spring (*Clarkia* spp.), 388
Farmer, J. F., 383
Farris, C. A., 75
Farris, K. L., 123, 272
FARSITE model, of fire behavior, 47–48, 51–52
Fasullo, J. T., 19
Fausch, K. D., 435
Feakins, S. J., 310
Federal Clean Air Act, 445
Federal Conformity Regulation, 447–48
Federal Endangered Species Act and the California Endangered Species Act, 520
Federal fire policy, 400
Federal Implementation Plan (FIP), 445, 447
Federal Land Assistance, Management, and Enhancement Act (FLAME Act), 189, 408, 511
Federal land management agencies, 406, 432, 448–49, 512–13
Fehmi, J. S., 287, 289
Fehmi, S., 289
Fellows, D. P., 471
Fenenga, G. L., 281
Fenner, R. L., 125
Fenn, M. E., 92, 98
Ferguson, D. E., 238
Fernandez, I., 92
Fernandez-Pello, A. C., 45
Ferrenberg, S. M.
Fettig, C. J., 123, 139, 272
Ffolliott, P. E., 88, 97
Ffolliott, P. F., 106
fiddle-necks (*Amsinckia* spp.), 111
Fiddler, G. O., 116
Fiedler, C., 272
Fiedler, C. E., 51
Field, C. B., 140
Fielder, C. E., 123
Figura, P. J., 236
Figurehead Mountain, 181*f*
filaree (*Erodium* spp.), 283
filbertworms (*Cydia latiferreana*), 387
Filipe, J. A. N., 164
Finch, D. M., 125, 134, 140
Finley, B. D., 441
Finney, M. A., 41, 45, 47, 49–52, 405, 408
fire. *See* wildland fires
Fire-Adapted Community Learning Network, 511
fire-adapted species, 104
fire-adapted vegetation, 3, 3*t*
fire behavior: atmospheric models of, 52; campfire analogy for studying, 47; coupled, 52; flaming front, 45–47; prediction systems, 43; simulations of, 53; surface fire spread, 47
fire behavior models: atmospheric models, 52; Behave model, 47, 53; BehavePlus model, 47–48, 51; FARSITE model, 47–48; FIRETEC model, 52; HIGRAD model, 52; hydrodynamics model, 52
fire, climate change caused by, 493; area burned, 499; California fire regimes, impacts, 496–97; climatic water deficit (CWD), 496, 498*f*; ecosystem change, changes in fire regimes, 502*f*; effects of, 501*f*; effects on vegetation, 500; fire ignitions, 500; fire intensity, 499–500; fire season, 500; fire severity, 499–500; future with more fire, 501–3; interactions, 493–94; precipitation, 496; predictions for California, 494; projected impacts due to, 500–501; snow accumulation, 496; Snow Water Equivalent, 497*f*; stress complex from Sierra Nevada forests, 503*f*; temperature, 494, 495*t*; vegetation shift, impacts on fire frequency/probability, 497–99
fire climax, 58
fire containment, 80
fire cycle, concept of, 64
fire danger rating, 43
Fire Danger Sub-Committee, 44
fire deficit, 212; compounding, 213
fire-dependent ecosystems, 69
fire-dependent species, 113–14
fire ecology, 7, 150, 228, 517; California pitcher plant, 186; Douglas-fir, 182; herbaceous alliance, 186; hydric vegetation types, 236; of knobcone pine, 207; in mid-montane zone, 203; for montane shrubs, 208; of quaking aspen, 206
fire effects, ecological ramifications of: on animals, 123; biomass consumption, 53; fire severity, 52; microclimate, 53; on nonvascular plants, 112; plant mortality, 52–53; tree crown scorch height, 52
Fire Effects Information System, 104
fire events, inferred from sedimentary charcoal record, 73*f*

fire exclusion, 256
fire exclusion-tree densification model, of vegetation change, 213
fire fighter and public safety, 423
fire-followers, 112
fire frequencies, 74; ecologically significance of, 75; estimation of, 75; impact of humans on, 6
fire hazards, 163
fire, in California, 381; to combat insects/diseases, 386–88; contrasting plant architectures, 384*f*; cultural burn, 394*f*; dead trunk of black oak, 395*f*; Eliza Coon, Pomo woman, 384*f*; fire-based management, 393*f*; fire-mediated human influence, extent/degree of, 391–92; food production, enhancing, 385–86; Freddie, a Hupa woman, 386*f*; hypothetical mosaic, 391*f*; indigenous burning, ecological impacts of, 388; individual organisms, 388; intermediate disturbance hypothesis, 389*f*; keeping the country open, 382–83; landscape, 390–91; management today, implications for, 392–93; managing/hunting wildlife, 383; mixed-conifer forest types, 390*f*; Native American, 382; plant communities, 388–90; plant material, quality/quantity of, 389*f*; plant materials for manufacturing baskets, 383–84; plant populations, 388; territories, California Indian language groups, 382*m*; Wahnomkot (Wukchumni Yokuts), 385*f*
fire in context of ecological theory: Clements' view of, 57–59; disturbance theory, 59–60; ecosystem theory, 59; hierarchical theory, 60; succession theory, 57–59
fire insulation, 104
fire intervals, in redwood forests, 160*t*
fire/invasive plants, 459; burned/invaded creosotebush scrub, 464*f*; in California, 461–62, 462*f*; coniferous forests, 463–64; desert shrublands/woodlands, 463; dilemma of controlling invasive species, 466–67*sb*; era of global change, 471–72; grass-fire cycle, schematic representation of, 461*f*; grasslands, 462, 469*t*, 470*t*; herbaceous species, control of, 465–69; inter-relationships, 461; managing invasive species, 465; mean species richness/absolute cover, 465*f*; mediterranean shrublands, 462–63; nonnative herbaceous species, abundance of, 468*f*; nonnative plants, 465*f*; patterns and processes, 460–61; riparian ecosystems, 464–65; situation in California, 461–62; studies published, 460*f*; woody invasive species, 469–70
fire-killed trees, harvesting of, 423
fireline intensity, 46–47; of grass fire, 66; high, 67; low, 67; measurement of, 66–67; moderate, 67; multiple, 67; patterns of, 52
fire management, 1, 7, 166; calaveras county, 419*f*; contemporary fire patterns, 414*sb*; cultural history of, 508–9; forest thinning, 415; fuel management phases, restoration/maintenance, 415–16; grazing/vegetation management programs, 415; institutional structure of, 512–15; mastication, 415; mega-fires, 422; methods used on complex landscapes, 416; post-wildfire management, 423; prescribed fire in California, 412; prescribed fire in Sierra Nevada mixed conifer forests, 413*f*; in red fir-Jeffrey pine forests, 413*f*; riparian ecosystems and, 417–18*sb*; wildfire for multiple objectives, mechanical treatments, 412–15; wildland-urban interface, 422–23
fire occurrence: data on, 77–79; dendrochronological reconstruction of, 74–75, 74*f*; fire scar-based methods of reconstructing, 75; in Klamath Mountains bioregion, 179; in Northeastern Plateaus bioregion, 221–22; patterns of, 71, 79; in red fir forests, 210; seasonality of, 71
fire-occurrence areas (FOA), 180, 180*f*
fire persister plants, 103
fire policy, 420
fire-prone: ecosystems, 518; environment, 140
fire records, completeness and precision of, 79
fire reestablishment, 435
fire regime condition class (FRCC), 82; fire history groups, 82*t*
fire regime-ecosystem changes, 518
fire regime–plant community interactions, 308; in coastal prairie zone, 155–56; in Douglas-fir-tanoak forests, 161; of foothill shrubland and woodland zone, 260–61; in lower Montane zone, 182–83, 229–30, 264–65; in mid-montane eastern, 207–8; in mid-montane western, 205–6; in mid-montane zone, 231–33; in mid to upper montane zone, 185–86; in Northern chaparral zone, 157; in Northern coastal pine forests zone, 157; in Northern coastal scrub, 155–56; in Oregon oak woodland zone, 162; in Redwood forest zone, 159–60; in Sagebrush Steppe zone, 226–28; in Sitka spruce forest zone, 158; in southwestern foothills, 203; in subalpine zone, 186–87, 268–69; in upper montane zone, 208–11, 234–36
fire regimes: animal migration and, 132*t*; archived written records of, 75–77; assessments of, 79; attribute-based, 62–63, 79; attributes for each vegetation type, 523–27; of California landscape, 4–7; characterizing chaparral, 72; comprehensive, 69; condition class, 82; defined, 61; departure from historical range of variability, 82–83; distributions for (*See* distributions curves, of fire regimes); in ecosystem management era, 7; in era of Euro-American and Asian-American settlement, 6–7; failure rates, 17; fire severity within, 62*f*; fire suppression era, 7; groups used for condition classes, 61*b*; Heinselman's, 61*b*; in Klamath Mountains bioregion, 179; magnitude attributes of, 66; during native American era, 6; paleoecological reconstructions of, 71–77; periodic perturbations in, 11; pre-Euro-American Fire Regime (PFR) groups, 82; presettlement, 77; previous, 60–62; prior to human settlement, 6; quantifying contemporary, 77–82; spatial attributes of, 65; temporal, 63; wildlife colonization and exploitation with changes in, 137*t*; wildlife reproduction and, 133*t*
fire-related mortality, 117
fire resistance, 103–8; buds, 104; crowns and foliage, 104–5; grasses and grass-like plants, 107–8; roots and underground tissues and organs, 106–7; stems and trunks, 105–6
fire-resistant species, 115
fire resister plants, 103
fire responses of important species: in coastal prairie zone, 155; in Douglas-fir-tanoak forests, 160–61; of foothill shrubland and woodland zone, 256–60; of Lower Montane forest, 261–64; in lower Montane zone, 182, 183*t*, 229, 230*t*; in mid-montane eastside, 206–7, 206*t*; in mid-montane westside, 203, 205*t*; in mid-montane zone, 231; in mid to upper montane zone, 184, 184*t*; in Northern chaparral zone, 156–57; in Northern coastal pine forests zone, 157; in Northern coastal scrub, 155; in Oregon oak woodland zone, 162; in Redwood forest zone, 158–59; in Sagebrush Steppe zone, 224–25; in Sitka spruce forest zone, 158; in southwestern foothills, 203; in subalpine zone, 186, 187*t*; in upper montane zone, 208, 234, 235*t*
fire restoration, 521
fire return interval, 63, 63*f*; animal reproduction, effect on, 131; consequences of, 114–15; fire regime distributions for, 64*f*; long, 65; medium, 65; plant communities and, 116–17; plant population, impact on, 114–15; short, 64–65; significance of, 64; and survival of a nonsprouting species, 64; tree ring-based estimates, 72; truncated long, 65; truncated medium, 65; truncated short, 64
fire return interval departure (FRID), 82, 124; analysis of, 82; estimations of, 82; in northern and southern regions of California, 82
fire return intervals (FRIs), 179, 185, 202, 209; distribution of, 185*f*
fire rotation: concept of, 64; historical archived datasets on, 78*t*; in lodgepole pine forests, 211
fire safe councils (FSC), 511
fire scars: fire-occurrence dates, 74; intra-ring locations for, 202*f*; reconstructions, 74; stand-origin dates, 74
fire scientists and ecologists, 480
fire season. *See* season of burning
fire severity, 52, 67–68, 124; change detection algorithms for mapping, 80; data on, 79–80; distributions and spatial complexity, 82; high, 68; historical archived datasets on, 78*t*; in Illilouette Creek Basin, 81*f*; impact on water quality, 96; in Klamath Mountains bioregion, 179, 180–81; in lodgepole pine, 211*f*; low, 68; mapping programs, 79–80, 79*t*; moderate, 68; monitoring trends in, 80; multiple, 68; normalized burn ratio (NBR) index, 79; in Southern Cascades bioregion, 202–3; very high, 68; in Yosemite National Park, 81*f*
fire size: large, 66; medium fire size, 66; small, 65–66; truncated small, 65
fire spread: crown, 49–51; direction of, 32; due to lighted embers, 51*f*; equation of, 43–44; rate of, 32, 47; Rothermel spread equation, 46–49, 48*b*; and slope wind, 32; spotting of, 51; surface, 47
fire suppression, 21, 23, 77, 206, 373, 517, 518; activities, 485; Clements' model, 508; effects on at-risk species, 485–87; fire regimes during era of, 7; in Klamath Mountains, 178; in North Coast bioregion, 154; in Southern Cascades bioregion, 201; in United States, 508
fire surrogates study (FFS), 214
FIRETEC model, of fire behavior, 52

fire tolerance, 108
fire-tolerant species, 115
fire treatments, 412
fire triangle, 39, 40*f*
fire types: crown fire (*See* crown fire); multiple, 69
fire weather: atmospheric inversions, 31; atmospheric moisture, 28–30; atmospheric temperature, 27–28, 30–31; California's dry season, 34–36; components of, 27–30; diurnal pattern of stability, 31; smoke dispersion, 33; "window" in vegetation, 36; winds, 31–34
fire whirls, travel path of, 34
firewood analogy, 43
Fischer, D. T., 300, 309
Fischer, R. A., 226, 242
Fischer, W. C., 268, 269
fisher (*Martes pennanti pacifica*), 139
Fisher, R. F., 92
Fisher, R. N., 134
Fitch, C. H., 75–76
Fites-Kaufman, J., 104, 254, 272
Fitzgerald, R. T., 5, 6
flame dimensions, for a wind-driven fire, 47*f*
flame lines, propagation of, 11
flame zone, 45; depth, 46
flaming front, 45–47; biomass consumed by, 53; equations for, 46*b*; fireline intensity, 46; flame zone depth, 46; rate of energy release, 46; reaction intensity, 46; residence time, 46
flaming phase, of combustion, 45
flaming zone area, 46
flammability: age-dependent, 20; thresholds, 20; of vegetation (*See* vegetation flammability)
flannelbush (*Fremontodendron californicum*), 203, 333
Flematti, G. R., 110
Fleming, M., 372
Flintham, S. J., 76
Florence, M. A., 313
floristic provinces, of California, 2
Florsheim, J. L., 96
Fluazifop-p-butyl (Fusilade®), 371
flying squirrel (*Glaucomys sabrinus*), 126
föhn effect, 251
föhn (foehn) winds, 33–34; Bernoulli effect, 34; diurnal pattern of, 34; fire spread caused by, 34; in northern Sacramento Valley, 34; Santa Ana wind, 33; velocity of, 34
Foin, T. C., 150, 155
foliar chlorophyll, 98
foliar damage, 104
foliar moisture content, 43, 105
foliar senescence, 98
Foltz, R. B., 433
Fonda, R. W., 159
Fons, W., 47
Fontaine, J. B., 111, 115, 118–19, 179
food availability, effects of fire on, 137
food storage, 104
foothill needle grass (*Stipa lepida*), 155
foothill pine (*Pinus sabiniana*), 260, 283, 303
foothill shrub, 259
Forbes, D. L., 140
forb species, 226
Ford, L. D., 285
Forero, L. C., 285
forest area safety taskforce (FAST), 514
forest canopy, 28
forest density, 206
forest ecosystems, dynamic nature of, 416
Forester, R. M., 221
forest fires aid the enemy, 508
forest fuels, heat contents of, 39
forest management, 141, 453
forests and rangelands, protection of, 7
Forest Service (FS), 453, 477
Forest Service Pacific Southwest Region, 81
Forest Vegetation Simulator, 51
Forrestel, A., 117
Forsberg, B., 439–40
Fortheringham, C. J., 230, 313, 336
Forthofer, J. M., 41, 49
Fosberg, M. A., 34–36, 43
fossil floras, 2
Fossum, H., 287
Foster, B. L., 112
Foster, G. M., 383
Fotheringham, C., 157
Fotheringham, C. J., 23, 72, 98, 110
foxtail pine (*Pinus balfouriana*), 65, 253
fragmentation, of plant communities, 463
Frakes, N., 363
Fralish, J. S., 93
Franciscan Formation, 149
Francis, R. C., 195
Franco, E., 257
Franco, G., 314
Franco-Vizcaíno, E., 15, 19, 20, 23, 75
Frandsen, W. H., 47
Frangioso, K. M., 165
Franklin, A. B., 187
Franklin, E. C., 88, 90, 92–94
Franklin, J., 264
Franklin, J. F., 78, 187, 408
Franklin, S. B., 93
Fredricksen, R. L., 88, 90, 92–94, 123, 126
Freed, T. J., 75
Frelich, L. E., 59, 60
fremont cottonwood (*Populus fremontii*), 289, 369
Fremont National Forest, 229
freshwater marsh vegetation, 291
Freudenberger, D. O.
fried chicken mushroom (*Lyophyllum decastes*), 386
Fried, J. S., 408
Friggens, M. M., 125, 134, 140
Fritze, H., 88
Froelich, R. C., 88, 90, 92–94
Fry, D. L., 74, 75, 77–78, 83, 156
fuel: crown fuel, 411; ladder fuel, 412; quantity/size, 411; surface fuel-surface fuel, 411
fuelbed: bulk density, 42, 49; compactness of, 41; components of, 44; packing ratio of, 42, 47
fuel characteristics classification system (FCCS), 44, 451
fuel consumption, remote sensing metrics of, 72
fuel continuity, 367
fuel energy: available, 47; total, 47
fuel load, 42
fuel management, 487; basics, 411–12; cost of not doing fuel treatments, 423; forest landscape-scale fuel treatment design, 420–21*sb*; fuel emergency, evolution of, 416–22; objectives, 411; risk of fuel management projects, 423; in twenty-first-century California, 422
fuel models, 43–44; crown fire initiation models, 44; limitations of, 44
fuel moisture, 165; reduction of, 501
fuel reduction, treatments for, 163*f*
fuels: available for combustion, 41*f*; in bristlecone/limber pine woodlands, 369; calculation of moisture content of, 43; calculation of surface area to volume ratio of, 42*b*; characteristics of, 41; day-to-day changes in moisture, 43; dead fuels, 43; dryness of, 35; duff fuel depth classes, 44*t*; firewood analogy, 43; fuelbed compactness, 41; fuel cycle, 16; heat content of, 43; heat transfer, 41; live fuel moisture, 42–43; management of, 7; mineral content of, 43; models of, 43–44; moisture, 29, 41; packing ratio, 41–42, 42*b*; particle size, 41; size classes, 43; surface area to volume ratio, 41; timelag classes of moisture, 43, 44*f*, 44*t*; woody fuel size, 44*t*
fuel stratum, 45
fuel treatments: common goals of, 411; types of, 412
Fulé, P. Z., 75, 408–9
Fuller, D. L., 18
Fuquay, D. M., 45
Fu, R., 314
Furman, W. A., 43
fur traders, 154

Gabel, K. O., 128
Gabrey, S. W., 291
Gabrielson, I. N., 401
Gagnon, P. R., 104
Gale, R. T., 405–6
galleta (*Hilaria jamesii*), 365
Gall, R. L., 15
gambel oak (*Quercus gambelii*), 368
Gamradt, S. C., 125, 130–31
Ganey, J. L., 123
Gannett, H., 195
Gara, R. I., 210
Garbelotto, M., 164
Gardali, T., 140
Garton, E. O., 125
Gately, C. K., 314
Gause, G. W., 324, 337
Gauthier, S., 111
Gautier, C. R., 112
Gavin, D. G., 73–74
Gay, T. E., Jr., 219
GCVTC recommendations, 447
Gebert, K. M., 405
Gedalof, Z., 69
Geil, K. L., 314
Geiser, L. H., 92
gelechiid moth (*Ghelechia* spp.), 263
Gemeno, C., 134
general winds, 32
Gentilcore, D., 363
geomorphic equilibrium, 95
Geospatial Technology and Applications Center (GTAC), 80
Gerlach, J., 283, 287
Ghisalberti, E. L., 110
giant chinquapin (*Chrysolepis chrysophylla*), 160, 175
giant reed (*Arundo donax*), 116, 290, 338, 370, 464
giant sequoia (*Sequoiadendron giganteum*), 20, 66, 254, 261–62, 439
giant sequoias, 119
giant wild-rye (*Elymus condensatus*), 331
Gill, A. M., 106, 402
Gille, C., 134
Gilles, J. K., 408
Gilligan, C. A., 164
Gill, S. J., 20
Gimeno, B. S., 92
Giovaninni, G., 88, 90, 92–93
girdling, 106
Gizinski, V., 162

glacial-interglacial transition, 73
Glass, D. W., 87, 92
Gleason, H. A., 58–59
Gleason, K. E., 93
Glendening, G. W., 12
Glenn-Lewis, D. C., 287
global circulation models (GCMs), 494
global positioning systems (GPS), 79
global temperature, rise in, 2
global warming, 19
global warming potential (GWP), 452
Glotfelty, G., 422
Glovinsky, M., 320
Glyphosate (Roundup®), 371
Godar, R. P., 155
Goebel, M. O., 90
Goforth, B. R., 78, 90
Goldammer, J. G., 97
Goldberg, P., 381
golden eagles (*Aquila chrysaetos*), 134
golden yarrow (*Eriophyllum confertiflorum*), 331
Golder, D., 422
Goldman, D. H., 2
Gold Rush, 5
gold rush mining period, 77
Gollner, M. J., 41, 49
Gómez-Dans, J., 6
Gonen, L., 74
Gonzale-Caban, A., 343
Gonzalez, A., 461
Gonzalez, P., 423
Goode, R. W., 393
Goodman, S. W., 406
goosenest adaptive management area (GAMA), 214
goose pens, 158
Gordon, D. T., 117
Gordon, T. R., 309
Gorham, D. J., 41, 49
Gotway-Crawford, C., 188
Goudey, C. B., 2, 3, 5
gowen cypress (*H. goveniana* subsp. *goveniana*), 308
Graber, D. M., 82
Grace, J. B., 6, 373
Graham, R. C., 14, 18, 89–90, 94
Grand Canyon Visibility Transport Commission (GCVTC), 446
grand fir (*Abies grandis*), 157
granivorous animals, food for, 137
grasses and grass-like plants, 107–8; effect of fire on, 108; fire responses, 107; perennial bunchgrasses, 107; perennial stoloniferous, 107; rhizomatous grasses, 107; tissue damage and mortality, 107; variation in time to kill cambium, 108*f*
grass-fire cycle, 114, 460
grassland ecosystems, 139
grasslands, of California, 400
Graumlich, L. J., 19
Gray, J., 158
Gray, M. E., 363
gray mule's ears (*Wyethia helenoides*), 388
gray pine (*Pinus sabiniana*), 175, 197
greasewood (*Sarcobatus vermiculatus*), 224
Great Basin Floristic Province, 2
Great Idaho Fire (1910), 401
Greely, W. B., 75, 77
greenbark ceanothus (*Ceanothus spinosus*), 334
greenhouse gases (GHGs), 97, 451–52; smoke and fire effects, 441; strategies for reduction, 454
greenleaf manzanita (*Arctostaphylos patula*), 175, 200, 231, 256, 339
Green, L. R., 72
Green, R., 422
Greenwood, M. C., 422
Gregg, M. A., 242
Gresswell, R. E., 128
Grob, I. J., 45
ground-dwelling fauna: habitat zone of, 127*f*; relative mobility and response to fire, 127*b*
Gruell, G. E., 77
Gude, P. H., 422
Gulf of Mexico, 252
Guries, R. P., 110, 111
Gutierrez, R. J., 123–24, 134, 141, 187
Gutzler, D. S., 175, 195
Guynn, Jr., D. C., 123, 128, 136

Haase, S. M., 51, 106, 123, 272
habitat alteration, degree of, 124
Hadley cell, 11
Hadlow, A. M., 43
Hadmer, J., 404
Hagmann, R. K., 78
Hahm, W. J., 250
Haines, D. F., 358
Haines Index, 196
Haire, S. L., 114
Hales, J. E., Jr., 15
Hall, A., 72
Hall, D. R., 134
Hallett, D. J., 73–74
Halofsky, J. E., 179
Halpern, C. B., 185
Halpert, M. S., 19
Halvorsen, R., 92
Hamilton, E. L., 97
Hamilton, L., 406
Hamilton, T. A., 214
Hammond, T. T., 314
Hamrick, J. L., 114
Hann, W. J., 62
Hanski, I., 114
Hanson, C. T., 77
Hardegree, S. P., 94–95
hardwoods, 106
Hardy, C. C., 61, 69, 79, 82
Hare, S. R., 195
Harlow, R. F., 123, 126
Harms, K. E., 104
Harrington, M. G., 51, 104
Harris, J. M., 325
Harrison, S., 71
Harrison, S. P., 74
Hartley, R., 162
Hart, S., 272
Hart, S. C., 93, 123
Hart, S. J., 117
Hartsough, B. R., 412, 415
Harty, F. M., 459
Harvey, M. D., 95–97
Haston, L., 2
Hatcher, T., 422
Hat Creek Valley, 196, 199, 207
Hauer, F. R., 128
Havrilla, C. A., 374
Hawk, G. M., 158
Hawk, K., 422
Hawkworth, D. L., 125, 134, 140
Hayes, S. A., 481
Hayes, T. P., 34
Hayfork fires (1987), 172
Hayhoe, K., 408
hazelnut (*Corylus cornuta*), 262, 383
Healey, S. P., 421
Healthy Forests Restoration Act, 80
heat capacity, of soil, 88
heat content, of fuels, 43
heat flux, 45
heat influx, from fire to aquatic ecosystems, 138
heat loss, from water vapor, 39
heat of combustion, 39
heat of preignition, 49
heat per unit of area, 46
heat scarification, 234
heat sink, 47, 47*f*; moisture as, 30; of plant water, 11
heat source, 47, 47*f*
heat transfer, 40–41, 45, 113; mechanisms of, 41; in soils, 88
heat yield, 39
Hebel, C. L., 88
Heede, B. H., 95–97
Heffner, M. S., 163
Heinselman, M. L., 57, 61–62, 64, 71, 74
Heinz, K. M., 371
Heisey, R. M., 264
Hektner, M. M., 150, 155
Helms, A. M., 163
Hemphill, M. L., 221
Hendricks, V., 320
herbaceous biomass, recycling of, 20
herbaceous fuels, 42, 179
herbaceous vegetation, 137
Herbert, T., 178
Hereford, R., 359
Hessburg, P. F., 128, 187
Heusser, L., 178
Heyerdahl, E. K., 108
Hidalgo, H. G., 79
hierarchical concept of ecosystem, 60
high-severity fire does, 423
HIGRAD model, of fire behavior, 52
Higuera, P. E., 74
hillslope: erosion, 93–95; hydrology, 93, 96; overland flow, 94; surface runoff, 94
Hinckley, T. M., 232
Historic Range of Variability (HRV), 187
Hiukka, R., 88
Hobbs, R. J., 188, 466
Hodge, W. C., 76
Hodgskiss, P. D., 175
Hoerling, M. P., 19
Hogue, T. S., 93, 96
Holden, Z. A., 80
hollyleafed cherry (*Prunus ilicifolia* subsp. *ilicifolia*), 322
hollyleaf redberry (*Rhamnus ilicifolia*), 182, 325
Holocene, 2
Holst, E., 435
Holt, R. D., 112, 461
Holyoak, M., 461
Holzman, B. A., 309–10
honey mesquite (*Prosopis glandulosa* var. *torreyana*), 369
Hood, F., 320
Hood, S. M., 53, 108
Hooker's manzanita (*A. hookeri* subsp. *hookeri*), 310
Hoopes, C. L., 178
Hoopes, M. F., 461
Hoover, R. F., 287
Hopkinson, P. J., 289
Hopper, S. D., 110
Hornbeck, J. W., 90, 93–96
Horn, L. H., 27, 30
horsebrush (*Tetradymia* spp.), 225
Horse Creek Watershed, 235; fire history of, 236*t*
Horton, D., 112
Horton, J. S., 97

Horton, R., 90
Hoshovsky, M., 288
Howard, J. L., 158
Howard, S., 80
Howard, W. E., 125
Howell, E. A., 7
Hu, A., 19
Huang, J., 93, 272, 335
Hubbert, K. R., 14, 18, 90, 92, 94
huckleberry oak (*Quercus vaccinifolia*), 175, 184, 200, 266
Huddleston, R., 186
Hudgeons, J. L., 371
Huffman, E. L., 90
Hu, F. S., 73–74
Hughes, M., 72
Huguet, K., 43
Hull, M., 320
Hultine, K., 371
human-caused fires, 179, 227
human-caused large forest fires, 403
human-fire relationships, 7
human settlement, fire regimes prior to, 6
Humboldt Redwoods State Park, 159
Humboldt-Toiyabe National Forest, 75
Hungerford, R. D., 88
Hunt, A. G., 19
Hunter, M. E., 6
Hunter, R. D., 164
Hu, Q., 314
Husari, S. J., 213, 422
Hussey, T., 282
Hustedt, N., 422
Huxman, T. E., 461
hydrodynamics model, of fire behavior, 52
hydrologic cycle, 93, 94*f*, 140
hydrology, 87; of hillslopes, 93
hydrophilic soils, 90
hydrophobic coatings, 90
hydrophobic soils, 90
Hylkema, M., 302

ice ages, 2, 178
Idaho fescue (*Festuca idahoensis*), 108, 155, 226, 229, 287
identification of fires, 75
ignition flux, 16, 45
ignition of fire: autoignition (spontaneous ignition), 45; critical mass flux, 45; by firebrand, 45; human-caused, 6; by lightning strike, 6, 45, 77; by Native Americans, 77; pilot flaming, 45; probability of, 45; smoldering, 45; sources of, 44–45
Iknayan, K. J., 314
Illilouette Creek Basin, 81*f*
Imeson, A. C., 90
Imper, D. K., 207
incense-cedar (*Calocedrus decurrens*), 98, 160, 175, 254, 313
indigenous burning practices, 6
industrialization of America, 7
infiltration, 93–94; on burned sites, 93; hydrologic cycle, 94*f*; postfire reduction in, 94; in soil, 93
insectivorous animals, food for, 137
intact fire regime, 519
Intergovernmental Panel on Climate Change (IPCC) Assessments, 494
Interior Coast Ranges subregion, 307
interior live oak (*Q. wislizeni*), 197, 203, 283
International System (SI) units, 52
iron oxides, 90
Irwin, L. L., 272
Isaacs, R. E., 75, 77–78, 83
Isaak, D. J., 481
Isbell, C. J., 80
Israelowitz, M., 134

Jabis, M. D., 314
jack pine (*Pinus banksiana*), 59
Jacks, P. J., 96
Jacobs, D. F., 331, 340
Jacobsen, A. L., 163
Jacobsen, H., 320
Jacobs, J. S., 465
Jacobson, A. L., 323
Jaffe, D. A., 441
Jain, T. B., 80
Jakober, M., 96, 123, 129
James, S., 109
Jamieson, L. P., 359
Jass, R. M. B., 74
Jeffrey pine (*Pinus jeffreyi*), 65, 68, 98, 104, 175, 185, 207, 221, 229, 253, 325, 368
Jenkins, M. J., 139
Jenness, J. S., 123
Jensen, A. E., 78
Jiang, X., 314
Johannsen, P. L., 78
Johnny-jump-up (*Viola pedunculata*), 284
Johnson, C. G., Jr., 229
Johnson, D. R., 175
Johnson, D. W., 87, 92
Johnson, E. A., 72, 106
Johnson, J. R., 115, 118–19, 286
Johnson, K., 125
Johnson, K. N., 78
Johnson, L. A., 287
Johnson, N., 314
Johnson, R. F., 92
Johnson, S., 284
Johnston, C. M., 42
Johnstone, J. F., 115, 118–19
Jolly, W. M., 43
Jones, B. E., 207
Jones, G., 123–24, 134, 141
Jones, J. A., 187
Jones, K., 422
Jones, M. B., 72
Jones, T. R., 123
Joseph, F., 452
Joshua tree (*Yucca brevifolia*), 359
Jovan, S., 92
Jubas, H., 188
Jurik, T. W., 287

Kaczynski, K. M., 464
Kahrl, W. L., 18
Kanamitsu, M., 314
kangaroo rats (*Dipodomys* spp.), 130
kangaroo rats (*Dipodomys stephensi*), 129
Karl, T., 439–40
Karnauskas, K. B., 314
Kauffman, J. B., 53, 109
Kayes, L. J., 179
Kay, R., 372
Keane, J. J., 187
Kearns, F., 404
Keeler-Wolf, T., 1, 2, 5, 62, 69
Keeley, J. E., 6, 23, 51, 72, 77, 79, 98, 103, 108–12, 114, 123, 157, 272, 230, 313, 336, 374
Keetch-Byram Drought Index, 500
Keeter, T. S., 6
Keifer, M. B., 117
Keiffer, R. J., 156
Keil, D. J., 2
Keith, D. A., 108–9, 111–12
Keller, E. A., 96, 324
Kelvin waves, 19
Kennedy, C., 288
Kephart, P., 150
Kershner, J. L., 128
Kessomkiat, W., 314
Ketterings, Q. M., 90
Keuler, N. S., 344
Key, C. H., 79–82
Keyes, C. R., 162
Kidder, P., 422
Kiernan, J. D., 140
Kilgore, B. M., 61, 117, 227
kilocalorie (kcal), 40
kilowatt hours (kW h), 40
Kimball, T. L., 401
Kimmins, J. P., 92
King, J. A., 35
Kingsly, K., 110
King, T., 3
Kiniry, J. R., 371
Kinney, J. J. R., 34
Kinoshita, A. M., 93, 96
Kinter, J., 314
Kirby, M. E., 310
Kirkham, M. B., 90
Kirkpatrick, R., 320
Kirtman, B., 314
Klamath Marsh, 224
Klamath Mountains bioregion, 2, 4, 68, 152, 195, 202, 205; area covered by, 171; average precipitation in, 173*t*; climate of, 171; ecological zones in, 175, 179–87; fire climate in, 174–75; fire history reconstructions of the last 500 years, 178; fire-occurrence areas (FOA), 180, 180*f*; fire regime–plant community interactions, 182–83, 185–86; fire regimes in, 179; fire responses of important species, 182, 184; fire severity in, 179, 180–81; fire size and total area burned, 179; floristic diversity and complexity in, 175; future directions in management of, 188–90; hardwood component of, 175; historic fire activity, 178–79; historic fire occurrence, 176–79; holocene fire, vegetation, and climate history, 176–78, 178*f*; human-caused fires, 179; lightning strikes in, 172–74, 179; lower and mid-montane zone, 175, 176*f*, 182; major fire episodes, 172; management issues in, 187–88; map of, 172*m*; maxima and minima temperatures in, 173*t*; mid to upper montane zone, 175–76, 183–86; Northwest Forest Plan, 187; physical geography of, 171; postfrontal conditions, 171; prefrontal conditions, 171–72; riparian zones, 181; smoke, presence of, 188; snowpack data, 174*t*; storm episode in, 174; subalpine woodlands, 176, 177*f*; subtropical High conditions, 172; temperature records in, 171; topography influencing fire regimes in, 179–80; twentieth-century fire activity, 179; warming climate, causes of, 189–90; weather systems of, 171–72; wildland-urban interface, 188; wildlife habitat in, 187–88; wildlife management in, 189
Klamath National Forest, 140
Klamath River, 171
Klein, J. G., 225
Kline, T. S., 402
Klinger, R., 363
Knapp, E. E., 80–81, 123, 172–73, 272
Kneitel, J. M., 108
knobcone pine (*Pinus attenuata*), 58, 150, 157, 180–81, 185, 206, 230, 256, 308, 312, 323
Knowd, T., 312

Knoxl, K. J., 104, 108
Knutson, A. E., 371
Knutson, K., 373
Kobziar, L. N., 412, 415, 418
Kolasa, J., 60
Kolb, K. J., 333
Kolb, T. E., 104
Kolden, C. A., 79
Koo, E., 156
Kosaka, Y., 19
Kotok, E. I., 75–77, 79
Kovacs, P., 140
Kraebel, C. J., 97
Kramer, S. H., 162
Krammes, J. S., 95
Krawchuk, M. J., 72
Krueger, D., 320
Kruger, L. M., 106
Kuljian, H., 117
Kumar, A., 19
Kumar, S., 314
Kummerow, J., 106
Kwon, J., 134

Laacke, R. J., 209
labrador tea (*Rhododendron columbianum*), 390
lacy phacelia (*Phacelia tanacetifolia*), 364
ladder-fuel effect, 104
ladder fuels, 49, 117
lady's-slipper (*Cypripedium montanum*), 232
Laguna Beach dudleya (*Dudleya stolonifera*), 484
Laird, J. R., 95–97
Lake, F. K., 162, 393
Lake Tahoe, 12–13, 27
Lake Tahoe Basin Management Unit, 75
Lambers, H., 110
Lamont, B. B., 104, 108, 111, 115
Lancaster, J. W., 43
landfire biophysical setting types, 369
Landis, R., 406
land management, 7, 71, 140, 243; Bureau of Land Management (BLM), 15, 430, 477; objectives of, 62; strategies for, 453
land managers working group, 453
LANDSAT data, 258
landscape trap, 188
landslides, ground-saturated, 96
Langenbrunner, B., 314
La Niña, 19, 299
Lantz, R. K., 106
La Purisima manzanita (*Arctostaphylos purissima*), 310
Larsen, C., 71, 73
Larson, A. J., 187
Larson, L. T., 75, 77
Lassen Volcanic National Park (LVNP), 207, 213; white fir/Jeffrey pine forests in, 209*f*
Latham, D. J., 45
Lathrop, E. W., 117
Latif, M., 19
Laude, H. M., 72
Laughrin, L. L., 111
Lau, K. A., 19
Launchbaugh, K., 372
laurel sumac (*Malosma laurina*), 330
Lava Beds National Monument: climate change, 241; postfire community development, 241*f*; sagebrush-conifer ecotone in, 240*f*; woodland development phases, 240*f*
Laven, R. D., 1
Lavin, M. J., 405
Lavoie, M., 74
Lawes, M. J., 104, 108
Law, R., 461
Laws, B. E., 179
laws protecting at-risk species, 477
lead isotope, 71
leaf blight (*Alternaria alternata*), 387
Ledig, F. T., 175
Lee, D. C., 128
Lefsky, M., 421
Lehmkuhl, J., 123–24, 272
Lehmkuhl, J. F., 123
Leiberg, J. B., 75–77
Leibold, M. A., 461
Leicester, B., 404
Lemmon's ceanothus (*Ceanothus lemmonii*), 203
Lemmon's needle grass (*Stipa lemmonii*), 226
lemonade berry (*Rhus integrifolia*), 330
Lenihan, J. M., 162, 182, 408
Leonard, J., 404
Leopold, L. B., 96
Leopold, S. A., 401
Lertzman, K. P., 73
Lettenmaier, D. P., 140
Levins, R., 112
Lewis, C. E., 88, 90, 92–94
Lewis, H. T., 77
Lewis, S. A., 80
Libby, W. J., 111
Libohova, Z., 128
lichens, 112
light burning, 77
Lightfoot, K. G., 154, 302
lightning strike densities (LSDs), 500
lightning strikes, in California, 15–17, 251, 320; by bioregion, 17*t*; cloud-to-ground, 45; density of, 17*m*; ecological time scales of, 16; frequency of, 354; ignition due to, 45, 77, 331; ignitions due to, 16; in Klamath Mountains, 172–74, 179; with low precipitation, 220; monthly distribution of fires caused by, 221*f*; network data, 300; in Northeastern plateaus bioregion, 221; in Southern Cascades bioregion, 196–97; variables contributing to, 17; variation in number of, 174*f*
lignotuber mortality, 72
lignotubers, 109
Likens, G. E., 188
Lile, D. F., 207
limber (*Pinus flexilis*), 361
Lindstrom, S., 1
Linhart, Y. B., 334
Linn, R. R., 52
Littell, J. S., 242
litter and duff, 92
Little Horse Peak Research Project, 214
Little Ice Age, 74
Liu, J., 302, 381
live fuel moisture, 42–43
livestock grazing, 155, 179, 207, 238
Lockwood, R. N., 96
lodgepole pine (*Pinus contorta*), 45, 58, 75, 110–11, 175, 200, 207, 268, 325; distribution of fire severity classes in, 211*f*; fire rotations, 211; fuelbeds, 269
Loeschke, M., 287
Loewen, M., 104
logging, 207
lomatium (*Lomatium*), 328
Lombardo, K. J., 72, 74, 314
Lompoc ceanothus (*C. cuneatus* var. *fascicularis*), 310
Longbotham, C. R., 98
Longbotham, G. J., 98
Long, C., 74
Long, C. J., 73–74
Long, D. G., 79
long-horned beetle (*Phymatodes nitidus*), 263
Long, J., 342
Long, J. W., 393
Long, L. N., 314
Longpre, C. I., 359
Lonsdale, W. M., 459
Loomis, J., 343
Lopez, A., 312
Loreau, M., 461
Losos, E., 459
Los Padres fire, 304
Lotan, J. E., 110–12
Lott, M. J., 325
Loudermilk, E. L., 174
Lovell, C., 20
Lovich, J. E., 371
lower montane zone: fire regime–plant community interactions, 182–83, 229–30; fire responses of important species, 182, 183*t*, 229, 230*t*; in Klamath Mountains bioregion, 182–83; in Northeastern plateaus bioregion, 228–30, 228*f*
low heat content, 39
Lucchesi, S., 88, 90, 92–93
Luce, C., 128
Luce, C. H., 481
Lukens, C. E., 250
Lu, R., 34–35
Lydersen, J. M., 75, 77–78, 83
Lyle, M., 178
Lynch, J. A., 73
Lyon, J. L., 123, 126

MacAulay, J. D., 110
MacDonald, L. H., 90, 95, 128
MacDougall, A. S., 471
Macnab cypress (*Hesperocyparis macnabiana*), 206–7
macrofossils: dating of, 71, 73; in lake, 254
Madro-Tertiary Flora, 2
Mahala mat (*Ceanothus prostratus*), 206, 233
Maher, S. P., 314
Mahlum, S. K., 96, 123, 129
Maier, L., 74
maintenance phase, 416
Malakoff, D. A., 110
Malamud, B. D., 17
Malechek, J. C., 373
Mallory, L., 320
Malmstrom, C.
Maloney, E. D., 314
managed wildfire, 416
Mandondo, A., 20
Mannio, D., 188
Mann, M. L., 314
Mantua, N. J., 195
many-stemmed sedge (*Carex multicaulis*), 264
manzanitas (*Arctostaphylos* spp.), 68, 103, 150, 184, 222, 256, 368, 385
maple (*Acer*), 384
marbled murrelet (*Brachyramphus marmoratus*), 479
Marble Mountain wilderness, 189
Marcot, B. G., 123
marigold navarretia (*Navarretia tagetina*), 288
marine air, 35
marine layer inversion, 31
Mariotti, A., 314
mariposa lilies (*Calochortus* spp.), 260
mariposa lily (*Calochortus*), 328
Markgraf, V., 19
Marlon, J., 71
Marlon, J. R., 74
Marquez, V. J., 331

Martin, B. D., 117
Martinez, A. Y., 370
Martin, R. E., 6, 53, 103
Martinson, E. J., 6
mass erosion, 96
Massey, B. M., 324
Mast, J. N., 114
Matchett, J. R., 363
Mauget, S. A., 314
Maurer, E. P., 314
Maxwell, R. S., 75, 77–78, 83
Mayle, F., 71
Maynard, T. B., 45
MC1 dynamic general vegetation model, 499
McAllister, S. S., 41, 45, 49
McBride, J. R., 162, 271, 412, 415, 418
McClaran, M. P., 285
McComb, B., 187
McCreedy, C., 358
McCutchan, M. H., 34
McDonald, P. M., 116
McDougald, N. K., 283, 285
McGarigal, K., 114
McGregor, K. F., 459, 465, 467
McGregor, S., 19
McGuirk, J. P., 19
McHugh, C. W., 53, 104
McIver, J. D., 51, 123, 272, 412, 415
McKelvey, K. S., 208, 213
McKenzie, D., 69, 80, 242
McKinley, R. A., 363
McMaster, G. S., 112
McNab, W. H., 3
McPhaden, M. J., 19
McPherson, E. G., 93
median fire return interval (MFRI), 227
Medica, P. A., 358
medieval droughts, 254
medieval warm period, 74
Mediterranean climate, 11
Mediterranean ecosystems, 1
mediterranean grass (*Schismus barbatus*), 365
mediterranean grass accentuate herbaceous fuel, 370
Mediterranean shrublands, 90
mediterranean split grass (*Schismus* spp.), 460
medusa head (*Elymus caput-medusae*), 224, 468
Meehl, G. A., 19
Meentemeyer, R. K., 164–65
Mees, R.
Megahan, W. F., 90, 93–96
megajoule (MJ), 40
Meixner, T., 92
Meko, D. M., 19
Melack, J. M., 96, 418
Menakis, J. P., 61, 69, 82
Menke, J. W., 287, 400
Mensing, S. A., 6, 72, 162, 310, 325
Merriam, K. E., 187–88
Merrill, A. G., 104
mesoscale convective systems, 15
Mesozoic sedimentary bedrock, 149
Metlen, K., 51
Metz, L. J., 123, 126
Metz, M. R., 165
Mexican cliffrose (*Purshia mexicana*), 357
Meyer, M. D., 83
Meyer, P., 188
Meyerson, J. E., 314
Michaelsen, J., 2, 6, 72
Michaletz, S. T., 106
Micheletty, P. R., 93
microbial biomass, 233
microburns, 20
microclimate, effect of fire on, 53
Midgley, J. J., 104, 106, 108
mid-montane zone, 231–33, 231*f*; fire-caused mortality in, 231; fire regime–plant community interactions, 231–33; fire responses of important species in, 231, 232*t*; fuels management challenges in, 231
Miles, L. J., 72
Miles, S. R., 2, 3, 5
Millar, C. I., 2, 3, 82, 119
Millar, J., 422
Miller, B. P., 111, 115
Miller, C., 114
Miller, D., 417
Miller, J. D., 68, 80, 81, 83, 172–73, 342
Miller, J. P., 96
Miller, M., 106–8, 370
Miller, M. E., 95
Miller, N. L., 314
Miller, P. C., 18
Miller, P. R., 34, 98
Miller, R. F., 225–26, 242
Miller, S. M., 90
Miller, W. W., 87, 92
Mills, J., 206, 214
Mills, T. J., 406
Millspaugh, S. H., 73
Minckley, T., 74
mineral damping coefficient, 48
mineral soil, 88, 117–18; factors influencing color of, 90
Miniat, C. F., 471
Minnich, R. A., 15–17, 19–24, 34–36, 75, 78, 90
Mitchell, R. J., 471
Mladenoff, D. J., 110–11
mobile emissions, 452
Mockrin, M. H., 344
moderate resolution image spectroradiometer (MODIS), 451
Modoc budworm (*Choristoneura viridis*), 232
Modoc National Forest, 222; number and area of fires recorded on, 222*t*
Moeur, M., 421
Moffet, C. A., 94–95
Moghaddas, E. Y., 92
Moghaddas, J., 272
MOGHADDAS, J. J., 51, 123, 412, 415, 421
"moist threshold" of combustion, 23
moisture damping coefficient, 48
moisture exchange, 29
moisture of extinction, 39
mojave (*Ceanothus vestitus*), 367
Mojave Desert, 35
Mo, K. C., 314
monitoring trends in burn severity (MTBS), 79, 79*t*, 80
monkey flower (*Mimulus aurantiacus*), 330
montane chaparral patches, 265
Monterey cypress (*Hesperocyparis macrocarpa*), 308
Monterey pine (*Pinus radiata*), 58
Montgomery, K. R., 116
Moody, T. J., 72, 83, 117, 408
Mooney, H. A., 466
Moore, J. W., 481
Moore, P. D., 113
Morais, M. E., 23, 301
Moratto, M. J., 77
Morein, G., 17
Moreno, J. M., 72
Morgan, P., 77, 79–80, 433
Moriarty, P., 20
Morikawa, Y., 232
Moritz, M. A., 59–60, 72, 117, 301, 314, 404, 408, 420
mormon tea (*Ephedra* spp.), 365
Morris, G. A., 51
Morro manzanita (*A. morroensis*), 310
Mortsch, L. D., 140
Mosley, J. C., 242
Mote, P., 69
Mott, J. A., 188
mountain alder (*Alnus incana*), 110, 261
mountain big sagebrush (*A. tridentata* subsp. *vaseyana*), 225*f*, 361
mountain dogwood (*Cornus nuttallii*), 113, 200, 261
mountain hemlock (*Tsuga mertensiana*), 65, 175, 200, 233, 253
mountain lions (*Puma concolor*), 134
mountain-mahogany (*Cercocarpus* spp.), 368
mountain misery (*Chamaebatia foliolosa*), 116, 199, 261
mountain snowberry (*Symphoricarpos oreophilus*), 231
mountain whitethorn (*Ceanothus cordulatus*), 200, 262
Mount Hamilton Range, 313
Mouquet, N., 461
Moyle, P., 140
Muick, P. C., 284
mule deer, 125
mule deer (*Odocoileus hemionus*), 124, 129, 222, 243
mule fat (*Baccharis salicifolia*), 291
Muller's oak (*Quercus cornelius-mulleri*), 367
multiple lighting strike fire complexes, 403
Murov, M. B., 358
Murphy, J. D., 87, 92
Murphy, J. F., 92
mustard (*Brassica* spp.), 283
Mutch, R. W., 116
Myers, J. A., 104

Nader, G. A., 344
Nagel, T. A., 75, 482
Naiman, R. J., 417–18
Nakamura, G. M., 344
Nakano, S., 482
Nakazato, T., 374
Napper, C., 80
Narog, M. G., 90, 94
Nason, J. D., 114
National Ambient Air Quality Standards (NAAQS), 439, 445–46
National Environmental Protection Act (NEPA), 480
National Fire Danger Rating System (NFDRS), 43–45, 451
National Fire Plan, 80, 487
National Fire Protection Association, 511
National Forest Reserves, 7
National Institute for Occupational Safety and Health (NIOSH), 441
National Park Service (NPS), 240, 256, 453, 477, 509
National Research Council (NRC), 499
National Study of Fire and Fire Surrogates (FFS), 412
National Wildfire Coordinating Group (NWCG), 44, 81; Fire Environment Committee of, 44
Native Americans: fire regimes during era of, 6; reasons for using fire, 77; reconstruction of fire patterns, 77
native birds, 480
Natural Community Conservation Planning program, California, 488
natural disturbance, theory of, 59–60

natural range of variation (NRV), 82–83
natural resource management, 7, 482
Natural Resources Conservation Service (NRCS), 433
Naveh, Z., 286
Neal, J. L., 90
Neary, D. G., 88, 95, 97, 106, 408
needle grass (*Stipa* spp.), 264
Neelin, J. D., 314
Ne'eman, G., 230, 313, 336
Neilson, R. P., 162, 188, 408
Neiman, P. J., 501
Nelson, D. C., 110
Nelson, R. M., 50
Nelson, S. K., 123
Nespor, S., 452
net fuel load, 48
nevada blue grass (*Poa secunda*), 286
Neville, H. M., 134, 143–44, 191
Newman, W., 309
Newton, C. W., 12
Newton, W. E., 471
Nicholson, S. E., 19
nighttime radiational cooling, 35
Nijhuis, M. J. M., 374
nitrifying bacteria, 88*b*, 92
nitrogen and nitrogen compounds, 92
nitrogen deposition, critical loads of, 447*m*
nitrogen dioxide (NO_2), 110, 445
nitrogen fixation, 91–92, 233
nitrogen oxides (NO_x), 97; photolysis of, 98
Nobble, I. R., 111
nodding wild buckwheat (*Eriogonum cinereum*), 330
Nolan, A. W., 93
nonnative tamarisk (*Tamarix* spp.), 369
nonvascular plants, 104; affect of fire on, 112; vulnerability to surface fires, 112
Noonan-Wright, E. K., 254, 272
Noon, B. R., 141
Nordquist, M. K., 15
Normalized Burn Ratio (NBR) index, 79
Norman, S. P., 20
North Africa grass (*Ventenata dubia*), 238
North American monsoon, 14–15; afternoon weather conditions, 16*f*; circulation models of, 15*f*
North coastal chaparral zone, 151*f*
North coastal pine forest zone, 151*f*
North Coastal Scrub, 151*f*
North Coast bioregion, 3, 150*m*; annual precipitation, 149; area burned by decade for fires, 155*f*; climatic patterns of, 149–50; coastal prairie zone, 154–55; Douglas-fir-tanoak forests, 153*f*, 160–61; ecological zones of, 150–52, 154–56; fire during current period, 154; fire during historic period, 154; fire during prehistoric period, 152–54; fire frequency in, 154*f*; fire records of, 154; fire regime–plant community interactions, 155–61; fire responses of important species, 155–57, 156*t*, 158–61; fire suppression in, 154; future direction and climate change, 162; historic fire occurrence in, 152–54; lightning fires, 153; management issues, 162–66; mean temperature, 149; Northern chaparral zone, 150, 156; Northern coastal pine forests zone, 151*f*, 157; Northern coastal scrub, 154–55; Oregon oak woodland zone, 153*f*, 161–62; pattern of fire frequencies, 153; physical geography of, 149; redwood forest zone, 152*f*, 158–60; Sitka spruce forests, 152*f*, 157–58; synoptic weather systems, 149
Northeastern Plateaus bioregion: altered fire regimes in, 238; atmospheric instability in, 220; climate patterns of, 219–21; ecological zones of, 221–38; fire climate variables of, 219–20; fire in current period, 222; fire in historic period, 222; fire in prehistoric period, 221–22; fire perimeter locations, 223*f*; fire regime–plant community interactions, 226--36; fire responses of important species, 224–26, 229, 231, 232*t*, 234, 235*t*; fire season of, 220; grazing, presence and level of, 238; historical fire occurrence in, 221–22; land management in, 243; lightning strikes in, 221; long-term fire effects monitoring in, 239*f*; long-term monitoring to support management in, 239–42; lower montane zone, 228–30, 228*f*; management issues in, 238–43; map of, 220*m*; meadows in, 236–37; mid-montane zone, 231–33, 231*f*; nonzonal vegetation, 236–38; physical geography of, 219; postfire community development in, 241*f*; prefrontal winds in, 220; Sagebrush Steppe zone, 222–28; strong subsidence/low relative humidity, 220; subalpine zone of, 236, 237*f*; thunderstorms in, 219, 220; timber harvesting in, 222; upper montane zone, 233–36; weather systems of, 220; wildlife management in, 238–43
Northern California Prescribed Fire Council (NCPFC), 513
northern coastal pine forests zone: fire regime–plant community interactions, 155–56; fire responses of important species, 157
northern coastal prairie, 155
northern coastal scrub zone, 154–55; fire regime–plant community interactions, 155–56; fire responses of important species, 155
northern flying squirrel (*Glaucomys sabrinus*), 485
northern goshawk (*Accipiter gentilis*), 136, 272
northern spotted owl (*Strix occidentalis caurina*), 479
North, M. P., 53, 264, 342, 408
North Pacific gyre, 19
North Pacific Ocean, 13
north-south latitudinal gradient, 149
northwestern foothills, 197
northwest forest plan, 187
Norton-Griffiths, M.
Noss, R. F., 123, 158
Noste, N. V., 110, 114–15
NPS FRID index, 82
nutrient availability, 103
nutrient, concept of, 59
nutrient cycling, 112, 140
nutrient volatilization, 91
nutritional value of plant materials, effect of fire on, 137

oak (*Q. lobata*), 150, 279, 283, 384, 399
Oakeshott, G. B., 171
oak, fire regimes, 339
Oakley, B., 264
oat grass (*Rytidosperma penicillatum*), 156
Oberbauer, A. T., 322
Oberhue, R. D., 42
obligate resprouters, 108
obligate seeders, 108; demography of, 333
obligate-seeding shrubs, 260
O'Brien, M. J., 308
O'Dell, C. A., 34–36
Odion, D. C., 72, 77, 110, 114
Odum, E. P., 59
Oechel, W. C., 72
Oertel, A., 320
Office of Management and Budget (OMB), 422
O'Hara, K. L., 158, 308, 435
Ohlson, M., 92
Ojeda, F., 104, 108, 111
Okland, T., 92
Oldershaw, D., 358
Oldfather, M. F., 314
Oliver, W. W., 208
Olson, D., 123
Olson, R. A., 242
Omi, P. N., 6
O'Neill, R. V., 60
oniongrass (*Melica bulbosa*), 264
onshore winds, 98
Oostindie, K., 90
Open Pine Forest sites (OPF), 207
Oppenheim, J., 125
Oregon oak (*Quercus garryana*), 64, 150, 160, 175, 199, 229
Oregon Oak Woodland Zone, 150, 153*f*, 161–62, 165; fire regime–plant community interactions, 162; fire responses of important species, 162
organic energy of plants, 11
organic matter: decomposition of, 91; of soil, 92; thermal conductivity of, 88
orographic precipitation, 13
Ortiz, N., 71
Otrosina, W., 123, 272
Ottmar, R. D., 44, 53, 98
Outcalt, K. W., 123, 272
Overby, S. T., 92
overgrazing, 329
overland flow, 94
overnight radiation cooling inversion, 31
oxidation, 39, 40, 53, 91, 439
ozone (O_3), 98, 439, 441, 445

Pacific Decadal Oscillation (PDO), 18, 19, 174, 195
Pacific fisher (*Martes pennanti pacifica*), 272
Pacific kangaroo rats (*Dipodomys agilis*), 130
Pacific madrone (*Arbutus menziesii*), 150, 160, 175, 256, 307
Pacific North American Pattern (PNA), 174, 195
Pacific starflower (*Trientalis latifolia*), 264
Pacific willow (*S. lasiandra*), 291
Pacific yew (*Taxus brevifolia*), 262
packing ratio, 42, 42*b*
Padgett, P. E., 35
Page-Dumroese, D., 433
Page, W. G., 139
Pagni, P., 156
Paine, C. E. T., 104
Pajares, J. A., 134
pajaro manzanita (*A. pajaroensis*), 310
Paleocene Epoch, 2
pallid manzanita (*Arctostaphylos pallida*), 485
Palmén, E., 12
Palmer, M. A., 140
Palomar Mountain, 14
Pan, Z., 314
Parker, M. S., 482
Parker, V. T., 72, 110, 113
Parmeter, J. R., Jr., 98
Parry's goldenbush (*Ericameria parryi*), 200
Parsons, A., 80
Parsons, D. J., 72, 107, 213
partial pressure of gas, 29

particle size distribution, effects of heating on, 90*t*
particulate matter (PM), 97, 440; health-protective and visibility range values, 441*t*; pollution caused by, 453
passive-active crown fire, 68
Passmore, H. A., 104
patch flammability, 36
patch mosaics, in chaparral, 21–24; burning of, 23*f*
Paterson Lake, 221
Patra, S., 221
Patterson, R., 2
Pattison, R. R., 371
Patz, J. A., 140
Pausas, J. G., 108–9, 111–12
Pavlik, B., 1, 284
Pearson, H. A., 123, 126
Peery, M. Z., 123–24, 134, 141
Peet, R. K., 110
Peinado, M., 309
Pellant, M., 372
Peng, R., 335
penstemons (*Penstemon* spp.), 264
percent FRID (PFRID), 82
peregrine falcons (*Falco peregrinus*), 134
perennial bunchgrasses, 107
perennial stoloniferous, 107
periodic perturbations, in fire regimes, 11
Perry, D. A., 111, 187
Perry, H. M., 92
Peters, I. D., 461
Peterson, A. T.
Peterson, C. J., 110
Peterson, D. H.
Peterson, D. L., 52, 69, 73, 104, 106, 179, 242
Peterson, J. L., 98
Peterson, P. M., 384, 397
Peterson, S. H., 301
petrochemicals, 39
Pettit, N. E., 429, 434
Pfaff, A., 72
Pfaff, A. H., 72
Pflugmacher, D., 421
Philander, S. G., 19
Phillips, A., 459
Philpot, C. W., 105, 405
pH of soil, 88*b*, 93
phosphorus (P), 92–93
photosynthesis, 39, 104, 112
phylica (*Phylica* spp.), 334
Pickett, S. T. A., 59, 60
Pierce, D. W., 314
Pierson, F. B., 94–95
Pietikainen, J., 88
pigweed (*Chenopodium* spp.), 385
pika (*Ochotona princeps*), 140
Pilliod, D., 373
pilot flaming ignition, 45
pine beetle (*Dendroctonus ponderosae*), 236
pine-interior live oak (*Pinus sabiniana-Quercus wislizeni*), 253
pine-mat manzanita (*Arctostaphylos nevadensis*), 200, 209, 266
pinnacles national monument, 305
Pinter, N., 74
pinyon deer mouse (*Peromyscus truei*), 140
pinyon-juniper (*Pinus-Juniperus*), 36
pinyon pine (*Pinus monophylla*), 357
Pitcairn, M. J., 286
pitch canker (*Fusarium circinatum*), 309
Pit River, 199
planetary heat transfer, 27
plant canopy, 93
plant communities: and animal interactions with fire, 136–38; components of, 115; defense against bark beetles, 108; disclimax for, 58; distribution of, 58*f*; fire and geographic or landscape patterns of, 119; fire and species composition, 116–17; fire and structure of, 117; fire intensity and, 117; fire return interval, 116–17; fire seasonality and, 116; food availability, 137; indirect environment and climate effects on, 117–19; interaction with fire, 115–17; and landscape position, 117; major aspects of, 115; organism concept of, 59; phenological patterns for two co-occurring, 116*f*; and reciprocal effects between fire and vegetation, 115; removal of CO_2 from atmosphere, 454; shift in, 137–38; tree-dominated, 117
plant death, 104, 106
plant density, 135
plant interactions with fire. *See also* fire regime–plant community interactions: algae growth and, 112; bark flammability, 106; carbohydrate storage, 109; cross-section of tree trunk showing protective bark, 107*f*; crown fire, 103; crowns and foliage, survival of, 104–5; death of plant cells, 104; degree of damage, 104; direct effects on individuals, 103–4; drought-stressed, 118; effects of fire, 103; environment and climate effects on, 117–19; factors associated with fire response, 105*t*; fire-dependent, 112; fire effects on nonvascular plants, 112; fire-enhanced, 112; on fire intensity and duration, 108; fire intensity and fire type, 115; fire persisters, 103; fire resistance, 103, 104–8; fire resisters, 103; fire return intervals, consequences of, 114–15; fire seasonality, 103; grasses and grass-like plants, 107–8; indirect and direct effects, 103; individual plants and plant vigor, 108; insulation against fire, 104; ladder-fuel effect, 104; long-term climate changes, impact of, 118–19; low-severity fire, 108; morphological and physiological traits, 103; morphological structures, 107*f*; nutritional value of plant materials and, 137; persistence and colonization, 108–11; plant community and, 115–17; plant population and, 112–15; plants structures and associated definitions, 105*t*; predicting species responses to fire, 111–12; rate of heat diffusion, 106; reproductive responses to fire, classification of, 109*t*; roots and underground tissues and organs, 106–7; seasonality of fire, 112–13; seeds, 110–11; short-term climate changes, impact of, 117–18; on size of fire, 113; spatial complexity of fire and, 113–14, 114*f*; species adaptation and evolution, 103; sprouting, 108–10; starch storage, 109; stems and trunks, 105–6; sunlight-blocking canopy, removal of, 203; tolerance to fire, 108; variation in fire response traits, 111; vegetation mortality, 117
plant invasions, 459
plant life-cycle events, 109
plant matter, combustion of, 71
plant microclimate, 119
plant mortality, due to fire, 52–53; predictor of, 52
plant population: age-class distribution, 113; characteristics of, 112; defined, 112; fire-dependent species, 113; fire intensity and fire type, impact of, 115; and fire regime characteristics, 112–15; fire return interval, consequences of, 114–15; fire size, effect of, 113; fire spatial complexity, 113–14, 114*f*; general patterns of, 113; importance of an individual, 113; interactions with fire, 112–15; metapopulation, 112; mortality and reduction of, 113; progeny recruitment, 113; recovery after fire, 113; recruitment from sprouts and seed banks, 115; season of fire, influence of, 112–13; seed banks for regeneration, 113
plant reproductive responses to fire, classification of, 109*t*
plant transpiration, 11
Platt, W. J., 104
Pleasant Valley mariposa lily (*Calochortus clavatus*), 113
Pliocene Epoch, 2
plum (*Prunus*), 384
Plumas National Forest, 75
Plumb, T. R., 106
plume-dominated fires, 50
Plummer, F. G., 162
Podhora, E., 140
Poff, N. L., 140
Point Reyes National Seashore, 157
poison oak (*Toxicodendron diversilobum*), 182
pollen, 71–74; in lake, 254
pollution transport, air–fire interactions on, 98
ponderosa mushroom (*Tricholoma magnivelare*), 386
Ponderosa pine (*Pinus ponderosa*), 45, 60, 68, 104, 106, 136, 160, 175, 197, 260, 325, 368, 439, 464, 517
Poole, D. K., 18
Popper, M., 284
population densities, 282
porcupine (*Erethizon dorsatum*), 126
pore size distribution, 89–90
porosity of soil, 89–90
Porter, D., 162
Port Orford cedar (*Chamaecyparis lawsoniana*), 157, 184–85
Portwood, K. A., 335
Postel, S. L., 140
postfire: community development, 241*f*; erosion, 95; flammability, 36; image acquisition, 80; management, 80, 226, 344, 372–74, 423, 461; restoration activities, 433; seeding, 110, 112, 119, 313, 373, 463, 480, 486
potential evapotranspiration (PET) rates, 15
Potter, D. A., 116
Potts, J., 156
Povak, N. A., 187
Powell, B., 406
Power, M. E., 482
Power, M. J., 71, 74
Powers, R. F., 92, 206, 214
Pratt, R. B., 163
precipitation, in California, 13–14; mean annual precipitation, 13, 14*m*; orographic lift, 13; postfrontal, 13; in Salmon and Siskiyou Ranges, 13; “spent” upwind, 13
pre-Euro-American Fire Regime (PFR) groups, 82
pre-Euro-American settlement fires, 164
prefire and postfire landscapes, 79
prefrontal winds, 220
Preisler, H., 423
Preisler, H. K., 90, 94
Prengaman, K. A., 463
Prentice, K., 363
prescribed fire, 77

prescribed fire information reporting system (PFIRS), 450
presettlement fire regime, 77
Preston, C. M., 92
prevailing surface winds, 34, 35*m*
prevention of significant deterioration (PSD) program, 446; regional haze and, 446–47
Prichard, S. J., 73
Prieto-Fernandez, A., 92
priority effects, 461
process-based hydrologic models, 494
professional forestry, 508
pronghorn antelope (*Antilocapra americana*), 125, 243
Prospect Peak, 212*f*
Provenza, F. D., 373
Psonis, A. A., 19
P-Torch, 51
public health, near fires, 33
public land survey system, 79
Purich, A., 19
purple needle grass (*Stipa pulchra*), 155, 284
purple sage (*Salvia leucophylla*), 309, 330
pygmy cypress (*Hesperocyparis pygmaea*), 150
pygmy-weed (*Crassula connata*), 288
Pyke, D., 372–73
Pyne, S. J., 1, 6–7
pyrodiversity, 517
pyrogenic vegetation, 1
pyrolysis, 11, 40, 45
Pysek, P., 459, 465, 467

quaking aspen (*Populus tremuloides*), 200, 223, 261, 266
quantification, of fire regimes: application of, 81; burned area emergency response (BAER), 80; characterizations of high-severity patches, 81; fire-occurrence data, 77–79; fire-severity data, 79–80; limitations of, 81–82; monitoring trends in burn severity (MTBS), 80; rapid assessment of vegetation condition (RAVG), 81
Quarles, S. L., 344
Quaternary Period, 2
Quayle, B., 80
Quinn-Davidson, L. N., 393
Quinney, D. L., 125
Quinn, J. F., 68
Quinones, R. M., 140
Quintana, M. A., 370

Radeloff, V. C., 110, 111, 344
radiant heat, 41
radiation, process of, 41
radiocarbon dates, 71
Radosevich, S. R., 233
Radtke, K. W.-H., 320
rain shadows, 13
rainstorm, 95
Ralph, F. M., 501
Ramage, B., 117
Ramage, B. S., 158
Rambo, T., 264
Ramirez, A. R., 163
Ramirez, C. M., 81
Randall, J., 288
Randall, P. J., 202
Rao, L. E., 92
Rapacciuolo, G., 314
Raphael, M., 271
rapid assessment of vegetation (RAVG), 79, 79*t*, 81
Rappold, A. G., 272
raptors, 126
Rasker, R., 422
Rasmussen, G. A., 370
Ratchford, J., 242
Rau, B. M., 92
Raymond, R., 158
reaction intensity, 46–47; equation for, 48*b*
Reardon, J. R., 418
reasonable progress goals (RPGs), 447
Rebain, S. A., 51
reconstruction, of fire regimes: anecdotal information on, 76–77; approaches for, 71–77; by archived written records, 75–77; datasets on, 77; dendrochronological, 74–75, 74*f*; fire scar-based, 74; limitations of, 77; map sheets surveyed by Fitch, 76*f*; paleoecological, 71–75; by published reports and articles, 75–76; sedimentary charcoal and pollen, 71–74; stand age, 74–75; tree ring-based, 74–75; United States Geological Survey (USGS), 75
Redak, R. A.
red alder (*Alnus rubra*), 150, 157
red brome (*Bromus madritensis* subsp. *rubens*), 366, 460
redbud (*Cercis occidentalis*), 182, 383
Redd, S. C., 188
red fescue (*Festuca rubra*), 264
red fir, 200
red-legged grasshoppers (*Melanoplus femurrubrum*), 383
red pine (*Pinus resinosa*), 61
redshank (*Adenostoma sparsifolium*), 107
redstem filaree (*Erodium cicutarium*), 36
red willow (*S. laevigata*), 291
redwood (*Sequoia sempervirens*), 150
redwood forest zone, 152*f*; climax status of, 159; fire intervals in, 160*t*; fire regime–plant community interactions, 159–60; fire responses of important species, 158–59; fire season in, 159
redwood mortality, 165
Redwood National Park, 162
Reed, L. R., 64
Reese, R., 320
Reeves, G., 417
Reeves, G. H., 128
Regelbrugge, J. C.
Rehfeldt, G. E., 238
Reible, D. D., 35
Reid, J., 439–40
Reifsnyder, W. E., 106
Reimann, L. F., 97
Reimer, J., 71
Reiner A. L., 254, 272
Reiner, R. J., 118–19
Reinhardt, E. D., 51–53, 106
Reisner, J., 52
Rejmanek, M., 461
relative humidity (RH), 28, 28*f*, 29, 35*m*, 220; relation with fuel moisture, 30
relativized version of the dNBR (RdNBR), 80
remote sensing metrics, of fuel consumption, 72
Renwick, J., 19
reseeders, 108
Resh, V., 418
Resh, V. H., 109
residence time, 46
resprouters, 108
restoration, 433
restoration phase, 415
restoring watershed ecological processes, 435
Reynolds, S., 288
rhizomatous grasses, 107
rhizomes, 106
Rice, K., 283, 287
Rice, R. M., 94
Rice, S. K., 72
Richard, P. J. H., 74
Richardson, D. M., 6
Rich, L. R., 95
Richter, R., 140
Rickman, T. H., 207
Riebe, C. S., 250
Rieck, H. J., 221
Riegel, G. M., 187
Rieman, B. E., 435, 482
Rieman, B. R., 128
Rieradevall, M., 418
Riggan, P. J., 96
Riley, W. J., 408
Riordan, E. C., 307
riparian ecosystem, 87
riparian systems, of California, 464
riparian zones, 338
Ritchie, M. W., 214
Ritsema, C. J., 90
Rivera-Huerta, H., 83
Rizvi, S. W. H., 134
Rizzo, D. M., 164, 165
Robards, T., 423
Robberecht, R., 287
Robertson, P. A., 93
Roberts, S. L., 123–24, 134, 136, 138–39, 141
Robichaud, P. R., 80, 90, 94–95, 433
Robinson, S., 406
Robock, A., 172, 181
Rocca, M. E., 471
Rochester, C. J., 134–35
Rock Creek Butte Research Natural Area (RCBRNA), 186
Rock, D. C., 272
Rock, S. L., 272
Rockwell, T., 358
Rocky Mountain lodgepole pine (*Pinus contorta* subsp. *latifolia*), 422
Rocky Mountains, 111
rodent communities: habitat condition, 130; peak abundance, 130; post-fire abundance of, 132*t*; post-fire survival of, 132*t*; response to fire, 130; time of appearance, 132*t*
Rojas, K., 370
Rollins, M. G., 79
Romme, W. H., 59, 60, 73–75
Root, R. R., 81
roots and underground tissues and organs, 106–7; absorption of water and nutrients from the soil, 106; death of, 106; effects of fire on, 106; feeder roots, 106; sensitivity to heat, 106; structural roots, 106; structural stability to the plant, 106; underground buds, 106; underground stems, 106
Ropelewski, C. F., 19
Roquefeuil, M. C., 283
Rorig, M. L., 173
Rosatti, T. J., 2
Rosburg, T., 287
rose leptosiphon (*Leptosiphon rosacea*), 287
Rosenkrance, L. K., 405
Rosenthal, J. S., 5–6
Rosentreter, R., 223
Rossi, A. M., 14, 18
Ross' sedge (*Carex rossii*), 229, 264
Rost, J., 471
Rotenberry, J. T., 362
Rothermel, R. C., 11, 41–50, 52, 116
Rothermel spread equation, 46–49, 48*b*, 50; limitations of, 49
Roth, L. F., 158
Rothstein, D., 459
Roth, T. R., 93

Roundtree, L., 287
Roundy, B. A., 262
Rowe, J. S., 111
Rowe, P. B., 93–94, 97
Rowlands, P. G., 365
Rowntree, R., 271
Royall, P. D., 282
Roy, B. A., 423
Royce, E. B., 18
Royle, J. A., 123
rubber and yellow rabbitbrush (*Ericameria nauseosa*), 225
rubber rabbitbrush (*E. nauseosa*), 200
Rudd, N., 186
Ruffolo, J., 431–32
Rummell, R. S., 139
Rundel, P. W., 72, 103, 114, 117, 307, 335
Running, S. W., 140
runoff, 11, 17–18, 20
Runyon, J. B., 139
rush (*Juncus* spp.), 291
Russell, K. R., 123, 128, 136
Russell, R. E., 123–24
Russell, W. H., 271
Russian olive (*Elaeagnus angustifolia*), 464
Russian thistle (*Salsola tragus*), 365
Rustigian-Romsos, H., 139
rusty popcornflower (*Plagiobothrys nothofulvus*), 284
Ryan, B. C., 34
Ryan, D. A., 18
Ryan, K. C., 52–53, 106, 418
rye grass (*Festuca perennis*), 460

Saab, V., 123–24
Saah, D., 423
sacaton-dominated grasslands (*Sporobolus airoides*), 286–87
Sackett, S. S., 106
Sacramento River, 184, 202, 219
Sado, Y., 207
Safford, H. D., 75, 77–83
sagebrush (*Artemisia tridentata*), 197, 223, 254, 368, 433
sagebrush sedge (*Carex filifolia* var. *erostrata*), 268
Sagebrush Steppe zone, 222–28, 224*f*, 237; fire regime–plant community interactions, 226–28; fire responses of important species, 224–26; ground cover of native herbaceous vegetation, 226; postfire management, 226; relative negative response to common prennial forbs in sagebrush biome to fire, 227*t*; seed crop production in, 225
sage-grouse (*Centrocercus urophasianus*), 242–43
Sahara mustard (*Brassica tournefortii*), 365
Saito, K., 41, 49
Sakai, H. F., 125
Saksa, P., 435
Sala, A., 108
salal (*Gaultheria shallon*), 150
Salazar, L. A., 154
saltbush (*Atriplex canescens*), 224, 281
Salter, R. B., 187
salt marsh bulrush (*Bolboschoenus maritimus*), 291
salvage logging, 80
Sampson, A. W., 72
Sampson, R. N., 61, 69, 82
San Bernardino larkspur (*Delphinium parryi*), 284
San Bernardino National Forest, 76
Sandberg, D. V., 44, 98
Sandeman, M, 422
sandmat manzanita (*A. pumila*), 310
sand mesa manzanita (*A. rudis*), 310
Sando, R. W., 117
sand rice grass (*Stipa hymenoides*), 365
Sands, A., 324
Sands, P. B., 283
Sandström, T., 439–40
San Francisco Bay-Delta, 281
San Gabriel Mountains, 124, 133
sanicles (*Sanicula* spp.), 264, 328
San Joaquin Valley, 14, 31, 34–36, 281, 287, 299
San Pedro Mártir mission, 20
Santa Ana Mountains, 14, 35, 324*f*
Santa Ana winds, 33, 320; weather maps of, 36*f*
Santa Barbara ceanothus (*Ceanothus impressus*), 310
Santa Clara River, 303
Santoso, A., 19
Sapsis, D. B., 6, 103
Sargent cypress (*Hesperocyparis sargentii*), 67, 308, 312
Sarna-Wojcicki, A. M., 221
Sauer, J. R., 123
Savage, M., 114
saw-toothed goldenbush (*Hazardia squarrosa*), 330
Sawyer, J. O., 1, 62, 69, 158
Scarface Fire (1977), 233
Schaefer, R. J., 238, 368
Schaffner, A. A., 72
Schandelmeier, L. A., 112
Scharf, E. A., 74
Schechter, C., 405
Scheller, R. M., 139, 174
Schenck, S. M., 385
Schiff, A. M., 154
Schlieter, J. A., 45
Schmid, S. M., 463
Schmidt, K. M., 61, 69, 82
Schmidt, M. W. I., 92
Schmoker, K., 482, 486
Schneider, A. C., 314
Schoenberg, F. P., 335
Schoenherr, A. A., 2, 5
Schoennagel, T., 117
Scholl, A. E., 75, 77–78
Schonher, T., 19
Schreier, W., 125
Schroeder, M. A., 242
Schroeder, M. J., 28–31, 33–36, 43–45, 320
Schultz, A. M., 59
Schultz, G. P., 331
Schuster, R. J., 406
Schwalbe, C. R., 374
Schweizer, D., 423
Schwilk, D. W., 104, 111, 123, 272, 412, 415
Schwind, B., 80
sclerophyllous shrubs, 48
Scoles-Sciulla, S. J., 363
Scoles, S. J., 358
scorch height: calculation of, 52; factors affecting, 52*f*; fireline intensity and, 52; tree crown, 52
Scotch broom (*Cytisus scoparius*), 288, 464
Scott, A. C., 74
Scott, J. H., 43–44, 48, 50–51, 90
Scott, M. J., 140
Scott, V. H., 90
Scouler's willow (*Salix scouleriana*), 234
scrub oak (*Quercus berberidifolia*), 203, 309, 322
Seabloom, E. W., 313
sea breeze, 32–34
Seager, R., 314
seaside woolly sunflower (*Eriophyllum staechadifolium*), 307
seasonal fire, dangers of, 43
seasonal severity rating (SSR), 500
season of burning, 43, 72; fire regime distributions for, 64*f*; late summer-fall fire season, 64; late summer-short fire season, 64; spring-summer-fall fire season, 63–64; summer-fall fire season, 64
sea surface temperatures, 18–19
Seavy, N. E., 140
Sebastian, J. R., 372
sedimentation, 96
sediment yield, 97
seed banks, 108, 110, 117–18, 208
seeders, 108
seedling recruitment, 336
seed mortality, 113
seeds, 110–11; chemical-induced germination of, 110; colonization of, 110–11; dispersal of, 110–11; establishment on burned sites, 110–11; heat tolerance of, 110; postfire germination of, 110; response to smoke and/or charcoal, 110; soil insulation, 110; soil-stored, 110
Seli, R. C., 53
senescent risk, 309
Sengupta, A., 19
Sequoia and Kings Canyon National Parks, 412
Sergius, L., 320
serotinous cones, 110
Serra, Y., 314
Seth, A., 314
severity mapping programs, 80
Sevink, J., 90
sexual regeneration: supported by recurring fire, 3
shadscale (*Atriplex confertifolia*), 224, 365
Shafer, D. S., 74
Shaffer, K. E., 482, 486
Shair, F. H., 35
Shakesby, R. A., 90
Shang-Ping, X., 19
Sharpe, S., 221
Shasta red fir (*A. magnifica*), 175
Shasta Valley, 197
Shaw, G., 423
Sheffield, J., 314
Shelton, M. L., 18
Sherlock, J. W., 80, 420
Sheu, P. J., 19
shore pine (*Pinus contorta* subsp. *contorta*), 150
short hair reedgrass (*Calamagrostis breweri*), 269
short-lived-pollutants (SLPs), 452
shortpod mustard (*Hirschfeldia incana*), 343
Show, S. B., 75–77, 79
showy milkweed (*Asclepias speciosa*), 387
shrub canopy recovery rate, 331
shrub communities, in California, 462
shrub forage kochia (*Kochia prostrata*), 373
shrublands, 137
shrub live oak (*Q. turbinella*), 367
shrub steppe communities, 242
shrub tanoak (*Notholithocarpus densiflorus* var. *echinoides*), 175
Shryock, D. F., 374
Shugart, H. H., 60
Shurin, J. B., 461
Sibold, J., 359
Siedinmyer, C., 439–40
Siefkin, N., 162
Sierra gooseberry (*Ribes roezlii*), 262

Sierra juniper (*Juniperus grandis*), 65, 253, 367
Sierra Nevada bioregion, 250, 250*m*; air quality, 272–73; alpine meadow, 253, 269; annual precipitation/temperature patterns, 251*f*; climate change, 272; climatic patterns, 250; current period, 256; description of, 249; eastside forest/woodland, 254, 270, 270*f*; eastside forest/woodland ecological zone, 271*t*; ecological zones, 249, 253, 256; elevation bands, 253*f*; fire climate variables, 250–51; fire/fuels management, 272; fire regime-plant community interactions, 260–61, 264--71; fire response, for important species, 261–64, 266, 268, 270; fire severity distribution, 258*f*; foothill shrubland/woodland, 253, 256, 259*t*; forest heterogeneity, 273; giant sequoias/fire, 263*sb*; historical fire occurrence, overview of, 254, 255*t*; historic period, 255–56; important species, fire responses of, 256–60; Jeffrey pine-mixed conifer forest, 257*f*; lightning strikes, by elevation, 252*f*; lightning strikes, distribution of, 252*m*; lower montane ecological zone, 262*t*; lower montane forest, 253, 261*f*; management issues, 271; physical geography, 249–50; post-burn forest restoration, 273; prehistoric period, 254–55; shrubland, 253; shrubland zone, 269; Sierra de San Pedro Mártir, 257*sb*; smoke dynamics, 273; species at risk, 272; subalpine forest, 268, 268*f*; subalpine forest ecological zone, 253, 269*t*; upper montane forest, 253, 265–66, 266*f*, 267*t*; urbanization, 271; water balance, for species, 252*f*; weather systems, 251–53; woodland zone, 253
Sierra Nevada bioregions, 2, 6
Sierra Nevada Ecosystem Project (SNEP), 202
Sierra Nevada forests, 418
Sierra Nevada lodgepole pine forests, 268
silk tassel bush (*Garrya* spp.), 334
silver sagebrush (*Artemisia cana*), 224
Simon, J-P., 111
Singer, F. J., 125, 137
Singer, M., 90
single-leaf pinyon pine (*Pinus monophylla*), 65, 254
single-leaf pinyon pine-Sierra juniper (*Pinus monophylla-J. grandis*), 222
Singleton, P. H., 187
Siskiyou National Forest, 186
Sitka spruce (*Picea sitchensis*), 65, 150
Sitka spruce forest zone, 152*f*, 157–58; fire regime–plant community interactions, 158; fire responses of important species, 158
sixweeks grass (*Festuca octoflora*), 364
Skalova, H., 459, 465, 467
Skelton, R., 106
Skinner, C. N., 20, 51, 75, 77–78, 80, 83, 123, 172–74, 187, 213–14, 265, 271–72, 393
slash, forest, 454
Slayback, D., 313
Slayter, R. O., 111
Sleeter, B. M., 302
slender-beaked sedge (*Carex athrostachya*), 291
slender bulrush (*Schoenoplectus heterochaetus*), 291
Sloan, L. C., 314
slope winds, 32
slurries of water, 96
small fescue (*F. microstachys*), 364
small-head clover (*Trifolium microcephalum*), 288
SMARTFIRE system, 451
Smith, A. M., 80
Smith, D. F., 45
Smith, E. M., 188
Smith, J. E., 88
Smith, M. M., 72
Smith, S. J., 6, 74, 119
Smith, S. L., 53
Smith, S. M., 110
smoke: asphyxiation due to, 134; chemicals generated by, 98; composition of, 439–40; dispersion caused by winds, 33; effects on ecosystems, 98–99; effects on photosynthetic efficiency, 98; emissions, calculation process, 440; emissions from fire, 87; emissions, fuel consumed by combustion phase, 444*t*; emissions, statewide emission inventory 2012, 443*t*; environment, 441; forecasting and messaging, 453; formaldehyde/acrolein, 441; greenhouse gases (GHGs), 441; human health, 440–41; impact on air quality, 87, 97; impact on human health, 87; in Klamath Mountains bioregion, 188; management of, 87; managing from fires in California, 441; plume dynamics for hot fire, 97*f*; from wildland fires, 439
smoke management program (SMP), 447, 449
smoke tree (*Psorothamnus spinosus*), 364
"Smokey Bear" campaign, 401
smoldering combustion, 106
smoldering ignition, 45
smoldering, on wildland fire, 40
Smythe, K. A., 339
Smythe, L., 358
Sneck, K. M., 74
Sneeuwjagt, R. J., 402
Sneva, F. A., 227
snow accumulation, 93
snowberry (*Symphoricarpos albus* var. *laevigatus*), 368
snow brush (*Ceanothus velutinus*), 110, 115, 270
snowdrop bush (*Styrax redivivus*), 182
snowfall, in California, 14
snow melt, 93
Snyder, M. A., 314
soap plant (*Chlorogalum pomeridianum*), 260
social interactions, 511
Sogard, S. M., 481
soil carbon, 233
soil chemistry, effect of fire on, 91–93
soil–fire interactions, 87; black carbon, 92; bulk density, 89–90; canopy and vegetation, 88*b*; on cation exchange capacity (CEC), 92; color of soil, 90; effect of slope steepness on, 88*b*; effect on chemical properties, 91–92; heat-stimulated germination, 110; hydrologic consequences of, 91*f*; impact on soil texture and structure, 88–89; on litter and duff, 92; on nitrogen, 92; on depth of the solum (soil profile) and water holding capacity, 88*b*; organic matter, 88*b*, 92; pH of soil, 93; on phosphorus, 92–93; on pore size distribution, 89–90; porosity, 89–90; on soil heating, 88; on soil pH, 88*b*; on soil temperature, 88*b*; on thermal conductivity, 88; water repellency, 90
soil heating, effects of, 89*f*
soil moisture, 104, 225
soil quality, 92
soils: aggregation, 89–90; black carbon, 92; bulk density of, 89–90; burn severity, 80; canopy and vegetation, 88*b*; cation exchange capacity (CEC), 93; changes caused by wildfire in, 92; chemical properties of, 91–93, 93*t*; color of, 90; depth of the solum (soil profile), 88*b*; erosion of, 89; fire interactions with (*See* soil–fire interactions); heat capacity of, 88; heating, 52, 72, 88; heat transfer in, 88; hydrophilic, 90; hydrophobic, 90; insulation effect, 89*f*; insulation of plants against fire, 104; litter and duff, 92; microflora of, 91; mineral soil, 88, 90; nitrogen accumulation, 87; nitrogen in, 92; organic matter, 88*b*; organic matter of, 92; permeability of, 90; pH of, 88*b*, 93; phosphorus, 92–93; physical properties, 88–90; pore size distribution, 89–90; porosity of, 89–90; quality, 87; recolonization of, 88; slope steepness, impact of, 88*b*; sterile, 88; structure of, 88–89; temperatures of, 88*b*; texture of, 88–89; thermal conductivity of, 88; water-holding capacity of, 88; water repellency of, 90, 91*f*
soil-stored seeds, 110
soil water holding capacity, 251
solar radiation, 117
Sosa-Ramirez, J., 15, 20, 75
Soulard, C. E., 302
sound vegetation management, 313
Soutar, A., 6
South Coast bioregion, 319; area burned percentage in foothills, 329*t*; arid sites, 333–34; bigcone Douglas-fir forests, 336–37, 336*f*, 337*t*; California black oak, 339; California sage scrub, 330–31; chaparral fire response, 331–32; chaparral fuels/fire, 335*sb*; climate, 319–21; closed-cone cypress, 335–36; coastal sage scrub, mosaic of, 324*f*; coastal valley/foothill zone, 322–25; current era, 327–28; ecological zones, 321–22, 328; fire for foothill, 328, 328*f*; fire regime-plant community interactions, 329, 331, 334–35, 337–38, 340–42; fire regimes, 327*sb*; fire response types for important species, 329*t*, 330*t*, 332*t*; foliar cover before and after fire, 333*f*; grasslands, 328–29; historical era, 325–27; historical fires, 325, 326*t*; lightning strike density, 321*f*; lodgepole pine forests, 342; low-elevation pines, 336; management issues, 342–44; mesic sites, 334; mixed-conifer forests, 341–42; montane coniferous forests, 340*t*; montane zone, 325; oak and walnut woodlands, 340*t*; oak/walnut woodlands, 339; Palmer drought severity index, decadal variation in, 322*f*; physical geography, 319; prehistorical era, 325; riparian woodland-shrubland, 338, 338*t*; southern oak woodland, landscape mosaic of, 324*f*; transverse ranges to Mexico, 320*m*; type conversion, 323*sb*; wildfire in the San Gabriel Mountains, 337*f*; yellow pine forests, 339–40
Southeastern Deserts bioregion, 2, 353–54; agency fire records, 359–61; annual plant fuels management, 370–72; burnable wildland area, 356*t*; climate, 353–54; climate effects, 357–59; current period, 359; desert montane, elevational characteristics of, 355*f*; desert montane woodland, 369*f*; desert montane zone, 368; desert riparian woodland/oasis, 370*f*; ecological sections and zones, 357*m*;

ecological sections/zones, 357*m*; ecological zones, 354, 363; extreme fire years of 2005/2006, 361–63; fire area/fire rotation, 360*t*; fire ecology/fire regime-plant community interactions, 366; fire effects, on desert tortoise, 374*sb*; fire management planning, 374–75; fire regime-plant community interactions, 364–65, 367–70; fire responses of important species, 364, 367–70; fire responses, of important species, 365–66; fire suppression, 373; Great Basin Desert, 354*m*; high-elevation desert zone, 367, 367*f*; historic fire occurrence, overview of, 356; historic period, 358; human effects, 358; lessons, from fire history information, 363; low-elevation desert shrubland, 364*f*; low-elevation desert zone, 364; management issues, 370; middle-elevation desert zone, 366*f*; nonnative annual grasses, 362–63*sb*; physical geography, 353; postfire management, 373–74; precipitation and fire-occurrence potential, 361*f*; prehistoric period, 356; Saltcedar, 371*sb*; woody perennial plant fuels management, 372–73
Southern bulrush (*Schoenoplectus californicus*), 291
Southern California black walnut woodlands, 339
Southern Cascades bioregion: annual precipitation in, 197*t*; Butte Valley, 208; changing stand structures of, 213; climate and fire issues, 195–96; compounding fire deficit, 213; current fire occurrence in, 201–2; ecological effects of fuels treatments, 213–14; ecological zones of, 197–201; fire patterns of, 202; fire regime–plant community interactions, 203, 205–8; fire responses of important species, 203, 204*t*, 206–7, 206*t*; fire severity, patterns of, 202–3; fire suppression in, 201; historic fire occurrence, 200–202; intra-ring locations for fire scars in, 202*f*; lightning strikes in, 196–97; lower elevations of the east side of, 199; lower montane eastside, 199*f*; management issues regarding, 212–13; map of, 196*m*; maxima and minima temperatures, 197*t*; mid-montane eastside, 199–200, 199*f*, 206–8; mid-montane westside, 199–200, 199*f*, 203–6; northwestern foothills of, 197, 198*f*, 203; physical geography of, 195; postfrontal conditions in, 196; prehistoric fire occurrence, 201; relative humidity in, 196; seasonality in, 202; snowpack data, 198*t*; southwestern foothills of, 197, 198*f*, 203; subalpine and alpine vegetation, 200, 201*f*, 211–12; topography of, 202; upper montane zone, 200, 200*f*, 208–11; weather conditions in, 196; wildland-urban interface, 212, 213; wildlife habitat, 212–13
South Warner Mountains, 228*f*
South Warner Wilderness Area, 235
southwestern foothills, 197; fire responses of important species in, 204*t*
Spaeth, K. E., 94–95
Spalt, K. W., 106
Sparks, R. E., 59–60
Spence, P., 19
Spencer, C. N., 128
Spencer, W. D., 139
Sperry, J. S., 335
Spies, T. A., 179, 187, 421
spiny hop-sage (*Grayia spinosa*), 224
spiny menodora (*Menodora spinescens*), 365
spiny redberry (*Rhamnus crocea*), 334
spot fire distance, model for calculating, 51
spotted owl (*Strix occidentalis*), 123; in fire-prone forests, 141
spotting, of fire spread, 51
sprouters, 108, 115
sprouting, 108–10; carbohydrate storage, 109; hormonal patterns governing, 109; phenology of, 109; starch storage, 109; from stolons or root crowns, 109
squirrel tail (*Elymus elymoides*), 108, 226, 284
SSPM wildfire, 258
Stahlschmidt, P. K., 406
Standiford, R. B., 285
stand-origin dates, 74
stand-replacing event, 75
stand-replacing fire, 75
Stanislaus National Forests, 75
Stanton, A., 174
starch storage, 109
Stark, N., 111
Starratt, S., 74
Starrs, C. F., 79
Statheropoulos, M., 97
Stednick, J. D., 90
steelhead trout (*Oncorhynchu mykiss*), 481
Steel, Z. L., 83
Steen, H. K., 29–30, 34
Steenhof, K., 125
Stein, E. D., 96
Steinhorst, R. K., 80
Stensrud, D. J., 15
Stephens, C. W., 74–75, 77–79, 92
Stephenson, N. L., 82
Stephens, S. L., 6, 20, 51, 74–75, 82–83, 96, 111, 123, 156, 187, 272, 402, 404, 408–9, 412, 415, 418, 420–21
Sterling, E. A., 76
Stewart, O. C., 6
Stewart, S. I., 344
Stewart, W., 435
Stewart, W. C., 79
Stires, J., 406
Stohlgren, T. J., 72, 108
Stoms, D. M., 313
Stone, E. C., 118
storage effects, 461
Storer, A. J., 309
storm flows, 95
Storms, N. K., 331
Strachan, S., 359
strawberry (*Fragaria* spp.), 385
streamflow hydrograph, 95*f*
Streeter, R., 405
stress-complex, 502
Striplen, C., 302
Strittholt, J. R., 139
Stromberg, M. R., 150
strong line thermals, 51
Strothmann, R. O., 160
Stuart, J. D., 210
Stuiver, M., 71
Stylinski, C. D., 331
subalpine fir (*Abies lasiocarpa*), 175
subalpine zone: fire regime–plant community interactions in, 186–87; fire responses of important species in, 186, 187*t*, 237*t*; of Klamath Mountains, 186–87; landscapes of, 186; of Northeastern plateaus bioregion, 236, 237*f*; of Southern Cascades bioregion, 211–12
subsidence inversion, 31
subterranean fauna, 125; community, 126*b*; habitat zone, 126*f*
subtropical high aloft type, 252
successional ecology theory, 57–59, 58*f*
Sudworth, G. B., 75–77
sugar pine (*Pinus lambertiana*), 75, 98, 160, 175, 200, 261, 390
Sugihara, N. G., 64, 342
Sulfometuron methyl (Oust®), 371
sundowners, 35
surface-dwelling fauna, 125
surface fires, 75, 115
surface fire spread, 47
surface-passive crown fire, 68
surface runoff, 94
surface winds, on mountain slopes, 35
Susfalk, R. B., 92
Svejcar, T., 226
Swanson, F. J., 187
Swartzendruber, P. C., 441
sweet tooth (*Hydnum repandum*), 386
sweet vernal grass (*Anthoxanthum odoratum*), 155
Swetnam, T. W., 20, 72, 74–75, 79, 119
Swezy, D. M., 106
Sydoriak, W. M., 269
Symons, J. N., 213–14
Symsted, A. J., 181
Syphard, A. D., 79, 139, 344
Syverson, C., 320
Szalay, F. A., 109

Tagestad, J., 358–59, 363, 375
Tahoe National Forest, 75
Takemoto, B. R., 98
Talley, S. N., 309
Talluto, M. V., 323
tamarisk (*Tamarix* spp.), 290, 371, 464
tanoak (*Notholithocarpus densiflorus*), 261, 307, 390, 512
Tansley, A. G., 59
tansy mustard (*Descurainia pinnata*), 364
Tappeiner, II, J. C., 161, 183, 229
tasmanian blue gum (*Eucalyptus globules*), 470
Tate, K. W., 207
Taubert, B. D., 123
Tausch, R. J., 242, 262
Tayler, K. E., 485
Taylor, A. H., 20, 74–75, 77–78, 83, 174, 187, 265, 271
Taylor, D. W., 150, 155
tecate cypress (*Hesperocyparis forbesii*), 322
Tempel, D. J., 123–24, 134, 141
temperatures, in California, 12–13; for January, 12*m*; for July, 13*m*
Teraoka, J. R., 162
terrestrial animals: aquatic habitat and ecosystems, 134; biotic and abiotic environments, 134; food availability, 137; habitat distribution due to fire, 136; habitat structure, 134; indirect effects of fire on, 134–38; plant community composition, 136–37; plant community shifts and, 137–38; plant density and size, 135; postfire changes to habitat, 134; snags and downed wood, use of, 135–36, 136*b*; soil and litter, 135; wildlife management, 134; woody plants and, 135
terrestrial ecosystems, 59; growth of, 87
terrestrial heat loss, 28
Teskey, R. O., 232
Tesky, J. L., 125
Thayer, D. J., 368
thermal balance, 32
thermal belt, 31
thermal conductivity, principles of, 88
thermal energy, 97
thermal inversions, 172, 181

thermochemical decomposition of wood. *See* pyrolysis
thermocouples, 72
Thibeault, J., 314
Thies, W. G., 104
thimbleberry (*Rubus parviflorus*), 262
Thode, A. E., 68, 80–81
Thomas, A. D., 90
Thomas, K. A., 358
Thompson, J. R., 179
Thompson, M. P., 405–8
Thompson, S., 420
Thomsen, C. D., 461
Thornburgh, D. A., 158
thunderstorms, 30, 33, 44–45, 174; in California, 15; cumulus stage, 33, 34*f*; development and decay of, 33; dissipating stage, 33; downdraft winds, 33, 34*f*; episodes of, 220; mature stage, 33; in Northeastern plateaus bioregion, 219; precipitation from high-based, 33; shifting winds associated with, 33; as source of fire ignitions, 33; three-stage evolution to, 33; *virga*, 33
Thurber's needle grass (*S. thurberiana*), 226
Tiedemann, A. R., 90, 93–96
Tilman, D., 461
timber: harvesting, 77, 213, 222; mortality due to forest fire, 75
Timbrook, J., 286
timelag, of fuel moisture, 43
Timmerman, K., 125
Timmermann, A., 19
Tinner, W., 74
Tisdale, E. W., 227–28
tobacco brush (*Ceanothus velutinus*), 175
Tobin, M. F., 163
Tonnesen, G. S., 92
Torching Index, 51
Torn, M., 408
torrey pine (*Pinus torreyana*), 323
total carbohydrate (TAC) levels, seasonal changes in, 109*f*
Townsend's chipmunk (*Tamias townsendii*), 135
toyon (*Heteromeles arbutifolia*), 110, 256, 310, 324
trade winds, 15
trail plant (*Adenocaulon bicolor*), 113, 264
transport winds, 33; and smoke dispersion, 33
Transverse Ranges, 2
Traugh, C. A., 163
tree basal area, 80
tree crown scorch height, 52
tree frogs, 126
tree mortality, 126
tree mortality, fire caused, 77
tree-of-heaven (*Ailanthus altissima*), 464
tree regeneration: episodic, 77; factors influencing, 75
tree-ring studies, 7, 74; datasets on, 77; uncertainty associated with, 75
tree squirrels, 126
Trenberth, K. E., 19
Trewartha, G. T., 27, 30
Trinity County, 137
Trinity River, 171, 178, 184
Trouet, V., 174
Truex, R. L., 272
Tubbesing, C. L., 164
Tubbs, A. M., 15
Tucker, C. M., 319, 461
tufted hair grass (*Deschampsia cespitosa*), 155, 307
tule elk (*Cervus canadensis nannodes*), 127*f*
Tule Lake, 224
tumble mustard (*Sisymbrium altissimum*), 224
turbidity, 96
Turco, R. P., 34–35
Turcotte, D. L., 17
Turhune, G., 213
Turkington, R., 471
Turner, M. G., 59–60, 110
Turner, S. R., 110
Tyler, C., 72, 108–9
Tyree, M., 314

Uchytil, R. J., 158
Uhrenholdt, B., 98
Ulery, A. L., 89
Ulrickson, B. L., 12
Umbanhowar, C. E., 74
Underwood, E. C., 272
Underwood, S., 162
ungulate migration, 131
unharvested high-severity burn patches, 434
United States Geological Survey (USGS), 75–76
unplanned fires, for resource benefits policy, 448
upper montane zone: fire regime–plant community interactions, 208–11, 234–36; fire responses of important species, 208, 208*t*, 234, 235*t*; fuel accumulation rates in, 234–35; of Northeastern Plateaus bioregion, 233–36; Owl Creek watershed, 234*f*; slopes dominated by shrubs and herbaceous plants, 234*f*; of Southern Cascades bioregion, 208–11
Urban, D., 415
Urness, P. J., 373
U.S. Fish and Wildlife Service (USFWS), 477
U.S. General Accounting Office (GAO), 433
Usner, D. J., 302, 304
Utah juniper (*Juniperus osteosperma*), 357
Utah service-berry (*Amelanchier utahensis*), 229
Uzoh, F. C. C., 207

Vaillant, N. M., 254, 272
Valachovic, Y., 344
Vale, T. R., 6, 45, 115, 117
valet tassels (*Castilleja attenuatus*), 287
valley oak (*Q. lobata*), 197, 307
valley winds, 32
Van Ballenberghe, V., 234
Vander Hagen, W. M.
Van der Ploeg, R. R., 90
VanDevender, T. R., 374
Van de Water, K. M., 79, 82
Van Horne, M. L., 75
van Lear, D. H., 123, 128, 136
Van Mullingen, E. J., 90
van Riper, III, C., 358
Van Wagner, C. E., 49–50, 52–53, 64van Wagtendonk, J., 405–6
van Wagtendonk, J. W., 15, 68, 81, 117, 213, 269
van Wagtendonk, K. A., 45, 81
van Wilgen, B. W., 64, 109, 111–12
vapor pressure, 29
vapor pressure deficit (VPD), 501
Varner III, J. M., 64, 117, 162
vascular cambium, 106
vascular plants, 104
Vazquez, A., 207
Veblen, T. T., 117
Vega, R., 123
vegetation: evolution of, 2; flammability, 11
vegetation change, fire exclusion-tree densification model of, 213
vegetation flammability, 11
vegetation management, to lower risk of fire: chemical and biological treatments, 483–84; mechanical treatments, 484; prescribed burning, 484–85
vegetation type map (VTM) survey, 368
velvet grass (*Holcus lanatus*), 155
Venable, D. L., 461
Verdú, M., 111
Verkaik, I., 418
Verner, J., 125
Verstraten, J. M., 90
Vesk, P. A., 108, 111
Viereck, L. A., 90, 93–96, 112
Viers, J. H., 83, 313
virga, 33
Virginia, R. A., 264
Vitousek, P. M., 36, 115
Vogl, R. J., 106, 157, 207, 230
volatile organic compounds (VOCs), 97, 441
von Schroeder, H. P., 134

Wade, D. D., 90, 93–96, 106, 108
Wagener, W. W., 105
Waide, J. B., 60
Wakeman, C. D., 95
Wake, T. A., 154
Wakimoto, R. H., 320
Walden, G. K., 314
Waldrop, T. A., 123, 272
Walker, L. R., 59
Walker, R. F., 87, 92
Wallace, C. S. A., 358
Wallace, J. M., 19, 195
Wallin, D. O., 187
walnut woodlands, fire regimes, 339
Walsh, M. K., 74
Walsh, R. E., 314
Walsh, R. P. D., 90
Walsh, T., 422
Walter, H. S., 480, 484–85
Wang, C., 314
Wangler, M., 367–68
Wan, S., 92
Warburton, A. D., 388
Ward, B. C., 139
Ward, D. E., 440
Ward, J. K., 325
Waring, R. H., 232
Warner Mountains, 220–21, 235, 237
Warner, P., 162
Warner, R., 140
Warton, D. I., 109, 111
water: hillslope hydrology, 93; infiltration, 93–94; interaction with fire (*See* water-fire interactions)
water balance, of burned landscapes, 93
water chickweed (*Montia fontana*), 288
water cycle. *See* hydrologic cycle
water diffusion, in dead fuel, 11
water–fire interactions: on catchment water yield, 97; channel erosion and sedimentation, 96; channel hydrology and, 95; effect on water quality, 96; effect on water quantity, 95–96; evapotranspiration and, 94; hillslope erosion, 94–95; hillslope hydrology, 93; infiltration, 93–94; on overland flow, 94–95; on sediment yield, 97; snow accumulation and, 93; snow melt and, 93; on water balance, 93; watershed hydrology and erosion, 96
water quality: effects of fire on, 87, 96, 127; fire severity, impact of, 96; nitrate loss, 96;

postfire water chemistry, 96; stream temperatures, effect of, 96; turbidity, 96
water quantity, effect of fire on, 95–96; base flows, 95; burned watersheds, 95; discharge rates, 95; flow timing, 96; on lag time to peak flows, 96; peak flows, 95; riparian vegetation, 95; storm flows, 95
water repellency, of soils, 90, 94; factors influencing, 90; hydrologic consequences of, 91*f*
watersheds, 93, 96, 202, 429; aquatic habitats, 429–30; Deer Creek watershed, 203; ecological/social values, 433–34; ecosystem, 429, 430*f*; erosion, 96, 429; existing condition, 434; hydrology, 96; influence of fire on, 429; integrating fire, 435–36; postfire rehabilitation/restoration, 433; postfire stabilization, 432–33; postfire stabilization, rehabilitation and restoration, 430–32; restoring fire, 434; riparian areas, treatment, 434–35; site *vs.* landscape level, 435; water quality laws and policies, 431–32*sb*
Waters, J. R., 125, 208
water vapor, net motion of, 29*f*
water yield, 97
Watt, A. S., 59
weather risk, role of, 24
Weatherspoon, C. P., 123, 213, 272
weather systems: of Central Coast bioregion, 300–301; of Central Valley bioregion, 280; of Klamath Mountains bioregion, 171–72; of Northeastern plateaus bioregion, 220; of Sierra Nevada bioregion, 251–53; of Southern Cascades bioregion, 196
Webb, R. H., 19, 358–59
Webster, J., 206, 214
Webster, S. R., 233
weevils (*Curculio* spp.), 387
Weide, D. L., 103
Weiner, S., 381
Weirich, F., 96
Weisberg, P. J., 79, 174
Weise, D. R., 156
Weiss, S. B., 92
Weitkamp, W. H., 283
Weixelman, D. A., 104, 234
Welling, R., 90
Wells, C. G. (1979), 88, 90, 92–94
Wells, W. G., II (1981), 95
Welsh, H. H., Jr., 123
Wenju, C., 19
Wennberg, P., 439–40
Wessel, C., 96
Westcott, V. C., 111, 115
West, D. C., 60
Westerling, A. L., 242
western blue flag (*Iris missouriensis*), 264
western choke-cherry (*Prunus virginiana* var. *demissa*), 385
western hemlock (*Tsuga heterophylla*), 157
western juniper (*Juniperus occidentalis*), 104, 175, 197, 200, 222, 254
western lily (*Lilium occidentale*), 393; US Fish and Wildlife Service Recovery Plan for, 393
western poison oak (*Toxicodendron diversilobum*), 284, 310
western redbud (*Cercis occidentalis*), 284
western red cedar (*Thuja plicata*), 157
western sycamore (*Platanus racemosa*), 289, 434
western white pine (*Pinus monticola*), 175, 253
western wild-rye (*E. glaucus*), 284
West, G. J., 158
Westlind, D. J., 104
West, N. E., 242
Westoby, M., 108–9, 111
Westrerling, A. L., 242
wet-bulb temperature, 28–29
Wheatley, M. J., 429
Wheeler, N. J. M., 34
Whelan, R. J., 282
whirlwinds, 34
Whisenant, S. G., 225, 228, 362, 368, 459, 460
whispering bells (*Emmenanthe penduliflora*), 110
white alder (*Alnus rhombifolia*), 237, 261
whitebark pine (*Pinus albicaulis*), 65, 119, 175, 200
white bursage (*Ambrosia dumosa*), 364
White, D. L., 230
white fir (*Abies concolor*), 65, 98, 104, 175, 200, 221, 253, 517
White, K. L., 157, 207
White, P. S., 59
White, R., 123–24
whiteleaf manzanita (*Arctostaphylos viscida*), 175, 203, 256, 283
whitethorn (*Ceanothus cordulatus*), 92
Whitley, M. G., 421
Whitlock, C., 71, 73–74
Whitmore, S. A., 123–24, 134, 141
Whitson, T. D., 242
Whittaker, R. H., 59
Whitworth, V., 325
Wickman, B. E., 232
Wickstrom, C. K. R., 282
Wiedenhoeft, A., 325
Wiedinmyer, C., 452
Wiene, C. L., 181
Wieslander, A. E., 78
Wigand, P. E., 221
Wijayratne, U. C., 225
Wikler, K., 309
Wilcove, D. S., 459
wildfire's ability, 484
Wildland Fire Leadership Council, 80
wildland fires: changes caused in soil due to, 92; cultural history of, 508–9; in early twentieth century, 58; ecological importance of, 2; as ecological process, 1–2; effects of atmospheric stability on, 30; impact of humans on frequency of, 6; impact on biological diversity, 1; influence acceptance of practice, 510–11; institutional structure of, 512–15; key policies and programs supporting public mitigation efforts, 511; largest fires by acres burned, 404*t*; management of, 1, 7; mountain area safety taskforce-case study, 514*sb*; national level, 511; Northern California prescribed fire council-case study, 513*sb*; public perceptions, 509–10; recurrences of, 23; smoke column, 28; smoldering on, 40; social dynamics in California, 507; state level efforts, 511–12; temperatures of, 89*f*; traditional use-case study, 512*sb*; and vegetation, 1
wildland-urban interface, 7, 69, 163; in Klamath Mountains bioregion, 188; in Southern Cascades bioregion, 212–13
wildlife colonization and exploitation, with changes in fire regimes, 137*t*
Wildlife Habitat Relationships (WHR) model, 3*t*
wildlife management: in Klamath Mountains bioregion, 189; in Northeastern plateaus bioregion, 238–43; of terrestrial animals, 134
wild mock orange (*Philadelphus lewisii*), 182
wild oats (*Avena fatua*), 36, 283, 330
Wilken, D. H., 2
Wilkin, K., 83
Williams, B., 185
Williams, C. J., 94
Williams, D. F., 125
Williams, G. W., 282
Williams, J., 422
Williams, J. W., 313
Williams, M. A., 78
Williams, M. P., 233
Williams, M. R., 96
Willig, M. R., 59
willow (*Salix* spp.), 110, 253
Wilson, B. C., 291
Wilson, E. O., 459
Wilson, M. V., 291
Wilson, R. B., 171
Wilson, T. P., 459
Wilson, T. S., 302
wind-driven fires, 51, 403; flame dimensions for, 47*f*
wind energy, 50
winds, 31–34; 6 m (20 ft) wind, 32; causes of, 32; cold-front winds, 33; direction of, 32; effects of surface friction on, 32; fire behavior, affects on, 32; fire spread due to, 32; föhn (foehn) winds, 33–34; fuel moisture, affects on, 32; general winds, 32; prefrontal, 220; Santa Ana wind, 33, 36*f*; sea breeze, 32–33; significant fire weather winds, 33; slope winds, 32; smoke dispersion caused by, 33; sundowners, 35; thunderstorms, 33; transport winds, 33; valley winds, 32; weather maps of, 36*f*; whirlwinds, 34
Wing, B. M., 214
winter fat (*Krascheninnikovia lanata*), 224, 365
Winterkamp, J., 52
Winters, D., 128
Winward, A. H., 224
Wirth, T. A., 373
Wirtz, II, W. O., 130–32, 134–36
Wise, E. K., 174, 195
Woche, S. K., 90
Wohlgemuth, P. M., 90, 94–95, 343
Wolfley, K. P., 225
wolf trees, 231
Wollum, A. G., 233
Wolman, M. G., 96
wolverine (*Gulo gulo*), 140
Woodbury, T. D., 75, 77
Wood, D. L., 308–9
woodland cup (*Peziza sylvestris*), 386
woodrats (*Neotoma* spp.), 125
Woods, S. W., 94
Wood, T. E., 374
woody fuel, 106
woody invasive species, 469
woody plants, 106, 135
woody vegetation, 36
Woolfenden, W., 119
Woolfenden, W. B., 254
woolly mule's ears (*Wyethia mollis*), 266
woolyleaf ceanothus (*C. tomentosa*), 333
Wormworth, J., 140
wound-wood formations, signs of, 75
Wrangham, R., 381
Wright, E., 90
Wright, H. A., 103–5
Wright, H. R., 7
Wright, J., 225
Wright, J. G., 225
Wrobleski, D. W., 242
Wrona, A. F., 264

Wuycheck, J. C., 481
Wylie, B., 372
Wyman, B., 314
wyoming big sagebrush (*Artemisia tridentata*), 361

Xiao, Q., 93
Xie, S.-P., 314
Xu, Q., 381
xylem and phloem, 104–5

Yadon, V., 150
Yang, J., 174
Yasuda, D., 342
Yasuda, D. A., 420
Yates, H. S., 78
Yaussy, D., 272
Yaussy, D. A., 123
Yelenik, S. G.
yellow pine (*P. jeffreyi* and *P. ponderosa*), 75
yellow pines (Diploxylon pines), 177
yellow star-thistle (*Centaurea solstitalis*), 283, 288, 462
Yensen, E., 125
yerba santa (*Eriodictyon californicum*), 256, 387
Yetman, D., 370
Yoder-Williams, M. P., 233
Yokelson, R., 439–40
Yool, S. R., 75
York, J. D., 74
Yosemite National Park, 68, 74*f*, 75, 81*f*, 124, 137, 227, 488
Yoshimura, K., 314
Young, A. B., 75, 77–78, 83
Youngberg, C. T., 233
Youngblood, A., 51, 123, 272
Younger-Dryas chronozone (YDC), 74
Young, M. A., 96
Young, M. K., 123, 129, 435
YPMC forests, 258
Yuan, F., 92
Yucca genus, 366

Zabel, C. J., 208
Zachmann, L. J., 363
Zack, J. A., 35
Zack, S., 272
Zac, S., 123
Zammit, C. A., 110
Zanner, W. C., 90
Zavaleta, E. S., 466
Zavitokovski, J., 333
Zavon, J. A., 289
Zedler, P. H., 110, 112, 114
Zhang, J., 206, 214
Zhang, Y., 195
Zhao, M., 314
Zhong, M., 19
Zhu, Z., 80
Zhu, Z.-L., 80
Ziegenhagen, L. L., 225
Zielinski, W., 123, 187
Ziemer, R. R., 429
Zimmerman, S. R., 162
Zimmerman, S. R. H., 310
Zinke, P. J., 150
Zink, T. A., 331
Ziogas, A. K., 90
Zobel, D. B., 158
Zouhar, K., 228
Zouhar, K. L., 155